HANDBOOK OF
TRANSPORTATION
ENGINEERING

HANDBOOK OF TRANSPORTATION ENGINEERING

Myer Kutz Editor

President, Myer Kutz Associates, Inc.

McGRAW-HILL

New York Chicago San Francisco Lisbon London Madrid
Mexico City Milan New Delhi San Juan Seoul
Singapore Sydney Toronto

The McGraw·Hill Companies

ISBN 0-07-139122-3

The sponsoring editor for this book was Kenneth P. McCombs and the production supervisor was Pamela A. Pelton. It was set in Times Roman by Pro-Image Corporation. The art director for the cover was Handel Low.

Printed and bound by RR Donnelley.

McGraw-Hill books are available at special quantity discounts to use as premiums and sales promotions, or for use in corporate training programs. For more information, please write to the Director of Special Sales, McGraw-Hill Professional, Two Penn Plaza, New York, NY 10121-2298. Or contact your local bookstore.

This book is printed on acid-free paper.

To my grandson, Jayden William Parry:
So many wonderful miles are ahead of you

CONTENTS

Part II   Traffic, Streets, and Highways

Chapter 19. Transportation Noise Issues *Judith L. Rochat*   **19.1**

Chapter 20. Transportation-Related Air Quality *Shauna L. Hallmark*   **20.1**

Part IV Non-Automobile Transportation

Chapter 21. Pedestrians *Ronald W. Eck*   **21.3**

Chapter 22. Bicycle Transportation *Lisa Aultman-Hall*   **22.1**

Chapter 28. Air Traffic Control System Design *Robert Britcher* **28.1**

Part V Operations and Economics

Chapter 29. Transportation Planning and Modeling *Konstadinos G. Goulias* **29.3**

Chapter 30. Transportation Economics *Anthony M. Pagano* **30.1**

Chapter 31. Innovative Information Technology Applications in Public Transportation *John Collura* **31.1**

Chapter 32. Parking Management *P. Buxton Williams and Jon Ross* **32.1**

Chapter 33. Trucking Operations   *Amelia Regan*　　　　　　33.1

Chapter 34. The Economics of Railroad Operations: Resurgence of a Declining Industry   *Wesley W. Wilson and Mark L. Burton*　　34.1

Chapter 35. Airline Management and Operations   *Saad Laraqui*　　35.1

Chapter 36. The Marine Transportation System   *James J. Corbett*　　36.1

Chapter 37. Freight Transportation Planning   *Kathleen Hancock*   37.1

Chapter 38. Transportation Management   *George L. Whaley*   38.1

CONTRIBUTORS

Michael S. Bronzini *Department of Civil, Environmental and Infrastructure Engineering, George Mason University, Fairfax, Virginia* (CHAPTER 1, NATIONAL TRANSPORTATION NETWORKS AND INTERMODAL SYSTEMS)

Ben Immers *Department of Civil Engineering, Transportation Planning and Highway Engineering, KU Leuven, Heverlee, Belgium* (CHAPTER 2, TRANSPORT NETWORK PLANNING: THEORETICAL NOTIONS)

Bart Egeter *Netherlands Organization for Applied Scientific Research, TNO Inro, Delft, The Netherlands* (CHAPTER 2, TRANSPORT NETWORK PLANNING: THEORETICAL NOTIONS)

Rob van Nes *Faculty of Civil Engineering and Geosciences, Transportation and Planning, Delft University of Technology, Delft, The Netherlands* (CHAPTER 2, TRANSPORT NETWORK PLANNING: THEORETICAL NOTIONS)

Andrew P. Tarko *School of Civil Engineering, Purdue University, West Lafayette, Indiana* (CHAPTER 3, TRANSPORTATION SYSTEMS MODELING AND EVALUATION)

Elena Shenk Prassas *Department of Civil Engineeering, Polytechnic University, Brooklyn, New York* (CHAPTER 4, SOFTWARE SYSTEMS AND SIMULATION FOR TRANSPORTATION APPLICATIONS)

Gary S. Spring *Department of Civil Engineering, Merrimack College, North Andover, Massachusetts* (CHAPTER 5, APPLICATIONS OF GIS IN TRANSPORTATION)

Baher Abdulhai *Department of Civil Engineering, University of Toronto, Toronto, Ontario, Canada* (CHAPTER 6, TRAFFIC ENGINEERING ANALYSIS)

Lina Kattan *Department of Civil Engineering, University of Toronto, Toronto, Ontario, Canada* (CHAPTER 6, TRAFFIC ENGINEERING ANALYSIS)

Arun Chatterjee *Department of Civil and Environmental Engineering, The University of Tennessee, Knoxville, Tennessee* (CHAPTER 7, TRAVEL DEMAND FORECASTING FOR URBAN TRANSPORTATION PLANNING)

Mohan M. Venigalla *Civil, Environmental, and Infrastructure Engineering Department, George Mason University, Fairfax, Virginia* (CHAPTER 7, TRAVEL DEMAND FORECASTING FOR URBAN TRANSPORTATION PLANNING)

Lily Elefteriadou *Department of Civil Engineering, The Pennsylvania State University, University Park, Pennsylvania* (CHAPTER 8, HIGHWAY CAPACITY)

Richard W. Denney, Jr. *Iteris Corporation, Sterling, Virginia* (CHAPTER 9, TRAFFIC CONTROL SYSTEMS: FREEWAY MANAGEMENT AND COMMUNICATIONS)

Richard W. Denney, Jr. *Iteris Corporation, Sterling, Virginia* (CHAPTER 10, TRAFFIC SIGNALS)

Philip M. Garvey *Pennsylvania Transportation Institute, Pennsylvania State University, University Park, Pennsylvania* (CHAPTER 11, HIGHWAY SIGN VISIBILITY)

Beverly T. Kuhn *System Management Division, Texas A&M, College Station, Texas* (CHAPTER 11, HIGHWAY SIGN VISIBILITY)

Kara Kockelman *Department of Civil Engineering, The University of Texas at Austin, Austin, Texas* (CHAPTER 12, TRAFFIC CONGESTION)

Brian Wolshon *Department of Civil and Environmental Engineering, Louisiana State University, Baton Rouge, Louisiana* (CHAPTER 13, GEOMETRIC DESIGN OF STREETS AND HIGHWAYS)

Joseph E. Hummer *Department of Civil Engineering, North Carolina State University, Raleigh, North Carolina* (CHAPTER 14, INTERSECTION AND INTERCHANGE DESIGN)

Yongqi Li *Arizona Department of Transportation, Phoenix, Arizona* (CHAPTER 15, PAVEMENT TESTING AND EVALUATION)

Rune Elvik *Institute of Transport Economics, Norwegian Center for Transportation Research, Oslo, Norway* (CHAPTER 16, TRAFFIC SAFETY)

Thomas J. Cova *Center for Natural and Technological Hazards, Department of Geography, University of Utah, Salt Lake City, Utah* (CHAPTER 17, TRANSPORTATION HAZARDS)

Steven M. Conger *Center for Natural and Technological Hazards, Department of Geography, University of Utah, Salt Lake City, Utah* (CHAPTER 17, TRANSPORTATION HAZARDS)

Ahmed Abdel-Rahim *Department of Civil Engineering, University of Idaho, Moscow, Idaho* (CHAPTER 18, INCIDENT MANAGEMENT)

Judith L. Rochat *U.S. Department of Transportation/Volpe Center, Cambridge, Massachusetts* (CHAPTER 19, TRANSPORTATION NOISE ISSUES)

Shauna L. Hallmark *Department of Civil, Construction, and Environmental Engineering, Iowa State University, Ames, Iowa* (CHAPTER 20, TRANSPORTATION-RELATED AIR QUALITY)

Ronald W. Eck *Department of Civil and Environmental Engineering, West Virginia University, Morgantown, West Virginia* (CHAPTER 21, PEDESTRIANS)

Lisa Aultman-Hall *Department of Civil and Environmental Engineering, University of Connecticut, Storrs, Connecticut* (CHAPTER 22, BICYCLE TRANSPORTATION)

Keith L. Hawthorne *Transportation Technology Center, Inc., Pueblo, Colorado* (CHAPTER 23, RAILWAY ENGINEERING)

V. Terrey Hawthorne *Newtowne Square, Pennsylvania* (CHAPTER 23, RAILWAY ENGINEERING)

(In collaboration with E. Thomas Harley, Charles M. Smith, and Robert B. Watson) (CHAPTER 23, RAILWAY ENGINEERING)

Ernest T. Selig *Department of Civil Engineering (Emeritus), University of Massachusetts, and Ernest T. Selig, Inc., Hadley, Massachusetts* (CHAPTER 24, RAILWAY TRACK DESIGN)

Sudhir Kumar *Tranergy Corporation, Bensenville, Illinois* (CHAPTER 25, IMPROVEMENT OF RAILROAD YARD OPERATIONS)

William H. Mason *Department of Aerospace and Ocean Engineering, Virginia Polytechnic Institute and State University, Blacksburg, Virginia* (CHAPTER 26, MODERN AIRCRAFT DESIGN TECHNIQUES)

William R. Graves *School of Aeronautics, Florida Institute of Technology, Melbourne, Florida* (CHAPTER 27, AIRPORT PLANNING AND DESIGN)

Ballard M. Barker *School of Aeronautics, Florida Institute of Technology, Melbourne, Florida* (CHAPTER 27, AIRPORT PLANNING AND DESIGN)

Robert Britcher *Montgomery Village, Maryland* (CHAPTER 28, AIR TRAFFIC CONTROL SYSTEM DESIGN)

Konstadinos G. Goulias *Department of Civil Engineering, The Pennsylvania State University, University Park, Pennsylvania* (CHAPTER 29, TRANSPORTATION PLANNING AND MODELING)

Anthony M. Pagano *Department of Managerial Studies, University of Illinois at Chicago, Chicago, Illinois* (CHAPTER 30, TRANSPORTATION ECONOMICS)

John Collura *Northern Virginia Center, Virginia Polytechnic and State University, Falls Church, Virginia* (CHAPTER 31, INNOVATIVE INFORMATION TECHNOLOGY APPLICATIONS IN PUBLIC TRANSPORTATION)

P. Buxton Williams *MPSA Partners, Oak Park, Illinois* (CHAPTER 32, PARKING MANAGEMENT)

Jon Ross *MPSA Partners, Chicago, Illinois* (CHAPTER 32, PARKING MANAGEMENT)

Amelia Regan *Computer Science and Civil and Environmental Engineering, University of California, Irvine, Irvine, California* (CHAPTER 33, TRUCKING OPERATIONS)

Wesley W. Wilson *Department of Economics, University of Oregon, Eugene, Oregon, Upper Great Plains Transportation Institute* (CHAPTER 34, THE ECONOMICS OF RAILROAD OPERATIONS: RESURGENCE OF A DECLINING INDUSTRY)

Mark L. Burton *Department of Economics, Marshall University, Huntington, West Virginia* (CHAPTER 34, THE ECONOMICS OF RAILROAD OPERATIONS: RESURGENCE OF A DECLINING INDUSTRY)

Saad Laraqui *Business Administration Department, Embry-Riddle Aeronautical University, Daytona Beach, Florida* (CHAPTER 35, AIRLINE MANAGEMENT AND OPERATIONS)

James J. Corbett *Department of Marine Studies, University of Delaware, Newark, Delaware* (CHAPTER 36, THE MARINE TRANSPORTATION SYSTEM)

Kathleen Hancock *Civil and Environmental Engineering Department, University of Massachusetts at Amherst, Amherst, Massachusetts* (CHAPTER 37, FREIGHT TRANSPORTATION PLANNING)

George L. Whaley *Organization and Management Department, College of Business, San Jose State University, San Jose, California* (CHAPTER 38, TRANSPORTATION MANAGEMENT)

PREFACE

Transportation engineering is a broad, multidisciplinary field. There are enough topics and issues for an editor to develop a handbook similar in style and format to a handbook devoted to one of the traditional engineering disciplines, such as civil engineering, even though transportation engineering is a sub-discipline of civil engineering. Transportation engineering involves the movement of people and goods by land, air, and sea, using human locomotion, bicycles, automobiles, buses, trucks, trains, aircraft, ships, and pipelines. It also involves the design of streets and highways, intersections, traffic control devices, mass transit systems, railroad tracks, and rail yards, as well as airport runways and air traffic control systems. Transportation engineers deal with a wide variety of issues, including highway sign visibility, travel demand, traffic flow, and the modeling, simulation, management, and economics of transportation systems. Among the problem areas transportation professionals have to deal with are traffic congestion, safety, hazards, incident management, noise, and air pollution, to name just a few.

To be sure, the transportation engineering literature is substantial. Plenty of textbooks, reference works, and manuals, as well as a modest number of journals, already exist. Because government agencies and legislative bodies are intimately involved with the whole wide range of transportation issues, professional associations and institutes, as well as government agencies themselves, publish materials, including handbooks and manuals, with essentially official status. Some of these publications represent, as one professor put it in a note to me, "the 'ONLY STOP' for information on their respective topics (unless, of course, the engineer wants to invite lawsuits)." But even with codification of design issues, practitioners and researchers continue to find new ways of planning, construction, maintaining, monitoring, and evaluating technologies, infrastructures, and systems. Also, existing professional-level publications tend to be highly specialized, directed only to the specific needs of particular groups of practitioners.

The audience for transportation engineering publications is as broad and multidisciplinary as the discipline itself. Not only transportation engineers, but also civil engineers, city and regional planners, public administrators, economists, social scientists, and urban geographers are among the professionals who are involved with transportation issues. They work in a variety of organizations, including design, construction, and consulting firms, as well as federal, state and local government agencies. A rationale for a general transportation engineering handbook is that every practitioner, researcher, and bureaucrat cannot be an expert on every topic, especially in such a broad and multidisciplinary field, and may need to read an authoritative summary on a professional level of a subject that he or she is not intimately familiar with but may need to know about for a number of different reasons.

My intention is that the *Handbook of Transportation Engineering* stand at the intersection of textbooks, research papers, and design manuals. For example, I want the handbook to help young engineers move from the college classroom to the professional office and laboratory where they may have to deal with issues and problems in areas they may not have studied extensively in school.

My endeavor has been to produce a practical reference for the transportation engineer who is seeking to solve a problem, reduce cost, or improve a system or facility. The handbook is not a research monograph. Its chapters offer design techniques, illustrate successful applications, or provide guidelines to improving the performance, the life expectancy, the effectiveness, or the usefulness, of systems, infrastructure, and facilities. The intent is to show the reader what options are available in a particular situation and which option he or she might choose to solve the problem at hand.

I want this handbook to serve as a source of practical advice to the reader. I would like the handbook to be the first information resource a practicing engineer reaches for when faced with a new problem or opportunity—a place to turn to even before turning to other print sources, even the officially sanctioned ones, or to sites on the Internet. So the handbook has to be more than a voluminous reference or collection of background readings. In each chapter, the reader should feel that he or she is in the hands of an experienced consultant (some of the authors actually are experienced consultants) who is providing sensible advice that can lead to beneficial action and results.

Can a single handbook cover this broad, interdisciplinary field? The *Handbook of Transportation Engineering* is designed as if it were serving as a core for an Internet-based information source. With voluminous references, most individual articles point readers to a web of information sources dealing with the subjects that the articles address. Furthermore, many of the articles, where appropriate, provide enough analytical techniques and data so that the reader can employ a preliminary approach to solving problems.

I have asked contributors to write, to the extent their backgrounds and capabilities make possible, in a style that will reflect practical discussion informed by real-world experience. I want readers to feel that they are in the presence of experienced teachers and consultants who know about the multiplicity of technical and societal issues that impinge on topics within transportation engineering. At the same time, the level is such that students and recent graduates can find the handbook as accessible as experienced engineers. Contributors have been mindful of the multidisciplinary nature of the transportation engineering audience.

I have divided the handbook into five parts. Part I, "Networks and Systems," which consists of five chapters, starting with a general consideration by Mike Bronzini of George Mason University of national transportation networks, corresponding network databases, modal networks, multimodal networks and intermodal connections in the United States, as well as a discussion of national and local applications of network databases for practical planning studies. Next, a team of researchers from The Netherlands, led by Ben Immers of the Netherlands Organization for Applied Scientific Research and the Department of Civil Engineering of KU Leven, and including Bart Egeter and Rob van Nes, uses detailed and profusely illustrated network designs in Hungary and Florida to illustrate original theories of network planning. Andrej Tarko of Purdue uses examples from a wide range of locations—Maine, Sacramento, Washington, DC, and Cincinnati—in presenting aspects of transportation system modeling, which he defines as "a formal description of the relationships between transportation system components and their operations." Elena Prassas of Brooklyn Poly provides information on the span of software systems and reference tools available to practicing engineers and planners, with much of the emphasis on traffic tools. And Gary Spring, formerly of the University of Missouri at Rolla and now at Merrimack College, defines and describes geographic information systems in the transportation industry (GIS-T), provides a review of GIS-T application areas, and presents issues involved in implementing these systems.

Part II, "Traffic, Streets, and Highways," which contains nine chapters, starts with a thorough analytical survey by Baher Abdulhai and Lina Kittan, of the University of Toronto, of important traffic engineering topics, many of which are covered from a more practical, design oriented perspective in the following chapters. Arun Chaterjee and Mohan Venigalla, of the University of Tennessee, write about forecasting travel demand in urban areas, which is essential for long-range transportation planning in order to determine strategies for accommodating future needs. In her chapter on highway capacity, Lily Elefteriadou of Penn State deals with the question, "How much traffic can a facility carry?" She provides an overview of the concept of highway capacity, discusses the factors that affect it, and supplies guidance on obtaining and using field capacity values. Rick Denney of Iteris Corporation is responsible for the next two chapters, which cover traffic control systems. In his first chapter, on freeway management and communications, he presents strategies for optimizing capacity and for managing demand, together with the required infrastructure. He describes freeway management activities in real time. His second chapter deals with the implementation of traffic signals at intersections. Phil Garvey of Penn State and Beverly Kuhn of Texas A&M discuss highway sign visibility, which they call "an imprecise term encompassing both sign detectability and sign legibility." Kara Kockelman of the University of Texas at Austin examines traffic congestion's defining characteristics, its consequences, and most importantly, its possible solutions. The next two chapters deal with roadway design issues. Brian Wolshon of LSU summarizes the geometric highway design process, as well as many of the details of the so-called Green Book, the most widely referenced and applied highway design resource in the United States, to make it more usable. Joe Hummer of NC State discusses current intersection and interchange design concepts, presenting basic design elements first, followed by configurations. Part II ends with a description by Yongqi Li of the Arizona Department of Transportation of the major tests and evaluation procedures commonly used to maintain highway pavements at a level that provides satisfactory riding comfort, structural integrity, and safe skid resistance.

The first chapter, by Rune Elvik of the Institute of Transport Economics in Oslo, in Part III, "Safety, Noise, and Air Quality" (which contains five chapters in total), covers essential topics in traffic safety analysis, including statistics, an empirical method for estimating safety, identifying and analyzing hazardous road locations, traffic engineering measures that affect road safety, methods of assessing the quality of road safety evaluation studies, and techniques that help policy makers choose the most effective road safety measures. Tom Cova and Steve Conger, of the University of Utah, discuss the many environmental hazards that commonly disrupt or damage transportation systems, transportation risks to proximal people and resources, and the role of transportation in emergency management. It is worth mentioning that the reference list for this chapter is unprecedented. Reducing the impact of incident-related congestion has become critical on roadway systems that are operating close to capacity under the best of conditions. Ahmed Abdel-Rahim of the University of Idaho reviews the latest developments in incident management. The last two chapters of Part III deal with environmental quality issues. Judy Rochat of the U.S. Department of Transportation at the Volpe Center in Cambridge, MA, discusses specific noise issues relating to different modes of transportation, including sources, prediction, metrics, and control of noise. And Shauna Hallmark of Iowa State provides up-to-date information on the effects that transportation has on air quality, a topic whose importance increases daily.

Part IV, "Non-Automobile Transportation," which contains eight chapters, starts with two discussions of alternative modes of transportation. The first of these chapters, by Ronald Eck of West Virginia University, takes a comprehensive look at incorporating pedestrians into the transportation system. Lisa Aultman-Hall of UConn talks about how bicycles should operate safely within the transportation system and provides an overview of the specific design knowledge available for designing bicycle infrastructure. Three chapters on rail transportation follow. Terrey and Keith Hawthorne update their chapter from the Marks Handbook on locomotives, freight cars, and passenger equipment. In writing about railway tracks, Ernie Selig, professor emeritus at UMass and now a busy consultant, focuses on a listing of design functions and a description of design methods. Sudhir Kumar, professor at IIT and principal at Tranergy Corporation, writes mainly about improving the operations of classification yards, which are used primarily to make new trains from the cars of arriving trains. Part IV ends with three chapters on air transportation. Bill Mason of Virginia Tech opens them with a comprehensive review (remarkable in a handbook-length chapter) of current design techniques for non-military aircraft. In his chapter on air traffic control systems, Bob Britcher, who works at Lockheed-Martin, emphasizes the design of systems to control commercial and general aviation in the United States. William Graves and Ballard Barker of the Florida Institute of Technology address the scope of airport-particular planning and design issues, including routine elements of airport planning and design, as well as issues, challenges, and approaches to help the transportation engineer address community needs. The central focus is on those elements of airport design typically performed by transportation engineers, rather than those commonly performed by aviation planners, architects, or such engineering specialties such as electrical or mechanical.

To open Part V, "Operations and Economics," which contains ten chapters, Kostas Goulias of Penn State provides an overview of transportation planning, modeling, and simulation, then defines methods of improving the design of simulation models and assesses the future of transportation policy analysis tools. Tony Pagano of the University of Illinois at Chicago starts his chapter on transportation economics with a discussion of the methodology of project appraisal, then moves on to transportation user benefits, intangible costs and benefits, and the externalities associated with transportation benefits. John Collura of the NoVA Graduate Center reviews the major areas in which transit operators in the U.S. have invested in information technology—public transportation management and operations, traveler information services, transit signal priority, and electronic payment and fare collection—and examines their experiences and lessons learned. In their chapter, Jon Ross and Buxton Williams, principals in MPSA Partners, provide a highly strategic, systems-oriented approach to parking management so municipalities can optimize revenues, efficiencies, and customer service, and leverage well-run parking management systems into benefits across other levels of government. Amelia Regan of the University of California, Irvine, provides an overview of the trucking industry, which, as she notes at the end of her chapter is "being impacted by emerging technologies and innovations in operating procedures. The past two decades have seen enormous productivity increases that are not likely to slow in the years ahead." Wes Wilson of the University of Oregon and Mark Burton of Marshall University collaborate on a chapter whose purpose is to carefully describe the railroad industry, its operations, the markets it serves, the current regulatory environment, and the specific economic conditions that motivate firm decision-making. Saad Laraqui of Embry-Riddle compares and contrasts successes and failures in airline operations in Europe and the US. He discusses the new and continuously changing airline business environment, suggesting that structural changes need to be met with a strategic paradigm that integrates strategic, financial and operations management into cohesive whole. In describing the role of the marine transportation system within a context of more familiar transportation modes, such as automobile, trucking, and rail, Jim Corbett of the University of Delaware provides some of the most interesting tables in the handbook. Kitty Hancock of UMass writes about the operations and planning of freight movement, which has evolved rapidly since World War II due to globalization, deregulation of transportation, changing government infrastructure, organic changes in business, and rapidly changing technology. The handbook closes with a chapter by George Whaley of San Jose State, who focuses on the management of people in transportation organizations and the special challenges managers in this industry face on a daily basis at three levels—the total organization, the department or team level and the individual level.

It is the individuals I've mentioned above—engineers and economists, mainly—who made this handbook a pleasure to develop. I am indebted to them. It's a not-so-minor miracle that they found the time to write their chapters so cogently and to provide illuminating and valuable tables and illustrations. And they all were pretty much on time. Thank you all! A word of thanks is also due Ken McCombs, my editor at McGraw-Hill, whose encouragement I appreciate tremendously. Thanks, too, to Lucy Luckenbaugh and the powers-that-be and staff at Pro-Image, who produced this volume speedily and without fuss. When I saw that first email from Lucy shortly after the manuscript had gone into production, I knew the project was in good hands. And thanks, finally, to Arlene, who's always here for me.

Myer Kutz

P · A · R · T · I

NETWORKS AND SYSTEMS

CHAPTER 1
NATIONAL TRANSPORTATION NETWORKS AND INTERMODAL SYSTEMS

Michael S. Bronzini
*Department of Civil, Environmental
and Infrastructure Engineering,
George Mason University, Fairfax, Virginia*

1.1   INTRODUCTION

Transportation systems of regional and national extent are composed of networks of interconnected facilities and services. It follows that nearly all transportation projects must be analyzed with due consideration for their position within a modal or intermodal network, and for their impacts on network performance. That is, the network context of a transportation project is usually very important. Thus, it is appropriate to begin a volume on transportation engineering with a chapter on national transportation networks.

The subject of national transportation networks may be approached from at least two different perspectives. One approach, common to most introductory transportation textbooks, describes the physical elements of the various transport modes and their classification into functional subsystems. A second approach focuses on the availability of national transportation network databases and their use for engineering planning and operations studies. The latter approach is emphasized in this chapter, with the aim of providing the reader with some guidance on obtaining and using such networks. In describing these network databases, however, some high-level descriptions of the physical networks are also provided.

The modal networks considered are highway, rail, waterway, and pipeline and their intermodal connections. Airports and airline service networks are deliberately excluded, as air transport is markedly different in character from the surface transportation modes. Likewise, urban highway networks and bus and rail public transportation networks are not covered, since the emphasis is on national and state-level applications. For reasons of space and focus, only transportation networks in the United States are included, although the general concepts presented apply to any national or regional transportation network.

The chapter begins with a general consideration of the characteristics and properties of national transportation networks and the corresponding network databases. The modal networks are then described, followed by a section on multimodal networks and intermodal connections. The concluding section discusses national and local applications of network databases for practical planning studies.

1.2   *NATIONAL TRANSPORTATION NETWORK DATABASES*

1.2.1   The U.S. Transportation Network

Table 1.1 indicates the broad extent of the U.S. surface transportation system. The national highway network (FHWA 2001) includes nearly 4 million miles of public roads, and total lane-miles are more than double that, at 8.2 million miles. The vast majority of the total highway mileage, 77.6 percent, is owned and operated by units of local government. States own 19.6 percent and the federal government owns only 3 percent. The interstate highway system, consisting of 46,677 miles, accounts for only 1.2 percent of total miles but carries 24 percent of annual vehicle-miles of travel. Another important subsystem is the National Highway System (NHS), a Congressionally designated system that includes the interstate highways and 114,511 miles of additional arterial roadways. The NHS includes about 4 percent of roadway miles and 7 percent of lane miles but carries over 44 percent of total vehicle-miles of travel. Highways are by far the dominant mode of passenger travel in the United States, and trucks operating on the vast highway system carry 29 percent of domestic freight ton-miles (BTS 2003).

The class I railroad network in the United States presently consists of 99,250 miles. This mileage has been decreasing over the past 40 years; in 1960 the class I railroads owned 207,334 miles of track (BTS 2002). Railroad mergers, rail line abandonment, and sales to short-line operators account for the decrease. While this mileage is limited, the rail mode continues to provide vital transportation services to the U.S. economy. For example, railroads carry 38 percent of domestic freight ton-miles, which exceeds total truck ton-miles, and Amtrak provides passenger service over 23,000 miles of track (BTS 2002).

The other modes of transportation listed in Table 1.1 are probably less familiar to the average citizen. The inland waterway system includes 26,000 miles of navigable channels. Of this total, about 11,000 miles are commercially significant shallow-draft waterways (BTS 2002), consisting primarily of the Mississippi River and its principal tributaries (notably the Ohio River system and the Gulf Intracoastal Waterway). To this could be added thousands of miles of coastal deep-draft shipping routes serving domestic intercoastal shipping (e.g., routes such as New York to Miami) and providing access to U.S. harbors by international marine shipping. Nearly totally hidden from view is the vast network of oil and gas pipelines. In fact, at 1.4 million miles, gas pipelines are second in extent only to the highway network. The water and oil pipeline modes each carry about 16 percent of domestic freight ton-miles (BTS 2002).

1.2.2   National Transportation Network Model Purposes and Uses

Motivating the development of national transportation network databases has been the need to consider broad national and regional policies and strategies, and projects for meeting

TABLE 1.1   U.S. Transportation Network

Transportation mode	Statute miles in the U.S. (2002)
Highways	3,936,229
Class I rail	99,250
Inland waterways	26,000
Crude petroleum pipeline	86,369[a]
Petroleum products pipeline	91,094[a]
Natural gas pipeline	1,400,386

[a]Data for year 2001.
Source: BTS 2002.

critical needs for mobility and economic development. Assessing the benefits of such projects often requires considering their role within the national transportation infrastructure. For example, consider the new highway bridge crossing the Potomac River on I-95, under construction near Washington, DC. When this project was nearing a critical funding decision, the question arose as to how much of the traffic using the existing bridge and other regional crossings was interstate truck traffic versus local traffic. Local modeling based on historical truck counts simply could not provide the requisite information. Answering this question (BTS 1998) required a regional or national network model of broad enough scope to capture a diverse set of commercial truck trips (BTS 1997).

Other examples of national network modeling are numerous. An early use of national rail networks was for analyzing the impacts of railroad mergers. The initial proposal to impose a diesel fuel tax on domestic inland waterway transportation was analyzed, in part, with a waterway system network model (Bronzini, Hawn, and Sharp 1978). Subsequent to the energy crisis of the mid-1970s, USDOT used national rail, water, highway, and pipeline networks to examine potential bottlenecks in the movement of energy products (USDOT/ USDOE 1980). The potential impacts of spent fuel shipments from nuclear power plants to the proposed waste repository in Nevada have been estimated with the aid of rail and highway network models (Bronzini, Middendorf, and Stammer 1987). Most recently, the Federal Highway Administration (FHWA) has developed the Freight Analysis Framework (FAF), which is a network-based tool for examining freight flows on the national transportation system. Information on the FAF may be found at http://www.ops.fhwa.dot.gov/freight/. Examples of state and local uses of network models are covered at the end of this chapter.

What these examples have in common is that the demand for using specific segments of the transportation system arises from a set of geographically dispersed travelers or shippers. Likewise, the impacts of improving or not improving critical pieces of the network are felt by that same set of diverse network users. Building network models for these types of applications used to be a daunting prospect, due to the lack of available network data. As will be seen later, much of this impediment has been overcome.

1.2.3   Characteristics of Large-Scale Transportation Networks

A network model of the transportation system has two basic analytic requirements: (1) it must be topologically faithful to the actual network; and (2) it must allow network flows along connected paths. A network model that included every mile of every mode would obviously be very unwieldy. Constructing the initial database would be very time-consuming, the quality of the data would likely be compromised, and maintaining and updating the model would be equally difficult. Hence, no such undertaking has yet been attempted, at least not for a model that fulfills both analytic requirements. Topographic databases, as used for mapmaking, do not satisfy the second requirement and hence are not entirely useful for computer-based transportation analyses.

Since the entire system cannot be directly represented in the network model, some judgment must be exercised in determining the model's level of detail. This is referred to as the granularity of the model, which is a relative property. A particular network model can only be characterized as coarser or finer than some other model of the same network, i.e., there is no accepted "granularity scale." Figure 1.1 displays two possible models of a simple highway intersection. In panel (a) the intersection is represented as four links, one for each leg of the intersection, meeting at one node. In panel (b) each direction of travel and each movement through the intersection is represented as a separate link. (In fact, many different types of detailed intersection network coding have been proposed.) The level of granularity adopted will depend upon whether the outcome of the analysis is affected by the details of the within-intersection traffic flows and upon the capabilities of the analytical software to be used in conjunction with the network database.

Related to network granularity is the granularity of the spatial units that contain the socioeconomic activity that generates transportation demands. It is customary to divide the

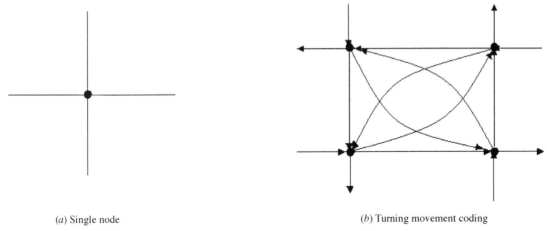

(*a*) Single node   (*b*) Turning movement coding

FIGURE 1.1   Representation of intersections in network models.

analysis area into zones or regions and to connect these regions with the transportation network model so as to allow analysis of the flows between the zones. For example, in a statewide model the spatial units could be counties and cities. Obviously, the zones and the network must have complementary degrees of granularity.

1.2.4   Typical Network Data Elements

Transportation networks inherently have a node and link structure, where the links represent linear features providing for movement, such as highways and rail lines, and the nodes represent intersections. Thus, the principal data content of a node is its name or number and location. Links usually have characteristics such as length, directionality, number of travel lanes, and functional class. Flow capacity, or some characteristics enabling ready estimation of the capacity, are also included. Of course, the whole assemblage of nodes and links will also be identified with a particular mode.

Another representational decision to be made is whether the network links will be straight lines or will have "shape points" depicting their true geography. Early network models were called "stick networks," which is topologically accurate but lacking in topographic accuracy. For many types of analyses this is of no concern; a software system that deals only with link-node incidences, paths, and network flows will yield the same answer whether or not the links have accurate shapes. For producing recognizable network maps and for certain types of proximity analysis, however, topographically accurate representations are needed (see Figure 1.2). Hence, most large-scale network models currently utilize shape points. This comes at a price, in that much more data storage is required, and plots or screen renderings are slowed. Fortunately, advances in computing power and geographic information systems (GIS) software have minimized these drawbacks to a large extent.

The idea of link capacity was mentioned above. In some networks this is stated directly for each link, in units such as vehicles per hour or tons per day. In others the functional class of a link points to an attribute table that has default capacity values. In the case of an oil pipeline, for example, the diameter of the pipe could be used to estimate flow capacity for various fluid properties. Nodes seldom are modeled as capacity-constrained, but in principal can be (and have been) treated in the same way as links.

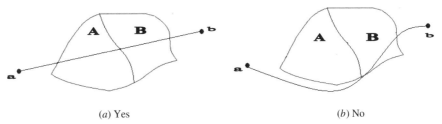

(a) Yes   (b) No

FIGURE 1.2   Does link *ab* enter region *A*?

1.3   EXAMPLES OF NATIONAL MODAL NETWORKS

The principal source of national transportation network data in the public domain is the National Transportation Atlas Database (NTAD), developed and distributed by the USDOT Bureau of Transportation Statistics (BTS). Information on the NTAD may be obtained at http://www.bts.gov/gis/. As stated there: "NTAD is a set of transportation-related geospatial data for the United States. The data consist of transportation networks, transportation facilities, and other spatial data used as geographic reference."

Figure 1.3 is a plot of a portion of the U.S. transportation system (excluding pipelines), centered on the state of Ohio, drawn from the NTAD. As could be seen by comparing this figure with state-level highway and rail maps, the NTAD does not contain data for the entire system. In particular, facilities that largely serve local traffic are not represented. Nonetheless, the facilities included carry the great bulk of intercity traffic, hence the networks have proven valuable for conducting national and regional planning studies.

1.3.1   Highway Networks

For the highway mode, the NTAD includes the National Highway Planning Network (NHPN), shown in Figure 1.4, which is a comprehensive network database of the nation's major highway system. Data for the NHPN are provided and maintained by the Federal Highway Administration (FHWA). The NHPN consists of over 400,000 miles of the nation's highways, including those classified as rural arterials, urban principal arterials, and all NHS routes. Functional classes below arterial vary on a state-by-state basis. The data set covers the 48 contiguous states plus the District of Columbia and Puerto Rico. The nominal scale of the data set is 1:100,000 with a maximal positional error of ±80 m. The NHPN is also used to keep a map-based record of the NHS and the Strategic Highway Corridor Network (STRAHNET), which is a subnetwork defined for military transportation purposes.

Highway nodes are labeled with an identification number and located by geographic coordinates, FIPS code, and other location identifiers. Links are designated by the nodes located at each end, a scheme common to all of the databases discussed in this section, and also have identifiers such as a link name or code, sign route, and street name. Other link attributes include length, direction of flow permitted, functional class, median type, surface type, access control, toll features, and any special subnetworks (such as the NHS) to which the link belongs. Each link also has a shape point file.

The NHPN originated at Oak Ridge National Laboratory (ORNL), which has gone on to develop further and maintain its own version of a national highway network database, the Oak Ridge National Highway Network. This is nearly identical in structure and content to the NHPN. For details see http://www-cta.ornl.gov/transnet/Highways.html. Like the NHPN, this database is in the public domain.

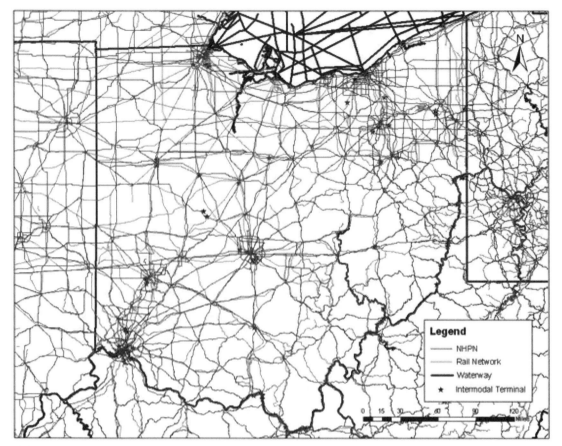

FIGURE 1.3 Extract from the National Transportation Atlas Database (2002).

1.3.2 Rail Networks

The Federal Railroad Administration (FRA) has developed and maintains a national rail network database. The BTS compiled and formatted the rail network data for release as part of NTAD 2002. The rail network (Figure 1.5) is a comprehensive data set of the nation's railway system at the 1:2,000,000 scale. The data set covers the 48 contiguous states plus the District of Columbia. Nodes and links are identified and located in the usual fashion. Link attributes include the names of all owning railroads and all other railroads that have trackage rights, number of main tracks, track class, type of signal system, traffic density class for the most recent year of record, type of passenger rail operations (e.g., Amtrak), and national defense status. FRA is also working on developing a 1:100,000 scale network, but that version has not yet been released.

As in the case of highways, ORNL also maintains and makes available its own version of the national railroad network database. This network is an extension of the Federal Railroad Administration's national rail network. In addition to the network attributes listed above, the ORNL rail network includes information on the location and ownership (including ancestry) of all rail routes that have been active since 1993, which allows the construction of

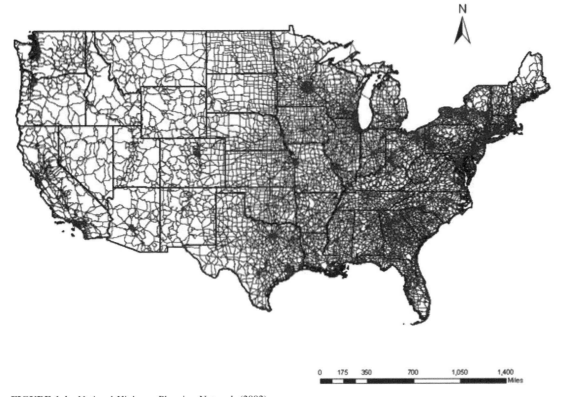

FIGURE 1.4   National Highway Planning Network (2002).

routable networks for any year since then. The geographic accuracy of this network is generally 100 m on active lines.

1.3.3   Waterway Network

The National Waterway Network is a comprehensive network database of the nation's navigable waterways. The data set covers the 48 contiguous states plus the District of Columbia, Puerto Rico, ocean routes for coastwise shipping, and links between domestic and international ocean routes and inland harbors. The majority of the information was taken from geographic sources at a scale of 1:100,000, with larger scales used in harbor/bay/port areas and smaller scales used in open waters. Figure 1.3 shows segments of the National Waterway Network database in and around the state of Ohio.

Links in the waterway network represent actual shipping lanes or serve as representative paths in open water where no defined shipping lanes exist. Nodes may represent physical entities such as river confluences, ports/facilities, and intermodal terminals, or may be inserted for analytical purposes. Approximately 224 ports defined and used by the U.S. Army Corps of Engineers (USACE) are geo-coded in the node database.

The National Waterway Network was created on behalf of the Bureau of Transportation Statistics, the USACE, the U.S. Census Bureau, and the U.S. Coast Guard by Vanderbilt

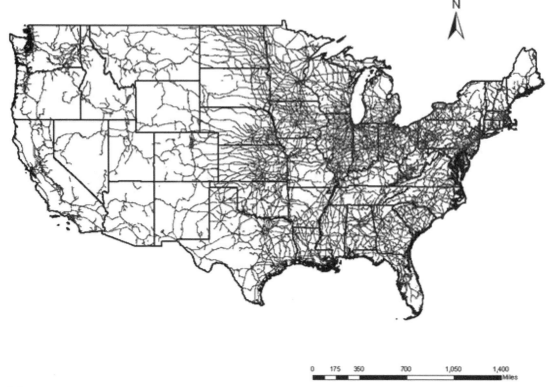

FIGURE 1.5 National rail network (2002).

University and Oak Ridge National Laboratory. Additional agencies with input into network development include Volpe National Transportation Systems Center, Maritime Administration, Military Traffic Management Command, Tennessee Valley Authority, U.S. Environmental Protection Agency, and the Federal Railroad Administration. In addition to its general uses, the network is used by the USACE to route waterway movements and compute waterborne commerce ton-miles for its *Waterborne Commerce* of the United States publication series.

1.3.4 Pipeline Networks

Pipeline network data are available from PennWell MAPSearch, an information provider to the oil, gas, electric, and related industries. Information is published as paper map and CD-ROM products, or licensed in either GIS or CAD formats. The oil and gas database provides pipeline logistical information, including diameter, owner/operator, direction of flow, storage terminals, gas processing facilities, refineries, truck loading/unloading, compressor/pump stations, marketing hubs and other facilities related to crude oil, LPG/NGL, natural gas, refined products, petrochemicals/olefins, and other petroleum-related commodities transported by pipeline. Further information is available at http://www.mapsearch.com/home.cfm.

The USDOT Office of Pipeline Safety (OPS) has underway a joint government-industry effort called the National Pipeline Mapping System. However, at this juncture it appears that the OPS project will not provide a public domain pipeline database, at least not in the near future.

1.4   MULTIMODAL NETWORKS AND INTERMODAL CONNECTORS

There are many applications of national network models that require consideration of traffic that uses more than one mode of transportation for travel between origin and destination areas. In most cases the exact routes and transfer locations of the individual movements are unknown, and hence a multimodal network model must be used to estimate these results. A good example is the processing system used to estimate ton-miles of traffic by commodity and mode for the national commodity flow surveys (CFS) conducted by the USDOT and the U.S. Census Bureau. The procedures used are described by Bronzini et al. (1996). The CFS collected information from shippers about specific intercity freight shipments, including the commodity, origin, destination, shipment size in tons, and the mode or modes of transportation used. Shipment distance by mode was not collected, so a multimodal network model was used to find routes through the U.S. freight transportation network, thereby allowing estimation of mileage by mode for each shipment in the survey. To allow for multimodal routings, the separate modal networks were connected at appropriate locations using intermodal transfer links.

Establishing analytically correct intermodal transfer links for a multimodal network is not a simple undertaking. To a first approximation, one could use GIS software to find nodes of different modes that are within some threshold distance of each other, and simply establish mode-to-mode connectors at all such locations. This, however, ignores the investment cost and special-purpose nature of intermodal transfer facilities, and tends to overestimate the number of intermodal connectors.

To assist with these types of applications, the NTAD includes a file called the Intermodal Terminal Facilities data set. The Oak Ridge National Laboratory developed the intermodal terminal facility data from which this database was derived. This database contains geographic data for trailer-on-flatcar (TOFC) and container-on-flatcar (COFC) highway-rail and rail-water transfer facilities in the United States. Attribute data specify the intermodal connections at each facility; i.e., the modes involved in the intermodal transfer, the AAR reporting marks of the railroad serving the facility, the type of cargo, and the direction of the transfer. These latter two attributes are extremely important. Even though two modes may have an intermodal connection at a given point, it does not follow that all commodities carried by the two modes can interchange there. Typically, each such connector handles only one commodity or type of commodity. For example, a coal terminal will not usually handle grain or petroleum products. Further, the transfer facility may serve flows only in one direction. A waterside coal transfer terminal, for example, may allow dumping from rail cars to barges but may not provide facilities for lifting coal from barges into rail cars. These examples illustrate why a simple proximity analysis method is unlikely to yield correct identification of intermodal connector links.

Attribute data for the Intermodal Terminal Facilities data set were extracted from the Intermodal Association of North America (IANA) 1997 Rail Intermodal Terminal Directory, the Official Railway Guide, the TTX Company Intermodal Directory, the Internet home pages of several railroads, the U.S. Army Corps of Engineers Port Series Reports, Containerization International Yearbook, the 1996 Directory of the American Association of Port Authorities (AAPA), and various transportation news sources, both in print and on the Internet. Attribute data reflect conditions at TOFC/COFC facilities during 1995–96 and are subject to frequent change. The database does not include TOFC/COFC and marine container facilities known to have been closed before or during 1996. However, because of the frequent turnover of this type of facility, some of the terminals included in the database may now be dormant or permanently closed.

The locations of TOFC/COFC facilities were determined using available facility address information and MapExpert, a commercial nationwide digital map database and software package, and recording the longitude/latitude of the approximate center of the facility. Facility locations are not bound to any current or previous highway, railway, or waterway network models. This is an advantage in that the facility locations in the database will be unaffected by changes in the other networks. Figure 1.3 shows some of the intermodal terminals that are included in the NTAD.

Further work for the CFS has validated the use of modal and multi-modal networks for national and regional commodity flow studies. A recent paper by Qureshi, Hwang, and Chin (2002) documents the advantages.

1.5 NETWORK MODEL APPLICATIONS

Section 1.2.2 briefly described use of transportation network models for national-level studies, an area of activity that dates back more than 20 years. Recent transportation studies carried out by states and Metropolitan Planning Organizations (MPOs), however, demonstrate that this type of analytical work is now within the reach of engineers and planners at those levels.

The prototypical use of network modeling at the state level is for statewide transportation planning. Horowitz and Farmer (1999) provide a good summary of the state-of-the-practice. Statewide passenger travel models tend to follow the urban transportation planning paradigm, using features such as separate trip generation and trip distribution models, and assignment of traffic to a statewide highway network. Michigan has one of the most well-developed statewide passenger models (KJS Associates, Inc. 1996). Statewide freight models also tend to follow this paradigm, with a focus on truck traffic on highways. Indiana (Black 1997) and Wisconsin (Huang and Smith 1999; Sorratini 2000) have mature statewide freight models, and Massachusetts (Krishnan and Hancock 1998) recently has done similar work.

Sivakumar and Bhat (2002) developed a model of interregional commodity flows in Texas. The model estimates the fraction of a commodity consumed at a destination that originates from each production zone for that commodity. The model includes the origin-destination distances by rail and truck, which were determined using the U.S. highway and rail networks that are included in TransCAD.

Work by List et al. (2002) to estimate truck trips for the New York City region is representative of freight network analysis activity at the MPO level. The model predicts link use by trucks based on a multiple-path traffic assignment to a regional highway network composed of 405 zones, 26,564 nodes, and 38,016 links. The model produced an excellent match between predicted and observed link truck volumes ($R^2 > 95\%$).

Switching back to the national level, Hwang et al. (2001) produced a risk assessment of moving certain classes of hazardous materials by rail and truck. They used national rail and highway network routing models to determine shipping routes and population densities along the routes for toxic-by-inhalation chemicals, liquid petroleum gas, gasoline, and explosives. Their work is fairly representative of network-based risk assessment methods. They assessed the routing results as follows: "Although the modeled routes might not represent actual routes precisely, they adequately represented the variations in accident probability, population density, and climate that characterize the commodity flow corridors for each hazardous material of interest." A similar statement could be made about most transportation network analysis results.

1.6 ACKNOWLEDGMENT

The figures in this chapter were prepared by Mr. Harshit Thaker.

1.7 REFERENCES

Black, W. R. 1997. *Transport Flows in the State of Indiana: Commodity Database Development and Traffic Assignment, Phase 2.* Transportation Research Center, Indiana University, Bloomington, IN, July.

Bronzini, M. S., S. Chin, C. Liu, D. P. Middendorf, and B. E. Peterson. 1996. *Methodology for Estimating Freight Shipment Distances for the 1993 Commodity Flow Survey.* Bureau of Transportation Statistics, U.S. Department of Transportation.

Bronzini, M. S., A. F. Hawn, and F. M. Sharp. 1978. "Impacts of Inland Waterway User Charges." *Transportation Research Record* 669:35–42.

Bronzini, M. S., D. P. Middendorf, and R. E. Stammer, Jr. 1987. "Analysis of the Transportation Elements of Alternative Logistics Concepts for Disposal of Spent Nuclear Fuel." *Journal of the Transportation Research Forum* 28(1):221–29.

Bureau of Transportation Statistics (BTS). 1997. *Truck Movements in America: Shipments From, To, Within, and Through States.* BTS/97-TS/1, Bureau of Transportation Statistics, U.S. Department of Transportation, Washington, DC, May.

———. 1998. *Truck Shipments Across the Woodrow Wilson Bridge: Value and Tonnage in 1993.* BTS/98-TS/3, Bureau of Transportation Statistics, U.S. Department of Transportation, Washington, DC, April.

———. 2002. *National Transportation Statistics 2002.* BTS02-08, Bureau of Transportation Statistics, U.S. Department of Transportation, Washington, DC.

———. 2003. *Pocket Guide to Transportation.* BTS03-01, Bureau of Transportation Statistics, U.S. Department of Transportation, Washington, DC.

Federal Highway Administration (FHWA). 2001. *Our Nation's Highways 2000.* FHWA-PL-01-1012, Federal Highway Administration, U.S. Department of Transportation, Washington, DC.

Horowitz, A. J., and D. D. Farmer. 1999. "Statewide Travel Forecasting Practice: A Critical Review." *Transportation Research Record* 1685:13–20.

Huang, W., and R. L. Smith, Jr. 1999. "Using Commodity Flow Survey Data to Develop a Truck Travel-Demand Model for Wisconsin." *Transportation Research Record* 1685:1–6.

Hwang, S. T., D. F. Brown, J. K. O'Steen, A. J. Policastro, and W. E. Dunn. 2001. "Risk Assessment for National Transportation of Selected Hazardous Materials." *Transportation Research Record* 1763:114–24.

KJS Associates, Inc. 1996. *Statewide Travel Demand Model Update and Calibration: Phase II.* Michigan Department of Transportation, Lansing, MI, April.

Krishnan, V., and K. Hancock. 1998. "Highway Freight Flow Assignment in Massachusetts Using Geographic Information Systems." 77th Annual Meeting, Transportation Research Board, Washington, DC, January.

List, G. F., L. A. Konieczny, C. L. Durnford, and V. Papayanoulis. 2002. "Best-Practice Truck-Flow Estimation Model for the New York City Region." *Transportation Research Record* 1790:97–103.

Qureshi, M. A., H. Hwang, and S. Chin. 2002. "Comparison of Distance Estimates for Commodity Flow Survey; Great Circle Distances versus Network-Based Distances." *Transportation Research Record* 1804:212–16.

Sivakumar, A., and C. Bhat. 2002. "Fractional Split-Distribution Model for Statewide Commodity Flow Analysis." *Transportation Research Record* 1790:80–88.

Sorratini, J. A. 2000. "Estimating Statewide Truck Trips Using Commodity Flows and Input-Output Coefficients." *Journal of Transportation and Statistics* 3(1):53–67.

USDOT/USDOE (1980). *National Energy Transportation Study.* U.S. Department of Transportation and U.S. Department of Energy, Washington, DC, July.

CHAPTER 2
TRANSPORT NETWORK PLANNING: THEORETICAL NOTIONS

Ben Immers
Department of Civil Engineering,
Transportation Planning and Highway Engineering,
KU Leuven, Heverlee, Belgium

Bart Egeter
Netherlands Organization for Applied Scientific Research,
TNO Inro, Delft, The Netherlands

Rob van Nes
Faculty of Civil Engineering and Geosciences,
Transportation and Planning, Delft University of
Technology, Delft, The Netherlands

2.1 INTRODUCTION

Mobility is undergoing constant change, in terms of both volume and spatial patterns. The traffic infrastructure has to respond to this continual process of change. Where bottlenecks emerge, improvements can be made from a whole palette of measures, varying from traffic management and pricing to the expansion of capacity in stretches of road and junctions.

This kind of bottleneck-oriented approach has offered some degree of solace for some time, but occasionally the need arises to completely review and rethink the whole structure of the network: Does the existing structure come to terms with changing mobility patterns? Are structural modifications necessary, such as a reconsideration of the categorizing of roads and the associated road design, expanding the robustness of the network, disentangling traffic flows, or changing the connective structure of urban areas? In other words: there is a need to *redesign* the network. The problem of network design that emerges then is a very complex one which requires a consideration to be made of the (vested) interests of various parties.

In The Netherlands a methodology has recently been developed for the integral design of the transport networks of different modalities.* In this the focus lies on networks on a

---

*The IRVS design method (Egeter et al. 2002), developed by the Netherlands Organization for Applied Scientific Research (TNO Inro) and sponsored by the Dutch Agency for Energy and the Environment (NOVEM).

regional scale. Parties that have worked with this methodology cite the following key features:

1. As a basis of the analysis, separate from the present infrastructure, an ''ideal network'' is designed.
2. Design occurs together with the stakeholders on the basis of clear, practicable steps.

By creating an ideal network separate from the network that is present, a very clear insight is gained into the structure of the network since it is not obscured by the existing situation which has emerged historically and therefore is not always ideal. Confronting this ideal situation with the existing situation will allow weaknesses in the structure to come to light. A second function of the ideal network is providing a long-term horizon within which short-term measures have to fit.

By reducing the theoretically highly complex design problem to a number of successive design steps or decisions, this methodology provides insight and is applicable in practical situations. What is important in this respect is that for each step there is commitment from the stakeholders before the next step is taken. It is, then, most effective when the methodology is used in a workshop-type situation whereby these parties themselves participate in the design process.

The result of the methodology is that stakeholders gain a clear picture of the crucial dilemmas and decisions. The methodology prevents thinking in terms of end solutions. Instead, the functions of the different parts of the network can be analyzed in terms of whether they actually fulfill the functions for which they were designed or to which they are now assigned. The function of a particular part of the network is thereby the leading factor for form and technique. Analysis may result in a whole palette of possible recommendations, from no action through traffic management, function adjustment coupled to modification of the road design and disentangling or expanding existing connections, to the construction of new junctions or new connections. This can be phased in, for instance by first applying traffic management and then in the longer term building new junctions or connections.

2.2 A FUNCTIONAL CLASSIFICATION OF TRANSPORT SYSTEMS

The approach described in this chapter is based on a classification of transport systems (ECMT 1998). This classification (see Table 2.1) is used to emphasize that what matters is the *quality* that is offered, not the modes and technologies used. It distinguishes five levels of scale (represented by their trip length) and two different types of organization (individual or collective transport). Roughly speaking, the term *individual systems* refers to road networks and the term *collective systems* refers to public transportation networks.

The design method focuses on the national (state) and regional level (I-3 and I-2, and C-3 and C-2), but is not limited to this level. Figure 2.1 shows the different subsystems, as well as the connections (the arrows in Figure 2.1) between different scale levels and between the individual and collective systems. The focus of this chapter is highlighted in gray.

2.3 KEY CHARACTERISTICS OF THE DESIGN METHOD

Designing successful transportation networks requires more than the application of the functional classification. In order to assist stakeholders in the design process, a step-by-step design process was set up. It is not a blueprint that tells stakeholders exactly what to do, merely a framework within which they make decisions. The stakeholders get to make the designs, but the method brings structure to the design process, by indicating which decisions

TABLE 2.1 Functional Classification of Transport Systems, by Scale Level and Organization Type

Scale level (trip length)	Individual, private transport	Collective, transport service supplied	Design speed	Accessibility; distance between access nodes
<1 km neighborhood	I-0 e.g., walking		0–10 km/h	
1–10 km district, medium-sized village, (part of) a town	I-1 e.g., bicycle, in-line skates, car on the local road network	C-1 e.g., local bus/tramway/ scheduled taxi service	10–30 km/h	0.2–1 km
10–50 km agglomeration, area, region	I-2 e.g., moped/scooter, urban car, car on regional road network (highway, expressway)	C-2 e.g., subway, light rail, commuter train service	30–80 km/h	2–5 km
50–300 km/h county, state	I-3 e.g., car on highway/ freeway network	C-3 e.g., long-distance train and bus services	80–200 km/h	10–30 km
>300 km/h state, interstate		C-4 e.g., high-speed train, airplane, Greyhound bus	>200 km/h	60–150 km

1 km = 0.62 mile; 1 mile = 1.61 km.

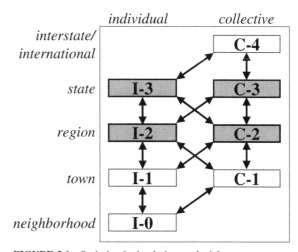

FIGURE 2.1 Scale levels the design method focuses on.

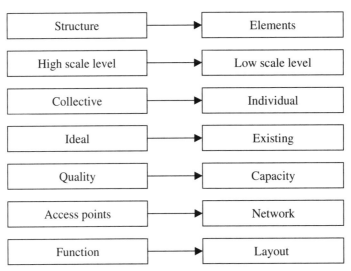

FIGURE 2.2   Main characteristics of the design methodology (in random order).

need to be made at what point in the process. It is based on a number of important characteristics, which are listed in random order in Figure 2.2.

2.3.1   First Structure, then Elements

First, a perspective on the complete *structure* of the network must be developed, such as which cities must be connected by the network, which scale levels are distinguished, etc. Only then can a decision be made about the *elements* (road sections, junctions, and routes/alignment). In practice, problems are usually solved at the element level: bottleneck by bottleneck. This kind of bottleneck-oriented approach has offered some degree of solace for some time, but occasionally the need arises to completely review and rethink the whole structure of the network: Does the existing structure come to terms with changing mobility patterns? Are structural modifications necessary, such as a reconsideration of the categorizing of roads and the associated road design, expanding the robustness of the network, disentangling traffic flows, or changing the connective structure of urban areas?

2.3.2   First the Higher Scale Level, then the Lower Scale Level

Networks for every scale level are designed independently, following a top-down approach: from the higher to the lower scale level, with a feedback loop bottom-up. Each network is designed to meet its functional requirements optimally. In order to achieve coherence between networks of different scale levels, access points of higher scale level are automatically included in the lower scale level.

2.3.3   First the Collective Networks, then the Individual Networks

Access to collective transport systems is much more cumbersome than access to individual transport, and therefore the situation of the access points of the collective system (public transport stops) requires more careful consideration than the situation of access points of the individual networks (e.g., highway and freeway entry points). This is because in the case of collective transport, unlike individual transport, access and egress by lower-level transport either requires physical transfers from and to other modes or takes place on foot. Therefore, important public transport nodes are preferably situated within a short distance of main origin and destination points. When integrating collective and individual networks (per scale level), the collective transport system receives priority in the design, for instance when it comes to the situation of intermodal transfer points.

2.3.4 First Ideal, then Existing

First, an ideal network is designed, ignoring the existing network. Subsequently, this ideal structure is confronted with the existing situation. The actions that need to be taken to change the existing situation into the ideal situation can then be prioritized. This way, improvements in the existing networks will be coherent; the ideal structure serves as a long-term perspective.

2.3.5 First Quality, then Capacity

The desired level-of-service, or *quality,* of the connections in the network needs to be defined clearly. Quality concerns characteristics such as speed, reliability, and comfort, but also pricing policies and traffic management strategies that are applied to the network. An acceptable volume-capacity ratio (*capacity*) is a prerequisite, but capacity should be considered separately from the desired quality. In practice, capacity is more often than not the primary aspect, which means that quality aspects receive less attention.

2.3.6 First Access Points, then the Network

A transport network serves to connect access points. Therefore, it is logical to define first which access points should be connected and then design the connections between these points (the network). In practice, it is often done the other way around. A well-known example in Europe is the discussion about which cities should get high-speed train stations on the line from Amsterdam to Paris. Whether or not the train was going to stop in The Hague, a decision that should have been made before a route was chosen, became dependent on the choice for one route or the other.

2.3.7 First Function, then Layout and Technique

Before the layout of the various components of the networks (access points, links, and junctions) is defined, it must be clear what the function of this component is. By gearing the layout to the functional requirements, it is more likely that this road will be used in accordance with the objectives set for this road. As a consequence, changing the function of a road (e.g., from national to regional) can lead to changing the layout (e.g., from highway to regional main road). The same principles apply to collective networks. For example, the choice between bus and rail should depend on the function; in some cases both techniques can meet the requirements.

2.4 DILEMMAS ENCOUNTERED IN DESIGNING A TRANSPORTATION NETWORK

A transportation system is made up of links, nodes, and a number of other design variables. Designing a transportation system is then a matter of assigning values to each variable. This sounds simple, but in practice, because of the different objectives set (by transport authorities, services providers and users), there will always be conflicting variables, resulting in so-called design dilemmas. The design method distinguishes four major design dilemmas:

1. The number of systems: differentiation versus cost reduction

2. Access point density: quality of a connection versus accessibility

3. Access structure: accessibility versus differentiation in use

4. Network density: quality of a connection versus cost reduction

These dilemmas are implicitly processed in the functional categorization used in transport systems.

2.4.1 Dilemma 1: The Number of Systems

Several subsystems make up the total transportation system (see Table 2.1). Having several subsystems makes it easier to fulfill the different functions a system may have. The more subsystems, the better their functions can be geared towards the needs of the traveler. Thus, offering more subsystems increases the user benefit. On the other hand, reducing the number of subsystems means reducing the investor costs, as this means the capacity offered can be used more efficiently. A practical example of this dilemma is the question of whether short- and long-distance travel should be combined on the same ring road: this means a high-quality road for short-distance travel, but disturbance of the long-distance traffic flow caused by the short distance between access points. In general, more subsystems can be offered in more urbanized areas, where the transport demand is higher.

2.4.2 Dilemma 2: Access Point Density

For any given subsystem, there is the question whether there should be few or many access points. The more access points, the better its accessibility. This means that a smaller part of the trip needs to be made on the lower-scale-level (and therefore slower) networks. On the other hand, the quality of connections (how fast, and how reliable from one access point to another) provided by the subsystem is higher when there are few access points. This dilemma plays a major role in the design of public transport networks, but it is also becoming more and more important in road networks. In many countries, long-distance traffic often encounters congestion near urbanized areas caused by regional or even local traffic entering and exiting the freeway and frequently causing disturbances in doing so. In general, higher-scale-level networks have fewer access points—this has to do with the fact that access points are usually found near cities, and fewer cities will be connected to the higher order networks.

2.4.3 Dilemma 3: Access Structure

Apart from defining the ideal structure of the connections between towns, there is the question of where to situate the access points: one access point in the middle (as is usual for train stations), or one or more at the edges of the built-up area (as is usual for through roads). The first option maximizes the accessibility of the system, but this often leads to misuse of the system by traffic that could use a lower-order network. It may affect livability in the area, and it undermines the intended differentiation in systems. Although this dilemma plays a role in individual as well as in collective systems, the outcome of the question is different for each type:

- In the collective systems, the access point is preferably situated in the center of the urban area. This is because changing from one collective system to another always involves a physical transfer (from one vehicle to the next). Transfers should be kept at a minimum, which means that it is desirable to concentrate access points of all collective subsystems in one location.

- In contrast, a transfer from one individual system to the next is almost seamless: passengers do not change vehicles. With livability issues in mind, access points are usually planned outside built-up areas. This also helps in fighting the undesired use of through roads (and sometimes congestion) by short-distance traffic.

2.4.4   Dilemma 4: Network Density

Once it has been established which cities need to be connected, it still has to be decided whether these cities should be connected by direct links or by way of another city. More links means higher-quality connections because there are fewer detours. In public transport, however, limiting the number of links makes higher frequencies possible. Obviously, more links mean higher costs, not only in infrastructure investments but also in the effects on the environment.

What network density will be acceptable depends chiefly on two factors:

- The amount of traffic: high volumes justify the need for extra infrastructure.
- The difference of quality between two subsystems: a greater difference (in design speed) between scale levels means that a greater detour is acceptable when using the higher-order system.

2.5   FEASIBILITY OF DESIGN

In practice there will be a trade-off between the ideal network design and the realistic network design. The difference between both networks is mainly related to the resources that are available to lay the new infrastructure. The term *feasibility of design* has to be interpreted, however, in relation to the gradual development of a network and the wish to have a long-term view. On the basis of such a view of the ideal structure of the network, the various investment steps can be better substantiated and the network coherence better guaranteed. The absence of a long-term view results in an incoherent bottleneck approach that poses questions. The risk is then considerable that all kinds of short-term utilization measures will form the basis of a long-term infrastructure policy.

2.6   THE DESIGN PROCESS

2.6.1   Rules of Thumb

Designing means making certain choices with regard to each dilemma. To help the designer, the design method includes a number of rules of thumb. Certain values to variables are proposed (different for each scale level), and the designer is free to use or discard these values. Per scale level we have defined what the optimal values are for:

- The number and size of the cities the network is meant to connect
- The expected travel distance over the network
- The desired distance between access points
- The desired distance between (center of) built-up area and access points
- The acceptable detour factor (the distance traveled over the network divided by the distance as the crow flies)

These variables determine, to a large extent, what the design is going to look like. Moreover, the design sessions held so far have shown that these variables are strongly interconnected within a scale level. Inconsistent combinations of values for these variables lead to inefficient networks. The optimal values (derived from the design speed for each scale level) depend on local circumstances.

2.6.2   The Design Method Step by Step

Applying the design method results in designs for the collective and individual networks for each scale level distinguished and the interchange points where the networks are connected. Every network at every scale level is designed independently, thereby ensuring that each network is optimally geared towards its function. Possibly, in a later stage of the design process, some of the connections from different scale levels will be combined on one route, or even on one road or railway line. In that case, however, it is a conscious choice, a trade-off between the advantages and disadvantages of combining functions on that particular connection. Because the situation of the access points for the collective systems is much more important than for the individual systems, the collective network for a scale level is always designed before the individual network.

Step 1: Distinguish Urbanization Levels (Urban/Rural).   The edges of urban areas provide good locations for intermodal transfer points, so the border between urban and rural area must be indicated on the map for later use.

Step 2: Define the Hierarchy of Cities and Towns.   In this step, the rule of thumb for the number and size of the nodes (cities and towns) (Figure 2.3) the network is meant to connect is used to define which towns should be accessible via the network, and in what order of importance. In doing so (for the scale level under consideration), first-, second-, and third-level nodes are selected and indicated on the map. Large cities are split up into several smaller units.

Step 3: Design Desired Connections.   The desired connections (heart-to-heart) are drawn on the map (Figure 2.4), according to the following rules:

- First connect first-level nodes.
- Add connections to second-level nodes.
- Include third-level nodes when they are close to an already included connection. When adjusting a connection to include a third-level node, one should check that this does not result in unacceptable detours in the network.

Step 4: Design the Ideal Network.   This is the most difficult and intuitive stage in the design method. The existing situation must be ignored. The desired connections must be translated into an efficient network with the right density. The access points must be put in the right place. Step by step this stage involves, for the individual network:

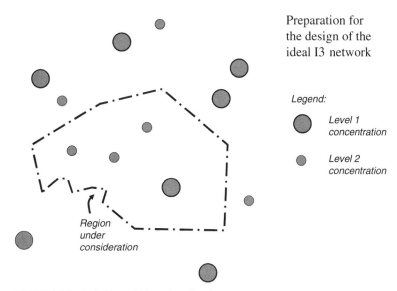

FIGURE 2.3 Definition of hierarchy of nodes.

1. For the super-regional scale levels: drawing circles around first- and second-level concentrations to indicate the desired distance between built-up area and through roads
2. Identifying main flow directions past first-level concentrations (at which side of town should the road pass)
3. Defining the optimal routes past concentrations (accessibility structures)

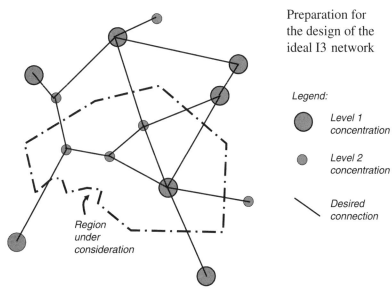

FIGURE 2.4 Drawing heart-to-heart connections.

4. Connecting the selected concentrations

5. Checking to see whether the network density is right and detours in the network are acceptable; if not, adding (or removing) connections

The result of sub-steps 1–3 is illustrated in Figure 2.5 (based on a design of an I3 or national road network for a province in The Netherlands).

Sub-steps 4 and 5 result in an ideal I3 network as depicted in Figure 2.6. It must be noted that many other designs are possible; the network in Figure 2.6, however, is the one that resulted for this region. This I3 (national) network formed the basis for the regional network that was subsequently designed.

The process is less complicated for the collective network because the stops should be as much in the center of the built-up area as possible.

Step 5: Assess Current Network.   The ideal network will differ from the existing network in several aspects:

- The connections that have been included
- The major traffic flows (which have implications for the layout of the interchanges of roads)

Step 5 has been included to assess how much of the existing network meets the requirements set by the method. This is done by looking at the existing connections that would most likely serve as a connection in the ideal network. The information gathered here can be used in a later stage, when it must be decided which part of the ideal network is given up in order to create a feasible network or to establish which parts of the network should be adapted first.

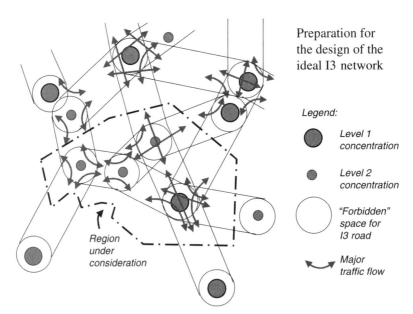

FIGURE 2.5   Sub-steps 1–3 in the design of the ideal network.

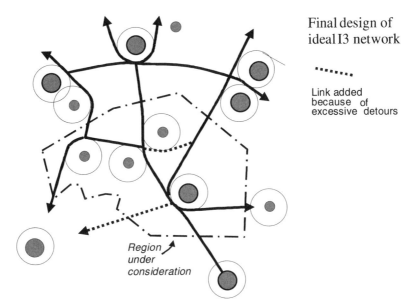

Final design of ideal I3 network

. . . .

Link added because of excessive detours

Region under consideration

FIGURE 2.6   Final design of ideal I3 network.

Design requirements to look at include:

• The distance between access points (too small?)
• The design speeds (too high? too low?)
• Requirements with respect to a logical layout of the network (do the through lanes at interchanges cater to the major flows?)

This step results in a map with connections on, over, or under the desired level of service and illogical points in the network.

Step 6: Design Realistic Network.   We now have an ideal network and an assessment of where the existing network falls short of the ideal network. It must now be decided what is an acceptable amount of new infrastructure. Also, the individual and collective networks must be connected to each other. Likewise, the networks of the different scale levels must be connected. This means:

• Selecting routes: following the ideal or existing network
• Choosing main flow directions (so illogical points will be avoided)
• Selecting access points for collective and individual networks of all scale levels and for connecting collective and individual networks

Depending on the time horizon chosen, a realistic network can be selected that is closer to either the existing or the ideal network. In our case two variants have been elaborated in this manner. Policy-makers were quite pleased with the design that stayed closer to the ideal network. It gave them many new ideas for their long-term plans. Interestingly, when the effects of these two designs were evaluated, it was found (with the help of an integrated land use and transport model) that the second design performed better in many respects (i.e., was more sustainable).

2.7 THEORETICAL BACKGROUND

This section discusses some theoretical issues related to the network design methodology, including the network design problem, hierarchical network structures, and some special issues.

2.7.1 Network Design Problem

A network consists of access nodes, nodes, and links connecting these nodes. In the case of transit networks, lines are included as well. The network design problem in its simplest form is to find a set of links that has an optimal performance given a specific objective. Basically, there are two kinds of network design problems:

1. Designing a new network, for instance a new higher-level network or a transit network
2. Improving an existing network, for instance increasing capacities or adding new roads

In this chapter the focus is on designing a new network.

The network design problem is known to be a very complicated problem, for three reasons. First of all, there is the combinatorial nature of the problem. Given a set of access nodes the number of possible link networks connecting all access nodes increases more than exponentially with the number of access nodes. Therefore, there are no efficient methods available for solving large-scale network design problems.

Second, the perspective on the design objectives might be very different. The key conflict is that between the network user, i.e., the traveler, and the investor or network builder. The traveler prefers direct connections between all origins and destinations, while the investor favors a minimal network in space (see Figure 2.7). There are three methods to reconcile these opposing perspectives:

1. Formulating an objective that combines the interests of both parties involved. Typical examples of such design objectives are maximizing social welfare and minimizing total costs.

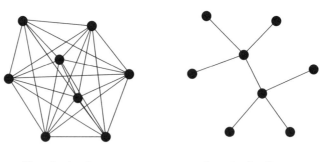

Travelers' optimum   Investors' optimum

FIGURE 2.7 Illustration of the difference in optimal network structures between the traveler's and the investor's point of view.

2. Focusing on the perspective of one of the parties, usually the traveler, while using the second perspective as a constraint, e.g., minimizing travel time given a fixed budget.
3. Again choosing a specific objective, in this case usually the investor's perspective, but at the same time taking into account the behavior of the other party involved, i.e., the traveler. An example of this approach is a transit operator maximizing profit while considering the fact that inadequate services will reduce patronage and thus revenues.

Third, there is a strong relationship between the demand for transport networks and transport networks themselves. Changes in transport networks lead to changes in travel behavior, and changes in travel behavior set requirements for the transport network. As such, the network design problem can be seen as a Stackelberg game in which one decision-maker, i.e., the network designer, has full knowledge of the decisions of the second decision-maker, the traveler, and uses this information to achieve his own objectives (see Figure 2.8).

These three complicating factors, combinatorial nature, conflicting perspectives, and relationship between transport network and transport demand, explain the huge amount of literature on transport network design. Most of the scientific research deals with mathematical models that can be used to solve the network design model. For transport planners, however, design methodologies such as presented in this chapter are more suitable.

2.7.2 Hierarchical Network Structures

Hierarchy as a Natural Phenomenon. It can easily be demonstrated that hierarchy is a common phenomenon in transport networks. Let us assume a perfectly square grid network where all origins and destinations are located at the crossings, all links being equal in length and travel time. The demand pattern is uniformly distributed, that is, at every origin the same number of trips start in all directions having the same trip length, leading to the same number of arrivals at all destinations coming from all directions. Since it is a grid network, the traveler may choose between a number of routes having the same length and travel time. In this hypothetical situation no hierarchies in demand or supply are assumed and at first sight no hierarchy in network usage results. However, if small deviations to these assumptions occur, a process is started that leads at least to a hierarchical use of the network. Examples of such small changes are:

- Travelers might prefer specific routes, even though all routes are equal in time and length from an objective point of view. Such a preference might be due to habit, to the traveler's own perception of the routes or perception regarding the crossings, or to information provided by other travelers.
- Link characteristics might differ slightly, leading to objective differences in route characteristics.

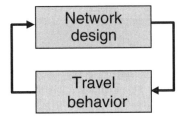

FIGURE 2.8 Network design problem as a Stackelberg game.

- Travelers might prefer to travel together, bringing in the stochastic element of travelers passing by and having an overlap with one of the possible routes.
- Some origins and destinations might be more attractive than others.

All of these deviations have the same effect regardless of the size of the change: namely, some routes will become more attractive than others. This effect is mainly caused by the demand side of the transport system. The higher usage of some routes, however, also influences the supply side of the transport system. In the long run the most intensively used routes will receive better facilities and become more attractive, while the less used routes will be neglected. The supply side of the transport system thus strengthens the hierarchy started by the demand side. In fact, the process described here is an example from economics based on increasing returns (see, e.g., Waldrop 1992; Arthur, Ermoliev, and Kaniovski 1987), which is a fundamental characteristic in all kind of evolutionary processes, be they in economics or in biology. The final result in this case is a hierarchical network structure consisting of two link types; in other words, a higher-level network is superimposed on the original lower-level network.

Hierarchy in settlements stimulates hierarchical network structures. Furthermore, the introduction of faster modes speeds up the processes leading to hierarchical networks. Similarly, hierarchical transport networks lead to concentration of flows, and if these flows are large enough they allow for more efficient transport, leading to lower travel costs per unit traveled (economies of scale), and reduce negative impact on the environment, which also stimulates the development of hierarchical network structures. Hierarchical networks are thus a natural phenomenon resulting from the interaction between demand and supply that, due to technological developments and modern decision processes focusing on environmental impact, are becoming more common in transport networks (see Figure 2.9).

Development of Hierarchical Network Structures.   The main process, that is, the interaction between demand and supply, might have self-organizing characteristics. Many networks, however, have been developed over a long period of time and are, therefore, influenced by

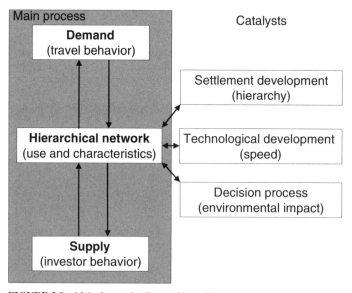

FIGURE 2.9   Main factors leading to hierarchical networks.

many factors. Hierarchy in spatial structure has always been such a factor. The importance of technology has substantially increased in the last two centuries. Rail networks were developed early in the 19th century and were a true accelerator for hierarchical network development in transport networks and spatial structures. The introduction of high-speed trains today will have a similar effect. The introduction of the private car at the beginning of the 20th century led to more ambiguous developments. Private cars improved space accessibility and thus had a reverse effect with respect to spatial structure. At the same time, however, the private car allowed substantially higher speeds given the quality of the infrastructure, and can thus be seen as an accelerator for hierarchical road network development. In the second half of the 20th century a strong focus on planning processes, especially with regard to environmental impact, and the concept of bundling of transport and thus of infrastructure became dominant issues. Hierarchical networks can therefore be seen as a result of a continuous interaction process between demand and supply, which has a strong correlation with spatial development and is influenced over time by other developments such as technological advances and decision processes.

Hierarchical Network Levels. A hierarchical network structure is a multilevel network in which the higher-level network is characterized by a coarse network, limited accessibility, and high speeds, and is especially suited for long-distance trips. The lower-level networks are intricate and have high accessibility and low speeds, making them suitable for short-distance trips and for accessing higher-level networks. It can be shown that the hierarchy in transport network levels is linked with the hierarchy in settlements (Van Nes 2002). Each network level then offers connections between cities of a specific rank and offers access to cities and networks of a higher rank. Figure 2.10 shows this concept as proposed for the German road network guidelines (FGSV 1988). Table 2.2 presents a classification for road networks as proposed by Van Nes (2002). Please note that presently no higher speeds are possible for the two highest network levels. These network levels will therefore need more attention with respect to directness and traffic quality, i.e., reliability. For transit networks, however, high-speed trains really make it possible to provide higher network levels.

Plausibility Scale Factor 3. A logical criterion for a higher-level network is that the lower-level network will not be considered as an alternative for a trip using the higher-level network. Another way of formulating this criterion is the elimination of shortcuts, which is a

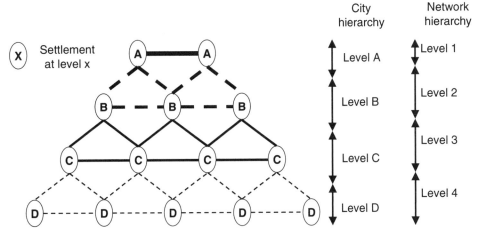

FIGURE 2.10 Road network structure according to Schönharting and Pischner (1983).

TABLE 2.2   Classification for Road Networks

Network level	Spatial level	Distance (km)	Road spacing (km)	Speed (km/h)
International	Metropolis	>300	300	*
National/Interstate	Agglomeration	100–300	100	*
Interregional/freeway	City	30–100	30	100–120
Regional highway	Town	10–30	10	60–70
Local highway	Village	3–10	3	35–40

*Theoretically these network levels should have higher speeds; however, these are not yet technically feasible.
Source: Van Nes (2002).

criterion that is primarily based on the traveler's perspective. A possible approach in this case is to look at the maximum detour for a shortcut in a single grid. This detour determines the necessary difference in travel speed between the network levels. The most realistic scale factor for road spacing can be found by calculating this travel speed ratio for a set of scale factors and selecting the scale factor resulting in the lowest travel speed ratio. The choice for the lowest value is based on the intuitive notion that the lower the travel speed ratio, the easier it will be to develop a higher-level network.

In this approach, the only assumption that is necessary is that the trip length be equal to or longer than the road spacing of the higher-level network. Within a grid network the trip having the maximum detour using the higher-level network can be defined as the trip between two nodes that are located at the middle of two opposing sides of the grid of the higher-level network (see Figure 2.11a). In the case that the scale factor, *sf*, for road spacing is uneven, this trip is located between two nodes as close to the middle as possible (Figure 2.11b). The trip distance using the lower-level network is always equal to the road spacing of the higher-level network.

If the scale factor, *sf*, for the road spacing is even, the trip distance using the higher-level network is twice as large, which implies that the travel speed for the higher-level network should be at least twice as high in order to have a shorter travel time using the higher-level network. This implies that in this case no choice for the most realistic scale factor can be made. In case *sf* is uneven, the trip distance for the higher-level network becomes $(2(sf - 1)/sf$ as large. In order to have a shorter travel time using the higher-level network, the travel speed should increase accordingly. It can easily be shown that the smallest increase of travel

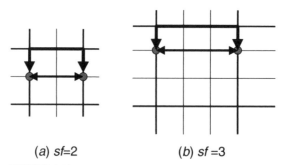

(a) sf=2   *(b) sf =3*

FIGURE 2.11   Maximum detours in a hierarchical grid network.

speed is found if the scale factor for road spacing *sf* equals 3: the speed of the higher level network then is 1.67 times the speed of the lower level network. As *sf* increases the necessary increase in travel speed converges to a factor 2. In both cases the maximum travel speed ratio is 2. Apparently it is not necessary to have larger travel speed ratios to avoid short cuts.

This analysis clearly shows that the existence of a scale factor 3 for the road spacing of hierarchical road networks can be explained using a simple and plausible mechanism based only on network characteristics. The corresponding scale factor for network speed is 1.67 and should not be larger than 2.

2.7.3 Special Issues

Steiner Nodes.   When building a network, planners usually consider only the nodes that have to be connected, i.e., cities or agglomerations. However, it might be an interesting option to introduce extra nodes that make it possible to reduce network length and thus investment costs. The impact of these so-called Steiner nodes is illustrated in Figure 2.12.

On the left-hand side we have four nodes that have to be connected. Using only these four nodes a grid network might be a proper solution. Introducing an additional node in or near the center, however, reduces the network length significantly (about minus 30 percent) while travel times are reduced in some cases and increased in other. The net effect on investment costs and travel times depend strongly on the demand pattern and the location of the additional node. Finally, it is possible to introduce an additional node where a specific road type ends, connected to the surrounding cities by links of lower-level networks.

Integrating Functions.   The notion of hierarchical transport networks is primarily functional. In urbanized areas, however, there is a strong tendency to integrate functionally different network levels within a single physical network. In urban areas the distance between access nodes for freeways, i.e., on- and off-ramps, is clearly shorter than in more rural areas. Integrating network levels might be attractive since they reduce the necessary investments. There is, however, an important pitfall for the quality of the transport network on the long run (Bovy 2001). Medium- and short-distance trips that theoretically would be served by lower-level networks experience a higher quality due to the higher accessibility and higher speed of the higher-level network. This higher quality influences all kinds of traveler choices, such as location, destination, mode, and route. The net result will be a relatively large increase of these medium- and short-distance trips using the freeway network, in quantity as well as in trip length. The resulting congestion reduces the transport quality for the long-distance trips for which the freeway network was originally designed. In some cases, the impact on location choice of individuals and companies might even limit the possibilities to increase the capacity in order to restore the required quality for long-distance trips. This unwanted impact of integrating functions requires special attention when planning higher-level networks in urbanized areas.

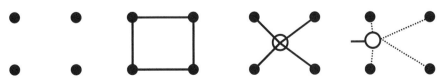

FIGURE 2.12 Application of Steiner nodes.

2.8 APPLICATION OF THE DESIGN METHOD

To illustrate the design method described, two concrete cases have been elaborated: design of the national network of Hungary and the federal network of the state of Florida. The following paragraphs present the results of the various design steps.

2.8.1 Hungary: Design of the National Road Network

See also Monigl (2002) and Buckwalter (2001).

Step 1: Hierarchy of Nodes. The first step in the design process is the decision how many (and which) nodes (cities) will have to be incorporated in the national network. Two approaches for this can be considered:

1. Based on distance classes (a quality approach)
2. Based on size of the various nodes (a user approach)

The national network is meant to be used for trips ranging from 50–300 km. Accommodating these trips adequately requires an optimal network density, and access point density and this should match the density of nodes. Example: if we apply the same density of nodes as used in The Netherlands, then we need to select approximately 40 nodes (30 in The Netherlands, as the size of the country is smaller). The consequence of this assumption is that we have to select nodes with a number of inhabitants of approximately 30,000.

In the second approach we assume that inclusion of a node in the national network is determined by the number of inhabitants. The number should be higher than 50,000. For the Hungarian situation this would result in the inclusion of 21 nodes.

Of course, the situation in Hungary differs from the situation in The Netherlands, e.g.,

* The size of the country (area): greater than The Netherlands
* The population density: lower than in The Netherlands
* The distribution of the population; quite unbalanced in Hungary, as the largest node (Budapest) has 1.8 million inhabitants and the second-largest node (Debrecen) only 210,000

It was decided to make a network design starting with 24 nodes (see Figure 2.13): minimum number of inhabitants per node is 40,000. The adopted approach is more or less a combination of the quality approach and user approach. The consequence of this is that the national network will not be used as intensively as in more densely populated countries like The Netherlands.

Step 2: International Connections. These connections have to be dealt with before designing the national network. Budapest is connected with the following large cities abroad (city and direction):

* Bratislava—direction Czech Republic and Slovakia, Poland
* Vienna—direction Austria, Germany
* Maribor—direction Slovenia
* Zagreb—direction Croatia/Slovenia/Italy
* Subotica—direction Serbia
* Arad/Timisoara—direction Romania

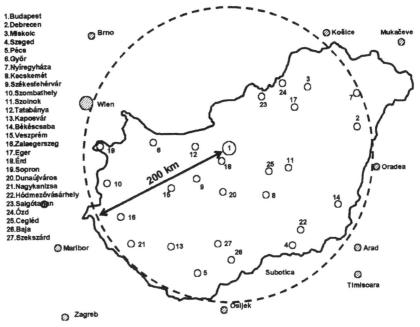

FIGURE 2.13 Nodes and ranking number.

- Oradea—direction Romania
- Mukaceve—direction Ukraine

Based on the criterion for accessibility, we also assume an international connection to the center of Slovakia (direction Zvolen). The result of this step is shown in Figure 2.14.

Step 3: Design of the Ideal National Network. The main structure of the national network is based on the international corridors and the 10 largest cities (central nodes) in Hungary (minimum of 80,000 inhabitants):

1. Budapest
2. Debrecen
3. Miskolc
4. Szeged
5. Pécs
6. Györ
7. Nyíregyháza
8. Kecskemét
9. Székesfehérvár
10. Szombathely

In addition to these nodes, Lake Balaton is also indicated as a rather important attraction node (especially during summertime). We further assume that all national connections are situated on national territory (no bypasses via neighboring countries).

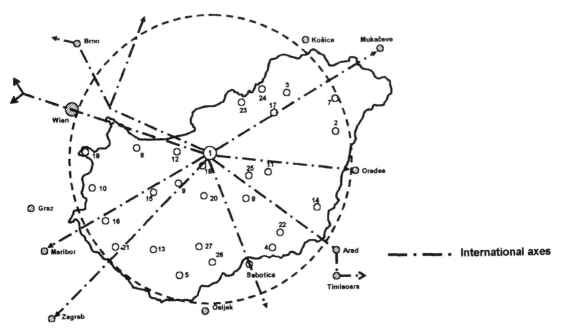

FIGURE 2.14 International axes.

Because of the dominant position of Budapest in the list of central nodes, the minimum spanning tree (connecting all cities) has a radial structure starting in Budapest.

Some remarks:

- Connection with Pécs: directly from Budapest (along river Danube via Szekszárd) or via Székesfehérvár. Because Szekszárd is a county capital, there is a strong preference for the link along the Danube.
- Only between Budapest and Oradea does a shortcut seem obvious; for all other connections the minimum spanning tree will do.

Now we add the nodes 11–24:

11. Szolnok

12. Tatabánya

13. Kaposvár

14. Békéscsaba

15. Veszprém

16. Zalaegerszeg

17. Eger

18. Érd

19. Sopron

20. Dunaújváros

21. Nagykanizsa

22. Hódmezovásárhely

23. Salgótarjan

24. Ózd

- Some of these nodes are already accessible via the main network structure.
- Additional links to nodes not yet connected are necessary in the western part (Sopron, Zalaegerszeg, and Nagykanizsa) and, if Pécs is connected via Székesfehérvár, we need to connect one node south of Budapest (Dunaújvaros). This further supports the realization of a direct connection with Pécs.
- In the northern part an additional link connects Budapest to Salgótarjan and Ózd (via Vác). In the eastern part of the country the international connection to Oradea goes via Cegled and Szolnok. Békéscsaba is connected to Kecskemét.
- In addition, we need to establish a few shortcuts:
  - South of Budapest: Gyor-Székesfehérvár- Dunaújvaros-Kecskemét
  - Along the southeast border: Pécs-Szeged-Hódmezovásárhely-Békéscsaba-Debrecen
  - Eger-Debrecen

The final result of this design step (the ideal national network) is presented in Figure 2.15.

Step 4: Analysis of (Comparison with) Existing Road Network (Including Roads under Construction). At present four motorway corridors are under construction (all starting in Budapest):

- Direction west: Gyor-Vienna and Gyor-Bratislava
- Direction northeast: Eger-Miskolc-Nyíregyháza-Debrecen

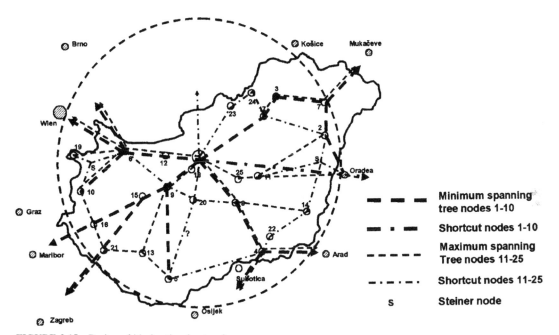

FIGURE 2.15 Design of ideal national network.

- Direction southeast: Kecsemét-Szeged-Serbia
- Direction southwest: Székesfehérvár-Balaton-Zagreb

This network still needs a few extensions:

- Direction northeast: extension in the direction Nyíregyháza-Debrecen (via Steiner node?)
- Direction southeast: extension to Szeged and Subotica (Serbia)
- Direction southwest: missing link along south bank of Lake Balaton to Croatian border via Nagykanizsa and to Slovenian border via Zalaegerszeg

After completion of this network the following regions are not yet connected:

- East corridor: Cegléd-Szolnok-Békéscsaba and possible extension to Arad/Timisoara (Romania)
- South corridor: direction Pécs (directly via Ráckeve and Dunaújvaros, or via Lake Balaton). This is important because Pécs is the fifth-largest city (pop. 170,000)
- The western part: Györ-Sopron and Györ-Szombathely (via Steiner node), Zalaerszeg
- The northern part: direction center of Slovakia and the nodes Salgótarjan and Ózd

Possible shortcuts:

- Szombathely-Zalaerszeg
- Along the southeast border: Pécs-Szeged-Hódmezovásárhely-Békéscsaba
- Szolnok-Debrecen-Oradea (via Steiner node)
- Kecskemét-Szolnok
- Ózd-Eger

Transit traffic Budapest:

- South ring road (connection between the motorways heading for Vienna, Lake Balaton, and Szeged): this part of the ring road already exists.
- East ring road (connection with motorway heading for Eger and further and the northern route towards Slovakia): this part does not exist yet.
- The availability of the two abovementioned ring roads, excludes (diminishes the necessity for) the construction of a ring road head to the south Györ-Székesfehérvár-Dunaújvaros-Kecskemét.

A realistic network design for the 10 largest nodes is shown in Figure 2.16.

National Roads. The question presents itself whether all suggested extensions to the existing motorway network and motorways under construction are of the same type. It does make sense to investigate whether some extensions/connections with low volumes (less than 1200 veh/hr/direction) could be constructed according to a lower standard, e.g., having the following characteristics:

- Dual carriageway (2*1) or one carriageway for both directions (with possibilities to overtake)
- Speed limit 90–110 km/hr
- Matching horizontal and vertical curve radius
- Limited number of intersections (access points), preferably grade-separated

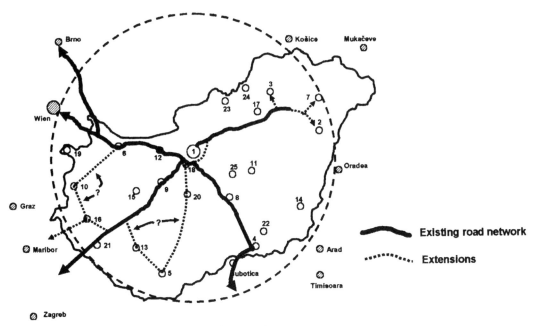

FIGURE 2.16 Realistic network design for 10 largest nodes.

Should the occasion arise, this type of road can combine the national and regional function. Figure 2.17 shows a further extension of the national network with lower order roads.

Step 5: Check on Opening-up Function (Is Not Applied). Besides the function of connecting economic centers, the transport system also needs to open up areas (provide access to as many travelers as possible this within an acceptable distance or timeframe). In this step additional access points are selected that contribute to this function. Possible criteria that can be used to find additional stops are:

- Ninety-five percent of the population should live within a specific distance (25 km for national network) of an access point; areas that are not yet served will get a connection if the area represents a least a specific number of inhabitants or departures/arrivals.
- Cities with a specific rank (e.g., county capital) lying outside the influence area of the network (e.g., 25 km or 30 minutes of travel from already existing access point).

Looking at the map and the network, there are a few areas (places) with poor access to the national road network, e.g., Jászberény, Esztergom and the northwest part of Lake Balaton. Some of these cities could be selected as a national node. Consequently, they should be given access to the network. This would result in a further extension of the national network.

2.8.2 Network Design for Florida

Step 1: Node Hierarchy. A summary of the municipalities with 100,000 inhabitants or more reveals that there are very few municipalities for level I3.

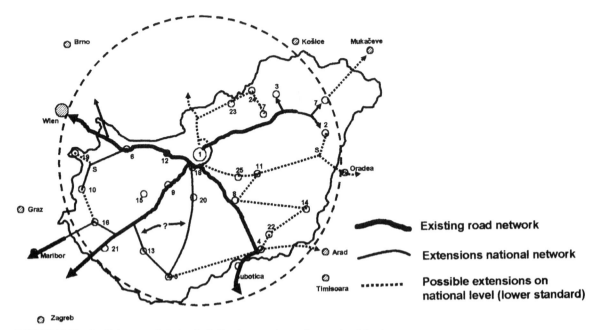

FIGURE 2.17 Realistic network design including lower-order roads at national level.

Municipality	Rank	Count
Jacksonville	1	735,617
Miami	2	362,470
Tampa	3	303,447
Saint Petersburg	4	248,232
Hialeah	5	226,419
Orlando	6	185,951
Fort Lauderdale	7	152,397
Tallahassee	8	150,624
Hollywood	9	139,357
Pembroke Pines	10	137,427
Coral Springs	11	117,549
Clearwater	12	108,787
Cape Coral	13	102,286

Source: Census 2000.

Analysis of the map also reveals the presence of clusters. This is particularly true of Miami (2, 5, 7, 9, 10, and 11) and the Tampa/Saint Petersburg/Clearwater cluster (3, 4, and 12). The number of nodes is considerably reduced by this, and so nodes with less than 100,000 inhabitants and that do not lie in clusters have also been examined. On this basis the following secondary nodes have been determined for the network:

14. Palm Bay/Melbourne (together 150,000)

15. Gainesville (95,000)

16. Port Saint Lucie (89,000)

17. West Palm Beach (82,000)

18. Lakeland (78,000)

19. Daytona Beach (64,000)

20. Pensacola (56,000)

With these nodes included a fair coverage of the area is achieved at the same time. It is also possible to include other criteria in selecting nodes, like employment or large tourist attractions (e.g., in Orlando). Figure 2.18 shows the selected nodes with ranking number.

Step 2: Interstate Connections.   These connections are focused on the largest (primary) nodes (Jacksonville (1), Miami (2), Tampa (3) and Orlando (4)) and the largest urban areas around Florida (New Orleans, Atlanta, and East Coast (towards Savannah)). A point of attention in the final design is the detour from New Orleans to southern Florida: it runs in this network via Jacksonville. The results of this step are shown in Figure 2.19.

Step 3: Ideal Network

Step 3a: Connections.   This step investigates how the other selected nodes can be incorporated in the ideal network by modifying the interstate connections or introducing new connections. Examples of possible network adaptations are Palm Bay/Melbourne (14) and Port Saint Lucie (16), Coral Springs (13), Lakeland (14), Tallahassee (8), and Pensacola (20).

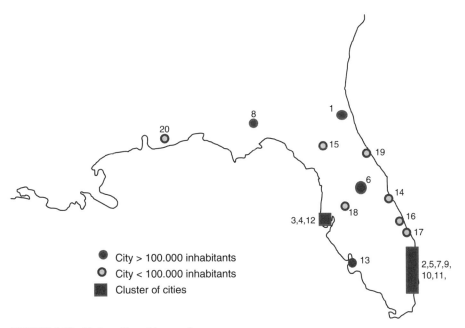

FIGURE 2.18   Nodes with ranking number.

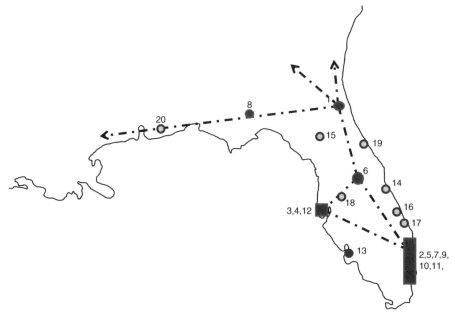

FIGURE 2.19   Interstate connections.

Extra connections are necessary to link up Gainesville (15). For the time being the preference is for the Gainesville–Jacksonville connection only. For Daytona Beach (19) possibly the connection with Orlando will suffice. The result is shown in Figure 2.20.

Step 3b: Accessibility of Nodes.   Analysis of the connections indicates quite clearly what the desired location of the I3 network at the nodes is:

• Jacksonville (1): westerly.
• Orlando (6): northerly from Orlando, splitting in a westerly branch (towards Tampa) and easterly branch (Miami). A point of attention is probably the accessibility of the tourist area southwest of Orlando in relation to Miami.
• Cluster Tampa/Saint Petersburg/Clearwater (3): easterly.
• Palm Bay/Melbourne (14): westerly.
• Miami (1): no through connection necessary, so only municipal access roads from the west and north, possibly westerly ring road.

Figure 2.21 presents the accessibility structure of the various nodes.

Step 3c: Ideal Network.   In the ideal network (see Figure 2.22) all nodes are included in the network, with the location around the nodes having no influence on the network. Discussion points/alternatives in respect of the ideal network are specifically the connections around Gainesville (15) and the Daytona Beach (19) and Jacksonville (1) connection.

Step 4: Confrontation of Existing Network.   In this step the relationship with the existing network and the acceptability of incorporating secondary nodes in this network are examined. In comparison with the existing work, there are three main points for discussion:

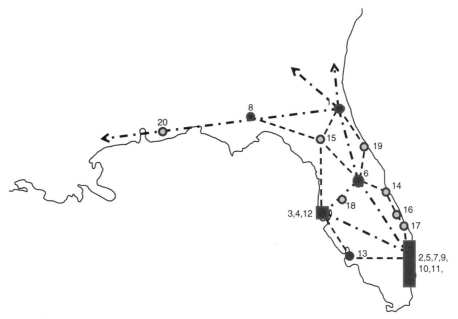

FIGURE 2.20   Ideal network: connections.

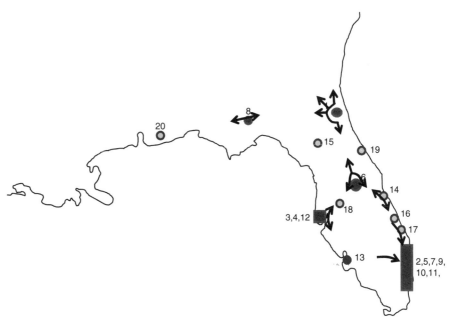

FIGURE 2.21   Accessibility structure of the various nodes.

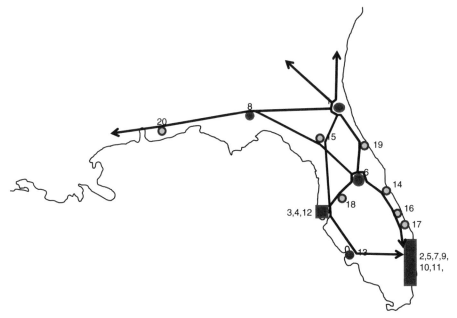

FIGURE 2.22 Ideal network.

- Position of Gainesville (15)
- Location of the I3 network at Orlando (6)
- Connection along the east coast

A direct connection from Orlando (6) to Atlanta via Gainesville (15) with a link from the Tampa cluster (3) reduces a number of detours, especially for the New Orleans–Tampa cluster and Tampa cluster–Atlanta connections, while no really new detours are created. In addition, this connection makes it possible to go from Jacksonville (1) via Daytona Beach (19) to Orlando (6), though the proposed Gainesville–Jacksonville connection does not have the required quality. This would need to be upgraded.

In Orlando (6) the I3 network lies relatively centrally. With the new structure northwards the branching structure proposed previously northwards from Orlando is less suitable. Keeping the existing structure would be suitable here if the number of on-/off-ramps in the neighborhood of Orlando could be restricted.

The Orlando (6)–Miami (2) connection can go via Port Saint Lucie (16). This does mean, however, that the Jacksonville (1)–Miami (2) connection has a clear detour and that the Palm Bay/Melbourne cluster (14) cannot be incorporated in the network. The alternative is that the Jacksonville (1)–Miami route runs completely along the coast with two special connections (municipal access roads) in the direction of Orlando from Daytona Beach (19) and Palm Bay/Melbourne (14). For the latter connection, the road to Cape Canaveral can be used.

Conclusion. The existing I3 network works well. The Interstate 75 Orlando–Atlanta in particular has a number of interesting advantages. Weak spots are the Gainesville–Jacksonville connection and the central location of the I3 network in Orlando. The final result (realistic network design) is shown in Figure 2.23.

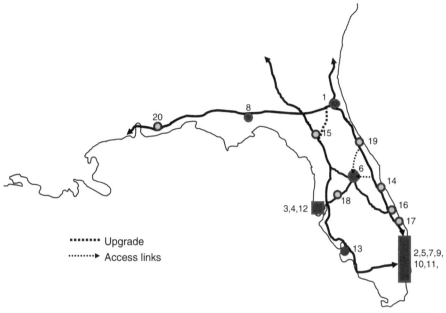

FIGURE 2.23 Realistic network design.

2.9 *EVALUATION OF THE NETWORK DESIGN*

Finally, this section provides additional information regarding the assessment of various effects of a new network design as well as the complexity of the policy process we are dealing with.

2.9.1 Costs–Benefits

In the design process the main emphasis lies on defining a network that connects and/or links up selected nodes (hierarchy) with a certain quality. It is self-evident that the accessibility of these nodes is thereby guaranteed. Of equal importance is the quality of the designed network as well as a verification of other sustainability criteria,* such as:

- Environment and livability
- Traffic safety
- Network accessibility
- Economy, accessibility
- Costs

In principle, such verification includes a cost–benefit ratio. Since the costs and benefits are spread over a longer period of time, they will have to be reduced to a net cash value with

---

*This list does not claim to include all relevant effects, only the most common criteria.

due regard to a discount rate and a time horizon. It is not intended to go into the technical details here of a cost–benefit analysis, but we will describe in brief what effects will have to be taken into account and how these effects can be calculated.

The cost–benefit analysis can be performed for a separate (newly designed) network, and it is also possible to compare certain network variants with each other.

Load on the Network.   To carry out a cost–benefit analysis one needs to know the load on the links and intersections from which the network is built up. For this purpose an origin–destination (O–D) matrix will have to be assigned to the network using a static or dynamic equilibrium assignment. The dynamic model is preferred because use of this model allows the congestion losses in the network to be more correctly calculated. And because the effects are also calculated for a future situation, it is better in principle to use O–D matrices for the longer term. If this approach becomes too complicated, an average growth percentage of the load on the network links will suffice.

For a more extensive explanation of traffic assignment models and traffic congestion, refer to chapters 7 and 12 of this Handbook.

Calculating Effects

Environment and Livability.   Effects that come under this heading are:

- Emissions of harmful substances (harmful to man, fauna, and flora)
- Noise nuisance
- Fragmenting the landscape

Traffic emits a number of harmful substances, such as CO, CO_2, C_xH_y, NO_x, Pb, and particulate matter. The size of these emissions depends on various factors, such as fuel usage, type of fuel, speed, driving cycle, and gradient. If we want to take account of all of these factors, this implies a highly complex method of calculation. This level of detail is undesirable at this stage in the design process and largely not feasible given the absence of required data. A global approach consists of a calculation of the emissions based on the kilometers covered per road type. The fuel (energy) consumption per km can be determined per road type (see Smokers et al. 1997):

Freeway: 2.64 MJ/km
Highway: 2.29 MJ/km

Making certain assumptions concerning the composition of the vehicle fleet, the speed of driving, and any acceleration and deceleration that may occur will allow these calculations to be refined.

To quantify noise nuisance one needs the following data:

- Load per stretch of road
- Composition of the traffic
- Speed of the traffic
- Distance of road axis from building facades
- Building density (number of premises/residents along the side of the road)

These data help determine the equivalent noise level which can be compared with the norm (e.g., 55 dB (A) threshold value). Subsequently the effect can be quantified by multiplying transgression of the norm per affected person by the cost (e.g., $21 per dB (A)) (ECMT 1998).

If one does not have access to the building density data, one can also use the number of vehicle kilometers (distinguishing between cars and trucks) or passengers and ton kilometers (road and rail) as a starting point.

The laying of new roads or widening of existing roads can have a significant impact on the landscape, such as:

- Slicing through the landscape, with the result that valuable landscapes and/or ecosystems are disturbed
- Dividing up areas such that they are subsequently too small to function as a habitat for a species
- Creating a barrier that hinders crossing for both man and animals

This effect is difficult to quantify generically; each case will have to be examined largely on its own merits. One way of calculating the effects is to determine which remediation measures (investments) are desirable to retain the original situation.

Traffic Safety. The safety of a trip in a network depends on many factors. To ascertain the effects of the proposed network structure on safety, the following procedure can be applied:

- The transfer kilometers* on the various types of road are multiplied by risk factors that indicate the possibility of an accident with (fatal) injury as a function of the distance covered. Per road type a distinction is made between the numbers of dead and injured.
- Depending on the type of road, risk factors** are used:

Deaths per 10^9 km: freeway = 2.250
Deaths per 10^9 km: highway = 6.617
Injured per 10^9 km: freeway = 12.631
Injured per 10^9 km: highway = 39.920

Accordingly, the costs of a fatal accident are estimated at $1.5 million, an accident with serious injury at $0.2 million, and an accident with slight injury at $0.03 million (ECMT 1998).

Network Accessibility. Network accessibility is related to the distance or trip time between origin and destination and a point of access in the network (system). A possible requirement for network accessibility could be that a certain percentage of the population lives within a radius of 10 km or a journey time of 15 minutes from an access point to the freeway network. Using a network accessibility requirement may result in additional access points having to be incorporated into a network.

Accessibility. In calculating the effects in respect of accessibility, a distinction can be made in terms of the components trip time, robustness, and average journey speed.

The trip times are calculated preferably on the basis of the resistance of the loaded road network, so it is necessary to assign an origin-destination matrix to the network. Use of a dynamic assignment model enables realistic trip times to be calculated.

The laying of new roads will tend to have a positive effect on the total trip time. The net cash value of the gain in trip time can be determined by the gain in hours (calculated on an annual basis) multiplied by the valuation of one hour's trip time.

The valuation of trip time differs per country (region), and it is essential to make a distinction in the valuation between the trip time losses that have an impact on normal

---

*These transfer kilometers are calculated using the equilibrium assignment model.

**No universal risk factors exist. The risk factors tend to be determined by country (state), year, and road type.

congestion (peak traffic) and trip time losses that have an impact on incidental congestion (e.g., incidents). In this latter case the trip time valuation will be higher.

To determine the robustness of the network, various possibilities are open, such as:

• Calculating the effects of fluctuations in demand and/or supply (e.g., as a result of the occurrence of an incident). Temporarily restricting the capacity of one or more links in a network allows the effects of this on the quality of the traffic flow to be calculated.
• A second factor that has a considerable influence on the robustness of a network concerns the traffic load of the arms of the intersections in the network. Especially the higher-order intersections can have a significant influence on the quality of the traffic flow in large parts of the network, even if the capacity of only one of the arms falls short. As a result of the "blocking-back" effect, the tail of a traffic jam can disturb the traffic flow in other parts of the network.

The robustness of a network can be investigated by varying both demand and supply. The network's capacity to accommodate varying load patterns can be determined on the basis of journey times (total and relative) and intersection loads (load on various parts such as slip roads, acceleration lanes, etc.).

The average trip speed indicates the average speed of transfers on the various networks and can also be determined relatively (in km/hr), thus showing the average speed between origin and destination. This last measurement is used mainly in comparing transfers across various distances.

2.9.2 Process

Proposals to modify the network are often sensitive issues given the numerous effects that would occur. We are increasingly being confronted by a situation that can be characterized as a complex policy problem, in which two dimensions exist:

• A *knowledge-intrinsic complexity* derived from the relationship between the various intrinsic themes—traffic and transport, spatial planning, economic development, environment, nature and landscape—that are interwoven. Any intervention in one theme has consequences for another. But all too often a swings-and-roundabouts mechanism occurs; as a solution is found for one theme, this causes a problem for another.
• A *process-related complexity* that is expressed in the involvement of many public and private actors at various levels of government and/or scales, who all want to see their vested interests covered in the problem definitions and solutions. The actors are very interdependent in the problem-solving and so form changing coalitions to this end.

In the design process, account will have to be taken of this complexity. This implies that:

• All stakeholders are invited to participate in the design process. This means having to invite other parties like environmental agencies, chambers of commerce, etc. in addition to the existing road authorities.
• The effects of the proposed measures (network variants) are fed back to the stakeholders. Ensuring that the feedback of effects is fast and clear to everyone stimulates the participants to enter into a structured discussion on potential solutions.

For such an approach to succeed, it is essential that all parties participate on the basis of equality, everyone has access to the same information, and the problem is not made more complicated than necessary.

2.10 ACKNOWLEDGMENT

We would like to thank Isabel Wilmink, Arno Hendriks, and Janos Monigl for their inspiring contribution in writing this chapter.

2.11 REFERENCES

Arthur, W. B., Y. M. Ermoliev, and Y. M. Kaniovski. 1987. "Path-Dependent Processes and the Emergence of Macro-Structure." *European Journal of Operational Research* 30:294–303.

Bovy, P. H. L. 2001. "Traffic Flooding the Low Countries: How the Dutch Cope with Motorway Congestion." *Transport Reviews* 21(1):89–116.

Buckwalter, D. W. 2001. "Complex Topology in the Highway Network of Hungary, 1990 and 1998." *Journal of Transport Geography* 9: 125–35.

European Conference of Ministers of Transport (ECMT). 1998. *Efficient Transport for Europe: Policies for Internalisation of External Costs.* Paris: OECD Publications Service.

Egeter, B., I. R. Wilmink, J. M. Schrijver, A. H. Hendriks, M. J. Martens, L. H. Immers, and H. J. M. Puylaert. 2002. *IRVS: Ontwerpmethodiek voor een integraal regionaal vervoersysteem.* Research project commissioned by NOVEM. TNO Inro, ref. 01 7N 200 71621. Delft, October.

Forschungsgesellschaft für Strassen und Verkehrswesen (FGSV). 1988. *Richtlinien für die Anlage von Strassen (RAS),* part *Leitfaden für die funktionale Gliederund des Strassennetzes (RAS-N).* Bonn: Kirschbaum Verlag.

Monigl, J. 2002. *Highway Network Planning in Hungary Including Tolled Motorways.* Budapest: TRANSMAN Consulting.

Schönharting, J., and T. Pischner. 1983. *Untersuchungen zur Ableitung von Strassenkategorieabhängigen Soll-Qualitäten aus Erreichbarkeitsanalysen.* Forschung Strassenbau und Strassenverkehrstechnik, Heft 399.

Smokers, R. T. M. et al. 1997. *Verkeer en vervoer in de 21e eeuw; Deelproject 2: Nieuwe aandrijfconcepten.* TNO Automotive commissioned by SEP; TNO report 97.OR.VM.089.1. TNO Automotive, December.

Van Nes, R. 2002. *Design of Multimodal Transport Networks: A Hierarchical Approach.* Trail Thesis Series T2002/5. Delft: DUP Science.

Waldrop, M. M. 1992. *Complexity: The Emerging Science at the Edge of Order and Chaos.* New York: Simon & Schuster.

CHAPTER 3
TRANSPORTATION SYSTEMS MODELING AND EVALUATION

Andrew P. Tarko
School of Civil Engineering, Purdue University,
West Lafayette, Indiana

3.1 INTRODUCTION

Transportation includes infrastructure, administration, vehicles, and users and can be viewed from various aspects, including engineering, economics, and societal issues. A *transportation system* can be defined narrowly as a single driver/vehicle with its second-by-second interactions with the road and other vehicles. The system can also be defined broadly as a regional transportation infrastructure with its year-by-year interactions with the regional economy, the community of transportation users and owners, and its control components such as transportation administration and legislature. These two extremes exemplify the range of transportation systems, with various intermediate scenarios possible.

Transportation models are a formal description of the relationships between transportation system components and their operations. Knowledge of these relationships allows for estimating or predicting unknown quantities (outputs), from quantities that are known (inputs). Because our knowledge of the transportation relationships is limited, transportation models are subsequently imperfect and selective. Awareness of the models' limitations facilitates using the models according to need, required accuracy, and budget.

Evaluation has two distinct meanings: "calculate approximately" and "form an opinion about." Both meanings are reflected in the two basic steps of transportation systems evaluation:

1. *Quantify* by applying a model
2. *Qualify* by applying evaluation criteria

The first step requires a valid model, while the second step uses preferences of decision-makers and transportation users. Modeling, in most cases, is a required part of transportation systems evaluation.

A transportation model is a simplification of transportation reality. It focuses only on what is essential at the level of detail appropriate for its application. If one wants to improve traffic at a specific location by redesigning signals, then optimal signal settings are the solution, which has a negligible economic effect on the regional economy and this variable therefore should not be considered in the model. The situation changes if one wants to program transportation improvements in the region that must compete with large-scale highway projects for funding. Then the economic impact of the decision is important and detail signal settings are not considered; instead, the overall effect of the typical control is represented in the analysis. These two cases require two distinct models that differ in scope and detail. A specific job requires a specific model. Understanding the basics of modeling in transportation engineering is helpful in selecting an adequate model, using it properly, and interpreting the results correctly.

This chapter aims to help decide whether a model is needed, select an adequate model, and use it effectively. The reader will find neither endorsements nor a complete overview of the existing modeling software packages, and specific references are mentioned for illustration of the points raised in the presentation without any intention either to compliment or criticize.

Although this chapter has been written with all the areas of transportation engineering in mind, examples are taken from surface transportation, which is the author's area of expertise. The author believes that this focus does not constrain the generality of the chapter.

3.2 TYPES OF TRANSPORTATION MODELS AND MODELING PARADIGMS

Classification of transportation models is challenging because there are a wide variety of transportation models and a considerable number of ways the models can be categorized. We will try to classify the models in a way that helps transportation engineers select models adequate to the job.

Transportation models are applied to individual highway facilities, groups of facilities, and entire transportation systems at the city, state, and national levels. Transportation models are also applied to time horizons, ranging from the present up to 20 and more years ahead. Depending on the use and scope, transportation models can focus on long-term prediction of demand for various transportation modes with adequately incorporated economic impacts; focus on routing and scheduling using choice models with properly represented connectivity between various network components; or focus on a faithful representation of traffic flows at various transportation facilities. From this perspective, transportation models can be divided into the following categories:

1. *Demand models* (econometric models, short-term traffic prediction, traffic generation, etc.)
2. *Network models* (modal split, traffic assignment, scheduling)
3. *Traffic models* (advanced traffic and control representation, interaction between vehicles)
4. *Performance models* (traffic quality perception, safety performance models, fuel consumption, air-pollution generation, noise generation, signal optimization, etc.)

3.2.1 Demand Models

Because the majority of trips are work or business-related, transportation demand depends strongly on the regional economy. On the other hand, economic growth in the area can be stimulated by the business-generating ability of transportation infrastructure or hindered by excessive transportation costs. *Econometric models* attempt to grasp these complex impacts through simultaneous statistical equations that represent the relationships among transportation, regional economy, land use, regional policies (laws, zoning, pricing, subsidies, etc.), and people travel preferences (example can be found in Johnson et al. 2001). These relationships are developed from the historical data for the region and then applied to predict future transportation demands. These models are typically highly aggregated and give total annual or daily numbers of trips between zones by various transportation modes. Although commercial packages provide a necessary modeling suite, models are developed to reflect properly local conditions, the objective of the analysis, and available data. The set of equations is developed on a case-by-case basis.

Econometric models are the high end of demand modeling; *extrapolation methods* including trends of growth rate method are the low end. According to the growth rate method, future volumes are predicted by multiplying current volumes with growth factors (growth method for highways is mentioned in Robertson 2000). Separate growth factors are developed for different regions, transportation modes, and types of transportation facilities (road category, airport category, etc.). This method is not able to reflect unusual changes either in the economy or in the land use.

The third category of demand models is *traffic generation models* (for example, in Meyer and Miller 1984). These models require future land use and household characteristics to be known because they link the number of trip ends in a traffic zone with particular zonal characteristics, such as the number of workplaces and the number of households. These models include the land use and the economic impact through zonal characteristics required as an input.

3.2.2 Network Models

Econometric models can and should consider all major transportation modes (air, rail, highway, pipes, and water). On the other hand, the network models have typically been developed to deal with certain transportation modes. The interactions between the transportation modes are presently attracting the interest of researchers who are trying to incorporate transportation intermodalism into regional planning. The primary focus of highway network modeling is on the connectivity of the transportation infrastructure and on the travelers' path choices across the network and of transportation modes (for example, highways with public transit routes). Modeling frameworks for travel choices include deterministic user equilibrium, stochastic user equilibrium, system optimum performance, and mixed flows (Sheffi 1985). Some models consider static equilibrium applied to hourly or daily network operations, while other models assign travelers to paths dynamically in short intervals (Peeta and Ziliaskopoulos 2001). Network models require travel demand to be known as a table of one-way flows between all possible pairs of network nodes or traffic zones.

Another large class of network models is logistic models that optimize transportation of commodities. These models are focused on routing commodities across a multimodal transportation network to minimize transportation costs and meet time constraints. Logistic models are of particular interest to private transportation companies and to large manufacturers that use supply-chain analysis to reduce transportation and storage costs. A good introduction to the logistics modeling and practice can be found in Fredendall, Hill, and Hill (2000).

3.2.3 Traffic Models

With a few exceptions, network models represent transportation facilities in a quite simplified manner through the analytical relationships between traffic demand and travel cost components; for example, the traditional BPR function (Bureau of Public Roads 1964). These simplifications do not allow for designing facilities and may raise concerns about the accuracy of the simplified demand-cost relationships. To allow design applications through detail analyses of traffic operations at these facilities, a group of models is available, here called traffic models. They incorporate the impact of facility geometric characteristics and traffic control on speed, delay, travel time, queue, etc. One of the best known depositories of traffic models for design and traffic analysis is the *Highway Capacity Manual* (Transportation Research Board 2000). These models are transportation facility-specific and require traffic demand, geometry, and control characteristics of the facility to be known.

3.2.4   Performance Models

Performance models quantify traffic characteristics (volume, speed, traffic density, travel time, delay, etc.) that are directly used to evaluate transportation systems and their components. Such quantities include costs, noise, air pollution, and users' perceptions of conditions. Widely known models include a model of vehicle emission MOBILE (Jack Faucett Associates 1994) and a model of aircraft noise generation and propagation INM (Volpe Center 2002). These and other performance models can accompany all the three transportation models already introduced, depending on the type of transportation analysis: planning, design, operations.

3.2.5   Modeling Paradigms

Figure 3.1 depicts the traditional modeling paradigm used in transportation engineering. Modeling is carried out in three separate modeling phases: demand modeling, network modeling, and traffic flow modeling. Each of the modeling phases provides input to performance models. The outcome of the demand modeling phase is used to model traffic assignment in the transportation networks, while the outcome of the network modeling provides necessary input to the traffic flow facility modeling at transportation facilities. Each phase can be performed by a different transportation engineering unit and the results transferred from one unit to another.

The traditional modeling paradigm was proven to provide a manageable modeling framework using data that are available and providing computational demands that are reasonable. This was accomplished in each phase through adequately defining and representing the transportation systems. Demand models represent the transportation and economy of the studied region with synthetic characteristics of sub-regions. Network models must be more detailed and present all important components of the transportation infrastructure. However, they represent only part of the region transportation infrastructure. Although they require more data, the task is distributed among multiple planning organizations. The geometric design and operational analysis of a transportation facility requires the most detailed information, but such an analysis is performed only for selected facilities. In the traditional modeling paradigm, the outputs from higher models are input to lower models and there is no two-way interaction.

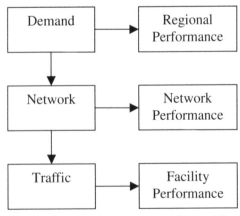

FIGURE 3.1 Traditional transportation modeling paradigm.

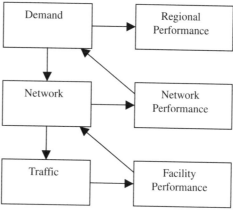

FIGURE 3.2 New transportation modeling paradigm.

Traditional transportation modeling has three phases, with a gradually growing level of detail and a narrowing geographical scope. The phases are executed sequentially and the outcome from the previous phase feeds the next phase. This modeling approach cannot properly treat many intelligent transportation systems (ITS) where traffic operations are important component of the ITS representation, while many strategic decisions about the system are made at the system level in real time. These decisions may even influence demand for traveling. A need for modeling such systems requires a new modeling paradigm where demand, network, and traffic are modeled simultaneously and dynamically in an integrated single phase. This paradigm is presented in Figure 3.2. The full interaction between different levels is obtained by adding feedback between the layers or developing models that truly integrate all the components into one model. It should be noted that the type of interaction between the demand and network layers in traditional transportation management is different than in the ITS-based dynamic management. In the traditional framework, the demand layer is used to predict aggregate average demand based on high-level policy and planning decisions. In the ITS-based management the changes in disaggregate demand are short-term and are caused by ITS operations.

3.3 EXAMPLE STUDIES AND MODELS

The number of available models and their computer versions is impressive. Most of the existing computer software for transportation modeling and evaluation includes multiple model types integrated in packages. Almost all such packages include performance models to produce results useful for evaluation and decision-making. Some of these packages integrate models and various levels of detail within the new modeling paradigm. This section presents selected models and studies to illustrate the model types and their applications.

3.3.1 Maine Statewide Travel Demand Model

The Maine DOT funded development of a modeling procedure to predict future travel demand in Maine (U.S. Department of Transportation 1998), based on future population and other socioeconomic data for traffic zones of the region. A gravity model was used to split the trip ends in a zone among all the zones. The traffic zones were defined inside and outside the study area to enable modeling internal trips with both ends inside the study area and external trips with one or both ends outside the study area. The model used current and assumed future characteristics of the transportation infrastructure. The model was calibrated for current data and demand predicted by changing the model inputs that represented the future. This model is an example of a traditional approach to long-term modeling of transportation demand where the future land use must be known. The Maine study is a good example of the use of GIS technologies to code data and visualize results.

3.3.2 Sacramento Area Travel Demand Model

The Sacramento area was modeled with econometric equations in the MEPLAN modeling framework (Johnston et al. 2001). The model includes the interaction between two parallel markets, the land market and the transportation market, as illustrated in Figure 3.3. In the land markets, price and cost affect decisions about the location of various activities. In the transportation markets, the costs of travel affect both mode and route selection. The interaction between the two markets was accomplished by including transportation costs in the attractiveness of zones for producing activities and including the producing activities in traffic demand factors. The generated trips were split by mode choice and assigned to routes based on stochastic user equilibrium with capacity restraint. The quasidynamic modeling was executed in time steps with time lag in the interaction between the two markets. This approach is an example of an advanced modeling of the two-way impact between land use and transportation where both the land use and the transportation demand are predicted.

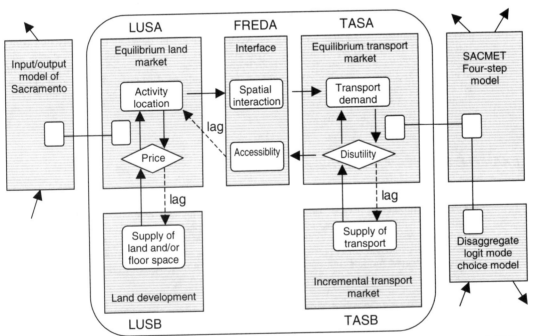

FIGURE 3.3   MEPLAN and other components of Sacramento model with interactions (based on Johnston et al. 2001).

3.3.3 ITS Evaluation for the Washington, DC, Metropolitan Area

In this study, the traffic flows for current and future years without ITS were predicted with traditional four-step demand and network modeling (Schintler and Farooque 2001). The four-step model is presented elsewhere in this Handbook. Several combinations of two sets of scenarios were examined for a selected portion of the arterial network in Washington, DC:

1. Present traffic and future traffic
2. Presence and absence of the SmarTraveler system

The traffic was simulated using a mesoscopic simulation package INTEGRATION for integrated arterials and freeways (Van Aerde et al. 1995). INTEGRATION models represent individual vehicles on freeways and on urban streets with traffic signals and ITS, while preserving the macroscopic properties of traffic flows. The program uses dynamic queuing-based traffic assignment, driver diversion, and rerouting during congested conditions. Travelers may behave according to their current experience or in response to information received via the Internet, from variable message signs, or from highway advisory radio. INTEGRATION models use advanced transportation management systems with real-time surveillance devices and adaptive traffic signal systems.

This study is a good example of integrating the traditional planning modeling with computer simulation where the three model types were integrated in a dynamic fashion to allow short-term demand adjustments to traffic conditions, network modeling, and rather sophisticated traffic modeling at the facility level.

3.3.4 *Highway Capacity Manual*

The *Highway Capacity Manual* is a leading reference for practicing highway transportation engineers who seek guidance in capacity and traffic quality modeling and evaluation for planning, design, and traffic operations applications (Transportation Research Board 2000). The *Highway Capacity Manual* focuses on traffic operations at various facilities, including freeway segments and ramp intersections, multilane and two-lane rural roads, urban streets, signalized and unsignalized intersections, pedestrian, bike, and public transit facilities. The *Highway Capacity Manual* includes performance models that convert traffic measures of effectiveness into levels of service that have become the standard procedure to qualify traffic performance and are an example of traffic models combined with performance models. The models are analytical and are implemented faithfully on a computer (McTrans 2002). The models predict neither traffic demand nor traffic diversion between routes. Traffic volume is among the required inputs and must be predicted with demand and network models.

3.3.5 Interactive Highway Safety Design Model

The Interactive Highway Safety Design Model (IHSDM) is road safety evaluation software being developed for highway planners and designers (visit http://www.fhwa.dot. gov/ihsdm/). It already includes two-lane rural highways and is ultimately intended to include multilane rural highways. IHSDM will consist of several components, including crash prediction and traffic analysis modules.

The crash prediction module will predict the expected number of crashes along the road segment given the basic geometric characteristics of the segment and traffic intensity (Harwood et al. 2000). The crash prediction models take the form of equations and are sometimes called safety performance functions.

The objective of the traffic analysis module is to provide practitioners with a tool to investigate the operational effects of alternative road designs, marking schemes, traffic controls, and vehicle size and performance characteristics for current and projected future traffic flow (Wade et al. 2000). TWOPAS, a traffic simulation model for two-lane rural highways, will form the basis for this module. The TWOPAS model simulates traffic operations on two-lane rural highways by reviewing the location, speed, and acceleration of each individual vehicle on the simulated road every second and advancing those vehicles along the road in a realistic manner. The model takes into account the effects of road geometry, driver characteristics and preferences, vehicle size and performance characteristics, and the presence of oncoming and same-direction vehicles that are in sight at any given time.

3.3.6   Microsimulation for Regional Planning—TRANSIMS

The Los Alamos National Laboratory is the leader in the development of a model called the Transportation Analysis and Simulation System (TRANSIMS) (Los Alamos National Laboratory 2002). TRANSIMS is an integrated microsimulation forecasting model designed to give transportation planners information on traffic impacts, congestion, and pollution in a metropolitan region. TRANSIMS models travelers' activities and the trips they take to carry out their activities and then builds a model of household and activity demand. Trips are planned to satisfy the individual's activity patterns. TRANSIMS simulates the movement of individuals across the transportation network, including their use of vehicles such as cars or buses, on a second-by-second basis. The model tries to forecast how changes in transportation policy or infrastructure might affect those activities and trips. The interactions of individual vehicles produce traffic dynamics from which the overall performance measures are estimated.

TRANSIMS is a courageous attempt to introduce detail-level modeling to regional planning. It is a departure from the widely accepted modeling framework presented in Figures 3.1 and 3.2 in that it represents traffic in the entire region at the level of individual travelers. To cope with the huge amount of data and computational demand, vehicles move in a simplified manner by jumping between road "cells." This modeling attempt may answer the question of whether data aggregation and representation simplicity in planning can be replaced with detail models that are traditionally applied to study traffic operations.

3.4   WHAT ARE THE USES OF TRANSPORTATION MODELS?

Modeling is a necessary component of transportation engineering if future traffic conditions are being analyzed or existing conditions are too expensive to observe. These circumstances are present in a majority of transportation studies. A practical approach to transportation studies is to solve transportation problems by first solving their models and then implementing the solutions to the real world. Modeling is applied in all the areas of transportation engineering: planning, design, and operations. Although planning is traditionally the most model-demanding, intelligent technologies and techniques applied to transportation have increased the demand for modeling in design and traffic operations areas.

3.4.1   Planning

Modeling for planning includes predicting future travel demands, identifying potential performance problems under the future travel demands, and proposing general solutions of the anticipated transportation problems. Modeling packages for long-range planning may include all the types of models presented and discussed previously: demand prediction, network modeling, traffic flows, and performance modeling (although traffic flows are modeled with much more simplification). From this perspective, long-range planning requires the most comprehensive modeling. Questions frequently addressed with modeling for planning applications include:

1. *Future travel demands.* Travel demand is typically defined as the volume of travelers willing to travel between two locations. A quadratic matrix of one-way traffic flows represents travel demand between multiple locations. Demand for travel strongly depends on land use and on economic conditions.

2. *Impact of regulations and policies on future travel demands.* Knowledge of this relationship is particularly useful if proper regulatory and economic measures are sought to keep travel demand at the desired level. The desired level of demand is determined by congestion management or environmental concerns.

3. *Identification of components of the existing infrastructure that will need improvements.* Performance of an existing system under future demands is frequently studied to identify the system bottlenecks—in other words, the infrastructure components that will cause excessive hazard, congestion, or environmental concerns. This analysis leads to the next one, whereby engineering solutions of the anticipated transportation-related problems are sought.

4. *Identification of projects needed to maintain acceptable performance of the system.* Once future transportation issues and weak components of the infrastructure are identified, adequate alternative solutions can be proposed and evaluated. Transportation models are run for various scenarios that represent various solutions, and consecutive evaluation of the scenarios is carried out to select the best one.

5. *Identification of improvements is needed to make the existing infrastructure more resilient to damage and more efficient in emergency situations.* Modeling of emergency conditions that follow natural or human-inflicted disasters (earthquake, hurricane, nuclear blast, biological/chemical attack) help identify critical components of a transportation system. A component of transportation infrastructure is critical if its failure brings severe deterioration of the system performance. Such components, once identified, can be structurally strengthened, better protected through enhanced security means, or supplemented with backup facilities. Scenarios of operations management can be prepared if a critical component is damaged.

3.4.2 Design

A well-designed transportation facility is economical in construction and maintenance, accommodates traffic demand during its lifetime, and does not expose its users to excessive danger. To meet these design criteria, a designer must be able to link design decisions about facility geometry and traffic control with facility performance (delays, speeds, safety, and costs). Although many design decisions related to safety and aesthetics are governed by design policies and guidelines, decisions that determine capacity (for example, the number of lanes on a freeway) require traffic modeling. Among many models used in design, two design tools have already been mentioned: the *Highway Capacity Manual* and the Interactive Highway Safety Design Model.

Evaluation of alternatives is a common approach to geometric design. More complex systems such as arterial streets or networks with coordinated signals may require quite advanced optimization tools for proper setting of signal parameters. Traffic volumes are assumed to be insensitive to design decisions and are provided by planning models.

3.4.3 Operations

Studies of existing systems may include direct observations of their performance. Although this is the most desirable method of evaluating existing systems, costs and measurement difficulties often make modeling desirable. Transportation analysts try to replicate the existing system through its model, and outcome from the model is used to evaluate the system's performance, detect operational problems, and evaluate possible solutions. Although this approach brings to mind planning studies, there is a considerable difference—not future but present traffic conditions are evaluated. Such studies are particularly justified if the considered improvements are low-cost and are implemented quickly. Signal retiming and reorganizing traffic circulation in the area are examples of such improvements.

Intelligent transportation systems require a new class of model applications. Advanced travel information systems and advanced traffic management systems use fast models run in real time to support real-time decisions about changes in schedules, signal retiming, and traveler rerouting in response to planned events and unplanned incidents. Real systems are simulated with a short time lag using real data. Short-term traffic prediction is desired to foresee the near future. Only then can optimal traffic control and management strategies be applied in a timely manner.

3.5 SELECTING A MODEL

Most of the available modeling tools are packages of models described in section 3.2. In this section, the term *model* stands for a model package. The variety of transportation studies and multiplicity of models make selection of a model difficult, and this section therefore is written with inexperienced transportation analysts in mind who seek general guidance for selecting a proper model.

A model is appropriate if it can do the job. Transportation analysts tend to seek a universal model that can do most of the jobs, but it is doubtful that such a model exists. A model is a simplification of reality and is developed as a result of a preassumed class of applications that determine which simplifications are acceptable and which are not. If the model is used for an application outside of the assumed class, then seemingly reasonable results may be in fact inaccurate or even useless. Selection of a model therefore must consider the job and, more precisely, the model's applicability for the job. For example, traditional long-term planning models estimate the travel times along links between nodes using the simplified BPR function. These travel time estimates are sufficient to model network flows but are useless if one wants to analyze, for instance, the performance of urban arterial streets during special events.

A modeling job is specified sufficiently for model selection if the following is known:

1. Studied transportation system
2. Purpose of the modeling
3. Required output scope and format
4. Required accuracy of results (model validation and calibration)
5. Resources available for the modeling

3.5.1 Output Scope and Format

Transportation models are used to produce specific results. If a model does not produce what is needed or the results are insufficient to calculate the needed outcome, then such a model cannot be used.

3.5.2 Model Validity

Another criterion that is often overlooked and is closely related to the validity of the model is the accuracy of the results, which is tested by inspecting the model error. The model is run a sufficient number of times using input collected in the field, and the obtained results are then compared with the measured values. The model is valid if the discrepancies between the calculated and observed results are acceptable. The validity of many available models is not sufficiently documented, part of which can be explained by the fact that validation of transportation models is not easy. Validation requires field-collected inputs and field-collected values corresponding to the model outputs. Not only are the high costs of data collection a problem, but sometimes so is the measurability of quantities used in the validated model. For example, a considerable number of microsimulation models require the percentages of drivers of certain personality types (aggressive, regular, or passive), intervehicle time gaps that drivers use when crossing a road, or desired speeds. Models that use a large number of input variables are difficult to validate because validation requires collection of all these variables simultaneously, which is often practically impossible.

Prediction models present a new category of validation issues. The only possible validation method is retrospection, where past observations are predicted using older observations. The weakness of this approach is that even good replication of the past data does not rule out failure in predicting the future.

Another way of evaluating model validity is to analyze the model's fundamentals and structure. Sufficient description of the model and explanation of its theoretical basis increase confidence in the model's validity, particularly for computer simulation if the component models are well documented in the literature and set together in a plausible manner. Proprietary computer codes unavailable to transportation analysts make indirect evaluation of these models difficult if not impossible. Preferable models are those well documented by developers and validated independently by research units.

3.5.3 Model Calibration

Even a valid model requires calibration if the local conditions differ from the ones for which the model was developed and validated. Frequently, a simulation model is developed from component models and the default values of model parameters are assumed using common sense or published research. Then the model is offered to clients with a presumption that users are responsible for model calibration to local conditions. For clarification, a parameter is an input whose value typically remains unchanged from one simulation to another.

There are two ways of setting model parameters. If a parameter has a meaning and can be measured in the field, then a measured value should replace the default one. If a parameter is not observable because it has no meaning or is difficult to measure, then it may serve as a "tuning knob." A calibration exercise is often understood as applying measured parameter values and changing the remaining parameters until the model returns results of satisfactory accuracy. Although the recipe seems simple, its application is not. The calibration difficulty caused by the large amount of data required was explained above in section 3.5.2.

The need for calibration of simulation models has been investigated at the University of Hawaii (Wang and Prevedouros 1998). Three microsimulation models, INTETRATION, CORSIM, and WATSim, were applied to a small congested network of two highway intersections and several segments. The simulated volumes and speeds were compared to the measured values, and the authors concluded that the results were reasonable after a large number of parameters were modified to replicate the real traffic conditions. In no case did the default parameters offer satisfactory results. Further, the authors concluded that the parameter calibration could be tedious and time-consuming and the effort for calibration multiplies with the size of the size of the transportation system and the number of model parameters.

Models, particularly microsimulation, may have a large number of parameters. Replicating real values in a small number of cases is always possible if the number of tuning knobs is sufficient. The problem is that a model calibrated for a small number of data sets may not perform well for new cases. This issue is similar to a regression model calibrated for a small amount of point data but with a large number of parameters. If the number of data points equals the number of parameters, it is always possible to find parameter values that allow a model to replicate the data exactly. Unfortunately, the predictive ability of the model is nonexistent.

Model validation and calibration require special attention, particularly when the models are complex. Given the difficulties with validation and calibration of complex models mentioned above, simple models sufficiently validated and with a limited number of parameters should be favored. Specific methods of planning models validation and calibration can be found in Barton-Aschman Associates and Cambridge Systematics (1997).

Numerous transportation models have been developed for particular analysis. For example, researchers from the University of Texas in Austin proposed a simulation model developed to evaluate the effect of introducing new large aircraft in 2006 (Chiu and Walton 2002). The method simulates the operating characteristics of a new large aircraft and its market share for various scenarios to address prediction uncertainty. The arrival passenger flows are modeled as a queuing network system composed of a series of passenger processing facilities. The model has been validated with survey data and statistics for international airports.

3.5.4 Scope of Input

A model is practical if it requires input that is feasible to obtain. Again, this condition should be considered in connection with a specific job for which a model is needed. In the planning phase, only major transportation characteristics are known or considered, and it is unusual, if not impossible, to consider the details of intersection geometry or the detail settings of a signal controller 20 years in advance. Not only the information is uncertain but also planners are not interested in such detail decisions. Models that require unavailable input lead to use of default values, and when default values may be used is in itself an important issue. The answer depends on the variability of the input and on the model sensitivity. If the input varies significantly from one case to another and the results are strongly dependant on the input, then use of a default value is questionable. A model that requires such an unknown input should not be used.

An excessively detailed model may impose difficulties of using default inputs, but an overly simplified model may defeat the purpose. For example, if a model is used to evaluate the introduction of a median on an arterial street, the model must incorporate the effect of the median on the capacity of the crossing and turning streams.

3.5.5 Modeling Costs

The costs of transportation analysis include gathering and formatting data, running a model, and documenting and reporting results. Data collection and formatting are frequently the primary component of the total costs, particularly when the model is complex and requires calibration. The costs can be significantly reduced if the data required by the model are routinely collected by a transportation agency and stored in well-maintained databases.

The costs can be further reduced if the amount of data required by the model is limited. The input required by models varies dramatically and depends on the level of modeling detail. Microsimulation requires the largest amount of data compared to other types of models. For example, the TRANSIMS microsimulation modeling tool was applied to the Portland area (Los Alamos National Laboratory 2002). Highway network representation included

100,000 nodes, 125,000 links, and 250,000 activity locations. In the introduction to the report presenting intermediate results after nearly two years of effort, the authors characterized collating the existing data (no special data were collected for this project) as a tremendous chore and a daunting task. The report makes the disclaimer that the purpose of the Portland study was to calibrate the model with the Portland historical data and that the stated objective was to allow the developers of the model to understand transportation planning methods and the model components. The Portland lesson indicates that the hurdle of using microsimulation for planning is the large effort required for data preparation and model calibration. Another issue that must be resolved is the massive computations needed. TRANSIMS was proposed for the modeling of the Switzerland transportation system, and in the interim report researchers mentioned the large computational load that required clusters of PCs and Internet connection (Raney et al. 2002).

3.5.6 Other Considerations

Client preferences must be considered when selecting a model and a modeling approach. For example, transportation administration may have its own standard models. For years, the *Highway Capacity Manual* was preferred by most state and local highway agencies. Simulation models were then approved by several states to be used concurrently or instead. In the private sector, modeling practice is more diversified and depends on the preferences of individual clients.

Modeling tools with a good visual presentation of the results, including microsimulation, are attractive because they facilitate communication between transportation analysts, decision-makers, and the public. The danger is that realistically displayed vehicles moving in a pseudorealistic transportation system, although not guaranteeing realistic results, may be convincing evidence of model validity to the layperson. Another dangerous myth is that the more detailed the representation of a transportation system is, the more accurate the results are. These two myths have continued to proliferate since microsimulation was introduced and modeling became an industry. Quality control, typically expensive and difficult in transportation modeling, may easily be lost in these transactions. The burden of quality assurance in modeling usually lies with clients that are public agencies.

3.6   TRANSPORTATION SYSTEMS EVALUATION

3.6.1 Steps in Evaluation

Transportation projects affect communities and are usually funded by public agencies. Evaluation of large projects that considerably impact whole communities requires public involvement. This specific requirement of transportation projects determines the method of their evaluation as the benefits and costs are estimated from the societal point of view. The primary objective of the evaluation effort is to select the best alternative solution, and typically no-build is one of the alternatives. Inclusion of the no-build scenario allows feasibility checking by comparing proposed solutions with the no-build scenario. The best alternative is selected from feasible solutions.

The first step of any evaluation is developing project alternatives. There are no specific guidelines on what and how alternatives should be developed. This part depends strongly on the transportation problem, existing infrastructure, and public and decision-makers' preferences. For example, the Central Area Loop study finished in 2001 (Driehaus 2001; Central Area Loop Study 2002) considered four alternatives for the Cincinnati area:

1. No-build
2. Enhanced shuttle bus service—increasing frequency and coverage, adding exclusive bus lanes and bus priority at traffic signals
3. Streetcar
4. Personal rapid transit, a novel automated transit system based on small cars

The second step is to select evaluation criteria. These criteria should be adequate to the evaluated alternatives and of course should include all aspects important to the decision-making body as well as to the community (travelers, business owners, and residents). In the Cincinnati case, 59 criteria were formulated shown in Table 3.1.

The third step is to evaluate alternatives using the developed criteria. This is the stage where demand and traffic flows are predicted and travel quality estimated. In addition, economic and environmental impacts are quantified. Transportation modeling is extensively used in this phase, and the outcome includes tables summarizing fulfillment of the criteria

TABLE 3.1 Evaluation Criteria in the Central Area Loop Study for Cincinnati, Ohio

Category	Evaluation Criterion
Cost-effectiveness	To what extent does this transportation system represent a cost-effective investment?
	Are there front-end costs and time associated with this transportation system to ready it for implementation?
	How severe are the secondary costs (utilities, street changes) due to placing this transportation system and its structures in likely locations?
	What is the technical life expectancy of this technology?
	To what extent does this transportation system imply a reasonable level of annual costs?
	Are there any extraordinary power requirements associated with this technology?
	Is this transportation system labor-intensive to operate and maintain?
	What relative degree of vehicle failure or downtime is likely with this technology?
	What level of vehicle spares seems indicated as prudent?
Equity	Will the transportation system distribute costs and benefits equally to all segments of the population within the service area?
	Will the transportation system serve a variety of populations?
	Will the transportation system provide affordable transportation to low-income individuals?
Safety/access	Is this transportation system ADA-compliant?
	Does the transportation system meet fire/life safety requirements?
	Is there a perception of personal safety within the vehicle and at the stations?
	Will there be difficulties meeting building code requirements?
	Does the transportation system provide convenient access to all users?
	Does the transportation system present a safety hazard to non-users?
Effectiveness	Does the transportation system have acceptable point-to-point travel times, including station dwell times?
	Does the system provide reliable service levels?
	Does the transportation system provide adequate service to the study area destinations in terms of frequency of service and geographic coverage?
	Will the transportation system adequately serve projected ridership and/or attract sufficient ridership to justify the investment?
	How reliable is the transportation system in maintaining schedule?
	Does this transportation system facilitate intermodal transfer movements among public transit service providers?

by the alternatives. Table 3.2 shows example results for the four alternatives for Cincinnati as published by the Sky Loop Committee (at http://www.skyloop.org/cals/cals-ending.htm). According to the revised values, personal rapid transit has the largest ridership and the lowest cost per passenger.

The fourth step is to select the best alternative. Although an evaluation analysis helps in making a selection, the final selection belongs to a decision-maker. In the Cincinnati case, the Ohio-Kentucky-Indiana Regional Council of Governments selected the improved bus service. The personal rapid transit alternative was rejected as not proven technologically (Driehaus 2001; Central Area Loop Study 2002). Other concerns included too-high capital costs, removing commuters from businesses at the street level, and poor aesthetics.

3.6.2 Conditions of Truthful Evaluation

The Central Area Loop Study for Cincinnati involved controversies around the modeling particulars. The draft final report prepared by a hired consultant was rebuked by one of the participating parties (Sky Loop Committee and Taxi 2000 Corporation 2001) as biased against one of the studied alternatives. This is not an unusual situation; parties involved in the evaluation process have their own preferences, opinions, and agenda. Therefore, it is important that the evaluation results be defendable. Several unquestioned technical conditions of truthful evaluation can be pointed out.

1. *Properly identified transportation problem.* Incorrectly identified transportation problems immediately defeat the purpose of the evaluation study. Current transportation problems are typically obvious and are identified through data routinely collected and analyzed by transportation agencies or through special studies triggered by travelers' complaints. Identification of future transportation problems requires modeling, is more challenging, and includes predicting future conditions for the no-build scenario. A transportation analyst must decide where prediction of future demand is needed with all or some of the constraints removed (for example, road capacity). Conditions 3–7 discussed below apply also here.

TABLE 3.2 Summary Results of the Cincinnati Central Loop Study

		Alternative		
Estimation items	No-build	TSM (enhanced bus service)	Streetcar	Personal rapid transit
Capital costs (with contingencies)	$0	$2,9000,000	$215,000,000	~~$450,000,000~~ $109,272,047
Design life (average for components)	10.0	12.5	31.6	34.3
Annualization factor	0.142	0.123	0.079	0.078
Annualized capital costs per year	$0	$350,000	$17,000,000	~~$35,000,000~~ $8,523,220
Operating and maintenance costs	$1,770,000	$5,710,000	$4,200,000	~~$13,900,000~~ $10,205,463
Passengers per year	468,000	1,812,600	2,098,200	7,951,200
Cost per passenger	$3.78	$3.34	$10.10	~~$6.15~~ $2.36

Source: As published at http://www.skyloop.org/cals/cals-ending.htm (last visited on December 26, 2002).

2. *Adequately and exhaustively developed alternatives.* Alternative projects are developed to solve the identified transportation problem. The proposed alternatives must cover a broad spectrum of possibilities.

3. *Input data of the best quality achievable.* The amount of data can be large, and its collection may be costly. It is difficult to explain why existing needed data are not used if they are available. Collection of new data depends on the budget and time frame.

4. *Honest assumptions where data are lacking.* It is not possible to have all required data available. In such cases assumption may be the only way, but these assumptions must be justified with analogous cases, with common knowledge, or at least with a consensus of the participating parties. The last option seems to be most difficult.

5. *Adequate model calibrated for local conditions.* Model selection and calibration issues are discussed above. It is obvious that an inadequate or inadequately calibrated model may generate inaccurate and even useless results. Model selection should consider resources required for model calibration to avoid a situation where no time or no funds are available to calibrate the selected model.

6. *Adequate and exhaustive evaluation criteria.* Evaluation criteria must reflect the preferences of the stakeholders and should consider all foreseeable benefits and losses caused by the alternatives. Therefore, input from the public is often sought if the subject transportation system is public or considerably influences the community.

7. *Proper documentation of the evaluation process and presentation of the results.* The final report must clearly state all the data sources, the assumptions made, and the tools used including the procedure for reaching the conclusions. The results should be presented in a friendly way to help laypersons understand the results, conclusions, and justification of the conclusions. All the critical results must be included. Modern presentation techniques allow for attractive layouts.

3.7 CLOSURE

The high costs of transportation investments require sound decisions. Evaluation of transportation systems is needed and required before a decision can be made, and this evaluation can be complex if its subject is complex. Modeling makes the evaluation task manageable.

Good modeling of transportation systems is resource-demanding. The costs can be reduced by using an adequate modeling tool that does not require redundant data and extensive calibration. Planning models and microsimulation are typically data-demanding. Combining these two creates large computational and input requirements. It should be kept in mind that an inappropriate model or one having incorrect calibration may produce inaccurate results.

Ethical issues are frequently present in transportation modeling and evaluation. The high costs of good modeling, the shortage of time and data, and the quality-control difficulties create conditions where the best modeling and evaluation practice may be replaced by a substitute process where the tools and outcomes are accepted because the results are timely available and seemingly usable.

This chapter discussed, in general terms, conditions of technically sound modeling and evaluation of transportation systems. More detailed presentation of various modeling and evaluation tools can be found in other chapters of this Handbook. The reader may visit the Online TDM Encyclopedia, created and maintained by the Victoria Transport Policy Institute, at http://www.vtpi.org/tdm/. This site offers a practical compendium of transportation demand management methods and issues. The United States Department of Transportation sponsors a clearinghouse for information on travel demand forecasting at http://tmip. tamu.edu/clearinghouse/. This site contains over 100 documents well organized and classified. Readers who would like to learn more about existing microsimulation models should visit http://www.its.leeds.ac.uk/projects/smartest/deliv3.html, which lists 32 existing microsimulation models and discusses various aspects of microsimulation.

3.8 REFERENCES

Barton-Aschman Associates, Inc., and Cambridge Systematics, Inc. 1997. *Model Validation and Reasonableness Checking Manual.* Prepared for Travel Model Improvement Program, Federal Highway Administration, http://tmip.tamu.edu/clearinghouse/docs/mvrcm/.

Bureau of Public Roads. 1964. *Traffic Assignment Manual.* Washington, DC: U.S. Department of Commerce, Urban Planning Division.

Central Area Loop Study. 2002. Final Recommendations. In *In the Loop*, The Official Newsletter of the Central Area Loop Study, no. 3. Published at http://www.oki.org/, January.

Chiu, C. Y., and C. M. Walton. 2002. "An Integrated Simulation Method to Evaluate the Impact of New Large Aircraft on Passenger Flows at Airport Terminals." Paper presented at the 2002 Annual Meeting of the Transportation Research Board, Washington, DC.

Driehaus, B. 2001. "OKI rejects 'sky loop' elevated rail system." *The Cincinnati Post, online edition,* http://www.cincypost.com/2001/sep/26/oki092601.html, September 26.

Fredendall, L. D., J. E. Hill, and E. Hill. 2000. *Basics of Supply Chain Management.* Boca Raton: Lewis.

Harwood, D. W., F. M. Council, E. Hauer, W. E. Hughes, and A. Vogt. *Prediction of the Expected Safety Performance of Rural Two-Lane Highways.* Federal Highway Administration, Report No. FHWA-RD-99-207, http://www.fhwa.dot.gov/ihsdm/, December.

Jack Faucett Associates. 1994. *Evaluation of MOBILE Vehicle Emission Model.* U.S. Department of Transportation, National Transportation Library, http://ntl.bts.gov/DOCS/mob.html.

Johnston, R. A., C. J. Rodier, J. E. Abraham, J. D. Hunt, and G. J. Tonkin. 2001. *Applying an Integrated Model to the Evaluation of Travel Demand Management Policies in the Sacramento Region.* MTI REPORT 01-03, The Mineta Transportation Institute, College of Business, San José State University, San José, CA. http://www.iistps.sjsu.edu/publications/SAC.htm, September.

Los Alamos National Laboratory. 2002. "TRansportation ANalysis SIMulation System (TRANSIMS)." *Portland Study Reports* 0-7, August.

McTrans. 2002. HCS2000 Version 4.1c. *On-line Products Catalog,* at http://mctrans.ce.ufl.edu/index.htm, last visited December 2002.

Meyer, M. D., and E. J. Miller. 1984. *Urban Transportation Planning: A Decision-Oriented Approach.* New York: McGraw-Hill.

Peeta, S., and A. Ziliaskopolous. 2001. "Foundations of Dynamic Traffic Assignment: The Past, the Present, and the Future." *Networks and Spatial Economics* 1:233–65.

Raney, B., A. Voellmy, N. Cetin, M. Vrtic, and K. Nagel. 2002. *Towards a Microscopic Traffic Simulation of All of Switzerland.* Department of Computer Science and Institute for Transportation Planning, ETH, Zürich, Switzerland, http://www.inf.ethz.ch/~nagel/papers/cse-adam-02/html/, last visited December 2002.

Robertson, H. D., ed. 2000. *Manual of Transportation Engineering Studies.* Washington, DC: Institute of Transportation Engineers.

Schintler, L. A., and M. A. Farooque. 2001. *Partners In Motion and Traffic Congestion in the Washington, D.C. Metropolitan Area.* Center for Transportation Policy and Logistics, School of Public Policy, George Mason University, prepared for Federal Highway Administration, Virginia Department of Transportation, School of Public Policy, George Mason University, Fairfax, Va, at http://www.itsdocs.fhwa.dot.gov/JPODOCS/REPTS_TE/13500.html.

Sheffi, Y. 1985. *Urban Transportation Networks: Equilibrium Analysis with Mathematical Programming Methods.* Englewood Cliffs: Prentice-Hall.

Sky Loop Committee and Taxi 2000 Corporation. 2001. *A Rebuttal to the Central Area Loop Study Draft Final Report*, Cincinnati, Ohio, at http://www.skyloop.org/cals/cals-ending.htm, September.

Transportation Research Board (TRB). 2000. *Highway Capacity Manual.* National Research Council, TRB, Washington, DC.

U.S. Department of Transportation. 1998. *Transportation Case Studies in GIS—Case Study 4: Maine Department of Transportation Statewide Travel Demand Model.* Federal Highway Administration, Office of Environment and Planning, Washington, DC, http://tmip.tamu.edu/clearinghouse/docs/gis/maine/maine.pdf, September.

Van Aerde and the Transportation Systems Research Group. 1995. *INTEGRATION—Release* 2, User's Guide, December.

Volpe Center. 2002. *Integrated Noise Model (INM) Version 6.0 Technical Manual.* U.S. Department of Transportation, Federal Aviation Administration, FAA-AEE-97-04.

Wade, A. R., D. Harwood, J. P. Chrstos, and W. D. Glauz. 2000. *The Capacity and Enhancement of VDANL and TWOPAS for Analyzing Vehicle Performance on Upgrades and Downgrades within IHSDM.* Federal Highway Administration, Report No. FHWA-RD-00-078, http://www.tfhrc.gov/safety/00-078.pdf, January.

Wang, Y., and P. D. Prevedouros. 1998. "Comparison of INTEGRATION, TSIS/CORSIM and WATSim in Replicating Volumes and Speeds on Three Small Networks." Paper presented at the 1998 Annual Meeting of the Transportation Research Board, Washington, DC, January.

CHAPTER 4
SOFTWARE SYSTEMS AND SIMULATION FOR TRANSPORTATION APPLICATIONS

Elena Shenk Prassas

*Department of Civil Engineering, Polytechnic University,
Brooklyn, New York*

Senior members of the profession speak of a time when signal optimization was done by physical time-space boards on which the intersection signal timing patterns were slid relative to each other (with an overlay of cords sloped for lead vehicle trajectories), and of a time when traffic assignment models were executed overnight on a mainframe computer, in a batch mode. Most data collection was by manual counts and classifications, or by roadside mechanical counters that contained paper rolls on which counts were printed.

Thanks to massive advances in computing and communications technology, we now live in an era in which much data can be downloaded electronically, transmitted wirelessly, input into spreadsheet programs, and presented in summary reports and visuals by automated processes. In traffic control, data availability in real time is a reality. Sophisticated computer programs aid computation, signal optimization, and network assignments. This includes dynamic traffic assignments, responsive to incidents and events. Networks are modeled and simulated to the microscopic level. The emphasis on intelligent transportation systems (ITS) has raised the level of both technological capability and public expectation. Global positioning systems (GPS), geographic information systems (GIS), pervasive cellular telephones, and electronic toll-collection systems have been added to the repertoire.

At the same time, much of the sensing is still done by road tubes, albeit linked to sophisticated devices. Integration of these varied technological capabilities into a seamless system of information is still a challenge. The very power and sophistication of some computational tools has led to a new generation of planning-level estimators, consistent with the newer operational tools.

The purpose of this chapter is to provide some information on the span of software systems and simulation tools available to practicing engineers and planners. Much of the emphasis is on traffic tools, but there seems to be a convergence of traffic and transportation tools—and thought—as the field develops.

4.1 COMPUTATION VERSUS SIMULATION

Some use the words "simulation" and "computation" rather interchangeably. It is better to make the clear distinction that if an equation or algorithm is well defined but being executed by computer for speed or efficiency, it is a *computational* tool; if subsystems are modeled, randomness introduced, and the subsystems linked, then it is generally "a *simulation* tool."

In this spirit, the *Highway Capacity Manual* (*HCM*) (TRB 2000) is a collection of computational procedures that are being realized in a set of computational tools, such as HCS and HCM/Cinema. A number of signal-optimization programs, including TRANSYT and PASSER, are also computational tools. Some tools that combine capacity and signal optimization, such as SIGNAL*2000*, SIG/Cinema, and SYNCHRO, also fit within this description.*

One of the earliest simulation tools was the UTCS-1 (FHWA 1974; see also Lieberman, Worrall, and Bruggeman 1972), prepared for the Federal Highway Administration (FHWA) as part of the Urban Traffic Control System testbed in Washington, DC, dating to the late 1960s. This was a true simulation, in that such subsystems as turning, speed selection and car-following, and lane selection were modeled stochastically, calibrated individually, and then incorporated into a system that simulated traffic movements on arterials and in networks at a microscopic (i.e., individual vehicle) level.

This tool later evolved into NETSIM (Andrews 1989) and was validated over the years at the macro level of overall network flows, speeds, and delay. An analogous tool was developed for traffic on freeways and aptly named FRESIM (FHWA 1994). These were integrated under the direction of FHWA into a corridor tool, CORSIM (Kaman Sciences Corporation 1996). Other corridor tools, such as WATSIM, have been produced as competing products, as have such traffic simulators as VISSIM and PARAMICS.**

The newer generations of transportation assignment tools are linked to such computational tools and/or to assignment models. Examples include EMME, AIMSUN, and TRANPLAN.†

One system for organizing the links between various computational and simulation models is TEAPAC*2000*.‡

4.1.1 Important Areas and Tools Not Covered

This chapter does not address a host of transportation computational tools available in the profession: transit scheduling and routing algorithms; commercial vehicle route selection or dynamic routing; emerging multimodal tools for corridor assessment; and several other areas. The selection for this chapter represents a judgment call on the set of tools and principles of interest to many practicing engineers and planners in their common practice and needs.

4.2 SENSING TRAFFIC BY VIRTUAL DETECTORS

The mainstay of traffic detection for decades has been magnetic sensing of vehicles by a loop installed into the road surface. While relatively reliable, complete intersection coverage is expensive, and the desired detail (vehicle classification, extent of queue, speed estimation) is lacking.

To a large extent, the newest tool is a standard or infrared video camera covering an area such as an intersection approach, coupled with software that creates "virtual detectors" that

---

*For HCS, TRANSYT, and PASSER, see the McTrans website, www.mctrans.ce.ufl.edu/. For HCM/CINEMA and SIG/CINEMA, see www.kldassociates.com/. For SIGNAL*2000*, see www.strongconcepts.com/. For SYNCHRO, see www.trafficware.com/.

**For WATSIM, see www.kldassociates.com. For VISSIM, see www.atacenter.org/tst/Vissim.html. For PARAMICS, see www.paramics-online.com/index2.htm.

†For EMME, see www.inro.ca/. For AIMSUN, see www.tss-bcn.com/news.html. For TRANPLAN, see www.uagworld.com/tranplan/index.html.

‡For TEAPAC*2000*, see www.strongconcepts.com/.

FIGURE 4.1   Software-based virtual detector, an illustration. (*Note:* five user-defined detectors are shown.)

are defined and located by the user, to observe counts at points, occupancy at points or areas (including queue extent), and speeds. Some capability to classify vehicles may also be included.

Using such a tool, the transportation professional can "locate" numerous "detectors" essentially by drawing them on top of the intersection image, and depending upon the software to process the data. Refer to Figure 4.1 for an illustration of virtual detector placement.

AutoScope is one of the pioneering video imaging systems; there are now a number of such products in the international ITS community.

The use of infrared imaging allows vehicles to be detected in a variety of weather conditions. The use of sophisticated algorithms based upon coverage of the underlying pavement image allows data to be collected from stationary traffic as well as moving traffic.

The literature now shows variants of these concepts, including laser systems for traffic detection, as well as a new generation of radar detectors.

4.3   TRAFFIC SOFTWARE SYSTEMS

There are national FHWA-designated software distribution centers, notably McTrans at the University of Florida (www.mctrans.ce.ufl.edu) and PC-TRANS at Kansas University (www.kutc.ku.edu/pctrans). Originally funded by FHWA during their startup phase, they now exist based upon software sales and related services.

To appreciate the range of software products available to the user community, note the categories listed on those two websites:

- Capacity analysis
- Construction management
- Demand modeling
- Environmental engineering
- General traffic
- Highway design
- Highway hydraulics
- Highway surveying
- Mapping and GIS
- Network assignment
- Paratransit operation
- Pavements/maintenance
- Planning data
- Project management
- Safety and accidents
- Signal timing/signal warrants
- Site analysis
- Structural engineering
- Traffic data
- Traffic maintenance
- Traffic simulation
- Transit operations
- Transit planning

plus categories named "Utilities (Computer)" at one site and "General Interest" and "McTrans Training" at the other.

From this list, the present chapter limits its attention primarily to a few categories: traffic simulation and, to some extent, signal timing and capacity analysis.

4.4 BASICS OF SIMULATION

Simulation is used extensively in manufacturing, traffic engineering, and other areas. The simulation models themselves can be classified into generic tools and application-specific models. Modern manufacturing simulations allow the user to define a production line or process by dragging icons of process steps into a logical order in a Windows environment, and to define the characteristics of each process step as well as the workload (demand) characteristics. Likewise, the report statistics may be defined by the user. The actual mechanisms of the simulation are rather invisible to the user.

Traffic-simulation models tend to be more application-specific; thus, NETSIM was tailored to only vehicular traffic on surface street networks. The network definition has become more user-friendly over the years, but is not yet the drag-and-click approach in generic manufacturing models. CORSIM—which includes both NETSIM and FRESIM—was intended to be more user-friendly. Several other packages (TEAPAC, AIMSUN) are designed to emphasize user-friendly links between models and/or easy network building.

4.4.1 Random Number Generation

The most basic feature of simulation is randomness, and yet digital computers are deterministic by their very nature. How then does one generate random numbers on a computer?

The answer is that one creates a computer code that generates a sequence of numbers that *appear* to be random. In fact, there is a pattern, but it repeats only after millions of numbers. Sequential numbers in the chain do not appear to be correlated to each other even with rather sophisticated statistical techniques. These "pseudo-random" number generators are now used extensively and are rather sophisticated in masking the underlying relation. Indeed, they are so routine that spreadsheets now incorporate random-number-generating capabilities.

In many applications, the user can specify a starting or "seed" number for the chain. By specifying a different number, the user picks up the chain in a different place for each choice.

The user can also specify the same number and be assured that exactly the same numbers are generated in exactly the same pattern. This is extremely useful when one wishes to see how two different control policies affect *exactly* the same traffic, something that is impossible in the real world.

From a statistical point of view, it also allows paired *t*-tests to be run on the performance data from a set of N replications under two different control policies, if the same set of N different seed numbers is used in the N runs for each policy.

4.4.2 Time-Based versus Event-Based Simulation

It is rather natural to think of moving ahead in discrete steps of time, say one second into the future. With knowledge of vehicle speeds and positions, a car-following relation, a lane-changing rule, and certain other rules, we can then simulate the next positions (and speeds) of individual vehicles. We can also estimate whether a new vehicle entered the system in each entry lane on each link. Further, we can gather information needed for the performance indices and vehicle trajectories. This approach is *discrete time-based* simulation.

Another approach is event-based simulation. Consider a single waiting line on which there are two important events: a customer service is completed, or a new customer arrives. Rather than estimating the probability of each event in the next second, we can generate the time t_1 to the next completion-of-service and the time t_2 to the next arrival, each generated from an appropriate distribution. We can then jump ahead by whichever is the smaller of t_1 and t_2 seconds and continue to jump ahead from one event to another.

Event-based simulation of neatly contained systems (such as the above-cited bank teller) generally involves fewer steps over the simulation period than a time-based simulation of the same process, and can therefore be much faster running and even require less computer memory. However, the events must be very well defined, and the actual simulation code tends to be built around the events. If the process mechanisms are redefined—one customer line, multiple servers or various types of customers and toll types—the number and definition of events can change, complicating the modeling, shortening the time between events, driving up the computation time and complexity. Nominal advantages over a time-based approach can thereby erode.

From the way many of us are educated, we find orderly discrete steps to be more natural in our thinking and planning. In addition, it is easier to think of a modular approach—the rules for car-following and for lane-changing are separable, even if they interact; desired free-flow speeds can be tagged to vehicles more easily; vehicles can be followed through the system.

The modular construct (that is, the construction by modules) is important, for we have certainly learned that traffic simulation models evolve as experience is gained and new needs are defined. Also, the power of modern desktop and notebook computers makes the older concerns for computational burden and memory efficiency less important than they once were.

4.4.3 Modeling the Mechanisms

The first issue is actually *enumerating* the mechanisms of the proposed simulation model and defining their interactions. This done, it is then necessary to model the individual mechanisms, link them, use them, and refine them in some iterative process that introduces reality. Consider the simple case of traffic traveling on an arterial. The primary mechanisms might be:

- Vehicle arrives.
- Vehicle travels in lane at desired speed.
- Vehicle interacts with others (car-following).
- Vehicle changes lane, or not (Overtaking? Turning? When? Why?).
- Signal indication influences vehicle.
- Vehicle decelerates.
- Vehicle accelerates.

This simplified list does not include the ways in which the vehicle may interact with pedestrians, geometrics (grade of road, for instance), weather, or sun glare. Nor does the list include the ways in which cross-traffic is considered, or many other factors.

Nonetheless, having defined these basic mechanisms, the immediate challenge is then to establish the details of the mechanisms, the links to other mechanisms (because they are generally interdependent), and the calibration/validation data needs.

Moreover, there must be a guiding principle that *anticipates the uses* to which the model will be put. This is often elusive, because users find new applications and because the state of the art changes. Nonetheless, the need exists.

Consider the simple decision to affix an identifying tag to each vehicle as it enters the network, which it retains as it moves through the network. This allows: (1) future applications to assign a route to that particular vehicle, and to update that route periodically; (2) desired speed and accel/decel patterns to be linked to that vehicle; (3) location data to be stored at each increment of time, allowing trajectories to be reconstructed. Without such an identity tag, the same vehicle might bounce through the network by Brownian motion, subject to a series of random turn decisions (with the turn percentage being a characteristic of the intersection), with no meaningful path information recoverable.

4.4.4 Calibrating the Mechanisms and/or the Model

If the mechanisms are constructed as linked modular entities, it should be possible to calibrate each mechanism individually.

Consider the left-turn mechanism under permissive signalization, which is basically a gap-acceptance rule that considers the size of the available gap, the speed of the approaching vehicle, and perhaps the type of opposing vehicle. It might also consider the type of left-turning vehicle, the "pressure" from queued vehicles behind the turner, the number of lanes to be crossed, and other factors.

The challenge is to define the mechanism in a way that is both realistic and capable of calibration with an affordable amount of data. The starting point is often the traffic theory literature. Once the data plan is constructed, it must be executed at sufficient sites to assure credibility.

At this level of detail, it may be that the model as an entity is not calibrated as such, but that it is run for reality checks on whether the overall performance is credible.

4.4.5 Validating the Mechanisms and/or the Model

Validation is a distinct operation from calibration and requires data reserved or collected for the purpose. Calibration data cannot be used to validate a model, simply because it is then a self-fulfilling prophecy.

Validation can be done on two levels, the microscopic checking of individual mechanisms and the macroscopic checking of aggregate performance measures. The latter category may verify that arterial travel times, average speeds, and delay conform to real-world observations when observed traffic data (volumes, composition, etc.) are fed into the model. Because this is a statistical test, usually operating with a null hypothesis that the model and real world results are the same, a validation may require a significant amount of data.

Microscopic validation of individual mechanisms is done less frequently, simply because of the data needs. However, it is logical that it be part of every research project in which particular mechanism(s) are pivotal in determining the aggregate results, or the specific performance is being investigated as part of the research.

A close look at the details of the simulation model may reveal items in which a new and detailed mechanism calibration is needed before the anticipated research investigation can be done. For instance, CORSIM has a mechanism by which vehicles can turn into or out of driveways. However, this mechanism does not include any details on the geometries of the driveways, and therefore CORSIM (as an example) could not be used to investigate such features.

4.5 ISSUES IN TRAFFIC SIMULATION

A number of issues in planning a traffic simulation deserve special attention.

4.5.1 Number of Replications

Assuming that the traffic simulation model is as random as the real world, at least in its internal mechanisms, the resulting performance measures will be *samples* or *observations* from a set of possible outcomes. That is, each performance measure is a random variable, with a mean and variance (among other properties).

Because of this, several runs (or "replications") of the situations are needed, each with its own set of seed random numbers. If one desires a certain confidence bound, a considerable number of replications might be required. If one can only afford a limited number of replications, then the resultant confidence bound might be disappointing.

Some practitioners cite a rule of thumb that three replications are the minimum. Many have translated that to mean that three replications will suffice. *This is without foundation, and confidence bounds must be considered in each case.*

Consider the following nine replications in which the average speed is read from simulation results:

Run	Average speed (mph)
1	45.2
2	52.5
3	43.7
4	48.4
5	47.3
6	53.2
7	46.7
8	42.9
9	50.1

From these results, one can estimate a mean of 47.8 mph and a standard deviation of 3.6 mph. Further, the 95 percent confidence bounds on the mean are ± 2.4 mph. If ± 1.0 mph were desired, then 51 replications would have been required, rather than 9.

If one could only afford 9 replications (or had time for only 9), then the 95 percent confidence bound on the mean would be unavoidably ± 2.4 mph.

Some situations are not as bleak as this particular example. There is also some advantage to be gained when two control policies are to be compared in terms of their effect on a performance measure (in this case, the mean speed).

Consider the case in which nine replications are done with each of two control policies, using the same nine sets of seed numbers for each policy (that is, the identical traffic, so to speak). This "pairs" specific sets of runs and allows us to construct the following:

Run	Control policy one, average speed (mph)	Control policy two, average speed (mph)	Paired difference (mph), CP2-CP1
1	45.2	48.1	2.9
2	52.5	54.3	1.8
3	43.7	44.3	0.6
4	48.4	51.4	3.0
5	47.3	50.2	2.9
6	53.2	55.8	2.6
7	46.7	48.3	1.6
8	42.9	45.2	2.3
9	50.1	53.5	3.4

If a test were done on this data without pairing, under the hypothesis of "the two means are the same," the variability in the data would lead us to *not reject* the hypothesis of equal means. *Even if such a difference existed, it could not be detected.*

However, by pairing the runs according to the seed numbers (i.e., the exact traffic patterns), the differences in each pair can be computed, as shown in the last column above.

In this particular case (which we arranged to be so dramatic), *all* of the differences are positive, with a mean of 2.3 mph and a standard deviation of 0.9 mph. Indeed, the 95 percent confidence bounds on the mean are ± 0.6 mph, and a hypothesis of "zero difference" is easily rejected. While not all cases are so dramatic, the reader must understand the advantages of a well-planned set of runs.

4.5.2 Length of Run

Early on in the practice of simulation, another rule of thumb emerged: runs should be (at least) 15 minutes long. Again, many users interpreted this to be that they can all be 15 minutes long.

Rather than follow such a rule of thumb, the user must focus on the defining events, determine how often they occur, and select the run duration so that a reasonable number of these cases occur during the observation period.

Consider the case in which the productivity of an approach is being considered and there are two defining events—the vehicle at the head of the queue is a left turner who traps everyone else, or it isn't. If the first situation occurs only 10 percent of the time, and by definition can occur only once every cycle length, then a 15-minute period with a 90-second cycle length will have *no* such blockages 35 percent of the runs and only one such blockage another 39 percent of the runs. Lengthening the run duration will dramatically lower these probabilities, if that is desired.

It is good practice to use runs of 1–2 hours of simulated time when a number of such "rare" events should be included in the typical period. Another situation that may lead to longer runs is taking the effect of buses into account, because the interarrival times on even a very busy route is often at most 3–5 minutes.

4.5.3 Detail of Specific Detailed Mechanisms

It was already noted that driveway geometrics were not included in the surface street part of CORSIM, limiting the analysis that can be done on the effect of driveway location/number/design on access management measures.

Consider another case, that of bus traffic in an urban area. If the simulation model specifies a dwell time distribution but does *not* link it to the interarrival time between buses, an important mechanism contributing to platooning of buses can be overlooked. This is because the delayed buses *in the model* do not have more people waiting for them, and early-arriving buses do not have fewer. If a user is looking at overall performance, this may not be a problem. But if the user is trying to study bus platooning specifically, this can be a major problem and the mechanism might have to be created and the model revised.

4.5.4 Avoiding Use of the Model beyond Its Limits

This is simply another aspect of the point just made, but it is a particular challenge when a user is trying a new application. A working knowledge of the model is needed, on a level that only the developers and a few others might have.

4.5.5 Selecting Performance Measures for the Model

The traffic engineering profession has a number of well-established measures (volume, flow rate, speed, delay) and a number of evolving ones, particularly as relate to multimodal considerations. Even so, the measures defined in some simulation models may not conform to the standard definitions and usage. For instance, there are many forms of "speed" used in both practice and in traffic models.

4.5.6 User-Friendly Input and Output

Whereas the preceding items focus on some of the important issues in applying models, the most important user issues tend to be in how user-friendly the input/output is and how efficiently it can link to existing databases, if at all. For instance, TEAPAC, AIMSUN, HCM/ Cinema, and to some extent CORSIM make major points on how clear the input process is and/or how much they link to other models.

4.6 VISUALIZATION

Two aspects of visualization deserve our attention: (1) the displays that tell system managers and the public the condition of the network, and (2) graphic representations of simulator outputs that provide powerful visual images of the situation.

4.6.1 Network Displays

Other sections of this Handbook and such references as Roess, Prassas, and McShane (2003) address the subjects of traffic signal optimization on surface streets and of freeway surveillance and control. These subjects are central themes in ITS applications, with greater emphasis placed on communication with the drivers and on the autonomous decisions then made by those drivers.

There has also been a growing emphasis on color-coded maps of the freeway system, available from traffic control centers and used by local TV stations. These graphically depict the trouble spots in the network at any given time.

Some TV and radio stations also depend heavily on a network of "web-cams" to observe conditions in the network and inform the public. Webcams were once the domain of the agencies operating the facilities and employing video cameras, but it is now relatively common for TV and radio to depend both on these and on privately sponsored web-cams. Some agencies feature such displays on their website or are used by TV and radio on their web sites.

Given the information flowing to drivers through such systems, through in-vehicle routing systems, and through traffic advisory services (often based upon cell phones), an interesting challenge emerges for the transportation professional: how are the drivers going to reroute themselves, and how does this affect the network loading and hence travel times?

This challenge can be stated as trying to estimate the actions of drivers based upon their current location, the origin-destination that underlies their initial route selection, and their (generally unknown) personal decision rules.

Therefore, the technical problems of (a) dynamic traffic reassignment and (b) estimation of O/D patterns from observed data take on greater importance, and in fact blend into the traffic optimization problem.

A related issue is the type of data available from which to estimate the underlying O/D pattern. Some work has assumed that only traffic count observations are available, with turn information. Other work has focused on "probe vehicles" that transmit information from which their exact routing is known. And, of course, with the rather pervasive use of cell phones—and subject to privacy issues related to accessing the data—the route information on many vehicles could be obtained from cell phone locations over time.

4.6.2 Visualization of Simulation Results

More and more, computer models are measured by their ability to effectively communicate the results, in two-dimensional or, preferably, three-dimensional representations.

FIGURE 4.2 VISSIM animation of output, with aerial photographs for context. (*Source:* VISSIM website, www.atacenter.org/tst/Vissim.html, used with permission.)

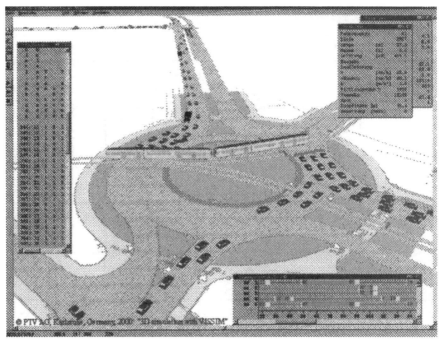

FIGURE 4.3 VISSIM 3-D animation of output, including an LRT line. (*Source:* VISSIM website, www.atacenter.org/tst/Vissim.html, used with permission.)

FIGURE 4.4 AIMSUN 3-D display of animated output. (*Source:* TSS website, www.inro.ca/, used with permission.)

FIGURE 4.5 WATSIM 3-D display, tailored to specific constrained urban setting. (*Source:* KLD Associates materials, www.kldassociates.com, used with permission.)

Figure 4.2 shows the output of the VISSIM model in which the two-dimensional simulation output is shown in the context of aerial photographs of the area; this is a standard feature available in VISSIM. Figure 4.3 shows another VISSIM output, with a three-dimensional perspective. The PARAMICS model also has three-dimensional displays.

Other models provide three-dimensional animations as part of the standard output, or as specialty displays for specific needs. Figure 4.4 shows an AIMSUN output, and Figure 4.5 shows a WATSIM output tailored to an acute operational problem in a very constrained urban environment.

4.7   FHWA NEXT GENERATION SIMULATION (NGSIM) PROGRAM

The Federal Highway Administration (FHWA) of the USDOT has been the pioneering force and sponsor behind the early development of traffic-simulation models, and to this date continues to maintain and support the CORSIM program, as part of its overall TSIS-5 system.

Having been the leader in establishing a viable market for traffic simulation tools, and several commercial products having emerged, FHWA has reviewed its role and redefined its role as a market facilitator. In this role, it will conduct and sponsor research on the building blocks for models (lane change logic, gap acceptance logic, and other fundamental behavioral models) *and* will ensure that the resultant core algorithms, validation data sets, and documentation are widely and openly available. FHWA will also facilitate the deployment and use of existing tools, by a combination of outreach, training, guidance, and technical support within its Traffic Analysis Tools Program.

In its work leading to this new role, FHWA conducted various polls of constituencies, one of which being a survey in which 40 state and local agencies responded to the question, "What are the major barriers to your use of traffic analysis tools?" FHWA reports the following top answers:*

- Lack of trained staff
- Lack of time
- Intensive data-gathering requirements
- Cost of software
- Lack of confidence in the results

To the question "What are the major enablers to your use of traffic analysis tools?" the top answers were reported as:

- Integrated training curriculum
- Short course on specific models
- Access to comprehensive technical assistance
- Face-to-face technical information exchange
- Overview course that introduces models

4.8   WHY SIMULATE? ADVANCED TRAFFIC CONTROL SYSTEMS, AND OTHER

The above user perceptions are interesting and represent significant barriers to general use and acceptance of simulation as a powerful and routine tool. At the same time, better support and information is identified as the solution path.

---

*See FHWA NGSIM Program website, ops.fhwa.dot.gov/Travel/Traffic_Analysis_Tools/ngsim_program.htm.

The most basic reasons for using simulation extensively are straightforward:

1. One can experiment with various control policies without disrupting real traffic and making that traffic part of an experiment.
2. The same can be said of alternative designs for remedies to existing or future conditions, such as lane additions, rerouting of traffic, changes in direction of links, and so forth.
3. Above all, these alternative control implementations and designs can be considered rapidly, without major capital investment.
4. Whenever desired, various alternatives can be considered with exactly the same traffic load and conditions, something that is simply not feasible in the real world.

Refer to Figure 4.6 for an illustration of a setting in which design and control changes may be useful in seeking a remedy.

The potential place of traffic simulators in real-time control is another important reason. Historically, simulators and signal optimization programs dependent upon simulators could really only be used off-line. Figure 4.7 shows that if one aspires to truly real-time control, the time period available for the computations is rather short. Therefore, the speed of the available computers and/or the simulator or computational tool is of critical importance.

Historically, the computer speed and computational burden has limited the options available for control. Table 4.1 shows the three planned generations of the FHWA Urban Traffic Control System (UTCS); these have to be viewed in the context of computing power available at the time.

Figure 4.8 shows the typical control loop, with emphasis on the logical place (and time) in which the computations and simulation has to be done. If it is not feasible to do so, the control interval has to be made longer (as in the early UTCS generations) or the aspiration deferred for faster models and/or faster computers.

FIGURE 4.6   3-D display of acute traffic problem, prepared to consider signal and design changes. (*Source:* KLD Associates materials, www.kldassociates.com, used with permission.)

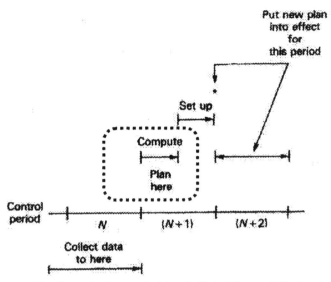

FIGURE 4.7 A timeline emphasizing the limited time available for true on-line control computations. (*Source:* Roess, Prassas, and McShane 2003, used with permission.)

4.9 TRAFFIC SIMULATORS

At the beginning of this chapter, a number of simulators and computational tools that embed or use simulations were enumerated. At the risk of overlooking a key model, it will simply be noted that the available and widely used tools (nationally and internationally) include CORSIM, WATSIM, VISSIM, PARAMICS, AIMSUN, and TEAPAC2000. There are other tools that incorporate some simulation, such as HCM/Cinema (for intersection animation),

TABLE 4.1 Three Generations of Control, Developed in the Context of Computing Speed and Algorithms

Feature	First generation	Second generation	Third generation
Optimization	Off-line	On-line	On-line
Frequency of update	15 Minutes	5 Minutes	3–6 Minutes
No. of timing patterns	Up to 40 (7 used)	Unlimited	Unlimited
Traffic prediction	No	Yes	Yes
Critical intersection control	Adjusts split	Adjusts split and offset	Adjusts split, offset, and cycle
Hierarchies of control	Pattern selection	Pattern computation	Congested and medium flow
Fixed cycle length	Within each section	Within variable groups of intersections	No fixed cycle length

Source: "The Urban Traffic: Control System in Washington, D.C.," U.S. Department of Transportation, Federal Highway Administration, 1974.

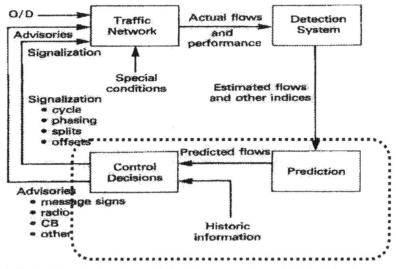

FIGURE 4.8   The place of control algorithm computations, and of simulation, in a true real-time control loop. (*Source:* Roess, Prassas, and McShane 2003, used with permission.)

and others that handle signal optimization and/or capacity, including SIGNAL*2000* and HCS, without simulation.

Figure 4.9 shows a short section of an arterial that is a representative small problem for a simulator. Figure 4.10 shows the corresponding link-node representation used in such tools as CORSIM and WATSIM. Some models, including VISSIM, emphasize a link-connector basis rather than a link-node structure; the advantage cited is flexibility in constructing complex intersections or lane alignments.

Table 4.2 shows a representative tabular output corresponding to the network just cited. The metrics by link include vehicle miles and trips, delay, time in queue, stopped time, and speed.

Figure 4.11 shows a diagram of the TEAPAC*2000* system and the relation to other tools.

Figure 4.12 shows an illustrative input screen for HCM/Cinema, which embeds a simulator to generate animations and certain statistics.

Because there are a number of models in the commercial market, each with its own satisfied user base, this chapter will not attempt to rank or rate the alternative models. Indeed, the literature contains little information on extensive *comparative* testing of various models relative to each other or to base cases with real field data.

Several of the models emphasize their suitability for ITS applications. One of the more interesting applications is the modeling of toll plazas with a mix of exact change, cash, and electronic toll collection (ETC). Because such systems are one of the most visible implementations of seamless regional transportation technology, and because ETC and "smart cards" can contribute important data, the next section addresses these topics.

As the use of these technologies spread even more widely, the potential for a new range— and quality—of network data becomes available. This shapes what simulators *can* do, and *how* they must be designed, to address problems that heretofore were deemed infeasible.

4.10   *ELECTRONIC TOLL COLLECTION AND SMART CARDS*

At the time of this writing, it is still appropriate to treat electronic toll collection (ETC) and the smart card as distinct topics, but the distinction is blurring as the field advances. The

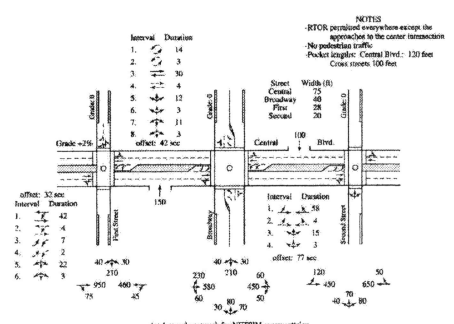

(a) A sample network for NETSIM representation

FIGURE 4.9 A sample arterial, to be coded into a simulation model. (*Source:* Roess, Prassas, and McShane 2003, used with permission.)

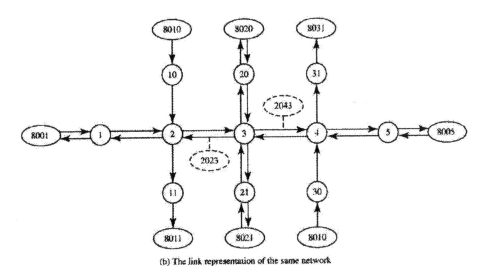

(b) The link representation of the same network

FIGURE 4.10 The same network, shown as nodes and links. (*Source:* Roess, Prassas, and McShane, used with permission.)

TABLE 4.2 Representative Summary Output of a Simulation Model

Part of an illustrative TRAF-NETSIM report

Cumulative NETSIM statistics at time 7:13:25

Elapsed time is 0:13:25 (805 seconds), time period 1 elapsed time is 805 seconds

Link	Vehicle		Vehicle minutes			Ratio move/total	Minutes/mile		Seconds/vehicle				Average values		
	Miles	Trips	Move time	Delay time	Total time		Total time	Delay time	Total time	Delay time	Queue time	Stop time	Stops (%)	Volume (vph)	Speed (mph)
(8010, 10)		314												1404	
(10, 1)	29.73	314	51.3	168.4	219.7	0.23	7.39	5.66	42.0	32.2	23.9	22.8	64	1404	8.1
(1, 2)	66.97	268	115.6	86.1	201.6	0.57	3.01	1.29	45.1	19.3	12.2	11.8	33	1198	19.9
(2, 3)	83.75	268	144.5	93.2	237.7	0.61	2.84	1.11	53.2	20.9	10.7	10.2	47	1198	21.1
(3, 4)	41.24	242	71.2	39.0	110.2	0.65	2.67	0.95	27.3	9.7	4.9	4.8	18	1082	22.5
(4, 5)	34.65	229	59.8	28.6	88.4	0.68	2.55	0.83	23.2	7.5	3.4	3.3	10	1024	23.5
(5, 6)	15.91	224	27.5	46.5	74.0	0.37	4.65	2.93	19.8	12.5	9.8	9.7	18	1001	12.9
(6, 20)	19.02	201	32.8	4.8	37.7	0.87	1.98	0.25	11.2	1.4	0.0	0.0	0	898	30.3
(8030, 30)		161												720	
(30, 1)	9.43	166	18.9	106.3	125.2	0.15	13.27	11.27	45.2	38.4	30.3	29.3	80	742	4.5
(1, 35)	10.68	190	21.4	4.4	25.8	0.83	2.41	0.41	8.1	1.4	0.0	0.0	0	849	24.9
(8035, 35)		161												720	

Data acquired using TRAF-NETSIM, U.S. Department of Transportation, Federal Highway Administration.
Source: Roess, Prassas and McShane 2003, used with permission.

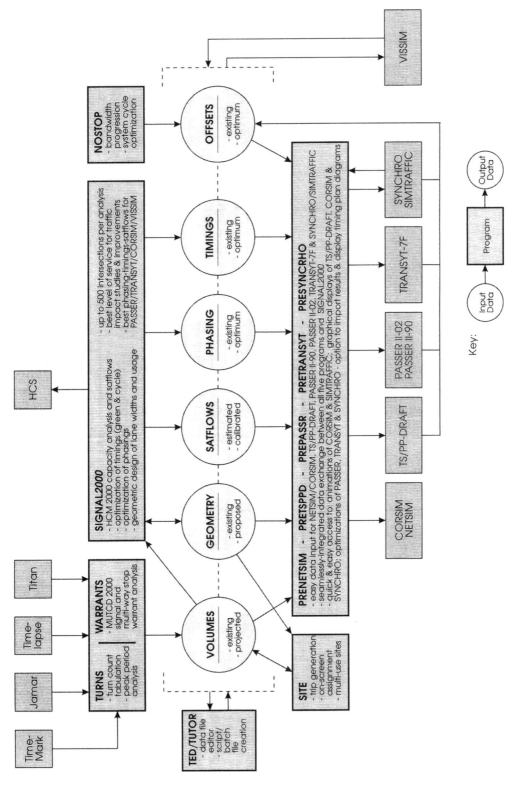

FIGURE 4.11 TEAPAC2000 and its linkages for an integrated system. (*Source:* Strong Concepts website, www.strongconcepts.com, used with permission.)

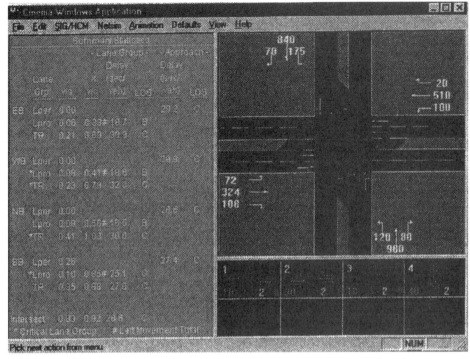

FIGURE 4.12   HCM/Cinema display. (*Source:* KLD Associates materials, www.kldassociates.com, used with permission.)

concept of one debit card to be used seamlessly on different transit services and also on toll roads is exceptionally attractive to the public and is evolving rapidly in the United States and elsewhere.

One of the highly visible systems in the United States is the E-Z Pass system used in a number of jurisdictions in New York, New Jersey, and more recently Delaware. The great appeal is that *one* in-vehicle device can be used on many different facilities, with the driver not concerned with which agency operates what facility.

Underlying the tag's popularity is of course special toll lanes that accept only the E-Z Pass and have much smaller queues (if any); refer to Figure 4.13. Another major feature is discounts for tag users, compared to tolls paid at the standard lanes.

At the time E-Z Pass was introduced, a great concern was whether the public would accept and use the system. In 2001, there were some 2 million E-Z Pass tags in use, and—in order to encourage reasonable speeds through the toll booths—the penalty that was presented for violations was suspension of the right to use E-Z Pass.

When the E-Z Pass system was introduced on the New Jersey Turnpike in 2000, the immediate effect was a 40 percent usage of the special lanes, thanks in good part to the extensive number of users in the region.

In Florida, a tag system known as E-Pass is in use. As of this writing, an extensive operational test of a public-private effort named the "Orlando Regional Alliance for Next Generation Electronic payment Systems" (ORANGES). This system is to create a seamless, multimodal electronic payment system that includes toll roads, bus transit riders, and parking services.

FIGURE 4.13 E-Z Pass booths in the New York metropolitan area. (*Source: Tolltrans* 2001, 45, used with permission.)

There are in fact a variety of ETC systems in use or planning throughout the world, including projects in Europe, Brazil, Canada, Australia, China, Malaysia, and Thailand.

Three issues/trends have been emerging since the initiation of the ETC systems:

1. Data can be collected from the tags to estimate travel times on sections of road and between toll stations. The same data can be used for noting routing patterns. At the present time, one of the uses of travel time data is providing the public with estimates of the condition of the system, a benign application. But the potential for using the same tags for speed enforcement exists, although it currently would be viewed as counterproductive in encouraging card use. The ability to establish routings and to know starting times, down to a personal level, also raises questions of privacy, an issue that is a companion to such an information-rich environment.

2. A new generation is emerging of toll stations that do not have (or need) the physical infrastructure of existing toll plazas, and can sense vehicles at prevailing speeds.

3. The pervasiveness of the tags enables congestion pricing to be considered as technically feasible. Indeed, various discounting options already in effect can lay the groundwork for public acceptance—and perhaps enthusiasm—for such systems.

Systems using smart cards have generally been of the variety of seamless travel using a common card on public transportation, thus making the trips easier. In New York City, the MetroCard (perhaps not fully enabled yet as a smart card) allows users free transfers between bus and subway if the transfer is made within a certain time. Other systems use variable pricing with distance and time of day.

As already cited, the ORANGES project is extending the use of smart cards from public transportation to parking and to toll roads.

One should look forward and consider at least four enticing issues with future generations of smart cards:

1. The opportunity to encourage seamless and efficient use of all transportation modes

2. The ability to introduce variable cost with time of day and with distance, thus moving to a wider use of the congestion pricing or road user cost concepts

3. The extensive database that is being created, and the potential to use that data to obtain both traffic statistics (volume, speeds, travel times) and O/D estimates for use in

- Building historical patterns, by season, weather, and other factors
- Observing trends and changes in the historical patterns
- Use in planning, in scheduling, and even in revising transit routes
- Real-time use of the data in traffic advisories and control, as well as response to incidents

4.11 PLANNING TOOLS

This chapter has focused on computational models and simulation models that are becoming more powerful over time and more capable of addressing sophisticated problems. This is proper, logical, and accurate. The profession is to some extent moving to a set of tools that simply cannot be executed by hand. For a document such as the *Highway Capacity Manual* to recognize and accept this principle for operational analyses is truly ground-breaking.

At the same time, there is a growing demand for *planning* techniques that are fast, simple, and easy to use and communicate. It is generally acknowledged that these must be consistent with operational techniques, but capable of giving preliminary guidance on the capability of facilities in terms of AADT and quality of flow. The Florida DOT (2002) has been a pioneer in this development of modern simplified tools to handle both the early stages of planning and some aggregate assessment of operational impacts.

This is given special note because these models contribute to the development of a spectrum of tools available to the professional, suited to the precision and data availability needed at different stages; in this continuum, it is the simulation models that often represent the most sophisticated of the approaches. And it is somewhat fascinating that the barriers and needs identified by FHWA for simulation tools (cited in this chapter) apply equally well to the planning tools of FDOT—training and technical support, in several forms.

4.12 RESOURCES FOR FURTHER INFORMATION

To a large extent, the references for this chapter have been listed as a set of websites, covering most of the well-used models and some software distribution centers.

4.13 REFERENCES

Andrews, B., E. Lieberman, and A. Santiago. 1989. "The NETSIM Graphics System." *Transportation Research Record* 1112.

Federal Highway Administration (FHWA). 1974. *The Urban Traffic Control System in Washington DC.* U.S. DOT, FHWA, Washington, DC, September.

———. 1994. *FRESIM User Guide, Version 4.5.* Turner Fairbank Highway Research Center, U.S. DOT, FHWA, Washington, DC, April.

Florida Department of Transportation. 2002. *Quality and Level of Service Handbook* Tallahassee.

Kaman Sciences Corporation. 1996. *TSIS User's Guide, Version 4.0 beta, CORSIM User's Guide Version 1.0 beta,* January.

Lieberman, E., et al. 1972. "Logical Design and Demonstration of UTCS-1 Network Simulation Model." *Transportation Research Record* 409.

Roess, R. P., W. R. McShane, and E. S. Prassas. 1998. *Traffic Engineering,* 2nd ed. Englewood Cliffs: Prentice-Hall.

Roess, R. P., E. S. Prassas, and W. R. McShane. 2003. *Traffic Engineering*, 3rd ed. Englewood Cliffs: Prentice-Hall.

Tolltrans—Traffic Technology International Supplement. 2001.

Transportation Research Board (TRB). 2000. *Highway Capacity Manual*. National Research Council, TRB, Washington, DC.

CHAPTER 5
APPLICATIONS OF GIS IN TRANSPORTATION

Gary S. Spring
Department of Civil Engineering, Merrimack College,
North Andover, Massachusetts

5.1 INTRODUCTION

The application of geographic information systems in the transportation industry (GIS-T) has become widespread in the past decade or so.* Indeed, as Miller and Shaw (2001) put it, GIS-T has arrived and is one of the most important applications of GIS.

In the United States, several key pieces of federal legislation passed in the 1990s— including the Intermodal Surface Transportation Efficiency Act of 1991, the Transportation Equity Act for the 21st Century of 1998 (Meyer 1999), the Clean Air Act Amendments of 1990 (CAAA), and legislation by states that mandate the development of transportation programs to reduce traffic impacts—contain explicit requirements for local and state governments to consider transportation systems through their interdependence with other natural, social, and economic systems. There exists a need, therefore, for enhanced approaches to store, manipulate, and analyze data spanning multiple themes—for example, highway infrastructure, traffic flow, transit characteristics, demographics, and air quality. Later in this chapter, several applications from each of these areas, and others, are described.

GIS offers a data management and modeling platform capable of integrating a vast array of data from various sources, captured at different resolutions, and on seemingly unrelated themes. The objectives of this monograph are to define and describe GIS-T, provide a review of GIS-T application areas, and present issues involved in implementing these systems. The discussion begins with some background to GIS and GIS-T, providing definitions and some discussion regarding the "why" of using them.

5.1.1 Background

The GIS is a computerized database management system that provides graphic access (capture, storage retrieval, analysis, and display) to spatial data. The most visually distinctive feature of GIS software is a map display that allows thematic mapping and graphic output overlain onto a map image. The key element that distinguishes GIS from other data systems is the manner in which geographic data are stored and accessed. The types of GIS packages

---

*The acronym GIS-T is often used when referring to GIS applied to transportation. For a full list of GIS acronyms and jargon, visit the National Center for Geographic Information and Analysis website (NCGIA, 1992) and download Report Number 92-13.

used for transportation applications store geographic data using topological data structures (objects' locations relative to other objects are explicitly stored and therefore are accessible) that allow analyses to be performed that are impossible using traditional data structures. The addition of this spatial dimension to the database system is, of course, the source of power of GIS. Without this dimension, the GIS is merely a database management engine. Linked with the spatial dimension, its database management features enable GIS to capture spatial and topological relationships among geo-referenced entities even when these relationships are not predefined. Standard GIS functions include thematic mapping, statistics, charting, matrix manipulation, decision support systems, modeling algorithms, and simultaneous access to several databases.

In contrast to other GIS applications, GIS-T has as its central focus a transportation network that has an intrinsic complexity due to its varying legal jurisdictions and multimodal nature. The digital representation of such a network is nontrivial and will be discussed in the following section.

5.2   BASIC PRINCIPLES OF GIS-T

5.2.1   GIS-T Data

Data and the modeling of those data are crucial elements of a successful system. The GIS derives information from raw data—one of its primary strengths. Although *data* and *information* are often used interchangeably, they can be quite different. Data consist of facts or numbers representing facts, whereas information derives from data and gives meaning to those facts. This section describes the nature, types, and sources of data and the modeling techniques and data models that are used to convert the raw data to information.

The Nature of Data.   There are two types of data in general: spatial data and attribute data. Spatial data describe the physical geography represented in a database, whereas attribute data, linked to the spatial data, describe the attributes of the spatial objects. For example, a point could represent a tree location and its attributes might be species, diameter, or height. If a line feature represents a road segment, attributes might include pavement type, number of lanes, or speed limit. Area attributes might include soil type, vegetation cover, or land use. An attribute is the generic descriptor for a feature, while each feature has a specific value. In the point example above, likely values for the species attribute would be "pine," "fir," and "aspen." Attributes may be described as the questions that would be asked and values as the answers. Attributes to be collected would be determined in the software planning process, while specific values would be entered in the field.

The two most common ways of representing spatial data are the vector and raster formats. A familiar analogy for these formats is bit-mapped photos (such as jpeg) corresponding to the raster format and vector-based ones (such as tiff) representing the vector format. Each of these approaches has advantages and disadvantages and the choice between them lies in the nature of the application for which the data are to be used. Data structure, that is, choice between raster and vector, is generally one of the first major decisions to be made in establishing a GIS. With regard to GIS-T, this decision is fairly straightforward, however. Transportation applications typically are concerned, as was pointed out earlier, with networks, political boundaries, and so on. These are all well-defined line systems and thus lend themselves better to the vector data structure. This notwithstanding, there are situations in which the raster format is used for transportation applications, and so the next few paragraphs provide a description of both structures, including some discussion of their relative advantages and disadvantages.

The Raster Data Structure.   Raster data consist of rows of uniform cells coded according to data values. For example, a landscape scene would be gridded and each cell in the grid would be given a single landscape identity, usually a code number that refers to a specific attribute measure (e.g., a particular land use or type of land cover). The number might also be an actual measurement value, such as an amount of rainfall. These cells are akin to pixels on a computer monitor. Indeed, carrying the analogy further, pixels have values associated with them as well—color, hue, and shades of grey—and the degree of resolution provided relates to the size of the pixel. Similarly, the degree of approximation in a landscape relates to the size of the raster cell. A grid composed of small cells will follow the true location of a boundary line more closely, for example. However, as with photo resolution, there is an overhead cost, namely the size of the raster file containing the data will increase as cell size decreases. In general, raster provides a simple data model and fast processing speed at the expense of the excellent precision provided by the vector model with its higher data needs. The nature of raster data lends itself to natural resources applications whose data elements tend to be continuous in nature—such as soil type.

The Vector Data Structure.   Vector data consist of points, lines, and closed polygons or areas, much the same as the drawing elements in computer-aided drafting (CAD) programs. The lines are continuous and are not broken into a grid structure. In the vector model, information about points, lines, and polygons is encoded and stored as a collection of x–y coordinates. The location of a point feature, such as a manhole, is described by a single x–y coordinate. Linear features, such as roads, are stored as a string of point coordinates. Area features, such as census tracts or traffic analysis zones, are stored as a closed-loop set of coordinates. The vector model represents discrete features like buildings better than continuous features like soil type. The vector format, in general, provides a more precise description of the location of map features, eliminates the redundancy afforded by the raster model and therefore reduces mass storage needs, and allows for network-based models. Vector models are, however, often computationally intensive—much more so than their raster counterparts.

Vector models are most often used for man-made infrastructure applications, such as transportation. Indeed, for transportation applications, vector-based GIS is by far the most commonly used data structure. GIS-T, in representing guideways (or roadways), uses a coordinate system to store and display primitive feature elements of points, lines, and areas. The roadway is represented as a line feature.

A combined collection of graphical links form a roadway network, but this representation alone is considered as having no intelligence—that is, connectivity (also known as topology) does not exist. Unlike purely graphical software applications, however, GIS can build and manage topology. A road network consists of a series of roadway line features that have the same defined attributes—for example, road name. This connectivity of line features is critical for use in routing and network data modeling. With respect to the latter, GIS network algorithms may be used to determine optimal travel routes, which roadway or other facility is closest, or what is near a particular site. Appropriate topology may be used most effectively if built into the GIS data.

Topology describes how graphical objects connect to one another by defining relative positions of points, lines and areas. For example, topology allows queries about which street lines are adjacent to a census tract area, or what intersection node points form the end points of a street segment (link). The latter information is essential for routing applications. Only vector-based data include topological information, which is in part why these data may have higher costs associated with them—the presence of topology implies the maintenance of additional databases whose purpose is to store information on the connectedness of points, lines, and areas in the database.

Scale and Accuracy.   Modeling the road network as a spatial or graphical layer in GIS must reflect the needs and requirements of an agency. Scale and accuracy are important data considerations, especially when using more than one data source, which is often the case in transportation applications. Generally, the source material and the standards of data development determine both the scale and precision of spatial datasets.

Simply defined, scale is the relationship between distance on the map and distance on the ground. A map scale usually is given as a fraction or a ratio—1/10,000 or 1:10,000. These "representative fraction" scales mean that 1 unit of measurement on the map—1 in. or 1 cm—represents 10,000 of the same units on the ground. Small-scale maps have smaller representative fractions than large-scale maps—for example, 1 in 500,000 versus 1 in 24,000, respectively. That is, large is small. As scale size increases, the ability to depict detail also increases. For example, a map with scale 1 in 500,000 would necessarily represent only main roads and only by centerline, whereas with a scale of 1 in 20,000, details such as ramps, collectors and direction of travel could be shown as well. See Figure 5.1 for examples. On much larger scales, such as 1 in 600 (1 in. = 50 ft), features such as pavement markings, actual lane designations, and specific design elements can be depicted. The latter scale is the level of resolution often used in roadway design work.

GIS employs a wide range of data sources reflecting the varied goals of the systems themselves. Since GIS-T may involve applications as varied as archeological analysis, marketing research, and urban planning, the source materials can be difficult to inventory and classify comprehensively. The fact that many different scales may be encountered adds to the complexity of the problem. Even within a single GIS project, the range of materials employed can be daunting. If multiple data sets are contemplated, and they do not have common scales, the process of conflation (the fusion or marrying together of data) may be used in some cases to spatially integrate the data sets to create a new master coverage with the best spatial and attribute qualities. In situations such as this, the analyst should use extreme care in the use of these conflated data.

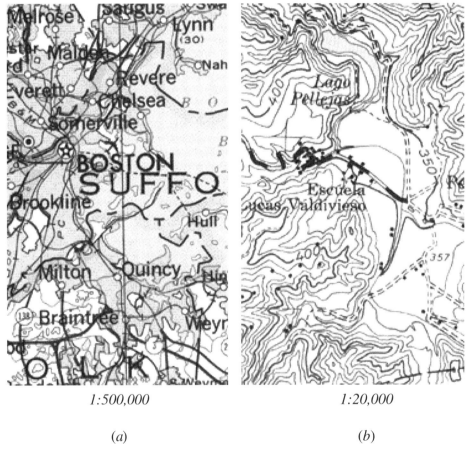

1:500,000   *1:20,000*

(a)   *(b)*

FIGURE 5.1   Effects of scale.

Geocoding. The linkage between attribute data and spatial data, as stated previously, is the key advantage of using GIS. The process by which this is accomplished is called geocoding. A great many tools exist to accomplish this process. Some involve matching a non-map-based database, such as a list of names and addresses, with a map-based database, such as one containing zip code centroids. Using a common field, and the relational model (described in the next section), coordinate information is attached to the name and address. Other common geocoding tools involve the use of distances along reference lines and offsets from these lines to place objects on a map, thus requiring some sort of linear referencing system (LRS), which will also be described in a later section. For example, going back to the non-map-based names and addresses database, one could use a street centerline file that includes address ranges on its links to geocode addresses by estimating the address locations based on the length of the segment and the address range assigned to the segment. For example, an address of 41 Elm Street would plot as shown in Figure 5.2.

Data Processing. For the processing and management of data the two paradigms most commonly used are the relational and object-oriented models. It should be noted that the database is the operation center of the GIS where much of the primary work is done. Graphics may or not be necessary for an application, but, almost without exception, the database is a key part of any analysis.

Relational database management systems (RDBMS) are well suited for ad hoc user queries, an important aspect of GIS analysis. The relational model uses tables of data arranged as columns (categories of data) and rows (each observation entry). Columns are called fields, and rows are called records, as shown in Figure 5.3. Consider, for example, the table of courses shown in Figure 5.4. Note that each course's attributes are read across, on a row. Queries may be made from these tables by specifying the table name, the fields of interest and conditions (for example, course credit hours greater than 3). The rows that meet those conditions are returned for display and analysis. The power of RDBMS lies in its ability to link tables together via a unique identifier, which is simply a field common to both tables and whose values appear only once. Thus, using the current example, one could, in addition

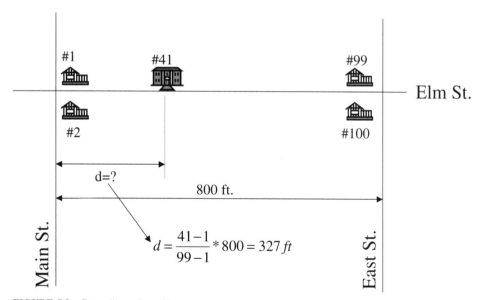

FIGURE 5.2 Geocoding using address ranges.

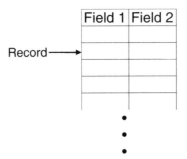

FIGURE 5.3 DBMS—the relational model.

to querying about courses from Figure 5.4, query about which students take courses from which professor, number of students in each class who hail from Missouri, geographic distribution of students, and so on. This is done via the common field, student ID, which links the two data tables, as shown in Figure 5.5. Within GIS, this capability to link tables is used to link spatial objects with tables containing information relating to those objects, as shown in Figure 5.6.

Object-oriented database management systems (ODBMS) offer the ability to integrate the GIS database with object-oriented programming languages such as Java and C++. Martin and Odell (1992) state that the object-oriented approach "models the world in terms of objects that have properties and behaviors, and events that trigger the operations that change the state of the objects. Objects interact formally with other objects." In short, the database has a set of objects with attached attributes. At this simplest level, ODBMS is analogous to the RDBMS model in that its objects represent the latter's "rows" in tables, and its attributes the RDBMS fields. The ODBMS model is much more powerful than this simple definition implies, however. Three concepts are crucial to understanding the ODBMS: abstraction, encapsulation, and inheritance.

Data abstraction is the process of distilling data down to its essentials. The level of abstraction indicates what level of detail is needed to accomplish a purpose. For example, on a small scale a road network may be presented as a series of line segments, but on a large scale the road network could include medians, edges of roadways, and roadway fixtures (Rumbaugh et al. 1991). Encapsulation includes procedures with the object data. In other words, code and data are packaged together. The code performs some behavior that an object can exhibit—example, calculating the present serviceability index of a section of road. Thus, encapsulation allows the representation of an object to be changed without affecting the applications that use it. Modern software programming provides a familiar example of this concept. The application Excel may be called as an object in a Visual Basic routine. The

Professor	Course	Hours	Course ID
Adams	CE341	3	44356
Burken	CE441	3	77877
Nanni	CE223	4	33645
Spring	CE210	3	65645
Zhang	CE446	4	48655

FIGURE 5.4 Table in the RDBMS.

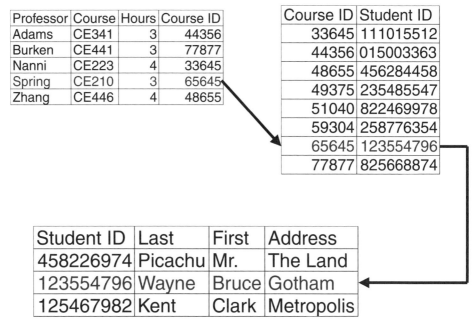

Professor	Course	Hours	Course ID
Adams	CE341	3	44356
Burken	CE441	3	77877
Nanni	CE223	4	33645
Spring	CE210	3	65645
Zhang	CE446	4	48655

Course ID	Student ID
33645	111015512
44356	015003363
48655	456284458
49375	235485547
51040	822469978
59304	258776354
65645	123554796
77877	825668874

Student ID	Last	First	Address
458226974	Picachu	Mr.	The Land
123554796	Wayne	Bruce	Gotham
125467982	Kent	Clark	Metropolis

FIGURE 5.5 Linking tables in the RDBMS.

routine would then have access to the application's full functionality through its attributes (built-in functions).

Finally, inheritance provides a means to define one class of object in terms of another. An object class represents a group of objects with common operations, attributes, and relationships (Fletcher, Henderson, and Espinoza 1995). An object is a specific instance of a class. Each object can have attributes and operations (or code, as explained above). For example, a conifer is a type of tree. There are certain characteristics that are true for all trees, yet there are specific characteristics for conifers. An attribute is a data value held by objects in a class (Rumbaugh et al. 1991)—for example, for the attribute "tree color" the value might be "green." An operation is a function that may be applied to or by objects in

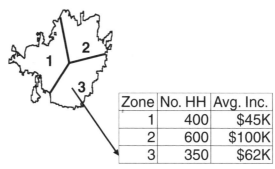

Zone	No. HH	Avg. Inc.
1	400	$45K
2	600	$100K
3	350	$62K

FIGURE 5.6 RDBMS in GIS.

a class. Objects and classes of objects may be connected to other objects. That is, they may have a structural relationship called an association with these other objects. The primary purpose for inheritance is software reuse and consistency (Lewis and Loftus 1998).

Although RDBMS continues to be the most commonly used database model for GIS applications, the power of ODBMS models with respect to their ability to interface with object-oriented programming languages, their ability to provide "smart" data, and their utility when applied to extremely complex databases all support their eventual preeminence in the GIS industry. Indeed, most GIS vendors are moving in this direction (Sutton 1996).

Linear Referencing Systems.   In attempting to maintain information on transportation infrastructure, many transportation agencies and planning organizations (e.g., U.S. DOT, state DOTs and MPOs) have developed linear referencing systems (LRS) for their transportation facilities. These are similar to the address range example mentioned previously but use transportation network-related measures, such as mile post, to locate objects. Baker and Blessing (1974) define LRS as "a set of office and field procedures that include a highway location reference method." Linear referencing is a means of identifying a location on a linear feature, such as a road or railroad. LRS support the storage and maintenance of information on events that occur within a transportation network, such as pavement quality, crashes, traffic flow, and appurtenances (traffic signs and signals for example). They provide the mechanism by which these data elements can be located in a spatial database. LRS has traditionally been accomplished using either references to the roadway or references to monuments along the roadway. This method is known as linear referencing. There exist several variations of linear location referencing systems, the most common of which fall into the five categories listed and described below (Adams, Koncz, and Vonderohe 2000).

- Route-milepost (RMP)
- Route-reference post (RRP)
- Link-node (LN)
- Route-street reference (RSR)
- Geographic coordinates

The Route-milepost system is perhaps the most common linear referencing method used, particularly at the state DOT level. It uses measured distance from a known point, such as a route's beginning or where it crosses a state or county line, to a referenced location. The distance is usually specified to the nearest hundredth of a mile, although in some cases other less precise measures may be used. The point of interest (such as the location of a crash) is always offset in a positive direction from the zero mile point and is not referenced to other intermediate points along the route. An advantage of this system is that reference posts or signs need not be maintained in the field. However, it requires that field personnel know where the route begins and the primary direction of the referencing system. It also means that with each realignment along the route, the referenced distances change.

The route-reference post (RRP) method uses signs posted in the field to indicate known locations. These signs, known as reference posts, may or may not reflect mileposts. All road feature data collected in the field are referenced to these markers in terms of distance and direction. The features can later be converted to corresponding mileposts using cross-referencing tables and maps. The advantage of this system over RMP is that the problems associated with changes in route length due to realignment are eliminated. This method overcomes the disadvantages of the mile point method but adds the burden of maintaining signage in the field.

In a link-node (LN) system, specific physical features, such as intersections, are identified as nodes. Each node is considered unique and is assigned some sort of unique identifier such as a node number. Links are the logical connection between nodes and may vary in length. Links also have unique identifiers that are often derived from the associated pair of node numbers. All features are measured as an offset distance from the nearest or lowest node number along a link.

The route-street reference (RSR) system is more commonly used in municipalities and relies on a local system of streets to locate physical features. In this system, an event is typically recorded as occurring on one street at a specified distance and direction from a reference street. A variation of this system is the use of two reference streets and no distance measurement. For example, a crash may be coded as occurring on street A between streets 22 and 23. This option results in a loss of detail with regard to precise location, but it still provides enough information for some uses, such as identifying road sections that have a high crash frequency.

The use of coordinates as a reference system is becoming more common with the advent of technologies such as global positioning systems (GPS). Cartesian coordinates use an x–y coordinate plane to represent location. Geographic coordinates use latitude and longitude that are measured in degrees along the axes of the sphere of the Earth. Transportation authorities at the state and local levels generally use the former system, in the form of state plane coordinates, to measure (in meters or feet) the distance east and west or north and south along a state origin or datum.

The need to manage, analyze, and understand transportation information in a linear context will persist because transportation systems are essentially linear phenomena. The current trend is to collect data using two- or three-dimensional devices, such as GPS, which allows for new economies in the field and the continued use of existing legacy while supporting network models, applications, and displays (Fletcher, Henderson, and Espinoza 1995). Advances in the use of GPS and their inexpensive data collection and navigation tools may someday replace the LRS (AASHTO 2002).

Data Sources.   Digital transportation data may be obtained in a variety of ways and from a large number of sources, all of which fall into one of the following four categories:

- Digitizing source materials
- Remote sensing
- GPS for road centerline data
- Existing digital sources

Digitizing Source Materials.   Digitizing source materials involves digitizing data elements from aerial photographs (considered to be primary source material) or from hardcopy maps (considered to be secondary source material). The process involves collecting the x–y coordinate values of the line features by tracing over each one using a digitizing tablet with a cursor or puck as the input device to locate and input map features into the computer from a paper map. This may also be done using a mouse with a digital raster image map. This latter approach is called ''heads up'' digitizing. This type of manual production requires planning, source material preparation, and production setup, in addition to postprocessing of data (Davis 1996). The costs for this type of data acquisition often represent the majority of system startup costs. Semiautomated methods using map scanning and line tracing technologies are sometimes used to lower the cost and improve the accuracy of the digitizing process.

Remote Sensing.   Remote sensing is the science (and to some extent, art) of acquiring information about the Earth's surface without actually being in contact with it. This is done by sensing and recording reflected or emitted energy and processing, analyzing, and applying that information. Remote sensing provides a convenient and efficient means of identifying and presenting planimetric data. Imagery is available in varying scales to meet the requirements of many different users. In general, there are two sources for remotely sensed data: photogrammetry, which uses cameras on board airplanes, and satellite imagery. The former produces photographs that may either be in paper form, in which case they must be digitized, as was described in the previous section, or in digital form, which may then be used directly as raster images. Cameras and their use for aerial photography are the simplest and oldest of sensors used for remote sensing of the Earth's surface. Cameras are framing systems that acquire a near-instantaneous snapshot of an area of the surface. Camera systems are passive optical sensors that use a lens (or system of lenses, collectively referred to as the optics) to form an image at the focal plane, the plane at which an image is sharply defined. Aerial photographs are most useful when fine spatial detail is more critical than spectral information, as their spectral resolution is generally coarse when compared to data captured with electronic sensing devices. The geometry of vertical photographs is well understood and it is possible to make very accurate measurements from them, for a variety of different applications (geology, forestry, mapping, etc.). The use of satellites involves an interaction between incident radiation and the targets of interest. Note, however that remote sensing also involves the sensing of emitted energy and the use of nonimaging sensors.

Global Positioning Systems.   Using GPS for collecting spatial data, specifically road centerline data, is gaining popularity with DOTs, but has limitations in technology and an overall high cost for statewide coverage (Spear 1998). The current trend is to collect data using two- or three-dimensional devices, such as GPS, and transform the coordinates into path-oriented or linear locations. These measurements are referenced to a standard ellipsoid model instead of base maps or linear field monuments. This allows for new economies in the field and the continued use of existing legacy, and it supports path and network models, applications, and displays.

Existing Digital Data from Other Sources.   A cost-effective alternative to digitizing is to acquire digital data from a third-party source, such as those described below.

In the recent past, most GIS projects have had to rely almost exclusively upon data available only in printed or "paper" form. Much of these data are now available in digital form while continuing to be published on paper. The ever-increasing pace of this transformation from paper to digital sources has many repercussions for GIS. Inexpensive, and in many cases free, access to high-quality data will enhance the use of GIS in the coming years. The Internet and the World Wide Web are being used more and more to distribute data and information, thus requiring users to know where to look and how to search networks.

All data sources have strengths and limitations. Digital sources are no different. It is important to understand their characteristics, costs, and benefits before using them. Learning a little about commonly employed digital formats will save much work in the long run. Although the types of materials will vary greatly from project to project, GIS practitioners know something of the characteristics and limitations of the most commonly available data sources. These are materials collected and published by a variety of government agencies and commercial interests and are used quite widely.

Local, state, and federal government agencies are major suppliers of digital data. Finding the data appropriate for a given system may often require significant on-line research. This is perhaps less the case at the federal level mainly because certain key agencies such as the Bureau of the Census, United State Geological Survey, Soil Conservation Service, National Aeronautics and Space Administration (NASA), and Federal Emergency Management Agency, provide standard sorts of information for the entire nation.

Key federal data sources and helpful indexes to those sources include the following.

The Bureau of Transportation Statistics provides links to state GIS resources, national transportation atlas data files, which consist of transportation facilities, networks, and services of national significance throughout the United States. The files are in shapefile format, which is the proprietary format for ESRI data. There are also links to DynaMap/1000 files that contain centerline data and features for nearly every street in the nation as well as to the National Highway Planning Network, which is a comprehensive network database of the nation's major highway system. It consists of over 400,000 miles of the nation's highways, composed of rural arterials, urban principal arterials, and all National Highway System routes. The data set covers the 48 contiguous States plus the District of Columbia, Alaska, Hawaii, and Puerto Rico. The nominal scale of the data set is 1:100,000 with a maximal positional error of ± 80 m. The network was developed for national level planning and analysis of the U.S. highway network. It is now used to keep a map-based record of the National Highway System and the Strategic Highway Corridor Network.

The United States Geological Survey (USGS) produces topographic maps for the nation, as well as land use and land cover maps, which include information about ownership and political boundaries, transportation, and hydrography. For access to these data as well as an extensive index of other federal on-line data sources, see the USGS publications website. The index includes links to the Global Land Information System and tends to concentrate on data created through the USGS.

The Bureau of the Census provides socioeconomic and demographic data, census tract boundary files and street centerline networks (TIGER files) for the entire nation. TIGER (Topologically Integrated Geographic Encoding and Referencing) was developed by the U.S. Census Bureau to support its mapping needs for the Decennial Census and other Bureau programs. The topological structure of the TIGER database defines the location and relationship of streets, rivers, railroads, and other features to each other and to the numerous geographic entities for which the Census Bureau tabulates data from its censuses and sample surveys. It has a scale of approximately 1 in = 100 ft and is designed to ensure no duplication of these features or areas.

The National Aeronautics and Space Administration provides remotely sensed data from all over the world. These data are typically in raster format and are 30-m resolution.

The National Technical Information Service (NTIS) established FedWorld, a clearinghouse for digital data, which is an excellent resource for on-line federal information of any kind.

Many states also provide extensive data on-line (for example, Missouri and New York). However, finding the data desired may be problematic since state and local government agencies organize themselves—and their data—in very different ways from place to place. Thus, discovering which department within an agency holds the desired information may require significant effort. In some cases, states and cities are required to collect certain types of information to comply with federal and state regulations—but again, the agencies that actually compile this information may vary from place to place. This means that finding information at the local and state levels can be time-consuming and may involve making phone calls, writing letters, asking questions, and visiting the agencies in person. Data may not be available in standard formats, may or may not come with adequate documentation and data quality reports, and may need to be checked as to origin and quality. Furthermore, the spatial data holdings themselves have limitations. As mentioned earlier, different data sources may have different scales associated with them and require conflation to integrate them. Conflation, however, may not yield the information required. For example, the most widely held data, the USGS 7.5-minute quadrangle data (1:24,000 scale), are increasingly becoming unsuitable for the uses of a transportation agency. For example, when collecting driveway data with a GPS unit at 1-m accuracy, centerlines based on 40-ft accuracy do not work well. Updating base maps with greater accuracy has the potential to be an expensive issue.

A primary advantage offered by government data sets is that most are in the public record and can be used for free or for a small processing fee. Some of the agencies that do not provide their data free of charge have found that this practice seriously impedes data sharing.

Many private sources of information also exist. Commercial mapmaking firms are among the largest providers, but other firms have for years supplied detailed demographic and economic information, such as data on retail trade and marketing trends. Some of this information can be quite expensive to purchase. Also, it is important to check on restrictions that might apply to the use of commercially provided data.

Many software vendors repackage and sell data in proprietary formats as well. These data are usually checked and corrected during the repackaging process. The use of these converted data sets can save time. Some firms will also build data sets to a user's specifications. Often termed "conversion" firms, these are usually contracted to build special-purpose data sets for utility companies and some government agencies. These data sets are often of such special purpose that they cannot be assembled from existing publicly available sources, say when an electric utility wishes to digitize its maps of its service area.

Data Quality.   It has been said that an undocumented data set is a worthless data set. If a data set's pedigree and quality are unknown, for example, the user must spend time and resources checking the data. Vendors should provide a data dictionary that provides a description of exactly what is in the file (data types and formats), how the information was compiled (and from what sources), and how the data were checked. The documentation for some products is quite extensive and much of the detailed information may be published separately, as it is for USGS digital products.

Some characteristics to consider when evaluating data sets are:

- Age
- Origin
- Areal coverage
- Map scale to which the data were digitized
- Projection, coordinate system, and datum used
- Accuracy of positional and attribute information
- Logic and consistency
- Format
- Reliability of the provider

In summary, sometimes the costs of using and converting publicly and commercially available digital files outweigh their value. No matter how much data becomes available publicly, there is no guarantee that exactly the sorts of information necessary for specific projects will be included.

Data Standards.   For a variety of reasons, GIS users often want to share or exchange data, usually having to do with overlapping geographic interests. For example, a municipality may want to share data with local utility companies and vice versa. GIS software developed by commercial companies store spatial data in a proprietary format protected by copyright laws. Although vendors often attempt to provide converters, in general, one company's product may not read the data stored in another's format. Applications of digital geospatial data vary greatly, but generally users have a recurring need for a few common themes of data. These themes include transportation, hydrography (rivers and lakes), geodetic control, digital imagery, government boundaries, elevation and bathymetry, and land ownership (or cadastral) information. A lack of investment, common standards, and coordination has created many situations in which these needs are not being met. As a result in some cases, important information may not be available; in others, data sets may be duplicated. A means to maintain and manage the common information being collected by the public and private sector does not exist. This results in increased costs and reduced efficiency for all involved.

Data standardization is therefore a fundamental consideration in developing GIS for integration with existing databases. All users of GIS data depend upon the establishment of standards that should address simple integration and processing of data within the GIS. Standards should include spatial data modeling and scale, accuracy, resolution, and generalization, and datum and projection mapping. As illustrated by the Internet and its interoperability, hardware platforms, operating systems, network environments, database systems, and applications software are generally *not* an issue. The standards for data definitions are much more important than these latter items for providing reliable and portable systems and applications. The GIS software standards, however, can add to the complexity of sharing data sets, since not all GIS share a common route system that is easily transferred from one vendor-specific application to another.

During the 1980s the USGS worked with academic, industrial, and federal, state, and local government users of computer mapping and GIS to develop a standard for transfer and exchange of spatial data. In 1992, after 12 years of developing, reviewing, revising, and testing, the resulting standard, SDTS, was approved as Federal Information Processing Standard (FIPS) Publication 173-1 (NIST 1994). The SDTS requires a two-step process for transferring data from one platform to another. The source data are exported by the first GIS to the SDTS format, then the second GIS imports the transfer file, creating a target data set in its own file format. This approach, while extremely useful in enabling data sharing, is cumbersome at best. Recognizing the criticality of data sharing, The National Spatial Data Infrastructure (NSDI) was established by President Clinton's Executive Order 12906 on April 11, 1994, to implement the recommendations of the National Performance Review published by his administration in the Fall of 1993 (superseded in 1998 by ANSI NCITS 320-1998 (ANSI 1998)).

The Order states that:

> Geographic Information is critical to promote economic development, improve our stewardship of natural resources, and protect the environment. . . . National Performance Review has recommended that the Executive Branch develop, in cooperation with state, local, tribal governments, and the *private sector,* a coordinated National Spatial Data Infrastructure to support public and private sector applications of geospatial data [emphasis added].

The concept of this infrastructure was developed by representatives of county, regional, state, federal, and other organizations under the auspices of the Federal Geographic Data Committee (FGDC). It will provide a base upon which to collect, register, or integrate information accurately.

The private sector is also actively engaged in building the National Spatial Data Infrastructure to meet marketplace needs defined as business opportunities. The Open GIS Consortium of over 200 private-sector, public-sector, and not-for-profit organizations as well as universities representing technology users and providers is addressing the issue of easy access to spatial information in mainstream computing. OGC is working to develop open software approaches that facilitate the development and use of location-dependent software applications using spatial data to increase farm productivity, identify disease and health threats, assist police and law enforcement in identifying crime patterns, and many more. In other words, GIS users will be able to access one another's spatial data across a network even if they are using different GIS software programs.

5.2.2 Spatial Modeling

The benefits of GIS-T are well established. It provides:

- The capability of storing and maintaining large data sets of spatial and tabular information
- Display and analytical capabilities that model the physical proximity of spatial features
- Flexibility in modeling spatial objects to suit the particular needs of the user or application
- Database integration
- Image overlay capabilities
- Network analyses (e.g., shortest-path routing)

These capabilities have developed as the technology has matured. Over the past 10 years, GIS-T has adapted to accommodate linear referenced data. Crash and roadway inventory data are examples of this type of linear data and can now be brought into GIS for display and analysis. This feature makes it easier to understand the spatial relationships within data that are not found in other information systems. In addition, GIS-T provides a programming environment that allows users to develop specific analysis programs or customize existing programs. All functions for display and analysis can be employed in a single-system design using common programming languages, such as Visual Basic, C++, and Java (Smith, Harkey, and Harris 2001). With the advent of object-oriented programming, GIS-T can be integrated into more mainstream enterprise applications, as well as Web-based client applications. GIS-T provides abilities broader than simply mapping data and includes several types of analytical capabilities, which can be broadly categorized into five groups:

1. Display/query analysis
2. Spatial analysis
3. Network analysis
4. Cell-based modeling
5. Dynamic segmentation

Display/Query Analysis.   The primary appeal of GIS to many is its graphical capabilities; a picture is indeed worth a thousand words. Maps are the pictures GIS uses to communicate complex spatial relationships that the human eye and mind are capable of understanding. The computer makes this possible, but it is still the GIS user who determines what data and spatial relationships will be analyzed and portrayed or how the data will be thematically presented to the intended audience. In short, the GIS allows analyses at a higher level of abstraction than do standard database tools. Using the database capabilities of GIS, the analyst can query the database and have the results displayed graphically. This query analysis, when spoken in everyday conversation, takes on the form of a "show me" question, such as "Can you show me sections of road that are in poor condition?" However, query analysis in GIS can also be used for other purposes, such as database automation, which might be used for error checking and quality control of coded data. As an example, the GIS roadway database could be queried automatically during the crash data entry process to verify the accuracy of speed limit and other crash report variables coded by an officer.

Spatial Analysis.   Several analytical techniques, grouped under the general heading "overlay analysis," are available in GIS for spatial analysis and data integration. GIS provides tools to combine data, identify overlaps across data, and join the attributes of data sets using feature location and extent as the selection criteria. For example, the number of acres of wetlands impacted by a proposed highway corridor could be obtained by this overlay process. Overlay techniques may also be used in combining data features by adding (or by applying some other function) one data set to another, or by updating or replacing portions of one data set with another, thus creating a new spatial data set. For example, the analyst could use these techniques to combine number of households and the average number of school-age children with pedestrian-related crashes in order to derive risk factors for the total number of pedestrian-related crashes relative to the total number of school-age children per road segment, for pedestrian-to-school safety analysis.

Proximity analysis represents the fundamental difference of GIS from all other information systems. Buffering is a means of performing this practical spatial query to determine the proximity of neighboring features. It is used to locate all features within a prescribed distance from a point, line, or area, such as determining the number of road crashes occurring within one half mile from an intersection, or the number of households that fall within 100 feet of a highway's layout line.

Network Analysis.   Unlike proximity analysis, which searches in all directions from a point, line, or area, network analysis is restricted to searching along a line, such as a route, or throughout a network of linear features, such as the road network. Network analysis can be used to define or identify route corridors and determine travel paths, travel distances, and response times. For example, network analysis may be used to assess the traffic volume impact of a road closure on adjacent roadways. GIS networking capabilities can also be used for the selection of optimal paths or routes—for example, finding shortest paths between zonal centroids. The network may include turning points, avoid improper turns onto one-way streets, represent posted traffic control restrictions, and include impedance factors to travel (such as mean travel speeds, number of travel lanes, and traffic volumes) to enhance the network analysis. It is this capability that one sees in the on-board navigation tools currently available in some models of automobile.

Cell-Based Modeling.   Cell-based modeling, also referred to as grid-based analysis, uses a grid or cells to aggregate spatial data for discrete distribution. Although similar to raster-based systems, these are vector-based. In cell-based modeling, the spatial data are developed as tiles of a given dimension, or points of a uniform distribution, as defined by the user, for display and analysis. Cell-based modeling is effective in displaying patterns over larger areas, such as representing the sum total of crashes that are located within a cell. This capability provides a quick means to view spatial clustering of crash data. Another example of this modeling approach is the representation of groundwater contamination levels using statistical models as the mathematical base. Since cell-based modeling aggregates data at a specified grid resolution, it would not be appropriate for site-specific spatial analysis. In cell-based modeling, special tools are available to merge grid data for overlay analysis. Cell-based overlay analysis is similar to the GIS overlay analysis previously discussed; however, the techniques and functions available in cell-based modeling are somewhat different. When the cells of different data sets have been developed using the same spatial dimensions, they can be merged on a cell-by-cell basis to produce a resulting data set. The functions and processes used in cell-based modeling to merge grid data are referred to as map algebra because the grid data sets in cell-based modeling are merged using arithmetic and Boolean operators called spatial operators.

Dynamic Segmentation.   Dynamic segmentation is a process that allows the association of multiple sets of attributes to a portion of a linear feature without having to modify feature geometry or topology. Although implementation of dynamic segmentation varies by GIS vendor, GIS uses dynamic segmentation to locate and display linear features along a route and/or to segment the route itself. The process consists of interpolating the distance along the measured line of the GIS route from the beginning measure to the ending measure of the line using attributes as the interpolation criteria. Consider, for example, a vector in a database representing a segment of road. As explained earlier, this line object is actively linked to a row in a table of attribute data, say road condition. The issue addressed by dynamic segmentation is the inability of the GIS to represent features that do not all begin at the beginning of the line and end at its end. Figure 5.7 depicts a situation where pavement condition and traffic volume levels are shown along the line segment and demonstrates the difficulty in determining various combinations of pavement condition and traffic volume using fixed segmentation. As the figure shows, dynamic segmentation allows representation of these nonuniform attributes without requiring the physical segmentation of the link—that is, no modification of geometry is necessary.

   This robust method to represent model links allows the dynamic management of many-to-one relationships between GIS segments attribute features. The method offers several advantages over fixed segmentation. First, the editing of the routes is preferable to editing the underlying base map. Second, the routes may be stored in simple correspondence tables that can be easily read by planning models, for example. Third, depicting the transportation network as linear features (or ''linear events'') in the GIS means that route tables and associated transit line tables can be managed automatically through dynamic segmentation. Dueker and Vrana (1992) describe dynamic segmentation in great detail.

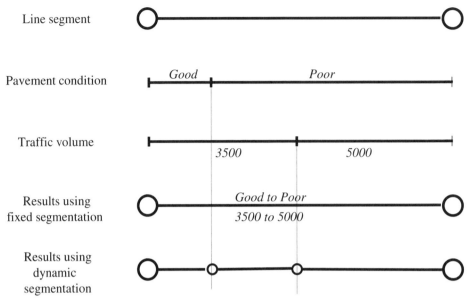

FIGURE 5.7   Dynamic segmentation example.

5.3   *GIS APPLICATIONS IN TRANSPORTATION*

Nearly all state departments of transportation (DOT) have now initiated some GIS activity. Budgets for GIS activities continue to grow, with nearly half of DOTs reporting annual GIS budgets in excess of one half million dollars. GIS use throughout the DOTs continues to grow rapidly, with the average number of end users tripling since the year 2000. There has been a significant increase in recent years in the number of state DOTs developing more accurate road base maps using digital imagery and kinematic GPS. There has also been significant growth in Web-based GIS applications and the use of GIS for managing road feature inventories (AASHTO 2002).

Certainly the legislation described at the beginning of this chapter has played a major role in the increasing use of GIS-T. But perhaps the more compelling reasons contributing to the advent of this technology are the technical ones, such as the increased ability to integrate diverse databases, better integration with modeling tools, and linkage with global positioning systems. As indicated earlier, transportation agencies across the nation have invested significant resources in developing GIS data sets and transportation models. Often, due to technological and other organizational constraints, these two activities have progressed in parallel, resulting in multiple transportation network files and databases. A variety of modeling techniques from different agencies or different sections within agencies are typically used for applications such as the modeling of traffic impacts, vehicle emissions, and transit planning. Data are commonly displayed on different map bases, different graphics programs being used to represent them. The fact that federal and state program requirements encourage modelers and planners to use more consistent methods of data integration and display has led to a growing interest in GIS-T (Sutton 1996).

The issue of data integration is a difficult one. Different database formats supporting diverse applications lead to inconsistencies, inaccuracies and duplication in updating. Dueker and Butler (2000) addressed this issue and proposed a framework for sharing transportation data that attempts to clarify roles among participants, data producers, data integrators, and data users. In a follow-up study, Dueker et al. (2001) proposed a clearinghouse approach to collecting and disseminating GIS-T data. The approach allows users to control the selection and segmentation of transportation features in updating existing GIS-T databases that meet the needs of various private and public applications.

While many GIS functions are also shared (database integration, data query, data management, use of buffers and overlays), certain of these activities are more specifically of concern to transportation. For example, network analysis is of key interest to transportation-related agencies. Its development within GIS is relatively new compared to the development of other forms of spatial analysis within GIS, but recently it has gained much more attention. Although GIS continues to be used principally for visualization and presentation, there do exist applications involving complex spatial or network analyses. This section describes current GIS-T applications to common transportation problems and provides examples in transportation planning and logistics, traffic engineering, management of infrastructure, ITS, and environmental assessment.

5.3.1 Transportation Planning

Decision support is the primary goal of transportation planning and consequently of GIS-T applied to transportation planning. Several models and data sets are available to planners for modeling trip generation for impact studies, land use projection, population and employment projection, four-step traffic projection for network traffic assignment and corridor studies, air-quality assessments of mobile sources of pollution, alternatives analyses, noise analysis, and energy studies.

Urban planning models have long been available to assist planners in recommending decisions regarding transportation investment. Today, many planning agencies also use GIS for inventory and data presentation, but few take full advantage of the GIS-T's modeling capabilities. Although many studies have suggested the utility of GIS in the context of transportation forecasting and scenario analysis, most agencies are not receiving the full benefits of integrating these two commonly used computer packages.

There exist a few commercially available transportation modeling programs that incorporate both modeling and GIS capabilities. For example, TransCAD (Caliper Corporation, http://www.caliper.com/) and UFOSNET (RST International Inc., http://www.ufosnet.com/) are GIS-based systems that include transportation forecasting procedures. In addition, the developers of TRANPLAN and QRSII are currently either beta testing or releasing GIS capable interfaces. However, while the commercial trend seems to be toward integration, many agencies continue to use standalone modeling packages or both GIS and modeling packages independently. Chief reasons are cost and prior investment in training and technology. In a 1996 Federal Highway Administration study, Anderson and Soulyrette developed a set of Windows-based programs that integrated urban planning models with desktop GIS packages. They demonstrated and evaluated the integration of Tranplan, a travel forecasting tool, and four common GIS tools and found that the main obstacle that may be faced by planning agencies wanting to implement GIS tools is the unwillingness of users to make the change.

Transportation planning applications involve network and non-network-based analyses. Applied to the latter, GIS-T serves primarily as a presentation tool. For example, GIS is well suited to store the data required by a typical site impact analysis. The analyst can perform innovative studies that more fully consider the relation between land use and transportation. Examples have included traffic impact studies at airports and analyses of hotel-casino trip generation that consider proximity of synergistic pedestrian facilities and land use (Souley-rette, Sathisan, and Parentela 1992). Additionally, Collins and Hartgen (1998) developed a GIS-T-based site impact analysis program, the objective of which was to develop a more efficient, yet understandable, means by which the traffic impacts of a new or expanded development can be identified and measured. The resulting tool was applied to proposed developments in Rock Hill, South Carolina. Traffic impact studies require knowledge of a proposed development's trip-generating characteristics, along with spatial information on surrounding streets and land use. Site impact analysis includes assessment of trip-generation characteristics, which depend on current and projected land use. It also includes the identification of trip-distribution splits, which in turn depend on intervening opportunities and existing travel patterns.

Applied to the former, namely network analysis, GIS can take several forms. Shortest path analysis can be used to find the minimum cost (travel time or distance, etc.) route between two points on a network defined to comply with certain criteria, such as lane width or the absence of highway rail at-grade crossings. This basic form of network analysis capability can be used in several ways, such as to support permit routing for oversize/overweight vehicles or to assist emergency vehicle response. As an example, Hu (2002) implemented a GIS-T framework for the dispatch and assignment of medical vehicles for emergency services via wireless communication. The framework combines distributed object computing GIS-T and a transportation logistics model in a single system.

GIS has not generally been used for full-scale network analysis in the form of urban transportation models. Several software packages, such as Tranplan and QRS II, are available for that specific purpose. The integration of these packages with GIS-T can occur through the exchange of network data and the calling of analytical routing across software packages. GIS-Trans (1996) proposed an approach to integrating various transportation databases that included planning, land use and engineering databases, and various transportation and GIS models as well, such as travel forecasting, dynamic segmentation, conflation, and others.

Network analysis may be extended, in shortest delivery route analysis, to incorporate several intermediate stops—a logistics application. These stops can be visited in a specific order and with specific delivery times. This functionality can be used by local transportation firms to plan delivery schedules and by transit agencies to develop bus routes or demand responsive paratransit routes and services. In the area of public transit, Saka (1998) used GIS to evaluate the viability of Baltimore's public transit system in providing employment transportation to the inner city low-income residents—this is also known as the "welfare-to-work" problem. He used demographic, socioeconomic, and employment data to map the inner city low-income residential areas, the location of new jobs, employment maps, and the public transit network. Using buffer analysis, a fundamental GIS-T application, the study found that the public transit system in Baltimore would not sustain the reverse commuting needs of the inner city's low-income residents.

Network analysis can also be used to find the nearest facility for a given purpose and the best route to follow. This technique can be used, for instance, to allocate students to the nearest elementary school or highway crash victims to the nearest hospital. Conversely, travel time or catchment areas around a facility can be used to find, for example, all areas within 10 minutes of a fire station, airport, or school. The analysis of catchment areas, and the characteristics of the populations within these areas, is very useful in assessing service regions and potential capacity problems. The surrounding population can be allocated to the most appropriate facility, subject to travel time and capacity constraints. One recent study (Da Silva et al. 2000) developed a tool, based on a traditional accessibility index, that will estimate the effects of proposed development on accessibility patterns in urban areas. The authors demonstrated how the tool could be used for selecting the best site for a new development among many alternatives.

Network analysis requires high-quality data and significant preparation before analysis can be conducted. Extensive and accurate information is needed on the connectivity of the network, turning options at intersections, network capacities, and demands on the system. The implications of poor-quality data, such as missing or incorrect line segments, incomplete node information, and inaccurate capacity and demand data, are perhaps more serious in transportation network analysis than in any other types of GIS activities.

5.3.2 Traffic Engineering

Application of GIS to traffic engineering is a fairly recent trend. GIS-T is especially well suited to maintaining, analyzing, and presenting city-wide spatial data useful in traffic engineering programs. The ability of GIS to integrate advanced technologies such as GPS, surveillance cameras, and vehicle location devices is now being applied to address traffic systems issues in large, congested urbanized areas (Easa and Chan 2000).

Safety management systems lend themselves extremely well to analysis using GIS tools. There have been a host of efforts to apply GIS to safety needs ranging from integrating GIS with existing safety analysis tools to the integration of computational intelligence tools with GIS. Vonderohe et al. (1993) documented the development of a conceptual structure, implementation plan, and management guidelines for a GIS-T capable of supplying the highway safety information requirements for most transportation agencies. Indeed, several agencies have begun to implement GIS-T within their safety management systems. Many have used global positioning systems in concert with GIS to develop point topology (Miller and Shaw 2001). Crash data, once geocoded, may be integrated with road attributes, sign, and signal inventories to enhance safety analyses using routing, dynamic segmentation, and buffering, all of which are built-in GIS functions. Much of the effort in such systems, though, has been focused on the very difficult problem of getting the data into the database, relying on built-in GIS functions for applications. In 1999, responding to the need for specialized tools specifically designed for GIS applications, FHWA released its GIS Safety Analysis Tools CD, which expands the analytical features of the Highway Safety Information System (HSIS) by integrating them with GIS functions. The tools provide practitioners with programs to perform spot/intersection analysis, cluster analysis, strip analysis, sliding-scale evaluations, and corridor analysis. The 2001 update includes additional pedestrian and bicycle safety tools to select safe routes to schools, assess the bicycle compatibility of roadways, and define high pedestrian crash zones (Smith, Harkey, and Harris 2001).

Given the ability to program within many GIS packages, researchers have attempted to expand GIS functionality by integrating advanced technologies. Spring (2002) developed a prototype knowledge-based GIS for the identification and analysis of hazardous highway locations. The system examined the use of these tools to address the problems of "false positives" intrinsic to most existing methods for identifying hazardous locations, and examined how knowledge-based systems could best interface with GIS to analyze these so-called black spots. An earlier study by Carreker and Bachman (2000) used GIS to address problems with crash data quality by providing better access to and improving multiple public and private road databases. Panchanathan and Faghri (1995) developed knowledge-based GIS for managing and analyzing safety-related information for rail-highway grade crossings. They developed an integrated system for the management of safety information. The system integrated rail-highway grade crossing safety data from the Federal Rail Administration and Delaware Department of Transportation (DelDOT), and an accident-prediction model developed by USDOT.

The GIS environment makes it possible to develop interrelated maps, databases, and traffic-management applications. Several studies have examined the use of GIS for incident management in particular. Siegfried and Vaidya (1993) developed a GIS-based automated incident-management plan for the Houston area. The study evaluates the use of GIS to relate incident locations with the transportation network and to make decisions and calculations for incident management. Their prototype system does alternate routing, incident response, and resource management, and it allows for network and incident information queries. Ozbay and Mukherjee (2000) implemented a Web-based real-time decision support system for wide-area incident management that can be used by all the involved agencies. The system is based upon an earlier effort described by Kachroo, Ozbay, and Zhang (1997) that developed an integrated expert system/GIS for this purpose. Ozbay implemented the system on the World Wide Web using a collection of Java-based tools and described implementation and development issues that they faced, such as communication speed, security issues, technological obsolescence, and use of heuristics to estimate key parameters (e.g., incident duration).

5.3.3 Infrastructure Management

The management of existing facilities rather than the construction of new ones has become the focus of most transportation agencies in the post-ISTEA era. Given the several million dollars worth of signs, signals, and markings maintained by the typical urban area, the hundreds of thousands of miles of paved roads under the care of most states as just two examples, this is not a trivial task. Many agencies have opted to use a GIS-based approach to collect, inventory, and maintain and manage their inventories. Spring (1992) developed a GIS-based sign inventory system for the city of Greensboro, North Carolina, in which he characterized the city's inventory activities and developed an interface that could be used in the field by city personnel to enter sign data. Functions included work plan development, ownership, performance reports, and maintaining sign histories and attributes. All are relatively straightforward database queries requiring no complex spatial queries such as buffering or polygon overlays. Zhang, Smith, and Hudson (2001) developed an implementation plan for using GIS to enhance pavement management practice in Texas. They proposed a three-stage implementation concept to assess the current practice, define the visionary system, and identify the intermediate solutions. Chen et al. (2000) successfully developed a GIS-based method for integrating decision support tools commonly used for the management of railway infrastructure. Harter (1998), once again in response to the ISTEA legislation, described a system that integrates GIS-T with existing transportation planning and analysis software used by the Florida Department of Transportation. The system manages, calculates, and stores roadway characteristics to estimate immediate and long-term infrastructure needs. It also calculates roadway level of service, identifies deficiencies, recommends improvements to meet operating standards, and, finally, calculates costs of the proposed improvement. Deficient segments, as well as road improvements and costs, are displayed thematically on GIS-generated maps.

5.3.4 Intelligent Transportation Systems

The integration of intelligent transportation systems (ITS) and GIS promises major economic and social benefits, including more efficient use of existing transportation infrastructure, increased highway safety, and better reliability, coverage, and resolution of real-time navigation systems for autos, buses, and trucks. Indeed, the relationship between ITS and GIS is increasingly being recognized for its importance. The National ITS Program Plan identifies 29 user services bundled into the primary ITS application areas (Roess, McShane, and Prassas 1998). Many transportation entities, such as highway agencies, railroad agencies, and emergency service agencies, are managed and operated as separate entities. To improve future ITS user service, it is important to integrate the transportation of people, movement of goods, and information services within the municipality. Chang, Ho, and Fei (1996) developed a system interface and information infrastructure design, within a GIS environment, that will integrate transportation-related information services, better satisfy transportation management functions, and support real-time community travel information service requirements.

5.3.5 Emergency Response / Logistics

Most urban areas have high-quality base maps with comprehensive, unique address ranges and street names. GIS is used for routing vehicles, allocating facilities, and placing vehicles in the most appropriate locations in anticipation of events (from the book). With respect to the last, Church and Cova (2000) report on the development of a GIS-based model that estimates evacuation risk during some extreme event, such as a wildfire or hazardous material spill. The model identifies network and demographic characteristics on transportation networks that may lead to significant problems in evacuation, such as neighborhoods with high ratios of population to exit capacity. Spring (1998) developed a prototype knowledge-based GIS for to serve as a decision aid in the scheduling and dispatching process for paratransit. The prototype makes insertions to paratransit skeleton schedules considering special customer needs, bus capacities, and slack time requirements and was used to minimize overall travel time.

5.4   IMPLEMENTATION ISSUES

Implementing GIS-T is not as simple as merely installing a new piece of software into an existing system (such as a pavement management system) and then operating it. In addition to the information technologies intrinsic to GIS-T, successful GIS implementation involves elements, such as personnel and their GIS skills, the organizational structure within which they work, and the institutional relationships that govern the management of information flow.

GIS is an enabling technology and serves as a platform for integrating various types of data, systems, and technologies. It may be applied to a host of different applications, as demonstrated by the preceding limited review of current applications. Among these are systems that allow for near real-time assessment of conditions, simultaneous sharing of large databases and integration, and enhancement of existing transportation models—all complex and challenging applications. The fact that the adoption rate of GIS technologies has grown exponentially makes successful implementation of these technologies even more challenging. Key areas that must be addressed when implementing GIS include data, people and organizations, and technology.

5.4.1   Data

The collection, maintenance, and use of data are challenging issues that must be addressed early in the implementation process. Who will collect and maintain data, who "owns" the data, and who will serve as the data's custodian are all questions that must be answered.

The main reason that these questions are important is that they are key in determining if one can take advantage of GIS's main strength, its ability to share data. That is, as stated above, GIS serves as a logical and consistent platform in a common location reference system and allows diverse databases to be integrated and shared among different divisions of a department, for example. Integration standards should therefore be established to integrate different databases, some spatial, some not. Answers to the questions also address barriers, such as institutional and organizational arrangements, to the implementation process.

Technical questions, such as what is the nature of the data, how is it to be spatially referenced, and what is its accuracy, also must be considered. To understand the various data elements in a complex system, metadata (information about data, such as when collected, by whom, scale) should be part of a system. Additionally, spatial data require establishment of some sort of linear referencing system, as described above.

All GIS data, spatial data, and associated attribute data suffer from inaccuracy, imprecision, and error to some extent. Data quality assurance and quality control rules ensure the delivery of high-quality data. Use of a data steward to collect, maintain, and disseminate GIS data would facilitate this.

5.4.2 Technology

While GIS applications no longer bear the stigma of new technology and have been accepted into the mainstream of professional practice, technological issues nevertheless exist that must be addressed when implementing them. Among these are identifying critical technologies to be used and keeping abreast of technological innovations. These include technologies that support interoperability (sharing of data and processes across application/system boundaries), Web-based GIS (given its significant growth in the past few years), and a rapidly expanding range of GIS applications. Of course, chosen technologies must match the architecture environment as well. The fact that industry standards are in their infancy exacerbates the problem of choosing the appropriate technologies for a given system. It is crucial to identify limiting technologies, statements by software vendors notwithstanding.

In making this decision, one should look to the future, but not too far into the future. One should also keep in mind that increasing pressure exists to make geo-referenced transportation data more accessible (and understandable) to the general public.

The introduction of new information technologies is necessarily accompanied by a change in organizational structures and institutional arrangements.

5.4.3 Organizations and People

Organizational and people issues are perhaps the most difficult to address. They continue to be more critical and more difficult to solve than technological issues. Personnel at both management and technical levels must be involved in implementing the system. Yet this has been identified consistently as a problem area—especially maintaining support of upper management. Convincing decision-makers to accept the idea of GIS is key to a system's success. A top-down, rather than bottom-up, GIS management strategy should be adopted for GIS planning and implementation.

Of primary interest to decision-makers are two questions: who pays (for the hardware, software, and personnel required) and from where will the resources be drawn? As with any investment decision, GIS must have economic justification. Hall, Kim, and Darter (2000) conducted a cost-benefit analysis of GIS implementation for the state of Illinois. They calculated a ratio of slightly less than 1, but also identified several intangible benefits for GIS project (shown in Figure 5.8) that indicate that GIS benefits outweigh implementation costs.

Champions (at both the management and grass-roots levels) are critical to a system's success as well. They can facilitate a positive decision to purchase and enhance the chances for system success.

With regard to the latter, qualified GIS personnel must be available. What training and education are necessary for minimum acceptable qualifications are based upon the level of GIS knowledge needed. There are likely to be three levels of GIS staff and users: local GIS users, local GIS specialists, and a GIS application/data steward. Training for the GIS support personnel is extremely important to the success of GIS.

Early implementation of GIS is often more dependent on vendor-supplied training. However, in the long term the GIS support group should develop specifications for in-house training. Several GIS entities (Zhang, Smith, and Hudson 2001) have proposed that a certification for GIS professionals be established to address this important issue. Getting and retaining adequately trained GIS staff continue to be problems. Indeed, there will be increased reliance on outside experts (consultants) for more complex GIS analyses, due to difficulty in finding and keeping expert staff on internal payroll.

Additional problems of which to be aware relate to the nature of GIS data, namely, that to be successful it must be shared. This leads to turf battles and questions about who should have access, and how much, to the GIS.

1. Integration of information for better decisions
2. Decreased risk of poor decisions due to incomplete data
3. Access to more accurate data
4. Reduction of duplicate data
5. Ability to visualize interrelationships of various data and projects quickly
6. Ability to develop public information maps
7. Assist with revenue initiatives
8. Quick response to internal and external queries
9. Easy access to information and maps
10. Significant time savings for information access
11. Unprecedented analysis of information
12. Identification of trends
13. Assist in public relations
14. Better cooperation with regulatory agencies
15. Increased safety for personnel
16. Ability to monitor department commitments

FIGURE 5.8 Intangible benefits of GIS.

5.5 CONCLUSIONS

GIS-T technologies hold much promise in improving how transportation does business and in making our transportation systems safer, cheaper, and more efficient. The transportation community has an unprecedented opportunity over the next few years to obtain, use, and distribute spatial data using these technologies. If used intelligently, the data will provide a wealth of information about transportation and its relationship to the quality of life on global and local scales.

This chapter has provided a brief overview of GIS-T, defined terms, reviewed general principles and analysis technologies, identified issues, and described major application areas within the GIS-T arena. This material can only serve as a superficial introduction to the applications of GIS to transport problems, but it is hoped that the reader will gain valuable insight into how these systems work, where they have been applied, and what difficulties GIS users face when implementing such systems. It is further hoped that, with the veil of ignorance removed, users will more readily consider this important class of tools for use.

5.6 REFERENCES

American Association of State Highway and Transportation Officials (AASHTO). 2002. Summary Report for 2002 Geographic Information Systems in Transportation Symposium, http://www.gis-t.org/yr2002/2002_GIST_Summary.htm.

Adams, T. M., N. A. Koncz, and A. P. Vonderohe. 2000. "Functional Requirements for a Comprehensive Transportation Location Referencing System." In *Proceedings of the North American Travel Monitoring Exhibition and Conference.*

Anderson, M. D., and R. Souleyrette. 1996. "A GIS-Based Transportation Forecast Model for Use in Smaller Urban and Rural Areas." *Journal of the Transportation Research Board* 1551.

American National Standards Institute (ANSI). 1998. Spatial Data Transfer Standard, ANSI NCITS 320-1998.

Baker, W., and W. Blessing. 1974. *Highway Linear Reference Methods, Synthesis of Highway Practice 21.* National Cooperative Highway Research Program, Washington, DC: National Academy Press.

Carreker, L. E., and W. Bachman. 2000. "Geographic Information System Procedures to Improve Speed and Accuracy in Locating Crashes." *Transportation Research Record* 1719:215–218.

Chang, E. C., K. K. Ho, and P. H. Fei. 1996. "GIS-T Design for Intelligent Transportation Systems (ITS) Applications." Paper delivered at Intelligent Transportation: Realizing the Future, Third World Congress on Intelligent Transport Systems.

Chen, F., L. Ferreira, J. Chen, and J. Zhang. 2000. "Railway Infrastructure Management Using a GIS based integrated approach." In *Proceedings of the Conference on Traffic and Transportation Studies, ICTTS 2000.* Reston, VA: American Society of Civil Engineers.

Church, R., L. Cova, and J. Thomas. 2000. *Transportation Research Part C: Emerging Technologies* 8(1):321–326.

Collins, C. E., and D. T. Hartgen. 1998. "A GIS-T Model for Incremental Site Traffic Impact Analysis." *ITE Journal* 68(11).

da Silva, L. R., A. N. R. da Silva, C. Y. Egami, and L. F. Zerbini. 2000. *Transportation Research Record* 1726, 8–15.

Davis, B. 1996. *GIS: A Visual Approach.* Santa Fe: OnWord Press.

Dueker, K., and J. A. Butler. 2000. "A Geographic Information System Framework for Transportation Data Sharing." *Transportation Research Part C: Emerging Technologies* 8(1):13–36.

Dueker, K., J. A. Butler, P. Bender, and J. Zhang. 2001. "Clearinghouse Approach to Sharing Geographic Information Systems—Transportation Data." *Transportation Research Record* 1768.

Dueker, K. J., and Vrana. 1992. "Dynamic Segmentation Revisited: A Milepoint Linear Data Model." *Journal of the Urban and Regional Information Systems Association* 4(2):94–105.

Easa, S., and Y. Chan, eds. 2000. *Urban Planning and Development Applications of GIS.* Reston, VA: American Society of Civil Engineers.

Fletcher, D. R. n.d. "Geographic Information Systems for Transportation: A Look Forward." White paper, Transportation Research Board.

Fletcher, D., T. Henderson, and J. Espinoza. 1995. *Geographic Information Systems—Transportation ISTEA Management Systems, Server—Net Prototype Pooled Fund Study Phase B Summary.* Sandia National Laboratory, Albuquerque, New Mexico.

GIS-Trans, Ltd. 1996. "Transportation Model—GIS Data Integration," White paper, http://www.gistrans.com/pub/model.pdf.

Hall, J. P., T. J. Kim, and M. I. Darter. 2000. "Cost-Benefit Analysis of Geographic Information System Implementation: Illinois Department of Transportation." *Transportation Research Record* 1719.

Harter, G. L. 1998. "An Integrated Geographic Information System Solution For Estimating Transportation Infrastructure Needs: A Florida Example." *Transportation Research Record* 1617.

Hu, T.-Y. 2002. "CORBA-Based Distributed GIS-T For Emergency Medical Services." National Research Council, 81st Transportation Research Board Annual Meeting, Washington, DC, Preprint.

Kachroo, P., K. Ozbay, and Y. Zhang. 1997. *Development of Wide Area Incident Management Support System.* FHWA Report DTHF71-DP86-VA-20.

Lewis, J., and W. Loftus. 1998. *Java Software Solutions: Foundations of Program Design.* Reading: Addison-Wesley.

Martin, J., and J. Odell. 1992. *Object-Oriented Analysis and Design.* Englewood Cliffs: Prentice-Hall.

Meyer, M. D. 1999. "Transportation Planning in the 21st Century." *TR News* 204.

Miller, H. J., and S. L. Shaw. 2001. *Geographic Information Systems for Transportation: Principles and Application.* New York: Oxford University Press.

National Institute of Standards and Technology (NIST). 1994. *Spatial Data Transfer Standard.* Federal Information Processing Standards Publication 173-1.

Ozbay, K., and S. Mukherjee. 2000. "Web-Based Expert Geographical Information System for Advanced Transportation Management Systems." *Transportation Research Record* 1719.

Padmanabhan, G., M. R. Leipnik, and J. Yoon. 1992. *A Glossary of GIS Terminology.* National Center for Geographic Information and Analysis, Report No. 92-13, ftp://ftp.ncgia.ucsb.edu/pub/Publications/Tech_Reports/92/.

Panchanathan, S., and A. Faghri. 1995. "Knowledge-Based Geographic Information System for Safety Analysis at Rail-Highway Grade Crossings." *Transportation Research Record* 1497.

Roess, R. P., W. R. McShane, and E. S. Prassas. 1998. *Traffic Engineering,* 2d ed. Englewood Cliffs: Prentice-Hall.

Rumbaugh, J., M. Blaha, W. Premerlani, F. Eddy, and W. Lorensen. 1991. *Object-Oriented Modeling and Design.* Englewood Cliffs: Prentice-Hall.

Saka, A. A. 1998. "Welfare to work: An application of GIS in assessing the role of public transit." In *Proceedings of the Congress on Computing in Civil Engineering, ASCE.* Reston, VA: American Society of Civil Engineers.

Siegfried, R. H., and N. Vaidya. 1993. *Automated Incident Management Plan Using Geographic Information Systems Technology for Traffic Management Centers.* Research Report 1928-1F, Texas Transportation Institute, College Station.

Smith, R. C., D. L. Harkey, and R. Harris. 2001. *Implementation of GIS-Based Highway Safety Analyses: Bridging the Gap.* FHWA-RD-01-039, Federal Highway Administration.

Souleyrette, R. R., S. K. Sathisan, and E. Parentela. 1992. "Hotel-Casino Trip Generation Analysis Using GIS." In *Proceedings of the Conference on Site Traffic Impact Assessment, Chicago, IL.* Reston, VA: American Society of Civil Engineers.

Spatial Technologies Industry Association (STIA). 2002. *Increasing Private Sector Awareness, and Enthusiastic Participation in the NSDI.* Report to the Federal Geographic Data Committee.

Spear, B. D. 1998. "Integrating the Transportation Business Using GIS." In *Proceedings of the 1998 Geographic Information Systems for Transportation (GIS-T) Symposium.*

Spring, G. S. 1998. "Development of a Knowledge-Based Approach to Automatic Scheduling and Dispatching Systems Using Automated Vehicle Location and Geographic Information System Technologies for Small Urban Areas," In *Proceedings of the 5th International Conference on Applications of Advanced Technologies in Transportation Engineering (AATT-5).*

Spring, G. S. 1992. "Geographic Information Systems for Traffic Inventory Management: A Case Study." In *Proceedings of ASCE's Computing in Civil Engineering Conference, Dallas, TX.* Reston, VA: American Society of Civil Engineers.

Spring, G. S. 2002. "Knowledge-Based GIS for the Identification, Analysis and Correction of Hazardous Highway Locations." In *Proceedings of the 7th International Conference on Applications of Advanced Technologies in Transportation Engineering (AATT-7).* Reston, VA: American Society of Civil Engineers.

Sutton, J. C. 1996. "Role of Geographic Information Systems in Regional Transportation Planning." *Transportation Research Record* 1518.

Thill, J. C. 2000. "Geographic Information Systems for Transportation in Perspective." In *Geographic Information Systems in Transportation Research,* ed. J. C. Thill, Oxford: Pergamon Press.

Vonderohe, A. 1993. *Adaptation of Geographic Information Systems for Transportation.* NCHRP Report 20-27, National Research Council, Washington, DC.

Zhang, Z., S. G. Smith, and R. W. Hudson. 2001. "Geographic Information System Implementation Plan for Pavement Management Information System: Texas Department of Transportation." *Transportation Research Record* 1769.

TRAFFIC, STREETS, AND HIGHWAYS

CHAPTER 6
TRAFFIC ENGINEERING ANALYSIS

Baher Abdulhai
Department of Civil Engineering,
University of Toronto, Toronto, Ontario, Canada

Lina Kattan
Department of Civil Engineering,
University of Toronto, Toronto, Ontario, Canada

6.1 TRAFFIC ENGINEERING PRIMER

6.1.1 Traffic Engineering

Traffic engineering, or, in more modern terms, traffic control and management, concerns itself with the provision of efficient mobility of people and goods while preserving safety and minimizing all harmful impacts on the environment. A broader look at traffic engineering might include a variety of engineering skills, including design, construction, operations, maintenance, and optimization of transportation systems. Practically speaking, however, traffic engineering focuses more on systems operations than on construction and maintenance activities.

6.1.2 Evolution of Current Transportation Systems and Problems

Automobile ownership and hence dependence and truck usage have been on the rise since the Second World War. In the United States, the Federal Aid Highway Act of 1956 authorized the National System of Interstate and Defense Highways. For a couple of decades thereafter, the prime focus was on the creation of this immense mesh of freeways, considered to be the largest public works project in the history of the planet. Very quickly, transportation professionals realized that the growth in automobile use and dependence is outpacing the growth in capacity building, not to mention other problems, such as lack of funds to maintain the giant infrastructure. Ever-rising congestion levels testify to this, and hence the sustainability of continued capacity creation came under the limelight, with fierce criticism by planners, and environmentalists alike. Recognition quickly crystallized that we need to move "smarter" and make intelligent use of existing capacity before any attempt to add more. The Intermodal Surface Transportation Efficiency Act (ISTEA, pronounced "ice tea") of 1991, marked the formal birth of intelligent transportation systems (ITS). Heavy emphasis was placed on the use of technology to utilize existing capacity efficiently instead of continued new construction, in addition to emphasizing intermodalism and modernization of public transport to curb automobile dependence. ISTEA dedicated $659 million dollars to research and development and experimental projects geared towards the intelligent use of the national transportation infrastructure. In 1998, the U.S. Congress passed the Transportation Equity Act for the 21st Century (TEA-21), which earmarked $1.2 billion dollars for mainstreaming ITS with emphasis on deployment. For the most part, similar initiatives took place all over the modern world, including Canada, Europe, Australia, and Japan. Hence, the modern transportation engineering field focused on, in addition to the basics, a number of key directions, such as intermodalism, using technology to improve transportation provisions under ITS, managing ever-rising congestion through supply control and demand management, and protecting the environment.

6.1.3 Transportation Systems: Mobility and Accessibility

Land transportation systems include all roadway and parking facilities dedicated to moving and storing private, public, and commercial vehicles. Those facilities serve two principal but contradicting functions: mobility and accessibility. Mobility is the common-sense objective of transportation, aiming at the fastest but safe movement of people or goods. Access to terminal points (homes, businesses) is also essential at trip ends. Mobility requires least friction with terminal points, while accessibility requires slow speeds and hence contradicts mobility. Fortunately, roads systems evolved in a hierarchical manner to serve both without conflict. For urban areas, for instance, the American Association for State Highway and Transportation Officials (AASHTO) defines the hierarchy of roads as follows:

1. Urban principal arterial system, including interstate highways, freeways, and other urban arterials, all have some level of access control to promote mobility; typified by high volumes and speeds.
2. Urban minor arterial street system, which augments the freeway system, emphasizes relatively high mobility while connecting freeways to collectors.
3. Urban collector street system, collecting traffic from local streets and streaming it onto arterials, with somewhat balanced emphasis on both mobility and accessibility.
4. Urban local street system, primarily provides access to terminal points, and hence deliberately discourages high mobility and emphasizes low volumes and speeds.

AASHTO has a somewhat similar classification for rural roads, defining the level of mobility versus accessibility provided by each class.

With the above hierarchical classification in mind, traffic control and management strategies must recognize and preserve the functional classification of the road at hand. For instance, improper provision of mobility on freeways and arterials might result in neighborhood infiltration by traffic, an undesirable and spreading phenomenon in today's congested urban areas.

6.1.4 Emerging Trends

Intelligent Transportation Systems (ITS). Intelligent transportation systems, an emerging global phenomenon, are a broad range of diverse technologies applied to transportation in an attempt to save lives, money and time. The range of technologies involved includes microelectronics, communications, and computer informatics, and cuts across disciplines such as transportation engineering, telecommunications, computer science, financing, electronics, commerce, and automobile manufacturing. ITS and the underlying technologies will soon put a computer (which has the potential to eliminate human error) in each car to guide us to our destinations, away from congestion, interact with the road, and even drive itself. Although ITS sounds futuristic, it is becoming reality at a very fast pace. Already, real systems, products, and services are at work all the over the world. Alot remains to be done, but the future of ITS is promising. Many aspects of our lives have been more pleasant and productive through the use of advanced technologies, and it is time for the transportation industry to catch up and benefit from technology. ITS can go beyond a transportation system whose primary controlling technology is the four-way traffic signal.

Some scattered computer-based solutions to transportation problems date back to the late 1950s and early 1960s. The first large-scale application of a computerized signal control system in the world took place in Metropolitan Toronto, Canada, during the early 1960s. However, the ITS field as we know it today started to mature only in the early 1990s, when it was known as intelligent vehicle and highway systems (IVHS). The name change to intelligent transportation systems reflects broadening to include all aspects of transportation. Several forces have driven the ITS field. As mentioned earlier, transportation practitioners and researchers alike realized that road building can never keep pace with the increasing demand for travel. Some countries, like the United States, invested billions of dollars in building road networks and infrastructure and are now faced with the challenge of revitalizing this huge network and making the best use of its already existing capacity before expanding further. Another set of driving forces is environment-related. Damage to the environment from traffic emissions rose to unprecedented alarming levels. In Canada for instance, transportation represents the single largest source of greenhouse gas emissions, accounting for 27 percent of the total emissions, which is estimated to increase to 42 percent by the year 2020. The problem is even greater in more car-dependent societies like the United States. Road safety, or the lack of, and escalating death tolls and injuries in traffic accidents each year are yet a third set of forces. For all these reasons, more road building is not always viable or desirable. High-tech computer, electronic, and communication technologies offer one attractive and promising approach, and hence the current appeal of ITS. A healthy ITS industry would also have other non-traffic-related societal benefits, including stimulation of new information technology-based industries and creation of new markets and jobs. Therefore, ITS is more than just intelligent solutions on the road. It is a new strategic direction for national and international economies. The market share of ITS is projected to expand over the next decade from an annual world market of $25 billion in 2001 to $90 billion in 2011. A projected $209 billion will be invested in ITS between now and 2011 (ITS-America). Access to this sizable market is vital to the transportation and related technology sectors.

One important attribute of this emerging new face of the transportation industry—shaped by ITS—is that it is no longer restricted to civil engineers or to a single department or agency. Given the broad range of technologies involved, the ITS field is multidepartmental, multiagency, and multijurisdictional, cutting across the public, private, and academic sectors. This broadness will certainly enhance potential, widen scope, and revolutionize the way we handle our transportation systems, but it will also pose institutional challenges that we must be aware of and prepared for.

ITS Subsystems.   Collectively, ITS aims to enhance the utilization of existing roadway capacity, as well as increase capacity itself. Enhancing the use of existing capacity is achievable through improved distribution of traffic, dynamically sending traffic away from congested hotspots to underutilized segments of the network, and the elimination of bottleneck-causing controls such as conventional toll plazas. Increasing the physical capacity itself is possible through automation of driving and elimination of the human behavior element altogether. This is the promise of automated highway systems that could potentially double or triple the number of vehicles a single lane can handle. From this perspective, ITS can be divided into two main categories of systems:

1. Advanced traffic management and traveler information systems (ATMS and ATIS, or combined as ATMIS)

2. Advanced vehicle control and automated highway systems (AVCS and AHS)

ATMIS provide extensive traffic surveillance, assessment of recurring congestion due to repetitive high demands, and detection of nonrecurring congestion due to incidents, traffic information and route guidance dissemination to drivers, and adaptive optimization of control systems such as traffic signals and ramp meters. Current and near-future trends in ATMIS tend to rely on centralized management in traffic management centers (TMCs). TMCs gauge traffic conditions by receiving information from vehicle detectors throughout the network as well as the vehicles themselves, as probes formulate control measures in the center and disseminate control to field devices as well as information and guidance to drivers. Newer trends of distributed control are emerging but have not crystallized yet. The main distinguishing characteristics of ATMIS are real-time operation and network-wide multijurisdictional implementation.

AVCS provides better control of the vehicle itself, either by assisting the driver or by automating the driving process in an auto-pilot-like fashion in order to increase capacity and enhance safety. Full automation (AHS) can result in higher speeds at lesser headways, and hence higher lane capacity. Automation can be applied to individual vehicles as free agents in a nonautomated mix of traffic or as fully automated lanes carrying platoons of electronically linked vehicles. Although AHS is technically promising, an array of unsettled issues remains, including legal liabilities in the event of incident due to any potential automatic controller failure, technical reliability issues, and social issues. Therefore, AHS is still "futuristic" at the current stage of ITS. The feasible alternative, however, is to use the technology to assist the driver, who remains in control of the vehicle—that is, to make the vehicle smarter. Such intelligent vehicles will detect obstacles on the road and in the blind spots and warn the driver accordingly, maintain constant distance from the vehicle ahead, and alert a sleepy driver who is going off the road. As technology improves further, the role of the intelligent vehicle can move from a simple warning to full intervention and accident prevention by applying the brakes or overriding faulty steering decisions.

The prime distinction between ATMIS and AVCS is that ATMIS focus on smoothing out traffic flow in the network by helping the driver make best route-choice decisions and optimizing the control systems in the network, while AVCS focus on the driver, the operation of the vehicle, and traffic maneuvers in the immediate vehicle vicinity. AVCS focus on enhancing the driver's awareness and perception, aiding decision-making by providing early warning and potentially initiating action, and eventually using sensory inputs and computer control in place of human sensory reactions and control.

ITS User Services.   Another way to look at the constituents of ITS is from the end-user perspective. In the United States, for instance, a collection of interrelated user services is defined and grouped into user-service bundles. As reported by the Intelligent Transportation Society of America (ITS-America), 29 user services have been defined to date as summarized in Table 6.1. These services and their definitions/descriptions are expected to evolve and undergo further refinements in time. User services are composed of multiple technological elements or functions, which may be in common with other services. For example, a single user service will usually require several technologies, such as advanced communications, mapping, and surveillance, which may be shared with other user services. This commonality of technological functions is one basis for the suggested bundling of services. In some other cases, the institutional perspectives of organizations that will deploy the services provided the rationale for the formation of a specific bundle. The users of this service or the ITS stakeholders include travelers using all modes of transportation, transportation management center operators, transit operators, metropolitan planning organizations (MPOs), commercial vehicle owners and operators, state and local governments, and many others who will benefit from deployment of ITS.

TABLE 6.1 ITS User Services

Bundle	User Services
1. Travel and Traffic Management	1. Pretrip Travel Information 2. En route Driver Information 3. Route Guidance 4. Ride Matching and Reservation 5. Traveler Services Information 6. Traffic Control 7. Incident Management 8. Travel Demand Management 9. Emissions Testing and Mitigation 10. Highway Rail Intersection
2. Public Transportation Management	1. Public Transportation Management 2. En-Route Transit Information 3. Personalized Public Transit 4. Public Travel Security
3. Electronic Payment	1. Electronic Payment Services
4. Commercial Vehicle Operations	1. Commercial Vehicle Electronic Clearance 2. Automated Roadside Safety Inspection 3. On-Board Safety Monitoring 4. Commercial Vehicle Administrative Processes 5. Hazardous Materials Incident Response 6. Commercial Fleet Management
5. Emergency Management	1. Emergency Notification and Personal Security 2. Emergency Vehicle Management
6. Advanced Vehicle Control and Safety Systems	1. Longitudinal Collision Avoidance 2. Lateral Collision Avoidance 3. Intersection Collision Avoidance 4. Vision Enhancement for Crash Avoidance 5. Safety Readiness 6. Pre-Crash Restraint Deployment 7. Automated Vehicle Operation
7. Information Management	1. Archived Data Function
8. Maintenance and Construction Management	1. Maintenance and Construction Operation

ITS Architecture.   We deal with and benefit from systems architectures almost every day, although we might not know what an architecture is or what it is for. For instance, one day you purchase a television set, and later you purchase a videocassette recorder from a different retailer and by a different manufacturer, but you never worry about whether they will work together. Similarly, you might buy a low-end radio receiver and a high-end compact disk player and again assume they will work together just fine. You travel with your FM radio receiver and it works everywhere. This seamless operation of different systems or components of a system has not come about by chance, thanks to a mature industry and widely adopted architectures and related standards that ensure such interoperability. The ITS industry, however, is still in its infancy. It is rapidly evolving in different places all over the world, and different groups are pursuing its development. Users are at risk of investing or adopting certain ITS equipment that works only locally. Similarly, if left without adequate guidance, stakeholders could easily develop systems solutions to their needs, which might be incompatible with those of regional neighbors. For instance, an in-vehicle navigation system purchased in California might not work in Nevada, or a system purchased in Ontario might not work once the user crosses the border to New York. Therefore, to ensure seamless ITS operation, some sort of global or at least national system architecture and related standards are needed. To maximize fully the potential of ITS technologies, system design solutions must be compatible at the system interface level in order to share data, provide coordinated, area-wide integrated operations, and support interoperable equipment and services where appropriate. An ITS architecture provides this overall guidance to ensure system, product, and service compatibility/interoperability without limiting the design options of the stakeholder. In this chapter we use the U.S. ITS National System architecture only as an example. We are confident that similar efforts are underway in almost every ITS-active country, producing architectures that are more or less similar to the American architecture.

In the United States, Congress directed the U.S. Department of Transportation (DOT) to promote nationwide compatibility of ITS. Spearheaded by the U.S. DOT and the Intelligent Transportation Society of America (ITS-America), four major teams were formed in 1993. The four teams proposed four different architectures. The two most promising architectures were selected, integrated, and refined in the form of the final architecture. A rich set of documents describing every detail of the final architecture can be found on the ITS-America website (www.itsa.org), which we summarize in the following section.

It is important to understand that the architecture is neither a system design nor a design concept. It is a framework around which multiple design approaches can be developed to meet the individual needs of the user while maintaining the benefits of a common architecture noted above. The architecture defines the *functions* (e.g., gather traffic information or request a route) that must be performed to implement a given user service; the *physical subsystems* where these functions reside (e.g., the roadside or the vehicle); the *interfaces*/information flows between the physical subsystems; and the *communication* requirements for the information flows (e.g., wireline or wireless). In addition, it identifies and specifies the requirements for the standards needed to support national and regional interoperability, as well as product standards needed to support economy of scale considerations in deployment. The function view of ITS is referred to as the logical architecture as shown in Figure 6.1. Functions such as "manage traffic," for instance, can be further divided into finer processes such as "detect pollution levels" and "process pollution data." The systems view is referred to as the physical architecture as shown in Figure 6.1. Figure 6.1 also shows communications requirements and information flows. The physical architecture partitions the functions defined by the logical architecture into systems and, at a lower level, subsystems, based on the functional similarity of the process specifications and the location where the functions are being performed. The physical architecture defines four systems, traveler, center, roadside, and vehicle, and nineteen subsystems. Subsystems are composed of equipment packages with specific functional attributes. Equipment packages are defined to support analyses and deployment. They represent the smallest units within a subsystem that might be purchased. In deployments, the character of a subsystem deployment is determined by the specific equipment packages chosen. For example, one municipal deployment of a traffic management subsystem may select collect traffic surveillance and basic signal control equipment packages, while a state traffic management center may select collect traffic surveillance and freeway control packages. In addition, subsystems may be deployed individually or in aggregations or combinations that will vary by geography and time based on local deployment choices. A traffic management center may include a traffic management subsystem, information provider subsystem, and emergency management subsystem, all within one building, while another traffic management center may concentrate only on the management of traffic with the traffic management subsystem.

The architecture has identified four communication media types to support the communication requirements between the nineteen subsystems: They are wireline (fixed-to-fixed), wide area wireless (fixed-to-mobile), dedicated short range communications (fixed-to-mobile), and vehicle-to-vehicle (mobile-to-mobile). Wireline technology, such as leased or owned twisted wire pairs, coaxial cable, or fiber optics, can be used by a traffic management

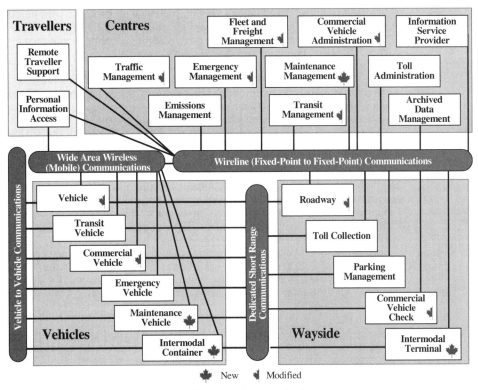

FIGURE 6.1   The Canadian physical architecture and communications connections.

center to gather information and monitor and control roadway subsystem equipment packages (e.g., traffic surveillance sensors, traffic signals, changeable message signs). Although wireless communications technologies can also be used in this case, they are used to provide fixed-to-fixed communications and consequently the architecture recognizes them as wireline communications media. One- or two-way wide area wireless (fixed-to-mobile) communications are suited for services and applications where information is disseminated to users who are not located near the source of transmission and who require seamless coverage, such as is the case for traveler information and route guidance dissemination. Short-range wireless is concerned with information transfer of localized interest. Two types of short-range wireless communication are identified by the architecture: vehicle-to-vehicle communication, which supports automated highway systems (AHS), and dedicated short-range communications (DSRC), used in applications such as automated toll collection.

In conclusion, the basic benefit of the architecture is to provide a structure that supports the development of open standards. This results in numerous benefits: the architecture makes integration of complex systems easier, ensures compatibility, and supports multiple ranges of functionality and designs.

ITS and Potential Economic Stimulation.   As mentioned above, ITS is more than just solutions to traffic problems. Investments in the ITS industry are actually large scale infrastructure investments that feature wide-spread application of high technology. The arguable question is whether ITS would promote national level economic growth. Information technology-related industries are increasingly becoming the heart of the economy of many industrialized nations worldwide. Directing information technology investments towards large-scale infrastructure developments has the potential to promote new industries that would have an impact on long-term economic growth. This might take place for several reasons (Transportation Infastructures 1995). First, such investments would create large economies of scale for new computing and communications products even before they could attain such scale economies in the marketplace. This would offer more rapid returns on investment to supporters of such industries. Second, as a consequence of such success, capital markets, which are usually risk-averse, might be more inclined to support such industries and make larger funds available for their expansion. Third, this would speed the adoption of new generations of communications and computing technology.

Governments can play critical roles in accelerating the growth of the ITS industry. Public investments could shape the setting of system architectures and standards and influence the development of applications that would encourage private sector investments by lowering the risk perceived by private investors. A national ITS mandate would likely reduce some of the risks and result in private companies taking a longer-range view of returns to their capital spending. The role of governmental leadership in the ITS industry can be easily appreciated if one contrasts the rapid growth of the ITS industry in a country like the United States with that in its neighbor Canada, for example. In the United States, ISTEA promoted and accelerated ITS research and development using federal funds, and resulted in large-scale involvement from the private sector. In Canada, on the other hand, the absence of a similar federal ITS mandate is severely crippling the growth of the ITS industry and the related job market, forcing Canadian talents and entrepreneurs in the ITS field to be export-oriented, shifting focus and effort toward the American market and the international market in general.

Sustainability. A 1987 United Nations-sponsored report entitled *Our Common Future* defined sustainable development as "development that meets the needs of the present without compromising the ability of future generations to meet their own needs." Thus, any economic or social development should improve rather than harm the environment. Sustainability has only recently begun to be applied to cities. With increasing environmental awareness, urban transportation planning process becomes more concerned with the air, land, and water and the likely ecological impact of transportation facilities.

Thus, sustainable transportation planning states that cities can become more livable, more humane, more healthy places, but they must learn how to achieve this by using fewer natural resources, creating less waste, and decreasing the impact on the environment.

Cities are increasingly involved in pursuing this sustainability agenda. For this purpose, specific indicators are defined to guide cities to move towards more livable communities while reducing their impact on the earth and the ecosystem. Examples of such indicators are taken from Newman and Kenworthy (1998):

- Energy and air quality (e.g., the reduction of energy use per capita, air pollutants, greenhouse gases)

- Water, materials, and waste (e.g., the reduction of total water user per capita, solid waste, consumption of building materials per capita)

- Land, green spaces, and biodiversity (e.g., preserve agricultural land and natural landscape and green space; increase proportion of urban redevelopment to new developments, increase density of population and employment in transit-oriented locations)

- Transportation (e.g., reduce auto use per capita; increase transit, walk, bike, and carpool; decrease parking spaces)

- Livability: human amenities and health (e.g., decrease infant mortality, increase educational attainment, decrease transport fatalities, increase proportion of cities allowing mixed-use and higher-density urban villages).

The above indicators are a scaled-down version of the original 150 indicators suggested by the World Bank and the UN Center for Human Settlements. They serve as evidence of how cities are contributing to global problems, namely greenhouse gases and oil depletion. Every particular city has to define the indicators that are applicable to its conditions, and consequently manage both local and global issues.

In order to meet the sustainability indicators, cities have to develop sustainability plans, known also as Local Agenda 21 plans, as stated in Agenda 21:

> Each Local authority should enter into a dialogue with its citizens, local organizations and private enterprises and adopt a "local" Agenda 21. Local authorities should learn from citizens and local, civic, community, business, and industrial organizations the information needed for formulating the best strategies. This process will also increase household awareness of sustainable development issues (Sitarz 1994, 177)

Sustainability plans require two central approaches, namely integrated planning as well as community participation. The integrated planning deals with the fusion of cities' physical and environmental planning with economic planning. On the other hand, community participation calls for public participation in planning. In other words, urban plans should be designed with local citizens to meet their local needs.

Auto-dependency is recognized as a great threat to sustainability. In fact, one of the central arguments for sustainable development is concerned with the critical impact of an automobile-based transportation system on a society. Thus, changes in travel behavior must occur in order to minimize transportation's impact on the environment (Newman and Kenworthy 1998). Three general approaches are to be implemented simultaneously to limit auto-dependency and consequently change cities over time to become more sustainable:

- *Automobile technological improvements:* the development of less-polluting cars to reduce air pollutants and emissions.
- *Economic instruments:* setting the right road user charges to meet the real cost of auto usage, such as pollution costs, health costs, road and parking costs, etc.
- *Planning mechanisms:* the need of a non-automobile-dependent planning. The New Urbanism trend encourages environment-friendly commuting modes, such as transit, cycling, and walking. This can be achieved by changing the urban fabric to become denser and mixed land use.

In addition, in order to reduce auto-dependency in cities, five policies should be followed. These policies bring together the processes of traffic calming, state-of-the-art transit, bicycle planning, and transit-oriented development, the neo-traditional urban design of streets for pedestrians, in particular the design of urban villages, growth management, as well as economic penalties for private transportation.

6.2 TRAFFIC STREAM PARAMETERS AND THEIR MEASUREMENT

6.2.1 Characteristics of Traffic Flow

Traffic flow can be divided into two primary types: interrupted flow and uninterrupted flow.

Uninterrupted flow occurs when vehicles traversing a length of roadway are not required to stop by any cause external to the traffic stream, such as traffic control devices. Uninterrupted flow is regulated by vehicle-vehicle interactions on one side and by the interactions between vehicles and the roadway environment and geometry on the other side. An instance of uninterrupted flow includes vehicles traveling on an interstate highway or on other limited access facilities where there are no traffic signals or signs to interrupt the traffic. Uninterrupted flow can also occur on long sections of rural surface highway between signalized intersections. Even when such facilities are experiencing congestion, breakdowns in the traffic stream are the results of internal rather than external interactions in the traffic stream.

Interrupted flow occurs when flow is periodically interrupted by external means, primarily traffic control devices such as stop and yield signs and traffic signals. Under interrupted flow conditions, traffic control devices play a primary role in defining the traffic flow, while vehicle-vehicle interactions and vehicle-roadway interactions play only a secondary role. For instance, traffic signals allow designated movements to occur only part of the time. In addition, because of the repeated stopping and restarting of traffic stream on such facilities, flow occurs in platoons.

6.2.2 Traffic Stream Parameters

Traffic stream parameters represent the engineer's quantitative measure for understanding and describing traffic flow. Traffic stream parameters fall into two broad categories: macroscopic parameters, which characterize the traffic stream as a whole, and microscopic parameters, which characterize the behavior of individual vehicles in the traffic stream with respect to each other.

The three macroscopic parameters that describe traffic stream are volume or rate of flow, speed, and density.

Volume and Flow.   Volume is simply the number of vehicles that pass a given point on the roadway or a given lane or direction of a highway in a specified period of time. The unit of volume is simply vehicles, although it is often expressed as annual, daily, hourly peak and off-peak. The subsequent sections explain the range of commonly used daily volumes, hourly volumes, and subhourly volumes.

Daily Volumes.   Daily volumes are frequently used as the basis for highway planning, for general trend observations, as well as for traffic volume projections. Four daily volume parameters are widely used: average annual daily traffic (AADT), average annual weekday traffic (AAWT), average daily traffic (ADT), and average weekday traffic (AWT).

- AADT is the average 24-hour traffic volume at a given location over a full year, that is, the total number of vehicles passing the site in a year divided by 365. AADT is normally obtained from permanent counting stations, typically bidirectional flow data rather than lane-specific flow data.

- AAWT is the average 24-hour traffic volume occurring on weekdays over a full year. AAWT is normally obtained by dividing the total weekday traffic for the year by the annual weekdays (usually 260 days). This volume is of particular importance since weekend traffic is usually low; thus, the average higher weekday volume over 365 days would hide the impact of the weekday traffic.

- ADT is the average 24-hour traffic volume at a given location for a period of time less than a year (e.g., summer, six months, a season, a month, a week). ADT is valid only for the period of time over which it was measured.

- AWT is the average 24-hour traffic volume occurring on weekdays at a given location for a period of time less than a year, such as a month or a season.

The unit describing all these volumes is vehicles per day (veh/day). Daily volumes are often not differentiated per lane or direction but rather are given as totals for an entire facility at a particular location.

Hourly Volumes.   As mentioned previously, daily volumes are used mainly for planning applications. They cannot be used alone for design and operational analysis. Hourly volumes are designed to reflect the variation of traffic over the different time period of a day. They are also used to identify single hour or period of highest volume in a day occurring during the morning and evening commute, that is, rush hours. The single hour of the day corresponding to the highest hourly volume is referred to as peak hour. The peak hour traffic volume is a critical input in the design and operational analysis of transportation facilities. The peak hour volume is usually a directional traffic, that is, the direction of flows is separated. Highway design as well as other operations analysis, such as signal design, must adequately serve the peak-hour flow corresponding to the peak direction.

Peak hour volumes can sometimes be estimated from AADT, as follows:

$$\text{DDHV} = \text{AADT} \times K \times D$$

where DDHV = directional design hourly volume (veh/hr)
  AADT = average annual daily traffic (24 hours) (veh/day)
    K = factor for proportion of daily traffic occurring at peak hour
    D = factor for proportion of traffic in peak direction

K and D values vary depending on the regional characteristics of the design facilities, namely, rural versus urban versus suburban. K often represents the AADT proportions occurring during the thirtieth or fiftieth highest peak hour of the year. K factor is inversely proportional to the density of development surrounding the highway. In design and analysis of rural areas, the thirtieth highest peak-hour volume is used, while in urbanized areas the fiftieth highest is used. The D factor depends on both the concentration of developments and the specific relationship between the design facility and the major traffic generators in the area. The *Highway Capacity Manual 2000* provides ranges for K and D factors depending on the facility types and the corresponding regional characteristics of the area.

Subhourly Volumes. Subhourly volumes represent traffic variation within the peak hour, i.e., short-term fluctuations in traffic demand. In fact, a facility design may be adequate for design hour, but breakdown may occur due to short-term fluctuations. Typical designs and operational analyses are based on 15-minute peak traffic within the peak hour (e.g., level of service analysis using *Highway Capacity Manual*).

The peak-hour factor (PHF) is calculated to relate the peak flow rate to hourly volumes. This relationship is estimated as follows:

$$\text{PHF} = \frac{V}{4 \times V_{15}}$$

where PHF = peak hour factor
    V = peak hour volume (veh/hr)
  V_{15} = volume for peak 15-min period (veh)

The PHF describes trip-generation characteristics. When PHF is known, it can be used to convert a peak-hour volume to an estimated peak rate of flow within an hour:

$$v = \frac{V}{\text{PHF}}$$

where v = peak rate of flow within hour (veh/hr)
    V = peak hourly volume (veh/hr)
  PHF = peak hour factor

Speed. The speed of a vehicle is defined as the distance it travels per unit of time. It is the inverse of the time taken by a vehicle to traverse a given distance. Most of the time, each vehicle on the roadway will have a speed that is somewhat different from the speed of the vehicles around it. In quantifying the traffic stream, the average speed of the traffic is the significant variable. The average speed, called the space mean speed, can be found by averaging the individual speeds of all of the vehicles in the study area.

Space Mean versus Time Mean Speed.   Two different ways of calculating the average speed of a set of vehicles are reported, namely the space mean speed and the time mean speed. This difference in computing the average speed leads to two different values with different physical significance. While the time mean speed (TMS) is defined as the average speed of all vehicles passing a point on a highway over a specified time period, the space mean speed (SMS) is defined as the average speed of all vehicles occupying a given section of a highway over a specified time period. TMS is a point measure and SMS is a measure relating to a length of highway or lane. TMS and SMS may be computed from a series of measured travel times over a measured distance. TMS takes the arithmetic mean of the observation. It is computed as:

$$\text{TMS} = \frac{\sum \frac{d}{t_i}}{n}$$

SMS could be calculated by taking the harmonic mean of speeds measured at a point over time. It is computed by dividing the distance by an average travel time, as shown below:

$$\text{SMS} = \frac{d}{\sum \frac{t_i}{n}} = \frac{nd}{\sum t_i}$$

where TMS = time mean speed (fps or mph)
      SMS = space mean speed (fps or mph)
         d = distance traversed (ft or mi)
         n = number of travel times observed
         t_i = travel time for the ith vehicles (sec or hr)

Density.   Density is the number of vehicles present on a given length of roadway or lane. Normally, density is reported in terms of vehicles per mile or per kilometer. High densities indicate that individual vehicles are very close to each other, while low densities imply greater distances between vehicles. Density is a difficult parameter to measure directly in the field. Direct measurements of density can be obtained through aerial photography, which is an expensive method, or it can be estimated from the density, flow, and speed relationship as explained in the paragraphs below.

Flow, Speed, Density Relationship.   Speed, flow, and density are all related to each other and are fundamental for measuring the operating performance and level of service of transportation facilities, such as freeway sections. Under uninterrupted flow conditions, speed, density, and flow are all related by the following equation:

$$\text{Flow} = \text{Density} \times \text{Speed: } v = S \times D$$

where v = flow (veh/hr)
         S = space mean (average running) speed (mph, km/hr)
         D = density (veh/mile, veh/hr)

The general form of relationships between speed, density, and flow is illustrated in Figure 6.2, also known as the fundamental diagrams of traffic flow. The relationship between speed and density is consistently decreasing. As density increases, speed decreases. This diagram as well as the above formula show that flow is zero under two different conditions:

• When density is zero: thus, there is no vehicle on the road
• When speed is zero: vehicles are at complete stop because of traffic congestion.

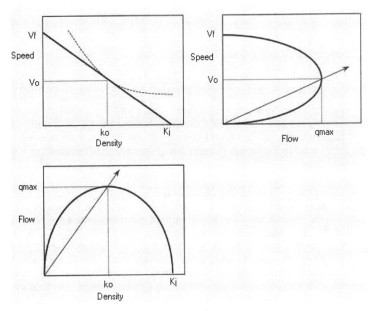

FIGURE 6.2 Fundamental flow-speed-density diagram.

In the first case, the speed corresponds to the theoretical maximum value: the free flow speed v_0, while in the second the density assumes the theoretical maximum value: the jam density, K_{jam}. The peak of the density flow curve (and speed-flow curve) occurs at the theoretical maximum flow (i.e., capacity) of the facility. The corresponding speed v_c and density k_c are referred to as the critical speed and the critical density at which maximum capacity occurs.

Density is the most important of the three traffic-stream parameters, since it is the measure most directly related to traffic demand and congestion levels. In fact, traffic is generated from various land uses, bringing trips on a highway segment. Generated trips produce traffic density, which in turn produces flow rate and speeds. Density also gives an indication of the quality of flow on the facilities. It is the measure of proximity of vehicles and is also the basis for LOS on uninterrupted facilities. In addition, density readings, in contrast to flow measurements, clearly distinguish between congested or uncongested conditions.

6.2.3 Other: Gap, Headway, and Occupancy

Flow, speed, and density are macroscopic parameters characterizing the traffic stream as a whole. Headway, gap, and occupancy are microscopic measures for describing the space between individual vehicles. These parameters are discussed in the paragraphs below.

Headway. Headway is a measure of the temporal space between two vehicles, or, more specifically, the time that elapses between the arrival of the leading vehicle and the following vehicle at the designated test point along the lane. Headway between two vehicles is measured by starting a chronograph when the front bumper of the first vehicle crosses the selected point and subsequently recording the time that the second vehicle's front bumper crosses over the designated point. Headway is usually reported in units of seconds.

Average value of headway is related to macroscopic parameters as follows:

$$\text{Average headway} = 1/\text{flow} \quad \text{or} \quad v = \frac{3600}{h_a}$$

where v = rate of flow
h_a = average headway

Gap. Gap is very similar to headway, except that it is a measure of the time that elapses between the departure of the first vehicle and the arrival of the second at the designated test point. Gap is a measure of the time between the rear bumper of the first vehicle and the front bumper of the second vehicle, where headway focuses on front-to-front times. Gap is also reported in units of seconds. Figure 6.3 illustrates the difference between gap and headway.

Occupancy. Occupancy denotes the proportion or percentage of time a point on the road is occupied by vehicles. It is measured, using loop detectors, as the fraction of time that vehicles are on the detector. Therefore, for a specific time interval T, occupancy is the sum of the time that vehicles cover the detector, divided by T. For each individual vehicle, the time spent on the detector is determined as function of the vehicle's speed, its headway, its length L, plus the length of the detector itself C. That is, the detector is affected by the vehicle from the time the front bumper crosses the start of the detection zone until the time the rear bumper clears the end of the detection zone. Occupancy is computed as follows:

$$\text{LO} = \frac{(L + C)/\text{speed}}{\text{headway}} = (L + C) \times \text{density} = k \times (L + C)$$

Assuming flow = density $\times$ speed

where LO = lane occupancy, i.e., percentage of time a lane is occupied with vehicles divided by total study time
K = density of flow
L = average vehicle length
C = length of detector

Therefore, if occupancy is measured as above, density can be estimated as:

$$k = \frac{\text{LO}}{(L + C)}$$

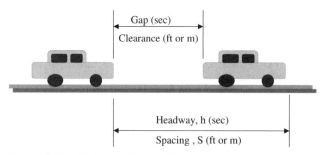

FIGURE 6.3 Illustration of gap and headway definition.

6.2.4 Loop Detector as Measuring Device

The inductive loop detector is by far the most common form of detector used for both traffic counting and traffic management purposes. It is used to measure traffic volume, flow rate, vehicle speed, and occupancy. Inductance loops are widely used detector systems and are known for their reliability in data measurement, flexibility in design, and relatively low cost.

The loop detectors' principal components (see Figure 6.4) include:

- One or more turns of insulated wire buried in a narrow, shallow saw-cut in the roadway
- Lead-in cable that connects the loop to the detector via a roadside pull-out box
- Detector unit (or amplifier) that interprets changes in the electrical properties of the loop when a vehicle passes over it

Data that can be determined from inductive loop detectors include lane occupancy, traffic densities, traffic composition, average and instantaneous vehicle velocities, presence of congestion, and length and duration of traffic jams. Depending on the technology used, these data can be directly or indirectly determined by the inductive loop detectors. Additional data include historical data, weather condition measurements, time of day (rush hour or otherwise), and type of day (weekday, weekend, public holiday).

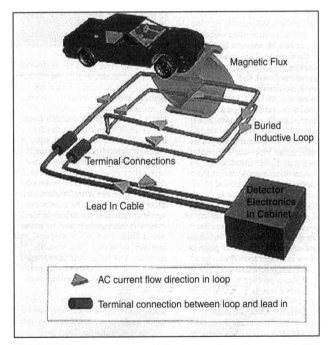

FIGURE 6.4 Car passing over inductive loop buried in pavement. The loop system becomes active when the detector unit sends an electric current through the cable, creating a magnetic field in the loop. When a vehicle passes over the loop, the metal of the vehicle disturbs the magnetic field over the loop, which causes a change in the loop's inductance. Inductance is an electrical property that is proportional to the magnetic field. The induced magnetic field increases the frequency of oscillation that is sensed by the detector unit. The loop sensor thus detects a vehicle.

Loop detectors are also necessary to measure the data that will be used to construct a traffic model and calibrate this model, as we will see in the next section (i.e., checking whether the behavior predicted by the model corresponds accurately enough to the real behavior of the system).

6.3 TRAFFIC FLOW THEORY

Knowledge of fundamental traffic flow characteristics (speed, volume, and density) and the related analytical techniques are essential requirements in planning, design, and operation of transportation systems. Fundamental traffic flow characteristics have been studied at the microscopic, mesoscopic, and macroscopic, levels. Existing traffic flow models are based on time headway, flow, time-space trajectory, speed, distance headway, and density. These models lead to the development of a range of analytical techniques, such as demand-supply analysis, capacity and level of service analysis, traffic stream modeling, shock wave analysis, queuing analysis, and simulation modeling (May 1990).

Traffic simulation models are also classified as microscopic, macroscopic, and mesoscopic models. Microscopic simulation models are based on car-following principles and are typically computationally intensive but accurate in representing traffic evolution. Macroscopic models are based on the movement of traffic as a whole by employing flow rate variables and other general descriptors representing flow at a high level of aggregation without distinguishing its parts. This aggregation improves computational performance but reduces the detail of representation. Mesoscopic models lie between the other two approaches and balance accuracy of representation and computational performance. They represent average movement of a group of vehicles (packets) on a link. Microscopic analysis may be selected for moderate-size systems where there is a need to study the behavior of individual units in the system. Macroscopic analysis may be selected for higher-density, large-scale systems in which a study of behavior of groups of units is adequate. Knowledge of traffic situations and the ability to select the more appropriate modeling technique is required for the specific problem. In addition, simulation models differ in the effort needed for the calibration process. Microscopic models are the most difficult to calibrate, followed by mesoscopic models. Macroscopic models are easily calibrated.

6.3.1 Traffic Flow Models

Microscopic traffic flow modeling is concerned with individual time and space headway between vehicles, while macroscopic modeling is concerned with macroscopic flow characteristics. The latter are expressed as flow rates with attention given to temporal, spatial, and modal flows (May 1990). This section describes the best-known macroscopic, mesosopic, and microscopic traffic flow models.

Macro Models.   In a macroscopic approach, the variables to be determined are:

- The flow $q(x,t)$ (or volume) corresponding to the number of vehicles passing a specific location x in a time unit and at time period t
- The space mean speed $v(x,t)$ corresponding to the instantaneous average speed of vehicles in a length increment
- The traffic density $k(x,t)$ corresponding to the number of vehicles per length unit

These macroscopic variables are defined by the well-known equation:

$$q(x,t) = k(x,t) \times v(x,t)$$

The static characteristics of the flow are completely defined by a fundamental diagram (as shown in Figure 6.5). The macroscopic approach considers traffic stream parameters and develops algorithms that relate flow to density and space mean speed. Various speed-density models have been developed and are shown also to fit experimental data. These models are explained below.

Greenshields Model. The first steady-state speed-density model was introduced by Greenshields, who proposed a linear relationship between speed and density as follows:

$$u = u_f - \left(\frac{u_f}{k_j}\right) \times k$$

where u = velocity at any time
u_f = free-flow speed
k = density at that instant
k_j = maximum density

As mentioned above, in these equations, as the flow increases, density increases and the speed decreases. At optimum density, flow becomes maximum (q_m) at $u = u_f/2$ and $k = k_j/2$.

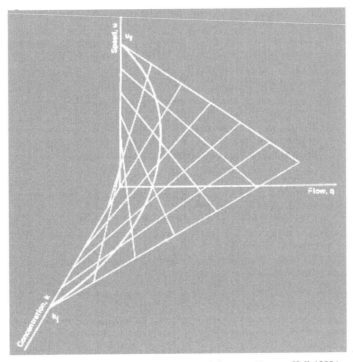

FIGURE 6.5 Three-dimensional fundamental diagram. (*Source:* Hall 1998.)

Greenberg Model. A second early model was suggested by Greenberg (1959), showing a logarithmic relationship as follows:

$$u = c \ln (k/k_j)$$

where u = velocity at any time
  c = a constant (optimum speed)
  k = density at that instant
  k_j = maximum density

Three-Dimensional Models. The idea of considering all three fundamental variables (q, k, v) simultaneously first appeared in TRB SR-165. The notion of a three-dimensional model appeared in the form of Figure 6.5, where $v = q/k$ represents the surface of admissible traffic stream models. The surface shown in Figure 6.5 is a continuous one; thus, by accepting that the $u = q/k$ relationship holds the entire range of traffic operations, one can reasonably conclude that it suffices to study traffic modeling as a two-dimensional problem (Hall 1998).

Lighthill and Whitham (LW) Model. The continuous-flow approach, proposed by Lighthill and Whitham (LW) (1955), represents the aggregate behavior of a large number of vehicles. This model is applicable to the distribution of traffic flow on long and crowded roads. The LW model reproduces qualitatively a remarkable amount of real traffic phenomena, such as the decreasing speeds with increasing densities and shock wave formation.

The LW model is derived from the physical law of incompressible fluid and is based on the following three fundamental principles (Cohen 1991):

1. *Continuous representation of variables:* It considers that at a given location x and time t, the traffic mean speed $u(x,t)$, the flow $q(x,t)$ and traffic density values $k(x,t)$ are continuous variables and satisfy the relation $u(x,t) = q(x,t)/k(x,t)$.

2. *The law of conservation of mass:* This is a basic speculation of the simple continuum model, which states that vehicles are not created or lost along the road. The law of the conservation of the number of vehicles leads to the continuity equation for the density $k(x,t)$:

$$\frac{\partial k(x,t)}{\partial t} + \frac{\partial q(x,t)}{\partial x} = 0$$

3. *The statement of fundamental diagrams:* The fundamental hypothesis of the theory is that at any point on the road, the speed u is a function of the density. In addition, speed is a decreasing function of concentration: $u = u(k)$.

Therefore, the law of traffic at a given section of the road during a given time period can be expressed in terms of an equation relating two out of the three variables flow, concentration, and speed (Cohen 1991).

For the macroscopic description of the theory the flow q(veh/hr), the density k (veh/km), and the mean speed u (km/hr) are considered as differentiable functions of time t and space x (Papageorgiou 1998).

From the continuity equation with the flow-density relation ($q = q(k)$) and the basic relation between traffic variables ($q = u \cdot k$), a differential equation of the density (k) is derived as follows:

$$\frac{\partial k(x,t)}{\partial t} + q'(k) \frac{\partial k(x,t)}{\partial x} = 0$$

The kinematic waves theory attempts to solve this partial differential equation to predict the concentration of flow at any point on the road at any time.

Figure 6.6 shows how the propagation speed of the shock wave corresponds to the slope of the tangent on the fundamental diagram. This hypothesis implies that slight changes in

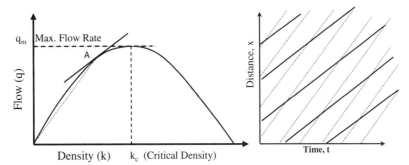

FIGURE 6.6 Speed and kinematic waves. (*Source:* Cohen 1991.)

traffic flow are propagated through the stream of vehicles along kinematic waves. Waves are propagated either:

- Forward, when density k is less than the critical density k_c, which corresponds to the uncongested region of the flow-density diagram, or
- Backward, when k is greater than critical density k_c, which corresponds to the congested region of the flow-density diagram.

This property leads to a distinction between two types of flow, namely uncongested flow (for $k < k_c$) and congested flow ($k > k_c$). In practice, once congestion occurs, disturbance propagates backward from downstream.

Under several assumptions and simplifications, the LW model is consistent with a class of car-following models. With regard to urban traffic flow in signalized networks, the LW model is more than sufficient because traffic flow dynamics are dominated by external events (red traffic lights) rather than by the inherent traffic flow dynamics (Papageorgiou 1998). For freeway traffic flow, the LW model achieves a certain degree of qualitative accuracy and is certainly an improvement over purely static approaches. However, the LW model includes a number of simplifications and fails to reproduce some real dynamic phenomena observed on freeways.

Shock Waves. Flow-speed-density states changes over space and time (May 1990). With the prompt occurrence of such change, a boundary is established that marks a discontinuity of flow and density from one side of the boundary in respect to the other. This discrepancy is explained by the generation of shock waves. Basically, a shock wave exists whenever the traffic conditions change abruptly. As such, shock waves can be generated by collisions, sudden increases in speed caused by entering free-flow conditions, or a number of other means. A shock, then represents a mathematical discontinuity (abrupt change) in k, q, or u.

Figure 6.7, from May (1990), shows two different densities, flows and speed of vehicles moving along a highway. The line separating these two flows represents the shock wave and is moving at a speed w_{AB}.

The propagation velocity of shock waves is

$$w_{AB} = (q_B - q_A)/(k_B - k_A)$$

where w_{AB} = propagation velocity of shock wave (mph or km/hr)
  q_B = flow prior to change in conditions (veh/hr)
  q_A = flow after change in conditions (veh/hr)
  k_B = traffic density prior to change in conditions (veh/mile or veh/km)
  k_A = traffic density after change in conditions (veh/mile or veh/km)

Thus, the shock wave separating the two flows travels at an intermediate speed. Since the

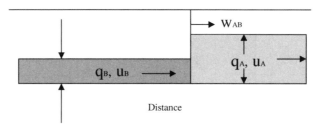

FIGURE 6.7   Shock wave analysis fundamentals. (*Source:* May 1990.)

shock wave in Figure 6.7 is moving with the direction of the traffic, it is a positive forward-moving shock wave. On the other hand, a backward-moving shock wave or negative shock wave travels upstream or against the traffic stream.

Figure 6.8, from Cohen (1991), demonstrates the use of traffic waves in identifying the occurrence of a shock wave and following its trajectory. The figure on the left represents a flow-concentration curve. The figure on the right represents the occurrence of a shock wave and following its trajectory. On the q–k curve, point A represents a situation where traffic travels at near capacity, implying that speed is well below the free-flow speed. Point B represents an uncongested condition where traffic travels at a higher speed because of the lower density. Tangents at points A and B represent the wave velocities of these two situations. The line connecting the two points on the q–k curve represents the velocity of the shock wave. In the space-time diagram the intersection of these two sets of waves has a slope equal to the slope of the line connecting the two points A and B on the q–k curve. This intersection represents the velocity of the shock wave.

Second-Order Model: Payne Model.   Payne (1971) proposed a method for relating macroscopic variables and car-following theories. Payne developed an extended continuum model that takes into consideration drivers reaction time and uses a dynamic speed equation as shown below:

$$\frac{dV}{dt} = \frac{2}{\tau}(V_e(\rho) - V) - \frac{D(\rho)}{\rho\tau}\frac{\partial\rho}{\partial x}$$

where the term $(V_e(\rho) - V)/\tau$ is denoted by the relaxation term and the term $(D(\rho)/\rho\tau)$ $(\delta\rho/\delta x)$ is denoted by the anticipation term.

The relaxation term allows for the delayed adjustment of the stream to a prespecified speed $V_e(\rho)$ as a result of reaction time τ and braking or acceleration procedures. The antic-

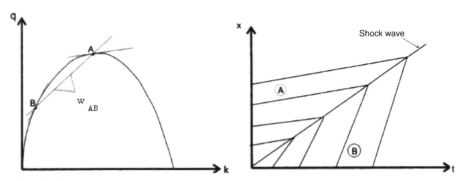

FIGURE 6.8   Shock wave diagrams.

ipation term allows the drivers to adjust their speeds in advance to changes in density lying ahead.

The 2nd-order model provides the possibility of a more realistic description of traffic flow (Kim 2002). The shock wave problem is alleviated through the application of the diffusion terms. Moreover, unstable congested states are derived by the interplay between anticipation and relaxation effects in the model. The 2nd-order model has a critical density, above which uniform flow conditions are unstable and the wave is oscillating with ever-increasing amplitude. In addition, the presence of oscillating waves, explains stop-and-go traffic conditions.

Microscopic Models. Much research has been devoted to the concept that traffic stream behavior can be analyzed at the microscopic level. At this level the behavior of individual drivers must be examined and modeled. Microscopic models use car following laws to describe the behavior of each driver-vehicle system in the traffic stream as well as their interaction. Examples of microscopic models include car-following models, General Motors models, and cell transmission and cellular automata models.

Car-Following. These models are based on supposed mechanisms describing the process of one vehicle following another, called follow-the-leader models (Lieberman and Rathi 1998). From the overall driving task, the subtask that is most relevant to traffic flow is the task of one vehicle following another on a single lane of roadway (car following). This particular driving subtask is relatively simple to describe by mathematical models as compared to other driving tasks. Car-following models describe the process of car-following in such a way as to approximate the macroscopic behavior of a single lane of traffic. Hence, car-following models form a bridge between individual car-following behavior and the macroscopic world of a line of vehicles and their corresponding flow and stability properties.

Pipes and Forbes Car-Following Models. Car-following theories were developed in the 1950s and 1960s. Early models employed simple rules for determining the distance gap between vehicles. For example, Pipes (1953) argued that the rule that drivers actually follow is the following, as suggested by the California Motor Vehicle Code: "The gap that a driver should maintain should be at least one car length for every 10 mph of speed at which he is traveling."

Using the notation shown in Figure 6.9 for the gap and the vehicle speed, the resulting distance headway d can be written as:

$$d_{\min} = [x_n(t) - x_{n+1}(t)]_{\min} = L_n \left[\frac{\dot{x}_{n+1}(t)}{(10)(1.47)} \right] + L_n$$

According to Pipes' car-following theory, the minimum safe distance headways increase linearly with distance.

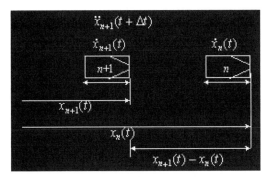

FIGURE 6.9 Car-following model.

General Motors Models. In reality, drivers conform to the behavior of the immediately leading vehicle. Under this notion, a stimulus response relationship exists that describes the control process of the driver-vehicle system. Researchers at General Motors (GM) have developed car-following models and tested these models using real-world data. The importance of these models lies in the discovery of the mathematical bridge linking the microscopic and macroscopic theories of traffic flow (May 1990).

The GM research team developed five generations of car-following models in terms of response-stimuli relationship. The general stimulus-response equation expresses the concept that a driver of a vehicle responds to a given stimulus according to a relation (May, 1990):

$$\text{Response} = \text{Function \{Sensitivity, Stimuli\}}$$

where Response = acceleration or deceleration of the vehicle, which is dependent on the sensitivity of the automobile and the driver himself
Sensitivity = ability of the driver to perceive and react to the stimuli
Stimuli = relative velocity of the lead and following vehicle.

The stimulus function may be composed of many factors: speed, relative speed, intervehicle spacing, accelerations, vehicle performance, driver thresholds, etc. The relative velocity is the most used term. It is generally assumed in car-following modeling that a driver attempts to (a) keep up with the vehicle ahead and (b) avoid collisions. The response is the reaction of the driver to the motion of the vehicle immediately in front of him/her. The response of successive drivers is to react (i.e., accelerate or decelerate) proportionally to the stimulus.

From the notation of Figure 6.10, assuming that the driver of the following vehicle will space himself/herself from the leading vehicle at a distance, such that in case the leading vehicle comes to an emergency stop he/she will be able to come to a rest without crashing. Thus, the spacing of the two vehicles at time *t* will be:

$$d(t) = [x_n(t) - x_{n+1}(t)] = \Delta T \cdot \dot{x}_{n+1}(t + \Delta t) + b_{n+1} + L - b_n$$

where *b* is the stopping distance of the vehicle.

Assuming equal braking distances for the two vehicles and differentiating with respect to time *t*, we obtain:

$$[\dot{x}_n(t) - \dot{x}_{n+1}(t + \Delta t)] = \Delta T \cdot \ddot{x}_{n+1}(t + \Delta t)$$

or

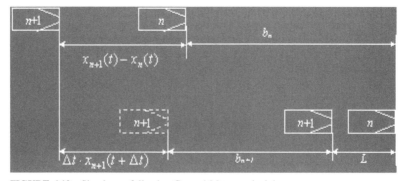

FIGURE 6.10   Simple car-following General Motors principle.

$$\ddot{x}_{n+1}(t + \Delta t) = \frac{1}{\Delta t} [\dot{x}_n(t) - \dot{x}_{n+1}(t + \Delta t)]$$

where $b =$ is the stopping distance of the vehicle, and
$\Delta t =$ reaction time (reciprocal of sensitivity).

The response of the following vehicle is to decelerate by an amount proportional to the difference of speeds. The measure of sensitivity is the reciprocal of the perception-reaction time of the driver. The response function is taken as the deceleration of the following vehicle. The response is lagged by the perception and reaction time of the following driver.

Another form in the simple car-following model is to distinguish the reaction time from the sensitivity by introducing a sensitivity term α as follows:

$$\ddot{x}_{n+1}(t + \Delta t) = \alpha[\dot{x}_n(t) - \dot{x}_{n+1}(t)]$$

The unit of the sensitivity term is $\sec^{-1}$. The stimuli term $[\dot{x}_n(t) - \dot{x}_{n+1}(t)]$ could be positive, negative, or zero, causing the response to be respectively either an acceleration, deceleration, or constant speed.

The two parameters α (sensibility term) and Δt (reaction time) must be selected in such a way that traffic behaves realistically. The choice of these terms is associated with the concept of stability, which is explained below.

Traffic Stability. There are two important types of stability in the car-following system: local stability and asymptotic stability. Local stability is concerned with the response of a following vehicle to a fluctuation in the motion of the vehicle directly in front of it, i.e., it is concerned with the localized behavior between pairs of vehicles. Asymptotic stability is concerned with the manner in which a fluctuation in the motion of any vehicle, say the lead vehicle of a platoon, is propagated through a line of vehicles. The analysis of traffic stability determines the range of the model parameters over which the traffic stream is stable.

Improvements over the First Generation of the General Motors Model. The first GM model was derived using a functional value for acceleration with the assumption that driver sensitivity is constant for all vehicles (May 1990). In a revised version of the GM model, discrepancy from field values indicated that the sensitivity of the driver was higher whenever the headway was less. Accordingly, the GM model was adjusted to account for this error.

Further improvements about the sensitivity were introduced by the speed difference, i.e., relative velocity, because as the speed difference increases, the sensitivity increases. Every system has a time lag to react to changes occurring ahead of it. This is accounted for by the term Δt, which represents the reaction time on the part of the following vehicle to accelerate and decelerate. Finally, the powers of the terms of speed and headway of the vehicle ahead were proposed and these constants were called speed component (m) and headway component (l). The resulting equation represents the fifth and final GM model and is stated as follows:

$$\ddot{x}_{n+1}(t + \Delta t) = \frac{\alpha_{l,m}\dot{x}_{n+1}^m(t + \Delta t)}{[x_n(t) - x_{n+1}(t)]^l} \cdot [\dot{x}_n(t) - \dot{x}_{n+1}(t)]$$

This is the generalized model, and all previous GM models can be considered a special case of this model (May 1990).

Macro-to-Micro Relationship. Gazis, Herman, and Potts (1959) studied the relationship between car-following models and macroscopic traffic stream models. They demonstrated that almost all macroscopic models were related to almost all car-following theory models (May 1990). Gazis, Herman, and Potts derived a generalized macroscopic model from the car-following models:

$$v^{1-m} = v_f^{1-m}[1 - (k/k_j)^{l-1}]$$

For instance, the Greenshields model lies within the following feasible range: when $m = 0$

and $l = 2$. Figure 6.11 shows the speed-density relationship for a number of cases with $m = 1$ and $l = 2.0$ to 3.0.

New Trends in Microscopic Traffic Flow Modeling. Cell-transmission models of highway traffic, developed by Daganzo (1994), are discrete versions of the simple continuum (kinematic wave) model of traffic flow that are convenient for computer implementation. They are in the Godunov family of finite difference approximation methods for partial differential equations. In cell-transmission models, the speed is calculated from the updated flow and density rather than being directly updated.

In the cell-transmission scheme the highway is partitioned into small sections (cells). The analyst then keeps track of the cell contents (number of vehicles) as time passes. The record is updated at closely spaced instants (clock ticks) by calculating the number of vehicles that cross the boundary separating each pair of adjoining cells during the corresponding clock interval. This average flow is the result of a comparison between the maximum number of vehicles that can be sent by the cell directly upstream of the boundary and those that can be received by the downstream cell.

The sending (receiving) flow is a simple function of the current traffic density in the upstream (downstream) cell. The particular form of the sending and receiving functions depends on the shape of the highway's flow-density relation, the proximity of junctions, and whether the highway has special lanes (e.g., turning lanes) for certain vehicles (e.g., exiting vehicles). Although the discrete and continuum models are equivalent in the limit of "vanishing" small cells and clock ticks, the need for practically sized cells and clock intervals generates numerical errors in actual applications.

The cell-transmission representation can be used to predict traffic's evolution over time and space, including transient phenomena such as the building, propagation, and dissipation of queues.

Recently there has been growing interest in studying traffic flow with cellular automata (CA) models. CA models are conceptually simple rules that can be used to simulate a

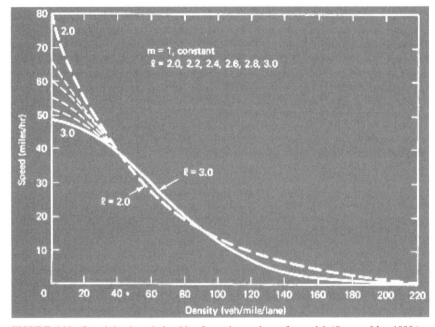

FIGURE 6.11 Speed-density relationships for various values of *m* and *l*. (*Source:* May 1990.)

complex physical process by considering a description at the level of the basic components of the system (Jiang). Only the essential features of the real interactions are taken into account in the evolution rules. Through the use of powerful computers, these models can capture the complexity of the real-world traffic behavior and produce clear physical patterns that are similar to real phenomenon.

Nagel and Schreckenberg (1992) introduced the CA model for traffic. The rationale of CA is not to try to describe a complex system from a global point of view, as it is described using for instance differential equations, but rather modeling this system starting from the elementary dynamics of its interacting parts. In other words, CA does not describe a complex system with complex equations; rather, it lets the complexity emerge by interaction of simple individuals following simple rules. These simple models have been shown to reproduce, at least qualitatively, the features of real traffic flow.

TRANSIMS microsimulation has adapted CA techniques for representing driving dynamics and simulating traffic in entire cities. In these models, the basic idea is to formulate a model in space and time. The space is the road divided into grid points or cells (typically 7.5-m length, which corresponds to the length that a car uses up in a jam). A cell is either empty or occupied by exactly one vehicle. In addition, car positions are updated synchronously, in successive iterations (discrete time steps) (Dupuis and Chopard 1998). During the motion, each car can be at rest or jump to the nearest neighbor site along the direction of motion. The rule is simply that a car moves only if its destination cell is empty. In essence, drivers do not know whether the car in front will move or is stuck by another car. That is, the state of a cell $s(t)$ at a given time depends only on its own state one time step previously, and the states of its nearby neighbors at the previous time step. This dynamic can be summarized by the following relation:

$$s_i(t + 1) = s(t)_{i-1}(1 - s(t)_i) + s(t)_i s(t)_{i+1}$$

where t is the discrete time step.

All cells are updated together. Movement takes place by *hopping* from one cell to another, using a 1-second time step, which agrees with the reaction-time arguments. Different vehicle speeds are represented by different hopping distances (Nagel and Rickert 2001). This implies for example that a hopping speed of 5 cells per time step corresponds to 135 km/hr. Accordingly, the rules for car following in the CA are:

1. If no car is ahead: linear acceleration occurs up to maximum speed.
2. If a car is ahead: velocity is adjusted so that it is proportional to the distance between the cars (constant time headway).
3. Sometimes vehicles are randomly slower than what would result from 1 and 2.

Lane changing is done as pure sideways movement in a sub-time step, before the forward movement of the vehicles, i.e., each time-step is subdivided into two sub-time steps (Nagel and Rickert 2001). The first sub-time-step is used for lane changing, while the second sub-time-step is used for forward motion. Lane-changing rules for TRANSIMS are symmetric and consist of two simple elements: decide that you want to change lanes, and check if there is enough gap to get in. A reason to change lanes is either that the other lane is faster or that the driver wants to make a turn at the end of the link and needs to get into the correct lane. In the latter case, the accepted gap decreases with decreasing distance to the intersection, that is, the driver becomes more and more desperate. In addition, details of the system, including lane changing, complex turns, and intersection configurations, are fully represented and each driver is given a destination and a preferred path.

In more advanced work, Nagel and Rickert (2001) proposed the parallel implementation of the TRANSIMS traffic microsimulation. In this parallelism, the road network is partitioned across many processors. This means that each CPU of the parallel computer is responsible for a different geographical area of the simulated region. The results show a significant speed-up in the computation efficiency.

Meso Models.   The mesoscopic models fall in between macroscopic and microscopic modeling. The mesoscopic traffic flow models describe the microscopic vehicle dynamics as a function of macroscopic fields. The gas-kinematic model, which is the most used mesoscopic traffic flow model, treats vehicles as a gas of interacting particles (Nagatani 2002). As such, when the number of vehicles is large, traffic flows is modeled in terms of one compressible gas. Prigogine and Herman have proposed the following Boltzmann equation for the traffic:

$$\frac{\partial f(x,v,t)}{\partial t} + v \frac{\partial f(x,v,t)}{\partial x} = - \frac{f(x,v,t) - \rho(x,t)F_{des}(v)}{\tau_{rel}} + \left(\frac{\partial f(x,v,t)}{\partial t} \right)_{int}$$

where the first-term on the right-hand side represents the relaxation of the velocity distribution function $f(x,v,t)$, to the desired velocity distribution $\rho(x,t)F_{des}(v)$, with the relaxation time τ_{rel}, in the absence of the interactions of vehicles. The second-term on the right-hand side takes into account the change arising from the interactions among vehicles.

Recently, the kinetic theories of a single-lane highway have been extended to two-dimensional flow for urban traffic and multilane traffic.

6.3.2   Traffic-Simulation Models

Computer simulation modeling has been a valuable tool for analyzing and designing complex transportation systems. Simulation models are designed to mimic the behavior of these systems and processes. These models predict system performance based on representations of the temporal and/or spatial interactions between system components (normally vehicles, events, control devices), often characterizing the stochastic nature of traffic flow. In general, the complex simultaneous interactions of large transportation system components cannot be adequately described in mathematical or logical forms. Properly designed models *integrate* these separate entity behaviors and interactions to produce a detailed, quantitative description of system performance.

In addition, simulation models are mathematical/logical representations (or abstractions) of real-world systems, which take the form of software executed on a digital computer in an experimental fashion (Lieberman and Rathi 1998). The inherent value of computer simulation is that it allows experimentation to take place off-line without having to go out in the real world to test or develop a solution. Specifically, simulation offers the benefits of being able to control input conditions, treat variables independently even though they may be coupled in real life, and, most importantly, repeat the experiment many times to test multiple alternative performance (Middleton and Cooner 1999). The user of traffic simulation software specifies a "scenario" (e.g., highway network configuration, traffic demand) as model inputs. The simulation model results describe system operations in two formats: (1) statistical and (2) graphical. The numerical results provide the analyst with detailed quantitative descriptions of what is likely to happen. Traffic simulation models may be classified according to the level of detail with which they represent the transportation performance, as well as flow representation, namely (see Table 6.2):

- In *microscopic* models, traffic is represented discretely (single vehicles); individual trajectories can be explicitly traced. Disaggregate performance measures are calculated based on explicit modeling of driver behavior.
- In *mesoscopic* models, traffic is represented discretely (vehicles or group of vehicles); individual trajectories can be explicitly traced as for microscopic models. However, aggregate performance measures are calculated as for macroscopic models.

TABLE 6.2 Classification and Examples of Traffic-Simulation Models

		Performance functions	
		Aggregate	Disaggregate
Flow representation	*Continuous*	MACROSCOPIC e.g., FREFLO, AUTOS, METANET	—
	Discrete	MESOSCOPIC e.g., DYNASMART, DYNAMIT, INTEGRATION	MICROSCOPIC e.g., INTRAS, CORSIM, PARAMICS, CORSIM, AIMSUN2, TRANSIMS, VISSIM, MITSIM

• In *macroscopic* models, traffic is represented continuously following the fluid approximation. Individual trajectories are not explicitly traced. Aggregate performance measures are calculated using relations derived from fluid-approximation models (Cascetta 2001).

Table 6.2 and Figure 6.12 illustrate the relationship between the type of simulation and the level of detail. In addition, Figure 6.12 illustrates, the relation between the type of simulation and the size of the transportation network that needs to be analyzed.

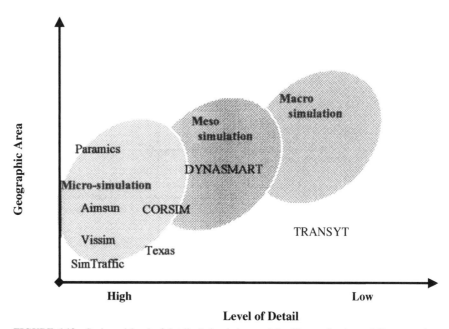

FIGURE 6.12 Scale and level of detail of simulation models. (*Source:* Institute of Transportation Studies, University of California, Irvine).

Microscopic Simulation Models.   A microscopic model describes both the system entities and their interactions at a high level of detail. Microscopic models simulate the journey of each single vehicle through explicit driving behavior models of speed adjustment (Cascetta 2001). These models contain processing logic, which describes how vehicles behave, including acceleration, deceleration, lane changes, passing maneuvers, turning movement execution, and gap acceptance. Typically, vehicles are input into the project section using a statistical arrival distribution (a stochastic process) and are tracked through the section on a second-by-second basis (Hoogendoorn and Bovy 2001).

Microscopic models describe vehicles (and often drivers) individually with varying characteristics and multiple classes. As well, these models describe the prevailing surrounding conditions in details such as traffic control logics (e.g., pretimed, actuated, adaptive). Their main trend has been to integrate as many traffic phenomena as possible. For instance, a lane-change maneuver at the micro level could use the car-following law for the subject vehicle with respect to its current leader. In addition, other detailed driving decision processes are reproduced, such as acceleration and braking; and driving behavior, such as in-route choice, lane change, gap acceptance, and overtaking. The duration of the lane-change maneuver can also be computed. Such models can be solved only by event-based or time-based simulation techniques.

Microscopic models provide very detailed traffic simulation on a small scale yet require a significant amount of data and effort for specification and calibration. Therefore, their application in on-line traffic control is limited due to the large computation time. For these reasons, microscopic models are used primarily for off-line traffic operations rather than transportation planning (Cascetta 2001). Microsimulators can prove quite useful to design intersections, analyze accurate emission and fuel consumption modeling, regulate traffic lights, or study the impact of variable message signs or of ramp metering and any other form of traffic control (Marchal 2001). Moreover, because of the large computing resources and expensive calibration procedures, this highly detailed approach is only practical for small networks (i.e., a few hundred links). Nevertheless, recent advances in cellular automata models, such as TRANSIMS, attempt to fill this gap with microsimulations of many thousand-link networks with the help of massive parallel computers.

Macroscopic Simulation Models.   Macroscopic models normally describe the movement of traffic as a whole by employing flow rate variables and other general descriptors representing it at a high level of aggregation as a flow, without distinguishing its parts. For instance, the traffic stream is represented in an aggregate manner using characteristics as flow-rate, density, and velocity. Macroscopic simulation takes place on a section-by-section basis rather than tracking individual vehicles. In addition, the choice of individual vehicle maneuvers, such as lane changes, is usually not explicitly represented (Schutter et al. 1999).

Macroscopic models are suited for large scale, network-wide applications, where macroscopic characteristics of the flow are of prime interest. Generally, calibration of macroscopic models is relatively simple compared to microscopic and mesoscopic models. However, macroscopic models are generally too coarse to describe correctly microscopic details and impacts caused by changes in roadway geometry (Hoogendoorn and Bovy 2001).

Mesoscopic Simulation Models.   Mesoscopic models fall in between macroscopic and microscopic models. Generally, they represent most entities at a high level of detail but describe their activities and interactions at a much lower level of detail than would a microscopic model. A mesoscopic model does not distinguish or trace individual vehicles; rather, it specifies the behavior of individuals in probabilistic terms. Individual vehicles with the same characteristics are represented by packets (small groups of vehicles). Mesoscopic models assume that packets of vehicles are moved together or that some "patterns" of decisions are modeled instead of individual decisions. Hence, each vehicle within a packet has the same origin and destination, the same route, and the same driver characteristics. In addition, vehicles on a link have the same speed, which is generally time-dependent. While the traffic is represented discretely by tracing the trips of single packets as characterized by a departure time, overtaking, lane-changing, and car-following behavior are not modeled microscopically. Instead, the aggregate speed-volume interaction of traffic (u–q relationship) is used on each link. Usually the smaller the size of the packet the more realistic the analysis. Mesoscopic models are able to simulate queue formation and spill-backs reasonably.

The computation time needed for the simulation is reduced compared to microscopic models (Schutter et al. 1999) but is still high compared to macroscopic models. Mesoscopic models can therefore handle medium-sized transportation networks. They also have the advantage of enabling description behaviors of individual vehicles without the need to describe their individual time-space behavior.

Model Calibration and Validation

Model Calibration.   In order to mimic real-world decisions, models must be calibrated. Model calibration is the process of quantifying model parameters using real-world data in the model logic so that the model can realistically represent the traffic environment being analyzed (Middleton and Cooner 1999).

Model calibration is conducted mainly by comparing user experimental conditions from simulation results with observed data from the real network (fields counts). The simulation is said to be accurate if the error between simulation results and the observed data is small enough. Thus, model parameters should be optimized to match (possibly site-specific) observed settings. However, finding the model parameters requires a decision on a data set and a decision on an objective function that can quantify the closeness of the simulation to observed data set. In general, calibration uses optimization techniques, such as generalized least square, to minimize the deviation between observed and corresponding simulated measurements.

Sensitivity analysis should be conducted to test the model robustness as well as to study the impacts of changes in the model parameters. Otherwise stated, the simulation is physically sound when a slight change in the experimental condition results in minimal oscillation in the simulated results. Sensitivity analysis is especially useful for complex microscopic models, in which effects of the parameters on the flow behavior are hard to analyze mathematically (Hoogendoorn and Bovy 2001). Sensitivity analysis is a time-consuming process because each parameter has to be individually analyzed. In fact, model transferability is dependent not only on an effect in calibration process but also on an in-depth understanding of the sensitivity of a model changes in parameters such as driver behavior (McDonald, Brackstone, and Jeffery 1994).

In general, vehicle characteristics and driver characteristics are the key parameters, which may be site-specific and require calibration (Taplin 1999). Polak and Axhausen (1990) identify three types of behavioral research needed to develop models:

- In-vehicle behavior in response to systems design
- Driving behavior (overtaking, gap acceptance, maneuver, signal behavior)
- Travel behavior, including route choice, compliance with ATIS advice, and responses to information from other sources.

All three types of research try to find how various factors contribute to choices in order to simulate a stochastic model based on each driver determining his or her driving and travel decision.

Calibration of Microscopic Models.   Calibrating a microscopic traffic model is a difficult issue. Both traffic data and knowledge about the traffic behavior are needed. At the micro level, the traffic throughput is decided by the driver behavior. The vehicle parameters are easily understood and possible to measure. Thus, calibration of microscopic models is to be conducted at both microscopic scale, with regard to vehicle-to-vehicle interactions (i.e., the calibration of the behavioral parameters) and macroscopic scale to ensure that the overall behavior is modeled correctly (McDonald, Brackstone, and Jeffery 1994). Therefore, reviewing the calibration of microscopic model also covers the calibration of macroscopic models.

Macroscopic models are relatively easy to calibrate using loop detector data (Cremer and Papageorgiou 1981). Data collection typically consists of minute-by-minute records of flows, average speeds/headways, and traffic composition which can be extracted from loop data or video recordings of the road (McDonald, Brackstone, and Jeffery 1994). Mostly, speed-density relations derived from observations are required. Kerner et al. (2000) show that traffic jam dynamics can be described and predicted using macroscopic models that feature only some characteristic variables, which are largely independent of roadway geometry, weather, and so forth. This implies that macroscopic models can describe jam propagation reliably without the need for in-depth model calibration.

Lind et al. (1999) suggest the collection of the following data to be used in the macroscopic calibration step:

- Flow and speed
- Travel time
- Headway
- Total queue time
- Maximum queue length in vehicle number
- Percentage stops
- Delay time

On the other hand, microscopic data are far more complex both to obtain and to calibrate. It is almost impossible to obtain data on all the behavioral parameters being modeled, not only because of the huge number of parameters required (microscopic models typically have 20 or more parameters), but also because many behavioral parameters are not directly related to easily observable measurements. Brackstone and McDonald (1998) recommend using suitable data sources (e.g., instrumented vehicles) to conduct such measurements. Examples include:

- A laser range-finder: capable of measuring the distances and relative speeds of immediately adjacent vehicles
- A laser speedometer: capable of accurately determining vehicles' speed and acceleration
- A video-audio monitor capable of providing permanent visual records

In addition, McDonald, Brackstone, and Jeffery (1994) propose other measurements relating driver behavior to road design, as well as observing driver eye position and duration.

Brackstone and McDonald (1998) also suggest disassembling the model, testing it in a step-by-step fashion, and adjusting of its distinctive entities individually. Whenever new behavioral rules are added to the model, they should be tested extensively, preferably in isolation.

In general, the lack of microscopic field data necessitates the analyst being confined to use macroscopic or mesoscopic calibration data. Moreover, calibration generally attempts to reproduce macroscopic quantities, such as speed-density curves, by changing parameters describing driving behavior. Unfortunately, this cannot produce the optimal parameters, since the number of degrees of freedom is far too large (Hoogendoorn and Bovy 2001).

In fact, only few microscopic simulation models have been extensively calibrated and validated. Ben Akiva et al. (2002) propose the calibration framework outlined in Figure 6.13 for calibrating the MITSIMLab microsimulation package in Stockholm. The calibration process is divided into two steps, aggregate and disaggregate calibration steps. First, using

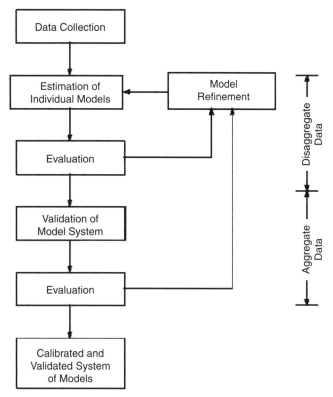

FIGURE 6.13 Calibration framework. (*Source:* Ben-Akiva et al. 2002.)

disaggregate data, individual models can be calibrated and estimated. Disaggregate data include information on detailed driver behavior, such as vehicle trajectories of the subject and surrounding vehicles.

Aggregate data (e.g., time headways, speeds, flows) is used to fine-tune parameters and estimate general parameters in the simulator. Aggregate calibration uses optimization techniques (such as generalized least square techniques) to minimize the deviation between observed and corresponding simulated measurements.

Since aggregate calibration uses simulation output, which is a result of the interaction among all these components, it is impossible to identify the effect of individual models on traffic flow.

In summary, microscopic model calibration is a tedious task, though the large number of sometimes unobservable parameters often plays a compromising role. Conversely, in macroscopic models, the number of parameters is relatively small and, more importantly, comparably easy to observe and measure.

Model Validation. Verifying a calibrated model by running the model on data different from the calibration data set is commonly called validation. In fact, calibration is useless without the validation step. The validation process establishes the credibility of the model by demonstrating its ability to replicate actual traffic patterns. Similar to calibration, validation is done on two levels, microscopic and macroscopic. Microscopic validation deals with the individual mechanism, while macroscopic validation deals with aggregate performance measures. Macroscopic validation verifies data such as arterial travel time, average speeds, and delays as compared to real-world observations. Microscopic validation deals with car-following, lane-changing, and route choice logic (McShane et al. 1998).

Similar to the calibration process, the validation of macroscopic models requires less effort than calibration of microscopic or mesoscopic models (Hoogendoorn and Bovy 2001). Accordingly, microscopic validation is carried out less frequently than macroscopic and mesoscopic models.

6.4 CAPACITY ANALYSIS

Capacity and level of service (LOS) are fundamental concepts that are used repeatedly in professional practice. Determination of the capacity of transportation systems and facilities is a major issue in the analysis of transportation flow. The *Highway Capacity Manual* defines the capacity of a facility as "the maximum hourly rate at which persons or vehicles can be reasonably expected to traverse a point or uniform segment of a lane or roadway during a given time period under prevailing roadway, traffic, and roadway conditions" (TRB 2000). Capacity analysis estimates the maximum number of people or vehicles that can be accommodated by a given facility in reasonable safety within a specified time period. Capacity depends on physical and environmental conditions, such as the geometric design of facilities or the weather. However, facilities are rarely planned to operate near capacity. Capacity analysis is only a mean to estimate traffic that can be accommodated by a facility under specific operational qualities.

The *Highway Capacity Manual* (*HCM 2000*), produced by the Transportation Research Board (TRB), is a collection of standards that define all the parameters related to capacity studies for transport infrastructure. The HCM presents operational, design, and planning capacity analysis techniques for a broad range of transportation facilities, including streets and highways, bus and rail transit, and pedestrian and bicycle facilities.

6.4.1   Capacity and Level of Service

Capacity and *level of service* (LOS) are closely related and can be easily confused. While capacity is a measure of the demand that a highway can potentially service, level of service is a qualitative measure of the highway's operating conditions under a given demand within a traffic stream and their perception by motorists and/or passengers. Thus, LOS intends to relate the quality of traffic service to given volumes (or flow rates) of traffic. The parameters selected to define LOS for each facility type are called measures of effectiveness (MOE). These parameters can be based on various criteria, such as travel times, speeds, total delay, probability of delay, comfort, and safety.

The *Highway Capacity Manual* defines six levels of service, designated A through F, with A being the highest level of service and F the lowest. The definitions of these levels of service vary depending on the type of roadway or roadway element under consideration. For instance, in the case of basic freeway sections, the levels of service are based on density. These are given in Table 6.3.

It is important to understand the concept of ideal condition of a facility, a term often used in the *HCM*. In principle, an ideal condition of a facility is one for which further improvement will not achieve any increase in capacity. Ideal conditions assume good weather, good pavement conditions, users familiar with the facility, and no incidents impeding traffic flow. In most capacity analyses, prevailing conditions are not ideal, and computations of capacity, service flow rate, or level of service must include adequate adjustments to reflect this absence of ideal conditions.

TABLE 6.3 Level of Service Definitions for Basic Freeway Segments

Level of service (LOS)	Density, pc/km/ln
A	0–7
B	7–11
C	11–16
D	16–22
E	22–28
F	>28

pc = passenger car, ln = lane.
Source: HCM 2000.

This section describes the capacity and LOS analysis of three types of facilities, two-lane highway, multilane highway, and weaving sections.

6.4.2 Two-Lane Highways

Two-lane highways differ from any other roadway in that the driver is required to share a lane with opposite traffic in order to pass a slow-moving vehicle. As traffic in one direction increases, passing maneuvers become more difficult. Hence, directional analysis is the foundation of capacity studies of two-lane highways. Actually, both capacity analysis and LOS of two-lane highways addresses the two-way capacity of the facility.

The methodology for calculating LOS is to start with the maximum volume at ideal conditions and then adjust the maximum as less-than-ideal conditions become apparent.

Two-lane rural highways are divided into two classes, with differing levels of service criteria:

* *Class I two-lane highways* are major intercity routes generally serving long-distance trips, such as major intercity routes, and primary arterials connecting major traffic generators, daily commuter routes, and primary links in state or national highway networks.

* *Class II two-lane highways* are usually access roads to class I facilities and serve shorter trips. Examples include access routes to class I facilities, scenic or recreational routes that are not primary arterials, and roads through rugged terrain on which drivers do not expect high speeds.

HCM 2000 determines the LOS of class I two-lane highways as function of percent-time-spent following (PTSF) and average travel speed (ATS). PTSF is defined as the average percent of travel time that vehicles must travel in platoons behind slower vehicles due to inability to pass. Simulation studies indicated that PTSF could be estimated as the percentage of vehicles traveling at headways of 3 seconds or less (Harwood et al. 1999). PTSF was assumed to describe traffic conditions better than density because density is less evenly distributed on two-lane highways than on multilane highways and freeways (Luttinen 2001).

LOS for class I and class II two-lane highways is determined as a function of percent-time-spent following. However, the ATS was selected as an auxiliary criterion for class I highways because ATS makes LOS sensitive to design speed. In addition, specific upgrades and downgrades can be analyzed by a directional-segment procedure.

Figure 6.14 and Table 6.4 illustrate the two-dimensional definition of level of service for respectively class I and class II two-lane highways, respectively (*HCM 2000*).

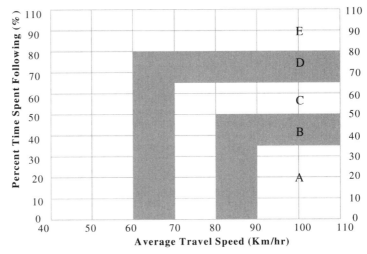

FIGURE 6.14   LOS graphical criteria for two-lane highways in class I. (*Source:* Exhibit 20-3 in *HCM 2000.*)

Analysis of Two-Way Segments

Capacity and LOS.   Under ideal conditions, total two-way capacity of a two-lane high-way is up to 3,200 veh/hr. For short distances a two-way capacity between 3,200 veh/hr and 3,400 veh/hr may be attained. Consequently, the capacity in major direction is 1,700 pc/hr until traffic flow in minor direction reaches 1,500 veh/hr. After that, the two-way capacity becomes the determining factor.

The objective of operational analysis is to determine the level of service (LOS) for a two-lane highway based on terrain and geometric as well as traffic conditions. The traffic data needed include the 2-hourly volume, the peak hour factor, and the directional distribution of traffic flow. Typically this methodology is applicable for highway section of at least 3 kilometers.

The analysis procedure for two-lane highways is outlined in the following steps:

1. Determine free-flow speed
2. Determine demand flow rate
3. Estimate average travel speed

TABLE 6.4   LOS Criteria for Two-Lane Highways in Class II

LOS	Percent time spent following
A	≤40
B	>40–55
C	>55–70
D	>70–85
E	>95

Source: *HCM 2000*, Exhibit 20.4.

4. Estimate percent-time-spent following

5. Determine LOS

6. Estimate other traffic performance parameters

Determination of Free-Flow Speed.   There are two options for determining free-flow speed (FFS) of a two-lane highway. The first option is through conducting field measurements and the other is to estimate FFS indirectly using *HCM* guidelines.

Using field measurement, FFS can be computed based on field data as below:

$$FFS = S_{FM} + 0.0125 \frac{V_f}{f_{HV}}$$

where FFS = estimate free-flow speed (km/hr)

S_{FM} = mean speed of traffic measured in the field (km/hr)

V_f = observed flow rate for the period when field data was obtained (veh/hr)

f_{HV} = heavy-vehicle adjustment factor (obtained from *HCM 2000*).

FFS can be estimated indirectly if field data are not available. To estimate FFS, the analyst must characterize the operating conditions of the facility in terms of a base free-flow speed (BFFS) reflecting traffic characteristics and facility alignment. BFFS can be estimated based on design speed and posted design speed of the facility. Then FFS is estimated based on BFFS adjusted for lane and shoulder width and access points as shown below:

$$FFS = BFFS - f_{LS} - f_A$$

where FFS = estimated FFS (km/hr)

BFFS = base FFS (km/hr)

f_{LS} = adjustment for lane width and shoulder width

f_A = adjustment for access points

Note that f_{LS} and f_A adjustments are found in Exhibits 20-5 and 20-6 in *HCM 2000*.

Determination of Demand Flow Rate.   Demand flow rate is obtained by applying three adjustments to hourly demand volumes: PHF, grade adjustment factor, and heavy vehicle adjustment factors. These adjustments are applied as follows:

$$v_p = \frac{V}{PHF * f_G * f_{HV}}$$

where v_p = passenger car equivalent flow rate for 15-min period (pc/hr)

V = demand volume for the full peak hour (veh/hr)

PHF = peak hour factor

f_G = grade adjustment factor, which accounts for the effect of grade on passenger car operation (f_G depends on type of terrain and is obtained from Exhibits 20-7 and 20-8 in *HCM 2000*)

f_{HV} = heavy vehicle adjustment factor applied to trucks and recreational vehicles (RVs). The heavy vehicle adjustment factor accounts for the effects of heavy vehicles. It is obtained as:

$$f_{HV} = \frac{1}{1 + P_T(E_T - 1) + P_R(E_R - 1)}$$

where P_T = proportion of trucks in the traffic stream, expressed as decimal

P_R = proportion of RVs in the traffic stream, expressed as decimal

E_T and E_R = passenger-car equivalent for respective trucks and RVs. E_T and E_R depend on type of terrain and are obtained from Exhibits 20-9 and 20-10 in *HCM 2000*

Estimation of Average Travel Speed.   The average travel speed is computed based on the FFS, the demand flow rate, and an adjustment factor for the percent no-passing zones. Average speed is determined from the following equation:

$$\text{ATS} = \text{FFS} - 0.0125v_p - f_{np}$$

where ATS = combined average travel speed for both directions of travel (km/hr)
f_{np} = adjustment for percentage of no-passing zones (Exhibit 20-11)
v_p = passenger-car equivalent flow rate for peak 15-min period (pc/hr)

Estimation of Percent-Time-Spent Following.   The percent-time-spent following is computed based on the directional split and no-passing zones. It is computed as follows:

$$\text{PTSF} = \text{BPTSF} + f_{d/np}$$

where PTSF = percent-time-spent following
BPTSF = base percent-time-spent following for both directions of travel combined, obtained using the equation

$$\text{BPTSF} = 100(1 - e^{-0.000879v_P})$$

where v_p = is the passenger car equivalent flow rate for 15-min period (pc/hr)
$f_{d/np}$ = adjustment of the percentage of no-passing zones on percent-time-spent following (obtained from Exhibit 20-11 in *HCM 2000*)

Determination of Level of Service.   LOS is obtained by comparing the passenger-car equivalent flow rate (v_p) to the two-way capacity of 3,200 pc/hr. If v_p is greater than capacity, the LOS is F (the lowest LOS) regardless of the speed or percent-time-spent following. Similarly, if the demand flow rate in either direction of travel is greater than 1,700 pc/hr, the level of service is also F. LOS of F corresponds to 100 percent of percent-time-spent following and highly variable speeds. In the case where v_p is less than capacity, then LOS criteria are applied for class I or class II. LOS for a two-way segment of a class I facility is determined as a function of both ATS and PTSF that corresponds to locating a point in Exhibit 20-3 in *HCM 2000* (see Figure 6.14). Alternatively, LOS for a class II highway is determined from PTSF alone, using the criteria in Table 6.4.

In addition to LOS analysis, *HCM 2000* provides procedures for computing volume/capacity ratio, total vehicle-miles (or vehicle-kilometers) of travel, and total vehicle-hours of travel for two-lane highways. For instance, volume to capacity ratio is computed as follows:

$$v/c = \frac{v_p}{c}$$

where v/c = is the volume capacity ratio
c = capacity taken as 3,200 pc/hr for a two-way segment and 1,700 for directional segment
v_p = passenger-car equivalent flow rate for peak minute period

Analysis of Directional Segments.   *HCM 2000* presents an operational analysis procedure for directional segments. The methodology for directional segments is similar to the two-way segment methodology, except that it analyzes traffic performance measures and LOS for one direction. However, the operational condition of one direction of travel is dependent on operational conditions not only for that direction of travel, but for the opposing direction of travel as well. *HCM 2000* methodology addresses three types of directional segments: extended directional segments, specific upgrades, and specific downgrades. The directional segment procedure incorporates the conceptual approach and revised factors discussed earlier in this chapter. Hence, the operational analysis procedure for directional segments includes the following five steps:

1. Determination of free-flow speed
2. Determination of demand flow rate
3. Determination of ATS
4. Determination of PTSF
5. Determination of LOS

Determination of Free-Flow Speed. The same procedure is followed as in the two-way segment analysis, taking each direction separately. As a result, demand flow rate for the studied direction is determined as follows:

$$v_d = \frac{V}{\text{PHF} * f_G * f_{HV}} \quad \text{and} \quad v_o = \frac{V_o}{\text{PHF} * f_G * f_{HV}}$$

where v_d and v_o = respectively the passenger-car equivalent flow rate for the peak 15-min period in respectively the direction analyzed and the opposing direction of travel

V = demand volume for the full peak hour (veh/hr) for the direction analyzed in the left-hand formula and for the opposing direction in the right-hand formula

PHF = peak hour factor

f_G = grade adjustment factor obtained from Exhibits 20-7 and 20-8 in *HCM 2000*, as explained previously for the two-way segment analysis

f_{HV} = heavy vehicle adjustment factor applied to trucks and RVs. f_{HV} is obtained as explained previously for the two-way segment analysis.

Determination of Demand Flow Rate. Similar to two-way segment analysis, demand flow rate is obtained by applying three adjustments to hourly demand volumes: PHF, grade adjustment factor, and heavy vehicle adjustment factors. These adjustments are applied much as in the two-way segment analysis. However, a specific grade procedure, i.e., specific upgrade or downgrade, is used if the downgrade exceeds 3 percent for a length of at least 1.0 km.

Specific upgrade: In the specific upgrade procedure, the directional segment procedure is followed with different tables for f_G and f_{HV} (Exhibits 20-13, 20-14, 20-15, 20-16, and 20-17 in *HCM 2000*).

Specific downgrade: The specific downgrade procedure differs from the extended segment procedure in considering heavy-vehicle effects. A special procedure is provided for locations where heavy trucks use crawl speeds on long, steep downgrades. f_{HV} for specific downgrade is given as follows:

$$f_{HV} = \frac{1}{1 + P_{TC} * P_T(E_{TC} - 1 + (1 - P_{TC})P_T(E_T - 1) + P_R(E_R - 1)}$$

where P_{TC} = proportion of all trucks in the traffic stream using crawl speeds on specific downgrade

E_{TC} = passenger-car equivalent for trucks using crawl speeds (obtained from Exhibit 20-18 in *HCM 2000*)

E_T and E_R = passenger-car-equivalent for trucks and RVs, respectively. E_T and E_R depend on type of terrain and are obtained from Exhibit 20-19 in *HCM 2000*.

Determination of Average Travel Speed.   Average travel speed (ATS) for directional analysis is computed from directional FFS using directional and opposing demand volumes, as well as adjustments for the percent no-passing zones. Average speed is determined from the following equation:

$$ATS_d = FFS_d - 0.0125(v_d + v_o) - f_{np}$$

where ATS_d = average travel speed in the analysis direction (km/hr)
$\quad\quad f_{np}$ = adjustment for percentage of no-passing zones in the analysis direction
$\quad\quad v_p, v_o$ = passenger-car equivalent flow rate for peak 15-min period (pc/hr) for the analysis direction and the opposing direction, respectively

Determination of PTSF.   The same procedure is followed as in the two-way segment analysis, with formula parameters considering directional $PTSF_d$ rather than two-way PTSF.

$$PTSF_d = BPTSF_d + f_{d/np}$$

where $PTSF_d$ = percent-time-spent following in the direction analyzed
$\quad\quad BPTSF_d$ = base percent time following for the direction analyzed obtained using the equation
$\quad\quad f_{d/np}$ = adjustment of the percentage of no-passing zones on percent-time-spent following (obtained from Exhibit 20-20 in *HCM 2000*)

$$BPTSF_d = 100(1 - e^{av_p^b})$$

where v_p = is the passenger-car equivalent flow rate for 15-min period (pc/hr)

The coefficients *a* and *b* are determined from Exhibit 20-21 in *HCM 2000*.

Determination of LOS.   LOS for directional analysis is obtained by comparing the passenger-car equivalent flow rate (v_p) to the one-way capacity of 1,700 pc/hr. If v_p is greater than capacity, the level of service is F. In the case where v_p is less than capacity, LOS criteria are applied to class I or class II, given by Table 6.4 and Figure 6.14, as in the two-way segment analysis.

Analysis of Directional Segments with Passing Lanes.   *HCM 2000* also presents an operational analysis procedure for directional segments containing passing lanes in level and rolling terrain. A passing lane is an added lane provided in one direction of travel on a two-lane highway in order to increase the availability of passing opportunities, which accordingly affects the LOS.

The procedure followed for analyzing this effect does not address climbing, lanes, which are added lanes in mountainous terrain or on specific upgrades. These are addressed separately.

The presence of passing lanes generally decreases PTSF and increases ATS on the roadway downstream of the passing lane; Figure 6.15 illustrates the effect of a passing lane on PTSF. The bold line shows the percent-time-spent following in the absence of a passing lane. The dotted line shows how the percent-time-spent following drops abruptly in the passing lane section and slowly returns to the non-passing lane values downstream from the terminus of the passing lane. The figure also shows that the effective length of a passing lane is actually greater than its actual length.

The operational analysis procedure followed for directional segments containing passing lanes incorporates the following five steps:

1. Application of the directional segment procedure without the passing lane in place
2. Division of the segment into regions
3. Determination of PTSF

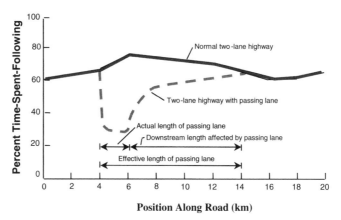

FIGURE 6.15 Operational effect of passing lane. (*Source:* Harwood and Hoban 1987.)

4. Determination of ATS

5. Determination of LOS

Application of the Directional Segment Procedure without Considering the Passing Lane in Place. The first step in a passing lane analysis is to apply the operational analysis procedure for the directional segment without considering the passing lane. The directional segment evaluated for a passing lane should be in level or rolling terrain. The results of the initial application of the directional segment procedure are estimates of PTSF and ATS for the normal two-lane cross-section.

Dividing the Segment into Regions.   The next step is to divide the analysis segments into four regions:

1. Upstream of the passing lane

2. Within the passing lane

3. Downstream of the passing lane but within its effective length

4. Downstream of a passing lane but beyond its effective length

These four lengths must, by definition, sum to the total length of the analysis segment. The analysis segments and their length will differ for estimation of PTSF and ATS because the downstream lengths for these measures differ. The length of the segment upstream of the passing lane (L_u) and the length of the passing lane itself (L_{pl}) are readily determined when the location (or proposed location) of the passing lane is known. The length of the downstream highway segment within the effective length of the passing lane (L_{de}) is determined from Exhibit 20-23 in *HCM 2000*. Once the lengths L_u, L_{pl}, and L_{de} are known, the length of the analysis segment downstream of the passing lane and beyond its effective length (L_d) can be determined as:

$$L_d = L_t - (L_u + L_{pl} + L_{de})$$

where L_t = total length of analysis segment
  L_u = length of two-lane highway upstream of the passing lane
  L_{pl} = length of passing lanes, including tapers
  L_{de} = downstream length of two-lane highway within the effective length of the passing lane. L_{de} is obtained from Exhibit 20-23 in *HCM 2000*.

Determination of Percent Time Spent Following (PTSF).   PTSF within lengths L_u and L_d is assumed to be equal to $PTSF_d$, as predicted by the directional segment procedure. Within the passing lane, PTSF is generally equal to 58–62 percent of its upstream value; this effect varies as a function of flow rate.

Within the downstream effective length of the passing lane, PTSF is assumed to increase linearly with distance from the within-passing-lane value to the normal upstream value. These assumptions result in the following equation for estimating PTSF for the analysis segment as a whole with the passing lane in place:

$$PTSF_{pl} = \frac{PTSF_d \left[L_u + L_d + f_{pl}L_{pl} + \left(\frac{1 + f_{pl}}{2} \right) L_{de} \right]}{L_t}$$

where $PTSF_{pl}$ = percent-time-spent following for the entire segment, including the passing lane
   $PTSF_d$ = percent-time-spent following for the entire segment without the passing lane
   f_{pl} = factor for the effect of a passing lane on percent-time-spent following. f_{pl} is obtained from Exhibit 20-24 in *HCM 2000.*

If the analysis section is interrupted by a town or a major intersection before the end of full effective length of the passing lane, then L_d is not used and the actual downstream length within the analysis segment L'_{de} is used. L'_{de} is less than the value of L_{de} as tabulated in Exhibit 20-23 in *HCM 2000.* The above equation is replaced by the equation:

$$PTSF_{pl} = \frac{PTSF_d \left[L_u + f_{pl}L_{pl} + f_{pl}L'_{de} + \left(\frac{1 - f_{pl}}{2} \right) \left(\frac{(L'_{de})^2}{L_{de}} \right) \right]}{L_t}$$

where L'_{de} = actual distance from end of passing lane to end of analysis segment $L'_{de} \leq L_{de}$ (L_{de} is obtained from Exhibit 20-23 in *HCM 2000*)

Determination of Average Travel Speed (ATS).   ATS with the passing lane in place is determined similarly to PTSF except that ATS is increased, rather than decreased, by the presence of the passing lane. ATS within lengths L_u and L_d is assumed to be equal to ATS_d, as predicted by the directional segment procedure. With the passing lane, ATS is generally equal to 8–11 percent higher than its upstream value. Within the downstream effective length of the passing lane, ATS is assumed to increase linearly with distance from the within-passing-lane value to the normal upstream value. These assumptions result in the following equation for estimating ATS for the analysis segment as a whole with the passing lane in place:

$$ATS_{pl} = \frac{ATS_d * L_t}{L_u + L_d + \dfrac{L_{pl}}{f_{pl}} + \dfrac{2L_{de}}{1 + f_{pl}}}$$

where ATS_{pl} = average travel speed for the entire segment, including the passing lane
   ATS_d = average travel speed for the entire segment without using the passing lane
   f_{pl} = factor for the effect of a passing lane on average travel speed

Determination of Level of Service.   LOS for a directional segment containing a passing lane is determined in a manner identical to a directional segment without a passing lane except that $PTSF_{pl}$ and ATS_{pl} are used in place of $PTSF_d$ and ATS_d.

Operational Analysis Procedure for Directional Segments Containing Climbing Lanes on Upgrades.   A climbing lane is a passing lane added on an upgrade to allow traffic to pass heavy vehicles with reduced speed. The operational analysis procedure for a directional segment containing a climbing lane on an upgrade is the same as the procedure for passing lanes, with the following modifications:

- In applying the directional segment procedure to the roadway without the added lane, the grade adjustment factor and the heavy vehicle adjustment factor should correspond to the specific upgrades. In cases where the added lane is not sufficiently long or steep, it should be analyzed as a passing lane rather than a climbing lane.
- PTSF and ATS adjustment factors for a climbing lane are based on different factors than the factors for passing lanes (adjustments factors are obtained from Exhibit 20-27 in *HCM 2000*).

6.4.3   Multilane Highways

Multilane highways usually have four to six lanes, often with physical medians or two-way left-turn lanes. Multilane highways, because they provide partial or full access to adjacent facilities, do not usually provide uninterrupted flow. In addition, traffic signals may be introduced with considerable distance in between. Due to this long distance, flow between these interruptions may operate similarly to freeways (uninterrupted flow) between two fixed interruption points (two signalized intersections at least 2 miles apart). In general, the posted speed limit in multilane highways ranges between 40 and 55 mph. This section discusses the methodologies for analyzing the capacity and level of service criteria for multilane highways based on *HCM 2000*.

Multilane Highway Basic Characteristics.   Multilane highways have the following characteristics:

- They generally have posted speed limits between 40 and 55 mph (65 km/hr to 90 km/hr).
- They usually have four or six lanes, often with physical medians or two-way left-turn-lane (TWLTL) medians, although they may also be undivided.
- They are typically located in suburban communities leading to central cities or along high-volume rural corridors that connect two cities or significant activities generating a substantial number of daily trips.
- Traffic signals may be found along such highways, although traffic signals spaced at 2.0 miles or less typically create urban arterial conditions.

Capacity and level of service analysis procedures for multilane uninterrupted flow begin with the calibration of a characteristic set of speed-flow-density relationships for highways operating under ideal conditions. "Ideal conditions" implies no heavy vehicles and only the presence of familiar drivers in the facility. The main parameters describing the HCM speed-flow models are free-flow speed, capacity, and density at capacity. *HCM 2000* has a standard family of curves for multilane highways under ideal conditions (see Figure 6.16). The general shapes of these curves are similar. When the flow rate exceeds a limit flow rate, speeds start decreasing below the free-flow speed with increasing flow rate towards the speed at capacity, due to an increasing level of interactions between vehicles. In addition, Figure 6.16 shows that the capacity varies with the free-flow speed of the facility.

LOS for multilane highways is defined on the basis of density-flow and speed-flow relationship. Thus, LOS criteria reflect the shape of the speed-flow and density-flow curves. Speed remains relatively constant across LOS A to D but is reduced as capacity is approached.

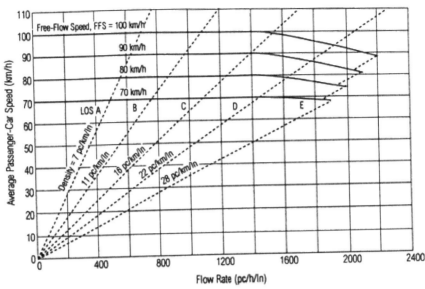

FIGURE 6.16 *HCM 2000* speed-flow models for four classes of multilane highways (based on functions given in *HCM 2000* Exhibit 21-3).

Capacity and Level of Service Analysis. The following section discusses the methodologies for analyzing the capacity and level of service criteria for multilane highways based on *HCM 2000*. The methodology is based on determining the reduction in travel speed that occurs for less than base conditions. The base conditions or ideal conditions for multilane highways imply the following:

- 12 ft (or 3.6 m) minimum lane width
- A minimum of 12 ft (or 3.6 m) total lateral clearance in the direction of travel
- No direct access point along the highway
- A divided highway
- FFS > 60 mph (95 km/hr)
- Traffic stream composed of passenger cars only

HCM 2000 procedure is composed of the following steps:

Step 1: Calculation of FFS
Step 2: Determination of flow rate
Step 3: Calculation of LOS

Calculation of FFS. FFS can be determined either from field measurement or by following HCM guidelines. If FFS is to be computed directly from field study, it is determined during periods of low to moderate flow, with an upper limit of 1400 pc/hr/lane.

Alternatively, FFS can be estimated indirectly from the *HCM 2000* formula. FFS decreases with decreasing lane width, decreasing lateral clearance, and increasing access point (intersection and driveway) density, and is reduced for undivided highways. The *HCM* formula is as follows:

$$FFS = BFFS - f_{LW} - f_{LC} - f_{M} - f_{A}$$

where BFFS = base free-flow speed
f_{LW} = adjustment factor for lane width (Exhibit 21-4 in *HCM 2000*)
f_{LC} = adjustment factor for lateral clearance (Exhibit 21-5 in *HCM 2000*)
f_{M} = adjustment factor for median type (Exhibit 21-6 in *HCM 2000*)
f_{A} = adjustment factor for access points (Exhibit 21-7 in *HCM 2000*)

Determination of Base Free-Flow Speed (BFFS). Base free-flow speed is based on the coded speed limit. It is equal to the following:

BFFS = Speed limit + 11 km/hr, for posted speed limits between 65 and 75 km/hr

= Speed limit + 8 km/hr, for posted speed limits between 80 and 90 km/hr

Adjustment Factor for Lane Width. The base condition for multilane highways requires lane widths of 3.6 meters. FFS should be adjusted for narrower lanes. Adjustment factors for lane widths are computed from Exhibit 21-4 in *HCM 2000*. Note that lanes larger than 3.6 meters are subject to no adjustment factors.

Adjustment Factor for Lateral Clearance (f_{LC}). The total lateral clearance is defined as the sum of the total lateral clearance from the right edge added to the total lateral clearance from the left side, as follows:

$$TLC = LC_{R} + LC_{L}$$

where LC_{R} = lateral clearance from the right edge of the travel lanes to roadside obstructions (maximum value of 1.8 m)
LC_{L} = lateral clearance from the left edge of the travel lanes to obstructions in the roadway median (maximum value of 1.8 m).

Lateral left clearance (LC_{L}) is computed for divided highways only. In all other cases, for instance if a continuous two-way left-turn lane exists or the facility is undivided, then LC_{L} is set to 1.8 meters. Facilities with one-way traffic operation are considered divided highways since there is no opposing flow to interfere with traffic.

Once TLC is computed, the appropriate reductions in FFS are obtained from Exhibit 21-5 in *HCM 2000*. Linear interpolation is used for intermediate values.

Adjustment Factor for Median Type. Exhibit 21-6 in *HCM 2000* states that the adjustment factor is taken as 2.6 if the highway is undivided and 0.0 if the highway is divided (including two-way left turning lanes).

Adjustment Factor for Access Points (f_{A}). Access points are intersections and driveways on the right side of the roadway in the direction of travel. The number of access points per kilometer is based directly on the values of the number of at-grade intersections with no traffic control devices and section length. A linear equation fits to the adjustment factor in form Exhibit 21-7 in *HCM 2000*. The linear equation indicates that for each access point per kilometer the estimated FFS decreases by approximately 0.4 km/hr.

Determination of Flow Rate. The flow rate is obtained by applying two adjustments to hourly demand volumes, PHF and heavy vehicle adjustment factors. Hence, the flow rate is obtained as follows:

$$v_{p} = \frac{V}{PHF * N * f_{HV} * f_{p}}$$

where v_{p} = passenger car equivalent flow rate for 15 min. period (pc/hr)
V = demand volume for the full peak hour (veh/hr)
PHF = peak hour factor

N = number of lanes
f_{HV} = heavy vehicle adjustment factor
f_p = driver population factor

PHF.   As stated previously, PHF represents the variation in traffic flow within an hour.

Adjustment Factor for Heavy Vehicles (f_{HV}).   The adjustment factor for heavy vehicles applies to three types of vehicles, trucks, recreational vehicles (RVs), and buses. Hence, the heavy vehicle adjustment factor is based on calculating passenger-car equivalents for trucks and buses and RVs as follows:

$$f_{HV} = \frac{1}{1 + P_T(E_T - 1) + P_R(E_R - 1)}$$

where P_T, P_R = proportion of trucks, buses, and RVs, respectively, expressed as a decimal (e.g., 0.15 for 15 percent)

E_T, E_R = passenger-car equivalents for trucks, buses, and RVs, respectively

f_{HV} = heavy vehicle adjustment factor for heavy vehicles

6.4.4 E_T and E_R

E_T and E_R are obtained from Exhibit 21-8 in *HCM 2000* for general terrain based on the type of terrain (i.e., level, rolling, or mountainous). Any grade of 3 percent or less that is longer than 1.6 kilometers or a grade greater than 3% for a length longer than 0.8 kilometers is treated as an isolated, specific grade.

Additionally, downgrades and upgrades are treated differently because the impact of heavy vehicles differs significantly in each case. In both downgrade and upgrade cases. E_T and E_R factors depend on both the percent grade and the length of the related section. Exhibits 21-9 and 21-10 in *HCM 2000* determine respectively the E_T and E_R factors for specific upgrades.

Similarly, Exhibit 21-11 in *HCM 2000* determines the E_T factor for specific downgrades greater than 4 percent. As for E_T factor for downgrades that are less than 4 percent and E_R factor for all downgrades, the same adjustment factors as for the level terrain are used (i.e., Exhibit 21-8 in *HCM 2000*).

Driver Population Factor (f_p).   The driver population factor reflects the effect of familiarity and unfamiliarity of drivers to the road. Recreational traffic can decrease the capacity by 20 percent. The driver population factor (f_p) ranges from 0.85 to 1.0. For urban highways, the driver population factor is set to 1.0 to indicate that drivers are familiar with roadway and traffic conditions (assuming that most of the traffic is composed of commuters). Thus, the analyst should consider f_p equal to 1.00, which corresponds to weekly commuter traffic, unless there is sufficient evidence that a lesser value reflecting more recreational trips should be used.

Calculation of LOS.   As stated above, Figure 6.16 illustrates the relationships between LOS, flow, and speed. The LOS criteria reflect the shape of the speed-flow and density-flow curves, particularly as speed remains relatively constant across LOS A to D but is reduced as capacity is approached. LOS can be directly determined from this figure by entering the corresponding free-flow speed and density values.

The density of the flow is determined as follows:

$$D = v_p / S$$

where D = density (pc/km/ln)
v_p = flow rate (pc/km/ln)
S = average passenger car travel speed

LOS can also be determined by comparing the computed density with the density ranges provided in Figure 6.16.

6.4.5 Weaving Sections

Weaving is defined as the crossing of two or more traffic streams traveling in the same general direction along a significant length of the roadway, without the aid of traffic control devices. Weaving areas are typically formed when a merging area is closely followed by a diverging area. Weaving sections are common design elements on freeway facilities such as near ramps and freeway-to-freeway connectors. The operation of freeway weaving areas is characterized by intense lane-changing maneuvers and is influenced by several geometric and traffic characteristics.

One example of freeway weaving section is a merge and a diverge in close proximity, combined with a change in the number of freeway lanes which requires either merging or diverging vehicles to execute a lane change. In such cases, traffic is subject to turbulence in excess of what is normally present on basic highway sections. Capacity is reduced in such weaving areas because drivers from two upstream lanes compete for space and merge into a single lane and then diverge into two different upstream lanes.

As lane changing is a critical component of the weaving areas, lane configuration (i.e., the number of entry lanes and exit lanes and relative placement) is one of the most vital geometric factors that need to be considered in computing the capacity of a weaving section. Other factors, such as speed, level of service, and volume distribution, are also taken into consideration.

General Consideration of Geometric Parameters.   The *Highway Capacity Manual* identifies three geometric variables that influence weaving segment operations: configuration, length, and width.

Configuration.   The configuration of the weaving segment (i.e., the relative placement of entry and exit lanes) has a major effect on the number of lane changes required of weaving vehicles to complete their maneuvers successfully.

In traffic engineering, three types of weaving sections are traditionally distinguished based on the minimum number of lane changes required for completing weaving maneuvers (TRB, 1994, 1997, 2000):

- *Type A weaving sections:* Every weaving vehicle (i.e., vehicle merging or diverging) must execute one lane change. The most common type A configuration is a pair of on-ramps and off-ramps connected by an auxiliary lane (see Figure 6.17).

- *Type B weaving sections:* One weaving movement can be made without making any lane change, while the other weaving movement requires at most one lane change. A common

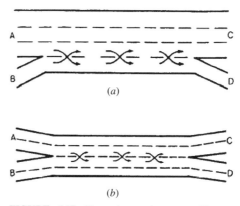

FIGURE 6.17 Type A weaving area. (*Source:* McShane et al. 1998.)

type B configuration has a lane added at an on-ramp. Merging traffic does not need to change lanes, but traffic diverging downstream must change onto this added lane to exit at the off-ramp (see Figure 6.18).

• *Type C* weaving sections: One weaving movement can be made without making any lane change, while the other weaving movement requires at least two lane changes (see Figure 6.19).

Weaving Length. The weaving length parameter is important because weaving vehicles must execute all the required lane changes for their maneuver within the weaving segment boundary from the entry gore to the exit gore. The length of the weaving segment constrains the time and space in which the driver must make all required lane-changing maneuvers. As the length of a weaving segment decreases (configuration and weaving flow being constant), the intensity of lane changing and the resulting turbulence increase. Similarly, as the length of the weaving area increases, capacity increases.

Weaving Width. The third geometric variable influencing the operation of the weaving segment is its width, which is defined as the total number of lanes between the entry and exit gore areas, including the auxiliary lane, if present. As the number of lanes increases, the throughput capacity increases.

Four different movements may exist in weaving. The two flows crossing each other's path within the section are called weaving flows, while those that do not cross are called nonweaving flows. The proportional use of the lanes by weaving and nonweaving traffic is another important factor in the analysis of weaving sections.

The following two variables are defined as follows:

N_w: number of lanes weaving vehicles must occupy to achieve balanced equilibrium operation with nonweaving vehicles

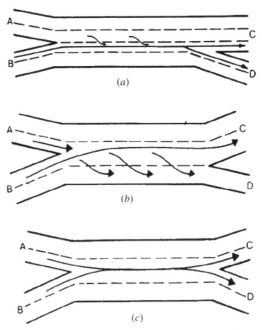

FIGURE 6.18   Type B weaving area. (*Source:* McShane et al. 1998.)

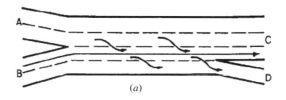

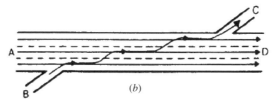

FIGURE 6.19 Type C weaving area. (*Source:* McShane et al. 1998.)

N_w (max): maximum number of lanes that can be occupied by weaving vehicles, based on geometric configuration

Accordingly, two distinguished types of operations are obtained as follows:

- *Constrained operation* ($N_w > N_w$ (max)): when weaving vehicles use a significantly smaller proportion of the available lanes and nonweaving vehicles use more lanes. Constrained operation results in larger difference in weaving and nonweaving vehicle speeds.
- *Unconstrained operation* ($N_w \leq N_w$ (max)): when configuration does not restrain weaving vehicles from occupying a balanced proportion of available lanes. Unconstrained operation results in a small difference in weaving and nonweaving vehicle speeds.

Analysis of Weaving Area:* HCM 2000 *Computational Methodology. The *HCM 2000* procedures for weaving sections involve computing the speeds of weaving and nonweaving vehicles, calculating densities, and then performing a table look-up to assign level of service. LOS criteria for weaving segments are found in Exhibit 24-2 in *HCM 2000* (Table 6.5). The

TABLE 6.5 LOS Criteria for Weaving Segments

	Density (pc/km/ln)	
LOS	Freeway weaving segment	Multilane and collector-distributor weaving segments
A	≤6.0	≤8.0
B	>6.0–12.0	>8.0–15.0
C	>12.0–17.0	>15.0–20.0
D	>17.0–22.0	>20.0–23.0
E	>22.0–27.0	>23.0–25.0
F	>27.0	>25.0

Source: *HCM 2000*, Exhibit 24.2.

geometric characteristics of the weaving section, the characteristics of vehicles by type, and their distribution over the traffic stream are important issues to be considered in the analysis. The procedure is outlined in the following steps:

1. Establish roadway and traffic condition types.
2. Convert all traffic flows into equivalent peak flow rates under ideal conditions, in pcph.
3. Construct a weaving segment diagram.
4. Compute unconstrained weaving and nonweaving speed.
5. Check for constrained operation.
6. Compute average (space mean) speed of all vehicles in the weaving area.
7. Compute average density of all vehicles in the weaving area.
8. Determine level of service and capacity.

Establish Roadway and Traffic Condition Types.   This necessitates the determination of the following:

• Weaving length and number of lanes
• Type of configuration (i.e., type A, type B, or type C)
• Traffic composition and movement types

Equivalent Peak Flow Rate Conversion.   The first computational step in a weaving analysis is the conversion of all demand volumes to peak flow rates under equivalent conditions:

$$v_i = \frac{V_i}{\text{PHF} - f_{\text{HV}} - f_p}$$

where v_i = peak flow rate under equivalent ideal conditions for movement i, pcph
$\quad V_i$ = hourly volume, veh/hr
$\quad$ PHF = peak-hour factor
$\quad f_{\text{HV}}$ = heavy vehicle factor (as in multilane highway methodology described above)
$\quad f_p$ = driver population factor (as in multilane highway methodology described above)

Weaving Segment Diagram.   After volumes are converted to flow rates, the weaving diagram is constructed to identify the traffic flow rates by type of movement, as follows:

• Weaving movements: ramp-freeway, freeway-ramp
• Nonweaving movements: ramp-ramp, freeway-freeway

Exhibit 24-4 in *HCM 2000* illustrates an example of the construction of such a diagram in Figure 6.20.

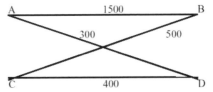

FIGURE 6.20   Weaving diagram. (*Source:* Exhibit 24-4 in *HCM 2000*.)

Unconstrained Weaving and Nonweaving Speed.   Compute unconstrained weaving speed (S_w) and nonweaving speed (S_{nw}) using the equation

$$S_i = S_{min} + \frac{S_{max} - S_{min}}{1 + W_i}$$

where S_i = mean speed of a weaving or nonweaving movement (denoted as S_w and S_{nw} for weaving and nonweaving speeds, respectively)

S_{max} = maximum speed expected in the section, assumed to be equal to S_{FF} + 5 mph

S_{min} = minimum speed expected in the section, normally assumed to be 15 mph

W_i = weaving intensity factor ($i = w$ for weaving and $i = nw$ for nonweaving)

Set S_{max} to free-flow speed (S_{FF}) added to 6 km/hr and S_{min} to 24 km/hr. Therefore, the equation is simplified to:

$$S_i = 24 + \frac{S_{FF} - 16}{1 + W_i}$$

Determination Weaving Intensity.   The weaving intensity factors (for weaving and non-weaving flows) are the measure of the influence of weaving activity on the average speed on both weaving and nonweaving vehicles. These factors are obtained as follows:

$$W_i = \frac{a(1 + VR)^b \left(\dfrac{v}{N}\right)^c}{(3.28L)^d}$$

where L = length of weaving area (ft)

VR = volume ratio, given as weaving flow rate/total flow rate

v = total adjusted rate of flow in the weaving section (pc/hr)

N = total number of lanes in the weaving section

a, b, c, d = constants that depend on weaving configuration: constrained or unconstrained operation and on weaving and nonweaving speed; a, b, c, and d factors obtained from Exhibit 24-6 in *HCM 2000*

Note that all predictions are conducted assuming that the type of operation is unconstrained.

Determinination of the Type of Operation.   As mentioned above, the type of operation depends on the proportional use of the lanes by weaving and nonweaving traffic. The type of operation is determined by determining the number of lanes N_w and N_w(max) required for weaving operation. N_w and N_w(max) are computed from Exhibit 24-7 in *HCM 2000*, depending on the configuration of the weaving segment (i.e., type A, B, or C).

- If $N_w \leq N_w$(max): unconstrained operation is valid, and speeds as computed previously.
- Otherwise, if $N_w > N_w$(max), operation is constrained. Thus, recompute speeds using a, b, c, d parameters corresponding to constrained operation.

Determination of Average (Space Mean) Speed.   Average speed is computed in *HCM 2000* as follows:

$$S = \frac{v_{nw} + v_w}{v_{nw}/S_{nw} + v_w/S_w}$$

where S_w = average speed of weaving vehicles (km/hr)

S_{nw} = average speed of nonweaving vehicles (km/hr)

v_w = weaving flow rate (pc/hr)
v_{nw} = nonweaving flow rate (pc/hr)

Determination of Average Density.   LOS criteria for uninterrupted weaving are based on density. Density is obtained from estimating average speed and flow for all vehicles in the weaving area. Accordingly, density is determined as:

$$D = \frac{v/N}{S} = \frac{v}{NS}; \qquad D \text{ is determined in pc/km/hr}$$

Determination of LOS and Weaving Segment Capacity.   LOS criteria for weaving areas are based on average density of all vehicles in the section as shown in Table 6.5.

The capacity of weaving segments is represented by any set of conditions resulting in LOS at the E/F boundary condition, i.e., an average density of 27 pc/km/ln for freeways, or 25.0 pc/km/ln for multilane highways. Thus, capacity depends on weaving configuration, length, number of lanes, free-flow speed, and volume ratio, and should never exceed capacity of similar basic freeway segments.

In addition, field studies suggest that capacities should not exceed 2,800 pcph for type A, 4,000 pcph for type B, and 3,500 pcph for type C. Furthermore, field studies indicate that there are other limitations related to the proportion of weaving flow VR, as follows:

VR = 1.00, 0.45, 0.35, or 0.20 for type A with two, three, four, or five lanes, respectively

VR = 0.80 for type B

VR = 0.50 for type C

Although stable operations may occur beyond these limitations, operation will be worse than predicted by methodology and failure is likely to occur.

Exhibit 24-8 in *HCM 2000* summarizes capacities under base conditions for a number of situations. These tabulated capacities represent maximum 15-minute flow rates under equivalent base conditions and are rounded to the nearest 10 pc/hr. The capacity of weaving section under prevailing conditions is computed as follows:

$$c = c_b \times f_{HV} \times f_p$$

where c = capacity as an hourly flow rate for peak 15 min of the hour
c_b = capacity under base conditions (Exhibit 24-8)
f_{HV} = heavy vehicle factor
f_p = driver population factor

If capacity in peak hourly volume is desired, then it is computed as

$$c_h = c \times \text{PHF}$$

where c_h = capacity under prevailing conditions (veh/hr)
PHF = peak hour factor

6.5   CONTROL

6.5.1   Objectives

The task of traffic control is to specify the control parameters, based on available measurements, estimations, or predictions, in response to forces/disturbances acting on the system, so as to achieve a prespecified goal or set of goals regarding the performance of the system.

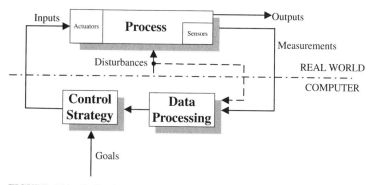

FIGURE 6.21 Basic elements of an automatic control system. (*Source:* Papageorgiou 1999.)

For instance, the task of ramp control is to specify the ramp metering rates, based on measured volumes and occupancies on the mainline and on the ramp, in response to varying traffic demand, so as to achieve minimum time spent in the network by travelers.

As we will see below, such control can come in many forms, from very simple to very advanced and hence complex. It can be manual or automatic, regulatory or optimal, open-loop or closed-loop. Furthermore, it is important to distinguish, particularly in our field, between dynamic (i.e., traffic-responsive control and adaptive control), as the two types are often confused. The following is a brief control theory primer for traffic applications. More elaborate treatments can be found in Papageorgiou (1999) and Dutton et al. (1997).

6.5.2 Control Theory Primer

If the control task described above is undertaken by a human operator, we have a manual control system. If the control task is undertaken by a computer program (a control strategy algorithm), we have an automatic control system. Hence, automatic control refers to theories, applications, and systems for enabling a system to perform certain prespecified tasks automatically while achieving a set of goals in the presence of continuous action of external forces (disturbances) on the system. Figure 6.21 shows the basic elements of an automatic control system, and Figure 6.22 shows an example of a traffic control system. In both figures, the process is either the real physical system, such as an urban traffic network and flows, or

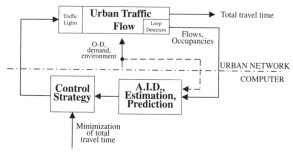

FIGURE 6.22 An example of a traffic control system. (*Source:* Papageorgiou 1999.)

a mathematical/computer model of the system. The control strategy is the brain that steers the system towards achieving its intended role by changing the control parameters or inputs such as traffic light timing plans. Disturbances are inputs to the systems that cannot be directly controlled, such as traffic demand or incidents. Process outputs are quantities that represent the behavior or performance of the system, such as travel times or delays, and can be inferred from direct measurements such as volumes and speeds. Those measurements can be fed directly into the control strategy or may have to be preprocessed for one reason or another, such as detecting incidents or forecasting near future measurements, based on current ones, if a form of proactive control is desired. In real-time applications, the control task is to specify in real time the process inputs, based on available measurements to steer the process out towards achieving a goal despite the varying impact of the disturbances. In the context of urban traffic control, the control strategy should calculate in real time the traffic light timing plans, the ramp metering rates, and traffic diversion plans among other control actions, based on measured traffic conditions, so as to achieve a goal such as minimizing travel times and maintaining densities at a certain desirable level, despite the variations in demand and the occurrence of incidents that continuously disturb the state of the system.

One should not confuse the process and its model with the control strategy. The process model, for instance a traffic simulator, is a mere replica of the real system that, given the inputs and disturbance, would accurately reproduce how the real system is likely to behave. The control strategy, on the other hand, is the algorithm that picks the proper control action given the system state and acting disturbances. Despite the continuous interaction between the process model and the control strategy, these are distinct and often separate components, the former to replicate the process behavior and the latter to make decisions on control action that will influence the process behavior.

Regulator versus Optimal and Open-Loop versus Closed-Loop Control.   System control strategies can be regulatory or optimal. A regulator's goal is to maintain the system's performance and output near an exogenously prespecified desired value. In a traffic environment, an example is freeway ramp metering, where the controller attempts to maintain traffic occupancy downstream of the ramp at or below critical occupancy (corresponding to capacity). If y is the process output (e.g., downstream occupancy), d is the disturbance (e.g., upstream demand, ramp demand, incidents), and u is the control vector (e.g., ramp metering rate), then the system output is given by

$$y = R(u,d)$$

The regulator guarantees that $y = y_d$ despite the presence of disturbances. The regulator R can be designed using standard control theory (Papageorgiou 1999; Dutton et al. 1997).

One could argue that in the above equation, if a desired output is prespecified y_d and the disturbance trajectory d is known, then one could solve for the control vector u that is required to yield the desired output, i.e.:

$$u = R^{-1}(y_d,d)$$

without having to measure the actual system output y, i.e., without feedback. In this case, elimination of the measurement of the actual system output turns the closed loop into an open loop as shown in Figure 6.23. The main advantage of such open-loop control is rapidity, since the control system need not wait for any feedback from the process. However, the real process state is never known, and hence if the process model is inaccurate or nonmeasurable disturbances are present, the control will be off and the process output will be far from y_d. For these reasons, closed-loop control would be preferable for most practical applications.

Recognizing that a desired output might not always be known, and even when known, there is no guarantee that the process performance would be best, a better control strategy would be an optimal one. In such a case, the control goal is to maximize or minimize quantity

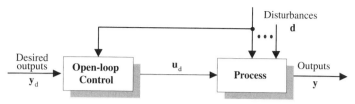

FIGURE 6.23   Open-loop regulator. (*Source:* Papageorgiou 1999.)

J, which is a function of the internal process (state) variables x and the applied control u, i.e., the control problem is an optimization problem of the form:

$$J(u,x) = y(u,x) \rightarrow \min$$

An urban traffic control example could be to minimize time spent in a freeway network using ramp metering. Note that in this case there is no need to specify desired values for the occupancies downstream of the ramps. Rather, the proper occupancy levels are obtained indirectly by applying the optimal control that minimizes time spent in the system by all travelers.

In the case of several competing optimization objectives (e.g., minimize travel time and reduce fuel consumption), the above equation would become:

$$J(u,x) = \alpha_1 y_1(u,x) + \alpha_2 y_2(u,x) + \cdots$$

where α_i are weighting parameters such that $\Sigma\ \alpha_i = 1$.

Optimal control can also be open-loop (aka P_1 problem) or closed-loop (aka P_2 problem). In open-loop optimal control the initial conditions and the disturbance trajectory in time are assumed known (forecasted) and the problem is to find the optimal control trajectory $u^*(t)$, as illustrated in Figure 6.24. This control trajectory u^* would be computed once at the beginning and applied afterwards without measuring any feedback from the process, i.e., open-loop. Hence, any unexpected disturbances could throw the control off since it was optimal only for the initial disturbance trajectory.

A better closed-loop control looks for an optimal control law or policy that is applicable at all times to compute the optimal control vector, i.e., find the function R

$$u^*(t) = R[x(t),t]$$

that minimizes the objective function J. Note that a full control vector trajectory is not obtained once at the beginning, as in open-loop optimal control. Rather, the control law or policy R is obtained that can be applied in real time based on the latest system state, i.e., closed-loop, and hence independent of the initial forecasted disturbance trajectory. This type of control is capable of recovering from the impact of any unexpected disturbances, as shown in Figure 6.24.

Both P_1 and P_2 problems can be solved using control theory ((Papageorgiou 1999; Dutton et al. 1997). P_1 is much simpler to solve and therefore is applicable to larger-sized practical problems. However, it is not accurate, as explained above. P_2 is more accurate, but suffers for the curse of dimensionality and is often infeasible for practical size problems. Many compromises are possible, including hierarchical control using a combination of P_1 and P_2 formulations, or repetitive optimization with rolling horizon, which essentially solves a P_1 problem several times over the control time horizon. New approaches for solving P_2 problems using artificial intelligence are emerging that are less computationally intensive than standard control theory approaches. A good example is the use of reinforcement learning (RL) for

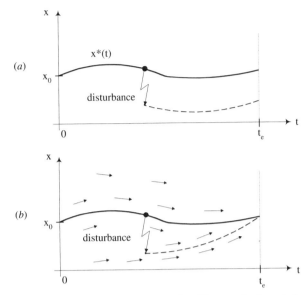

FIGURE 6.24 Open-loop versus closed-loop optimal control. (*Source:* Papageorgiou 1999.)

obtaining optimal control policies (see Sutton and Barto 1996; Abdulhai et al. 2002; and Abdulhai and Kattan 2002).

In closing this primer section, it is important to emphasize the difference between real-time control in general and adaptive control, as the two are widely confused in the traffic control field. Any control system is by default dynamic or real-time. When fed with varying disturbances over time, it responds with the proper (regulator or optimal) control as applicable. Most widely used traffic control systems are of this type, however often erroneously labeled as adaptive, because they adapt to traffic conditions. From a control theory perspective, the term *adaptive* means that the control law parameters themselves are adjusted in real time in order to account for possible time-varying process behavior. For example, the control system is capable of "learning" in a sense and it could respond to a given process state differently over time as the control policy evolves and adapts to time-varying process behavior. Driving and routing behavior, for instance, might vary widely from one city to another, and hence a generic signal control system might need to fine-tune itself according to local conditions. An attempt to develop such true adaptive control, applied to traffic signal control, can be found in Abdulhai et al. 2002.

6.5.3 Control Devices

Traffic control may be achieved by using various devices, including signs, traffic signals, ramp metering, islands, and pavement markings, to regulate, guide, inform and/or channel traffic.

The *Manual on Uniform Traffic Control Devices* (*MUTCD*) defines all policies and guidelines pertaining to traffic control devices and for determining where and whether a particular control type is suitable for a given location and/or intersection (Garber and Hoel 2002). The *MUTCD*, published by the Federal Highway Administration, can be found online at http://mutcd.fwa.dot.gov/.

The *MUTCD* defines the following five requirements for traffic control devices to be effective:

1. Fulfill a need
2. Command attention
3. Convey a clear simple meaning
4. Command the respect of road users
5. Give adequate time for proper response

These five requirements are governed mainly by five major factors: the design, placement, operation, maintenance, and uniformity of traffic devices. For instance, the traffic device should be designed with a combination of size, color, and shape that will be able to attract drivers' attention and clearly convey the intended message. In addition, the traffic device should be properly placed to fall within the field of vision of the driver. This will ensure that the driver has adequate time to respond while driving at normal speed.

6.5.4 Freeway Control: Ramp Metering

Freeway control is the application of control devices such as traffic signals, signing, and gates to regulate the number of vehicles entering or leaving the freeway to achieve some operational objective. Ramp metering has proven to be one of the most cost-effective techniques for improving traffic flow on freeways. Ramp meters can increase freeway speeds while providing increased safety in merging and reducing rear-end collisions on the ramps themselves. Although additional delays are incurred by the ramp traffic, mainline capacities are protected and the overall operational efficiency, usually measured in terms of travel time or speed, is improved (Kachroo and Krishen 2000).

Warrants for Ramp Metering.   The *MUTCD* provides general guidelines for the successful application of ramp control. These guidelines are mainly qualitative. Makigami (1991) suggests the following warrants for installation of freeway ramp-metering:

1. Expected reduction in delay to freeway traffic exceeds the sum of the expected delay to ramp users and the added travel time for diverted traffic on alternative routes
2. The presence of adequate storage space for the vehicles stopped at the ramp signal
3. The presence of alternative surface routes with adequate capacity to take the diverted traffic from the freeway ramps, and either:
   - Recurring congestion on the freeway due to traffic demand in excess of the capacity, or
   - Recurring congestion or a severe accident hazard at the freeway on-ramp because of inadequate ramp-merging area.

Ramp Metering Objectives.   Typically ramp meters are installed to address two primary objectives:

1. Control the number of vehicles that are allowed to enter the freeway
2. Break up the platoons of vehicles released from an upstream traffic signal

The purpose of the first objective is to balance demand and capacity of the freeway to maintain optimum freeway operation as well as ensure that the total traffic entering a freeway section remains below its operational capacity. The purpose of the second objective is to provide a safe merge operation at the freeway entrance (Chaudhary and Messer 2000). The primary safety problem of the merging operation is incidents of rear-ends and lane-change collisions caused by platoons of vehicles on the ramp competing for gaps in the freeway traffic stream. Metering is used to break up these platoons and enforce single-vehicle entry.

Advantages and Disadvantages of Ramp Metering. Implementation of ramp metering to control freeway traffic can bring both positive and negative impacts. Appropriate use of ramp meters can produce positive benefits, including increased freeway throughput, reduced travel times, improved safety, and reduced fuel consumption and emissions. Another important benefit is that short freeway trips may divert to adjacent underutilized arterial streets to avoid queues at the meters. If there is excess capacity on surface streets, it may be worthwhile to divert traffic from congested freeways to surface streets.

Inappropriate use of ramp meters can produce negative effects. Issues that have been receiving more concern with the implementation of ramp metering are (Wu 2001):

- Freeway trips may potentially divert to adjacent arterial streets to avoid queues at the meters. If insufficient capacity exists on arterial streets, metering can therefore result in adverse effects.
- Queues that back up onto adjacent arterial streets from entrance ramps can adversely affect the surface network.
- Local emissions near the ramp may increase due to stop-and-go conditions and vehicle queuing on the ramp.

The greatest challenge to a ramp metering policy is preventing freeway congestion without creating large on-ramp queue overflows. Fortunately, the formation of excessively long queues may be avoided. By preventing freeway congestion, optimal ramp metering policies are able to service a much larger number of vehicles in the freeway than when no metering policies are used (Horowitz et al. 2002).

The positive impact of ramp metering on both the freeway and the adjacent road network traffic conditions was confirmed in a specially designed field evaluation in the Corridor Peripherique in Paris (Papageorgiou and Kotsialos 2000). When an efficient control strategy is applied for ramp metering, the freeway throughput will generally increase. It is true that ramp metering at the beginning of the rush hour may lead to on-ramp queues in order to prevent congestion from forming on the freeway. However, queue formation may temporarily lead to diversion towards the urban network. But due to congestion avoidance or reduction, the freeway will be able to accommodate a higher throughput, thus attracting drivers from urban paths and leading to an improved overall network performance.

Ramp Metering Components. Ramp meters (Figure 6.25) are traffic signals placed on freeway entrance ramps to control the rate of vehicles entering the freeway so that demand stays below capacity.

Figure 6.25 shows the related ramp metering components:

1. Ramp metering signal and controller
2. Upstream and downstream mainline loop detectors
3. Presence sloop detectors
4. Passage loop detectors
5. Queue loop detectors

Metering Rate and Control Strategies. Ramp metering involves determination of a metering rate according to some criteria, such as measured freeway flow rates, speeds, or occupancies upstream and downstream of the entrance ramp. Maximum practical single-lane rate is generally set at 900 veh/hr, with a practical minimum of 240 veh/hr.

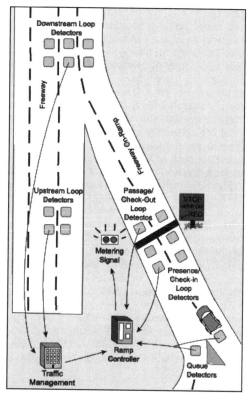

FIGURE 6.25   Ramp meter at freeway entrance from PATH. (*Source:* Partners for Advances Transit and Highways, http://www.path.berkeley.edu.)

When metering is to be used only as a means of improving the safety of the merging operation, the metering rate is simply set at a maximum consistent with merging conditions at the particular ramp. The metering rate selected must ensure that each vehicle has time to merge before the following vehicle approaches the merge area.

Metering rates may be fixed for certain periods based on historical data, or may be variable minute-by-minute (adaptive) based on measured traffic parameters. This later metering scheme may be based on regulator or optimal control strategies. In addition, ramp metering may operate either in isolation or in coordinated fashion. The coordinated ramp meters control strategy establishes metering rates for various ramps on the basis of total freeway conditions.

Pretimed Metering Stategies.   Pretimed ramp metering refers to a fixed metering rate, derived off-line, based on historical data. Thus, rates are not influenced by current real-time traffic conditions. Fixed-time ramp metering strategies are based on simple static models. They are derived for a freeway with several on-ramps and off-ramps by subdividing sections, each containing one on-ramp (Papageorgiou 2000). Pretimed strategy is typically used where traffic conditions are predictable. Its benefits are mainly associated with accident reductions from merging conflicts, but it is less effective in regulating mainline conditions. The main drawback of preset strategies is that they may result in overrestrictive metering rates if congestion dissipates sooner than anticipated, resulting in unnecessary ramp queuing and delays. Due to the absence of real-time observation and the reliance of historical observations, fixed ramp metering strategies are thus myopic. They lead either to overload of the mainline flow or to underutilization of the freeway.

Coordinated Ramp Metering Strategies. Coordinated metering refers to the application of metering to a series of entrance ramps rather than dealing with individual ramps independently. As such, coordinated ramp metering strategy makes use of all available mainstream measurements on a freeway section, to calculate simultaneously the ramp volume values, for all controllable ramps included in the same section. This requires the presence of detectors upstream and downstream of each ramp, as well as a communication medium and central computer linked to the ramps.

Coordinated ramp metering can be either pretimed or actuated. If it is pretimed, the metering rate for each of these ramps is determined independently from historical data pertaining to each freeway section. If it is actuated, the metering rate is based on real-time readings. Freeway traffic conditions are analyzed at a central control system and metering rates at the designated ramps are computed accordingly. This centralized configuration allows the metering rate at any ramp to be affected by conditions at other locations within the network. This provides potential improvements over local ramp metering because of more comprehensive information provision and because of coordinated control actions (Papageorgiou and Kotsialos 2000). In addition to recurring congestion, system-wide ramp metering can also manage freeway incidents, with more restrictive metering upstream and less restrictive metering downstream of the incident. Based on capacity, queue length, and demand conditions, linear programming is used to determine the set of integrated metering rates for each ramp, as shown below.

Example of Computation of Pretimed Coordinated Ramp Metering Strategies. Fixed ramp metering strategies are based on simple static models. The freeway section with several ramps and off-ramps is subdivided into sections, each containing one on-ramp. Mainline flow is given by the formula below (Papageorgiou and Kotsialos 2000):

$$q_j = \sum_{i=1}^{j} \alpha_{ij} r_i$$

where q_j = mainline flow section j
r_i = on-ramp volume (veh/hr) of section i
$\alpha_{ij} \in [0,1]$ = known portion of vehicles that enter the freeway section i and do not exit the freeway upstream of section j

To avoid congestion, the flow on freeway section j must be less than the capacity. Thus, $q_j \leq q_{\text{cap},j} \; \forall_j$, where $q_{\text{cap},j}$ is the capacity of section j.

On the other hand, the ramp metering rates are subject to the following constraints:

$$r_{j,\min} \leq r_j \leq \min \{r_{j,\max}, d_j\}$$

where d_j = demand on the ramp taken from historical data
$r_{j,\max}$ = ramp capacity at on-ramp j

Optimum metering rates are found by optimizing a given criterion, such as minimizing the total travel time in the network or minimizing the queue length on the ramps. The problem might be formulated as a quadratic programming or linear programming problem. For instance, when the objective function is to minimize queue length at ramps, the problem is formulated as follows:

$$Z(r_i) = \text{Arg min}_{r_i} \sum_{j} (d_j - r_j)^2$$

Subject to the following constraints:

$$q_j \leq q_{cap,j} \quad \forall j$$

$$q_j = \sum_{i=1}^{j} \alpha_{ij} r_i$$

$$r_{j,min} \leq r_j \leq \min \{r_{j,max}, d_j\}$$

This problem is easily solved by broadly available computer codes based on different algorithms, such as the gradient or conjugate gradient algorithms.

Actuated or Traffic-Responsive Ramp Metering Strategies.   In contrast to pretimed ramp control strategies, actuated ramp control strategies are based on real-time traffic measurements taken from mainline detectors placed in the immediate vicinity of the ramp to calculate suitable ramp metering values. Since occupancy is directly related to density, often occupancy data rather than flow are the commonly used parameter in actuated ramp metering. Occupancy readings, in contrast to flow measurements, distinguish between congested and uncongested conditions. Actuated traffic response strategies fall into two categories: regulator and optimal control. The difference between these two methodologies is detailed below.

Ramp Metering as Regulator Problem.   Regulator control strategies also fall into two types: reactive control and proactive control. Under reactive control, metering rates are the difference between the upstream flow measured in the previous period, usually 1 minute earlier, and the downstream capacity. Under proactive control, metering rates are based on occupancy measurements taken upstream of the ramp during the previous period, usually 1 minute prior.

Reactive Regulator.   Reactive ramp metering strategies are employed to keep freeway traffic at prespecified set values, based on real-time measurements (Papageorgiou and Kotsialos 2000). Metering rates are the difference between the upstream flow measured in the previous period, usually 1 minute earlier, and the downstream capacity. The algorithm determines the metering rate locally from input–output capacity considerations based on flow data as follows:

$$r(k) = \begin{cases} q_{cap} - q_{in}(k - 1) & \text{if } o_{out}(k) \leq o_{cr} \\ r_{min} & \text{else} \end{cases}$$

where $r(k)$ = number of vehicles allowed to enter in period k
    q_{cap} = capacity of freeway section
$q_{in}(k - 1)$ = upstream flow in period $k - 1$

The upstream flow, $q_{in}(k - 1)$ is measured by the loop detector, and the downstream capacity, q_{cap}, is a predetermined value.

The main criticism of these reactive algorithms is that they adjust their metering rates after the mainline congestion has already occurred. Traffic-predictive algorithms, such as feedback regulators, have been developed to operate in a proactive fashion, anticipating operational problems before they occur.

Feedback Regulator Algorithms: Proactive Regulators.   Feedback regulator algorithms use the occupancy rather than the flow in the upstream to determine the ramp metering rate for subsequent periods. One such algorithm is ALINEA (Asservissement Linéaire d'Entrée Autroutière) (Papageorgiou et al. 1991). ALINEA is a local-feedback control algorithm that adjusts the metering rate to keep the occupancy downstream of the on-ramp at a prespecified level, called the occupancy set point. The feedback control algorithm determines the ramp metering rate as a function of the following:

- The desired downstream occupancy
- The current downstream occupancy
- The ramp metering rate from the previous period

The number of vehicles allowed to enter the motorway is based on the mainline occupancy downstream of the ramp and is given by:

$$r(k) = r(k - 1) + k_R[\hat{o} - o_{out}(k)]$$

where $r(k)$ = number of vehicles allowed to enter in time period k
k_R = a regulator parameter
$\hat{o}$ = set of desired value for the downstream occupancy
$o_{out}(k)$ = downstream occupancy

While the reactive strategy reacts only to excessive occupancies o_{out}, proactive strategies such as ALINEA react smoothly even to slight differences $(\hat{o} - o_{out}(k))$. Proactive strategies may thus prevent congestion by stabilizing the traffic flow at a high throughput level.

Ramp Metering as an Optimal Control Problem.   Optimal control strategies have a better coordination level than regulator strategies. They tend to calculate in real time optimal and fair set values from a more proactive, strategic point of view (Papageorgiou and Kotsialos, 2000). Such a optimal approach takes into consideration the following:

- The current traffic condition on both the freeway and on the on-ramps
- Demand predictions over a sufficiently long time horizon
- The limited storage capacity of the on-ramps
- The ramp metering constraints, such as minimum rate
- The nonlinear traffic flow dynamics, including the infrastructure's limited capacity
- Any incidents currently present in the freeway network

The optimal control strategy is formulated as a nonlinear mathematical programming problem, with a parameter set to be adopted for the future time horizon, and the optimization is to be repeated after a specified time with a new data collection set. For instance, the mathematical problem could be minimizing an objective criterion, such as minimizing the total time spent in the whole network (including the on-ramps) while satisfying the set of constraints (e.g., queue length, ramp capacity). This problem is solved on-line with moderate computation time by use of suitable solution algorithms (Papageorgiou and Kotsialos 2000).

6.5.5 Surface Street Control

Warrants for Traffic Signals.   Traffic signals are one of the most effective means of surface street control. The *MUTCD* provides specific warrants for the use of traffic control signals. These warrants are much more detailed than any other control device. This is justified by the significant cost and negative impact of misapplication of traffic signals as compared to other control devices (McShane et al. 1998). Traffic volume represents the key factor in the *MUTCD* warrants. Other factors, such as pedestrian volume, accident data, and school crossing, also play a significant role (Garber and Hoel 2002).

Advantages and Disadvantages of Traffic Signals.   When properly placed, traffic signals offer the following advantages. They:

- Assign right-of-way
- Increase capacity
- Eliminate conflicts, thus reducing severity of accidents
- Allow for coordination plans at designated speeds
- Permit pedestrian movements
- Permit cross-street movements

Despite these advantages, when poorly designed, traffic signals or improperly operated traffic control signals can result in the following disadvantages:

- Volumes will increase.
- Signals will not function as safety devices; crashes will often increase.
- Delays will increase.
- Unjustified construction, operation, and maintenance costs will come about.

Traffic Signal Design. Traffic control systems are designed according to different kinds of control logic. They can be divided into three categories. The simplest is the pretimed category, which operates in a prespecified set of cycles, phases, and interval durations. The second type is the actuated signal, which establishes cycle and phase duration based on actual traffic conditions in the intersection approaches. Pretimed and actuated signals can operate either in isolated or coordinated mode. Each intersection, in isolated signals, is treated as a separate entity with its timing plan independent of its neighboring intersections. On the other hand, coordinated signals operate mostly on an arterial or network level, where a series of intersections along arterial streets are treated as a single system and their timing plans are developed together to provide good vehicle progression along the arterial or minimum travel time at the network level. Finally, the third type is the adaptive signal, where the system responds to inputs, such as vehicle actuation, future traffic prediction, and pattern matching, reflecting current traffic conditions.

The main objective of signal timing at an intersection is to reduce the travel delay of all vehicles, as well as traffic movement conflicts. This can be achieved by minimizing the possible conflict in assigning an orderly movement of vehicles coming from different approaches at different times (phases) (Garber and Hoel 2002). While providing more phases reduces the probability of accidents, it does not necessarily reduce the delay of vehicles, since each phase adds about 3 to 4 seconds of effective red. This means that the higher the number of phases, the more delay incurred. The analyst therefore has to find a trade-off between minimizing conflicting movements and vehicle delay.

In addition to the number of phases, designing signal timing plans require a complete specification of the following (McShane et al. 1998):

1. Phasing sequence
2. Timing of yellow and all red intervals for each phase
3. Determination of cycle length
4. Allocation of available effective green time to the various phases, often referred to as splitting the green
5. Checking pedestrian crossing requirements

Moreover, the phase plan must be consistent with the intersection geometry, lane use assignments, and volumes and speeds at the designated location.

Isolated Pretimed Traffic Signals. In pretimed traffic signals, the cycle length, phase, phase sequences, and intervals are predefined based on historical traffic data. However, the timing plan can change during the course of the day to respond to varying traffic demand patterns. Note that such intersections do not usually require the presence of loop detector sensors.

Signal Phasing.   Signal phasing is largely dependent on the treatment of left-turn movements, which fall into two different categories:

- *Protected left turns,* where the left turn has exclusive right-of-way (no conflict with through movement).
- *Permissive (permitted) left turns,* where the left turns are made in the presence of conflicting traffic (finding gaps); a sufficient gap to make a turn must be found, resulting in potential conflict.

The simplest phasing control for a four-leg intersection is a two-phase operation in which one phase allocates right-of-way to north-south street traffic and the other phase allocates right of way to east-west traffic. Such phasing provides separation in the major conflicting movements. However, the left turn must yield until an adequate gap is available.

Phases to protect left-turn movements are the most commonly added phases. However, adding phases may lead to longer cycle lengths, which lead to increased stop and delay to other approaches. Hence, a protected left turn is mainly justified for intersections with high left-turn volume. Nevertheless, other considerations, such as intersection geometry, accident experience, and high volume of opposing traffic, might also warrant left-turn protection.

Yellow Interval.   The objective of the yellow interval is to alert drivers to the fact that the green light is about to terminate and to allow vehicles that cannot stop to cross the intersection safely. A bad choice of yellow interval may lead to the creation of a dilemma zone, which is a distance in the vicinity of the intersection where the vehicle can neither stop safely before the intersection stop line nor continue through the intersection without accelerating before the light turns red. To eliminate the dilemma zone, a sufficient yellow interval should be provided.

It can be shown that this yellow interval τ_{min} is obtained by applying the following formula:

$$\tau_{min} = \delta + \frac{u_0}{2(a + Gg)} + \frac{(W + L)}{u_0}$$

where δ = perception plus reaction time (sec)
   u_0 = speed limit (mph)
   W = width of the intersection (ft)
   L = length of the intersection (ft)
   A = comfortable deceleration rate (ft/sec^2) = $0.27g$ (where g is the gravitational acceleration)
   G = grade of the approach

For safety considerations, the yellow interval is made not less than 3 seconds and no more than 5 seconds. If the calculation requires more than 5 seconds, the yellow interval is supplemented with all-red phase.

Cycle Length of Pretimed Signals.   Several methods have been developed to determine the optimum cycle length for an isolated intersection. The Webster method is described below.

According to the Webster method, the minimum intersection delay is obtained when the cycle length C_0 is given by:

$$C_0 = \frac{1.5L + 5}{1 - \sum_{i=1}^{\phi} Y_i}$$

where C_0 = optimum cycle length (sec)
  L = total lost time per cycle, it is usually 3.5 sec per distinct phase
  Y_i = maximum value of the ratios of approach flows (called critical movement) to saturation flows for all traffic streams using phase $i = V_{ij}/S_j$
  ϕ = number of phases
  V_{ij} = flow on lane j having the right of way during phase i
  S_j = saturation flow in one lane j, for a through lane

The computation of saturation flow rate is explained next.

Saturation Flow Rate. Saturation flow rate is the maximum number of vehicles from a lane group that would pass through the intersection in one hour under the prevailing traffic and roadway conditions if the lane group were given a continuous green signal for that hour. The saturation flow rate depends on roadway and traffic conditions, which can vary substantially from one region to another. Thus, the saturation flow rate is based on ideal saturation conditions, which are then adjusted for the prevailing traffic and geometric condition of the lane in consideration, as well as the area type. According to *HCM 2000,* the saturation flow rate is obtained as follows:

$$s = s_0 N f_w f_{HV} f_g f_p f_{bb} f_a f_{LU} f_{LT} f_{RT} f_{Lpb} f_{Rpb}$$

where s = saturation flow rate for the subject lane group
  s_0 = base saturation flow rate per lane, usually taken as 1900 or 2000 veh/hr/ln
  N = number of lanes in the group
  f_w = adjustment factor for lane width
  f_{HV} = adjustment for heavy vehicles in the lane
  f_g = adjustment for approach grade
  f_p = adjustment factor for the existence of parking lane and parking activity adjacent to the lane group
  f_{bb} = adjustment for the blocking effect of local buses that stop within the intersection area
  f_a = adjustment factor for area type (CBD or suburban)
  f_{LU} = adjustment factor for lane utilization
  f_{LT} = adjustment factor for left turns in the lane group
  f_{RT} = adjustment factor for right turns in the lane group
  f_{Lpb} = pedestrian adjustment factor for left-turn movements
  f_{Rpb} = pedestrian adjustment factor for right-turn movements

HCM 2000 also gives a procedure for computing the saturation rate from field measurements. If available, field measurements should be used instead of the analytical procedure given above.

Total Lost Time. The lost time is critical in the design of signalized intersection since it represents time not used by any vehicle in any phase. During the green and yellow interval some time is lost at the very beginning when the light turns green (i.e., start-up lost time) and at the very end of the yellow interval (clearance lost time). The start-up lost time represents the time lost when the traffic light turns green, before the vehicles start moving. The clearance lost time is the time between the last vehicle entering the intersection and the initiation of the green on the next phase. Thus, the lost time for a given phase i can be expressed as:

$$l_i = G_{ai} + \tau_i - G_{ei}$$

where l_i = lost time for phase i
  G_{ai} = actual green time for phase i (not including yellow time)
  τ_i = yellow time for phase i
  G_{ei} = effective green time for phase i

Accordingly, the total lost time for the cycle is given by:

$$L = \sum_{i=1}^{\phi} l_i + R$$

where R = total all-red during the cycle
l_i = lost time for phase i, where ϕ is the total number of phases.

Allocation of Green Time. In general, the total effective green time available per cycle is given by:

$$G_{te} = C - L = C - \left(\sum_{i=1}^{\phi} l_i + R \right)$$

where G_{te} = total effective green time per cycle
C = actual cycle length

The total effective green time is distributed among the different phases in proportion to their Y_i values to obtain the effective green time for each phase:

$$G_{ei} = \frac{Y_i}{Y_1 + Y_2 + \cdots + Y_\phi} G_{te}$$

Recall that $Y_i = V_{ij}/S_j$. Accordingly, the actual green time for each phase is obtained as

$$G_{si} = G_{ei} + l_i - t_i$$

Pedestrian Crossing Requirements. Signals should be designed to provide an adequate time for pedestrians to cross the intersection safely. Thus, the pedestrian crossing time serves as a constraint on the green time allocated to each phase of a cycle. Pedestrians can safely cross an intersection as long as there are no conflicting movements occurring at the same time. Depending on the intersection geometry, in some cases the length of green time interval may not be sufficient for the pedestrians to cross. The minimum green time can be determined using the HCM expressions given as:

$$G_p = 3.2 + \frac{L}{S_p} + \left[2.7 \frac{N_{\text{ped}}}{W_E} \right] \qquad \text{for } W_E > 10 \text{ ft}$$

$$G_p = 3.2 + \frac{L}{S_p} + [0.27 \, N_{\text{ped}}] \qquad \text{for } W_E \leq 10 \text{ ft}$$

where G_p = minimum green time (sec)
L = crosswalk length (ft)
S_p = average speed of pedestrians, usually taken as 4 ft/sec
W_E = effective crosswalk width
N_{ped} = number of pedestrian crossing during a given interval

Actuated and Semiactuated. In actuated signals, the phase lengths are adjusted in response to traffic flow as registered by the actuation of vehicle and/or pedestrian detectors. Signal actuation varies depending on the amount of vehicle detection used in the design. Full-actuated control requires the presence of detectors on all approaches. Semiactuated control operates with green time constantly allocated to major streets, except when there is a detector call from the minor street. Semiactuated signals are usually placed on minor streets with low traffic volumes and random peaks. Full-actuated signals adjust to traffic demand on all approaches. The green time allocation is based on calls from detector actuation as well as on minimum and maximum green time settings. In full-actuated signals, detectors are placed on all approaches. Full-actuated signals are usually placed in busy isolated intersections where demand fluctuations during the day make pretimed signals inefficient.

The main feature of actuated signals is the ability to adjust the signal's pretimed phase lengths in response to traffic flow. For instance, some pretimed phases could be skipped if the detector indicated no vehicles on the related approach. In addition, the green time for each approach is a function of the traffic flow and can be varied between minimum and maximum duration length, depending on flows. Moreover, cycle lengths and phases are adjusted at intervals set by vehicle actuation.

Actuated Signals: Signal Timing Parameters and Detector Placement

Signal Timing Parameters.   Each actuated phase has the following features:

Minimum green time intends to provide sufficient time for vehicles stored between detectors and stop line to enter the intersection.

Gap/passage time defines the maximum gap between vehicles arriving at the detector to retain a given green phase. It also allows a vehicle to travel from the detector to the stop line when a call is placed.

Maximum green time is the maximum green time allowed for the green phase. A phase reaches this maximum when there is sufficient demand.

Recall switch: Each actuated phase has one. When the switch is on, the green time is recalled from a terminating phase. When the switch is off, green is retained on the previous phase until a call is received.

Minimum Green Time.   The minimum green time is usually set to be equal to an initial interval allowing all vehicles potentially stored between the detectors and the stop line to enter the intersection. A start-up time of 4 seconds is also included, in addition to 2 seconds for each vehicle. Thus, the minimum green time is given by the formula

$$G_{min} = 4 + \left[2 \times \text{Integer} \left(\frac{d}{20} \right) \right]$$

where G_{min} = initial interval
d = distance between detector and stop line (ft)

The denominator of 20 denotes the distance between stored vehicles in ft.

Gap/Passage Time.   The passage time is set as the time required for one vehicle to move from the point of detection to the stop line, at its approach speed, and is given by:

$$P = \frac{d}{1.468S}$$

where P = passage time
d = distance between the detector and the stop line (ft)
S = approach vehicle speed (mph)

Maximum Green Time.   The maximum green time is generally set by working out an optimal cycle length and phase splits, as in the pretimed signal plan. The computed cycle length is usually increased by a factor of 0.23 or 1.5 to set the maximum green time.

Detector Location Strategies.   The optimal detector location is computed based on the approach speed and the desired minimum green time. Two different approaches are presented in what follows.

The first approach computes the detector location for an approach speed of 40 mph using the minimum green time G_{min}, as follows:

$$d = 10G_{min} - 40$$

The second approach places the detector back at a distance that is traveled by either 3 seconds or 4 seconds, as follows:

$$d = 1.468S(3) = 4.404S \text{ for a travel time of 3 sec}$$

$$d = 11.468S(4) = 5.872S \text{ for a travel time of 4 sec}$$

Pedestrian Times. Pedestrian crossing time must be coordinated with vehicle signal phasing. The minimum green value should be set to accommodate safe pedestrian crossing time as in the pretimed case. In addition, most actuated pedestrian signals must include a pedestrian push button and accordingly an actuated pedestrian phase.

Figure 6.26 illustrates the operation of an actuated phase based on the actuated setting. When a green indication is initiated, it will be retained for at least the minimum green period. If an actuation occurs during the minimum green period, the green time is extended by an interval equal to the passage time starting from the time of actuation. If a subsequent actuation occurs within the passage time interval, another extension is added, and this process continues. The green is terminated in two cases: if a passage time elapses without an additional actuation, or if the maximum time is reached and there is call on another phase.

Coordinated Signals. In an urban network with relatively closely spaced intersections, signals must operate in a coordinated fashion. Coordinated signalization has the following objectives:

- To coordinate green times of relatively close signals to move vehicles efficiently through the set of signals without having vehicle queues blocking the adjacent intersections
- To operate vehicles in a platoon, so that vehicles released from a signal maintain their grouping for more than 1,000 ft. This will also effectively regulate group speed.

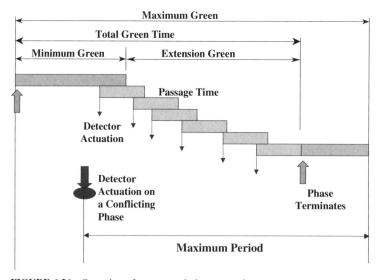

FIGURE 6.26 Operation of an actuated phase—maximum green.

Note that it is a common practice to coordinate signals less than half a mile apart on major streets and highways.

Advantages of Signal Coordination.   The primary benefit of signal coordination is reduction of vehicle stops and delays. This also results in conservation of energy and protection of the environment. Another benefit deals with speed regulation. Signals are set in such a way to incur more stops for speeds faster than the design speed. In addition, grouping vehicles into platoons (shorter headways) leads to more efficient use of intersections. Finally, stopping fewer vehicles is especially important in short blocks with heavy flows, which otherwise may overflow the available storage.

However, a number of factors may restrain the above benefits and make the implementation of coordinated signals rather difficult. These factors are related to the following (McShane et al. 1998):

- Inadequate roadway capacity
- Existence of substantial side frictions, including parking, loading, double parking, and multiple driveways
- Complicated intersections, multiphase control
- Wide range of traffic speed
- Very short signal spacing
- Heavy turn volumes, either into or out of the street

Moreover, easy coordination may not always be possible and some intersections may make exceptions in the coordination system. As an example, a very busy intersection may be located in an uncongested area. The engineer may not want to use the cycle time of the busy intersection as the common cycle time. Another example is the existence of a critical intersection that causes queue spillback. Such an intersection may have low capacity and may not handle the delivered vehicle volume. This situation may be addressed either by detaching this intersection from the coordination system or by building the coordination around it and delivering it in the volume that would not create storage problems.

Signal Coordination Features: Time-Space Diagram and Ideal Offsets.   The time-space diagram is simply the plot of signal indications as a function of time for two or more signals. The time-space diagram is usually scaled with respect to the distance. Figure 6.27 shows an example of a time-space diagram for two intersections.

The offset is the difference between the green initiation times at two adjacent intersections.

The offset is usually expressed as a positive number between zero and the cycle length.

Often the ideal offset is defined as the offset that will cause the minimum delay. Often the offset denotes the start of the green phase, i.e., the delay initiation green phase for downstream signal so that leading vehicles from upstream intersections arrive at downstream intersections just as the light turns green (see Figure 6.27). Under ideal conditions, offset is equal to travel time from the upstream intersection to the subject intersection, i.e.:

$$t(\text{ideal}) = \frac{L}{S}$$

where L = block length
$\quad\quad\ S$ = vehicle speed (mps)

Another important definition related to the coordinated signals is the bandwidth, defined as a window of green through which platoons of vehicles can move.

The time-space diagram shown in Figure 6.27 illustrates the concept of offset and bandwidth.

The design of progression depends on the type of road. For instance, progression is easier to maintain on a one-way street than on a two-way street. Furthermore, a two-way street

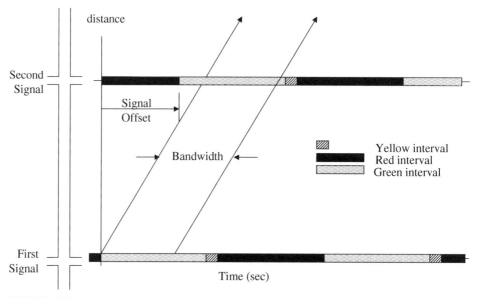

FIGURE 6.27 Time-space diagram for two intersections.

provides better progression than a network of intersection (grid). One-way street progression is easier to achieve because traffic signal phasing is less of a constraint. In two-way streets, finding good progression becomes easy if an appropriate combination of cycle length, block length, and platoon speed exist. Therefore, whenever possible, in new developments these appropriate combinations must be taken into account seriously.

Adaptive Signals. The use of simulation in conjunction with field controllers to provide simulation of optimization strategies has been increasingly used. TRANSYT is perhaps the most used model for optimizing signals off-line and has been quite successful in improving network performance when traffic demand is relatively constant (Stewart and Van Aerde 1998). However, such off-line signal coordination methods, as well as signal progression, are gradually giving way in larger cities to adaptive traffic responsive schemes such as SCOOT (Split, Cycle, Offset Optimization Technique) and SCATS (Sydney Coordinated Adaptive Traffic Control System).

Adaptive signal control systems help optimize and improve intersection signal timings by using real-time traffic information to formulate and implement the appropriate signal timings. The adaptive system is able to respond to inputs, such as vehicle actuation, future traffic prediction, and recurrent as well as current congestion. The intelligence of traffic signal control systems is an approach that has shown potential to improve the efficiency of traffic flow.

Adaptive control strategies use real-time data from detectors to perform constant optimizations on the signal timing plan parameters (i.e., the split, cycle length, and offset timing parameters) for an arterial or a network. Traffic adaptive control requires a large number of sensors that monitor traffic in real-time. As traffic changes, the splits, offsets, and cycle lengths are adjusted by small amounts, usually several seconds. The goal is to always have the traffic signals operating in a manner that optimizes a network-wide objective, such as minimizing queues or delays in the network.

6.6 REFERENCES

Abdulhai, B., and L. Kattan. 2003. "Reinforcement Learning Crystallized: Introduction to Theory and Potential for Transport Applications." Accepted, *CSCE Journal*, 2002.

Abdulhai, B., R. Pringle, and G. Karakoulas. 2003. "Reinforcement Learning and a Case Study on Adaptive Traffic Signal Control." *ASCE Journal of Transportation Engineering* 129(4).

Ben-Akiva, M., A. Davol, T. Toledo, H. T. Koutsopoulos, W. Burghout, I. Andréasson, T. Johansson, and C. Lundin. 2002. "Calibration and Evaluation of MITSIMLab in Stockholm." Submitted for Presentation and Publication at the 81st Transportation Research Board Meeting, January.

Brackstone, M., and M. McDonald. 1998. "Modeling of Motorway Operation." *Transportation Research Records* 1485.

Brundtland Commission (World Commission on Environment and Development). 1987. *Our Common Future*. New York: Oxford University Press.

Cascetta, E. 2001. *Transportation Systems Engineering Theory and Methods*. Dordrecht: Kluwer.

Chaudhary, N. A., and C. J. Messer. 2000. *Ramp Metering Technology and Practice*. Report No. FHWA/TX-00/2121-1, Texas Transportation Institute—The Texas A&M University System.

Cohen, S. 1991. "Kinematic Wave Theory." In *Concise Encyclopedia of Traffic and Transportation Systems,* ed. M. Papageorgiou. Oxford: Pergamon Press, 231–234.

Cremer, M., and M. Papageorgiou. 1981 "Parameter Identification for a Traffic Flow Model." *Automatica* 17:837–43.

Daganzo, F. C. 1994. "The Cell-Transmission Model: A Dynamic Representation of Highway Traffic Consistent with the Hydrodynamic Theory." *Transportation Research B* 28:269–87.

De Schutter, B., T. Bellemans, S. Logghe, J. Stada, B. De Moor, and B. Immers. 1999. "Advanced Traffic Control on Highways." *Journal A* 40(4):42–51.

Dupuis, A., and B. Chopard. 1998. "Parallel Simulation of Traffic in Geneva Using Cellular Automata." *Parallel and Distributed Computing Practices* 1(3):79–92.

Federal Highway Administration (FHWA). 1997. *Freeway Management Handbook*. U.S. Department of Transportation, FHWA, Washington, DC.

Gabart, J. F. 1991. "Car Following Models." In *Concise Encyclopedia of Traffic and Transportation Systems,* ed. M. Papageorgiou. Oxford: Pergamon Press, 231–34.

Garbacz, R. M. n.d. "Adaptive Signal Control: What to Expect." http://209.68.41.108/itslib/AB02H473.pdf.

Garber, N., and L. Hoel. 2002. *Traffic and Highway Engineering,* 3d ed. Pacific Grove, CA: Brooks/Cole Thomson Learning.

Gazis, D. C., R. Herman, and R. Potts. 1959. "Car-Following Theory of Steady-State Traffic Flow." *Operations Research* 9:545–95.

Hall, F. L. 1998. "Traffic Stream Characteristics." In *Traffic Flow Theory: A State-of the-Art Report.* U.S. Department of Transportation, Federal Highway Administration, Turner-Fairbank Highway Research Center, http://www.tfhrc.gov/its/tft/tft.htm.

Harwood, D. W., A. D. May, I. B. Anderson, L. Leiman, and A. R. Archilla. 1999. *Capacity and Quality of Service of Two-Lane Highways*. NCHRP Final Report 3-55(3), Midwest Research Institute, University of California—Berkeley.

Harwood, D. W., and C. J. Hoban. 1987. *Low-Cost Methods for Improving Traffic Operations on Two-Lane Highways—Informational Guide*. FHWA Report No. FHWA-IP-87-2, Federal Highway Administration.

Hoogendoorn, S. P., and P. H. L. Bovy. 2001. "State-of-the-Art of Vehicular Traffic Flow Modeling." Special Issue on Road Traffic Modeling and Control of the *Journal of Systems and Control Engineering*.

Horowitz, R., A. Skabardonis, P. Varaiya, and M. Papageorgiou. 2002. *Design, Field Implementation and Evaluation of Adaptive Ramp Metering Algorithms*. PATH RFP for 2001–2002.

Jiang, H. n.d. "Traffic Flow Simulation with Cellular Automata." http://sjsu.rudyrucker.com/~han.jiang/paper/.

Kachroo, P., and K. Krishen. 2000. "System Dynamics and Feedback Control Design Problem Formulations for Real-Time Ramp Metering." *Transactions of the Society for Design and Process Science (SDPS)* 4(1):37–54.

Kim, Y. 2002. "Online Traffic Flow Model Applying Dynamic Flow-Density Relations." Ph.D. Dissertation, University of Stuttgart, Germany.

Lieberman. E., and A. K. Rathi. 1998. "Traffic Simulation." In Traffic Flow Theory: A State-of the Art Report. U.S. Department of Transportation, Federal Highway Administration, Turner-Fairbank HIghway Research Center. http://www.tfhrc.gov/its/tft/tft.htm.

Lighthill, M. J., and G. B. Whitham. 1955. "On Kinematic Waves II: A Theory of Traffic Flow in Long Crowded Roads." In *Proceedings of the Royal Society* A 229:317–45.

Lind, G., K. Schmidt, H. Andersson, S. Algers, G. Canepari, C. Di Taranto, E. Bernauer, L. Bréheret, F.-F. Gabard, and K. Fox. 1999. *Best Practice Manual.* SMARTEST Project Deliverable D8 Smartest—Simulation Modeling Applied to Road Transport European Scheme Tests, http://www.its.leeds.ac.uk/smartest.

Luttinen, T. 2001. *Traffic Flow on Two-Lane Highways: An Overview.* TL Research Report 1/2001, TL Consulting Engineers, Ltd.

Makigami, Y. 2001. "On Ramp Control." In *Concise Encyclopedia of Traffic and Transportation Systems,* ed. M. Papageorgiou. Oxford: Pergamon Press, 285–89.

Marchal, F. 2001. "Contribution to Dynamic Transportation Models." Ph.D. Dissertation, Department of Economics, University of Cergy-Pontoise, France.

May, A. 1990. *Traffic Flow Fundamentals.* Englewood Cliffs: Prentice Hall.

McDonald, M., M. Brackstone, and D. Jefferey. 1994. "Data Requirements and Sources for the Calibration of Microscopic Motorway Simulation Models." In 1994 *IEEE Vehicle Navigation and Information Systems Conference Proceedings.*

McShane, W., R. Roess, and E. Prassas. 1998. *Traffic Engineering,* 2nd ed. Englewood Cliffs: Prentice Hall.

Middleton, M., and S. A. Cooner. 1999. *Simulation of Congested Dallas Freeways: Model Selection and Calibration.* Texas Transportation Institute, Report 3943-1- Project Number 7-3943. Sponsored by the Texas Department of Transportation, November.

Nagatani, T. 2002. "The Physics of Traffic Jams." *Physica* 62:1331–86.

Nagel, K., and M. Rickert. 2001. *Parallel Implementation of the TRANSIMS micro-simulation.* Dordrecht: Elsevier Science.

Nagel, K., and M. Schreckenberg. 1992. "Cellular Automaton Model for Freeway Traffic." *J. Physique I (Paris)* 2:2221.

Newman, P., and J. Kennworthy. 1998. *Sustainability and Cities: Overcoming Automobile Dependence.* Coreb, CA: Island Press.

Papageorgiou, M. 1998. "Some Remarks on Macroscopic Traffic Flow Modeling." *Transportation Research Part A* 32(5):323–29.

Papageorgiou, M. 1999. "Automatic Control Methods in Traffic and Transportation." In *Operations Research and Decision Aid Methodologies in Traffic and Transportation Management,* ed. P. Toint, M. Labbe, K. Tanczos, and G. Laporte. Berlin: Springer-Verlag, 46–83.

Papageorgiou, M., J.-M. Blosseville, and H. Hadj-Salem. 1989. "Macroscopic Modeling of Traffic Flow on the Boulevard Périphérique in Paris." *Transportation Research* B 28:29–47.

Papageorgiou, M., H. Salem, and J. Blosseville. 1991. "ALINEA: A Local Feedback Control Law for On-Ramp Metering." *Transportation Research Record* 1320.

Papageorgiou, M., and A. Kotsialos. 2000. "Freeway Ramp Metering: An Overview." In *2000 IEEE Intelligent Transportation Systems Conference Proceedings,* Dearborn, MI, October 1–3.

Payne, H. J. 1971. "Models of Freeway Traffic and Control, Simulation." In *Mathematical Models of Public Systems,* ed. G. A. Bekey. Simulation Council Proceedings Series 1(1):51–61.

Polak, J., and K. Axhausen, 1990. *Driving Simulators and Behavioural Travel Research.* TSU Ref. 559, University Transport Studies Unit, Oxford, U.K.

Sitarz, D. 1994. *Agenda 21.* Boulder: Earthpress.

Sutton, R., and A. Barto. 1996. *Reinforcement Learning: An Introduction.* Cambridge: The MIT Press.

Taplin, J. 1999. "Simulation Models of Traffic Flow." Paper presented at 34th Annual Conference of the Operational Research Society of New Zealand, University of Waikato, Hamilton, New Zealand, December 10–11.

Transportation Research Board (TRB). 2000. *Highway Capacity Manual.* Special Report 209, Transportation Research Board, National Research Council, Washington, DC.

Wu, J. 2001. *"Traffic Diversion Resulting from Ramp Metering."* M.S. thesis, University of Wisconsin—Milwaukee.

CHAPTER 7
TRAVEL DEMAND FORECASTING FOR URBAN TRANSPORTATION PLANNING

Arun Chatterjee
Department of Civil and Environmental Engineering,
The University of Tennessee, Knoxville, Tennessee

Mohan M. Venigalla
Civil, Environmental, and Infrastructure Engineering Department,
George Mason University, Fairfax, Virginia

7.1 INTRODUCTION

7.1.1 The Need for Determining Travel Demand: Existing and Future

The basic purpose of transportation planning and management is to match transportation supply with travel demand, which represents the need for transportation infrastructure. A thorough understanding of existing travel pattern is necessary for identifying and analyzing existing traffic-related problems. Detailed data on current travel pattern and traffic volumes are also needed for developing travel forecasting/prediction models. The prediction of future travel demand is an essential task of the long-range transportation planning process for determining strategies for accommodating future needs. These strategies may include land use policies, pricing programs, and expansion of transportation supply—highways and transit service.

7.1.2 Scope of Analysis and Levels of Planning

There are different levels of planning, directed to different types of problems. The terminology for these levels of planning and analysis varies according to the context. For example, the expressions "micro," "meso," and "macro" are sometimes used to describe the level of detail or the size of an area used for an analysis. Similarly, the expressions "site-specific," "corridor," and "area-wide" or "metropolitan" are used to describe variations in the scope of a problem. The approach and techniques for analyzing and forecasting travel would vary according to the level of analysis. Even for a particular level of analysis, the techniques may have to be adjusted to match the constraints of available data and human resources.

An example of a micro-level or site-specific analysis is the case of a congested road intersection. In this case traffic engineers would be interested in detailed traffic flow characteristics, including turning movements of vehicles along each approach and pedestrian volumes across each approach. Management strategies in this case would involve traffic operation and roadway design-oriented techniques. A corridor-level analysis, on the other hand, would cover a larger area, say 10 miles long and 2 miles wide. A major highway with severe congestion problem may require a corridor analysis. The origin and destination of trips and modal choice of travelers would be of interest in this case. Station-to-station movements of passengers may have to be estimated in the case of a rapid transit service along the corridor. At the macro level the concern may be total energy consumption by the transportation sector or the total emission of an air pollutant; for these cases, information on total vehicle-miles traveled (VMT) on each functional class of roads will be needed.

It is important to recognize that the nature of problems to be examined dictates the level of planning to be used as well as the technique for travel demand analysis. The discussion of this chapter will be oriented mostly to meso scale or area-wide travel demand analysis that is commonly performed in urban transportation planning studies. Even for this type of analysis for an urban area at the meso scale, the approach and details of techniques and models to be used would depend on the size of the area as well as the resources available for carrying out the work. For example, a small urban area may not have the manpower or funding needed for carrying out large-scale surveys and developing advanced mathematical models. The need for customizing the planning and modeling approaches based on specific situations was discussed in detail by Grecco et al. (1976).

7.2   CHARACTERISTICS OF TRAVEL

Certain special characteristics of travel demand require recognition for planning and design purposes. These are discussed below.

7.2.1   Spatial and Temporal Variations

The total magnitude of travel demand alone is not sufficient for detailed planning and management purposes. The spatial and temporal distributions of travel are also important items of information to be considered in determining supply strategies. The peaking of travel at certain time periods requires a level of transportation supply that is not needed at other times. However, due to the nature of supply, which cannot be adjusted easily, large investments have to be made to provide roadway or transit service capacities to accommodate peak period travel, and this capacity is not utilized efficiently at other times. An imbalance in the directional distribution of travel also creates similar inefficiencies.

The spatial orientation of trips has important influence on supply requirements and costs. A few typical spatial distribution patterns of trips in urban areas are listed below:

- Travel along dense corridors, which are usually radial connecting suburbs to central business district (CBD)
- Diffused travel pattern caused by urban sprawl
- Suburb to suburb or circumferential travel
- Travel within large activity centers in CBD and suburbs

Different modes of transportation may be needed to serve these different travel patterns. For example, fixed-route public transit service usually is efficient for concentrated travel along a dense corridor, but it is not ideally suited to serve a diffused travel pattern in a cost-effective manner.

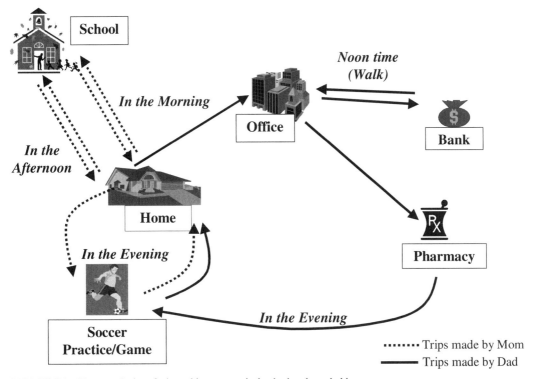

FIGURE 7.1   The complexity of trip making at a typical suburban household.

Choice of domicile and workplace, lifestyles and different travel needs of individuals and families make the comprehension of trip-making characteristics of a large metro area very complex. These complexities may be illustrated through trips made by a typical suburban U.S. household on a given weekday (Figure 7.1). Assume that this household has four members, including two children who go to grade school, and two cars. It can be seen that there are at least 11 trips made by this household at different times of day. Most of the trips are auto trips, and two trips are taken in the walk mode. Travel demand modeling attempts to capture such spatial and temporal variations in travel at an aggregate level, such as a zone, in which a number of households, businesses, and offices exist.

7.2.2   Classification of Travel by Trip Purpose and Market Segments

In addition to the spatial and temporal characteristics of travel demand, several other aspects of travel demand must be recognized. Trip purposes such as work, shopping, and social-recreation, and trip-maker's characteristics such as income and car ownership, are important factors influencing the elasticity of demand, reflecting its sensitivity with respect to travel time and cost. For example, work trips may be more likely to use public transit for a given level of service than trips of other trip purposes.

For a metropolitan study, it is useful to classify travel according to spatial orientation and trip purpose as shown in Figure 7.2. The concept of market segmentation is applicable to the classification of travel based on trip purpose, trip-makers' characteristics, and spatial-temporal concentration. This concept is used in the field of marketing for developing different types of consumer products targeted to match different tastes and preferences of potential

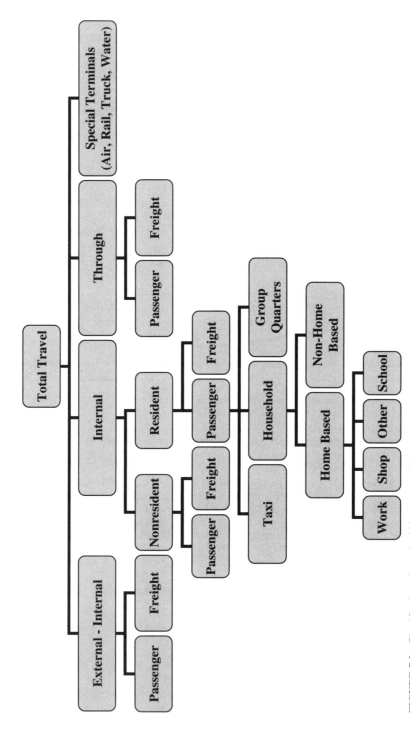

FIGURE 7.2 Classification of travel within a metropolitan area.

users/buyers of these products. The concept of market segmentation is applicable to public transportation planning. A single type of transit service is not suitable for all transit market segments. For example, express buses may be needed for a commuter market segment. Taxicabs serve a different market segment. Woodruff et al. (1981) examined this subject in depth.

7.3   UNITS FOR MEASURING TRAVEL DEMAND

Travel demand is measured and expressed in different ways for different types of analysis. Examples of different units of measurement are:

1. Trip (between two areas)
2. Trip end (in a given area)
2. Traffic volume (on a road segment)
4. Person trip and vehicle trip
5. Passenger vehicle and freight vehicle
6. Person-miles traveled and vehicle-miles traveled

The definition of each of these units should be understood clearly and an appropriate unit of measurement should be used to match the case being analyzed. For example, for a parking study, trip end is the appropriate unit for expressing parking demand. For estimating the number of lanes to be provided in a road segment, the demand should be expressed in terms of traffic volume. As pointed out earlier, the appropriate unit of travel for estimating fuel consumption and/or air pollution attributable to transportation is vehicle-miles traveled (VMT).

7.4   MEASURING EXISTING TRAVEL

Detailed information on existing travel is needed for two purposes: analyzing existing problems and developing mathematical models for forecasting travel. A variety of surveys can be performed for gathering information related to existing travel demand. However, travel surveys are expensive, and therefore care must be taken to identify the types of information that really would be useful for specific purposes, and then the most suitable procedures should be selected for gathering the information. Sampling techniques are useful, and adequate time and care must be devoted to developing sampling procedures. There are several different types of survey techniques, some of which are suitable for automobile travel, some for transit travel, and some for general passenger movement. Survey procedures for freight vehicles and commodity movements may be very different in certain respects from those of passenger travel. Good references for travel demand-related survey techniques are Dresser and Pearson (1994), Stopher and Metcalf (1996), and *Travel Survey Manual* (Cambridge Systematics, Inc. and Barton-Aschman Associates, Inc. 1996).

7.4.1   Time Frame for Travel Surveys

Since travel demand varies during a given year according to the season (or month of year) and day of week, a decision must be made carefully to select a specific time frame or reference for surveys. For urban transportation studies it is a common practice to develop travel demand information for an average weekday in the fall. However, the time can be different based on the nature of the problem to be analyzed. For example, in the case of a tourist-oriented urban area the major concern may be traffic problems during weekend days or holidays, and surveys may be done to capture information for one of those days. A few major types of surveys are discussed in the following sections.

7.4.2 Origin-Destination Surveys

The classification of trips into the three classes of internal, external-internal (and vice versa), and through trips is useful for meso-scale and metropolitan-level as well as small-area studies. This classification scheme is useful for developing forecasting procedures/models as well as policies and strategies for accommodating travel because strategies for each of these classes of travel would be different. For example, through trips may require a bypass facility. External-internal trips may have to be intercepted before they reach a heavily congested area such as the central business district (CBD).

The origins and destinations (O-D) of trips along with some other characteristics such as trip purpose and mode of travel can be determined in different ways:

1. Home interviews (for internal travel)
2. Roadside interviews at cordon stations (for external-internal and through trips)
3. On-board survey on transit vehicles

All three of these techniques involve sampling and require careful planning before their execution. The Federal Highway Administration (FHWA 1975b) and Urban Transportation Systems Associates (1972) developed detailed guidelines for O-D survey procedures. The reliability of the results of an O-D survey depends on its sampling scheme and sample size, and this issue was examined by Makowski, Chatterjee, and Sinha (1974).

Full-scale origin-destination surveys were widely used during the 1960s and 1970s to develop a variety of information, including "desire lines" of travel. Their use has decreased because of the cost and also due to the use of synthetic or borrowed disaggregate travel models, which require less survey data.

7.4.3 Traffic Volume and Passenger Counts

For determining the use of various roadway facilities and assessing their level of service, vehicle counts are taken at selected locations along roadways. Short-count techniques are useful provided appropriate expansion factors are developed based on previous or ongoing research on fluctuations of traffic by hour, weekday, and month. All state Departments of Transportation (DOTs) have extensive programs for gathering traffic volume data on an annual basis. These vehicle counts usually are taken with machines.

For urban transportation studies screen lines and cut-lines are established in the study area to select traffic count locations and take counts in an organized manner so that the major travel movements can be measured and analyzed. These counts are also used for checking the results of travel forecasting models. Similarly, traffic counts are taken at special traffic generators such as airports and large colleges and universities to capture their unique travel-generating characteristics.

For analyzing the use of a transit service, passenger counts are taken on-board transit vehicles and/or at selected stops or stations. These passenger counts usually are taken by observers who are assigned to specific transit vehicles and/or transit stops/stations according to a survey plan.

7.5 FORECASTING FUTURE DEMAND

The need for travel demand forecasts arises in various contexts of planning, short-range as well as long-range. Travel forecasting is one of the most important and difficult tasks of transportation planning. There are different types of travel prediction techniques, and the one to be used in a particular case must be compatible with the nature of the problem and scope of planning. Constraints of available time and resource also influence the selection of a technique.

7.5.1 Predicting Response to Service Changes Using Elasticity Coefficients

For short-range planning or a transportation systems management (TSM) study, it is often necessary to predict the effect of a proposed change in transportation service that can be implemented in the near future. For example, a planner may be asked to evaluate the impact on transit ridership of improving transit service in a travel corridor by providing increased frequency and/or limited stop service. The impact of changing the fare structure on transit ridership may also be of interest. In these cases demand elasticity coefficients, if available from past studies, would be useful. Typically an elasticity coefficient is developed with respect to a specific factor such as travel time or fare based on actual observation. The coefficient should exclude the effect of other factors that also may be influencing demand at the same time. Mayworm, Lago, and McEnroe (1980) give information on demand elasticity models. More information on the elasticity of transit use with respect fare may be found in Parody and Brand (1979) and Hamberger and Chatterjee (1987).

7.5.2 Stated Preference Surveys and Concept Tests for Forecasting

For transit planning it is sometimes necessary to ask people about their preferences and their likes and dislikes for various service characteristics. These surveys are used for determining how to improve an existing service and/or designing a new service, and also for forecasting ridership on a new service. These attitudinal and stated-preference surveys need sound statistical design for selecting the sample and analyzing the results. A discussion on stated preference survey may be found in the *Travel Survey Manual*. In the field of marketing, concept tests are performed for estimating the potential demand for a new consumer product, and this approach can be extended to ridership forecasts for new/innovative transit services.
Hartgen and Keck (1976) describe a survey-based method of forecasting ridership on a new dial-a-bus service. The interpretation of results of opinion-based surveys must be done carefully in order to account for any bias reflected in apparent results. Chatterjee, McAdams, and Wegmann (1983) present a case study involving noncommitment bias in public opinion on the anticipated usage of a new transit service.

7.5.3 Forecasting Future Travel on Road Segments and/or Transit Lines

A variety of forecasting procedures are available, ranging from the extrapolation of past trends to complex mathematical models involving several steps. A transportation planner must recognize the advantages and disadvantages of each procedure. Two procedures are examined for illustration.

Direct Estimation of Traffic Volume by Trend Analysis.   If traffic volume data are available for a road segment or a transit line of interest for several years in the past, the historical trend can be identified and extrapolated to estimate future volumes. This approach, of course, is appropriate if the trend is expected to continue, which commonly is true for short-range forecasts. Trend-based forecasts are appropriate also for aggregate values such as total VMT or transit rides in an urban area. However, major changes in the land development pattern and/or transportation network can cause substantial changes in the travel pattern, and if such changes are likely then trend extrapolation will not be appropriate. Therefore, for long-range forecasts of traffic volumes on individual segments of a road network or the number of passenger trips on individual transit routes, trend analysis is not used.

Stepwise/Sequential Procedure.   A widely used travel estimation procedure for long-range forecasts of traffic volumes on a highway network uses several steps in a sequence, as shown in the flowchart of Figure 7.3. Each step requires a particular type of model or procedure, and there are different choices of models at each step. One of the major advantages of this procedure is its ability to reflect several types of changes that may occur in the future:

1. Changes in trip-making rates
2. Changes in development pattern, resulting in altered travel pattern
3. Changes in transportation mode usage
4. Changes in transportation network

Another advantage of the stepwise, or sequential, procedure is that it generates several types of useful information at the end of various steps. The disadvantage of the procedure is that it needs a large amount of data for model development. It also requires a sound knowledge of one of the available computer software models that is specially designed for developing and applying these models. A great deal of research has been performed and is still being continued to improve this stepwise modeling procedure. It should be acknowledged that the staff of the former Bureau of Public Works and later the staff of the Federal Highway

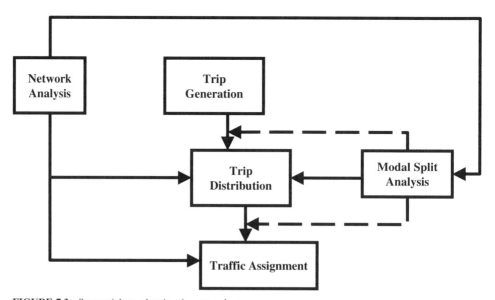

FIGURE 7.3 Sequential travel estimation procedure.

Administration and Urban Mass Transportation Administration made tremendous contributions to the development of various procedures and computer software. An historical overview of the development of planning and modeling procedures in the United States is presented by Weiner (1992).

The stepwise procedure is popularly known as the four-step modeling process because it includes four major steps: trip generation, trip distribution, mode choice, and traffic assignment. Additionally, network analysis must be done to develop a few types of information that are needed for the other steps. These steps and procedures involved with each are discussed in detail in the following sections.

7.6 TRIP GENERATION

Trip generation is a very important step because it sets up not only the framework for the following tasks but also some of the controlling values, such as the total number of trips generated in the study area by location and trip purpose. The commonly used units for trip generation analysis usually include a household, a dwelling unit (DU), and a business establishment. However, the results of a trip generation analysis for a study area are aggregated based on larger areas known as traffic zones.

A typical classification scheme of trips used for trip generation analysis is presented in Figure 7.2. A detailed discussion of this classification scheme is presented by Chatterjee, Martinson, and Sinha (1977). A thorough analysis of all these types of trips shown in the figure requires a large amount of data. These data are collected by using origin-destination (O-D) surveys, discussed briefly in an earlier section. This section will focus primarily on trip generation models for internal passenger trips made by households.

7.6.1 Models for Internal Passenger Trips: Aggregate and Disaggregate Models

The goal of trip generation models for internal passenger trips is to estimate the total number of trip ends for each purpose generated in each traffic zone based on socioeconomic and/or land use data for the respective zones. This task can be accomplished with either aggregate or disaggregate models. For aggregate models the total number of trips (trip ends) generated in a zone is used as the dependent variable, whereas for disaggregate models trips made by a household (or a business establishment) are used as the dependent variable. When disaggregate models are used, the trip ends generated by households and/or any other trip generating units such as business establishments in a zone are combined to produce the zonal (total) value. Both disaggregate and aggregate trip generation models are used in planning studies.

7.6.2 Trip Generation by Households

Household-generated trips comprise more than 80 percent of all trips in an urban area. Trips by nonresidents and a variety of other vehicles, including commercial vehicles such as taxis and trucks, and public utility and public service vehicles comprise the remaining portion of total travel. For the purpose of modeling, the trips generated by households are classified as home-based and non-home-based. Home-based trips have one end, either origin or destination, located at the home zone of the trip maker. If both ends of a trip are located in zones where the trip maker does not live, it is considered a non-home-based trip.

Definitions of Productions and Attractions and Trip Purpose. Because of the predominance of home-based trips in an urban area, the model development is simplified if it is assumed that the home end of a trip is a production (P) in the zone of the trip maker's residence irrespective of whether it represents the origin or destination of the trip. According to this approach, the non-home end of a home-based trip is considered to be attraction (A). For a non-home-based trip, which has neither its origin nor its destination at the trip maker's residence, production and attraction are synonymous with origin and destination, respectively. This definition of productions (P's) and attractions (A's) is depicted in Figure 7.4. It should be noted that for home-based trips the activity at the non-home end determines the trip purpose, and that non-home-based trips usually are not further stratified by purpose.

7.6.3 Cross-Classification or Category Models for Household Trip Generation

Household trip rates have been found to vary significantly according to certain socioeconomic characteristics and the size of a household. Household characteristics that have been

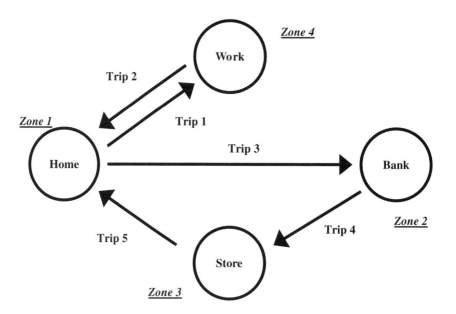

Zone	Production	Attraction (Purpose)
1	4 (2 work, 1 personal business, 1 shop)	0
2	1 (Non-home-based)	1 (Personal business)
3	0	2 (1 Non-home-based, 1 shop)
4	0	2 (work)
Total	5	5

FIGURE 7.4 Definitions of productions and attractions.

TABLE 7.1 A Hypothetical Category Model for Total Person Trips per Household

		Car ownership		
		No car	One car	Multicar
Household size	1 to 2	3.2	4.9	6.1
(persons per	3 to 4	5.2	7.1	8.7
household)	5 or more	7.2	9.6	12.0

found to be significant for trip generation and are commonly used in trip generation models are:

1. Household size
2. Auto ownership
3. Income

A hypothetical example of a trip generation model for households is presented in Table 7.1, which includes trip production rates for different types or categories of households defined in terms of different combinations of household size and auto ownership. This type of model is referred to as a cross-classification or category models, and such models are used widely for estimating trip production by households.

Alternative techniques for a statistical analysis of these models are discussed by Chatterjee and Khasnabis (1973). When the households in a traffic zone are cross-classified by size and auto ownership, total trips made by households in a zone for a specific purpose (P) = Summation of (No. of households of a category) $\times$ (Trip rate for households of that category and for that specific purpose). In mathematical notation this relationship is shown below:

$$P = \sum_{kl=1}^{n} (\text{HH}_{kl})(\text{TR}_{kl}) \tag{7.1}$$

where HH = households
    TR = trip rates
    kl = a particular combination of household size k and auto ownership l
    n = the number of combinations or categories

The choice of household characteristics to be used for developing the various categories for trip production rates may vary from one study to another. One advantage of disaggregate models is that for developing these models a full-scale O-D survey is not needed. A carefully selected small sample of households may be used for developing trip production rates as long as the number of cases for each category, or cell of the matrix, is statistically adequate.

7.6.4 Models for Trip Attractions

It is a common practice to use aggregate models in the form of regression equations for trip attractions. The dependent variable for these aggregate models is the total number of trip attractions for a specific trip purpose in a traffic zone. The independent variables usually are employment-related, and they represent zonal total values. Hypothetical examples of trip-attraction (A) models are presented below:

$$(\text{HBW A})_j = 1.5 \ (\text{Total employment})_j$$

$$(\text{HBNW A})_j = 8.5 \ (\text{Retail employment})_j + 1.0 \ (\text{Non-retail employment})_j$$
$$+ \ 0.9 \ (\text{Dwelling units})_j$$

$$(\text{NHB A})_j = 3.0 \ (\text{Retail employment})_j + 2.0 \ (\text{Non-retail employment})_j$$
$$+ \ 0.8 \ (\text{Dwelling units})_j$$

where $(\text{HBW A})_j$ = Home-based work attractions in zone j
$(\text{HBNW A})_j$ = Home-based non-work attractions in zone j
$(\text{NHB A})_j$ = Non-home-based attractions in zone j

The development of aggregate models usually requires a full-scale O-D survey. The co-efficients of the regression equations will vary from area to area. The choice of independent variables and trip-purpose categories also may vary from one study to another.

7.6.5 Balancing of Productions and Attractions

Due to the definition of productions and attractions, home-based productions in a zone may not be equal to the corresponding attractions in the same zone. Non-home-based productions in a zone should be equal to corresponding attractions in the same zone. However, area-wide (total) productions (P's) of any trip purpose—home-based or non-home-based—should be equal to the corresponding area-wide (total) attractions. Thus,

$$\Sigma(\text{HBW P})_i = \Sigma(\text{HBW A})_j \tag{7.2}$$

$$\Sigma(\text{HBNW P})_i = \Sigma(\text{HBNW A})_j \tag{7.3}$$

$$\Sigma(\text{NHB P})_i = \Sigma(\text{NHB P})_j \tag{7.4}$$

When synthetic or borrowed models are used, the estimated area-wide (total) productions would not be equal to the estimated area-wide (total) attractions. Therefore, to achieve a balance, zonal attractions are adjusted proportionately such that the adjusted area-wide attractions equal area-wide productions. Adjustment or scaling factors for attractions are calculated as follows:

$$\text{Adjustment factor for HBW A}_j\text{'s} = \frac{\Sigma(\text{HBW P})_i}{\Sigma(\text{HBW A})_j} \tag{7.5}$$

$$\text{Adjustment factor for NBNW A}_j\text{'s} = \frac{\Sigma(\text{HBNW P})_i}{\Sigma(\text{HBNW A})_j} \tag{7.6}$$

$$\text{Adjustment factor for NHB A}_j\text{'s} = \frac{\text{Total NHB productions}}{\Sigma(\text{NHB A})_j} \tag{7.7}$$

7.6.6 Commercial Vehicle Traffic in an Urban Area

It should be pointed out that although internal trips made by residents in passenger vehicles account for a large proportion of total trips in an urban area, the other categories of trips must not be overlooked. The classification scheme presented in Figure 7.2 shows the other categories. The proportion of each category of trips varies according to the size and other characteristics of an urban area. For example, the proportion of through trips is usually larger in smaller size areas. In some cases, trips of one or more of these other categories may be the cause of major problems and thus will require special attention. For example, through traffic may be the major issue in the case of a small or medium-sized urban area, and the planners may have to analyze these trips thoroughly. Similarly, the movement of large trucks may be of major interest in some urban areas. A comprehensive study should pay attention to travel demand of all categories, although the level of detail may vary.

The analysis of commercial vehicle travel has been neglected in most urban transportation studies. These vehicles are garaged in nonresidential locations and include trucks of all sizes, taxicabs, rental cars, service vehicles of plumbers and electricians, etc. There are a few useful references on how to estimate truck traffic in urban areas, including an article by Chatterjee et al. (1979), Ogden (1992), and Fischer (2001).

7.6.7 Forecasting Variables Used in Trip Generation Models

In developing trip generation models the availability of data for the independent variables of the models is an important issue that can influence the selection of a variable. Usually the availability of data for the base year is less problematic than that for future years. Of course, if data for an independent variable are not available for the base year, they cannot be used in model development. However, what the model developer must recognize before building and adopting a model is whether the independent variables used in the model can be forecast by the responsible planning agency, and if such forecasts would be very difficult then it may be desirable to avoid using those variables in the model. Sometimes transportation planners have to develop a procedure or model to be used for making such forecasts.

Usually aggregate values of socioeconomic parameters used in trip generation models are not very difficult to forecast with the existing state of the art. The difficulty usually involves the task of disaggregating socioeconomic data at the zonal level. For example, it may not be very difficult to predict the total number of households in each zone along with their average size and auto ownership. However, it would be difficult to cross-classify the predicted number of households in a zone according to specific categories based on household size and auto ownership. Similarly, predicting the average income of households in individual traffic zones may not be very difficult, but developing a breakdown of the households in every zone by income groups would be difficult. The disaggregate trip generation models thus present a challenge to planners for making detailed forecasts of socioeconomic characteristics for future years. In order to provide assistance for making forecasts in a disaggregate form, a few procedures have been developed, and examples of such household stratification models can be found in Chatterjee, Khasnabis, and Slade (1977) and FHWA (1975a).

7.7 *TRIP DISTRIBUTION*

The purpose of the trip distribution step of the stepwise travel-modeling procedure is to estimate zone-to-zone movements, i.e., trip interchanges. This step usually follows trip generation analysis. In some cases, but not commonly, trip distribution may come after trip generation and modal split analysis. The inputs to a trip distribution model are the zonal productions (P_i) and attractions (A_j). The model strives to link the productions and attractions based on certain hypotheses/concepts.

When the trip distribution phase precedes modal split analysis, productions and attractions include trips by all modes and the distribution model should be multimodal in nature. In actual practice, however, multimodal trip distribution models are uncommon, and in most cases highway-oriented models have been used to distribute trips of all modes. It should be noted that in the rare case where the trip distribution phase follows modal split analysis, mode-specific distribution models are needed. It is generally believed that ideally trip distribution should be combined with modal split analysis because decisions related to destination and travel mode are usually made simultaneously.

In this section a widely used trip distribution technique, the gravity model, will be discussed in detail, followed by a brief overview of other types of models. A good review of commonly used trip distribution models can be found in Easa (1993).

7.7.1   Formulation of a Gravity Model

The basic hypothesis underlying a gravity model is that the probability that a trip of a particular purpose k produced at zone i will be attracted to zone j is proportional to the attractiveness or pull of zone j, which depends on two factors: the magnitude of activities related to the trip purpose k in zone j, and the spatial separation of the zones i and j. The magnitude of activities related to trip purpose k in a zone j can be expressed by the number of zonal trip attractions of the same purpose, and the effect of spatial separation between zones i and j can be expressed by a friction factor, F_{ij}^k, which is inversely proportional to an appropriate measure of impedance, usually travel time. The attractiveness or pull of zone j with respect to zone i is proportional to $A_j^k F_{ij}^k$. The magnitude of trips of purpose k produced in zone i and attracted to zone j, T_{ij}^k, of course, also depends on the number of trips being produced at zone i, P_i^k. This can be expressed mathematically as follows:

$$T_{ij}^k = f(P_i^k, A_j^k, F_{ij}^k) \tag{7.8}$$

The above formulation is not sufficient for estimating the T_{ij}^k values because it yet does not reflect any considerations for other zones that are competing as alternative destinations for the trips P_i^k. Actually, the effective attractiveness of a zone is relative to others, and it can be expressed as the ratio of its own attractiveness with respect to the total. Thus the relative attractiveness of a zone for trips of purpose k being produced in zone i is expressed by the ratio $A_j^k F_{ij}^k / \sum_{j=1}^n A_j^k F_{ij}^k$. Dropping the subscript k, the trip distribution model can be written as follows:

$$T_{ij} = \frac{P_i A_j F_{ij}}{\displaystyle\sum_{j=i}^{n} A_j F_{ij}} \tag{7.9}$$

7.7.2   Application of Gravity Model Concept

The application of the gravity model concept for the trip distribution step of the stepwise travel forecasting procedure was introduced by Voorhees (1955). The classic example of an application of a gravity model as presented by Voorhees is shown in Figure 7.5. This application is based on the assumption that the effect of spatial separation with respect to trip making is proportional to the inverse of the square of travel time between the respective pairs of zones. The calculations presented below shows how the accessibility and attractiveness of zone 4 changed due to a new expressway resulting in an increased number of trip attractions. The number of shopping trips attracted to zone 4 from zone 1 was 28 without an expressway and increased to 80 as a result of a new expressway. The increase in trips attracted to zone 4 resulted in a decrease of trips attracted to the other zones.

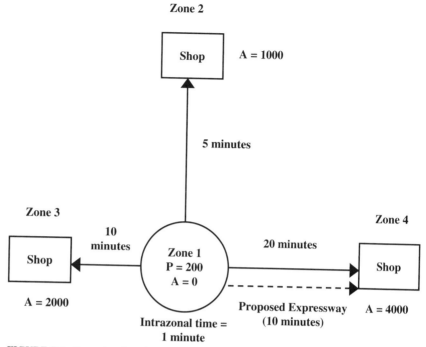

FIGURE 7.5 Example of gravity model concept as introduced by Alan M. Voorhees.

I. Situation without Expressway

Existing pulls from	Percent pull	No. of trips from zone 1 to
Zone 1 = $\dfrac{0}{1^2} = 0$	0/70 = 0%	200 × 0 = 0
Zone 2 = $\dfrac{1000}{5^2} = 40$	40/70 = 57%	200 × 0.57 = 114
Zone 3 = $\dfrac{2000}{10^2} = 20$	20/70 = 29%	200 × 0.29 = 58
Zone 4 = $\dfrac{4000}{20^2} = 10$	10/70 = 14%	200 × 0.14 = 28
Total pull = 70	100	200

II. Situation with Expressway

Existing pulls from	Percent pull	No. of trips from zone 1 to
Zone 1 $= \dfrac{0}{1^2} = 0$	$0/100 = 0\%$	$200 \times 0 = 0$
Zone 2 $= \dfrac{1000}{5^2} = 40$	$40/100 = 40\%$	$200 \times 0.40 = 80$
Zone 3 $= \dfrac{2000}{10^2} = 20$	$20/100 = 20\%$	$200 \times 0.20 = 40$
Zone 4 $= \dfrac{4000}{10^2} = 40$	$40/100 = 40\%$	$200 \times 0.40 = 80$
Total pull $= 100$	100	200

7.7.3 Calibration of Gravity Model Parameters

The three basic parameters of a gravity model are zonal trip productions and attractions, P_i's and A_j's, and friction factors F_{ij}'s. Whereas the P_i's and A_j's are estimated by trip generation models, friction factors must be derived as a part of the trip distribution phase. The basic concept of the gravity model implies the following form for friction factors:

$$F_{ij} = \frac{1}{(\text{Travel time between } i \text{ and } j)^x} \tag{7.10}$$

In the early application of the gravity concept, the exponent of the travel time was assumed to be a constant. However, further empirical analysis suggested that the exponent varies over the range of values for travel time. The actual values of friction factors are derived by a trial-and-error procedure and vary according to trip purpose.

For each trip purpose an arbitrary set of friction factors for a range of travel time values at an increment of 1 minute is assumed at the beginning of the calibration process, and the results of this initial application of the model are evaluated with respect to the actual trip distribution obtained from an O-D survey. The evaluation is made by comparing the trip-length frequencies generated by the model with those derived from the O-D survey, and if the results are not similar the friction factors are adjusted and the gravity model is applied again with the new factors. This trial-and-error procedure is continued until the trip-length frequencies of a model appear similar to those of the O-D survey. It may be noted that the absolute values of these factors have no special implications and that it is the relative weight with respect to each other that is important. The respective set of these frictions factors can be scaled up or down by a constant factor.

Balancing a Gravity Model. It must be pointed out that due to the basic nature of the gravity model formulation, the zonal productions obtained from the model application must equal the values of P_i's originally used as inputs to the model. However, the same is not true for the zonal attractions, and the model results must be compared with the original A_j's. In the cases when model-generated A_j's do not match closely with original A_j's, a balancing procedure is used by adjusting the input values of A_j's until the model results are satisfactory.

7.8 OTHER TYPES OF TRIP DISTRIBUTION MODELS

The gravity model is by far the most widely used trip distribution technique, but there are also other techniques used in urban transportation planning. One technique utilizes growth factors for each traffic zone and uses an iterative balancing procedure to project a base year trip matrix to the future year. The most popular of the growth factor techniques is that introduced by Fratar (1954) and known as the Fratar technique. The limitation of a growth factor procedure is that they are basically extrapolation techniques and cannot be used to synthesize movements between zone pairs if the base year trips are zero. However, the Fratar technique is utilized regularly for projecting through trips in an urban area and sometimes even for external-internal trips.

A somewhat complex trip distribution technique that was used by the Chicago Area Transportation Study, the transportation planning agency for Chicago, is the intervening opportunities model. The trip distribution theory underlying this model states that the probability that a trip originating in a zone i will find a destination in another zone j is proportional to possible destinations in zone j and also the possible destinations in zones closer to the origin of the trip. This model is rarely used by any agency today.

7.9 MODAL SPLIT

One widely researched step/phase of the sequential travel-modeling procedure for urban transportation planning is the modal split analysis, which involves the allocation of total person trips (by all modes) to the respective modes of travel, primarily automobile and public transit. It should be noted, however, that many studies for small and medium-sized urban areas omit this step by developing and using models for automobile trips only. This omission is justified in areas where transit trips constitute a very small fraction of total trips and are made primarily by captive riders.

Modal split models basically relate the probability of transit usage to explanatory variables or factors in a mathematical form. The empirical data necessary to develop these models usually are obtained from comprehensive O-D surveys in specific urban areas. In applying these models to predict the future transit usage, one must make the implicit assumption that the variables which explain the present level of transit usage will do so in much the same manner in the future.

7.9.1 Factors Affecting Mode Choice

Factors that may explain a trip maker's choosing a specific mode of transportation for a trip are commonly grouped as follows:

Trip Maker Characteristics

Income
Car ownership
Car availability
Age

Trip Characteristics

Trip purpose—work, shop, recreation, etc.

Destination orientation—CBD versus non-CBD

Trip length

Transportation Systems Characteristics

Waiting time

Speed

Cost

Comfort and convenience

Access to terminal or transfer location

7.9.2 Categories of Modal Split Models

The possible sequence of different types of modal split models with respect to the other steps of travel-modeling procedure is shown in Figure 7.3.

Predistribution (or Trip End) Models. This type of modal split model is used to separate the trip productions in each zone into the different modes to be distributed by mode-specific trip distribution models. The primary disadvantage of these models is that they cannot include variables related to transportation system characteristics. Predistribution models are not commonly used.

Postdistribution (or Trip Interchange) Models. This type of modal split model is very popular because it can include variables of all types. However, conceptually it requires the use of a multimodal trip distribution model and currently such distribution models are not used commonly. Figure 7.6 illustrates the sequence of application of a postdistribution model.

Simultaneous Trip Distribution and Modal Split Models. This type of model strives to estimate the number of trips between two zones by specific modes in one step directly following the trip generation phase. Conceptually and theoretically this type of a model has a sound basis, but it is not commonly used at this time.

7.9.3 Developing a Modal Split Model

Modal split models are developed from observed data on trip making available from home-interview surveys. The analysis involves the processing of a variety of data for both demand and supply.

Aggregate Model. Modal split models of the 1960s and early 1970s in most cases were based on an aggregate approach, which examined the mode choice of trip makers and their trips in groups based on similar socioeconomic and/or trip characteristics. These mode choice models usually involved two modes only: auto and transit. A detailed stratification scheme was used, and the share of each mode was determined for each stratified group of trips, which was then correlated with selected independent variables. The dependent variable was percent transit applicable to a group of trips of similar characteristics made by similar trip makers. Commonly used independent variables included the ratio of travel time by transit to that by automobile; the ratio of travel cost by transit to that by automobile; and the ratio of accessibility by transit to that by automobile. The relationship of the dependent variable,

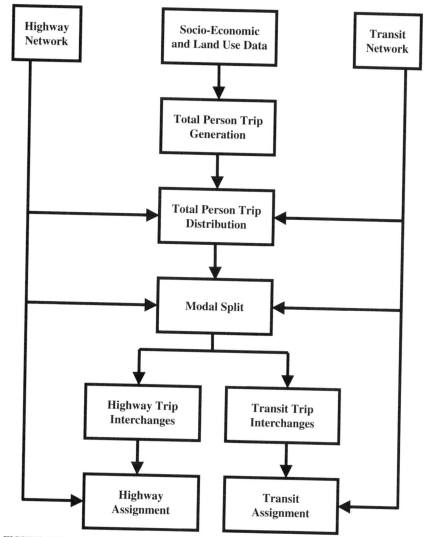

FIGURE 7.6   Travel simulation procedures with postdistribution modal split models.

percent transit, with the independent variable, say ratio of travel times, commonly was expressed by a set of curves. These curves sometimes were referred to as modal diversion curves.

The development of aggregate modal split models requires a large amount of data. Discussion of procedures used for developing different types of aggregate modal split models along with examples of these models can be found in Weiner (1969) and Chatterjee and Sinha (1975).

Disaggregate Behavioral Logit Models.   In late 1970s a new approach known as disaggregate behavioral method was developed and refined by a number of researchers. This approach recognized each individual's choice of mode for each trip instead of combining the trips in homogeneous groups. The underlying premise of this modeling approach is that an individual trip maker's choice of a mode of travel is based on a principle called utility maximization. Another premise is that the utility of using one mode of travel for a trip can be estimated using a mathematical function referred to as the utility function, which generates a numerical utility value/score based on several attributes of the mode (for the trip) as well as the characteristics of the trip maker. Examples of a mode's attributes for a trip include travel time and costs. The utilities of alternative modes can also be calculated in a similar manner. A trip maker chooses the mode from all alternatives that has the highest utility value for him or her.

A mathematical function that was used to represent the correlation of the probability of a trip maker's choosing a specific mode for a specific trip with a set of utility values is known as the logit function, and thus these models are also referred to as logit models. Binomial logit models deal with two modes, whereas multinomial logit models can deal with more than two modes. An example of the mathematical formulation of a multinomial logit model is given below:

$$p(k) = \frac{e^{U_k}}{\sum\limits_{x=1}^{n} e^{U_x}} \qquad (7.11)$$

where $p(k)$ = probability of using mode k
U_k = utility of using mode k
U_x = utility of using any particular mode x
n = number of modes to choose from

A special statistical procedure known as the maximum likelihood technique is used to derive an equation that combines different variable/factors in a meaningful way to calculate a utility (or disutility) value. The coefficients of each variable included in the utility (or disutility) function reflect certain behavioral aspects of a trip maker. Usually transportation-related variables used for a utility function include such items as access (or egress) time to (or from) transit stops/stations, wait time, line-haul time, and out-of pocket costs, and the coefficients of these variables are negative. Thus, the combined utility value comes out to be negative, which indicates disutility of using a mode. A trip maker's characteristics such as income are also built into the utility function.

One of the advantages of disaggregate mode choice models is that they do not need a full-scale O-D survey with household samples from every traffic zone. A carefully selected sample of 1,500 to 2,000 households would be adequate for developing these models. The mathematical theory related to multinomial logit models for mode choice analysis is fairly complex and beyond the scope of this chapter. Numerous articles and reports have been published on the subject of behavioral logit models, including Reichman and Stopher (1971) and McFadden (1978). Horowitz, Koppelman, and Lerman (1986) also have detailed information about disaggregate mode choice modeling.

7.10   TRAFFIC ASSIGNMENT

The task of the traffic assignment process is to develop the loadings, or user volumes, on each segment of a transportation network as well as the turning movements at intersections of the network. The user volumes may be the number of vehicles, the number of total persons, the number of transit riders, or any other units of travel demand that can be described by an origin and destination. For highway networks, user volumes are in terms of the number of vehicles, whereas for transit assignment the numbers of riders/passengers represent volumes. The relationship of the traffic assignment phase with respect to the other phases of the sequential travel simulation procedure is shown in Figure 7.3.

7.10.1 Inputs to Traffic Assignment Process

The two basic inputs to the assignment process are the transportation network and the zone-to-zone trip interchanges. The transportation network of automobiles, trucks, and taxis are analyzed separately from that of public transit systems, and usually traffic assignments are made separately for highway and transit systems. The typical inputs of a highway traffic assignment are shown in Figure 7.7. Transit network assignments are limited to internal person trips only.

7.10.2 Highway Network Analysis

For the purpose of computer analysis a highway network is represented by links and nodes and the traffic zones are represented by centroids, which are connected to the network. The characteristics of each link, such as the distance, speed, capacity, turn prohibitions, and functional classification, are coded. One of the primary tasks of network analysis is to determine the minimum time routes between each pair of centroids, and this task is performed utilizing Moore's algorithm. This algorithm does not require all possible routes between an

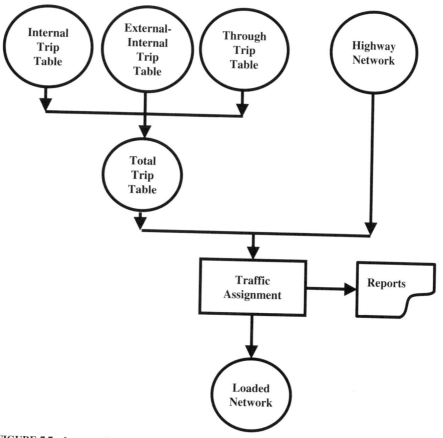

FIGURE 7.7 Inputs and outputs of traffic assignment.

origin and destination to be individually investigated to find the shortest route. Rather, a minimum tree is developed by fanning out from the origin to all other accessible nodes in increasing order of their impedance summation from the origin.

A tree is defined as the set of shortest routes from an origin to all other points in a network. An example of a path tree is shown in Figure 7.8. The travel time between a pair of zones is obtained by adding up the times on the individual links comprising the minimum time route, and this is repeated for every pair of zones. A skim tree usually refers to the interzonal travel time matrix.

It should be pointed out that the coding of a network for analysis with a computer-based algorithm requires a great deal of care and experience. There are many detailed issues and questions that come up with reference to such items as centroid connectors, representation of interchange ramps, whether to include certain roads or not, etc. Coding errors also can cause problems, and there are certain checks that can be done to minimize errors. Easa (1991) and Ismart (1990) discuss some of these issues and techniques for coding a network.

7.10.3 Alternative Techniques for Highway Traffic Assignment

A traffic assignment technique basically allocates the trips between each zone pair to the links comprising the most likely travel routes. The trips on each link are accumulated and

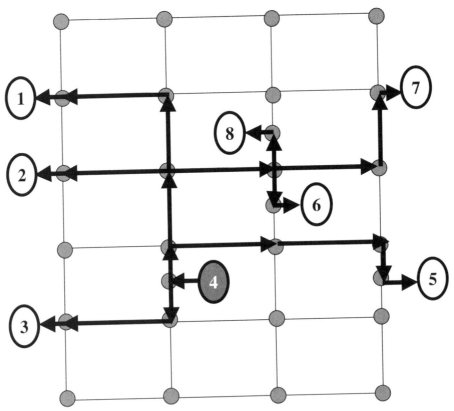

FIGURE 7.8 An example of a minimum path tree from home zone 4.

the total trips on each link are reported at the end of the assignment process. Alternative assignment techniques vary in terms of the criteria for route selection.

All-or-Nothing Assignment (AON).   This procedure assigns all trips between a zone pair to its minimum time route. This is the most commonly used technique, although the realism of its basic hypothesis is questionable. It should be noted that other, more advanced techniques make use of this technique as a part of their more involved procedure.

Diversion Techniques.   A diversion technique allocates the trips between a zone pair to more than one route. The most commonly used diversion technique considers two routes. One of these routes uses freeways and the other is the quickest alternative nonfreeway arterial route. The procedure assumes that a proportion of trips as determined by a diversion curve will be diverted from an arterial route to a freeway route based on the ratio of time via freeway with respect to time via quickest alternate arterial route. This procedure is documented in the *Traffic Assignment Manual* (1964) (Bureau of Public Roads 1964).

Diversion techniques were widely used in the early 1960s. Their advantage is in getting a spread of traffic between competing routes, and these techniques appear to be more realistic than the all-or-nothing assignment. With the introduction of the capacity restraint assignment procedure, diversion techniques are rarely used today for network assignments, although their usefulness should not be overlooked for corridor-type applications.

Capacity Restraint Assignment.   The capacity restraint procedure explicitly recognizes that as traffic flow increases the speed of traffic decreases. In this procedure several assignments are made based on the all-or-nothing concept. At the end of each assignment, however, the assigned volume on each link is compared with the respective capacity and the travel time is adjusted according to a given formula. A new set of minimum time routes is computed for the next assignment.

The original capacity restraint procedure developed by the then-Bureau of Public Roads, which is documented in the *Traffic Assignment Manual* (1964), assumed that the relationship between travel time and the volume peculiar to each link in a highway network can be expressed by the following equation:

$$t = t_0 \left[1 + 0.15 \left(\frac{V}{C_p} \right)^4 \right] \tag{7.12}$$

where t = travel time at which assigned volume can travel on the subject link
t_0 = base travel time at zero volume = travel time at practical capacity × 0.87
V = assigned volume
C_p = practical capacity

This process may be continued for as much iteration as desired. Usually four iterations are adequate. The analyst has the choice to accept the results of any single iteration. Sometimes the link volumes obtained from all iterations are averaged to produce the final result. This procedure strives to bring the assigned volume, the capacity of a facility, and the related speed into a proper balance.

Equilibrium Assignment.   Traffic assignment has been the subject of intense research for many years, and the research has resulted in several alternatives to the all-or-nothing and capacity restraint techniques, which were widely used during the 1960s and 1970s. One example of an assignment technique that was developed after the capacity restraint technique is the probabilistic multipath assignment technique, developed by Dial (1971). The most widely used procedure today is the user equilibrium assignment, which is based on the notion that traffic flows on network links are adjusted to an equilibrium state by the route-switching mechanism. That is, at equilibrium, the flows will be such that there is no incentive for route switching. As mentioned above, the travel time on each link changes with the flow, and therefore the travel time on several network paths changes as the link flows change. A stable condition is reached only when a traveler's travel times cannot be improved by unilaterally changing routes. This condition characterizes the user equilibrium (UE) condition.

The UE condition strives to optimize the utility of individual drivers. If the analysis is focused on optimizing a system-wide travel measure such as minimum aggregate travel time, then the problem is called a system optimal equilibrium (SOE) problem. Both UE and SOE problems rely on mathematical programming methods for developing the formulation and deriving a solution.

UE Problem Statement. Given a generalized function S_a that relates arc/link costs to traffic volumes, find the equilibrium traffic volumes on each arc/link of a directed graph, $G(N, A)$, with N nodes, A arcs, and a total number of origin-destination zones Z. This user equilibrium traffic assignment problem may be formulated in the following nonlinear optimization form.

$$\text{Minimize} \quad f(x) = \sum_a \int_0^{V_a} S_a(x) \, dx \tag{7.13}$$

$$\text{subject to} \quad V_a = \sum_i \sum_j \sum_k \delta_{ij}^{ak} x_{ij}^k \tag{7.14}$$

$$\sum_k x_{ij}^k = T_{ij} \tag{7.15}$$

$$x_{ij}^k \geq 0 \tag{7.16}$$

where, i = subscript indicates an origin zone/node $i \in Z$
$\quad\quad j$ = subscript indicates a destination zone/node $j \in Z$
$\quad\quad k$ = indicates a path between the origin zone (root) i and the destination zone j
$\quad\quad a$ = subscript for link/arc, $a \in A$
$\quad\quad u$ = subscript for volume category, $u \in U$
$\quad\quad T_{ij}$ = number of trips (all modes) originated at i and destined to j
$\quad\quad C_a$ = capacity of arc a
$\quad\quad V_a$ = total volume in category u on arc a in current solution
$\quad\quad W_a$ = all-or-nothing volume of u trips on arc a in current solution
$\quad\quad S_a(V_a)$ = generalized travel time (cost) function (also known as the link performance function) on link a which is determined by total flow V_a on each link
$\quad\quad x_{ij}^k$ = number of total trips from i to j assigned to path k
$\quad\quad \delta_{ij}^{ak}$ = 1 if link a belongs to path k from i to j, 0 otherwise

The most common form of the link performance function used for equilibrium traffic assignment problems is shown in equation (7.17).

$$t = t_0 \left[1 + a \left(\frac{V}{C_p} \right)^b \right] \tag{7.17}$$

where a and b are constants (note that when $a = 0.15$ and $b = 4$, equation (7.17) reduces to the form of BPR function shown in equation (7.12)).

Even for very small networks, it is very difficult to obtain a mathematical solution to this nonlinear optimization problem. Frank and Wolfe (1956) developed a heuristic solution that decomposes the problem into a number of steps. The solution is popularly known as Frank-Wolfe decomposition of the user equilibrium problem. Presented in Figure 7.9 is a schematic of computer implementation of Frank-Wolfe decomposition.

After several iterations, the heuristic reaches a situation where the UE condition is satisfied. As can be seen in Figure 7.9, the user equilibrium assignment technique utilizes a

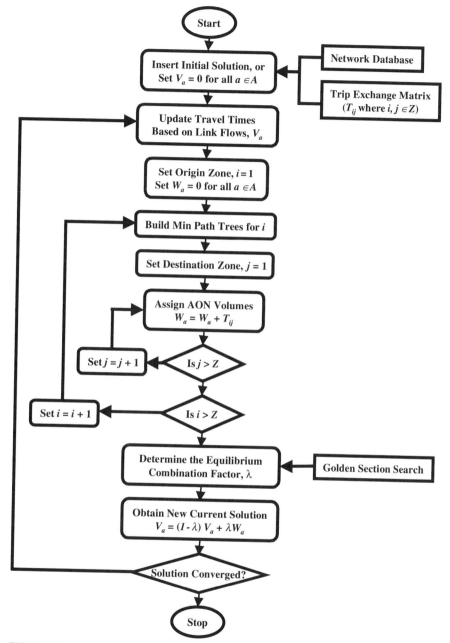

FIGURE 7.9 Frank-Wolfe decomposition of equilibrium traffic assignment problem.

convex combination method called the golden section search for direction finding. The mathematical concepts and optimization techniques underlying the user equilibrium assignment is fairly complex and beyond the scope of this chapter. More details about UE and other optimization solutions for network assignment problems can be found in Sheffi (1985).

7.11   MODEL ADJUSTMENTS AND USE OF SYNTHETIC/ BORROWED MODELS

When the models at different steps of the sequential modeling process are developed based on detailed data collected in the study area using large samples as prescribed for O-D and other survey procedures, the results at each step can be verified against survey-generated data that represent true values of the dependent variables of the models, such as trip ends (productions and attractions), trip-length frequencies, mode choice proportions, and traffic volumes on network links. The availability of detailed data allows model developers to make adjustments at respective steps as needed. For example, if during trip generation analysis it is found that the trip ends at a certain zone cannot be estimated closely by trip production or attraction models, a special investigation may be done, and an off-model procedure may be used for that zone. A special generator analysis for a college campus or an airport is an example of this type of a case. For gravity models, also, a special adjustment factor has been used in some cases to reflect the impact of certain physical features such as a river crossing on the attractiveness (or pull) of a group of traffic zones. Traffic counts taken at selected screen-lines are useful for comparing model-generated travel pattern with actual volumes of traffic crossing screen-lines and making adjustments, if needed.

At the traffic assignment step, network-related parameters such as travel speed and capacity of certain links may need adjustments in order to produce results that match closely existing traffic counts. For this purpose traffic counts taken at cut-lines are utilized for comparison with volumes generated by the traffic assignment model. For example, if it is found that along a freeway corridor the assigned volumes on freeway links are too high whereas those along a parallel arterial highway are too low in comparison to ground counts, the speeds and capacities of selected links along the corridor should be reexamined and adjusted. A good source of useful ideas for model adjustments including network coding is Ismart (1990).

Making adjustments to models to replicate the existing situation more closely is an important task of model development, and this requires some experience and sound understanding of how the models work. Adjustments to models assume a greater role in the case of synthetic or borrowed models. In this case a set of preexisting models for one area is transferred and adopted for another study area. Then the original values of key parameters such as trip generation rates and friction factors are adjusted, if necessary, to produce desired results. This process is referred to as model calibration. In the case of transferred/borrowed models, data from O-D survey are not available for checking the results at each step of the sequential modeling process, and the only data to check with are traffic volume counts on various road segments, which are to be replicated by the results of traffic assignment.

Synthetic models have been used widely, especially for small and medium-sized urban areas, in order to avoid or reduce the cost and time required for full-scale O-D surveys. Transportation planners in the North Carolina DOT's Planning and Research Branch have developed and used synthetic models for urban areas of a variety of sizes in North Carolina for many years. Chatterjee and Cribbins (1972), Bates (1974), Modlin (1974), Khasnabis and Poole (1975), and Chatterjee and Raja (1989) describe some these procedures. Sosslau et al. (1978) and Martin and McGuckin (1998) contain considerable information on the use of borrowed models and transferred parameters.

7.12   *TRAVEL DEMAND MODELING SOFTWARE PACKAGES*

The principles and the steps involved in the travel demand modeling process were first implemented in the form of computer software programs by the former Bureau of Public Roads and later refined by the Federal Highway Administration. This software package originally was called PLANPAC/BACKPAC, and it was primarily highway oriented. A report by FHWA (1974) contains the details of these programs. In early 1970s the Urban Transportation Planning System (UTPS) was developed by the former Urban Mass Transportation Administration to add the capability for transit planning. Later the scope of UTPS was expanded to include both highway and transit networks. UTPS was an IBM mainframe computer-based software system that has individual modules capable of performing a specific task. For example, the UMATRIX module performed matrix computations and the ASSIGN module performed AON or capacity restrained assignment. As with many computer programs in the 1960s and 1970s, the capabilities of early versions of UTPS were very limited. Performing a complete run of the four-step process took several days of work related to input preparation, debugging, and output analysis. Until the early 1990s several MPOs were still using the UTPS-based planning software.

With the advent and penetration of microcomputers in the early 1980s, different commercial versions of travel demand modeling (TDM) software were developed and marketed. Among the first was a software package called MINUTP, which was developed and marketed by COMSIS Corporation. MINUTP was an MS-DOS-based command-driven modeling package and similar to FHWA's PLANPAC software. Included among the command-driven TDM packages that were popular till the late 1990s and even in the early 2000s are TRANPLAN and QRS-II.

Since the advent of the Microsoft Windows operating system, travel demand modeling software landscape has changed even more dramatically. Current high-end TDM packages are not only capable of performing travel demand modeling, but are also compatible with geographic information systems (GIS). For example, TransCAD (Caliper Corporation) is a travel demand modeling package as well as a GIS software package. Other TDM packages include CUBE and TP+ by Citilabs, EMME/2, T-Model (Strong Concepts), and Saturn (UK).

7.13   *APPLICATIONS OF TRAVEL MODELS*

Depicted in Figure 7.10 is the traditional long-range planning process for a region (MPO) or sub-region. This process involves the identification of transportation-related problems followed by the determination of the future travel demand for a given situation. This in turn is followed by an attempt to find future transportation improvement that will meet the need of the future travel demand. Traditionally, in the planning process, the main criterion that is used objectively to evaluate alternative projects is congestion relief by capacity improvement, which typically involves building new highways, widening existing highways, and improving transit services. Land use-related alternatives also are examined. Travel forecasting models help assess the effectiveness of each alternative in reducing traffic congestion. Since traffic congestion of a serious nature usually occurs on major highways—primarily arterials—the travel forecasting procedure usually pays more attention to these highways, and this was reflected in network coding. Typically local roads and some minor collectors are not included in the network used for traffic assignment.

For the design of a new highway and/or the widening of an existing highway, the estimated traffic volume for the facility is the main item of interest that highway design engineers expect from travel forecasting models. For other related information, such as the proportion

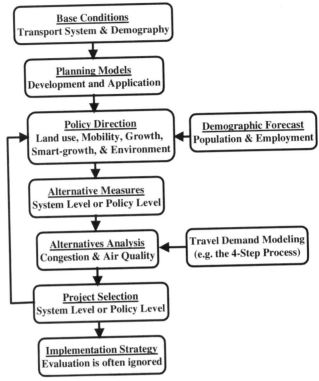

FIGURE 7.10   The long-range planning process.

of design hourly volume with respect to average daily traffic, directional split, and truck percentage, highway design engineers use other sources of information. However, this situation changed when more and more urban areas had to assess the air quality impact of highway networks as more detailed and accurate information related to travel was needed for air quality analysis.

7.14 TRANSPORTATION AND AIR QUALITY PLANNING

In recent years the planning process has given considerable emphasis on the assessment of the effect of transportation alternatives on the environmental consequences, especially air quality impacts. The United States Environmental Protection Agency (EPA) developed several versions of a model over time for estimating emission factors of air pollutants from mobile sources in terms of grams per mile. One of the recent versions of this model, MOBILE5, was used widely during the 1990s. The latest version, MOBILE6, is being used since early 2002. In the state of California a different emission factor model, called EMFAC, is used. These emission factor models need a variety of travel-related measures for the estimation of emissions from vehicular travel, and this need uncovered several deficiencies of the traditional travel forecasting models and led to various refinements and advancement of the modeling procedure. The integration of travel models with emission factors models is illustrated in Figure 7.11.

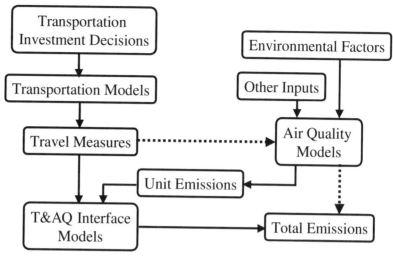

FIGURE 7.11   A schematic representation of TDM integrated with air quality modeling.

The travel-related inputs required for mobile source emissions estimation are discussed by Miller et al. (1992), and the deficiencies of the four-step models are examined by Stopher (1993). Another source of information on this subject is Chatterjee et al. (1997). A few examples of these weaknesses and refinements are discussed below.

7.14.1   Travel Speed

The amount of emissions released by vehicles when traveling varies considerably with speed, and therefore an accurate estimation of travel speed on each link of a highway network is important. However, the traffic assignment procedures usually focus on the accuracy of the assigned traffic volumes, and travel speeds are adjusted to produce better results for traffic volumes. In many cases the travel speeds generated by travel forecasting models are not accurate, and little effort was made in the past to improve the speed values because, as mentioned above, for the tasks of capacity-deficiency analysis and design of highway improvements, predicted traffic volume is the item of interest and there was no urgent need to improve the accuracy of speed estimates. During the 1990s, however, in response to the needs of emission factor models, considerable research was performed to improve the model-generated speed values. For this purpose feedback loops, as shown in Figure 7.12, and postprocessing procedures for speed calculations based on volume/capacity ratios were introduced, and these are widely used at this time. Alternative methods for feedback and iteration can be used, and there are also different convergence criteria to choose from. Boyce, Zhang, and Lupa (1994) discuss alternative methods of introducing feedback into the four-step procedure, and Dowling and Skabardonis (1992) provide good information on speed post-processing.

7.14.2   VMT

For the calculation of total emissions from mobile sources in an urban area, a reliable estimate of total VMT for each class of roads is needed. The estimation of VMT generated on

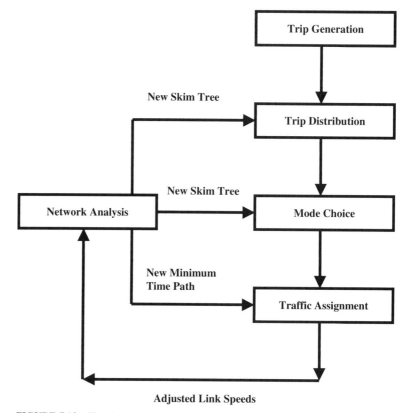

FIGURE 7.12   The planning process with feedback loops.

local and minor collector roads is important for this purpose. However, as discussed above, most of the local and minor roads usually are not included on the highway network used for travel demand modeling. Thus, network-based models are not capable of generating reliable estimates of VMT on local and minor collector roads. To fill this gap a few off-model procedures have been developed in recent years, and some of these are discussed by Chatterjee et al. (1997).

7.14.3   Vehicle Class Mix and Vehicle Age Distribution

For traditional transportation planning, which determines future needs of new and improved highways and transit services, detailed information regarding different types of vehicles operating on various highways is needed. Usually it is sufficient to use a rough estimate of the proportion of large trucks for capacity analysis. For air quality analysis, however, MOBILE5 provided separate emission rates for 8 different types of vehicles. The MOBILE6 moel requires input parameters for up to 28 different types of vehicles. The proportions of these classes of vehicles actually vary according to the functional class of roads and time of day. The data need becomes more demanding and complicated when it is recognized that emission rates depend not only on the size and weight of a vehicle but also on its age and mileage accumulation and that this information too has to be developed to take advantage of the MOBILE model's capabilities.

It is clear that commonly used travel forecasting models are not designed to develop detailed information on vehicle class and age and that other off-model procedures must be used for this purpose. However, the responsibility for developing this information lies with transportation planners who are working in the area of travel demand analysis.

7.14.4 Start versus Running Emissions

Earlier versions of MOBILE (version 5 and earlier) combined start emissions and running emissions into one composite emission factor. Owing to the realization that start emissions are much more pronounced than the running emission, the latest version of MOBILE (version 6) separates them from running emissions. Inputs on starts are provided in the form of start distributions by vehicle class.

7.14.5 Cold and Hot Soak Periods and Operating Modes

MOBILE 6 accounts for engine temperatures in the form of soak distributions (cold and hot soak distributions). The time lag between two successive trips of a vehicle determines the emissions at the beginning (start emissions) and end (evaporative emissions) of a trip. Inputs on this time lag are provided to the emission models in the form of soak period distributions. Venigalla and Pickrell (2002) describe procedures to obtain soak period inputs from large travel survey databases.

The operating modes of vehicles are related to engine temperature. Earlier versions of the MOBILE model classified operating modes into two broad categories—transient and hot stabilized modes. The transient mode is further categorized into two separate subcategories, cold start and hot start modes. EPA uses a few criteria based on engine soak period to determine the operating mode of a vehicle at a particular time. A few researchers used these time-based criteria in conjunction with travel surveys and innovative traffic assignment procedures to demonstrate how operating mode fractions can be estimated analytically (Venigalla, Miller, and Chatterjee 1995a, 1995b; Venigalla, Chatterjee, and Bronzini 1999; Chatterjee et al. 1996).

7.14.6 The Role of Transportation Planners in Air Quality Modeling

More and more the regulatory burdens of air quality modeling related to transporation projects are being shouldered by transportation planners. Air quality models are constantly being updated with new knowledge gained on transportation-related emissions. The current state of the art in emissions modeling requires more from the transportation planning community than ever before. For example, the concept of trip ends and trip chaining is easily extended to deriving travel-related inputs to the emission factor models (Chalumuri 2003). While model-improvement efforts undoubtedly improve the state of the practice, additional burdens are placed on the transportation modeling community to develop innovative methods to derive travel-related and vehicle activity inputs to the emissions model. In order to accommodate the needs of transportation-related air quality modeling in the foreseeable future, transportation planners are expected to develop new methods or adapt existing methods.

7.15 REFERENCES

Bates, J. W. 1974. "Synthetic Derivation of Internal Trips for Small Cities." *Transportation Research Record* 526, 93–103.

Boyce, D. E., Y. Zhang, and M. R. Lupa. 1994. "Introducing 'Feedback' into Four-Step Travel Forecasting Procedure Versus Equilibrium Solution of Combined Model." *Transportation Research Record* 1443, 65–74.

Bureau of Public Roads. 1964. *Traffic Assignment Manual.* U.S. Department of Commerce, Urban Planning Division, Washington, DC.

Cambridge Systematics, Inc., and Barton-Aschman Associates, Inc. 1996. *Travel Survey Manual.* Prepared for U.S. Department of Transportation and U.S. Environmental Protection Agency, Travel Model Improvement Program (TMIP), July.

Chalumuri, S. *Emersion Impacts of Personal Travel Variables.* A thesis submitted in partial fulfillment of Master of Science degree in Civil and Infrastructure Engineering Department, George Mason University, Fairfax, VA, August 2003.

Chatterjee, A., and P. D. Cribbins. 1972. "Forecasting Travel on Regional Highway Network." *Transportation Engineering Journal of ASCE* 98(TE2):209–24.

Chatterjee, A., and S. Khasnabis. 1973. "Category Models—A Case for Factorial Analysis." *Traffic Engineering,* October, 29–33.

Chatterjee, A., and M. Raja. "Synthetic Models for Through Trips in Small Urban Areas." *Journal of Transportation Engineering,* September, 537–55.

Chatterjee, A., and K. C. Sinha. 1975. "Mode Choice Estimation for Small Urban Areas." *Transportation Engineering Journal of ASCE* 101(TE2):265–78.

Chatterjee, A., S. Khasnabis, and L. J. Slade. 1977. "Household Stratification Models for Travel Estimation." *Transportation Engineering Journal of ASCE* 103(TE1):199–213.

Chatterjee, A., D. R. Martinson, and K. C. Sinha. 1977. "Trip Generation Analysis for Regional Studies." *Transportation Engineering Journal of ASCE* 103(TE6): 825–41.

Chatterjee, A., M. A. McAdams, and F. J. Wegmann. 1983. "Non-Commitment Bias in Public Opinion on Transit Usage." *Transportation* 11:347–60.

Chatterjee, A., T. L. Miller, J. W. Philpot, T. F. Wholley, R. Guensler, D. Hartgen, R. A. Margiotta, and P. R. Stopher. 1997. *Improving Transportation Data for Mobile Source Emission Estimates.* Transportation Research Board, National Research Council, National Cooperative Highway Research Program Report 394.

Chatterjee, A., P. M. Reddy, M. M. Venigalla, and T. M. Miller. 1996. "Operating Mode Fractions on Urban Roads Derived by Traffic Assignment." *Transportation Research Record* 1520, 97–103.

Chatterjee, A., F. J. Wegmann, J. D. Brogan, and K. Phiu-Nual. 1979. "Estimating Truck Traffic for Analyzing UGM Problems and Opportunities." *ITE Journal* 49(5):24–32.

Dial, R. B. 1971. "A Probabilistic Multipath Traffic Assignment Model Which Obviates Path Enumeration." *Transportation Research* 5(2):83–111.

Dowling, R., and A. Skabardonis. 1992. "Improving Average Travel Speeds Estimated by Planning Models." *Transportation Research Record* 1366, 68–74.

Dresser, G. B., and D. F. Pearson. 1994. *Evaluation of Urban Travel Survey Methodologies.* Texas Transportation Institute Research Report 1235-10, Texas Department of Transportation, Austin.

Easa, S. M. 1991. "Traffic Assignment in Practice: Overview and Guidelines for Users." *Journal of Transportation Engineering* 117(6):602–23

Easa, S. M. 1993. "Urban Trip distribution in Practice. I: Conventional Analysis." *Journal of Transportation Engineering* 119(6):793–815.

Eash R. 1993. "Equilibrium Traffic Assignment with High-Occupancy Vehicle Lanes." *Transportation Research Record* 1394, 42–48.

Eash, R. W., B. N. Janson, and D. E. Boyce. 1979. "Equilibrium Trip Assignment: Advantages and Implications for Practice." *Transportation Research Record* 728, 1–8.

Federal Highway Administration (FHWA). 1974. *FHWA Computer Programs for Urban Transportation Planning.* U.S. Department of Transportation, FHWA, July.

———. 1975a. *Trip Generation Analysis.* U.S. Department of Transportation, FHWA, August.

———. 1975b. *Urban Origin-Destination Surveys.* U.S. Department of Transportation, FHWA, reprinted July.

Fischer, M. J. 2001. *Truck Trip Generation Data.* NCHRP Synthesis 298, National Cooperative Highway Research Program, Transportation Research Board.

Frank, M., and P. Wolfe. 1956. "An Algorithm for Quadratic Programming." *Naval Research Logistics Quarterly* 3:95–110.

Fratar, T. J. 1954. "Forecasting Distribution of Inter-Zonal Vehicular Trips by Successive Approximations." *Proceedings of Highway Research Board* 33.

Gallo, G., and S. Pallottino. 1988. "Shortest Path Algorithms." *Annals of Operations Research* 13:3–45.

Grecco, W. L., J. Spencer, F. J. Wegmann, and A. Chatterjee. 1976. *Transportation Planning for Small Urban Areas*. National Cooperative Highway Research Program Report 167, Transportation Research Board.

Hamberger, C. B., and A. Chatterjee. 1987. "Effects of Fare and Other Factors on Express Bus Ridership in a Medium-Sized Urban Area." *Transportation Research Record* 1108, 53–59.

Horowitz, J. L., F. S. Koppelman, and S. R. Lerman. 1986. *A Self-Instructing Course in Disaggregate Mode Choice Modeling*. Technology Sharing Program, U.S. Department of Transportation, December.

Ismart, D. 1990. *Calibration and Adjustment of System Planning Models*. U.S. Department of Transportation, Federal Highway Administration, Washington, DC, December.

Khasnabis, S., and M. R. Poole. 1975. "Synthesizing Travel Patterns for a Small Urban Area." *Traffic Engineering,* August, 28–30.

Makowski, G. G., A. Chatterjee, and K. C. Sinha. 1974. "Reliability Analysis of Origin-Destination Surveys and Determination of Optimal Sample Size." *Proceedings of the Fifteenth Annual Meeting of Transportation Research Forum* 4(1):166–76. [Also published by UMTA through the National Technical Information Service, PB 236-986/AS.]

Martin, W. A., and N. A. McGuckin. 1998. *Travel Estimation Techniques for Urban Planning*. NCHRP Report 365, Transportation Research Board.

Mayworm, P., A. M. Lago, and J. M. McEnroe. 1980. *Patronage Impacts of Changes in Transit Fares and Services*. Ecsometrics Inc., Prepared for U.S. Department of Transportation, UMTA Report RR 135-1, Washington, DC, September.

McFadden, D. L. 1978. "The Theory and Practice of Disaggregate Demand Forecasting for Various Modes of Urban Transportation." In *Proceedings of the Seminar on Emerging Transportation Planning Methods,* Office of University Research, U.S. Department of Transportation, Washington, DC, August.

Miller, T., A. Chatterjee, J. Everett, and C. McIlvaine. 1992. "Estimation of Travel Related Inputs to Air Quality Models." In *Transportation Planning and Air Quality,* Proceedings of the National Conference of the American Society of Civil Engineers, 100–25.

Modlin, D. G. 1974. "Synthetic Through Trip Patterns." *Transportation Engineering Journal of ASCE* 100(TE2):363–78.

Moore, E. F. 1959. "The Shortest Path Through a Maze." In *Proceedings of the International Symposium on the Theory of Switching*. Harvard University Computation Laboratory Annals 30. Cambridge: Harvard University Press, 285–92.

Ogden, K. W. 1992. *Urban Goods Movement: A Guide to Policy and Planning*. Aldershot: Ashgate.

Parody, T. E., and D. Brand. 1979. "Forecasting Demand and Revenue for Transit Prepaid Pass and Fare Alternatives." *Transportation Research Record* 719, 35–41.

Reichman, S., and P. R. Stopher. 1971. "Disaggregate Stochastic Models of Travel-Mode Choice." *Highway Research Record* 369, 91–103.

Sheffi, Y. 1985. *Urban Transportation Networks: Equilibrium Analysis with Mathematical Programming Methods*. Englewood Cliffs: Prentice-Hall.

Sosslau, A. B., A. B. Hassan, M. M. Carter, and G. V. Wickstrom. 1978. *Quick Response Urban Travel Estimation Techniques and Transferable Parameters: User Guide*. NCHRP Report 187, Transportation Research Board, Washington, DC.

Stopher, P. R. 1993. "Deficiencies of Travel-Forecasting Methods Relative to Mobile Emissions." *Journal of Transportation Engineering* 119(5):723–41.

Stopher, P. R., and H. A. Metcalf. 1996. *Methods for Household Travel Surveys*. NCHRP Synthesis 236, Transportation Research Board. Washington, DC.

Urban Transportation Systems Associates, Inc. 1972. *Urban Mass Transportation Travel Surveys*. Prepared for U.S. Department of Transportation, August.

Venigalla, M. M, and D. H. Pickrell. 1997. "Implication of Transient Mode Duration for High Resolution Emission Inventory Studies." *Transportation Research Record* 1587.

Venigalla, M. M., and D. H. Pickrell. 2002. "Soak Distribution Inputs to Mobile Source Emissions Modeling: Measurement and Transferability." *Transportation Research Record* 1815.

Venigalla, M. M., A. Chatterjee, and M. S. Bronzini. 1999. "A Specialized Assignment Algorithm for Air Quality Modeling." *Transportation Research—D,* January.

Venigalla, M. M., T. Miller, and A. Chatterjee. 1995a. "Alternative Operating Mode Fractions to the FTP Mode Mix for Mobile Source Emissions Modeling." *Transportation Research Record* 1462, 35–44.

———. 1995b. "Start Modes of Trips for Mobile Source Emissions Modeling." *Transportation Research Record* 1472, 26–34.

Voorhees, A. M. 1955. "A General Theory of Traffic Movement." *Proceedings of the Institute of Traffic Engineers,* 46–56.

Weiner, E. 1969. "Modal Split Revisited." *Traffic Quarterly* 23(1):5–28.

———. 1992. *Urban Transportation Planning in the United States: An Historical Overview,* rev. ed. Washington, DC: Office of the Secretary of Transportation, November.

Woodruff, R. B., D. J. Barnaby, R. A. Mundy, and G. E. Hills. 1981. *Market Opportunity Analysis for Short-Range Public transportation Planning: Method and Demonstration.* National Cooperative Highway Research Program Report 212, Transportation Research Board, Washington, DC, September.

CHAPTER 8
HIGHWAY CAPACITY

Lily Elefteriadou
Department of Civil Engineering,
The Pennsylvania State University,
University Park, Pennsylvania

8.1 INTRODUCTION

How much traffic can a facility carry? This is one of the fundamental questions designers and traffic engineers have been asking since highways have been constructed. The term "capacity" has been used to quantify the traffic-carrying ability of transportation facilities. The value of capacity is used when designing or rehabilitating highway facilities, to obtain design elements such as the required number of lanes. It is also used in evaluating whether an existing facility can handle the traffic demand expected in the future.

The definition and value for highway capacity have evolved over time. The *Highway Capacity Manual* (*HCM 2000*) is the publication most often used to estimate capacity. The current version of the *HCM* defines the capacity of a facility as "the maximum hourly rate at which persons or vehicles reasonably can be expected to traverse a point or a uniform section of a lane or roadway during a given time period, under prevailing roadway, traffic and control conditions" (*HCM 2000*, 2-2). Specific values for capacity are given for various types of facilities. For example, for freeway facilities capacity values are given as 2,250 passenger cars per hour per lane (pc/hr/ln) for freeways with free-flow speeds of 55 miles per hour (mph), and 2,400 pc/hr/ln when the free-flow speed is 75 mph (ideal geometric and traffic conditions).

For a long time, traffic engineers have recognized the inadequacy and impracticality of the capacity definition. First, the expression "maximum . . . that can reasonably be expected" is not specific enough for obtaining an estimate of capacity from field data. Secondly, field data-collection efforts have shown that the maximum flow at a given facility varies from day to day, therefore a single value of capacity does not reflect real-world observations.

The main objective of this chapter is to provide transportation professionals with an understanding of highway capacity and the factors that affect it, and to provide guidance on obtaining and using field values of capacity. The next part of this chapter discusses the history and evolution of capacity estimation. The third part discusses the factors that affect capacity. The fourth part presents uninterrupted flow capacity issues, while the fifth part presents interrupted flow capacity issues. The last part presents a vision for the future of defining and estimating capacity.

8.2 CAPACITY DEFINITION AND ESTIMATION METHODS

8.2.1 The *Highway Capacity Manual*—A Historical Perspective

The *Highway Capacity Manual* is the publication most often used to estimate capacity. The first edition of the *Highway Capacity Manual* (1950) defined three levels of roadway capacity: basic capacity, possible capacity and practical capacity. Basic capacity was defined as "the maximum number of passenger cars that can pass a point on a lane or roadway during one hour under the most nearly ideal roadway and traffic conditions which can possibly be attained." Possible capacity was "the maximum number of vehicles that can pass a given point on a lane or roadway during one hour, under prevailing roadway and traffic conditions." Practical capacity was a lower volume chosen "without the traffic density being so great as to cause unreasonable delay, hazard, or restriction to the drivers' freedom to maneuver under prevailing roadway and traffic conditions."

The second edition of the *Highway Capacity Manual* (1965) defined a single capacity, similarly to the "possible capacity" of the *HCM 1950.* "Basic capacity" was replaced by "capacity under ideal conditions," while "practical capacity" was replaced by a series of "service volumes" to represent traffic operations at various levels of service. It is interesting to note that in the *HCM 1965,* the second chapter is titled "Definitions" and begins as follows: "The confusion that has existed regarding the meaning and shades of meaning of many terms . . . has contributed . . . to the wide differences of opinion regarding the capacity of various highway facilities. . . . In fact, the term which is perhaps the most widely misunderstood and improperly used . . . is the word 'capacity' itself." Thus, the definition of the term "capacity" allowed for various interpretations by different traffic analysts, and there was a desire to clarify the term. In the *HCM 1965,* the definition of capacity was revised to read as follows: "Capacity is the maximum number of vehicles which has a reasonable expectation of passing over a given section of a lane or a roadway in one direction (or in both directions for a two-lane or three-lane highway) during a given time period under prevailing roadway and traffic conditions." This definition includes the term "reasonable expectation," which indicates that there is variability in the numerical value of the maximum number of vehicles. Subsequent editions and updates of the *HCM* (1985, 1994, and 1997) define capacity in a similar manner, with the most recent definition (*HCM 2000*) as stated in the introduction of this chapter. This most recent definition indicates there is an expected variability in the maximum volumes, but it does not specify when, where, and how capacity should be measured, nor does it discuss the expected distribution, mean, and variance of capacity.

Capacity values provided in the *HCM* have increased over time. For example, the *HCM 1950* indicated that the capacity of a basic freeway segment lane is 2,000 pc/hr/ln, while the *HCM 2000* indicates that capacity may reach 2,400 pc/hr/ln for certain freeway facilities.

In addition to the definition of capacity, the HCM has historically provided (beginning with the *HCM 1965*) relationships between the primary traffic characteristics (speed, flow, and density) which have been the basis of highway capacity analysis procedures, particularly for uninterrupted flow facilities. Figure 8.1 presents a series of speed-flow curves that are provided in the *HCM 2000* and illustrate the relationship between speed and flow for basic freeway segments, and for various free-flow speeds (FFS), ranging from 55 to 75 mph. As shown in Figure 8.1, speed remains constant for low flows and begins to decrease as flow reaches 1,300–1,750 pc/hr/ln. The capacity for facilities with FFS at or above 70 mph is 2,400 pc/hr/ln and decreases with decreasing FFS. For example, the capacity of a basic freeway segment with FFS 55 mph is expected to be 2,250 pc/hr/ln.

Figure 8.2 provides the respective flow-density curves for basic freeway segments. Similarly to Figure 8.2, capacity values are shown to vary for varying free-flow speeds. This figure clearly illustrates the assumption used in the development of these curves that capacity is reached when density is 45 passenger cars per mile per lane (pc/hr/ln).

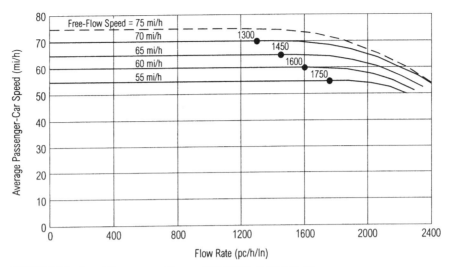

FIGURE 8.1   Speed-flow curves. (*Source: HCM 2000,* Exhibit 13-2.)

Both figures provide speed-flow-density relationships for undersaturated (i.e., noncongested) flow only. When demand exceeds the capacity of the facility, the facility will become oversaturated, with queues forming upstream of the bottleneck location. The *HCM 2000* does not provide speed-flow-density relationships for oversaturated conditions at freeways, because research has not been conclusive on this topic.

In summary, the definition of capacity within the *HCM* has evolved over time. There has been an implicit or, more recently, explicit effort to include the expected variability of maximum volumes in the capacity definition; however, there is no specific information in that document on where, when, and how capacity should be measured at a highway facility.

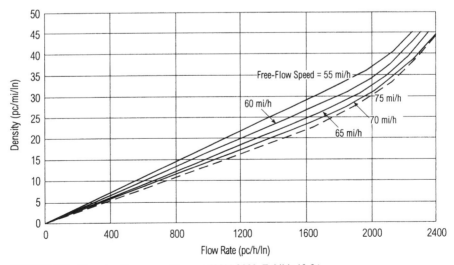

FIGURE 8.2   Flow-density curves. (*Source: HCM 2000,* Exhibit 13-3.)

8.2.2 Other Publications

For a long time, researchers have recognized the inadequacy and impracticality of this definition for freeway facilities. Field data collection of capacity estimates has shown that there is wide variability in the numerical values of capacity at a given site. This section summarizes the most recent literature findings regarding capacity definition and estimation.

Persaud and Hurdle (1991) discuss various definitions and measurement issues for capacity, including maximum flow definitions, mean flow definitions, and expected maximum flow definitions. They collected data at a three-lane freeway site over three days. In concluding, they recommend that the mean queue discharge flow is the most appropriate, partly due to the consistency the researchers observed in its day-to-day measurement.

Agyemang-Duah and Hall (1991) collected data over 52 days on peak periods to investigate the possibility of a drop in capacity as a queue forms, and to recommend a numerical value for capacity. They plotted prequeue peak flows and queue discharge flows in 15-minute intervals, which showed that the two distributions are similar, with the first one slightly more skewed toward higher flows. They recommended 2,300 pc/hr/ln as the capacity under stable flow and 2,200 pc/hr/ln for postbreakdown conditions, which corresponded to the mean value of the 15-minute maximum flows observed under the two conditions. The researchers recognized the difficulty in defining and measuring capacity, given the variability observed. Wemple et al. (1991) also collected near-capacity data at a freeway site and discuss various aspects of traffic flow characteristics. High flows (above 2,000 vehicles per hour per lane, vphpl) were identified, plotted, and fitted to a normal distribution, with a mean of 2315 vehicles per hour (vph) and a standard deviation of 66 vph.

Elefteriadou, Roess, and McShane (1995) developed a model for describing the process of breakdown at ramp-freeway junctions. Observation of field data showed that, at ramp merge junctions, breakdown may occur at flows lower than the maximum observed, or capacity flows. Furthermore, it was observed that, at the same site and for the same ramp and freeway flows, breakdown may or may not occur. The authors developed a probabilistic model for describing the process of breakdown at ramp-freeway junctions, which gives the probability that breakdown will occur at given ramp and freeway flows, and is based on ramp-vehicle cluster occurrence. Similarly to this research, Evans, Elefteriadou, and Natarajan (2001) also developed a model for predicting the probability of breakdown at ramp freeway junctions, which was based on Markov chains, and considered operations on the entire freeway cross-section, rather than the merge influence area.

Minderhoud et al. (1997) discuss and compare empirical capacity estimation methods for uninterrupted flow facilities and recommend the product limit method because of its sound theoretical framework. In this method, noncongested flow data are used to estimate the capacity distribution. The product-limit estimation method is based on the idea that each noncongested flow observation having a higher flow rate than the lowest observed capacity flow rate contributes to the capacity estimate, since this observation gives additional information about the location of the capacity value. The paper does not discuss transitions to congested flow, nor discharge flow measurements.

Lorenz and Elefteriadou (2001) conducted an extensive analysis of speed and flow data collected at two freeway-bottleneck locations in Toronto, Canada, to investigate whether the probabilistic models previously developed replicated reality. At each of the two sites, the freeway breakdown process was examined in detail for over 40 breakdown events occurring during the course of nearly 20 days. Examining the time-series speed plots for these two sites, the authors concluded that a speed boundary or threshold at approximately 90 km/hr existed between the noncongested and congested regions. When the freeway operated in a noncongested state, average speeds across all lanes generally remained above the 90 km/hr threshold at all times. Conversely, during congested conditions average speeds rarely exceeded 90 km/hr, and even then they were not maintained for any substantial length of time. This 90 km/hr threshold was observed to exist at both study sites and in all of the daily data samples evaluated as part of that research. Therefore, the 90 km/hr threshold was applied in the definition of breakdown for these sites. Since the traffic stream was observed to recover from small disturbances in most cases, only those disturbances that caused the average speed over all lanes to drop below 90 km/hr for a period of 5 minutes or more (15 consecutive 20-second intervals) were considered a true breakdown. The same criterion was used for recovery periods. The authors recorded the frequency of breakdown events at various demand levels. As expected, the probability of breakdown increases with increasing flow rate. Breakdown, however, may occur at a wide range of demands (i.e., 1,500–2,300 vphpl). The authors confirmed that the existing freeway capacity definition does not accurately address the transition from stable to unstable flow, nor the traffic-carrying ability of freeways under various conditions. Freeway capacity may be more adequately described by incorporating a probability of breakdown component in the definition. A suggested definition reads: "[T]he rate of flow (expressed in pc/hr/ln and specified for a particular time interval) along a uniform freeway segment corresponding to the expected probability of breakdown deemed acceptable under prevailing traffic and roadway conditions in a specified direction." The value of the probability component should correspond to the maximum breakdown risk deemed acceptable for a particular time period. A target value for the acceptable probability of breakdown (or "acceptable breakdown risk") for a freeway might initially be selected by the facility's design team and later revised by the operating agency or jurisdiction based on actual operating characteristics. With respect to the two-capacity phenomenon, the researchers observed that the magnitude of any flow drop following breakdown may be contingent upon the particular flow rate at which the facility breaks down. Flow rates may remain constant or even increase following breakdown. This may explain the fact that some researchers have observed the two-capacity phenomenon and others have not: it seems to depend on the specific combination of the breakdown flow and the queue discharge flow for the particular observation period. The paper does not discuss maximum prebreakdown flow, however, nor does it directly compare breakdown flows to maximum discharge flows for each observation day.

Elefteriadou and Lertworawanich (2002) examined freeway traffic data at two sites over a period of several days, focusing on transitions from noncongested to congested state, and developed suggested definitions for these terms. Three flow parameters were defined and examined at two freeway bottleneck sites: the breakdown flow, the maximum pre-breakdown flow, and the maximum discharge flow. Figure 8.3 illustrates these three values in a time series of flow-speed data at a given site.

It was concluded that:

- The numerical value of each of these three parameters varies and their range is relatively large, in the order of several hundred vphpl.

- The distributions of these parameters follow the normal distribution for both sites and both analysis intervals.

- The numerical value of breakdown flows is almost always lower than both the maximum prebreakdown flow and the maximum discharge flow.

- The maximum prebreakdown flow tends to be higher than the maximum discharge flow in one site, but the opposite is observed at the other site. A possible explanation for this difference may be that geometric characteristics and sight distance may result in different operations under high- and low-speed conditions.

In summary, several studies have shown that there is variability in the maximum sustained flows observed, in the range of several hundred vphpl. Three different flow parameters have been defined (maximum prebreakdown flow, breakdown flow, and maximum discharge flow), any of which could be used to define capacity for a highway facility. The maximum values for each of these are random variables, possibly normally distributed. Prebreakdown flow is often higher than the discharge flow. The transition from noncongested to congested flow is

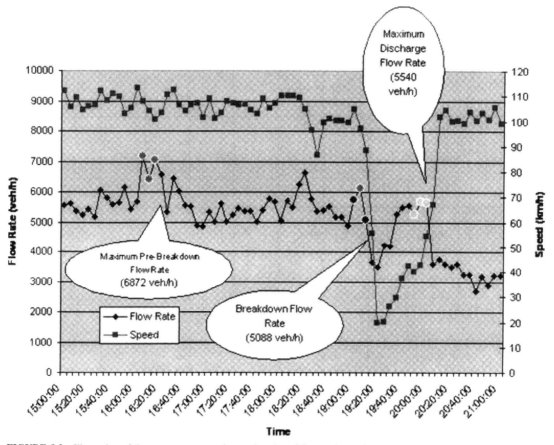

FIGURE 8.3   Illustration of three parameters on time-series plot of flow and speed.

probabilistic and may occur at various flow levels. The remainder of the chapter refers to these three as a group as "capacity," with references to a specific one when appropriate.

8.3   FUNDAMENTAL CHARACTERISTICS OF TRAFFIC FLOW

To understand the causes of variability in capacity observations, let us first review the fundamental characteristics of traffic flow. Figure 8.4 provides a time-space diagram with the trajectories of a platoon of five vehicles traveling along a freeway lane. The vertical axis shows the spacing (h_s, in ft) between each vehicle, while the horizontal axis shows their time headways (h_t, in sec). As shown, the spacing is different between each pair of vehicles if measured at different times (time 1 versus time 2). Similarly, the time headway between each pair of vehicles varies as they travel down the freeway (location 1 versus location 2). Flow can be expressed as:

$$\text{Flow} = 3600/\text{Average}\ (h_t) \tag{8.1}$$

When Average (h_t) is minimized, the flow is maximized (i.e., capacity is reached). Therefore,

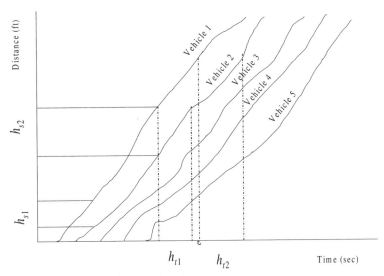

FIGURE 8.4 Vehicle platoon trajectories.

the distribution and values of h_t greatly affect the observed capacity of a facility. In Figure 8.4, the speed of each vehicle can be graphically obtained as the distance traveled divided by the respective time, or:

$$\text{Speed} = \text{Distance}/\text{Time} = h_s/h_t$$

$$h_t = h_s/\text{Speed} \tag{8.2}$$

Throughput is maximized when h_t is minimized, or as spacing decreases and speed increases. Therefore, spacing and speeds also have an impact on the maximum throughput observed.

In summary, the microscopic characteristics of traffic, i.e., the individual spacing, time headway, and speed of each vehicle in the traffic stream and their variability, result in variability in the field capacity observations. The remainder of this section discusses factors that affect these microscopic characteristics of traffic, and thus capacity.

8.4 THE THREE COMPONENTS OF THE TRAFFIC SYSTEM AND THEIR EFFECTS ON CAPACITY

The three components of the traffic system are the vehicle, the driver, and the highway environment. The vehicle characteristics, the driver characteristics, and the roadway infrastructure, as well as the manner in which the three components interact, affect the traffic operational quality and capacity of a highway facility. This section describes each of these three components and their specific aspects that affect capacity, along with their characteristics and interactions.

8.4.1 Vehicle Characteristics

As discussed above, highway capacity is a function of the speed, time headway, and spacing, which in turn are affected by the performance and size of the vehicles in the traffic stream.

The variability of these characteristics contributes in the variability in capacity observations. Vehicle characteristics that affect capacity include:

- *Wt/Hp (Weight-to-horsepower ratio):* The Wt/Hp provides a measure of the vehicle load to the engine power of the vehicle. It affects the maximum speed a vehicle can attain on steep upgrades (crawl speed), as well as its acceleration capabilities, both of which have an impact on capacity. Heavier and less powerful trucks generally operate at lower acceleration rates, particularly at steep upgrades. Slower-moving vehicles are particularly detrimental to capacity and traffic operational quality when there are minimal passing opportunities for other vehicles in the traffic stream.
- *Braking and deceleration capabilities:* The deceleration capability of a vehicle decreases with increasing size and weight.
- *Frontal area cross-section:* The aerodynamic drag affects the acceleration of the vehicle.
- *Width, length, and trailer-coupling:* The width of a vehicle may affect traffic operations at adjacent lanes by forcing other vehicles to slow down when passing. In addition, the width, length, and trailer coupling affects the off-tracking characteristics of a vehicle and the required lane widths, particularly along horizontal curves. The encroachment of heavy vehicles on adjacent lanes affects their usability by other vehicles and thus has an impact on capacity.
- *Vehicle height:* The vehicle height, even though not typically included in capacity analysis procedures, may affect the sight distance for following vehicles and thus may affect the resultant spacing and time-headways, and ultimately the capacity of a highway facility.

8.4.2 Driver Characteristics

Individual driver capabilities, personal preferences, and experience also affect highway capacity and contribute to the observed capacity variability. The driver characteristics that affect the capacity of a facility are:

- *Perception and reaction times:* These affect the car-following characteristics within the traffic stream. For example, these would affect the acceleration and deceleration patterns (and the trajectory) of a vehicle following another vehicle in a platoon. They also affect other driver actions such as lane changing and gap acceptance characteristics.
- *Selection of desired speeds:* The maximum speed at which each driver is comfortable driving at a given facility would affect the operation of the entire traffic stream. The effect of slower-moving vehicles in the traffic stream would be detrimental to capacity, particularly when high traffic demands are present.
- *Familiarity with the facility:* Commuter traffic is typically more efficient in using a facility than are drivers unfamiliar with the facility, or recreational drivers.

8.4.3 Roadway Design and Environment

The elements included under this category include horizontal and vertical alignment, cross-section, and traffic control devices.

- *Horizontal alignment and horizontal curves:* Vehicles typically decelerate when negotiating sharp horizontal curves. In modeling speeds for two-lane highways (Fitzpatrick et al. 1999), it has been shown that drivers decelerate at a rate that is proportional to the radius of the curve.

- *Vertical alignment and vertical curves:* Steep grades result in lower speeds, particularly for heavy trucks with low performance characteristics. Crawl speeds can be determined as a function of grade. Steep vertical crest curves would also affect sight distances and may act as local bottlenecks.
- *Cross-section:* The number and width of lanes, as well as the shoulder width, have been shown to affect speeds and thus the capacity of a highway facility. Provision of appropriate superelevation increases the speed and thus enhances the efficiency of a highway facility.
- *Traffic-control devices:* The clarity and appropriateness of traffic-control devices enhance the capacity of highway facilities.

Interactions between the three factors are also very important. For example, the effect of a steep upgrade on a heavy vehicles' performance is much more detrimental for capacity than generally level terrain. Also, the effect of challenging alignment would be much more detrimental to an unfamiliar driver than to a commuter.

8.5 CAPACITY OF UNINTERRUPTED FLOW FACILITIES

Uninterrupted flow facilities are defined as those where traffic is not interrupted by traffic signals or signs. These include freeway segments, weaving segments, ramp junctions, multi-lane highways, and two-lane highways. This section provides procedures for obtaining maximum throughput (i.e., capacity) estimates along uninterrupted flow facilities in field data collection, using microsimulation models. It provides guidance on observing and measuring maximum prebreakdown throughput, breakdown flow, and maximum discharge flow. The last part of the section discusses capacity estimation for uninterrupted flow facilities using the *HCM 2000*.

8.5.1 Field Data Collection

The four important elements that should be considered when observing breakdown and maximum throughput are site selection and measurement location, definition of breakdown, time interval, and sample size.

Regarding site selection, the site should be regularly experiencing congestion and break-down as a result of high demands and not as a result of a downstream bottleneck. For example, in Figure 8.5, which provides a sketch of a freeway facility with two consecutive

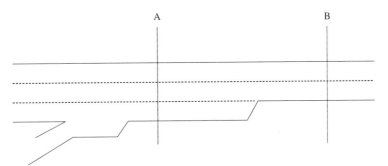

FIGURE 8.5 Freeway facility with two consecutive bottlenecks.

bottlenecks (one merge and one lane drop), the bottleneck at location B will result in queue backup into location A. The downstream bottleneck location (i.e., location B) should be the data-collection point.

To illustrate graphically the importance of site and location selection, Figure 8.6 provides the speed-flow relationships and the time-series of speed and flow for locations A and B across the freeway facility shown in Figure 8.5. Figure 8.6 illustrates the maximum throughput at location B is equivalent to its potential capacity, i.e., two-lane segment capacity. At location A the potential capacity is that of a three-lane segment, which, however, cannot be attained due to the downstream bottleneck. As soon as location B reaches capacity and breaks down, the queue created spills back into location A, which also becomes oversaturated (for additional discussion see May 1990). The speed time-series at location B shows the breakdown occurrence, which typically occurs with a relatively steep speed drop. At location A speed drops gradually as a result of the downstream breakdown.

The second important element when measuring maximum throughput is the definition and identification of breakdown. In a previous study Lorenz and Elefteriadou (2001) identified and defined breakdown at freeway merge areas as a speed drop below 90 km/hr, with duration of at least 15 minutes. Another study on weaving areas (Lertworawanich and Elefteriadou 2001) showed that the breakdown speed threshold exists at these sites as well, but has a different value (80 km/hr). Given that speed drop and its respective duration can uniquely identify the presence of breakdown, it is recommended that breakdown be quantitatively defined using these two parameters. As illustrated in Figure 8.6, at the bottleneck (location B, part (b)) the speed drop is typically sharp, and the breakdown can be clearly identified.

The third element that is important in clearly defining maximum throughput is the selection of an appropriate time interval. Time intervals that are typical in traffic operational analysis range between 5 and 15 minutes. Previous research has demonstrated that maximum throughput increases for smaller intervals, due to the general flow variability. Longer intervals result in averaging the flow over a longer time period, and thus the respective maximum throughput is lower.

The last element to be considered is the required sample size. Given the inherent variability of maximum throughput, it is important to observe an adequate number of breakdown

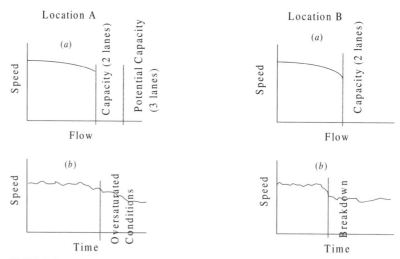

FIGURE 8.6   Speed-flow curves and time-series for freeway locations A and B.

events and the respective maximum prebreakdown and maximum discharge flows. Sample size-determination equations should be used to establish the required number of observations for the desired precision in the maximum throughput estimate.

Once these four elements are established, flow and speed data can be collected at the selected site(s), and time-series plots prepared (such as the one depicted in Figure 8.3) for each breakdown event. The breakdown can be identified based on the definition selected, and the respective breakdown flow can be obtained from the time series. Next, the maximum prebreakdown flows and the maximum discharge flows can be obtained for each breakdown event.

8.5.2 Additional Considerations for Obtaining Capacity from Microsimulation Models

The elements and procedure outlined above for field data collection can be followed when maximum throughput information is obtained through a microsimulation model. Additional considerations include simulation model selection and development of simulated demand patterns, which are discussed below.

Microsimulation model selection is a very broad topic and is only dealt here very broadly with regard to capacity and breakdown observations. The model selected should have the necessary stochastic elements and the capability to simulate breakdown as a random event. Stochastic elements include the vehicle and driver capabilities described above in section 8.4. Specifically, the acceleration and deceleration parameters for each vehicle, including the car-following models, should address and be calibrated for breakdown conditions. Similarly, the lane-changing algorithm of the model selected should consider and be calibrated for breakdown conditions.

The second consideration when using simulation modeling for capacity estimation is what demands to use and how to vary them so that breakdown is achieved. The most common technique is to run the model with incrementally higher demands, starting at a sufficiently low, below-capacity level. The analyst would need to load the network starting with relatively low demands and increasing them at constant intervals until breakdown is reached. Another, more complicated, technique is to develop random patterns of demand to simulate the demand patterns in the field. For both techniques the increments employed at each successive demand level would be a function of the desired interval in the capacity observations. Output data can then be collected on breakdown events and maximum throughput, similarly to the field data collection.

8.5.3 Capacity Estimates in the *HCM 2000*

The *HCM 2000* provides capacity estimates for basic freeway segments, ramp merge segments, weaving segments, multilane highways, and two-lane highways. The *HCM* provides, for each segment type and set of geometric conditions, a single value of capacity. For example, for freeway facilities, capacity values are given as 2,250 passenger cars per hour per lane (pc/hr/ln) for freeways with free-flow speeds of 55 mph, up to 2,400 pc/hr/ln when the free-flow speed is 75 mph (ideal geometric and traffic conditions). These values represent average conditions at similar sites around the United States, obtained based on general trends of maximum flows observed at various freeway locations. Note that the *HCM 2000* capacity definition is more closely aligned with the definition of maximum prebreakdown flow. The *HCM 2000* does not define breakdown flows and maximum discharge flows, nor does it provide estimates for these at various facility types.

8.6   THE CAPACITY OF INTERRUPTED FLOW FACILITIES

Interrupted flow facilities are those where traffic flow experiences regular interruptions due to traffic signs and signals. These include facilities such as signalized and unsignalized intersections, and roundabouts, all of which are discussed in this section.

8.6.1   Signalized Intersections

The capacity of a signalized intersection depends very much on the phasing and timing plan. Traffic flow on a signalized intersection approach is regularly interrupted to serve conflicting traffic. Thus, the capacity of the approach is a function of the amount of green given to the respective movements within a given time interval. For example, if the cycle length at a signalized intersection is 90 seconds and the eastbound traffic is given 45 seconds of green, then the total amount of time that the approach is given the right-of-way within an hour is:

Total green time = No. of cycles per hour × Green time

= 3600/90 × 45 = 1800 sec

This corresponds to 1800/3600 = 50 percent of the full hour.

In addition to this time restriction of right-of-way-availability, the time headways observed at the stop-line of a signalized intersection approach follow a different pattern as the traffic signal changes from green to yellow to red and then back to green. Figure 8.7 illustrates a series of consecutive time headways (also called discharge headways when referring to queued vehicles at signalized intersection approaches) observed as vehicles depart from a single-lane approach. The horizontal axis in the figure represents time (in seconds), while the vertical dashed lines represent events (signal changes and vehicle departures). At the beginning of the green there were 10 vehicles queued at the approach. The volume during this green interval is 16 vehicles. As shown, the first discharge headway, measured from the beginning of the green to the departure of the first vehicle in the queue, is also the largest among these first 10 queued vehicles. Subsequent discharge headways gradually decrease until they reach the "saturation headway" (s_h) level for the approach, which is the minimum time headway observed under conditions of continuous queuing—in this example 2 seconds. Next, saturation flow can be defined as the maximum throughput for the signalized intersection approach, if the approach were given the green for a full hour. The saturation flow for a single lane can be calculated as:

Saturation flow (vehicles per hour of green) = 3600/Average (s_h)

= 3600/2 = 1800 vehicles per hour of green per lane (vphgpl)   (8.3)

Considering that each approach does not have the green for the full hour, the maximum throughput that can be achieved at a signalized intersection approach depends on the percent of time the approach is given the green. Mathematically:

Max. throughput per lane = Percent eff. green × Saturation flow

$$= g/C \times 3600/s_h = (3600 \times g)/(s_h \times C)   (8.4)$$

where g is the duration of effective green for the approach and C is the cycle length for the intersection.

The effective green is defined here as the time the approach is effectively used by this movement, or the actual green time minus the lost time experienced due to start-up and acceleration of the first few vehicles (for discussion of lost time and its precise definition,

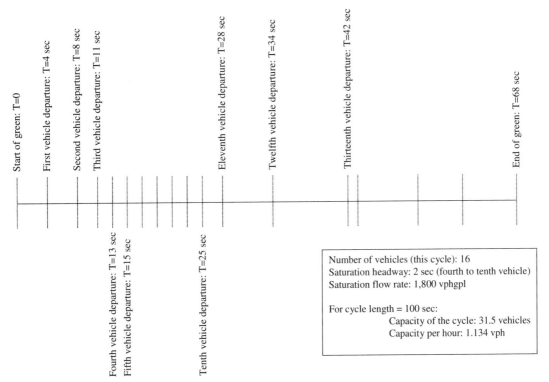

FIGURE 8.7   Time headways at a single-lane signalized approach.

consult the *HCM 2000*). In the example of Figure 8.7, the lost time is the extra time incurred by the first 10 vehicles after subtracting the saturation headways for each of these vehicles. Thus, the effective green time is:

$$\text{Effective green time} = \text{Actual green time} - \text{Lost time}$$

$$= 68 \text{ sec} - (25 \text{ sec} - 10 \text{ vehicles} \times 2 \text{ sec})$$

$$= 63 \text{ sec}$$

In the example of Figure 8.7, the maximum throughput (i.e., capacity), assuming that the duration of the effective green remains constant through the entire hour and that the cycle length is 100 seconds, is estimated from equation (8.4) as:

$$\text{Capacity} = (3600 \times 63 \text{ sec})/(2 \text{ sec} \times 100 \text{ sec}) = 1,134 \text{ vph}$$

An alternative method to estimate capacity within the hour is to estimate the capacity per cycle and multiply by the number of cycles within the hour. In the example of Figure 8.7, the capacity per cycle is:

$$\text{Capacity per cycle} = \text{Eff. green}/\text{Saturation headway}$$

$$= 63 \text{ sec}/2 \text{ sec} = 31.5 \text{ vehicles}/\text{cycle}$$

The total number of cycles in the hour is:

$$\text{Number of cycles} = 3600/100 \text{ sec} = 36 \text{ cycles/hr}$$

Thus, the capacity within an hour is:

$$\text{Capacity} = \text{Capacity per cycle} \times \text{Number of cycles}$$

$$= 31.5 \text{ vehicles per cycle} \times 36 \text{ cycles} = 1,134 \text{ vph}$$

8.6.2 Two-Way Stop-Controlled (TWSC) Intersections and Roundabouts

The operation of TWSC intersections and roundabouts (yield-controlled) is different than that of signalized intersections in that drivers approaching a stop or yield sign use their own judgment to proceed through the junction through conflicting traffic movements. Each driver of a stop- or yield-controlled approach must evaluate the size of gaps in the conflicting traffic streams, and judge whether he/she can safely enter the intersection or roundabout. The capacity of a stop- or yield-controlled movement is a function of the following parameters:

- The availability of gaps in the main (noncontrolled) traffic stream. Note that gap is defined in the *HCM 2000* as time headway; elsewhere in the literature, however, gap is defined as the time elapsing between the crossing of the lead vehicle's rear bumper and the crossing of the following vehicle's front bumper. In other words, the *HCM 2000* gap definition (which will be used in this chapter) includes the time corresponding to the crossing of each vehicle's length. The availability of gaps is a function of the arrival distribution of the main traffic stream.

- The gap acceptance characteristics and behavior of the drivers in the minor movements. The same gap may be accepted by some drivers and rejected by others. Also, when a driver has been waiting for a long time, he or she may accept a shorter gap, having rejected longer ones. The parameter most often used in gap acceptance is the critical gap, defined as the minimum time headway between successive major street vehicles, in which a minor-street vehicle can make a maneuver.

- The follow-up time of the subject movement queued vehicles. The follow-up time is the time headway between consecutive vehicles using the same gap under conditions of continuous queuing, and it is a function of the perception/reaction time of each driver.

- The utilization of gaps in the main traffic stream by movements of higher priority, which results in reduced opportunities for lower-priority movements (this is not applicable for roundabouts because there is only one minor movement).

The example provided below is a simplified illustration of the capacity estimation process for a stop- or yield-controlled movement. The capacity of the northbound (NB) minor through movement of Figure 8.8 will be estimated based on field measurements at the intersection. There is only one major traffic stream at the intersection (eastbound, EB) and one minor street movement (NB).

The critical gap was measured to be 4 seconds. It is assumed that any gap larger than 4 seconds will be accepted by every driver, while every gap smaller than 4 seconds will be rejected by every driver. The follow-up time was measured to be 3 seconds. Thus, for two vehicles to use a gap, it should be at least:

$$4 \text{ sec (first vehicle)} + 3 \text{ sec (second vehicle)} = 7 \text{ sec}$$

Conversely, the following equation can be used to estimate the maximum number of vehicles that can use a gap of size X:

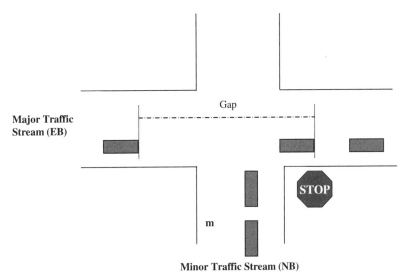

FIGURE 8.8 Capacity estimation for the NB through movement.

$$\text{Number of vehicles} = 1 + (\text{Gap Size } X - \text{Critical Gap})/\text{Follow-up time} \qquad (8.5)$$

Table 8.1 summarizes the field data and subsequent calculations for the intersection of Figure 8.8, and provides the capacity estimate. Column (1) of Table 8.1 provides the gaps measured in the field. Column (2) indicates whether the gap is usable, i.e., whether it is larger than the critical gap. Column (3) uses equation (8.5) to provide the number of vehicles

TABLE 8.1 Capacity Estimation for a Stop-Controlled Movement

	Gap size (sec) Col (1)	Usable Gap? (Y or N) Col (2)	Vehicles in NB movement that can use the gap (veh) Col (3)
	5	Y	1
	2	N	0
	7	Y	2
	10	Y	3
	4	Y	1
	2	N	0
	12	Y	3
	17	Y	5
	7	Y	2
	3	N	0
	2	N	0
	3	N	0
	13	Y	4
	16	Y	5
TOTAL	103 sec	9 usable gaps	26 vehicles

that can use each of the usable gaps. The total number of vehicles that can travel through the NB approach is the sum provided at the bottom of the column (3) (26 vehicles over 103 seconds), or (3600/103) × 26 = 909 vph, which is the capacity of the movement.

The *HCM 2000* methodology for estimating the capacity of TWSC intersections is based on the principles outlined above, using mathematical expressions of gap distributions and probability theory for establishing the use of gaps by higher-priority movements.

8.7 SUMMARY AND CLOSING REMARKS

This chapter provided an overview of the concept of highway capacity, discussed the factors that affect it, and provided guidance on obtaining and using field values of capacity.

The definition of capacity within the *HCM* has evolved over time. Even in the earliest publications related to highway capacity there is a general recognition that the "maximum throughput" of a highway facility is random, and thus there has been an effort to include the expected variability of maximum volumes in the capacity definition. Recently, several studies have used field data and have proven that there is variability in the maximum sustained flows observed, in the range of several hundred vphpl. The cause of this variability lies in the microscopic characteristics of traffic, i.e., the individual spacing, time headway, and speed of each vehicle in the traffic stream and their variability. This chapter also provided the fundamental principles of capacity estimation for uninterrupted and interrupted flow facilities, and provided guidance on obtaining capacity estimates in the field and using simulation.

The discussion provided in this chapter is not exhaustive and is only intended to provide the fundamental principles of capacity estimation for various highway facilities. Detailed methodologies for estimating highway capacity for various facilities are provided in the *HCM 2000,* as well as elsewhere in the literature.

8.8 REFERENCES

Agyemang-Duah, K., and F. L. Hall. "Some Issues Regarding the Numerical Value of Highway Capacity," Highway Capacity and Level of Service—International Symposium on Highway Capacity, Karlsruhe, July 1991, pp. 1–15.

Bureau of Public Roads. 1950. *Highway Capacity Manual.* U.S. Department of Commerce, Bureau of Public Roads, Washington, DC.

Elefteriadou, L., and P. Lertworawanich. 2002. "Defining, Measuring and Estimating Freeway Capacity." Submitted to the Transportation Research Board.

Elefteriadou, L., R. P. Roess, and W. R. McShane. 1995. "The Probabilistic Nature of Breakdown at Freeway-Merge Junctions." *Transportation Research Record* 1484, 80–89.

Evans, J., L. Elefteriadou, and G. Natarajan. 2001. "Determination of the Probability of Breakdown on a Freeway Based on Zonal Merging Probabilities." *Transportation Research B* 35:237–54.

Fitzpatrick, K., L. Elefteriadou, D. Harwood, R. Krammes, N. Irizzari, J. McFadden, K. Parma, and J. Collins. 1999. *Speed Prediction for Two-Lane Rural Highways.* Final Report, FHWA-99-171, 172, 173 and 174, June, www.tfhrc.gov/safety/ihsdm/pdfarea.htm.

Flannery, A., L. Elefteriadou, P. Koza, and J. McFadden. 1998. "Safety, Delay and Capacity of Single-Lane Roundabouts in the United States." *Transportation Research Record* 1646, 63–70.

Highway Research Board. 1965. *Highway Capacity Manual.* Highway Research Board, Special Report 87, National Academies of Science, National Research Council Publication 1328.

Lertworawanich, P., and L. Elefteriadou. 2001. "Capacity Estimations for Type B Weaving Areas Using Gap Acceptance." *Transportation Research Record* 1776, 24–34.

———. 2003. "A Methodology for Estimating Capacity at Ramp Weaves Based on Gap Acceptance and Linear Optimization." *Transportation Research B* 37(5):459–83.

Lorenz, M., and L. Elefteriadou. 2001. "Defining Highway Capacity as a Function of the Breakdown Probability." *Transportation Research Record* 1776, 43–51.

May, A. D. 1990. *Traffic Flow Fundamentals*. Englewood Cliffs: Prentice-Hall.

Minderhoud, M. M., H. Botma, and P. H. L. Bovy. "Assessment of Roadway Capacity Estimation Methods," Transportation Research Record 1572, National Academy Prem, 1997, pp. 59–67.

Persaud, B. N., and V. F. Hurdle. "Freeway Capacity: Definition and Measurement Issues," Highway Capacity and Level of Service—International Symposium on Highway Capacity, Karlsruhe, July 1991, pp. 289–308.

Transportation Research Board (TRB). 2000. *Highway Capacity Manual*. National Research Council, TRB, Washington DC.

Wemple, E. A., A. M. Morris, and A. D. May. "Freeway Capacity and Flow Relationships," Highway Capacity and Level of Service—International Symposium on Highway Capacity, Karlsruhe, July 1991, pp. 439–456.

CHAPTER 9
TRAFFIC CONTROL SYSTEMS: FREEWAY MANAGEMENT AND COMMUNICATIONS

Richard W. Denney, Jr.
Iteris Corporation, Sterling, Virginia

9.1 INTRODUCTION

The roadway system represents an asset, owned by the public or occasionally by private concerns, that provides a means of mobility for people, goods, and services in relatively small vehicles. Thus, the roadway network services a large number of disparate trips. Traffic is a fluid composed of particles, each of which is controlled by a human operator seeking to travel from an origin to a destination. Each traveler in a road network makes decisions about how to use that network based on a limited understanding, through experience or through communication from network observers, of the conditions in that network. These decisions encompass three basic choices: when to depart, what route to take, and whether to use a car or some other transportation mode.

Roadways require huge physical construction projects and are therefore one of the costliest of public works activities. They consume land, require reshaping that land, impose difficult drainage challenges, displace landowners and residents, create boundaries, and demand extensive and expensive maintenance. Consequently, public agencies are often pulled between competing influences: building more roads to relieve congestion or building fewer roads to avoid negative impact on quality of life. In this environment, agencies are strongly motivated to squeeze every possible bit of capacity a roadway network can offer. Doing so requires active traffic management, using a variety of techniques to manage demand and capacity.

In this chapter, the author will present freeway management as a congestion-relief activity, wholly tied to the demand and capacity relationship; strategies for optimizing capacity and for managing demand; and the infrastructure required to perform this management. The author will provide a description of freeway management activities in real time.

One of the most expensive and difficult components of real-time traffic management is establishing the ability to communicate with the management infrastructure in real time. Thus, the author will present a discussion of communications methods and technology as it relates to traffic management, to provide transportation engineers a basic understanding that will help them work effectively with communications professionals.

The author strongly advocates the *systems engineering approach*. In this approach, an engineer first defines user needs, which are written down in a concept of operations that describes how those needs are addressed in practice (not how the system will address them, but how humans address them). From those needs emerges a series of requirements that the system must meet. Finally, testing proves that the system meets the requirements. A fundamental principle of this approach is *traceability*. At each step of this process, the relevant features are explicitly traced back to the relevant features at the previous step. Thus, tests trace directly to requirements and requirements trace directly to user needs. Based on this concept, the author will continue to refer back to the primary goal of freeway management to modify either demand or capacity to reduce congestion.

9.2 FREEWAY MANAGEMENT

9.2.1 What Is Freeway Management?

Simply put, freeway management is a collection of activities with the goal of minimizing congestion in a freeway network. Congestion is the result of demand exceeding capacity, so strategies to achieve the goal include increasing capacity and decreasing demand. While every freeway management activity traces back to this goal through these strategies, this simple statement is still perhaps a little too simple.

Freeway managers have another goal that is not so often stated, and that is to help travelers reconcile themselves to unavoidable congestion and to manage their lives around the resulting delay. The doctor may not be able to effect a cure, but the fully informed patient can at least make plans.

Freeway managers are concerned with congestion that results from routine network usage, such as commuter demand, and congestion from unexpected changes in capacity as a result of traffic crashes, stalled vehicles, construction, and other incidents. Thus, freeway management includes both short-term strategies for incident management and long-term strategies for routine congestion.

9.2.2 Demand and Capacity: Congestion

Congestion is the natural result of demand exceeding capacity. A simple example is water pouring through a funnel. If more water pours through the funnel than the spout can pass, the excess water builds up in the funnel. Thus, the capacity of water flow through the funnel is controlled only by the diameter of the hole at its tip. A bigger funnel may store more water, but it will not move more water unless that hole is larger. If the objective is to keep the water from filling up the funnel, we must either enlarge the hole or pour in less water.

Capacity. The capacity of a freeway results from the operation of a fixed physical roadway. In the simplest case, a lane of a freeway carries at most a certain number of cars in an hour. In past years, it was assumed that, on average, drivers would not tolerate being closer together than 2-second headways. Headway is the time that passes between the front edge of one car passing a point and the front edge of the following car passing that point. The dynamics of the relationship between two cars is called *car-following theory*. Car-following rules are the subject of another chapter, but in this discussion it is enough to be able to characterize headways. As drivers become more aggressive, or more accustomed to congestion, their tolerance of close headways tightens. The traditional value of 2 seconds corresponds to a flow of 1,800 vehicles per hour (3,600 seconds/hour divided by 2 seconds/vehicle). Even in small cities, the tolerance is tighter than that, and the minimum capacity expected in most current situations is at least 2,000 vehicles per hour. The *Highway Capacity Manual* reveals that in certain circumstances one might see as many as 2,400 vehicles per hour, and higher flows than that for shorter periods.

Roadways, like rivers, are subject to constraints on capacity resulting from physical interruptions. For example, one of the busiest freeways in the United States, the Southwest Freeway in Houston, suffered a long-term capacity bottleneck where the freeway turned and rose over a bridge that spanned a railroad mainline. The combination of the upward slope (which only occurs at overpasses in flat Houston) and the horizontal curve created an obvious section of reduced capacity.

But most capacity constraints are less subtle. The typical capacity constraint occurs when a lane is dropped and the number of cars in each of the remaining lanes is higher than before the lane drop. This feature occurs in many places: in suburban areas where the freeway narrows at what was once the edge of urban development; in freeway interchanges where two-lane connectors merge into a single lane, at entrance ramps where two lanes merge into a single lane, and so on.

Most capacity problems occur downstream from merge points, where two roadways come together and are squeezed into a downstream roadway with less capacity than the total of the two entering roadways. For example, an entrance ramp of one lane merges into a four-lane freeway. The freeway still contains four lanes after the merge, but the capacity upstream from the merge was provided by five lanes.

The capacity constraints that most concern motorists and freeway managers alike are those that are not part of the physical features of the roadway. These include construction activities, traffic accidents, stalled vehicles, law-enforcement activities, and bad weather. Even a stalled vehicle parked on the shoulder can impose a slight reduction in capacity. These temporal reductions of capacity consume most of the attention of freeway managers because they are the easiest to manage and because they have an unexpected impact on travelers.

Demand. Travelers have a need to move from one place to another. Their trips may be nondiscretionary and demand a specific arrival time (such as going to work), or discretionary, with the option of going when it is most convenient (such as a shopping trip). Most traffic congestion results from nondiscretionary trips, because motorists seek to avoid wasting time if they can help it.

Travelers have other choices at their discretion, too. They may choose to travel by mass transit, for example. They may also choose to leave earlier or later.

When we sum up the choices made by all travelers, we get a profile of demand that varies as the day progresses. At times this profile may rise up over the limit of capacity, with the result that the traffic becomes congested. As with the funnel, the excess demand stacks up in a stream of slow-moving vehicles, or *queue*.

All traffic congestion is caused by excess demand. When the sum of the traffic on an entrance ramp and the traffic on the adjacent main lanes exceeds the capacity of the lanes downstream from the entrance ramp, a queue will form. The queue is quite sensitive to the amount of excess demand. For example, if the demand on a four-lane freeway is 8,400 vehicles per hour and the capacity in a bottleneck is also 8,400 vehicles per hour, we expect unstable operation with very little queue. If the demand rose to 8,600 vehicles per hour, which is an increase of only a little over 2 percent, 200 cars would stack up into a queue in an hour. Each lane's 50 cars of queue, if we assume maximum, or *jam*, density to be about 250 cars per mile, would extend not quite a quarter of a mile. The speed in that queue would be a little less than 10 mph on average (2,400 vehicles/hour at capacity divided by 250 cars/mile equals 9.6 mph). A 10 percent increase in demand would create a mile of additional queue. Therefore, demand only has to exceed capacity by a few percentage points to create severe congestion.

But the problem is worse than just the simple arithmetic of demand and capacity, because operational capacity is not constant. Queues are inefficient and reduce capacity somewhat just by their existence. The effect is subtle and not recognized by all experts, but only a small change can make a big difference.

In the case of a physical bottleneck, the only relief is a reduction in demand, but we can see that demand must be reduced to levels well below capacity to see much of a reduction in the queue. In our example with the flow of 8,600 and the capacity of 8,400 vehicles per hour, if the flow diminished to 8400 vehicles per hour the queue would not shrink because it is already there and it is being filled as fast as it is being emptied. Therefore, the demand must fall well below capacity. For example, if the demand diminished to 8,000 vehicles per hour, the 200 cars in the queue would take a half hour to dissipate.

Bottlenecks caused by accidents are usually more severe. Let's say that an accident blocks a lane and reduces capacity in our example to 6,400 vehicles per hour. When the accident is cleared and the capacity is restored, the front of the queue will be released. If the incident lasts an hour, the resulting queue will contain 2,000 cars. At a capacity flow of 8,400 vehicles per hour, those 2,000 cars will take a little less than 15 minutes to clear, during which time the arriving flow will add another 2,000 cars to the queue, which will take another 15 minutes to clear. Thus, the queue itself becomes a bottleneck that moves backward. Again, only a significant reduction in demand will allow the queue to clear completely.

From these examples, we can see that the smaller we can keep the queues, the easier it will be for them to dissipate when demand diminishes. Consequently, recognizing queues is one of the things we have to be able to do quickly to manage freeways effectively.

Reading a Freeway. Freeway managers need to be able to understand what they are seeing when they observe freeway flow, and they need to gain that understanding almost instinctively. In this section we will discuss what a freeway bottleneck looks like.

Most drivers connect the queue with the capacity constraint. But the queue grows upstream of the bottleneck, just as the water is stored in the funnel even though the bottleneck is the hole in its tip. Just as water moves quickly through the hole in the funnel while draining slowly from the wide hopper, so the freeway bottleneck is identified where slow queue traffic speeds up. As we have seen, the speeds in the queue are typically at around 10 mph or less. The speed of capacity flow is around 50 mph or a little more (or a little less on older freeways with lower design standards), and traffic in the bottleneck will be approaching this speed. Agencies often make the mistake of adding capacity where the queue is, but this does not solve the problem.

Brake lights are a good indicator of queue formation, even at its inception. When an observer sees brakes lights start to come on from groups of drivers, he or she can expect the queue to start forming quickly. But drivers do not apply their brakes because of a capacity constraint; they apply their brakes because the car in front of them is too close. So brake lights are a secondary effect of a bottleneck, just as the queue is a secondary effect.

We can now restate our goal of minimizing congestion by the strategies of reducing demand or increasing capacity. What are these strategies? And what are the tactics we need to implement them? Before we can answer these questions, we must first understand how freeway systems monitor traffic.

9.2.3 Surveillance

Freeway management systems use two different approaches to monitoring traffic: automatic and those requiring a human operator. The strategy is to know how many cars are on the facility, and how fast they are going or some other measure of their performance. The tactics are based on available technologies, and the author will introduce the currently available options in light of their strategic value.

Electronic Surveillance. The most basic form of surveillance in a freeway management system is point detection. Detection stations are constructed at regular intervals along a freeway, typically at half-mile or quarter-mile intervals. These stations use a detection technology to sense the presence of vehicles or to measure directly some aspect of their performance. These technologies can be broadly categorized into systems that measure the presence of a vehicle at a point and those that record the passage of particular vehicles at various points.

Point Detection. The oldest and still most widely used detection technology is the inductive loop detector. On freeways these are usually made by sawing a slot in the pavement in the form of a square 6 feet on a side, in the middle of each lane. Wires are laid into the slot so that they loop the square around its perimeter three or four times, forming a coil. The slot is then filled with a sealant to encapsulate the wires and to keep moisture out of the slot. When electricity is applied to the wire, the current through the coil creates a weak electromagnetic field. Passing cars change the loop's *inductance,* which is an electrical property of coils. This change in inductance is detected by the attached electronic device. Loop detectors are relatively inexpensive to install, especially if they can be installed in pavement already closed to traffic for some other reason, such as during its initial construction or subsequent reconstruction.

Loop detectors are simple and accurate in freeway applications. Most freeway detection stations use two loops in each lane, with about 10 feet of space between them. The small gap of time between the detection at the first loop and the detection at the second loop provides a direct measure of speed.

Loop detectors can therefore measure three things: the number of cars (volume), the percentage of time the loop is covered by cars (detector occupancy), and speed. *Occupancy* is a term widely used to describe the number of people in each vehicle, but in this application it means the percentage of time a detector is covered up. Occupancy is a surrogate for traffic density and speed. Of these three measures, occupancy is the most useful to freeway managers, and we will discuss why in a later section.

Other technologies are available that provide similar presence detection, such as magnetometers.

Several new technologies have emerged that allow presence detection without installing anything in the pavement. These are generally classed as *nonintrusive* detectors, and though their hardware usually costs more than loop detectors, they are far easier to install in pavement that is already carrying traffic. These technologies include acoustic detectors, infrared detectors, radio-frequency detectors, and certain examples of video-processing detectors.

Acoustic, infrared, and radio detectors work either by reflecting energy from vehicles being detected or by sensing the direct emissions of vehicles. These systems typically measure presence and some performance parameter such as speed, but some that use these technologies can also do simple classification of vehicles into trucks and cars.

The most glamorous of current detection technologies are the video-image-based systems. These systems digitize and evaluate video imagery by one of several means to find the vehicles and determine their performance. The key advantage to video-based detection systems is that the detection zone can be moved easily after installation, by the technician drawing the detection zone on a computer montior. This flexibility makes them ideal for maintaining detection capabilities within, for example, construction zones where the lanes will be shifted from time to time. Even so, video-based detection has found more enthusiastic application for detection associated with traffic signals, because it is at its worst when trying to count successive cars, especially in congested conditions and in conditions of fog or snow.

Vehicles as Probes. Point detection has a basic limitation: it can only provide data for one physical location. A common need within freeway management is the ability to measure travel time, and point detectors do not provide this capability. Several technologies that use vehicles as probes measure travel times directly and overcome this limitation. Travel-time information has not been used in automated incident detection, but it has found popular use for letting motorists know the condition of freeways. In most applications, however, probe-vehicle data collection grew from another capability and freeway managers have used the data as an opportunity.

For example, the Texas Department of Transportation developed the Houston area's entire automated surveillance capability in a remarkably short time by using existing electronic toll-collection systems. There were enough subscribers to the automatic toll-collection system for their two toll facilities to provide sufficient probe coverage throughout the network. The system in Houston uses automated toll readers on the freeways, at 2-mile intervals, in addition to their regular application at toll plazas. Studies have shown that as little as 3 or 4 percent of the traffic stream can provide reliable data about the traffic stream, so the system did not need many equipped vehicles.

Many emerging technologies promise easy access to vehicle probe information, but so far the costs have been too high in most cases to warrant the expenditure, or the technology has not lived up to its claims. Eventually, traffic management agencies will have access to reliable vehicle probe data, and at that time they may develop new uses for the data other than to inform motorists of conditions.

Human Surveillance. All freeway systems are built with an extensive capability for human surveillance using closed-circuit television. Humans have an ability to make judgments far beyond the skill of automated systems using electronic detection, but they are subject to loss of attention caused by boredom or distraction. Most systems combine electronic detection and human surveillance to optimize the strengths of each. Human operators might not be looking in the right place at the right time to see an incident occur, but once notified by an automated detection system, they can quickly determine what has happened and what should be done.

The major difficulty in establishing human surveillance is the infrastructure required to provide video to the location of the human operator. Because human operators are still more difficult to come by than infrastructure, however, we build control centers to bring all the surveillance information together. Control centers, in turn, require the surveillance information, including video, to be communicated from the field to the operator's workstation. The major cost of freeway management systems is therefore the communications infrastructure.

Video surveillance technology is improving at a rapid pace, and this text would quickly become dated if it attempted to document appropriate technology choices. Most cameras at the time of this writing use digital technologies for producing the image, convert them to analog signals to communicate them to a communications hub, convert them back to (differently formatted) digital signals to go over the communications infrastructure, and then convert them back to analog for switching and display in the control center. The trend toward digitizing all of these links will eventually bring about the replacement of analog technologies.

The typical arrangements of video cameras fall into three approaches: full redundant coverage, full coverage, and partial coverage. In the first category, cameras are placed frequently enough so that at least two cameras can see any given incident. This usually requires cameras at roughly half-mile spacing. Cameras can be placed at 1-mile intervals, but this requires cameras to be able to see half a mile, which stretches the capabilities of cameras and mounts unless the cameras include expensive image-stabilization technology. Typically, the traffic industry has avoided using such technologies in favor of simple cameras that are easier to maintain.

Partial coverage places surveillance cameras only at critical locations. Some traffic signal systems use partial coverage, but most freeway operators greatly prefer full coverage. The reason is that any section of freeway can become critical to the capacity of the facility should an incident occur there.

Cameras are usually mounted between 35 and 45 feet above the ground, with due care given to terrain, field of view, and lens flare caused by the sun. They are equipped with mounts that allow the cameras to be aimed from the control center, known as *pan-tilt-zoom* mounts. The sun is a major obstacle to clear viewing and will cause lens flare, blooming, and aperture errors when it is in the field of view. Most good designers, therefore, attempt to locate the cameras on the south side of east-west roadways, and alternating on the east and west sides of north-south roadways. The key advantage to full redundant coverage is that an incident can always be viewed by a camera that is facing away from the sun. When designing camera locations in the field, most engineers will resolve troublesome locations by elevating themselves to camera position in a bucket truck and videotaping the available field of view using a camcorder. These tapes can be reviewed by agency engineers and operators to select the location that provides the most effective view.

Camera technology has exploded in the last few years, but engineers who write specifications for these cameras are still concerned with durability and maintainability. Many new installations are equipped with lowering devices so the cameras can be lowered to ground level for maintenance. Some agencies use bucket trucks to reach cameras mounted at viewing height. A few agencies depend on ladder rungs and work platforms built into the poles, but these tend to be unusable for a variety of reasons by the normal maintenance technicians working for such agencies.

Most agencies build ground-level cabinets for their cameras, and most video manufacturers provide field equipment that allows a technician at the site to assume control of the camera from the central system for maintenance purposes.

9.2.4   Capacity Strategies

Freeway managers focus on two types of capacity constraints: queue formation and incidents. As we have seen, queues are themselves capacity constraints and linger after their cause has been removed. Minimizing queues resulting from physical bottlenecks is a major activity, but it will be discussed in more detail under demand strategies because improving the capacity within a physical bottleneck is not one of the capabilities of freeway management.

Capacity bottlenecks caused by incidents are another matter. Most freeway management systems are fully justified by the need to recognize, respond to, manage, and clear incidents as quickly as possible. There are many approaches for doing each of these.

Incident Recognition.   Unlike the sea, rivers move while their waves do not. For a given traffic flow, routine bottlenecks will produce expected queues and congestion at expected places. When this congestion violates expectations, an incident might be at fault. A number of algorithms have been developed over the years to automatically warn of potential incidents. None are even close to perfect, but agencies that operate large systems are motivated to refine their approach as much as possible. For example, the California Department of Transportation (Caltrans) reports managing as many as 60 incidents in a typical day in the Los Angeles area. If their incident-detection system flags 120 potential incidents, their operators will be overwhelmed by false calls. But if their algorithm flags only 30 incidents, the operators will not discover the remaining incidents as quickly.

The literature reports four different categories of automated incident detection technologies, including comparative, statistical, smoothing, and modeling methods. The comparative approach has been used most widely, and is the basis for the algorithms used by Caltrans. Caltrans uses a collection of incident-detection algorithms that work by comparing volume and occupancy to thresholds that define normal operation. When the measured conditions cross the thresholds, the algorithm assumes abnormal operation and warns the operator. These algorithms have been refined over many years, and the best of them report accuracy rates of around 75 percent and false alarm rates of around 1 percent. Agencies must spend a lot of time customizing the thresholds for normal operation, and they must aggressively maintain that customization as traffic conditions change over time.

The modeling approach has been used in Canada, based on successful research at McMaster University. The model predicts detector occupancy based on upstream data, and when measured occupancy and speed fall outside the predicted range, the algorithm warns of a possible incident. The modeling approach requires less tailoring to specific locations and is able to track trends in traffic over time, but it can be confused to some extent by the instability of the relationship between volume and congestion at the point of capacity. Nevertheless, the literature indicates that the McMaster method is at least as effective as the best of the Caltrans methods.

All of these approaches share a common flaw: they detect incidents based on secondary effects. For example, if two cars collide they will block at least one lane. The blocked lane will greatly reduce capacity in that freeway section, which will create a bottleneck. Traffic will queue up behind the bottleneck. Eventually, the queue will extend back and cover a freeway detector, causing the occupancy to shoot upwards, and that spike in occupancy, especially when coupled with the low occupancy at downstream locations, will flag the incident. The extending queue might not trigger the alarm for many minutes after the incident occurs. The best systems report a lag time of at least 3 minutes, but in many cases where the incident-induced congestion overlays existing congestion it will be much longer.

Many of these incident detection systems have become irrelevant in recent years with the explosion in the use of cellular telephones. Incidents have often been reported long before the queue builds back to the detector. Human operators also will often see the effects of incidents before the automated systems flag them, but many systems are operated with limited staff and operator fatigue is a critical issue in watching for incidents. Fortunately, cellular technology has solved the problem of identifying incidents. It has not solved the problem of getting proper information about the incidents, which is why we see no trend away from providing a full surveillance capability.

Incident Response. Incident response requires three elements: A notion of the effects of the incident at hand, a plan that addresses those effects, and a means of mobilizing that plan.

Considering the volume of traffic on the highway, incidents are rare events. Accident rates are typically measured in accidents per million miles of travel, which is an indication of how infrequent they are. The result of this relative rarity is that predicting incidents is nearly impossible and measuring their effects by observing the freeway is even more difficult.

When an incident occurs, it often starts with a catastrophic effect on freeway capacity—often closing down the entire facility for a few minutes until following drivers get over their shock and pick their way around the wreckage. The one or two lanes will get through for a while until vehicles not completely disabled can be moved to the side. More lanes may be closed by emergency personnel, and so on. The profile of closures for any one incident is therefore highly variable. This variability highlights the difficulty in characterizing incidents well enough to model their impact.

Here is where the philosophy of freeway management has a large effect on how incident management is conducted. Some freeway managers do not attempt to prepare response plans in advance, and the systems they develop focus on the ability to develop responses on the fly. Other freeway managers want the responses to be stored in the system so that they can be implemented on command from a control center. Other systems combine these two philosophies using an expert system to prepare elements of response plans on the fly from predesigned choices and decision trees.

For those systems that are designed around stored response plans, the development of those plans is a substantial task. The development itself follows a basic outline:

• List all potential incident scenarios (categorized by location, time of day, number of lanes closed, and duration of closure).
• Analyze conditions for each scenario.
• Design a response plan based on the analysis.
• Program the response plan into the freeway system's database.

For example, the TransGuide system in San Antonio, Texas, initially covered 27 miles of freeway. The Texas Department of Transportation developed approximately 125,000 scenarios for the initial implementation of the system.

Some systems avoid building a massive flat database by following a decision tree approach, sometimes called an expert system. In this approach, the possible responses are categorized into branches of decisions, where each of the categories listed above represent their own branches. Response elements common to an entire branch are applied when the decision to follow that branch is taken. Caltrans uses this approach in their District 12 freeway system, which covers Orange County, south of Los Angeles.

The key advantage of these systems is that the response plans can be implemented immediately, and in some cases automatically (though always with confirming approval from the system operator). The disadvantage is the massive task of defining the scenarios and designing responses for those scenarios and the maintenance required to keep them up to date. This approach may be overkill for incident management in a small city. In those cases, an experienced operator may be able to craft response plans on the fly.

Most agencies deploying incident response scenarios will employ analysis tools to help understand the effects of incidents, usually involving one of several simulation models.

Once incidents are modeled, freeway managers must determine how to respond to them. These responses fall into two categories: handling the emergency and managing traffic. Both are covered in the sections that follow. Agencies cannot implement these plans, however, without the means to do so. Jurisdictional boundaries must be crossed between traffic management agencies and emergency response agencies, and often across geographical jurisdictions as well. For example, the typical freeway system in a major metropolitan area will involve dozens of agencies. The state transportation department usually operates the freeway and will therefore usually be the lead agency. But the emergency response agencies, especially fire departments and emergency medical services, are usually attached to local governments, of which there may be many in a metropolitan region. State police may or may not exercise jurisdiction over freeways, and when they do not, local governmental police agencies will have to be involved. Freeway incidents often spill over into surrounding streets, and some systems are being designed with close coordination between state and local agencies because the latter usually operate the traffic signals. These jurisdictional issues can easily overwhelm the technical issues when implementing freeway management systems. Agreeing to incident response plans before the fact is one way to resolve these issues, by tying agencies together around a common understanding of the rules that will used when incidents occur. Often, however, these common approaches remain elusive, even when a system is being constructed, and often the needed cooperation results from system construction rather than guiding it.

Traffic Management during an Incident.   A number of different techniques help to restore capacity during an incident, all of which are based on the principle of opening lanes as quickly as possible. Most states have passed laws that require motorists to move vehicles out of the travel lanes if possible and if there are no injuries, though many motorists are reluctant to do so for fear that the facts of the accident will become obscured, making it difficult to assign responsibility. To reinforce the law, most police agencies refuse to investigate accidents when the vehicles can be removed from the travel lanes and there are no injuries.

One tactic to assist in clearing lanes after minor crashes involves creating protected sites for accident investigation outside the lanes of the freeway. This concept was popular among agencies because it could be implemented at low cost, but it was not popular with motorists because the accident investigation sites were not easy to find during the emergency.

But most techniques employed during an incident attempt to manage demand by informing motorists of the accident so that they can take an alternate route, choose a different time to make their trip, or make use of a more efficient travel mode. These strategies are discussed in the next section, on demand management.

9.2.5 Demand Strategies

Most freeway management strategies hope to modify the demand approaching an area of congestion. These strategies can be classified by the type of shift in demand they promote. At the beginning of this chapter, the author stated that drivers have one of three choices in using the network: the route that they take, the time that they depart, and the mode of travel that they use. Demand management seeks, therefore, to promote a spatial, temporal, or modal shift in demand.

Spatial Demand-Shift Methods.   All these approaches share the objective of encouraging (or directing) motorists to use a less-congested route in order to achieve a balanced overall use of the network capacity. These methods primarily include motorist information systems but also include lane-use signals.

Motorist Information Systems.   When congestion on the freeway creates much longer delays than motorists would face on alternate routes, the freeway managers will try to inform them of the delay to encourage a shift of demand to those relatively uncongested routes. Freeway managers have to be careful, however, not to invoke too strong a reaction. A lane of a freeway carries between 2,000 and 2,600 vehicles per hour at capacity, while a lane on a surface street might carry only 500 vehicles per hour. Any large diversion from a freeway to a city street system will overwhelm the latter if the demand is heavy. Most freeway managers are therefore unwilling to direct motorists to an alternate route, for fear of creating a worse problem. Most strategies hope to achieve as much as about 25 or 30 percent diversion, but no more, and a directive to follow a specific alternate route suggests total compliance. Agencies usually therefore avoid directives and allow motorists to choose their own response and thus be responsible for their choice.

Dynamic Message Signs.   When freeway managers first visualize a system, they typically think of video surveillance and overhead dynamic message signs. Therefore, fixed-location motorist information systems usually represent the most important elements of freeway systems. Dynamic message signs are the most popular fixed-location information systems. These large overhead guide signs provide real-time information to a passing stream of traffic.

Dynamic message signs can be divided into two broad categories: changeable message signs and variable message signs. These terms are often used interchangeably, but their meanings are distinct. Changeable message signs contain a fixed menu of messages that may be displayed, any one of which can be selected remotely by freeway system operators. Variable message signs provide the additional capability of editing and creating new messages remotely. Some display technologies fall into one category and not the other, but often this distinction is more imposed by the design of the system than by the hardware of the sign. The following list describes a few of the major sign technologies still in use:

- Rotating drums display lines of text by rotating drums mechanically within the sign. The technology is obsolete but is still in use in many places.
- Bulb matrix signs use long-life traffic signal bulbs in reflectors to make up the dots (or picture elements—*pixels*) that form letters on the sign. Bulb matrix signs provide the greatest brightness over the widest field of view, but they are difficult to maintain and extremely hungry for energy.
- Flip-disk signs display pixels by flipping a plastic disk so that its reflective face is shown to the traffic. They are efficient but require illumination and are subject to mechanical failure.
- Fiber-optic technology is usually used in conjunction with flip-disk technologies to provide night-time illumination of the pixels. The disks either reveal the continuously lighted pixel or hide it. One weakness of fiber-optic signs is that the light is focused in a relatively small field of view to make it bright enough, and the sign is only readable in a cone of about 20° facing the sign. Sign installers carefully aim the sign so that the illuminated field coincides with the motorists' positions.
- Light-emitting diodes, or LEDs, are mounted in clusters that combine to form pixels. Typically, 8 to 14 LEDs will be clustered in a 2-inch pixel, and the clusters are illuminated directly by applying electrical current to the LEDs. This approach shares the weakness of narrow field of illumination with fiber technologies, but it is by far the easiest technology to maintain. Most new freeway dynamic message signs currently being installed use this

approach. Most dynamic message signs for freeway management in the western hemisphere are a single color against a black background, but LED technology already provides a full-color capability that can be used to provide graphical information to motorists rather than plain text. As of this writing, freeway managers have only scratched the surface of this potential.

Dynamic message signs are typically installed upstream from decision points to provide information about the relative choices. In most places they are used to warn motorists of downstream incidents, but some agencies are using them for the display of useful non-incident real-time information.

One of the more heated debates within freeway management agencies is whether to display messages when there is no incident information to post. Some agencies insist that blank signs should be displayed in the absence of a compelling need to display a message. They argue that an illuminated sign, when compared with the routine blank sign, will strongly attract the attention of motorists. Other agencies argue that blank displays invoke complaints that the signs are not being used despite their cost.

The author resolves this issue with the definition of the sign as a *dynamic* message sign. More and more agencies are looking for information of real-time relevance whether or not it is related to incidents. For example, the Georgia NaviGAtor system displays travel time information to known landmarks downstream from the sign. This sort of real-time messaging is not incident-related, but it has proved quite popular in cities where it has been implemented. The author's conclusion is that dynamic message signs should be used to display dynamic information, even if it is not related to incidents.

Many agencies have instituted strict policies on what may be displayed on dynamic message signs to control liability. Agencies usually require that messages include descriptions of what is happening and wherever possible avoid dictating what the driver should do about it. Freeway systems are intended to empower motorists to make better decisions about how to use the network, and an attempt to make those decisions for motorists brings the possibility of directing them to do something that will cause them injury. Thus, most agencies are descriptive in their sign messages, not prescriptive.

Highway Advisory Radio.   Another well-established motorist information technology uses one of several protected bands of broadcast AM and FM radio to communicate to motorists. Because broadcast radio extends beyond specific facilities, these systems are used to provide wide-area information. Highway advisory radio (HAR) uses low-power transmissions (limited by the Federal Communication Commission to 10 watts with a range of about 5 miles) to provide messages using either looped recorded voice announcements or live announcers. The trend in major metropolitan areas is to use live announcers. For example, the New Jersey Department of Transportation uses HAR with live announcers extensively along suburban freeways around New York City. Most existing HAR systems use prerecorded voice loops, such as the systems along the New Jersey Turnpike. HAR facilities can be provided at frequent intervals using alternate frequencies to provide continuous information along the facility.

Temporal and Modal Demand-Shift Methods.   Once motorists are on the network, options for managing congestion are somewhat diminished because the car must go *somewhere*. Freeway control methods often seek to put motorists where they will do less harm to the demand-capacity relationship. A growing class of management techniques seeks to address traveler needs as opposed to motorist needs. The author will present control methods first.

Ramp Metering.   Ramp control has been one of the most controversial techniques employed in freeway systems in many areas. In some locations, agencies attempted to implement ramp control widely, only to be forced to retreat in the face of public outcry. In other places, established ramp control systems have had to justify their existence in the face of political challenges. Many agencies use ramp control with no problem, and other agencies will not even consider the alternative.

Ramp metering forces a temporal shift in demand by restraining traffic from entering a freeway for a brief period to keep the entering flow below a certain threshold. The idea is that if enough ramps are controlled to a moderate extent, the overall entering flow on the freeway can be held to a total that keeps demand below capacity at key bottlenecks.

Thus, ramp metering is best suited to freeways that have well-defined bottlenecks and a series of ramps feeding the bulk of traffic to that bottleneck.

Many motorists hate ramp meters. Minneapolis has established an aggressive use of ramp metering network-wide with careful refinement and maintenance of ramp metering strategies over many years. Even so, removal of the ramp meters became a political issue with the election of a new governor, and the compromise was a detailed study to determine the effectiveness of the ramp-metering program. Considering the political environment at the time, it speaks highly for the effectiveness of the program that after the test most ramp meters remained in operation. Other cities have not been able to demonstrate these benefits as clearly, and many had to remove most or all of their ramp meters after installation because of public pressure.

Thus, most agencies that are willing to consider ramp metering at all cite two critical principles in its successful application:

1. An aggressive, straightforward, and truthful public information campaign
2. Coupling ramp metering with popular freeway management services, such as effectively applied dynamic message signs and other motorist information systems

Traveler Information.   Another way to encourage both temporal shifts and modal shifts in demand is to provide good information about traffic conditions to travelers before their departure time and mode choices are made. As of this writing, the most popular method to providing traveler information for freeway management is by use of an Internet website. This has replaced many older approaches. Most current freeway management systems provide open access to much of the information available to the system operators. For example, the Georgia NaviGAtor website provides map displays showing currently congested sections of the freeway network, the messages being displayed on the various freeway dynamic message signs, locations of construction zones and incidents, and even snapshots from the video surveillance cameras. This information is also made available to media outlets, but often these provide limited information at best, and outdated and inaccurate information at worst.

If transit agencies cooperate with freeway management agencies and provide real-time transit information on their websites, then that information can help travelers make effective mode choices. Many commuters use their personal cars because transit modes are less convenient and sometimes more time-consuming. But if unexpected congestion because of a major incident will cause an extensive increase in travel time, the commuter may decide to use the transit service if it is easy to identify and use.

9.2.6   Operations

The operation of freeway management systems includes the manipulation of the various demand and capacity management strategies described above. But agencies must also maintain information about their effectiveness so that they can evaluate their operations. Thus, freeway management systems include software and hardware systems in the control centers to operate the field devices in the system and maintain a database of activities, incidents, and responses. These systems are complex and expensive and often consume a large share of the budget for the entire system. Because software systems change so frequently, describing them in detail is beyond the scope of this book. In general terms, however, software systems are designed in accordance with the system engineering approach outlined at the beginning of this chapter. The more closely agencies follow this approach, the more predictable their software development and acquisition will be. Many agencies now employ qualified systems engineers to manage their software acquisition, maintenance, and operation, and this trend is growing. The reader interested in further study on software development and acquisition approaches should review the systems engineering literature, particularly that published by the Institute of Electrical and Electronics Engineers.

Although this Handbook does not present software principles in detail, an understanding of good software management is critical to successful system deployments, and failures in the absence of good project management are unfortunately common. These failures have repeatedly resulted in cost overruns and blown schedules for system implementation.

9.3 COMMUNICATIONS

9.3.1 Why Learn Communications?

Communications infrastructure represents a huge portion of the cost of building traffic management systems. Most systems designers will say that at least two-thirds of the cost of systems will be used to establish communications with the field devices.

A review of systems projects around the country, particularly for traffic signal systems, reveals that the most common mistake in the designing and implementing these systems is a mismatch between the needs of the software systems in the control center and the services provided by the communications infrastructure. Managing implementation costs often point agencies to low-cost communications alternatives whose performance must be considered in the design of the software. Software providers, on the other hand, often do not see the requirements imposed by the communications system and design their software around unsupportable assumptions of communications performance. Only a good understanding of communications alternatives and architecture will prevent these mistakes from proliferating.

9.3.2 Architecture

The architecture of a communications network defines what kind of equipment will be used in the system. It does not necessarily define the medium; the same architecture can often be operated over different media, and different media are frequently used in the same system. In some cases, however, the selection of a particular architecture requires a specific medium, or at the very least allows or precludes classes of media.

In general, architectural approaches to communications systems fall into two camps. The first camp accommodates systems that *require* reliable real-time transmission of messages. The second camp does not. Control messages often have real-time requirements, while surveillance messages usually do not. Thus, the camps are generally demarked by those systems that do remote control versus those systems that delegate control to the field devices. Freeway systems usually fall into the former camp, and most (but not all) traffic signal systems fall into the latter camp. The need to accommodate real-time communications therefore becomes the single most important decision in the selection of communications architecture.

Real Time Defined. *Real time* is often defined as messages being passed within a specified amount of time. To most traffic engineers, this time frame is usually 1 second. Other branches of the computer industry, however, have very different standards. In industrial control systems, for example, real-time operation often requires response to a stimulus within 100 microseconds. Some process-control systems boast of response times as short as 6 microseconds. The use of the term *real time* to describe a desired response horizon is therefore not very specific.

A much more rigorous and useful definition of real time is that the system must respond to external stimuli predictably, that is, deterministically. In a traffic system that communicates control messages with the expectation of being implemented "immediately," for example, the very long 1-second event horizon might still be satisfactory if it can be provided predictably without fail. In any system that queues messages when message traffic exceeds communications capacity, this predictability is very hard to guarantee. The Internet provides a good example. Internet communications are entirely stochastic. The demand on the network for passing messages depends entirely on the behavior of a host of independent users. Therefore, demand on the network is a random variable that can be characterized by a bell curve of some sort.

Deterministic Architecture.   One way to guarantee predictable message delivery is to provide a path of fixed capacity that is reserved only for use between the system and a single field device. In these systems, the communications trunk lines are divided, either physically or logically (or both), into private channels with predetermined and fixed performance characteristics. If, for example, we wish to talk with 1,000 traffic signals and we must send a command message to each signal exactly once each second, then we will probably take the private channel approach. In such a system, we will group traffic signals into communications subgroups. Each subgroup will have its own physical path back to a hub site. If we have eight traffic signals on the path that serves that subgroup, then we will divide the time available on that path into eight time slices. Each time slice is a predetermined size, and each traffic signal will receive exactly that size portion of the physical path no matter what data is carried on it. Thus, each device gets a fixed and predictable communications rate. For example, traditional traffic signal systems employing private-channel architectures often allow eight controllers to share a single physical channel that has a capacity of 1,200 bits per second. The effective communications rate to each device is therefore 150 bits per second.

These subgroup paths are brought back to a hub and then combined onto a trunk line, which is divided in the same way. On the trunk line, therefore, we may have 64 or 128 traffic signals communicating, each on its predetermined time slice. The overall performance of the trunk line must be fast enough to accommodate all the fixed private channels. This process of devoting time slices to each of a large number of private channels is called *time-division multiplexing*.

The private-channel scheme was developed by the telephone industry. When a caller places a call, a private channel is established between the caller and another party. That channel may traverse a huge variety of distribution lines, trunk lines, and switching equipment, but the size and capacity of that channel is predetermined. Until the last few years, technologies that were developed within the telephone industry used time-division multiplexing, including the T-Carrier standards for twisted-pair wire and the Synchronous Optical Network (SONET) standards for fiber-optic cable.

The term *circuit switching* is often used to describe multiplexed, fixed-channel communications because each circuit is switched onto the trunk line for the duration of its time slice. In the past, circuit switching was the only way to guarantee communications performance, but it comes at a high cost with systems that typically only talk to a fraction of their field devices at any one time. Thus, the strong trend is towards systems that tolerate some lack of predictability, which opens the door to software-managed communications.

Stochastic Architecture.   When we allow devices to share channels, there is the possibility that more than one device will need the channel during any one instant. One device will therefore have to wait, which removes some degree of predictability. The more devices that want to talk at once, the more waiting will be imposed on some. Thus, the wait time is based on the demand, and we describe that wait time in terms of probability. In short, demand and wait time in such a network is stochastic.

To keep the wait from extending too long, each device will fragment its messages into small pieces. To allow reassembly at the other end, error-checking, and other communications services, each message will contain information about itself: its origin and destination, its composition and sequence number, an error-checking parameter, and any other information needed to manage the communications function. These messages are called *packets,* and because the information within each packet controls the use of the communications link, the management method is known as *packet-switching.* Unlike circuit-switching, which is handled by specialized hardware that is transparent to the data being moved, packet switching is managed by software in the same computers and devices that are doing the work of the system. This is the approach taken in most computer networks and internetworks, including the Internet. In fact, it was the computer networking industry that developed this approach.

Packet-switching can be very efficient because each field device has access to the full capacity of the channel and can send high-demand bursts of communications at high speed. Only a few devices can do this at any one time, however, so packet switching is most effective in networks where devices do not all want to talk at once. This is largely true for transportation systems, and packet-switching approaches are quickly sweeping away older circuit-switched techniques.

The disadvantage is that packet-switching requires specialized communications software in every device in the system in order to manage the communications network. Also, the demand in the network has to be carefully characterized to make sure that the performance can be kept to acceptable levels.

The Line Is Not So Clear.   Having established a sharp line between two camps, we must now render that line again with much less clarity. Both camps have been migrating towards one another for some years, and elements of each will be found in many modern systems. For example, it is quite common to see shared-channel communications carried over larger multiplexed deterministic communications lines leased from circuit-switched network owners such as the telephone company.

The Rise of Ethernet.   Ethernet is a technology originally developed for local computer networks using copper wire. It is a shared-channel approach that requires full software-managed packet-switching in all the computers that are part of the network. Traditionally, the transportation industry has had neither the software to manage communications within field devices nor powerful enough field devices to run such software. But this has changed, and it is now economical to provide Ethernet over a wide-area network to all field devices, even video cameras. As of this writing, many systems are being designed with fiber networks that carry Ethernet, Fast Ethernet, Gigabit Ethernet, and even 10-Gigabit Ethernet over fibers that until recently would only have carried SONET multiplexed communications.

Reliability and Redundancy.   Many communications systems enhance reliability by providing redundant elements. The most common redundancy scheme for communications is the self-healing ring, which is defined for fiber-optics systems that employ the SONET standards. In a self-healing ring, signals are transmitted simultaneously clockwise and counterclockwise around a ring of hubs. Each hub is therefore connected to the other hubs in two ways. If the cable is broken between two hubs, then all the hubs are still accessible from one or the other direction of the ring. The equipment in a SONET system is designed to switch automatically between directions to correct faults.

Some systems design self-healing rings along a linear path by squeezing the ring together until it forms a line. This does not provide protection as well as a true ring, because a physical interruption of the cable will likely affect both sides of the "ring" and still isolate some of the hubs. But this approach still allows the system to recover from equipment failures.

Within packet-switched networks, the grid is used more often than the ring, and network switches in the grid can find alternate paths when a path is lost. This capability is known as a *spanning tree* and is a feature in sophisticated Ethernet switch-routers.

Redundancy helps improve reliability, but it does not reduce maintenance requirements. Actually, ring topology and other redundancy schemes can increase maintenance requirements because the equipment is more complicated and, in some cases, duplicated (which means that more equipment must be maintained). Redundancy increases overall maintenance requirements but reduces the *urgency* of maintenance. When a link is damaged on a SONET, the system still functions, so operation of the system does not depend on immediate repairs. But the repairs must still be made quickly because the system no longer has redundancy. In many agencies, maintenance budgets are down to crisis levels, meaning that only those repairs that are critical to system functionality will be performed. In such circumstances, ring redundancy may be a problem rather than a solution because the redundant elements will not be repaired as they fail. Once the redundant elements have failed, the system no longer benefits from the enhanced reliability. But the agency has still borne the increased cost of redundancy, both in initial cost and in overall maintenance requirements. In such cases, agencies would be better served spending the money on systems that are easier to maintain and resist damage more effectively.

The technology of the communications should not outstrip the ability of the maintaining agency.

9.3.3 Guidelines and Standards

Circuit-Switching Standards. Private-channel architectures were initially developed by the telephone industry. Consequently, a host of standards is available for defining multiplexed digital communications. All these standards can be characterized in a single continuum known as the North American Electrical Digital Hierarchy, which is represented below:

Description	Total Data Transfer Rate	Composition
OC48	2.48832 Gb/s	48 DS3 (1344 DS0)
OC36	1.86624 Gb/s	36 DS3
OC24	1.24406 Gb/s	24 DS3
OC18	933.12 Mb/s	18 DS3
OC12	622.08 Mb/s	12 DS3
OC9	466.56 Mb/s	9 DS3
OC3	155.53 Mb/s	3 DS3
OC1	51.84 Mb/s	1 DS3 (plus SONET overhead)
DS3	44.736 Mb/s	28 DS1
DS2	6.312 Mb/s	4 DS1
DS1C	3.152 Mb/s	2 DS1
DS1 (T1)	1.544 Mb/s	24 DS0
DS0	64 Kb/s	Basic unit—one digital voice channel
DS0A	64 Kb/s	2.4, 4.8, 9.6, 19.2, or 56 Kb/s filling a single DS0
DS0B	64 Kb/s	20-2.4, 10-4.8, or 5-9.6 Kb/s subchannels on a single DS0

The hierarchy is divided generally into three sections. At the slowest speeds, we have the standards covering standard serial communications. These standards are defined by several agencies, including the Electronics Industries Association (for physical media) and the Consultative Committee for International Telephone and Telegraph (CCITT, for communications services above the physical media). Communications at this level range in speed from 110 bits per second to 115,000 bits per second, but standard multiplexing equipment is available only to combine the speeds shown for DS0A and DS0B.

The middle level of the electrical digital hierarchy covers high-capacity communications over copper-wire links. These are commonly known as the T1 or T-carrier standards, but are more specifically known as the digital signal types. "T1" is also used to describe DS1, or digital signal type 1. The T1 standards emerged from the telephone industry, and the American National Standards Institute (ANSI) now maintains them. The basic unit of T1 is a digital voice channel, defined nominally to be 64 kilobits per second, which is known as DS0. A DS0 can be further subdivided, but any signal larger than DS0, all the way up through SONET speeds on fiber networks, must first be broken down into DS0 channels, sent across however many channels are required, and reassembled on the other end. Higher standards are multiplexed combinations of multiple DS0 channels.

Packet-Switching Standards.   The rise of Ethernet has brought with it new standards, but they are mostly software protocols rather than signaling standards because most services are managed in software rather than hardware. The most common software protocol is now universally TCP/IP, which stands for Transmission Control Protocol/Internet Protocol, for the two main protocols used within the Internet. Packets (known here as datagrams) that include TCP/IP management information are now typically being carried over protocol-driven media, ranging from wireless digital packet protocols such as 802.11 (known as Wi-Fi) to conventional Ethernet. For hard-wired traffic systems, Ethernet is emerging as the technique of choice.

Ethernet performance is divided into the following standards:

Standard	Copper wire or fiber designation	Nominal data rate
Ethernet	10-BaseT (copper), 10-BaseF (fiber)	10 megabits/second
Fast Ethernet	100-BaseT (or F)	100 MB/s
Gigabit Ethernet	1000-BaseX (or GigE)	1 gigabit/s
10-Gigabit Ethernet	10-GigE	10 Gb/s

Ethernet networks are divided into sections that are connected using switches, routers, switch-routers, or bridges, depending on the nature of the connection.

To implement a wide-area Ethernet, all devices on the network must have an address that allows them to be identified by software-based protocols. This addressing capability is built into the TCP/IP protocol. TCP/IP-equipped communications interfaces can handle the connection to plain serial devices that do not have this capability, bringing Ethernet possibilities even to traditional and existing traffic control equipment. Devices intended for implementation on Ethernet packet-switched networks are often called "IP Addressable."

Polled and Non-polled Protocols.   Channel-sharing creates the possibility of two devices attempting to use the channel at the same time. This potential conflict requires some discipline in how devices can access the channel. The simplest scheme precludes remote devices from speaking unless they are individually commanded to speak by a master device. This approach is known as a *polled protocol.* The master device sends out a poll request message to a specific field device. That field device then sends a response. Other devices will hear the transaction (or at least the master's side of the transaction) but will keep silent. The polled protocol can be characterized by the phrase "speak only when spoken to." Most communications systems for transportation management applications use polled protocols, but this convention is giving way as field devices become more powerful and therefore more capable of processing complex protocols.

As of this writing, systems are emerging that allow field devices to report in to central systems without a previous poll request. The communications systems being designed for these systems have collision management built into them. For example, the 802.11 (Wi-Fi) standard for wide-area wireless communications has collision control as a basic service of systems supporting the standard.

Software Procotols.   Most of the previous discussion involves the specific protocols of the communications medium. All these standards are applied at the bottom layer of a communications system. The bottom layer is the physical layer, the part that directly manages the signaling on the medium in question. Once the signals are translated into bits and bytes, they will be processed by computer programs within the communications software on the device. Each message is encapsulated with a series of wrappers, each of which handles some communications service. These wrappers are called *layers*, and all such layers are called a multilayer protocol or *communications stack*. Above the physical layer are layers for handling transport issues such as breaking large messages apart and routing them through the network, layers for defining the protocol in use, and finally layers that understand traffic data.

A central principle of the multilayer model is that each layer need not know anything about the other layers. The application layer does not need to know anything about which protocol is used in the layers handling transport issues. Consequently, a profile of protocols can be defined, selecting the specific protocol for each layer as appropriate for the design of the system in question.

In transportation systems, the profile of protocols used conforms to the National Transportation Communications for ITS Protocol, or NTCIP. We will discuss NTCIP in a later section. NTCIP protocols represent a special case of the multilayer approach, using many of the Internet protocols but also providing a few that are specific to transportation.

9.3.4   Video Communications

The data transmission requirements for full-motion video are so huge that they deserve a separate discussion. The addition of video surveillance to a system design fundamentally changes the requirements of the system. For data only, the major decision concerns selecting deterministic digital communications or nondeterministic (stochastic) digital communications. When video is added, the major decision affecting cost is whether or not to use analog video transmission or digital video transmission.

Analog Video.   Full-motion video in analog form requires 6 MHz of bandwidth, of which 4.5 MHz is required by the color-burst, or picture, and the remainder for audio. Analog video is a continuous scan of electrons moving horizontally in rows down the screen. In North America, video signaling is defined by the National Television Standards Corporation (NTSC). NTSC video is defined as 525 lines of vertical resolution, or scan rows, and all rows are scanned 30 times a second. In many systems, including most consumer television systems, the signal strength and display equipment is not good enough to clearly transmit and display all 525 lines, and most consumers are more accustomed to seeing about half that. The standard is defined to allow this degradation without loss of synchronization.

Analog television may be transmitted over fiber very effectively using frequency modulation, or FM. In fact, in some systems with a single link this proves to be a cheaper and simpler method than digital video.

Analog video has the key advantage that it can be transmitted cheaply over poor-quality communications links and still be usable for traffic surveillance. The disadvantage is that it cannot be carried over a digital communications link, and increasingly *all* available communications links for transportation use are digital.

Digital Video. If we divide each horizontal scan line into 700 picture elements, or *pixels,* then each scanned screen image contains 367,000 pixels. If each pixel defines a level of red, green, and blue (the additive primary colors used in constructing a color video image), and if each color can be defined at 256 levels, then each pixel needs three bytes to describe it. The size, in bytes, of a single frame of broadcast-quality video is therefore a little over a megabyte. If we transmit 30 frames per second, then the data stream is going by at 265 megabits per second. In practice, an analog video image can be stream-digitized at less than broadcast quality at much lower rates. Typically, uncompressed full-motion video is expected to consume well over 100 megabits per second. Even at this rate, the bandwidth requirements of digital video are vast compared to data requirements for most transportation management systems. Consumer digital video is compressed to some extent, and consumes 25 megabits per second.

The standards for digital video compression are in continuous development by the Motion Picture Expert Group, which is a working group of the International Standards Organization. Compressed images conforming to their standards are called by their acronym, MPEG. Most systems can now cheaply use MPEG compression to reduce the data rate for high-quality video down to about 5 megabits per second, and less if some loss of picture quality can be tolerated. Even using TCP/IP protocols for managing streamed video, at least 15 channels of video can be transmitted at high quality over a single Fast Ethernet network. With SONET circuit-switched networks, compressed video of high quality can be transmitted over two T1 links.

9.3.5 Selection Techniques

Communications systems are complicated, but an understanding of their architecture and technologies can have a huge impact on the cost of a system. Most modern traffic management systems will spend about two-thirds of the construction budget on communications.

The first step is to determine the requirements. Two questions are critical:

1. Will the system transmit video images?

2. Does the system require mandatory real-time control communications, and if it does, what event horizon can be tolerated?

If the system will transmit video images, then several more questions must be answered:

1a. Will the video be transmitted over a leased communications service?

1b. What image quality is required?

1c. What is the maintenance capability of the agency?

If leased service is necessary, then digital video is probably the only recourse. If not, then the higher the image quality requirements, the more cost-effective analog becomes by comparison, at least within simple networks. Also, limited maintenance capabilities would suggest the simpler analog equipment, or the use of digital video over Ethernet networks, but would offer a warning with the use of SONET multiplexed networks.

If the system does not require mandatory real-time communications, then the reliability of the media can be reduced significantly and wireless media can be considered. Thus, the architecture of the system software has to be matched to the services provided by the communications plant. Also, systems that do not require real-time control can take advantage of the greater flexibility and interoperability offered by nondeterministic shared-channel networks that use packet-switching techniques. These systems are readily available for all communications media.

If the system does require mandatory real-time communications, then the selection is not so easy. A packet-switched system is still possible, though the system will have to impose the necessary discipline on the communications messages to ensure sufficiently deterministic performance. New Ethernet options make this much easier.

9.3.6   The National Transportation Communications for ITS Protocol

The NTCIP is an ongoing project resulting from a collaboration of the user community, represented by the Institute of Transportation Engineers (ITE) and the American Association of State Highway and Transportation Officials (AASHTO), and traffic equipment and software manufacturers, represented by the National Electrical Manufacturers Association (NEMA).

The NTCIP seeks to provide two fundamental capabilities: interchangeability and interoperability. Interchangeability has been defined as the ability to replace a device from one manufacturer with a similar device from another manufacturer without damaging the operation of the system. Interoperability refers to the ability of dissimilar devices to share a single communications channel.

Interoperability.   In order to allow a variety of systems and devices to share the same communications channel, each message transmitted on that channel must be formatted in a way that the software in all the systems and devices understands. Achieving this understanding requires that messages be constructed in consistent ways with consistent protocol management information. It is not necessary for a traffic signal controller to understand the meaning of a message intended for a dynamic message sign, but it must understand the format of the message sufficiently to know that it *is* intended for a dynamic message sign.

Systems for traffic management are designed around different system architectures, depending on the needs of the system. Thus, NTCIP provides many options to allow efficient application in varied circumstances.

Each possible configuration requires specific protocols used at each layer of the multilayer protocol. This list is called a *profile*. Figure 9.1 shows the NTCIP Standards Framework. The profiles represent the paths that ascend and descend through this diagram.

Many of the protocols identified in the figure are beyond the scope of this book to discuss, and interested readers should learn about them in the current version of the NTCIP Guide, which is available for download at http://www.ntcip.org. The "levels" in the NTCIP framework approximately correspond to the "layers" in multilayer communications protocols. Most of these protocols are taken directly from the Internet, with the result that NTCIP messages with the appropriate profile can be communicated across the Internet.

For example, a simple freeway management system might include traffic sensors, dynamic message signs, and a control system, with a simple single-channel multi-drop on agency-owned copper wire connecting the system to the devices. In this case, no routing is needed and all messages for these devices will be small enough to be contained in single packets. The profile, therefore, needs a protocol at the application level and at the sub-network level only. The protocols in the transport level provide services that are not needed by this simple network. The profile used in this example is shown highlighted in Figure 9.2.

A complex system will need a different solution. For example, a freeway management system might communicate with field devices using the Internet, in which case the network between the system and the field devices is highly complex. Each message will need routing and end-to-end error-checking and will require the User Datagram Protocol (for error checking) and the IP (for routing) contained in the transport level of the framework. A message with UDP and IP information in it can traverse the Internet.

In addition to managing the network, interoperability requires that messages make use a standard way of defining themselves in terms of content. We are not talking about the content

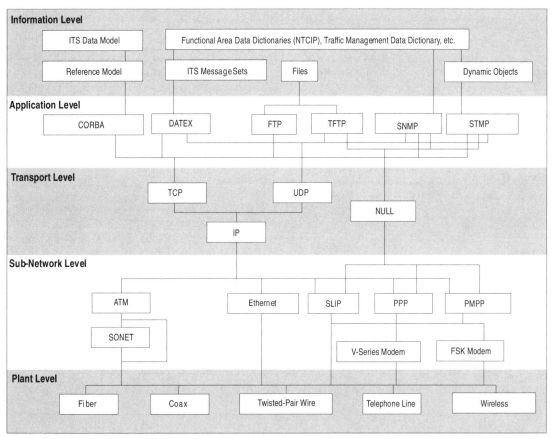

FIGURE 9.1 NTCIP Standards Framework.

itself, but rather how the content is identified. The Small Network Management Protocol (SNMP) in the application level of the NTCIP framework defines how message types are identified. Each of the messages supported by a particular device is stored in a management information base (MIB) that lists all the specific information needed by the protocol software to interpret the message. The SNMP protocol software incorporates a device's MIB and then uses that information to format messages as required by the system and device software. Each message object used by all devices in the transportation industry are defined within the standard, though manufacturers are allowed to create their own objects when the standard objects do not provide a specific feature they wish to support.

The content of each object determines the data it contains and how those data is defined. Thus, the profile information for interoperability defines *syntax,* while the object definitions in the MIB define *semantics.* Standardized semantics are required for interchangeability.

Interchangeability.   Interchangeability allows similar devices from different makers to be interchanged on the network without any differences in the resulting operation. Interchangeability has proved to be more difficult than interoperability because it defines the functionality of the system. Interchangeability requires the common definition of message object content.

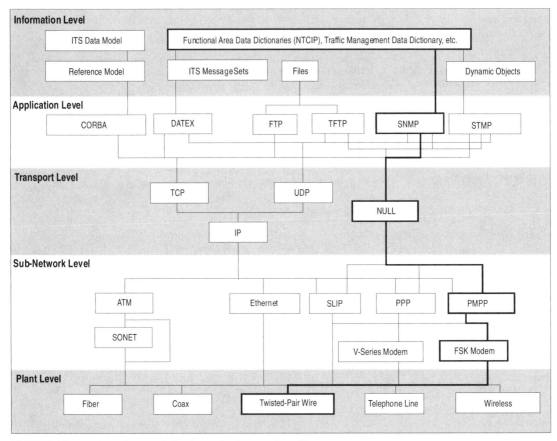

FIGURE 9.2 Standards framework showing a simple network profile.

Implementing NTCIP. For each system, the designer has a series of determinations to make. The first is to determine the required profile, or what software protocols provide the services needed in that system. The more protocols are used, the bigger each packet will be because of the additional data added by each protocol. Consequently, the designer must make sure that the data flow requirements of the system, including the overhead added by the NTCIP protocol, can be accommodated by the speed and capacity of the network being designed. In many systems, the traditional 1,200 bits per second will not nearly suffice if the system expects to see a polling rate that keeps map displays updated in approximate real time.

Once the profile and an appropriate network are designed, then the designer must determine which objects in the standard MIBs will be needed. Most of the objects in the MIB are optional, and they are grouped into mandatory and optional conformance groups that categorize the objects for a given device. All optional objects and conformance groups that are needed by the system must be specified during the design of the system.

The status of the ongoing standards development effort can be monitored on the web at http://www.ntcip.org. That site includes the downloadable *NTCIP Guide,* which provides the best available fully detailed explanation of NTCIP and how it is applied.

CHAPTER 10
TRAFFIC SIGNALS

Richard W. Denney, Jr.
Iteris Corporation, Sterling, Virginia

10.1 INTRODUCTION

10.1.1 Background

At the intersection of two streets with no traffic control, drivers are expected to decide for themselves who is to go and who is to wait. Their only guidelines are the rules of the road, as defined in the Uniform Vehicle Code and the laws of the state, and their courtesy and cooperation. For intersections where the traffic demand is very light and the conditions favorable, drivers do not have trouble assigning their own right-of-way in this manner, and their delay is very small. As soon as traffic begins to build, drivers are faced with too many decisions to make in order to safely assign their own right-of-way, and some help from the governing body is required to help them with these decisions. The simplest form of traffic control is a yield sign on one of the streets to let drivers know which movement has the right-of-way when a decision must be made. This, too, is only suitable for low traffic volumes. As traffic increases, drivers must be stopped so that they have time to make complex right-of-way decisions. A two-way stop, with stop signs on the minor street, serves this purpose.

Eventually the traffic will build to the point that assignment of right-of-way is not the only problem, and moving the traffic efficiently becomes a primary concern also. Stop signs are effective in the former task, but they force every driver to stop and start, which is inefficient in that a large proportion of the time is spent just waiting for the drivers to get going. At this point, traffic signals can assign right-of-way much more efficiently because the delay associated with starting up is only felt by the first two or three cars in the *platoon*.

The ideal isolated traffic signal assigns just enough green time to clear the queue of traffic and then move on to the next movement. When other signalized intersections are nearby, traffic engineers also try to arrange the timing so that the light will be green for platoons coming from the other signals when they arrive. These two goals sometimes compete. Current control equipment can provide a wide array of features, loosely organized into *fixed-time* and *traffic-actuated* capabilities. The former are usually used for calculating signal timings, especially in systems where those timings have to be consistent to maintain network flow, and actuated settings are used to allow the local controller to make adjustments based on the demand actually present at the intersection.

Before the actual discussion begins, the reader should be aware that though some very careful and scientific thought has gone into some of the principles, the traffic signal industry is very young and much more is not known than is known. Most experienced traffic professionals will happily depart from these methods when they do not seem to work on the street. But this freedom implies a warning: ***Do Not Break the Rules Before Learning Them.***

The reader should also be warned of the temptation to depend on computer programs without understanding their underlying methodologies. This strong temptation can be a downfall when traffic engineers are called by the public and their representatives to explain their actions.

10.1.2 Definitions

Because of the complexity of traffic signals, a new jargon has evolved to help signal professionals communicate efficiently. These definitions are intended to make clear exposition in this chapter possible, and thus are sometimes more precise or consistent than their usage within the industry might suggest. They are presented in logical order.

Traffic signal: Any power-operated device for warning or controlling traffic, except flashers, signs, and markings.

Legal authority: "Code" refers to the Uniform Vehicle Code (NCUTLO 2000), which is the basis for the motor vehicles laws in most states. The authors have assumed that the Uniform Vehicle Code is the controlling legal authority in this presentation. "*MUTCD*" refers to the *Manual on Uniform Traffic Control Devices* (FHWA 2001), as published by the Federal Highway Administration. Some states have their own versions of each of these documents. The federal *MUTCD* is available on-line at http://mutcd.fhwa.dot.gov.

Traffic movement: A flow of vehicles or pedestrians executing a particular movement.

Approach: The roadway section adjacent to an intersection that allows cars access to the intersection. An approach may serve several movements.

Major street: The street in the intersection that has greater importance or priority, usually (but not always) because it carries the greater traffic volume.

Minor street: The approaches that have the lesser importance, usually because they carry less traffic.

Right-of-way: The authority for a particular vehicle to complete its maneuver through the intersection.

Traffic signal system: A network of traffic signals coordinated to move traffic systematically. This system may be controlled at the individual intersections or by some central device or system of devices.

Signal indication: A particular message or symbol that is displayed by a traffic signal, such as a green ball or a WALK.

Green ball indication: Allows drivers to enter the intersection and go straight or turn right. If no left-turn signal is present and no sign prohibits it, left turns may be executed in appropriate gaps in the opposing traffic.

Green arrow indication: Allows drivers to complete the movement indicated by the arrow without conflict. Right turn arrows may conflict with a WALK indication, though this practice may cause problems at high pedestrian locations.

Yellow indication: Drivers are notified that right-of-way is ceasing and vehicles should stop, if practical.

Flashing yellow indication: Drivers should proceed through the intersection with more than usual caution.

Red indication: All vehicles must stop. Some turning movements are allowed after stopping, as provided in the Code.

Red arrow indication: A red signal applying to a particular movement.

Flashing red indication: All traffic must stop and then proceed when safe to do so. The meaning is the same as a stop sign.

WALK indication: Pedestrians may proceed to cross the intersection without conflict except from right-turning vehicles.

Flashing DONT WALK indication: A pedestrian clearance interval. Pedestrians in the crosswalk should complete their crossing, but no pedestrians should enter the crosswalk.

Steady DONT WALK indication: Pedestrians should wait at the curb; conflicting traffic has the right-of-way.

Signal section: A housing that contains the light source for one indication.

Signal head: An assembly containing all the sections for a particular traffic movement or movements, using an arrangement as defined in the *MUTCD.*

The following terms are purposely defined narrowly for clarity. The terms *phase* and *overlap* regularly cause confusion in traffic signal discussions, and these precise definitions help keeps things sorted out.

Traffic phase: A particular traffic movement or combination of nonconflicting movements that receive a green indication simultaneously.

A variety of coding and numbering schemes have been proposed over the years, with no hope of a consistent standard, until the National Electrical Manufacturers Association (NEMA) developed a specification for a traffic-actuated controller. In that standard, a numbering system was used that has now become a de facto standard. The NEMA phase numbering scheme for the movement in the intersection is as shown in Figure 10.1.

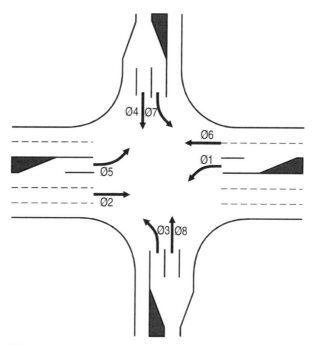

FIGURE 10.1   NEMA Phase numbering scheme.

Sometimes, when no protected left-turn movement is provided, the left-turners will be included in the straight-through phase number. For example, an intersection of two one-way streets may only use phases 2 and 4. A common convention is to orient phase 2 in the primary direction on the major street.

Some traditional literature uses numbering schemes that are now obsolete, but the NEMA scheme has become an established standard and we will use it in this chapter.

A traffic-actuated signal controller may operate two phases together. For example, phases 1 and 5 may operate at the same time for leading left turns. This is denoted by a plus sign, +, e.g., phase 1 + 5. An actuated controller usually allows separate actuation of each phase.

Phase sequence: The order in which traffic phases are presented to drivers. The following terms apply to specific phase sequences:

Leading left: A protected left-turn phase that comes up before or at the beginning of the through phase adjacent to it. For example, if phase 5 comes up before phase 2, or at the same time, then phase 5 is a leading left.

Lagging left: A protected left-turn phase that comes up at the end of after the adjacent through phase.

Dual leading left: When phases 1 and 5 come up simultaneously, followed by phases 2 and 6. All these terms can be equally applied to the side street phasing. Minor street dual leading left would mean that phases 3 and 7 come up first, followed by phases 4 and 8.

Dual lagging left: When phases 1 and 5 (or 3 and 7) come up and conclude simultaneously, preceded by phases 2 and 6 (or 4 and 8).

Lead-lag lefts: When the left turn for one approach is leading and the left turn for the opposing approach is lagging. For example, phases 2 + 5 (leading left), followed by phase 2 + 6 (through movements), followed by phase 1 + 6 (lagging left). The phase sequence 1 + 6, 2 + 6, 2 + 5 will sometimes be called lag-lead.

Split phase: When all the phases for one approach come up together, end together, and then are followed by all the phases for the opposing approach. For example, phase 2 + 5, followed directly by phase 1 + 6, with no intervening 2 + 6 interval.

Permitted left turns: When a green ball is displayed to left-turners, they may turn when it is safe to do so. If no left turn arrow is provided, the left turn is *permitted only.* If a green arrow is shown to the left-turners for part of the cycle and a green ball part of the time, then the operation is *protected-permitted.* When only a green arrow is presented to the left turners, or when so designated by a sign, then the operation is *protected only.*

Overlap phase: A special feature of actuated controllers. An overlap phase is a special phase that displays green at the same time as two or more regular traffic phases. For example, an intersection may allow a movement during two other traffic phases (which cannot be run together). The overlap phase goes green at the same time as either of these two *parent phases,* and stays green as long as either one of these phases is green. Do not confuse this hardware feature with the term *operational overlap.*

Operational overlap: Any interval when two normally distinct phases are green at the same time. One example is in the phase sequence 2 + 5, 2 + 6, 1 + 6. The 2 + 6 interval is called an operational overlap. Thus, split phasing is lead-lag phasing with no overlap.

Cycle length: The time required to service all the phases at an intersection. This time is measured from the start of green of a particular phase to the start of green of that same phase again.

Green split: Many times just called a *split.* The fraction of the cycle that is allotted to any one phase.

Traffic signal controller: The devices that directs the logic of signal timing. The traffic signal controller, or just *controller*, directs the phase order and interval length. The traffic signal controller is housed within a cabinet, as shown in Figure 10.2.

Master controller: A field device that controls a small number of intersections and that in some cases brokers communications with a signal system.

Local controller: The controllers in a signal system that receive coordination information either from master controller or from signal systems, or both.

Detectors: Equipment that detects the presence of traffic. A variety of equipment and hardware technologies exists for this purpose.

These definitions are not intended to be comprehensive, but rather to clear up the most commonly confused terms. Other terms will be described in the discussions that follow.

10.1.3 Advantages of Traffic Signals

Traffic signals are installed because they efficiently assign right-of-way to drivers approaching an intersection. Thus, signals have the following advantages over other traffic control schemes. These advantages can be negated if the traffic signals are installed when not needed.

- They provide for orderly traffic movement.
- They allow higher capacity than other methods.
- They can promote systematic traffic flow when interconnected into a system.

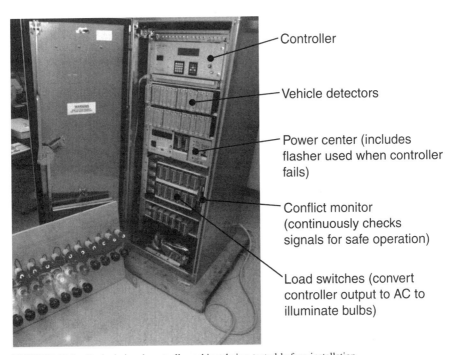

Controller

Vehicle detectors

Power center (includes flasher used when controller fails)

Conflict monitor (continuously checks signals for safe operation)

Load switches (convert controller output to AC to illuminate bulbs)

FIGURE 10.2 Typical signal controller cabinet being tested before installation.

- They provide gaps in heavy traffic flow that allow other drives to enter the traffic stream. Sometimes one signal can solve these problems on a long stretch of roadway.
- They can reduce the frequency of right-angle collisions (if such collisions are a problem).
- They are more economical and effective than manual control if properly designed.

10.1.4 Disadvantages of Traffic Signals

Traffic signals should never be considered as a panacea for all intersection problems. This misconception is common, especially among members of the public, but hard to explain given the large percentage of network delay that is caused by signals. In many urban areas, the majority of requests for signals concern locations where a lightly traveled minor street intersects a high-volume arterial highway. Because of volumes and speeds, the side-street drivers feel victimized, and their requests usually get political support. The political hierarchy will also complain about poor traffic flow on the major street, and never realize the paradox they present. Traffic signals work because they occasionally turn red and delay the major street traffic. This interruption in major street flow is inherently detrimental; if this was not so, the signal would not be needed, because adequate gaps would already exist. Traffic signals cannot add capacity to a free-flowing roadway; they can only reduce it.

Traffic signals do not always promote safety. Usually the number of rear-end and side-swipe collisions will increase, particularly on high-speed roadways.

The important concept is that signal installation is a trade-off between advantages and disadvantages. This trade-off must be evaluated, even if the decision will be made in the political arena, because the consequences need to be communicated to the decision-makers.

10.1.5 Classification of Signals

Traffic signals include a large variety of traffic control devices, including pedestrian cross-walks, emergency signals to serve occasional use by emergency vehicles, lane-use signals for traffic management, freeway ramp meters, railroad crossings, and so on. This chapter will be limited to signals at roadway intersections for use by regular traffic and pedestrians.

10.1.6 When to Install a Traffic Signal

Generally, two levels of criteria must be met before a traffic signal should be installed and operated. The first requirement is the legally defined minimum warranting conditions in the *MUTCD*, which is intended to assert when a traffic signal is *not* appropriate. The *MUTCD* warrants form a threshold, not a conclusion.

Secondly and more importantly, an engineering analysis that analyzes the trade-offs involved in signal installation must recommend a signal. The signal must solve more problems than it causes.

Both of these levels of criteria are required by the *MUTCD*. The Manual defines a traffic engineering study that includes the following data-collection requirements:

- Sixteen-hour volume count on each approach in 15-minute intervals
- Classification study to determine the relative quantities of trucks and cars
- Pedestrian count, where applicable
- Speed study on each approach
- Condition diagram to show the physical layout and features
- Accident history for the previous 2 years
- Operation data, such as delay, as necessary

Not all these data are collected at every location, because agencies do not usually have the resources to go into this much detail. Any time any one of these factors is significant, however, some field data should be collected. By "significant," the reader should understand that the critical factor is the one that motivated the request for the signal. If the request for the signal is based on perceived danger, accident statistics should be researched. If the request is motivated by excessive side-street delay, than that delay should be measured.

But it is critically important to understand that citizens who request a signal do not always explain their reasons in good faith. Most citizens requesting a signal will cite safety concerns, though in many cases the intersection in question shows no history of accidents. Often they will cite delay issues, and a delay study will show that a signal will worsen their plight. These study results will not persuade them, because they address the stated problem rather than the true problem. In many cases, the true problem is that motorists are uncomfortable making tight gap-choice decisions, and that discomfort has motivated the call. Drivers would prefer to wait longer at a signal, if it means they can enter the intersection without discomfort. Traffic engineers must separate a sense of discomfort from true risk, because accommodating the desire for comfort for side-street motorists may well cause considerable discomfort on the part of main-street drivers.

The *MUTCD* defines eight general conditions that suggest further consideration should be given. These can be grouped into several categories:

- Volume warrants are used to consider whether the general volume reaches a certain threshold, or whether the flow on the main street is high enough to likely present too few opportunities to enter the street. They are evaluated on the basis of the eight highest hours of the day, the four highest hours, or the peak hour, with the thresholds increasing for each. Pedestrian volumes can also be the basis for consideration.

- Safety warrants include school crossings and also intersections that display a specific and measurable crash history.

- Network and system warrants include those intersections that may not (yet) meet the volume warrants but that rise to a specific standard of importance in the network, or where a traffic signal might help keep vehicles platoon to improve the effectiveness of a signal system.

Readers seeking detailed explanations of the warrants should consult the *MUTCD* directly.

10.2 FIXED-TIME SIGNALS

10.2.1 Fixed-Time Signal Phasing and Timing

Even though nearly all signal controllers now used are traffic-actuated, understanding fixed-time signal operation provides the basis for understanding actuated control in coordinated systems.

10.2.2 Phasing

The first question to be answered in determining fixed-time operation is the order of phases. The primary question is the use of left-turn phases. It has been said that if left turns did not exist, signal timing engineers would be out of work. Thus, the decision to provide protected movements for left-turners will have an effect on all other operational decisions.

Left-Turn Phase Warrants.   Unfortunately, no widely accepted standards for when left turns should receive their own phase have emerged, and practitioners have resorted to various limited research findings, agency policies based on tradition, and rules of thumb. At least one committee of the Institution of Transportation Engineers has been charged with filling this gap, but so far no results have been published.

In the 1980s, research at the University of Texas at Austin presented a methodology for deciding when to use left-turn phases, and this method is presented here to give the reader at least a starting point. Other methods exist, and interested readers should review the literature and consider the methods used by their agencies.

The UT-Austin method assumes that signal timing is already known, or have been provisionally determined. The next step is to compare the left-turn volumes and green times against the formulas presented in the table below:

If the number of opposing lanes is . . .	. . . and the opposing traffic volume (Q_o) times the percentage of opposing green (G/C) is . . .	. . . then the left turn volume must exceed
One	Less than 1,000	$770\,(G/C) - 0.634Q_o$
	Between 1,000 and 1,350	$480\,(G/C) - 0.348Q_o$
Two	Less than 1,000	$855\,(G/C) - 0.500Q_o$
	Between 1,000 and 1,350	$680\,(G/C) - 0.353Q_o$
	Between 1,350 and 2,000	$390\,(G/C) - 0.167Q_o$
Three	Less than 1,000	$900\,(G/C) - 0.448Q_o$
	Between 1,000 and 1,350	$735\,(G/C) - 0.297Q_o$
	Between 1,350 and 2,400	$390\,(G/C) - 0.112Q_o$

G = green time displayed to opposing flow
C = cycle length
Q_o = opposing flow (veh/hr)

The formulas in this table were developed for intersections where adequate left-turn lanes exist.

A practitioner will filter this and all numerical methods through the judgment filter. Numerical methods imply precision where none exists.

Phase Sequence. Traditional phasing uses two opposing leading left-turn phases. This method works best when permitted left turns are used at isolated fixed-time signals, and it is nearly always used at isolated traffic-actuated signals. The use of permitted left turns varies substantially from city to city. Most cities are now using them whenever possible on newer installations, but they will usually have a number of intersections where protected-only operation is used. Other cities have converted their entire network to permitted operation, and still others do not use them at all. They should not be used in situations where speed, visibility, or geometric layout makes these maneuvers difficult.

One situation that should always be avoided is a lagging protected left turn that opposes a permitted left turn. Left-turners in the permitted direction see the through movement signal change to red and mistakenly assume that this change is occurring to the opposing through movement as well. But the opposing through movement is still green, coupled to its adjacent protected left turn. Drivers making a permitted left turn will assume that they can clear during the yellow interval and will be surprised by oncoming traffic that still has a green. This is called the left-turn trap, and several methods of addressing it have been proposed and are being studied in various research projects, with no nationally consistent method yet receiving wide recognition. Until a consistent means of addressing the problem emerges, cautious traffic engineers will avoid the left-turn trap altogether.

In signal systems, the phase sequence decision should be made with system operation in mind. Optimal coordination requires phase sequence flexibility.

10.2.3 Daily Patterns

Any intersection will experience a variety of different traffic demand patterns in any given day. Most new controllers provide many more pattern possibilities than any agency will use. One way to get a handle on the problem is to plot total intersection volume and major street volume on the same graph for a 24-hour period. Characteristics of the plot that would indicate the need for a new pattern would be when the total volume changes substantially (say, greater than 30 percent), or when the fraction of total volume on the major street changes significantly. Without a tremendous amount of analysis work, these decisions must be made based on judgment; no cookbook methods will suffice.

Most agencies, however, do not attempt to design patterns for individual intersections based on daily demand patterns. Rather, they design coordination patterns for networks of signals, and rely on the local intersection actuation capability to accommodate variations that do not affect network operation. Coordination patterns will be covered later in this chapter.

10.2.4 Critical Lane Volumes

Once phasing has been determined, the designer must review the traffic using each phase to determine the effective volume for that phase. In the past, methods used for capacity analysis have been applied to the signal-timing problem to adjust traffic volumes to passenger car equivalents. The designer must be careful in applying these corrections, however, because correction factors used in combination may be misleading and unrealistic.

Correction factors have also been used to convert turning traffic to through-traffic equivalents, and this approach is even more likely to lead one astray. Field experience is essential along with a healthy respect for variability. There is no sense in making signal timing decisions based on 5 percent effects when traffic will routinely vary by 20 percent or more from one cycle to the next or from one day to the next.

Here are some typical values to use as a starting point:

- One truck = 1.5 through passenger cars
- One right-turner = 1.4 through passenger cars (1.25 if the curb return radius exceeds 25 feet)
- One left-turner = 1.6 for nonprotected movement or left turns from a through lane, or 1.1 for protected movements from a left-turn bay

Once these equivalencies can be quantified, the designer can adjust the traffic volumes arriving at the intersection to equivalent through passenger cars and then assign them to lanes. Refer to the example in a later section to see how this is done.

10.2.5 Cycle Length

The next question to be answered is the cycle length to be used. In many cases, this value is already determined because the signal will be part of a coordinated system that has a cycle length already defined. When starting from scratch, however, the cycle length must be determined. As other timing parameters are developed, the cycle length may need to be adjusted, but a common mistake is to define all the other timing parameters and add them up to see what cycle length to use. This usually results in cycle lengths that are too long. Cycle length calculation is not standardized, and many different techniques are used by different practitioners and agencies. Among practitioners who know how to calculate cycle length manually, the two most popular and widely used methods are the Poisson method and Webster's method.

Poisson Method.   In this method, one uses the Poisson statistical distribution to guess at the near-maximum size of the arriving platoon based on average demand. The details of this method are beyond the scope of this book, but we mention it because it exemplifies a method that seeks to minimize *cycle failures,* or those green times that do not serve the demand on the approach at the time. Poisson assumes no vehicle interactions, so it only works at low-volume intersections.

Webster's Method.   In the 1960s, F. V. Webster of England's Road Research Laboratory took a different approach by observing actual operation. He observed the relationship shown in Figure 10.3.

For each situation, Webster found an optimum cycle length that gave the least delay. As cycle length increases, the average wait time increases geometrically, such that doubling the cycle length results in quadrupling the delay. When cycle lengths are too short, too much of the cycle is wasted in queue startup lost time and the number of cycle failures is so high that the queues build faster than they can be dissipated. Therefore, the optimum cycle length is the shortest cycle length that will provide the needed capacity.

Webster's formula is:

$$C = \frac{1.5L + 5}{1 - Y}$$

where C = the optimal cycle length
L = total lost time per cycle (calculated by multiplying the number of critical phases by 3.8 sec)
Y = the total saturation, which is the sum of the critical phases divided by the maximum saturation flow

The maximum saturation flow is subject to some debate, with values ranging from below 1,800 to as much as 2,200 in aggressive urban conditions. We will use 1,800 as the traditional value. This value is used only for cycle length calculation. The critical lane analysis method, when used to calculate capacity, uses a value of under 1,500 to provide some reserve capacity.

Webster's cycle length goes to infinity as demand approaches saturation flow. This is a flaw caused by the fact that he did not study congested intersections. Maximum cycle lengths are tremendously controversial in the traffic engineering community. Proponents of long

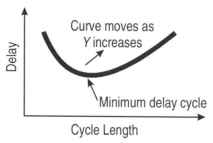

FIGURE 10.3 Webster's observed relationship between delay and cycle length.

cycle lengths argue that once the flow of cars is moving, it should not be stopped. Other long-cycle proponents try to reduce the percentage of the cycle time wasted in startup lost time. Some use long cycle times at intersections where the side street volumes are so low compared to the main street volumes that proportioned green times can only be obtained by having the main street green be extremely long.

Though these arguments can be eloquently voiced, some important points should be considered.

- Only the first 15 or so drivers may be tightly packed entering the intersection. Following drivers are responding to the car in front of them rather than the distant signal, and reaction times may increase departing headways.
- Some approaches are widened compared to the upstream roadways. Any green time longer than what it takes to empty the widened section will be fed by fewer lanes and the flow will go down.
- Driver attention spans are short, and drivers may stop paying attention to the signal after 30 seconds or so.

These experience-based observations suggest that excessive cycles might not gain the capacity increase that Webster (along with many traffic engineers) predicts, but the delay they induce is no illusion. Consequently, many experienced professionals establish a fixed maximum value. The author has used a maximum of 120 seconds in one major city, but practices vary widely.

10.2.6 Splits and Minimum Phase Times

Green splits are calculated by dividing the cycle length in proportion to the critical lane volumes. Phases serving light traffic demand at otherwise heavily traveled intersections may require a disproportionately large portion of the cycle to avoid being too short. What is too short?

Pedestrian Signal Timing.   The *MUTCD* requires that pedestrians be given adequate time to cross an intersection, whether or not pedestrian signal indications are provided. This requirement does not have a pedestrian demand component, and even those locations where no pedestrian traffic is expected are included. The principle conveyed in the *MUTCD* is that pedestrians can expect reasonable accommodation at every traffic signal.

When pedestrian signals are provided, two intervals are timed to accommodate pedestrians. The first is the WALK signal (or white walking person on international displays), which is used only to get waiting pedestrians started. The *MUTCD* suggests a minimum time for a WALK signal of 4 to 7 seconds, and different agencies will interpret this minimum differently, often with a stated policy that practitioners should follow. Pedestrian clearance intervals (flashing DONT WALK or orange hand indication) are calculated based on the length of the crosswalk. Sufficient time must be provided so that a pedestrian walking at 4 ft/sec, can cross the entire traveled width of the street before a conflicting signal is provided. Some of this time can be provided during the clearance interval of the nonconflicting cars being served at the same time, even though many controllers will not use flashing DONT WALK during that period.

At intersections with a protective median of at least 6 feet width, pedestrians can be provided sufficient time to cross only to and from the median. If this is the case, pedestrian signals (and pushbuttons if actuated) must be provided in the median.

At locations without pedestrian signals, the green interval that serves pedestrians and adjacent cars must still be long enough to provide sufficient pedestrian clearance before conflicting traffic is released. Without pedestrian signals, the use of pushbuttons does not relieve the agency of this requirement, because the pedestrian must know what to expect, using either pedestrian signals or sufficient green time.

Minimum Vehicular Movement Time.   Given that most intersections are controlled by actuated controllers, the minimum green time is usually controlled by the detection design. Most agencies have a policy for the minimum green time imposed by a coordination plan, and this usually ranges between 5 and 15 seconds.

Clearance Intervals.   Much discussion and research over the years has been devoted to clearance interval calculation. Most current methods apply equations of physics to stopping vehicles to determine clearance times. Some research has suggested a behavioral approach to try to determine how drivers react to the display of a yellow signal. Many methods have been devised, and three different representative approaches will be presented here.

The Traditional Rule of Thumb.   The traditional clearance interval is 1 second long for each 10 miles/hour of approach speed. This approach is not a method at all, but rather a rule based more on lore than science. The result is usually kept within the range of 3 to 5 seconds. This rule is based on the common (but not proved by research) assumption that faster cars require a greater clearance interval.

The Traditional Analytical Method.   The following equation derives from the physical laws of motion:

$$\text{Clearance interval} = t + \frac{v}{2d + gG} + \frac{l + w}{v}$$

where t = perception and reaction time of the driver, commonly considered 1 sec at intersections, though some research suggests slightly longer periods
v = approach speed, in ft/sec (1.47 times speed in mph)
d = deceleration rate, in ft/sec², normally taken to be 15 ft/sec² but may be closer to 10 ft/sec²
g = acceleration due to gravity (32.2 ft/sec²)
G = grade of approach (positive for uphill approaches, negative for downhill)
w = intersection crossing distance, in ft
l = vehicle length, in ft, typically 20 ft

The first term is the time required for the driver to see the yellow signal and apply the brake. The second term is the time required to traverse the distance that would be required to stop, and the third term is the time required to drive through the intersection.

The clearance interval includes yellow and red clearance (red clearance is the interval between the end of the yellow signal and the beginning of the conflicting green), though most methods (e.g., the publications of the Institute of Transportation Engineers) suggest using the $(l + w)/v$ term for the red clearance interval.

Some assumptions are built into this approach:

• The driver perception–reaction time is constant.

• The approach speed is constant.

Behavioral Research Approach.   Some researchers and practitioners dispute both of the assumptions listed above, having observed that reaction time actually diminishes with higher speeds and drivers who decide to go usually speed up. Their method is based on studies that have been conducted to measure the time required from the onset of yellow until last drivers who don't stop clear the intersection. The research that has considered this approach came to an interesting and convenient conclusion: Most (95 percent) drivers who do not stop take less than 4 seconds to reach the intersection, *no matter what their approaching speed.* The researchers and practitioners who implemented their work typically suggest a 4-second yellow clearance at all intersections plus a red clearance as calculated in the previous method.

These practitioners are in the definite minority, however, and the reader should understand the traditions and policies of the jurisdiction in question before deciding which method to use.

10.2.7 Detailed Example of Phasing and Timing: Northwest Boulevard at Woody Hill Drive

We will work through one typical example in detail. The example is a high-volume arterial street intersection at a relatively lightly traveled residential collector in the morning rush hour. It is perhaps a bit too realistic: the intersection is congested and many of the usual techniques fail. But it presents a scenario that is all too commonly seen by practitioners.

Woody Hill Drive is a low-to-moderate volume residential collector street that feeds a neighborhood on one side of Northwest Boulevard and a shopping center and apartment community on the other side. Northwest Boulevard is the major arterial street connecting the city and its northwest suburbs. This roadway carries about 90,000 vehicles a day at this intersection. The current traffic demand is substantially higher than can be accommodated by the intersection geometry, and the resulting operation is highly congested.

This example concerns the morning peak period. During this period, this intersection creates a queue of over a mile, with typical delays of 20 to 25 minutes. This intersection is a good example of congested intersection timing. Base turning movement counts and layout for this intersection are shown below in Figure 10.4.

Phasing. Let us assume in this case that the decision to use protected left turn north and south has already been made. A close look at the intersection will reveal that the width of the median opening is not enough to allow opposing two-lane left turns off of Woody Hill

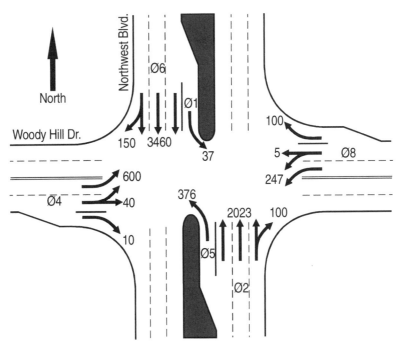

FIGURE 10.4 Example intersection layout.

Drive to run simultaneously, and therefore either lead-lag or split phasing must be used. The left-turn volume on the side street is greater than the through volume, which is common with intersections like this one, and split phasing is therefore the only remaining choice.

Adjusting Volumes.   The first step is to determine the truck correction factor. There are 10 percent trucks, and 1 truck is equal to 1.5 cars, so the correction factor looks like this:

$$(1 - 0.1) + (0.1)(1.5 \text{ cars/truck}) = 1.05$$

This value can be multiplied by the raw count to correct it to equivalent passenger cars. For movements in the same phase group, each volume is multiplied by the turning movement correction as follows:

Southbound Northwest Boulevard:

Right and through:

$$[(150)(1.25 \text{ right-turn factor}) + 3460](1.05) = 3,830 \text{ passenger car equivalents/hr (pceph)}$$

Left turn:

$$(37)(1.1 \text{ left turn factor})(1.05) = 43 \text{ pceph}$$

Eastbound Woody Hill Drive:

Right turns have an exclusive lane, so its volume is adjusted separately:

$$(10)(1.25 \text{ right turn factor})(1.05) = 14 \text{ pceph}$$

Through volumes are much less than the average left turns per lane, so the exclusive through lane does not overflow and the lane will contain both through and left-turn traffic:

$$[40 + (600)(1.1)](1.05) = 735 \text{ pceph}$$

Northbound Northwest Boulevard:

Right and through:

$$[(100)(1.25) + 2023](1.05) = 2255 \text{ pceph}$$

Left turn:

$$(376)(1.1)(1.05) = 435 \text{ pceph}$$

Westbound Woody Hill Drive:

Right turns:

$$(100)(1.25)(1.05) = 131 \text{ pceph}$$

Through volumes are much less than the average lane volume for the through and left lanes, and the exclusive through lane does not overflow.

$$[5 + (247)(1.1)](1.05) = 291 \text{ pceph}$$

We now must determine the average lane volume in each phase group. To do this, we just divide the total volume for that phase group by the number of lanes. When more than one movement uses a phase, the higher of the two average lane volumes is the critical lane volume for that phase.

Northwest Boulevard phases:

Phase 1: 43/1 lane = 43

Phase 2: 2,255/3 lanes = 752

Phase 5: 435/1 lane = 435

Phase 6: 3,830/3 lanes = 1277

Woody Hill Drive phases;

Phase 4: 735/2 lanes = 368

Phase 8:

Through and left: 291/2 lanes = 146

Right: 131/1 lane = 131

Because 146 > 131, 146 is the critical lane volume, and the through and left-turn critical lane volume controls this phase group.

The critical phase combinations must now be evaluated.

Because the left turns from Northwest Boulevard can run simultaneously with each other, or with nonconflicting through movements, their critical lane volumes must be combined. Of the four main street phases, only two combinations result in conflict. All other combinations of these four phases are legal (and will be used in a lead-lag situation). Phase 1 and phase 2 cannot run together without conflict, nor can phases 5 and 6. Consequently, the sequence 1-2 or 5-6 will be the critical sequence of phases on the main street. The critical lane volumes for each phase are added for these sequences to see which one must carry the higher traffic flow:

Phase 1 + Phase 2 = 43 + 752 = 795 pceph per lane

Phase 5 + Phase 6 = 435 + 1,277 = 1,712 pcephpl

Because 1,712 > 795, phases 5 and 6 are the critical phases for the main street. Because the minor street is using split phasing, both phases must run exclusively and are therefore both critical.

Cycle Length.   We will use Webster's method for calculating the cycle length. First we must determine the sum of the critical phase lane volumes:

Phases 5 + 6 + 4 + 8 = 435 + 1,277 + 146 + 368 = 2226 pcephpl

We can then use Webster's equation:

$$C = \frac{1.5L + 5}{1 - Y}$$

L is the total lost time, which is about 4 seconds for each critical phase. There are four critical phases in the cycle at this intersection, so the total lost time is 16 seconds.

$$C = \frac{1.5(16) + 5}{1 - 2{,}226/1{,}800}$$

$$= -121?$$

This bizarre result will always occur when the intersection is more heavily loaded than its capacity. The value of 1,800 is the saturation flow. As the total critical lane volume approaches this value, the resulting C increases to infinity. When 1,800 is exceeded, C becomes negative. Now we have seen the upper boundary of usefulness for Webster's method. Because the intersection is seriously beyond saturation, any cycle length will result in congestion, and the maximum cycle length allowed by policy should be used. In this case, we will assume that this maximum is 120 seconds. This situation is all too common at many locations in most cities, and it illustrates that traffic signals cannot add capacity; they can only distribute what capacity is provided by the available lanes. Traffic engineers are faced with situations repeatedly where they must rely on experience and judgment because reasonable design methods are overwhelmed.

Splits.   Once the cycle length is determined, the green times are prorated based on the ratio of the critical lane volume for that phase to the total critical lane volume. Clearance intervals are subtracted first because they are the same for all phases. Clearance times are set by policy in this jurisdiction at 4 seconds of yellow plus 1 second of red clearance.

Total available green time = 120 − 4 phases (5 sec clearance per phase) = 100 sec

Phase 5 = (435/2,226) (100 sec) = 20 sec green

Phase 6 = (1,277/2,226) (100 sec) = 56 sec

Phase 4 = (368/2,226) (100 sec) = 17 sec

Phase 8 = (146/2,226) (100 sec) =   7 sec

Total   100 sec

Minimum Intervals.   These phase times must be checked to ensure that minimum green times are accommodated. Left-turn minimum greens are assumed by policy to be 5 seconds, with 7-second minimums for through movements. These short minimums reflect timing policies in areas where the intersections are dramatically congested. All green times meet or exceed these values.

We must also check to make sure we have acceptable pedestrian crossing times. Here we may have a problem. While the main-street times are more than sufficient to cross the narrow side street, the side-street times may not be adequate to cross the wide main street. There is a median wide enough to provide pedestrian refuge, though there are no pedestrian signals provided and they will have to be installed. The crossing distance for pedestrians moving with phase 4 is 36 feet to the median and 48 feet from the median to the far side. At four ft/sec, a minimum green time of 12 seconds is required, and the 17-second green time is sufficient. For phase 8, however, things are tight. Pedestrians moving with phase 8 also have a 48-foot cross from the median to the far side, requiring 12 seconds of pedestrian clearance. The green time of 7 seconds is insufficient, but remember that we can also use the vehicle clearance interval because the pedestrians face no conflicting traffic during that interval. Thus, 7 seconds of green plus 5 seconds of clearance interval can all be used for pedestrian clearance, and we meet the 12-second requirement, but with no room to spare. Because of the highly congested nature of this intersection, and because its suburban location has few pedestrians, most practitioners would prefer to prohibit pedestrian crossings adjacent to the phase 8 movements, and require that pedestrians cross on the other side of the intersection.

Noncritical Phases.   Now that we know the critical phase green times, we can proportion the time to the noncritical phases, as follows:

Main street green time = Phase 5 + Phase 6 = 20 + 56 = 76 sec

Total noncritical lane volumes for main street:

Phase 1 + Phase 2 = 43 vphpl + 752 vphpl = 795 vphpl

The green times are determined as follows:

Phase 1 = (43/795)(76) = 4 sec

Phase 2 = (752/795)(76) = 72 sec

        Total   76 sec

Check minimum green times: Phase 1 is 4 sec, which is less than the minimum 5 sec, so set phase 1 at the minimum and assign the remainder to Phase 2.

The final interval timings, using the NEMA scheme for phase numbering, are as follows (Y is the yellow clearance and R is the red clearance for the phase in question.):

Phase	Length
1 + 5	5 sec
1Y + 5	4
1Y + 5R	1
2 + 5	10
2 + 5Y	4
2 + 5R	1
2 + 6	56
2Y + 6Y	4
2R + 6R	1
4	17
4Y	4
4R	1
8	7
8Y	4
8R	1
Total	120

This format explicitly spells out the entire operation at a glance and needs only to be accompanied by a diagram showing phase numbering to allow proper orientation on the street. Total phase length for any phase is easily obtained by adding the interval lengths for all intervals showing that phase number.

10.3   TRAFFIC-ACTUATED SIGNALS

Traffic-actuated signals vary their green time based on demand at the intersection as measured on detectors installed on the approach. These detectors vary in technology, but the most common is the inductive loop detector. A loop of wire is installed in the street and is charged with a slight electrical current, which creates an electrical field. As cars pass through the field, the inductance of the field changes, indicating the presence of a car.

Actuated traffic controllers use the same numbering scheme used in the previous section, and in reviewing the final signal timings of the example, we can see a pattern of operation emerge. This pattern is called ring operation and is the basis for intersection control in the United States. Figure 10.5 lays out the terms and the rules.

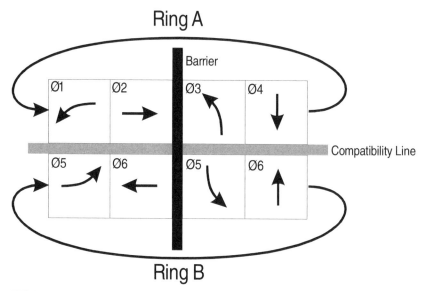

FIGURE 10.5   NEMA ring operation.

If you refer back to the phase-numbering scheme presented at the beginning of the chapter, you will notice right away that all the phases to the left of the barrier serve main-street movements and phases to the right of the barrier serve minor-street movements. You will also notice that any main-street phase in ring A can run simultaneously with any main-street phase in ring B, which is why the line that separates them is called the *compatibility line*. Ring rules state that phases in a ring are served sequentially and never simultaneously and that both rings must cross the barrier at the same time.

Within this context, each phase can either be served or skipped. The decision to skip a phase occurs when there is no car on the detector when the previous phase turns yellow. If a phase is served, it is served for a minimum period called the *minimum green* or *initial*. After the initial, the phase will rest in green until a car passes over a detector on a competing phase. At that point, the phase can be terminated by one of three processes:

- *Gap-out.* As soon as a competing phase has been called, the phase showing green will start a timer, called the extension timer, which counts down from the extension value to zero. The extension timer serves to extend the green for a short period to keep the signal green while tightly bunched cars are present. The extension is set to be long enough to allow closely spaced cars to keep being detected. With each new detection, the extension timer set again to the extension value. Once the gap in the traffic exceeds the extension timer, the timer will reach zero and phase will turn yellow.

- *Max-out.* If traffic is heavy, a sufficient gap to gap-out may not occur in a reasonable period. Thus the actuated controller provides a maximum time to prevent excessive cycle lengths. When this time is reached, the phase will turn yellow. Some controllers will not start the max timer until there is competing demand, and others will time the maximum time and let the controller dwell in green, terminating immediately upon a competing detection. The former method is more common. Maximum times are often arbitrarily set, but they should be set with some sensitivity to the demand. One useful approach is to calculate the fixed-time operation as outlined in the previous section and set the maximum greens to values perhaps 20 percent higher than that to allow some flexibility.

• *Force-off.* In coordinated signal systems, the network signal timing needs to impose discipline on the actuation process to keep the intersection in coordination with other intersections. Actuated controllers contain a cycle timer that starts with the end of the main-street green. When there are competing calls, the main-street green phase is held green until the time specified by the system, and then the controller can cycle through the other phases. Each of these phases may gap out or max out, but if they are extended beyond the time allowed by the system timings, they will be forced off. This force-off value is a setting in the controller that defines the point in the cycle timer at which this phase will be forced off. The relationship of this intersection with others is controlled by an *offset,* which is the time between the starting point of the cycle timer at this intersection and the starting point of an arbitrary system-wide cycle timer.

A whole chapter, or indeed an entire book, could be devoted to the individual settings and modifications that are possible to these basic timings, but they are beyond the scope of this book. Suffice to say that none of these values is constant. The minimum green (or initial), the extension, the maximum, and the force-off can all be modified by special features and processes based on demand conditions or design. In particular, the relationship of these values with the locations and dimensions of the detectors merits special study for those who are interested in deeper study.

10.4   *TRAFFIC SIGNAL SYSTEMS*

When traffic signals are close enough so that the platoons formed by one signal are still tightly formed at the next signal, traffic engineers can coordinate the intersections together so that the platoons avoid arriving at red lights as much as possible. This simply stated objective belies the tremendous complexity associated with this task, and indeed the signal coordination is far more important and complicated than fixed-time or actuated control. Coordination attempts to achieve some combination of the following objectives:

• Minimize delay
• Minimize stops
• Minimize fuel consumption
• Minimize pollution emissions
• Maximize smooth flow
• Maximize capacity
• Minimize the arrival of platoons at red lights

There are other more esoteric objectives than these dictated by changing social requirements. It should be noted that these objectives often compete with one another, and each traffic engineer has to find the mix of objectives that is most appropriate in the local jurisdiction. One should also note that verbalizing the objective is easier than attaining it, and sometimes traffic engineers claim different objectives than they truly seek. Most, for example, will talk about minimizing delay, but then they will calculate timings that provide better access to some side streets even though overall delay might be higher. The one objective that is a reality for nearly all agency practitioners is not stated in any of the research or computerized methods: Minimize citizen complaint calls

Achieving signal coordination requires two branches of knowledge. The first is signal timing on a network scale, and the second is understanding how traffic signals can be tied together into functioning signal systems. We will discuss each of these topics.

10.4.1 Signal Coordination

As was stated in the previous section, signals are coordinated by means of an offset value that defines how much the cycle at one intersection is delayed from an arbitrary system zero point. These offset values are calculated to provide the best network flow based on one of the objectives mentioned above. No matter what the objective, however, traffic engineers use time-space diagrams to visualize these timing patterns. A typical time-space diagram is shown in Figure 10.6.

The sequence of phases is shown extending horizontally and the distance along the road is shown vertically. Thus, a constant speed can be represented by the sloped lines. If a driver can maintain a constant speed through all intersections, we say that we have achieved *progression,* and his course through the time-space diagram is a continuous line. A *progression band* is the collection of all such lines.

Phase sequence is also an issue in progression. The time-space diagram in Figure 10.7 is the same example as above, but with a different phase sequence at the middle intersection. Notice how the progression bands have been widened as a result.

Notice also the resonance between signal spacing, progression speed, and cycle length. Figure 10.8 shows the same system and the same splits, but with a cycle length that prevents progression.

In past years, traffic engineers developed these diagrams by hand, optimizing them by hand or through the use of simple computer programs. Today, however, most traffic engineers use complex computer programs and simulations to develop signal timings, though most still visualize those timings using time-space diagrams.

10.4.2 System Design

Simply put, the objective of a traffic signal system is to implement signal timings in a way that can be efficiently operated and maintained. System designers devote a lot of energy,

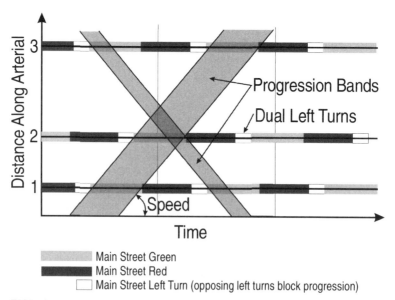

FIGURE 10.6 Time-space diagram.

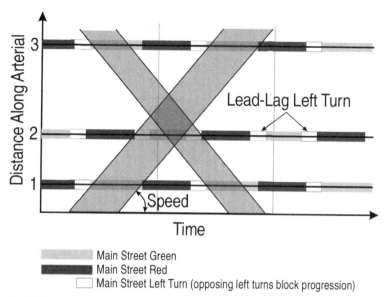

FIGURE 10.7   Time-space diagram showing more effective phase sequence.

however, to determining which timings are implemented at what times and whether conditions provide an opportunity to adjust those timings. To summarize these considerations, we will present three categories of signal systems: time-of-day operation, traffic-responsive operation, and adaptive control.

These systems are physically implemented in a variety of ways. At their simplest, the coordination settings may be installed in each controller, with accurate internal clocks syn-

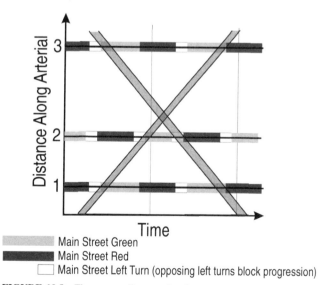

FIGURE 10.8   Time-space diagram showing unresonant cycle length.

chronized to maintain the relationship between intersections. This is known as time-based coordination and is subject to clock drift and other problems. Most agencies prefer to use it only as a short-term solution or as a backup when communications are lost. Small systems may use a device on the street to maintain coordination that can be dialed into from a computer at the traffic engineer's office. These *closed-loop* systems can also be tied together into larger systems, which is known as a distributed architecture. Centralized systems maintain central control of all the signal controllers. These days, the only thing preventing all systems from such central operation is the cost of the required communications infrastructure. That infrastructure was presented in Chapter 9 on freeway management systems, with which signal systems are increasingly integrated.

Time-of-Day Operation.   Most signal systems in the United States use signal timings that are designed based on historical traffic data and then stored in the system to be implemented according to a fixed, daily schedule. Time-of-day systems are still completely effective in many situations, especially considering the advanced features that traffic engineers can use to fine-tune them. These features include the ability to create exceptions for weekends, holidays, and other special-purpose plans. Some surveys show that time-of-day operation is used in as much as 95 percent of all signal systems. Some portray these findings as a waste of the traffic-responsive features in many of those same systems, but in fact there is no evidence to prove that traffic-responsive operation is really more effective in common situations.

Traffic-Responsive Operation.   Starting in the 1950s, some systems were able to use traffic detection to determine when preengineered patterns would be brought into operation. These methods were first established in common practice with the advent of the Urban Traffic Control System (UTCS), a federal research program in the early 1970s that created a public-domain signal system that was the mainstay of U.S. large-scale systems for the following 15 years. UTCS traffic-responsive operation has grown into two general approaches for responding to measured demand by the implementation of a preengineering signal timing plan.

The more common approach is intended for arterial networks that have a significant inbound and outbound flow pattern. These systems define system detectors for collecting data. System detectors are usually regular intersection detectors that have been assigned also as system detectors. As with freeway systems, they measure volume (number of cars) and detector occupancy (percentage of time detector is covered by a car). These systems combine these values into a single parameter that they compare first with an *offset-level threshold,* which indicates whether it is an inbound, outbound, or average pattern. Within each offset level, they then determine which cycle to use and then which collection of splits to use. As with all traffic-responsive systems, they use existing timing patterns that were designed off-line and stored in the system.

The other approach is better suited to grid networks where inbound and outbound influences might be harder to observe. It uses a pattern-recognition technology. Each signal-timing plan is associated with a collection of system-detector volume and occupancy values. The system compares observed detector values against the stored values associated with each plan. The plan whose detector values show the least difference from measured values is the one selected.

With both these approaches, detector data are averaged over a control period, which may range from several minutes to as much as half an hour. When a plan is selected, the controllers are instructed to implement the plan, which may take up to several minutes. Each controller will shorten or lengthen its force-off values to nudge the cycle into the new plan incrementally, with the transition process controlled by a series of parameters set by the traffic engineer. If the control period is too short, a large percentage of the time might be spent in transition, and signal coordination is poor during those periods. If the control period is too long, the condition to which the system is responding might have changed yet again by the time the response is in place.

To make a traffic-responsive system work better than a time-of-day system, traffic engineers much spend considerable time with experimentation and fine-tuning. For this reason, most agencies are sufficiently constrained by budget and time to make traffic-responsive operation not worth the difficulty.

Adaptive Control.   Some newer systems (and some not so new) have the ability to measure traffic conditions and develop new signal timings on the fly. These systems have been grouped under the heading of *adaptive control.* A number of technologies for adaptive control exist in theory, one or two of which have received wide implementation throughout the world. None have yet been fully implemented based on compatibility with U.S.-style controllers that follow ring rules. Thus, adaptive control systems have not been widely deployed in the United States.

Some adaptive control systems are based on incremental changes to the signal timing based on measured traffic flow on special system detectors, and others use those data to make predictions about traffic flow a few seconds or minutes into the future. All of these systems require extensive fine-tuning during implementation—far more than with traffic-responsive operation. Unlike systems that use canned signal timing plans, however, adaptive systems should be able to track long-term trends in traffic such that signal timings do not have to be calculated again every few years. This is often reported as the key advantage of adaptive systems.

Conceptually, there are reasons to be cautious in the implementation of adaptive control. The main concern is that traffic control experts cannot agree on the objectives and calculation of optimal signal timing plans using offline methods. Presumably, an off-line method has available nearly unlimited resources of time, data, and computing power compared with methods that must be computed in real time. Yet debate as to which off-line method is superior never ceases. If getting the absolute best out of any signal system requires the fully committed calculation and implementation of an off-line method, then an adaptive system cannot hope to marshal the resources necessary to achieve the same result online. Thus, a stable operation that is some fraction of the potential of the network is the best one can hope to achieve. On the other hand, many time-of-day and traffic-responsive systems are severely out of date, with their timings corrupted by years of incremental changes in response to a variety of influences. Consequently, few of those systems achieve anywhere near the potential for off-line optimization. Any agency considering their options must evaluate their own ability to implement a highly complex system requiring intensive fine-tuning versus their ongoing ability to monitor, tweak, and refine traditional signal timings.

10.5 REFERENCES

Federal Highway Administration (FHWA). 2001. *Manual on Uniform Traffic Control Devices, Millennium Edition.* U.S. Department of Transportation, FHWA, Washington, DC.

Gordon, R. L., R. Reiss, H. Haenel, R. Case, R. French, A. Mohaddes, and R. Wolcott. 1996. *Traffic Control Systems Handbook, 1995 Edition.* U.S. Department of Transportation, FHWA, Washington, DC.

Kell, J. H., and I. J. Fullerton. 1998. *Manual of Traffic Signal Design,* 2nd ed. Washington, DC: Institute of Transportation Engineers.

National Committee on Uniform Traffic Laws and Ordinances (NCUTLO). 2000. Uniform Vehicle Code. NCUTLO, Alexandria, VA.

Pline, J. L., ed. 1999. *Traffic Engineering Handbook,* 5th ed. Washington, DC: Institute of Transportation Engineers.

CHAPTER 11
HIGHWAY SIGN VISIBILITY

Philip M. Garvey
Pennsylvania Transportation Institute,
Pennsylvania State University, University Park, Pennsylvania

Beverly T. Kuhn
System Management Division,
Texas A&M, College Station, Texas

11.1 MEASURES OF SIGN EFFECTIVENESS

Sign visibility is an imprecise term encompassing both sign detectability and sign legibility. Sign detectability refers to the likelihood of a sign being found in the driving environment and is integrally associated with sign conspicuity. Sign conspicuity is a function of a sign's capacity to attract a driver's attention that depends on sign, environmental, and driver variables (Mace, Garvey, and Heckard 1994). Sign legibility describes the ease with which a sign's textual or symbolic content can be read. Sign legibility differs from sign recognizability in that the former refers to reading unfamiliar messages while the latter refers to identifying familiar sign copy. Sign legibility and recognizability in turn differ from sign comprehensibility in that the latter term implies understanding the message while the former merely involve the ability to discern critical visual elements.

While other measures of sign visibility such as blur tolerance (Kline et al. 1999) and comprehension speed (Ells and Dewar 1979) are sometimes used, sign visibility is most often assessed by determining threshold distance. The intent is to provide the sign's observer with the maximum time to read the sign, and to do that the observer must find it prior to or at its maximum reading distance. Therefore, in designing a sign it is desirable to achieve a detection distance that is equal to or greater than that sign's legibility distance.

To describe legibility distance across signs with different letter heights, the term legibility index is often used. Legibility index or (LI) refers to the legibility distance of a sign as a function of its text size. Measured in feet per inch of letter height (ft/in.), the LI is found by dividing the sign's legibility distance by its letter height. For instance, a sign with 10-inch letters legible at 400 feet has an LI of 40 ft/in. (400/10).

11.2 VISUAL PERCEPTION

In the initial stages of vision, light passes through the eye and is focused by the cornea and lens into an image on the retina. The retina is a complex network of nerve cells or light receptors composed of rods and cones. So named because of their basic shape, rods and cones perform different functions. Cones function best under high light intensities and are responsible for the perception of color and fine detail. In contrast, rods are more sensitive in low light but do not discriminate details or color. Rods and cones are distributed across the retina in varying densities.

The region of the central retina where a fixated image falls is called the fovea. The fovea has only cones for visual receptors and is about 1.5 to 2° in diameter. Beyond 2°, cone density rapidly declines reaching a stable low point at about 10°. Conversely, rod density rapidly increases beyond 2° and reaches a maximum at about 18° before dropping off (Boff, Kaufman, and Thomas 1986). From 18° outward toward the nose and ears, forehead, and chin, the number of rods decreases but still continues to be higher than the number of cones.

Functional detail vision extends to about 10°, worsening in the near periphery from about 10 to 18° and significantly deteriorating in the far periphery from about 18 to 100°. Occasionally the term *cone of vision* appears in the visibility literature. Although this expression is used freely, it is not well defined. In general, the cone of vision can be taken to refer to an area within the near periphery. The *Traffic Control Device Handbook* (FHWA 1983) section on driver's legibility needs states:

> When the eye is in a fixed position it is acutely sensitive within a 5 or 6 degree cone, but is satisfactorily sensitive up to a maximum cone of 20 degrees. It is generally accepted that all of the letters, words, and symbols on a sign should fall within a visual cone of 10 degrees for proper viewing and comprehension.

11.2.1 Visual Acuity

Visual acuity is the ability to discern fine objects or the details of objects. In the United States, performance is expressed in terms of what the observer can see at 20 feet, referenced to the distance at which a "normal" observer can see the same object. Thus, we have the Snellen expressions 20/20 and 20/40, where the test subject can see objects at the same and half the distance of the standard observer, respectively. Visual acuity can be measured either statically and/or with target motion, resulting in static visual acuity (SVA) and dynamic visual acuity (DVA). Although there are questions as to its relationship to driving ability, static visual acuity is the only visual performance measure used with regularity in driver screening.

11.2.2 Contrast Sensitivity

Contrast sensitivity differs from visual acuity in that it measures an individual's ability to detect contrast, or the difference between light and dark areas of a target. Contrast sensitivity is derived from an individual's contrast threshold. Contrast threshold is the minimum detectable luminance difference between the dark and light portions of a target. Contrast sensitivity is the reciprocal of the contrast threshold; that is, contrast sensitivity = 1/contrast threshold. Therefore, a high contrast threshold represents low contrast sensitivity—in other words, a large difference between the darkest and lightest portions of a target is necessary for target visibility.

11.2.3 Visual Field

A driver's field of vision or visual field is composed of foveal and peripheral vision, or literally everything the driver can see. Visual acuity is highest in the central fovea. Beyond the fovea, vision deteriorates rapidly. In the near periphery, individuals can see objects but color and detail discriminations are weak. The same holds true for far peripheral vision. Aside from these general visual field break points, no specific designations of intermediate vision exist.

The related concept of useful field of view (UFOV) has gained widespread acceptance as a potentially useful tool to describe vision in natural settings (Ball and Owsley 1991). Proponents of UFOV assert that while an individual's visual field is physically defined by anatomical characteristics of the eye and facial structure, it can be further restricted by emotional or cognitive states.

11.2.4   Glare

Another set of terms related directly to drivers' visual abilities concerns the adverse effects of light from external sources, which include, but are not limited to, signs, headlamps, and overhead lights. Four types of glare exist: direct, blinding, disability, and discomfort. Direct glare comes from bright light sources or the reflectance from such sources in the driver's field of view (FHWA 1978a). Blinding glare and disability glare differ only in degree. Blinding glare results in complete loss of vision for a brief period of time, whereas disability glare causes a temporary reduction in visual performance ranging from complete to minor. Viewing comfort is reduced by discomfort glare.

11.3   PHOTOMETRY

Photometry is the measurement of radiant energy in the visual spectrum; that is, the measurement of light. Photometric equipment is designed and calibrated to match the human eye's differential sensitivity to color and to daytime-versus-nighttime lighting. Therefore, photometry is the measurement of light as people see it. The four most important photometric measurements used to describe sign visibility are luminous intensity, illuminance, luminance, and reflectance.

Luminous intensity, expressed in candelas (cd), is a description of a light source itself and is therefore independent of distance. That is, no matter how far away an observer is from a lamp, that lamp always has the same intensity. Luminous intensity is the photometric measurement most often specified by lamp and LED manufacturers.

Illuminance, or incident light, is expressed in units of lux (lx) and is a measure of the amount of light that reaches a surface from a light source. Illuminance is affected by distance and is equal to luminous intensity divided by the distance squared ($lx = cd/d^2$). For example, if a source emits a luminous intensity of 18 cd, the illuminance level measured 3 meters away would be 2 lx ($18/3^2$). If the source's intensity is unknown, illuminance is measured with an illuminance or lx meter placed on the surface of interest facing the light source.

Luminance is expressed in candelas per square meter (cd/m^2). Luminance is the photometric that most closely depicts the psychological experience of "brightness." Luminance can refer to either the light that is emitted by or reflected from a surface, and is an expression of luminous intensity (cd) over an extended area (m^2). Like luminous intensity, a source's luminance is constant regardless of distance. To measure luminance, a luminance meter is placed at the observer's position, aimed at the target of interest, and a reading is taken.

Reflectance is the ratio of illuminance to luminance and, as such, reflectance describes the proportion of incident light that is absorbed and the proportion that is reflected by a surface. If, for example, 100 lx hits an object's surface and that surface has a luminance level of 5 cd/m^2, that surface has a reflectance of 5 percent (5/100).

A related term often found in traffic sign literature is retroreflection. Retroreflection describes material that, unlike normal matte (i.e., diffuse) or specular (i.e., mirror-like) surfaces, reflects most incident light directly back to the light source. With regard to signs and other vertical surfaces, retroreflection is represented by a coefficient of retroreflection, known as R_A, although it is also commonly referred to as the "specific intensity per unit area," or SIA (ASTM 2001). A material's SIA is the ratio of reflected light to incident light. SIA values for signs are commonly expressed in candelas per lux per square meter ($cd/lx/m^2$).

11.4   FEDERAL TRAFFIC SIGN REGULATIONS

The *Manual on Uniform Traffic Control Devices (MUTCD)* (FHWA 2000) is the legal document that governs traffic control device design, placement, and use in the United States. All local, county, and state transportation departments must use this document as a minimum standard. According to the document, a traffic control device is "any sign, signal, marking, or device placed on, over, or adjacent to a street or highway under the authority of a public body or official for the purpose of regulating, warning, or guiding traffic" (FHWA 2000).

11.4.1   Sign Height

The *MUTCD* requires that any sign placed at the side of a roadway have a minimum vertical height ranging from 5 to 7 feet, depending on the general location and the number of installed signs. For instance, a sign in a rural area or on freeways must have a vertical height of at least 5 feet, while a sign in a business, commercial, or residential area must have a vertical clearance of at least 7 feet (FHWA 2000). The reason behind this variation is that more vertical clearance is needed in areas where pedestrian traffic is heavier. To allow for the passage of the largest vehicles, overhead signs must have a vertical clearance of 17 feet.

11.4.2   Lateral Offset

As with sign height, lateral clearance requirements vary with sign location. For instance, the *MUTCD* states that traffic signs placed along a roadway should have a minimum lateral clearance of 6 feet from the edge of the shoulder. However in urban areas where lateral offset is often limited, offsets can be as small as 1 to 2 feet (FHWA 2000). These small lateral clearances reflect the lower vehicle speeds found in urban areas. The basic premise behind even the smallest minimum standards is a combination of visibility, safety, and practicality.

11.4.3   Reflectorization and Illumination

Regulatory, warning, and guide signs must be illuminated in such a manner that they have the same appearance at night as in daytime. The *MUTCD* provides general regulations regarding the illumination of traffic signs. For instance, signs may be illuminated in three ways: (1) by internal lighting (e.g., fluorescent tubes or neon lamps) that illuminates the main message or symbol through some type of translucent material; (2) by external lighting (including high-intensity discharge lamps or florescent lighting sources) that provides uniform illumination over the face of the sign; and (3) by some other method such as luminous tubing, fiber-optics, incandescent panels, or the arrangement of incandescent lamps (FHWA 2000).

On freeway guide signs, all copy (legends, borders, and symbols) must be retroreflectorized with the sign background either retroreflectorized or otherwise illuminated. On over-

FIGURE 11.1

head sign mounts, headlight illumination of retroreflective material may not be sufficient to provide visibility (FHWA 2000). In such situations, supplemental external illumination is necessary. Supplemental illumination should be reasonably uniform and sufficient to ensure sign visibility. Carlson and Hawkins (2002b) recently developed recommendations for minimum retroreflectivity for overhead guide signs and street name signs, and the FHWA is in the process of developing minimum retroreflectivity levels for all highway signs (FHWA 2000).

11.4.4   Message Design

Both the *MUTCD* and *Standard Highway Signs* (FHWA 2002) specify guide sign content characteristics. Although they allow some flexibility for overall sign layout, these two federal documents provide detailed specifications for message design.

11.4.5   Font

With regard to font, the federal specifications are very precise. The letter style used must be one of the Standard Highway Alphabet series (FHWA 2002). For mixed-case words (i.e., an initial capital followed by lower-case letters), the capital letter must be in Standard Highway Series E modified (Figure 11.1). Only one lower-case font is permitted (also E(M)). The FHWA is currently reviewing a new font called Clearview for use on all U.S. road signs and inclusion in Standard Highway Signs (Figure 11.2).

11.4.6   Case

Destination names on freeway guide signs must use mixed case. The upper-case letters should be about 1.33 times the lower-case loop height. Destination names on conventional road guide signs and word messages on street name signs may use either mixed or all upper-case lettering; however, all other lettering on conventional roadways must be upper-case.

FIGURE 11.2

11.4.7 Letter Height

As stated in *Standard Highway Signs* (FHWA 2002), the general rule for selecting letter height is based on a legibility index of 40 feet per inch of letter height (LI = 40). This is a 10 ft/in. reduction from the LI of 50 that has been the standard for almost 50 years (FHWA 1988). In addition to the general rule, the *MUTCD* (FHWA 2000) establishes 8 inches as the minimum letter height for freeways and expressway guide signs, although overhead signs and major guide signs require larger lettering. For example, on expressways the minimum height ranges from 20 inches/15 inches (upper-/lower-case) for a major roadway to 10.6 inches/8 inches for a road classified as minor, and for freeways 20 inches/15 inches to 13.3 inches/10 inches, respectively, for these two roadway types. *Standard Highway Signs* (FHWA 2002) states that the minimum letter height for the principal legend on conventional road guide signs on major routes is 6 inches; this drops to 4 inches for urban and less important rural roads. The long-held 4-inch minimum for street name signs found in previous versions of *Standard Highway Signs* (FHWA 1978b) has been changed to 6 inches with the 2002 edition.

11.4.8 Border

All signs are required to have a border. The border must be the color of the legend and, in general, should not be wider than the stroke width of the largest letter used on that sign. Guide sign borders should extend to the edge of the sign, while regulatory and warning sign borders should be set in from the edge.

11.4.9 Spacing

The spacing between the sign copy and the border should be equal to the height of the uppercase letters. The spacing between lines of text should be 75 percent of the upper-case letter height. Spacing between unique copy elements (for example, words and symbols) should be 1 to 1.5 times the upper-case letter height (FHWA 2002).

11.5 *SIGN VISIBILITY RESEARCH*

When reviewing the sign visibility literature, it is useful, if somewhat artificial, to divide the research by the type of visibility studied. The two main sign visibility research areas focus on sign detection and sign legibility. These areas are interdependent; one cannot read a sign that one cannot find, and there is little point in detecting a sign that cannot be read from the road. However, from a sign design and placement viewpoint, the characteristics that affect detection and legibility differ enough qualitatively to warrant separate consideration.

11.5.1 Detection

A sign's detectability is directly related to its conspicuity. If a sign is conspicuous, it will be detected at a greater distance than if it easily blends into its surroundings. While this relationship is fairly obvious, the variables that determine sign conspicuity may be less so. This section addresses conspicuity and the characteristics of the environment, sign, and driver that directly affect both the likelihood and distance of sign detection.

11.5.2 Sign-Placement Variables

One of the most important factors in sign detection has nothing to do with characteristics of the sign itself. That factor is sign placement. Because a sign that is not well placed cannot be seen and therefore cannot be read, a sign's positioning relative to its environment is key to its effectiveness. This section covers sign mounting height and offset and the immediate sign environment, or the sign's "surround."

Lateral and Vertical Offset. Careful placement of signs along the roadway ensures that a driver has sufficient time to detect the sign and take necessary action. Upchurch and Armstrong (1992) found placement of signs with respect to restricting geometric features of the roadway, such as hills and curves, to be important in maximizing detection distance. Mace and Pollack (1983) stated that as the distance between a target sign and "noise" items increases, the sign becomes more conspicuous, although this conspicuity is eroded as the sign becomes located further from the center of the driver's visual field. Claus and Claus (1975) quantified this when they wrote that signs should be placed within 30° of the driver's line of sight. Matson (1955) suggested "that a sign should fall within a visual cone of 10 to 12 deg on the horizontal axis and 5 to 8 deg on the vertical axis" (in Hanson and Waltman 1967). Jenkins and Cole (1986) took this statement a step further when they wrote, "[I]f a sign is to be noticed . . . it will be within 10 degrees of his line of sight. When the eccentricity . . . becomes greater than this, the sign is most unlikely to be noticed at all." Jenkins and Cole's statement is supported by Zwahlen's (1989) study of nighttime traffic sign conspicuity in the peripheral visual field. Zwahlen found that retroreflective signs placed in the foveal region resulted in twice the detection distances of those located 10° outside this central visual area. A further result from Zwahlen's study was that signs located 20 and 30° outside the fovea resulted in one-third and one-quarter the detection distances, respectively, of centrally located signs.

As indicated by Zwahlen's study, sign placement is particularly important for retroreflective signs. This is because the angle between the vehicle and the sign strongly influences sign brightness and therefore nighttime sign detection (McNees and Jones 1987). A retroreflective material returns light back to its source as a function of entrance angle which describes the relative positions of the headlamps and sign. As this angle increases, the amount of reflected light seen by the driver decreases (King and Lunenfeld 1971). Thus, when placing retroreflective signs it is important to obtain entrance angles as close to the manufacturer's recommendations as is practically possible.

Surround. Where a sign is placed in relation to other visual stimuli defines the visual complexity of the sign's surround. The factors that affect visual complexity include the number and overall density of noise in the driver's visual field and the density of noise items immediately adjacent to the sign (Mace and Pollack 1983). Research conducted on signs with various levels of retroreflectivity in different environments reveals that virtually any retroreflective sign can be seen at a reasonable distance in an environment that is not visually complex (Mace and Pollack 1983). In other words, if a sign does not have to compete with many other objects in a driver's cone of vision, it is conspicuous, even if its retroreflectivity is low. However, in an area that has more visual distractions, sign conspicuity becomes more a function of its retroreflectivity, size, color, and other variables (Mace and Pollack 1983; Mace, Garvey, and Heckard 1994). McNees and Jones (1987) supported Mace and his colleagues when they asserted that as the number of objects in the driver's cone of vision increases, the conspicuity of a sign decreases. However, when a sign is located in a visually complex environment, retroreflectivity may not be enough to ensure sign detection (Mace and Pollack 1983). Thus, in more complex environments, conspicuity boosters will be needed to achieve a desired detection distance. In such situations, additional lighting or sign redundancy may be necessary to provide adequate conspicuity to ensure timely sign detection.

11.5.3 Lighting Variables

The first step in visual perception is the detection of light. Differences in the quantity (e.g., luminance) and quality (e.g., color) of light are necessary to differentiate objects. A sign with the same luminance and color as its background is undetectable. Therefore, the term *lighting variables,* as used in this section, refers not only to illuminated nighttime sign display, but to all factors that fall within the category of photometric sign properties. While this category includes nighttime illumination techniques, it also covers daytime and nighttime sign luminance, sign color, and luminance and color contrast between the sign and surround.

External Luminance Contrast.   Sign contrast affects both conspicuity and legibility distance. In sign visibility research, there are two types of luminance contrast: external and internal. Detection distance is affected by external sign contrast, which is the ratio of the sign's average luminance and the luminance of the area directly surrounding the sign. Legibility distance is affected by internal sign contrast, defined by the ratio of the luminance of a sign's content and its background.

As a sign's external contrast ratio increases, so does the sign's conspicuity (Forbes et al. 1968a; Mace and Pollack 1983). Mace, Perchonok, and Pollack (1982) concluded that in low visual complexity locations, external contrast and sign size are the major determinants of sign detection. Cooper (1988) goes a step further in stating that external contrast plays a far greater role in sign conspicuity than does sign size. While no research provides optimal and minimum values for external contrast, McNees and Jones (1987) found high-intensity background sheeting (i.e., ASTM Type III, or encapsulated lens) with high-intensity copy, opaque background sheeting with button copy (i.e., letters embedded with "cat's eye" reflectors), and engineer-grade background sheeting (i.e., ASTM Type I, or enclosed lens) with button copy to provide acceptable freeway guide sign-detection distances.

Sign Luminance.   Mace and Pollack (1983) stated that sign conspicuity increases with higher sign luminance. Furthermore, Mace, Perchonok, and Pollack (1982) concluded that, with the exception of black-on-white signs, increasing sign luminance could even offset the detrimental effects of increased visual complexity. Pain (1969) stated "a higher [overall] brightness enhances a high brightness ratio [external contrast] by roughly 10 percent." Zwahlen (1989) buttressed these findings when he concluded that increasing retroreflective sign SIA values can offset the negative effects of peripheral location. Research conducted on various types of commonly used retroreflective background sheeting combined with reflective copy concurs, indicating that conspicuity increases as sign retroreflectivity increases (McNees and Jones 1987). Research on white-on-green signs (Mace, Garvey, and Heckard 1994) goes further in reporting that sign brightness can actually compensate for sign size. Mace et al. found that small (24-inch) diamond-grade (i.e., ASTM Type VII, or microprismatic) signs produced the same legibility distance as large (36-inch) engineer-grade signs.

Nighttime conspicuity research conducted by Mace, Garvey, and Heckard (1994) indicated that the relationship between sign brightness and detection distance is mediated by sign color. They found no difference in detection distance for either black-on-white or black-on-orange signs as a function of retroreflective material. However, higher-reflectance materials resulted in an improvement in detection distance for white-on-green signs at high and low visual complexity sites.

Color.   Forbes et al. (1968b) concluded that "relative brightness is of most importance, but hue [color] contrast enhances the brightness effects in some cases." Of the background sign colors black, light gray, and yellow, Cooper (1988) found yellow to be the most effective for sign detection. Mace, Garvey, and Heckard (1994) reported that black-on-orange and white-on-green signs were detected at greater distances than black-on-white signs. This is consistent with the research of Jenkins and Cole (1986), which found black-on-white signs to provide particularly poor conspicuity. Mace, Garvey, and Heckard (1994) concluded that the reason for this was that white signs were being confused with other white light sources and that it was necessary to get close enough to the sign to determine its shape before recognizing it as a sign. Mace et al.'s research punctuates the interaction between various sign characteristics (such as shape and color) in determining sign conspicuity. Zwahlen and Yu (1991) furthered the understanding of the role color plays in sign detection when they reported their findings that sign color recognition distance was twice that of shape recognition and that the combination of a highly saturated color and specific shape of a sign could double a sign's average recognition distance.

11.5.4   Sign Variables

In addition to environmental and photometric variables, there are a number of characteristics related to sign structure and content that have been found to affect sign detection. These characteristics include the size and shape of the sign and the message design.

Size and Shape.   The size and shape of a sign relative to other stimuli in the driver's field of vision play a role in determining the sign's conspicuity. Mace, Garvey, and Heckard (1994) found significant increases of around 20 percent in both nighttime and daytime detection distances with increases in sign size from 24 to 36 inches for black-on-white, black-on-orange, and white-on-green signs. In 1986, Jenkins and Cole conducted a study that provides corroborative evidence that size is a key factor in sign detection. Jenkins and Cole concluded that sign sizes between 15 and 35 inches are sufficient to ensure conspicuity, and that if signs this size or bigger are not detected, the problem is with external contrast or surround complexity. In addition to the effects of sign size, Mace and Pollack (1983) concluded that conspicuity also increases if the shape of the sign is unique compared to other signs in the area.

Display.   Forbes et al. (1968b) found green signs with high internal contrast to improve sign detection. In particular, these researchers found signs with bright characters on a dark background to have the highest conspicuity under light surround conditions and the reverse to be true for dark or nighttime environments. Hughes and Cole (1984) suggest that bold graphics and unique messages increase the likelihood of meaningful detection.

11.5.5   Legibility

Once a sign has been detected, the operator's task is to read its content; this is sign legibility. *Legibility* differs from *comprehensibility* in that *legibility* does not imply message understanding. Symbol signing provides a good example of this distinction. An observer could visually discern the various parts of a symbol and yet be unable to correctly report that symbol's meaning. The same is true for alphanumeric messages with confusing content. The problem with drawing a distinction between legibility and comprehension, however, is that familiar symbolic and textual messages are accurately reported at much greater distances than novel sign copy (Garvey, Pietrucha, and Meeker 1998). This well-documented phenomenon leads to the need to distinguish legibility from recognition.

Because recognition introduces cognitive factors, message recognition does not require the ability to discriminate all the copy elements—such as all the letters in a word, or all the strokes in a symbol—in order for correct identification to occur (Proffitt, Wade, and Lynn 1998). Familiar word or symbol recognition can be based on global features (Garvey, Pietrucha, and Meeker 1998). Sign copy recognition distances are, therefore, longer than would be predicted by either visual acuity or sign characteristics alone (Kuhn, Garvey, and Pietrucha 1998). In fact, one of the best ways to improve sign-reading distance is not through manipulation of sign characteristics, but rather by making the sign copy as familiar to the target audience as possible, a concept not lost on the advertising community.

However, modifications in sign design that improve sign legibility will enhance the reading distance for both novel and familiar content. The following sections provide an overview of more than 60 years of research on how to improve sign legibility. The research emphasizes the importance of sign characteristics such as photometric properties and symbol and textual size and shape.

11.5.6   Lighting Variables

The role of lighting variables in sign legibility is probably one of the best-researched areas in the sign visibility field. In this research, negative-contrast sign legibility (i.e., dark letters on a lighter background; e.g., regulatory and warning signs) is typically measured as a function of sign background luminance, and positive-contrast sign legibility (i.e., light letters on a darker background; e.g., guide signs) as a function of internal luminance contrast ratio.

Internal Contrast.   Sivak and Olson (1985) derived perhaps the most well-accepted optimum contrast value for sign legibility. These researchers reviewed the sign legibility literature pertaining to sign contrast and came up with a contrast ratio of 12:1 for "fully reflectorized" or positive contrast signs using the average of the results of six separate research efforts. This 12:1 ratio would, for example, result in a sign with a 24 cd/m^2 legend and a 2 cd/m^2 background. This single, optimal ratio was expanded in a 1995 synthesis report by Staplin (1995) that gave a range of acceptable internal contrast levels between 4:1 and 50:1.

McNees and Jones (1987) found that the selection of retroreflective background material has a significant effect on sign legibility. These researchers found four combinations of sheeting and text to provide acceptable legibility distances for freeway guide signs: button copy on super engineer-grade (ASTM Type II, or enclosed lens) background sheeting, high-intensity text on high-intensity background, high-intensity on super engineer, and high-intensity on engineer grade. Earlier research by Harmelink et al. (1985) concurs. These researchers found that observers favored high-intensity text on engineer-grade background, stating that this combination provided contrast ratios as good as those produced by high-intensity text on high-intensity background.

Sign Luminance.   Khavanin and Schwab (1991) and Colomb and Michaut (1986) both concluded that only small increases in nighttime legibility distance occur with increases in sign retroreflectivity, More recently, however, Carlson and Hawkins (2002a) found that signs made of microprismatic sheeting resulted in longer legibility distances than those using encapsulated materials. The research of McNees and Jones (1987), Mace (1988), and Garvey and Mace (1996) supports that of Carlson and Hawkins.

Based on a review of the literature, Sivak and Olson (1983) suggested an optimal nighttime sign legend luminance of 75 cd/m^2 and a minimum of 2.4 cd/m^2 for negative contrast signs. With positive contrast signs, Garvey and Mace (1996) found 30 cd/m^2 to provide maximum nighttime legibility distance. Again using positive contrast signs, Garvey and Mace (1996) found that daytime legibility distance continued to improve with increases in luminance up to 850 cd/m^2, after which performance leveled off.

Lighting Design.   Overall, the literature indicates that a sign's luminance and contrast have a greater impact on legibility than does the specific means used to achieve these levels. Jones and Raska (1987) found no significant differences in legibility distance between lighted and unlighted overhead-mounted retroreflective signs for a variety of sign materials (in McNees and Jones 1987). Other research extends this finding, indicating no significant difference in legibility distances for up to 10 different sign-lighting system types for freeway guide signs

(McNees and Jones 1987; Upchurch and Bordin 1987). However, in a study evaluating the effects of sign illumination type on storefront signs, Kuhn, Garvey, and Pietrucha (1999) found externally illuminated signs to perform worse at night than internally and neon-illuminated signs.

In a study of changeable message sign (CMS) visibility, Garvey and Mace (1996) found retroreflective and self-illuminated lighting design to provide equivalent legibility distances. Garvey and Mace did, however, find that the use of "black light" ultraviolet lighting severely reduced legibility. This was attributed to a reduction in internal luminance contrast and color contrast. Hussain, Arens, and Parsonson (1989) addressed this problem when they recommended the use of "white" fluorescent lamps for optimum color rendition and metal halide for overall performance (including color rendition) and cost-effectiveness.

11.5.7 Sign-Placement Variables

Lateral Sign Placement. Sign placement is as important to sign legibility as it is to detection. First, there is the obvious need to place signs so that traffic, pedestrians, buildings, and other signs do not block their messages. A less intuitive requirement for sign placement, however, involves the angle between the observer location and the sign (Prince 1958, in Claus and Claus 1974). Signs set off at large angles relative to the intended viewing location result in letter and symbol distortion. Prince recommended that the messages on signs at angles greater than 20° be manipulated (i.e., increased in height and/or width) to appear "normal" to the observer.

Longitudinal Sign Placement. The placement of directional signs must be far enough in advance of the location of the intended action so that the motorist can react and slow the vehicle or change lanes if necessary, after passing the sign and prior to reaching the appropriate crossroad or access road. The *MUTCD* (2000) states, "When used in high-speed areas, Destination signs [i.e., conventional road guide signs] should be located 200 ft or more in advance of the intersection. . . . In urban areas, shorter advance distances may be used." The distance of 200 feet at 65 mph translates to approximately 2.0 seconds.

This minimum distance should be increased on multilane roadway approaches to allow the motorist time to change lanes. The *MUTCD* (2000) states, "[W]here the road user must use extra time to adjust speed and change lanes in heavy traffic because of a complex driving situation" 4.5 seconds should be allotted for vehicle maneuvers.

The *MUTCD* recommendations can best be thought of as absolute minimums. To establish more conservative recommended distances for longitudinal sign placement for the National Park Service, a formula developed by Woods and Rowan (1970) was combined with deceleration rates from the AASHTO green book. Table 11.1 contains the results of that formula for single lane approaches. Adding 4.0 seconds to the single lane approaches results in the multilane recommendations.

In some cases, such as high-speed highways, two signs may be necessary. In fact, the *MUTCD* (2000) recommends, "[F]or major and intermediate interchanges, two and preferably three Advance Guide signs should be used. Placement should be 0.5 mi, 1 mi, and 2 mi in advance of the exit. At minor interchanges, only one Advance Guide sign should be used. It should be located 0.5 to 1 mi from the exit gore" (U.S. DOT 2000).

11.5.8 Sign Variables

Letter Height. If a response to a guide sign is required, the typical behavior is speed reduction and a turning maneuver at the appropriate crossroad or interchange. On multilane roadways, the motorist may also have to change lanes. For warning and regulatory signs the

TABLE 11.1 Recommended Reading Time, Letter Height, and Longitudinal Sign Placement for Various Operating Speeds and Number of Words on a Sign

Operating speed (mph)	Number of words	Reading time (sec)	Letter height (in.)	Longitudinal sign placement distance (ft/sec)	
				Single-lane approach	Multilane approach
25–40	1–3	3.0–4.5	4–6	375/6.4	600/10.4
	4–8	6.0	8		
41–50	1–3	3.5–4.5	6–8	500/6.8	800/10.8
	4–8	5.5–7.0	10–12		
51–60	1–3	4.0–5.0	8–10	650/7.4	1000/11.4
	4–8	5.5–7.0	12–14		
61–70	1–3	4.0	10	725/7.1	1100/11.1
	4–8	5.5	14		

motorist may have to reduce speed and will sometimes be required to change lanes or make a steering adjustment. Whether the sign is regulatory, warning, or for guidance, sign placement should allow sufficient time to comfortably react to the sign message after passing the sign. With the exception of corner-mounted street name signs, what occurs before the driver passes the sign should be limited to sign detection and reading for comprehension. Appropriate letter heights ensure sufficient time to accomplish the reading task.

Reading Speed. Proffitt, Wade, and Lynn (1998) reported that the average normal reading speed for adults is about 250 words per minute (wpm), or 4.2 words per second. Research evaluating optimum acuity reserve (the ratio between threshold acuity and optimal print size) has demonstrated that optimal reading speeds result from print size that may be as much as four times size threshold (Bowers and Reid 1997; Yager, Aquilante, and Plass 1998; Lovie-Kitchin, Bowers, and Woods 2000). In fact, Yager, Acvilante, and Plass (1998) reported 0.0 wpm reading speed at size threshold. This explains some of the disparity between "normal" reading speed of above size threshold text and the time it takes to read a sign, which often begins at acuity threshold.

Research on highway sign reading provides evidence that it takes drivers approximately 0.5 to 2.0 seconds to read and process each sign word. Dudek (1991) recommended a minimum exposure time of "one second per short word . . . or two seconds per unit of information" for unfamiliar drivers to read changeable message signs. In a study conducted by Mast and Balias (1976), average advance guide sign reading was 3.12 seconds and average exit direction sign reading was 2.28 seconds. Smiley et al. (1998) found that 2.5 seconds was sufficient for 94 percent of their subjects to read signs accurately that contained three destination names; however, this dropped to 87.5 percent when the signs displayed four or five names.

McNees and Messer (1982) mentioned two equations to determine reading time: $t = [N/3] + 1$ and $t = 0.31N + 1.94$ (where t is time in seconds and N is the number of familiar words). In a literature survey on sign comprehension time, Holder (1971) concluded that the second equation was appropriate if the sign was located within an angular displacement of 10°. In their own research, McNees and Messer (1982) found that the time it takes to read a sign depends, among other things, on how much time the driver has to read it; in other words, signs are read faster when it is necessary to do so. However, they also found that as reading speed increases, so do errors; an example of the well-documented speed-accuracy tradeoff. McNees and Messer (1982) concluded that, "a cut-off of approximately 4.0 sec to read any sign was critical for safe handling of a vehicle along urban freeways." If the 4.0 seconds are plugged back into the second equation, the number of familiar words on the sign would be 6.7, or 1.7 words per second.

While it is impractical to specify a single minimum reading time that will allow all drivers to read and understand all signs, the research on sign-reading speed indicates that signs with four to eight words could be comfortably read and comprehended in approximately 4.0 seconds and signs with one to three words in about 2.5 seconds.

Task Loading. In addition to sign reading, the driver must also watch the road and perform other driving tasks. Considering overhead guide signs, McNees and Messer (1982) estimated that a 4.5-second sign-reading time would actually require an 11.0-second and sign-legibility distance. This results from adding 2.0 seconds for sign-clearance time (when the vehicle is too close to the sign for the driver to read it) and dividing the remaining 9.0 seconds equally between sign reading and other driving tasks. In looking at shoulder-mounted signs, Smiley et al. (1998) provide more practical estimates. These researchers allowed for 0.5 seconds clearance time and a 0.5-second glance back at the road for every 2.5 seconds of sign reading (based on eye movement research by Bhise and Rockwell 1973). This would require a 5.0-second legibility distance for 4.0 second of sign reading and 3.0 seconds legibility distance for 2.5 seconds of sign reading. This is assuming that the driver begins to read the sign as soon as it becomes legible. Allowing an additional 1.0 seconds for sign acquisition after it becomes legible, appropriate legibility distance for signs displaying four to eight words would be 6.0 seconds and for signs with one to three words would be 4.0 seconds. Based on these calculations and assuming a legibility index of approximately 40 ft/in., Table 11.1 provides reading times and recommended letter heights as a function of the number of words on the sign and travel speeds.

Diminishing Returns. While research indicates that legibility distance increases with letter height, a point of diminishing return exists (Allen et al. 1967; Khavanin and Schwab 1991). For example, doubling letter height will increase, but will not double, sign reading distance. Mace, Perchonok, and Pollack (1994) and Garvey and Mace (1996) found that increases in letter height above about 8 inches resulted in nonproportional increases in legibility distance. Garvey and Mace (1996) found that a sign with 42-inch characters produced only 80 percent of the legibility index of the same sign with 18-inch characters. That is, the 42-inch character produced a legibility distance of approximately 1,350 feet (LI = 32 ft/in.) while the 18-inch characters resulted in a legibility distance of about 800 ft (LI = 44 ft/ in.).

Text versus Symbols. In a study of traffic sign comprehension speed, Ells and Dewar (1979) found symbolic signs to outperform those with textual messages. These researchers also discovered that symbolic signs were less susceptible than were text signs to visual degradation. In a 1975 visibility study, Jacobs, Johnston, and Cole assessed the legibility distance of almost 50 symbols and their textual counterparts. These researchers found that in the majority of cases the legibility distances for the symbols were twice that of the alphanumeric signs. This finding was replicated in Kline and Fuchs' (1993) research for a smaller set of symbols using young, middle-aged, and older observers. Kline and Fuchs' research also introduced a technique to optimize symbol legibility: recursive blurring, which results in symbols designed to "maximize contour size and contour separation." In other words, optimized symbols or logos will have elements that are large enough to be seen from a distance and spaces between the elements wide enough to reduce blurring between elements.

The literature clearly indicates that, from a visibility standpoint, symbols are superior to text. Symbols, however, require a different kind of comprehension than words. Symbol meaning is either understood intuitively or learned. Although traffic sign experts and traffic engineers agree that understandability is the most important factor in symbol design (Dewar 1988), other research has shown that what is intuitive to designers is not always intuitive to drivers, and that teaching observers the meaning of more abstract symbols is frequently unsuccessful. For example, in one study (Kline et al. 1990) even the relatively simple "HILL" symbol resulted in only 85 percent comprehension, while the "ROAD NARROWS" symbol accommodated only 52 percent of the respondents. In researching the Slow Moving Vehicle emblem, Garvey (2003) found correct symbol recognition to be approximately 30 percent for older and younger subjects under daytime and nighttime viewing conditions.

Upper-Case versus Mixed-Case.   Forbes, Moskowitz, and Morgan (1950) conducted perhaps the definitive study on the difference in sign legibility between text depicted in all upper-case letters and that shown in mixed-case. When upper- and mixed-case words subtended the same sign area these researchers found a significant improvement in legibility distance with the mixed-case words. Garvey, Pietrucha, and Meeker (1997) replicated this result with new sign materials, a different font, and older observers. They found a 12 to 15 percent increase in legibility distance with mixed-case text under daytime and nighttime conditions. It must be noted, however, that these results were obtained with a recognition task—that is, the observers knew what words they were looking for. In instances where the observer does not know the text, improvements with mixed-case are not evident (Forbes, Moskowitz, and Morgan 1950; Mace, Garvey, and Heckard 1994; and Garvey, Pietrucha, and Meeker 1997).

Font.   Assessing the effect of letter style on traffic signs has been limited by state and federal governments' desire to keep the font "clean," in other words, a sans serif alphabet that has a relatively constant stroke width. While sans serif letters are generally considered to provide greater legibility distance than serif letters (Prince 1957, in Claus and Claus 1974), a comparison of the sans serif Standard Highway Alphabet with Clarendon, the serif standard National Park Service font, revealed a slight improvement with the Clarendon font (Mace, Garvey, and Heckard 1994).

Currently, the only font allowed by FHWA on road signs is the Standard Highway Alphabet (U.S. DOT 2002). However, recent research on highway font legibility (Garvey, Pietrucha, and Meeker 1997, 1998; Hawkins et al. 1999; and Carlson and Brinkmeyer 2002) has led the FHWA to consider including a new font called Clearview in the next version of Standard Highway Signs (Figure 11.3). Clearview was designed to reduce halation (Figure 11.4) resulting from the use of high-brightness retroreflective materials (i.e., microprismatic sheeting) and improve letter legibility for older drivers. Related research for the National Park Service has led that agency to accept NPS Rawlinson Road as an alternate to the NPS's current Clarendon font (Figure 11.5). NPS Rawlinson Road has been shown to increase sign legibility by 10.5 percent while reducing word length by 11.5 percent (Garvey et al. 2001).

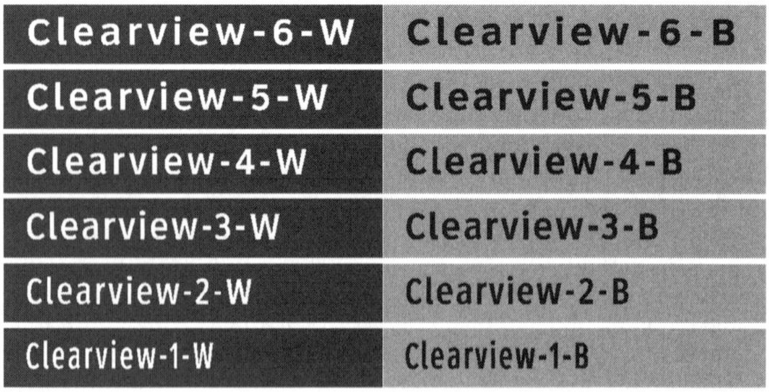

Clearview-6-W	Clearview-6-B
Clearview-5-W	Clearview-5-B
Clearview-4-W	Clearview-4-B
Clearview-3-W	Clearview-3-B
Clearview-2-W	Clearview-2-B
Clearview-1-W	Clearview-1-B

FIGURE 11.3

FIGURE 11.4

Stroke Width. Kuntz and Sleight (1950) concluded that the optimal stroke width-to-height ratio for both positive and negative-contrast letters was 1:5. Forbes et al. (1976) found increases in legibility distance of fully-reflectorized, positive-contrast letters and decreases in legibility for negative-contrast letters when the stroke width-to-height ratio was reduced from 1:5 to 1:7. That is, light letters on a darker background should have a thinner stroke and dark letters on a lighter background should have a bolder stroke. Improved legibility for fully reflectorized, white-on-green signs with thinner stroke width was also found by Mace, Garvey, and Heckard (1994) for very high contrast signs, and for mixed case text by Garvey, Pietrucha, and Meeker (1998).

Abbreviations. In a study of changeable message sign comprehension, Huchingson and Dudek (1983) developed several abbreviation strategies. These researchers recommended the technique of using only the first syllable for words having nine letters or more; for example, *Cond* for *Condition*. This technique should not, however, be used if the first syllable is in itself a new word. A second method using the key consonants was suggested for five- to seven-letter words; for example, *Frwy* for *Freeway*. Abbreviations, however, are to be used only as a last resort if limitations in sign size demand it, as they increase the possibility of incorrect sign interpretation. Alternative suggestions to deal with sign size limitations include selecting a synonym for the abbreviated word, reducing letter size, reducing message length, and increasing sign size.

Contrast Orientation. The research on this issue is clear; with the possible exception of tight inter-character spacing (Case et al. 1952), positive-contrast signs provide greater legibility distances than negative-contrast signs. As far back as 1955, laboratory research by Allen and Straub found that white-on-black signs (positive-contrast) provided longer legibility distances than black-on-white signs when the sign luminance was between 3 and 30 cd/m². Allen et al. (1967) replicated these results in the field. Garvey and Mace (1996) extended these results in their changeable message sign research with the addition of orange, yellow, and green signs. Positive-contrast signs resulted in improvements of about 30 percent over negative-contrast signs (Garvey and Mace 1996).

FIGURE 11.5

Color. Schnell et al. (2001) found small legibility improvements when comparing signs using fluorescent colors versus signs using matching nonfluorescent colors. However, in an evaluation of normal sign colors, Garvey and Mace (1996) found no difference in legibility distance that could not be accounted for by luminance, luminous contrast, or contrast orientation between signs using the following color combinations: white/green, black/white, black/orange, black/yellow, and black/red. This is also consistent with the findings of research on computer displays (Pastoor 1990). In general, the research indicates that if appropriate luminance contrast, color contrast, and luminance levels are maintained, the choice of specific colors for background and text does not affect legibility distance.

11.6 FINAL REMARKS

For any type of highway sign to be visually effective, it must be readable. While this seems like a fairly simple objective to achieve, the information presented herein indicates that the ability of a driver to detect and read a sign is a function of numerous human, environmental, and design factors with complex interrelationships. For example, visual acuity, contrast sensitivity, visual field, and glare can significantly impact the driver's ability to see the sign and read its message. Furthermore, the basic design and placement of the sign can greatly impact the detectibility and legibility of any highway sign. Features such as height, lateral offset, reflectorization, illumination, message design, font, case, letter height, border, and letter spacing are so critical that federal guidelines dictate these features to maximize the potential for a driver to see the sign within the highway environment. The photometric characteristics of the sign, including the internal contrast, luminance, and light design, can also directly impact how well a driver sees a sign. And if all of these factors are not enough, the location of a sign relative to the rest of the highway environment can either enhance its detectibility or force it to compete with other signs and objects for visibility. Thus, the task of designing detectible, legible, and understandable highway signs is a challenge that continues to be refined as we learn more about their role in the highway environment and their interaction with the driver. Ultimately, we hope to provide critical information to the traveler that they can use in a timely manner to navigate safely through the highway environment.

11.7 REFERENCES

Allen, T. M., F. N. Dyer, G. M. Smith, and M. H. Janson. 1967. "Luminance Requirements for Illuminated Signs." *Highway Research Record* 179, 16–37.

Allen, T. M., and Straub, A. L. 1955. "Sign Brightness and Legibility." *Highway Research Board Bulletin* 127, 1–14.

American Society for Testing and Materials (ASTM). 2001. *Standard Test Method for Coefficient of Retroreflection of Retroreflective Sheeting Utilizing the Coplanar Geometry.* Document Number ASTM E810-01, ASTM International.

Ball, K., and C. Owsley. 1991. "Identifying Correlates of Accident Involvement for the Older Driver." *Human Factors* 33(5):583–95.

Bhise, V. D., and T. H. Rockwell. 1973. *Development of a Methodology for Evaluating Road Signs.* Final Report, Ohio State University.

Boff, K. R., L. Kaufman, and J. P. Thomas. 1986. *Handbook of Perception and Human Performance,* vols. 1 and 2. New York: John Wiley & Sons.

Bowers, A. R., and V. M. Reid. 1997. "Eye Movement and Reading with Simulated Visual Impairment." *Ophthalmology and Physiological Optics* 17(5):492–02.

Carlson, P. J., and G. Brinkmeyer. 2002. "Evaluation of Clearview on Freeway Guide Signs with Microprismatic Sheeting." *Transportation Research Record* 1801, 27–38.

Carlson, P. J., and G. Hawkins. 2002a. "Legibility of Overhead Guide Signs Using Encapsulated versus Microprismatic Retroreflective Sheeting." Presented at the 16th Biennial Symposium on Visibility and Simulation, Iowa City, IA.

———. 2002b. "Minimum Retroreflectivity for Overhead Guide Signs and Street Name Signs." *Transportation Research Record* 1794, 38–48.

Case, H. W., J. L. Michael, G. E. Mount, and R. Brenner. 1952. "Analysis of Certain Variables Related to Sign Legibility." *Highway Research Board Bulletin* 60, 44–58.

Claus, K., and J. R. Claus. 1974. *Visual Communication through Signage,* vol. 1, *Perception of the Message.* Cincinnati: Signs of the Times.

Colomb, M., and G. Michaut. 1986. "Retroreflective Road Signs: Visibility at Night." *Transportation Research Record* 1093, 58–65.

Cooper, B. R. 1988. "A Comparison of Different Ways of Increasing Traffic Sign Conspicuity." *TRRL Report* 157.

Dewar, R. E. 1988. "Criteria for the Design and Evaluation of Traffic Sign Symbols." *Transportation Research Record* 1160, 1–6.

Dudek, C. L. 1991. *Guidelines on the Use of Changeable Message Signs.* Final Report, DTFH61-89-R-00053, U.S. DOT, Federal Highway Administration, Washington, DC.

Ells, J. G., and R. E. Dewar. 1979. "Rapid Comprehension of Verbal and Symbolic Traffic Sign Messages." *Human Factors* 21, 161–68.

Federal Highway Administration (FHWA). 1978a. *Roadway Lighting Handbook.* U.S. Department of Transportation, FHWA, Washington, DC.

———. 1978b. "Standard Highway Signs: As Specified in the *Manual of Uniform Traffic Control Devices.*" U.S. Department of Transportation, FHWA, Washington, DC.

———. 1988. *Manual on Uniform Traffic Control Devices for Streets and Highways.* U.S. Department of Transportation, FHWA, Washington, DC.

———. 2000. *Manual on Uniform Traffic Control Devices, Millennium Edition.* U.S. Department of Transportation, FHWA, Washington, DC, http://mutcd.fhwa.dot.gov/kno-millennium.htm.

———. 2002. "Standard Highway Signs: As Specified in the MUTCD Millennium Edition." U.S. Department of Transportation, FHWA, Washington, DC, http://mutcd.fhwa.dot.gov/.

Forbes, T. W., J. P. Fry, R. P. Joyce, and R. F. Pain. 1968a. "Letter and Sign Contrast, Brightness, and Size Effects on Visibility." *Highway Research Record* 216, 48–54.

Forbes, T. W., K. Moskowitz, and G. Morgan. 1950. "A Comparison of Lower Case and Capital Letters for Highway Signs." *Proceedings, Highway Research Board* 30, 355–73.

Forbes, T. W., R. F. Pain, R. P. Joyce, and J. P. Fry. 1968b. "Color and Brightness Factors in Simulated and Full-Scale Traffic Sign Visibility." *Highway Research Record* 216, 55–65.

Forbes, T. W., B. B. Saari, W. H. Greenwood, J. G. Goldblatt, and T. E. Hill. 1976. "Luminance and Contrast Requirements for Legibility and Visibility of Highway Signs." *Transportation Research Record* 562, 59–72.

Garvey, P. M. 2003. "Motorist Comprehension of the Slow Moving Vehicle (SMV) Emblem." *Journal of Agricultural Safety and Health* 9(2):159–169.

Garvey, P. M., and D. M. Mace. 1996. *Changeable Message Sign Visibility.* Publication No. FHWA-RD-94-077.

Garvey, P. M., M. T. Pietrucha, and D. Meeker. 1997. "Effects of Font and Capitalization on Legibility of Guide Signs." *Transportation Research Record* 1605, 73–79.

———. 1998. "Development of a New Guide Sign Alphabet." *Ergonomics in Design* 6(3)7–11.

Garvey, P. M., A. Z. Zineddin, M. T. Pietrucha, D. T. Meeker, and J. Montalbano. 2001. *Development and Testing of a New font for National Park Service Signs.* U.S. Department of the Interior, National Park Service Final Report.

Hanson, D. R., and H. L. Waltman. 1967. "Sign Backgrounds and Angular Position." *Highway Research Record* 170, 82–96.

Harmelink, M. D., G. Hemsley, D. Duncan, R. W. Kuhk, and T. Titishov. 1985. "Evaluation of Reflectorized Sign Sheeting for Nonilluminated Freeway Overhead Guide Signs." *Transportation Research Record* 1010, 80–84.

Hawkins, H. G., D. L. Picha, M. D. Wooldridge, F. K. Greene, and G. Brinkmeyer. 1999. "Performance Comparison of Three Freeway Guide Sign Alphabets." *Transportation Research Record* 1692, 9–16.

Holder, R. W. 1971. "Consideration of Comprehension Time in Designing Highway Signs." *Texas Transportation Researcher* 7(3):8–9.

Huchingson, R. D., and C. L. Dudek. 1983. "How to Abbreviate on Highway Signs." *Transportation Research Record* 904, 1–3.

Hughes, P. K., and B. L. Cole. 1984. "Search and Attention Conspicuity of Road Traffic Control Devices." *Australian Road Research Board* 14(1):1–9.

Hussain, S. F., J. B. Arens, and P. S. Aparsonson. 1989. "Effects of Light Sources on Highway Sign Color Recognition." *Transportation Research Record* 1213, 27–34.

Jacobs, R. J., A. W. Johnston, and B. L. Cole. 1975. "The Visibility of Alphabetic and Symbolic Traffic Signs." *Australian Road Research* 5(7):68–86.

Jenkins, S. E., and B. L. Cole. 1986. "Daytime Conspicuity of Road Traffic Control." *Transportation Research Record* 1093, 74–80.

Khavanin, M. R., and R. N. Schwab. 1991. "Traffic Sign Legibility and Conspicuity for the Older Drivers." In *1991 Compendium of Technical Papers.* Washington, DC: Institute of Transportation Engineers, 11–14.

Kuhn, B. T., P. M. Garvey, and M. T. Pietrucha. 1998. "The Impact of Color on Typical On-Premise Sign Font Visibility." Presented at TRB's 14th Biennial Symposium on Visibility, Washington, DC, April.

———. 1999. "On Premise Sign Legibility and Illumination." In *1999 Compendium of Technical Papers.* Washington, DC: Institute of Transportation Engineers.

King, G. F., and H. Lunenfeld. 1971. *Development of Information Requirements and Transmission Techniques for Highway Users.* NCHRP Report 123, National Cooperative Highway Research Program, Washington, DC.

Kline, D. W., and P. Fuchs. 1993. "The Visibility of Symbolic Highway Signs Can Be Increased among Drivers of All Ages." *Human Factors* 35(1):25–34.

Kline, D. W., K. Buck, Y. Sell, T. L. Bolan, and R. E. Dewar. 1999. "Older Observers' Tolerance of Optical Blur: Age Differences in the Identification of Defocused Test Signs." *Human Factors* 41(3): 356–64.

Kline, T. J. B., L. M. Ghali, D. W. Kline, and S. Brown. 1990. "Visibility Distance of Highway Signs among Young, Middle-Aged, and Older Observers: Icons Are Better than Text." *Human Factors* 32(5): 609–19.

Kuntz, J. E., and R. B. Sleight. 1950. "Legibility of Numerals: The Optimal Ratio of Height to Width of Stroke." *American Journal of Psychology* 63:567–75.

Lovie-Kitchin, J. E., A. R. Bowers, and R. L. Woods. 2000. "Oral and Silent Reading Performance with Macular Degeneration." *Ophthalmology and Physiological Optics* 20(5):360–70.

Mace, D. J. 1988. "Sign Legibility and Conspicuity." In *Transportation in an Aging Society.* Special Report 218, vol. 2. Washington, DC: National Research Council, Transportation Research Board, 270–93

Mace, D. J., and L. Pollack. 1983. "Visual Complexity and Sign Brightness in Detection and Recognition of Traffic Signs." *Transportation Research Record* 904, 33–41.

Mace, D. J., P. M. Garvey, and R. F. Heckard. 1994. *Relative Visibility of Increased Legend Size vs. Brighter Materials for Traffic Signs.* Report FHWA-RD-94-035, U.S. Department of Transportation, Federal Highway Administration, Washington, DC.

Mace, D., K. Perchonok, and L. Pollack. 1982. *Traffic Signs in Complex Visual Environments.* Report FHWA-RD-82-102, U.S. Department of Transportation, Federal Highway Administration, Washington, DC.

Mast, T. M., and J. A. Balias. 1976. "Diversionary Signing Content and Driver Behavior." *Transportation Research Record* 600, 14–19.

McNees, R. W., and H. D. Jones. 1987. "Legibility of Freeway Guide Signs as Determined by Sign Materials." *Transportation Research Record* 1149, 22–31.

McNees, R. W., and C. J. Messer. 1982. "Reading Time and Accuracy of Response to Simulated Urban Freeway Guide Signs." *Transportation Research Record* 844, 41–50.

Pain, R. F. 1969. "Brightness and Brightness Ratio as Factors in the Attention Value of Highway Signs." *Highway Research Record* 275, 32–40.

Pastoor, S. 1990. "Legibility and Subjective Preference for Color Combinations in Text." *Human Factors* 32(2):157–71.

Proffitt, D. R., M. M. Wade, and C. Lynn. 1998. *Creating Effective Variable Message Signs: Human Factors Issues.* Final Contract Report, Proj No. 9816-040-940, VTRC 98-CR31, Virginia Department of Transportation, Richmond, VA.

Schnell, T., K. Bentley, E. Hayes, and M. Rick. 2001. "Legibility Distances of Fluorescent Traffic Signs and Their Normal Color Counterparts." *Transportation Research Record* 1754, 31–41.

Sivak, M., and P. L. Olson. 1983. *Optimal and Replacement Luminances of Traffic Signs: A Review of Applied Legibility Research.* UMTRI-83-43, University of Michigan Transportation Research Institute, Ann Arbor, MI.

———. 1985. "Optimal and Minimal Luminance Characteristics for Retroreflective Highway Signs. *Transportation Research Record* 1027, 53–56.

Smiley, A., C. MacGregor, R. E. Dewar, and C. Blamey. 1998. "Evaluation of Prototype Tourist Signs for Ontario." *Transportation Research Record* 1628, 34–40.

Staplin, L. 1995. *Older Driver and Highway Safety Literature Review and Synthesis.* Working Paper in Progress. U.S. Department of Transportation, FHWA, Washington, DC.

———. 1983. *Traffic Control Device Handbook.* U.S. Department of Transportation, FHWA, Washington, DC.

Upchurch, J. E., and J. D. Armstrong. 1992. "A Human Factors Evaluation of Alternative Variable Message Sign Technologies." In *Vehicle Navigation and Information Systems.* Oslo: Norwegian Society of Chartered Engineers, 262–67.

Upchurch, J. E., and J. T. Bordin. 1987. "Evaluation of Alternative Sign-Lighting Systems to Reduce Operating and Maintenance Costs." *Transportation Research Record* 1111, 79–91.

Woods, D. L., and N. J. Rowan. 1970. "Street Name Signs for Arterial Streets." *Highway Research Record* 325, 54.

Yager, D., K. Aquilante, and R. Plass. 1998. "High and Low Luminance Letters, Acuity Reserve, and Font Effects on Reading Speed." *Vision Research* 38:2527–31.

Zwahlen, H. T. 1989. "Conspicuity of Suprathreshold Reflective Targets in a Driver's Peripheral Visual Field at Night." *Transportation Research Record* 1213, 35–46.

Zwahlen, H. T., and J. Yu. 1991. "Color and Shape Recognition of Reflectorized Targets under Automobile Low-Beam Illumination at Night." *Transportation Research Record* 1327, 1–7.

CHAPTER 12
TRAFFIC CONGESTION

Kara Kockelman
Department of Civil Engineering,
The University of Texas at Austin, Austin, Texas

12.1 INTRODUCTION

Congestion is everywhere. It arises in human activities of all kinds, and its consequences are usually negative. Peak demands for goods and services often exceed the rate at which those demands can be met, creating delay. That delay can take the form of supermarket checkout lines, long waits for a table at a popular restaurant, and after-work crowds at the gym. Yet the context in which we most often hear of congestion posing a serious problem, to ourselves and to our economy, is the movement of people and goods.

The average American reports traveling 78 minutes a day, over 80 percent of which is by automobile (FHWA 2001).* The Texas Transportation Institute (TTI) estimates that over 45 percent of peak-period travel or roughly one-third of total vehicle miles traveled occur under congested conditions in many U.S. metropolitan areas (Shrank and Lomax 2002). These include the predictable places like Los Angeles, Washington, DC, and Atlanta; they also include places like San Diego, California, Tacoma, Washington, and Charlotte, North Carolina. Though crime, education, taxes, and the economy certainly are key issues for voters and legislators, polls regularly report congestion to be the number one local issue (see, e.g., Scheibal 2002; Knickerbocker 2000; and Fimrite 2002).

Nonpersonal modes of transportation are certainly not immune to congestion, either. Intercity trucking carries almost 30 percent of freight ton-miles shipped in the United States every year (BTS 2002) and 72 percent of the value shipped (CFS 1997). These trucks are subjected to the same roadway delays, resulting in higher-priced goods, more idling emissions, and frayed nerves. The gates, runways, and traffic control systems of many popular airports are tested daily. And seaport berths, rail tracks, canals, and cables all have their limitations. As soon as demand exceeds supply, goods, people, and information must wait in queues that can become painfully long. Though not stuck in queues, others find themselves waiting at the destinations for expected shipments, friends, family members, and colleagues that fail to arrive on time.

Engineers, economists, operations researchers, and others have considered the problem of congestion for many years. The confluence of growing traveler frustration, technological

---

*NPTS-reported trip-making involves spatially very distinct locations, such as home and work, school and shopping center. In reality, we are moving much more regularly, between bedrooms, around offices, and along supermarket aisles. Such travel, while substantial, is probably much less impacted by congestion.

innovations, and inspirational traffic management policies from around the world provide added momentum for the modifications needed to moderate and, ideally, eliminate this recurring problem and loss. This chapter examines congestion's defining characteristics, its consequences, and possible solutions.

12.2   DEFINING CONGESTION

Notably, congestion is not always undesirable. Some "congested" experiences can be positive, and these tend to occur at one's destination, rather than en route. Myers and Dale (1992) point out that orchestrated congestion in public spaces, such as theater entry plazas, enhances public interaction, enables better land use mixing, and calms vehicular traffic. Taylor (2002) observes that congested city centers are often signs of vibrant city activity and prosperity. These forms of congestion are to some extent desirable and are not the concern of this chapter.

This chapter stresses instead the undesirable form of congestion: the kind that impedes travel between two points, effectively adding access costs to a desired destination. The travel itself is not enjoyable; it is instead a necessary expenditure.* This form of congestion is a slowing of service. Queues (or lines) of travelers will form if demand exceeds capacity. But these are not necessary for congestion to occur. All that is necessary is that the service speed be less than the "free-flow" or maximum speed, which exists when demand is light relative to capacity.

All transportation systems are limited by a capacity service rate. Operators at manual toll booths and transponder readers for electronic toll collection (ETC) cannot reliably serve more than a certain number of vehicles per hour. Commuter rail lines eventually fill up, along with their train cars. Port cranes exhibit functional limits, as do canals, runways, and pipelines. No system is immune; all physical pathways are constrained in some respect.

When systems slow down, delays arise. Delay generally is defined as the difference between actual travel time and travel time under uncongested or other acceptable conditions. The *Highway Capacity Manual 2000* (TRB 2000) defines signalized intersection delay as the sum of delay under uniform arrivals (adjusted for a progression factor), incremental delays (to account for randomness in arrival patterns), and any initial queue delays (to recognize spillovers from prior cycles). While vehicle detection and intersection automation can dramatically reduce signal delays, they cannot eliminate them. Any time two or more vehicles (users) wish to use the same space at the same time, delay—or else a crash—will result. One must cede that space to the other. The mechanism may be a signal, a queue, pricing, rationing, or other policies.

In general, then, congestion is the presence of delays along a physical pathway due to the presence of other users. Before discussing strategies to combat such delays, this chapter examines the general costs and consequences of congestion, its causes, and its quantification.

12.3   THE CONSEQUENCES OF CONGESTION

Automobile congestion has a myriad of impacts, from wasted fuel and added emissions to frayed nerves, more expensive goods, and elevated crash rates. Its clearest impact is delay,

---

*Mokhtarian and colleagues have been examining the extent to which travel may be desirable in and of itself, resulting in "excess travel" (Mokhtarian and Salomon 2001; Redmond and Mokhtarian 2001; Salomon and Mokhtarian 1998). Richardson (2003) found reasonably high proportions of travelers in Singapore to exhibit a zero value of travel time, causing him to conclude that, among other things, travel on the air-conditioned transit system in that humid city can be a relatively enjoyable (or at least refreshing) experience.

or lost time. Across the United States this may average 20 hours per year per person. In Los Angeles, it is estimated to exceed 60 hours, which translates to 10 minutes a day, or one-sixth of one's average travel time (Schrank and Lomax 2002). Presented this way, even Los Angeles's numbers may seem acceptable. Why, then, does this issue so consistently top opinion polls as our communities' number one policy issue? There are many reasons. One is that dense car traffic is more difficult to navigate, even if speeds stay high. Such travel is unpleasant and tiring in many ways. Another is that congestion tends to be unpredictable, even when it is recurring. As a result, peak-period travelers—including trucks and buses—regularly arrive early or late at their destinations, creating frustration. These suboptimal arrival times carry a cost: missed meetings and deliveries, loss of sleep, childcare fees for late pick-up, children waiting around for classes to begin, a supervisor's growing intolerance of missed work. Researchers have tried to quantify these additional costs.

Bates et al.'s (2001) review of travel time reliability research finds that every minute of *lateness* is regularly valued at two to five times a minute of travel time. And every minute *early* is valued at almost a minute of travel time (around 80 percent). In general, variation in travel time (as measured by standard deviation) is worth more to the typical traveler than the average travel time. So even if one's commute trip *averages* 25 minutes, if it exhibits a 10-minute standard deviation,* it typically is not preferred to a guaranteed travel time of 35 minutes. Using loop detector data for samples of freeway sections in cities across the United States, Lomax, Turner, and Margiotta (2003) estimated early-departure buffer times in order to ensure on-time (or early) arrivals in 95 percent of one's trip-making. For Austin, Texas, the required buffer was estimated to be 24 percent of the average travel time; in Los Angeles it was 44 percent. In Lomax et al.'s 21-city data set, congestion is highly correlated with high buffer times (and thus low reliability). Clearly, congestion is costly—in a variety of ways.

Another reason for society's impatience with congestion may relate to equity and an ability to prioritize consumption or purchases. On a "free" system of public roadways, every trip is treated the same. In other words, more valuable trips experience the same travel times as less valued trips. Persons and goods with very low values of time pay the same time-price as others, even though the monetary value of and/or willingness to pay for their trips can differ by orders of magnitude. Few options are available, and they can require significant adjustments: changes in one's home, work, school, or other locations; moved meeting times; and entirely forgone activities. In one exceptional circumstance, Silicon Valley pioneer Steve Jobs elected to purchase a helicopter to reduce his San Francisco Bay Area commute.

As mentioned above, some travelers find a moderate level of congestion acceptable, even desirable. Moreover, an evolution of in-vehicle amenities, from radios, air conditioning, and reclining seats to tape and CD players, stereo-quality sound systems, cellular phones, video players, and heated seats, plays a role in reducing the perceived costs of congestion. At the same time, increasing presence of two-worker couples, rising incomes, just-in-time manufacturing and delivery processes, and complication of activity patterns for adults and children have resulted in a variety of travel needs and time constraints that can make delays more costly and stressful. The market for scarce roadspace breaks down during peak travel hours in many places. So what are the costs?

The TTI studies estimate Los Angeles's congestion costs to exceed $14 billion each year, or more than $1,000 per resident. This figure is based on speed and flow estimates across the region's network of roads, where every hour of passenger-vehicle delay is valued at $12.85, every mile of congested truck travel at $2.95, and each gallon of gasoline consumed (while delayed) at $1.39 (Schrank and Lomax 2002).** Together with costs from 74 other

---

*The standard deviation is the square root of the expected value of squared differences between actual and mean travel times: $\sigma_t = \sqrt{E((t - \mu_t)^2)}$. For a normal or Gaussian distribution of travel times that averages 25 minutes, a standard deviation of 10 minutes implies a 16 percent probability that a trip will exceed 35 minutes.

**Passenger-vehicle occupancies were assumed to be 1.25 persons/vehicle. Five percent of congested-period vehicle-miles traveled (VMT) was assumed to be by trucks. Truck time delays were multiplied by congested speeds and $2.95 per mile traveled in order to provide truck delay costs, which are on the order of $100 per hour.

major U.S. regions (using the same unit-cost assumptions), the annual total reaches $70 billion. This is about 4¢ per vehicle mile traveled, or double U.S. gas taxes (which total roughly 40¢/gallon, or 2¢/mile). Essentially, the U.S. Highway Trust Fund could be trebled through the addition of these estimated costs. Of course, these neglect congestion in other travel modes and other U.S. locations, as well as environmental impacts, schedule delay, delivery difficulties, inventory effects, frustration, and other costs. Carbon monoxide and hydrocarbon emissions are roughly proportional to vehicle *hours* of travel (Dahlgren 1994); oxides of nitrogen also rise with slowed traffic, further worsening air quality (Beamon 1995). Congestion stymies supply chains and diminishes agglomeration economies (Weisbrod, Vary, and Treyz 2001). Taken all together, such costs may rival or even exceed the TTI 75-city cost estimates.

12.3.1 Congestion and Crashes

The consequences of congestion for crash frequency and severity are intriguing. The lowest crash rates (crashes per vehicle mile traveled per lane) tend to occur at intermediate levels of flow (for example, level of service C)* (see, e.g., Gwynn 1967 and Brodsky and Hakkert 1983). Controlling for traffic density, rather than flow, also is key, since low flows can occur under both uncongested (high-speed) and congested (low-speed) conditions. Garber and Subramanyan's recent work (2002) for weekday crashes on four highways indicates a steep reduction in police-reported crash rates (i.e., crashes per vehicle mile traveled) when densities are about half of critical density (where critical density corresponds to capacity flow rates).** Thus, crash rates generally appear to rise as congestion sets in. However, speeds tend to fall under such conditions, especially when demand exceeds capacity. And lower collision speeds mean less severe traffic crashes (see, e.g., Evans 1991; Kockelman and Kweon 2002).

Of course, crashes themselves generate congestion, by distracting other drivers and blocking lanes and shoulders. And drivers frustrated by congestion may take risks that offset some of the benefits of reduced speeds, such as tailgating, using shoulders as traffic lanes, cutting across dense oncoming traffic, and speeding up excessively when permitted. Such behaviors are typical of "road rage" and may worsen congestion—and result in crashes. Little is formally known regarding the magnitude and nature of such indirect safety effects of congestion.

12.4 QUANTIFYING CONGESTION

The explicit measurement or quantification of congestion has many uses. Such measures help communities identify and anticipate traffic problems by location, severity, and time of day. Their magnitude and ranking provide a basis for targeted investment and/or policy decisions. They also provide useful inputs for air quality models, which require travel speed and distance information. The following discussion provides a definition of congestion, along with methods for its estimation based on travel time formulae.

12.4.1 Congestion and the *Highway Capacity Manual*

For transmission of people, goods, or data, an important distinction exists between capacity (i.e., maximum-flow) speeds and free-flow speeds. In the case of roadways, congestion can

---

*Freeway level of service C implies conditions where speeds are near free-flow speeds but maneuverability is "noticeably restricted" (*Highway Capacity Manual 2000* [TRB 2000], 13–10).

**Many slight crashes may occur under congested conditions and go unreported to police. Thus, it is possible that total (reported and unreported) crash rates stay stable or even rise under congested conditions.

set in and speeds can fall well before capacity is reached. Chapter 23 of the *Highway Capacity Manual 2000* (TRB 2000) provides estimates of capacity and speeds for a variety of basic freeway segments. A traffic density of 45 passenger cars per lane-mile (pc/mi/ln) corresponds to (sustainable) capacity conditions, and it distinguishes levels of service E and F. On facilities with a 75 mph free-flow-speed (FFS), average speeds are expected to begin falling at relatively low densities (e.g., 18 pc/mi/ln). And at capacity conditions (2,400 pc/h/ln and 45 pc/mi/ln), the predicted prevailing speed is just 53.3 mph, well below the uncongested 75 mph FFS. On freeways exhibiting lower free-flow speeds, congestion is expected to set in later and speed reductions are less severe. For example, for FFS = 55 mph, small drops in prevailing speed arise at 30 pc/mi/ln densities, and capacity speeds are 50 mph (just 10 percent less than FFS). Travel on these lower-speed facilities will take longer, however, even if their conditions do not qualify as "congested," simply because their associated speeds are lower for every level of service.

Why do speeds fall as density increases? Because the smaller the spacing between vehicles, the more likely a conflict. So drivers slow down when spacings are tight. Safety also sets an upper limit on how fast people want to travel. Human and vehicle response times are limited: we take time to perceive threats, and our vehicles take time to slow down. Thus, drivers have maximum speed preferences, which govern when traffic is relatively light, and minimum spacing preferences, which govern when traffic is relatively heavy.* Driver spacing requirements go up with speed, so there is a natural limitation on how many can traverse a given road section in any given time. The *Highway Capacity Manual* predictions for freeway lanes are just one illustration of these safety-response phenomena.

On uncontrolled (nonfreeway) multilane highways, *Highway Capacity Manual* estimates of speeds and capacity flows are lower than those found on freeway lanes. (Driveways and other access points, left turns in the face of oncoming traffic, and other permitted behaviors necessitate more cautious driving.) However, the density values defining levels of service are almost the same as those for freeways, and the magnitudes of speed reduction leading to capacity conditions are minor (on the order of 1 to 3 mph).

On two-lane (undivided) highways, capacity flows are dramatically less (1,700 pc/hr/ln) and levels of service are defined by the percentage of time that vehicles follow slower vehicles (PTSF) and, in the case of class I facilities, by average travel speed. If passing is not permitted along a section, average travel speeds are predicted to fall by as much as 4.5 mph.

When travel is controlled by traffic signals, signal-related delay estimates define level of service. Speed calculations are not emphasized, though simulation software such as the FHWA's CORSIM (FHWA 1999) can generate estimates of trip start times, end times, and distances, thereby predicting operating speeds.

Significantly, *Highway Capacity Manual* methods permit no traffic speed estimation beyond level of service E, and Chapter 23's speed-flow curves and tables disappear. Beyond level of service E, traffic conditions can be characterized by speeds as high as 50 mph (on high-design freeways)—or by complete gridlock (0 mph). When demand exceeds capacity, a queue develops and straightforward speed models break down. At that point, it becomes more important to know when a traveler enters the queue than to know how many are entering it. And it is all characterized as level of service F.

Unfortunately, level of service F's oversaturated conditions are common in many regions and on many facilities. Local bottlenecks and incidents cause demand to exceed capacity, sending congestive shockwaves back upstream. Under these conditions, upstream speeds can fall well below those prevailing under capacity conditions, even to zero. While oversaturated traffic conditions are rather unstable and exhibit high variation, speeds can be approximated. The following two sections describe some applicable methods.

---

*Based on the traffic observations illustrated in Figures 12.1 and 12.2, Kockelman (2001) has estimated free-flow speed and spacing preferences for various freeway user classes.

12.4.2 Roadway Conditions: The Case of Interstate Highway 880

The *Highway Capacity Manual* offers traffic predictions based on empirical evidence for a variety of roadway types, designs, and locations. Yet conditions on specific facilities can differ rather dramatically. Loop detectors embedded in highway pavements offer continuous data-collection opportunities for key traffic variables.

Based on detector data from Interstate Highway (I.H.) 880's number-two lane, observed speeds and densities are plotted as Figure 12.1.* Speeds (measured in mph) fall as density (measured in vehicles per lane mile [veh/ln/mi]) increases. Density is inversely related to vehicle spacing;** and, as vehicles are added to a section of roadway, density rises and spacings fall. Drivers are inhibited by reduced spacings and the growing presence of others. For reasons of safety, speed choice, and reduced maneuverability, drivers choose lower speeds.

Since speed multiplied by density is flow,† Figure 12.1's information leads directly to Figure 12.2's speed versus flow values. Flow rates over these detectors may reach 3,000 vehicles per hour—or 1.2 seconds per vehicle, for a brief, 30-second interval. And average speeds appear to fall slightly throughout the range of flows, from roughly 65 to 55 mph. Travel times are impacted as more and more vehicles enter the lane, densifying the traffic stream. At some point downstream of this detector station, demand exceeds capacity or an

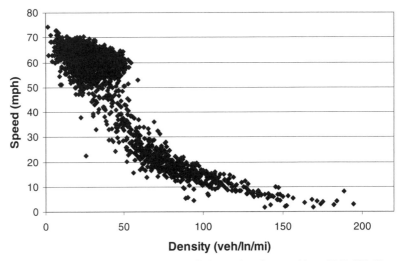

FIGURE 12.1 Speed versus density (lane 2 observations from northbound I.H. 880, Hayward, California).

---

*This northbound freeway section consists of five lanes near Hayward, California, and the number one (or leftmost) lane is an HOV lane. The plotted data come from a rainy day and a dry day in early 1993. The single-station dual inductive-loop detectors' 30-second data have been multiplied by 120 to represent equivalent hourly values. The data are indicative of this segment's general operations; however, these two days' data exhibit more congested points than typically observed during that period.

**Vehicles per unit distance (density) equals the inverse of distance per unit vehicle (spacing), where spacing is measured from the front of one vehicle to the front of the following vehicle (thereby including one vehicle's body).

†Flow is vehicles per unit time, speed is distance per unit time, and density is vehicles per unit distance. Under stationary traffic conditions, vehicles maintain constant speeds, the mix of these vehicles (and their speeds) is unchanging, and density times space-mean speed (rather than the commonly measured time-mean speed or average of spot speeds) equals flow.

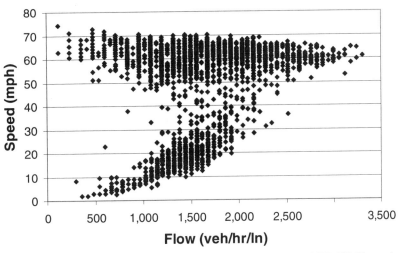

FIGURE 12.2   Speed versus flow (lane 2 observations from northbound I.H. 880, Hayward, California).

incident destabilizes traffic and shockwaves travel back upstream, forcing traffic into level of service F conditions. These oversaturated conditions correspond to speeds below 50 mph in this lane on this facility.

Figure 12.3 is a photograph of congested traffic that could have come from this same section of I.H. 880. Since passenger vehicles average 15 to 18 feet in length (*Ward's* 1999), the image suggests an average spacing of roughly 40 feet per vehicle (per lane). Such spacing translates to a density of 132 vehicles per lane-mile. Assuming that Figure 12.1's relationships are predictive of Figure 12.3's traffic behaviors, this density corresponds to average speeds between 5 and 15 mph. Since speed times density is flow, flows are likely to be around 1,300 veh/hr/ln, or about half of capacity.

As illustrated by Figure 12.2, a flow of just 1,300 veh/hr/ln could also correspond to a much higher speed (about 60 mph) and a much lower density (perhaps 25 veh/ln/mi, or a 211 ft/veh spacing). This contrast of two speeds (and two densities) for the same level of output (i.e., flow) is disturbing. Densities beyond critical (capacity-level) density and speeds below critical speed identify a loss: the restricted roadspace could be more fully utilized and traveler delays could be avoided, if only demand and supply were harmonized.

This "tragedy of the commons" (Hardin 1968) plays itself out regularly in our networks: heavy demand, downstream bottlenecks, and capacity-reducing incidents force upstream travelers into slow speeds. And facilities carry lower than capacity flows. In extreme cases, high-speed freeways as well as downtown networks become exasperating parking lots. To avoid underutilization of scarce resources, one must have a strong understanding of demand versus supply. And bottlenecks are to be avoided if the benefits can be shown to exceed the costs.

A chain is only as strong as its weakest link, and downstream restrictions can have dramatic impacts on upstream roadway utilization. Bridges are common bottlenecks; they are relatively costly to build* and expand and thus often are constructed with fewer, narrower lanes and limited shoulders. Given a trade-off in construction costs, travelers' willingness to

---

*Based on freeway construction-cost data from the Texas Department of Transportation, Kockelman et al. (2001) estimated bridge lanes to cost 5 to 10 times as much as regular lanes.

FIGURE 12.3   Image of congestion.

pay, and traveler delays, theoretically there is an optimal number of lanes to build in any section of roadway. Reliable information on demand (or willingness to pay) and associated link travel times (for estimates of delay) is necessary. Real-time roadway pricing, robust models of travel demand, and instrumentation of highways (for traffic detection) are key tools for communities aiming to make optimal investment—and pricing—decisions. The next section describes methods for estimating delays.

12.4.3   Link Performance Functions and Delay Estimation

Delay and the social value of this delay (such as the willingness of travelers to pay to avoid such delay) depend on the interplay of demand, supply, and willingness to pay. One must quantify congestion through link and network performance functions, and transform the resulting delays into dollars.

Travel time (per unit distance) is the inverse of speed. Thus, delays rise as speeds fall. And, as described earlier, speeds fall as density rises. Density rises as more and more users compete for limited roadspace, entering the facility and reducing inter-vehicle spacings, causing speeds to fall. To a certain extent, densities can rise fast enough to offset reduced speeds, so that their product (i.e., flow) rises. But flow can increase only so far: in general, it cannot exceed capacity. As soon as demand for travel across a section of roadway exceeds capacity, flow at the section "exit" will equal capacity and a queue will form upstream, causing average travel times across the congested section to rise. Unfortunately, these travel times tend to rise exponentially as a function of demand for the scarce roadspace. Thus, only moderate additions to demand can dramatically impact travel times. And if the resulting queuing impedes other system links, the delay impacts can be even more severe.

Referred to as the BPR (Bureau of Public Roads) formula (FHWA 1979), the following is a common travel-time assumption:

$$t(V) = t_f \left(1 + 0.15 \left(\frac{V}{C}\right)^4\right) \tag{12.1}$$

where $t(V)$ is actual travel time, as a function of demand volume V, t_f is free-flow travel time, and C is "practical capacity," corresponding to approximately 80 percent of true capacity.* Figure 12.4 illustrates the BPR relationship for a particular example where true capacity is 10,000 veh/hr, C is 8,000 veh/hr, and t_f is 1 minute (for example, the time to traverse a one-mile section at a free-flow speed 60 mph). If demand exceeds *practical* capacity by 30 percent ($V = 1.3C$), travel times will be twice as high as those under free-flow/uncongested conditions. The resulting 2 min/mi pace implies speeds of just 30 mph. It also implies one minute of delay for every mile of travel, with delay naturally defined as follows:

$$\text{Delay}(V) = t(V) - t_f$$

$$= 0.15t_f \left(\frac{V}{C}\right)^4 \tag{12.2}$$

Is a one-to-one correspondence of delay to free-flow travel time common? Shrank and Lomax (2002) estimate 54 seconds of delay for every minute of peak-hour travel in the Los Angeles

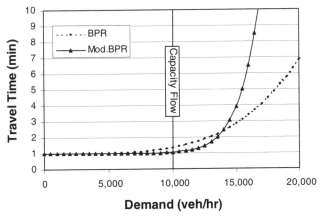

FIGURE 12.4   Travel time versus demand: BPR and modified-BPR formulae (capacity = 10,000, C = 8,000 veh/hr, t_f = 1 min).

---

*True capacity is understood to be the maximum service flow (MSF) under the *Highway Capacity Manual*'s level of service E. This practical capacity variable is a source of regular error and confusion in applications. Many (e.g., Garber and Hoel 2001) substitute a roadway's true capacity flow rate for C, resulting in an underprediction of travel times.

region. Kockelman and Kalmanje's (2003) survey results indicate that Austin, Texas, commuters perceive almost 60 seconds of delay for every 60 seconds of their commute travel. Thus, it may be in some regions that peak-period demands regularly exceed capacity by 30 percent or more.

Of course, this 30 percent figure for a doubling of travel times is based on the BPR formula. Researchers have proposed modifications to this formula. Horowitz (1991) suggested replacing the two constants (0.15 and 4) with 0.88 and 9.8 (for use with 70-mph-design speed freeways) and 0.56 and 3.6 (for 50-mph freeways). Dowling et al. (1997) recommended 0.05 and 10 (for freeways) and 0.20 and 10 (for arterials). For comparison purposes, Dowling et al.'s freeway formula is included in Figure 12.4, as the modified BPR curve. The two differ dramatically when demand exceeds capacity by more than 50 percent.

Beyond basic modifications to the BPR formula, there are other options. Akçelik's (1991) formula is wholly distinct and recognizes demand duration. The duration of queuing has important consequences for total queue lengths and thus overall delays, which depend on when one can expect to enter the queue.

Modifications in BPR factors and the underlying formulae can have dramatic impacts on travel time estimates, travel demand predictions, and policy implications.* Yet actual delay relationships remain poorly understood. Vehicles occupy space and roadway sections back up, spilling over onto other links in the network. The complexity of networks makes it difficult to measure travel times**—and even more difficult to ascertain "demand."

Delay Example: A Temporary Lane Loss.   Relying again on Figure 12.4 and equation (12.1)'s BPR formula, consider the impact of a loss of one lane. If capacity of 10,000 veh/hr corresponds to a four-lane high-design freeway, then the loss of one of these four lanes (by a crash or creation of a construction work zone, for example) results in an effective capacity of 7,500 veh/hr. At a demand of 13,000 veh/hr, travel times will jump by 115 percent, from 2.0 min/mi to 4.3 min/mi. This is now 330 percent longer than travel time under free-flow conditions. Under this dramatic situation, speeds would be just 14 mph—far less than the free-flow speed of 60 mph and well below the four-lane speed of 30 mph.

The reason that travel times rise so dramatically once demand exceeds capacity is that a roadway (like an airport or any other constrained facility) can accommodate no more flow. It behaves much like a funnel or pipe that can release only so much fluid per unit of time. Any additional users will be forced to form a slow-moving queue, backing up and impacting the rest of the system (by blocking off-ramps and on-ramps, or driveways and crossroads, upstream of the limiting section). This is a classic bottleneck situation, where demand exceeds supply. Capacity-reducing incidents can affect supply instantly, leading to essentially the same low-speed, high-delay conditions for which recurring bottlenecks are responsible. Unfortunately, there is no guarantee that congestion can be altogether avoided; supply disruptions, through incidents and the like, can occur at most any time.

12.4.4   Recurring and Nonrecurring Congestion

The above example of a temporary lane loss may be recurring or nonrecurring, predictable or unpredictable. Recurring congestion arises regularly, at approximately the same time of day and in the same location. It results from demand exceeding supply at a system bottleneck,

---

*Nakamura and Kockelman's (2002) welfare estimates for selective pricing on the San Francisco Bay Bridge were very dependent on the bridge's travel time function. Outputs of Krishnamurthy and Kockelman's (2003) integrated land use-transportation models of Austin, Texas, were most affected by the exponential term in the BPR formula.

**Lomax et al. (1997) recommend the use of probe vehicles to ascertain operating speeds. Loop detectors are presently only popular on freeway lanes and can assess only local conditions; frequent placement of loops is necessary to appreciate the extent of upstream queuing. Video cameras and sophisticated image processing techniques offer hope for future traffic data collection and robust travel-time estimation.

such as a bridge, tunnel, construction site, or traffic signal. Nonrecurring congestion results from unexpected, unpredictable incidents. These may be crashes, jackknifed trucks, packs of slow drivers, or foggy conditions.

Schrank and Lomax's (2002) extensive studies of regional data sets on travel, capacity, and speeds suggest that nonrecurring incidents account for roughly half of total delay across major U.S. regions. However, these percentages do vary. They depend on the levels of demand and supply, crash frequency, and incident response. For example, in the regularly congested San Francisco Bay Area, with its roving Freeway Service Patrols, incidents account for 48 percent of total delay. In the New York–Eastern New Jersey region, this estimate rises to 66 percent.

12.4.5 Evaluating the Marginal Costs of Travel

Whether a traveler opts to enter a facility that is congested for recurring or nonrecurring reasons, he or she pays a price (in travel time, schedule delay, and other costs) to use that facility. Because travel times rise with demand, his or her entry onto the facility (or at the back of the queue) also marginally increases the travel cost for others entering at the same time or just behind. This imposition of a cost, to be borne by others, is called a *negative externality*. Essentially, use of a congestible facility reduces the quality of service for others. This reduction in service quality is an external cost in the form of travel time penalties that others bear. Of course, all users bear it equally. So is it a problem?

Any time users "pay" a cost lower than society bears to permit the added consumption (in this case, use of a space-restricted facility), the good is overconsumed and society bears more cost than it should. Economists have rigorously shown that in almost any market goods should not be allocated beyond the point where marginal gain (or value to society) equals marginal cost to furnish the good. Marginal gains for most goods are well-specified by consumers' willingness to pay. And marginal costs are typically absorbed by suppliers of those goods. In the case of road use, costs arise in many forms—and they are absorbed by many parties: infrastructure provision and maintenance costs are absorbed by federal, state, and local agencies (and passed on through fuel and property taxes); travel time costs are absorbed by travelers; environmental damages are absorbed by humans and animals (on, off, and far from the facilities themselves); and crash losses are felt by a variety of individuals (through pain, suffering, delay, and EMS-related taxes).

The focus of this chapter is congestion, and therefore travel times in excess of free-flow travel times. A road's available space is fixed, and under congested conditions fellow travelers absorb the costs of delays arising from additional users. What are these marginal costs? Every link-performance function $t(V)$ implies these.

At a particular level of demand V, the marginal cost of an additional user $MC(V)$ is the change in total travel costs due to that added user. Total costs are average cost per user $AC(V)$, multiplied by the number of users V. And travel time (per user) $t(V)$ is the average cost. Using this logic, the standard BPR travel time function, and a little calculus (for continuous differentiation of the total-time formula), one has the following results:

$$AC(V) = t(V) = t_f \left(1 + 0.15 \left(\frac{V}{C} \right)^4 \right)$$

$$TC(V) = V \cdot t(V)$$

(12.3)

$$MC(V) = TC(V + 1) - TC(V) \approx \frac{\partial TC}{\partial V} = AC(V) + 0.6 t_f \left(\frac{V}{C} \right)^4$$

This last of three equations, the marginal cost of additional users, clearly includes the cost that the additional user experiences directly, $AC(V)$. But it also includes a second term: the

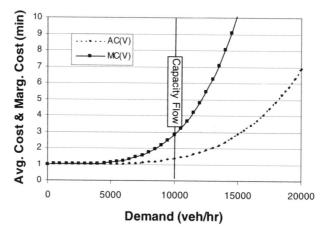

FIGURE 12.5   Average and marginal cost of demand (BPR formula, capacity = 10,000, C = 8,000 veh/hr, t_f = 1 min).

unpaid cost or negative externality that others endure in the form of higher travel times. In this BPR-based example, the externality also depends on the fourth power of the demand-to-practical capacity ratio; but it is amplified by a factor of 0.6, rather than 0.15. Thus, at certain levels of demand V, this second external cost will dominate the first average-cost term. It is this unpaid cost that is responsible for overconsumption of roadspace. Without assignment of ownership of the roads or some other method to ensure optimal use, excessive congestion results. If the roads are in heavy demand, the cost is more severe and the loss to society particularly striking.

Figure 12.5 plots the marginal cost curve above the standard BPR average cost curve for the FFS = 60 mph, 4-lane freeway example. The difference between these two is that which goes unpaid by the additional users. At capacity levels of demand, this *difference* exceeds average cost by 7 percent. At demand of 13,000 veh/hr, it constitutes more than double the average cost. Essentially, then, for optimal operations, perceived travel times or costs *should be* equivalent to 6.23 min/mi. Yet they are only 2.04 min/mi; the added travelers are enjoying an implicit subsidy of 4.18 min/mi at the expense of other travelers.

This situation may lead one to question: Who is the last, "marginal" driver and why (and how) should only he or she be penalized, when everyone should enjoy equal rights of access to the public right of way? The answer is that *all* drivers should weigh the true marginal cost of their trip before embarking—and then pay this cost (in the form of a toll) if they choose to make the trip. They will make the trip when their marginal gain exceeds the total cost of time and toll. This requirement is placed on buyers of any good in regular markets, even for items as basic as clothing, shelter, food, and health care.* Private providers are not asked to provide goods below their marginal cost. Such excessive production is unwise. In considering whether and how to combat congestion, we should ask whether society should provide space on roadways at prices below marginal cost. This question brings us to this chapter's final section.

---

*Education remains largely a public good. Regardless, subsidies targeting specific goods and consumer groups always can be provided (e.g., food stamps for low-income families and travel credits for welfare-to-work participants).

12.5 SOLUTIONS TO CONGESTION

Congestion may result from inadequate supply, imperfect information, flawed policies or any combination of these three. To address these possibilities, a variety of remedies have been proposed. Solutions may be supply- or demand-sided, long- or short-term, best suited for recurring or nonrecurring congestion. They may be demographically, temporally, and spatially extensive—or limited; they may be costly or inexpensive, mode-specific or multimodal. They may involve mode subsidies, special lanes, and/or pricing policies. Virtually all produce winners and losers.

In discussing various strategies for combating congestion, this section first details supply-side remedies, which generally aim to increase capacity by adding facilities or enhancing operations of existing facilities. On the demand side, the emphasis is on modification of prices and preferences.

12.5.1 Supply-Side Solutions

Traditionally, the solutions for congestion have been supply-sided. Engineers expected that roadway expansions and upgrades would relieve congestion. And as long as the latent demand for the facilities did not overtake the expansions (through, for example, Downs' [1962] "triple-convergence principle" of route, time of day, and mode choice adjustments), peak-hour speeds rose and travel times fell.

The same phenomenon holds true at airports, where gates and runways may be added and aircraft may be made larger (to add seating). It holds true on railways, where engines and track may be added and headways reduced (through added trains and track as well as better coordination of train schedules), and at shipyards, where container cranes may be added or upgraded and berths may be extended. Traffic actuation and synchronization of signals, rationalization of freight networks, targeted enhancements of bottlenecks, headway-reducing vehicle-guidance technologies, and other remedies also are very helpful for specific applications.

Points of recurring congestion are often well-suited for supply-side solutions. Returns on such investment will be more certain, as long as similarly sized, nearby bottlenecks do not negate the local expansion. (For example, a previously untested and thus undetected choke-point may lie just downstream and lead to similar backups.) In cases of nonrecurring congestion, solutions providing rapid detection (e.g., paired loop detectors, video processing, or transponder tag reidentification), rapid response (e.g., roaming freeway service patrols), and real-time information provision (e.g., variable message signs) will have the greatest impact.

Capacity Expansion.   There are a great many ways to expand the capacity of congestible systems. As long as increases in demand—through natural growth, changing preferences, and substitution (of origins, destinations, routes, modes, departure times, and other choices)—do not overtake these expansions, service times will fall. Yet, in many markets, capacity is so constrained and pent-up demand so significant that travel times and speeds on expanded sections of the network may not fall in any perceptible way. Investigations by Hansen and Huang (1997), Noland and Cowart (1999), Fulton et al. (2000), and Rodier et al. (2000) have resulted in high long-run elasticity predictions of demand (vehicle miles traveled) for roadspace (generally after controlling for population growth and income). Elasticity estimates of almost 1.0 (suggesting that new roadspace is almost precisely filled by new miles traveled) are not uncommon. However, several of these studies draw largely on California data, where congestion is relatively severe. In less congested regions, elasticities are expected to be lower. Even so, at roughly $1 million per added lane mile for freeway construction costs alone (Kockelman et al. 2001; Klein 2001),* funding constraints regularly preclude major supply-side solutions. And in regions not in attainment with air quality standards and/or wishing to limit sprawl and other features of long-distance driving, building one's way out of congestion may not be a viable option.*

Alternative Modes and Land Use.   To reduce roadway congestion, there are several supply-side enhancements of alternative modes that can cost less than new roadways while reducing driving and emissions. Improvements and expansions of bus, rail, ferry, and other services may qualify. However, transit is already heavily subsidized, per trip, in many countries. And even in downtown locations where its provision is reasonably extensive, transit ridership rates remain low in the United States. Thus, it is unlikely to attract many travelers, particularly for long trips in a U.S. context.**

Land use solutions have also been proposed as a way to increase the use of alternative modes and diminish congestion. Transit use and walking are highest in high-density, mixed-used areas (see, e.g., Pushkarev and Zupan 1977; Kockelman 1997; Cervero and Kockelman 1997). Transit- and walking-oriented New Urbanist designs strive to motivate mode shifts and reduce automobile reliance. But the resulting mode shifts are relatively weak, and neighborhood design—particularly in the form of higher development densities—is a poor instrument for combating roadway congestion (see, e.g., Boarnet and Crane 2001; Taylor 2002).

Managed Lanes.   Addition of managed lanes, both high-occupancy vehicle (HOV) and high-occupancy toll (HOT), is an intermediate option. HOT lanes help ensure against congestion for those whose trips are highly valued while facilitating full utilization of these special lanes (Peirce 2003). Fees can rise and fall (for example, up to 40 cents/mile) to keep the HOT lanes flowing smoothly, while carpoolers (HOV users) and transit buses ride free—and fast. Thanks to revenues generated, agencies can float bonds to help cover some of the construction and other costs, or spend the money on other services, such as increased transit service, roving freeway service patrols, and variable message signs with information on traffic conditions (Dahlgren 2002). However, without pricing of substitute routes and services, it is difficult if not impossible to raise sufficient revenues from the private sector. Public financing is still needed. And HOV/HOT lane construction costs generally exceed those of standard freeway lanes due to distinguishing features (such as longitudinal barriers and special access points).

Ramp Metering.   Another supply-side strategy is ramp metering (May 1964; Newman, Dunnet, and Mears 1969). In contrast to expansion of existing systems and services, the objective is *reduction* of ramp flows, to keep main freeway lanes moving safely and swiftly. This form of supply restriction can reduce travel times and improve safety (Chen, Jia, and Varaiya 2001; Klein 2001), but it also can penalize near-destination dwellers in favor of long-distance drivers, resulting in certain inequities (Levinson et al. 2002). Ramp metering aims to moderate the use of key links in a system, thus impacting route choice and demand, the subject of the next section.

---

*Right-of-way acquisition, traffic diversion delays during construction, and other features of major highway projects in congested areas will add further to the overall expense of such projects.

*Even though VMT may rise in proportion to expanded capacity, with congestion remaining high, there may be sufficient benefits accruing to warrant such expansions. For example, if households can afford better homes and enjoy more choice in stores, schools, jobs, and other activities, thanks to expanded travel options, those benefits should be recognized.

**There are a variety of reasons for this. Land use patterns (including dispersed, low-density origins and destinations), parking provision, and relatively low gas prices are just a few.

12.5.2 Demand-Side Solutions

Demand-side strategies seek to impact demand directly through policies and prices. Rather than expanding (or shrinking) existing services and facilities, one targets the relative prices of and/or access to these.

Parking policies offer valuable examples of demand-side strategies. Most parking is provided "free," at offices, shopping centers, schools, and elsewhere. When space is plentiful, attendants are not needed, operation costs are zero, and maintenance costs may be minimal. Largely for the sake of cost-collection efficiency, parking costs are borne indirectly by users, through, for example, reduced salaries (to employees) and higher goods prices (for shoppers). Everyone bears these indirect prices, however, so there is no price-based incentive for not parking. Preferential parking and other perks for carpoolers and others who reduce total driving and parking demands provide a way to impact driving demand; however, there is a cost to these policies. Shoup's (1992, 1994) cash-out policy, now in place in California,* requires that the cash-equivalent of parking expenses be given to those employees who do not use the parking. This form of clear remuneration makes good sense to those who agree that markets naturally clear at optimal levels when pricing and other signals are unambiguous and consistent. This argument raises the case for congestion pricing.

Congestion Pricing.   The objective of congestion pricing is efficient travel choices. It is a market-based policy where selfish pursuit of individual objectives results in maximization of net social benefits. Such laissez-faire capitalism is the guiding light behind Adam Smith's (1776) Invisible Hand. When market imperfections are removed (through pricing of negative externalities such as noise, emissions, and congestion; subsidy of positive externalities such as public schools for educating all community members; and provision of adequate information), private pursuit of goods is optimal.**

In 1952, Nobel Laureate William Vickrey proposed congestion pricing for the New York City subway through the imposition of higher fares during congested times of day. This proposal was followed by theoretical support from Walters (1954, 1961) and Beckmann, McGuire, and Winsten (1956). If peak-period pricing recognizes the negative externality (i.e., time-penalty equivalent) of each new, marginal user of a facility, optimal consumption and use decisions can result. In the case of Figure 12.5's standard-BPR curve for travel time, the value of the negative externality, or the optimal toll, is the difference between the average and marginal cost curves:

$$\text{Optimal toll} = 0.6t_f \left(\frac{V}{C}\right)^4 \text{VOTT}_{\text{avg}} \tag{12.4}$$

where VOTT_{avg} is the monetary value of travel time of the average marginal traveler (roughly \$5 to \$15 per hour, depending on the traveler and trip purpose).† Other variables are defined as for equation (12.1).

---

*Under California Health and Safety Code Section 43845, this policy applies to businesses with 50 or more employees in air quality nonattainment areas.

**Efficient travel choices will have effects on several related markets. For example, land use choices and wage decisions will become more efficient by recognition of the true costs of accessing goods, services, and jobs.

†The toll should equal the value of the additional travel time the last vehicle adds to the facility. This last vehicle adds $[dt(V)/dV]V$, where $V = \Sigma_i \, V_i(t,\text{toll})$ and i indexes the various classes of users demanding use of the road under that toll and travel time. The appropriate moneterarized value of travel time to interact with this added time is the demand-weighted average of VOTT_i's: $\text{VOTT}_{\text{avg}} = \Sigma_i \, \text{VOTT}_i \cdot V_i(t,\text{toll})/\Sigma_i \, V_i(t,\text{toll})$. Given demand functions sensitive to time and toll and a link's performance function, it is not difficult to solve for the optimal toll. However, these inputs are tricky to estimate; they will be based on sample data for use of an entire network, across times of day, and may involve significant error.

Figure 12.6 adds a downward-sloping demand or "willingness to pay" (WTP) curve to Figure 12.5's travel time "supply" functions. The untolled supply curve represents the approach currently defining roadway provision, while the higher, tolled curve incorporates all marginal costs. The intersection of the demand curve with these two defines the untolled (excessively congested) and tolled (less-congested) levels of use on this roadway. The first results in 13,000 veh/hr, travel times of 2.0 min/mi, and delays of 1.0 min/mi. The latter results in 10,450 veh/hr, travel times of 1.44 min/mi, delays of 0.44 min/mi, and (the equivalent of) a 1.75 min/mi toll. In this example, neither situation is uncongested. Demand exceeds capacity in each instance, so queues will build for as long as demand exceeds supply.

Imposition of the net-benefit-maximizing "optimal toll" equates marginal benefits (measured as willingness to pay, here in the form of time expenditures) and marginal costs. Under this 1.75 min/mi toll, realized demand is 10,450 veh/hr and toll revenues reach almost 18,300 min/hr (1.75 × 10,450). Moreover, traveler benefits—measured as WTP in excess of travel time plus toll—exceed 25,100 minutes.* If time is converted to money at a rate of $12 per vehicle-hour (for the marginal traveler), the toll is worth 35¢ per mile, revenues are $3,600 per hour per mile, and the traveler benefits are worth $5,000/hr/mi.

Without the toll, traveler benefits exceed 39,000 minutes ([8 − 2] × 13,000/2), or $7,800/hr/mi. But there are no toll revenues. Thus, the $8,600 of value derived under the optimally tolled situation exceeds the laissez faire approach by over 10 percent.

For a 10 percent addition in value during peak times of day on key roadway sections, should communities ask their public agencies to step in and start charging on the order of 35¢ per congested mile of roadway? There are many issues for consideration. First, the revenues should not go right back to those who bear the cost of congestion (i.e., the delayed travelers), since that will offset their recognition of the marginal costs and encourage prior levels of demand. But, if these revenues are not well spent, society loses. Second, if demand is highly inelastic (i.e., steeply sloped in Figure 12.6), realized demand will not change much under pricing; revenues simply will be transferred from traveler benefits to the collecting agent. Third, if the cost of implementing and enforcing congestion pricing is high, any improvements in added value of the policy may be overcome.

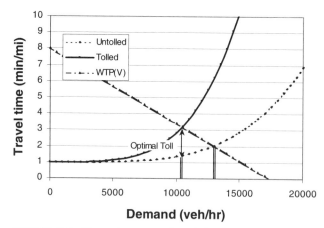

FIGURE 12.6 Demand versus supply: tolled and untolled cases (tolled and untolled equilibrium demand levels are 10,450 and 13,000 veh/hr; optimal toll is equivalent to 1.75 min/mi).

---

*Thanks to an assumed-linear demand function, this computation is relatively simple. It involves the triangular area under the demand curve, and above the 3.19-min generalized cost: (8 − 3.19) × 10,450/2 = 25,132 min (per hour of flow, on this mile-long section of roadway).

However, the benefits of pricing on highly congested links may well exceed 10 percent. While *average* travel times may rise by a factor of two during peak periods in congested areas like Los Angeles, certain sections of roadway (such as key bridges) may experience much more severe delays. Imagine the same example in Figure 12.6 with a peak-period demand line that still begins at 8 min/mi but lies flatter and crosses the $AC(V)$ curve at 17,500 veh/hr, or 75 percent beyond capacity. In this case, untolled travel times are about 4.5 min/mi and marginal costs lie 13.7 min/mi above that level (a significant externality, and implicit subsidy). The optimal toll would be worth 3.6 min/mi (or 72¢/mi), bringing demand down to 12,500 veh/hr. And tolled traveler benefits plus revenues would exceed untolled traveler benefits by a whopping 94 percent. Such a situation is probably a strong candidate for pricing.

Beyond increases in community benefits through removal of delay-related externalities, there are other advantages of congestion pricing. These include reduced emissions and gasoline consumption, since idling is reduced, closer destinations are chosen, and cleaner modes of transportation are selected. They result in healthier species, less crop and property damage, and diminished threat of global warming. Such benefits are difficult to value, but Small and Kazimi's work (1995) puts the human health costs of emissions close to 5¢/vehicle-mile in air basins like that of Los Angeles.

Another key benefit is the allocation of roadspace to the "highest and best users."* Many users with high values of travel time (including, for example, truck drivers delivering goods for just-in-time manufacturing processes) presently are doing whatever they can to avoid congested roadways. Those with little or no value of travel time (e.g., high school students traveling to a shopping mall) are taking up that scarce space. With the introduction of monetary prices, the types of travelers on our roadways would shift towards those with more money and less time.

Of course, high values of time often correspond to higher wages and wealth, so communities and policy-makers are understandably worried about the regressive impacts of congestion pricing—relative to the status quo. This is particularly true in the short term, when home, work, school, and other location choices are relatively fixed. Pricing will have land use effects that are difficult to predict; short-term activity location and timing inflexibilities suggest that restrained introduction of pricing will be necessary, and special cases of heavily impacted low-income travelers (for example, single parents with fixed work and childcare times that coincide with rush hour) may require credits and/or rebates. Toll revenues can be spent on alternative modes and other programs to benefit those most negatively impacted by such policies. For example, revenues from London's downtown cordon toll of £5 (roughly $8) go largely toward transit provision. This experiment seems to be succeeding: speeds have doubled (from 9.5 to 20 mph), bus ridership is up 14 percent, and only 5 percent of central London businesses claim to have experienced a negative impact (*Economist* 2003).

Equity Considerations: Minimum-Revenue Pricing, Rationing, and Reservations.   While various forms of congestion pricing already exist, in Southern California, Florida, Singapore, Milan, Rome, Trondheim, and now London, equity and revenue-distribution implications rouse public concern (Button and Verhoef 1998). These considerations also have spawned a number of creative policy proposals. They include Dial's (1999) "minimal revenue pricing" (where one route is kept essentially free between every origin-destination pair), DeCorla-Souza's (1995) "Fast and Intertwined Regular (FAIR) Lanes."** and Daganzo's (1995) alternating tolled days across users.

Penchina (Forthcoming) has demonstrated that demand must be relatively inelastic for Dial's (1999) proposal to have important advantages over marginal cost tolling (such as lower

---

*"Highest and best use" is a normative economic term traditionally used to describe a market-derived (top-bid) use. A social or community-derived highest and best use may differ.

**FAIR lanes involve demarcating congested freeway lanes into tolled fast lanes and untolled regular lanes. Regular lane drivers using electronic toll tags would receive credit for use of the facility; accumulated credits could be redeemed for use of the fast lanes.

tolls, more stable tolls, and fewer tolled links). Given the variety of travel-choice substitutes people face for many activities (e.g., time of day, destination, and mode), relatively inelastic cases may be unusual.

Nakamura and Kockelman (2002) examined Daganzo's idea in the context of the San Francisco-Oakland Bay Bridge to assess winners and losers under a variety of alternate pricing and link-performance scenarios. Net benefits were significant, but, as with almost any policy, they identified some losers—unless revenues are specially targeted and link-performance functions are favorable.

Another policy one might consider is reservation of scarce road capacity to allocate crossing of key network links (such as bridges) by time of day. This policy could involve a cap on free reservations and tolls for additional use. Airports grant (or sell) gate rights to airlines for exclusive use. And airlines sell specific seat assignments to passengers, effectively guaranteeing passage. Singapore rations vehicle ownership through a vehicle licensing system wherein only a certain number of 5-year licenses are auctioned off each year (through on-line sealed bids). Coupled with the most extensive road-pricing system in the world, this city-state is a clear leader in congestion-fighting policies.

Owing largely to budget constraints, U.S. transportation agencies are looking more and more at toll roads as a congestion-fighting and roadway-provision option. These may be variably priced throughout the day to recognize the premium service they provide when demand is high (yet tolled travel times remain reasonable). In addition, smooth-flowing high-occupancy toll (HOT) lanes encourage peak-period travelers to shift to buses and carpooling while allowing better use of such lanes through tolling of single-occupancy vehicles (SOVs).* Toll roads and HOT lanes may be the intermediate policies that spark widespread congestion pricing in the United States. Yet the issue of equity and redistributive impacts remains.

Ideally, any selected solution will be cost-effective and relatively equitable. There is one possible solution that has the potential for near-optimal returns while ensuring substantial equity and efficiency: "credit-based congestion pricing" (Kockelman and Kalmanje 2003).

Credit-Based Congestion Pricing.   Under a credit-based congestion-pricing plan, the tolling authority collects no *net* revenues (except perhaps to cover administrative costs), and travelers willing to reschedule their trips, share rides, or switch modes actually receive money by not exhausting allocated cash credits. All collected tolls are returned in the form of per-driver rebates to licensed drivers in a region or regular users of a corridor.

A credit system for roadway use that provides a base amount of "free" peak-period travel each month addresses public conerns to a much greater extent than pricing alone—and thus could generate considerably more support. Explicit credit banking and trading are becoming widespread in other domains. In 1999, the OECD found 9 programs involving tradable permit schemes in air pollution control, 5 in land use control, 5 in water pollution, 75 in fisheries management, and 3 in water resources (Tietenberg 2002). At present, many more such applications exist. And, as long as the cost of trading credits is not high, *any* distribution strategy theoretically will result in an efficient use of the constrained resource (Tietenberg 2002).

A National Academy of Science committee recently proposed a tradable credits system as an alternative to the current corporate average fuel economy standards (TRB 2002). For each gallon of fuel expected to be consumed by a new vehicle over its lifetime, the manufacturer would need a credit. The annual allocation of credits to a manufacturer would be based on the company's production level and the government's target for fuel consumption per vehicle.

Clearly, there is a distinction between industrial applications involving the trading of emissions, fuel economy credits, and the consumer-oriented credit-based congestion-pricing application considered here. In the case of roadway networks, where the temporal and spatial attributes of the congested resource are critical, appropriate link pricing must also exist.

---

*Toll collection can become an issue, however, since HOVs must be distinguished so that they are not charged.

Roadway pricing, therefore, is somewhat more complex than emissions or fisheries regulation; one cannot simply cap total vehicle miles traveled (VMT) and allow VMT credit use at any time of day or at any point in the network. Continual monitoring of traffic conditions and recognition of key links and times of day are needed. And prices, rather than credits, are charged. However, just as companies achieved the imposed reductions in SO_2 emissions more easily than had been foreseen initially, travelers also may have greater flexibility in their demand for peak-period travel than some anticipate. The potential for and presence of such flexibility is key.

Congestion Pricing Policy Caveats. While credit-based and other forms of pricing have the ability to reduce congestion to (and beyond) "optimal" levels, there are implementation issues that require careful consideration. First, the costs of the technology can pose serious hurdles. Many commercial vehicles carry GPS systems, but these presently cost hundreds of dollars. If the price of congestion is less than $100 per vehicle in a region per year (as it is in many regions), GPS systems probably will not make economic sense. If much cheaper local radio-frequency systems are used, their roadside readers (presently costing over ten thousand dollars) are likely to be selectively distributed, leaving much of the network unpriced and some locations still congested. In either case, private third-party distribution of identification codes and formal legal protections are probably needed to ensure privacy expectations are met. Second, policy administration is rife with special needs. If only a few regions per state merit pricing, visitors may need to buy or rent identification systems for travel through and around the priced systems.* For enforcement purposes, video-image processing of rear license plates is probably necessary at key network points in order to identify vehicles without transponders or system violators. Lists of license plate holders would then be necessary, and regions and states would have to share information (as they presently do for serious offenses).

Finally, much travel time is access time, at trip ends. The drive from one's home to the first arterial may have little to no travel time improvement available; similarly, the walk and elevator ride from one's parking space to one's office may already be minimized. These access times are not insignificant. Thus, even if congestion on all arterial roads is removed, total peak-period travel times may fall only 25 percent or less, for most travelers, even in highly congested regions.**

12.6 CONCLUSIONS

Congestion results from high demand for constrained systems and tends to manifest itself in the form of delay. A relative scarcity of roadspace and other forms of transportation capacity has led to substantial congestion losses in many travel corridors, across many regions, at many times of day. The delays may be severe or moderate; the congestion may be recurring and anticipated or nonrecurring and unpredictable.

Society pays for congestion not just through higher travel times and crash rates, uncertain and missed schedules, additional emissions, and personal frustration, but also through higher costs for goods and services. After all, commercial delivery services must confront the same traffic delays that personal vehicle occupants face. These delays translate to lowered productivity and more expensive deliveries and commutes, resulting in higher prices for everyone.

The relationship between demand and delay is not clear-cut. As demand approaches capacity, travel times tend to rise. When demand exceeds capacity, queues form and travel time impacts are pronounced. Small additions to demand can generate significant delays—

---

* Short-term visitors may be granted access without penalty, depending on time of stay and use of network.
** Taylor (2002) makes this point in his *Access* article "Rethinking Traffic Congestion."

and minor reductions can result in significant time savings. Removal of bottlenecks, toll road provision, and subsidy of alternative modes are proving popular mechanisms to combat congestion. But more effective solutions are needed in most cases. Individual travel choices will remain inefficient until travelers recognize and respond to the true costs of using constrained corridors and systems.

The good news is that new technologies are available to address the congestion issue. But robust estimates of individual demand functions and the marginal costs of additional users are needed to take congestion policies to the next level. Congestion pricing promises many benefits, and *credit-based* congestion pricing, as well as other strategies, may substantially offset the burdens on automobile-dependent low-income populations, particularly in the short term, when location and other requirements are relatively fixed for certain activity types.

Congestion is not just a roadway phenomenon. It affects nearly all pathways, including air, rail, water, and data ports. Fortunately, many of the strategies and technologies for resolving roadway congestion also apply to these other domains. Greater recognition of the negative externalities involved in oversaturation of our transportation systems will guide us to the most effective strategies for coping with congestion, with the promise of more efficient travel for everyone.

12.7 ACKNOWLEDGMENTS

The author is grateful for the excellent suggestions of Tim Lomax, Steve Rosen, Sukumar Kalmanje, and Annette Perrone.

12.8 REFERENCES

Akçelik, R. 1991. "Travel Time Functions for Transport Planning Purposes: Davidson's Function, Its Time-Dependent Form and an Alternative Travel Time Function." *Australian Road Research* 21(3):44–59.

Bates, J., J. Polak, P. Jones, and A. Cook. 2001. "The Valuation of Reliability for Personal Travel." *Transportation Research* 37E(2):191–229.

Beamon, B. 1995. "Quantifying the Effects of Road Pricing on Roadway Congestion and Automobile Emissions. Ph.D. Dissertation, Department of Civil and Environmental Engineering. Georgia Institute of Technology, Atlanta, GA.

Beckmann, M., C. B. McGuire, and C. B. Winsten. 1956. *Studies in the Economics of Transportation*, ch. 4. New Haven: Yale University Press.

Boarnet, M. G., and R. Crane. 2001. *Travel by Design: The Influence of Urban Form on Travel*. Oxford: Oxford University Press.

Brodsky, H., and A. S. Hakkert. 1983. "Highway Accident Rates and Rural Travel Densities." *Accident Analysis and Prevention* 15(1):73–84.

Bureau of Transportation Statistics (BTS). 2002. *National Transportation Statistics 2002*. U.S. Department of Transportation, BTS, Washington, DC.

Button, K. J., and E. T. Verhoef, eds. 1998. *Road Pricing, Traffic Congestion and the Environment*. Cheltenham, UK: Edward Elgar.

Cervero, R., and K. Kockelman. 1997. "Travel Demand and the Three Ds: Density, Diversity, and Design." *Transportation Research* 2D(3):199–219.

Chen, C., Z. Jia, and P. Varaiya. 2001. "Causes and Curves of Highway Congestion." *IEEE Control Systems Magazine* 21(4):26–33.

Commodity Flow Survey (CFS). 1997. U.S. Department of Transportation, Bureau of Transportation Statistics, and U.S. Department of Commerce, Economics and Statistics Administration, U.S. Census Bureau, Washington, DC.

Daganzo, C. 1995. "A Pareto Optimum Congestion Reduction Scheme." *Transportation Research B* 29: 139–54.

Dahlgren, J. 1994. "An Analysis of the Effectiveness of High Occupancy Vehicle Lanes." Ph.D. Dissertation, Department of Civil Engineering, University of California at Berkeley.

———. 2002. "High-Occupancy/Toll Lanes: Where Should They Be Implemented." *Transportation Research A* 36:239–55.

DeCorla-Souza, P. 1995 "Applying the Cashing Out Approach to Congestion Pricing." *Transportation Research Record* 1450, 34–37.

Dial, R. B. 1999. "Minimal Revenue Congestion Pricing Part I: A Fast Algorithm for the Single-Origin Case." *Transportation Research* 33B:189–202.

Dowling, R., W. Kittelson, J. Zegeer, and A. Skabardonis. 1997. "Planning Techniques to Estimate Speeds and Service Volumes for Planning Applications." NCHRP Report 387, National Research Council, Transportation Research Board, Washington, DC.

Downs, A. 1962. "The Law of Peak-Hour Expressway Congestion." *Traffic Quarterly* 16:393–409.

Economist, The. 2003. "Congestion Charge: Ken's Coup" March 22, 51.

Evans, L. 1991. *Traffic Safety and the Driver.* New York: Van Nostrand and Reinhold.

Federal Highway Administration (FHWA). 1979. *Urban Transportation Planning System (UTPS).* U.S. Department of Transportation, FHWA, Washington, DC.

———. 1999. CORSIM, Version 4.2. U.S. Department of Transportation, FHWA, Washington, DC (distributed through McTrans, University of Florida, Gainesville, FL).

———. 2001. National Household Travel Survey (NHTS). U.S. Department of Transportation, FHWA, Washington, DC.

Fimrite, P. 2002. "Traffic Tops List of Bay Area Banes, Weak Economy Is Number 2 Bane, Survey Shows." *San Francisco Chronicle,* http://sfgate.com/cgi-bin/article.cgi?file=/c/a/2002/12/05/MN51835.DTL#sections. Accessed on December 5, 2002.

Fulton, L., R. Noland, D. J. Meszler, and J. Thomas. 2000. "A Statistical Analysis of Induced Travel Effects in the U.S. Mid-Atlantic Region." *Journal of Transportation and Statistics* 3(1):1–14.

Garber, N. J., and L. A. Hoel. 2002. *Traffic and Highway Engineering,* 3rd ed. Pacific Grove, CA: Brooks-Cole.

Garber, N., and S. Subramanyan. 2002. *Feasibility of Incorporating Crash Risk in Developing Congestion Mitigation Measures for Interstate Highways: A Case Study of the Hampton Roads Area.* Virginia Transportation Research Council, Final Report VTRC 02-R17, Charlottesville, VA.

Gwynn, D. W. 1967. "Relationship of Accident Rates and Accident Involvements with Hourly Volumes." *Traffic Quarterly* 21(3):407–18.

Hansen, M., and Y. Huang. 1997. "Road Supply and Traffic in California Urban Areas." *Transportation Research A* 31(3):205–18.

Hardin, G. 1968. "The Tragedy of the Commons." *Science* 162:1243–48.

Horowitz, A. J. 1991. *Delay Volume Relations for Travel Forecasting Based on the 1985 Highway Capacity Manual.* Report FHWA-PD-92-015, U.S. Department of Transportaion, Federal Highway Administration, Washington, DC.

Klein, L.. 2001. *Sensor Technologies and Data Requirements for ITS.* Boston: Artech House.

Knickerbocker, B. 2000. "Forget Crime—But Please Fix the Traffic." *Christian Science Monitor,* February 16.

Kockelman, K. 1997. "Travel Behavior as a Function of Accessibility, Land Use Mixing, and Land Use Balance: Evidence from the San Francisco Bay Area." *Transportation Research Record* 1607, 116–25.

———. 2001. "Modeling Traffic's Flow-Density Relation: Accommodation of Multiple Flow Regimes and Traveler Types." *Transportation* 28(4):363–74.

Kockelman, K., and S. Kalmanje. 2003. "Credit-Based Congestion Pricing: A Policy Proposal and the Public's Response." Paper presented at the 10th International Conference on Travel Behaviour Research, Lucerne, Switzerland, August.

Kockelman, K., and Y.-J. Kweon. 2002. "Driver Injury Severity and Vehicle Type: An Application of Ordered Probit Models." *Accident Analysis and Prevention* 34(3):313–21.

Kockelman, K. K., R. Machemehl, A. Overman, M. Madi, J. Sesker, J.Peterman, and S. Handy. 2001. *Frontage Roads in Texas: A Comprehensive Assessment.* Report FHWA/TX-0-1873-2, University of Texas at Austin, Center for Transportation Research.

Krishnamurthy, S., and K. Kockelman. 2003. "Propagation of Uncertainty in Transportation-Land Use Models: An Investigation of DRAM-EMPAL and UTPP Predictions in Austin, Texas." *Transportation Research Record* No. 1831: 219–29.

Levinson, D., L. Zhang, S. Das, and A. Sheikh. 2002. "Ramp Meters on Trial: Evidence from the Twin Cities Ramp Meters Shut-off." Paper presented at the 81st Meeting of the Transportation Research Board, Washington, DC.

Lomax, T., S. Turner, and R. Margiotta. 2003. "Monitoring Urban Roadways in 2001: Examining Reliability and Mobility with Archived Data." Report FHWA-OP-02-029, U.S. Department of Transportation, Federal Highway Administration, Washington, DC.

Lomax, T., S. Turner, G. Shunk, H. S. Levinson, R. H. Pratt, P. N. Bya, and G. B. Douglas. 1997. *Quantifying Congestion: Volume 1, Final Report.* NCHRP Report 398, National Research Council, Transportation Research Board, Washington, DC.

May, A. D. 1964. "Experimentation with Manual and Automatic Ramp Control." *Highway Research Record* 59, 9–38.

Mokhtarian, P., and I. Salomon. 2001. "How Derived Is the Demand for Travel?" *Transportation Research* 35A:695–719.

Myers, B., and J. Dale. 1992. "Designing in Car-Oriented Cities: An Argument for Episodic Urban Congestion." In *The Car and the City: The Automobile, the Built Environment, and Daily Urban Life,* ed. Martin Wachs and Margaret Crawford. Ann Arbor: University of Michigan Press.

Nakamura, K., and K. Kockelman. 2002. "Congestion Pricing and Roadspace Rationing: An Application to the San Francisco Bay Bridge Corridor." *Transportation Research* 36A(5):403–17.

Newman, L., A. Dunnet, and J. Meirs. 1969. "Freeway Ramp Control: What It Can and Cannot Do." *Traffic Engineering* (June):14–25.

Noland, R. B., and W. A. Cowart. 2000. "Analysis of Metropolitan Highway Capacity and the Growth in Vehicle-Miles of Travel." Paper presented at the 79th Annual Meeting of the Transportation Research Board, Washington, DC.

Oak Ridge National Laboratory (ORNL). 2001. *1995 NPTS Databook.* ORNL/TM-2001/248, prepared for the U.S. Department of Transportation, Federal Highway Administration, Washington, DC.

Peirce, N. 2003. "Congestion Insurance: 'HOT' Lanes' Amazing Promise." *Washington Post,* March 5.

Penchina, C. M. Forthcoming. "Minimal-Revenue Congestion Pricing: Some More Good-News and Bad-News." Accepted for publication in *Transportation Research B.*

Pushkarev, B S., and J. M. Zupan. 1977. *Public Transportation and Land Use Policy.* Bloomington, IN: Indiana University Press.

Richardson, T. 2003. "Some Evidence of Travelers with Zero Value of Time." Paper presented at the 82nd Annual Meeting of the Transportation Research Board, Washington, DC.

Redmond, L. S., and P. L. Mokhtarian. 2001. "The Positive Utility of the Commute: Modeling Ideal Commute Time and Relative Desired Commute Amount." *Transportation* 28:179–205.

Rodier, C. J., J. E. Abraham, R. A. Johnston, and J. D. Hunt. 2000. "Anatomy of Induced Travel Using an Integrated Land Use and Transportation Model in the Sacramento Region." Paper presented at the 79th Annual Meeting of the Transportation Research Board, Washington, DC.

Salomon, I., and P. L. Mokhtarian. 1998. "What Happens when Mobility-Inclined Market Segments Face Accessibility-Enhancing Policies?" *Transportation Research* 3D(3):129–40.

Scheibal, S. 2002. "New Planning Group Kicks-off Effort with Survey on What Area Residents Want." *Austin American-Statesman,* August 26.

Schrank, D., and T. Lomax. 2002. *The 2002 Urban Mobility Report.* Texas A&M University, Texas Transportation Institute.

Shoup, D. 1992. *Cashing out Employer-Paid Parking.* Report No. FTA-CA-11-0035-92-1, U.S. Department of Transportation, Washington, DC.

———. 1994. "Cashing out Employer-Paid Parking: A Precedent for Congestion Pricing?" In *Curbing Gridlock: Peak-Period Fees to Relieve Traffic Congestion,* vol. 2. Washington, DC: National Academy Press, 152–200.

Small, K. A., and C. Kazimi. 1995. "On the Costs of Air Pollution from Motor Vehicles." *Journal of Transport Economics and Policy* 29(1):17–32.

Smith, A. 1776 *An Inquiry into the Nature and Causes of the Wealth of Nations.* London: W. Strahan and T. Cadell. Reprint, London: Dent & Sons, 1904.

Taylor, B. D. 2002. "Rethinking Traffic Congestion." *Access* 21:8–16.

Tietenberg, T. 2002. "The Tradable Permits Approach to Protecting the Commons: What Have We Learned?" Chapter 6 in *The Drama of the Commons.* Washington, DC.: National Research Council, National Academy Press.

Transportation Research Board (TRB). 2000. *Highway Capacity Manual 2000.* National Research Council, TRB, Washington, DC.

———. 2002, *Effectiveness and Impact of Corporate Average Economy (CAFE) Fuel Standards.* Washington, DC: National Academy Press.

Walters, A. A. 1954. "Track Costs and Motor Taxation," *Journal of Industrial Economics.*

———. 1961. "The Theory and Measurement of Private and Social Costs of Highway Congestion." *Econometrica* 29(4):676–699.

Ward's Automotive Yearbook 1998. 1999. Southfield, MI: Ward's Communications, Inc.

Weisbrod, G, D. Vary, and G. Treyz. 2001. *Economic Implications of Congestion,* NCHRP Report 463, National Cooperative Highway Research Program, Transportation Research Board., Washington, DC.

CHAPTER 13
GEOMETRIC DESIGN OF STREETS AND HIGHWAYS

Brian Wolshon
*Department of Civil and Environmental Engineering,
Louisiana State University, Baton Rouge, Louisiana*

13.1 INTRODUCTION

The geometric design of highways encompasses the design of features of the roadway associated with safe, efficient, and comfortable travel. The design process takes into account the range of different driver and vehicle types that are likely to use the facility, the topographic and land use characteristics of the area that surrounds it, and the costs involved in constructing and maintaining the highway. In addition, highway design also requires a balance between many overlapping and often competing functions and needs that will affect the safety, efficiency, and cost-effectiveness of the roadway.

The highway design process requires knowledge of most of the subdisciplines of civil engineering, including traffic and transportation engineering; geotechnical and materials engineering; structural engineering; hydraulic engineering; and surveying. It also requires an understanding of many aspects of the basic sciences, including physics, physiology, and psychology. This chapter provides an overview of both the theory and practice of highway design. It summarizes the geometric highway design process by highlighting the fundamental engineering and scientific principles that underlie it and summarizing the ways in which they are applied to the development of a final design.

Since the highway design process encompasses such a wide range of design specialties, it is obviously not possible to cover all of them within single book chapter. However, in recognition that the subject is considerably more complex and detailed than can be adequately addressed herein, the author has included references for additional resources that include a more detailed treatment of particular subjects. Readers are encouraged to review them as well as the related chapters of this book to gain a greater understanding of these specific topic areas.

The presentation of the material in this chapter approximates the coverage of similar subjects in the American Association of State Highway and Transportation Officials' *A Policy on Geometric Design of Highways and Streets* (AASHTO 2001). This manual, commonly referred to as the Green Book, is the most widely referenced and applied highway design resource in the United States and serves as the foremost source of highway design theory and application. It has evolved over several decades based on a consensus of state and local highway agency design methods, standards, and assumptions that are themselves based on years of testing and experience.

Since the Green Book represents a compilation of design policies, it also allows ample latitude for the use of engineering judgment and experience that may be appropriate to address specific design issues. However, its wide scope and, in some cases, lack of specificity can leave some of users overwhelmed with differing opinions, recommendations, and information when addressing specific design issues. Among the objectives of this chapter is an attempt to summarize some of the many of the details of the Green Book to make it more usable.

13.2 HIGHWAY FUNCTION AND DESIGN CONTROLS

The design of a highway needs to fit within a set of parameters that govern how it will be used, who will use it, and where it will be located. These parameters are a function of the traffic that it will serve and, often, the agency or entity that owns it. Thus, the classification, whether functional or jurisdictional, is a fundamental component that drives final design. Conversely, the classification can also impact the make-up of the traffic that will be using the roadway, including the their type (trucks, automobiles, buses, recreational vehicles, etc.), operating speeds, and, to some extent, the type of trips that they will make on it. The following sections summarize some of the differences between the functional and jurisdictional classifications of highways and describe some of the ways in which they impact their design.

Among the primary components that drive road design are its users. In addition to vehicles, users also include pedestrians and bicyclists. The composition of vehicles, commonly known as the *vehicle mix,* may itself include tractor semi-trailer trucks, small automobiles, motorcycles, and nearly every vehicle type in between. What is important about these vehicle categories is that each of them has different heights, widths, lengths, and weights, as well as varying capabilities to accelerate, brake, turn, and climb hills. Similarly, individual drivers of these vehicles also differ in terms of their ability to see, hear, read, comprehend, and react to stimulus. Despite the differences in vehicle characteristics and user abilities, only a single road can be constructed. Thus, it must be designed to accommodate this spectrum of vehicles and drivers safely and economically. The following sections will also discuss some of the key aspects of vehicle and driver characteristics and the way that they impact the design of the highway environment.

13.2.1 Roadway Functional Classification

A key factor that guides the development of a roadway design is its functional classification. The functional classification influences both the type of traffic (automobiles, heavy trucks, etc.) the road will carry as well as the trip characteristics (high-speed through route, low-speed local access route, etc.) of these vehicles. Generally speaking, roads serve two functions, providing mobility to drivers that seek to move quickly and efficiently from one location to another and providing access to properties adjacent to the road. These functions, while complementary and often overlapping, can also be conflicting. To serve the range of mobility and access demands, a hierarchy of roadway functionality has evolved to serve traffic, access adjacent properties, and guide the development of their design.

The three primary road classifications within this functional hierarchy are *arterial, collector/distributor,* and *local* roads and streets. There are also varying stages within each of these classifications. The basic function of arterial roadways is to accommodate through traffic over longer distances, on routes that allow comparatively high speeds, with a minimum of interruptions. An example of the highest level of arterial roadway is a freeway. Freeways tightly restrict access for entering and departing traffic, have no at-grade intersections, and allow traffic to operate at high speeds. At the opposite end of the functional classification spectrum are local roads, which are designed primarily to provide access to abutting lands. As a result, they often carry lower volumes of traffic than arterial roadways, are designed for low-speed traffic operation, and provide frequent points of access/egress. Residential neighborhood streets are the most common example of local roads. On them, posted speeds are typically in the range of 20 to 30 mph and access points (driveways) may be spaced as close as 50 to 100 feet apart. Collector/distributor roadways encompass the range of roads between the arterial and local categories. While they serve multiple purposes, including overlapping the arterial and local roles, their intended purpose is to serve as a transition between arterial and local streets by collecting traffic from local and arterial streets and distributing it up or down the functional hierarchy.

The distinction between the various roadway classifications and functions is not always clear and can become even more blurred as transportation needs and use land requirements change over time. An example of how this can occur is often seen in the vicinity of residential neighborhoods, when low-level collector distributor roads evolve (with or without conscious planning) to serve as minor arterial routes. An opposite case occurs when intensive land development takes place along arterial routes in areas that were once rural in character. In such cases, roads that once provided high mobility between distant locations must now adapt to provide access to new local developments. The functional conflicts illustrated in both of these examples also lead to traffic congestion and safety problems as the traffic mix changes to combine both through traffic seeking to travel at higher speeds and low-speed local traffic turning in and out of local businesses. These situations can, however, be addressed (or avoided altogether) though a combination of effective land use planning and travel forecasting analyses with design, rather than purely through design processes.

13.2.2 Design Parameters

In addition to operational considerations, a road's functional classification influences several other key design parameters, including the volume of traffic the road will carry, its composition (in terms on the percentage of trucks and other commercial and heavy vehicles), and its design speed. In turn, these parameters affect nearly all aspects of the geometric design as well many of the aspects of the pavement thickness requirements and culvert and bridge structural elements.

Design Vehicle Types. To simplify the variety of vehicles that may use a roadway, AASHTO has established a set of four general design vehicle groups based on representative sizes, weights, and operating characteristics. These vehicle classes include passenger automobiles, buses, trucks, and recreational vehicles and incorporate several specific subconfigurations within each. In addition to the specific dimension measurements that are shown in Table 13.1, each design vehicle also has a set of turning radii that also include off-tracking and vehicle overhang measurements for large vehicles.

Traffic Volume. The amount of traffic that is expected to use a roadway typically increases over time. Forecasts of traffic volume usually come from planning studies that take into account future land use development; improvements to the surrounding road network; population demographics and behavioral characteristics; and future economic trends. The traffic volume measures that are most useful in road design include the future *average daily traffic* (ADT) and the future *design hourly volume* (DHV).

DHV is closely related to the thirtieth highest hourly volume during the design year. This volume can also be used to assess the anticipated operational characteristics of planned roadways using the theories and computational procedures contained in the *Highway Capacity Manual* (*HCM*) (TRB 2000). ADT measures are typically more useful for the design of minor and low volume roads. It should also be noted that areas that experience widely

TABLE 13.1   Dimensions of AASHTO Design Vehicles

Design vehicle type	Symbol	Overall Height	Overall Width	Overall Length	Overhang Front	Overhang Rear	WB$_1$	WB$_2$	Typical kingpin to center of rear axle
		Dimension (ft)							
Passenger car	P	4.25	7.0	19.0	3.0	5.0	—	—	—
Single unit truck	SU	11–13.5 in.	8.0	30.0	4.0	6.0	20.0	—	—
Conventional school bus (65 pax)	S-BUS 36	10.5	8.0	35.8	2.5	12.0	21.3	—	—
Intermediate semi-trailer	WB-50	13.5	8.5	55.0	3.0	2.0	14.6	35.4	37.5
Motor home	MH	12.0	8.0	30.0	4.0	6.0	20.0	—	—

Source: AASHTO 2001.

fluctuating seasonal or morning and evening peak hour traffic volumes additional analyses of demand may find it necessary to provide adequate capacity for the anticipated demand. A more detailed discussion of the procedures used to acquire and compute both ADT and DHV can be found in Chapters 2 and 6 of this book.

Design Speed.   Roads must also be designed under an assumed set of driver and vehicle operating characteristics. One of the most influential of these is design speed. Design speed impacts most safety-related features of the design as well those associated with rideability and comfort. It can also impact the efficiency and capacity of a roadway. The selection of an appropriate design speed is based on a number of factors, including functional classification, the terrain and topography in the vicinity of the road, adjacent land use, and driver expectation.

In general, roads are engineered for design speeds between 15 and 75 mph. A design speed of 70 mph should be used for freeways, although depending on specific conditions (such as terrain, urban/rural conditions, etc.) actual design speeds may range from 60 to 80 mph. Most arterial roadways and streets are designed for between 30 and 60 mph. The design speed of collector/distributor roads is also in the lower area of this range and local streets typically 30 mph and under.

Common practice dictates that the posted speed limit on a road be somewhat less than the design speed, although this is not universally true (Krammes et al. 1996). There are several different methods of posting speeds. Two popular methods include posting at 5 to 10 mph under the design speed and the use of an 85th percentile running speed, which can be checked once the road is in operation.

Capacity and Level of Service.   Similar to fluid flow in a pipe, there is a limit to the amount of traffic volume that a highway can accommodate. The maximum amount of flow is commonly referred to as the capacity of the roadway. Highway capacity is affected by a variety of factors that influence a driver's ability to maintain a desired operating speed and maneuver in traffic, including the number of lanes, the sharpness and steepness of curves, the amount of heavy vehicles in the traffic stream, the quality of the signal coordination, and the location of lateral obstructions. The *HCM* contains standardized procedures to estimate the capacity

TABLE 13.2 Design Level of Service Based on Functional Classification

Roadway functional class	Appropriate level of service for combinations of area and terrain type			
	Rural level	Rural rolling	Rural mountainous	Urban and suburban
Freeway	B	B	C	C
Arterial	B	B	C	C
Collector/distributor	C	C	D	D
Local	D	D	D	D

Source: AASHTO (2001).

of highways in North America (TRB 2000). One of the fundamental aspects of highway design is the determination of how many lanes will be required to service the expected design traffic demand.

The *HCM* also contains procedures that allow alternative designs to be evaluated based on the anticipated quality of the operating conditions associated with each one. With them, the operational quality of service provided by a particular design can be designated on a level of service (LOS) rating scale of A to F, in which A is the highest quality of flow. On this scale LOS E describes the flow conditions at capacity and F the conditions when the traffic demand exceeds the capacity of the roadway.

Typically, different roadways are designed to operate at different LOSs. In general, higher LOSs are sought for roads of higher functional classification. The AASHTO LOS guidelines for various functional classifications are shown in Table 13.2. As shown in the table, there are several factors that can influence LOS assessments of specific roadways, including the location of the roadway (rural/urban), the terrain (flat/mountainous), and the costs associated with constructing the desired facility.

13.2.3 Sight Distance

Another fundamental requirement of a highway design is the need to maintain adequate sight lines for drivers. This can be accomplished in a number of different ways, including manual checks on design profiles and through the application of theoretical stopping and passing sight distance equations. At a minimum, all roads must be able to permit drivers operating vehicles at the design speed and within assumed conditions, to bring a vehicle to a controlled stop in the event that an obstacle is blocking the lane of travel. This must be true longitudinally (i.e., straight ahead), around horizontal curves, and over and under vertical curves. Where economically and practically feasible, designers may also wish to provide enough advanced visible distance to drivers such that they are able to perform a passing maneuvers in the lane of oncoming traffic.

The minimum lengths for sight distances are a function of a series of assumptions regarding the condition of the road and weather as well as the capabilities of a driver. The assumptions and resulting equations for determining minimum stopping and passing design requirements are described in the following sections.

Stopping Sight Distance. Design stopping sight distance (SDD) lengths are computed as the sum of two separate distances, the distance traveled during the perception–reaction process and the distance covered during the deceleration from the design speed to a stop as shown in equation (13.1) (AASHTO 2001).

$$\text{SSD} = 1.468 V_d t_{p-r} + \frac{V_d^2}{30 \left(\left(\dfrac{a}{32.2} \right) \pm G \right)} \tag{13.1}$$

where V_d = design speed (mph)
t_{p-r} = perception–reaction time (sec)
a = deceleration rate (ft/sec^2)
G = longitudinal grade of the road (%/100)

In the perception–reaction distance portion of the equation, it is assumed that a driver will require 2.5 seconds to identify and react to a hazard stimulus. This is a standard value that is used despite prior research showing that actual reaction times may range from less than a second to more than 7 seconds, depending on the level of expectation and complexity of the information that is given. The standard reaction time value does not account for drivers who may be momentarily distracted from the driving task or who may be in a less than fully lucid state, such as those who may be drowsy, taking medication, or affected by other impairments.

The second part of the equation accounts for the distance traveled while the vehicle is decelerating. This distance is affected by several factors, including the initial speed of the vehicle, the friction conditions that exist between the vehicle tires and the road surface, and the grade of the road. The friction relationship assumes that the tires on the design vehicle will be well worn with less than a full tread and that the road surface will be wet. These assumptions do not, however, take into account snow or icy conditions in which the level of friction between the tires and the road surface would be effectively zero. Thus, drivers are expected (and in most case required by law) to operate their vehicles in a manner "reasonable and prudent" for the apparent conditions, regardless of the design or posted speed limit. Road grade can also lengthen or shorten the required stopping distance, depending on whether the road profile in the direction of travel is oriented up- or downhill. The AASHTO stopping sight distances for flat grades are shown in Table 13.3.

Passing Sight Distance. Unlike stopping sight distance, which is required for all design, the provision for passing sight distance is an optional design feature. Typically, passing sight distance is provided where practicable and economically feasible. Since passing can increase the overall efficiency of a roadway, it is desirable to provide it on as high a proportion of

TABLE 13.3   Recommended Stopping Sight Distances

Design speed (mph)	Brake reaction distance (ft)	Braking distance on a flat grade (ft)	Stopping sight distance	
			Calculated (ft)	Rounded for design (ft)
20	73.5	38.4	111.9	115
30	110.3	86.4	196.7	200
40	147.0	153.6	300.6	305
50	183.8	240.0	423.8	425
60	220.5	345.5	566.0	570
70	257.3	470.3	727.6	730
80	294.0	614.3	908.3	910

Brake reaction distance assumes a 2.5-second driver perception–reaction time and a vehicle deceleration rate of 11.2 ft/sec^2.
Source: AASHTO (2001).

the roadway as feasible, particularly on low-volume roads. Where it is not practically feasible to provide passing sight distance in a design, the incorporation of passing lanes is suggested.

Similar to the stopping sight distance equation, the computation of passing sight distance is made up of multiple parts that account for the various physical processes that occur during the maneuver. The equation is based on the passing need for a two-lane highway in which a driver would need to see far enough ahead into the opposing lane of traffic that he can accelerate drive around the passed vehicle and return to his original lane while leaving an adequate safety distance between himself and the oncoming vehicle.

The first component distance of a passing action is the *initial maneuver*. This part includes the distance covered as the driver of the passing vehicle checks the downstream clearance distance and begins accelerating to overtake the vehicle to be passed. The equation for computing this distance is shown in equation (13.2) (AASHTO 2001).

$$d_1 = 1.468t_i + \left(v - m - \frac{at_i}{2}\right) \tag{13.2}$$

where t_i = time of initial maneuver (sec)
a = average acceleration (mph/sec)
v = average speed of the passing vehicle (mph)
m = difference in speed of passed and passing vehicles (mph)

One of the key assumptions in this equation is that the difference in speed between the passing and passed vehicles (the variable m in the equation) is 10 mph. In reality, this may or may not be the case.

The second distance component is that which is covered while the passing vehicle occupies the lane of oncoming traffic. This includes the distance in which any part of the passing vehicle encroaches into the lane of opposing traffic. The computation of this distance is based on the time that the passing vehicle is assumed to occupy the lane of opposing traffic as shown in equation (13.3) (AASHTO 2001).

$$d_2 = 1.468vt_2 \tag{13.3}$$

where t_2 = time passing vehicle occupies the left lane (sec)
v = average speed of the passing vehicle (mph)

The third component accounts for the distance covered by a vehicle heading toward the passing vehicle in the opposing lane. For design purposes, it is assumed to be two-thirds of the d_2 distance. The fourth and final distance is safety separation distance between the passing and opposing vehicles. This distance varies and is based on the design speed of the road.

The AASHTO passing sight distance ranges for various operating speed ranges are shown in Table 13.4. Of course, it must be understood that the assumptions both for stopping and passing conditions can vary significantly across any set of driver, vehicle, weather, and location conditions. However, an additional implied assumption is that it is unlikely that this a set of near-worst-case conditions will occur simultaneously at a single time in a single place. Thus, the available sight distances provided on most roads are more than adequate for the overwhelming majority of the time.

13.3 HORIZONTAL ALIGNMENT DESIGN

Although the construction of highways takes place in three-dimensional space, the procedures that govern their design have been developed within three separate sets of two-dimensional relationships that represent the *plan, profile,* and *cross-section* views. The plan view encompasses the *horizontal alignment* of the road in the *x–y* plane. In it, the straight and curved

TABLE 13.4 Recommended Passing Sight Distances

Design speed (mph)	Assumed vehicle speeds (mph)		Stopping sight distance	
	Passed vehicle	Passing vehicle	Calculated (ft)	Rounded for design (ft)
20	18	28	706	710
30	26	36	1,088	1,090
40	34	44	1,470	1,470
50	41	51	1,832	1,835
60	47	57	2,133	2,135
70	54	64	2,479	2,480
80	58	68	2,677	2,680

Source: AASHTO (2001).

sections of the road are described on a flat two-dimensional surface. The directions and lengths of straight or *tangent* segments of road are stated in terms of *bearings* and *distances* from one point to another and are referenced in a coordinate system of *northings* and *eastings,* which are analogous to *y* and *x* rectangular coordinates. The profile view, or *vertical alignment,* represents the roadway in terms of uphill and downhill slopes, or *grades,* that connect points of vertical *elevation.* Finally, the cross-section is a slice of the roadway through the *y–z* plane. The key components of the cross-section include the widths and cross-slopes of the lanes, shoulders, and embankment slopes as well as the depths or thicknesses of the pavement and embankment material layers.

Horizontal alignments can be thought of a series of straight lines, known as *tangents,* that meet one another, end-to-beginning, at *points of intersection* (POI). The goal of *horizontal curves* is to facilitate high-speed changes of travel direction between the various tangents. The sharpness of these curves dictates how abruptly a change in direction can be made at given design speed. The horizontal alignment must also be designed to provide adequate sight distances on these straight and curved sections of highway such that drivers will be able to operate their vehicles at the desire rate of speed without encountering any safety or efficiency obstacles.

The development of a horizontal alignment is based on a number of factors, the most important of which are safety, efficiency, access to adjacent properties, and cost. However, since both safety and efficiency can be designed into an alignment, the chief factors in most alignment selections are often cost and land access. The cost of an alignment may be affected by a number of factors, including right-of-way land acquisition, environmental considerations, and the cost of construction. Construction cost can be increased significantly by special features such as bridges and tunnels and from added earthwork manipulation such as that necessitated by unsuitable soil conditions.

13.3.1 Horizontal Curve Design

The design of horizontal curves must allow a driver to maintain control of a vehicle within a lane of travel, at a desired rate of speed, while permitting him to see an adequate distance ahead such that he would be able to take evasive action (stop, change lanes, etc.) if a hazard is present in his path. To accomplish these objectives two criteria have been developed: minimum radius curve design and lateral sight distance within a curve.

Minimum Radius.   According to Newton's first law, a body in motion tends to stay in motion unless acted on by an external force. This theory is applied to horizontal curve design from the perspective of a vehicle rounding a curve. The natural tendency of a vehicle is to go straight rather than turn. As such, vehicles are prone to skid out of a curve that is too sharp for the speed at which they are traveling. To account for this fact, a minimum curve radius formula shown in equation (13.4) has been developed (AASHTO 2001).

$$R_{min} = \frac{V_d^2}{15(0.01e_{max} + f_{max})} \qquad (13.4)$$

where R_{min} = minimum radius of curvature (ft)
V_d = design speed (mph)
e_{max} = maximum rate of superelevation (ft/ft)
f_{max} = coefficient of side friction

Equation (13.4) balances the constituent sideways, gravitational, and centripetal forces on a vehicle as it rounds a curve by effectively allowing all of the external forces to sum to zero (i.e., the vehicle will neither skid out of the curve nor slide down the banking) and allowing a driver to maintain a travel path lane position throughout the curve.

Curve Superelevation.   While equation (13.4) assumes that the curve will be superelevated throughout its length, in reality this is rarely, if ever, the case. In actual practice the lanes of tangent sections of road are designed with a crowned *cross-slope*. A crowned cross-slope facilitates the runoff of surface water across the lanes from a centerline high point to the outside edges of each lane. In a superelevated curve the road cross-slope is sloped continuously from one edge to the other. Thus, this roadway cross-slope change from a normal crown to a fully superelevated section must occur over a gradual transition distance. This transition distance can be accomplished in several different ways.

One method involves the design of a *spiral* or *transition curve* to transition the road alignment form a straight line to a curve. By definition, a spiral is a curve of constantly varying radius. When used to transition between a tangent and a curve, the spiral begins with an infinite radius (i.e., a straight line) and ends at the design radius of the curve. To transition from a normal crown to a fully superelevated section, the road surface cross-section is transitioned gradually over this length.

Another common method of transition is the application of a *maximum relative gradient* (MRG). The MRG numerically designates the rate at which the road cross-slope, and correspondingly the edge of pavement elevation, can change relative to the road centerline crown point. As shown in Table 13.5, the higher the design speed, the more gradual the transition. For example, at a design speed of 50 mph, the MRG is 0.50 percent or an elevation change of approximately 1 foot vertically for each 200 feet of longitudinal distance. This would mean that a two-lane highway of 24-foot width, with a 2 percent normal crown cross-slope, superelevated about the inside edge of pavement, at 6 percent, and designed with a 50-mph design speed, would require a total longitudinal distance of approximately 120 feet to transition from a normal crown to a fully superelevated cross-slope.

Each of the various methods of transition from a normal crown to fully superelevated cross-section (rotated about the centerline, inside edge of pavement, or outside edge of pavement) has subtle advantages and disadvantages. In south Louisiana, for example, it is common practice to design superelevated transitions by rotating about the inside edge of pavement because of high groundwater conditions. While this type of design may require a small increase in the required quantity of embankment fill, it permits designers to maintain the elevation of the inside edge of pavement and thereby not force it any lower than its original elevation. It should also be noted that the required length of transition also increases with wider pavement widths (i.e., more lanes).

TABLE 13.5 Maximum Relative Gradients

Design speed (mph)	Maximum relative gradient (%)	Equivalent maximum relative slope
20	0.74	1:135
30	0.66	1:152
40	0.58	1:172
50	0.50	1:200
60	0.45	1:222
70	0.40	1:250
80	0.35	1:286

Source: AASHTO (2001).

Horizontal Curve Sight Distance.   A key safety element of the design of horizontal curves is the provision of sight distance through the inside of the curve. Unlike a straight section of highway, the clear vision area on a curve may need to be maintained beyond the edges of the roadway. Although stopping sight distance is measured along the alignment, the direct line of vision to the end of the stopping distance includes areas within the inside of the curve. This relationship is also schematically illustrated in chapter 3 of the Green Book.

Using a special case the circular horizontal curve calculation procedure in which the middle ordinate is measured perpendicularly from the travel path around the innermost lane to a sight obstruction inside of the curve, an equation for calculating the minimum lateral clearance to provide adequate sight distance around a horizontal curve has been developed. The relationship, shown in equation (13.5), computes the middle ordinate distance from the center of innermost lane based on a corresponding curve radius and stopping sight distance requirement.

$$M = R \left(1 - \cos \frac{28.65S}{R} \right) \tag{13.5}$$

where S = stopping sight distance (ft)
  R = radius of the curve (ft)
  M = middle ordinate (ft)

Horizontal Alignment Layout.   The design of horizontal alignments also requires knowledge of several principles of curve layout that include the calculation of various parts of these curves. While a discussion of this process is outside the scope of this chapter, readers are encouraged to review the included sources of information on this subject (Meyer and Gibson 1980; Hickerson 1964; Crawford 1995; Garber and Hoel 2002; Schoon 2000).

13.4 VERTICAL ALIGNMENT DESIGN

Similar to horizontal alignments, profile design can also be regarded as a series of straight lines that meet one another at POIs. In the case of profiles, however, the tangents are uphill and downhill *grades* that meet at *points of vertical intersection* (PVI) that are connected by *vertical curves.* Another difference is that the vertical curves are based on parabolic, rather than circular, curve relationships. Although much like horizontal design, the sharpness of these curves is dictated by how abruptly a change in direction can be made at given design speed. Thus, the objective of a profile design is to provide a balance between vertical grade steepness and sight distance along the tangent and curved sections of highway so that drivers will be able to operate their vehicles at the desired rate of speed without encountering any significant efficiency or safety obstacles.

This section discusses the design considerations for the basic design and layout of vertical alignments. In recognition that vertical alignment design includes aspects that are considerably more complex than those presented here, readers are again urged to review the referenced resources for a more detailed treatment of the subject matter.

13.4.1 Maximum and Minimum Grades

Maximum road grades are based primarily on the ability of trucks and other heavy vehicles to maintain an efficient operating speed. A goal of any design is to permit uniform operating conditions throughout the length of the design segment. When assessing the impact of grade on operating conditions, both the steepness and the length of the grade should be considered. Most passenger cars are not affected by uphill or downhill grades equal to or less than 5 percent. However, heavy trucks have a comparatively lower weight-to-horsepower ratio that leads to a diminished ability to accelerate and maintain constant speeds on uphill grades.

In general, grades of 5 percent are considered the maximum for design speeds in the range of 70 mph. For lower-speed roads in the range of 25 to 35 mph, maximum grades in the range of 7 to 12 percent may be appropriate if the local terrain conditions are substantially rolling. These maximums should also be assessed in conjunction with the length of the sloped segment since it is also recommended that maximum grade be used infrequently and, where possible, for lengths of less than 500 feet.

Minimum grade requirements are based on the need to maintain surface runoff drainage. In general, a grade of 0.50 percent is considered to be the minimum to facilitate surface drainage on curbed roadway. However, grades of 0.30 percent may also be appropriate when pavement surfaces and cross-slopes are adequate to maintain surface flows. On uncurbed roadways, flat grades (0.0 percent slope) may also be used when the pavement cross-slope is consistent and adequate to laterally drain surface runoff.

13.4.2 Vertical Curve Design

Vertical curves are used to transition vehicle direction between various uphill and downhill grades. Depending on the magnitude and type of grade (positive or negative), vertical curves may be either crest or sag, although curves that connect positive-to-positive and negative-to-negative tangent grades are monotonic in that they do not have a turning point. Vertical curve designs seek to provide a design that is safe, comfortable, and aesthetically pleasing. This is accomplished by providing proper sight distances and adequate drainage and maintaining various vertical forces to within tolerable limits. The following sections present many of the key design equations and discuss both the theories that underlie them and the their application for design situations.

13.4.3 Designing for Sight Distance

A vertical curve may be designed for several different criteria. The most critical is sight distance. When designing for sight distance, the goal is to design a curve that permits a driver to see far enough ahead along the road profile to permit a deceleration to a stop or to allow passing maneuvers. To achieve this, a curve must be flat enough at a given set of design speed, driver eye height, object height, and road grades. However, the design of flatter curves also could require additional cutting and filling during construction. Thus, when designing for sight distance, an attempt is made to minimize the length of the vertical curve to minimize construction costs while still providing adequate sight lines.

In the case of crest curves, the apex of the rise must not interfere with the stopping and passing sight distance needs. This can be accomplished in several different ways, the most fundamental of which involves flattening the curve. To do this it is necessary to assume the height of the driver's eye in the design vehicle and the height of the target object. In the case of stopping sight distance conditions, the object is assumed to be the taillight of a preceding vehicle, and in the case of passing sight distance, the roof of an oncoming vehicle in the opposing lane.

In the case of sag curves the ability to provide stopping sight distance is not limited by any geometric obstructions, since a driver should be able to see across the curve when traveling on either of the approach grades. An exception to this would be at night when clear vision would be limited by darkness. Because of this, the limiting factor for sight distance on sag vertical curves is assumed to be the distance that vehicle headlights are able to cast light on the road ahead. And in general, this would equate to the stopping sight distance that would be computed in equation (13.1).

In both crest and sag vertical curve cases, two geometric relationships can occur between the curve length and the sight distance. The first is the case in which the stopping sight distance is contained within the length of the curve itself. Known commonly as the $S < L$ case, this occurs under combinations of lengthy curves or in shorter curves with lower design speeds. The second is the $S > L$ case, in which the stopping sight distance is longer than the curve itself and may extend prior to, after, or on both sides of the vertical curve. This is more common in shorter curves or when the *algebraic difference* between the right and left tangent grades is minimal.

The general equations for determining the minimum length of curve required to provide adequate sight distances on crest and sag curves for both the $S < L$ and $S > L$ cases are shown below.

Crest Curves.   When S is less than L,

$$L = \frac{AS^2}{100(\sqrt{2h_1} + \sqrt{2h_2})^2} \tag{13.6}$$

When S is greater than L,

$$L = 2S - \frac{200(\sqrt{h_1} + \sqrt{h_2})^2}{A} \tag{13.7}$$

where L = length of vertical curve (ft)
$\quad\ S$ = stopping sight distance (ft)
$\quad\ A$ = algebraic difference in grades (%)
$\quad h_1$ = height of eye above the road surface (ft)
$\quad h_2$ = height of object above the road surface (ft)

Sag Curves.   When S is less than L,

$$L = \frac{AS^2}{200[2.0 + S(\tan 1°)]} \tag{13.8}$$

When S is greater than L,

$$L = 2S - \frac{200[2.0 + S(\tan 1°)]}{A} \tag{13.9}$$

where L = length of vertical curve (ft)
$\quad\ \ S$ = light beam distance (ft)
$\quad\ \ A$ = algebraic difference in grades (%)

In all vertical curve design cases the standard design assumption is that the driver eye height will be 3.5 feet above the road surface. This represents the approximate height of a driver in a small sports car, for example. For stopping sight distance conditions the object height is assumed to be 2.0 feet, or the approximate height of a rear taillight. When these standard values are substituted in the general equations (13.6) through (13.9), the specific-case equations shown below result.

Crest Curves.   When S is less than L,

$$L = \frac{AS^2}{2158} \tag{13.10}$$

When S is greater than L,

$$L = 2S - \frac{2158}{A} \tag{13.11}$$

where L = length of vertical curve (ft)
$\quad\ \ S$ = stopping sight distance (ft)
$\quad\ \ A$ = algebraic difference in grades (%)

Sag Curves.   When S is less than L,

$$L = \frac{AS^2}{400 + 3.5S} \tag{13.12}$$

When S is greater than L,

$$L = 2S - \left(\frac{400 + 3.5S}{A}\right) \tag{13.13}$$

where L = length of vertical curve (ft)
$\quad\ \ S$ = light beam distance (ft)
$\quad\ \ A$ = algebraic difference in grades (%)

The stopping sight distance curve length requirements for crest and sag are also shown graphically in chapter 3 of the Green Book. In these figures curve lengths are represented as a function of design speed and the algebraic differences between the left and right tangents. Also apparent in these figures are the two regions of $S < L$ and $S > L$, which are divided by a dashed curved line in both tables.

Another area of note in the AASHTO figures is the use of K *values.* K values are computed by dividing the length of curve by the algebraic difference between the two tangent grades. Technically, K is a measure of the length of vertical curve per percent change in algebraic difference, or, in effect, the sharpness of the curve. K values are also a convenient method of quickly relating design speeds and algebraic differences to required curve lengths. The figures also shows a drainage maximum K of 167. This is to call attention to the fact that curves designed with K values greater than 167 will have relatively long flat longitudinal grades in the vicinity of the curve turning point. Thus, the need to maintain positive cross-slope drainage across the pavement surface will be of critical importance within the vicinity of the curve high or low points.

When computing crest curve length for the passing sight distance requirement, an object height of 3.5 feet is used to approximate the roof height of a small vehicle. The resulting specific-case equations are shown below (AASHTO 2001).

When S is less than L,

$$L = \frac{AS^2}{2800} \tag{13.14}$$

When S is greater than L,

$$L = 2S - \frac{2800}{A} \tag{13.15}$$

where L = length of vertical curve (ft)
S = stopping sight distance (ft)
A = algebraic difference in grades (%)

Although it is often difficult to design for passing sight distance in hilly areas because of the added expense of earthwork, the additional passing/climbing lane on the uphill grade segments into and on the curve can make provisions for passing in these areas a viable option. Since it is generally assumed that vehicles operating at night will utilize headlamp illumination, no equations have been developed for passing sight distance conditions on sag vertical curves because a passing vehicle would be able to see the headlamps of an oncoming vehicle in the opposing travel lane.

While at a minimum vertical curve lengths must be designed to provide minimum sight distances, curves are also often designed to satisfy other needs. These can include matching the elevation of an intersection or drainage feature, providing adequate vertical clearance above or below obstructions, limiting the amount of vertical forces on the driver, and co-ordinating with the associated horizontal alignment to provide safer and more distant perspectives of the highway ahead. Various procedures and equations have been developed specifically for these purposes and can be found in a number of related text references (Meyer and Gibson 1980; Hickerson 1964; Crawford 1995; Garber and Hoel 2002; Schoon 2002). Similar to the design of horizontal alignments, vertical alignments also require knowledge of several basic principals of survey layout that are outside the scope of this chapter. However, readers are encouraged to also review the same references for detailed information on this subject.

13.5 CROSS-SECTION DESIGN

The third dimension of highway design is cross-section. Roadway cross-section design encompasses elements such as the traveled way, shoulders, curbs, medians, embankments, slopes, ditches, and all other elements associated with the roadside area. While cross-section design deals primarily with the design of the widths and slopes of these elements, it also includes the design and analysis of safety-related aspects of roadside appurtenances such as sign supports, guardrails, and crash attenuators. Like horizontal and vertical alignment design, cross-section design is affected by both the design speed and other operational considerations. The following sections present the elements and design aspects of roadway cross-section features and discuss many of the philosophies that are used in the design of roadside features.

13.5.1 Cross-Section Elements

A highway section is subdivided into separate regions based mainly on their likelihood to carry traffic. The portion that carries the highest load traffic is the roadway itself. While this obviously includes the lanes of travel, it can also include the shoulders. And although shoulders are used less frequently by traffic, they still must be designed to support traffic stopping for emergencies or in some cases driving on the shoulder to avoid a blockage. As such, one of the key considerations in the development of road cross-section elements design standards is the amount and travel speed of the traffic it will carry.

Traveled Way.   The traveled way is the portion of the road used by vehicles under normal operation. At a minimum it includes the travel lanes. The width of travel lanes varies by design speed, traffic mix, and functional classification. They can vary in width from 9 to 13 feet. Although 11-foot lanes are common in areas with low volume, low operating speeds, and/or restricted right-of-way, 12-foot lanes are recommended for most roadways.

The cross-slope of the travel lanes may also vary. While attempts should be made to minimize the road cross-slope for operational purposes, the main purpose of cross-slopes is to facilitate drainage. Recommended cross-slopes for travel lanes can vary from a little as 1 percent for some roads to as much as 6 to 8 percent on others. A specific slope design is primarily a function of the surface type and need to remove surface runoff. For smooth pavement surfaces like asphalt and concrete, cross-slopes of 2 percent are recommended. Although in states of the Gulf Coast region of the United States where design rainfall intensities are significantly higher than those in other areas of the country, cross-slopes are commonly designed at 2.5 percent. For low-type surfaces and unpaved roadways recommended cross-slope increases to around 6 percent.

Shoulders.   Shoulders include the portion of the roadway immediately outside of the traveled way. Although they are intended to serve traffic in special cases, they are neither designed nor meant for high-speed operation or high traffic volume. The benefits gained from the use of shoulders include (AASHTO 2001):

1. Aiding drivers in the recovery of temporary loss of control or to provide room to perform emergency evasive action
2. Storing vehicles safely off the traveled way in emergency situations
3. Providing a safe means of accomplishing routine maintenance and navigational operations
4. Serving as a temporary traveled way during reconstruction or emergency operations
5. Serving as a primary clear area free of obstructions
6. Enabling greater horizontal sight distance in cut sections
7. Enhancing traffic flow and thereby capacity
8. Providing structural support to the pavement and traveled way

Shoulder design can also vary significantly, although it is based primarily on the function of the facility. Shoulders may be paved or unpaved (grass and gravel), and their width may range from 2 to 12 feet. On high-speed, high-volume roadways 12-foot shoulders are desirable, particularly in rural areas where wider rights-of-way are available. Since shoulders are not expected to be used for high-speed maneuvering, their cross-slopes are also higher than those of travel lanes, typically between 2 to 8 percent, although care must be taken to ensure that the algebraic difference in slope between the travel lane and shoulder is minimized so that severe slope breaks will not exist at the edges of the traveled way.

Medians.   In areas where it is economically feasible, highways that are planned to carry high volumes of traffic at high operating speeds are often designed to separate opposing streams with the use of a center median. The benefits of roadway medians include (ITE 1992):

1. Physically separating high-speed, opposing traffic, thereby minimizing the chances of serious head-on collisions

2. Providing a clear recovery area for inadvertent encroachments off of the traveled way

3. Providing a means of safely storing stopped or decelerating left-turning vehicles out of the higher-speed through lanes

4. Providing a means of safely storing vehicles turning left out of a minor street or driveway as they wait for an available gap

5. Providing safe storage for pedestrians crossing a high-speed or wide, divided highway

6. Restricting or regulating left turns to and from adjacent businesses except at designated locations

Like shoulder design, median design varies in terms of widths, slopes, and surface. Medians may be as narrow as 2 to 4 feet and as wide as 100 feet. In urban designs medians are typically 40 feet or less, although they are typically designed to accommodate shoulders and barriers (if warranted) and should allow adequate width for the storage of queued left-turning traffic and adequate turning radii for turning vehicles. On rural freeways 60-foot median widths are common to remove the need for barriers to separate traffic on opposing roadways.

13.5.2 Roadside Design

Years of experience and research have shown that the proper design of the area immediately outside of the traveled way is critical to reducing the potential for crash-related injuries and fatalities, particularly those involved in single-vehicle run-off-the-road accidents. The features that make up this area include embankment slopes (parallel and cross-slope), ditches, and recovery areas. The design of these elements needs to balance cost-efficiency and functional needs with safety so that drivers will be able "to recover safely from loss-of-control situations, or at least not suffer a serous injury or fatality should an accident occur" (AASHTO 1996).

The most widely applied reference for the design and analysis of these areas is the AASHTO *Roadside Design Guide* (AASHTO 1996). In addition to a serving as reference on geometric design issues, the *Guide* offers guidance on the design and application of barriers, sign and lighting supports, and crash-attenuating devices.

Roadway Clear Zone.   To minimize the chances for injuries in run-off-the-road accidents, the concept of a *clear recovery area* or *clear zone* was introduced to keep the area outside of the traveled way clear of hazardous obstructions and keep the cross-section slopes traversable for errant vehicles. A six-step hierarchy of design guidelines has been suggested to reduce the level of hazard in this area (AASHTO 1996):

1. Removing obstructions from the roadside area

2. Relocating obstructions to less hazardous areas of the roadside

3. Redesigning obstructions to make them more traversable (in the case of a shoulder embankment) or less of a hazard

4. Reducing the hazard potential of obstructions by making them breakaway (such as sign supports)

5. Shield obstacles from traffic by using barriers and crash attenuators

6. Delineating obstructions to make them more visible to drivers

Although the recommended clear zone depends on a number of factors, such as traffic volume, roadway design speeds, and slope steepness, the use of a 30-foot clear zone recovery area is recommended for most higher-classification facilities, including freeways. It should also be understood, however, that require clear zone for any particular location will vary based on the amount of available right-of-way and terrain features.

Embankments and Slopes.   Among the most common hazards in the roadside vicinity are embankment slopes resulting from earthwork cuts and fills. Embankment slopes are typically categorized into *recoverable* and *nonrecoverable*. Recoverable slopes include those in which drivers are assumed to be able to recover control of an errant vehicle. The most desirable slopes are those that are flatter than 6:1, although steeper slopes up to 4:1 are recommended when limiting the lateral extent of more significant cut and fill heights. Nonrecoverable slopes are those steeper than 4:1, in which there is a likelihood of vehicle rollover and other serious accidents. The hazard potential of slopes is also related to the total height of the cut or fill section (AASHTO 1996).

In general, fill slopes are considered to be somewhat more hazardous than cut slopes because of the increased risk from accelerating down an embankment. Protective design measures, such as the use of guardrails, are usually recommended in the vicinity of high and steep slopes.

Barriers.   In areas where it is costly or not economically feasible to construct a suitable clear zone, roadside barriers such as guardrails and crash cushions may be appropriate. Many different types of barrier designs may be used, based on the characteristics of particular location. However, the decision whether to use a barrier has to be made with the understanding that a barrier is itself a potential hazard and that there must be a judgment made to select and locate the barrier such that it protects drivers from obstacles while minimizing the potential for damage and injury when impacted.

In general, a barrier is warranted when the potential damage and injuries that could be sustained from the hazard are deemed to be greater than what that would be sustained from a collision with the barrier itself. The general nature of this definition gives considerable latitude to the designer in making safety assessments where fixed hazards are present. Examples of obvious hazards are steep shoulder drop-offs, watercourses, and other substantial fixed objects located within the desired clear zone. The *Guide* offers more specific guidance for the use of guardrails in the vicinity of fill embankments based on comparative risk associated with embankment fill slope steepness and height.

Another key element in barrier design is the length of need and the placement of the barrier relative to both the roadway edge and the hazard. Guidelines for placement are specified in the *Guide,* which details the design and use of other roadside reduction features, including crash attenuators and breakaway devices.

13.6   OTHER DESIGN CONSIDERATIONS

In addition to the general areas of geometric design principles presented in the preceding sections, a range of specialized areas of design are often required for the development of a final set of construction plans. These include the design of traffic control features, roundabouts, intersection channelization, railroad crossings, and roadway lighting, among many others. More specialized design principles are also required when attempting to modify existing designs to accommodate new development and/or reduce the occurrence or lessen the impact of cut-through traffic and high operating speeds. The following sections highlight some of the basic principles associated with the design of traffic control devices and intersections.

13.6.1 Traffic Control Devices

Traffic control is used to regulate and control the flow of traffic so that highways systems are able to achieve both safe and efficient operation. It is also a key element of highway design. Traffic control design elements are used to communicate information to the drivers—including where and when to stop, when to go, how fast to travel, which lane to turn from—and give advanced warning of potentially hazardous conditions that may exist within the highway environment. Traffic control devices (TCDs) include road signs, traffic signals, pavement markings, and other warning and communication devices.

The most widely accepted reference for the design, placement, and maintenance of TCDs is the *Manual on Uniform Traffic Control Devices* (*MUTCD*) (FHA 2000). The *MUTCD* was developed to provide guidance and establish national uniformity for the use of TCDs. It includes guidelines for colors, shapes, symbols, sizes, patterns, and use of text on traffic signs, signals, and pavement markings. It also gives guidance on their appropriate use under routine conditions and under special conditions such as school zones and construction work areas.

Although TCDs are used to control traffic, they must also be used only within the limits of there intended purpose. For some of the most critical devices, such as stop signs and traffic signals, the *MUTCD* includes a series of warrants that should be followed to establish whether one of these devices is truly needed. Traffic control devices should never be used indiscriminately and should never be used to alter or lessen the impacts of inadequate design and planning. One such example is the use of stop signs in residential areas as a speed control or traffic volume reduction measure. Inappropriate use of TCDs can lead to drivers placing little importance on or ignoring them.

A more detailed description of the design and application of the *MUTCD* and traffic control devices can be found in Chapter 9 of this book.

13.6.2 Intersections

When two highway alignments cross, an intersection is created. While some intersections can be avoided by creating a *grade-separated* intersection using a bridge or tunnel to achieve a vertical separation, the vast majority of intersections are *at-grade*. Because of the obvious impact that intersection can have on both the safe and efficient operation of a highway, special consideration must be given to their design and control.

The most obvious difference between an intersection and a road segment is the numerous points of conflict. A typical four-legged intersection has 32 *conflict points*. These are points where traffic streams intersect (16 of the 32), traffic from one stream merges with another (8 of the 32), and vehicles depart a traffic stream (8 of the 32). To minimize the safety impacts while maintaining acceptable levels of operational efficiency, four design objectives should be followed when designing intersections (ITE 1992):

1. Minimizing the number of conflict points
2. Simplifying conflict areas
3. Limiting conflict frequency
4. Minimizing conflict severity

The level hazard of many of these conflict points can be moved, reduced, and/or eliminated with various design measures such as traffic islands and exclusive turn lanes and traffic control devices such as stop signs and traffic signals. These points are discussed further in the following chapter.

It is most desirable to have roadways intersect at right angles. A 90° intersection allows drivers a greater field of view of oncoming traffic and facilitates turning movements for large trucks. When perpendicular intersections are not possible, other design procedures for re-aligning one or more approaches can be used to mitigate or correct the effects of skewed intersections. Additional techniques such as the use of channelization techniques and the redesign of corner radii can be used. Various techniques can also be used for intersecting roadways that cross at substantially varying grade profiles.

The provision of adequate sight distance is also a key element of intersection design. The evaluation and design of sight distances at intersections differs somewhat from that along roadway segments. As opposed to developing a design that provides stopping or passing sight distance along the mainline of a roadway, the design of sign distance at intersections seeks to provide an envelope of clear vision for drivers on a minor roadway attempting to cross or turn on to major roadway.

There are several different scenarios for which sight distance may be required at an intersection, including the no-control, yield control, stop control, and signal control cases. The no-control case requires clear vision areas on all approaches to the intersection so that vehicles approaching on any leg of the intersection can see approaching vehicles on the other approaching legs far enough in advance to adjust their speed to avoid a collision. The yield case is a special case of the stop control condition in which additional clear sight areas are required because vehicles are permitted to enter or cross the intersecting roadway if no conflicting vehicles are present on the major highway. The stop control case involves an analysis of the sight distance requirement of a vehicle on a stop-controlled minor street to make a left turn, right turn, or crossing maneuver. The required distance is affected by several factors, including the design vehicle, operating speed of the major road, grade of the minor street, and number of lanes/width of the median on the major road.

Required sight distances for right- and left-turning traffic at intersections are graphically presented in chapter 9 of the Green Book. The values represented within these figures, however, are for flat profile grades and passenger car vehicles, and thus they do not reflect the effect of grade or vehicle type. The additional distances required for these factors as well as more of the specifics for the location of the driver eye position and multilane highways can also be found in this reference.

13.6.3 Interchanges

Intersections that occur at freeways are based on the need to maintain access to and from these facilities. The specific location and configuration of freeway interchanges is primarily a function of the directional demand between the various intersecting freeways and roadways as well as other cost and right-of-way limitations. As with nonfreeway intersections, the objectives of freeway interchanges are to maximize the operational efficiency and safety aspects within the interchange vicinity. To achieve these objectives, eight interchange design principles have been suggested (ITE 1992):

1. Avoiding the incorporation of left-hand entrances and entrances
2. Striving for single exit designs from all approach directions
3. Placing exits in advance of (rather than after) the crossing roadway
4. Designing the interchange to permit the crossroad to pass over (rather than under) the freeway
5. Avoiding designs that require weaving maneuver within the interchange area
6. Providing decision sight distance in advance of the interchange on all approaches
7. Using auxiliary lanes, two-lane ramps, and special ramp designs to provide lane balance at all interchange ramp terminals
8. Designing interchanges to fit the principle of route continuity rather than forecast traffic patterns

Interchanges fall within two primary categories: *service interchanges,* which occur at the intersection of two freeway or other controlled access facilities, and *system interchanges,* which occur between a freeway and other noncontrolled access roadways, such as arterial and collector roadways. Among the differences between these two categories is the need to maintain (without stopping) traffic on system interchanges. Since interchange design is heavily influenced by the cost and availability of right-of-way, various loop and ramp configurations can be used to keep certain quadrants of the interchange clear of connecting roadways, such as those used in partial cloverleaf designs. These varied designs can also be used to facilitate left-turn movements at interchanges and can eliminate the need for stops at the crossroad.

13.7   CONCLUSION AND SUMMARY

The design of highway facilities is a process that requires an understanding of the theories that describe human behavior and the interactions of the physical world and the application of them in a practical and cost-effective manner. Highway designs must accommodate a range of drivers and vehicle types that encompass a wide range of characteristics and capabilities as well as a number of other external influences such as weather, lighting, and even local driving customs. Because of this there is no "one-size-fits-all" design. Each design must be evaluated within the specific location, objective, and constraints of the particular location.

While certainly a key component, a proper geometric design on its own will not ensure the safe and proper use of a highway. Effective design must be integrated with adequate and continuous enforcement and education efforts. Another consideration is that conditions can change over time, or the conditions assumed during the original design may turn out to be different once it is constructed and in operation. Thus, both safety and operational conditions must be monitored over time to minimize accidents and delays that may occur.

A design must be developed within the context of many related topics discussed in the various chapters of this book, including pedestrian facilities, roadway lighting, parking, and safety countermeasures. Designers also need to stay current on evolving standards and practices of highway design as well as the changing philosophies of design, including those associated with traffic calming, mixed land use developments, environmental controls (including those for air and noise pollution), and other socioeconomic and environmental impacts. Information can be found both through published sources and through involvement in professional organizations.

13.8   REFERENCES

American Association of State Highway and Transportation Officials (AASHTO). 1996. *Roadside Design Guide.* Washington, DC: AASHTO.

———. 2001. *A Policy on Geometric Design of Highways and Streets.* Washington, DC: AASHTO.

Crawford, W. 1995. *Construction Surveying and Layout,* 2nd ed. West Lafayette, IN: Creative Construction.

Ewing, R. 1999. *Traffic Calming: State of the Practice.* Washington, DC: Institute of Transportation Engineers.

Federal Highway Administration (FHWA). 2000. *Manual on Uniform Traffic Control Devices.* U.S. Department of Transportation, FHWA, Washington, DC.

Garber, N. J., and L. A. Hoel. 2002. *Traffic and Highway Engineering,* 3rd ed. Pacific Grove, CA: Brooks-Cole.

Hickerson, T. 1964. *Route Location and Design,* 5th ed. New York: McGraw-Hill.

Institute of Transportation Engineers (ITE). 1992. *Traffic Engineering Handbook.* Washington, DC: ITE.

Krammes, R. A., K. Fitzpatrick, J. D. Blaschke, and D. B. Fambro. 1996. Speed—Understanding Design, Operating, and Posted Speed. *Texas Transportation Institute Repor*t 1465-1, College Station, TX.

Meyer, C., and D. Gibson. 1980. *Route Surveying and Design,* 5th ed. New York: Harper-Collins.

Schoon, J. 2000. *Geometric Design Projects for Highways,* 2nd ed. New York: ASCE Press.

Transportation Research Board (TRB). 2000. *Highway Capacity Manual.* National Research Council, TRB, Washington, DC.

CHAPTER 14
INTERSECTION AND INTERCHANGE DESIGN

Joseph E. Hummer
Department of Civil Engineering,
North Carolina State University, Raleigh, North Carolina

14.1   INTRODUCTION

Intersections and interchanges are important parts of the highway system. They typically have much higher collision rates and cause much more delay than midblock segments. They are also particularly expensive parts of the highway system. The purpose of this chapter is to provide a summary of current intersection and interchange design concepts and to direct readers to sources of detailed information on those designs. The chapter discusses intersections first, and then interchanges. For each of these, basic design elements are presented first, followed by configurations.

14.2   BASIC INTERSECTION ELEMENTS

14.2.1   Spacing

For safe and efficient vehicular traffic, intersections should not be placed too close together. Drivers accelerating away from one intersection are not expecting to encounter traffic slowing for another intersection, for example. Also, a queue of vehicles from one intersection that blocks another intersection, called spill-back, will cause congestion to propagate and cause extra delay, as Figure 14.1 shows. Intersection spacing of at least 500 feet is typically desirable for vehicles. On the other hand, pedestrians and bicyclists enjoy shorter paths and greater mobility when intersections are more closely spaced, as in older central business districts of some U.S. cities. Intersection spacing as low as 300 feet works well for pedestrians and bicyclists.

Spacing between signalized intersections is critical. To ensure optimum progression in both directions on an arterial, signals should be spaced far enough apart that vehicles travel from one signal to the next in one-half the signal cycle length. For typical suburban speeds and cycle lengths, signal spacing around one-half mile provides for optimum two-way progression. Good two-way progression is often impossible with signal spacing from 500 to 2000 feet.

FIGURE 14.1   Traffic queuing at one intersection spills back to block another intersection, adding to delay and causing potential safety problems. (*Source:* Daniel L. Carter.)

14.2.2   Location

Intersections are safer in some locations than others. One important guideline is that intersections should not be on a horizontal curve if possible. Horizontal curves could restrict sight distances to the intersection and to traffic signals or signs near the intersection. A horizontal curve requires some drivers turning at the intersection to make complex reverse turn maneuvers. Also, horizontal curves are often superelevated, which makes the profile of the intersecting street very difficult to negotiate for turning or crossing motorists. Relative to the vertical alignment of a road, intersections should not be near a crest vertical curve, again due to restrictions on sight distance.

14.2.3   Angle

The angle of intersection is important to operations. As the angle between the two streets departs further from 90°, vehicle and pedestrian time in the area conflicting with other traffic streams increase so delays and collisions increase. In addition, the paved area—and therefore construction and maintenance cost—will increase. Tradition is that these effects increase dramatically with angles less than 60° or greater than 120°.

Several options exist for treating existing or proposed intersections with unfavorable angles. One or both streets could be realigned with horizontal curves to create a more favorable angle. In some cases, one intersection can be made into two, creating an offset intersection as described below. Another option is to use islands to guide drivers and pedestrians better through the intersection and reduce the paved area (see below).

14.2.4 Number of Approaches

All else being equal, an intersection with fewer approaches will be safer and more efficient than an intersection with more approaches. The reason for this is the number of conflict points in the intersection—points where one traffic stream crosses, merges with, or diverges from, another traffic stream. A standard three-legged (T) intersection has 9 vehicle conflict points, a standard intersection between a two-way street and a one-way street has 13 vehicle conflict points, and a standard four-approach intersection has 32 vehicle conflict points. Among other effects, fewer conflict points mean fewer phases are needed at a traffic signal, which in turn means less lost time and lower delays.

There is a lively ongoing debate about whether it is safer and more efficient to create one four-legged intersection or two three-legged intersections separated by several hundred feet, an *offset intersection*. It appears that for some combinations of traffic volumes and spacing the offset intersection is a better choice than one standard four-legged intersection. Analysis tools like the *Highway Capacity Manual* (TRB 2000) and CORSIM (ITT 1995–2001) are good ways to examine the efficiency of the offset design compared to a standard four-legged design.

Intersections with five or more approaches are particularly inefficient and difficult for road users to negotiate. The treatments for such cases include terminating one or more legs before they reach the intersection, using horizontal curves to redirect one or more legs (likely creating another intersection), or making one or more legs one-way moving away from the intersection.

14.2.5 Turn Bays

Turn bays provide room for left-turning or right-turning vehicles to decelerate before their turns and/or to queue while waiting to turn. Left-turn bays are particularly effective at reducing delay and collisions by getting those vehicles out of the way of through vehicles. At busy signalized intersections, dual and triple left-turn lanes are used effectively to reduce the time that those vehicles need the right-of-way. Dual right-turn lanes are also used at some intersections. The drawbacks to using turn bays include higher right-of-way costs and longer crossing distances for pedestrians.

Through the years, many criteria have been published for left-turn and right-turn bays. The criteria are typically different for unsignalized and signalized intersections. For signalized intersections, one well-known set of turn bay criteria is provided in the *Highway Capacity Manual*. The *Manual* recommends:

- A single left-turn bay for peak hour left-turn volumes of 100 veh/hr or more
- A dual left-turn bay for peak hour left-turn volumes of 300 veh/hr or more
- A single right-turn bay for peak hour left-turn volumes of 300 veh/hr or more

The *Manual* also recommends additional through lanes for each 450 veh/hr of through volume. Many agencies use a triple left-turn bay for peak hour left turn volumes of 500 veh/ hr or more and a dual right-turn bay for peak hour right-turn volumes of 600 veh/hr or more.

Many other authors have provided turn bay criteria for signalized and unsignalized intersections. Pline (1996) synthesized information on left-turn treatments in the mid-1990s. Fitzpatrick and Wooldridge (2001) included this topic in their 2001 synthesis of recent research findings in geometric design.

Once a designer has decided to provide turn bays, he or she must decide on bay length. AASHTO (2001) recommends that overall turn bay length should be the sum of the taper length, deceleration length, and storage length. Taper rates into a turn bay are typically between 8:1 and 15:1, with lower rates for bays on urban roads with lower speeds. Taper rates into dual left-turn bays are sometimes even lower than 8:1 to maximize storage area. It is desirable to allow vehicles to decelerate fully after having departed a through lane, although this is sometimes impractical in urban areas. AASHTO (2001) states that typically lengths needed to decelerate from 45, 50, and 55 mph speeds to a full stop are 430, 550, and 680 feet, respectively, on grades of less than 3 percent. Storage lengths needed for unsignalized intersections are typically short, accommodating only a couple of vehicles. At signalized intersections, storage lengths typically must be much longer. AASHTO (2001) passes along traditional guidance that the turn bay should be able to store 1.5 to 2 times the average number of vehicles desiring storage per cycle. However, more sophisticated methods are available that consider more factors, including the possibility that a queue of through vehicles could block the entrance to the turn bay (Kickuchi, Chakroborty, and Vukadinovic 1993).

One recent innovation increasing the safety of turn bays that serve permissive left turns is providing a positive offset. A positive offset moves the intersection end of the left turn bay as far to the left as possible, as shown in Figure 14.2, so that a left-turning driver's view of opposing through and right-turning traffic is not blocked by the opposing left-turn queue.

14.2.6 Islands

Providing islands to separate traffic streams, also called channelization, has a number of benefits and a few drawbacks. The benefits generally include (AASHTO 2001):

- Separation of conflicts
- Control of angle of conflict

FIGURE 14.2   Positive offset left-turn bay. (*Source:* Daniel L. Carter.)

- Reduction in excessive pavement areas
- Regulation of traffic and indication of proper use of intersection
- Arrangements to favor a predominant turning movement
- Protection of pedestrians
- Protection and storage of turning and crossing vehicles
- Location of traffic control devices
- Aesthetics

Separating and controlling the angle of conflicts are fundamental principles of intersection design that should lead to tangible collision reductions. However, designers should be cautious about applying islands everywhere, because in some places they do have drawbacks. Islands should help guide the motorist through the intersection using a natural path that meets their expectations, but islands that place unexpected obstacles in motorists' paths will likely cause collisions. In addition, very small islands are difficult for motorists to see and may pose a hazard. The AASHTO *Policy on Geometric Design of Highways and Streets* (AASHTO 2001) recommends that the smallest size for a triangular turning island should be 50 square feet in an urban area and 75 square feet in a rural area.

One final note on choosing islands is that they tend to make the intersection look large and discouraging for pedestrians. Islands provide refuge for pedestrians, but the multiple-stage crossing can be slow and confusing for pedestrians. In places where the context calls for encouraging pedestrians, designers generally try to avoid wide channelized intersections.

Once a designer has decided to provide an island, he or she must decide on the type of surface, shape, and offset to travel lanes, among other details. Islands may have pavement flush with the surrounding travel lanes (separated only by pavement markings), may have a raised concrete surface, or may be landscaped with or without curbs. In deciding upon a surface, designers should consider the visual target value, the possibility of motorist violations, the protection afforded pedestrians using it as a refuge, maintainability, drainage, and other factors. Designers should also remember that any fixed objects placed on an island may be struck by errant motorists and may block drivers' or pedestrians' lines of sight. Designers should only use curbed islands at intersections with fixed-source lighting or where the curbs are delineated well. The shape of islands is typically dictated by the swept paths of the design vehicles chosen for the intersection and the widths of turning roadways (see below). The edge of an island is typically offset by 2 to 6 feet from the edges of the travel lanes or turning roadways surrounding the island, with a larger offset at points where driving paths diverge from each other. Island corners typically have one-foot to two-foot radii.

14.2.7 Curb Radii

The choice of curb radii is critical to optimum intersection function. Curb radii that are too small will lead to large vehicles slowing dramatically and/or encroaching over curbs and lane lines, causing delays and possibly collisions. Curb radii that are too large waste construction and right-of-way funds and cause longer intersection crossing times for pedestrians, which also causes delays and leads to collisions. Larger curb radii also decrease the distance from the curb to the right-of-way edge when the right-of-way edges are not curved (i.e., corner clearance), which could restrict pedestrian queuing area, among other effects. As noted above in the discussion of islands, choice of a design vehicle and the context of the intersection are critical in this aspect of design. In an exurban area with higher speeds and higher truck volumes larger curb radii (30 feet or more) are appropriate, while in a denser urban area with fewer large vehicles where pedestrians are encouraged smaller curb radii (15 to 25 feet) are appropriate. Parking lanes, with parking appropriately restricted near the intersection, help reduce the radius that would otherwise be needed.

One way a designer can compromise between the needs of large vehicles and pedestrians is by choosing the best type of intersection curb design. AASHTO (2001) provides for three types of intersection curb designs: a constant radius; a combination of taper, constant radius, and second taper; and a three-center compound curve. A constant radius is the simplest of the three choices to lay out and construct and is therefore the typical choice where neither large vehicle volumes nor pedestrian volumes are high, such as an intersection between two local residential streets. Since the tapered or compound curve designs more closely approximate modern vehicle turning paths, they use less space while accommodating larger vehicles at higher speeds with fewer encroachments. Tables in AASHTO (2001) recommend taper rates and curve radii given an intersection angle and a design vehicle.

14.2.8   Turning Roadway Widths

AASHTO (2001) recommends widths of turning roadways at intersections based on several factors, including the radius of the inner edge of the turn, the number of lanes, design vehicles, and the type of curb or shoulder. Recommended widths range from 12 feet for a 500-foot radius turn with one lane serving predominantly passenger cars to 45 feet for a 50-foot radius turn with two lanes serving large numbers of large semi-trailer trucks.

14.2.9   Medians and Openings

Medians at intersections generally provide many of the same benefits as islands, described above, including separating opposing traffic directions (making head-on conflicts rarer), providing crossing pedestrians and vehicles refuge, providing a place for traffic control devices and light fixtures, and reducing pavement area. Since medians restrict left turns, median opening spacing can be a terrific access control mechanism.

The median width at an intersection is a major decision for a designer. Wider medians typically mean higher right-of-way costs, encourage vehicles turning left from the cross-street to "lock up" with each other, increase wrong-way movement potential, and mean higher minimum green times and lost times at signals, so engineers usually try not to design them wider than needed. Some of the key features of various median widths at intersections are:

• Narrower than 4 feet: difficult for drivers to see so the nose may be struck more often.

• Four feet or wider: provides a pedestrian refuge.

• Sixteen feet or wider: provides a pedestrian refuge and room for a left-turn bay.

• Twenty feet or wider: provides refuge for a crossing passenger car. According to the *Highway Capacity Manual,* at unsignalized intersections this refuge typically means a healthy improvement in level of service for the minor street.

• Twenty-eight feet or wider: provides a pedestrian refuge and room for a dual left-turn bay.

• Forty-nine to 59 feet: minimum median widths to allow U-turns by various trucks into the outer lane of a four-lane highway (AASHTO 2001). Allow U-turning vehicles to encroach on the shoulder reduces these widths.

Several median nose designs are available. Designers typically design the median nose and the median opening width to conform to the turning paths of design vehicles, making sure that the paths of vehicles that are supposed to move simultaneously do not encroach on each other. AASHTO (2001) provides recommendations on minimum median opening widths based on design vehicle, median opening, nose shape, and intersection angle.

14.2.10 Access Points

To increase traffic safety and efficiency, driveways and minor streets should be separated from major intersections as far as possible. Some of the safety threats from driveways near major intersections include drivers accelerating away from an intersection being surprised by a driver ahead slowing to turn into a driveway, and a "double-threat" when a driver in one lane stops to let someone in or out of a driveway while a driver in an adjacent lane does not stop. Many agencies have programs to try to close existing driveways and minor street intersections near major street intersections or prohibit such driveways opening in the first place. Gluck, Levinson, and Stover reviewed corner clearance policies from many agencies across the United States and provided application guidelines. The mid-range corner clearance called for in guidelines around the United States is 100 to 200 feet, with some agencies requiring up to 300 feet. One popular treatment to a problem of existing access points near a major intersection is installation of a narrow raised median to prohibit left turns into or out of the driveway, as Figure 14.3 shows.

14.2.11 Designing for Pedestrians and Bicycles

In recent years intersection designers have been taking the needs of nonmotorist road users, particularly pedestrians and bicyclists, into account more and more. In some dense urban areas, pedestrians and bicyclists are the dominant users, while even in suburban fringe areas a surprisingly high number of pedestrians and bicyclists need to be considered as they attempt

FIGURE 14.3 A narrow, raised median helps treat a safety problem created by a driveway near a major intersection. (*Source:* Daniel L. Carter.)

to cross safely. Principles designers should adopt regarding pedestrians at intersections include:

- Provide as short a crossing as possible. Keep curb radii as short as possible and provide curb "bulb-outs" (sidewalk extensions into parking lanes) if possible.
- For wider crossings, consider medians and/or islands to provide pedestrian refuge.
- Provide a crossing path that is as direct as possible, since many pedestrians will attempt to walk on the shortest path anyway.
- Consider using a different or rougher pavement surface for the crosswalk to help alert and slow drivers. Figure 14.4 shows a "speed table" that raises the whole intersection surface and has been effective in some places by slowing vehicles appropriately in the presence of pedestrians.
- Provide appropriate curb ramps for wheelchairs, strollers, skaters, and other sidewalk users with wheels.
- Consult the *Highway Capacity Manual* for procedures to calculate the level of service for pedestrians at signalized or unsignalized intersections. Levels of service criteria are based upon delay and available queuing space for signalized intersections and upon delay for unsignalized intersections. Design treatments for intersections with poor pedestrian levels of service include medians, islands, larger corner queuing areas, and anything that would reduce the signal cycle length.

One of the main decisions facing designers regarding bicycles is how to route bicycle lanes through intersections. If there is an exclusive right-turn lane, the bicycle lane is typi-

FIGURE 14.4 A speed table can help slow vehicles, providing for a safer pedestrian crossing. (*Source:* Daniel L. Carter.)

cally placed between the turn lane and the through lanes, which means that bicycles must weave across right-turning vehicles at the beginning of the right turn lane. If there is no exclusive turn lane, the bicycle lane will stay to the right of the through lanes and bicycles will conflict with right-turning vehicles at the intersection. In either case, clear sight lines and good traffic control devices will increase safety. AASHTO has published a *Guide for the Development of Bicycle Facilities* (AASHTO 1999) that provides many useful details on designing for bicycles at intersections.

14.2.12   Grades

Steep grades hamper traffic operations at intersections. Steep downgrades on an intersection approach increase stopping distances and make turning more difficult. Steep upgrades on an intersection approach make idling difficult for vehicles with manual transmissions and make acceleration slower for all vehicles, which in turn increases necessary gap sizes and sight distances for crossing and turning movements. Generally, grades under 2 percent do not cause many operational problems, grades from 2 to 4 percent begin to introduce noticeable problems, and grades over 4 percent should be avoided where practical.

14.2.13   Intersection Sight Distances

Intersection designers should ensure that approaching drivers can see all applicable traffic control devices and conflicting traffic soon enough to take appropriate actions. Sight may be blocked by the alignment of the road the driver is on, the alignment of the intersecting road, objects or terrain in the "sight triangle" near the corner, or objects or terrain in the median. Given the alignments and roadside obstacles, the designer needs to compute the intersection sight distance provided and compare that to the minimum sight distance recommended in AASHTO (2001). The AASHTO *Policy* helps the designer do this for six different situations, labeled case A through case F.

- Case A—intersections with no traffic control. Because vehicles typically slow down considerably on approach to this type of intersection, these intersection sight distances are lower than the stopping sight distance for any given design speed.

- Case B—intersections with stop control on the minor road. The sight distance of interest in this case is that for a vehicle that has stopped at the intersection and must be able to see the traffic on the major street far enough away to make a good decision about when to proceed. The calculation is different for vehicles turning left, turning right, and crossing the major road. The choice of design vehicle and the presence of a median (making a left turn of a crossing maneuver into a two-stage process) affect the calculation greatly. The computation of case B intersection sight distances has been simplified in AASHTO (2001) compared to that outlined in earlier editions.

- Case C—intersections with yield control on the minor road. The computation in this case differs for vehicles that are crossing the major road from vehicles that are turning onto the major road. In both instances, intersection sight distances in case C exceed those for case B.

- Case D—intersections with traffic signal control. For this case, the AASHTO *Policy* states that the only minimal sight distances, allowing a vehicle at one stop bar to see vehicles at the other stop bars, are necessary, with three exceptions. First, permissive left-turning vehicles must be able to see gaps in oncoming through and right-turning traffic. Second, if flashing signal operation is allowed during off-peak times, the appropriate case B sight distances are recommended. Third, if right turns on red are allowed, the appropriate case B sight distances are recommended.

- Case E—intersections with all-way stop control. Like case D, the AASHTO *Policy* states that the only minimal sight distances, allowing a vehicle at one stop bar to see vehicles at the other stop bars, are necessary, but this time with no exceptions. All-way stop control is therefore an excellent treatment option for an intersection with sight distance problems.

- Case F—left turns from the major road. This should generally be provided at every point along a road where permissive left turns may be made now or in the future (i.e., designers should consider future intersection and driveway locations). If stopping sight distance has been provided along a road, and case B and C sight distances have been provided for each minor road intersecting the major road, the AASHTO (2001) states that no separate check for case F intersection sight distances may be needed.

Designers finding that an existing or proposed intersection does not provide the recommended intersection sight distance have a number of options to improve the situation, including realigning a road, installing a wide median, removing sight obstacles on the roadside, changing the type of traffic control, reducing speed limits, or installing warning signs.

14.3 INTERSECTION CONFIGURATIONS

14.3.1 Need for Alternatives

A basic configuration serves well at the vast majority of intersections. However, many intersections with a basic configuration are no longer serving today's higher traffic demand levels very well from a capacity or a safety point of view. Once good signal technology and some of the features discussed above, like turn bays and islands, have been installed, engineers have often exhausted the range of conventional improvements. Additional through lanes, bypasses, and structures are expensive and environmentally disruptive; alternative modes of travel and demand management remove few cars and trucks from the road; and intelligent transportation systems are years from making a major impact at most intersections. In this atmosphere, there is a great need for alternative intersection configurations as a practical way to improve safety and efficiency. The next few sections discuss some of those alternative configurations that designers should consider to improve intersection operations.

14.3.2 One-Way Approaches

An intersection involving one or two one-way streets operates much more efficiently and probably more safely than a similar conventional intersection between two two-way streets. Reasons for this greater efficiency and safety include:

- Signals at intersections involving one-way streets require fewer phases, reducing lost time. At an intersection between a two-way street and a one-way street, only one left-turning movement is opposed by vehicular traffic.

- It is possible to establish perfect signal progression along a one-way street with any signal spacing at any speed.

- The number of conflicts between traffic streams in the intersection is greatly reduced.

- Crossing pedestrians face fewer directions of conflicting vehicular traffic.

One-way streets also have some well-known drawbacks, including greater travel distances and potential for dangerous wrong-way movements. On the balance, however, converting conventional intersections into intersections involving one-way streets will improve traffic operations.

14.3.3 Traffic Calming

At some intersections, the primary goal is to discourage high vehicular volumes and/or speeds. *Traffic calming,* as it is termed, is typically appropriate on collector or local streets in residential areas and is often aimed at through traffic. Many measures have been used successfully at intersections to decrease vehicular volumes and speeds. Curb bulb-outs, textured crosswalks, and raised speed tables were mentioned above as design measures that assist crossing pedestrians, and these might be considered traffic-calming measures. Other design measures that engineers use to calm traffic at intersections include:

- Small traffic circles and roundabouts (see section below)
- Chokers (narrowing lanes with curbs and/or landscaping)
- Semi-diverters (allowing one-way in or out of the intersection only on a particular approach)
- Forced-turn diverters (allowing no through movements for one street)
- Diagonal diverters (forcing two approaches to turn only left and the other two approaches to turn only right)
- Vehicular cul de sacs (usually allowing nonmotorized and emergency users to get through)

If done well, traffic calming at intersections can improve the overall quality of service and safety for road users and will improve the quality of the environment surrounding the intersections. If done poorly, though, traffic calming can cause collisions, increase travel times, increase frustrations, and harm the nearby environment. Some of the important issues designers need to examine when considering traffic calming at intersections include visibility to the design feature, forgiveness of the design feature if struck, driver expectations, aesthetics of the design feature, access for emergency vehicles, and whether problems mitigated at one intersection will simply migrate to another location. Ewing (1999) has written an excellent resource on traffic calming that will help designers negotiate some of these issues.

14.3.4 Roundabouts

A modern roundabout, as shown in Figure 14.5, offers greater safety and efficiency than a conventional intersection if designed and operated well. The idea of modern roundabouts came to the United States from Europe and Australia in the late 1980s, and modern roundabouts have since moved beyond the experimental stage to be a part of standard engineering practice in many U.S. agencies. Modern roundabouts differ from earlier, often unsuccessful, traffic circles because they have the following features:

- Yield control upon entry to the circle (traffic in the circle always has the right-of-way)
- Low design speeds for circulating traffic
- All traffic diverts from a straight-line path through the intersection
- No parking in or near the circle
- No pedestrians in the circle

Single-lane roundabouts have been proven in recent U.S. research to be safer than the conventional intersections they replaced (Persuad et al. 2001). Traffic-analysis software also typically shows that single-lane roundabouts that remain below capacity reduce delays compared to signalized intersections handling the same volumes. The capacity of single-lane roundabouts is typically about 30,000 vehicles per day, meaning that single-lane roundabouts are a good solution for intersections between two collector streets or minor arterials. Firm

FIGURE 14.5 Modern roundabout. (*Source:* Daniel L. Carter.)

results on the safety and efficiency of roundabouts with two or more circulating lanes in the United States are not available to this point, however.

Roundabout designers face many decisions, including design speed, circle diameter, circle roadway width, approach flares, and splitter island size and shape. The FHWA has assembled an excellent information source for these and other roundabout design issues (FHWA 2000).

14.3.5 Auxiliary Through Lanes

Designers have sometimes successfully used short auxiliary through lanes—added before an intersection and dropped after the intersection—to increase capacity and reduce delay at the intersection. Frequently, such auxiliary lanes are an interim step in a general road-widening project. The main question designers have in considering auxiliary lanes is how long they have to be to allow safe diverges and merges and to entice enough motorists into the auxiliary lane to make a difference. If the lanes are too short, drivers will not choose to use them and the extra capacity provided on paper will not be realized in the field. Hurley (1997) has provided the best research available to this point guiding designers on auxiliary through lane lengths.

14.3.6 Median U-Turns

Reducing and separating conflicts between traffic streams, especially those involving left turns, often creates safer and more efficient intersections. In recent years designers have been able to choose from a menu of unconventional intersection designs that attempt, one way or

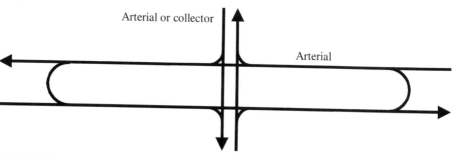

FIGURE 14.6 Median U-turn design.

another, to achieve that conflict reduction and separation. On top of that menu is the median U-turn, shown in Figure 14.6. Median U-turns have been in place for 40 years in Michigan, where over 1,000 miles are in service, as well as other states. They work by rerouting all left-turning drivers away from the main intersection to one of the U-turn roads typically located about 600 feet from the main intersection. The main intersection then requires only a simple two-phase signal.

Research has shown that the median U-turn typically produces lower overall travel times than a comparably sized conventional intersection or other unconventional intersection designs carrying the same traffic volumes (Reid and Hummer 2001). Arterials with median u-turn intersections also have lower average collision rates than arterials with conventional intersections (Castronovo, Dorothy, and Maleck 1998). Other advantages of median U-turns over conventional intersections include easier progression for through traffic and fewer threats to crossing pedestrians. Disadvantages of the median U-turn design relative to a conventional intersection design include potential driver confusion, increased travel time for left-turning vehicles, wider rights-of-way along the arterial, and longer minimum green times for cross-streets to accommodate crossing pedestrians.

14.3.7 Other Unconventional Intersection Designs

Besides the median U-turn described above, there are 11 other unconventional intersection designs on the menu that offer advantages in certain situations. Figure 14.7 shows 6 of these other designs, with a note on how each works and the main advantage each offers. All 12 unconventional designs have been discussed in a peer-reviewed publication, and most have been placed in operation somewhere in the United States. Hummer and Reid (2002) recently provided a summary of the state-of-the-art for all 12 unconventional designs.

14.4 BASIC INTERCHANGE ELEMENTS

Interchanges are junctions between two or more roads where the conflicting traffic streams cross over or under each other using structures and connections between the roads are made using ramps. Interchanges serve two main purposes. First, one or more of the roads is a freeway that cannot have at-grade intersections. Second, an interchange is a safe way to add capacity and reduce delay at an intersection. The next few sections describe the basic design elements of an interchange.

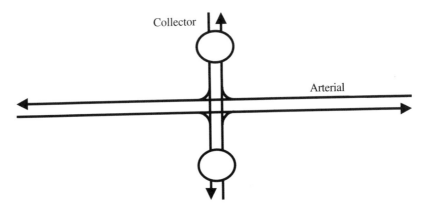

Bowtie:
- All left turns are routed to one of the roundabouts on the minor street
- Works like a median U-turn, but does not need a wide median on the major street

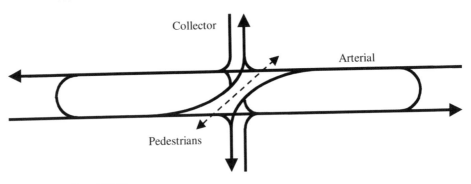

Superstreet:
- All minor street through and left turn traffic uses a median crossover
- Allows perfect progression with any signal spacing on the major street

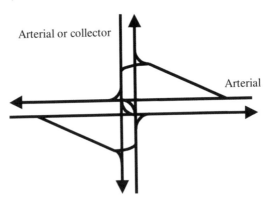

Jughandle:
- Left turns from the major street use a right-side ramp to the minor street
- Without left-turn bays or U-turns, the major street median can be narrow

FIGURE 14.7 A diagram and note on six other unconventional intersection designs.

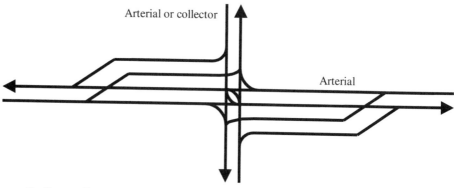

Continuous Flow Intersection:
- Left turns cross opposing traffic prior to the main intersection
- The main intersection has a two-phase signal with unopposed left turns

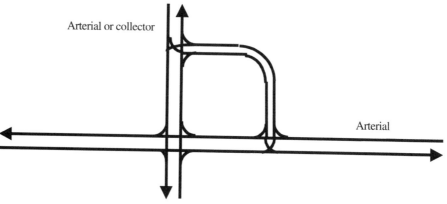

Single Quadrant Intersection:
- All left turns use the connecting roadway
- This generally vies with the median U-turn as most efficient unconventional design

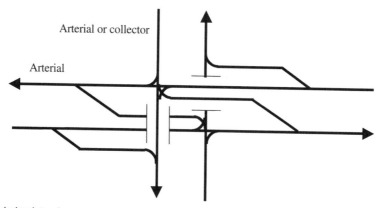

Echelon Interchange:
- Half of each roadway meets on the top, while the other halves meet on the bottom
- Two-phase signals with unopposed left turns mean great efficiency

FIGURE 14.7   (*Continued*)

14.4.1 Interchange Spacing

In general, interchanges should be spaced as far apart as possible. Interchanges that are too close force traffic streams to weave across each other, causing delays and collisions. Interchanges that are too close also make signing difficult, forcing drivers to divert too much attention to reading too many signs and causing driver confusion. Closely spaced interchanges encourage short trips on the freeway and reduce the mobility function of the facility. AASHTO (2001) states that a general rule of thumb on minimum interchange spacing is 1 mile in urban areas and 2 miles in rural areas.

14.4.2 Route Continuity, Lane Continuity, and Lane Balance

Route continuity means drivers should not have to use a ramp to follow an important interstate route through an interchange, even if the route turns at the interchange. Figure 14.8 shows how a designer should employ route continuity at an interchange where the main interstate route turns. Route continuity is important to meet driver expectations, avoid driver confusion, and avoid sudden speed changes by large groups of drivers.

Lane continuity means that drivers should not have to change lanes to follow a certain main interstate route. For long stretches along a main interstate route, the freeway should contain the same basic number of through lanes. If auxiliary lanes are needed to boost capacity at some points along the route, they should be added (typically) to the right side and then terminated where they are no longer needed. Like route continuity, lane continuity helps meet driver expectations, avoids driver confusion, and ultimately eliminates collisions.

Lane balance is a related concept that helps designers decide how many freeway lanes are necessary before and after ramp junctions to satisfy driver expectations and increase efficiency. AASHTO (2001) states that:

1. At entrances, the number of lanes beyond the merging of two traffic streams should not be less than the sum of all traffic lanes on the merging roadways minus one, but may be equal to the sum of all traffic lanes on the merging roadway.
2. At exits, the number of approach lanes on the highway should be equal to the number of lanes on the highway beyond the exit, plus the number of lanes on the exit, minus one. . . .
3. The traveled way of the highway should be reduced by not more that one traffic lane at a time.

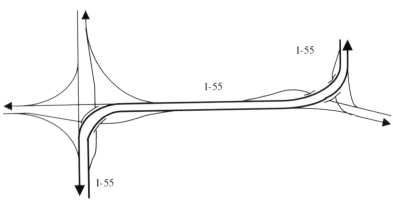

FIGURE 14.8 Interchange design with route continuity.

There are exceptions to point 2 above for cloverleaf interchanges and where interchanges are closely spaced. Lane continuity and lane balance conflict at some interchanges, but the use of auxiliary lanes can remove the conflict.

14.4.3 Overpass or Underpass

Sometimes the alignments of the freeway and crossroad dictate whether the freeway must be over or under the crossroad. Where the designer has a choice, though, it is an important one. Some of the positive aspects of placing the crossroad over the freeway include:

- A single structure, probably smaller than for the freeway going over the crossroad
- Off-ramps go uphill and on-ramps go downhill, allowing gravity to aid deceleration and acceleration
- Good visibility on the freeway to the crossroad, providing an early alert to exiting drivers
- Less noise impact from the freeway

Some of the positive aspects of placing the freeway over the crossroad include:

- Shorter structural spans
- No height restrictions on the freeway
- Less disturbance to the crossroad during construction
- Easier to drain the freeway

14.4.4 Types of Ramps

There are three basic types of ramps, as shown in Figure 14.9. Direct ramps are very common and accommodate most right-turn movements at interchanges. Direct ramps are inexpensive and can have higher design speeds—which drivers expect on freeway-to-freeway movements—but can accommodate left turns only if there is a stop sign or traffic signal at the crossroad. Indirect ramps are often used for left-turn movements. Indirect ramps can also allow higher design speeds, but they have high structural costs. Left turns not accommodated by indirect ramps are usually made with loop ramps. Loop ramps can accommodate left turns without additional structures, but they usually have relatively low design speeds, require large land areas, and cannot accommodate right turns without a stop sign or signal at the crossroad.

14.4.5 Consistent Ramp Pattern

Drivers expect a particular pattern of off-ramps and on-ramps as they negotiate an interchange, and the interchange will be safer if designers satisfy that expectation. Signing and visibility to decision points are also improved if the pattern is provided. In particular, the pattern is:

- All ramps merge into or diverge from the right side of the freeway.
- All off-ramps depart the freeway prior to the crossroad.
- All on-ramps join the freeway beyond the crossroad.

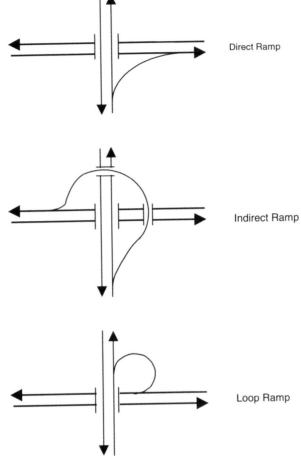

FIGURE 14.9   Three basic types of ramps.

Left-side ramps obviously violate this pattern. Agencies rarely build new left-side ramps on freeway mainlines and often try to remove those that currently exist. Loop ramps often violate this pattern by merging into or diverging from the wrong side of the crossroad. Designers can make loop ramps fit the pattern by extending the ramp parallel to the freeway until the gore (the point where the ramp left edge line meets the freeway right edge line) is on the expected side of the crossroad.

14.4.6 Ramp Widths

AASHTO (2001) recommends ramp widths, like intersection turning roadway widths, at interchanges based on several factors, including the radius of the inner edge of the turn, design vehicles, whether passing a stalled vehicle is possible, and the type of curb or shoulder. Recommended widths for one-lane ramps range from 12 feet for a 500-foot radius turn serving predominantly passenger cars with no provision for passing a stalled vehicle, to 30 feet for a 50-foot radius turn with provision for passing a stalled vehicle serving large numbers of large semi-trailer trucks. Recommended two-lane ramp widths range from 25 feet to 45 feet for those same conditions. For long ramps, designers sometimes choose two-lane ramps where a one-lane ramp would have been adequate because the difference in widths is not that great.

14.4.7 Ramp Acceleration and Deceleration Lengths

Ramp acceleration and deceleration lengths may be of either the tapered or the parallel type, as Figure 14.10 shows. Agencies across the United States are fairly evenly split on using tapered or parallel designs, and many believe that both designs provide safe and efficient merging and diverging if properly designed.

The acceleration or deceleration length provided between the end of the ramp curve and the end of the taper (i.e., where normal freeway lanes resume) is a critical design choice. Longer acceleration or deceleration areas mean better levels of service and fewer collisions, but if the areas are too long they may confuse drivers and they will waste pavement and right-of-way. AASHTO (2001) recommends minimum lengths for flat grades (less than 2 percent) ranging from 340 feet for deceleration from a 70-mph freeway design speed to a 50-mph ramp curve design to 1,620 feet for acceleration from a stop condition to a 70-mph freeway design speed. Grades have a substantial effect on acceleration or deceleration lengths needed. For example:

- Downgrades of 3 to 4 percent increase minimum deceleration lengths by a factor of 1.2.
- Downgrades of 5 to 6 percent increase minimum deceleration lengths by a factor of 1.35.
- Upgrades of 3 to 4 percent increase minimum acceleration lengths by a factor of 1.3 to 1.8.
- Upgrades of 5 to 6 percent increase minimum acceleration lengths by a factor of 1.5 to 3.0.

14.4.8 Weaving Areas

Weaving areas are created where an entrance ramp is followed within 2,500 feet by an exit ramp, where the ramps are connected by an auxiliary lane, and where the conflicting traffic streams are not controlled by stop signs or traffic signals. Weaving areas are usually found

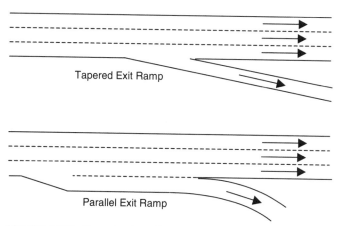

Tapered Exit Ramp

Parallel Exit Ramp

FIGURE 14.10 Tapered and parallel ramp terminals.

in the middle of cloverleaf interchanges or where two interchanges are close together. Weaving areas are relatively inefficient and unsafe because of the conflicting traffic streams. Currently, many highway agencies are not building new weaving areas on freeways and are, in fact, engaged in eliminating existing ones from their systems. Methods of eliminating weaving areas from freeways shown in Figure 14.11 include collector-distributor roads, braided ramps, and reversal of ramp direction (conversion to an X-interchange) where there are continuous one-way frontage roads.

14.4.9 Sign Placement

Guide sign placement is a critical decision for freeway designers because drivers moving at freeway speeds need long distances to read and respond properly to the sign messages. Five important human factors concepts that come into play in interchange sign placement are overload, spreading, primacy, repetition, and redundancy. Overload is providing too much information to drivers at one time and can result in drivers making poor decisions or neglecting more important driving tasks. Spreading is a way to avoid overload by spacing signs for an interchange further apart. The *Manual on Uniform Traffic Control Devices* (*MUTCD*) (FHWA 2001) mandates that freeway guide signs not be closer than 800 feet apart, and ideally they should be spaced farther apart if possible. Primacy means providing the most important information, such as where precisely the gore is located, closest to the interchange. Designers place less important information, like guide signs for attractions and services, farther from the interchange. Repetition and redundancy mean providing important information more than once—repetition in the same form, and redundancy in different forms. This helps make sure drivers do not miss important information, especially in a complex

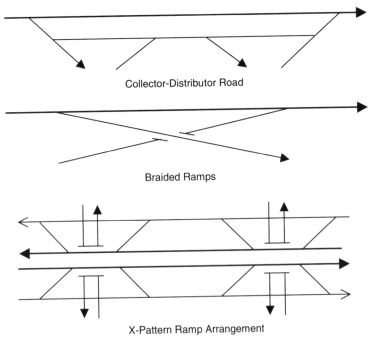

Collector-Distributor Road

Braided Ramps

X-Pattern Ramp Arrangement

FIGURE 14.11 Methods to eliminate weaving areas on freeways.

environment with many sight restrictions. The *MUTCD* provides much more detailed suggestions and requirements for interchange signing.

14.4.10 Decision Sight Distances

When designing mainline and ramp alignments near interchanges, designers may use decision sight distance as the control instead of stopping sight distance. Decision sight distance allows drivers to take more time before deciding on a course of action when presented with a choice in a complex area. AASHTO (2001) provides decision sight distance values for five avoidance maneuvers:

A: Stop on rural road

B: Stop on urban road

C: Speed, path, and/or direction change on rural road

D: Speed, path, and/or direction change on suburban road

E: Speed, path, and/or direction change on urban road

Decision sight distance values provided by AASHTO (2001) range from just slightly greater than stopping sight distance values at the same design speed for maneuver A to three times as high for maneuver E at low design speeds. Using minimum decision sight distances in alignment design will mean much flatter curves than using minimum stopping sight distances.

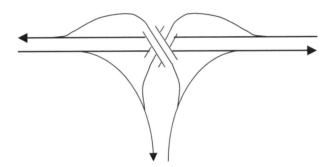

Three-level advantage: can fit in small right-of-way or can have high ramp speeds

Disadvantage: high bridge costs

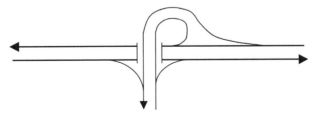

Trumpet advantage: one bridge

Disadvantages: high right-of-way, loop has low speed and capacity

FIGURE 14.12 Three-level and trumpet three-legged interchange designs.

14.5 *INTERCHANGE CONFIGURATION*

Interchange designers have many factors to consider, but fortunately they have many configurations to choose from. The following paragraphs describe some of the basic interchange configurations and briefly describe their advantages and disadvantages.

14.5.1 Three-Legged

Many three-legged interchanges connect one freeway to another. There are two basic configurations of this type of interchange, the trumpet and the three-level, as Figure 14.12 shows. Trumpets have one left turn on a higher-speed indirect ramp and one left turn on a lower-

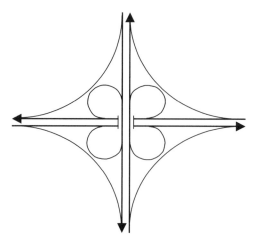

Cloverleaf advantages: one bridge, drivers expect this configuration

Disadvantages: large right-of-way, weaving areas break down easily

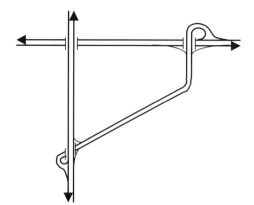

Single quadrant advantages: can use any quadrant, easy toll collection

Disadvantages: some long travel distances, violates expected ramp pattern

FIGURE 14.13   Common four-legged freeway-to-freeway designs.

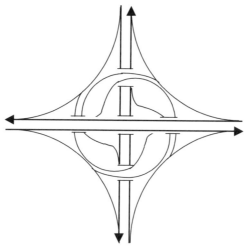

Pinwheel advantages: small right-of-way, all right-side indirect and direct ramps

Disadvantages: high bridge costs, low speed ramps

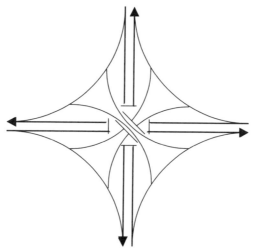

Four-level advantages: meets expected ramp pattern, efficient with high volumes

Disadvantages: very high bridge costs, steep ramp grades

FIGURE 14.13   (*Continued*)

speed loop ramp, so the lower volume left-turn movement or the left-turn movement from the road that has the lower functional class is typically relegated to the loop. Considering the relative advantages provided in Figure 14.12, designers typically use trumpets in lower-volume rural areas and three-level designs in higher-volume urban areas.

The trumpet is a very efficient and relatively low-cost interchange and thus designers often use it to connect freeways to other roads. Other common designs for three-legged freeway-to-surface street interchanges include the diamond (see below) and a modified diamond with a roundabout on the top of the "T."

14.5.2 Four-Legged Freeway-to-Freeway

There is a wide variety of four-legged freeway-to-freeway interchanges in service. Four common designs that are still considered for new interchanges include the cloverleaf, the single quadrant, the pinwheel, and the four-level. Figure 14.13 shows these designs and the relative advantages of each. Considering those relative advantages, designers typically use cloverleaf designs in lower-volume rural areas, single-quadrant designs on a toll facility, and pinwheel and four-level designs in higher-volume urban areas.

14.5.3 Four-Legged Freeway-to-Surface Street

Most interchanges in the United States have four legs and connect freeways to surface streets. Figure 14.14 shows the most common configurations for this type of interchange. All of the four-legged freeway-to-freeway designs could also be employed in this category, but they are generally not, due to high costs and safety concerns with traffic coming off high-speed ramps onto surface streets. The diamond is the most common configuration of this type, but its drawbacks have become more exposed in recent years and other configurations are built now where diamonds would have been in the past. The single-point is probably the fastest growing type of interchange, especially in urban and suburban areas, because of its great efficiency in a compact area. Raindrops are common in Europe and are gaining acceptance in the United States due to great efficiency with a narrow bridge. Partial cloverleaf designs are common and quite efficient, particularly where the loops accommodate the left-turn movement from the freeway to the surface street as shown in Figure 14.14. Finally, the median U-turn interchange is a version of the median U-turn intersection described earlier in the chapter and is efficient because no left turns are made where the ramps meet the surface street.

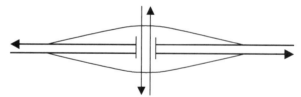

Diamond advantages: inexpensive, drivers expect this configuration

Disadvantages: high travel time, may need wide bridge

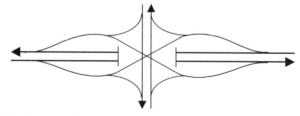

Single-point advantages: low travel time, small right-of-way

Disadvantages: long bridge, difficult pedestrian crossing

FIGURE 14.14 Common four-legged freeway-to-surface street interchange designs.

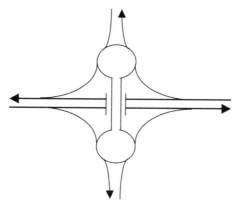

Raindrop advantages: narrow bridge, low travel times with moderate volumes

Disadvantages: large right-of-way at ramp terminals, driver confusion

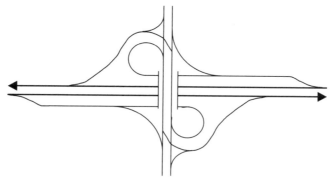

Partial cloverleaf advantages: low travel time, loops can be in any quadrant

Disadvantages: large right-of-way in loop quadrants, difficult pedestrian crossing

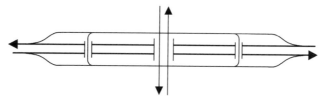

Median U-turn advantages: low travel time, narrow right-of-way

Disadvantages: driver confusion, three bridges needed

FIGURE 14.14   (*Continued*)

14.5.4 Collector-Distributor Roads

Collector-distributor roads are one-way roads built to freeway standards running parallel to the freeway mainline that are meant to shield the mainline from multiple ramp merges and diverges. As noted earlier, collector-distributor roads are a good solution to at least move weaving areas off the mainline. Collector-distributor roads are also beneficial at other places where there is a high density of ramps.

14.5.5 Frontage Roads

Frontage roads are one-way or two-way roads built parallel to a freeway but not to freeway standards. Frontage roads typically serve land uses along the freeway and may keep some shorter trips off the freeway. Continuous one-way frontage roads are a common feature in Texas and other states and provide great efficiency, particularly with an X-pattern of ramps (shown earlier in Figure 14.11). Continuous one-way frontage roads are very expensive, however, particularly because they add another level at freeway-to-freeway interchanges. Two-way frontage roads sometimes pose problems for designers because they intersect the cross-street at an interchange very close to the ramp terminals.

14.6 ACKNOWLEDGMENT

The photos and drawings for this chapter were provided by Daniel L. Carter.

14.7 REFERENCES

American Association of State Highway and Transportation Officials (AASHTO). 1999. *Guide for the Development of Bicycle Facilities,* 3rd ed. Washington, DC: AASHTO.

———. 2001. *A Policy on Geometric Design of Highways and Streets,* 4th ed. Washington, DC: AASHTO.

Castronovo, S., P. W. Dorothy, and T. L. Maleck. 1998. "Investigation of the Effectiveness of Boulevard Roadways." *Transportation Research Record* 1635.

Ewing, R. 1999. *Traffic Calming: State of the Practice,* Washington, DC: Federal Highway Administration and Institute of Transportation Engineers.

Federal Highway Administration (FHWA). 2000. *Roundabouts: An Informational Guide.* FHWA-RD-00-067, U.S. Department of Transportation, Washington, DC.

———. 2001. *Manual on Uniform Traffic Control Devices, Millennium Edition.* Washington, DC: U.S. Department of Transportation, FHWA.

Fitzpatrick, K., and M. Wooldridge. 2001. *Recent Geometric Design Research for Improved Safety and Operations.* NCHRP Synthesis 299, National Research Council, Transportation Research Board, Washington, DC.

Gluck, J., H. S. Levinson, and V. Stover. 1999. *Impacts of Access Management Techniques.* NCHRP Report 420, National Research Council, Transportation Research Board, Washington, DC.

Hummer, J. E., and J. D. Reid. 2002. "Access Management and Unconventional Arterial Designs: How Well Do the Various Design Accommodate Driveways?" In *Proceedings, Fifth National Conference on Access Management,* Transportation Research Board, Austin, TX, June 24.

Hurley, J. W., 1997. "Utilization of Auxiliary Through Lanes at Signalized Intersections with Downstream Lane Reductions." *Transportation Research Record* 1572.

ITT Industries, Inc. 1995–2001. Traffic Software Integrated System Version 5.0, Users' Manual. ITT Industries, Inc., Colorado Springs, CO.

Kickuchi, S., P. Chakroborty, and K. Vukadinovic. 1993. "Lengths of Left-Turn Lanes at Signalized Intersections." *Transportation Research Record* 1385,

Persaud, B. N., R. A. Retting, P. E. Garder, and D. Lord, "Safety Effects of Roundabout Conversions in the United States: Empirical Bayes Observational Before-After Study," *Transportation Research Record* 1751.

Pline, J. L. 1996. *Left-Turn Treatments at Intersections.* NCHRP Synthesis 225, National Research Council, Transportation Research Board, Washington, DC.

Reid, J. D., and J. E. Hummer. "Travel Time Comparisons Between Seven Unconventional Arterial Intersection Designs." *Transportation Research Record* 1751.

Transportation Research Board (TRB). 2000. *Highway Capacity Manual.* National Research Council, TRB, Washington, DC.

CHAPTER 15
PAVEMENT TESTING AND EVALUATION

Yongqi Li
Arizona Department of Transportation, Phoenix, Arizona

15.1 INTRODUCTION

A "good" pavement provides satisfactory riding comfort, structural integrity, and safe skid resistance. This chapter describes the major tests and evaluation procedures commonly used to assist pavement engineers to maintain pavements on their highway systems at such a level. These procedures evaluate pavement structural capacity, roughness, surface friction, and distress conditions.

Pavement structural capacity is an engineering concept to indicate the ability of pavement to carry the designed traffic loads adequately. It can be defined in terms of either the mechanical response (displacement, stress and strain) of the pavement under simulated wheel-loading conditions or an index that characterizes the load-bearing abilities of pavement material layers such as structural number (AASHTO 1993). The evaluation of existing pavement structural capacity is required for the determination of pavement maintenance and rehabilitation (M&R) strategy and design. Pavement roughness reflects the traveling public's perception of the quality of the highways and is related to vehicle operating costs. It is defined as the distortion of the road surface, which contributes to an undesirable or uncomfortable and unsafe ride. Uses of roughness information include determining the need for pavement improvements from the user's perspective, identifying severely rough sections as M&R projects, and implementing a smoothness-based specification to measure the initial quality of the contractor's work. Skid resistance is the pavement surface's ability to resist sliding when braking forces are applied to the vehicle tires. Without adequate skid friction, the driver may not be able to retain directional control and stopping ability. The major reason for collecting skid resistance data is to identify and repair pavement sections with low level of friction and thus to prevent or reduce accidents. The friction data are a major factor affecting the prioritization of pavement M&R projects and the determination of the appropriate M&R strategies.

Pavement distress surveys are another important part of pavement evaluation. They record visible distresses, such as cracking, flushing, and patching. Distress surveys are needed for assessing the maintenance measures needed to prevent accelerated, future distress, or the rehabilitation strategies needed to improve the pavement. The distresses survey data, together with the test results of pavement structural capacity, roughness, and skid resistance, constitute the basic database of a pavement management system, which assists decision-makers of a highway agency in finding optimum strategies for providing and maintaining pavements in a satisfactory condition.

There have been many pavement-testing devices and evaluation methods, the majority of which are in the forms of ASTM or other organization's standards. This chapter describes only the basic principles and applications of those most widely used by highway agencies or major research facilities. For the detailed testing procedures readers should consult the published standards (ASTM 1999; AASHTO 1990; ISO 1998).

15.2   EVALUATION OF PAVEMENT STRUCTURAL CAPACITY

The testing techniques for pavement structural capacity include nondestructive deflection testing (NDT), full-scale accelerated pavement testing (APT), and destructive methods. In practice, NDT is more widely used than the other two approaches for many reasons, including convenience, efficiency, low cost, fewer traffic interruptions, less pavement damage, high measurement accuracy, and high reliability. NDT results are usually used to formulate pavement M&R strategies; determine overlay design thickness, load limits, and load transfer across joints (concrete pavements); and detect void and remaining structural life. Destructive tests are restricted to pavements showing severe evidence of distress (Hudson and Zaniewski 1994). APT is mainly used in academic circles (Highway Research Board 1962; Hass and Metcalf 1996).

15.2.1   Nondestructive Deflection Testing and Evaluation

Based on loading mode, deflection-measuring devices are categorized as static, steady-state dynamic, and impulse; impulse devices are the most widely used. Fewer highway agencies use steady-state dynamic devices. Automated static deflection-measurement equipment is used only in Europe (Croney and Croney 1991). The well-known Benkelman beam, a static device that played an important role worldwide in pavement research, evaluation, and design in the last four decades, now is mainly used for pavement research.

Static Equipment.   The Benkelman beam is typical of this type. It is a simple hand-operated deflection measurement device (Figure 15.1). It consists of a lever arm supported by an aluminum frame. It is used by placing the tip of the lever arm between the dual tires of a loaded truck at the point where deflection is to be measured. As the loaded vehicle moves away from the beam, the rebound or upward displacement of the pavement is recorded by the dial gauge or an LVDT displacement transducer, which is installed at the rear end of the beam. When it is used with a pavement accelerated loading facility (ALF), the beam tip can be positioned at one side of the dual tires of the ALF. Thus, the beam will not block the movement of the wheels and the pavement deflections can be measured at any loading speeds. This is particularly useful for the investigation of the effect of loading speeds on the pavement response.

Because Benkelman beam testing is very slow and labor intensive some special vehicles mounted with automated deflection beam devices were developed based on the same principles as the Benkelman beam. The most common one of this type is the La Croix deflectograph, which is manufactured in France and has been widely used in Europe. The testing speed is approximately 2 to 4 km/hr. Neither the Benkelman beam nor the deflectograph measures the absolute deflection of the pavement because the beam supports in both cases are to some extent within the influence of the truck axles during testing.

Steady-State Dynamic Deflection Equipment.   This type of equipment induces a steady-state vibration to the pavement with a dynamic force generator, which can be either electro-mechanical or electrohydraulic. Pavement deflections are measured with velocity transducers.

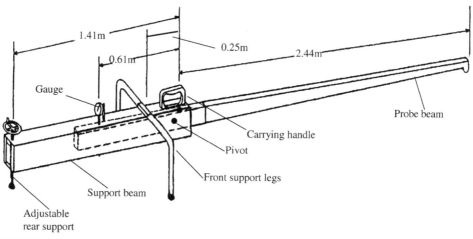

FIGURE 15.1   Sketch of basic components of the Benkelman beam.

Figure 15.2 shows a typical force output. The dynaflect was one of the first devices of this kind available in the market (Smith and Lytton 1984). It is trailer-mounted and can be towed by a standard vehicle. Its electromechanical force generator has a pair of counter-eccentric masses, rotating in opposite directions at 8 Hz, to produce a cyclic force in the vertical direction. The static weight of the unbalanced mass is normally 907 kg and the produced peak-to-peak dynamic force of 4536 kN at 8 Hz is distributed between two rigid load wheels. The resulting deflection basin is measured by five geophones that are mounted on a placing bar at 0.3-meter intervals. Other devices in this category include Road Rater and the U.S. Army Engineer Waterways Experiment Station vibrator. The steady-state dynamic deflection equipment is highly reliable. Their shortcoming is that they require a relatively large static preload and pavement resonance may affect deflection measurements.

Impulse Deflection Equipment.   Impulse deflection devices deliver a transient impulse load (Figure 15.3) to the pavement and the resulting pavement deflections are measured using either geophones or seismometer. The force is generated by a one-mass or two-mass (falling

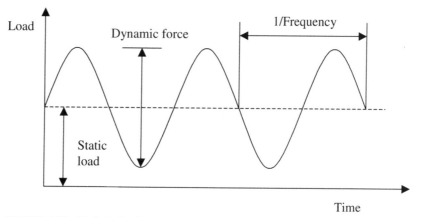

FIGURE 15.2   Typical vibrating steady-state force by Dynaflex.

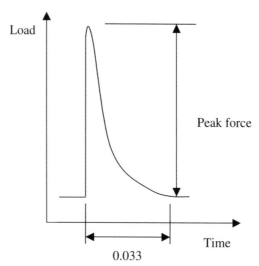

FIGURE 15.3 Typical load pulse produced by FWD.

weight) system (Figure 15.4), which is raised to one or more predetermined heights and dropped with a guide system. These devices are typically referred to as falling weight deflectometers (FWD). The force magnitudes are mainly determined by the drop height and the weight of the mass and can vary from 13,620 kN to over 227,000 kN depending on the model used. FWD devices have relatively low static preloads, which eliminates the negative effects of a high preload usually associated with steady-state dynamic deflection equipment. The loading plate, mass system, type, and precision of deflection and load sensors, the location where the falling weight system is mounted at the vehicle, and other features of FWD vary from one manufacturer to another. The sensor spacing is adjustable. Typically, seven deflection sensors are located to measure the deflections at the center of the loading plate, 0.3, 0.6, 0.9, 1.2, 1.5, and 1.8 meters from the center. The primary impulse-deflection equipment commercially available includes the Dynatest FWD, the JILS FWD, the Phonix FWD, and the KUAB FWD (Smith and Lytton 1984). To reduce the individual equipment error and the between-equipment error of the measurement, four regional FWD calibration centers have been established for the load cell and displacement sensors of a FWD to be calibrated regularly using the Strategic Highway Research Program SHRP procedure as the reference (FHWA 1993).

Comparison between FWD and Other Deflection Devices. Theoretically, it is impossible to compare the deflections measured from different devices due to the difference of loading modes between FWD and other deflection devices and the intrinsic complexity of pavement systems' response to loading of different modes. From a practical standpoint, however, there may be situations where there are no alternatives but to develop the statistical correlation between different devices. In these cases, such correlations should be used cautiously. It should be kept in mind that the correlation equation is a function of pavement type, time of testing, material properties, and many other variables under which it was developed and it is impossible to estimate its accuracy on any given pavement. It has been suggested that the literature that documented the correlation equation should be tracked down to check the R^2 and the conditions under which it was derived to make the judgment whether the equation is suitable to the pavement and the conditions under which it will be used (FHWA 1994).

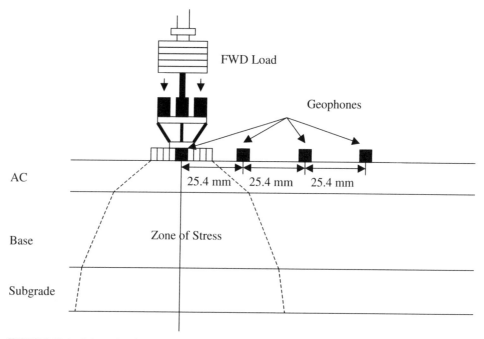

FIGURE 15.4 Schematic of stress zone within pavement structure under the FWD load. (*Source:* AASHTO 1993.)

Influencing Factors of Measured Deflection.   Load is the primary factor affecting measured deflection. The deflections measured from different types of devices inevitably differ when the peak load magnitude applied is equal. The same type of equipment also provides significantly different results at the same load magnitude if the load pulse shape and duration are different. The peak values of the center deflection can vary as much as 10 to 20 percent for an FWD (Royal Institute of Technology 1980).

Deflection measures the overall mechanical response of pavement system to load. Any factor affecting the mechanical properties of pavement layers has a direct impact on the measured deflection. For example, water content and high temperature can cause a dramatic reduction of the moduli of subgrade soil and asphalt layer, respectively, resulting in high measured deflection. Due to the limited pavement dimensions and variation of pavement distress, different testing locations in the same section can cause a significant difference in deflection measurement. For flexible pavement, deflections measured near cracks are normally much higher than the measurements in nondistressed areas (FHWA 1998). Similarly, deflection measurements near longitudinal joints, transverse joints, or corners are higher than those measured at mid-slab for concrete pavements. Thermal and moisture gradient in the vertical direction of concrete slabs cause curling and warping and have a significant influence on deflection measurements. Measurements taken at night or in the early morning are considerably different from these obtained in the afternoon.

In cold areas that experience freeze-thaw, the influence of season on deflection measurements shows a clear pattern (Figure 15.5; Scrivner et al. 1969). In the period of deep frost, the measured deflection is the lowest due to the frozen subgrade. The pavement is the weakest and the deflection increases rapidly in the thaw period, during which the frost begins to disappear and the subgrade becomes soft. When the excess of free water from the melting

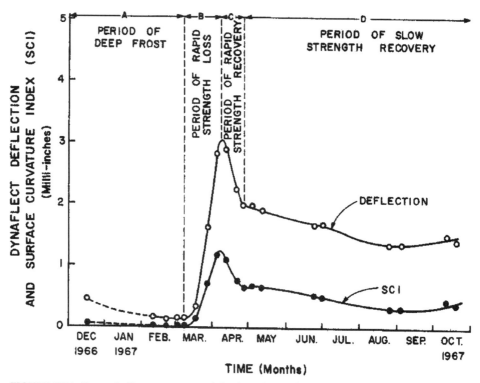

FIGURE 15.5 Seasonal effects on pavement deflection. (*Source:* Scrivner et al. 1969.)

frost leaves the pavement, the soil begins to gain strength, rapidly at first and slowly afterwards.

Because deflection measurement is influenced by so many factors, it is important for an agency to develop a standard deflection test procedure. Ideally, deflection measurements should be conducted at approximately the same temperature and in the same season. If possible, adjustment factors should be applied to account for temperature and moisture variations. In cold areas, it is desirable to measure deflection during the spring thaw period, when the deflection measurements are at the maximum. It is particularly important to make sure that the measured deflections at different conditions are adjusted to the standard condition especially when pavement overlay design is based on measured deflections. In any cases, along with deflection measurements, the measurement locations, pavement condition, and temperature should be recorded.

Use of Measured Deflection. The measured deflections are mainly used to determine overlay thickness design, material properties of pavement layers, load transfer across joints (concrete pavement), void detection, and layer bonding.

Overlay Thickness Design. The commonly used methodology is to establish a regression relationship between the actual traffic loads and the pavement structural capability indices, which are represented by deflection basin parameters, and other factors under a defined criterion of the final condition of the pavement. The procedure used to design overlays in Arizona is typical of this methodology. Through a statistical analysis of the pavement performance data of several historical overlay projects, the following regression equation was established (Way et al. 1984):

$$\text{Log } L = 0.0587T(2.6 + 32.0\ D5)^{0.333} - 0.104\ \text{SVF} + 0.000578P_0 + 0.0653\ \text{SIB} + 3.255$$

$$(15.1)$$

where L = design 18 kip ESALs
T = overlay thickness, in.
SVF = seasonal variation factor
P_0 = roughness before overlay, in./mi
SIB = spreadability index before overlay. For the FWD, SIB = 2.7 (FWD SI)$^{0.82}$
$D5$ = #5 Dynaflect sensor reading, mils. For the FWD, $D5$ = 0.16 (FWD $D7$)$^{1.115}$
SI = $(D_0 + D_1 + D_2 + D_3 + D_4 + D_5 + D_6) * 100/D_0$

The thickness is determined from the above regression equation as follows:

$$T = \frac{(\log L - 3.255) + 0.104\ \text{SVF} + 0.000578P_0 - 0.0653\ \text{SIB}}{0.0587(2.6 + 32.0\ D5)^{0.333}} \qquad (15.2)$$

Like any other regression relationships, the validity of a design equation of this type is limited to the specific conditions and pavement types under which the equation was established. Usually the R^2 of the regression equation is very low.

Pavement Layer Moduli Back-Calculation. Back-calculation is the process of estimating the fundamental engineering properties (elastic moduli or Poisson's ratio) of pavement layers and underlying subgrade soil from measured pavement surface deflections. This process can be illustrated using the point load case of Boussinesq's problem. Assuming that a semi-infinite medium is homogeneous, isotropic and linear elastic, the surface deflection D_0 at the center line of the circular uniformly distributed load p, is calculated from:

$$D_0 = \frac{2ap(1 - \mu^2)}{E} \qquad (15.3)$$

where a = radius of the circular area
E = elastic modulus
μ = Poisson's ratio

Given the values of a, p, μ and measured D_0, E can be back-calculated by substituting the known values into equation (15.3).

For a typical pavement, which is usually a multilayer system (Figure 15.4), the equation for calculating pavement surface deflections is closed form. However, there is no closed-form solution to the back-calculation of the layer moduli from the measured deflections and other parameters as in Boussinesq's problem. The existing back-calculation programs determine pavement layer moduli by an iterative procedure (Figure 15.6; Lytton 1989). They search for the set of layer moduli from which the calculated deflections agree with the measured within a given tolerance limit. Different programs may use different moduli inter-active procedures and tolerance limits. Examples of these programs include MODULUS, BISDEF, ELSDEF, and CHEVDEV. Elastic layered programs BISAR, ELSYM5, CIRCLY, and CHEVRON are generally used to calculate deflections. ILLL_BACK is a back-calculation program for rigid pavements that uses Westergard's equation to calculate deflections.

It is very important to note that the back-calculated moduli do not necessarily reflect the fundamental property of the layer materials as the moduli determined in the laboratory do (Ullidtz and Coetzee 1995). A comparison study showed that not only are laboratory-determined moduli not identical in values to back-calculated moduli statistically but there is literally no correlation between the two at all with the R^2 of 0.013 (Figure 15.7; Mamlouk et al. 1988). The significant difference between the two moduli may be partly caused by the factors that affect the accuracy of laboratory-determined moduli, such as disturbance of the samples and nonrepresentive stress condition. However, the major source of this discrepancy

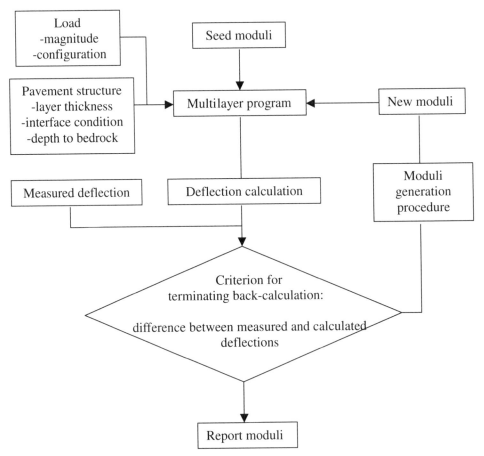

FIGURE 15.6 Common procedure of back-calculation programs.

is related to the back-calculation procedure itself. First, the elastic layered model used in back-calculation to calculate deflections is based on the assumption that the pavement materials conform to an idealized condition of being linearly elastic, uniform, and continuous. This assumption ignores discontinuities (cracks, voids), nonlinearity and plasticity of materials, material variability, the effects of temperature gradients, and many other factors. This inevitably causes errors of the back-calculated moduli. Furthermore, the analytical solution of deflection itself does not guarantee that there is a unique solution of layer moduli for a given set of deflections. In other words, there is no solid theoretical base for the back-calculation of moduli from measured deflections for a typical pavement.

Therefore, it is critically important to notice that back-calculation procedure can by no means prove its accuracy itself. A set of back-calculated moduli that provide a deflection basin perfectly matching the measured one does not necessarily mean that they are correct or accurate. In fact, the only way to evaluate the reliability of back-calculated moduli is to use the directly measured moduli as the reference.

Joint Load Transfer. When a load is applied near a transverse joint, both the loaded and the unloaded sides deflect because a portion of the load that is applied to the loaded side is carried by the other side through load transfer at the joint. Efficient load transfer can reduce the edge stresses and deflections, thus reducing fatigue damage and minimizing pumping.

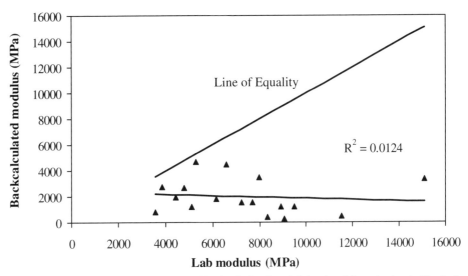

FIGURE 15.7   Correlation between lab and back-calculated moduli (produced from the data in Mamlouk et al. 1988.)

Therefore, load transfer at the joint or crack is essential for satisfactory performance of rigid pavement. Deflection testing can be used to measure the load transfer efficiency. The test can be conducted by using the FWD, with the loading plate positioned adjacent to the joint and a deflection sensor located across the joint (Figure 15.8). The load transfer efficiency (LTE) can be calculated as follows (FHWA 1998):

$$LTE = (\delta_u / \delta_l) * 100 \qquad (15.4)$$

where δ_u is deflection on unloaded side of joint and δ_l is deflection on loaded side of joint. The deflection load transfer of the joint can be evaluated approximately by the following scale (FHWA 1998):

Good: greater than 75 percent
Fair: 50–75 percent
Poor: less than 50 percent

The test result of load transfer is greatly affected by temperature. When the temperature increases, the adjacent concrete slabs expand and more contact between them occurs at the joint, causing the load transfer to increase. When the temperature decreases, the slabs contract and the joint opens, causing the load transfer between slabs to decrease. Studies show that on a summer day when the pavement temperature rises substantially, load transfer at the same joint may increase from 50 percent in the morning to 90 percent in the afternoon. Testing is usually recommended to be performed when the joint is open. It should be noticed that the LTE calculated from equation (15.4) is less than 100 percent in most cases but can be greater than 100 percent in some conditions.

Detection of Voids.   Pumping and the subsequent voids beneath the slab are one of the major causes of deterioration of concrete pavements. Deflection testing can be conducted to detect voids below slab corners. The common techniques include corner deflection profile and variable load corner deflection analysis (AASHTO 1993).

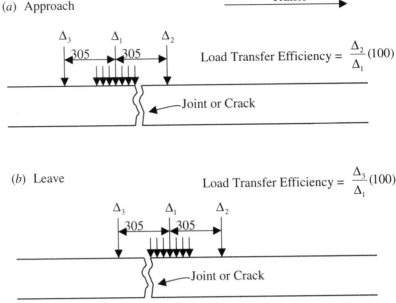

FIGURE 15.8 Arrangement of deflection sensors for determining load transfer efficiency (after FHWA 1994, 5-29).

In the first method, the deflection at the approach and leave corners are measured and plotted. The corners that exhibit the lowest deflections are expected to have full support value. Therefore, if the deflection at a corner is significantly greater than the lowest value, it can be concluded that voids exist at the corner. In the second method, the corner deflections are measured at three load levels and the load-deflection curve is plotted, (Figure 15.9). Studies show that for the locations with no voids, the intercept at the deflection axis is very near the origin (less than 50 μm). Thus, if the intercept at the deflection axis is significantly greater than 50 μm it can be suspected that there are voids at the corner. Because slab curling caused by temperature gradient and moisture substantially affect the deflection measurement, the deflection testing to detect voids should be conducted at the appropriate time to eliminate or reduce the influence of slab curling.

Detection of Loss of Bonding.   Consider a composite pavement system as illustrated in Figure 15.10. A wheel load $p(r)$ is acted upon this system. It is assumed that the two Poisson ratios are equal in magnitude. Using the plate theory on composite section, the equivalent flexural rigidity can be computed as follows (Li and Li):

$$D_e = \frac{E_1 h_e^3}{12(1 - \mu^2)} \tag{15.5}$$

where h_e is the equivalent thickness with respect to the type of interface.
For an unbonded interface, the equivalent thickness is expressed as

$$h_e = h_1 \left(1 + \frac{E_2}{E_1} \left(\frac{h_2}{h_1}\right)^3\right)^{1/3} \tag{15.6}$$

where h and E are the thickness and elastic modulus of the corresponding layer material, respectively, as shown in Figure 15.11a.

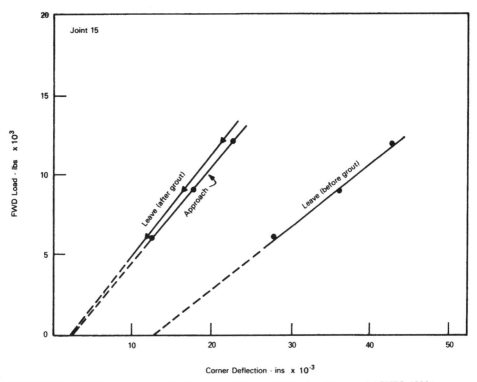

FIGURE 15.9 Variable load corner deflection analysis for void detection. (*Source:* AASHTO 1993.)

For a fully bonded interface, the equivalent thickness is computed below:

$$h_e = h_1 \left[1 + \frac{E_2}{E_1} \left(\frac{h_2}{h_1} \right)^3 + \frac{3 \frac{E_2}{E_1} \frac{h_2}{h_1} \left(1 + \frac{h_2}{h_1} \right)^2}{\left(1 + \frac{E_2}{E_1} \frac{h_2}{h_1} \right)} \right]^{1/3} \tag{15.7}$$

where all variables are as defined earlier.

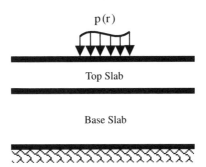

FIGURE 15.10 Composite pavement structure. (*Source:* Li and Li.)

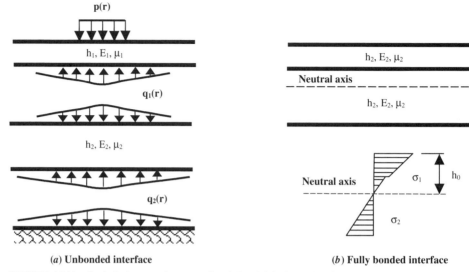

(*a*) **Unbonded interface**   (*b*) **Fully bonded interface**

FIGURE 15.11 Analytical approaches to unobonded and fully bonded slabs. (*Source:* Li and Li.)

Examination of the two equations indicates that the determination of equivalent thickness includes two same terms, i.e., $\{1 + (E_2/E_1)(h_2/h_1)^3\}$, for both the unbonded and fully bonded interfaces. The only difference is the third term on the right side in equation (15.7), which accounts for the effect of the bonding across the interface. The contribution of a fully bonded interface is an increase in the equivalent thickness in terms of the third term in equation (15.7). Based on this observation, the coefficient of bonding K is introduced as follows:

$$h_e = h_1\left[1 + \frac{E_2}{E_1}\left(\frac{h_2}{h_1}\right)^3 + K\frac{3\dfrac{E_2}{E_1}\dfrac{h_2}{h_1}\left(1 + \dfrac{h_2}{h_1}\right)^2}{\left(1 + \dfrac{E_2}{E_1}\dfrac{h_2}{h_1}\right)}\right]^{1/3} \tag{15.8}$$

For an unbonded interface, $K = 0$. For a fully bonded interface, $K = 1.0$. K is extended to characterize a partially bonded interface by assuming that K is a number varying from zero to one. By substituting equation (15.8) into equation (15.5) and subsequently substituting equation (15.5) into Westergard's equation, the deflection can be related to the coefficient of bonding K.

15.2.2 Destructive Structural Evaluation

The techniques used for destructive pavement evaluation generally involve taking cores or cutting trenches transversely across pavements. The objectives are usually to remove samples for inspection and testing and to diagnose the causes and mechanisms of pavement failures. Because of the extreme complexity and variability of pavement failures, sometimes postmortem is the only way to review what really happens inside pavements. Such evaluation procedures are used mostly on test roads and occasionally on in-service pavements.

Trenches were cut at the AASHO test road and other APT test sites to investigate the permanent deformation accumulated at the top of each of the structural layers (Highway

Research Board 1962; Sharp 1991; Li et al. 1999). At the AASHO test road, it was found that rutting was mainly due to decrease in thickness of the pavement layers, which was caused by lateral movement of the materials. Postmortem was conducted on the asphalt pavements with cement-treated base (CTB) at Australia's Accelerated Loading Facility (ALF). It was found that the failure mechanism of the CTB pavements was debonding of the CTB layers under the repeated loading, followed by the penetration of water and subsequent erosion at the layer interfaces, eventually causing the break-up of the top CTB layer (Figure 15.12; Sharp 1991).

15.2.3 Full-Scale Accelerated Pavement Testing (APT)

"Full-scale accelerated pavement testing is defined as the controlled applications of a prototype wheel loading, at or above the appropriate legal load limit to a prototype or actual, layered, structural pavement system to determine pavement response and performance under a controlled, accelerated, accumulation of damage in a compressed time period" (Metcalf 1996). The acceleration of pavement damage process is achieved by means of increased load magnitude and frequency, imposed climatic conditions, thinner than standard pavements, or a combination of these factors. Based on the length of tested pavements and loading facilities, full-scale APT can be broadly classified into two groups: test roads and test tracks (Metcalf 1996). Test roads usually have a long loop of tested pavement (several miles) with several sections; loading of traffic is accomplished by either directed actual traffic or calibrated and controlled test trucks. An example of this group includes the well-known Road Test conducted between 1958 and 1960 by the AASHO (Highway Research Board 1962). Other more recent ones include MnRoad (Minnesota), WesTrack (Nevada), and NCAT Track (Alabama) (Epps et al. 1999; Harris, Buth, and Van Deusen 1994; Brown and Powell 2001). Test tracks have a shorter test pavement and a specially designed loading facility, which is of circular, linear, or free-form layout. Linear tracks have been widely used in the United States, including the Australian-designed Accelerated Loading Facility (ALF) (Figure 15.13) and the South African-designed and manufactured Heavy Vehicle Simulator (HVS) (Sharp 1991; Harvey, Prozzi, and Long 1991).

APT is an essential part of pavement research strategy. It also has a specific application to modifying existing design for conventional materials to heavier traffic, evaluating new pavement materials and structural designs, investigating pavement failure mechanisms, estimating remaining pavement life, and evaluating environmental effects. A successful appli-

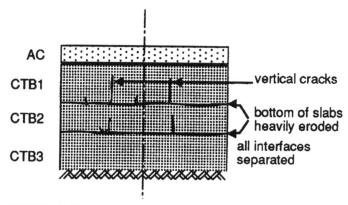

FIGURE 15.12 Typical failure mode of CTB pavements. (*Source:* Sharp 1991.)

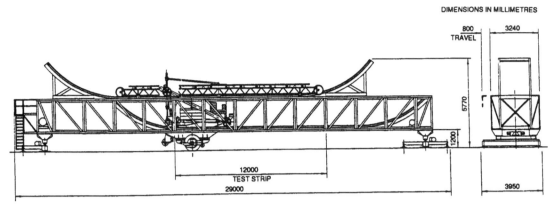

FIGURE 15.13   Schematic diagram of ALF. (*Source:* Sharp 1991.)

cation of APT was the AASHO test road, which resulted in the AASHTO Pavement Design Guide (AASHTO 1993). Most of the recent APT programs in the United States have focused on evaluating the SHRP Superpave mix design, innovative mix design, and the relationship between pavement roughness and fuel consumption.

APT facilities usually have instrumented test pavements to measure pavement responses under accelerated loading, such as the horizontal strain at layer interfaces, vertical pressures at the surface of subgrade, and displacement at multiple depths. Climate data are also collected. The survival rate and reliability of the instruments (strain gauges, pressure cells, displacement gauges, and moisture gauges) are usually very low due to the debonding between the instruments and measured pavement materials. It is also a challenging task to analyze and interpret the data from these gauges.

15.3   EVALUATION OF PAVEMENT ROUGHNESS

Roughness is an indicator of a car driver's perception of pavement riding comfort and safety. The measurement of roughness depends primarily on the vertical acceleration experienced by a driver. Vertical acceleration in turn depends on three factors: the pavement profile, the vehicle mass and suspension parameters, and the travel speed. Based on the ways to record and characterize vertical acceleration, roughness measurement methods are classified as response-type and profile-type. In a response-type system, a transducer installed on a passenger vehicle or a special trailer measures and accumulates deflections of the vehicle suspension as it travels on the road. The vertical movement of accumulated suspension stroke, normalized by the distance traveled, is used to quantify roughness. On the other hand, instead of recording the actual vertical movements of a real vehicle's suspension directly, a profile-type roughness measurement system records the pavement longitudinal profile. It then runs a mathematical model, which simulates the suspension response of a real vehicle moving on the road, to calculate the simulated suspension motion and normalize it by the traveled distance to give profile index. This profile index is used to quantify pavement roughness. Profile-type roughness measurement systems have many advantages over response-type systems and are the most widely used nowadays.

15.3.1   Response-Type Road Roughness Measuring Systems (RTRRMS)

The Bureau of Public Roads (BPR) Roughometer (Figure 15.14, was the earliest mechanical RTRRMS (Hass, Hudson, and Zaniewski 1994). It is a trailer-mounted system. The trailer,

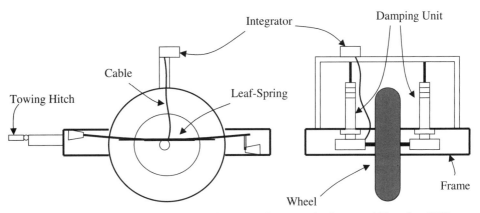

FIGURE 15.14   Sketch of basic components of BPR Roughometer (after Sayers and Karamihas 1998).

with a heavy rectangular chassis, is supported on a central wheel. The wheel is supported by two single-leaf springs positioned one on each side of the wheel. Two dashpot-damping assemblies are installed between the chassis and the wheel axle. With the appropriately selected parameters of the mass, dashpot damping unit, and spring, the trailer approximately represents one-quarter of a passenger car. In operation, the differential movement between the wheel axle and the chassis is summed by a mechanical integrator unit installed on the chassis. The roughness index is given as the integrated vertical movements divided by the traveled distance, with a unit of m/km. Other RTRRMS systems following the concept of the BPR Roughometer include the TRRL bump integrator, Australia's NASRA meter, and the Mays Ride Meter. The last was the most popular in the United States before profilers were widely adopted.

The other type of RTRRMS uses an accelerometer as the primary motion sensor. The accelerometer can be mounted on either a special trailer or the vehicle body or axle. The output of the accelerometer can be integrated twice to obtain the vertical movement. The integration process can magnify the effect of noise signal. The axle-mounted accelerometers are not as sensitive to the vehicle parameters as the mechanical RTRRMS. Generally, the accelerometer-based RTRRMS uses the summarized root mean square value of acceleration (RMSA) to quantify pavement roughness (Hass, Hudson, and Zaniewski 1994).

Because each and every factor that influences vehicle response to the pavement surface impacts measures from response-type systems, these systems are host vehicle-dependent and inevitably have two problems. First, they are not stable with time. Measures made at different times cannot be compared with confidence. Second, they are not transportable, meaning that measures from different equipments on the same pavement are seldom reproducible. In practice, response-type systems have to be regularly calibrated by the measures from a valid profiler (Gillespie, Sayers, and Segel 1980).

15.3.2   Profile-Type Road Roughness Measuring Systems

Using a profiler to measure pavement roughness involves two steps. The first step is to measure the pavement longitudinal profile. The second step is to run a mathematical model against the profile to simulate a vehicle's suspension response to the actual pavement. Mathematical models simulating different roughness measurement devices result in different roughness indices. The most widely used one is International Roughness Index (IRI), which comes from a mathematical model simulating BPR-like devices (Sayers, Gillespie, and Paterson 1986; Sayers 1995).

Profile Measurement.   The rod and level, familiar to most civil engineers, can be used to measure pavement profile. However, the accuracy requirements for measuring a profile valid for computing roughness are much more stringent than is normal for road surveying. The elevations must be taken at close intervals of 0.3 meters or less, and the individual height measures must be accurate to 0.5 millimeters or less. Therefore, using the rod- and-level method to measure a profile for roughness is extremely slow and labor-intensive. One alternative to road and level equipment is Dipstick, which is faster than road and level for measuring profiles suited for roughness analysis. When it is "walked" along the line being profiled, a precision inclinometer installed in the device measures the differential height between the two supports, normally spaced 0.3 meters apart. The elevation at the second support is calculated from that at the first support.

The rod-and-level method and Dipstick measure the real profile and are usually used as a reference to check the accuracy of other profiler. The most widely used equipment for routine pavement profiling is the inertial profiler (Figure 15.15; Karamihas and Gillespie 1999). As the vehicle moves at a certain speed, an accelerometer mounted in the vehicle to represent the vertical axis of the vehicle measures vertical acceleration of the vehicle body. The accelerations are integrated twice to obtain the vertical movement d_1 of the vehicle body. A transducer is used for measuring the distance between the pavement surface and the vehicle d_2. The elevations of the pavement surface are obtained by adding d_2 to d_1. A distance odometer measures the distance along the pavement. The on-board data acquisition system records the data to obtain the profile files for real time or postroughness analysis. The inertial profilers not only works at normal highway speeds, it requires a certain speed (usually >15 km/hr) to function.

The profile from an inertial profiler does not look like one measured statically from either the road and level or Dipstick. For example, Figure 15.16 shows profiles measured from the Dipstick and two inertial profilers. The different appearances are mainly due to the fact that neither of these devices accurately measures the grade and long undulations, which, for roughness analysis, we are not interested in. In order to make practical use of a profile, we have to remove the components of the grade and long undulations from the measured profile. In fact, after the grade and long undulations have been removed, the profiles from different devices match very closely to each other. They all present the pavement unevenness in such a scale that it causes severe disturbance to the public driving over it. Figure 15.17 shows the same profiles in Figure 15.16 after the road grade and long waves have been removed.

Removal of the grade and long undulations involves filtering techniques. A moving average is a simple filter commonly used in roughness analysis of profiles. It replaces each profile point with the average of several adjacent points. For a profile that has been sampled as interval ΔX, the filter is defined as:

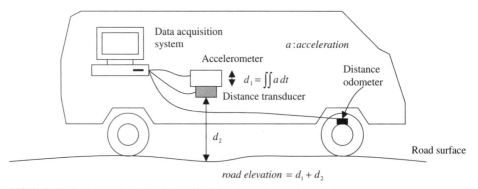

road elevation $= d_1 + d_2$

FIGURE 15.15   Schematic of inertial profiler (after Karamihas and Gillespie 1999).

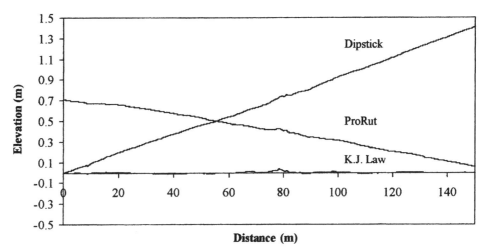

FIGURE 15.16   Unfiltered profiles.

$$p_{fL}(i) = \frac{1}{2n+1} \sum_{j=i-n}^{i+n} p_{(i)} \qquad (15.9)$$

where $p_{(i)}$ and $p_{fL(i)}$ are the values of the original and filtered profile point at ith location, respectively, B is the base length of the moving average, and $B = 2n\ \Delta X$. The effect of a moving average filter is to smooth the profile by averaging out the point-by-point fluctuations. More importantly, it can also be used to filter out the smoothed components from the original profile to obtain the deviations from the smoothed profile. The filter to achieve this is to subtract the smoothed profile from the original:

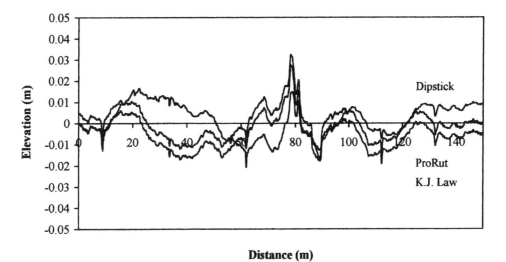

FIGURE 15.17   Profiles filtered with a 91-meter high-pass filter.

$$p_{fH(i)} = p_{(i)} - p_{fL(i)} \qquad (15.10)$$

The filter $p_{fH(i)}$ is usually called a high-pass filter, while the filter $p_{fL(i)}$, which is used to smooth a profile, is usually called a low-pass filter. When the sampling interval ΔX is very small (<25.4 millimeters) a low-pass filter with a small B (about 0.3 meters) is usually applied to the original profile to average out the fluctuation within the small range in which we are not interested. Then a high-pass filter with a relatively large B (30 to 90 meters) is applied to the original or smoothed profile. Figure 15.17 is the resulting profiles after a high-pass filter with a base length of 90 meters is applied to the profiles shown in Figure 15.16. A high-pass filter is mandatory for roughness analysis, but a low-pass filter is not. It is the high-pass filtered profile that is used to calculate profile indices.

IRI. The IRI is a profile index based on a mathematical model called a quarter-car (Figure 15.18; Karamihas and Gillespie 1999). The quarter-car model calculates the suspension deflection of a simulated mechanical system with a response-to-pavement profile similar to that of a passenger car. The IRI is defined as the accumulated suspension motion from the quarter-car with a set of specified parameters moving on an appropriately filtered profile divided by the distance traveled. It is also expressed in unit of m/km. The computer program to calculate the IRI from a profile can be found elsewhere (Sayers, Gillespie, and Querioz 1986).

The IRI originally came out of the effort of the World Bank in 1982 to search for a calibration standard for roughness measurements, especially from response-type devices (Sayers, Gillespie, and Querioz 1986). Since the IRI is essentially a computer-based virtual response-type system, it has a good correlation with the roughness indices from response-type systems. In addition, the IRI is defined as a property of a profile and independent of profilers as long as they provide valid profiles. It is reproducible, portable, and stable with

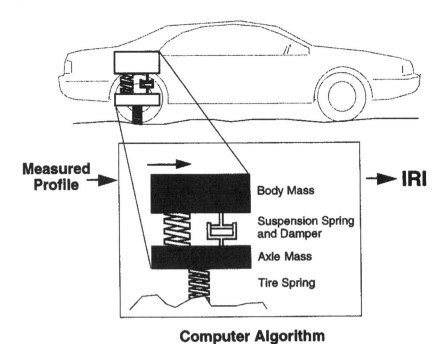

FIGURE 15.18 Quarter-car model for IRI. (*Source:* Karamihas and Gillespie 1999.)

time. The IRI is an ideal pavement condition indicator. It captures the roughness qualities of a pavement profile that impact vehicle response. Figure 15.19 shows IRI ranges represented by different classes of road (Sayers and Karamihas 1998). Currently most states in the United States use inertial profiler as the profile measurement equipment and IRI as roughness index. Other profile indices are not in broad use because there is little reason to use them when IRI can be conveniently calculated.

It should be noted that the accuracy of an IRI measurement depends solely on the accuracy of profile measurement. A profile measurement used for IRI calculation has to be particularly accurate in a range of wavelengths from 1.3 to 30 meters. The guidelines for measuring a longitudinal pavement profile to use in computing IRI can be found elsewhere (Karamihas and Gillespie 1999).

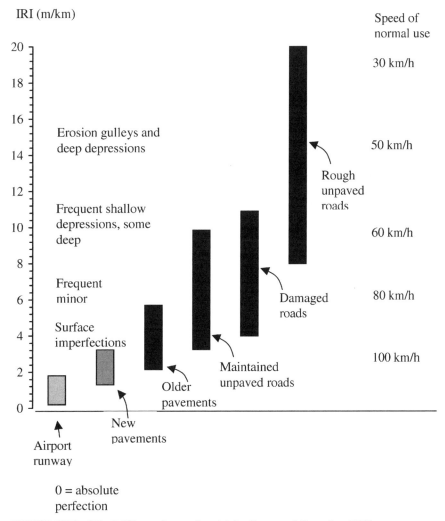

FIGURE 15.19   IRI of different classes of road (after Sayers and Karamihas 1998).

15.3.3 Profilograph

Profilographs have been widely used for evaluating the smoothness of concrete pavements during construction. The design of these devices is all based on the same principle, as shown in Figure 15.20. These devices consist of bogie wheel sets at front and rear, a recording wheel at the center, and a recorder to record the vertical movement of the recording wheel. A profilograph actually performs the similar function as a high-pass moving average filter does. The average is established by the bogie wheels, and deviations are measured relative to the average, which is not for points on the same path as in the case of profiles. The profilogram from a profilograph is usually recorded on a scale of 1:300 longitudinally and 1:1 vertically. In the following, the Arizona method to determine the profile index (PrI) and locate individual high areas that need to be grinded from the profilogram is described (Arizona DOT 1999).

The device for determining the PrI is a plastic scale 1.7 inches wide and 21.12 inches long, representing a pavement length of one tenth of a mile. Near the center of the scale is an opaque band 0.2 inch wide extending the entire length. On either side of this band are scribed lines 0.1 inch apart parallel to the opaque band. These lines, called scallops, serve as a scale to measure deviations of the graph above or below the blanking band. The plastic scale should be placed over the profile in such a way as to blank out as much of the profile as possible. When it is impossible to blank out the central portion of the trace, the profile should be broken in short sections (Figure 15.21). The height of all the scallops appearing both above and below the blanking band should be measured and totaled to the nearest 0.05 inch. Only the scallops ≥0.03 inch high and ≥0.08 inch long (2 feet pavement length) are included in the count. The PrI is defined as the total count of tenths of an inch divided by 10 times the miles of profile.

The device to determine scallops in excess of 0.3 inch is a plastic template having a line 1 inch long scribed on one face with a small hole at either end and a slot 0.3 inch from the parallel to the scribed line (Figure 15.22). The 1-inch line corresponds to a pavement distance of 25 feet. At each prominent peak on the profile trace, the template should be placed so that the small holes intersect the profile trace to form a chord across the base of the peak or indicated bump. The line on the template need not be horizontal. A line should be drawn using the narrow slot as a guide. Any portion of the trace extending above this line indicates the length and height of the deviation in excess of 0.3 inch. When the distance between easily recognizable low points is less than one inch, a shorter chord length should be used in making the scribed line on the template tangent to the trace at the low points. The baseline for measuring the height of bumps should be as near 25 feet (1 inch) as possible, but in no case should exceed this value.

15.4 EVALUATION OF SKID RESISTANCE

The coefficient of friction is obtained by dividing the frictional force in the plane of interface by the force normal to the plane. For pavements, the coefficient of friction measured when the tire is sliding on the wet pavement is usually called skid resistance. Skid resistance is very important to driving safety. Without adequate skid resistance the driver may not be able to retain directional control and stopping ability and an accident may occur. Skid resistance measurement is a major pavement test.

It has been demonstrated that the skid resistance decreases as the velocity of the tire surface relative to pavement surface increases (Moyer 1934). This relative velocity is called the slip speed. There are four types of pavement skid resistance measuring devices: locked wheel, side force, fixed slip, and variable slip. They measure the skid resistances at different slip speeds. The locked-wheel method simulates emergency braking without an antilocking system, the side-force method measures the ability to maintain control in curves, and the

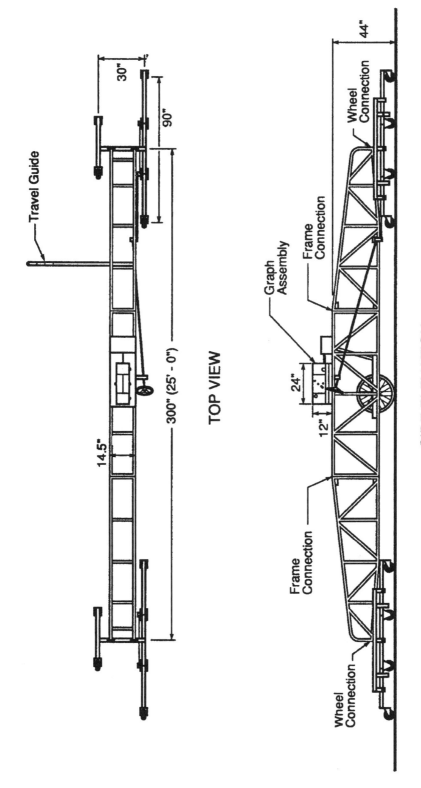

FIGURE 15.20 Schematic of a profilograph.

30"

90"

Travel Guide

TOP VIEW

300" (25' - 0")

14.5"

44"

Wheel Connection

Frame Connection

Graph Assembly

24"

12"

Frame Connection

Wheel Connection

SIDE ELEVATION

15.21

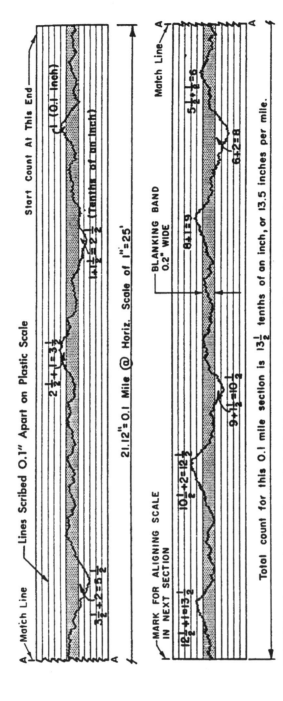

FIGURE 15.21 Example of showing method of deriving profile index from profilograms. (*Source:* Arizona Department of Transportation 1999.)

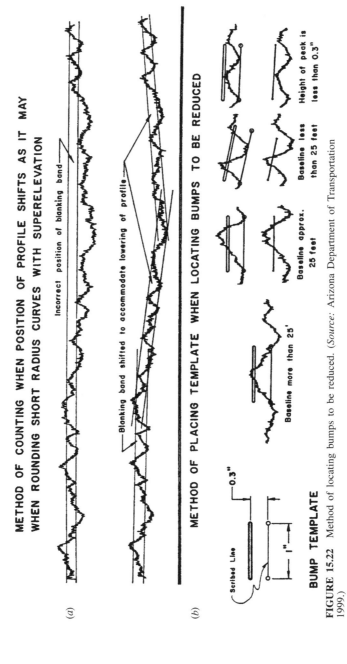

METHOD OF COUNTING WHEN POSITION OF PROFILE SHIFTS AS IT MAY
WHEN ROUNDING SHORT RADIUS CURVES WITH SUPERELEVATION

Incorrect position of blanking band

Blanking band shifted to accommodate lowering of profile

(a)

METHOD OF PLACING TEMPLATE WHEN LOCATING BUMPS TO BE REDUCED

(b)

Baseline more than 25'

Baseline approx. 25 feet

Baseline less than 25 feet

Height of peak is less than 0.3"

Scribed Line

0.3"

1"

BUMP TEMPLATE

FIGURE 15.22 Method of locating bumps to be reduced. (*Source:* Arizona Department of Transportation 1999.)

15.23

fixed-slip and variable-slip methods concern braking with antilock systems. The locked-wheel method is most widely used in North America. Outside the United States, side-force and fixed-slip methods are commonly used.

The locked-wheel system measures friction at a 100 percent slip condition in which the slip speed is equal to the vehicle speed. After the test wheel is locked and has been sliding for a suitable distance, the frictional force is measured and the skid number, SN, is calculated by multiplying the friction value by 100. Fixed-slip devices measure skid resistance at a constant slip, usually between 10 and 20 percent. The test wheel is driven at a lower angular velocity than its free-rolling velocity. These devices measure low-speed skid resistance. Variable-slip devices measure skid resistance at a predetermined set of slip ratios. Some locked-wheel testers can operate in a mode in which they measure the friction as the test wheel changes from free-rolling to the fully locked condition (0 to 100 percent slip). Side-force devices measure skid resistance in the yaw mode, where the wheels are turned at some angle to the direction of travel. The sideways frictional force varies with the magnitude of yaw angle, and it is desirable to use an angle that is relatively insensitive to small variations. The yaw angle is relatively small and the slip speed for these devices, approximately $V \sin \alpha$ (where V is the vehicle speed and α is yaw angle), is therefore also small even though the vehicle speed is high. A fairly simple and widely used version of a yaw mode device is the Mu-Meter.

Another machine widely used to assess skid resistance is the Pendulum Tester, which was developed by the British Transport and Road Research Laboratory (Biles, Sabey, and Cardew 1962). It involves dropping a spring-loaded rubber shoe attached to a pendulum. The results are reported as British Pendulum Numbers (BPN). It is mainly used for laboratory testing and areas such as intersections. Also, because the shoe contacts the pavement at a relatively low speed, the results do not correlate well with locked-wheel trailer test results conducted at 40 mph.

Correlations among various full-scale skid measurement devices are usually very poor, largely because each device measures a different aspect of the frictional interaction between vehicle and pavement (Hass, Hudson, and Zaniewski 1994). For example, side-force devices and fixed-slip devices are low-slip-speed systems and therefore are sensitive to microtexture but insensitive to macrotexture, whereas locked-wheel testers are high-slip systems and mostly reflect the effect of macrotexture (Leu and Henry 1983).

Having recognized the dependence of friction measurement on slip speed researchers have recommended that macrotexture depth be measured with skid resistance. In 1992, the World Road Association (formerly PIARC) conducted extensive pavement friction and texture tests with various types of devices. As a result of these tests, the International Friction Index (IFI) was proposed (PIARC 1995). The IFI is composed of two numbers that jointly describe the full feature of the skid resistance of a pavement: the speed constant (S_p) and the friction number (F60). The speed constant reflects the macrotexture characteristics and is linearly related to a macrotexture measurement (TX). The friction number (F60) is determined from a measurement of friction by:

$$F60 = A + B \text{ FRS } e^{S-60/S_p} + C \text{ TX} \tag{15.11}$$

where FRS = measurement of friction by a device operating at a slip speed (S)
   A, B, C = constants determined for a specific friction device.

The value of C is zero for a smooth-tread tire because the measured friction already reflects the effect of macrotexture. However, the term of C TX is needed for ribbed or patterned test tires because they are relatively insensitive to macrotexture. The values of these constants for the devices in the experiment are reported in ASTM Standard Practice E-1960. One advantage of the IFI is that F60 and S_p are sufficient to describe the friction as a function of slip speed. Another advantage is that F60 will be the same regardless of the slip speed. This allows any device operate at its safe speeds in different conditions and reported on a common scale.

The traditional device to measure pavement macrotexture is the sandpatch method (ASTM Standard Practice E-965). It requires spreading a specified volume of Ottawa sand or glass spheres of a specific sieve range in a circular motion. The mean texture depth (MTD) is obtained by dividing the volume by the area of the circular patch. Another device for characterizing macrotexture is the outflow meter (Henry and Hegmon 1975). This is a transparent cylinder that rests on a rubber annulus placed on the pavement. When the measurement is conducted, the cylinder is filled with water and the valve at the bottom of the cylinder is then opened. The time for the water level to fall by a fixed amount is measured. The time is reported in seconds as the outflow time (OFT). The OFT reportedly has a good correlation with MTD (Henry 2000). The most recent method is to evaluate pavement macrotexture from pavement profile measured by laser-based profilers. The measured pavement profile is used to calculate the mean profile depth (MPD) (ASTM Standard Practice E-1845; ISO Standard 13472). It was found that the MPD is the best macrotexture measure for the prediction of skid resistance (PIARC 1995).

In practice, skid-resistance evaluation, especially for the purpose of assessing pavement future rehabilitation needs, should consider changes on a time and/or traffic basis, as well as on a climatic effect basis. The latter can involve both short and longer periods of time (i.e., rainfall or icing of a short duration versus seasonal climatic changes). On a short-term basis, skid-resistance changes can occur rapidly, usually because of rainfall. On a somewhat longer, seasonal basis, skid resistance may fluctuate. On a still longer basis of, say, several years or several million vehicle passes, most pavements show a continual decrease of skid resistance.

15.5 PAVEMENT DISTRESS SURVEY

Distress surveys measure the type, severity, and density, or extent and location, of various pavement surface distresses. For asphalt pavements, the survey usually includes cracking, patching, potholes, shoving, bleeding, rutting and raveling, etc. The survey for concrete pavements measures cracking, joint deficiencies, various surface defects, and miscellaneous distresses. A pavement distress survey must be conducted following a well-developed survey manual to ensure the consistency and uniformity of the survey data. A manual includes a clear definition of the distress type and a definite procedure to determine or calculate the severity and density of each distress (Shabin and Kohn 1979; Strategic Highway Research Program 1993). Currently, most pavement distress surveys are performed walking along the road or from a moving vehicle. Walking surveys provide the most precise data of the distresses and are generally used for project review. Driving along the road, usually on the shoulder, at slow speed is suitable for pavement network survey. In recent years, several types of automated distress survey equipment based on film, video, or laser technology have been developed. Some agencies have started using these kinds of automated survey equipment (Hass, Hudson, and Zaniewski 1994).

15.6 SUMMARY

Almost all pavement tests are empirically based, and the results produced by the various devices and test methods are affected by many factors. Thus, it is important to follow the standard methods for the calibration and use of the devices. However, the discussion in this chapter focuses primarily on the principles of the test methods; it is not intended to be used as a substitute for reading and following the specifications of the standard test methods. Readers should consult officially published standards, such as the *Book of ASTM Standards*, for detailed information.

15.7 REFERENCES

American Association of State Highway and Transportation Officials (AASHTO). 1990. *Standard Specifications for Transportation Materials and Methods of Sampling and Testing.* Washington, DC: AASHTO.

———. 1993. *AASHTO Guide for Design of Pavement Structures.* Washington, DC: AASHTO.

American Society for Testing and Materials (ASTM). 1999. *Annual Book of ASTM Standards.* West Conshohocken, PA: ASTM.

Arizona Department of Transportation. 1999. *ADOT Materials Testing Manual.* Materials group, Intermodal Transportation Division, Arizona Department of Transportation, Phoenix.

ASTM Standard Test Method E-965. "Measuring Pavement Macrotexture Depth Using a Volumetric Technique." *Book of ASTM Standards,* vol. 04.03, ASTM, West Conshohocken, PA.

ASTM Standard Practice E-1845. "Calculation Pavement Macrotexture Profile Depth." *Book of ASTM Standards,* vol. 04.03. ASTM, West Conshohocken, PA.

ASTM Standard Practice E-1960. "Calculating International Friction Index of a Pavement Surface." *Book of ASTM Standards,* vol. 0403, ASTM, West Conshohocken, PA.

Biles, C. G., B. E. Sabey, and K. H. Cardew. 1962. *Development and Performance of the Portable Skid Resistance Tester.* ASTM Special Technical Publication No. 326.

Brown, E. R., and R. B. Powell. 2001. "A General Overview of Research Efforts at the NCAT Pavement Test Track." Paper presented at International Symposium on Maintenance and Rehabilitation of Pavements and Technological Control, Auburn, AL, July 29–August 1.

Croney D., and P. Croney. 1991. *The Design and Performance of Road Pavements.* New York: McGraw-Hill.

Epps, J. A., R. B. Leahy, T. Mitchell, C. Ashmorem, S. Seeds, S. Alavi, and C. L. Monismith. 1999. "Westrack—the Road to Performance-Related Specifications." Paper presented at First International Conference on Accelerated Pavement Testing, Reno, NV, October 18–20.

Federal Highway Administration (FHWA). 1993. *SHRP FWD Calibration Protocol.* Long-Term Pavement Performance Group, FHWA, McLean, VA, April.

———. 1994. *Pavement Deflection Analysis.* National Highway Institute Course No. 13127, FHWA-HI-94-021, U.S. Department of Transportation, FHWA, Washington, DC, February.

Gillespie, T. D., M. W. Sayers, and L. Segel. 1980. *Calibration of Response-Type Road Roughness Measuring Systems."* NCHRP Report 228, National Research Council, Washington, DC.

Harris, B., M. Buth, and D. Van Deusen. 1994. *Minnesota Road Research Project: Load Response Instrumentation Installation and Testing Procedures.* Mn/Pr-94/01, Minnesota Department of Transportation, Office of Materials Research and Engineering, Maplewood, MN.

Harvey, J. T., J. A. Prozzi, and F. Long. 1999. "Application of CAL/APT Results to Long Life Flexible Pavement Reconstruction." Paper Presented at First International Conference on Accelerated Pavement Testing, Reno, NV, October.

Hass, R., W. R. Hudson, and J. Zaniewski. 1994. *Modern Pavement Management.* Malabar, FL: Malabar.

Henry, J. J. 2000. *Evaluation of the Italgrip Systems.* Highway Innovative Technology Evaluation Center, Civil Engineering Research Foundation, Washington, DC.

Henry, J. J., and R. R. Hegmon, 1975. Pavement Texture Measurement and Evaluation. ASTM Special Technical Publication No. 583. West Conshohocken, PA: American Society for Testing and Materials.

Highway Research Board (HRB). 1962. *The AASHO Road Test: Report 5—Pavement Research.* HRB Special Report 61-E.

International Standards Organization (ISO). 1998. ISO Standards. Geneva, ISO.

ISO Standard 13473. 1998. "Characterization of Pavement Texture by Use of Surface Profiles—Part 1: Determination of Mean Profile Depth." ISO, Geneva.

Karamihas, S. M., and T. D. Gillespie. 1999. *Guideline for Longitudinal Pavement Profile Measurement.* NCHRP Report 434, National Research Council, Washington, DC.

Leu, M. C., and J. J. Henry. 1983. "Prediction of Skid Resistance as a Function of Speed from Pavement Texture." *Transportation Research Record* 946.

Li, Y., J. B. Metcalf, S. A. Romanoschhi, and M. Rasoulian. 1999. "The Performance and Failure Modes of Louisiana Asphalt Pavements with Soil Cement Bases under Full-Scale Accelerated Loading." *Transportation Research Record* 1673, 9–15.

Li, S., and Y. Li. "Mechanistic-Empirical Characterization of Bonding between Ultra-Thin White-Topping and Asphalt Pavement." Paper submitted to ASTM.

Lytton, R. L. 1989. *Backcalculation of Pavement Layer Pavements and Backcalculation of Moduli.* ASTM STP 1026, American Society for Testing and Materials, West Conshohocken, PA, 7–38.

Mamlouk, M. S., W. N. Houston, S. L. Houston, and J. P. Zaniewski. 1988. *Rational Characterization of Pavement Structures Using Deflection Analysis,* vol. 1. report FHWA-AZ88-254, Arizona Department of Transportation, Phoenix.

Metcalf, J. B. 1996. *Application of Full-Scale Accelerated Pavement Testing.* NCHRP Report 235, National Research Council, Washington, DC.

Moyer, R. A. 1934. *Skidding Characteristics of Automobile Tires on Roadway Surfaces and Their Relation to Highway Safety.* Bulletin 120, Iowa Engineering Experiment Station, Ames.

PIARC. 1995. *International PIARC Experiment to Compare and Harmonize Texture and Skid Resistance Measurements.* PIARC Report 01.04.T, The World Road Association, Paris.

Royal Institute of Technology. 1980. *Bulletin 1980: 8 Testing Different FWD Loading Times.* Royal Institute of Technology, Department of Highway Engineering, Stockholm, Sweden.

———. 1998. *Techniques for Pavement Rehabilitation—Reference Manual (Sixth Edition).* National Highway Institute Course No. 13108, FHWA-HI-98-033, U.S. Department of Transportation, Federal Highway Administration, Washington, DC, August.

Sayers, M. W. 1995. "On the Calculation of International Roughness Index from Longitudinal Road Profile." *Transportation Research Record* 1501, 1–12.

Sayers, M. S., and S. M. Karamihas. 1998. *The Little Book of Profiling—Basic Information about Measuring and Interpreting Road Profiles.* Ann Arbor: University of Michigan Transportation Research Institute.

Sayers, M. W., T. O. Gillespie, and W. D. O. Paterson. 1986. *Guideline for Conducting and Calibrating Road Roughness Measurements.* World Bank Technical Paper Number 46, World Bank, Washington, DC.

Sayers, M. W., T. D. Gillespie, and C. A. Que011. 1986. *The International Road Roughness Experiment—Establishing Correlation and Calibration Standard for Measurements.* World Bank Technical Paper Number 45, World Bank, Washington, DC.

Scrivner, F. H., R. Peohl, W. M. Moore, and M. B. Phillips. 1969. *Detecting Seasonal Changes in Load-Carrying Capabilities of Flexible Pavements.* National Cooperative Highway Research Program Report 76, National Research Council, Highway Research Board, Washington, DC.

Shahin, M. Y., and S. D. Kohn. 1979. *Development of Pavement Condition Rating Procedures for Roads, Streets, and Parking Lots—Volume I Condition Rating Procedure.* Technical Report M-268, Construction Engineering Research Laboratory, U.S. Corps of Engineers.

Sharp, K. G. 1991. "Australian Experience in Full-Scale Pavement Testing Using the Accelerated Loading Facility." *Journal of the Australian Road Research Board* 21(3):23–32.

Smith, R. E., and R. L. Lytton. 1984. Synthesis Study of Nondestructive Testing Devices for Use in Overlay Thickness Design of Flexible Pavements. Report No. FHWA/RD-83/097, U.S. Department of Transportation, Federal Highway Administration, Washington, DC, April.

Strategic Highway Research Program. 1993. *Distress Identification Manual for the Long-Term Pavement Performance Project.* SHRP-P-338, National Research Council, The Strategic Highway Research Program, Washington, DC.

Ullidtz, P., and N. F. Coetzee. 1995. "Analytical Procedures in Nondestructive Testing Pavement Evaluation." *Transportation Research Record* 1482, 61–66.

Way, G. B., J. F. Eisenberg, J. Delton, and J. Lawson. 1984. Structural Overlay Design Method for Arizona." Paper presented at AAPT annual conference, AAPT, Scottsdale, AZ, April 9–11.

P · A · R · T · III

SAFETY, NOISE, AND AIR QUALITY

CHAPTER 16
TRAFFIC SAFETY

Rune Elvik
Institute of Transport Economics, Norwegian Center for
Transportation Research, Oslo, Norway

16.1 INTRODUCTION

This chapter presents essential elements of traffic safety analysis. The topics covered in this chapter include:

1. Basic concepts of road accident statistics
2. Safety performance functions and the empirical Bayes method for estimating safety
3. Identifying and analyzing hazardous road locations
4. The effects on road safety of some common traffic engineering measures
5. How to assess the quality of road safety evaluation studies
6. Formal techniques for setting priorities for safety treatments

16.2 BASIC CONCEPTS OF ROAD ACCIDENT STATISTICS

The basic concepts of road accident statistics were developed by the French mathematician Simeon Denis Poisson more than 150 years ago. Poisson investigated the properties of binomial trials. A binomial trial is an experiment that has two possible outcomes: success or failure. The probability of success is the same at each trial. The outcome of a trial is independent of other trials. Repeated tosses of a coin are an example of a sequence of binomial trials. If a coin is tossed four times, the outcomes can be zero heads, heads once, heads twice, heads three times, or heads all four times. At each trial, the probability of heads (p) is 50 percent. When a coin is tossed four times, the probability of getting heads n times ($n = 0, 1, \ldots, 4$) is:

Heads 0 times:	0.0625
Heads 1 time:	0.2500
Heads 2 times:	0.3750
Heads 3 times:	0.2500
Heads 4 times:	0.0625

The expected number of heads in N trials is $N \cdot p$. Since $p = 0.5$, the expected number of heads when $N = 4$ is 2. Poisson studied what happened to the binomial probability distribution when the number of trials N became very large while at the same time the probability of failure p became very low. Denote the expected value in N trials by λ. Poisson found that the probability of x failures in N trials could be adequately described by the following probability function, which bears his name:

$$P\,(X = x) = \frac{\lambda^x e^\lambda}{x!} \qquad (16.1)$$

The parameter lambda (λ) indicates the expected value of the random variable X, x is a specific value of this variable, e is the base of the natural logarithms ($e = 2.71828$), and $x!$ is the number of permutations of x. If, for example, $x = 3$, then $x! = 1 \cdot 2 \cdot 3$. If $x = 0$, then $x! = 1$. $\lambda = N \cdot p$, when N is very large and p is very small. A random variable is a variable that represents the possible outcomes of a chance process. Translating these abstract terms to a language more familiar to traffic engineers gives:

Expected number of accidents (λ) = Exposure (N) $\cdot$ Accident rate (p)

Accident rate is traditionally defined as the number of accidents per unit of exposure:

$$\text{Accident rate} = \frac{\text{Number of accidents}}{\text{Unit of exposure}}$$

Exposure denotes the number of trials, and a commonly used unit of exposure in traffic engineering is one kilometer of travel (Hakkert and Braimaister 2002). The idea, deeply rooted in probability theory, that the expected number of accidents depends on exposure and accident rate, is perhaps the source of the assumption traditionally made in traffic engineering that one can account for the effects of traffic volume on accidents by using accident rates. However, as will be discussed in the next section, this assumption is no longer tenable.

Definitions of some key concepts are given in Figure 16.1.

It is essential to keep in mind that accidents are subject to random variation. The variation observed in counts of accidents will nearly always be a mixture of pure random variation and systematic variation. Two questions immediately come to mind:

1. How can we know if the variation in the count of accidents in a set of study units is not just random?

Expected number of accidents

The mean number of accidents expected to occur in the long run for a given combination of exposure and accident rate.

Random variation in the number of accidents—regression-to-the-mean

Variation in the recorded number of accidents around a given expected number of accidents. Regression-to-the-mean is the return of an abnormally high or low recorded number of accidents to figures closer to the expected number.

Systematic variation in the number of accidents

Variation in the expected number of accidents (across time, space, modes of transportation, etc.).

Exposure

The volume of an activity exposed to risk. Risk is the product of the probability of an unwanted event (in our case an accident) and the consequences of that event (in our case accident severity).

FIGURE 16.1 Definition of key concepts of accident statistics.

2. If accidents are found to occur at random, is it still possible to reduce the number of accidents?

Table 16.1 presents data that can be used to answer both of these questions. The table show the distribution of highway-railroad grade crossings on public roads in Norway by the number of accidents recorded during the 10 years 1959–1968 (Amundsen and Christensen 1973). There are two categories of grade crossings: those protected by signals only, and those protected by automatic gates. For each category, the first column shows the actual distribution of crossings by number of accidents. There were, for example, 48 crossings protected by signals that had 1 accident during the years 1959–1968. The mean number of accidents per crossing was 0.515, and the variance was 0.545. Next to the actual distribution of accidents is shown the distribution that would be expected if accidents occurred entirely at random. This distribution, the Poisson distribution, was estimated by means of equation (16.1) by inserting the value 0.515 for the mean (λ). It can be seen that the Poisson distribution is very similar to the actual distribution of accidents. Whether the two distributions really differ can be tested by means of a chi-square test. This statistical test is described in any elementary textbook in statistics. For grade crossings protected by signals, the value of chi-square is 1.395. There are two degrees of freedom (the number of categories (4) minus 1, minus the number of parameters of the Poisson distribution). The exact P-value is 0.498, which means that the actual distribution of accidents does not differ significantly from the Poisson distribution. An identical analysis for grade crossings protected by automatic gates leads to the same conclusion.

It is fair to conclude on the basis of these tests that accidents occurred at random in grade crossings protected by signals only. Yet had all these grade crossings been protected by automatic gates, the mean number of accidents would have been reduced by more than 50 percent. The mean number of accidents for grade crossings protected by automatic gates was 0.241, which is less than half the mean number of accidents for grade crossings protected by signals only (0.515). A random distribution of accidents in a sample of study units does not necessarily mean that it is impossible to reduce the mean number of accidents. It just

TABLE 16.1   Accident Data for Highway-Railroad Grade Crossings in Norway 1959–1968

Count of accidents	Highway-railroad grade crossings by type of protective device			
	Signals only		Automatic gates	
	Actual distribution	Poisson distribution	Actual distribution	Poisson distribution
0	108	105	71	68
1	48	54	13	16
2	17	14	1	2
3	3	3	2	1
N	176	176	87	87
Mean	0.517		0.241	
Variance	0.545		0.344	
Chi-square		1.395		2.194
Degrees of freedom		2		2
Exact P-value		0.498		0.334

Source: Amundsen and Christensen (1973).

means that it is impossible to identify one of the study units as having a higher expected number of accidents than another.

In the Poisson distribution, the variance is equal to the mean. Looking at Table 16.1, the variance for grade crossings protected by signals only was 0.545, which is close to the mean number of accidents (0.515). For the grade crossings protected by automatic gates, however, the variance (0.344) is somewhat larger than the mean (0.241). Since, by definition, the size of the purely random variation in the count of accidents equals the mean number of accidents, the total variation in the count of accidents found in a sample of study units can be decomposed into random variation and systematic variation (Hauer 1997):

$$\text{Total variation} = \text{Random variation} + \text{Systematic variation} \tag{16.2}$$

There is systematic variation in number of accidents whenever the variance exceeds the mean. This is usually referred to as overdispersion. The amount of overdispersion found in a data set can be described in terms of the overdispersion parameter, which is defined as follows:

$$\text{Var}(x) = \lambda \cdot (1 + \mu\lambda) \tag{16.3}$$

Solving this with respect to the overdispersion parameter gives:

$$\mu = \frac{\dfrac{\text{Var}(x)}{\lambda} - 1}{\lambda} \tag{16.4}$$

For the distributions listed in Table 16.1, the overdispersion parameter can be estimated to 0.105 for grade crossings protected by signals and 1.773 for grade crossings protected by gates. For grade crossings protected by signals only, the total variance consists of 95 percent random variation and 5 percent systematic variation. For grade crossings protected by gates, there is 70 percent random variation in accidents, 30 percent systematic variation.

A stepwise approach to the statistical analysis of accident occurrence within a system for which a traffic engineer is professionally responsible is suggested in Figure 16.2. The first

---

Step 1: Define suitable sets of study units

Examples of study units used in traffic engineering studies include road sections of a given length, intersections, driveways, horizontal curves, highway-railroad grade crossings, bridges, tunnels.

Step 2: Analyze distribution of accidents in each set of study units

For each set of study units defined, the distribution of accidents should be analyzed with respect to the mean number of accidents and the variance.

Step 3: Identify the safety performance function in each set of study units

A safety performance function is an equation that describes the sources of systematic variation in accidents, fitted by means of appropriate multivariate techniques of analysis.

Step 4: Estimate safety for each study unit using the empirical Bayes method

The empirical Bayes method combines information from two clues to safety and can be used to estimate the expected number of accidents for each study unit.

Step 5: Define hazardous road locations and identify them statistically

A hazardous road location is any study unit for which the expected number of accidents is abnormally high.

FIGURE 16.2 Stepwise approach to statistical analysis of accidents.

step of analysis is to define sets of study units. These sets are often categories of roadway elements, such as road sections, intersections, or curves. In a city there may, for example, be a few thousand intersections. It is important that the study units be identically defined and can be counted.

For each set of study units, the distribution of accidents should be analyzed as illustrated above for grade crossings (step 2 in Figure 16.2). The count of accidents should represent at least a few years of data. The objective of the analysis is to determine the amount of systematic variation in the number of accidents. If there is very little systematic variation, as was the case for grade crossings protected by signals only, there is little point in continuing to the next steps of analysis indicated in Figure 16.2. If, on the other hand, the number of accidents is found to contain significant systematic variation, the next step of analysis is to identify sources of systematic variation (step 3 in Figure 16.2). The effects of the variables that explain systematic variation in the number of accidents are usually summarized in terms of a safety performance function, whose general form will be discussed in the next section. Once a safety performance function has been estimated, the information that this function gives about safety can be combined with the accident records for each study unit to estimate the expected number of accidents for each study unit (step 4 in Figure 16.2). Combining these two sources of information about safety is the essential feature of the empirical Bayes method to the estimation of road safety, which is illustrated in the next section.

When the expected number of accident has been estimated for each study unit, a distribution of study units according to the expected number of accidents can be formed and hazardous road locations can be identified (step 5 in Figure 16.2). Several definitions of hazardous road locations can be imagined; common to all definitions is that they are intended to help identify statistically road locations with a high expected number of accidents.

16.3   SAFETY PERFORMANCE FUNCTIONS: THE EMPIRICAL BAYES METHOD FOR ESTIMATING ROAD SAFETY

A safety performance function is any mathematical function that relates the normal, expected number of accidents to a set of explanatory variables. The following form for the safety performance function is widely applied:

$$E(\lambda) = \alpha Q^{\beta} \, e^{\sum \gamma_i x_i} \tag{16.5}$$

The estimated expected number of accidents $E(\lambda)$ is a function of traffic volume Q and a set of risk factors X_i ($i = 1, 2, 3, \ldots n$). The effect of traffic volume on accidents is modeled in terms of an elasticity, that is a power β to which traffic volume is raised (Hauer 1995). This elasticity shows the percentage change of the expected number of accidents, which is associated with a 1 percent change in traffic volume. If the value of β is 1.0, the number of accidents is proportional to traffic volume, as traditionally assumed when using accident rates in road safety analysis. If the value of β is less than 1, the number of accidents increases by a smaller percentage than traffic volume. If the value of β is greater than 1, the number of accidents increases by a greater percentage than traffic volume.

A number of recent studies (Fridstrøm et al. 1995; Persaud and Mucsi 1995; Mountain, Fawaz, and Jarrett 1996; Fridstrøm 1999) have estimated safety performance functions for total accidents, fatal and injury accidents, and specific types of accidents (such as single-vehicle and multiple-vehicle). It is typically found that the value of β is less than 1 for fatal accidents, close to 1 for injury accidents and in some cases greater than 1 for property damage-only accidents. Figure 16.3 shows functions based on a value for β of 0.7 for fatal accidents, 0.9 for injury accidents, and 1.1 for property-damage-only accidents.

The effects of various risk factors that influence the probability of accidents, given exposure, are generally modeled as an exponential function, that is, as e (the base of natural

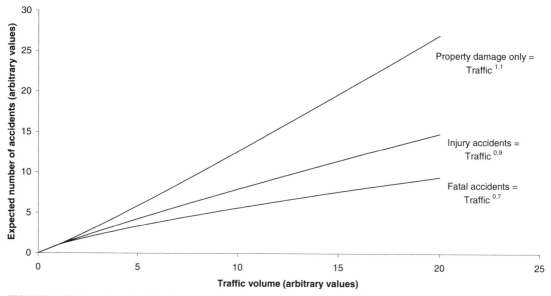

FIGURE 16.3 Typical relationships between traffic volume and the expected number of accidents.

logarithms, see section 16.2) raised to a sum of the product of coefficients γ_i and values of the variables x_i, denoting risk factors. A model of the form shown in equation (16.5) can be fitted with several commercially available computer software packages. When estimating a safety performance function, it is important to specify the distribution of the residual terms correctly. The residual term of a model is the part of systematic variation in accident counts, which is not explained by the model. If a model explains all the systematic variation in accident counts there is in a data set, the residuals will by definition contain random variation only and can be specified as Poisson distributed. Usually, however, a model will not be able to explain all systematic variation in accident counts. The residuals will then contain some overdispersion, which can usually be adequately described by the negative binomial distribution.

Two questions need to be answered with respect to the use of safety performance functions:

1. How should the variables to be included in a model be selected?

2. How can one determine the success of a model in explaining accidents?

The choice of variables to be included in a safety performance function will often be dictated by the availability of data. In traffic engineering analyses, variables referring to highway design and traffic control would normally be included. Such variables include number of lanes, lane width, number of intersections or driveways per kilometer of road, type of roadside terrain, horizontal curvature, and speed limit. The success of a model in explaining accidents can be evaluated by comparing the overdispersion parameter of a fitted model to the overdispersion parameter in the original data set. Table 16.2 presents a data set for national highways in Norway, showing the number of fatalities per kilometer of road during 1993–2000 (Ragnøy, Christensen, and Elvik 2002). The mean number of road accident fatalities per kilometer of road was 0.0646, and variance was 0.0976. The overdispersion parameter can be estimated to 7.91. A multivariate accident model was fitted to the data, assuming a negative binomial distribution for the residuals. The coefficients estimated for

TABLE 16.2 Number of Road Accident Fatalities per Kilometer of Road, National Highways, Norway 1993–2000.

Number of fatalities	Distribution of road sections by number of fatalities			Safety performance function	
	Actual	Poisson	Negative binomial	Explanatory variables	Coefficient
0	19,957	19,728	19,974	Constant	−7.154
1	895	1,274	854	Ln(AADT)	0.842
2	135	41	163	Speed limit 50 km/hr	Reference category
3	43	1	39	Speed limit 60 km/hr	−0.020
4	9	0	10	Speed limit 70 km/hr	0.385
5	3	0	3	Speed limit 80 km/hr	0.172
6	1	0	1	Speed limit 90 km/hr, rural road	0.090
7	0	0	0	Speed limit 90 km/hr, class B road	0.610
8	1	0	0	Speed limit 90 km/hr, class A road	0.879
N	21,044	21,044	21,044	Number of lanes	−1.967
				Number of intersections/km	0.082
				Dummy for trunk road	0.255
Mean	0.0646				
Variance	0.0976			Estimated variance	0.0745
Over-dispersion parameter	7.91			Over-dispersion parameter	2.39

Source: Ragnøy, Christensen, and Elvik (2002).

this model are listed in Table 16.2. The overdispersion parameter for the model was 2.39. Inserting this into equation (16.3) gives an estimated variance of 0.0745. The contributions of various factors to the observed variance in accident counts can be determined as follows:

Random variation = 0.0646/0.0976 = 0.662 = 66.2% of all variance

Systematic variation = (0.0976 − 0.0646)/0.0976 = 0.338 = 33.8%

Systematic variation explained by safety performance function = (0.0976 − 0.0745)/0.0976 = 0.237 = 23.7%

Remaining systematic variation not explained by safety performance function = (0.0745 − 0.0646)/0.0976 = 0.101 = 10.1%

The safety performance function explains 0.237/0.338 = 0.701 = 70.1% of all systematic variation found in this data set.

The coefficients given in the safety performance functions show the effects of each variable, controlling for the other variables included in the model. By inserting these coefficients into equation (16.5), the normal expected number of fatalities during eight years for a road section of 1 kilometer can be estimated for any conceivable combination of values observed for the explanatory variables.

What is the best estimate of the long-term expected number of accidents or accident victims for a given roadway element, given the fact that we know some but not all of the factors affecting accident occurrence? According to the empirical Bayes method (Hauer 1997), the best estimate of safety is obtained by combining two sources of information: (1) the accident record for a given site, and (2) a safety performance function, showing how various factors affect accident occurrence. Denote by R the recorded number of accidents and by λ the normal, expected number of accidents as estimated by a safety performance function. The best estimate of the expected number of accidents for a given site is then:

$$E(\lambda/r) = \alpha \cdot \lambda + (1 - \alpha) \cdot r \qquad (16.6)$$

The parameter α determines the weight given to the estimated normal number of accidents for similar sites when combining it with the recorded number of accidents in order to estimate the expected number of accidents for a particular site. The best estimate of α is:

$$\alpha = \frac{1}{1 + \dfrac{\lambda}{k}} \qquad (16.7)$$

λ is the normal expected number of accidents for this site, estimated by means of a safety performance function, and k is the inverse value of the overdispersion parameter of this function, that is, $1/\mu$. To illustrate the use of the empirical Bayes method, suppose that the normal expected number of accidents for a 1-kilometer road section during a period of eight years has been estimated by means of a safety performance function to be 3.73. The overdispersion parameter for the safety performance function is 0.3345; hence k is 2.99. The weight to be given to the estimate based on the safety performance function thus becomes $1/[1 + (3.73/2.99)] = 0.445$. Seven accidents were recorded. The long-term expected number of accidents is estimated as:

$$E(\lambda|r) = 0.445 \cdot 3.73 + (1 - 0.445) \cdot 7 = 5.54.$$

The interpretation of the three different estimates of safety can be explained as follows: 3.73 is the number of accidents one would normally expect to occur at a similar site, that is, one that has the same traffic volume, the same speed limit, the same number of lanes, etc., as the site we are considering. Seven accidents were recorded. Part of the difference between the recorded and normal number of accidents for this type of site is due to random variation. An abnormally high number of accidents due to chance cannot be expected to continue; a certain regression-to-the-mean must be expected. In the example given above, the regression-to-the-mean expected to occur in a subsequent 8-year period is $(7 - 5.54)/7 = 0.209 = 20.9\%$. The difference between the site-specific expected number of accidents (5.54) and the normal, expected number of accidents for similar sites (3.73) can be interpreted as an effect of local risk factors for the site, causing it to have a higher expected number of accidents than similar sites.

Applying the empirical Bayes method to a site that had 0 recorded accidents, but was otherwise identical to the site used as an example above, gives a site-specific expected number of accidents during 8 years of 1.66 (the normal, expected number of accidents for similar sites was 3.73). For this site, the difference between the site-specific expected number of accidents and the normal, expected number of accidents for similar sites can be interpreted as the effect of local safety factors, that is, factors causing the site to be safer than otherwise similar sites.

16.4 IDENTIFYING AND ANALYZING HAZARDOUS ROAD LOCATIONS

No standard definition exists of a hazardous road location. The following definition conforms to common usage of the term (Elvik 1988): A hazardous road location is any site at which the site-specific expected number of accidents is higher than for similar sites, due to local risk factors present at the site. Hazardous road locations should be identified in terms of the expected number of accidents, not the recorded number of accidents. An abnormally high recorded number of accidents could to a large extent be the result of random variation and does not necessarily mean that a site has a high expected number of accidents.

The empirical Bayes method illustrated in section 16.3 offers an ideal way of identifying hazardous road locations. For each road location, one would estimate the normal, expected number of accidents by means of a safety performance function. This would then be combined with the accident record for each site, yielding an estimate of the site-specific expected number of accidents. A road location would be considered hazardous if the site-specific expected number of accidents was (substantially) higher than the normal, expected number of accidents for similar sites. This approach to the identification of hazardous road locations conforms to the definition of the concept given above and utilizes all information that gives clues to safety.

Once hazardous road locations have been identified statistically, the task of finding remedies to reduce the hazards can be started. The traditional approach to this task has been to perform a detailed analysis of accidents, producing collision diagrams, analyzing the distribution of accidents by time of the day, road surface condition, type of accident, and so on. In addition to analyzing accidents, elements of the traditional approach to black spot treatment include site visits, detailed traffic counts, filming the site, and driving through it from all directions. Although this kind of detailed examination can no doubt give valuable information, there are inherent limitations of the approach that render it less than perfectly reliable as a means to test hypotheses about why there is a concentration of accidents at a particular site.

Accident analysis at hazardous road locations amounts to proposing hypotheses based on an analysis of known data, which means that the data that generated the hypotheses cannot also be used for testing them. Take the analysis presented in Table 16.3 as an example. Eight accidents were recorded at a hazardous road location. The analysis shows that five of the eight accidents were pedestrian accidents, whereas one would normally expect one in eight accidents to involve pedestrians. If the distribution of accidents by type is modeled as a binomial trial (each accident is either a pedestrian accident or nonpedestrian accident), it is found that recording five pedestrian accidents in a total of eight accidents is a highly unlikely outcome, given that the initial probability of a pedestrian accidents is 0.125 (the probability of a nonpedestrian accident is 0.875).

On the whole, the predominant accident pattern found in Table 16.3, pedestrian accidents occurring at night on a wet road surface, suggests that local risk factors related to the amount of pedestrian traffic, road surface friction, and visual obstructions may be present at the location. Yet a more careful investigation would be needed in order to determine whether the factors suggested are actually responsible for the abnormally high number of pedestrian accidents at this particular location.

Convincing evidence showing that current methods for analyzing accidents at hazardous road locations are able to separate the true black spots from the false ones does not exist. In a very interesting paper, Jarrett, Abbess, and Wright (1988) estimated the regression-to-the-mean effect for two samples of road accident black spots. These black spots had been identified according to the recorded number of accidents only, not the expected number of accidents. One of the samples consisted of black spots that had been selected for treatment but for which the treatment had, for various reasons, been deferred or given up. Accident analyses had been made for these candidate sites, with a view to identifying the dominant accident pattern for the purpose of selecting an appropriate treatment. If these analyses were really able to tell the true black spots from the false ones, one would expect the regression-to-the-mean effect to be smaller for the candidate sites than for black spots in general. But Jarrett, Abbess, and Wright found no evidence for this. The number of accidents regressed to a lower mean value by almost the same percentage at the candidate sites as for black spots in general. In other words, accident analyses did not correctly identify those sites that had a higher long-term expected number of accidents.

These results are supported indirectly in a study reported by Elvik (1997). This study investigated the effects attributed to black spot treatment depending on which confounding factors evaluation studies controlled for. It was found that the effect of black spot treatment was zero in the best-controlled studies. Belief in the effectiveness of black spot treatment

TABLE 16.3 Hypothetical Results of Accident Analysis at a Hazardous Road Location

Accident number	Type of accident	Time of day	Road surface	Vehicles involved	Alcohol involved	Excessive speed	Failure to see
1	Pedestrian	11 PM	Wet	Car	Yes, pedestrian	Yes	Yes
2	Rear-end	10 AM	Wet	Truck	No	No	No
3	Rear-end	5 PM	Dry	Car	No	No	No
4	Pedestrian	8 PM	Dry	Car	No	Yes	No
5	Pedestrian	9 PM	Wet	Car	Yes, pedestrian	No	Yes
6	Pedestrian	11 AM	Wet	Car	No	Yes	Yes
7	Overturning	1 PM	Dry	Motorcycle	No	Yes	No
8	Pedestrian	11 PM	Wet	Truck	Yes, pedestrian	No	Yes
Key finding Normal value	5 pedestrian 1 pedestrian	4 in evening 2 in evening	5 on wet road 2 on wet road	Nothing abnormal	3 with alcohol 1 with alcohol	4 speeders 3 speeders	4 did not see 2 did not see
P-value (binomial)	0.0011	0.0865	0.0231	Not tested	0.0561	0.2112	0.0865
Predominant accident pattern	The predominant type of accident is a pedestrian accident at night on a wet road surface, in which the parties did not see each other. Some over-involvement of alcohol among pedestrians.						

seems at least in part to have rested on an uncritical acceptance of poor evaluation studies whose findings may to a great extent reflect the effects of uncontrolled confounding factors.

A better way to analyze hazardous road locations for the purpose of identifying contributing risk factors and propose treatment is by means of a blinded matched-pair comparison. The essential elements of this approach can be stated as follows:

1. For each hazardous road location, find a safer-than-average comparison location, matched as closely to the hazardous road location as possible with respect to variables included in the safety performance function.

2. For each matched pair of sites, search for local risk factors or safety factors from a list of factors drawn up on the basis of the analysis of accidents at the hazardous road location.

3. Blind analysts to accident records. Analysts should not know which site was hazardous and which site was safer than average.

The use of this approach is shown in Table 16.4. Hazardous and safe sites are matched in pairs according to the values observed for the variables included in the safety performance function. Two matched pairs are shown in Table 16.4. Once the pairs have been formed, each site is inspected and data collected regarding local risk factors. A sample of such data, not necessarily exhaustive, is shown in Table 16.4.

In the first pair of sites it was found that road surface friction was significantly worse, there were more pedestrians crossing the road, and there were more sources of visual obstruction at the hazardous site than at the safe site. This information confirms the hypotheses regarding contributing factors proposed on the basis of the analysis of accidents. The analysis has therefore successfully identified local risk factors. Keep in mind that the analysts identifying risk factors should be blinded to accident records to prevent their knowledge of accident records from biasing their observations.

TABLE 16.4 Verification of Traditional Accident Analysis by Identification of Risk Factors Contributing to Accidents

	Case 1: Local risk factors successfully identified		Case 2: Local risk factors not identified	
	Hazardous	Safe	Hazardous	Safe
Matching variables	AADT, speed limit, number of lanes, number of intersections, trunk road status = nearly same values observed for both sites		AADT, speed limit, number of lanes, number of intersections, trunk road status = nearly same values observed for both sites	
Local risk factors (sample):				
Road surface friction—dry	0.70	0.82	0.78	0.77
Road surface friction—wet	0.25	0.48	0.47	0.49
Pedestrians crossing per day	2,500	1,000	1,200	1,250
Sources of visual obstruction	5	2	3	3
Minimum sight distance (m)	100	155	110	115
Driveways per km of road	2	2	0	0
Public bar nearby	Yes	No	No	No
Accident records:	8 accidents in total, of which 5 involving pedestrians on a wet road surface	0 accidents	7 accidents in total, but no clear pattern	0 accidents

The other case shown in Table 16.4 was less successful. It turned out that there were no differences between the hazardous and the safe site with respect to the risk factors surveyed. Hence, accidents must be attributed to other factors, for example a widespread violation of speed limits or other traffic control devices.

The implications for safety treatment of a successful analysis of local risk factors at a hazardous road location are rarely obvious. For the successful case presented in Table 16.4, one can imagine many treatments that might be effective, including lowering the speed limit to allow pedestrians and cars more time for observation, removing obstacles to vision, improving road surface friction, providing or upgrading road lighting, upgrading pedestrian crossing facilities, conducting random breath testing (there was a bar nearby), or stepping up police enforcement in general. It is not obvious which of these measures would be the most cost-effective; choice of treatment should rely on a detailed analysis of the cost-effectiveness of alternative treatments.

16.5 EFFECTS ON ROAD SAFETY OF SELECTED TRAFFIC ENGINEERING MEASURES

A very large body of information exists concerning the effects on road safety of traffic engineering measures. Unfortunately, as pointed out by Elvik (2002b) and Hauer (2002), not all of this information can be trusted; in fact, some of it deserves to be labeled disinformation. The next section of this chapter attempts to explain why this is so and provides some advice to traffic engineers about how to read evaluation studies critically and how to extract information from such studies. Information is given on the effects of selected highway traffic engineering and traffic control measures (Table 16.5) for which meta-analyses summarizing evidence from several studies have been reported.

Amundsen and Elvik (2002) summarized evidence on the road safety effects of constructing or upgrading urban arterial roads. All the studies included in their review controlled for regression-to-the-mean and long-term trends in the number of accidents. It was found that the construction of new urban arterial roads tends to induce new traffic, which offsets the reduction of the accident rate, resulting in very small changes in the number of accidents. Upgrading urban arterial roads by means of lane additions and a median is associated with a large reduction of the number of accidents. Bypass roads around small towns have been found to reduce the number of accidents by about 25 percent (Elvik, Amundsen, and Hofset 2001).

The conversion of intersections to roundabouts (Elvik 2003), the installation of road lighting (Elvik 1995a) and the provision of guardrails (Elvik 1995b) reduce both the number and severity of accidents. The effects of area-wide urban traffic calming schemes (Bunn et al. 2003) and of treatment of horizontal curves (Elvik 2002a) are somewhat more uncertain. The effects of speed limits can be appraised in terms of the effects on accidents of certain percentage changes in the mean speed of traffic. On the average, a 10 km/hr change in the speed limit can be expected to result in a change of 3–4 km/hr in the mean speed of traffic (Elvik, Mysen, and Vaa 1997). Nearly all studies evaluating the effects on accidents of changes in speed have found that increases in speed are associated with increases in the number of fatal and injury accidents, whereas reductions in speed are associated with reductions in the number of fatal and injury accidents. The effect of changes in speed on property-damage-only accidents is less well known.

16.6 HOW TO ASSESS THE QUALITY OF ROAD SAFETY EVALUATION STUDIES

As noted above, an enormous number of road safety evaluation studies exist. How can users of this research assess whether or not to trust the findings reported by these studies? The

TABLE 16.5 Best Estimates of the Effects on Accidents of Selected Highway and Traffic Engineering Measures

Measure	Accidents affected	Fatal accidents		Injury accidents		Property-damage-only accidents	
		Best estimate	95% CI	Best estimate	95% CI	Best estimate	95% CI
New urban arterial roads	All accidents	−14	(−50, +50)	−1	(−9, +8)	Not available	Not available
Lane additions and median	All accidents	−71	(−91, +35)	−51	(−65, −32)	+15	(+8, +22)
Bypass roads	All accidents	−17	(−56, +57)	−25	(−33, −16)	−27	(−38, −13)
Roundabouts (yield on approaches)	At intersections	−81	(−93, −50)	−47	(−56, −34)	+1	(−19, +25)
Road lighting (previously unlit)	In darkness	−64	(−74, −50)	−30	(−35, −25)	−18	(−22, −13)
Upgrading road lighting	In darkness	−50	(−79, +15)	−32	(−39, −25)	−17	(−22, −11)
Guard rails along roadside	Run-off-road	−44	(−56, −32)	−47	(−52, −41)	−7	(−35, +33)
Median guard rails	Median crossings	−43	(−53, −31)	−30	(−36, −23)	+24	(+21, +27)
Area wide urban traffic calming	All accidents	−30	(−78, +128)	−11	(−21, −1)	−5	(−19, +11)
Treatment of hazardous curves	In curves	Not available	Not available	−16	(−35, +9)	−18	(−44, +21)
Speed limit (mean speed − 5%)	All accidents	−6	(−11, −2)	−10	(−11, −9)	Not available	Not available
Speed limit (mean speed −10%)	All accidents	−24	(−32, −15)	−17	(−19, −15)	Not available	Not available
Speed limit (mean speed +5%)	All accidents	+17	(+13, +22)	+19	(+16, +21)	Not available	Not available
Speed limit (mean speed + 10%)	All accidents	Not available	Not available	+8	(−2, +20)	Not available	Not available

Source: See text, section 16.5.

first thing a reader should look for is a description of study design. If a study employed an experimental design involving the random assignment of safety treatment to study units, it is in most cases safe to assume that the study shows the effects of the safety treatment only and has, by way of randomization, controlled for all confounding factors. Most road safety evaluation studies do not employ an experimental study design. Various versions of before-and-after designs are common. Hauer (1997) gives a comprehensive guide to the critical assessment of before-and-after studies.

One may assess the credibility of before-and-after studies by checking whether the studies have controlled for important confounding factors such as (Elvik 2002a):

1. Regression-to-the-mean
2. Long-term trends affecting the number of accidents or injured road users (during several before periods)
3. General changes of the number of accidents from before to after the road safety measure is introduced (from one before period to one after period)
4. Changes in traffic volume
5. Any other specific events introduced at the same time as the road safety measure

A confounding variable is any exogenous (i.e., not influenced by the road safety measure itself) variable affecting the number of accidents or injuries whose effects, if not estimated, can be mixed up with effects of the measure being evaluated. Two main approaches can be taken to control these variables: (1) estimate the effects of a confounding factor statistically, and (2) use a comparison group. In many cases, both approaches will be used in the same study. Regression-to-the-mean is usually controlled for by means of statistical estimation. The effects on accidents of changes in traffic volume are often also estimated statistically. A comparison group is used to control for all confounding factors whose effects cannot be estimated statistically.

The more confounding factors an evaluation study has controlled for, the less likely are the results to have been caused by confounding factors not controlled for by the study. The results of before-and-after studies of road safety measures can be greatly influenced by the approach taken to controlling for confounding factors in such studies. There is a tendency for studies that do not control for any confounding factors to exaggerate the effects of the road safety measure being evaluated. In some cases this exaggeration can be substantial, amounting to a difference of some 20–30 percent in the size of the effect attributed to the road safety measure.

Readers of observational before-and-after studies of road safety measures should always pay very careful attention to what these studies say about control of confounding factors. If a study does not state explicitly that it controlled for a certain confounding factor, one is almost always right in believing that the study did not control for that factor. Simple before-and-after studies, which do not control for any confounding factors, should never be trusted.

Provided a study has controlled adequately, but perhaps not perfectly, for important confounding factors, the most important aspect of study quality is the size and diversity of the accident sample. The larger the accident sample, the more precise are the estimates of the effects of the safety measure. Moreover, a large accident sample may permit estimates to be made of the effect of a measure for different types of accidents, different levels of accident severity, and different types of traffic environment. A common problem in many studies evaluating traffic engineering safety treatments is that little or no information is given concerning how the sample of sites was obtained. This is regrettable, as it is then difficult to know the generality of study findings.

16.7 FORMAL TECHNIQUES FOR PRIORITY SETTING OF ROAD SAFETY MEASURES

Broadly speaking, three formal techniques have been developed to help policy-makers choose the most effective road safety measures from a set of potentially effective measures:

1. Cost-effectiveness analysis, which seeks to identify those road safety measures that give the largest reductions in accidents or injuries per dollar spent to implement the measures.

2. Cost-benefit analysis, which seeks to identify those road safety measures for which the benefits are greater than the costs. All benefits are converted into monetary terms.

3. Multiattribute utility analysis, which seeks to maximize the overall attainment of a set of goals, for each of which the utility function of the responsible decision-makers is determined.

The simplest of these techniques is cost-effectiveness analysis. It does not require a monetary valuation of accidents or injuries, nor of any other relevant policy objective. A cost-effectiveness analysis simply estimates the cost-effectiveness ratio, which can be defined as:

$$\text{Cost-effectiveness ratio} = \frac{\text{Number of accidents prevented}}{\text{Cost of measure}}$$

The number of accidents prevented forms the numerator, consistent with the idea that one wants to maximize the cost-effectiveness ratio. Cost refers to the direct costs of implementing the measure. There are three limitations of cost-effectiveness analysis:

1. The concept of cost-effectiveness becomes a problem if accidents of different severities are to be considered. It may then be necessary to estimate a cost-effectiveness ratio for each level of accident severity and then compare ratios across levels of severity.

2. Cost-effectiveness analysis does not include a criterion stating when a certain measure should be regarded as cost-ineffective, that is, as giving too small safety benefits compared to the costs of the measure. Cost-effectiveness analysis can only be used to rank order measures by cost-effectiveness.

3. Cost-effectiveness analysis cannot be used to make tradeoffs against other policy objectives. It seeks to maximize a single objective only, that of preventing accidents or injuries.

Cost-benefit analysis seeks to overcome these limitations of cost-effectiveness analysis. Accidents or injuries of different severities are made comparable by estimating the benefits to society, stated in monetary terms, of preventing them. Measures are rejected as inefficient if benefits are smaller than costs. Trade-offs against other policy objectives are made possible by converting all policy objectives to monetary terms. As far as road safety policy is concerned, the most important potentially conflicting policy objectives are those related to travel time, costs of transport (vehicle operating costs), and quality of the environment (noise, air pollution).

Estimates of road accident costs to be used in cost-benefit analysis have been made in most highly motorized countries. For the United States, the National Safety Council recommends the following comprehensive costs of traffic injury for use in cost-benefit analyses (National Safety Council 2002; U.S. dollars 2000 prices):

Injury severity	Cost per injured person (US$ 2000)
Fatal injury	3,214,290
Incapacitating injury	159,449
Nonincapacitating evident injury	41,027
Possible injury	19,528
No injury	1,861

These cost estimates refer to the so-called KABCO scale for injury severity, used by the police in reporting accidents to state and federal highway agencies.

Multiattribute utility analysis is rarely applied in formal analyses of road safety measures. It is a complex technique, conceptually closely related to cost-benefit analysis but differing from it by not requiring relevant policy impacts to be converted to monetary terms. Each policy impact is expressed in "natural" units (number of accidents, hours or minutes of travel time, vehicle operating costs in dollars, noise in decibels, and so on), but a preference function (utility function) is defined for each impact and for the pairwise trade-offs between sets of impacts. Readers will find an excellent introduction to the technique in a book by Keeney and Raiffa (1976).

Regardless of which of the three formal techniques for assessing the effectiveness of road safety measures a policy-maker wants to rely on, the first step of any formal analysis is to survey potentially effective road safety measures. Figure 16.4 lists the essential steps of the application to road safety of formal techniques of priority setting.

It is essential to conduct a broad survey of potentially effective road safety measures, to make sure that one does not miss the most efficient measures (step 1 in Figure 16.4). Once a list of potentially effective road safety measures has been made, the scale of use of each measure should be assessed (step 2 in Figure 16.4). By scale of use is meant the number of locations where it is, in principle, possible to introduce the measure. This could refer to the number of intersections that can be converted to roundabouts, the length of road for which public lighting can be provided, and so on. Based on an assessment of the scale of use of

---

Step 1: Survey potentially effective road safety measures

A measure is potentially effective if (1) credible evaluation studies have found that it improves safety, or (2) it favorably affects one or more risk factors known to contribute to accidents or injuries, and (3) it has not yet been fully implemented.

Step 2: Estimate costs and effects of each potentially effective measure

Assess the scale of use conceivable for each potentially effective road safety measure and estimate the attendant costs and likely effects on safety.

Step 3: Choose a formal criterion of efficiency for ranking measures

There are three possible criteria: (1) cost-effectiveness, (2) benefit-cost ratio, and (3) multiattributive utility value.

Step 4: Do a marginal analysis of efficiency for each measure

The objective of a marginal analysis is to determine the marginal costs and marginal benefits of each measure.

Step 5: Choose those measures that maximize overall efficiency

Determine the mix of measures that maximizes overall efficiency according to the criterion chosen (cost-effectiveness, benefit-cost ratio, or overall utility).

FIGURE 16.4 Steps in the application of formal techniques of priority setting for road safety measures.

each measure, it is possible to develop alternatives for the use of each measure, such as: convert 10 intersections to roundabouts, convert 20, convert 30, and so on. A formal criterion of efficiency by which to compare all measures should be chosen (step 3 in Figure 16.4). The most frequently used criterion is probably the benefit-cost ratio.

A marginal analysis of each road safety measure is essential (step 4 in Figure 16.4). By a marginal analysis is meant an analysis of the additional costs and additional benefits of increasing the use of a measure by one unit. This refers, for example, to the additional costs and benefits of the eleventh conversion of an intersection to a roundabout, when the first 10 conversions are ranked in order of declining benefit-cost ratio. A marginal analysis is needed if one wants to put together an optimal mix of safety measures, that is, a mix that gives the largest benefits for a given cost. Once the marginal benefits and marginal costs of all measures are known, an optimal mix is obtained by using each measure up to the point where marginal benefits equal marginal costs (step 5 in Figure 16.4).

16.8 REFERENCES

Amundsen, A. H., and R. Elvik. 2002. *Evaluering av hovedvegomlegginger i Oslo.* TØI Report 553, Institute of Transport Economics, Oslo.

Amundsen, F. H., and P. Christensen. 1973. *Statistisk opplegg og bearbeiding av trafikktekniske effektmålinger.* TØI Report, Institute of Transport Economics, Oslo.

Bunn, F., T. Collier, C. Frost, K. Ker, I. Roberts, and R. Wentz. 2003. "Area-Wide Traffic Calming for Preventing Traffic Related Injuries." Cochrane Collaboration systematic review. *The Cochrane Library* 3.

Elvik, R. 1988. "Ambiguities in the Definition and Identification of Accident Black Spots." In *Proceedings of International Symposium on Traffic Safety Theory and Research Methods,* Amsterdam, April 26–28, Session 1, Context and scope of traffic-safety theory.

———. 1995a. "A Meta-Analysis of Evaluations of Public Lighting as an Accident Countermeasure." *Transportation Research Record* 1485, 112–23.

———. 1995b. "The Safety Value of Guardrails and Crash Cushions: A Meta-analysis of Evidence from Evaluation Studies." *Accident Analysis and Prevention* 27:523–49.

———. 1997. "Evaluations of Road Accident Blackspot Treatment: A Case of the Iron Law of Evaluation Studies?" *Accident Analysis and Prevention* 29:191–99.

———. 2002a. "The Importance of Confounding in Observational Before-and-After Studies of Road Safety Measures. *Accident Analysis and Prevention* 34:631–35.

———. 2002b. "Measuring the Quality of Road Safety Evaluation Studies: Mission Impossible?" Paper presented at Transportation Research Board annual meeting, Washington, DC, January, special session 539. Available on request.

———. 2003. "Effects on Road Safety of Converting Intersections to Roundabouts: A Review of Evidence from Non-US studies." Paper 03-2106, Transportation Research Board, annual meeting, Washington, DC, January.

Elvik, R., F. H. Amundsen, and F. Hofset. 2001. "Road Safety Effects of Bypasses." *Transportation Research Record* 1758, 13–20.

Elvik, R., A. B. Mysen, and T. Vaa. 1997. *Trafikksikkerhetshåndbok,* 3rd ed. Oslo: Institute of Transport Economics.

Fridstrøm, L. 1999. *Econometric Models of Road Use, Accidents, and Road Investment Decisions,* vol. 2. Report 457, Institute of Transport Economics, Oslo.

Fridstrøm, L., J. Ifver, S. Ingebrigtsen, R. Kulmala, and L. Krogsgård Thomsen. 1995. "Meauring the Contribution of Randomness, Exposure, Weather, and Daylight to the Variation in Road Accident Counts." *Accident Analysis and Prevention* 27:1–20.

Hakkert, A. S., and L. Braimaister. 2002. *The Uses of Exposure and Risk in Road Safety Studies.* Report R-2002-12, SWOV Institute for Road Safety Research, Leidschendam, The Netherlands.

Hauer, E. 1995. "On Exposure and Accident Rate." *Traffic Engineering and Control* 36:134–38.

———. 1997. *Observational Before-After Studies in Road Safety.* Oxford: Pergamon Press.

————. 2002. "Fishing for Safety Information in the Murky Waters of Research Reports." Paper presented at Transportation Research Board annual meeting, Washington, DC, January, special session 539. Available at www.roadsafetyresearch.com.

Jarrett, D. F., C. R. Abbess, and C. C. Wright. 1988. "Empirical Estimation of the Regression-to-Mean Effect Associated with Road Accident Remedial Treatment." In *Proceedings of International Symposium on Traffic Safety Theory and Research Methods,* Amsterdam, April 26–28, Session 2, Models for evaluation.

Keeney, R. L., and H. Raiffa. 1976. *Decisions with Multiple Objectives: Preferences and Value Trade-offs.* New York: John Wiley & Sons.

Mountain, L., B. Fawaz, and D. Jarrett. 1996. "Accident Prediction Models for Roads with Minor Junctions." *Accident Analysis and Prevention* 28:695–707.

National Safety Council. 2000. "Estimating the Cost of Unintentional Injuries. Paper accessible at www.nsc.org. Accessed November 7, 2002.

Persaud, B. N., and K. Mucsi. 1995. "Microscopic Accident Prediction Models for Two-Lane Rural Roads." *Transportation Research Record* 1485, 134–39.

Ragnøy, A., P. Christensen, and R. Elvik. 2002. *Skadegradstetthet. Et nytt mål på hvor Yarlig en vegstrekning er.* TØI-report 618, Institute of Transport Economics, Oslo.

CHAPTER 17
TRANSPORTATION HAZARDS

Thomas J. Cova
Center for Natural and Technological Hazards,
Department of Geography, University of Utah,
Salt Lake City, Utah

Steven M. Conger
Center for Natural and Technological Hazards,
Department of Geography, University of Utah,
Salt Lake City, Utah

17.1 INTRODUCTION

Transportation systems are designed to move people, goods, and services efficiently, economically, and safely from one point on the earth's surface to another. Despite this broad goal, there are many environmental hazards that commonly disrupt or damage these systems at a variety of spatial and temporal scales. Whereas road-curve geometry and other engineered hazards can be addressed through design (Persaud, Retting, and Lyon 2000), hazards such as extreme weather, landslides, and earthquakes are much more difficult to predict, manage, and mitigate. These adverse events can dramatically reduce network serviceability, increase costs, and decrease safety. The economic livelihood of many individuals, firms, and nations depends on efficient transportation, and this is embodied in 20th-century innovations like just-in-time manufacturing and overnight shipping. As the movement of people, goods, and services increases at all scales due to population growth, technological innovation, and globalization (Janelle and Beuthe 1997), the systematic study of these events becomes increasingly important.

Research in the area of transportation hazards aids governments in allocating scarce resources to the four phases of emergency management: mitigation, preparedness, response, and recovery. New fields of study are emerging to address this need, as in the case of *Highway Meteorology,* which focuses on the adverse effects of extreme weather on transportation systems (Perry and Symons 1991). The growing importance of this particular field in the United States can be seen in the recent publication of *Weather Information for Surface Transportation—National Needs Assessment Report* (OFCM 2002). Some transportation agencies organize special teams to manage and mitigate the effects of one or more of these hazards. Recurrence intervals for an event span from daily to centuries, while the associated consequences range from inconvenient to catastrophic. In some cases one event may cause another–torrential rain can trigger a landslide that blocks a road. Some occur unexpectedly, while others arrive with significant warning, but all are amenable to some level of prediction and mitigation.

Transportation systems also create hazards. Accelerated movement comes with risks, and the corresponding accidents that occur disrupt lives and transportation systems daily. Vehicles collide, trains derail, boats capsize, and airplanes crash often enough to keep emergency managers and news reporters busy. The transportation of hazardous materials (HazMat) is a controversial example in this regard because it places substantial involuntary risks on proximal people and the environment. From the *Lusitania* to the World Trade Center, we are occasionally reminded that transportation disasters can be intentional acts. Lesser-known transportation hazards include elevated irrigation canals, gas pipelines, and electrical transmission lines. Intramodal risks are present in many transportation systems, as in wake turbulence behind large aircraft (Gerz, Holzapfel, and Darracq 2002; Harris et al. 2002), but intermodal risks are also a significant factor—a train might collide with a truck at an at-grade crossing (Austin and Carson 2002; Panchaanathan and Faghri 1995), or a river barge might bump a bridge, leading to the derailment of a train.

Transportation systems that are disrupted by a hazardous event also play a critical role in emergency management. Transportation lifelines are generally considered the most important in an emergency because of their vital role in the restoration of all other lifelines. Emergency managers must route personnel to an accident site, restore lifelines, relocate threatened populations, and provide relief, all of which rely on transportation. Research in this area is increasing, and there are many methods and tools to aid in addressing problems in this domain. The 2000 Cerro Grande Fire in Los Alamos, New Mexico, is a case where a low-capacity transportation network was partially disabled yet successfully used to manage a large fire and safely relocate more than 10 thousand residents.

This chapter reviews recent research and practice in three areas related to transportation and hazards: environmental hazards to transportation systems, transportation risks to proximal people and resources, and the role of transportation in emergency management.

17.2 HAZARD, VULNERABILITY, AND RISK

The study of adverse transportation events can be broadly divided into transportation *hazard analysis, vulnerability analysis,* and *risk analysis.* The focus in hazard analysis is identifying threats to a transportation system, its users, and surrounding people and resources. This is also referred to as *hazard identification.* The term *hazard* is often used to refer to environmental threats like fog, wind, and floods, but transportation hazards exist at all scales, from a sidewalk curb that might trip a pedestrian to the potential for sea-level rise to flood a coastal highway. In the most general sense, a hazard is simply a threat to people and things they value. Vulnerability analysis focuses on variation in the susceptibility to loss from hazardous events. Vulnerability can be viewed as the inverse of resilience, as resiliency implies less susceptibility to shocks. Risk analysis incorporates the likelihood of an event and its consequences, where an event can range from a minor road accident to a dam break that inundates an urban area. For example, identifying the lifelines in a given area that might be compromised by a landslide would be transportation hazard analysis. The loss of a lifeline to a landslide, or a reduction in its service, will have varying consequences depending on the design of the lifeline, its importance in the system, and the spatial economic consequences to the region. Analyzing this variation would constitute vulnerability analysis. In risk analysis, the likelihood of a landslide and its associated consequences would both be incorporated, often with the goal of identifying potential landslides that represent an "unacceptable" risk. The following sections review these three areas in greater depth.

17.2.1 Hazard Analysis

There are many questions that drive transportation hazard analysis. In the simplest case, we could assemble a list of the potential hazards that might affect transportation systems in a region. This could be accomplished by creating a hazard matrix (hazard against travel mode) that indicates whether a given hazard threatens a mode. The next level would be to identify where and when these events might occur. This is typically approached from two perspectives. In one case, we might map the potential for each hazard in a region and overlay areas of high hazard with road, rail, pipeline, and transmission networks to identify points where the two coincide. In the second case, we could select a link and inventory its potential hazards. The first approach requires a method for hazard mapping. This can be further divided into deductive and inductive modeling approaches to hazards mapping (Wadge, Wislocki, and Pearson 1993). In a deductive approach, an analyst builds a physical process model using governing equations. For example, if landslides are the hazard in question, one could use slope instability equations to determine landslide hazard along a road. In an inductive approach to landslide hazard mapping, an empirical study is undertaken to map past events to determine the conditions that lead to their occurrence. Areas with similar characteristics are then identified, often with techniques in map overlay, because they may also be hazardous. The line between inductive and deductive approaches should not be drawn too sharply, because most hazard analyses rely on both. For example, past events may be studied to help build a deductive process model.

There are a number of important dimensions in transportation hazard analysis, most notably the spatial and temporal scales. The spatial scale includes both the extent of the study and the resolution or detail. The spatial extent might be global, national, regional, local, or an individual link in a network. Detail and spatial extent are correlated, but as computer storage continues to increases, this is weakening, and we may soon see national (or larger) studies with very fine spatial and temporal detail. The temporal extent and resolution are also important. A central question is the time-horizon of the study, which can range from a single time period (cross-sectional) to any duration (longitudinal). Time is also important because of the many cycles that affect the potential for hazards. Road icing is most common at night in the winter, and thus it varies seasonally and diurnally. Landslides occur more often during the rainy season, avalanches occur in the winter, and fires occur during the dry season. Figure 17.1 depicts the changing likelihood of hazardous events over time, and this becomes more important in risk analysis.

17.2.2 Vulnerability Analysis

Vulnerability is an increasing focus in researching threats to transportation systems (Berdica 2002; Lleras-Echeverri and Sanchez-Silva 2001; Menoni et al. 2002). There are many definitions of hazard vulnerability in the research literature (Cutter 1996). As noted, vulnerability in a transportation context recognizes that susceptibility is not uniform across people, vehicles, traffic flow, infrastructure, and the environment. Vulnerability can refer to the physical vulnerability of the users or the potential for an incident to decrease the serviceability of the transportation system. Vulnerability in a transportation context can also be approached from the point of view of network *reliability,* as a reliable network is less vulnerable, and Berdica (2002) links these two concepts. For an example of differing road network vulnerability, a road accident in a two-way tunnel may temporarily cripple a regional transportation system, leading to significant delays, but a system with a separate tunnel in each direction would be less vulnerable to an incident halting traffic in both directions. People and environmental resources in proximity to a transportation corridor are also vulnerable to adverse events. For example, in transporting hazardous materials along a populated corridor, vulnerability along the corridor may vary significantly from point to point, and two potential incidents a few miles apart can have very different outcomes. There are also regional economic vulnerabilities because adverse events can disrupt commerce. Individuals can miss meetings, retail outlets can lose customers, commodities can be delayed, and tourism can be adversely impacted, all of which have economic consequences.

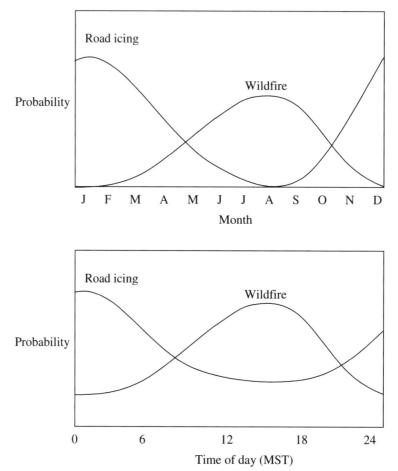

FIGURE 17.1   Seasonal and diurnal variation in the probability for two North American transportation hazards.

17.2.3 Risk Analysis

The most common definition of risk incorporates both the likelihood of an event and its consequences. It is not possible to avoid all risks, only to choose from risk-benefit trade-offs (Starr 1969). Kaplan and Garrick (1981) define risk as a set of triplets:

$$(s, p, c) \tag{17.1}$$

where s is a scenario, p its probability, and c its consequences. Risk analysis can be viewed as the process of enumerating all triplets of interest within a spatial and temporal envelope. The probability of a scenario varies inversely with its consequences, which is embodied in the concept of a risk curve (Figure 17.2). In Kaplan and Garrick's framework, the definition of a scenario can be arbitrarily precise. For example, one scenario might be an intoxicated driver crashing on a wet road at night, while another might be an earthquake-induced land-slide above a town. The concept of vulnerability enters the triplet through the consequence

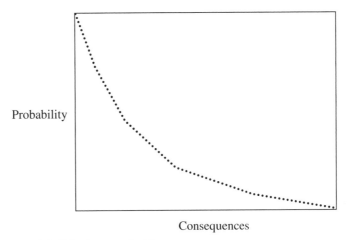

FIGURE 17.2   An example risk curve.

term, which varies as a function of the unique vulnerabilities of the scenario elements. In accident analysis, the consequence term can be held constant for comparison purposes, as in a *road casualty*. This effectively removes the c term, which allows an analyst to focus on estimating p for different scenarios and levels of risk exposure (Thorpe 1964; Chapman 1973; Wolfe 1982). An example would be comparing the probability of a daytime versus nighttime road casualty. It is difficult to estimating p for extreme events with little historical data, and Bier et al. (1999) provide an excellent survey of current methods to address this problem.

A thought experiment might help convey the related concepts of hazard, vulnerability, and risk in a transportation context. Imagine two motorcyclists riding in adjacent lanes with the hazard in question being a crash. All characteristics of the drivers, vehicles, and the environment are equal. We would say that the two face the same hazard, vulnerability, and risk because the likelihood and consequences of either motorcyclist crashing are equal. To understand vulnerability, place a helmet on one rider. The likelihood of a crash has not been altered, but both the vulnerability and the risk of the rider with the helmet have decreased. Now imagine that both riders are wearing a helmet, but the surface of one lane is wet. The vulnerability of both drivers is equal, but the likelihood of a crash is higher (as is the risk) for the rider in the wet lane. To make it tricky, imagine that the rider in the wet lane is wearing a helmet but the rider in the dry lane is not. One has a greater likelihood of a crash and the other a greater vulnerability to a crash, but which rider is at greater risk? An empirical approach to this problem would be to compare the casualty rate for motorcyclists wearing a helmet in rainy conditions with the rate for riders without a helmet in dry conditions, attempting to control for all other variables.

Despite the challenges presented by quantitative risk assessment and its many assumptions, risk analysis has many benefits that outweigh the drawbacks. Evans (1997) reviews risk assessment practices by transport organizations for accidents and notes that the benefits of quantitative risk assessment include:

1. It makes possible the prioritization of safety measures when resources are scarce, or where there are different approaches to achieving the same end.

2. It makes possible the design of systems (engineering or management) aimed at achieving specified safety targets or tolerability limits.

3. It facilitates proactive rather than just reactive safety regulation.

4. It provides a basis for arguing against safety measures whose benefits are small compared with their costs, and for justifying such decisions on a rational basis.

An overarching goal in quantitative risk assessment is to determine if a given transportation risk is "acceptable." If it is not, mitigation actions are in order. One approach to this problem is to compare the given risk with commonly accepted risks. Thus, a rock fall study along a highway might compare the results with other risks like air travel, drowning, lightning, or structural failure to determine if the risk of a rock fall fatality is significantly greater than other risks (Bunce, Cruden, and Morgenstern 1997). Another approach is to compare the risk of two scenarios to compute their relative risk using a risk ratio. For example, if there were 10 road accidents on rainy weekends on average and 5 on dry weekends, then the risk ratio of rainy-day weekend driving to fair-weather weekend driving would be $10/5 = 2$, or twice as risky, assuming that the amount of driving (aggregate exposure) was roughly the same from weekend to weekend.

17.3 HAZARDS TO TRANSPORTATION SYSTEMS

There are many environmental hazards that may damage or disrupt transportation systems, and we review only the more common ones here. For example, Figure 17.3 depicts familiar road hazards grouped by their principal effect along with some of their causal relationships. In general, road hazards can: (1) compromise the quality of the surface, (2) block or damage infrastructure, (3) compromise user visibility, (4) compromise steering, (5) create a temporary obstacle, or (6) some combination of the prior five. From the figure, it is clear that rain, wind, and earthquakes have causal links with many other hazards. Rain and earthquakes can both induce a flood, landslide, rock fall, or debris flow. Earthquakes can also start a fire or result in a toxic release. Extreme wind can kick up dust, start a fire, drive smoke from a fire, blow trees and debris into the roadway, or redeposit snow, leading to an avalanche. This

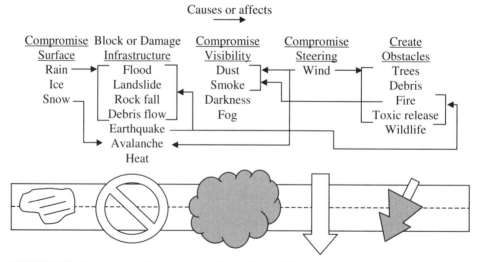

FIGURE 17.3 Example road hazards grouped by their principle effect, including some of their causal relationships.

is only a sample of the many hazards and relationships that might exist. Hazards can also coincide, as in a nighttime earthquake in severe rain. This section reviews recent research in the analysis of many of these hazards, but it should not be considered comprehensive. The review is multimodal and driven primarily by these questions:

- What is the hazard?
- What has been done to address the hazard in research and practice?
- What travel modes does the hazard affect?
- How well can we predict the hazard in space and time?
- What are the consequences of the hazard and how are they defined and measured?
- What mitigation actions exist and what might be developed?

17.3.1 Avalanches

An avalanche is a sudden transfer of potential energy inherent in a snow pack into kinetic energy. The principal contributing factors include snow, topographic effects, and wind which can redeposit snow. *Snow structure* refers to the composition of its vertical profile, which can become unstable as new layers are added. An avalanche occurs when the strength of the snowpack no longer exceeds the internal and external stresses. Avalanches are typically divided into dry or wet and loose or slab avalanches. Dry slab avalanches accelerate rapidly and can reach speeds in excess of 120 mph, but wet avalanches move much more slowly.

The systematic study of avalanches in North America dates back to the 1950s in Alta, Utah. Figure 17.4 depicts the most active and damaging slide paths in Alta. The most useful, general reference is McClung and Schaerer's (1993) avalanche handbook. Avalanches typically reduce the serviceability of a road, but they can also damage infrastructure and cause injury or death. Other modes affected include rail, pipelines, and transmission lines. The science of predicting the timing of avalanches is called *forecasting* (Schweizer, Jamieson, and Skjonsberg 1998), and it has improved significantly over the last 50 years. Snow pits, weather instrumentation, field observation, and remote sensing are combined to forecast avalanches. The corridors that receive the greatest attention are those with high traffic volume and a documented avalanche history. Avalanche path identification using terrain and vegetation is also a common task in areas where historical records may not be available.

Three challenges that transportation agencies face in avalanche control are (1) selecting paths where mitigation would be most beneficial, (2) evaluating mitigation measures, and (3) comparing the risks of different roads. The avalanche-hazard index (Schaerer 1989) combines forecasting with traffic flow volumes to address these needs. The index includes the likelihood of vehicles being impacted by an avalanche along a road as well as the potential consequences. It also incorporates the observation that loss of life can occur when a neighboring avalanche overcomes traffic halted by another slide. The composite avalanche-hazard index for a road is:

$$I = \sum_i \sum_j w_j (P_{mij} + P_{wij}) \qquad (17.2)$$

where P_{mij} = the likelihood that *moving* traffic might be hit by an avalanche of class j at path i

P_{wij} = the likelihood that *waiting* traffic might be hit by an avalanche of class j at path i

w_j = the consequence of an avalanche of class j

The index can also be calculated separately for each avalanche path along a road to determine where mitigation would make the largest contribution to overall hazard reduction.

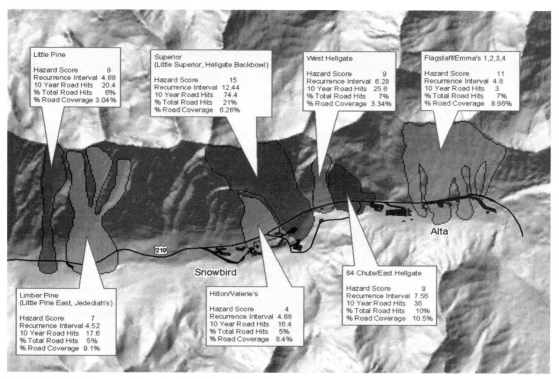

FIGURE 17.4 The most active avalanche slide paths in Little Cottonwood Canyon, Utah shown with their mean recurrence interval (years), the number of road hits in the last 10 years, the percentage of times they hit the road, and the percentage of the length of the road they covered. (*Source:* William Naisbitt, Alex Hogle, and Wendy Bates.)

A number of avalanche risk case studies for transportation corridors have been performed, including at Glacier National Park (Schweizer, Jamieson, and Skjonsberg 1998), the Colorado Front Range (Rayback 1998), and the Himalayas (De Scally and Gardner 1994). Avalanche-mitigation options, increasing in cost, include explosives, snow sheds, and deflection dams. Rice et al. (2000) provide an example of system for automatically detecting avalanches on rural roads.

17.3.2 Earthquakes

The study of earthquakes and seismic risk spans many fields in the sciences and social sciences. They are widely researched by transportation engineers from a variety of perspectives because they can severely damage and disrupt transportation systems. A devastating earthquake epitomizes a low-probability, high-consequence event in risk analysis. The recurrence interval for a large earthquake in a region can be centuries, varying inversely with magnitude, yet devastating earthquakes occur almost every year somewhere in the world. For many populated areas without a history of severe earthquake loss, the likelihood of facing an earthquake that damages transportation lifelines is a near certainty because the geologic record reveals past large earthquakes (Clague 2002). No major transport mode is exempt from the adverse affects of an earthquake. Roadways, railways, pipelines, transmission lines, and airports and seaports can all be damaged, with tremendous economic costs

(Cho, Shinozuka, and Chang 2001). Earthquakes can also start fires, trigger landslides (Refice and Capolongo 2002), release toxic chemicals (Lindell and Perry 1996), cause dam failures, and create sudden earthen dams via landslides leading to inevitable flooding (Schuster 1986).

Preimpact earthquake research in transportation engineering focuses on vulnerable structures like bridges (Malik 2000), tunnels (Hashash et al. 2001), and water-delivery systems (Chang, Svekla, and Shinozuka 2002). The central problem is estimating the response characteristics of these structures to ground shaking and liquefaction (Price et al. 2000; Selcuk and Yüceman 2000; Romero, Rix, and French 2000). Werner (1997) notes that earthquake losses to highway systems depend not only on the response characteristics of the highway components, but also on the nature of the overall highway system's configuration, redundancy, capacity, and traffic demand (see also Basoz and Kiremidjian 1996). For example, two bridges may be equally susceptible to ground shaking, but one may be much more important in serving the daily travel demand to an important destination. Retrofitting is typically in high order when a bridge highly susceptible to the effects of an earthquake is also essential in serving a large volume of travel demand.

Postimpact earthquake research focuses on immediate damage assessment (Park, Cudney, and Inman 2001), the performance of the transportation system (Chang 2000), and the lifeline restoration process (Isumi, Nomura, and Shibuya 1985; Opricovic and Tzeng 2002). Chang (2000) examines postearthquake port performance following the Kobe quake in 1995 and frames the economic loss (and thus vulnerability) in terms of three types of traffic:

1. Cargo originating from or destined to the immediate hinterland
2. Cargo from/to the rest of Japan
3. Foreign transshipment cargo

By examining the pre- and post-conditions of these cargo types, Chang concludes that 2 and 3 suffered the most, resulting in both short-term loss of revenue and long-term loss of competitive position. Economic impacts may last beyond the point where the infrastructure has been repaired. Kobe demonstrates that 3 is especially important, and the central port vulnerability question can be framed as the percentage of a port's revenue tied to transshipment cargo.

17.3.3 Floods and Dam Breaks

Floods cause the greatest loss in many countries because they occur frequently and their severity is compounded by dense development along many rivers. The National Weather Service (NWS) in the United States estimates that greater than half of all flood-related deaths occur in vehicles at low-water crossings. Flood damage to transportation systems represents one of the largest losses in the public sector. Intense rainfall is the chief cause of floods, but hurricanes also hold the potential to cause a significant amount of storm surge inundation. Dam breaks are included here as a special type of technologically-induced flood. This includes earthen dam breaks caused by earthquake-induced landslides (Schuster 1986). The modeling of dam breaks has increased in recent years because agencies such as the U.S. Bureau of Reclamation (USBR) are required to submit a report and associated inundation animations of potential dam breaks to local emergency managers downstream from all dams for emergency planning purposes.

Figure 17.5 depicts an example of modeling flooding across a transportation network. The depth of the flood is shown in meters, with the direction and velocity of the flood depicted using a vector field. This example is output from the MIKE 21 flood simulation system for modeling two-dimensional free surface flows. The system can model many conditions that occur in a floodplain, including flooding and drainage of floodplains, embankment overtopping, flow through hydraulic structures, tidal forces, and storm surge. MIKE 21

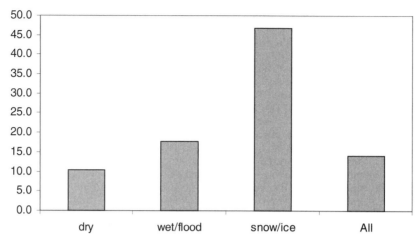

FIGURE 17.5 Example skidding rates (the percentage of accidents where skidding is a factor) for Great Britain (adapted from Perry and Symons 1991).

is an excellent example of a deductive process-oriented hazard mapping approach because the system solves nonlinear equations of continuity and conservation of momentum for flooding.

Flooding is a serious problem in many areas because of its ability to degage the serviceability of a transportation network at various points rapidly. Ferrante, Napolitano, and Ubertini (2000) combine a numerical model for flood propagation in urban areas with a network path-finding algorithm to identify least-flood-risk paths for rescuing people as well as providing relief. They use Dijkstra's (1959) shortest-path algorithm, but calculate cost of a link in a very novel manner using the flood flow depth and velocity across the road. In this way, the "cost" of traversing a link is a function of both the length of the road and its flood characteristics:

$$c_{ij} = \frac{L_{ij}}{\alpha_h \alpha_v}$$

(17.3)

where c_{ij} = the cost/risk of traversing the link
L_{ij} = the length of the link
α_h = a parameter (0–1) related to flood height
α_v = a parameter related to water velocity (0–1)

Each alpha parameter *decreases* as flood height or velocity increases, respectively, until the maximum allowable flood height (e.g., 0.3 meters) or velocity (e.g., 1 m/sec) is reached, whereby they become 0. At this point, link cost is infinite and is no longer traversable. Thus, the travel cost of a link without flooding is its length, but as flood height and velocity across the link increase, its cost and risk quickly increase. This example links a hazard process model with a network algorithm, which points to a valuable opportunity for analysts, as many hazards reduce the serviceability of network links. Real-time path finding in a network degraded by a hazard is a very valuable application. The challenge is to develop a means for acquiring accurate, timely information on the hazard as well as to manage and convey the uncertainty in the results.

17.3.4 Fog, Dust, Smoke, Sunlight, and Darkness

Fog, dust, smoke, sunlight, and darkness are transportation hazards that compromise the visibility of system users. This hazard category does not apply to pipeline networks, transmission lines, and other networks where visibility is not an issue. From a roadway perspective, Perry and Symons (1991) provide an excellent source on these hazards. Musk (1991) thoroughly covers the fog hazard, and Brazel (1991) describes a dust storm case study for Arizona. Although smoke from wildfires routinely disrupts roadways and inhibits operations at airports each summer, it appears to be an underresearched topic in transportation hazards. Darkness also has an understandably adverse effect on road safety, especially when combined with fog, smoke, or dust.

Fog can cause spectacular road accidents involving hundreds of vehicles on a roadway. Musk (1991) describes the fog potential index (FPI), which expresses the susceptibility of a location p on a road to thick radiation fog on a scale from 0 to 100. The values of two locations are comparative in that a value of 30 at location A and a value 20 at location B means that location A should experience 50 percent more hours of thick radiation fog than location B. The index is of the form:

$$I_p = 10d + 10t + 2s + 3e \qquad (17.4)$$

where d = the distance of the location p from standing surface water
   t = a function of the local topography at p (e.g. hill or valley)
   s = a function of the road site topography (e.g., bridge or embankment)
   e = the incorporation of other environmental features likely to affect the formation of radiation fog (e.g., proximity to power station cooling)

The index coefficients are weights that affect the relative importance of the variables. This index can be applied at any linear resolution, but 1 kilometer is common. The index can then be tested against in situ observations of visibility.

17.3.5 Rain, Snow, and Ice

Rain, snow, and ice are common hazards that compromise visibility and the quality of a road, rail or airport surface (Benedetto 2002; Andrey 1990). All road users are familiar with road signs like "slippery when wet" or "bridges may be icy" (Carson and Mannering 2001). Ice is also a hazard for aircraft because of its effect on lift, as well as sea travel because it creates obstacles (Tangborn and Post 1998). In a road network context, skidding is the most common explanation for accidents that occur in the context of these hazards. The *skidding rate* is the statistic used to quantify this factor, which is the percentage of accidents where one or more vehicles are reported to have skidded (Perry and Symons 1991). Example skidding rates for cars are given in Figure 17.6 for Great Britain in 1987. This figure shows that rain roughly doubles the percentage of accidents where skidding is a factor over dry conditions, and snow and ice quadruple the rate over dry conditions. The overall skidding rate for cars for all road conditions is about 14 percent.

The question of how rain, snow, and ice affect the total number of road accidents is not straightforward. Palutikof (1983) found that people drive more carefully in snow or simply postpone or cancel journeys. This leads to reduction in the total number of accidents over that which would be expected. Rain does not seem to have the same effect on travel decision making, and Brodsky and Hakkert (1988) found that the number of accidents increases in wet conditions. Al Hassan and Barker (1999) found a slightly greater drop in traffic activity owed to inclement weather on the weekend (>4 percent) than on weekdays (<3 percent). In a case study in Chicago, Bertness (1980) found that rain roughly doubled the number of

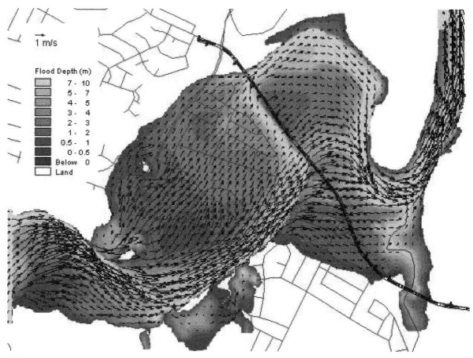

FIGURE 17.6   A floodplain inundation map depicting flood depth (m) and velocity (m/sec) over a transportation network. (*Source:* DHI Water & Environment and Camp Dresser & McKee, Inc. ©.)

road accidents, with the greatest effect in rural areas. It is important to keep in mind that rain, snow, and ice studies tend to underestimate the risk because road accidents are typically underreported.

Hazards that affect the road surface represent the most costly maintenance function for many cities, counties, and state transportation departments. Salt is the most common road de-icer, with about 10,000,000 tons applied each year in the United States (Perry and Symons 1991). This is expensive and comes with environmental side effects. Eriksson and Norman (2001) note that road weather information systems have a very high benefit–cost ratio in reducing weather-related risk. The widespread adoption of Doppler radar has greatly improved the reporting of precipitation, and some systems can now report rain intensity to levels as detailed as an individual street segment. There is much work in developing and installing in situ road sensors to automatically detect poor road conditions. This can greatly improve road maintenance procedures because managers can apply mitigation measures like salt where it is most needed.

17.3.6   Landslides, Rock Fall, and Debris Flow

Many miles of roads, rail, and pipeline travel through areas with rock faces and steep slopes in mountainous terrain. Geomorphic hazards that commonly affect transportation corridors include landslides, rock fall, and debris flow. A debris flow is essentially a fast-moving landslide. These hazards can damage or reduce the serviceability of infrastructure, crush or bury vehicles, and result in death. In some cases they occur without little or no warning, but they are typically preceded by intense rain (Al Homoud, Prior, and Awad 1999). They can also be earthquake- or volcanically induced (Dalziell and Nicholson 2001) and create sudden earthen dams that lead to flooding (Schuster, Wieczorek, and Hope 1998). An excellent general source on landslides and debris flows is the Transportation Research Board (TRB) report on landslides edited by Turner and Schuster (1996). In terms of case studies, Marchi, Arattano, and Deganutti (2002) examine 10 years of debris flows in the Italian Alps, Evans and Savigny (1994) examine landslides in Canada; He, Ma, and Cui (2002) looked at debris flows along the China-Nepal Highway; Budetta (2002) conducted a risk assessment for a 1-kilometer stretch of road subject to debris flows in Italy; and Petley (1998) examined geomorphic road hazards along a stretch of road in Taiwan. Fish and Lane (2002) discuss a rock-cut management system, and Franklin and Senior (1997) describe a rock fall hazard rating system.

Bunce, Cruden, and Morgenstern (1996) provide an excellent example method for assessing the risk of loss of life from rock fall along a highway. They used rock fall impact-mark mapping supplemented by documented rock fall records to establish a rock fall frequency for the Argillite Cut on Highway 99 in British Columbia. The method relies on separate calculations for the risk of a rock hitting a stationary vehicle versus a moving vehicle, as well as a moving vehicle hitting a rock on the road. The probability that one or more vehicles will be hit is given as:

$$P(S) = 1 - (1 - P(S\,|\,H))^N \qquad (17.5)$$

where $P(S\,|\,H)$ is the probability that a vehicle occupies the portion of the road affected by a rockfall and N is the number of rocks that fall. This equation states that the probability of a vehicle being hit is one minus the probability that a vehicle will not be hit. With a series of assumptions, they estimate the risk of death due to rockfall for a one-time road user and daily commuter at 0.00000006 and 0.00003 per year, respectively.

17.3.7 Wind, Tornadoes and Hurricanes

Wind is a significant hazard to road, rail, sea, and air transport (Perry and Symons 1994). Gusts, eddies, lulls, and changes in wind direction are often greatest near the ground in extreme wind episodes. In these episodes, the majority of fatalities are generally transport-related. It is difficult to summarize the effects of wind on road and rail transport because little data exists, although it is generally viewed as less of a hazard than ice, snow, and rain. Figure 17.7 depicts wind that is blowing smoke across an interstate and blocking traffic. Perry and Symons (1994) divide the wind hazard into three categories: direct interference with a vehicle, obstructions, and indirect effects. Direct interference includes its effects on vehicle steering, which may push one vehicle into another or run a vehicle off the road. Extreme winds can overturn high-profile trucks and trains when the wind vector is orthogonal to the direction of travel because the force of the wind is proportional to the vehicle area presented (Baker 1988). Wind can impede transport by blowing dust or smoke across a road, which can reduce visibility. It can also blow trees and other debris onto a road or railway and create temporary obstacles. Indirect effects include the redeposition of snow leading to an avalanche, as well as its adverse effect on bridges and air and sea-based termini. Overall, wind can impede transport operation or damage vehicles and infrastructure, all of which can result in economic impacts, injuries, and fatalities.

Air transport faces the greatest hazard from wind. A violent downdraft from a thunderstorm (microburst) on takeoff or landing is one example, but any exceptionally large local wind gradient (wind shear) can affect lift adversely at low altitudes (Vorobtsov 2002; Goh and Wiegmann 2002). In many air disasters, wind is considered the primary contributing factor. Small aircraft are much more vulnerable to in-flight storms and are often warned to avoid storms completely. Measures to reduce wind hazard include permanent wind breaks,

FIGURE 17.7 A wildfire adjacent to an interstate blocking traffic. (*Source:* http://www.commanderchuck. com).

warnings, road closures, and low-level wind shear alert systems. An airport wind-warning system generally consists of a set of anemometers that are analyzed by computer. A warning is issued when levels differ by some threshold. Automated wind-warning systems for individual roads may appear soon because of advances in weather instrumentation. The finest level that wind warnings are commonly issued is at a county scale. Improved weather forecasting is generally viewed as the principal means for reducing the hazard (Perry and Symons 1994).

Hurricanes and tornadoes represent special cases of extreme winds. Due to satellite, radar, and other in situ sensor networks, their prediction has greatly increased in recent years. Much of the transportation research in this area focuses on evacuation. Wolshon (2001) reviews the problems and prospects for contraflow freeway operations to reduce the vulnerability of coastal communities by reversing lanes to increase freeway capacities in directions favorable for evacuation. This problem is simple conceptually but represents a significant challenge for both traffic engineers and emergency managers.

17.3.8 Wildlife

Wildlife is a familiar hazard to most drivers because of the many warning signs along roadways. Wildlife accidents typically result in vehicle damage, but they can also result in injury or death. Two common examples of wildlife hazards include the threat that ungulates such as moose (Joyce and Mahoney 2001) present to vehicles and the threat that birds present to aircraft. The number of these collisions is staggering, and it is estimated that in 1991 greater than half a million deer were killed by vehicles in the United States (Romin and Bissonette 1996). Lehnert and Bissonette (1997) review research on deer-vehicle collisions and describe a field experiment on the effectiveness of highway crosswalk-structures as a means of mitigation. The crosswalk system evaluated forces deer to cross at specific areas that are well marked for motorists. Although deer fatalities decreased by 42 percent following the installation of the crosswalks, they were unable to attribute this reduction to the crosswalks because there was an 11 percent probability that it might have occurred by chance.

Bird hazards to aircraft are also a significant concern, and Lovell and Dolbeer (1999) provide a recent review with a study to validate the results of the U.S. Air Force (USAF) bird avoidance model (BAM). BAM provides information to pilots regarding elevated bird activity based on refuge surveys, migration dates, and routes. Lovell and Dolbeer note that since 1986 birds have caused 33 fatalities and almost $500 million in damage to USAF aircraft alone. On average, USAF aircraft incur 2,500 bird strikes a year, with most occurring in the fall and spring migration. Waterfowl and raptors account for 69 percent of the damaging strikes to low-level flying military aircraft. Lovell and Dolbeer found that BAM predicted significantly higher hazard for routes where bird strikes have occurred in the past and thus can assist in minimizing strikes.

17.4 TRANSPORTATION AS HAZARD

In addition to the many environmental hazards that threaten transportation systems, transportation itself presents hazards to people, property, and the environment. Road traffic accidents are the most common example, accounting for the majority of transportation casualties in most countries. The contributing factors for road accidents are typically classified into those associated with the driver, vehicle, and the environment. Contributing factors associated with the driver include error, speeding, experience, and blood-alcohol level. Factors associated with the vehicle include its type, condition, and center of gravity. Environmental factors include the quality of the infrastructure, weather, and obstacles. The majority of road accidents are attributed to driver factors (Evans 1991), and this holds for many other modes such as boats (Bob-Manuel 2002), bicycles (Cherington 2000), snowmobiles (Osterom and Eriksson 2002), and all-terrain vehicles (Rogers 1993). Taken together, this implies that most transportation casualties in the world are road accidents chiefly attributed to the driver. Not surprisingly, research on driver factors represents the largest area of transportation hazards research (see the journal *Accident Analysis and Prevention*). Transportation accidents have severe effects on those directly involved, as well as side effects on others. Other effects might include severe traffic delays leading to missed meetings, lost sales to businesses, delayed commodity shipments, and increased insurance costs. Research in accident analysis spans all modes and typically focuses on assessing the role of various driver, vehicle, and environmental factors as well as methods for mitigating accidents. (See chapter 18 on incident management.)

In addition to common traffic incidents, there are also low-probability, high-consequence transportation events that place risks on people and environmental resources in proximity to transportation corridors and ports. Rail, road, pipeline, and marine HazMat transport is the prime example in this area because it places considerable involuntary risks on people (and resources) who do not perceive much benefit to the transport of hazardous materials. HazMat has been studied from a number of perspectives for many materials and modes, so there are numerous frameworks for analysis (Bonvicini, Leonelli, and Spadoni 1998; Cassini 1998; Erkut and Ingolfsson 2000; Fabiano et al. 2002; Helander and Melachrinoudis 1997; Jacobs and Warmerdam 1994; Klein 1991). Aldrich (2002) provides an historic perspective on rail HazMat shipments from 1833 to 1930, and Cutter (1997) reviews recent trends in hazardous material spills. Caputo and Pelagagge (2002) present a system for monitoring pipeline

HazMat shipments. Singh, Shonhardt, and Terezopoulos (2002) examine spontaneous coal combustion in sea transport. Raj and Pritchard (2000) present a risk-analysis tool used by the Federal Railway Administration. Hwang et al. (2001) present a comprehensive risk analysis approach for all modes that includes 90 percent of the dangerous chemicals. Abkowitz, Cheng, and Lepofsky (1990) describe a method for evaluating the economic consequences of HazMat trucking. Verter and Kara (2001) review HazMat truck routing in Canada. Dobbins and Abkowitz (2002) look at inland marine HazMat shipments. Saccommano and Haastrup (2002) focus on HazMat risks in tunnels. Marianov, ReVelle, and Shih (2002) propose that proximal communities receive a tax reduction to offset the risk of proximal HazMat shipments.

Following the events of September 11, 2001, *transportation security* has become a national research priority led by the Transportation Security Administration (TSA), recently reorganized under the Department of Homeland Security. Transportation terrorism has not been a focus of transportation hazards researchers in the past, so there is little to review at this point. However, reports and proposals are beginning to surface that indicate that this will be one of the largest areas of transportation hazards research for many years.

17.5   TRANSPORTATION IN EMERGENCY MANAGEMENT

Transportation lifelines are vital during an emergency and play an important role in all four phases of emergency management: mitigation, preparedness, response, and recovery. The concern in the mitigation phase is reducing the likelihood of an event, its consequences, or both. The focus of the preparedness phase is improving operational capabilities to respond to an emergency such as training emergency personnel, installing notification systems, and redeploying resources to maximize readiness (Sorensen 2001). The mitigation and preparedness phases both help reduce the impact of hazardous events. The response phase begins immediately following an event, and this is when plans devised in the preparedness phase as well on-the-fly plans are activated. Common concerns include evacuating and sheltering victims, providing medical care, containing the hazard, and protecting property and the environment. The recovery phase addresses longer-term projects like damage assessment and rebuilding, which feeds back into the mitigation phase because this phase presents an opportunity to rethink hazardous areas.

Mitigation strategies for specific hazards and assets were discussed in the prior section on hazards to transportation systems. The overarching challenge in the mitigation phase is identifying and prioritizing mitigation projects in a region and allocating scarce resources to their completion. Benefit-cost analysis is a valuable method in this regard, but it must be preceded by risk assessments for all potential hazards. The effectiveness of the mitigation strategy is also important, and this can be considered part of the benefit.

Research in the preparedness and response phase has been fueled by new technologies. Enhanced 911 (E-911) is a significant relatively recent innovation, and this is covered in chapter 18 on incident management. Relevant topics that are actively researched in this phase include optimally locating emergency teams (List and Turnquist 1998), locating and stocking road maintenance stations, finding optimal fire station location for urban areas and airports (Revelle 1991; Tzeng and Chen 1999), and installing hazard-specific warning systems. Evacuation planning in this phase focuses on delimiting emergency planning zones (Sorensen, Carnes, and Rogers 1992), designing and simulating evacuations (Sinuany-Stern and Stern 1993; Southworth 1991; Cova and Johnson 2002), developing and testing evacuation routing schemes (Dunn 1993; Yamada 1996; Cova and Johnson 2003), and identifying potential evacuation bottlenecks (Cova and Church 1997). Reverse 911 systems that allow police to call evacuees are becoming increasingly important in dealing with notification. State-of-the-art systems allow emergency managers to send custom messages with departure timing and routing instructions to zones defined on the fly with a mouse. Other research in the preparedness and response phase includes methods for keeping roadways open following an earthquake or landslide (Santi, Neuner, and Anderson 2002).

One problem that complicates emergency planning by transportation agencies is the increasing amount of development in many hazardous areas. This is nearly universal as populations increase in floodplains, coastal areas subject to hurricanes, fire-prone wildlands, areas near toxic facilities (Johnson and Zeigler 1986), regions at risk to seismic activity, and so on. This presents a problem because in many of these areas (and at many scales) the transportation system is not being improved to deal with these increasing populations. This means that evacuating threatened populations is becoming increasingly difficulty at all scales as new development occurs. In other words, vulnerability to environmental hazards is continually increasing due to the fact that populations in hazardous areas are increasing at the same time that the ability of emergency managers to invoke protective actions such as evacuation are decreasing.

Figure 17.8 is two maps from Cova and Johnson (2002) that show the effect of a new road on household evacuation times for a community at risk to wildfire near Salt Lake City. Before the construction of the new road, homes in the back of the canyon had the greatest evacuation times, as the sole road out of the canyon would get congested. In this scenario, the average vehicle departure-time following notification to evacuate was 10 minutes and the average number of vehicles per household was 2.5, so it can be considered a reasonably urgent evacuation when most residents are home. Note that houses in the back of the canyon stand to gain much more from the construction of the new exit because their evacuation times decrease much more than those for homes near the original exit from this community. Also, all evacuation times become more consistent because the second exit reduces the delay caused by everyone using one exit. Viewed another way, the new exit reduces the number of households per exiting road from $250/1 = 250$ to $250/2 = 125$.

17.6   NEW TECHNOLOGIES

There are many new technologies that hold promise to aid transportation agencies in reducing the effects of transportation hazards. Weather instrumentation is prime example that is improving both in terms of breadth of measurement, as well as the number of installed road weather stations. The suite of geospatial technologies including the global positioning system (GPS), geographic information systems (GIS), and remote sensing also hold much promise to improve the amount of information available to transportation users, planners, and emergency responders. The recent formation of the National Consortia for Remote Sensing in Transportation (NCRST) is dedicated to this task (Gomez 2002). The consortia are divided into four themes: hazards, environment, infrastructure, and flow. NCRST Hazards (NCRST-H) is the most relevant in the context of applying geospatial technologies to monitoring and mitigating transportation hazards.

A simple benefit of GPS in accident analysis is that it is an inexpensive means for greatly improving the locational component of crash data (Graettinger 2001). Remote sensing can be used to detect and monitor fires, volcanoes (Oppenheimer 1998), landslides, avalanches, and many other hazards. One technology that is having a significant effect on the study of transportation hazards is GIS. For a comprehensive review of GIS in transportation (GIS-T) see Miller and Shaw (2001). GIS is being used in transportation applications such as mapping collision data (Arthur and Waters 1997; Austin, Tight, and Kirby 1997), routing HazMat shipments (Brainard, Lovett, and Parfitt 1996; Lepofsky, Abkowitz, and Cheng 1993), identifying hazardous highway locations (Spring and Hummer 1995), and modeling the vulnerability of populations to toxic spills (Chakraborty and Armstrong, 1996), among many other transportation hazard applications.

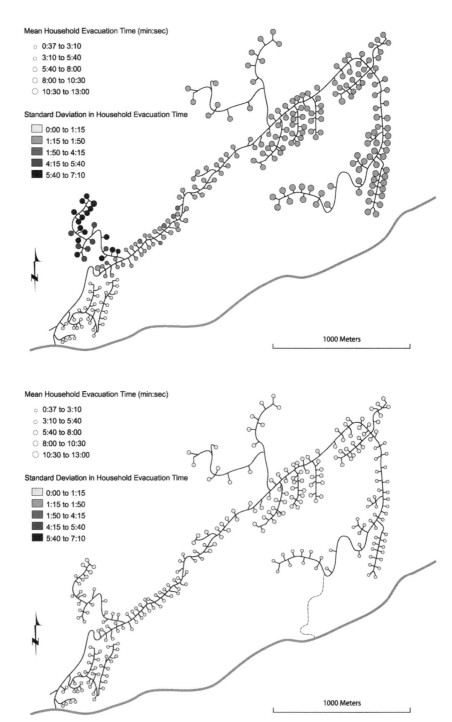

FIGURE 17.8   The effect of the construction of a second access road (dashed line) on household evacuation times in a fire-prone canyon east of Salt Lake City, Utah.

17.7 CONCLUSION

Transportation and hazards is a growing field in terms of research and practice. New methods for predicting and mitigating hazards are continually being developed, and researchers are linking these methods and models to tasks in transportation planning and mitigation. Globalization is increasing our dependence on transportation systems at all scales. For this reason, disruptions will only become more costly and important to mitigate. New information technologies are converging that promise to change this field drastically in the coming years. These technologies are emerging during a shifting emphasis toward transportation security. This will bring information-based research on reducing the effects of transportation terrorism to the forefront of this research area. Finally, development in hazardous areas is increasing along with corresponding increases in traffic volumes along many lifelines at risk to hazards. This will present substantial challenges to transportation researchers and analysts for the foreseeable future.

17.8 ACKNOWLEDGMENTS

The authors would like to thank Justin P. Johnson for his time in creating figures. The lead author was supported through the U.S. DOT Research and Special Programs Administration (RSPA) as part of the National Center for Remote Sensing in Transportation Hazards Consortium (NCRST-H) administered as Separate Transactions Agreement #DTRS56-00-T-003.

17.9 REFERENCES

Abkowitz, M., P. D. M. Cheng, and M. Lepofsky. 1990. "Use of Geographic Information Systems in Managing Hazardous Materials Shipments." *Transportation Research Record* 1261, 35–43.

Aldrich, M. 2002. "Regulating Transportation of Hazardous Substances: Railroads and Reform, 1883–1930." *Business History Review* 76(2):267–97.

Al Hassan, Y., and D. J. Barker. 1999. "The Impact of Unseasonable or Extreme Weather on Traffic Activity Within Lothian Region, Scotland." *Journal of Transport Geography* 7(3):209–13.

Al Homoud, A. S., G. Prior, and A. Awad. 1999. "Modeling the Effect of Rainfall on Instabilities of Slopes along Highways." *Environmental Geology* 37:317–25.

Andrey, J. 1990. "Relationships Between Weather and Traffic Safety: Past, Present, and Future Directions." *Climatological Bulletin* 24:124–36.

Arthur, R., and N. Waters. 1997. "Formal Scientific Research of Traffic Collision Data Utilizing a GIS." *Transportation Planning and Technology* 21:121–37.

Austin, K., M. Tight, and H. Kirby. 1997. "The Use of Geographical Information Systems to Enhance Road Safety Analysis." *Transportation Planning and Technology* 20:249–66.

Austin, R. D., and J. L. Carson. 2002. "An Alternative Accident Prediction Model for Highway-Rail Interfaces." *Accident Analysis and Prevention* 34:31–42.

Baker, C. J. 1988. "High-Sided Articulated Lorries in Strong Cross Winds." *Journal of Wind Energy and Industrial Aerodynamics* 31:67–85.

Basoz, N., and A. Kiremidjian. 1996. *Risk Assessment for Highway Transportation Systems*. J. A. Blume Earthquake Engineering Center, Department of Civil Engineering, Stanford University, Report No. 118.

Benedetto, A. 2002. "A Decision Support System for the Safety of Airport Runways: The Case of Heavy Rainstorms." *Transportation Research A* 8:665–82.

Berdica, K. 2002. "An Introduction to Road Vulnerability: What Has Been Done, Is Done and Should Be Done." *Transport Policy* 9:117–27.

Bertness, J. 1980. "Rain Related Impacts on Selected Transportation Activities and Utility Services in the Chicago Area." *Journal of Applied Meteorology* 19:545–56.

Bier, V. M., R. Zimmerman, Y. Y. Haimes, J. H. Lambert, and N. C. Matalas. 1999. "A Survey of Approaches for Assessing and Managing the Risk of Extremes." *Risk Analysis* 19:83–94.

Bob-Manuel, K. D. H. 2002. "Probabilistic Prediction of Capsize Applied to Small High-Speed Craft." *Ocean Engineering* 29:1841–51.

Bonvicini, S., P. Leonelli, and G. Spadoni. 1998. "Risk Analysis of Hazardous Materials Transportation: Evaluating Uncertainty by Means of Fuzzy Logic." *Journal of Hazardous Materials* 62:59–74.

Brainard, J., A. Lovett, and J. Parfitt. 1996. "Assessing Hazardous Waste Transport Using a GIS." *International Journal of Geographical Information Science* 10:831–49.

Brazel, A. I. 1991. "Blowing Dust and Highways: The Case of Arizona, USA." In *Highway Meteorology,* ed. A. H. Perry and L. J. Symons. New York: E. & F. N. Spon, 131–61.

Brodsky, H., and A. S. Hakkert. 1988. "Risk of a Road Accident in Rainy Weather." *Accident Analysis and Prevention* 17:147–54.

Budetta, P. 2002. "Risk Assessment from Debris Flows in Pyroclastic Deposits along a Motorway, Italy." *Bulletin of Engineering Geology and the Environment* 61:293–301.

Bunce, C. M., D. M. Cruden, and N. R. Morgenstern. 1997. "Assessment of the Hazard from Rock Fall on a Highway." *Canadian Geotech Journal* 34:344–56.

Caputo, A. C., and P. M. Pelagagge. 2002. "An Inverse Approach for Piping Networks Monitoring." *Journal of Loss Prevention in the Process Industries* 15:497–505.

Carson, J., and F. Mannering. 2001. "The Effect of Ice Warning Signs on Ice-Accident Frequencies and Severities." *Accident Analysis and Prevention* 33:99–109.

Cassini, P. 1998. "Road Transportation of Dangerous Goods: Quantitative Risk Assessment and Route Comparison." *Journal of Hazardous Materials* 61:133–38.

Chakraborty, J., M. P. Armstrong. 1996. "Using Geographic Plume Analysis to Assess Community Vulnerability to Hazardous Accidents." *Computers, Environment, and Urban Systems* 19:341–56.

Chang, S. E. 2000. "Disasters and Transport Systems: Loss, Recovery and Competition at the Port of Kobe after the 1995 Earthquake." *Journal of Transport Geography* 8:53–65.

Chang, S. E., W. D. Svekla, and M. Shinozuka. 2002. "Linking Infrastructure and Urban Economy: Simulation of Water-Disruption Impacts in Earthquakes." *Environment and Planning B* 29(2):281–301.

Changnon, S. A. 1999. "Record Flood-Producing Rainstorms of 17–18 July 1996 in the Chicago Metropolitan Area. Part III: Impacts and Responses to the Flash Flooding." *Journal of Applied Meteorology* 38:273–80.

Chapman, R. 1973. "The Concept of Exposure." *Accident Analysis and Prevention* 5:95–110.

Cherington, M. 2000. "Hazards of Bicycling: From Handlebars to Lightning." Seminars in Neurology 20(2):247–53.

Cho, S., M. Shinozuka, and S. Chang. 2001. "Integrating Transportation Network and Regional Economic Models to Estimate the Costs of a Large Urban Earthquake." *Journal of Regional Science* 41: 39–65.

Clague, J. J. 2002. "The Earthquake Threat in Southwestern British Columbia: A Geologic Perspective." *Natural Hazards* 26:7–34.

Collins-Garcia, H., M. Tia, R. Roque, and B. Choubane. 2000. "Alternative Solvent for Reducing Health and Environmental Hazards in Extracting Asphalt—An Evaluation." Construction 2000—*Transportation Research Record* 1712, 79–85.

Cowen, D. J., J. R. Jensen, C. Hendrix, M. E. Hodgson, and S. R. Schill. 2000. "A GIS-Assisted Rail Construction Econometric Model That Incorporates LIDAR Data." *Photogrammetric Engineering and Remote Sensing* 66:1323–28.

Cova, T. J. 1999. "GIS in Emergency Management." In *Geographical Information Systems: Principles, Techniques, Applications, and Management,* ed. P. A. Longley, M. F. Goodchild, D. J. Maguire, and D. W. Rhind. New York: John Wiley & Sons, 845–58.

Cova, T. J., and R. L. Church. 1997. "Modelling Community Evacuation Vulnerability Using GIS." *International Journal of Geographical Information Science* 11(8):763–84.

Cova, T. J., and J. P. Johnson. 2002. "Microsimulation of Neighborhood Evacuations in the Urban-Wildland Interface." *Environment and Planning A* 34(12):2211–29.

———. 2003. "A Network Flow Model for Lane-Based Evacuation Routing." *Transportation Research A*, 37:579–604.

Cutter, S. 1996. "Vulnerability to Environmental Hazards." *Progress in Human Geography* 20:529–39

———. 1997. "Trends in U.S. Hazardous Materials Spills." *Professional Geographer* 49:318–31.

Dalziell, E., and A. Nicholson. 2001. "Risk and Impact of Natural Hazards on a Road Network." *Journal of Transportation Engineering* 127:159–66.

De Scally, F. A., and J. S. Gardner. 1994. "Characteristics and Mitigation of the Snow Avalanche Hazard in Kaghan Valley, Pakistan Himalaya." *Natural Hazards* 9:197–213.

Dobbins, J. P., and M. D. Abkowitz. 2002. "Development of an Inland Marine Transportation Risk Management Information System." Marine Transportation and Port Operations—*Transportation Research Record* 1782, 31–39.

Dunn, C. E., and D. Newton. 1992. "Optimal Routes in GIS and Emergency Planning Applications." *Area* 24:259–67.

Eriksson, M., and J. Norrman. 2001. "Analysis of Station Locations in a Road Weather Information System." *Meteorological Applications* 8:437–48.

Erkut, E., and A. Ingolfsson. 2000. "Catastrophe Avoidance Models for Hazardous Materials Route Planning." *Transportation Science* 34:165–79.

Evans, S. G., and K. W. Savigny. 1994. "Landslides in the Vancouver-Fraser Valley-Whistler Region." *Bulletin—Geological Survey of Canada* 481:251–86.

Evans, A. W. 1997. "Risk Assessment by Highway Organizations." *Transport Reviews* 2:145–63.

Evans, L. 1991. *Traffic Safety and the Driver.* New York: Van Nostrand Reinhold.

Fabiano, B., F. Curro, E. Palazzi, and R. Pastorino. 2002. "A Framework for Risk Assessment and Decision-Making Strategies in Dangerous Good Transportation." *Journal of Hazardous Materials* 93: 1–15.

Ferrante, M., F. Napolitano, and L. Ubertini. 2000. "Optimization of Transportation Networks during Urban Flooding." *Journal of the American Water Resources Association* 36:1115–20.

Fish, M., and R. Lane. 2002. "Linking New Hampshire's Rock Cut Management System with a Geographic Information System." *Transportation Research Record* 1786, 51–59.

Franklin, J. A., and S. A. Senior. 1997. "The Ontario Rockfall Hazard Rating System." In *Proceedings of the International Association of Engineering Geologists Conference, Engineering Geology and the Environment,* vol. 1, 647–56.

Fridstrom, L., J. Ifver, S. Ingebrigtsen, R. Kulmala, and L. K. Thomsen. 1995. "Measuring the Contribution of Randomness, Exposure, Weather, and Daylight to the Variation in Road Accident Counts." *Accident Analysis and Prevention* 27:1–20.

Gerz, T., F. Holzapfel, and D. Darracq. 2002. "Commercial Aircraft Wake Vortices." *Progress in Aerospace Sciences* 38(3):181–208.

Goh, J., and D. Wiegmann. 2002. "Human Factors Analysis of Accidents Involving Visual Flight Rules Flight into Adverse Weather." *Aviation Space and Environmental Medicine* 73:817–22.

Gomez, R. B. 2002. "Hyperspectral Imaging: A Useful Technology for Transportation Analysis. *Optical Engineering* 41:2137–43.

Graettinger, A. J., T. W. Rushing, and J. McFadden. 2001. "Evaluation of Inexpensive Global Positioning System Units to Improve Crash Location Data." Highway Safety: Modelling, Analysis, Management, Statistical Methods, and Crash Location—*Transportation Research Record* 1746:94–101.

Harris, M., R. I. Young, F. Kopp, A. Dolfi, and J. P. Cariou. 2002. "Wake Vortex Detection and Monitoring. *Aerospace Science and Technology* 6:325–31.

Hashash, Y. M. A., J. J. Hook, B. Schmidt, and J. I. C. Yao. 2001. "Seismic Design and Analysis of Underground Structures." *Tunnelling and Underground Space Technology* 16(4):247–93.

He, Y., D. Ma, and P. Cui. 2002. "Debris Flows along the China-Nepal Highway." *Acta Geographica Sinica* 57:275–83.

Helander, M. E., and E. Melachrinoudis. 1997. "Facility Location and Reliable Route Planning in Hazardous Material Transportation." *Transportation Science* 31:216–26.

Hwang, S. T., D. F. Brown, J. K. O'Steen, A. J. Policastro, and W. E. Dunn. 2001. "Risk Assessment for National Transportation of Selected Hazardous Materials." Multimodal and Marine Freight Transportation Issues, *Transportation Research Record* 1763, 114–24.

Isumi, N., N. Nomura, and T. Shibuya. 1985. "Simulation of Post-Earthquake Restoration for Lifeline Systems." *International Journal of Mass Emergencies and Disasters* 87–105.

Jacobs, T. L., and J. M. Warmerdam. 1994. "Simultaneous Routing and Siting for Hazardous Waste Operations." *Journal of Urban Planning and Development* 120:115–31.

Janelle, D. G., and M. Beuthe. 1997. "Globalization and Research Issues in Transportation." *Journal of Transport Geography* 5:199–206.

Johnson, J. H., and D. J. Zeigler. 1986. "Evacuation Planning for Technological Hazards: An Emerging Imperative." *Cities* (May):148–56.

Joyce, T. L., and S. P. Mahoney. 2001. "Spatial and Temporal Distributions of Moose-Vehicle Collisions in Newfoundland." *Wildlife Society Bulletin* 29:281–91.

Kaplan, S., and B. J. Garrick. 1981. "On the Quantitative Definition of Risk." *Risk Analysis* 1:11–27.

Kim, K., and N. Levine. 1996. "Using GIS to Improve Highway Safety." *Computers, Environment, and Urban Systems* 20:289–302.

Klein, C. M. 1991. "A Model for the Transportation of Hazardous Waste." *Decision Sciences* 22:1091–1108.

Lehnert, M. E., and J. A. Bissonette. 1997. "Effectiveness of Highway Crosswalk Structures at Reducing Deer-Vehicle Collisions." *Wildlife Society Bulletin* 25:809–18.

Lepofsky, M., M. Abkowitz, and P. Cheng. 1993. "Transportation Hazard Analysis in Integrated GIS Environment." *Journal of Transportation Engineering* 119:239–54.

Levine, N., and K. E. Kim. 1998. "The Location of Motor Vehicle Crashes in Honolulu: A Methodology for Geocoding Intersections." *Computers, Environment, and Urban Systems* 22:557–76.

Lindell, M. K., and R. W. Perry. 1996. "Addressing Gaps in Environmental Emergency Planning: Hazardous Materials Releases During Earthquakes." *Journal of Environmental Planning and Management* 39:529–43.

List, G. F., and M. A. Turnquist. 1998. "Routing and Emergency-Response-Team Siting for High Level Radioactive Waste Shipments." *IEEE Transactions on Engineering Management* 45:141–52.

Lleras-Echeverri, G., and M. Sanchez-Silva. 2001. "Vulnerability Analysis of Highway Networks: Methodology and Case Study." *Proceedings of the Institution of Civil Engineers—Transport* 147:223–30.

Lovell, C. D., and R. A. Dolbeer. 1999. "Validation of the United States Air Force Bird Avoidance Model." *Wildlife Society Bulletin* 27:167–71.

Malik, A. H. 2000. "Seismic Hazard Study for New York City Area Bridges." Fifth International Bridge Engineering Conference, vols. 1 and 2. *Transportation Research Record* 1696, 224–28.

Marchi, L., M. Arattano, and A. M. Deganutti. 2002. "Ten Years of Debris-Flow Monitoring in the Moscardo Torrent (Italian Alps)." *Geomorphology* 46:1–17.

Marianov, V., C. ReVelle, and S. Shih. 2002. "Anticoverage Models for Obnoxious Material Transportation." *Environment and Planning B* 29:141–50.

McClung, D., and P. Schaerer. 1993. *The Avalanche Handbook.* Seattle: The Mountaineers.

Menoni, S., V. Petrini, F. Pergalani, and M. P. Boni. 2002. "Lifelines Earthquake Vulnerability Assessment: A Systemic Approach." *Soil Dynamics and Earthquake Engineering* 22:1199–1208.

Miller, H. J., and S. L. Shaw. 2001. *Geographic Information Systems for Transportation: Principles and Applications.* New York: Oxford University Press.

Musk, L. F. 1991. "Climate as a Factor in the Planning and Design of New Roads and Motorways." In *Highway Meteorology,* ed. A. H. Perry and L. J. Symons. New York: E. & F. N. Spon, 1–25.

Office of the Federal Coordinator for Meteorological Services and Supporting Research (OFCM). 2002. *Weather Information for Surface Transportation.* National Needs Assessment Report, FCM-R18-2002, U.S. Department of Commerce, National Oceanic and Atmospheric Administration (NOAA).

Oppenheimer, C. 1998. "Volcanological Applications of Meterological Satellites." *International Journal of Remote Sensing* 19:2829–64.

Opricovic, S., and G. H. Tzeng. 2002. "Multicriteria Planning of Post-Earthquake Sustainable Reconstruction." *Computer-Aided Civil and Infrastructure Engineering* 17:211–20.

Osterom, M., and A. Eriksson. 2002. "Snow Mobile Fatalities: Aspects on Preventive Measures from a 25-Year Review." *Accident Analysis and Prevention* 34:563–68.

Panchaanathan, S., and A. Faghri. 1995. "Knowledge-Based Geographic Information System for Safety Analysis at Rail-Highway Grade Crossings." *Transportation Research Record* 1497, 91–100.

Park, G., H. H. Cudney, and D. J. Inman. 2001. "Feasibility of Using Impedance-based Damage Assessment for Pipeline Structures." *Earthquake Engineering and Structural Dynamics* 30:1463–74.

Perry, A. H., and L. J. Symons. 1991. *Highway Meteorology.* New York: E. & F. N. Spon.

———. 1994. "The Wind Hazard in the British Isles and Its Effects on Transportation." *Journal of Transport Geography* 2:122–30.

Persaud, B., R. A. Retting, and C. Lyon. 2000. "Guidelines for Identification of Hazardous Highway Curves, Highway and Traffic Safety: Crash Data, Analysis Tools, and Statistical Methods." *Transportation Research Record* 1717, 14–18.

Petley, D. N. 1998. "Geomorphological Mapping for Hazard Assessment in a Neotectonic Terrain." *Geographical Review* 164:183–201.

Price, B. E., M. Stilson, M. Hansen, J. Bischoff, and T. L. Youd. 2000. "Liquefaction and Lateral Spread Evaluation and Mitigation for Highway Overpass Structure—Cherry Hill Interchange, Davis County, Utah." Soil Mechanics 2000—*Transportation Research Record* 1736:119–26.

Raj, P. K., and E. W. Pritchard. 2000. "Hazardous Materials Transportation on US Railroads: Application of Risk Analysis Methods to Decision Making in Development of Regulations." *Transportation Research Record* 1707:22–26.

Rayback, S. A. 1998. "A Dendrogeomorphological Analysis of Snow Avalanches in the Colorado Front Range." *Physical Geography* 19:502–15.

Refice, A., and D. Capolongo. 2002. "Probabilistic Modeling of Uncertainties in Earthquake-Induced Landslide Hazard Assessment." *Computers and Geosciences* 28:735–49.

ReVelle, C. 1991. "Siting Ambulances and Fire Companies: New Tools for Planners." *Journal of the American Planning Association* 57:471–84.

Rice, R., R. Decker, N. Jensen, R. Patterson, and S. Singer. 2000. "Rural Intelligent Transportation System for Snow Avalanche Detection and Warning." *Transportation Research Record* 1700:17–23.

Rogers, G. B. 1993. "All Terrain Vehicle Injury Risks and the Effects of Regulation." *Accident Analysis and Prevention* 25:335–46.

Romero, S., G. J. Rix, and S. P. French. 2000. "Identification of Transportation Routes in Soils Susceptible to Ground Motion Amplification in the New Madrid Seismic Zone." *Transportation Research Record* 1736, 127–33.

Romin, L. A., and J. A. Bissonette. 1996. "Deer-Vehicle Collisions: Status of State Monitoring Activities and Mitigation Efforts." *Wildlife Society Bulletin* 24:276–83.

Saccomanno, F., and P. Haastrup. 2002. "Influence of Safety Measures on the Risks of Transporting Dangerous Goods Through Road Tunnels." *Risk Analysis* 22:1059–69.

Santi, P. M., E. J. Neuner, and N. L. Anderson. 2002. "Preliminary Evaluation of Seismic Hazards for Emergency Rescue Route, U.S. 60, Missouri." *Environmental and Engineering Geoscience* 8:261–77.

Schaerer, P. 1989. "The Avalanche-Hazard Index." *Annals of Glaciology* 13:241–47.

Schuster, R. L., G. F. Wieczorek, and D. G. Hope II. 1998. "Landslide Dams in Santa Cruz County, California Resulting from the Earthquake." U.S. Geological Survey Professional Paper 1551-C, 51–70.

Schweizer, J., B. Jamieson, and D. Skjonsberg. 1998. "Avalanche Forecasting for Transportation Corridor and Backcountry in Glacier National Park (BC, Canada)." In *Proceedings of 25 Years of Snow Avalanche Research,* ed. E. Hestnes. Publication 203, Norwegian Geotechnical Institute, 238–44.

Selcuk, A. S., and M. S. Yücemen. 2000. "Reliability of Lifeline Networks with Multiple Sources under Seismic Hazard." *Natural Hazards* 21:1–18.

Singh, R. N., J. A. Shonhardt, and N. Terezopoulos. 2002. "A New Dimension to Studies of Spontaneous Combustion of Coal." *Mineral Resource Engineering* 11:147–63.

Sinuany-Stern, Z., and E. Stern. 1993. "Simulating the Evacuation of a Small City: The Effects of Traffic Factors." *Socio-Economic Planning Sciences* 27:97–108.

Sorensen, P. A. 2001. "Locating Resources for the Provision of Emergency Medical Services." Ph.D. dissertation, Department of Geography, University of California Santa Barbara.

Sorensen, J., S. Carnes, and G. Rogers. 1992. "An Approach for Deriving Emergency Planning Zones for Chemical Stockpile Emergencies." *Journal of Hazardous Materials* 30:223–42.

Southworth, F. 1991. *Regional Evacuation Modeling: A State-of-the-Art Review.* ORNL/TM-11740, Oak Ridge National Laboratory.

Spring, G. S., and J. Hummer. 1995. "Identification of Hazardous Highway Locations Using Knowledged-Based GIS: A Case Study." *Transportation Research Record* 1497, 83–90.

Starr, C. 1969. "Societal Benefit versus Technological Risk." *Science* 165:1232–38.

Tangborn, W., and A. Post. 1998. "Iceberg Prediction Model to Reduce Navigation Hazards." In *Ice in Surface Waters,* ed. H. T. Shen. Rotterdam: A. A. Balkema, 231–36.

Thorpe, J. 1964. "Calculating Relative Involvements Rates in Accidents without Determining Exposure." *Australian Road Research* 2:25–36.

Turner, A. K., and R. L. Schuster. 1996. *Landslides: Investigation and Mitigation.* Transportation Research Board Special Report 247, National Academy Press, Washington, DC.

Tzeng, G.-H., and Y.-W. Chen. 1999. "The Optimal Location of Airport Fire Stations: A Fuzzy Multi-objective Programming and Revised Genetic Algorithm Approach." *Transportation Planning and Technology* 23:37–55.

Ullman, G. L. 2000. "Special Flashing Warning Lights for Construction, Maintenance, and Service Vehicles—Are Amber Beacons Always Enough?" Work Zone Safety; Pavement Marking Retroreflectivity—*Transportation Research Record* 1715, 43–50.

Verter, V., and B. Y. Kara. 2001. "A GIS-Based Framework for Hazardous Materials Transport Risk Assessment." *Risk Analysis* 21:1109–20.

Vorobtsov, S. N. 2002. "Estimation of the Hazard of Aircraft Flight in Conditions of Shears of a Three-Dimensional Wind." *Journal of Computer and Systems Science International* 41:703–15.

Wadge, G., A. Wislocki, and E. J. Pearson. 1993. "Spatial Analysis in GIS for Natural Hazard Assessment." In *Environmental Modelling with GIS,* ed. M. F. Goodchild, B. O. Parks, and L. T. Steyaert. Oxford: Oxford University Press, 332–38.

Werner, S., C. E. Taylor, and J. E. Moore. 1997. "Loss Estimation Due to Seismic Risks to Highway Systems." *Earthquake Spectra* 13:585–604.

Wolfe, A. 1982. "The Concept of Exposure and the Risk of a Road Traffic Accident and an Overview of Exposure Data Collection Methods." *Accident Analysis and Prevention* 14:337–40.

Wolshon, B. 2001. "One-Way-Out: Contraflow Freeway Operation for Hurricane Evacuation." *Natural Hazards Review* 2(3):105–112.

Yamada, T. 1996. "A Network Flow Approach to a City Emergency Evacuation Planning." *International Journal of Systems Science* 27:931–36.

CHAPTER 18
INCIDENT MANAGEMENT

Ahmed Abdel-Rahim
*Department of Civil Engineering, University of Idaho,
Moscow, Idaho*

18.1 INTRODUCTION

Incident management has become an important component of the activities of departments of transportation nationwide. With much of the nation's roadway system operating very close to capacity under the best of conditions, the need to reduce the impact of incident-related congestion has become critical. Traffic incidents are causing thousands of hours of congestion and delay annually. Incident-related delay accounts for between 50 and 60 percent of total congestion delay. In smaller urban areas, it can account for an even larger proportion (FHWA 2000b). Incidents also increase the risk of secondary crashes and pose safety risks for incident responders on the incident scene and elsewhere. While incidents on roadways cannot be predicted or prevented entirely, the implementation of an effective incident management system can mitigate the impacts of the incidents.

The success of an incident management program depends largely on coordination and collaboration between different agencies involved in the incident management operations. The emergence of intelligent transportation system (ITS) technologies in the 1990s has brought opportunities to increase the timeliness, effectiveness, and efficiency of incident management operations. The evolution of ITS promises not only new tools for real-time communication, but also tools for incident detection and verification, integrated network-wide management, and the provision of up-to-date traveler information.

This chapter contains a review of the latest developments in incident management programs, with the intention of providing an incident management practical reference and guide to transportation professionals and also to other professionals who might be involved in the incident management process. The topics in this chapter are not covered in great detail, as the breadth of the traffic incident management activities is too wide for this Handbook. More extensive references can be consulted (FHWA 2000a,b; Koehne, Mannering, and Hallenbeck 1995; Carvell et al. 1997).

18.2 CHARACTERISTICS OF TRAFFIC INCIDENTS

18.2.1 Incident Definition

Traffic incidents can be defined as nonrecurring events that cause congestion and delay by restricting normal traffic flow. Traffic incidents can result from either a reduction in the roadway capacity or an increase in the traffic demand. Based on this definition, incidents can be classified into two groups: planned and random events. Planned events include:

- Highway maintenance and reconstruction projects
- Special nonemergency events (e.g., sports activities, concerts, or any other event that significantly affects roadway operations)

Random events include:

- Traffic crashes
- Disabled vehicles on the road
- Spilled cargo
- Natural or man-made disasters

The incident management activities covered in this chapter are focused primarily on random events incidents.

18.2.2   Incident Types

Incident types and frequencies are location-dependent. Differences are due to factors such as road geometry, level of traffic demand, weather, grade, and shoulder availability. The reported relative frequencies of incident types are quite diverse. However, the majority of the reported incidents were found to involve minor events, such as disabled vehicles on the shoulder and other incidents that have little impact on the roadway capacity. An example of a typical composite profile of reported freeway incidents is presented in Figure 18.1, which illustrates the distribution of freeway incidents by incident type, duration, and incident-related delay (ATA and Cambridge Systematics, Inc. 1997).

18.3   INCIDENT IMPACTS

The magnitude of incident-related problems is severe. Traffic incidents pose three primary concerns:

1. Reduction in the operational efficiency of the transportation network, causing significant delay to motorists, including delay to emergency responders
2. Increased risk of secondary crashes
3. Safety risks to incident responders due to increased danger at the incident scene

Problems associated with incidents are not limited to these primary areas; they also include (FHWA 2000b):

- Increased response time by police, fire, and emergency medical services due to roadway delay and reduction in manpower
- Lost time and a reduction in productivity and increased cost of goods and services.
- Increased fuel consumption and reduced air quality and other adverse environmental impacts
- Negative public image of public agencies involved in incident management activities
- Traveler frustration and road rage

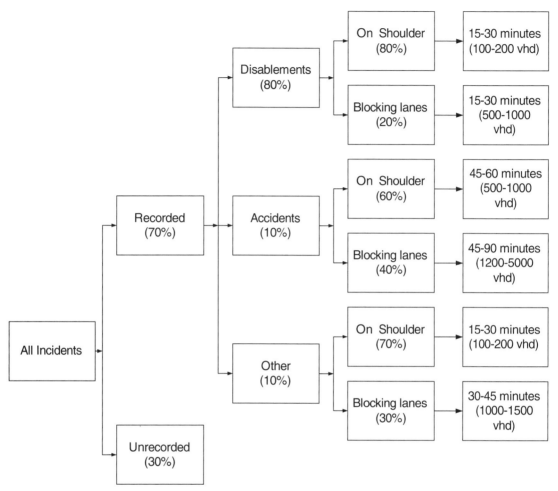

FIGURE 18.1   Incident types. (*Source:* Cambridge Systematics, Inc.)

18.3.1   Incident-Related Congestion

The amount of incident-related congestion for the 10 most congested urban areas in the United States in 1998 ranged from 218,000 to 1,295,000 person-hours of delay. This is equivalent to an annual cost to of $140/driver to $291/driver due to the additional fuel consumed (FHWA 2000b). The incident-related delay is caused by the reduction of roadway capacity that results from the incident. The *Highway Capacity Manual* (*HCM 2000*) (TRB 2000), provides guidelines on the expected reduction in the proportion of freeway segment capacity as a function of number of freeway lanes and the type of incident. Incident types included in the guidelines range from shoulder disablement to full lane closure, as shown in Table 18.1.

Figure 18.2 is a graphical representation of traffic behavior upstream and downstream of an incident location when the incident partially or totally blocks the roadway. Upstream of the incident, the characteristics represent traffic conditions moving at normal speeds and normal density. The area located immediately upstream of the incident represents the high-

TABLE 18.1   Proportion of Freeway Segment Capacity Available under Incident Conditions

Number of freeway lanes by direction	Shoulder disablement	Shoulder accident	One lane blocked	Two lanes blocked	Three lanes blocked
2	0.95	0.81	0.35	0.00	N/A
3	0.99	0.83	0.49	0.71	0.00
4	0.99	0.85	0.58	0.25	0.13
5	0.99	0.87	0.65	0.40	0.20
6	0.99	0.89	0.71	0.50	0.26
7	0.99	0.91	0.75	0.57	0.36
8	0.99	0.93	0.78	0.63	0.41

Source: *Highway Capacity Manual 2000.*

density congested area where vehicles are queuing and traveling at low speeds. The region immediately downstream of the incident reflects traffic flowing at a metered rate with low density and slightly higher speed than normal flow. The far downstream area from the incident represents traffic flow similar to the area far upstream of the incident, with normal traffic flow and normal density. This behavior continues until the incident is completely removed and the queued vehicles upstream of the incident are completely discharged. The length of the congested area, and thus the total incident-related delay, will depend on the incident severity (number of lanes blocked) and the incident duration. Figure 18.3 illustrates incident-based delay represented by cumulative arrivals and departures diagrams throughout the incident duration. Figure 18.4 illustrates possible reduction in incident-based delay with an incident management system. Reducing incident detection, verification, response, and clearance times and diverting some of the freeway traffic to alternate routes can significantly reduce the incident-based delay and queues, as illustrated in the figure.

18.3.2   Increased Risk of Secondary Crashes

The severity of secondary crashes is often greater than that of the original incident. A study conducted in Minnesota found that 13 percent of all peak-period crashes were secondary

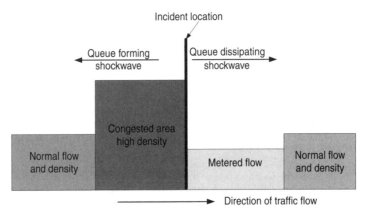

FIGURE 18.2   Behavior of traffic under an incident.

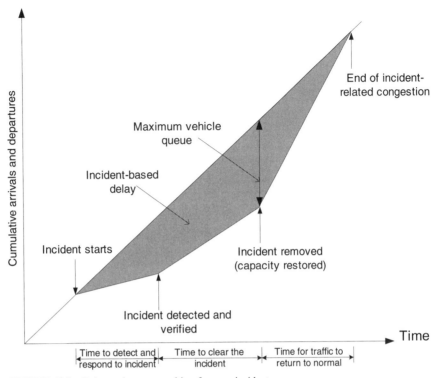

FIGURE 18.3 Delay and queue resulting from an incident.

crashes (Minnesota DOT 1982). Similar results were found in a study by the Washington State Department of Transportation. The study found that 3,165 shoulder crashes had occurred on interstate, limited access, or other state highways during a seven-year period. These collisions caused a total of 40 deaths and 1,774 injuries. The study reported that 41 percent of all shoulder collisions involved injuries (IACP 1996). An analysis of accident statistics in California revealed that secondary crashes represent an increase in collision risk of over 600 percent (Volpe 1995).

18.3.3 Safety Risks to Incident Responders

Hazards at the incident scene put emergency personnel who respond to the incident at risk of being struck by passing vehicles. In 1997, nearly 40 percent of all law enforcement officers who died in the line of duty died in traffic (*The Police Chief* 1998). In addition to the hazards of the incident scene, emergency responders traveling to and from the incident scene are also at risk because of incidents. In 1998, there were 143 fatalities in the United States involving emergency vehicles, 77 of which occurred when the vehicle was responding to an emergency (NHTSA 1998).

18.4 THE INCIDENT MANAGEMENT PROCESS

Incident management consists of a series of activities involving personnel from a variety of response agencies and organizations. These activities are not necessarily performed sequen-

FIGURE 18.4 Possible reduction in delay and queue with an incident management system.

tially. The activities include incident detection, incident verification, incident response, site management and clearance, and motorist information, as shown in Figure 18.5. Table 18.2 lists the roles and responsibilities of different agencies involved in the incident management activities. The following sections discuss each of these components of the incident management process.

18.4.1 Incident Detection

Incident detection is the process by which an incident is identified and brought to the attention of the responsible agency. Some of the methods commonly used to detect incidents include (FHWA 2000b):

- Cellular telephone calls from motorists
- Closed circuit television (CCTV) cameras viewed by operators in traffic management centers
- Automatic incident detection algorithms (AID)
- Motorist aid telephones or call boxes
- Police and service patrol vehicles
- Fleet vehicles (transit and trucking)

With the widespread use of wireless phones, most incidents are being reported by multiple wireless calls to 911-emergency numbers or 511-travel numbers (Figure 18.6).

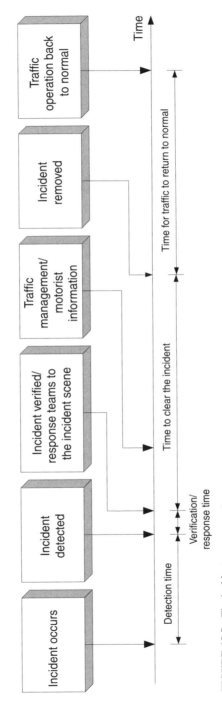

FIGURE 18.5 The incident management process.

TABLE 18.2 Role and Responsibilities of Different Agencies Involved in the Incident Management Process

Agency	Role and Responsibility
Police department	• Assist in incident detection • Secure the incident scene • Assist disabled motorists • Provide emergency medical aid until help arrives • Direct traffic • Conduct accident investigations • Serve as incident commander • Safeguard personal property • Supervise scene clearance
Fire and rescue	• Protect the incident scene • Provide traffic control until police or DOT arrival • Provide emergency medical care • Provide initial HazMat response and containment • Perform fire suppression • Rescue crash victims from wrecked vehicles • Arrange transportation for the injured • Serve as incident commander • Assist in incident clearance
Hazardous materials (HazMat)	• Rescue crash victims from contaminated environments • Clean up and dispose of toxic or hazardous materials.
Emergency medical service (EMS)	• Provide advanced emergency medical care • Determine of destination and transportation requirements for the injured • Coordinate evacuation with fire, police, and ambulance or airlift • Serve as incident commander for medical emergencies • Determine approximate cause of injuries for the trauma center • Remove medical waste from incident scene
Transportation agencies	• Assist in incident detection and verification • Initiate traffic management strategies on incident impacted facilities • Protect the incident scene • Initiate emergency medical assistance until help arrives • Provide traffic control • Assist motorists with disabled vehicles • Provide motorist information • Determine incident clearance and roadway repair needs • Establish and operate alternate routes • Coordinate clearance and repair resources • Serve as incident commander for clearance and repair functions • Repair transportation infrastructure
Towing and recovery	• Remove vehicles from incident scene • Protect victims' property and vehicles • Remove debris from the roadway • Provide transportation for uninjured vehicle occupants • Serve as incident commander for recovery operations
Media and information service providers	• Report traffic incidents • Broadcast information on delays • Provide alternate route information • Update incident status frequently • Provide video or photography services

FIGURE 18.6 911-emergency cellular phone calls help expedite incident detection.

The percentages of incidents detected by cell phone callers to 911 or other dedicated phone numbers are as high as 90 percent, with most incidents being reported within 2 minutes of their occurrence. When multiple calls are received for the same incident, it is important to identify accurately the location of the callers to determine the exact location of the incident. Some agencies use 0.1 mileposts to aid callers in identifying their locations. Cellular phones with GIS-based systems will allow agencies to accurately identify the caller's location and hence expedite the incident detection and verification process.

Automated incident detection (AID) algorithms have been the focus of much research and have been implemented and tested in many systems throughout North America. AIDs can be divided into four major categories:

- Comparative (or pattern recognition) algorithms that compare traffic parameters at a single detector station or between two detector stations against thresholds that define the incident conditions. This category includes 10 modified California algorithms, as well as the pattern recognition (PATREG) algorithm.
- Statistical algorithms that use statistical techniques to determine whether observed detectors' data differ statistically from historical or defined conditions.
- Time-series/smoothing algorithms that compare short-term predictions of traffic conditions to measured traffic conditions.
- Modeling algorithms that use standard traffic flow theories to model expected traffic conditions on the basis of current traffic measurements. Dynamic model and McMaster algorithms fall into this category.

The effectiveness of incident detection algorithms depends mainly on the detection system configuration and the prevailing traffic flow profiles. Successful implementation of incident detection algorithms require a dense network of detectors capable of transmitting real-time speed and occupancy data. The performance of incident detection algorithms can be measured using the following three factors:

- Percentage of incidents detected
- Time required to detect an incident
- False alarm rate

Algorithms with high detection sensitivity and short detection time will typically yield high false alarms. Agencies must decide what is an acceptable balance between detection sensitivity and false alarm rates for their detection system. False alarms can be tolerated in order to achieve a higher detection sensitivity. Research indicates that the best performing algorithms detect 70 to 85 percent of all incidents, with a false alarm rate of 1 percent or less (Hellinga and Knapp 2000; Mahamassani et al. 1999; Chang 1998; Barton-Aschman Associates, Inc. 1993; FHWA 1997.) The state-of-the-practice is to use AID algorithms to flag operators to any change in traffic flow patterns, then use CCTV cameras to verify the nature of problem.

18.4.2   Incident Verification

Incident verification is the process of confirming that an incident has occurred, determining its exact location, and obtaining as many relevant details about the incident as possible before dispatching responders to the incident scene. Verification includes gathering enough information to dispatch the proper initial response. Methods of incident verification include the following:

- Operators viewing CCTV cameras
- Dispatching field units such as police or service patrols to the incident site
- Combining information from multiple cellular phone calls

CCTV cameras can be used to provide the traffic operation centers with valuable information about the specific location, nature, and extent of incidents. If an incident has occurred within the CCTV coverage area, operators can monitor the incident from the operation center. Agencies with CCTV cameras covering more than 90 percent of their freeway system can verify incidents reported by cell phones in less than 2 minutes from the time they are reported. This can reduce the incident response and clearance time significantly.

18.4.3   Incident Response

Incident response includes dispatching the appropriate personnel and equipment to the incident scene. It also includes activating the preestablished communication links between different agencies and the motorist information system. The response should start as soon as the incident is verified. Effective incident response requires preparedness on the part of each responding agency. This is achieved through planning and personnel training, both on an individual basis and collectively with other response agencies.

Two ITS technologies can facilitate better incident response: computer-aided dispatch (CAD) and traffic signal preemption. CAD combines computer and communications technologies to manage communications better among emergency responders and each group's dispatch center. Computer-aided dispatch systems are in place in thousands of fire, police, and other emergency service agencies throughout North America. CAD functions assist dispatchers in tracking the status of field units and in assigning units to respond to an incident based on the severity of the incident, the unit's location, and the equipment required at the incident scene. They also provide police and fire personnel with access to multiple databases containing information on issues such as hazardous materials identification and handling procedures. In addition, CAD systems archive incident response activities, which can be used to conduct postincident review and evaluations.

Traffic signal preemption provides emergency vehicles with green lights, accelerating their move through signalized intersections en route to the incident scene. Signal preemption systems consist of an emitter-equipped vehicle, a detector, and a signal phase selector. Each emitter's communication signal is encoded with a unique identifier and user authorization code. The detector receives the emitter's message and transfers it to the signal controller. The phase selector validates the signal and requests priority control from the traffic controller. Command priority results in the system providing a green light to authorized vehicles after appropriate intersection timing is complete, usually within a few seconds. However, emergency vehicle preemption benefit comes with considerable disutility cost. Every emergency vehicle preemption causes disruption to the corridor operation and may cause considerable delay to motorists.

18.4.4 Site Management

Site management is defined as the process of managing the incident site and coordinating on-scene resources. The primary objective of site management is to ensure the safety of response personnel, incident victims, and other motorists on or near the incident scene (Figure 18.7). When multiple agencies respond, the unified command structure concept has helped solve some of the coordination and communication problems. The unified command structure provides a management tool to facilitate cooperative participation by representatives from many agencies and/or jurisdictions. It encourages the incident commanders of the major agencies to work together in a unified command post. Communication to the unified command post from the traffic management center is of key importance.

18.4.5 Incident Clearance

Incident clearance is the process of removing vehicles, wreckage, debris, spilled material, and other items to return roadway capacity to normal levels. Clearance is the most critical step in managing major incidents, due to the length of time required to restore traffic flow. The objectives of incident clearance are to:

- Restore the roadway to its preincident capacity as quickly and safely as possible to minimize motorist delays.
- Make effective use of all clearance resources.

FIGURE 18.7 Site management for an incident on a multilane freeway section—safety and operational challenges.

- Enhance the safety of responders and motorists.
- Protect the roadway system and private property from unnecessary damage during the removal process.

Planning for efficient incident clearance involves scheduling and deploying of response personnel and selecting the appropriate methods to remove all types of incidents. It also involves identifying the resources available and obtaining agreements for the use of resources. Training for response personnel and postincident reviews can significantly improve the incident-clearance operation.

Tow trucks, typically operated by private companies, are the most common resource used in incident-clearance activities. Before dispatching tow trucks to the incident scene, tow companies need to be provided with specific information on the incident type so they can dispatch the appropriate equipment to the incident location. For example, heavy-duty wreckers are used for removal or recovery of large trucks, and the availability of heavy recovery equipment is the primary factor in the clearance of major truck-related incidents (Figure 18.8).

18.4.6 Accident Investigations

Accident investigations are part of any incident and play a significant role in incident clearance time. Evidence preservation and determining collision dynamics require specific procedures and take a significant amount of time. When collision scenes become crime scenes, priority is given to thorough investigations. Incident scenes are thus closed and access is limited to authorized personnel only. The priorities for thorough investigation often conflict with the desire of transportation agencies to reopen the road as quickly as possible. Investigators can often prioritize their work to facilitate movement of traffic. They may complete the portions that allow partial opening of the roadway first, provided there is adequate traffic control available to maintain their safety while finishing other investigative tasks. Shoulder investigation for minor accidents minimizes traffic disruption and reduces incident-based delay (Figure 18.9).

FIGURE 18.8 Dispatching the right equipment to the incident scene reduces incident clearance time.

FIGURE 18.9 Shoulder accident investigation reduces traffic disruption and improves network operation.

Accident investigations require precise and detailed measurements of the crash scene. The survey needed to take these measurements often lengthens the closure time. New techniques and technology advances, such as total station surveying systems and photogrammetry, have significantly reduced the time required for the on-site surveys. Total station surveying systems expedite the on-scene measurements and can require fewer lane closures. Photogrammetry is quickly emerging as an alternative to total station measuring. Digital photographs are taken at different angles and locations. A software program has been designed to determine accurate measurements from the photographs. The Oregon, California, Arizona, Washington, and Utah State Police are now using this system (FHWA 2000b).

18.5 INCIDENT TRAFFIC MANAGEMENT

Traffic management, in the context of an incident, is the implementation of temporary traffic control measures at the incident site and on facilities affected by the incident to minimize traffic disruption while maintaining a safe workplace for responders. Traffic control measures can be categorized into two groups: measures intended to improve traffic flow past the incident scene and measures intended to improve traffic flow in affected areas and alternate routes.

Measures to improve flow past the incident site include:

- Establishing point traffic control at the incident scene and deploying appropriate personnel to assist in managing traffic
- Effectively managing the incident scene space and taking necessary measures to block lanes and other areas to ensure the safety of the responders. This also includes staging and parking emergency vehicles and equipment in a way that minimizes the impact on traffic flow and ensures maximum safety

Measures to improve traffic flow in areas affected by the incident and on alternate routes include:

- Real-time management of traffic control devices (including ramp meters, lane control signs, and traffic signals) in the areas where traffic flow is affected by the incident
- Designating, developing, and operating alternate routes

18.5.1 Traffic Management and Incident Site Operation

After an incident has occurred, it is critical to establish a safe worksite as quickly as possible. Two primary concepts that should be the focus of traffic management personnel are establishing point traffic control at the scene and managing the roadway space.

Establish Point Traffic Control at the Scene.   In order for traffic to move smoothly and safely past the incident, traffic control needs to be established at the incident scene. If lanes or roadways will be closed, traffic needs to be clearly marked to merge into lanes or shoulders that will remain open to traffic. Portable signs that warn motorists of the lane closure transition should be placed upstream from the incident site. Shoulders and even the oncoming lanes can be utilized to provide a lane of travel around the incident scene (Figure 18.10). Responders who will set up traffic control at the incident scene should be familiar with the requirements of the *Manual on Uniform Traffic Control Devices (MUTCD)*.

Manage the Roadway Space at the Incident Location.   The primary concepts in managing the roadway space are twofold: (1) to close only those lanes that are absolutely essential for protection of the incident responders and victims, and (2) to minimize the time that those lanes are closed.

The number of lanes that must be closed may change with the stage of the incident management and clearance efforts. This means that traffic control may be established and then changed several times. The incident responders responsible for traffic management will need to stay informed about the planned sequence of work to clear the incident. They should continually assess the impacts of the incident on traffic flows and monitor the extent of the queues. This information should be communicated to the traffic management center staff, who, in turn, can pass the information on to the motorist information functions. Information

FIGURE 18.10   Point traffic management at the incident scene to ensure responders and motorist safety.

can also be passed on from the center to those responsible. This information should be used to adjust and modify the site traffic control plan.

Methods for reducing the duration of incidents or the number of lanes closed may include (FHWA 2000b):

- Immediately move any vehicle which can move under its own power.
- Equip responders with push-bumpers to facilitate expedient clearance.
- Open individual lanes as soon as they are cleared. In some cases this opportunity occurs when one of the response vehicles leaves the scene.
- Clear incident scenes from left to right, gradually yet systematically shrinking the scene and moving toward the right shoulder and the nearest freeway exit.
- Encourage the first responder from any agency to remove accident debris or a small portion of a spilled load to clear one or two lanes.
- Pour oil dry or sand onto small spills to restart traffic flow in one lane.
- Require the first arriving wrecker to clear travel lanes first, aggressively and in a safe manner.
- Do not allow elaborate rigging, cargo off-loading, vehicle repairs, or the loading and se-curing of cars on roll back (flatbed) wreckers until all the travel lanes are cleared.
- Make sure that appropriate type and number of tow units is requested very early in the incident.
- Consider including in towing and recovery agreements the requirement that all wreckers that respond to freeway incidents have push-bumpers installed. This includes heavy-duty units and flatbeds.

A high priority must also be given to the vehicles trapped between the diversion point, usually the last upstream exit, and the incident site. In some cases these motorists can be allowed to proceed past the scene by opening a shoulder or even a portion of one lane for a brief period then securing the scene and continuing the incident management activities.

18.5.2 Improve Traffic Flow on Alternate Routes—Integrated Incident Management Systems

Many states and metropolitan areas have already initiated integrated incident management systems to ensure optimal utilization of the existing freeway and arterial system networks. These often consist of an extensive network of incident detection, surveillance, and information-dissemination devices strategically located along the freeway and arterial net-works and connected to a local or regional traffic management center (Figure 18.11). Careful analysis of diversion strategies, which includes examination of the real-time operational characteristics of both the freeway and alternate routes, can lead to more efficient and ef-fective incident management strategies.

Some studies suggest that the implementation of dynamic traffic assignment algorithms, which provide analytical tools to develop optimal integrated incident management plans, can significantly improve the flow of traffic on the network during incident situation (Messmer, Albert, and Papageorgiou 1995; Al-Deck and Kanafani 1991; Reis, Gartner, and Cohen 1991). The developments in dynamic algorithms and in sensing and communication tech-nologies, however, have not been accompanied by corresponding developments in evaluation and decision-making tools that can take advantage of real-time data to produce more efficient control strategies for managing freeway incidents. The state-of-the-practice is often limited to having diversion plans that cannot effectively account for the stochasticity of demand,

FIGURE 18.11 Traffic management center operation (courtesy ADA County High District, Boise, Idaho).

incident severity, and system disturbances such as drivers' compliance with the provided control.

Alternate Route Planning. When a major incident occurs that severely limits roadway capacity, motorists will naturally find ways to divert around the incident. Some regions have chosen to formally establish alternate routes to direct traffic to the routes that are best suited to handle this increased traffic demand. If a region chooses to implement alternate routes, it is critical that all agencies affected by the implementation and operation of alternate routes be involved in every step of the planning and operation of those diversion plans. Diversion practices are also discussed in detail in the National Cooperative Highway Research Program Synthesis 279, "Roadway Incident Division Practices" (NCHRP 2000).

Alternate Route Operations and Real-Time Control. When an incident occurs that results in traffic diverting to other routes, whether as a result of using a planned diversion route or because motorists chose to continue their trip on a less congested route, measures should be taken to manage the increased traffic on alternate routes.

The existing traffic control plan should be adjusted to accommodate the diverted traffic. This may include the following (Figure 18.12):

- Use variable message signs and highway advisory radio to disseminate information on the incident or to direct traffic to the most appropriate alternate route.
- Adjust ramp metering rates to account for changes in traffic flows.
- Adjust signal timing to account for changing traffic conditions.

Since freeways and arterial streets are typically operated by different agencies, personnel from the agencies responsible for operating these traffic control devices will need to imple-

FIGURE 18.12 Example of alternate route plan (courtesy ADA County High District, Boise, Idaho).

ment the changes. Close cooperation and collaboration in these situations is usually facilitated by early planning efforts and ongoing interaction.

The operational procedures of implementing alternate routes must be developed to meet the needs of individual regions. The typical chronology of establishing alternate routes include the following steps (FHWA 2000b):

- Determination that an alternate route should be implemented
- Identification of most applicable preplanned alternate route
- Notification of noninvolved agencies that will be affected by the alternate routes
- Modification of traffic signal timings on alternate routes
- Activation of traffic control devices and traveler information sources
- Monitoring of the alternate routes
- Communication with noninvolved agencies affected by alternate routes that the use is about to be terminated
- Termination of the use of the alternate routes

Uncertainties Associated with Alternate Route Planning and Operation.   There are several uncertainties associated with incidents that affect the incident management and alternate route planning and operations, including:

- The actual duration of the incident cannot be accurately predicted.
- There is no effective way in practice to achieve a theoretical optimum diversion from an open freeway. When drivers are allowed to make their own choice based on the advisory

messages disseminated to them through traveler information systems (uncontrolled and elective), diversion transportation professionals have very little control over how many drivers actually divert.

• The routing of vehicles that are diverted cannot be accurately predicted. Specifically, not all diverted vehicles will return to the freeway, as some drivers will choose to complete their trip on the arterial network once they have been diverted from the freeway. Many of the drivers who are willing to divert will pursue routes other than the designated diversion route to complete their trip.

18.6   MOTORIST INFORMATION

Motorist information involves activating various means of disseminating incident-related information to affected motorists. Media used to disseminate motorist information include:

• Commercial radio broadcasts
• Highway advisory radio (HAR)
• Dynamic message signs (DMS)
• Telephone information systems
• In-vehicle or personal data assistant information or route guidance systems
• Commercial and public television traffic reports
• Internet/on-line services
• A variety of dissemination mechanisms provided by information service providers

   Motorist information includes the dissemination of incident-related information to two different motorist groups. The first group are motorists who are at or approaching the scene of the incident and receive motorist information en route. The second group are motorists who plan to travel using a route that passes through the incident location and receive pre-trip motorist information.

   Motorist information should be provided as early in the incident management process as possible and should continue until the incident has been cleared completely and the traffic queues resulting from the incident have completely dissipated. Motorist information reduces incident-based delay and has been shown to help incident response and clearance activities. It does this by reducing traffic demand at and approaching the incident scene, reducing secondary incidents, and improving responder safety on-scene.

18.6.1   En Route Motorist Information

Dynamic message signs (DMS), also known as variable message signs (VMS) or changeable message signs (CMS), can be used to disseminate travel information on a real-time basis. Fixed-location and portable signs can be used to support incident management functions. DMSs are most often used to:

• Inform motorists of varying traffic, roadway, and environmental conditions (including variable speed limits in adverse conditions).

• Provide specific information regarding the location and expected duration of incident-related delays.

• Suggest alternate routes because of construction or a roadway closure.

- Redirect diverted drivers back onto the freeway.
- Inform drivers when passage along the shoulder is permissible (e.g., in the event of a major incident, where this can help restore the traffic flow safely).
- Display amber alerts.
- Provide information on special events that may affect freeway traffic (concerts, football games, etc.).
- Provide specific travel time information to freeway users regarding expected travel time to specific destination.

Highway Advisory Radio (HAR) information is communicated to drivers via their vehicles' AM radio receivers (Figure 18.13). With signs and flashing lights placed upstream of the incident scene, drivers are instructed to tune into a specific frequency to hear a message about upcoming congestion. Message transmission can be controlled either on-site or from a remote location. Information disseminated by HAR is similar to that provided by VMS. An advantage of HAR, according to research, is that since the information is audible, the driver experiences less sensory overload than when having to read and process a written message, which also requires taking his or her eyes off the road. Another advantage of HAR over VMS is that longer, more complex messages can be communicated.

HAR does have disadvantages: drivers who do not have functioning radios cannot access the information and drivers who tune in midway through the message have to listen through another cycle. It also requires specific action on the part of the driver: that he or she pay attention to the advance signal and then tune in to the specified station.

With the spread of cellular telephone usage, transportation agencies have a new and quite powerful way to communicate with travelers who are already on the road. Wireless phone hotlines can be used by drivers to access construction, congestion, and incident-related travel information on the route or routes they are interested in traveling, via touch-tone menus. Commercial radio has long provided traffic information to commuters and travelers. Its clear advantage is that it is well understood and utilized by millions of drivers.

FIGURE 18.13 Highway advisory radio (HAR) effectively disseminate information to motorists at and near the incident scene.

18.6.2 Pre-Trip Motorist Information

The increasing number of Internet users at home, in the workplace, and in schools has made traffic websites relatively inexpensive and effective tools to disseminate pre-trip traffic information to motorists. Many agencies maintain websites with easy-to-read real-time traffic information such as system maps that are color-coded to indicate travel speeds and congestion on given links.

Commercial television and radio stations in most major cities provide traffic reports to indicate incident locations, other congestion, malfunctioning traffic signals, and other traffic-related information. To get around the limited time that commercial TV will typically devote to traffic information, another possibility is through public access TV. Travel and traffic information stations, which may take the form of video monitors mounted in a wall or on a countertop at fixed public locations, can also be used to provide incident-related information to travelers such as expected travel times, roadway closures, and alternate routes.

Alphanumeric pagers can be used to provide real-time traffic information for specified routes. This technology has already been tested as part of ITS programs in Seattle, Minneapolis, and New Jersey. A limitation of this technology is the small number of characters that can be broadcast in a given message. Personal data assistants (PDAs) allow users to interact directly with travel information systems. This interaction allows users to obtain route-planning assistance, traffic information broadcasts, and other information.

18.7 ASSESSING INCIDENT MANAGEMENT PROGRAM BENEFITS

The greatest benefits of an effective incident management program are achieved through the reduction of incident duration. Reducing the duration of an incident can be achieved by:

- Reducing the time to detect incidents
- Initiating an expedient and appropriate incident response
- Reducing incident clearance time

Substantial reductions in response and clearance of incidents can be achieved through the implementation of policies and procedures that are understood and agreed upon by each player in the incident management process. Benefits resulting from an effective incident management program can be characterized as both quantitative and qualitative.

18.7.1 Quantitative Benefits

No consistent standard has been identified that can be uniformly applied to evaluate the quantifiable benefits of an effective incident management program. In part, this results from the relatively diverse structure and operations of incident management programs (FHWA 2000b). Each program is developed to meet the unique identified needs of the given region. Incident management programs are also generally developed to fit within the existing institutional framework. In addition, baseline data against which to measure a new program's benefits (e.g., incident response times) are rarely available. The Federal Highway Administration *Traffic Incident Management (TIM) Self-Assessment Guide* (FHWA 2002) provides general guidelines that can help agencies assess the effectiveness of their traffic incident management programs.

Quantifiable benefits generally associated with an effective incident management program include:

- Increased survival rates of crash victims
- Reduced delay
- Improved response time
- Improved air quality
- Reduced occurrence of secondary incidents
- Improved safety of responders, crash victims and other motorists

18.7.2 Qualitative Benefits

Qualitative benefits generally associated with an effective incident management program include:

- Improved public perception of agency operations
- Reduced driver frustration
- Reduced travel time and improved air quality
- Improved coordination and cooperation of response agencies

18.8 INCIDENT MANAGEMENT WITHIN NATIONAL ITS ARCHITECTURE

One of the market packages defined in the National ITS Architecture (USDOT 1998, 2001) is the Incident Management market package (ATMS08—Incident Management System), which includes incident-detection capabilities through roadside surveillance devices and regional coordination with other traffic management centers. Information from these diverse sources is collected and correlated by this market package to detect and verify incidents and implement an appropriate response.

ATMS08 also supports traffic operations personnel in developing an appropriate regional response in coordination with emergency and other incident response personnel to confirmed incidents. Incident response also includes dissemination of information to affected motorists using the Traffic Information Dissemination market package (ATMS06) and dissemination of incident information to travelers through the Broadcast Traveler Information (ATIS1) or Interactive Traveler Information (ATIS2) market packages. Other market packages specific to incident management are HazMat management, emergency response, and emergency routing.

18.9 PLANNING AN EFFECTIVE INCIDENT MANAGEMENT PROGRAM

18.9.1 Define Mission, Goals, and Objectives

The process of defining mission, goals, and objectives and identifying existing problems or limitations helps determine incident management program needs. The incident management program mission is ultimately to create a safe and reliable transportation system. Goals are the desired effects of an effort. They provide ways of defining the mission in terms of specific achievements. Common goals of traffic incident management are to reduce delay and congestion caused by traffic incidents on freeways and reduce the number and severity of secondary crashes. Visible outcomes help define opportunities for system improvement and specific results to be achieved. A clear set of quantitative objectives must be defined at the early stages of the planning incident management planning process. An example of an objective might be a 50 percent reduction in average detection and clearance time for minor traffic incidents. Another important component of any effective incident management planning process is the identification of problems that limit the ability to meet stated goals and objectives.

18.9.2 Compare Alternatives and Define Performance Measures

Through each step of the incident management process, many alternatives that address specific objectives and deliver results should be considered. This generally includes operational and procedural alternatives combined in a comprehensive system to support the program mission, goals and objectives. Performance measures are used to evaluate how well various alternatives meet program objectives. Performance measures are most clearly applied in terms of quantifying an objective, but they can be measured in less quantitative ways. Responder observation and public feedback have been used as qualitative performance measures by many agencies.

18.10 BEST PRACTICE IN INCIDENT MANAGEMENT PROGRAMS

The following represent the best practice in incident management programs (FHWA 2000b).

18.10.1 Transportation Agencies

- Develop predesigned response plans for freeway closures, which include diversion routes and traffic control. These response plans have to be developed and reviewed in coordination with police, fire, and other local officials.
- Deploy service patrol vehicles to remove debris from travel lanes and assist motorists broken down on the freeway shoulder or in travel lanes. These service vehicles are critically important during peak periods and must be equipped with arrow boards to assist with traffic control for incidents.
- Create video links from traffic management centers to share with law enforcement and fire/rescue agencies.
- Participate in the incident command system on the incident scene and communicate with fire and police agencies for the prompt clearance of the scene.
- Set up safe traffic control around the crash scene, divert traffic upstream of an incident through the use of changeable message signs, and provide traffic information to the media and general public.
- Install reference markers at 2/10th-mile increments, which will allow cellular phone callers to report incident locations accurately.

18.10.2 Law Enforcement Agencies

- Coordinate with transportation, fire, and transportation agencies to develop incident response plans.

- Within the unified incident command system at the incident scene, communicate with transportation agencies to establish traffic management plans/detours and direct a partial or complete reopening of the roadway as quickly as possible.
- For accident investigations, efficiently collect evidence and survey scene using advanced tools such as total station equipment or aerial surveying.
- For minor noninjury crashes that involve property damage only, have dispatchers provide guidance to drivers on local policy for moving vehicles from travel lanes, and exchanging information as per state law.

18.10.3 Fire and Emergency Medical Agencies

- Dispatch the minimum amount of equipment necessary, to reduce the exposure of personnel at the scene.
- Provide for effective training in the identification of hazardous materials, to avoid lengthy lane closures for material that does not pose a threat to people or the environment.
- Provide effective training in temporary traffic control around incidents.
- Set up an effective communication system as part of the incident command system, so that partner response agencies are aware of progress in rescue efforts and can make correct decisions regarding traffic management, and provide traveler information to local media.

18.10.4 Towing and Recovery

- Prequalify towing companies so the towing company called to the incident scene has the capability needed to handle the vehicles involved.
- Train law enforcement personnel, to ensure that responders can request the correct equipment be dispatched to the incident.
- Weigh the cost–benefit of calling in third-party recovery teams, and consider whether their distance/time of travel will have excessive impact on the amount of time lanes remain closed.
- Move commercial vehicles or trailers to the roadside or shoulder to restore as many travel lanes as possible, as soon as possible, then perform any necessary salvage operations after the peak hour.

18.11 REFERENCES

Al-Deek, H., and A. Kanafani. 1991. "Incident Management with Advanced Traveler Information Systems." In *Proceedings of the Vehicle Navigation and Information Systems Conference,* Warrendale, PA, 563–76.

American Trucking Association (ATA) and Cambridge Systematics, Inc. 1997. *Incident Management: Challenges, Strategies and Solutions for Advancing Safety and Roadway Efficiency.* Final Technical Report, ATA and Cambridge Systematics, Inc., February.

Barton-Aschman Associates, Inc. 1993. *Incident Detection and Response System in North Dallas County.* Prepared for the Texas Department of Transportation, October.

Carvell, J. D., K. Balke, J. Ullman, K. Fitzpatrick, L. Nowlin, and C. Brehmer. 1997. *Freeway Management Handbook.* U.S. Department of Transportation, Federal Highway Administration, August.

Chang, E. C.-P. 1998. "Operational Sensitivity Evaluation of a Speed-Based Incident Detection Algorithm." In *Transportation Technology for Tomorrow: Conference Proceedings,* ITS America 8th Meeting, Detroit.

Federal Highway Administration (FHWA). 1997. *Development and Testing of Operational Incident Detection Algorithms.* Executive Summary, September.

——. 2000a. *Incident Management Successful Practices—A Cross-Cutting Study: Improving Mobility and Saving Lives.* Publication No. FHWA-JPO-99-018, U.S. Department of Transportation, FHWA, ITS Joint Program Office, April.

——. 2000b. *Traffic Incident Management Handbook.* Publication No. DOT-T-01-01, U.S. Department of Transportation, FHWA, Office of Traffic Management. Prepared by PB Farradyne, November.

——. 2002. *Traffic Incident Management (TIM)* Self-Assessment Guide. U.S. Department of Transportation, FHWA Office of Operation, November.

Hellinga, B., and G. Knapp. 2000. "Automatic Vehicle Identification Technology-Based Freeway Incident Detection." *Transportation Research Record* 1727, 142–53.

International Association of Chiefs of Police (IACP). 1996. *The Highway Safety Desk Book.* IACP, April.

Koehne, J., F. L. Mannering, and M. E. Hallenbeck. 1995. *Framework for Developing Incident Management Systems.* U.S. Department of Transportation, Federal Highway Administration, October.

Lomax, T., S. Turner, H. L. Levinson, and R. H. Pratt. 1996. *Quantifying Congestion.* Phase III Final Report, National Research Council, Transportation Research Board, National Cooperative Highway Research Program, September.

Mahamassani, H. S., C. Haas, J. Peterman, and S. Zhou. 1999. *Evaluation of Incident Detection Methodologies.* Report FHWA/TX-00-1795-S, Project Summary, Federal Highway Administration/Texas Department of Transportation/University of Texas, Austin, October.

Messmer, A., and M. Papageorgiou. 1995. "Route Diversion Control in Motorway Networks via Nonlinear Optimization." *IEEE Transactions on Control Systems Technology* 3(1):144–54.

Minnesota Department of Transportation. 1982. *I-35 Incident Management and the Impact of Incidents on Freeway Operation,* January.

National Highway Traffic Safety Administration (NHTSA). 1998. *Fatality Analysis Reporting System.* U.S. Department of Transportation, NHTSA.

Police Chief, The. 1998. "National Police Week Observed." International Association of Chiefs of Police, May.

National Cooperative Highway Research Program (NCHRP). 2000. *Roadway Incident Division Practices.* Synthesis 279, National Research Council, Transportation Research Board, NCHRP, Washington, DC.

Reis, R. A., N. H. Gartner, and S. L. Cohen. 1991. "Dynamic Control and Traffic Performance in a Freeway Corridor: A Simulation Study." *Transportation Research A* 25A(5):267–76.

Transportation Research Board (TRB). 2000. *Highway Capacity Manual.* National Research Council, TRB, Washington, DC.

U.S. Department of Transportation (USDOT). 1998. *Developing Freeway and Incident Management Systems Using the National ITS Architecture.* USDOT, Intelligent Transportation System Joint Program Office, August.

——. 2001. Version 4.0 of the National ITS Architecture.

Volpe National Transportation Systems Center. 1995. *Intelligent Transportation Systems Impact Assessment Framework.* Final Report, September.

CHAPTER 19
TRANSPORTATION NOISE ISSUES

Judith L. Rochat
U.S. Department of Transportation/Volpe Center,
Cambridge, Massachusetts

19.1   INTRODUCTION

Noise is a serious issue that should be considered in all stages of transportation system projects, from original design and construction to modifications. Transportation-related noise affects millions of people and, in many cases, requires local, state, and federal governments to provide noise abatement to help improve or restore their quality of life. The impact of noise on the quality of life can be substantial, especially with expanding transportation systems. This is why there is an increasing need for noise control, and why the field of transportation-related noise is thriving.

There are three modes of transportation where noise issues are typically addressed: highway traffic, aircraft, and rail. Although less common, other transportation-related noise concerns can also warrant noise control or at least consideration; these include: construction noise, noise inside vehicles, recreational vehicle noise, and underwater noise.

This text provides an overview of noise issues, followed by discussions of the specific issues relating to different modes of transportation, including noise sources, noise prediction, noise metrics, noise control, and relevant resources.

19.2   SOUND AND NOISE

19.2.1   Sound

Sound is a vibratory disturbance created by a moving or vibrating source. Examples of transportation sound sources include steady traffic on a highway, construction equipment used to build a highway or bridge, and a jet flying overhead. Each sound source can be described by its associated spectrum, amplitude, and time history. The *spectrum* reveals the frequency content of a sound. Figures 19.1 and 19.2 show two examples of typical spectra corresponding to transportation noise sources. Figure 19.1 shows an example spectrum of steady highway traffic; as can be seen, the most dominant frequency range is between 200 and 2000 Hz. Figure 19.2 shows an example of a propeller-driven aircraft in flight; for this spectrum, it is seen that there are some tonal components (perceived as a pitch by humans) in the lower frequencies, a noticeable tone being at 100 Hz. Both of these sound sources are broadband in nature—they are complex sounds containing multiple frequencies.

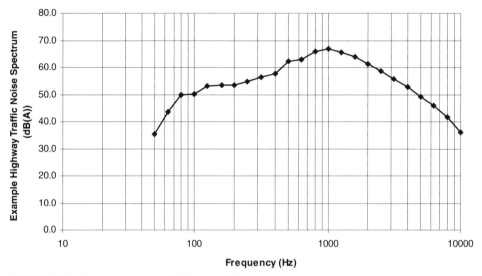

FIGURE 19.1   Example spectrum of highway traffic noise.

The *amplitude* of a sound indicates its volume; in general, higher amplitudes indicate louder sounds. The amplitude of a sound can be measured as a sound pressure level with a microphone/sound level meter or spectrum analyzer system; for guidance on instrumentation, please refer to section 19.3. Time-averaged sound levels or maximum sound levels are metrics often used to quantify sound, the associated quantities presented in units of decibels (dB). A decibel level is a logarithmic representation of sound energy and is calculated using

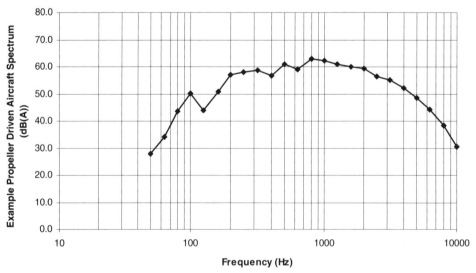

FIGURE 19.2   Example spectrum of a propeller-driven aircraft.

TABLE 19.1 Perceptions of Loudness

Sound level change	Descriptive change in perception
+20 dB	Four times as loud
+10 dB	Twice as loud
+5 dB	Readily perceptible increase
+3 dB	Barely perceptible increase
0 dB	Reference
−3 dB	Barely perceptible reduction
−5 dB	Readily perceptible reduction
−10 dB	Half as loud
−20 dB	One quarter as loud

a medium-dependent (e.g., air or water) reference energy level. The reference level is based on human perception such that the sound energy associated with the threshold of unimpaired human hearing will correspond to 0 dB. In general, a sound that is 10 dB higher than another sound with the same spectral characteristics is said to be twice as loud; please refer to Table 19.1 for different perceptions of loudness. The previously presented spectral data (Figures 19.1 and 19.2) show the amplitude for each third-octave frequency band; the band levels logarithmically added together equate to the overall amplitude.

The *time history* of a sound reveals its amplitude variation over time. The perception of a sound (to be discussed further in the noise section that follows) is also dependent on its time-varying characteristics. Sound with little variation over time is a continuous sound; sound that exists only for a brief time is a transient sound. Figures 19.3 and 19.4 show two examples of typical time histories: Figure 19.3 shows the time history for steady highway traffic, a relatively continuous sound, and Figure 19.4 shows the time history for a jet flying overhead, a transient sound. The onset of the transient sound in Figure 19.4 is relatively

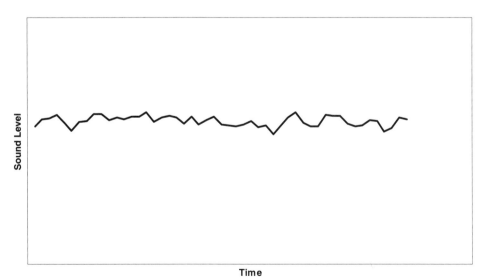

FIGURE 19.3 Example time history of continuous highway traffic.

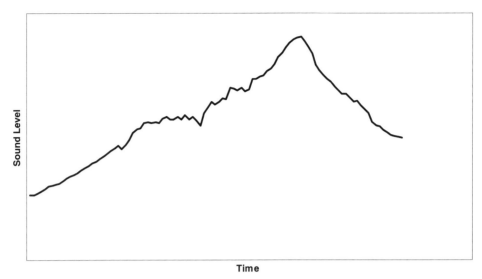

FIGURE 19.4 Example time history of a jet flyover.

gradual; in general, a commercial jet can be heard approaching, flying overhead, and then departing. Many transient sounds have more rapid onset times (sometimes referred to as impulsive sounds), where the amplitude increases substantially within a fraction of a second.

19.2.2 Noise

Certain sounds are considered to be pleasant and therefore found to be acceptable to a listener; some sounds, on the other hand, are unwanted and therefore considered to be noise. Noise is defined in different ways, depending on the listener; it is dependent on the spectrum, amplitude, and time history of the sound. However, some types of noise are understood to be objectionable. Many types of transportation noise sources are often included in this category. Figure 19.5 shows differing sound levels for typical noise sources, including transportation noise sources. As can be seen, a train horn 200 feet (60 meters) away from the source reaches a level of about 105 dB(A); the train horn noise is a loud burst that is tonal and startling. A jet flying overhead at an altitude of 1,000 feet (305 meters) reaches a maximum level on the ground of about 90 dB(A); the jet noise has a longer onset time than the train horn and is less startling, with broader frequency content. Highway traffic noise at a distance of about 300 feet (90 meters) from the highway has an almost continuous noise level of about 60 dB(A); often highway traffic noise is a continuous broadband sound.

It is straightforward to measure sound pressure levels in order to quantify sound; however, describing noise involves quantifying its perception. Quantifying perception accounts for the hearing abilities of the listener. Humans do not hear equally well at all frequencies; they ideally hear frequencies from 20 to 20,000 Hz (although sensitivity to higher frequencies decreases with age), where the hearing is most sensitive from about 1,000 to 6,300 Hz. To describe sound levels in a manner that closely approximates normal human hearing, the actual sound levels are typically modified by applying *A weighting*. This is a response function that spans the audible frequency range, emphasizing frequencies in the most sensitive range and deemphasizing frequencies out of the sensitive range. Although the A-weighted sound level is the most widely used measure of environmental noise and is internationally

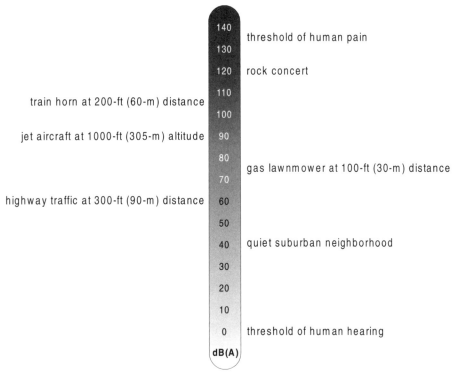

FIGURE 19.5 Example noise sources.

accepted, there are other frequency weighting functions that may be more suitable to specific noise analyses.

19.2.3 Effects of Noise

The effects of noise on people and wildlife can be substantial; as a result, serious consideration should be given to transportation noise impacts. Effects of noise on people range from annoyance to stress-related illnesses. Noise can interfere with communication and concentration, an example being that noise diminishes the ability of students to learn in a classroom environment (Seep et al. 2000). Noise can cause sleep deprivation and stress, contributing to high blood pressure and heart disease. For wildlife, there are serious concerns about noise interfering with communication, migration, and reproduction. For a review on noise effects see Sandberg and Ejsmont (2002), FRA (1998), and Lee and Fleming (2001), and for specifics on marine life, refer to Richardson et al. (1995) and Rochat (1998).

19.2.4 Noise Metrics

The metric applied to quantify noise depends on the noise source and purpose for the noise measurement. There are hundreds of metrics defined (please refer to Shultz 1982). Table 19.2 lists noise metrics that are commonly used in transportation noise engineering. Some

TABLE 19.2 Summary of Commonly Used Transportation Noise Metrics

Metric name	Metric symbol	Metric abbreviation	Metric description
A-weighted equivalent sound level	L_{AeqT} (T = time increment)	LAEQ	Sound level associated with the sound energy averaged over a specified time period.
Day-night average sound level	L_{dn}	DNL	L_{Aeq24h} with a 10 dB(A) penalty between the hours of 10 p.m. and 7 a.m. (sleeping hours).
Community noise equivalent level	L_{den}	CNEL	L_{Aeq24h} with a 10 dB(A) penalty between the hours of 10 p.m. and 7 a.m. (sleeping hours) and a 5 dB(A) penalty between the hours of 7 p.m. and 10 p.m. (relaxation, conversation hours).
A-weighted maximum sound level	L_{Amax}	LAMAX	A-weighted maximum sound level during a noise event or specified time period.
Percent exceeded sound level	L_X (e.g., L_{10})	—	The sound level exceeded X percentage of the time during a specified time period.
A-weighted sound exposure level	L_{AE}	SEL	The time integral of sound level over the course of a single event.
Effective perceived noise level	L_{EPN}	EPNL	A sound level based on human perception and accounting for tonal components and event duration, used primarily for assessing aircraft noise.
Time above/percent time above	TA, %TA	TA, %TA	Time or percentage of time that the A-weighted noise level is above a user-specified sound level during a specified time period.
Detectability level	D′L	D′L	A measure of the ability of detecting a particular sound in the presence of other noise. A function of the signal-to-noise ratio.

of these metrics require additional calculations beyond sound level measurements, but noise prediction software usually calculates the desired metric for the user. With each of these metrics, different frequency weightings, as previously mentioned, can also be applied. See the appropriate noise policy to find out which metrics are acceptable in a particular situation. In later sections of this chapter, when addressing specific modes of transportation, typical noise metrics will be listed.

19.2.5 Vibratory Disturbances

Vibratory disturbances other than sound are also generated by transportation systems. Acoustic waves traveling through the ground or air cause structures such as windows, walls, and floors to vibrate. These vibrations can adversely affect building inhabitants because of structural noise, uncomfortable sensations, or interference with delicate procedures where vibration isolation is mandatory (e.g., surgical procedures).

Vibrations can also adversely affect the structures themselves, potentially causing structural damage (e.g., cracks). Example sources of such vibrations are trains and aircraft. Vibrations from trains extend beyond the track, the waves traveling through the ground into nearby communities. Low-altitude or fast-flying aircraft can generate loud sounds that cause structures to vibrate, sometimes violently. (Sonic booms from supersonic aircraft have, on rare occasions, shattered windows in nearby buildings.)

For guidance on vibration vocabulary, please refer to ISO 2041. For guidance on the measurement and evaluation of vibration in buildings, please refer to ANSI (1990). For guidance on methods for analysis and presentation of vibrational data, refer to ANSI (1971). References for the appropriate mode of transportation will provide guidance to sources of vibration, measuring vibration, vibratory analysis, vibration control, and related policy issues.

19.2.6 General Noise Resources

References are listed by topic. See the References at the end of this chapter for full citations.

Overall: Beranek and Vér (1992), Harris (1991)

Transportation noise: Federal government website on transportation noise, http://ostpxweb.dot.gov/policy/noise/noise.htm; Fleming et al. (2000)

Noise measurements: Beranek (1988)

Noise metrics: Shultz (1982)

Noise effects: Seep et al. (2000), Sandberg and Ejsmont (2002), FRA (1998), Lee and Fleming (2001), Richardson et al. (1995), Rochat (1998)

19.3 *INSTRUMENTATION FOR MEASURING NOISE*

It is common to measure noise using a microphone/sound level meter or spectrum analyzer system. The choice of field instrumentation and settings is highly dependent upon the type of noise being measured and the purpose of the measurements. For specific guidance on instrumentation requirements, please refer to references cited in this section.

19.3.1 Microphones

Microphone choice is largely based on sensitivity, frequency response, directivity, and environmental performance. A few examples will be given for the four elements listed:

Microphone sensitivity: Larger-diameter microphones are used for measuring quieter sounds and smaller-diameter microphones for louder sounds; for typical community noise measurements, one-half-inch diameter microphones are most commonly used.

Frequency response: Smaller microphones generally exhibit better high-frequency response; for community noise measurements, microphones with good frequency response over the range of 50 to 10,000 Hz will convey information required for most transportation noise measurements.

Microphone directivity: Directional microphones give best results when pointed directly at the sound source (normal incidence), whereas pressure-response microphones give uniform response to sound sources in a plane perpendicular to the axis of the microphone (grazing incidence). (See Figure 19.6 for an illustration.) Microphone orientation requirements are determined by regulatory and practical concerns.

Environmental performance: Electret condensor microphones provide improved reliability in a humid environment over standard condensor microphones, which combine a flat frequency response and high sensitivity.

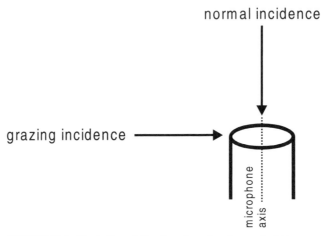

FIGURE 19.6   Illustration of directions for microphone orientation.

For more information on microphones, please refer to Beranek (1988) and also microphone manufacturers' documentation.

19.3.2 Sound Level Instrumentation

Sound level instrumentation choice is dependent on the required noise metric (spectrum analyzer when frequency content is required, integrating capability for time-averaged sound levels, desired frequency weighting capabilities, etc.). It is also important to consider requirements for portability and environmental reliability/durability for a particular set of measurements. In addition, regulations and standards for particular noise measurements recommend or require specific types of sound level instrumentation. Sound level meters meeting the type 1 specifications of ANSI S1.43-1997 (ANSI 1997) are usually recommended/required for transportation noise measurements.

For resources on sound level instrumentation refer to ANSI (1983, 1986, 1997) and Beranek (1988).

For the latest on available instrumentation, refer to current issues of *Sound and Vibration Magazine* or *Physics Today Magazine.* Also, many vendors attend noise-related conferences where product literature and demonstrations are available.

Conferences include Inter-Noise (International Institute of Noise Control Engineering, http://www.i-ince.org), Noise-Con (Institute of Noise Control Engineering of the USA, http://www.inceusa.org), Acoustical Society of America (http://asa.aip.org), and Transportation Research Board (http://www.nas.edu/trb/).

19.4 *PREDICTION OF TRANSPORTATION NOISE*

There are several computer programs available for the prediction of a specific mode of transportation noise. These models help to determine which residences/communities are impacted by transportation noise and allow the user to design some noise-abatement strategies. The computer programs will be listed under the corresponding mode of transportation.

19.5 HIGHWAY TRAFFIC NOISE

Highway traffic noise is something many of us encounter on a daily basis, although those most affected are people who live, work, or attend school next to a busy highway or roadway. Highways dominate our ground transportation systems. In the United States the road network is over 3.9 million miles (6.3 million kilometers) in length. Millions of people drive on the highways, including those driving automobiles commuting to work (where a single automobile is typically the quietest vehicle type on the road) and those driving tractor-trailers hauling freight across the country (where a single heavy truck is typically the loudest vehicle type on the road). All of the vehicles on the highway contribute to the noise that affects adjacent communities.

19.5.1 Highway Traffic Noise Sources

Highway noise is caused by tire/pavement interaction, aerodynamic sources (turbulent airflow around and partly through the vehicle), and the vehicle itself (the power unit noise created by the engine, exhaust, transmission, etc.). At highway speeds, tire/pavement interaction is generally the most dominant source. Figure 19.7 shows a generalized plot of highway noise levels as a function of speed for five vehicle types and average pavement. (The sound levels are averaged over different types of pavement.)

The interaction between tires and pavement is quite complex, the generated noise level being highly dependent on the road surface and the tire tread pattern and construction.

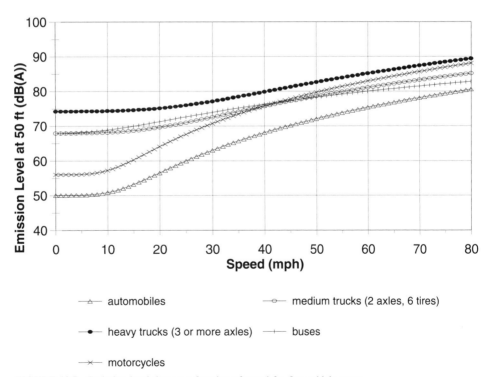

FIGURE 19.7 Emission level data as a function of speed for five vehicle types.

A brief overview of different pavement types will be described here; see Sandberg and Ejsmont (2002) for more details.

Two broad categories of pavements are asphalt and cement. Asphaltic concrete (AC) can vary in chipping or stone size (the aggregate), filler, and binder; the final formula and the construction affect its acoustical properties. Dense-graded asphaltic concrete (DGAC or DAC) is less porous, and open-graded asphaltic concrete (OGAC or PAC for porous AC) is more porous, the latter allowing some noise to be absorbed. In general, OGAC is "quieter" than DGAC; also, in general, increasing the maximum chipping size increases the noise. Portland cement concrete (PCC) is paved with a smooth surface, after which a surface treatment is applied for reasons of safety. An example of a PCC surface treatment is tining. Tining (small grooves in the pavement) is done longitudinally (in the direction of traffic flow) or transversely (perpendicular to the direction of traffic flow); longitudinal tining is "quieter" than transverse tining. In general, DGAC and OGAC surfaces are "quieter" than PCC surfaces.

19.5.2 Highway Traffic Noise Prediction

Several computer software packages are available for obtaining highway traffic noise predictions. These include:

1. The Federal Highway Administration's Traffic Noise Model (TNM) (Anderson et al. 1998; Menge et al. 1998)

2. The Federal Highway Administration's Standard Method in Noise Analysis (Stamina) (Barry and Regan 1978)

3. The University of Florida's Community Noise Model (CNM) (Wayson and MacDonald 1998)

4. SoundPLAN (SoundPLAN)

TNM is the official United States Federal Government model and it (or its predecessor, Stamina, which will soon be phased out) must be used in order for state departments of transportation to receive federal funding for highway noise-abatement projects.

19.5.3 Highway Traffic Noise Metrics

There are some generally accepted practices for quantifying highway traffic noise. For quantifying highway traffic noise in communities, the A-weighted equivalent sound level (L_{AeqT} or LAEQ), the day-night average sound level (DNL or L_{dn}), the community noise equivalent level (L_{den} or CNEL), and the percent exceeded sound level (L_X, e.g., L_{10} for 10 percent) are commonly applied. For analyzing pavements, the A-weighted maximum sound level (L_{Amax} or LAMAX) is applied; it is also useful to look at spectral information. For a single number to characterize pavement, the Statistical Pass-By Method (SPB) and/or the Close-Proximity Method (CPX) (Sandberg and Ejsmont 2002) can be applied, where L_{Amax} is measured. For most highway noise measurements, an A-weighted metric is usually applied. Refer to Table 19.2 for metric descriptions. Also, see the appropriate federal, state, or local noise policy to find out which metrics are acceptable for highway noise analysis in a particular area.

19.5.4 Highway Traffic Noise Control

Communities with residences, schools, parks, etc. that meet the criteria for noise abatement are eligible to receive some form of noise control. The qualifications for noise abatement can be found by accessing the appropriate noise policy; refer to Knauer et al. (2000) for U.S. policies by state, or contact the state noise representative—contact information can be found at www.thewalljournal.com/a1f04/states.htm.

The most typical highway noise-abatement tool is a noise wall or wall type noise barrier; many are designed to meet a goal of 10 dB(A) reduction. These walls are constructed alongside a highway, blocking the line of sight between the vehicles and people in the community. Blocking the line of sight typically reduces the sound by 5 dB; extending the wall above the line of sight helps to reduce the noise further. There are many variations in noise wall material and construction, where considerations are made for the available space, community acceptance, and durability, among other items. Refer to Knauer et al. (2000) for more details.

In some cases, noise walls are not feasible or effective, and buildings next to a highway may need to be altered (e.g., double-paned windows installed). An acoustically absorptive type of pavement on the highway surface may also be used to reduce highway noise; such pavement would require periodic examination to determine the potential need for repaving or cleaning.

19.5.5 Highway Traffic Noise Resources

References are listed by topic. See the references at the end of this chapter for full citations.

Overview: Sandberg and Ejsmont (2002); Rochat and Fleming (1999); FHWA (2000)

Measurement: ANSI (1995, 1998); Lee and Fleming (1996); Sandberg and Ejsmont (2002)

Prediction software: Menge et al. (1998); Barry and Regan (1978); Wayson and Mac-Donald (1998); SoundPLAN

Noise-abatement policy: Knauer et al. (2000); State DOTs (www.thewalljournal.com/a1f04/states.htm); FHWA (1995); 23 CFR Part 772

Tires/roadways: Sandberg and Ejsmont (2002)

Noise barriers: Knauer et al. (2000)

19.6 AIRCRAFT NOISE

There are over 14,000 airports in the United States, some of the major airports being the busiest in the world. In the United States and other countries around the world, noise from aircraft is affecting millions of people daily, people who live, work, go to school, etc. near the airports or helipads and under the flight paths.

19.6.1 Aircraft Noise Sources

The noise generated from aircraft is attributed to the engines and aerodynamic noise sources. In general, each type of aircraft generates a distinctive type of sound that is dependent on its construction, mode of propulsion (e.g., jet or propeller), and state of operation (e.g., departure, approach, cruise, etc.). For fixed-wing aircraft, jets in general can be heard on the ground as a broadband noise source, with some having higher-frequency tones (>1000 Hz); the sound of propeller-driven aircraft in general is more biased toward lower frequencies, usually with a prominent low-frequency tone (<500 Hz). For rotorcraft, helicopters can be heard on the ground sometimes as noise similar to that of jets and sometimes similar to that of propeller-driven aircraft, often with a repetitive impulsive quality; tiltrotor aircraft (a multi-modal aircraft that uses both wings and rotors to generate lift) can sound like helicopters or propeller-driven aircraft depending on the mode of flight.

For each aircraft the generated noise is affected by the aircraft's operations. Three operational states are:

1. Taking off from a runway or helipad
2. Approaching a runway or helipad
3. En route flight (cruise)

As part of the operations, the physical aerodynamic configuration of the aircraft (flaps, landing gear, etc.), engine settings (thrust, propeller speed, etc.), and relative motion of the aircraft (airspeed, attitude, etc.) can all affect the noise. The specific design of the aircraft (shape, size, structural configuration, etc.) also affects the noise. Lastly, the total weight of the aircraft (including fuel, passengers, and cargo) affects the noise; this variable is dependent on both the design limitations and the aircraft's operations.

Some aircraft operations can cause structural vibrations. A sonic boom, for example, created by an aircraft flying faster than the speed of sound, is impulsive and can be very loud. The received noise can reveal itself as window and wall vibrations (sometimes causing structural damage).

As a note, aircraft noise received on the ground can be substantially affected by the atmosphere since the sound is traveling over great distances when the aircraft is in flight. Wind, turbulence, and temperature gradients can affect the received sound level and the spectral content.

19.6.2 Aircraft Noise Prediction

Several computer software packages are available for obtaining aircraft noise predictions. These include:

1. The Federal Aviation Administration's Integrated Noise Model (INM) (Gulding et al. 1999)
2. The Federal Aviation Administration's Helicopter Noise Model (HNM) (Fleming and Rickley 1994)
3. The Federal Aviation Administration's Area Equivalent Method (AEM) (FAA 2001)
4. The Federal Aviation Administration's Air Traffic Noise Screening Model (ATNS) (FAA 1999)
5. The Federal Aviation Administration's Noise Integrated Routing System (NIRS) (FAA NIRS)
6. The U.S. Air Force's NOISEMAP (Galloway 1974)
7. NASA Langley Research Center's Rotorcraft Noise Model (RNM) (Lucas 1998)

INM and HNM are official U.S. federal government models, and they must be used in order to receive federal funding for aircraft noise-abatement projects.

19.6.3 Aircraft Noise Metrics

There are some generally accepted practices for quantifying aircraft noise. For aircraft noise in communities, the day-night average sound level (DNL or L_{dn}) is often applied. For noise certification of aircraft, the effective perceived noise level (L_{EPN} or EPNL) and A-weighted sound exposure level (L_{AE} or SEL) are applied, the metric dependent on the type of aircraft;

the noise certification process is overseen by the Federal Aviation Administration and is necessary to certify that new aircraft or new configurations of aircraft generate acceptable noise levels (CFR 14 Part 36). For noise in national parks, possible supplementary metrics that can be applied are Time above/Percent time above (TA/%TA) and detectability level (D'L) (Fidell et al. 1994, Reddingius 1994); the study of aircraft noise in wilderness areas is important to preserving the natural quiet environment. Refer to Table 19.2 for metric descriptions. Also see the appropriate noise policy to find out which metrics are acceptable for aircraft noise analysis in a particular area.

19.6.4 Aircraft Noise Control

For those who meet the criteria to receive noise abatement, some measures can be taken to help reduce the noise. Noise walls can be constructed to help reduce idle and take-off noise in the nearby communities. Where there is no option to block the line of sight between the source and the receiver, buildings can be better insulated to help the reduction; for example, double-paned windows can be installed or air conditioning units can be purchased to allow windows to be closed.

19.6.5 Aircraft Noise Resources

References are listed by topic. See the references at the end of this chapter for full citations.

Overview: Hubbard (1995); FAA, Environmental Network (www.faa.gov/programs/en/noise/); FAA, Office of Environment and Energy (AEE) (www.aee.faa.gov/Noise/)

Measurement: ANSI (1995); Fleming (1996); SAE-ARP-4821

Prediction software: Gulding et al. (1990); Fleming and Rickley (1994); FAA (1999, 2001); FAA NIRS; Galloway (1974); Lucas (1998)

Noise-abatement policy: 14 CFR Part 36; 14 CFR Part 150; FAA, Environmental Network (www.faa.gov/programs/en/noise/); FAA, Office of Environment and Energy (AEE) (www.aee.faa.gov/Noise/); FAA Order 1050

Sonic booms: Plotkin and Sutherland (1990); Rochat (1998); Hubbard 1995

19.7 RAIL NOISE

The total length of the rail network in the United States extends over 145,000 miles (225,000 kilometers). The noise and/or vibrations from high-speed trains, lower-speed trains, and train horns all affect communities near the tracks or guideways.

19.7.1 Rail Noise Sources

The sources creating the noise from rail operations are: engine noise (propulsion machinery); mechanical noise (wheel/rail interactions or guidance vibrations); aerodynamic noise (for speeds greater than 160 mph); and warning device noise (horns).

For safety purposes, train horns are sounded at a minimum of 96 dB measured at a distance of 100 feet (30 meters) away.

19.7.2 Rail Noise Predictions

Most rail noise predictions are accomplished through mathematical calculations rather than software packages. The program FTANOISE (see www.fta.dot.gov/library/planning/enviro/noise/ftanoise.html) uses a spreadsheet program that allows users to predict rail noise.

19.7.3 Rail Noise Metrics

There are some generally accepted practices for quantifying rail noise. For single events, such as a train passing through a community, the A-weighted maximum sound level (L_{Amax} or LAMAX) or the sound exposure level (L_{AE} or SEL) is applied. For the cumulation of multiple events, the A-weighted equivalent sound level (L_{AeqT} or LAEQ) or the day-night average sound level (DNL or L_{dn}) is applied. For train horns, L_{Amax} is applied. Refer to Table 19.2 for metric descriptions. Also, see the appropriate noise policy to find out what metrics are acceptable for rail noise analysis in a particular area.

19.7.4 Rail Noise Control

Abatement of rail noise can be addressed in multiple ways. First, the noise source can be modified (modifications to the cars or rails). To help reduce the noise before it reaches the community, noise walls can be installed and the buffer zone between the rail and community can be increased. Also, improved sound insulation in the buildings will help reduce the noise level.

Because train horns are an important element of highway-rail grade crossing safety, their sounds often go unmitigated. There are, however, some noise-control measures that have been implemented or are proposed. These include:

1. Removing the safety hazard at the crossing to remove the need for the horn (e.g., by constructing a bridge over the tracks, by introducing retractable solid barriers to block vehicles from entering the track area during train operations, etc.)

2. Limiting the maximum sound level

3. Limiting the direction of the sound

19.7.5 Rail Noise Resources

References are listed by topic. See the references at the end of this chapter for full citations.

Overview: FRA (1998) (high-speed rail); FRA (1999) (horns); Stusnick et al. (1982) (rail); FTA (1995) (rail)

Measurement: FRA (1998) (high-speed rail); FRA (1999) (horns); Stusnick et al. (1982) (rail); FTA (1995) (rail)

Prediction software: FTA (information and download of FTANOISE, www.fta.dot.gov/library/planning/enviro/noise/ftanoise.html)

Noise-abatement policy: 40 CFR Part 201; 49 CFR Part 210; FRA (1998) (high-speed rail); FRA (1999) (horns); Stusnick et al. (1982) (rail); FTA (1995) (rail)

19.8 OTHER TRANSPORTATION-RELATED NOISE

Although the three modes of transportation discussed in sections 19.5, 19.6, and 19.7 are those most commonly referred to in the noise-abatement literature, other types of transportation-related noise should also be mentioned.

The construction of our transportation systems also causes noise. The noise from machinery used to build our roadways, for example, varies in frequency content, amplitude, and time history—the noise generated from jackhammers to bulldozers to back-up alarms. Although the machinery itself has noise limitations, an entire construction site as a whole can be quite disturbing to its neighbors. Some noise abatement is temporarily applied in these situations; usually noise barriers are placed between the construction sites and the communities and restricted hours of operation are enforced. Resources for construction noise are Reagan and Grant (1976) and Schexnayder and Ernzen (1999); some of the references for specific modes of transportation also discuss construction noise (e.g., FRA 1998).

In addition to the communities affected by transportation vehicle noise, it should also be noted that those people operating the vehicles are also affected. The noise in the cab of a train, for example, can be quite loud; strict noise limits must be adhered to, and in some cases hearing protection is essential. See the previous section on rail noise for resources.

Recreational vehicle noise is also a source of concern. Open wilderness areas such as national parks attract vehicles like snowmobiles and all-terrain vehicles (ATVs), contributing to noise in areas trying to maintain a natural quiet environment and also to noise in neighboring communities. As an example, the noise from snowmobiles is an issue at Yellowstone National Park. For resources on recreational vehicle noise, refer to the specific area or park in question.

Underwater noise can be created by several transportation sources: shipping traffic, bridge construction, underwater naval operations, and aircraft. All must be considered in order to quantify the underwater sound levels. In some cases, noise control is warranted in order to reduce adverse effects on marine life. Resources for underwater noise are Richardson et al. (1995) and Rochat (1998).

19.9   OTHER NOISE INFORMATION

The elements required for a successful noise study include:

1. Proper identification of the noise source(s)
2. Proper noise metric selection
3. Proper instrumentation and field measurements
4. Proper noise prediction
5. Proper data analysis and interpretation

Additionally, for successful noise control projects, other items include:

1. Adequate public participation
2. Proper application of federal, state, and local requirements and limitations

19.10   ACKNOWLEDGMENTS

The author wishes to thank Dave Read for his valuable contributions. The author also wishes to thank other members of the John A. Volpe National Transportation Systems Center, Acoustics Facility, particularly Gregg Fleming, Christopher Roof, Amanda Rapoza, and Eric Boeker, for their input.

19.11 REFERENCES

American National Standards Institute (ANSI) and the Acoustical Society of America Standards. 1971. *Methods for Analysis and Presentation of Shock and Vibration Data.* ANSI S2.10-1971 (R2001). Acoustical Society of America, New York, NY.

———. 1983. *Specification for Sound Level Meters.* ANSI S1.4-1983 (R2001). Acoustical Society of America, New York, NY.

———. 1986. *Specification for Octave-Band and Fractional Octave-Band Analog and Digital Filters.* ANSI S1.11-1986 (R1998). Acoustical Society of America, New York, NY.

———. 1990. *Vibration of Buildings—Guidelines for the Measurement of Vibration and Evaluation of Their Effects on Buildings.* ANSI S2.47-1990 (R2001). Acoustical Society of America, New York, NY.

———. 1994. *Acoustical Terminology.* ANSI S1.1-1994 (R1999). Acoustical Society of America, New York, NY.

———. 1995. *Measurement of Sound Pressure Levels in Air.* ANSI S1.13-1995 (R1999). Acoustical Society of America, New York, NY.

———. 1997. *Specifications for Integrating Averaging Sound Level Meters.* ANSI S1.43-1997 (R2002). Acoustical Society of America, New York, NY.

———. 1998. *Methods for Determining the Insertion Loss of Outdoor Noise Barriers.* ANSI S12.8-1998. Acoustical Society of America, New York, NY.

Anderson, G. S., C. S. Y. Lee, G. G. Fleming, and C. W. Menge. 1998. *FHWA Traffic Noise Model, Version 1.0: User's Guide.* Reports No. FHWA-PD-96-009 and DOT-VNTSC-FHWA-98-1, U.S. Department of Transportation, Volpe National Transportation Systems Center, Acoustics Facility, Cambridge, MA.

Barry, T. M., and J. Regan. 1978. *FHWA Traffic Noise Prediction Model.* Report No. FHWA-RD-77-108, U.S. Department of Transportation, Federal Highway Administration, Washington, DC.

Beranek, L. L. 1988. *Acoustical Measurements.* Woodbury, NY: Acoustical Society of America.

Beranek, L. L., and I. L. Vér, eds. 1992. *Noise and Vibration Control Engineering: Principles and Applications.* New York, NY: John Wiley & Sons.

14 CFR Part 36. Federal Aviation Administration—Code of Federal Regulations, 14 CFR Part 36, *Noise Standards: Aircraft Type and Airworthiness Certification.* U.S. Department of Transportation, Federal Aviation Administration, Washington, DC. www.aee.faa.gov/Noise/.

14 CFR Part 150. Federal Aviation Administration—Code of Federal Regulations, 14 CFR Part 150, *Airport Noise Compatibility Planning.* U.S. Department of Transportation, Federal Aviation Administration, Washington, DC. www.aee.faa.gov/Noise/.

23 CFR Part 772. Federal Highway Administration—Code of Federal Regulations, 23 CFR Part 772, *Procedures for Abatement of Highway Traffic Noise and Construction Noise.* U.S. Department of Transportation, Federal Highway Administration, Washington, DC.

40 CFR Part 201. Federal Railroad Administration—Code of Federal Regulations, 40 CFR Ch. 1 (7-1-01 Edition) Part 201, *Noise Emission Standards for Transportation Equipment: Interstate Rail Carriers.* U.S. Department of Transportation, Federal Railroad Administration, Washington, DC.

49 CFR Part 210. Federal Railroad Administration—Code of Federal Regulations, 49 CFR Ch. 11 (10-1-00 Edition) Part 210, *Railroad Noise Emission Compliance Regulations.* U.S. Department of Transportation, Federal Railroad Administration, Washington, DC. www.fra.dot.gov/rdv/environment/index.html.

Federal Aviation Administration (FAA). 1999. *Air Traffic Noise Screening Module (ATNS) Version 2.0 User Manual.* U.S. Department of Transportation, FAA, Washington, DC. www.faa.gov/programs/en/noise/.

———. 2000. "Aviation Noise Abatement Policy 2000." *Federal Register* 65, no. 136. U.S. Department of Transportation, FAA, Washington, DC. www.aee.faa.gov/Noise/ or www.faa.gov/programs/en/noise/.

———. 2001. *Area Equivalent Method (AEM) Version 6.0c User's Guide.* U.S. Department of Transportation, FAA, Office of Environment and Energy, Washington, DC. www.faa.gov/programs/en/noise/.

———. *Noise Integrated Routing Systems* (FAA NIRS). U.S. Department of Transportation, FAA, Washington, DC. www.faa.gov/programs/en/noise/.

———. FAA Order 1050, *Policies and Procedures for Considering Environmental Impacts.* U.S. Department of Transportation, Federal Aviation Administration, Washington, DC. www.faa.gov/programs/en/noise/.

Federal Highway Administration (FHWA). 1995. *Highway Traffic Noise Analysis and Abatement Policy and Guidance.* U.S. Department of Transportation, FHWA, Office of Environment and Planning, Noise and Air Quality Branch, Washington, DC. www.fhwa.dot.gov/environment/noise/index.htm.

———. 2000. *Highway Traffic Noise in the United States—Problem and Response.* U.S. Department of Transportation, FHWA, Office of Environment and Planning, Noise and Air Quality Branch, Washington, DC. www.fhwa.dot.gov/environment/noise/index.htm.

Federal Railroad Administration (FRA). 1998. *High-Speed Ground Transportation Noise and Vibration Impact Assessment.* U.S. Department of Transportation, FRA, Office of Railroad Development, Washington, DC. www.fra.dot.gov/rdv/environmental_impact_assesment/guidance.htm.

———. 1999. *Proposed Rule for the Use of Locomotive Horns at Highway-Rail Grade Crossings.* Draft Environmental Impact Statement, U.S. Department of Transportation, FRA, Office of Railroad Development, Washington, DC.

Federal Transit Administration (FTA). 1995. *Transit Noise and Vibration Impact.* Report No. DOT-T-95-16, U.S. Department of Transportation, FTA, Washington, DC.

Fleming, G. G. 1996. *Aircraft Noise Measurement—Instrumentation and Techniques,* Letter Report No. DTS-75-FA653-LR5, U.S. Department of Transportation. Volpe National Transportation Systems Center, Acoustics Facility, Cambridge, MA.

Fidell, S., K. Pearsons, M. Sneedon. 1994. Evaluation of the Effectiveness of SSAR 50-2 in Restoring Natural Quiet to Grand Canyon National Park. BBN Report No. 7197.

Fleming, G. G., and E. J. Rickley. 1994. *HNM Helicopter Noise Model, Version 2.2, User's Guide.* Reports No. DOT-FAA-EE-94-01 and DOT-VNTSC-FAA-94-3, U.S. Department of Transportation, Volpe National Transportation Systems Center, Acoustics Facility, Cambridge, MA.

Fleming, G. G., R. E. Armstrong, E. Stusnick, K. D. Polcak, and W. Lindeman. 2000. *Transportation-Related Noise in the United States.* TRB A1F04, Committee on Transportation-Related Noise and Vibration, Millennium Publication, National Research Council,Transportation Research Board, Washington, DC.

Galloway, W. J. 1974. *Community Noise Exposure Resulting from Aircraft Operations: Technical Review.* Report No. AMRL-TR-73-106, Wright-Patterson Air Force Base, Aerospace Medical Research Laboratory.

Gulding, J., et al. 1999. *Integrated Noise Model (INM) Version 6.0 User's Guide.* Report No. FAA-AEE-99-03, U.S. Department of Transportation, Federal Aviation Administration, Office of Environment and Energy, Washington, DC.

Harris, C. M., ed. 1991. *Handbook of Acoustical Measurements and Noise Control.* New York, NY: McGraw-Hill.

Hendriks, R. W. 1998. *Technical Noise Supplement: A Technical Supplement to the Traffic Noise Analysis Protocol.* California Department of Transportation, Division of Environmental Analysis, Sacramento, CA. www.dot.ca.gov/hq/env/noise/index.htm.

Hubbard, H. H., ed. 1995. *Aeroacoustics of Flight Vehicles: Theory and Practice,* vols. 1 and 2. Woodbury, NY: Acoustical Society of America.

ISO 2041. *Vibration and Shock—Vocabulary.* International Organization for Standardization (ISO),1990. www.iso.ch.

Knauer, H. S., S. Pedersen, C. S. Y. Lee, and G. G. Fleming. 2000. *FHWA Highway Noise Barrier Design Handbook.* Reports No. FHWA-EP-00-05 and DOT-VNTSC-FHWA-00-01, U.S. Department of Transportation, Volpe National Transportation Systems Center, Acoustics Facility, Cambridge, MA.

Lee, C. S. Y., and G. G. Fleming. 1996. *Measurement of Highway Related Noise.* Reports No. FHWA-PD-96-046 and DOT-VNTSC-FHWA-96-5, U.S. Department of Transportation, Volpe National Transportation Systems Center, Acoustics Facility, Cambridge, MA.

———. 2001. *General Health Effects of Transportation Noise.* Letter Report No. DTS-34-RR297-LR2, U.S. Department of Transportation, Volpe National Transportation Systems Center, Acoustics Facility, Cambridge, MA.

Lucas, M. J. 1998. *Rotorcraft Noise Model Manual.* Report No. WR 98-21, Wyle Research Center Arlington, VA, for the Boeing Company.

Menge, C. W., C. F. Rossano, G. S. Anderson, and C. J. Bajdek. 1998. *FHWA Traffic Noise Model, Version 1.0: Technical Manual.* Reports No. FHWA-PD-96-010 and DOT-VNTSC-FHWA-98-2, U.S. Department of Transportation, Volpe National Transportation Systems Center, Acoustics Facility, Cambridge, MA.

National Park Service. 1994. *User's Manual for the National Park Service Overflight Decision Support System,* BBN Report 7984, prepared by BBN Systems and Technologies. U.S. Department of Interior, National Park Service, Denver, CO.

Olmstead, J. R., et al. *Integrated Noise Model (INM) Version 6.0 Technical Manual.* Report No. FAA-AEE-02-01, U.S. Department of Transportation, Federal Aviation Administration, Office of Environment and Energy, Washington, DC.

Plotkin, K. J., and L. C. Sutherland. 1990. *Sonic Boom: Prediction and Effects.* Tallahassee, FL: AIAA Professional Studies Series.

Reagan, J. A., and C. A. Grant. 1976. *Highway Construction Noise: Measurement, Prediction and Mitigation.* U.S. Department of Transportation, Federal Highway Administration, Office of Environmental Policy, Washington, DC.

Reddingius, N. H. 1994. Users Manual for the National Park Service Overflight Decision Support System. BBN Report No. 7984.

Richardson, W. J., C. R. Greene, Jr., C. I. Malme, and D. H. Thomson. 1995. *Marine Mammals and Noise.* San Diego, CA: Academic Press.

Rochat, J. L. 1998. "Effects of Realistic Ocean Features on Sonic Boom Noise Penetration into the Ocean: A Computational Analysis." Ph.D. thesis, Pennsylvania State University.

Rochat, J. L., and G. G. Fleming. 1999. *Acoustics and Your Environment—The Basics of Sound and Highway Traffic Noise.* Video Production and Letter Report No. DTS-34-HW966-LR1, U.S. Department of Transportation, Volpe National Transportation Systems Center, Acoustics Facility, Cambridge, MA.

Sandberg, U., and J. A. Ejsmont. 2002. *Tyre/Road Noise Reference Book.* Kisa, Sweden: INFORMEX. www.informex.info.

SAE-ARP-4721. Society of Automotive Engineers (SAE). *Monitoring Noise from Aircraft Operations in the Vicinity of Airports.* Forthcoming.

Schexnayder, C. J., and J. Ernzen. 1999. *Mitigation of Nighttime Construction Noise, Vibration, and Other Nuisances.* NCHRP Synthesis of Highway Practice 218, National Research Council, Transportation Research Board, Washington, DC.

Seep, B., R. Glosemeyer, E. Hulce, M. Linn, P. Aytar, and R. Coffeen. 2000. *Classroom Acoustics.* Melville, NY: Acoustical Society of America. http://asa.aip.org/map_publications.html.

Schultz, T. J. 1982. *Community Noise Rating,* 2nd ed. New York, NY: Applied Science.

SoundPLAN LLC. SoundPLAN (a software package). Braunstein + Berndt GmbH, Shelton, WA.

Stusnick, E., M. L. Montroll, K. J. Plotkin, and V. K. Kohli. 1982. *Handbook for the Measurement, Analysis, and Abatement of Railroad Noise.* Report No. DOT/FRA/ORD-82/02H, U.S. Department of Transportation, Federal Railroad Administration, Office of Research and Development, Washington, DC.

Wayson, R. L., and J. M. MacDonald. 1998. *The AMAA Community Noise Model, Version 5.0, User Guide.* Orlando, FL: University of Central Florida.

CHAPTER 20
TRANSPORTATION-RELATED AIR QUALITY

Shauna L. Hallmark
Department of Civil, Construction, and Environmental Engineering,
Iowa State University, Ames, Iowa

20.1 INTRODUCTION

Air pollution is the presence of undesirable material in the air in sufficient quantities to pose health risks to humans, damage vegetation, negatively impact ecological systems, reduce visibility, or damage property (de Nevers 2000). A number of naturally occurring processes contribute to air pollution, such as dust, forest fires, and volcanoes. In urban areas, the majority of pollution originates from human activity and includes stationary sources, such as factories or power plants; area sources including facilities such as dry cleaners; and mobile sources (USEPA 2000). Mobile sources include automobiles, trucks, buses, aircraft, trains, marine activity, farming equipment, construction equipment, lawn mowers, etc. Mobile sources contribute approximately 60 percent of carbon monoxide (CO), 50 percent of nitrogen oxides (NO_x), and 40 percent of volatile organic compounds (VOC) in urban areas, the majority of which is attributed to highway vehicles (USEPA 2001). In certain urban areas, CO emissions from mobiles sources have been estimated as high as 95 percent. Relative contributions from the transportation sector for CO, VOC, and NO_x are provided in Figures 20.1 to 20.3.

20.2 NATIONAL AMBIENT AIR QUALITY STANDARDS

The Clean Air Act (CAA) was passed in 1963 and subsequently amended in 1970, 1977, and 1990. The Clean Air Act and Amendments (CAAA) provide the legal basis for air pollution laws in the United States. The role of the United States Environmental Protection Agency (USEPA) is to formulate and publish regulations demonstrating how air quality laws should be applied. Regulations set forth by the USEPA are subject to public hearings, approval by the Office of Management and Budget (OMB), and in some cases litigation, before they become they have the force of law (de Nevers 2000).

The Clean Air Act set National Ambient Air Quality Standards (NAAQS). Primary standards were instituted at levels to protect the public health of the most sensitive members of the population: children, the elderly, and asthmatics. Secondary standards were also set with

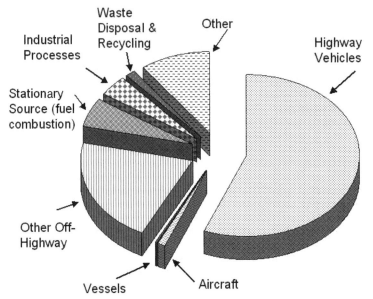

FIGURE 20.1 Major sources of carbon monoxide. (*Source:* USEPA 2001.)

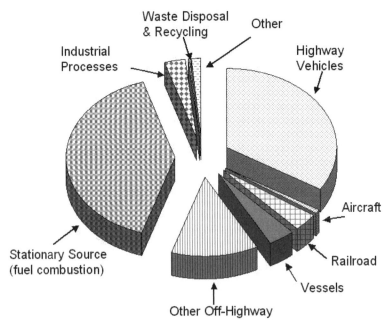

FIGURE 20.2 Major sources of NO_x. (*Source:* USEPA 2001.)

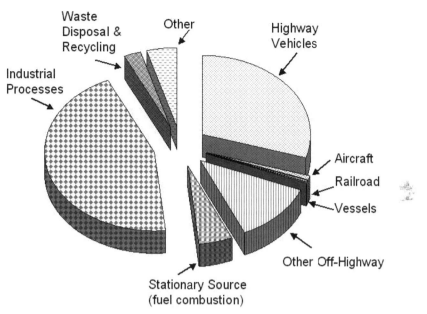

FIGURE 20.3 Major sources of VOC. (*Source:* USEPA 2001.)

the intent to protect the public welfare and the environment including damage to crops, ecosystems, vegetation, buildings, and decreased visibility. The six criteria pollutants are ozone, carbon monoxide, nitrogen dioxide, lead (Pb), sulfur dioxide (SO_2), and particulate matter (PM). An estimated 125 million people in the United States live in counties that are not in compliance with the primary NAAQS standards for at least of the one of the criteria in 1999 (USEPA 2001).

The Clean Air Act Amendments require nonattainment areas to reduce emissions to meet NAAQS. Areas that are not in compliance with NAAQS for ozone, CO, NO_x, SO_2, and PM must develop Statewide Implementation Plans (SIPs) to demonstrate conformity. Areas in nonattainment for the criteria pollutants are required to show compliance by projecting mobile source inventories for SIPs and estimating differences between alternatives (including "do-nothing") for transportation plans, programs, and projects. Emissions from alternatives must be less than "do-nothing" alternatives in order for transportation plans and programs to proceed. "Do-nothing" is defined as the current transportation system including future projects that receive environmental approvals under National Environmental Policy Acts (NEPA) (Chatterjee et al. 1997). Table 20.1 shows the primary standards for NAAQS.

NAAQS include two types of pollutants. Primary pollutants are those emitted directly into the atmosphere. Secondary pollutants are those formed indirectly. Primary pollutants include carbon dioxide (CO), sulfur dioxide (SO_2), and lead (PB). Ozone is a secondary pollutant and is a byproduct of a photochemical reaction in the atmosphere between volatile organic compounds (VOC) and nitrogen oxides (NO_x). Airborne PM is a combination of primary and secondary pollutants (TRB 2000). Mobile sources contribute a significant amount of pollution for only four of the six criteria pollutants; VOC, CO, NO_x, and particulates. Only minor proportions of lead and SO_2 are released by mobile sources, and consequently they are typically not evaluated.

TABLE 20.1 Primary Standards for NAAQS as of December 2000 (USEPA 2001)

Pollutant	Type of average	Concentration
CO	8-hour	9 ppm
	1-hour	35 ppm
Pb	Maximum quarterly average	1.5 μg/m^3
NO$_2$	Annual	0.053 ppm
O$_3$	Maximum daily 1-hr average	0.12 ppm
	4th Maximum daily 8-hour average	0.08 ppm
PM$_{10}$	Annual arithmetic mean	50 μg/m^3
	24-hour	150 μg/m^3
PM$_{2.5}$	Annual arithmetic mean	15 μg/m^3
	24-hour	65 μg/m^3
SO$_2$	Annual arithmetic mean	0.03 ppm
	24-hour	0.14 ppm

20.3 TRANSPORTATION-RELATED POLLUTANTS

20.3.1 Carbon Monoxide

Carbon monoxide is a colorless, odorless gas that can be lethal when inhaled at high concentrations. It also causes headaches and fatigue. Epidemiological studies have indicated that a relationship may exist between elevated levels of CO and cardiovascular problems (Godish 1997). In the body, hemoglobin binds with CO rather than oxygen and the body is deprived of oxygen (Wark, Warner, and Davis 1998). As shown in Table 20.1, the 1-hour standard for CO is 35 ppm and the 8-hour standard is 9 ppm. Neither standard can be exceeded annually. Carbon monoxide is typically evaluated on a localized basis because of its immediate health effects.

The amount of CO produced depends on the air to fuel ratio (A/F) in the combustion process. Under rich combustion (lower A/F) CO production is increased (de Nevers 2000). Fuel-rich conditions occur with cold starts or under engine loading, such as acceleration, high speeds, or operation on steep grades when cars are not tuned properly, and at higher altitudes (USEPA 2001). Lean fuel mixtures (higher A/F) occur when cruising or decelerating (Homburger et al. 1996).

20.3.2 Oxides of Nitrogen

Two main oxides of nitrogen are formed during the combustion process. Nitric oxide (NO) is formed during combustion from naturally occurring N$_2$ and O$_2$ under high temperatures. Vehicle engines emit a mixture composed of primarily NO with some nitrogen dioxide (NO$_2$). NO oxidizes to NO$_2$ in the atmosphere and is a major contributor to photochemical smog. The concentrations that are produced depend on the temperature of combustion and spark ignition timing (Homburger et al. 1996). Oxides of nitrogen are one of the principal components in smog and react with VOCs in the presence of sunlight to form ozone.

20.3.3 Ozone

Ozone is a severe irritant. It occurs naturally in the upper atmosphere and with NO$_2$ is a major component of smog. Ozone is a secondary pollutant and is a byproduct of a photochemical reaction in the atmosphere between VOCs and nitrogen oxides. The USEPA requires reductions for both NO_x and VOC to control ozone formation but focuses on reduction of VOCs as the most effective strategy. Ozone damages materials, causes eye and respiratory irritations, decreases pulmonary function, and decreases heart rate and oxygen intake. Effects include sore throat, chest pain, cough and headache. The effects are more adverse for the young and old. Ozone also affects plant growth and causes crop and vegetation damage (Wark, Warner, and Davis 1998). The 1-hour standard for ozone is 0.12 parts per million (ppm). The standards cannot be exceeded more than three times in a continuous 3-year period.

20.3.4 Volatile Organic Compounds

VOCs are not a criteria pollutant but react with NO_x to form ozone. Hydrocarbon (HC) emissions, a form of VOCs, are a result of incomplete combustion or fuel evaporation as shown in Figure 20.4. About half of HC emissions are tailpipe emissions (de Nevers 2000). Hydrocarbons are evaporated during vehicle operation as well as while the vehicle is parked. Evaporative emissions make up a significant portion of HC emissions and can be categorized as (USEPA 1994):

> *Diurnal* emissions are a result of evaporation that occurs as the fuel tank heats and cools throughout the day due to changes in ambient temperature.

> *Running losses* occur during vehicle operation. Additional evaporation occurs due to fuel being in contact with a hot engine and exhaust system. Evaporative emissions escape while the vehicle is operating. Emissions depend on ambient temperature, fuel vapor pressure, and driving cycle (TRB 2000).

> *Hot soak* emissions result from evaporation during the first hour following engine shutdown. While the engine is still hot, fuel is heated above ambient temperatures and evaporation occurs primarily from the fuel tank and carburetor bowl in carburetor vehicles. Hot soak emissions depend on a number of factors, including ambient temperature, fuel vapor pressure, proportion of fuel in the fuel tank, and vehicle technology (TRB 2000). The majority of hot-soak emissions occur in the first 10 minutes following engine shutdown (USEPA 1999).

FIGURE 20.4 Schematic of vehicle emissions. (*Source:* USEPA 1994.)

Refueling During refueling, fuel vapors are forced from the fuel tank contributing to HC emissions.

20.3.5 Particulates

Particulates are fine particles of material suspended in the air. Particulate matter is produced by industry, natural sources, motor vehicles, agricultural activities, mining and quarrying, and wind erosion. Particles emitted from vehicles consist primarily of lead compounds, motor oil, and carbon particles (Homburger et al. 1996). Particulates are classified by size. PM_{10} refers to all particles with a diameter of ≤ 10 micrometers. Sources include road dust and windblown dust. PM_{10} settles rapidly and consequently its range of influence is limited to immediate area. $PM_{2.5}$ are particles with a diameter ≤ 2.5 micrometers. It is formed from fuel combustion, fireplaces, and wood stoves. $PM_{2.5}$ is much smaller than PM_{10} and settles less rapidly. Consequently, it is transported farther distances in the atmosphere and is more uniformly dispersed in urban areas (USEPA 2001).

Both PM_{10} and $PM_{2.5}$ are inhaled and accumulate in the respiratory system. Coarser particulate matter aggravates respiratory conditions such as asthma, while fine particulate matter results in decreased lung function and increased hospital admissions and emergency room visits. Particulate matter is a major cause of reduced visibility, including in national parks. It also affects the acidity balance in land or water systems, and deposition on plants can corrode plants or interfere with plant metabolism (USEPA 2001).

20.4 ESTIMATING TRANSPORTATION-RELATED EMISSIONS

The Clean Air Act Amendments and corresponding provisions in the Intermodal Surface Transportation Efficiency Act (ISTEA) set forth a number of requirements for nonattainment areas to work towards meeting NAAQS. For transportation activities, estimates of both area-wide and project-level transportation emissions are necessary to meet the various requirements established by the legislation (TRB 1998). Mobile source emissions (typically in grams) are estimated by:

$$\text{Total emissions} = \text{vehicle activity} \times \text{emission rate} \qquad (20.1)$$

Vehicle activity estimates vary by agency and type of analysis but are frequently provided in miles of vehicle activity. Emission rates are almost exclusively provided by either the USEPA's MOBILE model or, in California, the EMFAC model. The MOBILE model is the primary emission factor tool to estimate on-road emissions. It is used by national, state, and local agencies to estimate emissions, make policy decisions, measure environmental impacts, and create national, regional, and urban emission inventories (TRB 2000). In California, the EMFAC model is the primary emission factor model. Estimates of mobile source emission are used to determine both the current and projected future contributions of mobile sources.

20.4.1 MOBILE6

The USEPA recently released emission factor model MOBILE6, which estimates fleet-average, in-use fleet emission rates for hydrocarbons (HC), carbon monoxide (CO), and oxides of nitrogen (NO_x). Twenty-eight individual vehicle types can be modeled, including gas, diesel, and natural gas-fueled passenger vehicles, heavy trucks, buses, and motorcycles for calendar years 1952 to 2050. Vehicle classes included in MOBILE6 are shown in Table 20.2. Emissions can be modeled at different average speeds from 2.5 to 65 miles per hour

TABLE 20.2 Vehicle Classes in MOBILE6

Abbreviation	Vehicle type	Weight (lb)
LDGV	Light-duty gasoline vehicles (passenger cars)	
LDGT1	Light-duty gasoline trucks 1	0 to 6,000 GVWR; 0 to 3,750 LVW
LDGT 2	Light-duty gasoline trucks 2	0 to 6,000 GVWR; 3,751 to 5,750 LVW
LDGT 3	Light-duty gasoline trucks 3	6,001 to 8,500 GVWR; 0 to 5,750 ALVW
LDGT 4	Light-duty gasoline trucks 4	6,001 to 8,500 GVWR; >5,750 ALVW
HDGV2b	Class 2b heavy-duty gasoline vehicles	8,501 to 10,000 GVWR
HDGV3	Class 3 heavy-duty gasoline vehicles	10,001 to 14,000 GVWR
HDGV4	Class 4 heavy-duty gasoline vehicles	14,001 to 16,000 GVWR
HDGV5	Class 5 heavy-duty gasoline vehicles	16,001 to 19,500 GVWR
HDGV6	Class 6 heavy-duty gasoline vehicles	19,501 to 26,000 GVWR
HDGV7	Class 7 heavy-duty gasoline vehicles	26,001 to 33,000 GVWR
HDGV8a	Class 8a heavy-duty gasoline vehicles	33,001 to 60,000 GVWR
HDGV8b	Class 8b heavy-duty gasoline vehicles	>60,000 GVWR
LDDV	Light-duty diesel vehicles (passenger cars)	
LDDT12	Light-duty diesel trucks 1 and 2	0 to 6,000 GVWR
HDDV2b	Class 2b heavy-duty diesel vehicles	8,501 to 10,000 GVWR
HDDV3	Class 3 heavy-duty diesel vehicles	10,001 to 14,000 GVWR
HDDV4	Class 4 heavy-duty diesel vehicles	14,001 to 16,000 GVWR
HDDV5	Class 5 heavy-duty diesel vehicles	16,001 to 19,500 GVWR
HDDV6	Class 6 heavy-duty diesel vehicles	19,501 to 26,000 GVWR
HDDV7	Class 7 heavy-duty diesel vehicles	26,001 to 33,000 GVWR
HDDV8a	Class 8a heavy-duty diesel vehicles	33,001 to 60,000 GVWR
HDDV8b	Class 8b heavy-duty diesel vehicles	>60,000 GVWR
MC	Gasoline motorcycles	
HDGB	Gasoline buses (school, transit, and urban)	
HDDBT	Diesel transit and urban buses	
HDDBS	Diesel school buses	
LDDT34	Light-duty diesel trucks 3 and 4	6,001 to 8,500 GVWR

(mph). However, the user-specified average speed applies to all vehicle types. Emission rates can also be allocated by four roadway categories: (1) freeways, (2) arterials (includes both arterials and collectors), (3) local roads, and (4) freeway on- and off-ramps (USEPA 2002).

Basic emission rates were developed for MOBILE from emission tests that were conducted under standard conditions, such as driving cycle, fuel type, and ambient temperature. Adjustments are made to the basic emission rates to represent conditions that differ from those standard conditions. Adjustments to base emission rates are calculated for a number of variables, including average speed by roadway type, environmental conditions, air conditioning use, fuel type, off-cycle driving, etc. (USEPA 2002).

Emission rates represent national averages. Users can tailor model output to reflect local conditions by customizing input. Parameters that can be tailored to local conditions include fuel characteristics, environmental conditions, vehicle age distribution, average speed distribution by hour of the day and roadway, hot soak duration, vehicle activity by vehicle class, etc. (USEPA 2002).

20.4.2 Average Vehicle Speeds

Emission factors are correlated to average vehicle speeds. Depending on the pollutant, emissions rates are generally higher at lower average speeds, less sensitive for mid-range speeds,

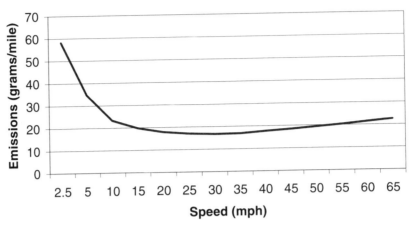

FIGURE 20.5 CO emissions versus average speed.

and higher as speeds increase. Typically, average speeds are output for a roadway link or facility type from travel demand forecasting models and a single average speed input to MOBILE to represent all vehicle types. Figures 20.5 to 20.7 illustrate emission rates by average speed output from MOBILE6. As shown, emission rates are highly speed-dependent. Emission rates for all three pollutants are highest at low speeds and then decrease toward mid-speed ranges. Carbon monoxide begins to increase gradually and NO_x rates increase sharply again at higher speeds. Consequently, average speed is input to emission factor models and speed-specific rates used to estimate emissions. Average speeds are estimated by segment or facility and represent all vehicle types unless specified otherwise. Heavy vehicles, including trucks and buses, are assumed to operate at the same speeds as passenger vehicles (TRB 1998).

The input speed used for MOBILE6 is average link speed and includes the time a vehicle spends in cruise, idle, acceleration, and deceleration along a roadway section. Average travel speed over a segment of roadway is given by:

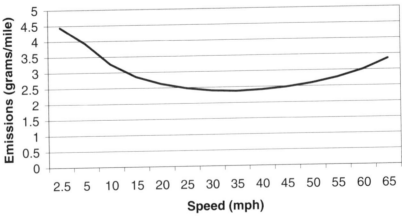

FIGURE 20.6 NO_x emissions versus average speed.

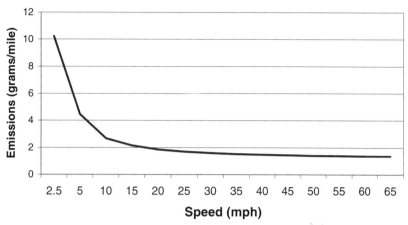

FIGURE 20.7   VOC emissions versus average speed.

$$v_s = \frac{L}{t_t} \tag{20.2}$$

where v_s = average travel speed (mph)
  L = length of segment (miles)
  t_t = total time to traverse the segment including delay (hours)

Average speeds on road are related to the capacity of the roadway and traffic volumes. As volumes increase, level of service decreases and vehicle speeds also decrease. Accordingly, average speeds are modeled at different times of the day (Chatterjee et al. 1997). Average speeds are frequently estimated by using output from travel demand forecasting models, which usually base average speeds on speed-flow curves (typically BPR curves). The speed-volume relationship employed in BPR curves is given by:

$$v_c = \frac{v_{ff}}{[1 + 0.15(V/C)^4]}$$

where v_c = congested average travel speed for segment (mph)
  v_{ff} = free-flow average travel speed for segment (mph)
  V = volume for segment
  C = capacity of segment

20.4.3 Estimates of Vehicle Activity

Average speeds are input to emission factor models and speed-specific emission rates output. Emission rates are multiplied by vehicle activity to produce the total amount of pollution produced. Speed estimates may vary by type of analysis. For regional emission estimation, vehicle miles traveled (VMT) is used to represent vehicle activity for the area. For nonattainment areas, the CAAA and Conformity Rule require base year and forecasts of future VMT. VMT estimates are required by vehicle mix and functional class. Acceptable methods to estimate VMT for areawide emissions modeling are either the Highway Performance Monitoring System (HPMS) method or network based travel demand models. The HPMS calculates VMT for different for different functional classes. The estimates are based on sampling and statistical extrapolation methods approved by the Federal Highway Administration (FHWA) (Chatterjee et al., 1997). Traffic volumes, which are subsequently converted to VMT, are also output from the traffic assignment step of travel demand models.

20.5 DISPERSION MODELING

The total amount of pollution produced is important. Health effects, however, are measured in concentrations of pollution, so NAAQS represent amounts of ambient pollution. Dispersion models are used to model how pollutants are transported and dispersed once they are emitted from vehicles or sources. Dispersion of pollutants depends on a number of factors, including:

• Chemical and physical characteristics of the pollutants
• Meteorological conditions such as atmospheric stability, wind speed and direction, and ambient temperature
• Topography
• Distance from the source

Models predict the concentrations of pollutants present at a specific location. The urban airshed model (UAM) is the most common regional dispersion model. It uses a three-dimensional photochemical grid to simulate the atmosphere mathematically and calculate pollutant concentrations based on physical and chemical processes of the atmosphere. Typical input requirements include hourly gridded emission for NO_x and VOCs, hourly estimates of the height of the mixed layer, ambient temperature, ambient humidity, solar radiation, cloud cover, atmospheric pressure, and hourly 3-D wind fields. Typical model outputs include hourly average concentrations by grid square for pollutants of interest, instantaneous concentrations of pollutants by grid square at the beginning of the averaging period, calculation of summer ozone levels, concentrations of winter CO, and projects of future emission scenarios (Chatterjee et al. 1997).

Microscale analysis is usually performed to measure whether local violations of NAAQS for carbon monoxide may occur. Microscale analysis is frequently used to model concentrations at intersections where vehicles may spend significant amounts of time queuing and idling. Locations of queuing and idling are often characterized by elevated emission rates. CAL3QHC is the microscale dispersion model required by the USEPA. In California, CALINE-4 is the microscale dispersion model used. Both are Gaussian plume models. For microscale analysis, concentrations are typically measured at specific receptor locations and measured during peak hours.

20.6 CONTROL OF AIR POLLUTION FROM VEHICLES

The majority of improvements in reduction of motor vehicle emissions, with the exception of lead, have been technological improvements to the vehicles themselves. One type of improvements is engine design features such as fuel injectors with computer-controlled fuel and air injection rates that optimize the combustion mixture, chamber design, compression ratio, spark timing, and exhaust gas recirculation. Add-on pollution control technology has also contributed to emission reduction. The most significant add-on pollution control improvement is the catalytic converter, which treats engine exhaust by reducing NO_x to N_2 and O_2 and oxidizing VOC and CO to H_2O and CO_2. Another technological improvement is carbon canisters, absorb evaporative VOC emissions from the hot engine once the vehicle has been turned off. When the vehicle is started again, intake air is circulated through the canister and absorbed VOCs are sent back into the cylinders for combustion. Another device, positive crankcase ventilation valves, routes air from the engine crankcase back into the cylinders for burning (Cooper and Alley 2002). Reduction of lead from motor vehicles came about almost entirely by removal of lead from gasolines.

20.6.1 Fuels

Gasoline is a mixture of olefinic, aromatic, and paraffinic HC compounds. Composition varies from refinery to refinery and even from one geographic area to another. The composition of fuel is an important factor in the amount of emissions that are produced by both evaporation and combustion (Godish 1997). One improvement in fuels was reductions in the maximum allowable Reid vapor pressure (RVP) of gasoline, resulting in reductions in VOCs from vehicles as well as in evaporative emissions at gas stations and gasoline storage areas (Cooper 2002).

Oxygenated fuels are another improvement in fuels. Oxygenated compounds include MTBE, ETBE, tertiary amyl methyl ether, and ethanol. MTBE is the most frequently used oxygenated compound. Oxygenated fuels have one or more oxygen atoms embedded in the fuel molecule. They are commonly added during winter months to reduce CO emissions. A 3 percent oxygen by weight blend is estimated to reduce CO emission by 30 percent. HC emissions are slightly decreased as well, although NO_x emissions may increase slightly (Cooper 2002). Several alternative fuels are in use or have been evaluated as replacements for gasoline. They include compressed natural gas, liquefied petroleum gas, pure methanol, pure ethanol, and alcohol-gasoline blends. Reductions in emission vary by fuel type. Natural gas and propane produce significantly less CO and lower emissions of VOCs. Evaporative emissions from ethanol and methanol are less reactive.

20.6.2 Transportation Control Measures

Although technological controls on vehicles have led to significant reductions in the amount of pollutant released by each vehicle, increases in the total number of vehicles on the road and in the number of vehicle miles traveled annually by each vehicle threaten to outpace technological advances. As a result, methods to reduce overall travel and reduce the amount of time vehicles spend in modes where emissions are higher, such as idling, are necessary. The Clean Air Act Amendments identify transportation control measures (TCMs) that are expected to decrease motor vehicle use and emissions. TCMs are listed in Section 108(f) of the Clean Air Act Amendments of 1990. They are shown in Table 20.3.

TCMs are strategies intended both to reduce the total number of vehicle miles traveled (VMT) and to make that travel more efficient (Cambridge Systematics, Inc. 1991). Transportation control measures are required to help reduce the amount of pollution released by the transportation sector to improve air quality and meet federal requirements. Although various definitions exist for TCMs, a general description is that they are actions designed to change travel demand or vehicle operating characteristics to reduce motor vehicle emissions, energy consumption, and congestion. TCMs include transportation supply improvement strategies and transportation demand management strategies. Transportation supply improvement strategies either change the physical infrastructure or implement actions for more efficient use of existing facilities to improve traffic flow and decrease stop-and-go movement. Supply improvement strategies take the form of bottleneck relief, construction improvements, improved signal timing, ramp metering, applications of intelligent transportation system technology, and alterations to land use patterns. Demand management measures attempt to change driver behavior to reduce the frequency and length of automobile trips. Demand management measures include, but are not limited to, no-drive days, employer-based trip-

TABLE 20.3   TCMs Listed in the CAAA

Public transit
High-occupancy vehicle facilities
Employer-based transportation management plans
Trip-reduction ordinances
Traffic flow improvements
Park and ride/fringe parking
Vehicle-use limitations or restrictions in downtown areas during peak periods
Limitations on certain roads for use by pedestrians and nonmotorized vehicles
Ride-share and high-occupancy vehicle (HOV) programs
Bike lanes and storage facilities
Control of extended vehicle idling
Reduction of cold starts
Flexible work schedules
Programs to encourage non-automobile travel and reduce the need for single-occupant vehicle travel
   (includes provision for special events and major activity centers)
Pedestrian and nonmotorized vehicle paths
Scrappage of older vehicles

reduction programs, parking management, park-and-ride programs, work schedule changes, transit fare subsidies, ride-sharing, and public awareness programs (Guensler 1998).

20.7   REFERENCES

Cambridge Systematics, Inc. 1991. *Transportation Control Measure Information Documents.* Prepared for the U.S. Environmental Protection Agency Office of Mobile Sources, Cambridge, MA.

Chatterjee, A., T. L. Miller, J. W. Philpot, T. F. Wholley, Jr., R. Guensler, D. Hartgen, R. A. Margiotta, and P. R. Stopher. 1997. *Improving Transportation Data for Mobile Source Emission Estimates.* NCHRP Report 394, National Research Council, Transportation Research Board, Washington, DC.

Cooper, D. C., and F. C. Alley. 2002. *Air Pollution Control: A Design Approach,* 3rd ed. Prospect Heights, IL: Waveland Press.

de Nevers, N. 2000. *Air Pollution Control Engineering,* 2nd ed. Boston: McGraw-Hill.

Godish, T. 1997. *Air Quality.* Boca Raton: Lewis.

Guensler, R. 1998. "Increasing Vehicle Occupancy in the United States." In *L'avenir des déplacements en ville* (The future of urban travel), ed. O. Andan et al., vol. 2, 127–55. Lyon, France: Laboratoire d'Economie des Transports.

Homburger, W. S., J. W. Hall, R. C. Loutzenheiser, and W. R. Reilly. 1996. *Fundamentals of Traffic Engineering,* 14th ed. Berkeley: University of California, Berkeley, Institute of Transportation Studies, May.

Transportation Research Board (TRB). 1998. *NCHRP Research Results Digest: Number 230.* National Research Council, Transportation Research Board, Washington, DC, August.

———. 2000. *Modeling Mobile-Source Emissions.* National Research Council, Transportation Research Board, Washington, DC.

U.S. Environmental Protection Agency (USEPA). 1994. *Automobile Emissions: An Overview.* Fact Sheet OMS-5, EPA 400-F-92-007, USEPA, August.

———. 1999. *Hot Soak Emissions as a Function of Soak Time.* EPA420-P-98-018, M6.EVP.007, USEPA, Office of Mobile Sources.

———. 2000. *Latest Findings on National Air Quality: 1999 Status and Trends.* EPA-454/F-00-002, USEPA, Office of Air Quality Planning and Standards, Research Triangle Park, NC, August.

———. 2001. *National Air Quality and Emissions Trends Report, 1999.* EPA 454/R-01-004, USEPA, Office of Air Quality Planning and Standards. March.

———. 2002. *User's Guide to MOBILE6.0: Mobile Source Emission Factor Model.* EPA420-R-02-001, USEPA, Office of Air and Radiation.

Wark, K., C. F. Warner, and W. T. Davis. 1998. *Air Pollution: Its Origin and Control,* 3rd ed. Menlo Park, CA: Addison-Wesley.

P · A · R · T · IV

NON-AUTOMOBILE TRANSPORTATION

CHAPTER 21
PEDESTRIANS

Ronald W. Eck
*Department of Civil and Environmental Engineering,
West Virginia University, Morgantown, West Virginia*

21.1 INTRODUCTION AND SCOPE

A pedestrian is defined as any person on foot. While everyone is a pedestrian at one time or another, in the United States walking is viewed primarily as a recreational activity. However, for relatively short trips, walking can be an efficient and inexpensive mode of transportation.

There are many factors that influence choice of travel mode and, specifically, the decision to walk. The National Bicycling and Walking Study (FHWA 1992) showed that there is a three-tiered hierarchy of factors:

1. *Initial considerations.* Many people rely on their automobile to go virtually anywhere and never seriously consider the option of walking. An individual's attitudes and values also play a role—e.g., walking may be considered as "not cool." Perceptions are also important in the decision to walk—e.g., safety concerns about traveling at night. Finally, there are situational constraints that, if they do not preclude the decision to walk, they do require additional planning and effort. Examples include needing a car at work or having to pick up children from soccer practice.

2. *Trip barriers.* Concern for safety in traffic is a frequently cited reason for not walking. This is particularly true where there are no alternatives to walking along high-speed, high-volume roadways. There may be problems with access and linkage, e.g., lack of connections between neighborhoods and shopping areas or parks. Environmental factors such as rugged topography or extremes in weather can also be considered as barriers.

3. *Destination barriers.* Lack of support from employers or coworkers can act as a barrier, e.g., relaxing the dress code or establishing a policy of flextime.

Increased levels of walking can result in benefits in terms of health and physical fitness, the environment, and transportation. Studies have demonstrated that even low to moderate levels of exercise, such as regular walking, can reduce the risk of coronary heart disease, stroke, and other chronic diseases, help reduce healthcare costs, and contribute to greater functional independence in later years and improve the quality of life at every stage. Replacing automobile trips with walking trips could result in significant economic benefits. Since it is nonmotorized and nonpolluting, walking would result in reduced emissions, energy consumption, and congestion. Facilitating walking means additional travel options for those unable to drive or who choose not to drive for some trips. Where there is a truly intermodal transportation system in which walking is an important component, the livability of communities is enhanced.

Passage of the Intermodal Surface Transportation Efficiency Act of 1991 (ISTEA) helped focus new attention on walking as an important nonmotorized mode of transportation. ISTEA requires each state and metropolitan planning organization (MPO) to include nonmotorized elements in its transportation plans.

ISTEA has opened up new sources of potential funding for nonmotorized transportation improvements, planning, and programs as well as for recreational trails and intermodal linkages. With these funding programs and the planning requirements established by ISTEA, there is increased interest in pedestrian transportation.

This chapter takes a comprehensive look at incorporating pedestrians into the transportation system. While the chapter provides an overview, it also directs readers to sources of more detailed technical information.

21.2   CHARACTERISTICS OF PEDESTRIANS

21.2.1   Ambulation

In normal walking, the lead foot swings forward and, as it approaches the end of the stride, the heel comes down on the surface gently as the ankle allows the foot to rotate forward at a controlled rate until the sole also contacts the floor. There is little horizontal force applied during the heel-strike phase, but if the heel edge slips at this point, the leading foot slides forward. Therefore, the characteristic slip pattern is that the victim falls backwards. This type of fall usually results in the most severe injury. Forward slips may occur when the sole of the foot slides in the retreating portion of the step. This tends to produce a forward fall where the injury is often less severe. Many slips result from the more complex activities of turning and changing direction. In these instances, the fall pattern is less predictable.

A trip occurs when a pedestrian's foot is impeded in the walking process by striking against some obstacle in the path of travel. If the interruption of motion is great enough, the pedestrian will fall forward. The person's reflex action is to extend the upper arms to absorb the energy of impact. Falling onto the extremities in this fashion can cause injury to wrist, forearm, elbow, upper arm and shoulder. One or both of the knees may also strike the ground.

Pedestrians tend to look ahead to their objective; the normal line of sight is about 15° below horizontal relative to the eyes. Pedestrians do not usually look down deliberately unless something attracts their attention. Consequently, even small changes in surface elevation or characteristics are not always seen. However, even if someone is looking down at the surface, a problem may not be perceived. Color, texture, low light levels, and glare can obscure changes in walking surfaces.

Good design takes this human factor into account in two ways. First, many problems can be avoided by giving thoughtful attention to safety in facility planning, design, and construction. Second, where something cannot be avoided, e.g., a single step, the designer must build in visual and tactile cues that will cause the pedestrian to look down and see the hazard before tripping over it or slipping on it.

Another characteristic of pedestrians is that they will always take the least energy route (i.e., the shortest distance and flattest path) between two points. They tend to cross streets at the most convenient locations rather than at designated crossings. If it is easier to cross at mid-block than to walk a long distance to a corner crosswalk, pedestrians will take the most direct route and cut across the street between intersections. Where a choice of routes is available, substantial barriers may be needed to prevent pedestrians from taking the direct route where that route includes hazards (e.g., at-grade crossing of a busy arterial roadway).

According to the American Association of State Highway and Transportation Officials (AASHTO) roadway design policy (2001), pedestrians usually do not walk over 1 mile to work or over 0.5 mile to a transit stop. About 80 percent of the distances traveled by pedestrians will be less than 0.5 mile. Pedestrian volumes are influenced by conditions such as weather, advertised sales, or special events.

Effective pedestrian facilities are designed to accommodate the full range of users. There is no such thing as a standard pedestrian. The stature, travel speeds, endurance limits, physical strength, and judgment abilities of pedestrians vary greatly. Users include children, older adults, families, and people with and without disabilities. Pedestrians may be carrying packages or luggage, walking a dog, pushing children in strollers, or pulling delivery dollies. Pedestrians transporting items cannot react as quickly to potential hazards because they are more physically taxed and distracted.

21.2.2 Seeing and Being Seen

Visibility of a pedestrian in the traffic stream is influenced by environmental conditions, behavior, and attire. The key factor is the degree of contrast between the pedestrian and his or her environment.

Environmental factors can affect the visibility of pedestrians to motorists. Rain, snow, fog, shadows, and glare all reduce visual range and acuity. The National Highway Institute's (NHI) *Participant Workbook* (NHI 1996) notes that vehicles themselves are equally important. Dirty or cracked windshields not only reduce vision but magnify the effects of glare.

There are also "visual screens" in the driving environment. Moving vehicles, particularly buses and commercial vehicles can block pedestrians' and motorists' view of one another. Stationary features such as parked vehicles, shrubs, structures, and traffic signal controller boxes can have the same effect.

About one-half of fatal pedestrian crashes occur in low-light or dark conditions. At night, pedestrians are frequently difficult to see because they lack conspicuity. All of the factors affecting conspicuity become increasingly critical during times of reduced light or darkness. For example, according to Federal Highway Administration (FHWA) data, at night the average driver of a vehicle operating with low-beam headlights will see a pedestrian in dark clothing in the roadway about 80 feet in front of the vehicle. At speeds greater than about 20 to 25 miles per hour, drivers do not have enough time to perceive the pedestrian, identify him or her, and make a decision to stop or swerve in time to avoid striking him or her.

Pedestrians tend to overestimate their visibility to motorists. Motorists who have been involved in nighttime crashes with pedestrians often remark, "I don't know what I hit—I thought I struck an animal." This is an indication they did not see the pedestrian until it was too late to react.

21.2.3 Groups of Particular Concern

Older Adults.   By 2020, it is estimated that 17 percent of the U.S. population will be older than 65. Although aging itself is not a disability, most persons aged 75 or older have a disability. Many of the characteristics commonly associated with aging can limit mobility. The aging process often causes a general deterioration of physical, cognitive, and sensory abilities. Characteristics may include:

• Vision problems such as degraded acuity

• Reduced range of joint motion

• Reduced ability to detect, localize, and differentiate sounds

• Limited attention span, memory, and cognitive abilities

- Reduced endurance
- Decreased agility, balance, and stability
- Slower reflexes
- Impaired judgment, confidence, and decision-making abilities

Older adults generally need more frequent resting places and prefer more sheltered environments. The FHWA publication (1999) on designing pedestrian facilities for access notes that many have increased fears for personal safety. Statistics confirm these fears indicating that older pedestrians appear to be at increased risk for crime and crashes at places with no sidewalks, sidewalks on only one side, and places with no streetlights. Older pedestrians would benefit from accessible paths that are well-lit and policed.

Ambulation of older adults is affected by their reduced strength. Travel over changes in levels, such as high curbs, can be difficult or impossible for older adults.

Because older people tend to move more slowly than younger pedestrians, they require more time to get across the street than many other sidewalk users. For many years, engineers have designed traffic signals based on an average pedestrian walking speed of 4 feet per second. Using a walking speed of 2.8 feet per second better accommodates older pedestrians.

The reduced visual acuity of older people can make it difficult for them to read signs or detect curbs. Older people are more dependent on high contrast between sign backgrounds and lettering. Contrast resolution losses can cause them to have difficulty seeing small changes in level, causing trips and falls on irregular surfaces.

Children.   Children have fewer capabilities than adults due to their developmental immaturity and lack of experience. Compared to adults, children tend to exhibit the following characteristics:

- One-third less peripheral vision
- Less accuracy in judging speeds and distances
- Difficulty in localizing the direction of sounds
- Overconfidence
- Inability to read or comprehend warning signs and traffic signals
- Unpredictable or impulsive actions
- Trust that others will protect them
- Inability to understand complex situations

Disabled.   According to the 1990 Census, one in every five Americans has a disability. In fact, 85 percent of Americans living to their full life expectancy will suffer a permanent disability. People with disabilities are also more likely to be pedestrians than other adults since some physical limitations can make driving difficult. According to the FHWA publication on providing access (1999), disabilities can be divided into three categories: mobility, sensory, and cognitive.

People with mobility impairments include those who use wheelchairs, crutches, canes, walkers, orthotics, and prosthetic limbs. However, there are many people with mobility impairments who do not use assistive devices. Characteristics common to people with mobility limitations include substantially altered space requirements to accommodate assistive device use, difficulty in negotiating soft surfaces, and difficulty negotiating surfaces that are not level.

Although sensory disabilities are more commonly thought of as total blindness or deafness, partial hearing or vision loss is much more common. Other types of sensory disabilities can affect touch, balance, or the ability to detect the position of one's own body in space. Color blindness is considered a sensory defect.

Cognition is the ability to perceive, recognize, understand, interpret, and respond to information. It relies on complex processes such as thinking, knowing, memory, learning, and recognition. Cognitive disabilities can hinder the ability to think, learn, respond, and perform coordinated motor skills. Such individuals might have difficulty navigating through complex environments such as city streets and might become lost more easily than other people.

The aforementioned FHWA publication (1999) on designing for access presents an excellent overview of the characteristics associated with different types of disabilities. The publication also reviews design approaches for accommodating specific categories of disabilities.

21.3   PLANNING

21.3.1   Conceptual Planning

Conceptual planning is relatively simple, consisting of determining the general direction that walkways should take. Focus should be on pedestrian generators such as schools, shops, cultural attractions, and work and play places. First, look at routes that presently exist before establishing new ones. Privacy, views, access, and local character must be understood and incorporated into the planning.

Initially, planners should determine where people want or need to travel, the routes they might travel, and who these people are. The most likely users of improved pedestrian facilities are:

- Children who must be driven to school, play, and other activities.
- Parents who have to drive children and would appreciate safe walking routes so their children can move around the community by themselves.
- Older people who may not drive but have time to walk may be able to carry out some of their daily chores, enjoy the outdoors, and exercise all on the same trip.
- Commuters who may be able to walk to bus or carpool stops.
- Recreational users, especially those who jog or walk regularly, would benefit from improved routes and separation from fast-moving traffic.

Simple pedestrian volume counts usually do not yield enough information about where people are going or coming from, trip purpose, and any special pedestrian needs that should be met. Such data are best obtained through an origin/destination survey that should include the following information:

1. Location of major pedestrian generators such as parking facilities, transit stations, and major residential developments
2. Location of significant pedestrian attractions such as shopping centers, office and public buildings, theaters, colleges, hospitals, and sports arenas
3. Existing and potential pedestrian routes between major destinations
4. Time periods in which major pedestrian flow occurs

Some questions to consider include:

1. Do existing routes satisfy the heaviest travel demand? Can a need for new routes be clearly identified?
2. Do existing routes require improvement to resolve circulation problems?

3. Which areas seem to be preferred locations for development of new activities to generate pedestrian movement?

21.3.2 Access and Linkages

There is no question that transportation modes have influenced the way cities have grown and the forms they have taken. Before the advent of the automobile, cities were more compact in terms of area and population. However, in the United States the automobile clearly is the dominant transportation mode. One manifestation of this is the phenomenon of suburbanization that occurred after World War II. Characteristics of suburbs include:

1. Suburban land use planning encourages low density and separation of land use types.
2. Street design standards typically require wide streets that encourage high-speed traffic and sometimes do not require sidewalks.
3. It is not easy to use public transportation in suburban locations.
4. Barriers to walking are created unintentionally.

Consequently, suburban activities essentially require the use of a car and generate large amounts of vehicular traffic.

Pedestrian travel is often an afterthought in the development process. The results are impassable barriers to walking both within and between developments. For example, early suburban communities had no sidewalks. Later, some communities required developers to install sidewalks. Consequently, in most suburbs there is a patchwork of sidewalks that start and stop but often are not linked.

Suburban neighborhood design can be modified to encourage walking. A pedestrian-oriented neighborhood should include the following characteristics:

• Streets that are laid out in well-connected patterns on a pedestrian scale so that there are alternative automobile and pedestrian routes to every destination.
• A well-designed street environment that encourages intermodal transportation. These streets should include pedestrian scale lighting, trees, sidewalks, and buildings that are within close walking distance to the sidewalk.
• Residential and internal commercial streets that are relatively narrow to discourage high-speed automobile traffic.
• On-street parallel parking is recommended where it can be used as a buffer between pedestrians and motor vehicle traffic. Parked cars also serve to slow down passing traffic.
• Building uses that are often interspersed, i.e., small homes, large homes, outbuildings, small apartment buildings, corner stores, restaurants and offices.
• In addition to streets, public open spaces, around which are larger shops and offices as well as apartments.

Local zoning ordinances can be revised to require more attention to the needs of pedestrians. Some examples are discussed below.

Residential subdivision layout should provide safe, convenient, and direct pedestrian access to nearby (within one-half mile) and adjacent residential areas, bus stops, and neighborhood activity centers such as schools, parks, commercial and industrial areas, and office parks. Cul-de-sacs have proven to be effective in restricting automobile through traffic. However, they can also have the effect of restricting pedestrian mobility unless public accessways are provided to connect the cul-de-sacs with adjacent streets. Trail connections between cul-de-sacs and adjacent streets (shown in Figure 21.1) should be provided wherever possible

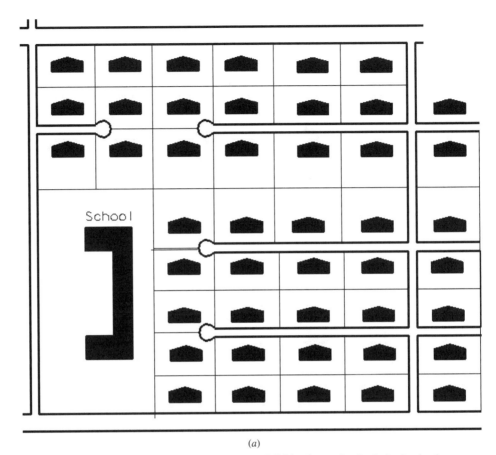

(a)

FIGURE 21.1 Subdivision layouts: (a) Conventional subdivision layout showing lack of pedestrian connectivity between streets and other land uses; (b) subdivision layout that provides pedestrian connections (indicated by the heavy lines) between streets and other land uses.

to improve access for pedestrians. Pedestrian facilities should be designed to meet local and statewide design standards.

In some high-density residential areas, regulations require off-street parking and reduced lot frontage. This results in homefronts that consist largely of garage doors. Ordinances should be modified to allow for rear-lot access (alleys) or other innovative solutions in these areas. Parking codes can be modified to allow for a reduced parking option for developments located on bus routes and which provide facilities that encourage biking and walking.

One of the most important factors in a person's decision to walk is the proximity of goods and services to homes and workplaces. The most conducive land use for pedestrian activity is one with a higher density mix of housing, offices, and retail. Major pedestrian improvements will occur as land use changes reduce the distances between daily activities. Such land-use changes include increasing density and mixing land use. While converting suburban locations to accommodate pedestrians is more difficult than downtown, such low-density development offers opportunities not possible in built-up neighborhoods. A strategy of (1) linking internal spaces where possible and (2) making the street usable for pedestrians will enhance suburban living for many people.

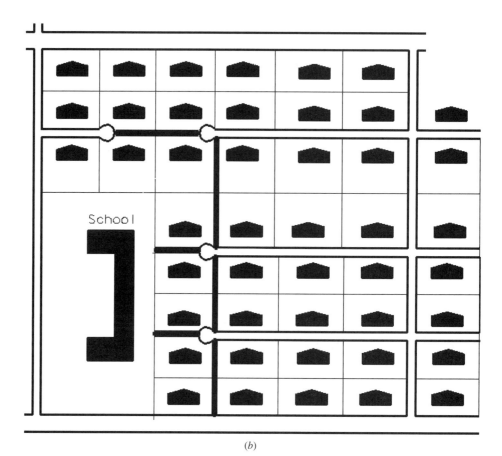

(b)

FIGURE 21.1 (*Continued*)

Extending and/or improving pedestrian access within a suburban community may eliminate some need for a car, allowing increased flexibility for those who have to drive. Most walkways should be planned in conjunction with roads so pedestrians can reach all developments that are located along the road.

Sidewalk width should vary to adjust to physical conditions and pedestrian volumes. Sidewalks near stores and schools should be wider to accommodate more people. Where there is a view, the sidewalk should be widened and a bench and landscaping added. Locations of commuter bus stops should be noted. Opportunities for short-cuts that make access easier should be identified. The success of suburban transit depends partly on the adequacy of the sidewalks and ease with which people can walk to bus stops.

Entrances to many commercial and retail centers are oriented toward automobile travel. Pedestrian access to storefronts is not only difficult and awkward, but often unsafe. A typical shopping center or strip mall is separated from the roadway by a wide parking lot. There are often no pathways linking store entrances to the sidewalks along the street, and in fact there may be no sidewalks along the street to be linked. Parking lots with multiple access points allow traffic circulation in different directions, creating hazards and confusion for walkers. Figure 21.2 illustrates a location that is not pedestrian friendly.

FIGURE 21.2 Lack of pedestrian connections to and within a retail center.

Such locations can be redeveloped to serve pedestrians better. As older commercial/retail locations undergo renovations, they should be redesigned to serve customers who arrive via automobile, transit, and bicycle and on foot. Specific methods include:

- Maximize pedestrian and transit access to the site from adjacent land uses.
- Provide comfortable transit stops and shelters with pedestrian connections to the main buildings.
- Locate transit stops and pedestrian drop-offs with reasonable proximity to building entrances.
- Provide attractive pedestrian walkways between the stores and the adjacent sites.
- Ensure that fencing and landscaping do not create barriers to pedestrian mobility.
- Rework entrances and orient buildings toward pedestrians and transit facilities instead of parking lots.
- Connect all buildings on site to each other via attractive pedestrian walkways, with landscaping and pedestrian-scale lighting. Provide covered walkways between buildings if possible.
- Minimize pedestrian-automobile conflicts by consolidating auto entrances into parking lots.

While many transit agencies in the United States have expended significant planning and design efforts to meet the needs of pedestrians in transit stations, relatively little attention has been devoted to the pedestrian environment to and from the stations. This hinders consideration of the door-to-door experience of using public transportation. It is not unusual for several different entities to maintain independent control over the various facilities that are used by someone walking to/from a single transit stop. State and local governments with responsibility for constructing and maintaining pedestrian facilities should cooperate with transit agencies, the private sector, and interested citizens in developing action programs to reduce barriers to pedestrian access to transit.

When given sidewalks and traffic-calmed streets to walk along, safe and convenient ways to cross streets, and a comfortable and attractive environment, most people are willing to walk farther to reach public transportation. Unfortunately, matters beyond the boundaries of the transit station have typically not received much attention in the United States. For example, park-and-ride lots are often located near freeways and/or shopping areas where residential housing is quite far away and there are no paths or sidewalks nearby.

However, the experience of some transit agencies shows there is growing awareness of the need to look at the larger environment that surrounds and leads to transit stations and bus stops. Examples of enhancing pedestrian access include placing pedestrian signals and detectors at key intersections, wider sidewalks, pedestrian shortcuts to key destinations, and trees.

21.3.3 Pedestrian Level of Service

The *Highway Capacity Manual* (TRB 2000) suggests that the quality of a pedestrian's experience is similar to an automobile-based measure, i.e., vehicles that are moving at slow speed or stopped due to congestion have low levels of service. They acknowledge that environmental factors also influence pedestrian activity. While speed, space, and delay are one set of measures of quality, there are other indicators of a good walking environment. Burden (1999) indicates that high pedestrian densities may indicate success. He notes that there are a number of attractions throughout the world where large crowds give excitement and security to a place.

Unlike high-speed motorists, whose travel is facilitated when the environment lacks detail, pedestrian travel is slow and interactive with the block, street, or neighborhood. The slower pace requires attention to detail. In fact, as Berkovitz (2001) points out, pedestrians are often intimidated by empty sidewalks and long travel distances, particularly along high-speed roads.

Landis et al. (2002) point out that evaluating the performance of a roadway section for walking is much more complex in comparison to that of the motor vehicle mode. While motor vehicle operators are largely insulated in their travel environment and are influenced by relatively few factors, pedestrians are relatively unprotected and are subject to a variety of environmental conditions. They note that there is not yet consensus among planners and engineers on which features of a roadway environment have statistically reliable significance to pedestrians. Several "walkability audits" have been developed that include a large number of features of the entire roadway corridor environment. Landis et al. (2002) present a list of factors on which it is generally agreed that pedestrians' sense of comfort and safety within a roadway corridor is based. These include: personal safety (i.e., threat of crashes), personal security (i.e., threat of assault), architectural interest, pathway or sidewalk shade, pedestrian-scale lighting and amenities, presence of other pedestrians, and conditions at intersections.

21.4 PEDESTRIAN SAFETY

The engineering literature on pedestrian safety tends to focus on pedestrian crash types, i.e., incidents where a pedestrian comes into contact with a motor vehicle. This is logical since many of these "crashes" are reported through the uniform crash reporting systems in each state. However, reliance on such data significantly understates the magnitude of the pedestrian safety problem since it typically ignores crashes that occur on private property such as parking lots or driveways. More importantly, incidents where pedestrians slip, trip, or fall on a sidewalk, stair, ramp, or other facility are not included. As described below, available data suggest that these non-motor-vehicle-involved incidents occur more frequently than motor-vehicle-involved crashes. This section examines both types of incidents by discussing common crash types and/or injury mechanisms. Tort liability issues are outlined along with risk management strategies for enhancing pedestrian safety.

21.4.1 Pedestrian Crash Types

According to Federal Highway Administration (FHWA) data, approximately 6,500 pedestrians are killed each year as a result of collisions with motor vehicles. In some large urban areas, pedestrians account for as much as 40 to 50 percent of traffic fatalities. Many more injuries are probably not reported to record-keeping agencies.

A FHWA study (Hunter, Stutts, and Pein, 1997) found that compared to their representation in the overall U.S. population, young persons (under 25 years of age) were overrepresented in pedestrian-motor vehicle crashes. Older adults (ages 25 to 44) and the elderly (age 65+) were underrepresented. However, elderly pedestrians in crashes were more than twice as likely to be killed (15 percent versus 6 percent) compared to young persons.

In general, pedestrian crashes occurred most frequently during the late afternoon and early evening hours. These are times when exposure is probably highest and visibility may be a problem. Alcohol or drug use was noted in about 15 percent of the pedestrian crashes overall. This increased to 31 percent for pedestrians in the 25–44 age group. Alcohol/drug crashes were also more frequent on weekends and during hours of darkness.

Forty-one percent of crashes occurred at roadway intersections, and an additional 8 percent occurred in driveways or alley intersections. About two-thirds of crashes were categorized as urban. Fifteen percent of the pedestrian crashes reported occurred on private property, primarily in commercial or other parking lots. The elderly were overrepresented in commercial parking lot crashes, young adults in non-commercial parking lot crashes, and children under age 10 in collisions involving driveways, alleys, or yards.

The eight most common pedestrian crash types identified by Hunter, Stutts and Pein (1997) are outlined below:

- Vehicle turn/merge: Pedestrian and vehicle collided while the vehicle was preparing to turn, in the process of turning, or had just completed a turn or merge.
- Mid-block dash: At a mid-block location, the pedestrian was struck while running and the motorist's view of the pedestrian was not obstructed.
- Not in roadway: The pedestrian was struck when not in the roadway. Areas included parking lots, driveways, private roads, sidewalks, service stations, and yards.
- Walking along roadway: Pedestrian was struck while walking (or running) along a road without sidewalks. The pedestrian may have been hitchhiking, walking with traffic and struck from behind or from the front, walking against traffic and struck from behind or from the front, or walking along a road but the details are unknown.
- Intersection dash: Pedestrian was struck while running through an intersection and/or the motorist's view of the pedestrian was blocked until an instant before impact.
- Intersection—other: Crash occurred at an intersection but does not conform to any of the specified crash types.
- Backing vehicle—Pedestrian was struck by a vehicle that was backing.
- Mid-block—other: The crash occurred at mid-block but does not conform to any of the specified crash types.

21.4.2 Non-Motor-Vehicle-Involved Incidents

The number of "same-level" falls is unknown. As Hyde et al. (2002) point out, falls that do not cause injury are not recorded. Therefore, researchers are limited to studying the incidence of such falls as reported by hospitals and similar institutions. Even with a significant number of injury and non-injury incidents going unreported, clearly the magnitude of the fall problem is significant. Data show that falls are the leading cause of injury reported to an emergency room. According to *The Injury Fact Book* (Baker et al. 1992), accidental falls are the second-leading cause of unintentional death, the second-leading cause of both spinal cord and brain injury, and the most common cause of hospital admission for trauma. While data exist on reported falls, i.e., the more serious falls, there is not good information about the circumstances surrounding the fall, i.e., did the individual slip on ice, trip on a raised section of sidewalk or fall on a stairway? In this regard, the author (Eck and Simpson 1996) attempted to use hospital records to obtain detailed information about falls to assist in countermeasure development. Results pointed out the importance of surface conditions to pedestrian safety. Two general types of surface condition problems were identified: slippery surfaces due to accumulation of ice and snow and holes or openings in the surface itself, e.g., missing, ill-fitting, or defective grates. While the former problem is one of maintenance, the latter issue has facility design, construction, and maintenance implications.

21.4.3 Tort Liability

As discussed above, planners and roadway designers and engineers must consider the needs of the pedestrian. Design, construction, maintenance, and operation of roads and streets, bridges, and surface condition must recognize the pedestrian. Roadway and recreational facilities that fail to incorporate the needs of all users increase the likelihood of claims against facility owners/managers.

Liability is an increasingly important issue for both public agencies and private entities. Implementing an aggressive and well-publicized risk management program can help head off these problems. It is useful to look at some of the design, construction, maintenance, and operational problems that commonly are cited in claims or lawsuits involving pedestrians.

- Open drainage grates in travel way
- Inadequate utility box covers (raised/depressed, poor skid resistance, or structural problems)
- Paths that suddenly end at "bad" locations with no transition or escape route provided
- Long-term, severe surface irregularities (e.g., broken pavement, potholes)
- Foreign substances on travel surface (e.g., water, sealants, oil/grease, loose stones or gravel)
- Vertical elevation differences in walkways
- Bridges that are hazardous to pedestrians
- Design errors, e.g., at curb ramps
- Wheelstops in parking lots
- Lack of railings between path and adjacent slope (and conditions at bottom)

Figure 21.3 shows wheelstops intruding into a walkway, creating a tripping hazard for pedestrians.

21.4.4 Risk Management Strategies

Outlined below are elements of an effective risk management strategy. Conscientiously implementing these elements should result in safer pedestrian facilities with fewer injuries, and therefore fewer claims, and should increase the likelihood of a successful defense for those claims that proceed to trial.

FIGURE 21.3 Wheelstops from parking lot intruding onto and near sidewalk, creating tripping hazard for pedestrians.

Perhaps the overarching strategy is to follow common sense principles for a defensible program. This means providing immediate response to the risks identified, including signing and warning for those conditions that cannot be immediately changed and funding spot improvements for those that can be changed. Listen to the public. If a parent calls, an editorial is written, or any other input from a "customer" suggests action, the situation should be evaluated and action taken as quickly and as intelligently as possible. All agencies, departments, and other parties whose duties are affected need to be involved. The risk-reduction effort takes cooperation and ongoing coordination.

Other specific actions that can be taken include:

- Incorporate accepted standards and guidelines.
- Use established engineering, planning, and design principles.
- Consider all potential users.
- Do it right!
- Promote community involvement and awareness.

21.5 PEDESTRIAN FACILITY DESIGN

21.5.1 Walkways, Sidewalks, and Public Spaces

A successful urban sidewalk should have the following characteristics:

- Adequate width
- Buffer from travel lane
- Gentle cross-slope (2 percent or less)

- Buffer to private properties
- Adequate sight distances around corners and at driveways
- Shy distances to walls and other structures
- Continuity
- Clear path of travel free of street furniture
- Well-maintained condition
- Ramps at corners and flat areas across driveways
- Sufficient storage capacity at corners

Sidewalks require a minimum width of 5 feet if set back from the curb or 6 feet if at the curb face. Walking is a social activity. For two people to walk together, 5 feet of space is the bare minimum needed. In some areas, such as near schools, sporting complexes, some parks, and many shopping districts, the minimum width for a sidewalk is 8 feet.

The desirable width for a sidewalk is often much greater. Some shopping districts require 12, 20, 30, or even 40 feet of width to handle the volume of pedestrian traffic they encounter. Pennsylvania Avenue in Washington, DC, has 30-foot sidewalk sections to handle tour bus operations. The *Highway Capacity Manual* (TRB 2000) covers the topics of sidewalk width and pedestrian level of service.

It is important to determine the commercial need for outdoor cafes, kiosks, corner gathering spots, and other social needs for a sidewalk. In commercial areas, designers should consult property owners, chambers of commerce, downtown merchants associations, and landscape architects to ascertain if the desired width is realistic. Corner or mid-block bulb-outs can be used to advantage for creating both storage space and for roadway crossing and social space. Figure 21.4 shows a corner bulb-out on an arterial street.

Most sidewalks are made of concrete due to its long life, distinct pattern, and lighter color. In some cases, asphalt can provide a useful surface.

FIGURE 21.4   Corner bulb-out on arterial street.

Paver stones can also be used. These colorful brick, stone, or ceramic tiles are often used to define corners or crosswalks (as illustrated in Figure 21.4), to create a mood for a block or commercial district, or to help those with visual impairments. The blocks need to be set on a concrete pad for maximum life and stability.

Certain cautions are in order relative to paver stones, bricks, and similar materials. The FHWA publication (2001) on best practices in designing trails for access points out that decorative surfaces may create a vibrating, bumpy ride that can be uncomfortable and painful for those in wheelchairs. Pavers or bricks can settle or buckle, creating changes in level. This creates a tripping hazard for people with vision impairments and for ambulatory pedestrians with visibility impairments. Finally, decorative surface materials can make it more difficult for pedestrians with vision impairments to identify detectable warnings that provide critical information about the transition from sidewalk to the street. For these reasons, paver, brick, or cobblestone sidewalks are not recommended. Concrete sidewalks with paver or brick trim preserve the decorative qualities but present an easier surface to negotiate.

Desirably, a border area should be provided along streets for the safety of motorists and pedestrians as well as for aesthetic reasons. The border area between the roadway and the right-of-way line should be wide enough to serve multiple purposes, including provision of buffer space between pedestrians and vehicular traffic, snow storage, an area for placement of underground utilities, and an area for maintainable landscaping. The border may be a minimum of 5 feet but desirably it should be 10 feet.

Nature strips, particularly in downtown areas, are a good location to use paver stones for easy and affordable access to underground utilities. In downtown areas, nature strips are also a convenient location for the swing width of a door and for placement of parking meters, hydrants, lampposts, and other furniture.

On-street parking has two distinct advantages for pedestrians. First, it creates desired physical separation from motor vehicle traffic. Second, it has been shown to reduce motorist travel speeds, thus creating a safer environment for street crossings.

On the back side of sidewalks, a minimum width buffer of 1 to 3 feet is essential. Without such a buffer, vegetation, walls, buildings, and other objects encroach on the usable sidewalk space.

Pedestrians require a shy distance from fixed objects, such as walls, fences, shrubs, buildings, parked cars, and other features. The desired shy distance for a pedestrian is 2 feet. Allowance for this shy distance must be made in determining the functional width of a sidewalk.

The literature points out that attractive windows in shopping districts cause curious pedestrians to stop momentarily. This is a desired element of a successful street. These window-watchers take up about 18 to 24 inches of space.

Because of its relatively slow pace, walking is more detail-oriented than driving. Walkers are attracted to locations with amenities, interesting storefronts, and outdoor cafes. On the other hand, pedestrians will avoid sidewalks lined with walls or that lack interesting details. Compare Figure 21.5, with its amenities and interesting details, with the wall effect shown in Figure 21.6. Where would you rather walk?

Newspaper racks, mailboxes, and other street furniture should not encroach into walking space. The items can be placed in a nature strip, a separate storage area behind the sidewalk, or in a corner or mid-block bulb-out. These items should be bolted in place.

AASHTO's highway design policy (2001) states that landscaping should be provided for aesthetic and erosion-control purposes in keeping with the character of the street and its environment. Landscaping should be arranged to permit sufficiently wide, clear, and safe pedestrian walkways. Combinations of turf, shrubs, and trees are desirable in border areas along roadways. However, care must be taken to ensure that sight distances and clearances to obstructions are preserved, particularly at intersections.

Landscaping can also be used to control, partially or fully, crossing points of pedestrians. Low shrubs in commercial areas and near schools are often desirable to channel pedestrians to crosswalks or crossing areas.

FIGURE 21.5   Sidewalk café and other amenities create an inviting sidewalk.

FIGURE 21.6   Empty wall (due to drawn blinds) and lack of amenities do not attract pedestrians.

Management of corner space is critical to the success of a commercial street. This small public space enhances the corner sight triangle, permits underground piping of drainage so that street water can be captured on both sides of the crossing, provides a resting place, stores pedestrians waiting to cross the roadway, and offers a location for pedestrian amenities. Well-designed corners, particularly in a downtown or village-like shopping district, can become a focal point for an area. Benches, telephones, newspaper racks, mailboxes, bike racks, and other features help to enliven this area. Corners are often the most secure places on a street.

Parking structures in commercial districts should ideally be placed away from popular walking streets. If this is not possible, driveway and curb radii should be kept tight to maximize safety and to minimize discomfort to pedestrians.

If possible, grades should be kept to no more than 5 percent; terrain permitting, grades steeper than 8 percent should be avoided. Where this is not possible, railings and other aids should be considered to assist older adults. The Americans with Disabilities Act (ADA) does not require designers to change topography but only to work within its limitations and constraints. Do not create a constructed grade that exceeds 8 percent.

Stairs should be avoided where possible since they are a barrier to accessibility and falls are common on poorly designed stairs. It is critical that stairs be well-constructed and maintained, easily detectable, and slip-resistant. The following principles apply; consult local building codes for additional details. Minimum stairway width is 42 inches, to allow two people to pass. Stairs require railings on at least one side. Railings must extend 18 inches beyond the top and bottom stair. For wide stairs, such as might be present at passenger transportation terminals, there should be railings on both sides and one or more in mid-stair areas. Open risers should be avoided. The stair should have a uniform grade with constant tread and rise along the stair. For exterior stairs, the tread should have a forward slope of 1 percent to drain water. Stairs should be illuminated at night.

Sidewalks are recommended on both sides of all urban arterials, collectors and on most local roadways. Codes should require sidewalks for new construction. Lack of sidewalks on a road or street means conflicts with vehicles are maximized. Children, older adults, and people with disabilities may not have mobility under these circumstances. When prioritizing missing sidewalks, the following factors should be considered: schools, transit stops, parks, shopping districts and commercial areas, medical complexes and hospitals, retirement homes, and public buildings

Experience has shown that the features summarized below are desirable to achieve robust commercial activity and to encourage added walking versus single-occupant motor vehicle trips. Sucher's excellent book (1995) on "city comforts" is recommended for more detailed information.

Trees. FHWA text (n.d.) states that the most charming streets are those with trees gracing both sides of a walkway. The canopy effect is attractive to pedestrians. Trees should be set back 4 feet from the curb.

Awnings. Retail shops should be encouraged to provide protective awnings to create shade, provide protection from rain and snow, and add color and attractiveness to the street. Awnings are especially important in warm climates on the sunny side of the street.

Outdoor Cafés. Careful regulation of street vendors, outdoor cafés, and other commercial activity help enliven a place—the more activity, the better. One successful outdoor café helps create more activity, and in time an entire area can be helped back to life. When outdoor cafés are present, it is essential to maintain a reasonable walking passageway. Elimination of two or three parking spaces in the street and addition of a bulb-out area can often provide the necessary extra space when café seating is needed.

Alleys and Narrow Streets.   Alleys can be cleaned up and made attractive for walking. Properly planned and lit, they can be secure and inviting. Some communities have covered over alleys and made them into access points for a number of shops. Alleys can become attractive places for outdoor cafés, kiosks, and small shops.

Gateways.   Gateways identify a place by defining boundaries. They create a sense of welcome and transition.

Kiosks.   Small tourist centers, navigational kiosks, and attractive outlets for other information can be handled through small-scale or large-scale kiosks. Well-located interpretive kiosks, plaques, and other instructional or historic place markers are essential to visitors. These areas can also serve as safe places for people to meet and can generally help with navigation.

Fountains, Play Areas, and Public Art.   Public play areas and interactive art can enliven a corner or central plaza. Project for Public Spaces, Inc. (2001) points out that for such amenities to work, they must respond to the needs of a location, to the activities that take place there, and to people's pattern of use.

Pedestrian Streets, Transit Streets, and Pedestrian Malls.   Many cities throughout the world have successfully converted streets to transit and pedestrian streets. These conversions need to be made with a master plan so that traffic flow and pedestrian movements are provided for fully.

21.5.2 Pedestrian Plazas

Pedestrian plazas are defined as places of abundant vegetation, artwork, seating, and perhaps fountains, which are intended not only as quiet spots for rest and contemplation but as centers where communities can come together to socialize and take part in a variety of activities (Project for Public Spaces, Inc. 2001). Unfortunately, many recently constructed plazas serve more to enhance the image of the building on the lot in that they are too large and uncomfortable for pedestrians. Problems with plazas include:

Some are windswept.

Some are on the shady side of buildings.

Some break the continuity of shopping streets.

Some are inaccessible because of grade changes.

Most are without benches, planters, cover, shops, or other pedestrian comforts.

To be comfortable, large spaces should be divided into smaller ones. Landscaping, benches, and wind and rain protection should be provided, and shopping and eating should be made accessible. Encourage the use of bandstands, public display areas, outdoor dining space, skating rinks, and other features that attract crowds.

FHWA (n.d.) indicates that no extra room should be provided. It is usually better to be a bit crowded than too open and to provide many smaller spaces instead of a few large ones. It is better to have places to sit, planters, and other conveniences for pedestrians than to have clean, simple, and architectural space. It is better to have windows for browsing and stores adjacent to the plaza space, with cross-circulation between different uses, than to have the plaza serve one use. It is better to have retailers rather than offices border the plaza. It is better for the plaza to be part of the sidewalk instead of separated from the sidewalk by walls. The popular downtown plaza in Montreal, shown in Figure 21.7, has most of these attributes.

FIGURE 21.7   Popular plaza in downtown Montreal.

Ideally, plazas should be located to provide good sun exposure and little wind exposure in places that are protected from traffic noise and in areas that are easily accessible from streets and shops. Planners should inventory the area for spaces that can be used for plazas, especially small ones. Appropriate spaces include locations where buildings may be demolished and new ones constructed, vacant land, or streets that may be closed to traffic or may connect to parking. The Project for Public Spaces, Inc. (2002) handbook on creating successful public spaces is an excellent resource on this topic.

21.5.3 Intersections

Intersections are locations where the paths of vehicles and pedestrians come together. They can be the most challenging part of negotiating the pedestrian network. If pedestrians cannot cross the street safely, then mobility is severely limited, access is denied, and walking as a mode of travel is discouraged. In designing and operating intersections that are attentive to the needs of pedestrians, the following considerations should be addressed:

* Enhancing visibility of pedestrians through painted crosswalks, moving pedestrians out from behind parked vehicles by using bulb-outs, and increasing sight distances by removing obstructions such as vegetation and street furniture
* Minimizing time and distance pedestrians need to cross roadway
* Making pedestrian movements more predictable through the use of crosswalks and signalization
* Using curb ramps to provide transition from walkway to street

The following features of intersections should be designed from the pedestrian as well as motor vehicle standpoint:

Crosswalks. One way to shorten the crossing distance for pedestrians on streets where parking is permitted is to install curb bulbs or curb extensions. As shown in Figure 21.4, curb bulbs project into the street usually for a distance equal to the depth of a typical parallel parking space, thereby making it easier for pedestrians to see approaching traffic and giving motorists a better view of pedestrians. When designing curb bulbs at intersections where there is low truck traffic, the corner radius should be as small as possible to have the effect of slowing down right-turning traffic.

Signal Timing, Indications, and Detection. Pedestrians are often confused by pedestrian phase signal timing and pushbuttons since these seem to vary not only from place to place but also from intersection to intersection. The timing of WALK and DONT WALK phases appears arbitrary. Many pedestrians do not know that the flashing DONT WALK is intentionally displayed before the average pedestrian can get completely across the street. Or the signal timing may be too fast for slow walkers such as older pedestrians or people with disabilities. FHWA (n.d.) recommends that agencies develop policies regarding pedestrian signal timing and pushbutton actuation to ensure fair treatment for pedestrians. Signal timing should be calculated based on a walking speed of 2.8 feet per second, as opposed to the average walking speed of 4 feet per second. This will accommodate older pedestrians and those with mobility impairments. The pushbutton should be placed at the top of and as near as possible to the curb ramp and clearly in line with the direction of travel. This is especially important for pedestrians with low vision. The pushbutton box should provide a visible acknowledgment that the crossing request has been received.

Refuge Islands. Pedestrian refuge islands are the areas within an intersection or between lanes of travel where pedestrians may safely wait until vehicular traffic clears, allowing them to cross a street. Such islands are commonly found on wide, multilane streets where adequate pedestrian crossing times cannot be provided without adversely affecting traffic flow. They also provide a resting place for those pedestrians (elderly, wheelchair-bound, or others) unable to cross the intersection completely within the allotted time. Pedestrian refuge islands may be installed at intersections or mid-block locations as determined by engineering studies. They must be designed in accordance with the AASHTO design policy (2001) and *MUTCD* (FHWA 2000) requirements. Raised curb islands need cut-through ramps at pavement level or curb ramps for wheelchair users. The island should be at least 6 feet wide from face of curb to face of curb (minimum width shall not be less than 4 feet). The island should not be less than 12 feet long or the width of the crosswalk, whichever is greater. Minimum island size should be 50 square feet. There should be no obstructions to visibility, e.g., barriers, vegetation, or benches.

21.5.4 Mid-Block Crossings

Designers often assume that pedestrians will cross roadways at established intersections. However, pedestrians routinely cross the street at mid-block locations. Pedestrians will rarely go out of their way to cross at an intersection unless they are rewarded with a much-improved crossing. As noted earlier, pedestrians will take the most direct route, even it means crossing several lanes of high-speed traffic. Pedestrians crossing at random and unpredictable locations create confusion and increase risks to both the pedestrians and drivers. Well-designed and properly located mid-block crossings can actually provide safety benefits for pedestrians. Two primary ways to facilitate non-intersection crossings are medians and mid-block crossings. Figure 21.8 shows a mid-block crossing which incorporates a median and traffic calming elements. The crossing connects a parking structure with a retail mall.

A median or refuge island is a raised longitudinal space separating opposing traffic directions. A pedestrian faced with crossing one or more lanes in each direction must determine a safe gap in two, four, or six lanes at a time—a complex task. Younger and older drivers have reduced gap-acceptance skills compared to pedestrians in other age groups. Pedestrian

FIGURE 21.8 Mid-block crossing connecting parking structure with downtown retail mall.

gap-assessment skills are particularly poor at night. A median allows pedestrians to separate the crossing into two tasks, negotiating one direction of traffic at a time. Crossing times are reduced, making the walk across the street much safer.

Desirably, a median should be at least 8 feet wide to allow a pedestrian to wait comfortably in the center, 4 feet from moving traffic. Normally there will be an open flat cut rather than a ramp, due to the short width.

Mid-block crossings are located and placed according to a number of factors, including roadway width, traffic volume, traffic speed and type, desire lines for pedestrian movement, and adjacent land use. Due to their low speed and volume, local roads generally do not have median treatments. However, there may be exceptions, particularly around schools or hospitals where traffic calming is desired.

The design of mid-block crossings uses warrants similar to those used for conventional intersections, e.g., stopping sight distance, effects of grade, need for lighting, and other factors. The designer should recognize that pedestrians have a strong desire to continue their intended path of travel. Natural patterns should be identified. For example, a parking lot on one side of the street connecting a large office complex or shopping center on another establishes the desired crossing location. Grade-separated mid-block crossings have been effective in a few isolated locations. Due to their cost and potentially low use, engineering studies should be conducted. Given a choice, on most roadways pedestrians generally prefer to cross at grade.

21.5.5 Accessibility

According to the FHWA publication (1999) on designing facilities for access, there are 49 million people in the United States with disabilities. At one time or another, virtually all of them are pedestrians. Anyone can experience a temporary or permanent disability at any time due to age, illness, or injury. It is estimated that 85 percent of Americans living to their full life expectancy will suffer a permanent disability. Good design is important so that these individuals are not restricted in their mobility.

The Americans with Disabilities Act (ADA) was enacted in 1990 to ensure that a disabled person has access to all public facilities in the United States. The law has specific requirements for pedestrian facilities on public and private property. This section provides an overview of basic accessibility requirements that are relevant to designing pedestrian facilities. The complete set of standards can be found in the Americans with Disabilities Act Accessibility Guidelines (ADAAG) developed by the Access Board. Note that the rules are updated from time to time.

Sidewalks.   Wheelchairs require a 3-foot minimum width for continuous passage. Therefore, sidewalks should have a minimum clearance width of 5 feet. Sidewalks should be surfaced with a smooth, durable, and slip-resistant material. They should be kept in good condition, free from debris, cracks, and rough surfaces. Sidewalks should have the minimum cross-slope necessary for proper drainage. The maximum cross-slope is 2 percent (1:50). A person using crutches or a wheelchair has to exert significantly more effort to maintain a straight course on a sloped surface than on a level surface. Driveway slopes should not encroach on the sidewalk.

Ramps.   Ramps are locations where the grade exceeds 5 percent along an accessible path. Longitudinal grades on sidewalks should be limited to 5 percent but may be a maximum of 1:12 if necessary. Long, steep grades should have level areas every 30 feet since traversing a steep slope with crutches or artificial limbs or in a wheelchair is difficult and level areas are needed for the pedestrian to stop and rest.

Street Furniture.   Street furniture includes things such as benches, newspaper boxes, trash receptacles, and bus shelters. To accommodate the disabled, street furniture should be out of the normal travel path as much as possible. For greater conspicuity, high-contrast colors such as red, yellow, and black are preferable. Guidelines for street furniture include the following:

- No protruding object should reduce the clear width of a sidewalk or walkway path to less than 3 feet.
- No object mounted on a wall or post or freestanding should have a clear open area under it higher than 2.3 feet off the ground.
- No object higher than 2.3 feet attached to a wall should protrude from that wall more than 4 inches.

Pedestrian Signals.   Some individuals have difficulty operating conventional pedestrian pushbuttons. There may be a need to install a larger pushbutton or to change the placement of the pushbutton. Pedestrian pushbuttons should always be easily accessible to individuals in wheelchairs and should be no more than 42 inches above the sidewalk. The force required to activate the button should be no greater than 5 pounds.

Accessible pedestrian signals provide audible and/or vibrotactile information coinciding with visual pedestrian signals to let blind or low-vision pedestrians know precisely when the WALK interval begins. Bentzen (1998) identifies intersections that may require evaluation for accessible pedestrian signal installation. These include very wide crossings, secondary streets having little traffic, nonorthogonal or skewed crossings, T-intersections, high volumes of turning vehicles, split-phase signal timing, and noisy locations. Where these conditions occur, it may be impossible for a pedestrian who is blind to determine the onset of parallel traffic or to obtain usable orientation and directional information about the crossing from the cues that are available.

Curb Cut Ramps.   The single most important design consideration for persons in wheelchairs is to provide curb cuts. New and rebuilt streets with sidewalks should have curb cuts at all crosswalks. It is desirable to provide two curb cuts per corner (single ramps located in the center of a corner are less desirable). Separate ramps provide greater information to visually impaired pedestrians in street crossings, especially if the ramp is designed to be parallel to the crosswalk. These also benefit others with mobility limitations, elderly pedestrians, and persons pushing strollers or carts.

ADAAG specifies that curb ramps should be at least 36 inches wide, not including the width of the flared sides. According to ADAAG, the slope of a curb ramp should not exceed 8.33 percent; the cross-slope should not exceed 2 percent. If the landing (level area of sidewalk at the top of a curb ramp) width is less than 48 inches, then the slope of the flares at the curb face should not exceed 8.33 percent. If the landing width is greater than 48 inches, a 10 percent slope is acceptable.

Ramps should be checked periodically to make sure large gaps do not develop between the gutter and street surface. Drainage is very important with curb cuts. Standing water can obscure a drop-off or pothole at the base of a ramp and makes the crossing messy. Storm drain inlets should be clear of the crosswalk.

21.5.6 Multiuse Trails

Multiuse or off-road trails provide environments for walking and other non-motorized users that, as shown in Figure 21.9, are separate from motor vehicle traffic. Such trails are often extremely popular facilities that are in high demand among in-line skaters, bicyclists, joggers, people walking dogs, and a variety of other users. These different trail users have different objectives. The resulting mix of objectives and volume of non-motorized traffic can create problems that should be anticipated during design by understanding the needs of these users and accommodating expected levels and types of use.

There are many types of trails, including:

FIGURE 21.9 Multiuse riverfront trail in Reno, Nevada.

- Urban trails and pathways
- Rail trails
- Trails in greenways
- Interpretive trails
- Historical/heritage trails
- Primitive trails

All of these can be designed for use by pedestrians (including joggers, casual strollers, hikers, in-line skaters, and others), people with disabilities, and bicyclists. What distinguishes one type of trail from another is its context.

The National Highway Institute (NHI) workbook (1996) cautions that design alone cannot generally restrict use. For example, if bicycles are to be prohibited on a path, education and enforcement programs should be implemented to back up the prohibition. Design can passively discourage certain users from traveling on a given trail but only to a limited extent. For example, "road" bicyclists are not likely to ride on an unpaved surface.

It may be easier to separate types of trail users. Several different separation techniques are available, including:

- Parallel pathways—one for "wheels" and one for "heels" (as shown in Figure 21.10)
- Striped lanes—pedestrians and bikes in separate lanes
- Directional separation—traffic in one direction stays to the right; traffic in the other direction stays to the left
- Signing and enforcement, e.g., "EQUESTRIAN USE ONLY"

National guidelines for the design of multiuse trails are provide by AASHTO's (1999) *Guide for the Development of Bicycle Facilities*. There are a number of other excellent guides

FIGURE 21.10   Separate pathways for pedestrians and bicycles.

as well, covering a variety of trail types (e.g., Steinholtz and Vachowski 2001). Schwarz (1993) presents a thorough discussion of greenway trail design including types of trail layouts. Figure 21.11 shows a stone-surface multiuse trail in steep terrain that incorporates a timber safety railing.

The minimum width for two-directional trails is 10 feet, although 12- to 14-foot widths are preferred where heavy traffic is expected. Centerline stripes should be considered for paths that generate substantial amounts of pedestrian traffic. Appropriate speed limit and warning signs should be posted. Trail etiquette signs should clearly state that bicycles should give an audible warning before passing other trail users.

The surfacing material on a trail significantly affects which user groups will be capable of negotiating the path. Soft surfaces such as sand and gravel are more difficult for all users to negotiate. They present special difficulties for those using wheeled devices such as road bicycles, strollers, and wheelchairs.

The FHWA guide to access (1999) notes that local conditions determine the choice of trail surface. Recreational trail surfaces are often composed of naturally occurring soil. Surfaces ranging from concrete to wood chips may be used depending on the designated user types, the expected volume of traffic, the climate, and the surrounding environment.

Trail/roadway intersections can become areas of conflict if not carefully designed. For at-grade intersections, the following characteristics should be included:

1. Position the crossing at a logical and visible location.

2. Warn motorists of the approaching crossing. Warning signs and pavement markings used to alert motorists of trail crossings should be used in accordance with the *MUTCD* (FHWA 2000).

3. Maintain visibility between trail users and motorists. Vegetation, signs, and other objects in the right-of-way should be removed or relocated so that trail users can observe traffic conditions and motorists can see approaching trail users.

FIGURE 21.11 Multiuse trail in steep terrain.

4. Inform trail users of the upcoming intersection. Signs and pavement markings on the trail can provide advance warning of the intersection, especially in areas where the intersection is not clearly visible.

The need for parking should be anticipated during the trail-planning process. Adequate parking at trailheads is necessary so that trail users do not park on the shoulder of the road near intersections, blocking sightlines of both motorists and trail users.

Unauthorized motor vehicle access is a problem at some trail/roadway intersections. Trail bollards are the most effective method of restricting motor vehicle traffic. However, care must be taken in their use because they present an obstacle when located in the travel path of pedestrians and bicycles.

Bollards should be painted a bright color and permanently reflectorized to maintain their visibility. They should be 3 feet tall and can be constructed of a variety of materials. Commercial manufacturers offer bollards that can be unlocked and removed to allow emergency vehicle and maintenance access.

The NHI workbook (1996) points out that trails can be used to provide connections between transportation modes. Intermodal linkage possibilities include:

1. To shopping, schools, work, and transit

2. Between parking and transit drop-off points

3. To ports, rivers, and scenic areas

4. To ferry or bridge connections

Access to trails is a key consideration. A *trailhead* is a location where people can access a trail. The NHI workbook (1996) indicates that a trailhead can be as simple as a trail marker and few parking stalls or a virtual visitor's center with vending machines, snack bar, and interactive trail guide information.

Good trails attract people who need places to park, rest, get trail information, use restrooms, dispose of trash, and get a drink of water. These are the basic considerations. Trailhead design should consider:

- Easy access from public streets
- Provision of adequate parking
- Location relative to transit facilities
- Potential joint use, e.g., can also serve as picnic area
- Rest rooms and trash disposal
- Weather protection
- Potential for interpretation of area's historic, cultural, and natural features
- Location relative to on-site concessions and ancillary facilities such as bike rentals

21.6   OPERATIONS AND MAINTENANCE

21.6.1   Pedestrian Signs and Pavement Markings

Traffic engineers use a wide variety of signs and pavements markings relative to pedestrians. Some are used to alert motorists to pedestrian activity and others to direct pedestrians to defined crossings.

Signing is governed by the *MUTCD* (FHWA 2000), which provides guidelines for the design and placement of traffic control devices installed within public rights-of-way. It must be noted that signs are often ineffective in modifying driver behavior and overuse of signs breeds disrespect.

Colors for signs and markings should conform to the color schedule recommended by the *MUTCD* to promote uniformity and understanding from jurisdiction to jurisdiction. The Millennium Edition of the *MUTCD* includes a new color (in addition to standard yellow) for pedestrian warning signs—fluorescent yellow-green. However, the *MUTCD* cautions that the mixing of standard yellow and fluorescent yellow-green backgrounds within a selected site area should be avoided.

The discussion below presents an overview of key issues related to traffic control devices for pedestrians. The discussion is organized by type of device.

Regulatory Signs.   The NO TURN ON RED sign may be used in some situations to fa- cilitate pedestrian movements. Due to conflicting research results, there has been considerable controversy regarding pedestrian safety and right-turn-on-red. Use of NO TURN ON RED signs at intersections should be evaluated on a case-by-case basis using engineering judg- ment. In August 2002, the Institute of Transportation Engineers was considering guidelines stating that prohibition of turns on red should only be considered after the need has been fully established and less restrictive methods have been reviewed or tried. Part-time prohi- bitions should be discouraged; however, they are preferable to full-time prohibitions when the actual need occurs for only short periods of time. ITE recommends that less restrictive alternatives be considered in lieu of prohibiting all turns on red. A supplemental plate reading WHEN PEDESTRIANS ARE PRESENT may aid the pedestrian without unduly restricting vehicular traffic flow. Education and enforcement play important roles in the benefits and safety of right-turn-on-red. Enforcement is important relative to turns being made only after stopping and yielding to other road users and that the necessary prohibitions are being observed.

Other signs include the pedestrian pushbutton signs or other signs at signals directing pedestrians to cross only on the green light or WALK signal. Pedestrian pushbutton signs should be used at all pedestrian-actuated signals. It is helpful to provide guidance to indicate which street the button is for. The signs should be located adjacent to the pushbutton and pushbuttons should be accessible to pedestrians with disabilities.

Educational plaques may be used for pedestrians at traffic signals to define the meaning of the WALK, DONT WALK, and flashing DONT WALK signal indications. The decision to use these signs is strictly engineering judgment and is primarily for educational purposes (to improve pedestrian understanding of pedestrian indications).

Warning Signs.   The pedestrian crossing sign (W11-2, as shown in Figure 21.12*a*) may be used to alert road users to locations where unexpected entries into the roadway by pedestrians might occur. The crossing sign should be used adjacent to the crossing location. If the crossing location is not delineated by crosswalk pavement markings, the crossing sign shall be supplemented with a diagonal downward-pointing arrow plaque showing the location of the crossing. If the crossing location is delineated by crosswalk pavement markings, the diagonal downward-pointing arrow plaque shall not be required. To avoid information over- load, this sign should not be mounted with another warning or regulatory sign.

It is important to note that overuse of warning signs breeds disrespect and should be avoided. Care should be taken in sign placement in relation to other signs to avoid sign clutter and allow adequate motorist response.

The playground sign (W15-1, as shown in Figure 21.12*b*) may be used to give advance warning of a designated children's playground that is located adjacent to the road. This sign is not intended for use on local or residential streets where children are expected. CAUTION CHILDREN AT PLAY or SLOW CHILDREN signs should not be used. They may encour- age children to play in the street and parents to be less vigilant. Such signs provide no guidance to motorists in terms of safe speed, and the sign has no legal basis for determining what a motorist should do. Furthermore, motorists should expect children to be ''at play'' in

FIGURE 21.12   (*a*) Pedestrian crossing sign (W11-2); (*b*) playground sign (W15-1).

all residential areas; the lack of signing on some streets may indicate otherwise. Use of these nonstandard signs may also imply that the involved jurisdiction approves of streets as playgrounds, which may result in the agency being vulnerable to tort liability.

Informational Signs.   Guide or directional signs for pedestrians are intended to assist unfamiliar pedestrians or those who may not know the most direct route to a destination by foot. Use distances meaningful to pedestrians, such as the number of blocks or average walking time.

Crosswalk Markings.   Crosswalk markings provide guidance for pedestrians who are crossing roadways by defining and delineating paths on approaches to and within signalized intersections and on approaches to other intersections where traffic stops. Crosswalk markings also serve to alert road users of a pedestrian crossing point across roadways not controlled by traffic signals or stop signs. At non-intersection locations, crosswalk markings legally establish the crosswalk. The standard in the *MUTCD* is that when crosswalk lines are used, they shall consist of solid white lines that mark the crosswalk. Crosswalk lines should not be used indiscriminately. An engineering study should be performed before they are installed at locations away from traffic signals or stop signs. Typical types of crosswalk markings are shown in Figure 21.13. Lalani and the ITE Pedestrian and Bicycle Task Force (2001) present an excellent summary of various treatments used by local agencies in the United States, Canada, Europe, New Zealand, and Australia to improve crossing safety for pedestrians, including mid-block locations and intersections.

21.6.2   Pedestrians and Work Zones

When construction or maintenance activities take place on or near sidewalks or crosswalks, pedestrians may be exposed to a variety of hazards, including detours that are difficult to navigate or force pedestrians into the street, uneven walking surfaces, walking surfaces contaminated with foreign substances, restricted sight distances, and conflicts with vehicles and equipment. In addition, the mobility of persons with disabilities may be adversely affected. It is important to develop and implement temporary traffic control zone policies that minimize these problems. All parties involved should be made aware of the needs of pedestrians and made responsible for providing safe and continuous passage.

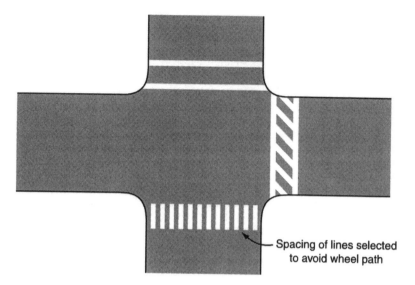

Spacing of lines selected
to avoid wheel path

FIGURE 21.13 Crosswalk markings as presented in *MUTCD*.

Developing a workable policy for pedestrian access through work areas requires the cooperation of traffic engineers, construction inspectors, crew chiefs, contractors, and advocates. The policy should apply whenever construction or maintenance work affects pedestrian access, whether the work is done by private firms or city, county, or state personnel.

Permits required for street construction or construction projects that encroach upon sidewalks or crosswalks should be contingent on meeting pedestrian access policies. Contractors should be given copies of the standards when they apply for a permit. Crew chiefs and crews should be trained so that they understand and follow the policy. The Millennium Edition of the *MUTCD* (FHWA 2000) provides more specific guidance than previous editions on pedestrian access around work areas.

21.6.3 Facility Maintenance

Walkways are subject to debris accumulation and surface deterioration and require maintenance to function safely and efficiently. Poorly maintained facilities become unusable and a liability as users risk injury.

As noted earlier, while walking a person typically looks ahead and around without noticing cracks and other discontinuities in the walking surface. A smooth, level surface is critical for young, elderly and disabled pedestrians. Pedestrians also depend on motorists respecting traffic signs and signals. These must also be properly maintained for pedestrian safety.

A walkway maintenance program is necessary to ensure adequate maintenance of facilities. The program should establish maintenance standards and a schedule for regular maintenance and inspection as outlined below. Recommended maintenance practices include:

1. Sweeping: Loose gravel, snow/ice control abrasives, broken glass, and other debris are not only unattractive but can cause slip-and-fall hazards for pedestrians, particularly when the walkway is on a gradient. A periodic inspection and maintenance program should be implemented so that such loose materials are regularly picked up or swept. Debris from the roadway should not be swept onto sidewalks.

2. Surface repairs: A smooth walkway surface free of cracks, potholes, bumps, and other physical problems should be provided and maintained. Surfaces should be inspected regularly and potentially hazardous conditions repaired as soon as possible.

3. Vegetation: Vegetation encroaching onto walkways is both a nuisance and a problem. Roots should be controlled to prevent break-up of the surface. Raised sidewalk slabs are a significant cause of trip-and-fall accidents, particularly for the elderly. It is also important that adequate clearances and sight distances be maintained at driveways and intersections. Pedestrians must be visible to approaching motorists and not hidden by overgrown shrubs or low-hanging branches. Local ordinances should allow authorities to control vegetation that originates from private property.

4. Traffic control devices: New pedestrian-related signs and pavement markings (e.g., crosswalks) are highly visible, but over time they weather and become harder to see, especially at night. Retroreflectivity of signs and markings should be inspected at night and defective devices replaced.

5. Drainage: New drainage devices function well, but they deteriorate over time. Repair or relocate faulty drainage at intersections where water backs up onto the curb cut or into the crosswalk.

6. Utility cuts: These can leave a rough surface if not constructed carefully. Sidewalk cuts should be finished as smooth as a new sidewalk.

7. Snow removal: Snow should be cleared from publicly owned sidewalks. Sidewalks are not appropriate for snow storage. Ordinances regarding removal of snow from private sidewalks should be publicized and enforced.

21.7 REFERENCES

American Association of State Highway and Transportation Officials (AASHTO). 1999. *Guide for the Development of Bicycle Facilities.* AASHTO Task Force on Geometric Design, Washington, DC.

———. 2001. *A Policy on Geometric Design of Highways and Streets.* AASHTO, Washington, DC.

Architectural and Transportation Barriers Compliance Board (Access Board). *Americans with Disabilities Act Accessibility Guidelines* (ADAAG). Access Board, Washington, DC, current edition.

Baker, S. P., B. O'Neill, M. J. Ginsburg, and G. Li. 1992. *The Injury Fact Book,* 2nd ed. New York: Oxford University Press.

Bentzen, B. L. 1998. *Accessible Pedestrian Signals.* U.S. Access Board, Washington, DC, August 4.

Berkovitz, A.. 2001. "The Marriage of Safety and Land-Use Planning: A Fresh Look at Local Roadways." *Public Roads* (September/October): 7–19.

Burden, D. 1999. *Pennsylvania Pedestrian and Bicyclist Safety and Accommodation.* Assembled for Pennsylvania Department of Transportation.

Eck, R. W., and E. D. Simpson. 1996. "Using Medical Records in Non-Motor-Vehicle Pedestrian Accident Identification and Countermeasure Development." *Transportation Research Record* 1538, 54–60.

Federal Highway Administration (FHWA). 1992. *The National Bicycling and Walking Study Case Study No. 1: Reasons Why Bicycling and Walking Are and Are Not Being Used Extensively as Travel Modes.* FHWA-PD-93-041, U.S. Department of Transportation, FHWA, Washington, DC.

———. 1999. *Designing Sidewalks and Trails for Access, Part I: Review of Existing Guidelines and Practices.* U.S. Department of Transportation, FHWA, Washington, DC, July.

———. 2000. *Manual on Uniform Traffic Control Devices, Millennium Edition.* U.S. Department of Transportation, FHWA, Washington, DC, December.

————. 2001. *Designing Sidewalks and Trails for Access, Part II: Best Practices Design Guide*. U.S. Department of Transportation, FHWA, Washington, DC.

————. N.d. *FHWA Course on Bicycle and Pedestrian Transportation: Student Workbook*. U.S. Department of Transportation, FHWA, Washington, DC.

Hunter, W. W., J. C. Stutts, and W. E. Pein. 1997. *Pedestrian Crash Types: A 1990's Informational Guide*. FHWA-RD-96-163, Department of Transportation, Federal Highway Administration, McLean, VA, April.

Hyde, A. H., G. M. Bakken, J. R. Abele, H. H. Cohen, and C. C. La Rue. 2002. *Falls and Related Injuries: Slips, Trips, Missteps and Their Consequences*. Tucson: Lawyers & Judges.

Lalani, N., and ITE Pedestrian and Bicycle Task Force. 2001. *Alternative Treatments for At-Grade Pedestrian Crossings*. Informational Report, Institute of Transportation Engineers, Washington, DC.

Landis, B. W., V. R. Vattikuti, R. M. Ottenberg, D. S. McLeod, and M. Guttenplan. 2001. "Modeling the Roadside Walking Environment—Pedestrian Level of Service," *Transportation Research Record* 1773, pp. 82–88.

National Highway Institute (NHI). 1996. *Pedestrian and Bicyclist Safety and Accommodation—Participant Workbook*. FHWA-HI-96-028, U.S. Department of Transportation, Federal Highway Administration, Washington, DC, May.

Project for Public Spaces. 2000. *How to Turn a Place Around: A Handbook for Creating Successful Public Spaces*.

————. 2001. *Getting Back to Place: Using Streets to Rebuild Communities*. New York.

Schwarz, L. L., ed. 1993. *Greenways: A Guide to Planning, Design, and Development*. Washington, DC: Island Press.

Steinholtz, R. T., and B. Vachowski. 2001. *Wetland Trail Design and Construction*. Technology and Development Program, USDA Forest Service, Missoula, MT, September.

Sucher, D. 1995. *City Comforts—How to Build An Urban Village*. Seattle: City Comforts Press.

Transportation Research Board (TRB). 2000. *Highway Capacity Manual*. Special Report 209, National Research Council, TRB, Washington, DC.

CHAPTER 22
BICYCLE TRANSPORTATION

Lisa Aultman-Hall
*Department of Civil and Environmental Engineering,
University of Connecticut, Storrs, Connecticut*

22.1 ARE BICYCLES REALLY TRANSPORTATION IN AMERICA TODAY?

In an automobile-dominated society, it is often difficult for transportation professionals, as well as private citizens, to remember that alternative modes of transportation exist and that it is in everyone's interest to promote their use. A diversified transportation system allows for flexibility and choices. The bicycle represents a mode of transportation that is relatively inexpensive, space-efficient, and accessible to almost all members of society. There are many virtuous reasons for promoting the bicycle, including the reduction of air pollution impacts, parking needs, and user costs. But benefits related to physical health, economic development, and tourism have also been shown to increase when more people in a community bicycle or when bicyclists are attracted to a community for bicycling.

Moreover, from a purely traffic engineering point of view, the use of bicycles for more utilitarian or purposeful trips (as opposed to purely recreation) also represents one marginal or incremental solution to the traffic-congestion problems plaguing our transportation systems. For example, while at low traffic volume conditions removing one automobile trip from the system might not be very beneficial, at higher volumes every trip removed represents travel time savings to all other motorized vehicles in the system. The overall widespread benefits of small or marginal reductions in demand on the system when it is operating near or above capacity are what traffic engineers are striving to accomplish today. Changes in traffic signal timings, small increases in ridesharing, and promotion of telecommuting are small efforts to accomplish marginal decreases in demand near or above capacity. It might be tempting to assume that the bicycle is a relatively specialized marginal mode of transportation and that because relatively few people use it, it offers relatively little in terms of benefits to the overall system. However, like traffic signaling timings or other smaller management solutions, a small number of purposeful bicycle trips at congested times represents important travel time savings to all users of the system. For this reason, the bicycle should not be considered an unimportant mode even if few people use it, but rather a mode that contributes to important marginal decreases in travel demand near or above capacity. It is one of many important incremental solutions to our current traffic congestion crisis. Bicycles are worth promoting even if they will not be used by most travelers.

The bicycle has the further benefit of offering relatively independent transportation to many who do not have other options. In 1995, 8 million households did not own an automobile (BTS 1995). Up to 37 percent of a state's population is too young to drive an

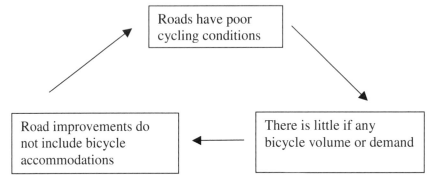

FIGURE 22.1   A transportation planning cycle driven by existing demand.

automobile or is elderly or disabled and may not be able to drive (FHWA 1994). However, despite this potential demand for bicycle use, in many places promoting bicycling has become lost in a self-defeating cycle. Often, as planners and engineers, we look at a portion of the system and see no one biking and therefore we assume there is no need to design for or accommodate cyclists on that route. However, the route may well be a major traffic artery with no place for bicycles that people perceive as safe or pleasant. Therefore, no one bicycles and with no existing bicycle demand, a transportation planning system that primarily builds to meet current or future demand does not make accommodations to create the bicycle demand. This cycle is shown in Figure 22.1. One can argue that we need transportation planning that is smarter than simply building to meet a demand. We need transportation planning that designs and builds systems that create the type of demand we want; travel demand that produces more optimal use of space, better communities, and fewer environmental impacts. Instead of building to meet the demand, we need to design the demand with the land use patterns and transportation infrastructure we provide. To help accomplish this goal, this chapter covers many of the aspects of how bicycles should operate safely within our transportation system, as well as an overview of the specific design knowledge that is available today for designing bicycle infrastructure. This chapter is intended to challenge engineers and planners to become informed about the latest information and standards in bicycle transportation in order to become part of educating cyclists, engineers, planners, and the public to make the best use of the bicycle as one of many solutions to our transportation dilemmas.

22.2   STANDARDS AND REGULATIONS FOR BICYCLE TRANSPORTATION

Many standards and regulations at the federal level, and in most states, support the reality that bicycles are vehicles in our transportation system. They are officially considered such in the Uniform Vehicle Code. Both the Intermodal Surface Transportation Equity Act (ISTEA) of 1991 and the Transportation Efficiency Act for the 21st Century (TEA-21) in 1998 contained programs and policies that include the bicycle as a mode of transportation in the multimodal and intermodal transportation system envisioned by these two federal acts. One of the most influential requirements of these acts was that for a bicycle and pedestrian coordinator in each state department of transportation. In some states, this position consists of only a portion of one employee's job. Yet in many other states, several professionals work within a bicycle and pedestrian program. In 2002, the American Association of State Highway and Transportation Officials (AASHTO) established a task force composed of 16 of the state bicycle pedestrian coordinators as a permanent part of its highway design committee. This is one more indication that accommodation of bicycles within the transportation system is no longer just a factor in "special" bicycle-friendly communities but a factor throughout the United States. As a result of ISTEA and TEA-21, federal spending on bicycles and pedestrians increased from $0.49 million in 1988 to $416 million in 2002.

In 2000, the United States Department of Transportation adopted a policy statement *Accommodating Bicycle and Pedestrian Travel: A Recommended Approach* (FHWA 2000) in response to the requirements in TEA-21. The policy requires that "bicycling and walking facilities will be incorporated into all transportation projects unless exceptional circumstances exist." The document, which outlines suggested approaches and guidance to accommodating bicycles in different circumstances, was the product of a national task force on which many groups were represented. However, the implementation of the policy was left to individual states themselves. Some states already had design manuals and regulations in place to specifically address bicycles and pedestrians. Others, such as Kentucky, formed their own task force, drafted design guidance, and adopted state-specific design guidelines in response to the FHWA guidance. Still other states have not yet responded or changed their design process in response to the 2000 FHWA document.

From the above discussion, it is clear that our laws and transportation agencies regard bicycling as a form of transportation. However, one controversial item remains: is there a difference between bicycle transportation and bicycle recreation? Or in other words, which types of bicycling are transportation? The most recent laws, especially TEA-21, have clearly targeted most programs and efforts towards utilitarian or purposeful bicycle trips. These trips are those that might reasonably replace a motorized vehicle trip. This argument suggests that those who bicycle for recreation, sightseeing, or exercise are not undertaking a utilitarian trip and are therefore not transportation. Others argue that all forms of movement between two points within our urban or rural areas are transportation, no matter what mode is used to accomplish this movement. Proponents of this argument cite those who use their cars to travel to the park, purchase ice cream on a hot summer day, or tour through decorative lights in December, and they question whether or not these motorized vehicle trips are transportation. In these cases we still strive as transportation professionals to provide safe travel and good levels of service on our transportation system. By extension, one can argue that all bicycle trips are transportation no matter what purpose motivates an individual to bicycle between two points. Similarly, within this chapter all bicycle trips are assumed to be transportation and the design guidance discussed applies to bicycle facilities for cyclists of all skills and ages and with different travel purposes.

The majority of the discussion of bicycle facility design guidance found in this chapter is based on the 1999 AASHTO *Guide to the Design of Bicycle Facilities,* but other references to design guides are also provided. The *Manual on Uniform Traffic Control Devices* (*MUTCD*) provides standard signs and lane markings that are used for both dedicated and shared bicycle facilities. The Americans with Disabilities Act is the basis for the *Designing Sidewalks and Trails for Access Guide* (Beneficial Designs Inc. 1999 and 2001), which contains important considerations that are not necessarily well known within the wider transportation design community. Each state and many cities may have a bicycle plan or bicycle design guide that deviates from national recommendations and should be consulted for projects within a given jurisdiction.

22.3 TYPES OF BICYCLE FACILITIES

Oftentimes we think of off-road *shared-use paths** as being the only type of bicycle facility. A shared-use path is an important facility dedicated to nonmotorized transportation because

---

*Note that these facilities are not called *bike paths,* the term sometimes used by popular media as well as engineers. The term *trail* has recently seen less use and is used for less formal, narrow, often dirt paths such as one might see through a wooden area. Paths as described in this chapter are also not those used in less developed areas for mountain biking. Readers are also cautioned that the term *path* has a much different meaning in Europe than it does in the United States and Canada.

it appeals to new cyclists who may not be able or willing to bicycle with heavy motor vehicle traffic. Even those who are able to bike with traffic sometimes prefer the quieter slower pace along a shared-use path. In addition to providing recreational transportation, shared-use paths often form important backbones of regional bicycle networks, sometimes allowing bypassing of major traffic arterials. Many older shared-use path facilities are not designed according to the solid design parameters that were documented throughout the 1990s. These paths might require upgrading or retrofitting. This is particularly important given the safety issues involved when bicycles, pedestrians, and other nonmotorized modes such as bladers are mixed together on relatively narrow facilities. The different operating characteristics of these users, especially speed, make safe design and operation of shared-use paths particularly challenging. Research has shown that bicyclists on paths have similar crash and injury rates to those found on roadways (Doherty, Aultman-Hall, and Swaynos 1997).

Bicycle lanes along roadways are a dedicated bicycle facility that has been growing in mileage in the United States. The design parameters for bicycle lanes have also been well documented, but the relative safety rates along bicycle lanes versus roads, as well as their impact on ridership levels in terms of attracting new riders, have not been conclusively determined at this time. Preliminary findings suggest bicycle lanes have improved bicycle safety (Clarke and Tracy 1995; Moritz 1998).

Many jurisdictions have been opting for *wide curb lanes* as a bicycle facility. These facilities allow the bicycle more room the travel with motorized traffic than is available in a standard-width travel lane, but do not require as much space as is needed for a bicycle lane. Some believe these facilities avoid some confusion on the part of drivers and cyclists about the use of bicycle lanes because it is clear the cyclists should act and be treated as a vehicle in the right-hand lane.

In most cases in our cities, towns and rural areas a bicycle facility consists of a *shared roadway.* Shared roadways are simply roads where cycling is not prohibited but dedicated facilities have not been provided. The quality of cycling and the safety of cycling on different shared roadways varies, and not all routes are realistically acceptable for all levels of cyclists.

A *signed bicycle route* is a combination of shared roadways, routes with wide curb lanes or bicycle lanes, and shared-use paths. It is delineated by green bicycle route signs (Figure 22.2) and leads from an origin to a destination. The 1999 AASHTO guide states that a signed bicycle route should offer some advantage over the alternatives. In some states, the bicycle routes are numbered or named in the same ways highways are. In the past, bicycle routes were sometimes established simply along roads where conditions were thought to be good or ideal for cycling. These routes started and ended at points that were not necessarily destinations but rather the location where the ideal bicycle conditions ended. In recent years, the signed bicycle route has been taken very seriously as part of a continuous system or bicycle network connecting areas of a city together. These routes might consist of several types of roads and facilities, but the new focus has been on ensuring continuity and directness. Many studies have shown that cyclists are not willing to travel significant extra distance between points even if the bicycling conditions are perceived to be much better on the longer route. While local and collector roads might seem ideal for inclusion in a bicycle route

FIGURE 22.2 Bicycle route sign.

system, the transportation planner should note that the curvy and disconnected patterns of these streets, particularly in newer suburban areas, often leave cyclists lost, and direction/route signs are most desperately needed in these areas.

Readers should specifically note that *sidewalks* are not on the list of bicycle facilities provided here. Numerous studies (Aultman-Hall and Kaltenecker 1999; Aultman-Hall and Hall 1998; Moritz 1997; and Wachtel and Lewiston 1994) have shown that sidewalks are a more dangerous place for bicycles when compared to either roads or shared-use paths. Sidewalks are often narrow, have proximate objects in the clear zone, intersect numerous driveways, and have uneven surfaces. These characteristics make them dangerous for cycling. Child cyclists are often going slower speeds, similar to pedestrians, and therefore they might be the only appropriate cyclists for sidewalks. In some locations special ramps onto a sidewalk have been provided to allow cyclists to use the sidewalk on bridges, especially long bridges or causeways. These sidewalks should have low pedestrian volumes, good bicycle-friendly railings (at least 42 inches) and sufficient width. Care is required to ensure safe access to the sidewalk at both ends without requiring the cyclist to break traffic rules, such as by wrong-way riding.

Selecting the right combination of facilities to create a comprehensive and connective network within a town or city is not straightforward. The process requires significant public input, especially from cyclists who use the routes. Specific guidance on where to use which type of facility has been documented and is available to engineers, planners, and the public (Wilkinson et al. 1994).

22.4   THE BICYCLE AS A DESIGN VEHICLE—DEFINING PARAMETERS

Four main characteristics define the bicycle as a design vehicle: physical dimensions, speed, stopping distance, and climbing ability. The uninformed designer might unknowingly make incorrect assumptions about the operating characteristics of the bicycle. For example, it is common for bicycles to be assumed slower than they can actually travel or narrower than they actually are. *The Guide for the Design of Bicycle Facilities* (AASHTO 1999) delineates these four operating characteristics for bicycles.

Although the tires of a bicycles are very narrow, the effective width of the vehicle plus rider and the width required for safe operation of a bicycle are much greater. Including a 5-inch (0.125-meter) buffer on both the right and the left, the overall vehicle (bicycle plus rider) is 40 inches (1 meter) wide. Given that bicycles cannot be driven, even by the most experienced cyclist, in a perfectly straight line and that some room for lateral shifts is needed, more than 1 meter of space is required for operation of a bicycle. AASHTO further recommends that a design speed of 30 km/hr or 20 mph be used for bicycles on paved surfaces, while 25 km/hr or 15 mph be used for unpaved surfaces. This translates into up to 220 feet of required stopping sight distance on a level, paved surface. Climbing ability varies greatly with bicycle type and the individual bicyclist; however, grades of less than 5 percent are recommended and AASHTO provides recommended grades based on grade length for shared-use paths. Designers should also note that on downgrades speeds are often very high and that ice, debris, or rumble strips are particularly dangerous in these situations.

22.5   HOW BICYCLES SHOULD OPERATE AS VEHICLES

There is a great deal of controversy over how bicycles should operate in the transportation system. Even some experienced cyclists disagree on how best to make left-hand turns or whether they ride with or against traffic. The engineer or planner working with bicycle transportation projects who is not a cyclist is handicapped in meaningfully participating in this debate. Furthermore, the noncyclist is in danger of making incorrect assumptions about how bicycles should operate safely as vehicles. This section of the chapter describes the current philosophy among bicycle safety education professionals as to how bicycles should operate in hopes that designers will implement designs that *promote* this type of riding and that transportation professionals might promote safe cycling.

The League of American Bicyclists promotes an education program entitled *Effective Cycling*. A main tenet of this program is the idea that "Bicyclists fare best when they act and are treated as drivers of vehicles." Acting and being treated as vehicles might also be considered simply following the rules of the road. We have rules and hierarchies of right-of-way within our transportation system in order to increase predictability and minimize ambiguity. This creates safety because users know what to expect from other users in different circumstances. Accomplishing this requires not only that cyclists follow rules and that designers provide appropriate facilities, but also that law enforcement is informed and enforces laws. In most but not all states, the bicycle is granted all the rights and responsibilities of other vehicles.

In 1994, the Federal Highway Administration established a system of categorizing cyclists by their skills and needs. Type A cyclists or advanced cyclists are confident and experienced in most traffic conditions. Type B cyclists or basic cyclists are casual, new, or less confident riders. A child cyclist or type C is self-explanatory. Clearly not all cyclists can operate as vehicles on all types of roads. A type B or C cyclist could not operate in the curb lane of an urban multilane arterial roadway. In these situations, a cyclist might elect to ride on the sidewalk or the wrong way against traffic. Safety research has shown that some of these tactics are simply not safe. A good rule of thumb is that if a cyclist does not feel comfortable riding with the traffic on a given road, then he or she should not be on that road. This creates a challenge for engineers and planners as they seek to provide a comprehensive bicycle network within a region: if all types of cyclists are to be served, bi-passes around major arteries must be sought.

Lane positioning is an important consideration for the cyclist who is riding with traffic. Oftentimes we assume the cyclist will be very narrow and as close to the right hand curb as possible. This is not recommended. The edge of the roadway contains many hazards: the curb face, drainage grates, potholes, or debris. Educators recommend between 1 and 3 feet be placed between the cyclist and the road edge or curb. Furthermore, if parallel parked cars are present, the cyclist should consider traveling far enough from the cars that an open door would not strike him or her. This is perfectly within the cyclist's rights.

It is not appropriate for the cyclists always to be in the rightmost lane, as is often assumed. The League of American Bicyclists promotes the slogan that a cyclist should be "in the rightmost lane that leads to their destination." So, for example, a cyclist making a left-hand turn would not do so from the rightmost lane but rather from the center of the leftmost lane (or the center of the rightmost left-turn lane if multiple left-turn lanes exist). Staying on the absolute right edge would create a conflict point between the bicycle and the through motor vehicle movements, as shown in Figure 22.3. This should make sense to traffic engineers because with all intersection control strategies we seek to minimize conflicting traffic streams in our attempts to maximize safety. Similarly, if a cyclist approaches an intersection where there is a right-hand turn lane and is going straight through, he or she should merge into the through lane and not stay on the rightmost edge of the rightmost lane. Staying in the right lane would create a conflict between the cyclist and any right-turning vehicles, as shown in Figure 22.4. In a shared right-turn through lane, the cyclists should merge to the center of that lane to proceed through the intersection to avoid conflicts with the right-turning vehicles.

These guidelines for lane usage simply follow the premise that bicycles should operate as vehicles. However, they also illustrate why not all cyclists may be able to operate on the roadway as vehicles. These cyclists may chose to use lower-volume single-lane roadways or off-road shared-use paths. It is not recommended that the cyclists improperly use the road-

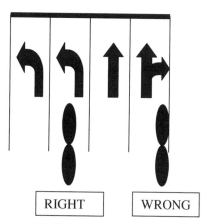

FIGURE 22.3 Bicycle positioning for left turns.

ways where they are not comfortable acting as vehicles. Designers should assume all bicycles will operate as vehicles as illustrated in these examples.

22.6 DESIGNING BICYCLE LANES

Although bicycle lanes are relatively new in terms of widespread use, the AASHTO Design Guide provides explicit guidance for lane placement, dimensions, and signage based on decades of use in California and other places. AASHTO defines a bicycle lane as "a portion of the roadway which has been designated by striping, signing and pavement markings for the preferential or exclusive use of bicycles." In special or unique circumstances, alternative solutions and designs are still being tried by various jurisdictions. Readers are directed to use the Web or the state bicycle pedestrian coordinators' network to ensure the most up-to-

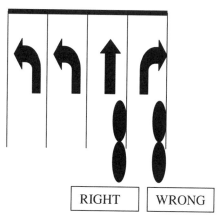

FIGURE 22.4 Bicycle positioning for through movement with a right-turn lane.

date options are considered. Bicycle lane designs continue to change and be improved. Innovative design strategies for bicycle lanes and other bicycle facilities have been documented by various groups, including the Institute for Transportation Engineers (ITE) (Nabti and Ridgway 2002).

The typical bicycle lane configuration is shown in Figure 22.5. If a roadway does not have a curb and gutter, a minimum of 4 feet is recommended by AASHTO for a bicycle lane. If a gutter is present, 5 feet including the gutter is recommended and no more than 2 of the 5 feet can be within the gutter. Caution should be used if areas wider than 7 feet are to be used as a bicycle lane. Motor vehicles can fit in this space and will be tempted to use the lanes to bypass traffic, particular for making right turns. This is a very undesirable situation. A 6-inch solid white stripe is used to delineate the bicycle lane from the adjacent travel lane. If a rumble strip* is present, the bicycle lane dimensions do not include the rumble strip, but there is not widespread agreement on whether the bicycle lane should be inside or outside the rumble strip. When parallel parking is present on the roadway, a minimum of 5 feet is recommended for the bicycle lane and a 6-inch solid white strip is placed on both sides of the bicycle lane: one between the bicycle lane and the parked vehicles and one between the bicycle lane and the right-most travel lane. Bicycle lanes with diagonal parking are not recommended. Note that when the bicycle lane approaches an intersection, a transition to a broken white line indicating that turning vehicles and bicycles may be crossing paths is required.

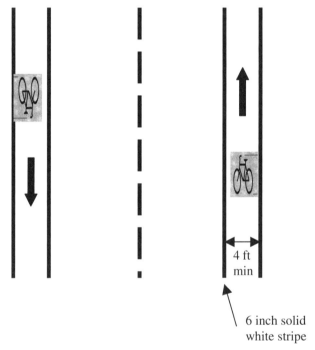

FIGURE 22.5 Typical bicycle lane configuration.

---

*Readers are cautioned that rumble strips, while shown to improve run-off-the-road motor vehicle safety, are dangerous to cyclists, particularly on downgrades, where cyclists can lose control. Several states have tested and adopted more bicycle-friendly rumble strips that include gaps every 60 to 100 feet to allow cyclists to exit the shoulder for movements such as left turns. The type and size of the rumble strip is also critical for bicycle safety.

When an existing right-of-way is being restriped for bicycle lanes, a reduction in the width of motorized vehicular lanes may be considered. The AASHTO guide recommends that restriping can reduce lanes to 10 or 10.5 feet if travel speeds are less than or equal to 25 mph. When speeds are between 30 and 40 mph, 11-foot lanes are suggested, with any two-way left-turn lanes being 12 feet wide. If speed is above 45 mph, then standard 12-foot lanes should be provided with the restriping. Sometimes the extra space needed for bicycle lanes can be found by reducing the parking lanes to 7 feet. The city of Chicago installed 75 miles of bicycle lanes between 2000 and 2002. Their detailed design drawings for bicycle lane configurations on streets with different curb-to-curb widths ranging from 44 to 60 feet are contained in a new *Bike Lane Design Guide* (City of Chicago 2002). The drawings contain configurations that include both one-way and two-way streets, bus stops, and parking. These excellent drawings are useful and also illustrate a departure from AASHTO standards: the city of Chicago uses 5 feet as the minimum width for a bicycle lane instead of 4 feet.

The *MUTCD* provides the templates for black-and-white signs indicating the start and end of bicycle lanes. Note that these signs are different than the green signs that indicate bicycle routes and also different from yellow warning signs targeted at motorists to advise caution in certain areas (such as where a shared-use path crosses a roadway).

Two-way bicycle lanes are no longer recommended and are rarely used today. They contradict the notion described in the previous section that bicycles fare best when they act and are treated as vehicles. In a two-way bicycle lane scenario (such as shown in Figure 22.6*a*) the number of opposing traffic steams that are adjacent to each other increases. The predictability and visibility of the wrong-way rider is compromised at driveways and intersections. In some cases two-way bicycle lanes are provided on one-way streets, as shown in Figure 22.6*b*. This type of design can provide an important link in the bicycle network. In cases where space on the one-way street will only allow for one bicycle lane, a contra flow lane can be provided for bicycles as shown in Figure 22.6*c*. In this case, the bicycles traveling the same way as the motorized vehicular traffic use the traffic lane. This is confusing to some inexperienced cyclists who try to use the bicycle lane in the wrong direction. Arrows (templates provided in the *MUTCD*) should be used to indicate where bicycles should travel. In some cases, bicycle lanes may be placed on the left edge of a one-way street if the right edge is being used for bus stops or another conflicting activity.

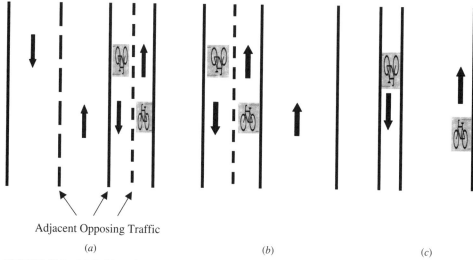

Adjacent Opposing Traffic

(a)   (b)   (c)

FIGURE 22.6 (*a*) Problematic bicycle lane configuration; (*b*) One-way street; (*c*) Contra-flow bicycle lane.

In addition to signage and direction arrows, a painted bicycle logo is recommended for placement in bicycle lanes. In previous years, different jurisdictions have had various ways to indicate bicycle lanes, including the use of the words "bike lane." There is now widespread support for the bicycle template as a common nationwide symbol.

The space at intersections is often limited, and therefore striping of bicycle lanes on intersection approaches and through intersections is not straightforward. Some jurisdictions elect to discontinue the bicycle lane at or before intersections. Others stripe the lane through using broken stripes to indicate where right-turning vehicles can cross the bicycle lane. In some places, a shared right-turn bicycle lane (shown in Figure 22.7) has been tested (Hunter 2001), but this design is not yet included in the AASHTO guide.

The road diet transformation illustrated in Figure 22.8 is increasing in popularity throughout the country. In many cases, it is the bicyclists or planners who put forth this idea when a repaving project is considered or when left-turn traffic volumes become problematic. When at least 48 feet of curb-to-curb distance is available on the originally four-lane road, the option of five narrow motorized traffic lanes is also considered (four through lanes and one two-way center left-turn lane). Instead, the road diet consists of replacing the four lanes with three standard-width travel lanes with two bicycle lanes. To the public and media eye, the transformation often seems to consist of only removing a traffic lane and adding bicycle lanes to the street. But many benefits beyond simply providing good standard bicycle lanes come with the road diet. It removes motorized traffic from proximate fixed objects on the edges of the roadway, provides standard wide travel lanes for trucks and buses, and increases turning radii at intersections and driveways. It also provides a buffer for pedestrians if the sidewalk is immediately behind the curb. Admittedly, the road diet also has both the pros and cons of adding a two-way left-turn lane. The road diet does not necessarily reduce the capacity of the roadway. Capacity changes depend on the volume of turns, particularly left turns. In some cases, these transformed roadways carry over 20,000 vehicles per day.

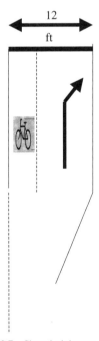

FIGURE 22.7 Shared right-turn and bicycle lane.

Before

After

Bike lanes or paved shoulders

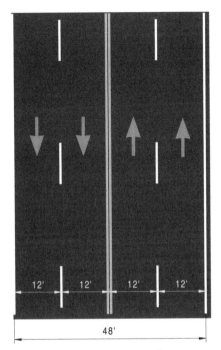

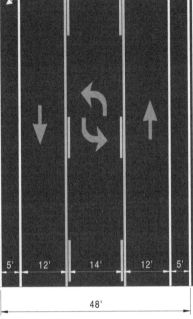

FIGURE 22.8 Road diet example.

22.7 DESIGNING SHARED ROADWAYS

Bicycle lanes are, at the current time, widely considered the desirable means to accommodate cyclists on roadways. They make bicycles a visible part of the transportation system even when bicycles are not present due to the lane markings and signage. Bicycle lanes fit better with the rules and conventions of the road network, avoiding the complications often encountered when connecting shared-use paths to the road network. Preliminary safety findings (Clarke and Tracy 1995; Moritz 1998) are positive. However, it is unreasonable to expect that all roads can or should have bicycle lanes, particularly in existing developments, older cities, or completely rural areas. This section of the chapter discusses the use of shoulder bikeways and wide curb lanes as a compromise between true bicycle lanes and no special treatment on roads. The subsequent subsection describes road features or attributes that can make a road good or bad for bicycling even when no special facilities are placed on a given road. The final subsection describes bicycle suitability measures that can gauge quality of service for a bicycle on a given road.

22.7.1 Shoulder Bikeways

Shoulder bikeways are particularly useful in rural areas where shoulders often exist and can be used for biking. Obviously a cyclist might choose to use a shoulder at any time, but designating a shoulder as bikeway requires that signage or another designation be provided. Care needs to be used when rumble strips are present, as noted in the previous section of this chapter. A higher level of responsibility is required for sweeping and maintaining a shoulder bikeway as the debris from the road is often swept by traffic air currents onto the shoulder. Ideally at least 4 feet is recommended by the AASHTO guide for a shoulder bikeway. This width should be increased to at least 5 feet if a guardrail* or other roadside barrier is present. When the shoulder area decreases in width due to bridges or other barriers, warning signs for the cyclists should be considered. A shoulder bikeway must be paved. However, even when it is paved, designers should note that some cyclists prefer to travel on the edge of the travel lane rather than the shoulder and are within their rights to do so. Debris in the shoulder can be a safety hazard, and on some racing or road bikes with smaller tires the risk of tire damage is great.

Shoulder bikeways are even used in some states on freeways or interstates. This requires appropriate consideration of rumble strips and interchanges but has been particularly successful in providing access in places were no other route exists for cyclists such as over rivers. In some cases, access to the freeway shoulder is provided immediately before and after the bridge, avoiding the need for cyclists to negotiate an interchange.

22.7.2 Wide Curb Lanes

Sometimes when restriping is considered for bicycle lanes there is simply not sufficient right-of-way to provide the required space for the bicycle lane but there is space to widen the curb lane. This wide curb lane could also be accomplished in combination with reducing the width of other travel lanes on a multilane roadway. At least 14 feet, and ideally 15 feet, is recommended for a wide curb lane (not including the gutter pan). This bicycle facility is often useful only to experienced cyclists who ride with traffic, but it is consistent with the philosophy of accommodating bicycles as vehicles within the road network and is therefore considered very positive by many bicycle advocates.

22.7.3 Road Attributes Requiring Special Attention for Bicycles

Whether shoulder bikeways or wide curb lanes are provided or not, certain road attributes require special attention on shared roadways to ensure safe and quality service is provided to cyclists. These issues are particularly important where a road is designated as part of a signed bicycle route. Many of these seemingly small details have been shown important to accommodating cyclists. A planner would want to consider these factors carefully for bicycle route networks that include shared roadways as one of the types of facilities connecting origins and destinations throughout an area.

Drainage grates are hazardous to bicycle tires and can cause damage or even crashes. While most jurisdictions have moved to replace grates, some have not, and one can argue that bicycle routes should have priority for replacement. Openings should not only be small but should be crossways to the direction of travel. The potholes and pavement problems that occur at drainage grates should also be considered when bicycle traffic is expected or encouraged.

Similar to drainage grates, railway crossings represent a hazard to cyclists because of the potential for a tire to be caught in the rail, causing the cyclist to fall or crash. Special textured mats exist to reduce the size of gaps where tires might get caught, but these are often only

---

*Designers should consider that guard rails or other barriers can present a hazard to cyclists if they crash into them. Cyclists have a high center of gravity and can flip over guardrails, meaning the conditions on the other side of the barrier should be considered.

useful when the bicycle crosses the rails at a 90° angle. When non-90° crossings are required, special pavement widening should be considered as shown in Figure 22.9. Otherwise the cyclist may stray into vehicular traffic in order to try to cross the rails at 90°. Similar consideration should be made when shared-use paths cross rail lines.

The pavement surface quality and edge condition of the roadway are particularly important for cyclists. Uneven surfaces slow cyclists, damage bicycles, and can cause crashes. Cyclists are known to divert to alternative routes based on paving quality and surface conditions. If the edge of the roadway has a sharp drop-off, the effective area for cyclists is greatly reduced. Similar hazards are present on bridge decks and at bridge joints.

Actuated traffic signals can be a great frustration for cyclists who in many cases cannot trigger the signal and may be motivated to ignore it. Detection levels can be altered along bicycle routes or special bicycle detectors can be used. In many cities special paint markings indicate the location of extra wire coils in the inductive loops that can detect cyclists. Many new video and radar detectors do not have these limitations as long as they are aimed at the right edge of the road, where a cyclist is likely to be stopped.

One of the best ways to determine the safety hazards along roads or routes within an area is to hold public input meetings. The cyclists who use a route are the best source of infor-

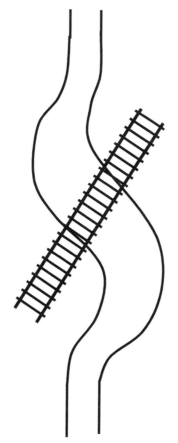

FIGURE 22.9 Widening for angled railway crossings.

mation on needed changes. The designer or planner can then apply solid design principles to address the problems. In some cases, field inspections (Figure 22.10) can point planners to locations where routes are needed.

22.7.4 Bikeability Measures

One of the ways to evaluate the suitability of a road for bicycling is to use one of the formal evaluation tools developed over the last decade by researchers who were motivated by the need to move beyond pure subjective measurement of shared roadways for bicycle traffic. These objective measurement tools are important because planning processes and project prioritization often require solid grounds as a basis for decision making.

The bicycle level of service (BLOS) (Landis, Vattikuti, and Brannick 1997) was developed by having cyclists bike real-world routes and evaluate them. The characteristics or attributes of the route sections were recorded and used as potential explanatory variables. The resulting regression models include motor vehicle traffic volume per lane, motor vehicle speed, traffic mix, cross-traffic levels, pavement surface, and width for cycling. Given the prevalence of city geographic information systems (GIS) containing much of this information on a block-by-block basis, it has become common in some cities to evaluate the whole arterial and collector road network. Sometimes volunteers are used to fill data gaps. The results are used to plot color-coded maps of the suitability of roads for biking based on the BLOS and can be useful in planning connective routes. The BLOS is an interesting contrast to the LOS measures in the *Highway Capacity Manual* (TRB 2000), which are used for motorized traffic on all types of roads. These are typically based on road capacity and travel time or delay. The BLOS reflects the reality that the quality of a ride for a cyclist is based on perceived safety and comfort level; in most places bicycle capacity is not yet an issue.

FIGURE 22.10 Finding route locations by field inspections.

A second objective measure is the bicycle compatibility index (BCI). This method was originally proposed by Sorton and Walsh (1994) and has been expanded using a larger sample of cyclists (Harkey, Reinfurt, and Knuiman 1998). In this case cyclists view road segments on videotape and evaluate the comfort level. The resultant model bases compatibility on the presence of a bicycle lane, its width, curb lane width, curb lane traffic volume, other lane traffic volume, traffic speed, parking, type of roadside development and adjustments for trucks, parking turnover, and right-turn volume. This measure can be used in the same way discussed above to label and plot block-by-block bicycle compatibility. Both models have been programmed into straightforward Excel-format spreadsheets.

22.8   DESIGNING SHARED-USE PATHS

Shared-use paths are important and popular facilities in a community not only for cyclists but also for other users, including pedestrians and bladers. Paths are particularly important in that they are a low-stress traffic environment, a place for children and novice riders, and can fill missing links in the bicycle network. The mix of users often requires establishment and promotion of trail user rules. However, as Morris (2002) summarizes, greenway paths or trails have been documented to offer much more to the community: healthy lifestyle promotion, economic development, historic preservation, utility corridors, and social interaction.

The basic segments of shared-use path design are straightforward, but the intersections, relationship to roadways, use of structures, and maintenance procedures can create challenges for the transportation or park agency. Many times a path is to be located along an old rail corridor* that affords adequate lateral space, limited grades, and existing infrastructure.

22.8.1   Basic Features

A basic shared-use path segment should be a minimum of 10 feet wide with 2 feet of graded shoulder (usually turf) and a minimum of 3 feet of clear zone to proximate objects such as signs, rocks, trees, or fences (AASHTO). A 2 percent minimum cross-slope for drainage is also recommended.

Use of a striped centerline varies from region to region and is sometimes only used on horizontal curves, where it helps keep cyclists on the right side of the path. If possible, a center stripe should be applied to help with user rules (such as walk and ride on the right, warn before you pass). AASHTO recommends broken stripes where passing sight distance is sufficient; however, this author finds that the labor required to determine passing sight distances for the range of user speeds is not worthwhile. Regardless of whether a stripe is present, pedestrians should walk on the right to ensure the appropriate user is in the correct position to yield the right of way when meeting. Figure 22.11 illustrates this principle: when walkers facing traffic (bicycles riding on the right) see a bicycle approaching, their tendency is to step to the other side of the trail and get out of the way of the bicycle they *see* approaching (bicycle B) (Figure 22.11*a*). Unfortunately, if they do not look behind them, they may step into the path of another cyclist (A) (or other higher-speed user such as a skater). All trail users are quiet, especially on a paved trail, and this situation is likely. The reality that addresses this situation is for all users to be on the right-hand side of the trail in

---

*Transportation planners might consider doing a background check on the legal requirements for rail banking or maintaining rail corridors for possible future use. In many jurisdictions the ownership and laws regarding abandoned rail lines complicate this process. The grass roots organization the Rails to Trail Conservancy can be of assistance in many cases.

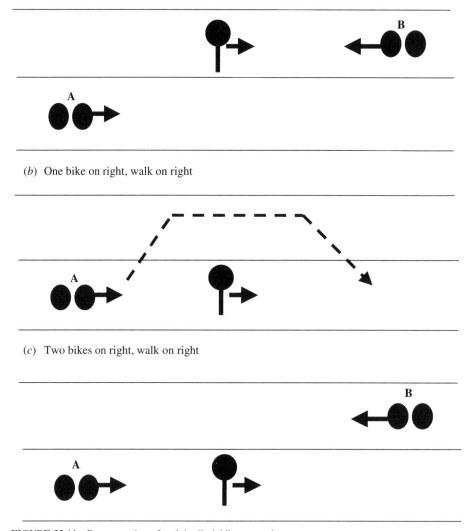

(*a*) Bike on right, walk on left

(*b*) One bike on right, walk on right

(*c*) Two bikes on right, walk on right

FIGURE 22.11   Demonstration of path/trail yielding scenarios.

their direction of travel as in the second and third pictures of Figure 22.11. In these cases, the bicycle behind the pedestrian sees and must yield to the oncoming bicycle.

The horizontal curve design recommended by AASHTO for paths is based on a 15° lean angle for the bicycle. Bicycles can lean inward to counter the centrifugal force that pulls the bicycle or any vehicle outwards on a curve. To some extent this means bicycles can turn sharper turns than other vehicles, but the extent of leaning does depend on the skill level of the cyclist. Ultimately the limiting factor is pedal height. For a 20-mph design speed (as one might expect on a paved path) the minimum curve radius is 100 feet (27 meters). Note that path speed limits might be posted substantially lower, between 10 and 16 mph. Superelevation can be used in the same way it is on highways to reduce the required radius at the same design speed. However, designers should note that ADA requires that superelevation cannot exceed 2–3 percent. This can reduce the radius at 20 mph to 30 feet.

Grade or vertical curvature must be minimized on paths whenever possible in order to accommodate cyclists of all skill levels. Dismounting and weaving at slow speeds can cause path operational problems. The length of grade affects the maximum percent grade that can be used in the same way it affects highway grades for trucks. Grades of less than 5 percent can be used without length limitation and are recommended as the maximum for paved paths. A grade of 5 percent can only be used for a maximum of 800 feet and 10 percent for 100 feet (AASHTO). In many cases, the downgrade and stopping distance is a more important consideration, particularly if horizontal curves or intersections are located at the bottom of the grade. Cyclist dismount signs can be used in some cases. Stopping sight distance may still be a controlling criterion for the length of vertical curves, as can the lateral clearance on horizontal curves. These factors must be checked in the same manner as for highways. On a level surface the stopping sight distance can be up to 225 feet (75 meters).

22.8.2 Paths and Roads

Inevitably paths have to start at or cross roads, and design of these points is not straightforward. These points often require innovative design treatments as documented in several reports (e.g., Ridgway and Nabti 2002). There are simply too many different scenarios for standard design guidance to exist for all cases, and even then, new treatments are always being tested and evaluated. The basic rule of thumb is to design the path/road intersections in such a way that (1) cyclists are encouraged to follow traffic rules, (2) both cyclists and motorists are warned and anticipate the intersection, and (3) the unambiguous and consistent right-of-way is clearly communicated using signs and pavement markings. When it is appropriate to combine traffic calming into a project, raised path crossings are sometimes considered. The path crossing is typically striped across the roadway. In some places, modeling after European efforts, blue bicycle crossing areas have been tried. Whenever possible, paths should cross roads at a 90° angle. Sometimes curves can be added to the path before the road to ensure this design standard. Mid-block crossings should be away from adjacent major intersections. A refuge area between opposing traffic direction is often included, especially on busier roads, but there must be enough space for several queued bicycles. Plantings and vegetation should be selected to ensure sight distance is not obstructed. Ideally crossings will be away from grades. Special care should be used to ensure cyclists are not required to stop too quickly following a downgrade or that they do not start on a significant upgrade. Traffic control, stop signs, or traffic signals are used for both the path and the road, as appropriate.

Paths that are parallel to roadways are particularly challenging, especially when driveways are involved. AASHTO recommends at least 5 feet between the adjacent road edge and the path, but this value is far less than is ideal for a two-way path. On a two-way path the wrong-way bicycle traffic is sometimes trapped between oncoming bicycles on the path and oncoming motor vehicles on the road, increasing the potential for head-on crashes. Furthermore, drivers exiting driveways or on intersecting streets are not expecting faster-moving traffic such as bicycles on their right (see Figure 22.12). Two-way paths adjacent to roads should only be used when driveways and intersecting roads are at a minimum. If cross-streets have even moderate traffic, the path can be pulled back a significant distance from the intersection (Figure 22.13). This simplifies bicycle turning movements and allows cross-bicycle traffic to move without interacting with intersection vehicular traffic. The path should either be close enough to the roadway intersection to be controlled and operated in an integral way or set back far enough that it can operate independently. Additional operational concerns arise when the entrance or start of these paths is located at intersections. The design is often difficult and sometimes requires odd or complex turns on the part of the cyclist (see Figure 22.14).

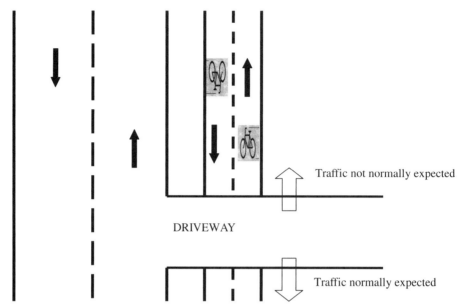

FIGURE 22.12   Entering driver expectations for intersecting traffic.

22.8.3 Maintenance, Surface, and Structures

The paving system and structures for a shared-use path must be designed for an emergency vehicle and maintenance trucks. Failure to do so can mean lack of access or premature pavement failure as shown in Figure 22.15. Similarly, any barriers placed at the entrance to paths must be removable for emergency or maintenance access. Posts are often used with a minimum separation of 5 feet for bicycle safety and a center post that is locked but removable. Caution should be used to ensure the posts do not constitute a fixed-object danger for cyclists. Some plastic options reduce this risk. Paint or thermal plastic warnings can be used. The separation used must be sufficient for bicycle trailers.

Bicycles are particularly sensitive to maintenance due to the lack of suspension systems and the high-pressure tires. Removal of the glass and debris that creates the risk of tire damage can be a frequent and labor-intensive task. The remote location of some trails makes them attractive for parties or loitering that lead to increased litter. Proper trash receptacles can help alleviate the problem. A well-populated path also helps keep "eyes" on the trail to discourage littering.

Another maintenance threat is creeping vegetation on the path edges that can narrow the path over time and affect safety. Many jurisdictions have programs to allow cyclists to report maintenance issues or other hazards along routes, including both paths and roads.

The decision whether to pave a path requires maintenance considerations as well. Unpaved paths have higher maintenance costs, including drainage and erosion problems. They do come with the added benefit of slowing bicycle traffic. Bladers are usually not found on unpaved paths, meaning that a surface treatment choice can actually impact demand and operating characteristics. Mopeds and horses, which make a poor mix with bicycles, might be found in different numbers on paved versus unpaved paths.

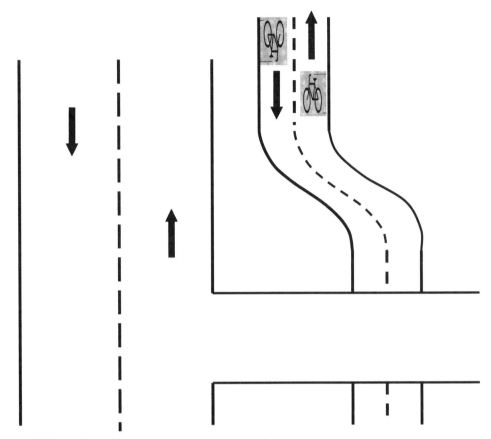

FIGURE 22.13 Shifting of shared-use path away from intersection.

22.9 *BICYCLE PARKING AND INTERMODALISM*

Studies have shown that parking is important for encouraging bicycle use, especially for utilitarian or transportation trips (Goldsmith 1992). Parking areas must be lighted and visible to counter theft and promote safety for lone cyclists at night. Indoor parking is very desirable for the same reasons it is for motor vehicles: vehicles are protected from the elements and to some degree theft. In some locations, security is provided via video camera or personal attendant. The "Cadillac" of individual bicycle parking is the bicycle locker, which provides personal guaranteed parking for an individual at a given location and consists of a bicycle-sized garage-like shelter. In many areas these lockers have proven to be moneymakers.

Bicycle racks are the most simple and common form of bicycle parking. Many undesirable types of parking racks have been used over the years. The Association of Pedestrian and Bicycle Professionals has recently published a guide for bicycle parking (APBP 2002). They recommend invert U or "post-and-loop" racks and discourage older-style racks where only a portion of one wheel can be locked and holds the bike up. The APBP recommends that bicycles be considered effectively 6 feet in length and that 4 feet be provided between rows of parked bicycles (not the racks). The APBP further recommends that bicycle parking be

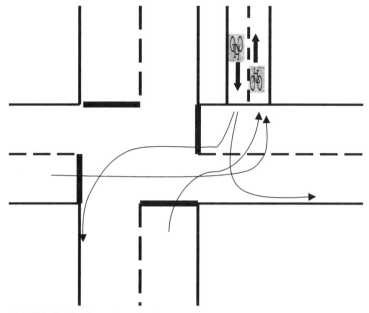

FIGURE 22.14 Examples of awkward bicycle movements needed at path-street junctions.

FIGURE 22.15 Rutting from maintenance vehicles on a shared-use path with substandard paving structure.

placed within 50 feet or 30 seconds of a building entrance. In the past, many bicycle parking areas were located so remotely that they were not known to cyclists and therefore not used. Care must be used when placing bicycle parking along sidewalks. Not only does this situation encourage sidewalk riding for access, it often creates negative space issues and blockage on the sidewalk for pedestrians.

Provision of adequate and secure bicycle parking at transit stations has been shown effective in promoting bicycle intermodalism: the use of the bicycle in conjunction with other modes for a complete trip. Formal bike-and-ride programs are becoming more widespread. The use of bicycle racks on buses and the relaxing of rules regarding bicycles on trains and subways has resulted in the bicycle being a common part of intermodal passenger transportation. In some cities insufficient capacity for bicycles on trains or bus racks has become a problem. More people want to bike and ride than can be accommodated.

22.10 CONCLUSION

In summary, the bicycle is a flexible mode of transportation that can be used for a variety of trips within the intermodal transportation system by a wide range of people. Creative designs have been required in many cases to retrofit our auto-dominated road system to promote and encourage bicycling. However, over the last decade strong progress has been made throughout the United States to reintroduce well-engineered bicycle transportation facilities into cities, towns, and rural areas. A wide range of guidance is available for designers who may not be cyclists or bicycle facility designers to ensure safe, appropriate, and consistent bicycle routes, signs, and operations can continue to spread throughout the country.

22.11 REFERENCES

American Association of State Highway and Transportation Officials (AASHTO). 1999. *Guide for the Development of Bicycle Facilities.* AASHTO Task Force on Geometric Design, Washington, DC.

Association of Pedestrian and Bicycle Professionals (APBP). 2002. *Bicycle Parking Guidelines.* Washington DC: APBP.

Aultman-Hall, L., and F. L. Hall. 1998. "Ottawa-Carleton Commuter Cyclist On- and Off-Road Incident Rates." *Accident Analysis and Prevention* 30(1):29–43.

Aultman-Hall, L., and K. G. Kaltenecker. 1999. "Toronto Bicycle Commuter Safety Rates." *Accident Analysis and Prevention* 31:675–686.

Beneficial Designs Inc. 1999. *Designing Sidewalks and Trails for Access: Part 1 of 2 Review of Existing Guidelines and Practices.* Washington, DC: U.S. Department of Transportation.

———. 2001. *Designing Sidewalks and Trails for Access: Part 2 of 2 Best Practices Design Guide.* Washington, DC: U.S. Department of Transportation.

Bureau of Transportation Statistics (BTS). 1995. *Our Nation's Travel: 1995 NPTS Early Results Report.* U.S. Department of Transportation, BTS, Washington, DC.

City of Chicago. 2002. *Bike Lane Design Guide.* Pedestrian and Bicycle Information Center, City of Chicago, Chicagoland Bicycle Federation and Association of Pedestrian and Bicycle Professionals.

Clarke, A., and L. Tracy. 1995. *Bicycle Safety-Related Research Synthesis.* Report 94-062, U.S. Department of Transportation, Federal Highway Administration, Washington, DC, April.

Doherty, S., L. Aultman-Hall, and J. Swaynos. 2000. "Commuter Cyclist Accident Patterns in Toronto and Ottawa, Canada." *Journal of Transportation Engineering* 126(1):26–27.

Federal Highway Administration (FHWA). 1994. *The National Bicycling and Walking Study.* FHWA-PD-94-023, U.S. Department of Transportation, FHWA, Washington, DC.

———. 1995. *Bicycle Safety-Related Research.* Synthesis, U.S. Department of Transportation, FHWA, Washington, DC.

————. 2000. *Manual of Uniform Traffic Control Devices, Millennium Edition.* U.S. Department of Transportation, FHWA, Washington, DC.

Goldsmith, S. A. 1992. *Reasons Why Bicycling and Walking Are Not Being Used More Extensively as Travel Modes.* Case Study Number 1, National Bicycling and Walking Study, U.S. Department of Transportation, Federal Highway Administration, Washington, DC.

Harkey, D., D. Reinfurt, and M. Knuiman. 1998. *Development of the Bicycle Compatibility Index: A Level of Service Concept.* FHWA-RD-98-072, U.S. Department of Transportation, Federal Highway Administration, Washington, DC.

Hunter, W. 2002. "Evaluation of a Combined Bicycle Lane/Right-Turn Lane in Eugene, Oregon." In *CD Proceedings of the Transportation Research Board Annual Meeting.* Washington, DC: National Academy of Science, January.

Landis, B. W., V. R. Vattikuti, and M. T. Brannick. 1997. "Real-Time Human Perceptions: Toward a Bicycle Level of Service." *Transportation Research Record* 1578, 119–126.

Moritz, W. E. 1997. "A Survey of North American Bicycle Commuters: Design and Aggregate Results." *Transportation Research Record* 1578, 91–101.

Morris, H. 2002. *Trails and Greenways: Advancing the Smart Growth Agenda.* Washginton, DC: Rails-to-Trails Conservancy.

Ridgway, M., and J. Nabti. 2002. *Innovative Bicycle Treatments.* Washington, DC: Institute for Transportation Engineering.

Sorton, A., and T. Walsh. 1994. "Bicycle Stress Level as a Tool to Evaluate Urban and Suburban Bicycle Compatibility." *Transportation Research Record* 1438, 17–24.

Transportation Research Board (TRB). 2000. *Highway Capacity Manual.* National Research Council,TRB,Washington, DC.

Wachtel, A., and D. Lewiston. "Risk Factors for Bicycle-Motor Vehicle Collisions at Intersections." *ITE Journal* (September): 30–35.

Wilkinson, W. C., A. Clarke, B. Epperson, and R. Knoblauch. 1994. *Selecting Roadway Design Treatments to Accommodate Bicycles.* Report No. FHWA-RD-92-073, U.S. Department of Transportation, Federal Highway Administration, Washington, DC.

CHAPTER 23
RAILWAY ENGINEERING

Keith L. Hawthorne
Transportation Technology Center, Inc., Pueblo, Colorado

V. Terrey Hawthorne
Newtowne Square, Pennsylvania

**(In collaboration with E. Thomas Harley,
Charles M. Smith, and Robert B. Watson)**

23.1 DIESEL-ELECTRIC LOCOMOTIVES

Diesel-electric locomotives and electric locomotives are classified by wheel arrangement; letters represent the number of adjacent driving axles in a rigid truck (A for one axle, B for two axles, C for three axles, etc.). Idler axles between drivers are designated by numerals. A plus sign indicates articulated trucks or motive power units. A minus sign indicates separate nonarticulated trucks. This nomenclature is fully explained in RP-5523, issued by the Association of American Railroads (AAR). Virtually all modern locomotives are of either B-B or C-C configuration.

The high efficiency of the diesel engine is an important factor in its selection as a prime mover for traction. This efficiency at full or partial load makes it ideally suited to the variable service requirements of routine railroad operations. The diesel engine is a constant-torque machine that cannot be started under load and hence requires a variably coupled transmission arrangement. The electric transmission system allows it to make use of its full rated power output at low track speeds for starting as well as for efficient hauling of heavy trains at all speeds. Examples of the most common diesel-electric locomotive types in service are shown in Table 23.1. A typical diesel-electric locomotive is shown in Figure 23.1.

Most diesel-electric locomotives have a dc generator or rectified alternator coupled directly to the diesel engine crankshaft. The generator/alternator is electrically connected to dc series traction motors having nose suspension mountings. Many recent locomotives utilize gate turn-off inverters and ac traction motors to obtain the benefits of increased adhesion and higher tractive effort. The gear ratio for the axle-mounted bull gears to the motor pinions which they engage is determined by the locomotive speed range, which is related to the type of service. A high ratio is used for freight service where high tractive effort and low speeds are common, whereas high-speed passenger locomotives have a lower ratio.

23.1.1 Diesel Engines

Most new diesel-electric locomotives are equipped with either V-type, two strokes per cycle, or V-type, four strokes per cycle, engines. Engines range from 8 to 20 cylinders each. Output

TABLE 23.1 Locomotives in Service in North America

| Locomotive[a] | | | | Weight min./max. (1,000 lb) | Tractive effort | | Number of cylinders | Horsepower rating (r/min) |
Builder	Model	Service	Arrangement		Starting[b] for min./max. weight (1,000 lb)	At continuous[c] speed (mph)		
Bombardier	HR-412	General purpose	B-B	240/280	60/70	60,400 (10.5)	12	2,700 (1,050)
Bombardier	HR-616	General purpose	C-C	380/420	95/105	90,600 (10.0)	16	3,450 (1,000)
Bombardier	LRC	Passenger	B-B	252 nominal	63	19,200 (42.5)	16	3,725 (1,050)
EMD	SW-1001	Switching	B-B	230/240	58/60	41,700 (6.7)	8	1,000 (900)
EMD	MP15	Multipurpose	B-B	248/278	62/69	46,822 (9.3)	12	1,500 (900)
EMD	GP40-2	General purpose	B-B	256/278	64/69	54,700 (11.1)	16	3,000 (900)
EMD	SD40-2	General purpose	C-C	368/420	92/105	82,100 (11.0)	16	3,000 (900)
EMD	GP50	General purpose	B-B	260/278	65/69	64,200 (9.8)	16	3,500 (950)
EMD	SD50	General purpose	C-C	368/420	92/105	96,300 (9.8)	16	3,500 (950)
EMD	F40PH-2	Passenger	B-B	260 nominal	65	38,240 (16.1)	16	3,000 (900)
GE	B18-7	General purpose	B-B	231/268	58/67	61,000 (8.4)	8	1,800 (1,050)
GE	B30-7A	General purpose	B-B	253/280	63/70	64,600 (12.0)	12	3,000 (1,050)
GE	C30-7A	General purpose	C-C	359/420	90/105	96,900 (8.8)	12	3,000 (1,050)
GE	B36-7	General purpose	B-B	260/280	65/70	64,600 (12.0)	16	3,600 (1,050)
GE	C36-7	General purpose	C-C	367/420	92/105	96,900 (11.0)	16	3,600 (1,050)
EMD	AEM-7	Passenger	B-B	201 nominal	50	33,500 (10.0)	NA[d]	7,000[e]
EMD	GM6C	Freight	C-C	365 nominal	91	88,000 (11.0)	NA	6,000[e]
EMD	GM10B	Freight	B-B-B	390 nominal	97	100,000 (5.0)	NA	10,000[e]
GE	E60C	General purpose	C-C	364 nominal	91	82,000 (22.0)	NA	6,000[e]
GE	E60CP	Passenger	C-C	366 nominal	91	34,000 (55.0)	NA	6,000[e]
GE	E25B	Freight	B-B	280 nominal	70	55,000 (15.0)	NA	2,500[e]

[a] Engines: Bombardier—model 251, 4-cycle, V-type, 9 × 10½ in. cylinders.
EMD—model 645E, 2-cycle, V-type, 9⅟₁₆ × 10 in. cylinders.
GE—model 7FDL, 4-cycle, V-type, 9 × 10½ in. cylinders.
[b] Starting tractive effort at 25% adhesion.
[c] Continuous tractive effort for smallest pinion (maximum).
[d] Electric locomotive horsepower expressed as diesel-electric equivalent (input to generator).
[e] Not applicable.

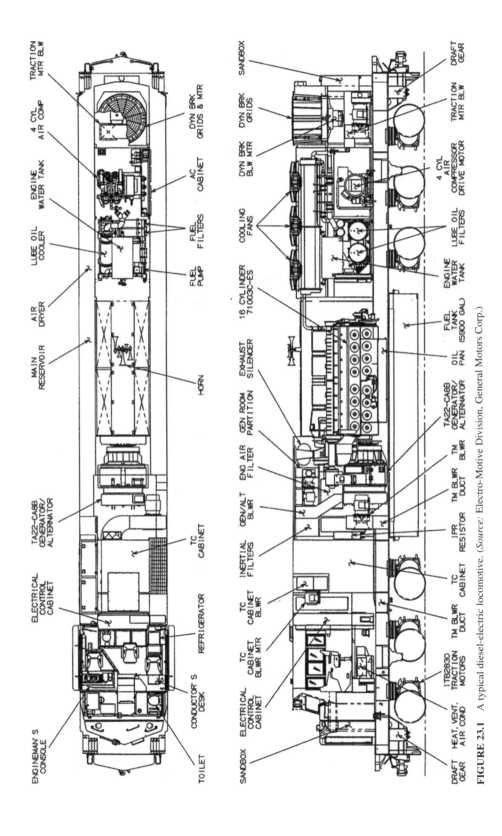

FIGURE 23.1 A typical diesel-electric locomotive. (*Source:* Electro-Motive Division, General Motors Corp.)

23.3

power ranges from 1,000 hp to over 6,000 hp for a single engine application. These medium-speed diesel engines range from 560 to over 1,000 cubic inches per cylinder.

Two-cycle engines are aspirated by either a gear-driven blower or a turbocharger. Because these engines lack an intake stroke for natural aspiration, the turbocharger is gear-driven at low engine speeds. At higher engine speeds, when the exhaust gases contain enough energy to drive the turbocharger, an overriding clutch disengages the gear train. Free-running turbochargers are used on four-cycle engines, as at lower speeds the engines are aspirated by the intake stroke.

The engine control governor is an electro-hydraulic or electronic device used to regulate the speed and power of the diesel engine.

Electro-hydraulic governors are self-contained units mounted on the engine and driven from one of the engine camshafts. They have integral oil supplies and pressure pumps. They utilize four solenoids, which are actuated individually or in combination from the 74-V auxiliary generator/battery supply by a series of switches actuated by the engineer's throttle. There are eight power positions of the throttle, each corresponding to a specific value of engine speed and horsepower. The governor maintains the predetermined engine speed through a mechanical linkage to the engine fuel racks, which control the amount of fuel metered to the cylinders.

Computer engine control systems utilize electronic sensors to monitor the engine's vital functions, providing both electronic engine speed control and fuel management. The locomotive throttle is interfaced with an on-board computer that sends corresponding commands to the engine speed control system. These commands are compared to input from timing and engine speed sensors. The pulse width and timing of the engine's fuel injectors are adjusted to attain the desired engine speed and optimize engine performance.

One or more centrifugal pumps, gear-driven from the engine crankshaft, force water through passages in the cylinder heads and liners to provide cooling for the engine. The water temperature is automatically controlled by regulating shutter and fan operation, which in turn controls the passage of air through the cooling radiators, or by bypassing the water around the radiators. The fans (one to four per engine) may be motor-driven or mechanically driven by the engine crankshaft. If mechanical drive is used, current practice is to drive the fans through a clutch, since it is wasteful of energy to operate the fans when cooling is not required.

The lubricating oil system supplies clean oil at the proper temperature and pressure to the various bearing surfaces of the engine, such as the crankshaft, camshaft, wrist pins, and cylinder walls. It also provides oil internally to the heads of the pistons to remove excess heat. One or more gear-type pumps driven from the crankshaft are used to move the oil from the crankcase through filters and strainers to the bearings and the piston cooling passages, after which the oil flows by gravity back to the crankcase. A heat exchanger is part of the system; all of the oil passes through it at some time during a complete cycle. The oil is cooled by engine-cooling water on the other side of the exchanger. Paper element cartridge filters in series with the oil flow remove fine impurities in the oil before it enters the engine.

23.1.2 Electric Transmission Equipment

A dc main generator or three-phase ac alternator is directly coupled to the diesel engine crankshaft. Alternator output is rectified through a full-wave bridge to keep ripple to a level acceptable for operation of series field dc motors or for input to a solid-state inverter system for ac motors.

Generator power output is controlled by (1) varying engine speed through movement of the engineer's controller and (2) controlling the flow of current in its battery field or in the field of a separate exciter generator. Shunt and differential fields (if used) are designed to maintain constant generator power output for a given engine speed as the load and voltage vary. The fields do not completely accomplish this, thus the battery field or separately excited field must be controlled by a load regulator to provide the final adjustment in excitation to load the engine properly. This field can also be automatically deenergized to reduce or remove the load, when certain undesirable conditions occur, to prevent damage to the power plant or other traction equipment.

The engine main generator or alternator power plant functions at any throttle setting as a constant-horsepower source of energy. Therefore, the main generator voltage must be controlled to provide constant power output for each specific throttle position under the varying conditions of train speed, train resistance, atmospheric pressure, and quality of fuel. The load regulator, which is an integral part of the governor, accomplishes this within the maximum safe values of main-generator voltage and current. For example, when the locomotive experiences an increase in track gradient with a consequent reduction in speed, traction-motor counter-emf decreases, causing a change in traction-motor and main-generator current. Because this alters the load demand on the engine, the speed of the engine tends to change to compensate. As the speed changes, the governor begins to reposition the fuel racks, but at the same time a pilot valve in the governor directs hydraulic pressure into a load regulator vane motor, which changes the resistance value of the load regulator rheostat in series with the main-generator excitation circuit. This alters the main-generator excitation current and consequently main-generator voltage and returns the value of main-generator power output to normal. Engine fuel racks return to normal, consistent with constant values of engine speed.

The load regulator is effective within maximum and minimum limit values of the rheostat. Beyond these limits the power output of the engine is reduced. However, protective devices in the main generator excitation circuit limit the voltage output to ensure that values of current and voltage in the traction motor circuits are within safe limits.

23.1.3 Auxiliary Generating Apparatus

DC power for battery charging, lighting, control, and cab heaters is provided by a separate generator, geared to the main engine. Voltage output is regulated within 1 percent of 74 V over the full range of engine speeds. Auxiliary alternators with full-wave bridge rectifiers are also utilized for this application. Modern locomotives with on-board computer systems also have power supplies to provide "clean" dc for sensors and computer systems. These are typically 24 V systems.

Traction motor blowers are mounted above the locomotive underframe. Air from the centrifugal blower housings is carried through the underframe and into the motor housings through flexible ducts. Other designs have been developed to vary the air output with cooling requirements to conserve parasitic energy demands. The main generator/alternators are cooled in a similar manner.

Traction motors are nose-suspended from the truck frame and bearing-suspended from the axle (Figure 23.2). The traction motors employ series-exciting (main) and commutating field poles. The current in the series field is reversed to change locomotive direction and may be partially shunted through resistors to reduce counter-emf as locomotive speed is

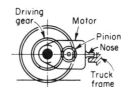

FIGURE 23.2 Axle-hung traction motor for diesel-electric and electric locomotives.

increased. Newer locomotives dispense with field shunting to improve commutation. Early locomotive designs also required motor connection changes (series, series/parallel, parallel) referred to as transition, to maintain motor current as speed increased.

AC traction motors employ a variable-frequency supply derived from a computer-controlled solid-state inverter system, fed from the rectified alternator output. Locomotive direction is controlled by reversing the sequence of the three-phase supply.

The development of the modern traction alternator with its high output current has resulted in a trend toward dc traction motors that are permanently connected in parallel. DC motor armature shafts are equipped with grease-lubricated roller bearings, while ac traction-motor shafts have grease-lubricated roller bearings at the free end and an oil-lubricated bearing at the pinion drive end. Traction-motor support bearings are usually of the plain sleeve type with lubricant wells and spring-loaded felt wicks that maintain constant contact with the axle surface. However, many new passenger locomotives have roller support bearings.

23.1.4 Electrical Controls

In the conventional dc propulsion system, electropneumatic or electromagnetic contactors are employed to make and break the circuits between the traction motors and the main generator. They are equipped with interlocks for various control-circuit functions. Similar contactors are used for other power and excitation circuits of lower power (current). An electropneumatic or electric-motor-operated cam switch, consisting of a two-position drum with copper segments moving between spring-loaded fingers, is generally used to reverse traction-motor field current ("reverser") or to set up the circuits for dynamic braking. This switch is not designed to operate under load. On some dc locomotives these functions have been accomplished with a system of contactors. In the ac propulsion systems, the power-control contactors are totally eliminated since their function is performed by the solid-state switching devices in the inverters.

Locomotives with ac traction motors utilize inverters to provide phase-controlled electrical power to their traction motors. Two basic arrangements of inverter control for ac traction motors have emerged. One arrangement uses an individual inverter for each axle. The other uses one inverter per axle truck. Inverters are typically GTO (gate turn off) or IGBT (insulated gate bipolar transistor) devices.

23.1.5 Cabs

In order to promote uniformity and safety, the AAR has issued standards for many locomotive cab features, RP-5104. Locomotive cab noise standards are prescribed by CFR 49 § 229.121.

Propulsion control circuits transmit the engineer's movements of the throttle lever, reverse lever, and transition or dynamic-brake control lever in the controlling unit to the power-producing equipment of each unit operating in multiple in the locomotive consist. Before power is applied, all reversers must move to provide the proper motor connections for the direction of movement desired. Power contactors complete the circuits between generators and traction motors. For dc propulsion systems excitation circuits then function to provide the proper main-generator field current while the engine speed increases to correspond to the engineer's throttle position. In ac propulsion systems, all power circuits are controlled by computerized switching of the inverter.

To provide for multiple-unit operation, the control circuits of each locomotive unit are connected by jumper cables. The AAR has issued Standard S-512 covering standard dimensions and contact identification for 27-point control jumpers used between diesel-electric locomotive units.

Wheel slip is detected by sensing equipment connected either electrically to the motor circuits or mechanically to the axles. When slipping occurs on some units, relays automatically reduce main generator excitation until slipping ceases, whereupon power is gradually reapplied. On newer units, an electronic system senses small changes in motor current and reduces motor current before a slip occurs. An advanced system recently introduced adjusts wheel creep to maximize wheel-to-rail adhesion. Wheel speed is compared to ground speed, which is accurately measured by radar. A warning light and/or buzzer in the operating cab alerts the engineer, who must notch back on the throttle if the slip condition persists.

23.1.6 Batteries

Lead-acid storage batteries of 280 or 420 ampere-hour capacity are usually used for starting the diesel engine. Thirty-two cells on each locomotive unit are used to provide 64 V to the system. (See also Avallone and Baumeister 1996, Sec. 15.) The batteries are charged from the 74-V power supply.

23.1.7 Air Brake System

The "independent" brake valve handle at the engineer's position controls air pressure supplied from the locomotive reservoirs to the brake cylinders on only the locomotive itself. The "automatic" brake valve handle controls the air pressure in the brake pipe to the train (Figure 23.3). On more recent locomotives, purely pneumatic braking systems have been supplanted by electro-pneumatic systems. These systems allow for more uniform brake applications, enhancing brake pipe pressure control, enabling better train handling. The AAR has issued Standard S-5529, *Multiple Unit Pneumatic Brake Equipment for Locomotives.*

Compressed air for braking and for various pneumatic controls on the locomotive is usually supplied by a two-stage three-cylinder compressor, usually connected directly or through a clutch to the engine crankshaft. An unloader or the clutch is activated to maintain a pressure of approximately 130 to 140 psi (896 to 965 kPa) in the main reservoirs. When charging an empty trainline with the locomotive at standstill (maximum compressor demand), the engineer may increase engine (and compressor) speed without loading the traction motors.

23.1.8 Dynamic Braking

On most locomotives, dynamic brakes supplement the air brake system. The traction motors are used as generators to convert the kinetic energy of the locomotive and train into electrical energy, which is dissipated through resistance grids located near the locomotive roof. Motor-driven blowers, designed to utilize some of this braking energy, force cooler outside air over

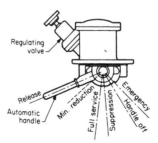

FIGURE 23.3 Automatic-brake valve-handle positions for 26-L brake equipment.

the grids and out through roof hatches. By directing a generous and evenly distributed air stream over the grids, their physical size is reduced in keeping with the relatively small space available in the locomotive. On some locomotives, resistor-grid cooling is accomplished by an engine-driven radiator/braking fan, but energy conservation is causing this arrangement to be replaced by motor-driven fans, which can be energized in response to need using the parasitic power generated by dynamic braking itself.

By means of a cam-switch reverser, the traction motors are connected to the resistance grids. The motor fields are usually connected in series across the main generator to supply the necessary high excitation current. The magnitude of the braking force is set by controlling the traction motor excitation and the resistance of the grids. Conventional dynamic braking is not usually effective below 10 mph (16 km/hr), but it is very useful at 20 to 30 mph (32 to 48 km/hr). Some locomotives are equipped with "extended range" dynamic braking which enables dynamic braking to be used at speeds as low as 3 mph (5 km/hr) by shunting out grid resistance (both conventional and extended range are shown on Figure 23.4). Dynamic braking is now controlled according to the "tapered" system, although the "flat" system has been used in the past. Dynamic braking control requirements are specified by AAR Standard S-5018. Dynamic braking is especially advantageous on long grades where accelerated brake shoe wear and the potential for thermal damage to wheels could otherwise be problems. The other advantages are smoother control of train speed and less concern for keeping the pneumatic trainline charged. Dynamic brake grids can also be used for a self-contained load-test feature which permits a standing locomotive to be tested for power output. On locomotives equipped with ac traction motors a constant dynamic braking (flat-top) force can be achieved from the horsepower limit down to 2 mph (3 km/hr).

23.1.9   Performance

Engine-Indicated Horsepower.   The power delivered at the diesel locomotive drawbar is the end result of a series of subtractions from the original indicated horsepower of the engine, which take into account the efficiency of transmission equipment and the losses due to the power requirements of various auxiliaries. The formula for the engine's indicated horsepower (ihp) is:

$$\text{ihp} = PLAN/33{,}000 \tag{23.1}$$

where P = mean effective pressure in the cylinder (psi)
L = length of piston stroke (ft)
A = piston area (in.^{2})
N = total number of cycles completed per minute

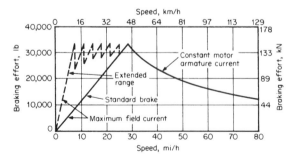

FIGURE 23.4   Dynamic-braking effort versus speed. (*Source:* Electro-Motive Division, General Motors Corp.)

Factor P is governed by the overall condition of the engine, quality of fuel, rate of fuel injection, completeness of combustion, compression ratio, etc. Factors L and A are fixed with design of engine. Factor N is a function of engine speed, number of working chambers, and strokes needed to complete a cycle.

Engine Brake Horsepower. In order to calculate the horsepower delivered by the crankshaft coupling to the main generator, frictional losses in bearings and gears must be subtracted from the *indicated horsepower* (ihp). Some power is also used to drive lubricating-oil pumps, governor, water pump, scavenging blower, and other auxiliary devices. The resultant horsepower at the coupling is *brake horsepower* (bhp).

Rail Horsepower. A portion of the engine bhp is transmitted mechanically via couplings or gears to operate the traction motor blowers, air compressor, auxiliary generator, and radiator cooling fan generator or alternator. Part of the auxiliary generator electrical output is used to run some of the auxiliaries. The remainder of the engine bhp transmitted to the main generator or main alternator for traction purposes must be multiplied by generator efficiency (usually about 91 percent), and the result again multiplied by the efficiency of the traction motors (including power circuits) and gearing to develop rail horsepower. Power output of the main generator for traction may be expressed as

$$\text{Watts}_{\text{traction}} = E_g \times I_m \tag{23.2}$$

where E_g is the main-generator voltage and I_m is the traction motor current in amperes, multiplied by the number of parallel paths or the dc link current in the case of an ac traction system.

Rail horsepower may be expressed as

$$\text{hp}_{\text{rail}} = V \times \text{TE}/375 \tag{23.3}$$

where V = velocity (mph)
    TE = tractive effort at the rail (lb)

Thermal Efficiency. The thermal efficiency of the diesel engine at the crankshaft, or the ratio of bhp output to the rate at which energy of the fuel is delivered to the engine, is about 33 percent. Thermal efficiency at the rail is about 26 percent.

Drawbar Horsepower. The drawbar horsepower represents power available at the rear of the locomotive to move the cars and may be expressed as

$$\text{hp}_{\text{drawbar}} = \text{hp}_{\text{rail}} - \text{Locomotive running resistance} \times V/375 \tag{23.4}$$

where V is the speed in mph. Train resistance calculations are discussed in section 23.5. Theoretically, therefore, drawbar horsepower available is power output of the diesel engine less the parasitic losses and losses described above.

Speed-Tractive Effort. At full throttle the losses vary somewhat at different values of speed and tractive effort, but a curve of tractive effort plotted against speed is nearly hyperbolic. Figure 23.5 is a typical speed-tractive effort curve for a 3,500 hp (2,600 kW) freight locomotive. The diesel-electric locomotive has full horsepower available over the entire speed range (within the limits of adhesion described below). The reduction in power as continuous speed is approached is known as "power matching." This allows multiple operation of locomotives of different ratings at the same continuous speed.

Adhesion. In Figure 23.6 the maximum value of tractive effort represents the level usually achievable just before the wheels slip under average rail conditions. Adhesion is usually expressed as a percentage of vehicle weight on drivers, with the nominal level being 25

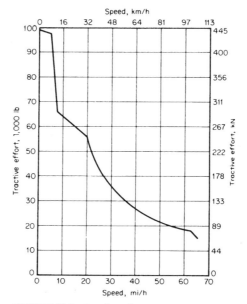

FIGURE 23.5 Tractive effort versus speed.

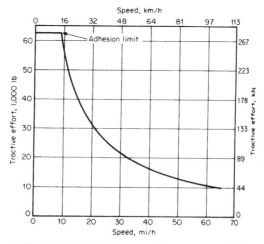

FIGURE 23.6 Typical tractive effort versus speed characteristics. (*Source:* Electro-Motive Division, General Motors Corp.)

percent. This means that a force equal to 25 percent of the total locomotive weight on drivers is available as tractive effort. Actually, at the point of wheel slip, adhesion will vary widely with rail conditions, from as low as 5 percent to as high as 35 percent or more. Adhesion is severely reduced by lubricants, which spread as thin films in the presence of moisture on running surfaces. Adhesion can be increased with sand applied to the rails from the loco-motive sanding system. More recent wheel slip systems permit wheel creep (very slow con-trolled slip) to achieve greater levels of tractive effort. Even higher adhesion levels are available from ac traction motors; for example, 45 percent at start-up and low speed with a nominal value of 35 percent.

Traction Motor Characteristics.   Motor torque is a function of armature current and field flux (which is a function of field current). Since the traction motors are series connected, armature and field current are the same (except when field shunting circuits are introduced), and therefore tractive effort is solely a function of motor current. Figure 23.7 presents a group of traction motor characteristic curves with tractive effort, speed, and efficiency plotted against motor current for full field (FF) and at 35 (FS1) and 55 (FS2) percent field shunting. Wheel diameter and gear ratio must be specified when plotting torque in terms of tractive effort. (See also Avallone and Baumeister 1996, Sec. 15.)

Traction motors are usually rated in terms of their maximum continuous current. This represents the current at which the heating due to electrical losses in the armature and field windings is sufficient to raise the temperature of the motor to its maximum safe limit when cooling air at maximum expected ambient temperature is forced through it at the prescribed rate by the blowers. Continuous operation at this current level ideally allows the motor to operate at its maximum safe power level, with waste heat generated equal to heat dissipated. The tractive effort corresponding to this current is usually somewhat lower than that allowed by adhesion at very low speeds. Higher current values may be permitted for short periods of time (as when starting). These ratings are specified in time intervals of time (minutes) and are posted on or near the load meter (ammeter) in the cab.

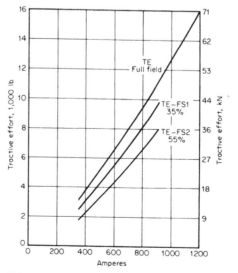

FIGURE 23.7 Traction-motor   characteristics. (*Source:* Electro-Motive Division, General Motors Corp.)

Maximum Speed.   Traction motors are also rated in terms of their maximum safe speed in r/min, which in turn limits locomotive speed. The gear ratio and wheel diameter are directly related to speed as well as the maximum tractive effort and the minimum speed at which full horsepower can be developed at the continuous rating of the motors. Maximum locomotive speed may be expressed as follows:

$$(mph)_{max} = \frac{\text{Wheel diameter in.} \times \text{Maximum motor r/min}}{\text{Gear ratio} \times 336} \qquad (23.5)$$

where the gear ratio is the number of teeth on the gear mounted on the axle divided by the number of teeth on the pinion mounted on the armature shaft.

Locomotive Compatibility.   The AAR has developed two standards in an effort to improve compatibility between locomotives of different model, manufacture, and ownership: a standard 27-point control system (Standard S-512, Table 23.2) and a standard control stand (RP-5132). The control stand has been supplanted by a control console in many road locomotives (Figure 23.8).

TABLE 23.2 Standard Dimensions and Contact Identification of 27-Point Control Plug and Receptacle for Diesel-Electric Locomotives

Receptacle point	Function	Code	Wire size, AWG
1	Power reduction setup, if used	(PRS)	14
2	Alarm signal	SG	14
3	Engine speed	DV	14
4*	Negative	N	14 or 10
5	Emergency sanding	ES	14
6	Generator field	GF	12
7	Engine speed	CV	14
8	Forward	FO	12
9	Reverse	RE	12
10	Wheel slip	WS	14
11	Spare		14
12	Engine speed	BV	14
13	Positive control	PC	12
14	Spare		14
15	Engine speed	AV	14
16	Engine run	ER	14
17	Dynamic brake	B	14
18	Unit selector circuit	US	12
19	2d negative, if used	(NN)	12
20	Brake warning light	BW	14
21	Dynamic brake	BG	14
22	Compressor	CC	14
23	Sanding	SA	14
24	Brake control/power reduction control	BC/PRC	14
25	Headlight	HL	12
26	Separator blowdown/remote reset	SV/RR	14
27	Boiler shutdown	BS	14

*Receptacle point 4—AWG wire size 12 is "standard" and AWG wire size 10 is "Alternate standard" at customer's request. A dab of white paint in the cover latch cavity must be added for ready identification of a no. 10 wire present in a no. 4 cavity. From *Mark's Standard Handbook for Mechanical Engineers*, 10th ed.

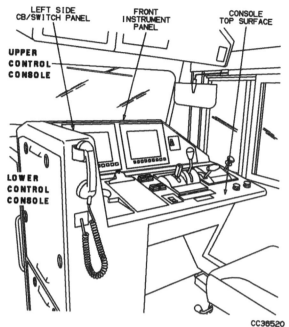

FIGURE 23.8 Locomotive control console. (*Source:* Electro-Motive Division, General Motors Corp.)

Energy Conservation. Efforts to improve efficiency and fuel economy have resulted in major changes in the prime movers, including more efficient turbocharging, fuel injection, and combustion. The auxiliary (parasitic) power demands have also been reduced with improvements including fans and blowers, that only move the air required by the immediate demand, air compressors that declutch when unloaded, and a selective low-speed engine idle. Fuel-saver switches permit dropping trailing locomotive units off the line when less than maximum power is required, while allowing the remaining units to operate at maximum efficiency.

Emissions. The U.S. Environmental Protection Agency has promulgated regulations aimed at reducing diesel locomotive emissions, especially oxides of nitrogen (NO_x). These standards also include emissions reductions for hydrocarbons (HC), carbon monoxide (CO), particulate matter (PM), and smoke. These standards are executed in three tiers. Tier 0 goes into effect in 2000, tier 1 in 2002, and tier 3 in 2005. EPA locomotive emissions standards are published under CFR 40 §§ 85, 89, and 92.

23.2 ELECTRIC LOCOMOTIVES

Electric locomotives are presently in very limited use in North America. Freight locomotives are in dedicated service primarily for coal or mineral hauling. Electric passenger locomotives are used in high-density service in the northeastern United States.

Electric locomotives draw power from overhead catenary or third-rail systems. While earlier systems used either direct current up to 3,000 V or single-phase alternating current at 11,000 V, 25 Hz, the newer systems in North America use 25,000 or 50,000 V at 60 Hz.

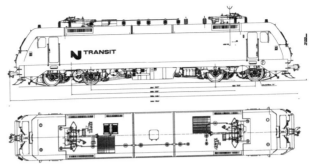

FIGURE 23.9   A modern electric high-speed locomotive. (*Courtesy of* Bombardier Transportation.)

The higher voltage levels can only be used where clearances permit. A three-phase power supply was tried briefly in this country and overseas many years ago, but was abandoned because of the complexity of the required double catenary.

While the older dc locomotives used resistance control, ac locomotives have used a variety of systems, including Scott-connected transformers; series ac motors; motor generators and dc motors; ignitrons and dc motors; silicon thyristors and dc motors; and, more recently, chopper control. Examples of the various electric locomotives in service are shown in Table 23.1.

An electric locomotive used in high-speed passenger service is shown in Fig. 23.9. High short-time ratings (Figures 23.10 and 23.11) render electric locomotives suitable for passenger service where high acceleration rates and high speeds are combined to meet demanding schedules.

The modern electric locomotive in Figure 23.9 obtains power for the main circuit and motor control from the catenary through a pantograph. A motor-operated switch provides for the transformer change from series to parallel connection of the primary windings to give a constant secondary voltage with either 25 kV or 12.5 kV primary supply. The converters

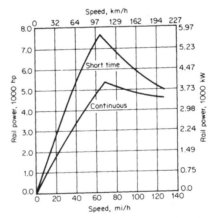

FIGURE 23.10   Speed-horsepower characteristics (half-worn wheels). [Gear ratio = 85/36; wheel diameter = 50 in (1,270 mm); ambient temperature = 60°F (15.5°C).] (*Courtesy of* M. Ephraim, ASME, Rail Transportation Division.)

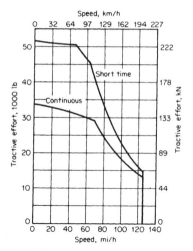

FIGURE 23.11 Tractive effort versus speed (half-worn wheels). [Voltage = 11.0 kV at 25 Hz; gear ratio = 85/36; wheel diameter = 50 in (1,270 mm).] (*Courtesy of* M. Ephraim, ASME, Rail Transportation Division.)

for armature current consist of two asymmetric type bridges for each motor. The traction-motor fields are each separately fed from a one-way connected-field converter.

Identical control modules separate the control of motors on each truck. Motor sets are therefore connected to the same transformer winding. Wheel slip correction is also modularized, utilizing one module for each two-motor truck set. This correction is made with a complementary wheel-slip detection and correction system. A magnetic pickup speed signal is used for the basic wheel slip. Correction is enhanced by a magnetoelastic transducer used to measure force swings in the traction-motor reaction rods. This system provides the final limit correction.

To optimize the utilization of available adhesion, the wheel-slip control modules operate independently to allow the motor modules to receive different current references depending on their respective adhesion conditions.

All auxiliary machines, air compressor, traction motor blower, cooling fans, etc., are driven by three-phase 400-V, 60-Hz induction motors powered by a static inverter that has a rating of 175 kVA at a 0.8 power factor. When cooling requirements are reduced, the control system automatically reduces the voltage and frequency supplied to the blower motors to the required level. As a backup, the system can be powered by the static converter used for the head-end power requirements of the passenger cars. This converter has a 500-kW, 480-V, three-phase, 60-Hz output capacity and has a built-in overload capacity of 10 percent for half of any 1-hour period.

Dynamic brake resistors are roof-mounted and are cooled by ambient airflow induced by locomotive motion. The dynamic brake capacity relative to train speed is shown in Figure 23.12. Regenerative braking can be utilized with electric locomotives by returning braking energy to the distribution system.

Many electric locomotives utilize traction motors identical to those used on diesel-electric locomotives. Some, however, use frame suspended motors with quill-drive systems (Figure 23.13). In these systems, torque is transmitted from the traction motor by a splined coupling to a quill shaft and rubber coupling to the gear unit. This transmission allows for greater relative movement between the traction motor (truck frame) and gear (wheel axle) and reduces unsprung weight.

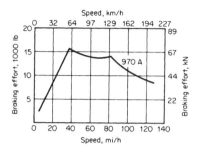

FIGURE 23.12 Dynamic-brake performance. (*Courtesy of* M. Ephraim, ASME, Rail Transportation Division.)

23.3 FREIGHT CARS

23.3.1 Freight Car Types

Freight cars are designed and constructed for either general service or specific ladings. The Association of American Railroads (AAR) has established design and maintenance criteria to assure the interchangeability of freight cars on all North American railroads. Freight cars that do not conform to AAR standards have been placed in service by special agreement of the railroads over which the cars operate. The AAR *Manual of Standards and Recommended Practices* specifies dimensional limits, weights, and other design criteria for cars that may be freely interchanged between North American railroads. This manual is revised annually by the AAR. Many of the standards are reproduced in the *Car and Locomotive Cyclopedia,* which is revised periodically. Safety appliances (such as end ladders, sill steps, and uncoupling levers), braking equipment, and certain car design features and maintenance practices must comply with the Safety Appliances Act, the Power Brake Law, and the Freight Car Safety Standards covered by the Federal Railroad Administration's (FRA) Code of Federal Regulations (CFR Title 49). Maintenance practices are set forth in the *Field Manual* of the AAR Interchange Rules and repair pricing information is provided in the companion *Office Manual.*

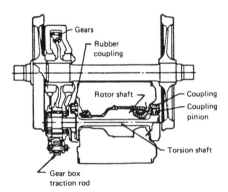

FIGURE 23.13 Frame-suspended locomotive transmission. (*Courtesy of* M. Ephraim, ASME, Rail Transportation Division.)

The AAR identifies most cars by nominal capacity and type, and in some cases there are restrictions as to type of load (i.e., food, automobile parts, coil steel, etc.). Most modern cars have nominal capacities of either 70 or 100 tons.* A 100-ton car is limited to a maximum gross rail load of 263,000 lb (119.3 ton) with 6½ by 12 in. (165 by 305 mm) axle journals, four pairs of 36-in. (915-mm) wheels, and a 70-in. (1,778 mm) rigid truck wheelbase. Some 100-ton cars are rated at 286,000 lb and are handled by individual railroads on specific routes or by mutual agreement between the handing railroads. A 70-ton car is limited to a maximum gross rail load of 220,000 lb (99.8 ton) with 6 by 11 in. (152 by 280 mm) axle journals, four pairs of 33-in (838-mm) wheels with trucks having a 66-in. (1,676-mm) rigid wheelbase.

On some special cars where height limitations are critical, 28-in. (711-mm) wheels are used with 70-ton axles and bearings. In these cases wheel loads are restricted to 22,400 lb (10.2 ton). Some special cars are equipped with two 125-ton two axle trucks with 38-in. (965-mm) wheels in four wheel trucks having 7 by 14 in. (178 by 356 mm) axle journals and 72-in. (1,829-mm) truck wheelbase. This application is prevalent in articulated double-stack (two-high) container cars. Interchange of these very heavy cars is by mutual agreement between the operating railroads involved.

The following are the most common car types in service in North America. Dimensions given are for typical cars; actual measurements may vary.

Boxcars.   There are six popular boxcar types (Figure 23.14*a*):

1. Standard boxcars may have either sliding or plug doors. Plug doors provide a tight seal from weather and a smooth interior. Unequipped box cars are usually of 70-ton capacity. These cars have tongue-and-groove or plywood lining on the interior sides and ends with nailable floors (either wood or steel with special grooves for locking nails). The cars carry typical general merchandise lading: packaged, canned, or bottled foodstuffs; finished lumber; bagged or boxed bulk commodities; or, in the past when equipped with temporary door fillers, bulk commodities such as grain.**

2. Specially equipped boxcars usually have the same dimensions as standard cars but include special interior devices to protect lading from impacts and over-the-road vibrations. Specially equipped cars may have hydraulic-cushion units to dampen longitudinal shock at the couplers.

3. Insulated boxcars have plug doors and special insulation. These cars can carry foodstuffs such as unpasteurized beer, produce, and dairy products. These cars may be precooled by the shipper and maintain a heat loss rate equivalent to 1°F (0.55°C) per day. They can also protect loads from freezing when operating at low ambient temperatures.

4. Refrigerated boxcars are used where transit times are longer. These cars are equipped with diesel-powered refrigeration units and are primarily used to carry fresh produce and meat. They are often 100-ton cars.†

5. "All-door" boxcars have doors that open the full length of the car for loading package lumber products such as plywood and gypsum board.

6. High-cubic-capacity boxcars with an inside volume of 10,000 ft³ (283 m³) have been designed for light-density lading, such as some automobile parts and low-density paper products.

Boxcar door widths vary from 6 to 10 ft for single-door cars and 16 to 20 ft (4.9 to 6.1 m) for double-door cars. All-door cars have clear doorway openings in excess of 25 ft

---

*1 ton = 1 short ton = 2,000 lb; 1 ton = 1 metric ton = 1,000 kg = 2,205 lb
**70 ton: L = 50 ft 6 in. (15.4 m), H = 11 ft 0 in. (3.4 m), W = 9 ft 6 in. (2.9 m), truck centers = 40 ft 10 in. (12.4 m).
†L = 52 ft 6 in. (16.0 m), H = 10 ft 6 in. (3.2 m), truck centers = 42 ft 11 in. (13.1 m)

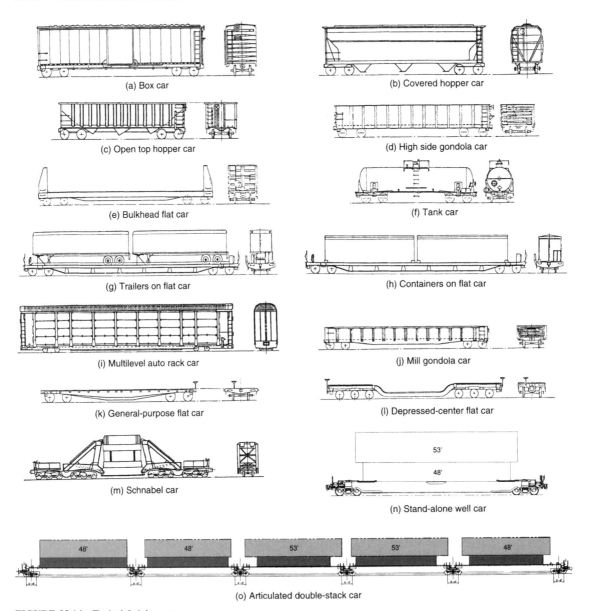

FIGURE 23.14 Typical freight cars.

(7.6 m). The floor height above rail for an empty uninsulated boxcar is approximately 44 in. (1,120 mm) and for an empty insulated boxcar approximately 48 in. (1,220 mm). The floor height of a loaded car can be as much as 3 in. (76 mm) lower than the empty car.

Covered hopper cars (Figure 23.14*b*) are used to haul bulk commodities that must be protected from the environment. Modern covered hopper cars are typically 100-ton cars with roof hatches for loading and from two to six bottom outlets for discharge. Cars used for dense commodities, such as fertilizer or cement, have two bottom outlets, round roof hatches, and volumes of 3,000 to 4,000 ft³ (84.9 to 113.2 m³).* Cars used for grain service (corn, wheat, rye, etc.) have three or four bottom outlets, longitudinal trough roof hatches, and volumes of from 4,000 to 5,000 ft³ (113 to 142 m³). Cars used for hauling plastic pellets have four to six bottom outlets (for pneumatic unloading with a vacuum system), round roof hatches, and volumes of from 5,000 to 6,000 ft³**(142 to 170 m³).

Open-top hopper cars (Figure 23.14*c*) are used for hauling bulk commodities such as coal, ore, or wood chips. A typical 100-ton coal hopper car will vary in volume depending on the light weight of the car and density of the coal to be hauled. Volumes range from 3,900 to 4,800 ft³ (110 to 136 m³). Cars may have three or four manually operated or a group of automatically operated bottom doors. Some cars are equipped with rotating couplers on one end to allow rotary dumping without uncoupling.† Cars intended for aggregate or ore service have smaller volumes for the more dense commodity. These cars typically have two manual bottom outlets.‡ Hopper cars used for woodchip service are configured for low-density loads. Volumes range from 6,500 to 7,500 ft³ (184 to 212 m³).

High-side gondola cars (Figure 23.14*d*) are open-top cars typically used to haul coal or wood chips. These cars are similar to open-top hopper cars in volume but require a rotary coupler on one end for rotary dumping to discharge lading since they do not have bottom outlets. Rotary-dump coal gondolas are usually used in dedicated, unit-train service between a coal mine and an electric power plant. The length over coupler pulling faces is approximately 53 ft 1 in. (16.2 m) to suit the standard coal dumper.§ Woodchip cars are used to haul chips from sawmills to paper mills or particle-board manufacturers. These high-volume cars are either rotary-dumped or, when equipped with end doors, end-dumped.¶ Rotary-dump aggregate or ore cars, called "ore jennies," have smaller volumes for the high-density load.

Bulkhead flat cars (Figure 23.14*e*) are used for hauling such commodities as packaged finished lumber, pipe, or, with special inward canted floors, pulpwood. Both 70- and 100-ton bulkhead flats are used. Typical deck heights are approximately 50 in. (1,270 mm).¶¶ Special, center beam, bulkhead flat cars designed for pulpwood and lumber service have a full-height, longitudinal divider from bulkhead to bulkhead.

Tank cars (Figure 23.14*f*) are used for liquids, compressed gases, and other ladings, such as sulfur, that can be loaded and unloaded in a molten state. Nonhazardous liquids such as corn syrup, crude oil, and mineral spring water are carried in nonpressure cars. Cars used to haul hazardous substances such as liquefied petroleum gas (LPG), vinyl chloride, and anhydrous ammonia are regulated by the U.S. Department of Transportation. Newer and earlier-built retrofitted cars equipped for hazardous commodities have safety features including safety valves, specially designed top and bottom "shelf" couplers, which increase the interlocking effect between couplers and decrease the danger of disengagement due to derailment, head shields on the ends of the tank to prevent puncturing, bottom-outlet protection if bottom outlets are used, and thermal insulation and jackets to reduce the risk of rupturing in a fire. These features resulted from industry-government studies in the RPI-AAR Tank Car Safety Research and Test Program.

---

* 100 ton: L = 39 ft 3 in. (12.0 m). H = 14 ft 10 in. (4.5 m), truck centers = 26 ft 2 in. (8.0 m).

** 100 ton: L = 65 ft 7 in. (20.0 m), H = 15 ft 5 in. (4.7 m), truck centers = 54 ft 0 in. (16.5 m).

† 100 ton: L = 53 ft ½ in. (16.2 m), H = 12 ft 8½ in. (3.9 m), truck centers = 40 ft 6 in. (12.3 m).

‡ 100 ton: L = 40 ft 8 in. (12.4 m), H = 11 ft 10⅞ in. (3.6 m), truck centers = 29 ft 9 in. (9.1 m).

§ 100 ton: L = 50 ft 5½ in. (15.4 m), H = 11 ft 9 in. (3.6 m), W = 10 ft 5⅜ in. (3.2 m), truck centers = 50 ft 4 in. (15.3 m).

¶ 100 ton: L = 62 ft 3 in. (19.0 m), H = 12 ft 7 in. (3.8 m), W = 10 ft 5 ⅜ in. (3.2 m), truck centers = 50 ft 4 in. (15.3 m).

¶¶ 100 ton: L = 61 ft ¾ in (18.6 m), W = 10 ft 1 in. (3.1 m), H = 11 ft 0 in. (3.4 m), truck centers = 55 ft 0 in (16.8 m).

Cars for asphalt, sulfur, and other viscous-liquid service have heating coils on the shell so that steam may be used to liquefy the lading for discharge.*

Intermodal Cars.   Conventional 89-ft 4-in. (27.2-m) intermodal flat cars are equipped to haul one 45-ft (13.7-m) and one 40-ft (12.2-m) trailer with or without end-mounted refrigeration units, two 45-ft (13.7-m) trailers (dry vans), or combinations of containers from 20 to 40 ft (6.1 to 12.2 m) in length. Hitches to support trailer fifth wheels may be fixed for trailer-only cars or retractable for conversion to haul containers or to facilitate driving trailers onto the cars in the rare event where "circus" loading is still required. Trailer hauling service (Figure 23.14*g*) is called TOFC (trailer on flat car). Container service (Figure 23.14*h*) is called COFC (container on flat car).**

Introduction of larger trailers in highway service has led to the development of alternative TOFC and COFC cars. One technique involves articulation of skeletonized or well-type units into multiunit cars for lift-on loading and lift-off unloading (standalone well car, Figure 23.14*n*, and articulated well car, Figure 23.14*o*). These cars are typically composed of from 3 to 10 units. Well cars consist of a center well for double-stacked containers.†

Another approach to hauling larger trailers is the two-axle skeletonized car. These cars, used either singly or in multiple combinations, can haul a single trailer from 40 to 48 ft long (12.2 to 14.6 m) with nose-mounted refrigeration unit and 36- or 42-in. (914- to 1,070-mm) spacing between the kingpin and the front of the trailer.

Bilevel and trilevel *auto rack cars* (Figure 23.14*i*) are used to haul finished automobiles and other vehicles. Most recent designs of these cars feature fully enclosed racks to provide security against theft and vandalism.‡

Mill gondolas (Figure 23.14*j*) are 70-ton or 100-ton open-top cars principally used to haul pipe, structural steel, scrap metal, and, when specially equipped, coils of aluminum or tinplate and other steel materials.§

General-purpose or *machinery flat cars* (Figure 23.14*k*) are 70-ton or 100-ton cars used to haul machinery such as farm equipment and highway tractors. These cars usually have wood decks for nailing lading-restraint dunnage. Some heavy-duty six-axle cars are used for hauling off-highway vehicles such as army tanks and mining machinery.¶

Depressed-center flat cars (Figure 23.14*l*) are used for hauling transformers and other heavy, large materials that require special clearance considerations. Depressed-center flat cars may have four-, six-, or dual four-axle trucks with span bolsters, depending on weight requirements.

Schnabel cars (Figure 23.14*m*) are special cars for transformers and nuclear power plant components. With these cars the load itself provides the center section of the car structure during shipment. Some Schnabel cars are equipped with hydraulic controls to lower the load for height restrictions and shift the load laterally for wayside restrictions.¶¶ Schnabel cars must be operated in special trains.

---

*Pressure cars, 100 ton: volume = 20,000 gal (75.7 m³), L = 59 ft 11¾ in (18.3 m), truck centers = 49 ft 0¼ in. (14.9 m). Nonpressure, 100 ton, volume = 21,000 gal (79.5 m³), L = 51 ft 3¼ in. (15.6 m), truck centers = 38 ft 11¼ in. (11.9 m).

**70 ton: L = 89 ft 4 in. (27.1 m), W = 10 ft 3 in. (3.1 m), truck centers = 64 ft 0 in. (19.5 m).

†10-unit, skeletonized car: L = 46 ft 6⅜ in. (14.2 m) per end unit, L = 465 ft 3½ in. (141.8 m).

‡70 ton: L = 89 ft 4 in. (27.2 m), H = 18 ft 11 in. (5.8 m), W = 10 ft 7 in. (3.2 m), truck centers = 64 ft 0 in. (19.5 m).

§100 ton: L = 52 ft 6 in. (16.0 m), W = 9 ft 6 in. (2.9 m), H = 4 ft 6 in. (1.4 m), truck centers = 43 ft 6 in. (13.3 m).

¶100 ton, four axle: L = 60 ft 0 in. (18.3 m), H = 3 ft 9 in. (1.1 m), truck centers = 42 ft 6 in. (13.0 m). 200 ton, 8 axle: L = 44 ft 4 in. (13.5 m), H = 4 ft 0 in. (1.2 m), truck centers = 33 ft 9 in. (10.3 m).

¶¶472 ton: L = 22 ft 10 in. to 37 ft 10 in. (7.0 to 11.5 m), truck centers = 55 ft 6 in. to 70 ft 6 in. (16.9 to 21.5 m).

23.3.2 Freight Car Design

The AAR provides specifications to cover minimum requirements for design and construction of new freight cars. Experience has demonstrated that the AAR Specifications alone do not ensure an adequate car design for all service conditions. The designer must be familiar with the specific service and increase the design criteria for the particular car above the minimum criteria provided by the AAR. The AAR requirements include stress calculations for the load-carrying members of the car and physical tests that may be required at the option of the AAR committee approving the car design. In some cases, it is advisable to operate an instrumented prototype car in service to detect problems that might result from unexpected track or train-handling input forces. The car design must comply with width and height restrictions shown in AAR clearance plates furnished in the specifications (Figure 23.15). In addition, there are limitations on the height of the center of gravity of the loaded car and on the vertical and horizontal curving capability allowed by the clearance provided at the coupler. The AAR provides a method of calculating the minimum radius curve which the car design can negotiate when coupled to another car of the same type or to a standard AAR base car. In the case of horizontal curves, the requirements are based on the length of the car over the pulling faces of the couplers.

In the application for approval of a new or untried type of car, the Equipment Engineering Committee of the AAR may require either additional calculations or tests to assess the design's ability to meet the AAR minimum requirements. These tests might consist of a static compression test of 1,000,000 lb (4.4 MN), a static vertical test applied at the coupler, and impact tests simulating yard impact conditions.

Freight cars are designed to withstand single-ended impact or coupling loads based upon the type of cushioning provided in the car design. Conventional friction, elastomer, or com-

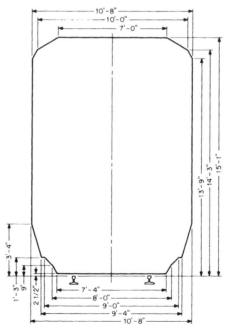

FIGURE 23.15 AAR plate B equipment-clearance diagram.

bination draft gears or short-travel hydraulic cushion units that provide less than 6 in (152 mm) of travel require a structure capable of withstanding a 1,250,000-lb (5.56-MN) impact load. For cars with hydraulic units that provide greater than 14 in. (356 mm) of travel, the required design impact load is 600,000 lb (2.7 MN). In all cases, the structural connections to the car must be capable of withstanding a static compressive (squeeze) end load of 1,000,000 lb (4.44 MN) or a dynamic (impact) compressive load of 1,250,000 lb (5.56 MN).

The AAR has adopted requirements for unit trains of high-utilization cars to be designed for 3,000,000 mi (4.8 Gm) of service based upon fatigue life estimates. General-interchange cars that accumulate less mileage in their life should be designed for 1,000,000 mi (1.6 Gm) of service. Road environment spectra for various locations within the car are being developed for different car designs for use in this analysis. The fatigue strengths of various welded connections are provided in the AAR *Manual of Standards and Recommended Practices,* Sec. C, Part II.

Many of the design equations and procedures are available from the AAR. Important information on car design and approval testing is contained in AAR *Manual of Standards and Recommended Practices,* Sec. C-II M-1001, Chap. XI.

23.3.3 Freight Car Suspension

Most freight cars are equipped with standard three-piece trucks (Figure 23.16) consisting of two side-frame castings and one bolster casting. Side-frame and bolster designs are subjected to both static and fatigue test requirements specified by the AAR. The bolster casting is equipped with a female centerplate bowl upon which the car body rests and with side bearings located (generally 25 in. (635 mm)) each side of the centerline. In most cases, the side bearings have clearance to the car body and are equipped with either flat sliding plates or rollers. In some cases, constant-contact side bearings provide a resilient material between the car body and the truck bolster.

The centerplate arrangement consists of various styles of wear plates or fiction materials and a vertical loose or locked pin between the truck centerplate and the car body.

Truck springs nested into the bottom of the side-frame opening support the end of the truck bolster. Requirements for spring designs and the grouping of springs are generally specified by the AAR. Historically, the damping provided within the spring group has utilized a combination of springs and friction wedges. In addition to friction wedges, in more recent years some cars have been equipped with hydraulic damping devices that parallel the spring group.

A few trucks have a "steering" feature, which includes an interconnection between axles to increase the lateral interaxle stiffness and decrease the interaxle yaw stiffness. Increased lateral stiffness improves the lateral stability and decreased yaw stiffness improves the curving characteristics.

23.3.4 Freight Car Wheel-Set Design

A freight car wheel set consists of wheels, axle, and bearings. Cast- and wrought-steel wheels are used on freight cars in North America (AAR *Manual of Standards and Recommended Practices,* Sec. G). Freight car wheels are subjected to thermal loads from braking, as well as mechanical loads at the wheel-rail interface. Experience with thermal damage to wheels has led to the introduction of "low-stress" or curved plate wheels (Figure 23.17). These wheels are less susceptible to the development of circumferential residual tensile stresses, which render the wheel vulnerable to sudden failure if a flange or rim crack occurs. New wheel designs introduced for interchange service must be evaluated using a finite-element technique employing both thermal and mechanical loads (AAR S-660).

Freight car wheels range in diameter from 28 to 38 in. (711 to 965 mm) depending on car weight (Table 23.3). The old AAR standard tread profile (Figure 23.18a) has been re-

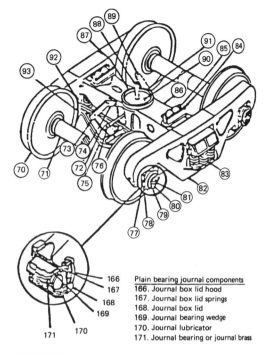

Plain bearing journal components
166. Journal box lid hood
167. Journal box lid springs
168. Journal box lid
169. Journal bearing wedge
170. Journal lubricator
171. Journal bearing or journal brass

Unit beam roller bearing truck components

70. Wheel	82. Truck side frame
71. Axle	83. Truck springs
72. Truck dead lever	84. Truck side bearing
73. Dead lever fulcrum	85. Side bearing roller
74. Dead lever fulcrum bracket	86. Truck bolster
75. Brake beam	87. Truck center plate cast
76. Bottom rod	88. Truck live lever
77. Roller bearing adapter	89. Center pin
78. Roller bearing assembly	90. Horizontal wear plate
79. End cap	91. Vertical wear plate
80. End cap retaining bolt	92. Brake shoe key
81. Locking plate	93. Brake shoe

FIGURE 23.16   Unit-beam roller-bearing truck with inset showing plain-bearing journal. (*Source:* AAR Research and Test Department.)

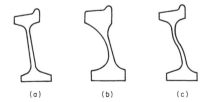

(a)   (b)   (c)

FIGURE 23.17   Wheel-plate designs. (*a*) Flat plate; (*b*) parabolic plate; (*c*) S-curved plate.

TABLE 23.3 Wheel and Journal Sizes of Eight-Wheel Cars

Nominal car capacity, ton	Maximum gross weight, lb	Journal (bearing) size, in.	Wheel diameter, in.
50	177,000	$5\frac{1}{2} \times 10$	33
[a]	179,200	6×11	28
70	220,000	6×11	33
100	263,000[c]	$6\frac{1}{2} \times 12$	36
125[b]	315,000	7×12	38

[a] Limited by wheel rating.
[b] Not approved for free interchange.
[c] 286,000 in special cases.

placed with the AAR-1B (Fig. 23.18c) profile, which represents a worn profile to minimize early tread loss due to wear and provides a stable profile over the life of the tread. Several variant tread profiles, including the AAR-1B, were developed from the basic Heumann design (Figure 23.18b). One of these, for application in Canada, provided increasing conicity into the throat of the flange, similiar to the Heumann profile. This reduces curving resistance and extends wheel life.

Wheels are also specified by chemistry and heat treatment. Low-stress wheel designs of classes B and C are required for freight cars. Class B wheels have a carbon content of 0.57 to 0.67 percent and are rim-quenched. Class C wheels have a carbon content of 0.67 to 0.77 percent and are also rim-quenched. Rim-quenching provides a hardened running surface for a long wear life. Lower carbon levels than those in Class B may be used where thermal cracking is experienced, but freight car equipment generally does not require their use.

Axles used in interchange service are solid steel forgings with raised wheel seats. Axles are specified by journal size for different car capacities (Table 23.3).

Most freight car journal bearings are grease-lubricated, tapered-roller bearings (see Avallone and Baumeister 1996, Sec. 8). Current bearing designs eliminate the need for periodic field lubrication.

Wheels are mounted and secured on axles with an interference fit. Bearings are mounted with an interference fit and retained by an end cap bolted to the end of the axle. Wheels and bearings for cars in interchange service must be mounted by an AAR-inspected and approved facility.

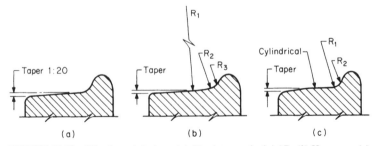

FIGURE 23.18 Wheel-tread designs. (a) Obsolete standard AAR; (b) Heumann; (c) new AAR-1B.

23.3.5 Special Features

Many components are available to enhance the usefulness of freight cars. In most cases, the design or performance of the component is specified by the AAR.

Coupler Cushioning.   Switching of cars in a classification yard can result in relatively high coupler forces at the time of the impact between the moving and standing cars. Nominal coupling speeds of 4 mph (6.4 km/hr) or less are sometimes exceeded, with lading damage a possible result. Conventional cars are equipped with an AAR-approved draft gear, usually a friction-spring energy-absorbing device, mounted between the coupler and the car body. The rated capacity of draft gears ranges between 20,000 ft-lb (27.1 kJ) for earlier units to over 65,000 ft-lb (88.1 kJ) for later designs. Impact forces of 1,250,000 lb (5.56 MN) can be expected when a moving 100-ton car strikes a string of standing cars at 8 to 10 mph (12.8 to 16 km/hr). Hydraulic cushioning devices are available to reduce the impact force to 500,000 lb (2.22 MN) at impact speeds of 12 to 14 mph (19 to 22 km/hr). These devices may be mounted either at each end of the car (end-of-car devices) or in a long beam that extends from coupler to coupler (sliding centersill devices).

Lading Restraint.   Many forms of lading restraint are available, from tie-down chains for automobiles on rack cars to movable bulkheads for boxcars. Most load-restraining devices are specified by the AAR *Manual of Standards and Recommended Practices* and approved car loading arrangements are specified in the AAR *Loading Rules,* a multivolume publication for enclosed, open-top, and TOFC and COFC cars.

Covered Hopper Car Discharge Gates.   The majority of covered hopper cars are equipped with rack-and-pinion-operated sliding gates that allow the lading to discharge by gravity between the rails. These gates can be operated manually, with a simple bar or a torque-multiplying wrench, or mechanically with an impact or hydraulic wrench. Many special covered hopper cars have discharge gates with nozzles and metering devices for vacuum or pneumatic unloading.

Coupling Systems.   The majority of freight cars are connected with AAR standard couplers. A specification has been developed to permit the use of alternative coupling systems such as articulated connectors, drawbars, and rotary-dump couplers.

23.3.6 Freight Train Braking

The retarding forces acting on a railway train are rolling and mechanical resistance, aero-dynamic drag, curvature, and grade, plus that force resulting from friction of the brake shoes rubbing the wheel treads. On locomotives so equipped, dynamic or rheostatic brakes using the traction motors as generators can provide all or a portion of the retarding force to control train speed.

Quick-action automatic air brakes of the type specified by the AAR are the common standard in North America. With the automatic air brake system, the brake pipe extends through every vehicle in the train, connected by hoses between each locomotive unit and car. The front and rear end brake pipe angle cocks are closed.

Air pressure is provided by compressors on the locomotive units to the main reservoirs, usually at 130 to 150 psi (900 to 965 kPa). (Pressure values are gage pressures.) The engineer's automatic brake valve, in "release" position, provides air to the brake pipe on freight trains at reduced pressure, usually at 75, 80, 85, or 90 psi (520, 550, 585, or 620 kPa) depending on the type of service, train weight, grades, and speeds at which a train will operate. In passenger service, brake pipe pressure is usually 90 or 110 psi (620 to 836 kPa).

When brake pipe pressure is increased, the control valve allows the reservoir capacity on each car and locomotive to be charged and at the same time connects the brake cylinders to exhaust. Brake pipe pressure is reduced when the engineer's brake valve is placed in a "service" position and the control valve cuts off the charging function and allows the reservoir air on each car to flow into the brake cylinder. This moves the piston and, through a system of levers and rods, pushes the brake shoes against the wheel treads.

When the engineer's automatic brake valve is placed in the emergency position, the brake pipe pressure (BP) is reduced very rapidly. The control valves on each car move to the emergency-application position and rapidly open a large vent valve, exhausting brake pipe pressure to atmosphere. This will serially propagate the emergency application through the train at from 900 to 950 ft/sec (280 to 290 m/sec). With the control valve in the emergency position both auxiliary- and emergency-reservoir volumes (pressures) equalize with the brake cylinder and higher brake cylinder pressure (BCP) results, building up at a faster rate than in service applications.

The foregoing briefly describes the functions of the fundamental automatic air brake based on the functions of the control valve. AAR-approved brake equipment is required on all freight cars used in interchange service. The functions of the control valve have been refined to permit the handling of longer trains by more uniform brake performance. Important improvements in this design have been (1) reduction of the time required to apply the brakes on the last car of a train, (2) more uniform and faster release of the brakes, and (3) availability of emergency application with brake pipe pressure greater than 40 psi (275 kPa).

The braking ratio of a car is defined as the ratio of brake shoe (normal) force to the car's rated gross weight. Two types of brake shoes, high-friction composition and high-phosphorus cast iron, are used in interchange service. Because these shoes have very different friction characteristics, different braking ratios are required to assure uniform train braking performance (Table 23.4). Actual or net shoe forces are measured with calibrated devices. The calculated braking ratio R (nominal) is determined from the equation

$$R = PLANE \times 100/W \tag{23.6}$$

where P = brake cylinder pressure, 50 psi gage
L = mechanical ratio of brake levers
A = brake cylinder area (in^2)
N = number of brake cylinders
E = brake rigging efficiency = $E_r \times E_b \times E_c$
W = car weight (lb)

To estimate rigging efficiency, consider each pinned joint and horizontal sliding joint as a 0.01 loss of efficiency; i.e., in a system with 20 pinned and horizontal sliding joints, $E_r = 0.80$. For unit-type (hangerless) brake beams $E_b = 0.90$, and for the brake cylinder $E_c = 0.95$, giving the overall efficiency of 0.684 or 68.4 percent.

The total retarding force in pounds per ton may be taken as:

$$F = (PLef/W) = F_g G \tag{23.7}$$

where P = total brake-cylinder piston force (lbf)
L = multiplying ratio of the leverage between cylinder pistons and wheel treads
ef = product of the coefficient of brake shoe friction and brake rigging efficiency
W = loaded weight of vehicle (tons)
F_g = force of gravity, 20 lb/ton/percent grade
G = ascending grade (%)

Stopping distance can be found by adding the distance covered during the time the brakes are fully applied to the distance covered during the equivalent instantaneous application time.

$$S = \frac{0.0334V_2^2}{\left[\dfrac{W_n B_n (p_a/p_n)ef}{W_a}\right] + \left(\dfrac{R}{2000}\right) \pm (G)} + 1.467t_1 \left[V_1 - \left(\frac{R + 2000G}{91.1}\right)\frac{t_1}{2}\right] \tag{23.8}$$

TABLE 23.4 Braking Ratios, AAR Standard S-401

Type of brake rigging and shoes	With 50 lb/in² brake cylinder pressure			Hand brake[a]
	Percent of gross rail load		Maximum percent of light weight	Minimum percent of gross rail load
	Min.	Max.		
Conventional body-mounted brake rigging or truck-mounted brake rigging using levers to transmit brake cylinder force to the brake shoes				
Cars equipped with cast iron brake shoes	13	20	53	13
Cars equipped with high-friction composition brake shoes	6.5	10	30	11
Direct-acting brake cylinders not using levers to transmit brake cylinder force to the brake shoes				
Cars equipped with cast iron brake shoes				
Cars equipped with high-friction composition brake shoes	6.5	10	33	11
Cabooses[b]				
Cabooses equipped with cast iron brake shoes			35–45	
Cabooses equipped with high-friction composition brake shoes			18–23	

[a] Hand brake force applied at the horizontal hand brake chain with AAR certified or AAR approved hand brake.
[b] Effective for cabooses ordered new after July 1, 1982, hand brake ratios for cabooses to the same as lightweight ratios for cabooses.
Note: Above braking ratios also apply to cars equipped with empty and load brake equipment.
From *Marks' Standard Handbook for Mechanical Engineers*, 10th ed.

where S = stopping distance (ft)
V_1 = initial speed when brake applied (mph)
V_2 = speed at time t_1
W_n = weight on which braking ratio B_n is based (lb) (see the table below for values of W_n for freight cars); (for passenger cars and locomotives, W_n is based on empty or ready-to-run weight)
B_n = braking ratio (total brake shoe force at stated brake cylinder (psi), divided by W_n)
P_n = brake cylinder pressure on which B_n is based, usually 50 psi
P_a = full brake cylinder pressure, t_1 to stop
e = overall rigging and cylinder efficiency, decimal
f = typical friction of brake shoes (see below)
R = total resistance, mechanical plus aerodynamic and curve resistance (lb/ton)
G = grade in decimal, + upgrade, − downgrade
t_1 = equivalent instantaneous application time, s

Capacity, ton	W_n, 1,000 lb
50	177
70	220
100	263
125	315

Equivalent instantaneous application time is that time on a curve of average brake cylinder buildup versus time for a train or car where the area above the buildup curve is equal to the area below the curve. A straight-line buildup curve starting at zero time would have a t_1 of half the total buildup time.

The friction coefficient f varies with the speed; it is usually lower at high speed. To a lesser extent, it varies with brake shoe force and with the material of the wheel and shoe. For stops below 60 mph (97 km/hr), a conservative figure for a high-friction composition brake shoe on steel wheels is approximately

$$ef = 0.30 \qquad (23.9)$$

In the case of high-phosphorus iron shoes, this figure must be reduced by approximately 50 percent.

P_n is based on 50 psi (345 kPa) air pressure in the cylinder; 80 psi is a typical value for the brake pipe pressure of a fully charged right train. This will give a 50-psi (345 kPa) brake cylinder pressure during a full-service application on AB equipment and a 60-psi (kPa) brake cylinder pressure with an emergency application.

To prevent wheel sliding, $F_R \leq \phi W$, where F_R = retarding force at wheel rims resisting rotation of any pair of connected wheels (lb), ϕ = coefficient of wheel-rail adhesion or friction (a decimal), and W = weight upon a pair of wheels (lb). Actual or adhesive weight on wheels when the vehicle is in motion is affected by weight transfer (force transmitted to the trucks and axles by the inertia of the car body through the truck center plates), center of gravity, and vertical oscillation of body weight upon truck springs. The value of ϕ varies with speed as shown in Figure 23.19.

The relationship between the required coefficient ϕ of wheel-rail adhesion to prevent wheel sliding and rate of retardation A in mph/sec may be expressed by $A = 21.95\phi_1$.

There has been some encouraging work to develop an electric brake that may eventually make obsolete the present pneumatic brake systems.

Test Devices.   Special devices have been developed for testing brake components and cars on a repair track.

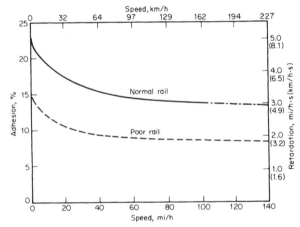

FIGURE 23.19 Typical wheel-rail adhesion. Track has jointed rails. (*Source:* Air Brake Association.)

End-of-Train Devices.   To eliminate the requirement for a caboose crew car at the end of the train, special electronic devices have been developed to transmit the end-of-train brake-pipe pressure to the locomotive operator by telemetry.

23.4 *PASSENGER EQUIPMENT*

During the past two decades most main-line or long-haul passenger service in North America has become a function of government agencies, i.e., Amtrak in the United States and Via Rail in Canada. Equipment for intraurban service is divided into three major categories: commuter rail, heavy rail rapid transit, and light rail transit, depending upon the characteristics of the service. Commuter rail equipment operates on conventional railroad rights-of-way, usually intermixed with other long-haul passenger and freight traffic. Heavy rail rapid transit operates on a dedicated right-of-way, which is commonly in subways or on elevated structures. Light rail transit (LRT) utilizing light rail vehicles (LRV) has evolved from the older trolley or streetcar concepts and may operate in any combination of surface, subway, or elevated dedicated rights-of-way, semireserved surface rights-of-way with grade crossing, or intermixed with other traffic on surface streets. In a few cases LRT shares the trackage with freight operations, which are separated to comply with FRA regulations.

Since main-line and commuter rail equipment operates over conventional railroad rights-of-way, the structural design is heavy to provide the FRA-required crashworthiness of vehicles in the event of accidents (collisions) with other trains or at grade-crossings with automotive vehicles. Although HRT vehicles are designed to stringent structural criteria, the requirements are somewhat less severe since the operation is separate from freight equipment and there are usually no highway grade crossings. Minimum weight is particularly important for transit vehicles so that demanding schedules over lines with close station spacing can be met with minimum energy consumption.

23.4.1 Main-Line Passenger Equipment

There are four primary passenger train sets in operation across the world; diesel locomotive hauled, diesel multiple unit (DMU), electric locomotive hauled, and electric MU (EMU). In recent years the design of main-line passenger equipment has been controlled by specifications provided by APTA and the operating authority. Most of the newer cars provided for Amtrak have had stainless steel structural components. These cars have been designed to be locomotive-hauled and to use a separate 480-V three-phase power supply for heating, ventilation, air conditioning, food car services, and other control and auxiliary power requirements. Figure 23.20 shows an Amtrak coach car of the Superliner class.

Trucks for passenger equipment are designed to provide a superior ride as compared to freight-car trucks. As a result, passenger trucks include a form of "primary" suspension to isolate the wheel set from the frame of the truck. A softer "secondary" suspension is provided to isolate the truck from the car body. In most cases, the primary suspension uses either coil springs, elliptical springs, or elastomeric components. The secondary suspension generally utilizes either large coil rings or pneumatic springs with special leveling valves to control the height of the car body. Hydraulic dampers are also applied to improve the vertical and lateral ride quality.

23.4.2 Commuter Rail Passenger Equipment

Commuter rail equipment can be either locomotive-hauled or self-propelled (Figure 23.21). Some locomotive-hauled equipment is arranged for "push-pull" service. This configuration

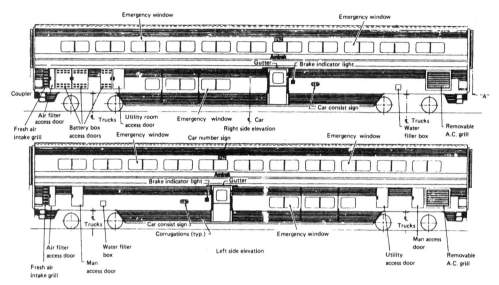

FIGURE 23.20 Main-line passenger car.

permits the train to be operated with the locomotive either pushing or pulling the train. For push-pull service some of the passenger cars must be equipped with cabs to allow the engineer to operate the train from the end opposite the locomotive during the push operation. All cars must have control trainlines to connect the lead (cab) car to the trailing locomotive. Locomotive-hauled commuter rail cars use AAR-type H couplers. Most self-propelled (single- or multiple-unit) cars use other automatic coupler designs that can be mechanically coupled to an AAR coupler with an adaptor.

23.4.3 Heavy Rail Rapid Transit Equipment

This equipment is used on traditional subway/elevated properties in such cities as Boston, New York, Philadelphia (Figure 23.22), and Chicago in a semiautomatic mode of operation over dedicated rights-of-way that are constrained by limiting civil features. State-of-the-art subway-elevated properties include such cities as Washington, Atlanta, Miami, and San Francisco, where the equipment provides highly automated modes of operation on rights-of-way with generous civil alignments.

The cars can operate bidirectionally in multiple with as many as 12 or more cars controlled from the leading cab. They are electrically propelled, usually from a dc third rail which makes contact with a shoe insulated from and supported by the frame of the truck. Occasionally, roof-mounted pantographs are used. Voltages range from 600 to 1,500 V dc.

The cars range from 48 to 75 ft (14.6 to 22.9 m) over the anticlimbers, the longer cars being used on the newer properties. Passenger seating varies from 40 to 80 seats per car, depending upon length and the local policy (or preference) regarding seated to standee ratio. Older properties require negotiation of curves as sharp as 50 ft (15.2 m) minimum radius with speeds up to only 50 mph (80 km/hr) on tangent track, while newer properties usually have no less than 125 ft (38.1 m) minimum radius curves with speeds up to 75 mph (120 km/hr) on tangent track.

All North American properties operate on standard-gage track, with the exception of the 5 ft 6 in. (1.7 m) San Francisco Bay Area Rapid Transit (BART), the 4 ft 10⅞ in. (1.5 m)

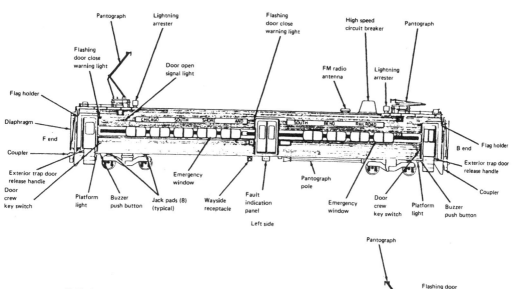

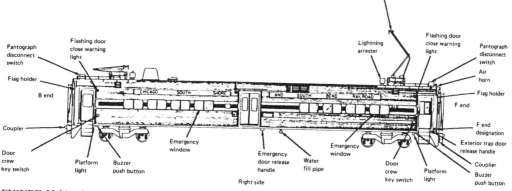

FIGURE 23.21 Commuter rail car.

Toronto Transit Subways, and the 5 ft 2½ in. (1.6 m) Philadelphia Southeastern Pennsylvania Transportation Authority (SEPTA) Market-Frankford line. Grades seldom exceed 3 percent, and 1.5 to 2.0 percent is the desired maximum.

Typically, newer properties require maximum acceleration rates of between 2.5 and 3.0 mph/sec (4.0–4.8 km/hr/sec) as nearly independent of passenger loads as possible from 0 to approximately 20 mph (32 km/hr). Depending upon the selection of motors and gearing, this rate falls off as speed is increased, but rates can generally be controlled at a variety of levels between zero and maximum.

Deceleration is typically accomplished by a blended dynamic and electro-pneumatic friction tread or disk brake, although a few properties use an electrically controlled hydraulic friction brake. Either of these systems usually provides a maximum braking rate of between 3.0 and 3.5 mph/sec (4.8–5.6 km/hr/sec) and is made as independent of passenger loads as possible by a load-weighing system that adjusts braking effort to suit passenger loads. Some employ regerative braking to supplement the other brake systems.

Dynamic braking is generally used as the primary stopping mode and is effective from maximum speed down to 10 mph (16 km/hr) with friction braking supplementation as the characteristic dynamic fade occurs. The friction brakes provide the final stopping forces. Emergency braking rates depend upon line constraints, car subsystems, and other factors,

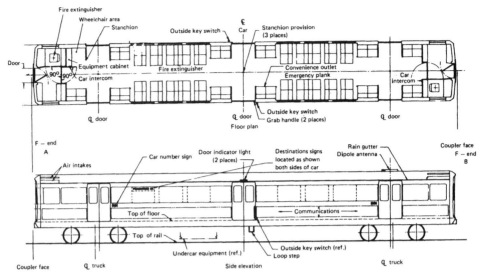

FIGURE 23.22   Full-scale transit car.

but generally rely on the maximum retardation force that can be provided by the dynamic and friction brakes within the limits of available wheel-to-rail adhesion.

Acceleration and braking on modern properties are usually controlled by a single master controller handle that has power positions in one direction and a coasting or neutral center position and braking positions in the opposite direction. A few properties use foot-pedal control with a "deadman" pedal operated by the left foot, a brake pedal operated by the right foot, and an accelerator (power) pedal also operated by the right foot. In either case, the number of positions depends upon property policy and control subsystems on the car. Control elements include motor-current sensors and some or all of the following: speed sensors, rate sensors, and load-weighing sensors. Signals from these sensors are processed by an electronic control unit (ECU), which provides control functions to the propulsion and braking systems. The propulsion systems currently include pilot-motor-operated cams, which actuate switches or electronically controlled unit switches to control resistance steps, or chopper or inverter systems, which electronically provide the desired voltages and currents at the motors.

In some applications, dynamic braking utilizes the traction motors as generators to dissipate energy through the on-board resistors also used in acceleration. It is expected that regenerative braking will become more common since energy can be returned to the line for use by other cars. Theoretically, 35 to 50 percent of the energy can be returned, but the present state of the art is limited to a practical 20 percent on properties with large numbers of cars on close headways.

Car bodies are made of welded stainless steel or low-alloy high-tensile (LAHT) steel of a design that carries structural loads. Earlier problems with aluminum, primarily electrolytic action among dissimilar metals and welding techniques, have been resolved and aluminum is also used on a significant number of new cars.

Trucks may be cast or fabricated steel with frames and journal bearings either inside or outside the wheels. Axles are carried in roller-bearing journals connected to the frames so as to be able to move vertically against a variety of types of primary spring restraint. Metal springs or airbags are used as the secondary suspension between the trucks and the car body. Most wheels are solid-cast or wrought steel. Resilient wheels have been tested but are not in general service on heavy rail transit equipment.

All heavy rail rapid transit systems use high-level loading platforms to speed passenger flow. This adds to the civil construction costs but is necessary to achieve the required level of service.

23.4.4 Light Rail Transit Equipment

The cars are called light rail vehicles (LRVs; Figure 23.23 shows typical LRVs) and are used on a few remaining streetcar systems such as those in Boston, Philadelphia, and Toronto on city streets and on state-of-the-art subway, surface, and elevated systems such as those in Edmonton, Calgary, San Diego, and Portland, Oregon, in semiautomated modes over partially or wholly reserved rights-of-way.

As a practical matter, LRVs' track, signal systems, and power systems utilize the same subsystems as heavy rail rapid transit and are not lighter in a physical sense.

LRVs are designed to operate bidirectionally in multiple with up to four cars controlled from the leading cab. LRVs are electrically propelled from an overhead contact wire, often of catenary design, which makes contact with a pantograph or pole on the roof. For reasons of wayside safety, third-rail pickup is not used. Voltages range from 550 to 750 V dc.

The cars range from 60 to 65 ft (18.3 to 19.8 m) over the anticlimbers for single cars and from 70 to 90 ft (21.3 to 27.4 m) for articulated cars. The choice is determined by passenger volumes and civil constraints.

Articulated cars have been used in railroad and transit applications since the 1920s. They have found favor in light rail applications because of their ability to increase passenger loads in a longer car that can negotiate relatively tight-radius curves. These cars require the additional mechanical complexity of the articulated connection and a third truck between two car-body sections.

Passenger seating varies from 50 to 80 or more seats per car, depending upon length, placement of seats in the articulation, and the policy regarding seated/standee ratio. Older systems require negotiation of curves down to 30 ft (9.1 m) minimum radius with speeds up to only 40 mph (65 km/hr) on tangent track, while newer systems usually have no less

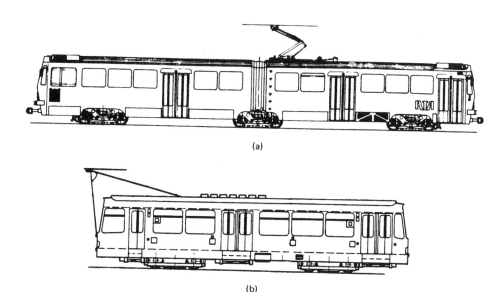

(a)

(b)

FIGURE 23.23 Light-rail vehicle (LRV). (*a*) Articulated; (*b*) nonarticulated.

than 75 ft (22.9 m) minimum radius curves with speeds up to 65 mph (104 km/hr) on tangent track.

The newer properties all use standard-gage track; however, existing older systems include 4 ft 10⅞ in. (1,495 mm) and 5 ft 2½ in. (1,587 mm) gages. Grades have reached 12 percent, but 6 percent is now considered the maximum and 5 percent is preferred.

Typically, newer properties require maximum acceleration rates of between 3.0 to 3.5 mph/sec (4.8–5.6 km/hr/sec) as nearly independent of passenger load as possible from 0 to approximately 20 mph (32 km/hr). Depending upon the selection of motors and gearing, this rate falls off as speed is increased, but the rates can generally be controlled at a variety of levels.

Unlike most heavy rail rapid transit cars, LRVs incorporate three braking modes: dynamic, friction, and track brake, which typically provide maximum service braking at between 3.0 and 3.5 mph/sec (4.8–5.6 km/hr/sec) and 6.0 mph/sec (9.6 km/hr/sec) maximum emergency braking rates. The dynamic and friction brakes are usually blended, but a variety of techniques exist. The track brake is intended to be used primarily for emergency conditions and may or may not be controlled with the other braking systems.

The friction brakes are almost exclusively disc brakes since LRVs use resilient wheels that can be damaged by tread-brake heat buildup. No single consistent pattern exists for the actuation mechanism. All-electric, all-pneumatic, electro-pneumatic, electro-hydraulic, and electro-pneumatic over hydraulic are in common use.

Dynamic braking is generally used as the primary braking mode and is effective from maximum speed down to about 5 mph (8 km/hr) with friction braking supplementation as the characteristic dynamic fade occurs.

As with heavy rail rapid transit, the emergency braking rates depend upon line constraints, car-control subsystems selected, and other factors, but the use of track brakes means that higher braking rates can be achieved because the wheel-to-rail adhesion is not the limiting factor.

Acceleration and braking on modern properties are usually controlled by a single master controller handle that has power positions in one direction and a coasting or neutral center position and braking positions in the opposite direction. A few properties use foot-pedal control with a deadman pedal operated by the left foot, a brake pedal operated by the right foot, and an accelerator (power) pedal also operated by the right foot. In either case, the number of positions depends upon property policy and control subsystems on the car.

Control elements include motor-current sensors and some or all of the following: speed sensors, rate sensors, and load-weighing sensors. Signals from these sensors are processed by an electronic control unit (ECU) that provides control functions to the propulsion and braking systems.

The propulsion systems currently include pilot-motor-operated cams that actuate switches or electronically controlled unit switches to control resistance steps. Chopper and inverter systems that electronically provide the desired voltages and currents at the motors are also used.

Most modern LRVs are equipped with two powered trucks. In two-section articulated designs (Figure 23.23a), the third (center) truck may be left unpowered but usually has friction and track brake capability. Some European designs use three powered trucks, but the additional cost and complexity have not been found necessary in North America.

Unlike heavy rail rapid transit, there are three major dc-motor configurations in use: the traditional series-wound motors used in bimotor trucks, the European-derived monomotor, and a hybrid monomotor with a separately excited field—the last in chopper-control version only.

The bimotor designs are rated between 100 and 125 shaft hp per motor at between 300 and 750 V dc, depending upon line voltage and series or series-parallel control schemes (electronic or electromechanical control). The monomotor designs are rated between 225 and 250 shaft hp per motor at between 300 and 750 V dc (electronic or electromechanical control).

The motors, gear units (right angle or parallel), and axles are joined variously through flexible couplings. In the case of the monomotor, it is supported in the center of the truck, and right-angle gearboxes are mounted on either end of the motor. Commonly, the axle goes through the gearbox and connection is made with a flexible coupling arrangement. Electronic inverter control drives with ac motors have been applied in recent conversions and new equipment.

Dynamic braking is achieved in the same manner as with heavy rail rapid transit.

Unlike heavy rail rapid transit, LRV bodies are usually made only of welded LAHT steel and are of a load-bearing design. Because of the semireserved right-of-way, the risk of collision damage with automotive vehicles is greater than with heavy rail rapid transit and the LAHT steel has been found to be easier to repair than stainless steel or aluminum. Although LAHT steel requires painting, this can be an asset since the painting can be performed in a highly decorative manner pleasing to the public and appropriate to themes desired by the cities.

Trucks may be cast or fabricated steel with either inside or outside frames. Axles are carried in roller-bearing journals, which are usually resiliently coupled to the frames with elastomeric springs as a primary suspension. Both vertical and a limited amount of horizontal movement occur. Since tight curve radii are common, the frames are usually connected to concentric circular ball-bearing rings, which in turn are connected to the car body. Air bags, solid elastomeric springs, or metal springs are used as a secondary suspension. Resilient wheels are used on virtually all LRVs.

Newer LRVs have low-level loading doors and steps, which minimize station platform costs.

23.5 VEHICLE-TRACK INTERACTION

23.5.1 Train Resistance

The resistance to a train in motion along the track is of prime interest, as it is reflected directly in locomotive energy requirements. This resistance is expressed in terms of pounds per ton of train weight. *Gross train resistance* is that force that must be overcome by the locomotives at the driving-wheel-rail interface. *Trailing train resistance* must be overcome at the rear drawbar of the locomotive.

There are two classes of resistance that must be overcome: *inherent* and *incidental*. Inherent resistance includes the rolling resistance of bearings and wheels and aerodynamic resistance due to motion through still air. It may be considered equal to the force necessary to maintain motion at constant speed on level tangent track in still air. Incidental resistance includes resistance due to grade, curvature, wind, and vehicle dynamics.

23.5.2 Inherent Resistance

Of the elements of inherent resistance, at low speeds rolling resistance is dominant but at high speeds aerodynamic resistance is the predominant factor. Attempts to differentiate and evaluate the various elements through the speed range are a continuing part of industry research programs to reduce train resistance. At very high speeds, the effect of air resistance can be approximated as an aid to studies in its reduction by means of cowling and fairing. The residence of a car moving in still air on straight, level track increases parabolically with speed. Because the aerodynamic resistance is independent of car weight, the resistance in pounds per ton decreases as the weight of the car increases. The total resistance in pounds per ton of a 100-ton car is much less than twice as great as that of a 50-ton car under similar conditions. With known conditions of speed and car weight, inherent resistance can be predicted with reasonable accuracy. Knowledge of track conditions will permit further refining of the estimate, but for very rough track or extremely cold ambient temperatures, generous allowances must be made. Under such conditions, normal resistance may be doubled. A formula proposed by Davis (1926) and revised by Tuthill (1948) has been used extensively for inherent freight-train resistances at speeds up to 40 mph:

$$R = 1.3W + 29_n + 0.045WV + 0.0005AV^2 \qquad (23.10)$$

where R = train resistance (lb/car)
W = weight per car (tons)
V = speed (mph)
n = total number of axles
A = cross-sectional area (ft²)

With freight-train speeds of 50 to 70 mph (80 to 112 km/hr), it has been found that actual resistance values fall considerably below calculations based on the above formula. Several modifications of the Davis equation have been developed for more specific applications. All of these equations apply to cars trailing locomotives.

1. Davis equation as modified by Tuthill (1948):

$$R = 1.3W + 29_n + 0.045WV + 0.045V^2 \qquad (23.11)$$

Note: In the Totten modification, the equation is augmented by a matrix of coefficients when the velocity exceeds 40 mph.

2. Davis equation as modified by the Canadian National Railway:

$$R = 0.6W + 20_n + 0.01WV + 0.07V^2 \qquad (23.12)$$

3. Davis equation as modified by the Canadian National Railway and Erie-Lackawanna Railroad for trailers and containers on flat cars:

$$R = 0.6W + 20_n + 0.01WV + 0.2V^2 \qquad (23.13)$$

Other modifications of the Davis equation have been developed for passenger cars by Totten (1937). These formulas are for passenger cars pulled by a locomotive and do not include head-end air resistance.

1. Davis equations modified by Totten for streamlined passenger cars:

$$R = 1.3W + 29_n + 0.045WV + [0.00005 + 0.060725(L/100)^{0.88}]V^2 \quad (23.14)$$

2. Davis equations modified by Totten for non-streamlined passenger cars

$$R = 1.3W + 29_n + 0.045WV + [0.00005 + 0.1085(L/100)^{0.7}]V^2 \qquad (23.15)$$

where L = car length in ft

23.5.3 Aerodynamic and Wind Resistance

Wind-tunnel testing has indicated a significant effect on freight train resistance resulting from vehicle spacing, open tops of hopper and gondola cars, open boxcar doors, vertical side reinforcements on railway cars and intermodal trailers, and protruding appurtenances on cars. These effects can cause significant increases in train resistance at higher speeds. For example, the spacing of intermodal trailers or containers greater than approximately 6 ft can result in a new frontal area to be considered in determining train resistance. Frontal or cornering ambient wind conditions can also have an adverse effect on train resistance which is increased with discontinuities along the length of the train.

23.5.4 Curve Resistance

Train resistance due to track curvature varies with speed and degree of curvature. The behavior of rail vehicles in curve negotiation is the subject of several ongoing AAR studies. Lubrication of the rail gage face or wheel flanges has become common practice for reducing friction and the resulting wheel and rail wear. Recent studies indicate that flange and/or gage face lubrication can significantly reduce train resistance on tangent track as well (Allen, "Conference on the Economics and Performance of Freight Car Trucks," October 1983). In addition, a variety of special trucks (wheel assemblies) that reduce curve resistance by allowing axles to steer toward a radial position in curves have been developed. For general estimates of car resistance and locomotive hauling capacity on dry (unlubricated) rail with conventional trucks, speed and gage relief may be ignored and a figure of 0.8 lb/ton per degree of curvature used.

23.5.5 Grade Resistance

Grade resistance depends only on the angle of ascent or descent and relates only to the gravitational forces acting on the vehicle. It equates to 20 lb/ton for each "percent of grade" or 0.379 lb/ton for each foot per mile rise.

23.5.6 Acceleration Resistance

The force (tractive effort) required to accelerate the train is the sum of the forces required for linear acceleration and that required for rotational acceleration of the wheels about their axle centers. A linear acceleration of 1 mph/sec (km/hr/sec) is produced by a force of 91.1 lb/ton. The rotary acceleration requirement adds 6 to 12 percent, so that the total is nearly 100 lb/ton (the figure commonly used) for each mile per hour per second. If greater accuracy is required, the following expression is used:

$$R_a = A(91.05W + 36.36n) \qquad (23.16)$$

where R_a = the total accelerating force (lb)
A = acceleration (mph/sec)
W = weight of train (tons)
n = number of axles

23.5.7 Acceleration and Distance

If in a distance of S ft the speed of a car or train changes from V_1, to V_2 mph, the force required to produce acceleration (or deceleration if the speed is reduced) is

$$R_a = 74(V^2/2 - V^2/1)/S \qquad (23.17)$$

The coefficient, 74, corresponds to the use of 100 lb/ton. This formula is useful in the calculation of the energy required to climb a grade with the assistance of stored energy. In any train-resistance calculation or analysis, assumptions with regard to acceleration will generally submerge all other variables; e.g., an acceleration of 0.1 mph/sec (0.16 km/hr/

sec) requires more tractive force than that required to overcome inherent resistance for any car at moderate speeds.

23.5.8 Starting Resistance

Most railway cars are equipped with roller bearings requiring a starting force of 5 or 6 lb/ton.

23.5.9 Vehicle Suspension Design

The primary consideration in the design of the vehicle suspension system is to isolate track input forces from the vehicle car body and lading. In addition, there are a few specific areas of instability that railway suspension systems must address. See AAR *Manual of Standards and Recommended Practice,* Sec. C-II-M-1001, Chap. XI.

Harmonic roll is the tendency of a freight car with a high center of gravity to rotate about its longitudinal axis (parallel to the track). This instability is excited by passing over staggered low rail joints at a speed that causes the frequency of the input for each joint to match the natural roll frequency of the car. Unfortunately, in many car designs this occurs for loaded cars at 12 to 18 mph (19.2 to 28.8 km/hr), a common speed for trains moving in yards or on branch lines where tracks are not well maintained. Many freight operations avoid continuous operation in this speed range. This adverse behavior is more noticeable in cars with truck centers approximately the same as the rail length. The effect of harmonic roll can be mitigated by improved track surface and by damping in the truck suspension.

Pitch and *bounce* are the tendencies of the vehicle to either translate vertically up and down (bounce), or rotate (pitch) about a horizontal axis perpendicular to the centerline of track. This response is also excited by low track joints and can be relieved by increased truck damping.

Yaw is the tendency of the car to rotate about its axis vertical to the centerline of track. Yaw responses are usually related to *Truck hunting*. Truck hunting is an instability inherent in the design of the truck and dependent on the stiffness parameters of the truck and on wheel conicity (tread profile). The instability is observed as a "parallelogramming" of the truck components at a frequency of 2 to 3 Hz, causing the car body to yaw or translate laterally. This response is excited by the effect of the natural frequency of the gravitational stiffness of the wheel set when the speed of the vehicle approaches the kinematic velocity of the wheel set. This problem is discussed in analytic work available from the AAR Research and Test Department.

23.5.10 Superelevation

As a train passes around a curve, there is a tendency for the cars to tip toward the outside of the curve in response to centrifugal force acting on the center of gravity of the car body (Figure 23.24*a*). To compensate for this effect, the outside rail is superelevated, or raised, relative to the inside rail (Figure 23.24*b*). The amount of superelevation for a particular curve is based upon the radius of the curve and the operating speed of the train. The balance or equilibrium speed for a given curve is that speed at which the centrifugal force on the car matches the component of gravity force resulting from the superelevation between the amount required for high-speed trains and the amount required for slower-operating trains. The FRA allows a railroad to operate with 3 in. of unbalance, or at the speed at which equilibrium would exist if the superelevation were 3 in. greater. The maximum superelevation is usually 6 in. but may be lower if freight operation is used exclusively.

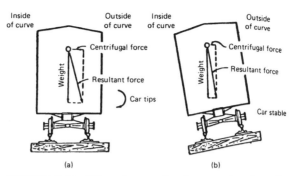

FIGURE 23.24   Effect of super relevation on center of gravity of car body.

23.5.11 Longitudinal Train Action

Longitudinal train (slack) action is a term associated with the dynamic action between individual cars in a train. An example would be the effect of starting a long train in which the couplers between each car had been compressed (i.e., bunched up). As the locomotive begins to pull the train, the slack between the locomotive and the first car must be traversed before the first car begins to accelerate. Next the slack between the first and second car must traversed before the second car beings to accelerate, and so on. Before the last car in a long train begins to move, the locomotive and the moving cars may be traveling at a rate of several miles per hour This effect can result in coupler forces sufficient to cause the train to break in two.

Longitudinal train action is also induced by serial braking, undulating grades, or braking on varying grades. The Track-Train Dynamics Program has published guidelines titled *Track-Train Dynamics to Improve Freight Train Performance* that explain the causes of undesirable train action and how to minimize the effects. Analysis of the forces developed by longitudinal train action requires the application of the Davis equation to represent the resistance of each vehicle based upon its velocity and location on a grade or curve. Also, the longitudinal stiffness of each car and the tractive effort of the locomotive must be considered in equations that model the kinematic response of each vehicle in the train. Computer programs are available from the AAR to assist in the analysis of longitudinal train action.

23.6 REFERENCES*

Allen, R. A. 1983. In Conference on the Economics and Performance of Freight Car Trucks, October.

American Public Transportation Association (APTA). *Standard for the Design and Construction of PassengerRailroad Rolling Stock.* APTA, Washington, DC. www.apta.com.

American Railway Engineering and Maintenance-of-Way Association (AREMA). *Manual for Railway Engineering.* Landover, MD: AREMA. www.arema.com.

American Society of Mechanical Engineers (ASME). *Proceedings.* New York: ASME. www.asme.org.

Association of American Railroads (AAR). *Field Manual of the AAR Interchange Rules.* AAR, Washington, DC. www.aar.com.

---

*In addition to the works cited here, see also the publications of the Association of American Railroads (AAR) Research and Test Department.

———. *Manual of Recommended Standards and Recommended Practices.* Mechanical Division, AAR, Washington, DC. www.aar.com.

———. *Office Manual of the AAR Interchange Rules.* AAR, Washington, DC. www.aar.com.

———. Multiple Unit Pneumatic Brake Equipment for Locomotives. Standard S-5529. AAR, Washington, DC. www.aar.com.

Avallone, E. A., and T. Baumeister III, eds. 1996. *Marks' Standard Handbook for Mechanical Engineers,* 10th ed. New York: McGraw-Hill.

Davis, W. J. 1926. "Tractive Resistance of Electric Locomotive and Cars." *General Electric Review* 29: 685–708.

Kratville, W. M. 1997. *The Car and Locomotive Cyclopedia of American Practice.* Omaha: Simmons-Boardman.

The Official Railway Equipment Register—Freight Connections and Freight Cars Operated by Railroads and Private Car Companies of North America. East Windsor, NJ: Commonwealth Business Media.

Railway Line Clearances. East Windsor, NJ: Commonwealth Business Media.

Totten, A. I. 1937. "Resistance of Light Weight Passenger Trains." *Railway Age* 103 (July).

Tuthill, J. K. 1948. "High Speed Freight Train Resistance." *University of Illinois Engineering Bulletin* 376.

CHAPTER 24
RAILWAY TRACK DESIGN

Ernest T. Selig
Department of Civil Engineering, (Emeritus),
University of Massachusetts, and Ernest T. Selig, Inc.,
Hadley, Massachusetts

24.1 INTRODUCTION

Railway track as it is considered in this chapter consists of a superstructure and a substructure. The superstructure is composed of steel rails fastened to crossties. The rails are designed to support and guide flanged steel wheels through their prescribed position in space. The superstructure is placed on a substructure. The substructure is composed of a layered system of materials known as ballast, subballast and subgrade. These track components are illustrated in Figure 24.1 (Selig and Waters 1994).

Special track components are added to perform needed functions. These include switches to divert trains from one track to another, crossing diamonds to permit one track to cross another, level grade crossings to permit roads to cross over the train track at the same elevation, types of warning devices such as hot bearing detectors, and dragging equipment detectors. The last example is rail attached directly to a reinforced concrete slab in a tunnel or to a bridge structure. The substructure incorporates a drainage system to remove water from the track.

The track design needs to consider soil and rock conditions, weather conditions (precipitation, temperature), traffic requirements (wheel loads, total annual tonnage), and maintenance costs for the designed track.

This chapter will provide a listing of design functions, a description of design methods, and references to sources of information on design details.

24.2 FUNCTIONS OF TRACK COMPONENTS

For each of the main track components the functions are the following (Selig and Waters 1994; Agarwal 1998; Hay 1982).

24.2.1 Rails

1. Guide the flanged wheels in the vertical, lateral, and longitudinal directions.
2. Provide a smooth running surface.

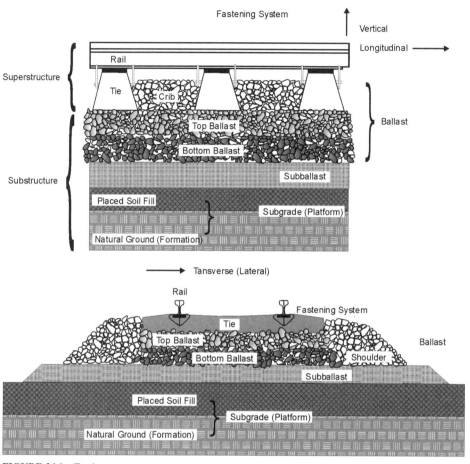

FIGURE 24.1   Track structure components.

3. Transfer wheel loads to spaced ties without large deflection.

4. Resist tension failure from longitudinal tensile force caused by rail temperature reduction.

5. Help resist buckling from longitudinal compression force caused by rail temperature increase.

6. Resist fatigue cracking from repeated wheel loads.

7. Provide strong bolted or welded joints.

8. Limit rail impact by maintaining track geometry and truing wheels to limit "false flange" wear on wheel and rail and reduce wheel defects such as engine burns, corrugations, and flat spots.

9. Permit tracks to cross over each other and permit trains to switch from one track to another.

24.2.2 Fastening Systems

1. Restrain the rail in the vertical, longitudinal, and transverse directions.

2. Resist overturning of rail from lateral wheel force.

3. Connect sections of rail to permit safe and smooth train operation.

4. Create a canted (inclined) surface to provide proper wheel/rail contact—wood ties.

5. Spread the rail seat force over a larger part of the tie surface to reduce tie damage—wood ties.

6. Provide resiliency under the vertical wheel load—concrete ties.

7. Reduce tie abrasion at rail seat—concrete ties.

8. Provide damping of the high frequency wheel-induced vibrations—concrete ties.

24.2.3 Crossties

1. Transfer the vertical wheel load from the rail through the rail seat to the bottom of the ties to provide an acceptable level or stress for the ties and ballast.

2. Hold the fastening system so that it can restrain the rails at the proper vertical, lateral, and longitudinal position and maintain the required gage.

3. Provide a canted (inclined) surface for proper wheel/rail contact—concrete ties.

24.2.4 Ballast

1. Restrain the ties against vertical, lateral, and longitudinal forces from the rails.

2. Reduce the pressure from the tie-bearing area to a level that is acceptable for the underlying materials.

3. Provide the ability to adjust track geometry by rearranging the ballast particles by tamping and lining.

4. Assist in drainage of water from the track.

5. Provide sufficient voids between particles to allow an efficient migration of unwanted fine particles from the ballast section.

6. Provide some resiliency to the track to decrease rail, rail component, and wheel wear.

24.2.5 Subballast

1. Maintain separation between the ballast and subgrade particles.

2. Prevent attrition of the hard subgrade surface by the ballast.

3. Reduce pressure from the ballast to values that can be sustained by the subgrade without adverse effects.

4. Intercept water from the ballast and direct it to the track drainage system.

5. Provide drainage of water flowing upward from the subgrade.

6. Provide some insulation to the subgrade to prevent freezing.

7. Provide some resiliency to the track.

24.2.6 Subgrade

1. Provide a stable platform on which to construct the track.

2. Limit progressive settlement from repeated traffic loading.

3. Limit consolidation settlement.

4. Prevent massive slope failure.

5. Restrict swelling or shrinking from water content change.

24.2.7 Drainage

Drainage is the single most important factor governing the performance of track substructure. A properly functioning drainage system provides the following:

1. Intersects the water seeping up from the subgrade

2. Diverts the surface water flowing toward the track

3. Removes water falling onto the track

4. Carry off stone dust, sand, and other debris that otherwise could foul the track.

24.3 TRACK FORCES

The forces applied to the track are vertical, lateral (parallel to the ties), and longitudinal (parallel to the rails). These forces are affected by train travel speed. An important point to recognize is that track design involves many force repetitions, not just one load, as in building foundation design. Thus, allowable forces must be considerably smaller than the failure forces in a single load test in order to perform satisfactorily over a period of time. This has long been recognized in the field of material fatigue.

24.3.1 Vertical

The main vertical force is the repetitive downward action of the wheel load. In addition, this wheel/rail interaction produces a corresponding lift-up force on the ties away from the wheel load points.

The nominal vertical wheel force, also called the static force, is equal to the gross weight of the railway car divided by the number of wheels. This force ranges from about 12,000 lb (53 kN) for light rail passenger cars to 39,000 lb (174 kN) for heavy freight cars. However, due to the inertial effects of the moving train traveling over varying geometry on a track with defects, the vertical load can be much greater or much smaller than the nominal value. The largest forces are produced from the impact of the wheel on the rail, which is accompanied by vibration that often can be felt at considerable distance from the track. The passage of trains over the track causes the initial track geometry to deform. The inertia of the moving train causes the vertical wheel force to vary above and below the nominal wheel load. The vertical impact dynamic load has two components, a short-duration larger force and a longer-duration smaller force. The first is expected to be more harmful to the rails and ties, while the second does more damage to the ballast and track geometry.

The major factors affecting the magnitude of the dynamic vertical forces are:

• Nominal wheel load
• Train speed
• Wheel diameter
• Vehicle unsprung mass
• Smoothness of the rail and wheel surfaces

- Track geometry
- Track modulus or vertical track stiffness

The traditional approach for representing the geometry-driven dynamic wheel load is to multiply the nominal wheel load by an impact factor that is greater than 1.

The impact factor recommended by AREMA (2003, chap. 16) is a function of the train travel speed and the wheel diameter. The actual maximum dynamic force on any track may be much different from that obtained from this approach.

These dynamic wheel forces increase the rate of track component deterioration. For example, studies in Europe (Esveld 1989) have indicated that the maintenance cost ratio is represented by the force ratio to the power n. Values of n from European work are 1 for rail fatigue, 3 for track geometry deterioration, and 3.5 for rail surface defects.

24.3.2 Lateral

One type of lateral force applied to the rail is the wheel force transmitted through friction between the wheel and top of the rail and by the wheel flange acting against the inside face of the rail head, particularly on curves. Another lateral force is the rail buckling resistance force.

The design lateral wheel force depends upon a number of factors, including:

- Vehicle speed
- Track geometry
- Elevation difference between the two rails at the same cross section
- Transverse hunting movement due to the train-track dynamics

As the train speed increases, the lateral force outward on the outside rail of curves increases, and simultaneously the lateral force on the inside rail decreases.

When the field joints in the track are removed and the rails are welded, long lengths of track result, which are subjected to considerable changes in longitudinal stress due to rail temperature changes. Temperature decrease relative to the temperature at the time of welding causes tensile force parallel to the rail, which can result in rail pull-apart, while temperature increase causes compressive force, which can result in track buckling.

The wheels on a railway vehicle are tapered so that the diameter decreases from the inside to the outside. This helps center the wheels on straight track and compensates in part for the greater distance that the outer wheels travel on a curve. Because the wheels are fixed to the axle, both wheels must turn together. Thus, wheel slip is required to the extent that the circumference of the wheels does not compensate for the difference in the inside and outside rail length in a curve. The vehicle wheels on a fixed axle may take a longer time than the curve spiral allows to become oriented to the curvature of the rail. This causes additional stress on the gage or wheel climb on the high rail, and rollover possibilities on the low rail.

A flange on the inside face of each wheel limits the lateral movement of the wheels to the distance between the wheel flange and the inside face (gage) of the rail. The combination of the wheel and railhead shapes, the inclination of the rails (cant), and the difference in elevation between the inside and outside rail in a curve serve to guide the train wheels along the intended alignment.

A spiral is a transition between the tangent track (straight) and the full radius curve. In a curve the outer rail is at a higher elevation than the inner rail so that the resultant of the weight of the train, the load balance in the car, and the centrifugal force is designed to be perpendicular to the track. This is a function of train speed, so if the train is operating above the design speed there will be a transverse force causing the flanges of the wheel to move against the outer rail. In cases of lower than design speed, the transverse force will move against the inner rail.

To achieve the desired alignment, the geometric components of a track are tangents, spirals (transition between straight track and constant radius curves), and constant radius curves in the horizontal plane, and gradients and vertical curves in the vertical plane.

24.3.3 Longitudinal

Sources of longitudinal rail forces are:

• Speed
• Locomotive traction
• Locomotive and car braking
• Expansion and contraction of the rails from temperature change
• Track grade
• Special track, i.e., turnouts, at grade crossings, rail crossings, dragging equipment, hot bearing detectors

The ratio of lateral to vertical force (L/V) is also important because it can cause loss of alignment and even track buckling.

24.4 TRACK SYSTEM CHARACTERISTICS

Track system performance is a function of the composite response of the track components under the action of the train loads. Two response characteristics are important to consider in track design: vertical track stiffness and lateral track stability.

24.4.1 Stiffness

The vertical response model is illustrated in Figure 24.2. The vertical track stiffness k is the vertical load on one rail divided by the vertical deflection at the loaded point. The track modulus u is the composite vertical support stiffness of the rails consisting of the fasteners, ties, ballast, subballast, and subgrade. Track modulus cannot be measured directly, but is calculated from track stiffness using the bending stiffness of the rail (Selig and Waters 1994).

Comparable models for horizontal (longitudinal and lateral) track response are not available.

The subgrade is the component that has the greatest influence on the track stiffness. It is also the component with the most variation and the most uncertainty about its property values. The track should be designed to have a stiffness that is neither too high nor too low. Both extremes will shorten the life of the components.

24.4.2 Lateral Stability

Track buckling is a result of increasing longitudinal rail force from increasing temperature. Buckling occurs in the lateral direction. Because this is the least stable direction, the lateral resistance is greatest directly under the wheel loads because the weight of the train increases

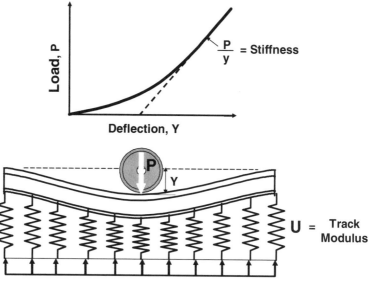

FIGURE 24.2 Track stiffness.

the lateral restraint provided by the ties. Buckling is most likely to occur on track disturbed by maintenance, when high temperatures develop in the rails, or where rail has been changed at a low temperature so that a large temperature increase can occur and is not adjusted when possible to the laying temperature. Additional information is available from Kish and Samavedam (1991).

Increasing resistance to buckling can be achieved by such means as increasing the ballast shoulder width. Tests have shown that little or no benefit is derived beyond 18 inches (460 millimeters), and most railroads only go to 12 to 16 inches (300 to 400 millimeters). In addition, increasing buckling resistance can be achieved through: (1) dynamic compacting of the crib and shoulder ballast after maintenance or out-of-face tamping, (2) using concrete ties, which are heavier than timber ties, and (3) avoiding tamping in extremely hot weather. The problem is that tamping involves lifting the track and rearranging the ballast to fill the space below the ties. This puts the track in its least stable condition until dynamic compaction is completed and/or further traffic, at decreased speeds, stabilizes the ballast.

Compaction and development of residual stresses in the ballast around the ties from train traffic greatly increases the lateral track stability. When the geometry deteriorates to the point when the track has to be lifted and tamped, the lateral track stability diminishes considerably. Consequently, it is common practice to run trains slowly at first after tamping until the lateral stability has increased from the train traffic. *Dynamic stabilizing* is used to reduce speed restrictions to only the first train over a stabilized area, whereas 10 heavy trains are commonly used for track that has not been stabilized.

24.5 RAILS

The goal of rail cross-section design should be to select the shape, size, material, and rail hardness to provide the most economical rail with the required strength and ductility for wear durability (Agarwal 1998; Marich, Mutton, and Tew 1991a). The rails are produced in the factory by hot-rolling the steel, and then cooled and straightened to form finished lengths of 80 feet (24 meters). Fixed welding plants, where the customer can send rail to be welded, are located all over North America. Mobile welding plants will come to a railroad, and in-track-flash butt welding machines will go on their track and weld jointed rails together. Most railroads still use thermite welds for rail change-outs and other short jobs in the field. These lengths can be electric flash-butt-welded (EFW) to form longer lengths, typically 1,440 feet (439 meters), that can be transported to the field for installation.

The rail may be heat-treated at the mill for increased hardness to increase the wearing resistance. Varying the steel composition can also give increased hardness with a slight loss of ductility. Overall rail life increases as the rail weight increases largely due to the ability to maintain profile over the thicker head with increased maintenance grinding. Rail properties to consider when choosing the size and type of rail are:

1. Wearability

2. Hardness

3. Ductility

4. Manufacture defects in the rail material

5. Rail straightness

The rail sections are connected in the field by either bolted joints or welding. The locations of the bolted joints are high-maintenance areas because of the impact of the wheels passing the rail end gap at the center of the joints. The combination of the impact load and reduced rail stiffness of the supporting joint bars causes greater stress on the fasteners, ties, ballast, and subgrade. This in turn causes fastener looseness, plate cutting (wood ties), pad deterioration and concrete tie seat abrasion (concrete ties) and more rapid track settlement and geometry deterioration. This problem can be reduced by eliminating the joints by field welding. This approach is preferred on high-speed and heavy axle load (HAL) lines.

The spacing of the rails is standardized at a value termed *gage*. The gage is the distance between the inside faces of the rail at $\frac{5}{8}$ in. (14 mm) below the top of the railhead. Gauge limitations and excesses are defined by the U.S. Federal Railroad Administration (FRA) Track Safety Standards (FRA 1998) based on the class of track. For example, in North America the gage is 4 ft $8\frac{1}{2}$ in. (1435 mm) with various tolerances as defined by the FRA. A range of standardized rail cross-sections is available for the designer to choose from, for example, AREMA Manual of Engineering (AREMA 2003, chap. 4). Properties for a light rail and a heavy rail are given in Table 24.1 as an indication of representative values.

TABLE 24.1 Example of Rail and Tie Properties

Rail weight	Size	Mass lb/yd (kg/m)	Area in.² (cm²)	Moment of inertia horizontal axis in.⁴ (cm⁴)
Light	115 RE	115 (56.9)	11 (72.6)	66 (2730)
Heavy	136 RE	136 (67.6)	13 (86.1)	95 (3950)
Tie material	Base width in. (mm)	Mass lb (kg)	Length ft (mm)	Spacing in. (mm)
Wood	9 (229)	200 (91)	8.5 (2590)	19.5 (495)
Concrete	11 (286)	800 (360)	8.6 (2629)	14 (610)

24.6 TIES

Concrete ties are both prestressed and reinforced. AREMA (2003, chap. 10) recommends that the average ballast pressure at the base of concrete ties not exceed 85 psi (590 kPa) for high-quality abrasion-resistant ballast. AREMA (2003, chap. 16) recommends a limit of 65 psi (450 kPa). The pressure would be reduced for lower quality ballast. The limits should also consider the durability of the tie bottoms, but this is not a part of the AREMA consideration for the maximum pressure. The reason it should be considered is because the abrasion resistance of the cement in concrete is less than the resistance of much of the rock currently used for ballast.

Timber ties are both hardwood and softwood. Natural wood used for timber ties will have defects such as knots, splits, checks, and shakes. Specifications exist (for example, see AREMA (2003, chap. 3)) for the maximum size of allowed defects. Wood ties are treated with a preservative for protection against deterioration from bacteria, insects, and fire. The performance of the track can help project the need for maintenance. The upper curve represents a low-quality track because of the rapid increase in roughness with time.

Timber is the most common material used for the manufacture of crossties in North America. Next most common is prestressed/reinforced concrete. A small percentage of crossties are manufactured from other materials such as steel and cast iron. Some new materials are being introduced such as glued wood laminates and recycled plastic. Representative values of concrete and timber tie properties are given in Table 24.1. Concrete ties, at approximately 800 pounds (360 kilograms), are heavier than timber ties, at approximately 200 pounds (91 kilograms), so concrete resists track buckling better but timber ties are easier to handle. Concrete ties generally have more secure fastening systems than timber, so concrete holds the rails better. Timber ties have natural resiliency, whereas concrete ties require compressible pads for some resiliency.

One design consideration is the bending stresses in the ties caused by the wheel loads moving over the tie. These bending stresses are significantly affected by the pressure distribution of the ballast along the bottom of the ties. When the track is lifted and tamped to smooth the geometry, a gap is produced under the middle of the ties to cause the tie-bearing area to be limited to the tamp zone on both sides of the rail. With traffic the track will settle, eventually bringing the center of the tie into contact with the ballast. This condition, called center binding, will greatly increase the bending stresses in the ties. Because it is not possible to predict the exact pressure distribution along the bottom of the tie, some simplified assumptions are commonly used (Marich, Mutton, and Tew 1991b).

Analysis of the pressure distribution using the vertical track model shows that the distribution is dependent upon the flexibility of the tie, the contact-bearing area between the ballast and the tie, the compactness of the ballast under the tie, and stiffness of the subballast, ballast, and subgrade. The peak values of pressure distribution are also a function of the tie base dimensions and the center-to-center tie spacing.

The vertical track model, if available, can determine the maximum rail seat loads. The rail seat loads can also be estimated using the beam on elastic foundation model, which requires an estimate of the track modulus and the bending characteristics of the rail (Selig and Waters 1994; Hay 1982).

The maximum tie-bending moments depend on all the same factors as the maximum contact pressure at the base of the tie. The maximum bending moments are at the rail seat and the center of the tie length. Because of the difficulty of accurately predicting the maximum bending moments and bending stresses as well as contact pressures, it is quite common to select the ties based on experience in track, in the environment, and under the loading for the design conditions. In this regard, both the maximum magnitude and the number of repetitions must characterize the load. The latter affects the durability requirement (e.g., ballast crushing, tie abrasion, and fatigue life).

To complete the design based on flexural considerations, the maximum allowable bending stress needs to be determined. This can be calculated from the maximum bending moment. The maximum bending stress depends upon the material from which the tie is constructed. The use of tie plates between the rails and the ties will spread the rail seat load and therefore further reduce the bending moment.

AREMA (2003, chap. 3) indicates that an estimate of the maximum allowable bending stress in the timber ties under repeated wheel loading could be taken as 28 percent of the modulus of rupture in bending test to failure. Accordingly, values were reported as 1 ksi (7 MPa) for softwood and 1.3 ksi (9 MPa) for hardwood. Similar methods have been developed for reinforced concrete and steel ties. The designer must ensure that the appropriate values of the maximum bending stress under repeated loading are obtained for the ties being considered.

A less conservative assumption that the ballast bearing pressure under concrete ties is uniformly distributed may be appropriate for these materials because their properties are better controlled and the ties are more expensive.

24.7 FASTENING SYSTEMS

A rail joint is desired to be as stiff and strong as the rail itself. Welded joints approach this condition, but bolted joints do not (Talbot 1933). The bolted joints have bars that fit within the railhead and base fillets and against the web of the rail. Holes are drilled through the rail concentric with the holes in the joint bars. Insulation can be placed between the bars and the rail to electrically isolate the signal circuits.

For timber ties, steel tie plates are fit to the rail base with a ¼-inch (6.4-millimeter) shoulder on either side of the base for line-spiking and up to 18 inches (460 millimeters) in length secured to the timber tie with 6-inch (150-millimeter) cut spikes or screw spikes. The plates are placed between the rail and the tie surface to spread the rail seat load. The plates work together with a variety of rail anchors or other elastic fasteners for horizontal and lateral restraint of the track. The tie plates provide the cant to the rail. Tie plates are available in a variety of sizes (AREMA 2003, chap. 5).

Tie plates come with four holes. Cut spikes are used in a variety of patterns depending on the geometry of the track, tangent, or curve. At least one spike is driven through the hole immediately outside the shoulder on each side of the rail for line stability. Again, at least one spike is driven through the opposing corner at each end of the plate to hold the plate in position on the tie. The primary function of the spike is to hold the plate to the tie and provide line stability for the rail as it fits within the shoulders of the tie plate. The heads of the spikes are driven down to a ⅛-inch (3-millimeter) height above the top of the rail base, but through time and the natural plate cut that occurs from the flexing and uplift of the rail, spikes are lifted up somewhat while still providing stable line. Specifications and proper spike driving patterns are given in AREMA (2003, chap. 5).

For concrete ties, spring clips, known as elastic fasteners, are connected to the top of tie and press down on the top of the rail base and against the web of the rail. These same fasteners come secured within the concrete tie pour, which makes fewer parts and more stable holding ability. There are several other elastic fastener designs on the market, all of which are designed to secure the rail in vertical, lateral, and longitudinal directions. A pad is placed between the bottom of the tie and the rail seat to provide resiliency and insulation for signal conductivity and help prevent rail seat and tie abrasion due to the L/V forces resulting from the load.

24.8 BALLAST

The ballast component of track shown in Figure 24.1 is subdivided into four zones:

1. Crib—material between the ties
2. Shoulder—material beyond the tie ends down to the bottom of the ballast layer
3. Top ballast—upper portion of supporting ballast layer that is disturbed by tamping
4. Bottom ballast—lower portion of supporting ballast layer, which is not disturbed by tamping and generally is the more fouled

The mechanical properties of the ballast layer result from a combination of the physical properties of the individual particles and the degree of fouling together with the in-place density of the assembly of particles. Fouling refers to the small particles that infiltrate the space between the ballast particles. The main factors producing the density are tamping, and train traffic. Tamping involves the insertion of tools into the ballast to rearrange the particles to fill the space under the ties resulting from track lift. This leaves the ballast in a relatively loose state. The many load cycles from the trains produce most of the compaction. Most of the major freight and passenger railroads use a combination of measured dynamic stabilizing and restricted speed over disturbed track.

24.8.1 Ballast Particle Requirements

Index tests have been established for characterizing the ballast properties. These cover mechanical strength, shape, water absorption, specific gravity, surface texture, particle size, and breakdown from cycles of freezing and thawing. Each railroad has a set of ballast specifications that stipulates limits for the values from the index tests. These specifications are known to be insufficient for ensuring satisfactory performance. One major limitation is that no correlation exists between index tests so that trade-offs can be established between two ballast materials, which differ in the values of the individual index properties. Petrographic analysis of the parent rock is a valuable aid in assessing ballast suitability. This information should be supplemented by observations of performance in track.

For ballast to perform its intended functions (section 24.2.4), it should consist of the following characteristics:

1. Most particles in the 0.8- to 2.5-inch (19- to 64-millimeter) size range
2. Produced by crushing hard, durable rock
3. Planar fractured faces intersecting at sharp corners (to give angularity)
4. Particles with a maximum ratio of 3:1 for largest to smallest dimensions
5. Rough surface texture preferred
6. Low water absorption

The relatively small range of particle size limits segregation when the particles are rearranged during tamping and also minimizes the loosening effect of the tamping process. The large size of particles creates large void spaces to permit migration and holding of fine particles while delaying the time when the ballast performance is significantly degraded by accumulation of the fine particles. The condition also permits rapid flow of water through the ballast layer. The fractured faces, with rough texture and high angularity together with restrictions on the amount of flat and elongated particles, provide high strength and stability for the assembly of ballast particles. Hard, durable rock is needed to reduce the particle breakage caused by the repeated train loading and from the tamping action during maintenance to smooth the track geometry. The low water absorption indicates stronger particles and reduces breakdown from water expansion during freezing temperatures. The stress-reduction function depends of the above characteristics and the layer thickness.

The optimum choice of particle characteristics depends on the magnitude of axle load and number of repetitions, together with the cost to deliver the ballast. Lower-quality ballast can be more cost-effective than higher-quality ballast on low-traffic lines, especially when the lower-quality ballast is closer.

24.8.2 Ballast Fouling

Over a period of time in track the ballast gradation typically becomes broader and finer than the initial condition because the larger ballast particles will break into smaller particles and additional smaller particles from a variety of sources will infiltrate the voids between the ballast particles. This process is known as fouling. Five categories of fouling material have been identified:

1. Particles entering from the surface such as wind-blown sand or coal fines falling out of cars
2. Products of wood or concrete tie wear
3. Breakage and abrasion of the ballast particles by train loading
4. Particles migrating upward from the granular layer underlying the ballast
5. Migration of particles from the subgrade

The main causes of ballast fouling should be identified so that proper steps can be taken to reduce the rate of fouling. The most frequent cause of ballast fouling is ballast breakdown, but there are individual situations in which each one of the other categories dominates. Geotextiles (filter fabrics) generally have not been found to be useful in solving ballast fouling problems (Selig and Waters 1994). A proper subballast layer is the best cure for fouling from the underlying granular layer and from the subgrade. When subgrade is the source of fouling material one of two main mechanisms usually is present: (1) abrasion of the subgrade surface by ballast particles in contact with the subgrade, or (2) crack pumping resulting from hydraulic erosion of water-filled cracks in the subgrade subjected to repeated train loading.

Most commonly observed fouling problems are restrictions of drainage and interference with track maintenance. However, as the voids become completely filled with fines, the ballast begins to take on the characteristics of the fines, with the ballast particles acting as filler. Soaked fines represent mud and hence the ballast becomes soft and deformable. When wet fouled ballast becomes frozen the resiliency is lost. When the fines become dried (but still moist), they act as a stiff binding agent for the crushed rock particles. This also causes loss of resiliency. All of these conditions prevent proper track surfacing.

The term *cemented ballast* is frequently used in the railroad industry to represent a condition in which the ballast particles are bound together. Although this term has not been officially defined, in most cases it appears to be used to represent dried fouled ballast. However the word *cemented* has led to the notion that a chemical bonding is involved, such as in the case of portland cement, a derivative of limestone rock. This is one of the reasons given by the railroad industry for preferring not to use limestone ballast.

A thorough examination of cemented ballast conditions is needed to determine the cause. Such a study could very well show that chemical bonding as in cement is not the main bonding mechanism in cemented ballast because it is not normally the type of bonding in dried fouled ballast.

24.8.3 Petrographic Analysis

The value of petrographic analysis as a means of assessing and/or predicting behavior of an aggregate has been long recognized by the concrete industry. Techniques for evaluating aggregate for use in concrete and for examining hardened concrete have been established by ASTM in standards C295 for aggregate and C856 for hardened concrete. The purposes of this petrographic examination are

1. To determine the physical and chemical properties of the material that will have a bearing on the quality of the material for the intended purpose
2. To describe and classify the constituents of the sample
3. To determine the relative amounts of the constituents of the sample, which is essential for the proper evaluation of the sample, especially where the properties of the constituents vary significantly

The value of the petrographic analysis depends to a large extent on the ability of the petrographer to correlate data provided on the source and proposed use of the material with the findings of the petrographic examination.

Petrographic analysis is very helpful in the selection of a suitable quarry for ballast and also for prediction of the shape and character of the components of future ballast breakdown (i.e., the fines generated by breakage and abrasion of the ballast). An experienced petrographer can estimate the relative mechanical properties, including hardness, shape, type of fracture, and durability in track.

24.8.4 Ballast Compaction

At the time when surfacing is required to correct track geometry irregularities, the ballast is in a dense state, particularly beneath the tie-bearing areas. When the rail and tie are raised to the desired elevation, tamping tines are inserted in the crib next to the rail to displace the ballast into the voids under the tie that were created by the raise. This tamping process disturbs the compact state of the ballast and leaves it loosened (Figure 24.3). The more fouled the ballast is, and the greater the raise, the looser the ballast is after tamping.

The loosened ballast beneath the tie results in renewed settlement as the traffic, or track equipment made for this purpose, stabilizes the ballast. The loosened crib ballast results in a significant reduction in lateral buckling resistance of the rail in the unloaded state. Crib surface vibratory compactors can be used to compact the crib ballast immediately after tamping, but not the ballast under the tie.

Traffic is the most effective means of compacting ballast under the tie, but this takes time and results in nonuniform track settlement. Traffic also causes crib ballast to stabilize.

In addition to increasing density, there is evidence that both the traffic and the crib compactor produce residual horizontal stresses in the ballast. These residual stresses may be one of the most important factors influencing ballast performance in track. Fouled ballast in the crib will reduce densification of crib ballast by traffic after tamping and hence diminish any tendency for the development of lateral residual stress against the sides of the ties.

At present no adequate correlation exists between ballast index tests as a group and ballast performance in track. What is needed to select ballast is a method that takes into account the effect of differences in ballast gradation and particle composition and, in addition, simulates field service conditions such as ballast depth, subgrade characteristics, traffic loading, and track parameters.

24.8.5 Ballast Layer Thickness

The thickness of the ballast layer beneath the ties should generally be the minimum that is required for the ballast to perform its intended functions. Typically this will be 9 to 18 inches

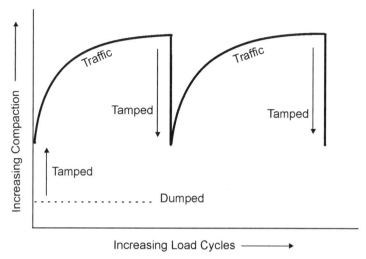

FIGURE 24.3   Effect of tamping and traffic on ballast compaction.

(230 to 460 millimeters). The minimum subballast layer thickness beneath the ballast should be 6 inches (150 millimeters). Thus, the minimum granular layer thickness beneath the ties would be 15 to 24 inches (380 to 600 millimeters). A check then needs to be made to determine whether this is enough granular layer thickness to prevent overstressing the subgrade (Li and Selig 1998). If not, then the granular layer thickness needs to be increased. Increasing the subballast thickness, not the ballast thickness, should do this. A thick ballast layer, when unconfined, can be prone to settlement from particle movement, particularly in the presence of vibration.

24.8.6 Track Stiffness

Clean ballast provides only a small amount of the total track stiffness, unless the subgrade is rock or a thick concrete slab.

24.9 *SUBBALLAST*

Subballast is a very important but not adequately recognized track substructure component. It serves some of the same functions as ballast, but it also has some unique functions. Like ballast, it provides thermal insulation and resiliency. Also, like ballast, it is a structural material that further reduces pressure on the subgrade. Water falling on the track will enter the ballast and flow downward to the subballast. Generally the water will enter the subballast, and this will be diverted to the sides of the track by the subgrade surface unless the subgrade is similar to the subballast in permeability. If the subgrade surface is not properly sloped for drainage, then some of the water will be retained under the track, where it will weaken the substructure. The remainder of the water will be shed by the subballast. The subballast also allows water flowing up from the subgrade to discharge without eroding the subgrade soil, which ballast cannot do because ballast is too coarse.

One unique function of subballast is to prevent the fine subgrade particles from migrating into the ballast voids, whether from repeated train loading or from flow of water. Finally, a particularly important function of subballast is to prevent the ballast particles from coming into contact with the subgrade soil, where they abrade or grind away the subgrade surface (subgrade attrition). The fine soil particles produced then mix with water and form mud that squeezes into the ballast voids. This is mainly a problem with hard subgrade. Inserting a 6-inch (150-millimeter) layer of properly graded and durable subballast between the ballast and the subgrade solves the problem.

Crushing durable rock to form sand- and gravel-sized particles forms subballast. Suitable subballast materials are commonly found in natural deposits. The aggregate must be resistant to breakdown from cycles of freezing and from repeated cycles of train loading. However, the durability requirements are not as severe as for ballast because the subballast particles are smaller and the stresses are lower. The finest particles less than 0.003 inch (0.075 millimeter) must be nonplastic. Depending on the permeability requirements for drainage, the fine particles must not exceed 5 to 10 percent by weight and may be less than 0 to 2 percent in some cases. Subballast materials satisfying these requirements, when placed and compacted, will satisfy the structural requirements of pressure reduction to the subgrade and resiliency.

Subballast must be well drained so that it is not saturated during repeated train loading, particularly dynamic loading from impact forces. Saturated and undrained subballast materials can deform significantly during train loading and even liquefy.

To provide separation between the ballast and subgrade particles, the subballast gradation must satisfy the requirement in Figure 24.4. This provides that the finest subballast particles are smaller than the largest subgrade particles, and correspondingly the largest subballast particles must be larger than smallest ballast particles. There must also not be gaps in the gradation of the subballast. Subballast satisfying these requirements will also be satisfactory for preventing attrition on the hard subgrade surface by the ballast.

The subballast must also be permeable enough to serve the drainage functions discussed more fully in section 24.11.

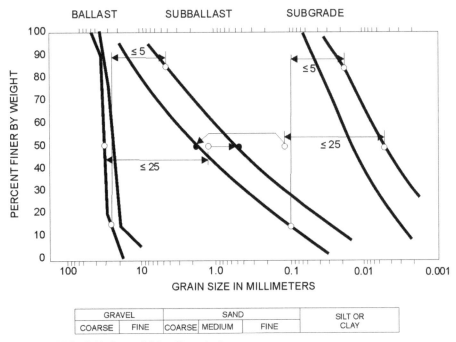

FIGURE 24.4 Subballast satisfying filter criteria.

To provide freezing protection of the subgrade and not contribute to frost heave-thaw softening problems, the subballast must be well drained and contain less than 5 percent fines (silt and clay-sized particles).

24.10 SUBGRADE

The subgrade is the platform upon which the track structure is constructed. Its main function is to provide a stable foundation for the subballast and ballast layers. The influence of the traffic-induced stresses extends downward as much as 5 meters below the bottom of the ties. This is considerably beyond the depth of the ballast and subballast. Hence, the subgrade is a very important substructure component that has a significant influence on track performance and maintenance. For example, subgrade is a major component of the superstructure support resiliency and hence contributes substantially to the elastic deflection of the rail under wheel loading. In addition, the subgrade stiffness magnitude is believed to influence ballast, rail, and sleeper deterioration. Subgrade also is a source of rail differential settlement due to movement of the subgrade from various causes.

The various types of subgrade problems are listed in Table 24.2 together with their causes and features.

The subgrade may be divided into two categories (Figure 24.1): (1) natural ground (formation) and (2) placed soil (fill). Anything other than soils existing locally are generally uneconomical to use for the subgrade. Existing ground should be used without disturbance as much as possible. However, techniques are available to improve soil formations in place if they are inadequate. Often some of the formation must be removed to construct the track at its required elevation, which is below the existing ground surface. This puts the track in a cut with the ground surface sloping downward toward the track. If the excavation intercepts the water table, slope erosion or failure can occur, carrying soil onto the track. Placed fill is used either to replace the upper portion of unsuitable existing ground or to raise the subgrade surface to the required elevation for the superstructure and the remainder of the substructure.

The subgrade is often the weakest substructure layer. Thus, a combined ballast and subballast thickness is required that will reduce the pressure on the subgrade to a level that produces an acceptably small deformation from the repeated train loading for the desired design life. The design method must consider the type and strength of the subgrade soil, the distribution of dynamic wheel loads and number of repetitions, and the substructure layer resilient moduli. The various levels of wheel loads and their corresponding numbers of cycles are converted to a single representative design load and equivalent number of cycles.

The design method is described in detail in Li and Selig (1998). Two analyses are performed:

• Limiting the cumulative plastic strain on the subgrade surface accompanying progressive shear to an acceptably small value over the life of the track to restrict the subgrade squeeze

• Limiting the cumulative plastic settlement of the compressible subgrade to prevent forming a "bathtub" depression in the subgrade that traps water

24.10.1 Limiting Strain Method

The steps in the method are:

1. Select the allowable strain limit based on design life desired.

2. Determine the equivalent number of design load cycles.

3. Estimate the static compressive strength of the subgrade soil.

4. From Figure 24.5, calculate the allowable cyclic stress.

TABLE 24.2 Major Subgrade Problems and Features

Type	Causes	Features
(1) Progressive shear failure	• Repeated over stressing subgrade • Fine-grained soils • High water content	• Squeezing of subgrade into ballast shoulder • Heaves in crib and/or shoulder • depression under ties trapping water
(2) Excessive plastic deformation (ballast pocket)	• Repeated loading of subgrade • Soft or loose soils	• Differential subgrade settlement • ballast pockets
(3) Subgrade attrition with mud pumping	• Repeated loading of subgrade stiff hard soil • Contact between ballast and subgrade • Clay-rich rocks or soils • Water presence	• Muddy ballast • Inadequate subballast
(4) Softening subgrade surface under subballast	• Dispersive clay • Water accumulation at soil surface • Repeated train loading	• Reduces sliding resistance of subgrade soil surface
(5) Liquefaction	• Repeated dynamic loading • Saturated silt and fine sand • Loose state	• Large track settlement • More severe with vibration • Can happen in subballast
(6) Massive shear failure (slope stability)	• Weight of train, track, and subgrade • Inadequate soil strength	• Steep embankment and cut slope • Often triggered by increase in water content
(7) Consolidation settlement	• Embankment weight • Saturated fine-grained soils	• Increased static soil stress as from weight of newly constructed embankment • Fill settles over time
(8) Frost action (heave and softening)	• Periodic freezing temperature • Free water • Frost-susceptible soils	• Occurs in winter/spring period • Heave from ice lens formation • Weakens from excess water content on thawing • Rough track surface
(9) Swelling/shrinkage	• Highly plastic or expansive soils • Changing moisture content	• Rough track surface • Soil expands as water content increases • Soil changes as water content decreases
(10) Slope erosion	• Surface and subsurface water movement • Wind	• Soil washed or blown away • Flow onto track fouls ballast • Flows away from track can undermine track
(11) Slope collapse	• Water inundation of very loose soil deposits	• Ground settlement
(12) Sliding of side hill fills	• Fills placed across hillsides • Inadequate sliding resistance • Water seeping out of hill or down slope is major factor	• Transverse movement of track

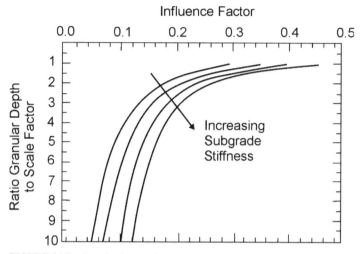

FIGURE 24.5   Granular layer thickness chart.

5. Calculate the strain influence factor.

6. From Figure 24.6, determine the required minimum granular layer thickness.

24.10.2   Limiting Deformation Method

The steps in the method are:

1. Select the allowable deformation limit.

2. Determine the equivalent number of design load cycles (same as method 1).

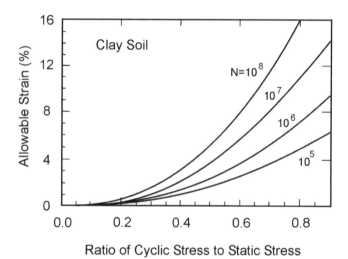

FIGURE 24.6   Allowable deviator stress.

3. Estimate the static compressive strength of the subgrade soil (same as method 1).

4. Calculate the deformation influence factor.

5. From a figure similar to Figure 24.6, determine the required minimum granular layer thickness.

Sections 24.8 and 24.9 indicate the minimum ballast and subballast thickness. The design granular layer thickness is the greater of (1) the combined ballast and subballast layer minimum thickness and (2) the thickness required for the subgrade protection.

Some alternatives are available to reduce the settlement in cases where the subgrade is overstressed. Reducing the wheel load and the total annual traffic million gross tons are assumed to be unacceptable alternatives in most cases. Two general categories may be designated: (1) with the track in place and (2) with the track removed (this would include new construction).

24.10.3 Track in Place

With the track remaining in place, there are several options for improving the subgrade performance:

1. Improve drainage.

2. Increase the granular layer thickness.

3. Add tensile reinforcement in the subballast (such as geogrid, geoweb).

4. Use special on-track machines that can renew substructure conditions while working beneath the track.

24.10.4 Track Removed

With the track removed or not yet placed, additional options become available:

1. Install proper drainage.

2. Remove soft soils and replace with compacted suitable soils.

3. Place impermeable membrane to prevent water from coming into contact with the soil.

4. Lime or cement stabilization of soils by mechanical mixing.

5. Insert hot mix asphalt concrete layer on subgrade.

Clearly, designing and installing the substructure to meet the track needs without the track in place is easier and more effective. Obviously many reasons exist why this is not done.

AREMA (2003, chap. 16) recommends a method for determining ballast depth to limit wheel load-induced stress on top of subgrade so that the subgrade will not fail. The method involves determining the depth for a given track modulus and wheel load that results in an allowable pressure of 25 psi. This value applies to all soils. The number of wheel load repetitions is not considered in the AREMA method. This is another major deficiency in this manual. The correct allowable stress at top of subgrade is not constant but depends on the soil conditions and number of wheel load repetitions. The allowable stress on the top of subgrade for good track performance is determined by cumulative deformation (settlement) rather than by bearing capacity. For a mix of traffic the heaviest loads mainly cause the deformation.

The following are a few examples of subgrade remedial treatment methods to fix the problems in Table 24.2:

1. *Grouting:* Some grouts penetrate the voids of the soils and strengthen them or reduce water seepage. Other grouts compact and reinforce the soils to strengthen them or displace the soils to compensate for settlement. Jet grouting mixes cement with soil to form columns of strengthened soil.

2. *Soil mixing:* This is a process in which soil is mixed with augers and paddles to create a mixture of soil and cement based grout. Soil mixing creates a column of strengthened soil for compression and shear reinforcement.

3. *Modification of clay properties with lime:* There are several alternatives: quick lime is placed in boreholes to strengthen the soil; lime is mechanically mixed with soil to form columns of material with increased strength; lime and water mixed to form slurry is injected into clay soil under pressure with the expectation of improving the clay properties. This last is a common but not usually effective treatment with undesirable side effects. It fractures the clay instead of penetrating the voids and also solidifies ballast.

4. *Reconstruction:* Compaction of existing soils in layers at proper water content or substitutions of better soils will give improved subgrade. Chemicals such as cement or quick hydrated lime, mechanically mixed with the soils in layers before compaction, will form a stronger or less reactive soil after compaction. The chemistry of the soils should be checked or tests performed to verify the effectiveness of the treatment, because some combinations can be harmful. All of these methods generally require removal of the track.

5. *Reinforcement:* Various plastic grids, metal strips, or cellular materials placed in the soils give tensile reinforcement. Alternatively, steel reinforcing can be installed in grout-filled boreholes.

6. *Stress reduction:* Increasing the thickness of the ballast and subballast will reduce the pressure on the weaker subgrade caused by the train loading. Contrary to the AREMA engineering manual, the allowable pressure is not constant but varies widely and must be determined in each case for correct design. The correct strength considers the magnitude of the repeated loading from the trains and the number of repetitions. For a given axle load, a high-tonnage line would have a much lower apparent strength than a low-tonnage line. Thus, the high-tonnage line needs a greater ballast/subballast thickness for the same subgrade properties.

24.11   *TRACK DRAINAGE*

Drainage of railway tracks is essential to achieve acceptable track performance. Water in the track substructure originates from three potential sources (Figure 24.7):

1. Precipitation onto the track
2. Surface flow from areas adjacent to the track
3. Groundwater flow

A complete drainage system must include provisions for handling water from all three sources (Heyns 2000).

24.11.1   Drainage of Precipitation Falling on the Track

Precipitation onto the track will enter the ballast, unless the ballast is highly fouled. The water will then flow laterally out of the ballast into the trackside drainage system or enter the subballast. The water entering the subballast will either drain laterally out of the subballast or continue downward onto the subgrade. The ability of water to drain laterally

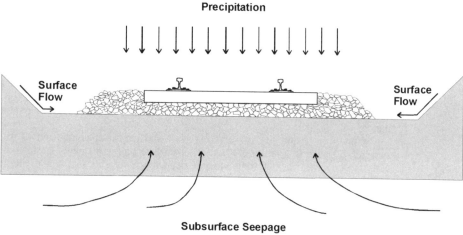

FIGURE 24.7   Sources of water entering the track.

requires that the drainage paths at the edge of the ballast and subballast layers not be blocked. Two conditions need to be met to achieve this requirement: (1) the ballast shoulder and the edge of the subballast must be free-draining and (2) discharged water must be able to flow away from the track. The lowest point in the granular layer system (bottom of subballast) is most critical, assuming low subgrade permeability. Surface ditch drains can collect water from the ballast and subballast. The type of rainstorm event, as well as the permeability of the materials, will affect the amount of water entering the subballast.

The degree of fouling generally controls the form of seepage through the ballast. Under clean ballast condition, the ballast will have a high void ratio and water will drain freely. As fouling of the ballast progresses, voids are filled with fouling material and the permeability and flow velocities are reduced. Due to the reduced flow velocities, the subballast is exposed for a longer time to the precipitation draining out of the ballast, resulting in a higher subballast infiltration rate. When the permeability of the subballast is low, there is less infiltration and more water discharge laterally out of the ballast. Also, the greater the lateral slope of the subballast towards the side, the less subballast infiltration occurs.

In a similar manner, for a given ballast degree of fouling and a given subballast permeability, low-intensity storms of long duration cause more seepage into the subballast than high-intensity storms of short duration. Storms of high intensity and short duration cause a higher water level within the ballast and hence a high-energy head, and so water is discharged quickly out of the ballast. The subballast is therefore exposed to the precipitation for a relatively short period of time. During a storm of low intensity but long duration, however, water build-up in the ballast is less and so the energy head is low. The subballast is therefore exposed to the precipitation for a longer time period.

Particularly difficult is drainage of water from tracks surrounded on one or both sides by other tracks. Not only is the drainage path to the side longer than for a single track, but a suitable drainage path is more difficult to maintain. Cross-drains under the outer tracks or longitudinal drains between tracks may be needed.

24.11.2 Surface Drainage Systems

A proper surface drainage system is necessary to remove surface water in the right-of-way. Sources of this water are seepage out of the ballast and subballast, runoff from the cut slope, and surface runoff from areas adjacent to the track. Open ditches parallel to the track are the most common component of a surface drainage system. In cuts, ditches parallel to the track usually drain water discharged from the cut face as well as lateral discharge out of the tracks.

Ditches on top of cuts should intercept water from drainage basins adjacent to the cut before it reaches the cut slopes and divert it to a drainage inlet structure or to a natural watercourse nearby. This reduces slope erosion problems and also reduces the required capacity of the trackside ditches. Because the cutoff ditches are placed on top of cuts and are usually not visible from the tracks, they are often overlooked during regular track maintenance.

To predict the quantity of surface runoff and lateral discharge out of the tracks that the system will need to handle, the rainfall at the site being evaluated needs to be characterized. This is done in terms of frequency of occurrence of a storm of a particular duration and intensity. Intensity-duration-frequency curves can be developed from available meteorological records for the site. Then an appropriate storm return period needs to be selected. As the return period becomes longer, the maximum storm intensity likely to be encountered will increase. Then the larger the selected return period for design, the smaller the risk of the ditch overflowing and causing damage to the tracks, but the higher the cost of the system to accommodate the larger quantity of water. A design return period of 5 to 10 years is typically appropriate for ditch design.

24.11.3 Considerations for Ditch Design

Ditches parallel to the track are usually unlined and require a high level of maintenance to remove vegetation and sedimentation and to restore the ditch side-slopes where they have eroded. For the ditches to remain functional and to avoid deposition of sedimentation, a longitudinal minimum slope of 0.5 percent is recommended (Heyns 2000). Ditches placed in long cuts or on flat terrain with the track profile less than the recommended minimum ditch slope therefore become deep (and hence far bigger than the required flow capacity) as they drain towards the outlet. As an alternative, the ditch may be lined to allow placement at a grade shallower than 0.5 percent. For example, smooth-lined concrete ditches can typically be placed at a minimum grade of 0.25 percent.

Stabilization against erosion of ditches is necessary for severe hydraulic conditions. Stabilization measures include rigid linings, such as concrete, or flexible linings, such as vegetation or riprap. Rigid linings are impermeable and are useful in flow zones where high shear stress or nonuniform flow conditions exist, such as at a cut-to-fill transition where the ditch outlets onto a high fill. Although rigid linings are nonerodible, they are susceptible to structural failure. The major causes of such failures are underlying soil movement from freeze-thaw cycles, swelling-shrinking cycles, and undermining by water.

24.11.4 Subsurface Drainage Systems

Groundwater flow into the subballast from the subgrade is a problem only in cuts where water can enter the ground from higher elevations. This water needs to be intercepted by subsurface drains to prevent it from weakening the subgrade. However, groundwater also can be a problem in level ground when the water table in the track subgrade is within the zone of influence of the train loading 13 feet (4 meters). Subsurface drains can be used to lower this water table and may also be needed to help drain the subballast. An exploration should be made to estimate the extent and nature of the groundwater in order to design the subsurface drainage system properly.

A subsurface drainage system is an underground means of collecting gravitational or free water from the track substructure. Provided that a proper surface drainage system is present, gravitational or free water in the track substructure comes from both precipitation onto the track and groundwater flow. If a proper surface drainage system is not present, surface flow also could be a source of subsurface water and must be considered in the design of the drainage system.

The flow rate of water out of the subballast to a drainage ditch may not be fast enough to keep the subballast from saturating. Factors causing this include long seepage distance, low permeability of the subballast, and settlement of the subgrade surface causing a depression. In these cases, a subsurface drainage system is appropriate for removing the water trapped within the subballast.

Subsurface drains that run laterally across the track are classified as transverse drains and are commonly located at right angles to the track centerline. If the ground water flow tends to be parallel to the track, transverse drains can be more effective than longitudinal drains in intercepting and/or drawing down the water table. Also, where ballast pockets exist in the subgrade, transverse drains can be an effective way to drain water from the low location in the ballast pocket. Transverse systems usually connect to the longitudinal subsurface system or the surface drainage system, such as a ditch.

In a multitrack system the tracks in the center should have lateral subballast slopes that match (or are higher than) the lateral slopes of the outside tracks to allow water to discharge under the outside tracks. Where the center tracks are lower than the outside tracks, the granular layer thickness of the outside tracks can be increased to allow continuous lateral drainage or a longitudinal drainage system should be placed between the tracks. In either case the seepage distance is long, resulting in slow drainage. An alternative would be to install transverse drains to carry water from the inside tracks under the outside tracks to a discharge point.

In a multitrack system the path for water to flow to the surface drainage system may become very long, even where a proper subballast lateral slope exists. For example, in a four-track system a water particle falling between the center tracks has to drain two track widths; thus, drainage may be inadequate. Therefore, it may always be desirable practice to drain water with a proper drainage system between the tracks.

24.12   MAINTENANCE IMPLICATIONS

The decisions made during design and construction of new track have a major effect on the cost of track maintenance. Special attention should be paid to subballast and subgrade drainage from under the track because they are very difficult to fix after the track is in service. Cutting construction costs on these important components may result in large maintenance cost for years afterwards to compensate for the construction shortcomings.

24.13   ACKNOWLEDGMENTS

Vincent R. Terrill is acknowledged for sharing his extensive railway experience with the writer over many years, and in particular for his willingness to review the manuscript of this chapter and provide many valuable suggestions.

24.14   REFERENCES

Agarwal, M. M. 1998. *Indian Railway Track,* 12th ed. New Delhi: Prabha & Co.

American Railway Engineering and Maintenance-of-Way Association (AREMA). 2003. *Manual for Railway Engineering.* Landover, MD: AREMA.

ASTM C295. "Standard Practice for Petrographic Examination of Aggregates for Concrete." *ASTM Annual Book of Standards,* Section 4, *Construction,* vol. 04.02, *Concrete Mineral Aggregates.*

ASTM C856. "Standard Recommended Practice for Petrographic Examination of Hardened Concrete." *ASTM Annual Book of Standards,* Section 4, *Construction,* vol. 04.02, *Concrete and Mineral Aggregates.*

Esveld, C. 1989. *Modern Railway Track.* Duisburg: MRT-Productions.

Federal Railroad Administration (FRA). 1998. *Track Safety Standards,* Part 213, Subpart A to F, Class of Track 1 to 5 and Subpart G for Class of Track 6 and higher. U.S. Department of Transportation, FRA, Washington, DC.

Hay, W. W. 1982. *Railroad Engineering,* 2nd ed. New York: John Wiley & Sons.

Heyns, F. J. 2000. "Railway Track Drainage Design Techniques." Ph.D. dissertation, University of Massachusetts, Department of Civil and Environmental Engineering, May.

Kish, A., and G. Samavedam. 1991. "Dynamic Buckling of Continuous Welded Rail Track: Theory, Tests, and Safety Concepts." Rail-Lateral Track Stability, 1991—*Transportation Research Record* 1289.

Li, D., and E. T. Selig. 1998. "Method for Railroad Track Foundation Design: Development" and "Method for Railroad Track Foundation Design: Applications." *Journal of Geotechnical and Geo-environmental Engineering, ASCE* 124(4):316–22 and 323–29.

Marich, S., P. J. Mutton, and G. P. Tew. 1991a. *A Review of Track Design Procedures,* vol. 1, *Rails.* Melbourne: BHP Research-Melbourne Labs.

———. 1991b. *A Review of Track Design Procedures,* vol. 2, *Sleepers and Ballast.* Melbourne: BHP Research-Melbourne Labs.

Selig, E. T., and J. M. Waters. 1994. *Track Geotechnology and Substructure Management.* London: Thomas Telford.

Talbot, A. N. 1933. "Sixth Progress Report of the Special Committee on Stresses in Track." *Bulletin* 358.

CHAPTER 25
IMPROVEMENT OF RAILROAD YARD OPERATIONS

Sudhir Kumar
Tranergy Corporation, Bensenville, Illinois

25.1 INTRODUCTION

25.1.1 The Importance of Railroad Yards

Railroad yards play an important role in railroad operations. Yards consist of a large number of tracks grouped for the purpose of disassembling, sorting, and assembling cars in a train. Trains are brought into a receiving yard and sent on to a classification yard where cars are sorted and assembled into new trains, which are then dispatched to their new destinations. Railroads assemble a new train of cars based on the delivery requirements of their customers. (For a general discussion of yards and terminals see Petracek et al. 1997; *Railway Age* 2001; Wong et al. 1978; Christianson et al. 1979; AREMA 2001.)

Figure 25.1 shows a simplified diagram of car movement of which the classification yard process is a major component. Cars arrive in a yard in one of three ways:

1. From over the road trains
2. From an interchange yard where the cars of different railroad companies are interchanged and dispatched as regrouped cars
3. Loaded by a shipper and moved by a switch engine from an industry siding or yard to be dispatched to a destination specified by the shipper

After arriving in a receiving yard, these cars are moved into an adjacent yard called the classification yard. Here they are classified by being moved into different tracks and regrouped to make up various trains. These trains are finally moved into a departure yard and become one of the three types of train—over the road train or an interchange yard or delivered to an industry yard or siding via a switch engine as the final destination of that car.

For a railroad, the yard process is an essential but non-revenue-producing component. It has been estimated that nearly one-fourth of a railroad's expense is yard-related. In a DOT study (Petracek et al. 1997) it was reported that a typical freight car spends an average of 62 percent of its time in either terminal yards or intermediate yards. By comparison, the car spends only 14 percent of its time in line haul operations and, of this, only 6.6 percent in revenue producing service. This means that in that time a car spends only one in 15 days in revenue producing operation and more than 9 days in different yard processes. The im-

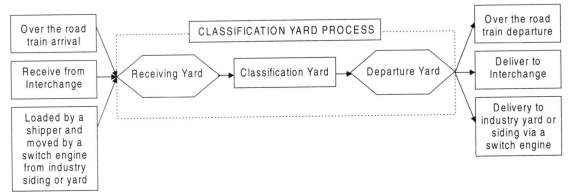

FIGURE 25.1   A simplified car movement diagram showing the classification yard process.

portance of yard operations was noted in an interview with Hunter Harrison (*Railway Age* 2001) regarding the seven steps he proposes for superior railroad service and a better bottom line. Of the seven steps, four concerned yard operations:

1. Minimize car dwell time in yards.
2. Minimize classifications in yards.
3. Use multiple traffic outlets between yards to keep traffic moving.
4. Space trains to support a steady workload flow through various yard processes.

25.1.2 Types of Railroad Yards

The present chapter is focused largely on railroad classification yards. There are several other types of yards, including industrial yards, passenger terminals, TOFC/COFC yards, storage yards, and interchange yards.

Classification Yards.   As stated earlier, classification yards are used primarily to make new trains from the cars of arriving trains.

Industrial Yards.   Industrial yards are used to collect and distribute freight cars to local industries. Cars are received from a classification yard to a local industrial yard for final delivery. In these yards, a resorting of cars takes place and local or industrial switch trains are made up, which are then sent to industrial sidings using switch engines.

Passenger Terminals.   Passenger terminals can be thought of as yards in which the passengers travel to different trains on different tracks. The trains are not sorted and stay on one track as an integral unit.

Storage Yards.   As the name implies, storage yards are used to store rolling stock, which often consists of empty cars. Depending on their demand, surplus cars are kept at such locations.

Interchange Yards.   These yards are used as points of interchange for cars between connecting railroads. Some yards are used exclusively for interchange movements and operations.

TOFC/COFC Yards. These yards handle both the trailer-on-flat-car (TOFC) or container-on-flat-car (COFC) loadings and unloadings. Their design can vary considerably based on the movement of cars and the loading and unloading processes. They also generally carry high-speed cars, which are classified as intermodal traffic cars.

25.1.3 Classification Yards

All classification yards are designed to sort and reassemble cars into new trains. The type of classification yard used depends on the volume of cars to be handled per day. Other factors that influence the type of yard suitable are the character of traffic to be handled and train schedules. An economic study is therefore necessary to determine how to minimize the costs of classifying each car.

Most classification yards provide several supporting railroad operations other than classification. These include locomotive and freight car service and repair, inspection of cars for mechanical defects, and weighing of cars for revenue purposes. The inclusion of all these activities makes the classification yard a major center of railroad operations.

There is no fixed number of cars to be handled by each type of yard. However, the following numbers give an idea of the type of yard to be used, which is based on, among other things, the volume of car traffic that it handles.

1. Hump yards or gravity yards handle the largest volumes of cars, with most of them humping over 1,000 cars per day and some more than 2,000 cars per day.
2. Mini-hump yards and ladder track yards, as the name implies, are smaller versions of regular hump yards and handle 600–1,500 cars per day.
3. Double flat yards handle between 300 and 600 cars per day.
4. Single flat yards handle the smallest volumes, up to 300 cars per day.

Hump Classification Yards. Hump yards, or gravity yards, are the largest car volume carriers of the railroads. There are presently over 50 hump yards in the United States. On average, each hump yard classifies about 1,500 cars a day. This translates to a nationwide total of 75,000 cars per day and 27,000,000 cars per year classified in these yards.

Figure 25.2 shows the general arrangement of a hump classification yard with inspection, maintenance, and servicing facilities. Inbound trains enter the receiving yard, and when the hump yard is ready to receive them the cars are moved on to the main hump. One or more cars are cut from the train and released on a hilltop whose incline varies based on the speed required to take the car(s) down to the classification track. On the way to their final classification, cars may pass through master, intermediate, and group retarders. Only three groups of classification tracks are shown in Figure 25.2. There may be up to eight groups in a yard. There may also be intermediate retarders (not shown) before the cars reach the group retarder. Once all the cars for a particular track have been classified, they are moved to the outbound or departure yard. Here they are coupled with one or more locomotives and the final train leaves the departure yard for the main line. Depending on the number of groups and tracks in a yard, there are generally two or more towers. One tower controls and oversees the classification process while another oversees the trim and departure process. Controllers in the towers have a wide view of the yard and control the cars through retardation and switches to classify them and make up a train. There may also be an inspection and running repair shop, a light car repair shop, a heavy-car repair track, locomotive servicing shop, and inspection and fuel tracks. The figure shows a nearly full service yard for a railroad.

Figure 25.3 is a condensed view of the profile of a typical hump yard. As the name implies, the cars are pushed to a hump by a locomotive, a point that is usually 10 feet or more than the average elevation of the yard. Most hump yards are shaped like a bowl, with cars entering the yard with a downhill (negative) grade and leaving the yard in the trim area

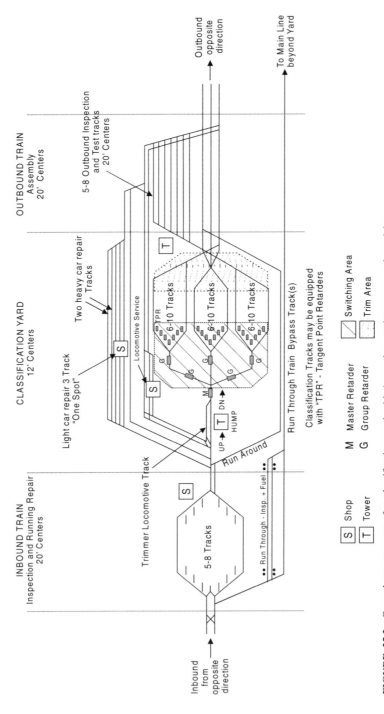

FIGURE 25.2 General arrangement of a classification yard with inspection, maintenance, and servicing facilities (based on a drawing by the late David G. Blaine).

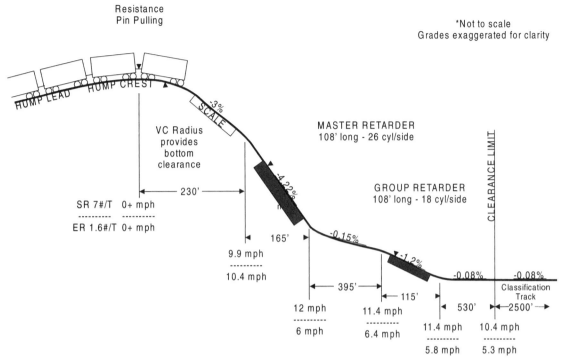

FIGURE 25.3 Profile of a typical hump yard (based on a drawing by the late David G. Blaine).

with a slight uphill or positive grade. The hump crest follows the hump lead. At the crest the resistance pin is pulled from the coupler of the cars, releasing one or more cars on a downhill slope track at about 3 percent grade. While in motion, the car is weighed by an electronic weighing scale. The car enters the master retarder at a speed of 9.9–10.4 mph (the grade of the master retarder is often over 4 percent) and exits at 6–12 mph. The car(s) then enter the group retarder at 6–11 mph and exit at about 5 mph. Using this method, a car enters the classification track and couples at a speed lower than the exit speed from the group retarder. The ideal coupling speed of approximately 4 mph minimizes damage to the car and the contents inside. Railroads prefer lowering exit speeds from the group retarder to make cars roll and couple consistently at about 4 mph. This is often not possible largely due to continually changing weather-dependent wheel-rail rolling friction. Many yard controllers therefore allow cars to exit the group retarder at higher speeds so that even the slowest rolling car reaches its destination and couples with the car ahead. As a result, most cars couple at speeds much higher than 4 mph, and some collide and derail. This can also cause the track to become damaged. The reasons for this and a solution to this problem will be discussed later.

Group retarders primarily control car speed. However, techniques developed recently combine different types of retarders for more accurate car speed control. The distances shown in Figure 25.3 of the different grades and slopes in the yards are for a small hump yard. These distances can be considerably larger for a bigger hump yard. The number of classification tracks shown in Figure 25.3 is only between 20 and 30. In modern larger hump yards, the number of tracks can exceed 60. Large-classification hump yards are expensive to operate and maintain, causing many railroads and industries with their own railroads to use a smaller yard, commonly known as a mini-hump yard.

Mini-Hump Yards.   The mini-hump yard is a variation of the conventional large hump yard. It can handle a reasonably large traffic volume of 500–1,500 cars per day ("Pint Sized Gravity Yards" 1975). It has a smaller hump crest 3–10 feet in height and may not use any retarders. The distance between the crest and clear point is also less, about 500 feet, and it has fewer classification tracks, generally less than 20. Some of these yards have only one master retarder, and some may have group retarders as well. Some of these yards use hydraulic retarders called Dowty retarders on tangent track (to be discussed below). The control of these retarders may be fully automatic, semiautomatic, or manual. Because of their smaller size, these yards cost less than conventional hump yards and often do not exceed the cost of flat yards. They can achieve humping rates comparable to conventional yards, which are about 2–3 cars per minute. The flat yards discussed below generally have much smaller rates of classification.

Flat Yards.   Most of the railroad yards in the United States are flat yards. These yards are designed to handle a smaller volume of cars. A good classification rate for such yards is 200–300 cars per day. Figure 25.4 shows a typical flat yard for switching from either end. A train approaches the yard backwards (the locomotive pushes the cars into the yard). As the cars to be uncoupled and classified approach the entry point, the locomotive accelerates the train to between 4 and 10 mph and then brakes. With the resistance pin at the coupling of the car pulled, the cars become uncoupled by inertia and continue to travel at the speed at which they became uncoupled. The rest of the train slows to a stop due to the applied brakes. The uncoupled car(s) enter the flat yard and go on track AB or CD depending on the side that the train enters the flat yard. The switch of the desired track in the yard is closed so that that car or group of cars can then roll into the classification track, where they are intended to make up the new train. The point of entry A may have flat track as shown in the diagram or may be a minor hump of a few feet. The mini-hump used in certain yards has the advantage of providing the potential energy needed by the cars to keep rolling on AB or CD. In any case, a typical yard has a 300-feet-wide strip at a grade of approximately 0.3 percent downhill so that cars keep moving to enter the switch without slowing down or getting hung up. If a car stalls anywhere along the track and fails to reach its destination, the entire train enters the yard at point A and along track AB or CD to push the car into

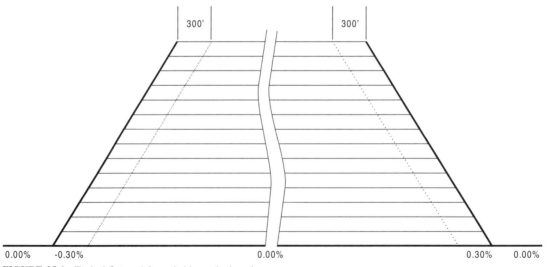

FIGURE 25.4 Typical flat yard for switching at both ends.

place. This "trimming" operation can become time-consuming if many cars stall during classification. Stalls occur in spite of train engineer experience because cars roll differently based on the weather and time of day. The inconsistent rollability and high energy consumption in going through a switch are serious problems that have been solved only recently; they will be discussed later. In a well-designed flat yard, cars may enter from either side, thus allowing a train to kick cars into the yard from point A or point C. The central section of the yard EFGH has zero slope to ensure that car rollability is similar for cars entering from either direction.

Ladder Track Yards.   Mini-hump yards with ladder track are capable of handling larger volumes of cars than flat yards. The ladder track or mini-hump yards take advantage of gravity and occasionally, retarders. Both the clasp-type friction retarders and the hydraulic Dowty retarders are used. This allows the train to back into the mini-hump and release cars one by one or in a group so that they reach the ladder track and eventually the individual track in which they have to be switched. Figure 25.5 is an example of a mini-hump yard with ladder tracks. The hump may only be a few feet above the level of the yard. If it is located at position O in the figure, the car accelerates for a couple of hundred feet. This part of the track, known as the acceleration grade, can be at a grade of anywhere between −1.5 and 0.5 percent. Before reaching the king switch, a clasp-type friction retarder may be located at point A. The ladder lead tracks, which essentially carry cars to individual tracks, are spread from B to C and D to E. The grade for these ladder lead tracks can be between −0.3 and −0.5 percent to keep cars rolling at a speed of approximately 6 mph. Dowty retarders may be located along the ladder lead track to slow faster rolling cars to the desired speed.

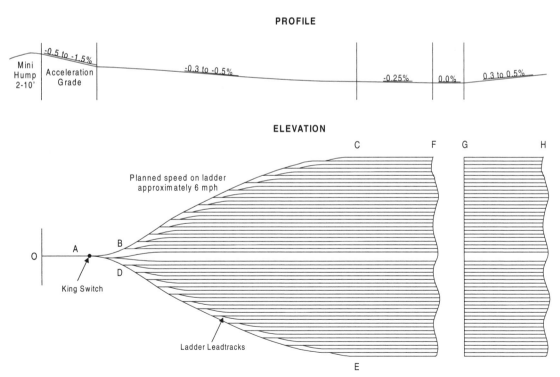

FIGURE 25.5   Typical mini-hump yard with ladder track.

The ladder lead tracks may stretch for several hundred feet, eventually coming to a steady grade tangent track of a few hundred feet length in the region C-F and E-I. The center of the yard at F-G and I-J is relatively flat, almost 0 percent grade. Towards the end of the yard there is an upward grade between 0.3 and 0.5 percent that slows cars if they are traveling at higher speeds. This section can also be a few hundred feet in length.

25.1.4 Retarders and Car Control

Cars move through yards due either to gravity (hump yard) or a locomotive kick (flat yards). That is, cars move with their own energy, derived either through potential or kinetic means. It is therefore necessary to have a good method to control car motion, particularly speed. Devices known as retarders are used to control car speed. Cars are coupled through impact at low speed. Railroads try to maintain that speed at approximately 4 mph. In practice, good coupling speeds are between 4 and 6 mph, preferably closer to 4 mph. Retarders are used to help achieve this coupling speed. Another reason that retarders are necessary to control the speed of free-rolling freight cars is to maintain a certain separation in cars as they are either humped or kicked. This separation is necessary to throw the different switches along the way in time for the track to receive the car. In practice, cars are switched in a 6–10-second interval and a minimum distance of 50 feet must be maintained between cars sent down the hump. The majority of retarders are clasp-type friction-based retarders. The Dowty system, which is a series of hydraulic cylinders bolted to the gage side of the rail, is also in use.

Friction-Type Retarders.   These retarders use a series of clasps with braking shoes that apply a predetermined force to the rims of car wheels a short distance above the rail to develop braking friction. The magnitude of braking force applied is based on the speed desired for the car. The braking shoes may be a composite material or a long beam that pushes against the wheels in their motion through the retarder. Retarder brake shoes should be kept free of grease, oil, and other lubricants in order to avoid compromising their retarding ability. When the wheels are clasped, an ear-piercing squeal radiates from the shoe-wheel contact. Some railroads set up 8-foot tall sound barriers on both sides of the retarder to mitigate this sound. The most common retarder mechanism for wheel braking is electro-pneumatic air cylinders. Other retarders use electrically actuated hydraulic power or electrically actuated spring power for the same purpose. Some other retarders are all-electric models suitable for heavy-duty jobs and amenable to automatic control. Finally, some retarders use the weight of the car wheel to determine the retardation force. These are often hydraulically, and sometimes mechanically, actuated. The degree of retardation in this case is determined by the speed and weight of the car, which needs to be slowed to a desired speed. Friction is the most utilized method of speed retardation because it is the most economical and efficient way to dissipate car energy.

Dowty Retarders.   These retarders are made up of a series of relatively small hydraulic cylinders mounted to the gage side of both rails for a certain length. Figure 25.6 shows a cross-section of one such Dowty cylinder unit (Petracek et al. 1997; Bick 1984; Melhuish 1983). As a car wheel flange rolls over the cylinder, the sliding piston is depressed by the weight of the car and oil or hydraulic fluid is moved from the chamber below the piston to the chamber above the speed-control valve. This motion of hydraulic fluid through an orifice dissipates car energy. A hydraulic oil pressure cylinder is mounted sequentially along the rail, and as the car moves forward it depresses each successive cylinder. Car speeds are maintained at a relatively low range of 4–6 mph in this process. Dowty cylinders function moderately well at low speeds but are prone to frequent failure. When car speeds are high,

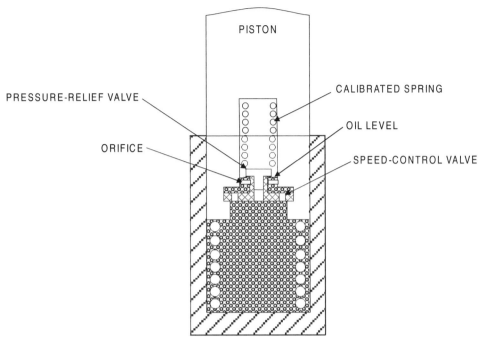

FIGURE 25.6   Dowty cylinder unit. (*Source:* Petracek et al. 1997.)

they do not respond well, resulting in car wheels riding on the cylinder and wheel tread temporarily losing contact with the head of the rail. Heavy, faster cars cause a larger degree of damage to the cylinders. These oil pressure units have an advantage in that they do not produce wheel "squeal" but produce instead a "ringing bell" type of sound as each cylinder is depressed. They are also not affected by wheel contamination that may result from leakage of freight car contents such as oil, grease, molasses, etc.

Other types of retarders, including electrodynamic, pure hydraulic, and linear induction types, have been experimented with, but none have been effective or efficient enough to warrant their cost and be adopted by the industry.

Skates.   Skates are devices installed on top of the railhead to block the movement of a wheel in a given direction. They consist of a steel frame with a blocking head on top of the railhead. They stop a car from moving unless the lateral force becomes excessive, at which point they slide. The friction force between the skates and the rail produces retardation. Although still used by some yards, skates are considered obsolete and have been replaced by spring-loaded retarders and other devices.

25.2   RECENT IMPROVEMENTS IN YARD EQUIPMENT AND OPERATIONS

Yards have traditionally been manpower-intensive. In the last several decades, therefore, the most important development has been computer control of the retarders, switches, and hump engine speed. Nearly all hump yards in the United States use computers for such operations. One or more manned hump towers are used to observe and control yard operations. The tower, or an adjacent building, houses computers and offices for the operation of the yard. The computers control the degree of retardation and switches in the classification section of the yard. Switches are operated by an electrically powered switch machine. Computers also control hump engine speed. A desired speed is maintained during humping to facilitate higher hump rates and complete humping of a train in one operation called a "cut." This is somewhat simpler than using variable humping speeds, which can also be computed and utilized to maximize the output of the yard. Thus, the computer control extends to switch operation, retarder control and operation, hump engine speed control, and overall traffic process control in the yard. The theory of operations advanced significantly in the 20th century, particularly during and after World War II. Yard operations have benefited from this theory as well, including processes such as multistage switching, sorting by block sequence, initial sorting by outbound trains, triangular switching, geometrical switching, and preblocking (Landow 1972; Daganzo, Dowling, and Hall 1982; Kraft 2000; Kubala and Raney 1983; Mundy, Heide, and Tubman 1992; Finian 1994). No attempt will be made here to explain any detail of these operations. There have been many additions to the yard operation that have made yard classification more efficient. Electronic devices mounted on cars called transponders identify the cars as they pass by. These are passive devices that respond to a radiated signal and bounce back the identification codes of the cars. In addition, there are devices like hot-box detectors that detect bad bearings on a car, dragging equipment detectors, broken wheel flange detectors, and loose wheel detectors. All these devices are used to identify bad cars, which are then diverted to a car maintenance shop in the yard to correct the defects. Each car is weighed in motion, within a certain accuracy, when it rolls over rail that is mounted on a scale.

Car computer control utilizes an important input that is measured by a device called a distance-to-couple measurement system. This measurement is derived from track circuits installed on the classification tracks. One method is to measure the impedance of a classification track from the clearance point to the nearest axle of a car present on the track; the axle acts as a shunt across the track. Because the impedance of the rails is proportionate to the distance from the circuit origin to the nearest shunt, it becomes possible to correlate impedance with the distance to couple. Other variations and methods also exist to measure the distance to couple. This is an important measurement needed to determine proper car coupling at a defined speed.

Rail lubrication, by wayside greasers on the gage side of the rail or through a hole in the railhead, is another addition that has developed to reduce car rolling resistance on sharp curves. Unfortunately, these greasers create a terrible mess in their surroundings. A long soaking pad between the rails absorbs excess grease and must occasionally be replaced—a process that can be filthy, expensive, and manpower-intensive.

25.3 THEORETICAL CONSIDERATIONS FOR CLASSIFICATION YARD DESIGN AND OPERATION

Railroad classification yards are designed to disassemble and sort cars without the use of locomotives. In hump yards, cars travel with the help of gravity and switches to a desired classification track and couple with the car ahead at speed of 4–6 mph. In flat yards, cars are kicked at a certain speed to a desired track and couple with the car ahead at speeds of 4–6 mph. In both cases, cars roll to their destination on their own, generally without additional motive power. The majority of their energy is used in overcoming wheel rail friction through sharp curves, switches, and tangent track. In this section, therefore, wheel rail friction will be discussed before discussing the mechanics of a yard.

25.3.1 Wheel Rail Rolling Friction

Wheel rail rolling friction coefficient μ, also known as adhesion coefficient, is defined as

$$\mu = \frac{\text{Tangential force in wheel rail contact}}{\text{Normal force on the wheel}} = \frac{\mu N}{N}$$

(25.1)

where N is the normal force.

In Figure 25.7, T is the torque applied to the wheel, and V is the forward velocity, ω is the angular velocity, and d is the wheel diameter.

$$T = (\mu N d)/2 \tag{25.2}$$

Rolling friction is different from Coulomb friction in that there is only a microslip in the wheel rail contact called creep (ξ), which is not visible. Coulomb friction involves a macroslip. The microslip increases as torque T increases with μ until it reaches a maximum value, μ_{max}, for a certain microslip. If more power is applied, T or μ will not increase. Only wheel slip or creep will increase, and for dry rail μ will actually decrease.

The wheel microslip or creep ξ is defined as

$$\xi = \frac{\text{Microslip velocity}}{\text{Forward velocity}} \tag{25.3}$$

or

$$\%\xi = \frac{(\omega D/2) - V}{V} \times 100 \tag{25.4}$$

For a freely rolling wheel on tangent track, creep in the rolling direction is a small fraction of 1 percent. On curves, this value increases because the rolling radii of the two conical wheels mounted on a rigid axle are different, forcing a microslip of one or both wheels. Thus, the friction in the rolling direction (longitudinal adhesion) and the microslip (longitudinal creep) increase significantly on sharp curves. Other factors that also come into play on curves will be discussed below. (Considerable detail on wheel-rail friction adhesion and

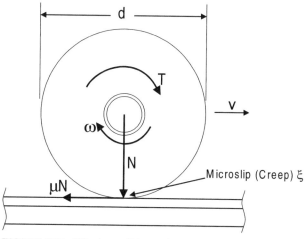

FIGURE 25.7 Wheel on rail in tractive mode.

creep characteristics is provided by Alzoubi 1998; Kumar 1995; TCRP 1997; Kumar and Mangasahayam 1980; Kumar, Rajkumar, and Sciammarella 1980; Kumar, Krishnamoorthy, and Rao 1986.)

Figure 25.8 shows two typical adhesion creep plots of wheel and rail. The highest friction is produced under normal clean dry rail conditions, with μ reaching values of 0.6 or higher for steel wheel and rail. This benefits locomotive wheels, which are designed to produce traction through good friction levels, but is not so good for rolling car wheels, for which the rolling resistance is increased. It is particularly not beneficial for car wheels in a yard where the rolling of a car to its destination depends on a fixed amount of potential energy stored from a hump or kinetic energy transferred from a locomotive push. Adhesion levels on the rail, varying anywhere from 0.6 to 0.1 depending on rail contamination, the weather, and time of day, further exacerbate the problem. This makes it difficult for computer controls to function efficiently without continuous manual adjustment and leads to car collisions and/ or stalls before they reach their destination.

In addition to friction in the rolling or longitudinal direction, wheel sets experience a much larger friction force in the lateral direction while negotiating a sharp curve, switch, or frog. Figure 25.9 shows a truck about to navigate a turnout. The wheels are forced to change direction by an angle θ, called the angle of attack. This results in a lateral slip of the wheel on the tread contact.

$$\text{Lateral creep } \xi_L = \frac{\text{Lateral sliding velocity}}{\text{Forward velocity}}$$

$$\xi_L = \frac{V \sin \theta}{V} = \text{Sin } \theta \cong \theta$$

(25.5)

or

$$\xi_L = \theta \text{ (radians)}$$

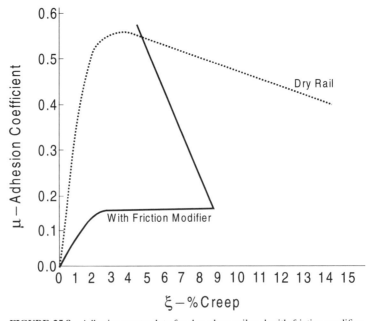

FIGURE 25.8 Adhesion-creep plots for dry, clean rail and with friction modifier.

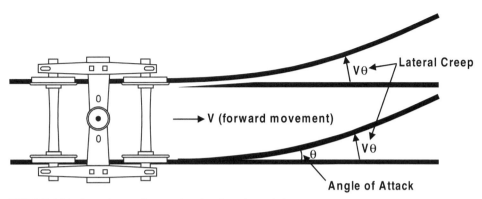

FIGURE 25.9 Lateral creep for a truck going through a switch or curve.

With the sign convention, a positive angle of attack is associated with positive force on the wheel.

Figure 25.10 shows a wheel set sitting on rail through the region of contact, going over a curve, switch point, or frog. The wheel flange on the wheel opposite the turning direction hits the rail gage corner with a large force (up to 20,000 lb). This is responsible for wheel flange wear and the large gage-side wear of the high rail. It can sometimes result in derailments due to tipped rail or wheel climb on the rail. What is generally not recognized is that this large flange force is produced by lateral creep wheel friction F_1 and F_2 on top of both rails, which is produced by the lateral creep-related friction.

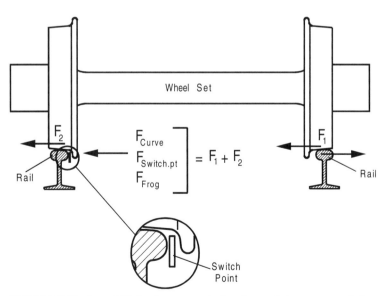

FIGURE 25.10 Lateral friction forces on a single axle negotiating a curve, switch or frog.

$$\mu_L = \frac{\text{Lateral creep force}}{\text{Normal force}}$$

$$\mu_L = \frac{\xi_L \cdot k \cdot n}{N} = k\xi_L \tag{25.6}$$

for the linear part of the μ, ξ relation. In equation (25.6), k is a proportionality constant that changes with rail surface contamination. It is highest in value for dry, clean rail and lowest for lubricated top of rail. As lateral creep increases, so does the lateral force.

Lateral creep is unavoidable on a curve because it is geometry-related, but the high lateral force associated with it is avoidable. This is achieved by using a suitable friction modifier on the wheel tread and the top of the rail, as discussed below.

In addition to the longitudinal and lateral creep, spin creep is experienced by the throat of the wheel flange and tread when this part of the wheel contacts the rail gage corner, such as in a conformal contact. The effect of this is not significant on the factors being discussed here and is therefore not further elaborated.

25.3.2 Mechanics of Yard Operation and Design

A car entering a classification yard has a given amount of energy that it can use to get to its destination. In the case of hump yards it has a certain amount of potential energy or energy head, and in the case of flat yards it has a certain amount of kinetic energy or velocity head, determined by the speed at which it is kicked. With this available energy it needs to overcome:

• Rolling friction (resistance) on tangent track
• Rolling friction (resistance) on curved track
• Rolling friction on turnout, switch, and frog
• Air and wind resistance

Yard classification tracks are designed with downward or negative grades such that the speed of a car is generally maintained constant. This does not always happen, because the rolling resistance (lb/ton) of lighter axle load cars such as empty cars is considerably higher than the rolling resistance of heavy axle load cars. Thus, heavy cars tend to roll much faster and collide with cars ahead while light cars may stop short on the curve and need to be trimmed. This is further complicated by the change of rail friction coefficient with time of day, weather, etc., which will be discussed below.

Figure 25.11 shows the profile and grades of a hump yard track without retarders.

Let V_0, V_1, V_2, V_3 be car speeds at 0, 1, 2 and 3 in ft/sec

m be the mass of the car in lb

h_1 h_2 be car heights between 0 and 1

D_1, D_2, D_3 be distances in ft from 0 to 1, 1 to 2, and 2 to 3

G_1, G_2 be gradient coefficients between 0,1 and 1,2

$$G_1 = h_1/D_1, \qquad G_2 = h_2/D_2$$

R_T be total resistance in lb per lb in car motion

H_0, H_1, H_2 be velocity heads at 0, 1, and 2

g be acceleration due to gravity

w be car weight (lb)

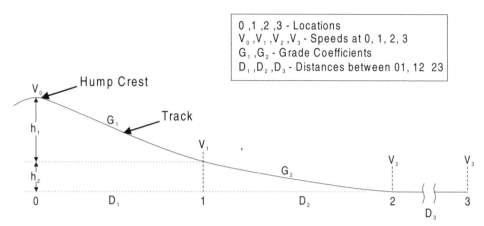

FIGURE 25.11 Profile of track grade in a hump yard.

For a car traveling from 0 to 1,

Potential energy $\quad$ Its kinetic $\quad$ Energy used to overcome
(based on energy $= mgh_1 =$ energy $\quad +$ all resistance (work done)
head of h_1 in ft) $\qquad$ $\frac{1}{2}mV_1^2 \qquad R_T \times D_1 \times W/2,000$

or $\qquad\qquad\qquad mgh_1 = \frac{1}{2}mV_1^2 + [(R_T \cdot mg)/2000] \times D_1$

or $\qquad\qquad\qquad 2gh_1 = V_1^2 + [(R_T \times D_1 \times g)/1,000] \qquad\qquad (25.7)$

Similar relations can be written when the car travels from 1 to 2 and 2 to 3:

$$V_1 = \sqrt{[2gh_1 - (R_T D_1 g/1,000)]} \qquad\qquad (25.8)$$

Rollability and Rollability Coefficient. Cars have good rollability if they have low rolling resistance. Rolling resistance is expressed as pound force per ton needed to move the car at constant speed on level tangent track. A rollability coefficient C is defined as the resistance force in lb per lb weight of the car:

$$C = \text{Resistance force (lb)/lb weight of car} \qquad\qquad (25.9)$$

$$\text{Resistance force} = [R_T \text{ lb/ton}] \times [W(\text{lb})/2,000 \text{ lb/ton}] = [R_T W/2,000]\text{lb}$$

and so

$$C = [R_T W/2,000]/W = R_T/2,000 \qquad\qquad (25.10)$$

If, for example, the force resisting the motion of the car is 5 lb/ton

$$C = 5 \text{ lb}/2,000 \text{ lb} = 0.0025 \text{ lb force/lb weight} = 0.25\%$$

This car will travel at constant speed on a -0.25% grade tangent track. The higher the value of the rollability coefficient, the harder it is for that car to roll.

Velocity Head Consideration. Velocity head at 0

$$H_0 = 0$$

Coming down to point 1 has a drop of height h_1 and

$$h_1 = D_1 G_1$$

Then the velocity head at 1 is

$$H_1 = D_1 G_1 - \text{loss due to rolling resistance}$$

$$= D_1 G_1 - D_1 C_1$$

Similarly, velocity head at point 2:

$$H_2 = D_1 G_1 - D_1 C_1 - D_2 C_2, \text{ and so on}$$

The car resistance on one track consists of:

Resistance on tangent track (lb/ton)
Resistance on curved track (lb/ton/degree of curve)
Aerodynamic and wind resistance (lb/ton at a given car speed)

Accordingly, the total rollability coefficient C_T is a sum of four elements.

$$C_T = C_0 + C_D + C_A + C_W \tag{25.11}$$

where typical values of these coefficients are given below:

C_0 = Rollability coefficient on tangent track = 0.0005 to 0.006 lb/lb

C_D = Rollability coefficient on curve of degree D = 0.00045 lb/lb/degree

C_A = Rollability coefficient due to aerodynamics = 0.00016 lb/lb/ft/sec car speed

C_W = Rollability coefficient due to wind = 0.0001 lb/lb/ft/sec of wind velocity

The rollability of a car is the inverse of its rollability coefficient. In other words, the smaller the rollability coefficient of a car, the better its rollability and vice versa.

The values of rollability coefficients given above are approximate and can be estimated in yards where cars operate at generally low speeds. In fact, these values change significantly with axle load, car shape, weather, and speed.

Example

Rolling coefficient of a car:

$$\text{Car weight} = 100 \text{ tons} = 200,000 \text{ lb}$$

$$\text{On a good tangent track } C_0 \cong 0.001$$

If the car is operating on a 15° curve in the yard, then

$$C_D = 0.00045 \times 15 = 0.00675$$

Assuming a car speed of 6 mph or 8.8 ft/sec

$$C_A = 0.00016 \times 8.8 = 0.00141$$

If there is no wind, $C_W = 0$

Thus, the total rollability coefficient is:

$$C_A = 0.001 + 0.00675 + 0.00141 = 0.00916 \text{ lb/lb}$$

It should be noted that in this example the curve rollability coefficient of the car is by far the largest. In addition, there are large energy losses when a car negotiates a switch, which are not included in the above calculation.

There is no good estimate available for rollability coefficient on a switch. It is a well-known fact that the cars in a yard lose considerable speed when they go through a switch. This is a major problem that has been solved only recently.

The rolling resistance of cars on curves and tangent track is quite high and varies with time and weather. For efficient car control, it is necessary to reduce it and keep it consistent. This will be discussed in the next section.

Rolling Resistance and Loss in Speed with Wheel Tread and Top of Rail Friction Modification.   It has been well accepted in the rail industry that top of rail friction modification (TOR-FM) reduces the rolling resistance and the lateral forces produced on a wheel on a curve (Kumar 1999; Davis, Kumar, and Sedelmeier 2000; FRA 2000; Clapper et al. 2001; Kumar, Yu, and Witte 1995; Reiff, Gage, and Robeda 2001; Reiff and Davis 2003; Reiff 2000). Yards are good candidates for the application of this phenomenon. This section presents a brief analysis of such an application. Technical details of achieving this will be discussed later.

The rolling resistance of a car can be expressed as:

$$R(\text{lb/ton}) = a + bV + cV^2$$

where a, b, and c are constants and V is car speed in mph. In a yard where car speeds are small (≤ 10 mph), the cV^2 term is small enough to be neglected.

So car resistance in a yard can be expressed as:

$$R = a + bV \qquad (25.12)$$

Using this expression and starting with a car arriving at point 1 (Figure 25.11), let us use two scenarios for analysis. There are two conditions of operation, A and B. For condition A, the rail head and wheel treads are dry with high friction, and for condition B the top of rail and wheel treads are coated with a friction modifier. The friction modifier used acts as a lubricant for rolling car wheels but does not compromise the traction and braking of locomotive wheels.

Loss of speed from 1 to 2 in dry condition $= \Delta V_A$

Loss of speed from 1 to 2 in friction modified condition $= \Delta V_B$

$$\Delta V_A = (V_1 - V_2)_A$$

$$\Delta V_B = (V_1 - V_2)_B$$

As a good first approximation, it can be assumed that the loss of speed by a car is affected only by the force of its rolling resistance. The grade and curve are the same for both conditions.

Kinetic energy of the car at location 1 is

$$= \frac{1}{2}mV_1^2$$

While rolling to location 2, some of this energy is used up in doing work against the rolling resistance R. Kinetic energy of the car at location 2 is therefore

$$= \tfrac{1}{2}mV_2^2 = \tfrac{1}{2}mV_1^2 - mRD_2$$

or $\qquad \tfrac{1}{2}(V_1^2 - V_2^2) = RD \quad$ and $\quad \Delta V = V_1 - V_2 = 2RD/(V_1 + V_2)$

so $\qquad \Delta V_A = \dfrac{2R_A D_2}{V_1 + (V_2)_A}$ $\hfill (25.13)$

and $\qquad \Delta V_B = \dfrac{2R_B D_2}{V_1 + (V_2)\ \mathrm{B}}$ $\hfill (25.14)$

Percentage change in speed as a result of friction modification is

$$= 100 \times \frac{\Delta V_B - \Delta V_A}{\Delta V_B}$$

$$= 100 \times \frac{2D_2[R_B/(V_1 + (V_2)_B) - R_A/(V_1 + (V_2)_A)]}{2D_2[R_B/(V_1 + (V_2)_B)]}$$

$$= 100 \times \left[1 - \frac{R_A}{R_B} \times \frac{V_1 + (V_2)_B}{V_1 + (V_2)_A} \right]$$

For most practical values of speed, the term $V_1 + (V_2)_B/V_1 + (V_2)_A$ is very close to 1 and therefore, % change in loss of speed from 1 to 2

$$= 100\,(1 - R_A/R_B) \hfill (25.15)$$

due to the friction modifier. This is the same as percentage change in rolling resistance due to friction modification. Therefore, we conclude that

% change in loss of speed = % change in rolling resistance

This means that the % change in loss of velocity is a very meaningful number that correlates directly to the % change in rolling resistance of the car. It is also a measure of the improvement in rollability.

Example

   Car exit speed at 1 = 8 mph

   For case A—dry rail

   Speed at point 2 = 6 mph

   (Loss of speed)$_A$ = 2 mph

   For case B—with friction modifier

   Speed at point 2 = 7 mph

   (Loss of speed)$_B$ = 1 mph

   % change in speed due to use of friction modifier $= \dfrac{2 - 1}{2} \times 100$

   $= 50\%$

So improvement in rollability = 50%

25.4   *YARD FRICTION MODIFIERS SOLVE MANY PROBLEMS*

In spite of considerable modernization and automation, yards continue to face serious problems in the areas of safety, lost productivity, loss and damage, increased track maintenance, and environmental concerns.

25.4.1   Problems in Yard Operations

Yard Safety Issues.   Railroad yards continue to face the hazards of car collisions and derailments. In addition to being safety concerns, these events can cost thousands of dollars in damage and lost productivity. Some of the problems associated with car derailments are tipped rail, crossed couplers, short stops, worn-out rail and switch points, and damaged and misaligned track.

Car collisions are not uncommon in a yard. They are caused mainly by rollout, insufficient speed control, inefficient skates due to excessive grease on the rails, and worn, out of maintenance retarders.

Loss and Damage.   The number of derailments and collisions that take place in a yard is a significant portion of the total derailments or collisions that occur in a railroad. Some of these are due to human error, but a significantly large percentage are due to insufficient control of car speeds. When collisions or derailments take place, there is significant damage to the car and the freight inside, as well as the track, which in turn increases the track maintenance cost.

Loss of Productivity.   Collisions, derailments, and cars stalled before their destination cause a loss of productivity in a yard because classification is suspended while these problems are remedied. In theory, 10 or more cars can be humped every minute and one or more cars can be kicked in a flat yard in the same time. In reality, however, output is considerably less, about one-tenth of the maximum. One of the factors that contribute to this is stalled cars that have to be trimmed by locomotives.

Main Problem.   At present it is not possible to achieve consistent rolling speeds in hump yards or flat yards. In hump yards, computer-controlled retarders cannot be set once for ever-changing friction conditions. In flat yards, engineers have to make intelligent guesses on prevailing friction conditions and then estimate kicking speed. Both scenarios often result in either car collisions or short stops. Sharply curved track sometimes produces such high lateral forces that it tips the rail and causes derailments. High lateral forces combined with cross-coupler impact at high speeds are a recipe for derailments.

Wayside Greasers.   Until recently, many yards used wayside greasers to reduce the rolling resistance of cars on curves. Greasers are not generally effective in solving the problem and in some cases make things worse. In order to make rolling easier, it is a common practice to set greasers at high rates of grease application. Excessive grease makes the surroundings messy by forming grease pools under the applicator. A coating of grease also forms between the rails for a stretch of track beyond the application site, making the area a safety hazard. Thus, the effective overall cost of these greasers is quite high considering the large amounts of grease used and the manpower required to maintain them and clean up excess grease.

25.4.2   The New Yard Friction Modifiers

The newly developed yard friction modifiers solve many of the problems stated above. Many yard personnel use the abbreviated form ''modifiers'' to describe these units. This system sprays a clean, environmentally safe friction modifier fluid on the wheels of a car as it enters its classification track. The application is in the form of microburst lubricant jets, which hit both wheels of the lead axle of an approaching truck of a car. Figure 25.12 shows how the modifier works. It is located just before the main switch of a group of tracks into which the classification is intended. In the case of a hump yard it is located right after the retarder. There are two sensors mounted on the rail PS and FS as shown in the figure. There are also two nozzle holder units marked N on each of the rails. As the car wheelset trips sensor PS, a signal is sent to start a pump in the main unit C to develop a desired pressure in the system. When the axle trips sensor FS, a predetermined small quantity of friction modifier fluid is fired on the wheels by both nozzle units. The shot is fired in such a way that both the flange and tread of the wheel receive some of the friction modifier (FM). The box C contains a microprocessor and software that controls the entire process. Each modifier unit is set to apply a specified amount of FM to each group of tracks based on its requirements. The system also does not fire when it detects locomotive wheels. The unit can also be set to skip a desired number of cars before firing again if so desired. The amount of FM dispensed for each shot can also be adjusted within certain limits. Such systems have replaced greasers on a large number of hump yards and some flat yards in the United States. The modifiers offer a clean and cost-effective solution to yard problems and lead to increased productivity and improved performance. They are currently produced by Tranergy Corporation in the United States.

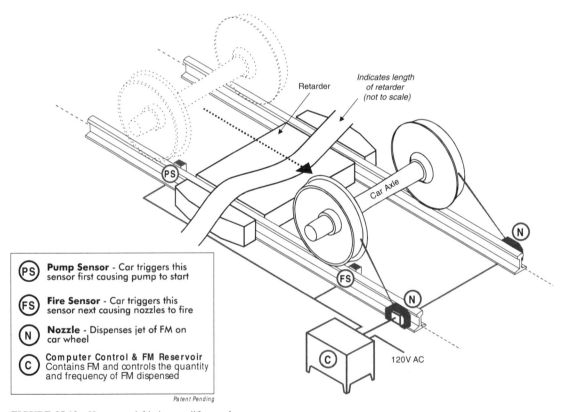

FIGURE 25.12   How a yard friction modifier works.

25.4.3   Friction Modifier Fluid

Friction modifier fluid is a thin, synthetic polymer liquid, and not a grease. It is safe to handle and biodegradable. It contains no solids such as graphite or molybdenum disulphide and dissipates as trains pass over it, leaving little or no buildup on rails or wheels. It is effective for a wide range of temperatures, from −20°F to 150°F, ensuring smooth delivery flow throughout the year. In the United States, it is currently produced for Tranergy Corporation by Shell Oil Co.

25.4.4   How Many Modifier Units Does a Yard Need?

The number of units needed by a yard is determined by the groups of tracks present in the yard. Figure 25.13 shows the layout of a hump yard equipped with modifiers. It has five groups of tracks, five corresponding retarders, and one primary or master retarder. The modifier units needed are shown in the diagram. In this example there are five modifiers present, corresponding to the five groups, and a possible sixth unit placed right after the master retarder. At the minimum, a yard needs one modifier unit after the final retarder of each group of tracks. The friction developed on each group of tracks is different and hence each unit is tuned to its group. The unit after the master retarder provides consistent rolling and entry speeds to the group retarders further down the track. It also reduces wear and tear on the track that follows it.

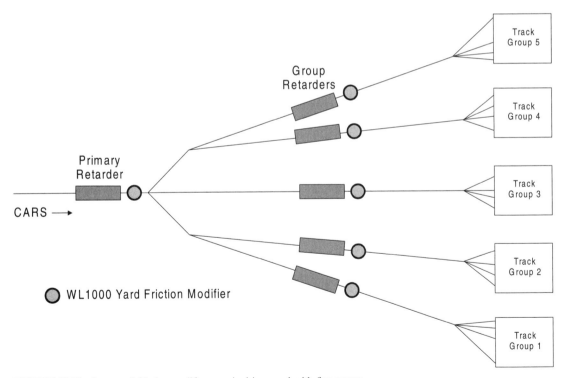

FIGURE 25.13   Layout of friction modifiers required in a yard with five groups.

25.4.5   Proven Improvements in Car Rollability and Reduction of Short Stops

A considerable amount of initial testing was conducted with the modifiers. It is generally accepted in the United States that in addition to being very clean, modifiers improve car rollability and reduce short stops in a yard. Data were first gathered by the Union Pacific Railroad Neff Yard in Kansas City, Kansas. Testing was conducted over a period of one month in the year 2000. The yard classified about 950 cars a day before the installation of the modifier units. After the installation of the modifier units and computer control adjustment of the retarder, the yard classified over 1,200 cars per day. While the modifiers were not solely responsible for this increase, they helped to improve consistent car rolling, which enabled the yard to be more productive. The test data in Figure 25.14 shows that car rollability improved anywhere between 30 and 57 percent for cars of various weights.

The Burlington Northern and Santa Fe Railroad also gathered data in their yard in Galesburg, Illinois. The rollability test data (not presented here) were similar to data gathered in the Union Pacific Yard. The Galesburg yard also gathered data for short stops in the yard for a day on one group of tracks (Figure 25.15). The data clearly indicate that the number of short stops was reduced by over 60 percent. To date, the use of modifiers in over a dozen yards in the United States has shown the following:

- Reduced derailments
- Reduced short stops and needed locomotive trim

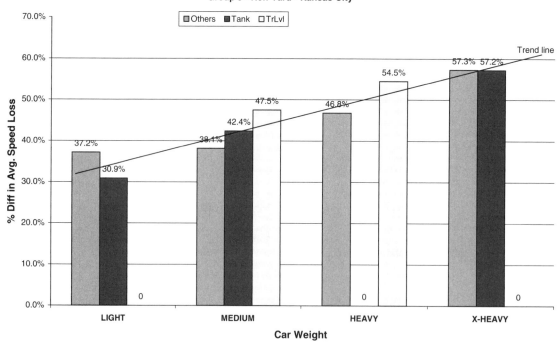

FIGURE 25.14   Improved rollability of cars with friction modifier (WL1000) opposed to greasers.

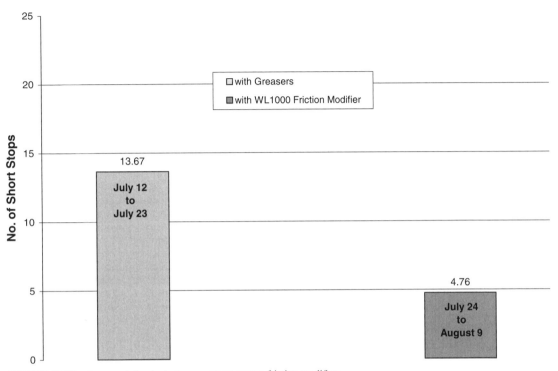

FIGURE 25.15 Average daily short stops: greasers versus friction modifiers.

- Reduced exit speeds from group retarders
- Reduced car-coupling speeds
- Eliminated tipped rail
- Consistent car-rolling speeds
- Reduced lateral forces on sharp curves
- Reduced rollouts and collisions
- Reduced track maintenance
- Increased yard output per day

25.5 NEW RAIL SWITCH ENHANCER IMPROVES YARD OPERATION AND SAFETY

A new electro-hydraulic system called the rail switch enhancer (RSE) has been developed to enhance yard operations. It automatically lubricates rail switches and the top of switch rail using computer-controlled quantities of lubricant to improve switch life, performance, and safety. For the benefit of the general reader unfamiliar with rail switches, a brief description of switches is presented before RSE is discussed.

25.5.1 General Background

Switches, found extensively in railroad industrial yards, are used in turnouts and crossovers to divert trains to other tracks. A typical hump yard may have 200–400 switches. Mini-hump yards and flat yards have 50–150 switches. Switches are a high-maintenance item for a railroad engineering department. It is estimated that nearly 30 percent of the engineering maintenance budget is used for switches and frogs. Enhancements in switch performance and life can thus lead to major savings and profitability for the railroad. Figure 25.16 shows the essential elements of a rail switch. The stock rail remains straight while turnout rail turns away from the main track. The switch in the diagram is in a position that would make a car travel straight. The switch point shown is moved with the help of switch rods and when moved into the turnout position causes the wheels to go on the turnout. The switch point, a rail with a sharp knife-like end, pivots about the switch heel. Only one rod is shown in the diagram, but there may be as many as five, depending on the length of the switch point. The switch point slides over switch plates in order to get into one of two positions—one on stock rail and the other on the turnout rail. Rail switches are used in conjunction with frogs and sometimes guardrails. All such railroad track components experience serious impact and wear depending on the sharpness of the turns in which they are located. It is therefore necessary to maintain these components regularly. In spite of maintenance, however, engineering departments find it necessary to replace switch rails and other moving components much more frequently than other rail. Switches are also a major factor in derailments. Studies by the U.S. Federal Railroad Administration have shown that the majority of derailments take place within 200–300 yards of a switch. As a car enters a switch, a sudden change in direction results in a lateral impact force on the wheels of the car (Figure 25.10). This force is produced by a sudden change in the lateral creep force on top of the switch rail and the other two rails. This contributes to the dynamic instability of the car, which can lead to derailments under certain conditions. In a yard, rolling cars often stall at or near a switch, indicating that considerable car energy is taken away by the switch, which is also due to the lateral creep

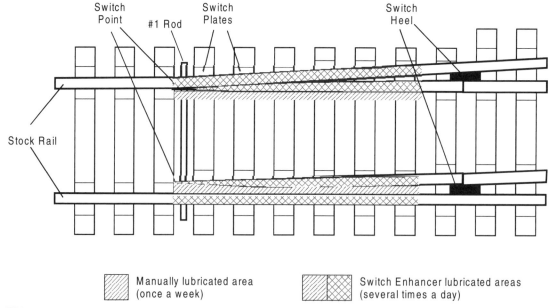

FIGURE 25.16 Typical rail switch (split switch) on freight tracks in the United States.

forces mentioned above. It has been reported by certain rail transit systems that a large percentage of rail fractures occur near switches. It is theorized that these fractures are also related to the lateral creep impact force mentioned above. The current practice of maintaining switches involves using a lubricant or grease to lubricate the sliding plates of the switch. This reduces friction and makes the switch move more easily, but it does not provide any reduction of lateral force on top of the rails. The same is true for rail frogs, which are present at all rail turnouts and crossovers along with switches. Frogs, like switches, are affected by the creep force impact. At present there is no consistent protection or performance enhancement available for either rail switches or frogs. The current lubrication practice is manpower-intensive and leads to irregular maintenance of the switch, especially in remote areas. Switches are operated either manually or by a powered switch machine. When switches become difficult to throw, the life of the switch machine is reduced. If it is a manual switch, then an injury hazard develops for the person throwing the switch. The present practice of maintaining and operating a switch involves high costs. The highest among these is derailment cost. Another large cost component is disruption of traffic. In a yard, several principal switches are essential to yard operation. The failure of these switches can lead to severe disruption of traffic. These switches are prime candidates for performance enhancement and automatic maintenance. On the main line, disruption of traffic costs can be even higher. The maintenance and replacement of points is also a major cost. Other costs include wear on the power machine and/or injury to the person throwing the switch. Thus, an avoidance of all these costs and a new approach to improving the performance, operation, and safety of rail switches has been badly needed. Such a new approach is now available and is discussed in the next section.

25.5.2 The New Rail Switch Enhancer

The rail switch enhancer is an electro-hydraulic system that automatically lubricates rail switches and enhances their performance. Figure 25.17 shows a RSE placed on the switch of a rail turnout. The stock rail turnout rails along with frog and guardrails are shown in the

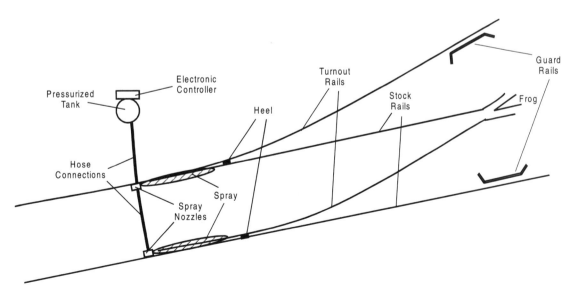

FIGURE 25.17 Schematic of a switch enhancer installed on a switch.

diagram. The switch rails pivot on the switch heels. The switch enhancer components include a set of check valves and nozzles in a nozzle holder on brackets attached to each rail. These nozzles are supplied with lubricant under pressure by a hose or pipe from a pressurized tank containing the lubricant. The controller, located in a box, is mounted on or near the tank. The tank and the box may be enclosed in another box for security. The nozzle spray occurs at a specified frequency, dispensing a specified amount of lubricant to cover the switch, plates, switch rods, base of the stock rails, and top of the switch and the stock rails, providing protection and enhancing switch performance. On main line switches with heavy traffic, it applies a lubricant shot prior to every train going through the turnout. This is triggered by the signal from the switch controller that opens the switch for the turnout. The lubricant sprayed on top of the stock and switch rails is carried forward by the train wheels and lubricates the top of all four rails, reducing the lateral creep impact forces and thus protecting the frog and the guard rails as well. Figure 25.18 shows a cross section of the switch point rail and the stock rail. Figure 25.19 shows a cross-sectional frontal view of the stock rail and the switch rail. The stock rail is mounted on a tie plate. The switch rail is supported on its own switch plate. The current occasional practice of manual lubrication provides lubricant coverage on the upper surface of the switch plate and the stock rail base. Lubrication is actually required for the upper surfaces of the rails in addition to the lubrication of the switch plate and the stock rail base. These are shown as crosshatched surfaces that are lubricated by the switch enhancer. Figure 25.19 shows a plan view of the switch enhancer spraying the lubricant jets. The spray nozzle brackets and the nozzle holders are mounted on the stock rail. The nozzle holders each have two separate orifices. In order to increase the coverage area laterally, a nozzle with a wider spray, nozzle 1, is used in addition to another nozzle with a long range of spray application, nozzle 2. This nozzle spray may reach the heel, if so desired. This arrangement provides lubrication to all the surfaces discussed earlier as well as the top of the rails of the switch point and stock rail.

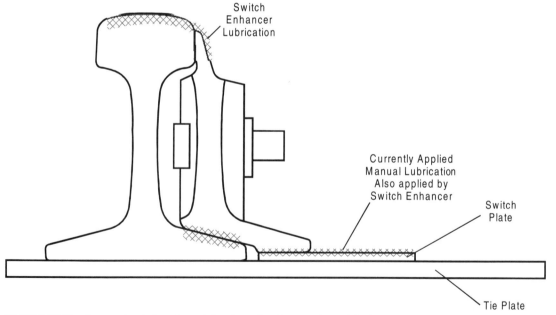

FIGURE 25.18   Cross-sectional view of a switch point and surfaces needing lubrication.

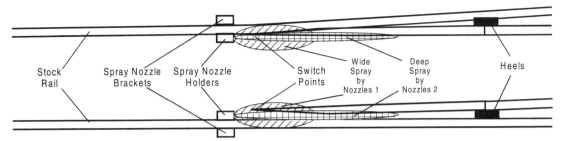

FIGURE 25.19 Plan view of a switch enhancer showing dual nozzle sprays covering the switch and the stock rail (cross-hatched).

In this way the rail switch enhancer continues to make an automatic lubricant application several times a day to switch components and the top of the rails at a defined time and quantity with one lubricant charge in the tank. The regularity of a small lubricant application reduces switch wear and improves train rolling on the switch, thereby enhancing switch operation.

25.6 ACKNOWLEDGMENTS

The author is most thankful to Mr. Suneet Cherian of Tranergy for his great support throughout the article, in transcribing the text, creating all figures, and editing the manuscript through each cycle of revision. The author also thanks Mr. Nikul Patel and Mr. Naved Tirmizi of Tranergy for transcribing the first version of the theoretical section of the text. Finally, he also thanks Mr. Saud Al Dajah, his Ph.D. student at I.I.T. Chicago, for participating in the Adhesion-Creep subsection.

25.7 REFERENCES

Alzoubi, M. 1998. "Adhesion-Creepage Characteristics of Wheel-Rail System under Dry and Contaminated Rail Surfaces." Ph.D. thesis, Illinois Institute of Technology, Chicago, IL.

American Railway Engineering and Maintenance-of-Way Association (AREMA). 2001. *Manual for Railway Engineering,* vol. 3, ch. 14, "Yards and Terminals." Landover, MD: AREMA.

Bick, D. E. 1984. "A History of Dowty Marshalling Yard Wagon Control System." *Proceedings of the Institute of Mechanical Engineers* 198B(2):19–26.

Christianson, H. B., et al. 1979. "Committee 14, Yards and Terminals,Report on Assignment #7 Yard System Design for Two-Stage Switching." *American Railway Engineering Association Bulletin* 81(11): 145–55.

Clapper, J., V. Dyavanapalli, S. Kumar, and E. Wolf. 2001. "Top of Rail Lubrication Pilot Project Results on Wheeling & Lake Erie Railway." In *Proceedings of 2001 ASME International Mechanical Engineering Congress and Exposition,* New York, November 11–16.

Daganzo, C. F., R. G. Dowling, and R. W. Hall. 1982. "Railroad Classification Yard Throughput: The Case of Multistage Triangular Sorting." *Transportation Research A* 17A(2):95–106.

Davis, K., S. Kumar, and G. Sedelmeier. 2000. "Further Developments in Top of Rail Lubrication Testing." In *Proceedings of the 62nd Annual Meeting, Locomotive Maintenance Officers Association,* September 18–20.

Federal Railroad Administration (FRA). 2000. *Evaluation of a Top of Rail Lubrication System.* Research results, RR00-01, U.S. Department of Transportation, FRA, Washington, DC.

Finian, D. H. 1994. *Current Methods for Optimizing Rail Marshalling Yard Operations.* NTIS#ADA289371, Kansas State University, Manhattan, KS.

Kraft, E. R. 2000. "A Hump Sequencing Algorithm for Real Time Management of Train Connection Reliability." *Journal of Transportation Research Forum* 39(4):95–115, published jointly with *Transportation Quarterly* 54(4).

Kubala, R., and D. Raney. 1983. "A Modular Approach to Classification Yard Control." *Transportation Research Record* 927, 62–67.

Kumar, S. 1995. "Wheel/Rail Adhesion Wear Investigation Using a Quarter Scale Laboratory Testing Facility." Paper delivered at 1995 ASME International Mechanical Engineering Congress and Exposition, San Francisco, CA, November 15.

Kumar, S. 1999. "Top of Rail Lubrication System for Energy Reduction in Freight Transportation by Rail." Society of Automotive Engineers Technical Paper Series, 1999-01-2236.

Kumar, S., and R. Mangasahayam. 1980. "A Parametric and Experimental Analysis of Friction, Creep and Wear of Wheel and Rail on Tangent Track." *Proceedings of ASME Symposium on the General Problem of Rolling Contact,* AMD 40:139–55.

Kumar, S., P. K. Krishnamoorthy, and D. L. P. Rao. 1986. "Influence of Car Tonnage and Wheel Adhesion on Rail and Wheel Wear: A Laboratory Study." *ASME Transactions, Journal of Engineering for Industry* 108(1):48–58.

Kumar, S., B. R. Rajkumar, and C. Sciammarella. 1980. "Experimental Investigation of Dry Frictional Behavior in Rolling Contact under Traction and Braking Conditions." Proceedings of 7th Leeds-Lyon Symposium on Tribology, Leeds, UK, Sept. 1980, *Friction and Traction.* Mechanical Engineering Publications, September, 207–19.

Kumar, S., G. Yu, and A. C. Witte. 1995. "Wheel-Rail Resistance and Energy Consumption Analysis of Cars on Tangent track with Different Lubrication Strategies." Paper delivered at IEEE/ASME Joint Railroad Conference, Baltimore, April 4–6.

Landow, H. T. 1972. "Yard Switching with Multiple Pass Logic." *Railway Management Review Quarterly* 72(1):11–23.

Melhuish, A. W. 1983. "Developments in the Application of the Dowty Continuous-Control Method." *Transportation Research Record* 927, 32–38.

Mundy, R. A., R. Heide, and C. Tubman. 1992. "Applying Statistical Process Control Methods in Railroad Freight Classification Yards." *Transportation Research Record* 1341, 53–62.

Petracek, S. J., A. E. Moon, R. L. Kiang, M. W. Siddiquee. 1997. "Railroad Classification Yard Technology: A Survey and Assessment." FRA/ORD 76/304, Stanford Research Institute, Menlo Park, CA, January.

"Pint Sized Gravity Yards Keep the Traffic Rolling." 1975. *Southern Pacific Magazine* (January/February): 10–11.

Railway Age. 2001. Interview with Hunter Harrison. (May): 28.

Reiff, R. P. 2000. *Top of Rail Lubrication Implementation Issues.* Report# RS-00-001, Transportation Technology Center Inc., July.

Reiff, R. P., and K. Davis. 2003. "Implementing Locomotive-Based Top-of-Rail Friction Control for Improving Network Efficiency." Paper delivered at International Heavy Haul Association, 2003 Specialist Technical Session, Dallas, 5.87, 5.95, May 5–9.

Reiff, R. P., S. E. Gage, and J. A. Robeda. 2001. "Alternative Methods for Rail/Wheel Friction Control." In *Proceedings of 7th International Heavy Haul Conference,* Brisbane, Australia, June, 559–62.

Transit Cooperative Research Program (TCRP). 1997. "Improved Methods for Increasing Wheel/Rail Adhesion in the Presence of Natural Contaminants." *Research Results Digest* 17.

Wong, P. J., C. V. Elliot, R. L. Kiang, M. Sakasita, and W. A. Stock. 1978. Railroad Classification Yard Technology, Design Methodology Study. Federal Railroad Administration, Washington, DC, September.

CHAPTER 26
MODERN AIRCRAFT DESIGN TECHNIQUES

William H. Mason

Department of Aerospace and Ocean Engineering,
Virginia Polytechnic Institute and State University,
Blacksburg, Virginia

26.1 INTRODUCTION TO AIRCRAFT DESIGN

This chapter describes transport aircraft design. We discuss the key issues facing aircraft designers, followed by a review of the physical principles underlying aircraft design. Next we discuss some of the considerations and requirements that designers must satisfy and the configuration options available to the designer. Finally, we describe the airplane design process in some detail and illustrate the process with some examples. The modern commercial transport airplane is a highly integrated system. Thus, the designer has to have an understanding of a number of aspects of engineering, the economics of air transport, and the regulatory issues.

Large transports are currently manufactured by two fiercely competitive companies: Boeing in the United States and Airbus in Europe. Smaller "regional jets" are manufactured by several companies, with the key manufacturers being Bombardier of Canada and Embraer of Brazil. Any new airplane designs must offer an advantage over the products currently produced by these manufacturers (known as "airframers"). Key characteristics of current designs can be found in the annual issue of *Aviation Week and Space Technology,* the *Sourcebook.* The other standard reference is *Jane's All the World's Aircraft* (Jackson 2002). An electronic appendix to Jenkinson, Simpkin, and Rhodes (1999) provides an especially complete summary. Also, essentially all new transport aircraft use turbofan engines for propulsion, although there are a number of smaller turboprop airplanes currently in service.

In picking the basis for a new aircraft design, the manufacturer defines the airplane in terms of range, payload, cruise speed, and takeoff and landing distance. These are selected based on marketing studies and in consultation with potential customers. Two examples of decisions that need to be made are aircraft size and speed. The air traffic system operates near saturation. The hub-and-spoke system means that many passengers take several flights to get to their destination. Often this involves traveling on a regional jet carrying from 50–70 passengers to a major hub, and then taking a much larger airplane to their destination. They may even have to transfer once again to a small airplane to get to their final destination. From an airport operations standpoint, this is inefficient. Compounding the problem, the small regional jets require the same airspace resources as a large plane carrying perhaps 10 times the number of passengers. Thus, the designer needs to decide what size is best for both large and small passenger transports. At one time United Airlines operated two sections of a flight between Denver and Washington's Dulles Airport using wide-body aircraft. Frequently, both aircraft were completely full. Thus, there is a need for even larger aircraft from the airspace operations viewpoint even though it may not be desirable on the basis of the operation of a single plane. Based on the system demands, Airbus has chosen to develop a very large airplane, the A380. Alternatively, Boeing predicts that the current hub-and-spoke system will be partially replaced by more point-to-point operations, leading to the major market for new aircraft being for aircraft about the size of B767. They reached this conclusion based on the experience of the North Atlantic routes, where B747s have been replaced by more frequent flights using the smaller B767s.

Another consideration is speed. Designers select the cruise speed in terms of the Mach number M, the speed of the plane relative to the speed of sound. Because the drag of the airplane rises rapidly as shock waves start to emerge in the flow over the airplane, the economical speed for a particular configuration is limited by the extra drag produced by these shock waves. The speed where the drag starts to increase rapidly is known as the drag divergence Mach number M_{DD}. Depending on the configuration shape, the drag divergence Mach number may occur between $M = 0.76$ and $M = 0.88$. It is extremely difficult to design an airplane to fly economically at faster speeds, as evidenced by the decision to withdraw the Concorde from service. Numerous supersonic transport design studies since the introduction of the Concorde have failed to produce a viable successor. In addition to the aerodynamic penalties, the sonic boom restriction for supersonic flight over land and the difficulty of achieving low-enough noise around airports (so-called community noise) makes the challenge especially severe.

The choice of design characteristics in terms of size and speed is at least as important as the detailed execution of the design. Selecting the right combination of performance and payload characteristics is known as the "you bet your company" decision. The small number of manufacturers building commercial transports today provides the proof of this statement.

The starting point for any vehicle system design work is to have information about current systems. In this chapter, we will use 15 recent transports as examples of current designs. We have divided them into three categories: narrow-body transports, which have a single aisle; wide-body aircraft, which have two aisles; and regional jets, which are small narrow-body aircraft. Table 26.1 provides a summary of the key characteristics of these airplanes. The values shown are the design values, and the range and payload and associated takeoff and landing distances can vary significantly. Detailed performance data can be found for Boeing airplanes on their website: www.boeing.com. Other airframers may provide similar information.

While this section provides an overview, numerous books have been written on airplane design. Two that emphasize commercial transport design are Jenkinson, Simpkin, and Rhodes (1999) in the United Kingdom and Schaufele (2000) in the United States. Paul Simpkin had a long career at Rolls Royce, and the book he coauthored includes excellent insight into propulsion system considerations. Roger Schaufele was involved in numerous Douglas Aircraft Company transport programs. Two other key design books are by Raymer (1999) and Roskam (1987–1990) (an eight-volume set).

26.2 ESSENTIAL PHYSICS AND TECHNOLOGY OF AIRCRAFT FLIGHT

Aircraft fly by exploiting the laws of nature. Essentially, lift produced by the wing has to equal the weight of the airplane, and the thrust of the engines must counter the drag. The goal is to use principles of physics to achieve efficient flight. A successful design requires

TABLE 26.1 Key Current Transport Aircraft

Aircraft	TOGW (lb)	Empty weight (lb)	Wingspan (ft)	Number of passengers	Range (nm)	Cruise (mach)	Takeoff distance (ft)	Landing distance (ft)
				Narrow body				
A320-200	169,800	92,000	111.8	150	3,500	0.78	5,900	4,800
B717-200	121,000	68,500	93.3	106	2,371	0.76	5,750	5,000
B737-600	143,500	81,000	112.6	110	3,511	0.782	5,900	4,400
B757-300	273,000	141,690	124.8	243	3,908	0.80	8,650	5,750
				Wide body				
A330-300	513,670	274,650	197.8	440	6,450	0.82	8,700	5,873
A340-500	811,300	376,800	208.2	375	9,960	0.83	10,450	6,601
A380-800	1,234,600	611,000	261.8	555	9,200	0.85	9,350	6,200
B747-400	875,000	398,800	211.4	416	8,356	0.85	9,950	7,150
B747-400ER	911,000	406,900	211.4	416	8,828	0.85	10,900	7,150
B767-300	345,000	196,000	156.1	218	5,450	0.80	7,550	5,200
B777-300	660,000	342,900	199.9	368	6,854	0.84	12,150	6,050
B777-300ER	750,000	372,800	212.6	365	8,258	0.84	10,700	6,300
				Regional jets				
CRJ200(ER)	51,000	30,500	69.7	50	1,895	0.74	5,800	4,850
CRJ700(ER)	75,000	43,500	76.3	70	2,284	0.78	5,500	4,850
ERJ135ER	41,888	25,069	65.8	37	1,530	0.76	5,052	4,363
ERJ145ER	54,415	26,270	65.8	50	1,220	0.76	5,839	4,495

TOGW, takeoff gross weight.

the careful integration of a number of different disciplines. To understand the basic issues, we need to establish the terminology and fundamentals associated with the key flight disciplines. These include:

- Aerodynamics
- Propulsion
- Control and stability
- Structures/materials
- Avionics and systems

Shevell (1989) describes these disciplines as related to airplane design, together with methods used to compute airplane performance.

To understand how to balance these technologies, designers use weight. The lightest airplane that does the job is considered the best. The real metric should be some form of cost, but this is difficult to estimate. Traditionally, designers have used weight as a surrogate for cost. For designs using similar technology and sophistication, the lightest airplane costs least. One airplane designer has said that airplanes are like hamburger, you buy them by the pound. A study carried out at Boeing (Jensen, Rettie, and Barber 1981) showed that an airplane designed to do a given mission at minimum takeoff weight was a good design for a wide range of operating conditions compared to an airplane designed for minimum fuel use or minimum empty weight.

We can break the weight of the airplane up into various components. For our purposes, we will consider the weight to be:

$$W_{TO} = W_{empty} + W_{fuel} + W_{payload} \tag{26.1}$$

where W_{TO} = takeoff weight
W_{empty} = empty weight, mainly the structure and the propulsion system
W_{fuel} = fuel weight
$W_{payload}$ = payload weight, which for commercial transports is passengers and freight

Very crudely, W_{empty} is related to the cost to build the airplane and W_{fuel} is the cost to operate the airplane. The benefit of a new technology is assessed by examining its effect on weight.

Example. Weight is critically important in aircraft design. This example illustrates why. If

$$W_{TO} = W_{struct} + W_{prop} + W_{fuel} + \underbrace{W_{payload} + W_{systems}}_{W_{fixed}}$$

$$= W_{TO} \left(\frac{W_{struct}}{W_{TO}} + \frac{W_{prop}}{W_{TO}} + \frac{W_{fuel}}{W_{TO}} \right) + W_{fixed}$$

or

$$\left[1 - \left(\frac{W_{struct}}{W_{TO}} + \frac{W_{prop}}{W_{TO}} + \frac{W_{fuel}}{W_{TO}} \right) \right] W_{TO} = W_{fixed}$$

and

$$W_{TO} = \frac{W_{fixed}}{\left[1 - \left(\dfrac{W_{struct}}{W_{TO}} + \dfrac{W_{prop}}{W_{TO}} + \dfrac{W_{fuel}}{W_{TO}} \right) \right]}$$

Using weight fractions, which is a typical way to view the design, the stuctural fraction could be 0.25, the propulsion fraction 0.1, and the fuel fraction 0.40. Thus:

$$W_{TO} = \frac{W_{fixed}}{(1 - 0.75)} = 4 \cdot W_{fixed}$$

Here 4 is the *growth factor,* so that for each pound of increased fixed weight, the airplane weight increases by 4 pounds to fly the same distance. Also, note that the denominator can approach zero if the problem is too difficult. This is an essential issue for aerospace systems. Weight control and accurate estimation in design are very important.

The weight will be found for the airplane carrying the design payload over the design range. To connect the range and payload to the weight, we use the equation for the range R, known as the Brequet range equation:

$$R = \frac{V(L/D)}{sfc} \ln \left(\frac{W_i}{W_f} \right) \tag{26.2}$$

where R = range of the airplane (usually given in the design requirement)
V = airplane speed
L = lift of the airplane (assumed equal to the weight of the airplane, W)
D = drag

Since the weight of the plane varies as fuel is used, the values inside the log term correspond to the initial weight W_i and the final weight W_f. The specific fuel consumption (sfc) is the fuel used per pound of thrust per hour. The aerodynamic efficiency is measured by the lift to drag ratio L/D, the propulsive efficiency is given by sfc, and the structural efficiency is given by the empty weight of the plane as a fraction of the takeoff weight.

26.2.1 Aerodynamics

The airplane must generate enough lift to support its weight, with a low drag so that the L/D ratio is high. For a long-range transport this ratio should approach 20. The lift also has to be distributed around the center of gravity so that the longitudinal (pitching) moment about the center of gravity can be set to zero through the use of controls without causing extra drag. This requirement is referred to as *trim*. Extra drag arising from this requirement is *trim drag*.

Drag arises from several sources. The viscosity of the air causes friction on the surface exposed to the airstream. This "wetted" area should be held to a minimum. To account for other drag associated with surface irregularities, the drag includes contributions from the various antennas, fairings, and manufacturing gaps. Taken together, this drag is generally known as *parasite drag*. The other major contribution to drag arises from the physics of the generation of lift and is thus known as *drag due to lift*. When the wing generates lift, the flowfield is deflected down, causing an induced angle over the wing. This induced angle leads to an induced drag. The size of the induced angle depends on the span loading of the wing and can be reduced if the span of the wing is large. The other contribution to drag arises due to the presence of shock waves. Shock waves start to appear as the plane's speed approaches the speed of sound, and the sudden increase in drag once caused an engineer to describe this "drag rise" as a "sound barrier."

To quantify the aerodynamic characteristics, designers present the aerodynamic characteristics in coefficient form, removing most of the size effects and making the speed effects more clear. Typical coefficients are the lift, drag, and pitching moment coefficients, which are:

$$C_L = \frac{L}{\frac{1}{2}\rho V_\infty^2 S_{\text{ref}}}, \qquad C_D = \frac{D}{\frac{1}{2}\rho V_\infty^2 S_{\text{ref}}}, \qquad C_m = \frac{M}{\frac{1}{2}\rho V_\infty^2 \bar{c} S_{\text{ref}}} \qquad (26.3)$$

where L, D, and M are the lift, drag, and pitching moment, respectively. These values are normalized by the dynamic pressure q and a reference area S_{ref} and length scale c, as appropriate. The dynamic pressure is defined as $q = 1/2\rho V^2$. Here ρ is the atmospheric density. The subscript infinity refers to the freestream values. One other nondimensional quantity also frequently arises, know as the aspect ratio, $\text{AR} = b^2/S_{\text{ref}}$.

In particular, the drag coefficient is given approximately as a function of lift coefficient by the relation

$$C_D = C_{D_0} + \frac{C_L^2}{\pi A R E} \qquad (26.4)$$

where C_{D_0} is the parasite drag and the second term is the drag due to lift term mentioned above. E is the airplane efficiency factor, usually around 0.9. Many variations on this formula are available, and in particular, when the airplane starts to approach the speed of sound and wave drag starts to arise, the formula needs to include an extra term, $C_{D_{\text{wave}}} (M, C_L)$. Assuming that wave drag is small and that the airplane is designed to avoid flow separation at its maximum efficiency, the drag relation given above can be used to find the maximum value of L/D (which occurs when the parasite and induced drag are equal) and the corresponding C_L:

$$\left(\frac{L}{D}\right)_{\text{max}} = \frac{1}{2} \sqrt{\frac{\pi ARE}{C_{D_0}}} \tag{26.5}$$

and

$$(C_L)_{L/D_{\text{max}}} = \sqrt{\pi AREC_{D_0}} \tag{26.6}$$

These relations show the importance of streamlining to achieve a low C_{D_0}. They also seem to suggest that the aspect ratio should be large. However, the coefficient form is misleading here, and the way to reduce the induced drag D_i is actually best shown by the dimensional form:

$$D_i = \frac{1}{q\pi E}\left(\frac{W}{b}\right)^2 \tag{26.7}$$

where b is the span of the wing.

Finally, to delay the onset of drag arising from the presence of shock waves, the wings are swept. We will present a table below containing the values of wing sweep for current transports.

The other critical aspect of aerodynamic design is the ability to generate a high enough lift coefficient to be able to land at an acceptable speed. This is characterized by the value of $C_{L_{\text{max}}}$ for a particular configuration. The so-called stalling speed (V_{stall}) of the airplane is the slowest possible speed at which the airplane can sustain level flight, and can be found using the definition of the lift coefficient as:

$$V_{\text{stall}} = \sqrt{\frac{2(W/S)}{\rho C_{L_{\text{max}}}}} \tag{26.8}$$

and to achieve a low stall speed we need either a low wing loading, W/S, or a high $C_{L_{\text{max}}}$. Typically, an efficient wing loading for cruise leads to a requirement for a high value of $C_{L_{\text{max}}}$, meaning that high-lift systems are required. High-lift systems consist of leading and trailing edge devices such as single-, double-, and even triple-slotted flaps on the rear of the wing, and possibly slats on the leading edge of the wing. The higher the lift requirement, the more complicated and costly the high lift system has to be. In any event, mechanical high-lift systems have a $C_{L_{\text{max}}}$ limit of about 3. Table 26.2 provides an example of the $C_{L_{\text{max}}}$ values for various Boeing airplanes. These values are cited by Brune and McMasters (1990). A good recent survey of high-lift systems and design methodology is van Dam (2002).

TABLE 26.2   Values of $C_{L_{\text{max}}}$ for Some Boeing Airplanes

Model	$C_{L_{\text{max}}}$	Device type
B-47/B-52	1.8	Single-slotted Fowler flap
367-80/KC-135	1.78	Double-slotted flap
707-320/E-3A	2.2	Double-slotted flap and Kreuger leading edge flap
727	2.79	Variable camber Kreuger and triple-slotted flap
747/E-4A	2.45	Variable camber Kreuger and triple-slotted flap
767	2.45	Slot and single-slotted flap

26.2.2 Propulsion

Virtually all modern transport aircraft use high-bypass-ratio turbofan engines. These engines are much quieter and more fuel efficient than the original turbojet engines. The turbofan engine has a core flow that passes through a compressor and then enters the combustor and drives a turbine. This is known as the hot airstream. The turbine also drives a compressor that accelerates a large mass of air that does not pass through the combustor and is know as the cold flow. The ratio of the cold air to the hot air is the bypass ratio. From an airplane design standpoint, the key considerations are the engine weight per pound of thrust and the fuel consumption.

$$W_{eng} = \frac{T}{(T/W)_{eng}} \tag{26.9}$$

where the engine thrust is given by T. Typical values of the T/W of a high-bypass-ratio engine are around 6–7. The fuel flow is given as:

$$sfc = \frac{\dot{w}_f}{T} \tag{26.10}$$

where $\dot{w}_f$ is the fuel flow in lb/hr and the thrust is given in pounds. Thus, the units for sfc are per hour. There can be some confusion in units because the sfc is sometimes described as a mass flow. But in the United States the quoted values of sfc are as a weight flow. Table 26.3 provides the characteristics of the engines used in the aircraft listed in Table 26.1.

Values for thrust and fuel flow of an engine are quoted for sea-level static conditions. Both the maximum thrust and fuel flow vary with speed and altitude. In general, the thrust decreases with altitude, and with speed at sea level, but remains roughly constant with speed at altitude. The sfc increases with speed and decreases with altitude. Examples of the variations can be found in Appendix E of Raymer (1999). More details on engines related to airplane design can be found in Cumpsty (1998).

26.2.3 Control and Stability

Safety plays a key role in defining the requirements for ensuring that the airplane is controllable in all flight conditions. Stability of motion is obtained either through the basic airframe stability characteristics or by the use of an electronic control system providing apparent stability to the pilot or autopilot. Originally airplane controls used simple cable systems to move the surfaces. When airplanes became large and fast, the control forces using these types of controls became too large for the pilots to be able to move surfaces and hydraulic systems were incorporated. Now some airplanes are using electric actuation. Traditionally, controls are required to pitch, roll, and yaw the airplane. Pitch stability is provided by the horizontal stabilizer, which has an elevator for control. Similarly, directional stability is provided by the vertical stabilizer, which incorporates a rudder for directional control. Roll control is provided by ailerons, which are located on the wing of the airplane. In some cases one control surface may be required to perform several functions, and in some cases multiple surfaces are used simultaneously to achieve the desired control. A good reference for control and stability is Nelson (1997).

Critical situations defining the size of the required controls include engine-out conditions, crosswind takeoff and landing, and roll response. Longitudinal control requirements are dictated by the ability to rotate the airplane nose up at takeoff and generate enough lift when the airplane slows down to land. These conditions have to be met under all flight and center-of-gravity location conditions.

TABLE 26.3   Engines for Current Transport Aircraft

Aircraft	Engine	Thrust (lb)	Weight (lb)	sfc	T/W_{eng}
		Narrow body			
A320-200	IAE V2527-A5	26,500	5,230	0.36	5.1
B717-200	RR BR 715	21,000	4,597	0.37	4.6
B737-600	CFM56-7B	20,600	5,234	0.36	3.9
B757-300	PW 2040	41,700	7,300	0.345	5.7
		Wide body			
A330-300	Trent 768	71,100	10,467	0.56	6.8
A340-500	Trent 553	53,000	10,660	0.54	5.0
A380-800	Trent 970	70,000	—	0.51	—
B747-400	GE CF6-80C2	58,000	9,790	0.323	5.9
B747-400ER	GE CF6-80C2	58,000	9,790	0.323	5.9
B767-300	GE CF6-80C2	58,100	9,790	0.317	5.9
B777-300	RR Trent 892	95,000	13,100	0.56	7.25
B777-300ER	GE90-115	115,000	18,260	—	6.3
		Regional jets			
CRJ200(ER)	GE CF34-3B1	9,220	1,670	0.346	5.5
CRJ700(ER)	GE CF34-8C1	13,790	2,350	0.37	5.9
ERJ135ER	AE3007-A3	8,917	1,586	0.63	5.6
ERJ145ER	AE3007-A1/1	8,917	1,586	0.63	5.6

26.2.4   Structures/Materials

Aluminum has been the primary material used in commercial transports. However, composite materials have now reached a stage of development that allows them to be widely used, providing the required strength at a much lighter weight. The structure is designed for an extremely wide range of loads, including taxiing and ground handling (bump, touchdown, etc.) and flight loads for both sustained maneuvers and gusts.

Typically, transport aircraft consist of a constant cross-section pressurized fuselage that is essentially round and a wing that is essentially a cantilever beam. The constant cross-section of the fuselage allows the airplane to be stretched to various sizes by adding additional frames, some in front of and some behind the wing, to allow the plane to be properly balanced. However, if the airplane becomes too long, the tail will scrape the ground when the airplane rotates for takeoff. The wing typically consists of spars running along the length of the wing and ribs running between the front and back of the wing. The wing is designed so that fuel is carried between the front and rear spars. Fuel is also carried in the fuselage, where the wing carry-through structure is located. Carrying fuel in the wing as well as the wing support of pylon-mounted engines helps reduce the structural weight required by counteracting the load due to the wing lift. Because the wing is a type of cantilever beam, the wing weight is reduced by increasing the depth of the beam, which increases the so-called thickness-to-chord ratio (t/c). This increases the aerodynamic drag. Thus, the proper choice of t/c requires a system-level trade-off. An excellent book illustrating the structural design of transport aircraft is Niu (1998).

26.2.5 Avionics and Systems

Modern aircraft incorporate many sophisticated systems to allow them to operate efficiently and safely. The electronic systems are constantly changing, and current periodicals such as *Aviation Week* should be read to find out about the latest trends. The survey by Kayton (2003) provides an excellent overview of the electronics systems used on transports. Advances in the various systems allowed modern transports airplanes to use two-man crews. Fielding (1999) has a good summary of the systems use on transport aircraft. The basic systems are:

Avionics Systems

Communications
Navigation
Radar
Auto pilot
Flight control system

Other Systems

Air conditioning and pressurization
Anti-icing
Electrical power system
Hydraulic system
Fuel system
Auxiliary power unit (APU)
Landing gear

Each of these systems, listed in a single line, is associated with entire companies dedicated to providing safe, economical components for the aircraft industry.

26.3 *TRANSPORT AIRCRAFT DESIGN CONSIDERATIONS AND REQUIREMENTS*

26.3.1 The Current Environment and Key Issues for Aircraft Designers

In addition to the overall selection of the number of passengers and design range, described above, the designer has to consider a number of other issues. One key issue has been the selection of the seat width and distance between seats, the pitch. The seating arrangements are closely associated with the choice of the fuselage diameter. This has been a key design issue since the selection of the fuselage diameter for the DC-8 and B-707s, the first modern jet transports. This can be a key selling point of the aircraft. For example, currently Boeing uses the same fuselage diameter for its 737 and 757 transports: 148 inches. The comparable Airbus product, the A320, uses a fuselage diameter of 155 inches. Because of the details of the interior arrangements, both companies argue that they have superior passenger comfort. Typically, in economy class the aisles are 18 inches wide and the seats are approximately 17.5 to 19 inches wide, depending on how they are measured (whether the armrest is considered). In general, the wider the aircraft, the more options are available, and the airlines can select the seating arrangement.

The distance between rows, known as the pitch, can be selected by the airline and is not as critical to the design process. Airplanes can be lengthened or shortened relatively cheaply. The fuselage diameter essentially cannot be changed once the airplane goes into production. Typically, the pitch for economy class is 30 inches, increasing for business and first class seating.

Emergency exits (which are dictated by regulatory agencies), overhead bins, and lavatories are also key considerations. In addition, access for service vehicles has to be considered. In some cases, enough ground clearance must be included that carts can pass underneath the airplane.

In addition to passengers, transport airlines depend on freight for a significant portion of their revenue. Thus, the room for baggage and freight also requires attention. There are a number of standardized shipping containers, and the fuselage must be designed to accommodate them. The most common container, known as an LD-3, can be fit two abreast in a B777. The LD-3 is 64 inches high and 60.4 inches deep. The cross section is 79 inches wide at the top and 61.5 inches wide at the bottom, the edge being clipped off at approximately a 45° angle to allow it to fit efficiently within the near circular fuselage cross section. This container has a volume of 158 cubic feet and can carry up to 2,830 pounds.

The modern transports turn out to have about the right volume available as a natural consequence of the near-circular cylindrical fuselage and the single passenger deck seating. Regional jets, which have smaller fuselage diameters, frequently cannot fit all the passenger luggage in the plane. When you are told that "The baggage didn't make the flight," it probably actually means it didn't fit on the plane. A similar problem exists with large double-deck transports, where some of the main deck may be required to be used for baggage and freight.

Details of passenger cabin layout are generally available from the manufacturer's website. Boeing and Embraer are particularly good. Texts such as Jenkinson, Simpkin, and Rhodes (1999) provide more details.

26.3.2   Regulatory Requirements

The aircraft designer has to accommodate numerous requirements. Safety is of paramount importance and is associated with numerous regulatory considerations. Environmental considerations are also important, with noise and emissions becoming increasingly critical, especially in Europe. In addition, security has become an important consideration. These requirements arise independently of the aircraft economics, passenger comfort, and performance characteristics of the introduction of a successful new airplane.

In the United States the Federal Aviation Administration (FAA) must certify aircraft. The requirements are given in Federal Airworthiness Regulations (FARs). In Europe the regulations are Joint Airworthiness Requirements (JARs). Commercial aircraft are generally governed by:

- Regulatory design requirements:
  - FAR Pt 25: the design of the aircraft
  - FAR Pt 121: the operation of the aircraft
  - FAR Pt 36: noise requirements
  - Security
  - Airport requirements
  - Icing
  - Extended-range twin-engine operations (ETOPS)

An airplane design has to be consistent with the airports it is expected to use. Details of airport design for different size airplanes can be found in Ashford and Wright (1992). Table 26.4 defines the basic characteristics. The FAA sets standards and defines airplanes within six categories, related to the airplane wingspan. A key consideration for new large airplanes

TABLE 26.4 FAA Airplane Design Groups for Geometric Design of Airports

Airplane design group	Wingspan (ft)	Runway width (ft)	Runway centerline to taxiway centerline (ft)
I	up to 49	100	400
II	49–79	100	400
III	79–118	100	400
IV	118–171	150	400
V	171–197	150	varies
VI	197–262	200	600

is the maximum wingspan on the class VI airport of 262 feet, the so-called 80-meter gatebox limit. The new Airbus A380, listed in Table 26.1, is constrained in span to meet this requirement. Because we have shown that the wingspan is a key to low induced drag, it is clear that the A380 will be sacrificing aerodynamic efficiency to meet this requirement.

Another issue for airplane designers is the thickness of the runway required. If too much weight is placed on a tire, the runway may be damaged. Thus, you see fuselage-mounted gears on a B747, and the B-777 has a six-wheel bogey instead of the usual four-wheel bogey. This general area is known as flotation analysis. Because of the weight concentrated on each tire, the pavement thickness requirements can be considerable. The DC-10 makes the greatest demands on pavements. Typically, it might require asphalt pavements to be around 30 inches thick and concrete pavements to be 13 inches thick. An overview of landing gear design issues is available in Chai and Mason (1996).

26.4 VEHICLE OPTIONS: DRIVING CONCEPTS—WHAT DOES IT LOOK LIKE?

26.4.1 The Basic Configuration Arrangement

The current typical external configuration of both large and small commercial transport airplanes is similar, having evolved from the configuration originally chosen by Boeing for the Boeing B-47 medium-range bomber shortly after World War II. This configuration arose following the development of the jet engine by Frank Whittle in Britain and Hans von Ohain in Germany, which allowed for a significant increase in speed (Gunston 1995). The discovery of the German aerodynamics development work on swept wings to delay the rapid increase in drag with speed during World War II was incorporated into several new jet engine designs, such as the B-47, immediately after the war. Finally, Boeing engineers found that jet engines could be placed on pylons below the wing without excessive drag. This defined the classic commercial transport configuration. The technical evolution of the commercial transport has been described by Cook (1991), who was an active participant. A broader view of the development, including business, financial, and political aspects of commercial transports, has been given by Irving (1993). The other key source of insight into the development of these configurations is by Loftin (1980).

So where do we start when considering the layout of an airplane? In general, form follows function. We decide on candidate configurations based on what the airplane is supposed to do. Generally, this starts with a decision on the type of payload and the mission the airplane is supposed to carry out with this payload. This is expressed generally in terms of:

• What does it carry?
• How far does it go?

- How fast is it supposed to fly?
- What are the field requirements? (How short is the runway?)
- Are there any maneuvering and/or acceleration requirements?

Another consideration is the specific safety-related requirements that must be satisfied. As described above, for commercial aircraft this means satisfying the Federal Air Regulations (FARs) and JARs for Europe. Satisfying these requirements defines the takeoff and landing distances, engine-out performance requirements, noise limits, icing performance, and emergency evacuation, among many others.

With this start, the designer develops a concept architecture and shape that responds to the mission. At the outset, the following list describes the considerations associated with defining a configuration concept. At this stage we begin to see that configuration design resembles putting a puzzle together. These components all have to be completely integrated.

- Configuration concept:
  - Lifting surface arrangement
  - Control surface(s) location
  - Propulsion system selection
  - Payload
  - Landing gear

The components listed above must be coordinated in such a fashion that the airplane satisfies the requirements given in the following list. The configuration designer works to satisfy these requirements with input from the various team members. To be successful, the following criteria must be met:

- Good aircraft:
  - Aerodynamically efficient, including propulsion integration (streamlining)
  - Must balance near stability level for minimum drag
  - Landing gear must be located relative to *cg* to allow rotation at takeoff
  - Adequate control authority must be available throughout the flight envelope
  - Design to build easily (cheaply) and have low maintenance costs
  - Today, commercial airplanes must be quiet and nonpolluting

Two books do an especially good job of covering the aerodynamic layout issues: Whitford (1987) and Abzug and Larrabee (1997). The titles of both these works are slightly misleading. Further discussion of configuration options can be found in Raymer (1999) and Roskam (1987–1990).

We can translate these desirable properties into specific aerodynamic characteristics. Essentially, they can be given as:

- *Design for performance*
  - Reduce minimum drag:

      Minimize the wetted area to reduce skin friction

      Streamline to reduce flow separation (pressure drag)

      Distribute area smoothly, especially for supersonic aircraft (area ruling)

      Consider laminar flow

      Emphasize clean design/manufacture with few protuberances, steps or gaps
  - Reduce drag due to lift:

      Maximize span (must be traded against wing weight)

Tailor spanload to get good span *e* (twist)

Distribute lifting load longitudinally to reduce wave drag due to lift (a supersonic requirement, note R. T. Jones's oblique wing idea)

Camber as well as twist to integrate airfoil, maintain good two-dimensional characteristics

- Key constraints:

At cruise: buffet and overspeed constraints on the wing

Adequate high lift for field performance (simpler is cheaper)

Alpha tailscrape, $C_{L\alpha}$ goes down with sweep

- *Design for handling qualities*
  - Adequate control power is essential

Nose-up pitching moment for stable vehicles

Nose-down pitching moment for unstable vehicles

Yawing moment, especially for flying wings and fighters at high angle of attack

Consider the full range of *cg*'s.

Implies: must balance the configuration around the *cg* properly

- *FAA and military requirements*
  - Safety: for the aerodynamic configuration this means safe flying qualities

FAR Part 25 and some of Part 121 for commercial transports

MIL STD 1797 for military airplanes

Noise: community noise, FAR Part 36, no sonic booms over land (high *L/D* in the takeoff configuration reduces thrust requirements, makes plane quieter)

To start considering the various configuration concepts, we use the successful transonic commercial transport as a starting point. This configuration is mature. New commercial transports have almost uniformly adopted this configuration, and variations are minor. An interesting comparison of two different transport configuration development philosophies is available in the papers describing the development of the original Douglas DC-9 (Shevell and Schaufele 1966) and Boeing 737 (Olason and Norton 1966) designs. Advances in performance and reduction in cost are currently obtained by improvements in the contributing technologies. After we establish the baseline, we will examine other configuration component concepts that are often considered. We give a summary of the major options. Many, many other innovations have been tried, and we make no attempt to be comprehensive.

The Boeing 747 layout is shown in Figure 26.1. It meets the criteria cited above. The cylindrical fuselage carries the passengers and freight. The payload is distributed around the *cg*. Longitudinal stability and control power comes from the horizontal tail and elevator, which has a very useful moment arm. The vertical tail provides directional stability, using the rudder for directional control. The swept wing/fuselage/landing gear setup allows the wing to provide its lift near the center of gravity and positions the landing gear so that the airplane can rotate at takeoff speed and also provides for adequate rotation without scraping the tail (approximately 10 percent of the weight is carried by the nose gear). The wing has a number of high-lift devices. This arrangement also results in low trimmed drag. The engines are located on pylons below the wing. This arrangement allows the engine weight to counteract the wing lift, reducing the wing root bending moment, resulting in a lighter wing.

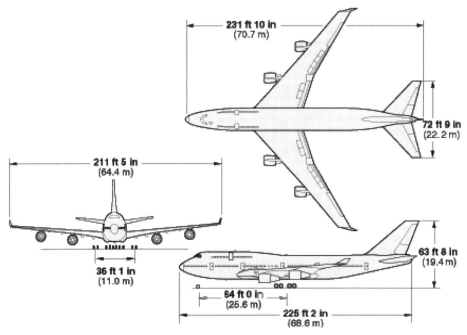

FIGURE 26.1 The classic commercial transport, the Boeing 747 (www.boeing.com).

This engine location can also be designed so that there is essentially no adverse aerodynamic interference.

26.4.2 Configuration Architecture Options

Many another arrangements are possible, and here we list a few typical examples. All require attention to detail to achieve the claimed benefits.

- *Forward swept wings:* reduced drag for severe transonic maneuvering conditions
- *Canards:* possibly safety, also possibly reduced trim drag, and supersonic flight
- *Flying wings:* elimination of wetted area by eliminating fuselage and tail surfaces
- *Three-surface configurations:* trim over wide *cg* range
- *Slender wings:* supersonic flight
- *Variable sweep wings:* good low speed, low altitude penetration, and supersonic flight
- *Winglets:* reduced induced drag without span increase

Improvements to current designs can occur in two ways. One way is to retain the classic configuration and improve the component technologies. This has been the recent choice for new aircraft, which are mainly derivatives of existing aircraft, using refined technology, e.g., improved aerodynamics, propulsion, and materials. The other possibility for improved designs is to look for another arrangement.

Because of the long evolution of the current transport configuration, the hope is that it is possible to obtain significantly improved aircraft through new configuration concepts. Studies looking at other configurations as a means of obtaining an aircraft that costs less to build and operate are being conducted. Two concepts have received attention recently. One integrates the wing and the fuselage into a blended wing body concept (Liebeck 2002), and a second uses strut bracing to allow for increased wingspan without increasing the wing weight (Gundlach et al. 2000).

26.4.3 The Blended Wing Body

The blended wing body concept (BWB) combines the fuselage and wing into a concept that offers the potential of obtaining the aerodynamic advantages of the flying wing while providing the volume required for commercial transportation. Figure 26.2 shows the concept. This configuration offers the potential for a large increase in L/D and an associated large reduction in fuel use and maximum takeoff gross weight (TOGW). The major overview is given by Liebeck (2002), who predicted that the BWB would have an 18 percent reduction in TOGW and 32 percent in fuel burn per seat compared to the proposed A380-700.

Because the BWB does not have large moment arms for generating control moments, and also requires a nontraditional passenger compartment, the design is more difficult than traditional designs and requires the use of multidisciplinary design optimization methods to obtain the predicted benefits (Wakayama 1998). Recently the concept has been shown to be able to provide a significant speed advantage over current commercial transports (Roman, Gilmore, and Wakayama 2003). Because of the advantages of this concept, it has been studied by other design groups.

26.4.4 The Strut-Braced Wing

Werner Pfenninger suggested the strut-braced wing concept around 1954. His motivation was actually associated with the need to reduce the induced drag to balance his work in reducing

FIGURE 26.2   The blended wing body concept (courtesy Boeing).

FIGURE 26.3 The strut-braced wing concept.

parasite drag by using active laminar flow control to maintain laminar flow and reduce skin friction drag. Since the maximum L/D occurs when the induced and parasite drag are equal, the induced drag had to be reduced also. The key issues are:

- Once again, the tight coupling between structures and aerodynamics requires the use of MDO (multidisciplinary design optimization) (see section 26.7) to make it work.
- The strut allows a thinner wing without a weight penalty and also a higher aspect ratio and less induced drag.
- Reduced t/c allows less sweep without a wave drag penalty.
- Reduced sweep leads to *even lower* wing weight.
- Reduced sweep allows for some natural laminar flow and thus reduced skin friction drag.

The benefits of this concept are similar to the benefits cited above for the BWB configuration. The advantage of this concept is that it does not have to be used on a large airplane. The key issue is the need to provide a mechanism to relieve the compression load on the strut under negative g loads. Work on this concept was done at Virginia Tech (Grasmeyer et al. 1998; Gundlach et al. 2000). Figure 26.3 shows the result of a joint Virginia Tech-Lockheed Martin study.

There are numerous options for the shape of the aircraft. Other possibilities exist, and there is plenty of room for imagination. See Whitford (1987) for further discussion of configuration options.

26.5 VEHICLE SIZING—HOW BIG IS IT?

Once a specific concept is selected, the next task is to determine how big the airplane is, which essentially means how much it weighs. Typically, for a given set of technologies the maximum takeoff gross weight is used as a surrogate for cost. The lighter the airplane, the less it costs, both to buy and operate. Some procedures are available to estimate the size of the airplane. This provides a starting point for more detailed design and sizing and is a critical element of the design. The initial "back-of-the-envelope" sizing is done using a database of existing aircraft and developing an airplane that can carry the required fuel and passengers to do the desired mission. This usually means acquiring data similar to the data presented in Tables 26.1 and 26.2 and doing some preliminary analysis to obtain an idea of the wing area required in terms of the wing loading W/S and the thrust to weight ratio T/W, as shown in Table 26.5.

Following Nicolai (1975), consider the TOGW, called here W_{TO}, to be:

$$W_{TO} = W_{fuel} + W_{fixed} + W_{empty} \tag{26.11}$$

where the fixed weight includes a nonexpendable part, which consists of the crew and equipment, and an expendable part, which consists of the passengers and baggage or freight. W_{empty} includes all weights except the fixed weight and the fuel. The question becomes: for a given (assumed) TOGW, is the weight left enough to build an airplane when we subtract the fuel and payload? We state this question in mathematical terms by equating the available and required empty weight:

$$W_{Empty\,Avail} = W_{Empty\,Reqd} \tag{26.12}$$

where $W_{Empty\,Reqd}$ comes from the following relation:

$$W_{Empty\,Reqd} = KS \cdot A \cdot TOGW^B \tag{26.13}$$

and KS is a structural technology factor and A and B come from the data gathered from

TABLE 26.5 Derived Characteristics of Current Transport Aircraft

Aircraft	TOGW (lb)	Empty weight (lb)	Wing area (ft²)	Sweep (quarter chord)	Aspect ratio	W/S	W/b	T/W
Narrow body								
A320-200	169,800	92,000	1,320	25.0	9.47	129	1,519	0.312
B717-200	121,000	68,500	1,001	24.5	8.70	121	1,297	0.347
B737-600	143,500	81,000	1,341	25.0	9.45	107	1,274	0.287
B757-300	273,000	141,690	1,951	25.0	7.98	140	2,188	0.305
Wide body								
A330-300	513,670	274,650	3,890	30.0	10.06	132	2,597	0.272
A340-500	811,300	376,800	4,707	30.0	9.21	172	3,897	0.261
A380-800	1,234,600	611,000	9,095	33.5	7.54	136	4,716	0.227
B747-400	875,000	398,800	5,650	37.5	7.91	155	4,139	0.265
B747-400ER	911,000	406,900	5,650	37.5	7.91	161	4,309	0.255
B767-300	345,000	196,000	3,050	31.5	7.99	113	2,210	0.337
B777-300	660,000	342,900	4,605	31.6	8.68	143	3,302	0.278
B777-300ER	750,000	372,800	4,694	31.6	9.63	160	3,528	0.307
Regional jets								
CRJ200(ER)	51,000	30,500	520	26.0	9.34	98	732	0.36
CRJ700(ER)	75,000	43,500	739	26.8	7.88	102	983	0.37
ERJ135ER	41,888	25,069	551	20.3	7.86	76	637	0.43
ERJ145ER	54,415	26,270	551	20.3	7.86	82	690	0.39

information in Table 26.5. Now, the difference between the takeoff and landing weight is due to fuel used (the mission fuel). Figure 26.4 shows how this relation is found from the data. Note that KS is very powerful and should not be much less than 1 without a very good reason.

Next we define the mission in terms of segments and compute the fuel used for each segment. Figure 26.5 defines the segments used in a typical sizing program. Note that the mission is often defined in terms of a radius (an obvious military heritage). Transport designers simply use one-half of the desired range as the radius. At this level of sizing, reserve fuel is included as an additional range, often taken to be 500 nm. To use the least fuel, the airplane should be operated at its best cruise Mach number (BCM), and it best cruise altitude (BCA). Often, air traffic control or weather conditions may prevent being able to fly at these conditions in actual operation.

Mission segment definitions for Figure 26.5:

1–2	engine start and takeoff
2–3	accelerate to subsonic cruise velocity and altitude
3–4	subsonic cruise out
4–5	accel to high speed (supersonic) dash/cruise
5–5+	supersonic cruise out
	combat (use fuel, expend weapons)

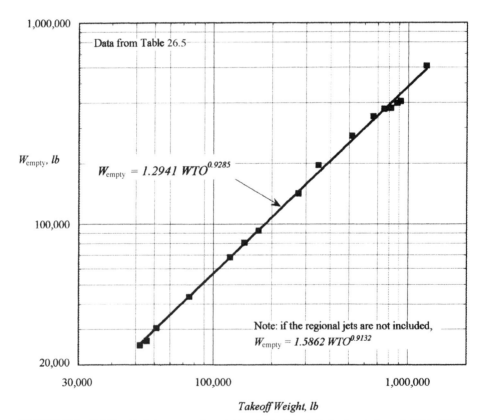

FIGURE 26.4 Relationship between empty weight and takeoff weight for the airplanes in Table 26.5.

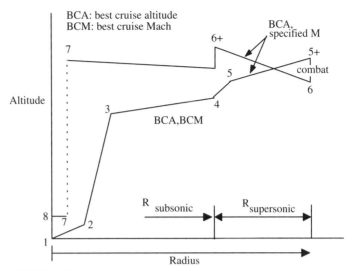

FIGURE 26.5 Mission definition.

6–6+   supersonic cruise back
6+–7   subsonic cruise back
7–8   loiter
8   land

To get the empty weight available, compute the fuel fraction for each mission segment. For the fuel fraction required for the range, invert the Brequet range equation given above:

$$\frac{W_{i+1}}{W_i} = e^{-R \cdot \text{sfc}/(V \cdot L/D)} \tag{26.14}$$

and for loiter:

$$\frac{W_{i+1}}{W_i} = e^{-R \cdot \text{sfc}/(L/D)} \tag{26.15}$$

The values of the cruise L/D and sfc have to be estimated, and the velocity for best range also has to be estimated, so it takes some experience to obtain these values. Note that this approach can also be used to establish the values of L/D and sfc required to perform a desired mission at a desired weight. Values for takeoff and climb are typically estimated and can be computed for more accuracy. However, for a transport aircraft the range requirement tends to dominate the fuel fraction calculation, with the rest of the fuel fractions being near unity. Therefore, we compute the mission weight fraction as:

$$\frac{W_{\text{final}}}{W_{\text{TO}}} = \frac{W_8}{W_1} = \underbrace{\frac{W_2}{W_1} \cdot \frac{W_3}{W_2} \cdot \frac{W_4}{W_3} \cdots \frac{W_8}{W_7}}_{\text{fuel fraction for each segment}} \tag{26.16}$$

and solve for the fuel weight in equation (26.16) as:

$$W_{\text{fuel}} = \left(1 + \frac{W_{\text{reserve}}}{W_{\text{TO}}} + \frac{W_{\text{trapped}}}{W_{\text{TO}}}\right)\left(1 - \frac{W_8}{W_1}\right) W_{\text{TO}}$$

$$= \left(1 + \frac{W_{\text{reserve}}}{W_{\text{TO}}} + \frac{W_{\text{trapped}}}{W_{\text{TO}}}\right)(W_{\text{TO}} - W_{\text{landing}})$$

(26.17)

so that we can compute $W_{\text{Empty Avail}}$ from:

$$W_{\text{Empty Avail}} = W_{\text{TO}} - W_{\text{fuel}} - W_{\text{fixed}}$$

(26.18)

The value of W_{TO} that solves the problem is the one for which $W_{\text{Empty Avail}}$ is equal to the value of $W_{\text{Empty Reqd}}$, which comes from the statistical representation for this class of aircraft. An iterative procedure is often used to find this value. The results of this estimate are used as a starting point for the design using more detailed analysis. A small program that makes these calculations is available on the Web (Mason n.d.).

We illustrate this approach with an example, also from Nicolai (1975). The example is for a C-5. In this case we pick:

Range: 6000 nm

Payload: 100,000 lb

sfc: 0.60 @ M = 0.8

h = 36,000 ft altitude

L/D = 17

Figure 26.6 shows how the empty and available weight relations intersect, defining the weight of the airplane, which is in reasonable agreement with a C-5A. Note that as the requirements become more severe, the lines will start to become parallel, the intersection weight will increase, and the uncertainty will increase because of the shallow intersection.

Note that the author's students have used this approach to model many commercial transport aircraft and it has worked well, establishing a baseline size very nearly equal to existing aircraft and providing a means of studying the impact of advanced technology on the aircraft size.

Once the weight is estimated, the engine size and wing size are picked considering constraints on the design. Typical constraints include the takeoff and landing distances and the cruise condition. Takeoff and landing distance include allowances for problems. The takeoff distance is computed such that in case of engine failure at the decision speed the airplane can either stop or continue the takeoff safely, and includes the distance required to clear a 35-foot obstacle. The landing distance is quoted including a 50-foot obstacle, and it includes an additional runway distance. Other constraints that may affect the design include the missed approach condition, the second segment climb (the ability to climb if an engine fails at a prescribed rate between 35 and 400 feet altitude), and the top-of-climb rate of climb. Jenkinson, Simpkins, and Rhodes (1999) have an excellent discussion of these constraints for transport aircraft. Figure 26.7 shows a notional constraint diagram for T/W and W/S. Typically, the engines of long-range airplanes are sized by the top-of-climb requirement and the engines of twin-engine airplanes are sized by the second-segment-climb requirement (Jenkinson, Simpkin, and Rhodes 1999).

26.6 CURRENT TYPICAL DESIGN PROCESS

The airplane design process is fairly well established. It starts with a conceptual stage, where a few engineers use the sizing approaches slightly more elaborate than those described above

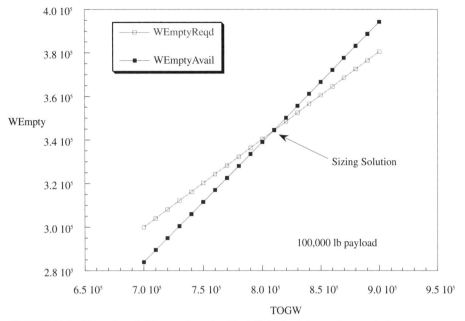

FIGURE 26.6 Illustration of sizing results using Nicolai's back-of-the-envelope method.

to investigate new concepts. Engine manufacturers also provide information on new engine possibilities or respond to requests from the airframer. If the design looks promising, it progresses to the next stage: preliminary design. At this point the characteristics of the airplane are defined and offered to customers. Since the manufacturer cannot afford to build the airplane without a customer, various performance guarantees are made, even though the

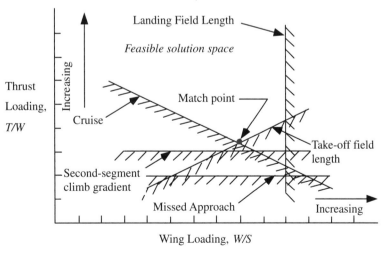

FIGURE 26.7 Typical constraint diagram (after Loftin 1980).

airplane has not been built yet. This is risky. If the guarantees are too conservative, you may lose the sale to the competition. If the guarantees are too optimistic, a heavy penalty will be incurred.

If the airplane is actually going to be built, it progresses to detail design. The following, from John McMasters of Boeing, describes the progression.

- *Conceptual design (1% of the people)*:
  - Competing concepts evaluated   What drives the design?
  - Performance goals established   Will it work/meet requirement?
  - Preferred concept selected   What does it look like?

- *Preliminary design (9% of the people)*:
  - Refined sizing of preferred concept   Start using big codes
  - Design examined/establish confidence   Do some wind tunnel tests
  - Some changes allowed   Make actual cost estimate (you bet your company)

- *Detail design (90% of the people)*:
  - Final detail design   Certification process
  - Drawings released   Component/systems tests
  - Detailed performance   Manufacturing (earlier now)
  - Only "tweaking" of design allowed   Flight control system design

26.7 MDO—THE MODERN COMPUTATIONAL DESIGN APPROACH

With the increase of computer power, new methods for carrying out the design of the aircraft have been developed. In particular, the interest is in using high-fidelity computational simulations of the various disciplines at the very early stages of the design process. The desire is to use the high-fidelity analyses with numerical optimization tools to produce better designs. Here the high-quality analysis and optimization can have an important effect on the airplane design early in the design cycle. Currently, high-fidelity analyses are used only after the configuration shape has been frozen. At that point it is extremely difficult to make significant changes. If the best tools can be used early, risk will be reduced, as will the design time. Recent efforts have also focused on means of using large-scale parallel processing to reduce the design cycle time. These various elements, taken all together, are generally known as multidisciplinary design optimization (MDO). One collection of papers has been published on the subject (Alexandrov and Hussaini 1997), and there is a major conference on MDO every other year sponsored by the AIAA, ISSMO, and other societies. Perhaps the best survey of our view of MDO for aircraft design is Giunta et al. (1996). We will outline the MDO process and issues based on these and other recent publications.

Our current view of MDO is that high-fidelity codes cannot be directly coupled into one major program. There are several reasons. Even with advanced computing, the computer resources required are too large to perform an optimization with a large number of design variables. For 30 or so design variables, with perhaps 100 constraints, hundreds of thousands of analyses of the high-fidelity codes are required. In addition, the results of the analyses are invariably noisy (Giunta et al. 1994), so that gradient-based optimizers have difficulty in producing meaningful results. In addition to the artificial noise causing trouble, the design space is nonconvex and many local optima exist (Baker 2002). Finally, the software integration issues are complex, and it is unlikely that major computational aerodynamic and structures codes can be combined. Thus, innovative methods are required to incorporate MDO into the early stages of airplane design.

Instead of a brute-force approach, MDO should be performed using surrogates for the high-fidelity analyses. This means that for each design problem, a design space should be constructed that uses a parametric model of the airplane in terms of design variables such as wingspan and chords, etc. The ranges of values of these design variable are defined, and a database of analyses for combinations of the design variables should be constructed. Because the number of combinations will quickly become extremely large, design of experiments theory will need to be used to reduce the number of cases that need to be computed. Because these cases can be evaluated independently of each other, this process can exploit coarse-grain parallel computing to speed the process. Once the database is constructed, it must be interpolated. In statistical jargon, this means constructing a response surface approximation. Typically, second-order polynomials are used. This process automatically filters out the noise from the analyses of the different designs. These polynomials are then used in the optimization process in place of the actual high-fidelity codes. This allows for repeated investigations of the design space with an affordable computational cost. A more thorough explanation of how to use advanced aerodynamics methods in MDO, including examples of trades between aerodynamics and structures, has been presented by Mason et al. (1998).

Current issues of interest in MDO also include the consideration of the effects of uncertainty of computed results and efficient geometric representation of aircraft. MDO is an active research area and will be a key to improving future aircraft design.

26.8 REFERENCES

Abzug, M. A., and E. E. Larrabee. 1997. *Airplane Stability and Control*. Cambridge: Cambridge University Press.

Alexandrov, N. M., and M. Y. Hussaini. 1997. *Multidisciplinary Design Optimization*. Philadelphia: SIAM.

Ashford, N., and P. H. Wright. 1992. *Airport Engineering*, 3rd ed. New York: John Wiley & Sons.

Aviation Week and Space Technology. 2003. *Sourcebook*. New York: McGraw-Hill, January 13.

Baker, C. A., B. Grossman, R. T. Haftka, W. H. Mason, and L. T. Watson. 2002. "High-Speed Civil Transport Design Space Exploration Using Aerodynamic Response Surface Approximations." *Journal of Aircraft* 39(2):215–20.

Brune, G. W., and J. H. McMasters. 1990. "Computational Aerodynamics Applied to High Lift Systems." In *Applied Computational Aerodynamics*, ed. P. Henne. Progress in Astronautics and Aeronautics 125. Washington, DC: AIAA, 389–413.

Chai, S. T., and W. H. Mason. 1996. *Landing Gear Integration in Aircraft Conceptual Design*. MAD Center Report MAD 96-09-01, September 1996, revised March 1997. http://www.aoe.vt.edu/~mason/Mason_f/MRNR96.html.

Cook, W. H. 1991. *The Road to the 707*. Bellevue: TYC.

Cumpsty, N. A. 1998. *Jet Propulsion: A Simple Guide to the Aerodynamic and Thermodynamic Design and Performance of Jet Engines*. Cambridge: Cambridge University Press.

Fielding, J. P. 1999. *Introduction to Aircraft Design*. Cambridge: Cambridge University Press.

Giunta, A. A., J. M. Dudley, B. Grossman, R. T. Haftka, W. H. Mason, and L. T. Watson. 1994. "Noisy Aerodynamic Response and Smooth Approximations in HSCT Design." AIAA Paper 94-4376, Panama City, FL, September.

Giunta, A. A., O. Golovidov, D. L. Knill, B. Grossman, W. H. Mason, L. T. Watson, and R. T. Haftka. 1996. "Multidisciplinary Design Optimization of Advanced Aircraft Configurations." In *Proceedings of the 15th International Conference on Numerical Methods in Fluid Dynamics*, ed. P. Kutler, J. Flores, and J.-J. Chattot. Lecture Notes in Physics 490. Berlin: Springer-Verlag, 14–34.

Grasmeyer, J. M., A. Naghshineh, P.-A. Tetrault, B. Grossman, R. T. Haftka, R. K. Kapania, W. H. Mason, and J. A. Schetz. 1998. *Multidisciplinary Design Optimization of a Strut-Braced Wing Aircraft with Tip-Mounted Engines.* MAD Center Report MAD 98-01-01, January.

Gundlach, J. F., IV, P.-A. Tetrault, F. H. Gern, A. H. Naghshineh-Pour, A. Ko, J. A. Schetz, W. H. Mason, R. K. Kapania, B. Grossman, and R. T. Haftka. 2000. "Conceptual Design Studies of a Strut-Braced Wing Transonic Transport." *Journal of Aircraft* 37(6):976–83.

Gunston, B. 1995. *The Development of Jet and Turbine Aero Engines.* Sparkford: Patrick Stevens.

Irving, C. 1993. *Wide-Body: The Triumph of the 747.* New York: William Morrow & Co.

Jackson, P., ed. 2002–2003. *Jane's All the World's Aircraft.* Surrey: Jane's Information Group.

Jenkinson, L. R., P. Simpkin, and D. Rhodes. 1999. *Civil Jet Aircraft Design.* London: Arnold; Washington, DC: AIAA. Electronic Appendix at http://www.bh.com/companions/034074152C/appendices/data-a/default.htm.

Jensen, S. C., I. H. Rettie, and E. A. Barber. 1981. "Role of Figures of Merit in Design Optimization and Technology Assessment." *Journal of Aircraft* 18(2):76–81.

Kayton, M. 2003. "One Hundred Years of Aircraft Electronics." *Journal of Guidance, Control, and Dynamics* 26(2):193–213.

Liebeck, R. H. 2002. "The Design of the Blended-Wing-Body Subsonic Transport." AIAA Paper 2002-0002, January.

Loftin, L. K. 1980. *Subsonic Aircraft: Evolution and the Matching of Size to Performance.* NASA Reference Publication 1060, August.

Mason, W. H. ACsize, Software for Aerodynamics and Aircraft Design. http://www.aoe.vt.edu/~mason/Mason_f/MRsoft.html#Nicolai.

Mason, W. H., D. L. Knill, A. A. Giunta, B. Grossman, R. T. Haftka, and L. T. Watson. 1998. "Getting the Full Benefits of CFD in Conceptual Design." Paper delivered at AIAA 16th Applied Aerodynamics Conference, Albuquerque, NM, June. AIAA Paper 98-2513.

Nelson, R. C. 1997. *Flight Stability and Automatic Control,* 2nd ed. New York: McGraw-Hill.

Nicolai, L. M. 1975. *Fundamentals of Aircraft Design.* San José, CA: METS.

Niu, M. C. Y. 1999. *Airframe Structural Design,* 2nd ed. Hong Kong: Conmilit Press. [Available through ADASO/ADASTRA Engineering Center, P.O. Box 3552, Granada Hills, CA, 91394, Fax: (888) 735-8859, http://www.adairframe.com/~adairframe/.]

Olason, M. L., and D. A. Norton. "Aerodynamic Design Philosophy of the Boeing 737." *Journal of Aircraft* 3(6):524–28.

Raymer, D. P. 1999. *Aircraft Design: A Conceptual Approach,* 3rd ed. Washington, DC: AIAA.

Roman, D., R. Gilmore, and S. Wakayama. 2003. "Aerodynamics of High-Subsonic Blended-Wing-Body Configurations." AIAA Paper 2003-0554, January.

Roskam, J. *Airplane Design.* 1987–1990. 8 vols. Lawrence, KS: DARcorporation. http://www.darcorp.com/Textbooks/textbook.htm.

Schaufele, R. D. 2000. *The Elements of Aircraft Preliminary Design.* Santa Ana: Aries.

Shevell, R. S. 1989. *Fundamentals of Flight,* 2nd ed. Upper Saddle River, NJ: Prentice Hall.

Shevell, R. S., and R. D. Schaufele. 1966. "Aerodynamic Design Features of the DC-9." *Journal of Aircraft* 3(6):515–23.

Van Dam, C. P. 2002. "The Aerodynamic Design of Multi-element High-Lift Systems for Transport Airplanes" *Progress in Aerospace Sciences* 38:101–44.

Wakayama, S. 1998. "Multidisciplinary Design Optimization of the Blended-Wing-Body." AIAA Paper 98-4938, September.

Whitford, R. 1987. *Design for Air Combat.* Surrey: Jane's Information Group.

CHAPTER 27
AIRPORT PLANNING AND DESIGN

William R. Graves
School of Aeronautics, Florida Institute of Technology,
Melbourne, Florida

Ballard M. Barker
School of Aeronautics, Florida Institute of Technology,
Melbourne, Florida

27.1 INTRODUCTION

Transportation engineering is a broad and somewhat synthetic field, involving a myriad of public interests and multiple professional disciplines. The majority of engineers working on transportation issues were probably educated as civil engineers and are graduates of the many ABET-accredited programs in the United States that carefully prescribe the essentials of engineering civil works, which must be imparted in a four-year curriculum. Graduate degrees in civil engineering, as in most disciplines, tend to develop more depth than scope of expertise in the field. As a consequence, many competent and experienced engineers newly employed on transportation engineering works have little formal training or direct experience with transportation matters. Airport planning and design, a subset of transportation engineering, gets even less attention in civil engineering education. The majority of engineers probably receive their first education in airport matters in on-the-job training after being employed by a firm doing airport projects. Indeed, a persistent lament of many airport planning firms and architectural and engineering firms involved in airport projects has been that engineers generally lack necessary education and knowledge of airport-related matters.

It is the goal of this chapter to make the reader aware of the scope of airport-specific planning and design issues and to point toward the modest body of literature that addresses routine elements of airport planning and design, as well as issues, challenges, and approaches to help the transportation engineer address community needs. The central focus, however, is on those elements of airport design typically performed by transportation engineers, rather than those commonly performed by aviation planners, architects, or narrower engineering specialties such as electrical or mechanical.

27.2 AIRPORT PLANNING

Airport *planning* is a prerequisite of proper airport *design* activities. The need for thorough airport planning in the context of addressing a region's transportation, socio-economic, and environmental concerns is globally recognized. The International Civil Aviation Organization

(ICAO), an operating agency of the United Nations with nearly 190 contracting states, addresses the need for guidance material to assist states with planning for construction and expansion of international airports in its Annex 14 to the convention on civil aviation. ICAO publishes and periodically revises its Airport Planning Manual (ICAO Doc. 9184-AN/902), which details a model airport planning process for airside and landside development, operations and support facilities, and land use and environmental controls, and provides generic guidance for the selection of airport consultants and construction services.

In the United States, there is a Federal Aviation Administration (FAA)-driven, four-tiered bureaucratic hierarchy of products dealing with airport planning in one way or another. It is FAA-driven in the sense that since implementation of the Airport Development Aid Program in 1970, any of the nearly 4,000 national interest airports in the United States that wish to apply for federal airport development grants under the Airport Improvement Program, or receive FAA approval of local Passenger Facility Charges for airport development, must participate in the FAA-steered process.

At the local level, airport operators develop *airport master plans*. The airport master plan is asubstantial document designed to reflect the airport's, and hopefully the community's, realistic vision of the airport's missions, roles, and facilities 5, 10, and 20 years into the future. It lays out phased airport development based upon forecast needs, alternative assessments, and financial planning for accomplishment. The airport master plan is the one planning document commonly used by transportation engineers as a basis for gross design guidance. The form and process of the airport master plan will be discussed in more detail in a later section.

It is common for major metropolitan areas and self-defined regions within or between states to develop general plans to address regional airport and other transportation needs. These *regional/metropolitan airport systems plans* incorporate the visions of local airport master plans and seek to harmonize them while accommodating present and future transportation needs in the subject area. Adequacy of air service to the region and coordination of the roles of the various airports in the area are typical elements. Each state transportation agency also compiles a *statewide integrated airport system plan* that is similar in level of detail and scope of concerns to the regional/metropolitan plans. Statewide plans are typically designed to identify and coordinate airport development needs for enhancement of state economies, and sometimes assist in prioritization of state planning and development aid to airports and communities.

The National Plan of Integrated Airport Systems (NPIAS) incorporates much of the information on future airport development needs from the states' plans and complements it with FAA forecasts of airport terminal area and national aviation activity. The NPIAS is required by the Airport and Airways Development Act of 1970 to be published biennially, ostensibly to aid in the appropriation and programming of AIP funds by the federal government. It is quite literally nothing more than a compiled summary of the estimated cost of desired airport development projects at each of the nation's national interest airports for running 10-year periods, and is of little practical value airport to airport planners and engineers.

27.2.1 The Airport Master Plan

As stated earlier, the airport master plan is the airport's concept for future development of the airport to meet community needs and concerns. It is the planning document with the most relevance and utility to engineers working on airport projects. Its preparation is the airport's responsibility, but at most airports it is developed by aviation consulting firms working with key members of airport staff. Broad guidance on the process and content of master plans for U.S. airports is presented in FAA Advisory Circular 150/5070-6A, *Airport Master Plans*. Some states, such as Florida, also publish excellent guidelines for airport master planning. These state guides are typically intended to give practical guidance for the development of useful and cost-effective planning products for general aviation and small commercial service airports.

The airport master plan has several objectives in addition to presenting a technically, economically, and environmentally sound development concept. Primary is development of a meaningful *capital improvement program* (CIP) that addresses at least the subsequent 5-year period in detail. The CIP should be accompanied by a corresponding conceptual financial plan that will allow for the phased implementation of the CIP as projected development needs become actual needs. The airport community's actual and forecast aeronautical demands are documented in association with the community's role and priorities for airport development. The airport master plan process should also provide ample opportunity for public involvement in plan development, to include review by regional and state agencies. Last, the master plan should become the foundation and framework for responsive and effective evolution of the airport to meet future needs of the community.

The elements of airport master planning vary with each airport's particular situation and may include:

1. An inventory of all relevant airport infrastructure, land uses, operational activities, and development issues.

2. Forecasts of aeronautical demand on airport infrastructure for subsequent 5-, 10-, and 20-year periods. This is arguably the most critical step of the planning process because of its effect on all subsequent planning actions. This step should also include realistic predictions of future aircraft types that are likely to serve the airport.

3. Determination of future airport requirements and alternative concepts for demand satisfaction. This step involves comparison of present capabilities with forecast demand at 5-year intervals, and the development of concepts to best satisfy different infrastructure needs. Various simulations may be employed to assist analysis in complex situations.

4. Site location and situation of new, relocated, or expanded infrastructure.

5. Environmental analyses and procedural compliance. These may include environmental assessments, archeological studies, and documentation of the situation.

6. Airport plan drawings. A series of large-scale drawings of overall airport layout (airport layout plan), terminal area plan, land use plan, surface transportation access plan, and the airspace plan. The airport layout plan (ALP) is the single most important and commonly used drawing because it includes all minimally required information and is an FAA prerequisite for any federally assisted airport development.

General guidelines and useful references for completing each of the several airport master plan elements are provided in FAA Advisory Circular 150/5070-6A, *Airport Master Plans,* and will not be repeated in this text.

27.3 AIRPORT DESIGN

27.3.1 Airport Layout Design

Once master planning has determined the desired roles of the airport, the needed capacity, and the critical aircraft for which the airport should be designed, airport layout (also referred to as airport geometry) can be determined using the appropriate ICAO or FAA standards. Airport geometry refers to the two-dimensional horizontal layout of airport surfaces as they are depicted on the airport layout plan drawing contained in the master plan. A principal goal of the airport layout should be to allow for future growth and change in the aircraft movement area and other airside facilities.

The airport needs to be designed to the standards associated with the most demanding airplane(s) to be accommodated during the planning window. The most demanding airplane for runway length may not be the same airplane used for determining the appropriate airplane design group or for pavement strength. If the airport is to have two or more runways and associated movement areas, it is highly desirable to design all airport elements to the standards associated with the most demanding airplane(s). However, it may be more cost-effective for some airport elements, e.g., a secondary runway and its associated taxiway, to be designed to standards associated with less demanding airplanes that will use the airport.

Runway Location and Orientation.   Runway location and orientation are paramount to aviation safety, airport efficiency, airline operating economies, and environmental compatibility. The weight given to each of the following runway location and orientation factors depends, in part, on: (a) the characteristics of airplanes expected to use each runway; (b) the volume of air traffic expected on each runway; (c) the meteorological conditions to be accommodated; and (d) the nature of the airport environment.

Runway location and orientation factors include:

1. *Area winds.* A wind analysis must be performed to determine the optimum runway orientation for purposes of wind coverage and to determine the necessity for a crosswind runway. Appendix 1 of AC 150/5300-13, *Airport Design,* provides information on wind data analysis for airport planning and design.

2. *Airspace availability.* Existing and planned instrument approach procedures, missed approach procedures, departure procedures, control zones, special-use airspace, restricted airspace, and other traffic patterns should be carefully considered in the development of new airport layouts and locations.

3. *Environmental factors.* Environmental studies should be made to ensure that runway development will be as compatible as possible with the airport environs. These studies should include analyses of the impact upon air and water quality, wildlife, existing and proposed land use, and historical/archeological factors.

4. *Obstructions to air navigation.* An obstruction survey should be conducted to identify those objects that might affect airplane operations. Approaches free of obstructions are desirable and encouraged, but, as a minimum, runways require a location that will ensure that the approach areas associated with the ultimate development of the airport can be maintained clear of airport hazards.

5. *Land consideration.* The location and size of the site, with respect to the airport's geometry, should be such that all of the planned airport elements, including the runway clear zone, are located on airport property. If it is anticipated that the airport will need to purchase additional land to accommodate development, AC 150/5100-17, *Land Acquisition and Relocation Assistance for Airport Improvement Program Assisted Projects,* should be referenced.

6. *Topography.* Topography affects the amount of grading and drainage work required to construct a runway and its associated movement areas. In determining runway orientation, the costs of both the initial work and ultimate airport development should be considered. For guidance see AC 150/5320-5, *Airport Drainage.*

7. *Airport facilities.* The relative position of a runway to associated facilities such as other runways, the terminal, hangar areas, taxiways, aprons, fire stations, navigational aids, and the airport traffic control tower, will affect the safety and efficiency of operations at the airport. A general overview of the siting requirements for navigational aids located on, or in close proximity to, the airport, including references to other appropriate technical publications, is presented in AC 150/5300-2, *Airport Design Standards—Site Requirements for Terminal Navigational Facilities.*

Additional Runways.   Many airports may have conditions that indicate one or more additional runways are necessary to accommodate the local circumstances such as:

1. *Wind conditions.* When a single runway or set of parallel runways cannot be oriented to provide 95 percent wind coverage, an additional runway (or runways), oriented in a manner to raise coverage to at least that value, should be provided. The 95 percent wind coverage is computed on the basis of the crosswind not exceeding 10.5 knots for Airport Reference Codes A-I and B-I, 13 knots for Airport Reference Codes A-II and B-II, 16 knots for Codes A-III, B-III, and C-I through D-III, and 20 knots for Codes A-IV through D-VI (AC 150/5300-13, App. 1).

2. *Environmental considerations.* New runways may be justified to reduce adverse impacts on areas around the airport during certain seasons or hours of the day or for certain types of aircraft operations. An example might be an additional runway that would be preferential for diverting night air cargo operations from overflight of dense residential areas.

3. *Operational demands.* An additional runway or runways of the configurations listed below may be warranted when the traffic volume exceeds the existing runway capacity. With rare exceptions, capacity-justified runways should be oriented parallel to the primary runway. In addition, runways will have different centerline separation and threshold offset distances depending on the type of operations to be accommodated.

Additional runways may have any of the following combinations of intended use and configuration:

Parallel Runways Designed for Simultaneous Use During Instrument Flight Rule (IFR) Operations (AC 150/5300-13).   When more than one of the listed conditions applies, the largest separation is required as the minimum. When centerline spacing of less than 2,500 feet (750 meters) is involved, wake turbulence avoidance procedures must be observed by aircraft and air traffic managers. Additionally, runways may be separated farther than the minimum distance to allow for placement of terminal facilities between runways in order to minimize taxi times and runway crossings. Types of simultaneous operations are:

1. *Simultaneous approaches.* For operations under instrument meteorological conditions (IMC), specific electronic navigational aids and monitoring equipment, air traffic control, and approach procedures are required. Simultaneous precision approaches for parallel runways require centerline separation of at least 4,300 feet.

2. *Simultaneous departures: non-radar environment.* May be conducted from parallel runways whose centerlines are separated by at least 3,500 feet (1,000 meters).

3. *Simultaneous departures: radar environment.* Departures may be conducted from parallel runways whose centerlines are separated by at least 2,500 feet (750 meters).

4. *Simultaneous approach and departure: radar environment.* Simultaneous, radar-controlled approach and departure may be conducted on parallel runways whose centerlines are separated as follows:

   • Thresholds are not staggered—a separation distance between runway centerlines of 2,500 feet (750 meters) is required.
   • Thresholds are staggered and the approach is to the nearer threshold—the 2,500 feet (750 meters) separation may be reduced by 100 feet (30 meters) for each 500 feet (150 meters) of threshold stagger to a minimum limiting separation of 1,000 feet (300 meters). For Airplane Design Group V, however, a minimum centerline separation of 1,200 feet (360 meters) is recommended.
   • When the thresholds are staggered and the approach is to the threshold—the 2,500 feet (750 meters) separation must be increased by 100 feet (30 meters) for every 500 feet (150 meters) of threshold stagger.

Parallel Runway Separation—Simultaneous VFR Operations. For simultaneous landings and takeoffs using visual flight rules (VFR), the standard minimum separation between centerlines of parallel runways is 700 feet (210 meters). The minimum runway centerline separation recommended for Airplane Design Group V is 1,200 feet (360 meters). Wake turbulence separation standards apply for runways separated by less than 2,500 feet. Separation standards are presented in Tables 27.1 and 27.2 as extracted from AC 150/5300-13.

Aircraft Weight Limitations

1. Maximum takeoff weight (MTW) and maximum landing weight (MLW) are structural limitations established by the manufacturer.
2. Maximum allowable takeoff weight (MATW) and maximum allowable landing weight (MALW) are climb performance limitations.
3. Desired takeoff weight (DTW) and desired landing weight (DLW) are operational requirements.

TABLE 27.1 Runway Separation Standards for Aircraft Approach Categories C & D

Item	DIM[a]	Airplane Design Group					
		I	II	III	IV	V	VI
Non-precision instrument and visual runway centerline to:							
Parallel runway centerline	H	Refer to discussion in section 27.2.1 on simultaneous operations					
Hold line[b]		250 ft 75 m	250 ft 75 m	250 ft 75 m	250 ft 75 m	250 ft 75 m	250 ft 75 m
Taxiway/taxilane centerline[b]	D	300 ft 90 m	300 ft 90 m	400 ft 120 m	400 ft 120 m	c c	600 ft 180 m
Aircraft parking area	G	400 ft 120 m	400 ft 120 m	500 ft 150 m	500 ft 150 m	500 ft 150 m	500 ft 150 m
Helicopter touchdown		Refer to Advisory Circular 50/5390					
Precision instrument runway centerline to:							
Parallel runway centerline	H	Refer to section 27.2.1					
Hold line[b]		250 ft 75 m	250 ft 75 m	250 ft 75 m	250 ft 75 m	280 ft 85 m	325 ft 98 m
Taxiway/taxilane centerline[b]	D	400 ft 120 m	400 ft 120 m	400 ft 120 m	400 ft 120 m	c c	600 ft 180 m
Aircraft parking area	G	500 ft 150 m	500 ft 150 m	500 ft 150 m	500 ft 150 m	500 ft 150 m	500 ft 150 m
Helicopter touchdown pad		Refer to Advisory Circular 150/5390-2					

[a] Letters correspond to the dimensions on Figure 2-1 (AC 150/5300-13 change 4) page 12.

[b] The separation distance satisfies the requirement that no part of an aircraft (tail tip, wing tip) at the holding location or on a taxiway centerline is within the runway safety area or penetrates the obstacle free zone (OFZ). Accordingly, at higher elevations, an increase to these separation distances may be needed to achieve this result.

[c] For Airplane Design Group V, the standard runway centerline to parallel taxiway centerline separation distance is 400 feet (120 meters) for airports at or below an elevation of 1,345 feet (410 meters); 450 feet (135 meters) for airports between elevations of 1,345 feet (410 meters) and 6,560 feet (2,000 meters); and 500 feet (150 meters) for airports above an elevation of 6,560 feet (2,000 meters).

TABLE 27.2 Taxiway and Taxilane Separation Standards

ITEM	DIM[a]	\multicolumn{6}{c}{Airplane Design Group}
		I	II	III	IV	V	VI
Taxiway centerline to:							
Parallel taxiway/taxilane centerline	J	69 ft 21 m	105 ft 32 m	152 ft 46.5 m	215 ft 65.5 m	267 ft 81 m	324 ft 99 m
Fixed or movable object[b,c]	K	44.5 ft 13.5 m	65.5 ft 20 m	93 ft 28.5 m	129.5 ft 39.5 m	160 ft 39.5 m	193 ft 59 m
Taxilane centerline to:							
Parallel taxilane centerline		64 ft 19.5 m	97 ft 29.5 m	140 ft 42.5 m	198 ft 60 m	245 ft 74.5 m	298 ft 91 m
Fixed or movable object[b,c]		39.5 ft 12 m	57.5 ft 17.5 m	81 ft 24.5 m	112.5 ft 34 m	138 ft 42 m	167 ft 51 m

[a] Letters correspond to the dimensions on RPZ designations.
[b] This value also applies to the edge of service and maintenance roads.
[c] Consideration of the engine exhaust wake impacted from turning aircraft should be given to objects located near runway/taxiway/taxi-lane intersections.
[d] The values obtained from the following equations are acceptable in lieu of the standard dimensions shown in Table 27.2:
- Taxiway centerline to parallel taxiway/taxilane centerline equals 1.2 times airplane wingspan plus 10 feet.
- Taxiway centerline to fixed or movable object equals 0.7 times airplane wingspan plus 10 feet.
- Taxilane centerline to parallel taxilane centerline equals 1.1 times airplane wingspan plus 10 feet.
- Taxilane centerline to fixed or movable object equals 0.6 times airplane wingspan plus 10 feet (3 meters).

27.3.2 Runway Length Design

Runway design length is computed using as its basis the takeoff and landing requirements of the most demanding aircraft that will use the runway. FAA worksheets such as the one used below are typical of the type used to determine full-strength pavement requirements. Landing distance is defined as the horizontal distance necessary to land and come to a complete stop from a point 50 feet above the landing surface.

Runway Length Calculation Sheet

Design Conditions

Airplane: Boeing 737-200 C (JT8D-15 Eng.) 1 + 15 reserve

Mean daily maximum temperature (°F):	50°F
Airport elevation (ft):	1,500 ft
Effective runway gradient (percent):	1.0%
Length of haul (statute miles):	400
Payload (lb):	31,930

Landing Weight Limitations. Landing gross weight must not exceed the lowest maximum weights allowed for:

1. Compliance with runway length requirements (landing distance must be equal to or less than 0.6 times the landing distance available). The landing distance set forth in the *Airplane Flight Manual* may not exceed 60 percent of the landing distance available.

2. Compliance with approach requirements (Landing weight ≤ MALW). An aircraft on approach to landing must be able to execute a missed approach under the following conditions. Maximum allowable landing weight (MALW) is the highest weight at which the aircraft can meet this requirement with existing temperature and pressure altitude.

Approach Climb Requirement. In the approach configuration corresponding to the normal all-engine-operating procedure in which V_S for this configuration does not exceed 110 percent of the V_S for the related landing configuration. The steady gradient of climb may not be less than 2.1 percent for two-engine airplanes, 2.4 percent for three-engine airplanes, and 2.7 percent for four-engine air planes, with:

1. The critical engine inoperative, the remaining engines at the available takeoff power or thrust
2. The maximum allowable landing weight
3. A climb speed established in connection with normal landing procedures, but not exceeding 1.5 V_S

Landing Climb Requirement. In the landing configuration, the steady gradient of climb may not be less than 3.2 percent, with

1. All engines operating
2. The engines at the power or thrust that is available 8 seconds after initiation of movement of the power or thrust controls from the minimum flight idle to the takeoff position
3. A climb speed of not more than 1.3 V_S
4. Structural limit of the airplane (Landing weight ≤ MLW)

Desired Landing Weight	
Typical operating empty weight plus reserve fuel:	70,138
Payload:	+31,930
Desired landing weight:	102,068
Landing Runway Length (flaps 40°, Table 27.3)	
Temperature:	50°F
Airport elevation:	1,500 ft
Maximum landing weight: MLW = MALW = 103.0 > DLW	103,000
Landing runway length:	5,544 ft

Using Table 27.3, runway length versus weight and pressure altitude, find 5.33 thousand feet at 100,000 pounds and 1,000 feet elevation, and 5.47 at 2,000 feet elevation. The maximum allowable weight for landing is 103,000 pounds, so we must interpolate between 100,000 and 105,000 pounds and 1,000 and 2,000 feet elevation. The numbers below are selected from Table 27.3 as described above.

The answer of 5.544 below is multiplied by 1,000 to determine the runway length required for landing.

TABLE 27.3 Aircraft Performance, Landing (Boeing 737-200 Series) JT8D-15 Engine, 40° FLAPS Maximum Allowable Landing Weight (1,000 lb)

Temp °F	Airport elevation (ft)								
	0	1,000	2,000	3,000	4,000	5,000	6,000	7,000	8,000
50	103.0	103.0	103.0	103.0	103.0	103.0	102.7	98.8	95.0
55	103.0	103.0	103.0	103.0	103.0	103.0	102.7	98.8	95.0
60	103.0	103.0	103.0	103.0	103.0	103.0	102.7	98.8	95.0
65	103.0	103.0	103.0	103.0	103.0	103.0	102.7	98.8	95.0
70	103.0	103.0	103.0	103.0	103.0	103.0	101.9	98.0	94.0
75	103.0	103.0	103.0	103.0	103.0	103.0	100.8	97.1	93.2
80	103.0	103.0	103.0	103.0	103.0	103.0	99.6	95.9	92.2
85	103.0	103.0	103.0	103.0	103.0	101.8	98.1	94.5	91.0
90	103.0	103.0	103.0	103.0	103.0	100.1	96.5	93.0	89.5
95	103.0	103.0	103.0	103.0	101.9	98.2	94.7	91.2	87.8
100	103.0	103.0	103.0	103.0	99.8	96.2	92.7	89.2	85.9
105	103.0	103.0	103.0	101.1	97.5	93.9	90.5	87.1	83.8
110	103.0	103.0	102.1	98.5	94.9	91.5	88.1	84.7	81.5

****

Weight 1,000 lb	Runway length (1,000 ft) Airport elevation (ft)								
	0	1,000	2,000	3,000	4,000	5,000	6,000	7,000	8,000
70	3.95	4.05	4.14	4.23	4.31	4.40	4.50	4.59	4.70
75	4.15	4.25	4.35	4.44	4.54	4.64	4.75	4.86	4.98
80	4.35	4.46	4.56	4.66	4.77	4.88	5.00	5.12	5.25
85	4.56	4.67	4.78	4.89	5.01	5.13	5.25	5.39	5.53
90	4.77	4.88	5.00	5.12	5.25	5.38	5.51	5.65	5.80
95	4.98	5.11	5.23	5.36	5.49	5.63	5.77	5.92	6.07
100	5.20	5.33	5.47	5.60	5.74	5.88	6.03	6.18	6.34
105	5.42	5.57	5.71	5.85	5.99	6.14	6.29	6.44	6.61

****

Airplane characteristics	Measure	Unit of advanced options	
		200	200C
Maximum takeoff weight	lb	109,000	115,500
Maximum landing weight			
Flaps 30°	lb	98,000	103,000
Flaps 40°	lb	89,700	103,000
Typical operating empty	lb	67,238	70,138[a]
Weight plus reserve fuel	lb	71,480	74,380[b]
Average fuel consumption	lb/mile	15	15
Typical maximum passenger			
Load @200 lb/passenger	lb	26,000	26,000
Maximum structural payload	lb	34,830	31,930

[a] Based on 1.25 hours of reserve fuel.
[b] Based on 2.00 hours of reserve fuel.

$$
\begin{array}{ccc}
 & 1000 & 1500 \quad 2000 \\
\end{array}
$$

$$
\begin{bmatrix} 100 \\ 103 \\ 5 \\ 105 \end{bmatrix} 3 \quad x \begin{bmatrix} 5.33 \\ \underline{0.144} \\ 5.474 \\ 5.57 \end{bmatrix} 0.24 \quad y \begin{bmatrix} 5.47 \\ \underline{0.144} \\ 5.614 \\ 5.71 \end{bmatrix} 0.24 \quad \begin{array}{l} 5.614 \\ \underline{5.474} \\ 11.088 \div 2 = 5.544 \times 1{,}000 = 5{,}544 \end{array}
$$

$$3/5 = x/0.24 \qquad\qquad 3/5 = y/0.24$$
$$x = (0.24)(0.6) = 0.144 \qquad y = (0.25)(0.60) = 0.144$$

Desired Takeoff Weight	
Length of haul:	400
Average fuel consumption:	× 15
Haul fuel:	6,000
Typical operating empty weight plus reserve fuel:	+ 70,138
Weight, no payload:	76,138
Payload:	+ 31,930
Desired takeoff weight:	108,068

Takeoff Weight Limitations. Takeoff gross weight must not exceed the lowest the maximum weights allowed for:

1. Compliance with runway length requirements. [takeoff distance (TOD) ≤ takeoff distance available (TODA), accelerate stop distance (ASD) ≤ accelerate stop distance available (ASDA)]

In determining the allowable gross weight for takeoff for any given runway, the performance of the airplane must be related to the dimensions of the airport; that is, the required takeoff distance and accelerate-stop distance for the gross weight must not exceed the runway length available.

2. (Takeoff weight ≤ MATW)
3. Compliance with en route performance requirements.
4. Compliance with maximum landing weight, considering normal fuel burnout en route.
5. Structural limit of the airplane (Takeoff weight ≤ MTW)

Takeoff Runway Length (Flaps 5°, Table 27.4)	
Temperature:	50°F
Airport evaluation:	1,500
Maximum takeoff weight: MATW $= \dfrac{115.5 + 114.1}{2} = 114.8 >$ DTW	114,800
Reference factor "R":	51.25
Limiting weight: 103,000 + 6,000 > DTW = 108,068	108,068
Runway length:	5,967
Gradient correction: 5967 (R/WL) × 0.10 × 1.0 (ERG) =	597
Corrected runway length:	6,564

TABLE 27.4 Aircraft Performance, Takeoff (Boeing 737-200 Series) JT8D-15 Engine, 5° FLAPS Maximum Allowable Takeoff Weight (1,000 lb)

Temp °F	Airport Elevation (ft)								
	0	1,000	2,000	3,000	4,000	5,000	6,000	7,000	8,000
50	115.5	115.5	114.1	110.1	106.0	102.0	98.1	94.4	91.0
55	115.5	115.5	114.1	110.1	106.0	102.0	98.1	94.4	91.0
60	115.5	115.5	114.1	110.1	106.0	102.0	98.1	94.4	91.0
65	115.5	115.5	114.1	110.1	106.0	102.0	98.1	94.4	91.0
70	115.5	115.5	113.5	109.6	105.6	101.5	97.4	93.6	90.0
75	115.5	115.1	111.6	107.7	103.8	99.8	95.8	92.1	88.6
80	115.5	113.3	109.7	105.8	101.9	98.0	94.2	90.6	87.2
85	114.8	111.4	107.7	104.0	100.1	96.3	92.6	89.1	85.8
90	113.1	109.5	105.8	102.1	98.3	94.6	91.0	87.5	84.3
95	111.3	107.6	103.9	100.2	96.5	92.9	89.4	86.0	82.9
100	109.5	105.7	102.0	98.3	94.7	91.1	87.3	84.5	81.5
105	107.8	103.9	100.1	96.4	92.8	89.4	86.1	83.0	80.0
110	106.0	102.0	98.1	94.5	91.0	87.7	84.5	81.5	78.6

****

Reference factor "R"
Airport elevation (ft)

Temp °F	0	1,000	2,000	3,000	4,000	5,000	6,000	7,000	8,000
50	48.0	49.5	53.0	57.5	62.2	67.1	72.4	78.1	84.5
55	48.1	50.1	53.3	57.8	62.4	67.4	72.7	78.5	84.8
60	48.2	50.8	53.8	58.3	62.9	67.9	73.2	79.1	85.5
65	48.3	51.4	54.5	59.0	63.6	68.7	74.1	80.0	86.5
70	48.5	51.3	55.5	59.9	64.6	69.7	75.2	81.3	87.8
75	48.6	52.4	56.6	61.0	65.8	71.0	76.7	82.8	89.3
80	50.0	53.7	57.9	62.4	67.3	72.6	78.4	84.6	91.4
85	51.4	55.2	59.4	63.9	69.0	74.4	80.4	86.8	93.7
90	53.0	56.8	61.0	65.7	70.9	76.5	82.6	89.2	96.3
95	54.7	58.6	62.9	67.7	73.1	78.9	85.2	91.9	99.2
100	56.5	60.5	65.0	70.0	75.5	81.5	88.0	95.0	102.5
105	58.4	62.6	67.2	72.4	78.1	84.4	91.1	98.3	106.1
110	60.4	64.8	69.7	75.1	81.0	87.5	94.5	102.0	110.0

****

Runway length (1,000 ft)
Reference factor "R"

Weight 1,000 lb	50	60	70	80	90	100	110
70	2.61	3.12	3.56	3.96	4.36	4.78	5.25
75	2.93	3.49	4.00	4.50	5.01	5.57	6.21
80	3.28	3.90	4.49	5.08	5.70	6.39	7.16
85	3.67	4.36	5.04	5.72	6.44	7.23	8.10
90	4.06	4.86	5.64	6.42	7.23	8.09	9.04
95	4.51	5.41	6.29	7.17	8.06	8.89	9.96
100	4.98	6.00	7.00	7.98	8.94	9.91	10.87
105	5.48	6.64	7.76	8.84	9.87	10.85	11.78
110	6.02	7.32	8.57	9.75	10.84	11.82	12.67
115	6.59	8.05	9.44	10.77	11.86	12.82	

TABLE 27.5 Minimum Positive Climb Gradient—One Engine Inoperative

1st segment	Two-engine aircraft	Positive
	Three-engine aircraft	0.3%
	Four-engine aircraft	0.5%
2nd segment	Two-engine aircraft	2.4%
	Three-engine aircraft	2.7%
	Four-engine aircraft	3.0%
3rd segment	Two-engine aircraft	1.2%
	Three-engine aircraft	1.5%
	Four-engine aircraft	1.7%

Net takeoff flight path

Takeoff flight path reduced by:		
	Two-engine aircraft	0.8%
	Three-engine aircraft	0.9%
	Four-engine aircraft	1.0%

Summary

Landing weight:	103,000
Takeoff weight:	108,068
Landing runway length:	5,544
Takeoff runway length:	6,564
Design runway length:	6,600

$$3.068/5 = x/0.54$$
$$x = (0.61)(0.54) = 0.331$$

WT   50   51.25   60

$$\begin{bmatrix} 105 \\[2pt] \overset{\textbf{1}}{108.068} \\[4pt] 5 \\ 110 \end{bmatrix}\!\!\overset{\textbf{2}}{\Big]}3.068 \quad x\begin{bmatrix} 5.48 \\ .331 \\ \overline{5.811} \\[6pt] 6.02 \end{bmatrix}\!\!\overset{\textbf{3}}{}\Big]0.54 \quad y\begin{bmatrix} 6.64 \\ .417 \\ \overline{7.057} \\[6pt] 7.32 \end{bmatrix}\!\!\overset{\textbf{4}}{}\Big]0.68 \;\; \textbf{5}$$

$$3.068/5 = y/0.68$$
$$y = (0.61)(0.68) = 0.417$$

$$1.25/10 = z/1.246$$

$$z = (0.125)(1.246) = 0.156 + 5.811 = 5.967 \times 1,000 = 5,967'$$

Runway Threshold Placement. The landing threshold identifies (by markings and lighting) the beginning of that portion of the full-strength runway surface that is available for landing. The threshold should normally be located at the beginning of the full-strength runway surface, but it may be displaced down the runway length when an object that obstructs the airspace required for landing airplanes is beyond the airport authority's power to remove, relocate, or lower. The new location is the *displaced threshold* and is treated as the new end of the runway for landing on that end. The runway pavement preceding the displaced threshold is available for takeoff in either direction and for landing rollout from the opposite direction. The following alternatives should be considered if an object penetrates an approach surface as defined in the preceding section:

1. Remove or lower the object so that it will not penetrate the applicable surface.
2. Apply a less demanding surface; e.g. convert the runway from a precision ornon-precision approach to a VFR-only approach.
3. Displace the threshold so that the object will not penetrate the applicable surface, and accept a shorter landing surface. Relevant factors to be evaluated include:
   - Types of airplanes that will use the runway and their performance characteristics.
   - Operational disadvantages associated with accepting higher landing minima.
   - Cost of removing, relocating, or lowering the object.
   - Effect of the reduced available landing length when the runway surface is adversely affected by precipitation or snow and ice accumulations.
   - Cost of extending the runway if insufficient runway would remain as a result of displacing the threshold. The environmental and public acceptance aspects of a runway extension must also be evaluated under this consideration.
   - Cost and feasibility of relocating visual and electronic approach aids such as threshold lights, visual approach slope indicator, runway end identification lights, localizer, glide slope (to provide a threshold crossing height of not more then 60 feet), approach lighting system, and runway markings.
   - Effect of the threshold change on aircraft noise-abatement procedures.

Displacing the Threshold. If, after consideration of alternatives, the decision is made to displace the threshold, the required displacement distance can be determined in a three-step process. Given: (a) type of approach to that runway end, (b) obstacle location, and (c) obstacle height above the ground; one calculates (d) approach surface height at the obstacle, (e) difference in height between obstacle and approach surface, and (e) obstacle displacement distance required.

Clearways and Stopways. There are certain runways where there is inadequate full-strength runway length to accommodate the full takeoff distance (TOD) of certain airplanes. Federal Aviation Regulations permit the use of a *clearway* to provide part of the takeoff distance required for turbine-powered airplanes. The clearway is defined as a plane, above which no object protrudes, extending from the end of the runway with an upward slope not exceeding 1.25 percent, not less than 500 feet wide, centrally located about the extended centerline of the runway, and under the control of the airport authorities. Threshold lights may protrude above the clearway plane, however, if their height above the end of the runway is 26 inches or less and if they are located to each side of the runway. Although the use of a clearway is a technique that permits higher allowable operating weights without an increase in runway length, the runway length recommended without use of a clearway (or stopway—see paragraph below) for the most demanding airplane should be provided.

A runway is normally designed with a critical aircraft and the maximum performance mission requirement in mind. The clearway should serve only as a means of accommodating the takeoff distance requirements for that occasional heavy operation requiring a greater takeoff distance than the most demanding airplane for which the runway length is designed. When the frequency of this "occasional" operation increases to a certain point, a new "most demanding" airplane for runway design length exists, and additional runway length should be provided. An airport owner interested in providing a clearway should be aware of the requirement that the clearway be under his control, although not necessarily by direct ownership. The purpose of such control is to ensure that no takeoff operation intending to use a clearway is initiated unless it has been absolutely determined that no fixed or movable object will penetrate the clearway plane during that operation.

A *stopway* is an area beyond the runway used for takeoff that is designated by the airport authority for use in decelerating an airplane during an aborted takeoff. A stopway is at least as wide as the runway it serves and is centered on the extended centerline of the runway. It should be able to support an airplane during an aborted takeoff without causing structural damage to the airplane.

Declared Distances.   Introduction of stopways and clearways and the use of displaced thresholds on runways have created a need for accurate information concerning the distances available and suitable for the landing and takeoff of airplanes. There are also situations when additional runway length is needed but the construction of additional full-strength runway is not feasible or practical. The concept of *declared distances* has been developed to accommodate certain operations in those circumstances. The declared distances that must be calculated for each runway direction are: (a) takeoff run available (TORA), (b) takeoff distance available (TODA), (c) accelerate stop distance available (ASDA), and (d) landing distance available (LDA). When a runway is not provided with either a stopway or clearway, and the threshold is located at the extremity of the runway, the four declared distances should normally be equal to the length of the runway, as shown in Figure 27.1*a*. When a runway is provided with a clearway, then the TODA will include the length of clearway, as shown in Figure 27.1*b*. Where a runway is provided with a stopway, the ASDA will include the length of stopway, as shown in Figure 27.1*c*. The LDA will be reduced by the length of the threshold displacement when a runway has a displaced threshold as shown in Figure 27.1*c*. A displaced threshold affects only the LDA for approaches made to that threshold; declared distances for landings in the opposite direction and takeoffs in either direction are unaffected. Figures 27.1*b* through 27.1*d* illustrate a runway provided with a clearway or a stopway or having a displaced threshold. More than one of the declared distances will be modified when more than one of these features exists, but the modification will follow the same principle illustrated. An example of situation where all these features exist is shown in Figure 27.1*d*. If a runway direction cannot be used for takeoff or landing or both because it is operationally forbidden, then this should be declared and the words "not usable" or the abbreviation "NU" entered.

27.3.3   Airport Approach and Departure Design

The preceding sections on airport layout design and runway length design focused on development of safe and efficient airport surfaces containing aircraft movement areas and the balance of the airport operating area. This section will address three-dimensional design standards for the protection of aircraft when operating in the immediate airport airspace, such as during approaches and departures. These standards are prescribed in three regulations: FAR Part 77, *Civil Airport Imaginary Surfaces;* FAR Part 25, *Airworthiness Standards: Transport Category Airplanes;* and FAR Part 121, *Operating Requirements: Domestic, Flag, and Supplemental Operators.*

FAR Part 77 was developed to help protect airspace in the vicinity of airports by defining imaginary surfaces that objects should not penetrate and thus constitute hazards or obstructions to air traffic. A *hazard* to air navigation is defined as a fixed or mobile object of a greater height than any of the heights or surfaces presented in Subpart C of FAR Part 77 that has not been properly charted and marked or lighted. An *obstruction* is an equivalent object that has been so marked, lighted, and charted. One of the surface sets referred to in Subpart C is the set of civil airport imaginary surfaces that is defined in FAR Part 77.25. The various imaginary surfaces affect operational procedures and threshold placement at all runways. The following civil airport imaginary surfaces (see Figure 27.2) are established with relation to the airport and to each runway. The size of each such imaginary surface is based on the category of each runway according to the type of approach available or planned for that runway. The slope and dimensions of the approach surface applied to each end of

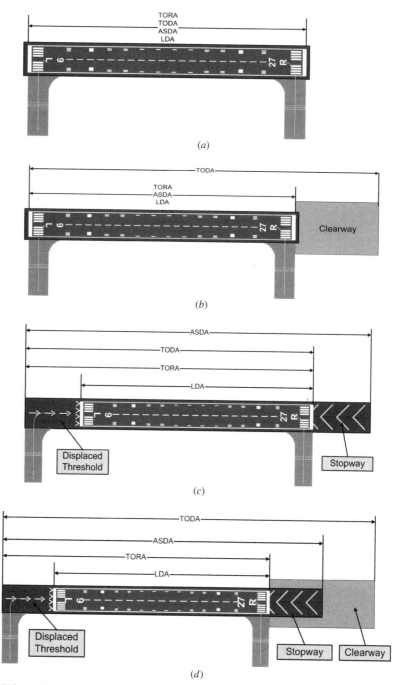

FIGURE 27.1 Declared distances *a*, *b*, *c*, and *d* show runway operations from left to right.

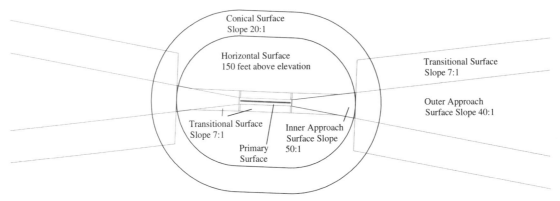

FIGURE 27.2   FAR Part 77 imaginary surfaces.

a runway are determined by the most precise approach existing or planned for that runway end:

Horizontal Surface.   A horizontal plane 150 feet above the established airport elevation, the perimeter of which is constructed by swinging arcs of specified radii form the center of each end of the primary surface of each runway and connecting the adjacent arcs by lines tangent to those arcs. The radius of each arc is:

* 5,000 feet for all runways designated as utility or visual
* 10,000 feet for all other runways

The radius of the arc specified for each end of a runway will have the same arithmetical value. That value will be the highest determined for either end of the runway. When a 5,000-foot arc is encompassed by tangents connecting two adjacent 10,000-foot arcs, the 5,000-foot arc shall be disregarded on the construction of the perimeter of the horizontal surface.

Conical Surface.   A conical surface is extended outward and upward from the periphery of the horizontal surface at a slope of 20:1 for a horizontal distance of 4,000 feet.

Primary Surface.   A surface longitudinally centered on a runway. When the runway has a specially prepared hard surface, the primary surface extends 200 feet beyond each end of that runway; but when the runway has no specially prepared hard surface or planned hard-surface, the primary surface ends at each end of that runway. The elevation of any point on the primary surface is the same as the elevation of the nearest point on the runway centerline. The width of a primary surface will be the width prescribed for the most precise existing or planned approach to either end of that runway:

* 250 feet for utility runways having only visual approaches
* 500 feet for utility runways having non-precision instrument approaches
* For other than utility runways:
  * (i) 500 feet for visual runways having only visual approaches
  * 500 feet for non-precision instrument runways having visibility minima greater than three-fourths statute mile
  * 1,000 feet for a non-precision instrument runway having visibility minima as low as three-fourths of a statute mile, and for all precision instrument runways

Approach Surface.   A surface longitudinally centered on the extended runway centerline and extending outward and upward from each end of the primary surface. An approach surface is applied to each end of each runway based upon the type of approach available or planned for that runway end.

- The inner edge of the approach surface is the same width as the primary surface, and it expands uniformly to a width of:
  - 1,250 feet for that end of a utility runway with only visual approaches
  - 1,500 feet of a runway other than a utility runway with visual approaches
  - 2,000 feet for that end of a utility runway with a non-precision approach
  - 3,500 feet for that end of a non-precision instrument runway other than utility, having visibility minimums greater than three-fourths of a statute mile
  - 4,000 feet for that end of a non-precision instrument runway, other than utility, having a non-precision instrument approach with visibility minimums as low as three-fourths statute mile
  - 16,000 feet for all precision instrument runways
- The approach surface extends for a horizontal distance of:
  - 5,000 feet at a slope of 20:1 for all utility and visual runways
  - 10,000 feet at a slope of 34:1 for all non-precision instrument runways other than utility
  - 10,000 feet at a slope of 50:1 with an additional 40,000 feet at a slope of 40:1 for all precision instrument runways
  - The outer width of an approach surface to an end of a runway will be that width prescribed in this subsection for the most precise approach existing or planned for that runway end

Transitional Surface. These surfaces extend outward and upward at right angles to the runway centerline and the runway centerline extended at a slope of 7:1 from the sides of the primary surface and from the sides of the approach surfaces Transitional surfaces for those portions of the precision approach surface that project through and beyond the limits of the conical surface extend a distance of 5,000 feet measured horizontally from the edge of the approach surface and at right angles to the runway centerline.

Takeoff Path Requirements.   FAR Part 25 prescribes airworthiness standards for certification of transport category airplanes, and FAR Part 121 prescribes rules governing air carrier flight operations. The following excerpts from those regulations demonstrate the concept of climb gradient and obstacle clearance that must be accommodated in airport design. Major portions of the regulations have been omitted or edited in the interest of brevity and clarity. For example, because of the inherent differences in the performance characteristics of reciprocating and turbine-powered engines, the certification and operating standards for these airplane types differ. In this text, only the regulations applicable to turbine-powered airplanes are summarized; however, if one understands the concepts presented in this text, there should be no difficulty addressing situations involving reciprocating engine airplanes.

Takeoff Path (FAR Part 25.111)

1. The takeoff path extends from a standing start to a point in the takeoff at which the airplane is 1,500 feet above the take-off surface, or at which the transition from the aircraft's takeoff to the enroute configuration is completed, whichever point is higher. In addition:

   a. The airplane must be accelerated on the ground to V_{EF}, at which point the critical engine must be made inoperative and remain inoperative for the rest of the takeoff
   b. After reaching V_{EF}, the airplane must be accelerated to V_2.

2. During the acceleration to speed V_2, the nose gear may be raised off the ground at a speed V_R. Landing gear retraction begins once airborne.

3. During the takeoff path determination:

    a. The slope of the airborne part of the takeoff path must be positive at each point;

    b. The airplane must reach V_2 before it is 35 feet above the takeoff surface and must continue at a speed as close as practical to, but not less than V_2, until it is 400 feet above the takeoff surface;

    c. At each point along the takeoff path, starting at the point at which the Airplane reaches 400 feet above the take-off surface, the available gradient of climb may not be less than:

        i. 1.2 percent for two-engine airplanes

        ii. 1.5 percent for three-engine airplanes

        iii. 1.7 percent for four-engine airplanes

    d. Except for gear retraction and propeller feathering, the airplane configuration may not be changed, and no change in power or thrust that requires action by the pilot may be made, until the airplane is 400 feet above the takeoff surface.

Takeoff Distance and Takeoff Run (FAR Part 25.113)

1. Takeoff distance is the greater of:

    a. The horizontal distance along the takeoff path from the start of the takeoff to the point at which the airplane is 35 feet above the takeoff surface, as determined under Part 25.111, or

    b. 115 percent of the horizontal distance along the takeoff path, with all engines operating, from the start of the takeoff to the point at which the airplane is 35 feet above the takeoff surface, as determined by a procedure consistent with Part 25.111.

Takeoff Flight Path (FAR Part 25.115)

1. The takeoff flight path begins 35 feet above the takeoff surface at the end of the takeoff distance determined in accordance with 25.113 (a).

2. The net takeoff flight path data must be determined so that they represent the actual takeoff flight paths (determined in accordance with 25.111 and with paragraph (1) of this section) reduced at each point by a gradient of climb equal to:

    a. 0.8 percent for two-engine airplanes;

    b. 0.9 percent for three-engine airplanes; and

    c. 1.0 percent for four-engine airplanes.

Climb: One-Engine-Inoperative (FAR Part 25.121)

1. Takeoff with landing gear extended. Critical takeoff configuration exists along the flight path between the points at which the airplane reaches V_{LOF} and the point at which the landing gear is fully retracted. The configuration used in 25.111, but without ground effect, the steady gradient of climb must be positive for two-engine airplanes, and not less than 0.3 percent for three-engine airplanes of 0.5 percent for four-engine airplanes, at V_{LOF} and with:

    a. The critical engine inoperative and the remaining engines at the power or thrust available when retraction of the landing gear. is begun in accordance with Part 25.111 unless there is a more critical power operating condition existing later along the flight path but before the point at which the landing gear is fully retracted; and

    b. The weight equal to the weight existing when retraction of the landing gear is begun, determined under Part 25.111.

2. Takeoff with landing gear retracted. The takeoff configuration exists at the point of the flight path at which the landing gear is fully retracted. The configuration used in Part 25.111 but without ground effect, the steady gradient of climb may not be less than 2.4 percent for two-engine airplanes, 2.7 percent for three-engine airplanes, and 3.0 percent for four-engine airplanes, at V_2 and with:

a. The critical engine inoperative, the remaining engines at the takeoff power or thrust available at the time the landing gear is fully retracted, determined under 25.111, unless there is a more critical power operating condition existing later along the flight path but before the point where the airplane reaches a height of 400 feet above the takeoff surface; and

b. The weight equal to the weight existing when the airplane's landing gear is fully retracted determined under Part 25.111.

Transport Category Airplanes: Turbine Engine Powered Takeoff Limitations (FAR Part 121.189)

An airplane certified after September 30, 1958 is allowed a net takeoff flight path that clears all obstacles either by a height of at least 35 feet vertically, or by at least 200 feet horizontally within the airport boundaries and by at least 300 feet horizontally after passing airport boundaries. The following information is provided to further clarify the climb gradient and obstacle clearance requirements:

1. In order to achieve compliance with this regulation, the takeoff gross weight for any given flight must not exceed the lowest of the maximum weights allowed for:

  a. Compliance with runway length and obstacle clearance requirements
  b. Compliance with take-off climb requirements
  c. Compliance with en route performance requirements
  d. Compliance with maximum landing weight taking into account normal fuel burnout en route
  e. Structural limit of the airplane

2. Takeoff climb requirement. The maximum weight for takeoff may not exceed that weight that will allow airplane performance equal to the climb gradients specified in Table 3. The takeoff path may be considered as the trajectory, or elevation profile made good on a takeoff with an engine failure occurring at V_1 speed. The path is considered as extending from the standing start to a point in the takeoff where a height of 1,500′ above the takeoff surface is reached, or to a point in the takeoff where transition from takeoff to en route configuration is complete, whichever is higher. The airborne part of the takeoff is comprised of the following parts:

  a. First Segment. Starts at liftoff and ends when gear retraction is complete.
  b. Second Segment. Starts at gear retraction and ends when at 400 feet above the runway.
  c. Third Segment. Starts at 400 feet above the runway, and continues to 1,500 feet above the takeoff surface, or until transition to the aircraft's enroute configuration is completed, which ever occurs last.
  d. Net Takeoff Flight Path. A profile starting at the end of the takeoff distance, having a gradient specified in Table 3, below the takeoff flight path. The "net" flight path must clear all obstacles by 35 feet vertically or 300 feet horizontally. As the actual flight path altitude is greater than the "net" flight path, the airplane will have clearance above obstacles in the flight path.

A climb gradient profile is depicted in Figure 27.3. The climb segments are located relative to the departure end of the runway. Adherence to the manufacturer's limits on maximum allowable takeoff weight and to the appropriate speed schedules will ensure compliance with takeoff climb requirements. During our classroom sessions we will be locating the climb segments based on typical takeoff data (liftoff distance, gear retraction distance and takeoff distance). Once the climb segments are located, the net takeoff flight path can be plotted to determine compliance with obstacle clearance requirements.

27.3.4 Airport Pavement Design

Airport traffic forecasts for airports encompass a variety of aircraft having different landing gear geometry and weights, the effects of which must be identified in terms of the critical

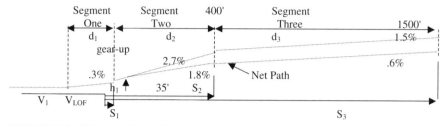

FIGURE 27.3   Climb gradient profile.

aircraft for that design feature. Each aircraft must be converted to an equivalency of the landing gear type represented by the design aircraft, as discussed in AC 150/5320-6D. The factors below should be used to make that conversion:

Aircraft in question landing gear wheel configuration	Design aircraft landing gear wheel configuration	Multiply departures by
Single	Dual	0.8
Single	Dual tandem	0.5
Dual	Dual tandem	0.6
Double dual tandem	Dual tandem	1.0
Dual tandem	Single	2.0
Dual tandem	Dual	1.7
Dual	Single	1.3
Double dual tandem	Dual	1.7

The conversion to equivalent annual departures of the design aircraft continues using the following formula:

$$R_1 = R_2 \sqrt{W_2/W_1}$$

where R_1 = equivalent annual departures by the design aircraft
 R_2 = annual departures expressed in design aircraft landing gear
 W_1 = wheel load of the design aircraft
 W_2 = wheel load of the aircraft in question

Ninety-five percent of the gross weight of an aircraft rests on the main landing gear. Wide-body aircraft have radically different landing gear assemblies from narrow-bodied airplanes, and therefore special consideration is given to maintain the relative effects. Each wide-body is treated as if it were a 300,000 pound dual-tandem wheel aircraft when computing equivalent annual departures. Remember, the factor multiplier is based on the actual gear configuration for wide-body aircraft, but the R_1, the equivalent annual departures, are determined using the modification described above. The number of equivalent annual departures for each aircraft is computed and the aircraft population is added to define the total equivalent annual departures (TEAD).

27.4 CONCLUSION

This chapter has introduced and explained just a few of the key concepts in airport design. Airport design involves a broad array of airside and landside elements and often requires a multidisciplinary approach. It definitely requires familiarization with, and reference to, the sizable body of ICAO and FAA documents as a foundation. There is also a growing body of literature in technical and scholarly journals dealing with airport design issues. The most important thing in the end is that the transportation engineer contribute to airport developments that are effective, economical to build and operate, flexible in design, and capable of expansion and change without undue stress on the host community and airport users.

27.5 REFERENCES

Ashford, N., and P. Wright. 1991. *Airport Engineering*, 3d ed. New York: John Wiley & Sons.

Caves, R., and G. Gosling. 1999. *Strategic Airport Planning*. London: Pergamon.

De Neufville, R., and A. Odoni. 2003. *Airport Systems: Planning, Design, and Management*. New York: McGraw-Hill.

Federal Aviation Administration (FAA). 1970. *Planning the Metropolitan Airport System*. Advisory Circular 150/5070-5, FAA, Washington, DC.

———. 1975. *The Continuous Airport System Planning Process*. Advisory Circular 150/5050-5, FAA, Washington, DC.

———. 1983a. *Airport Capacity and Delay*. Advisory Circular 150/5060-5, FAA, Washington, DC.

———. 1983b. *Noise Control and Compatibility Planning for Airports*. Advisory Circular 150/5020-1, FAA, Washington, DC.

———. 1985. *Airport Master Plans*. Advisory Circular 150/5070-6A, FAA, Washington, DC.

———. 1987. *A Model Zoning Ordinance to Limit Height of Objects Around Airports*. Advisory Circular 150/5190-4A, FAA, Washington, DC.

———. 1988. *Planning and Design Guidelines for Airport Terminal Facilities*. Advisory Circular 150/5360-13, FAA, Washington, DC.

———. 1989a. *Planning the State Aviation System*. Advisory Circular 150/5050-3B, FAA, Washington, DC.

———. 1989b. *Standards for Specifying Construction Standards on Airports*. Advisory Circular 150/5370-10A, FAA, Washington, DC.

———. 1990. *Runway Length Requirements for Airport Design*. Advisory Circular 150/5325-4A, FAA, Washington, DC.

———. 1994. *Architectural, Engineering, and Planning Consultant Services for Airport Grant Projects*. Advisory Circular 150/5100-14C, FAA, Washington, DC.

———. 1996a. *Airport Pavement Design and Evaluation*. Advisory Circular 150/5320-6D, FAA, Washington, DC.

———. 1996b. *Land Acquisition and Relocation Assistance for Airport Improvement Program Assisted Projects*. Advisory Circular 150/5100-17, FAA, Washington, DC.

———. 1996c. *Proposed Construction or Alteration of Objects That May Affect the Navigable Airspace*. Advisory Circular 70/7460-2J, FAA, Washington, DC.

———. 1999. *The National Plan of Integrated Airport Systems (NPIAS) 1998–2002*. FAA, Washington, DC.

———. 2002a. *Airport Design* (with Change 7). Advisory Circular 150/5300-13, FAA, Washington, DC.

———. 2002b. *Operational Safety on Airports During Construction*. Advisory Circular 150/5370-2D, FAA, Washington, DC.

———. *Airport Environmental Handbook*. Order 5050.4A, FAA, Washington, DC.

———. Federal Aviation Regulations Part 25, *Airworthiness Standards: Transport Category Airplanes*. FAA, Washington, DC.

————. Federal Aviation Regulations Part 77, *Objects Affecting Navigable Airspace.* FAA, Washington, DC.

————. Federal Aviation Regulations Part 121, *Operating Requirements: Domestic, Flag, and Supplemental Operations.* FAA, Washington, DC.

————. Federal Aviation Regulations Part 150, *Airport Noise Compatibility Planning.* FAA, Washington, DC.

————. Federal Aviation Regulations Part 157, *Notice of Construction, Alteration, Activation and Deactivation of Airports.* FAA, Washington, DC.

Florida Department of Transportation Aviation Office. *Guidebook for Airport Master Planning.* Florida Department of Transportation, Tallahassee, FL.

Horonjeff, R., and F. McKelvey. 1994. *Planning and Design of Airports,* 4th ed. New York: McGraw-Hill.

International Air Transport Association (IATA). 1995. *Airport Development Reference Manual,* 8th ed. Montreal: IATA.

International Civil Aviation Organization (ICAO). 1983. *Airport Planning Manual, Part 3: Guidelines for Consultant/Construction Services,* 1st ed. (Doc 9184-AN/902). Montreal: ICAO.

————. 1985. *Airport Planning Manual, Part 2: Land Use and Environmental Control,* 2d ed. (Doc 9184-AN/902). Montreal: ICAO.

————. 1987. *Airport Planning Manual, Part 1: Master Planning,* 2d ed. (Doc 9184-AN/902). Montreal: ICAO.

————. 1988. *Annex 16 to the Convention on International Civil Aviation, Environmental Protection,* vol. 1, *Aircraft Noise.* Montreal: ICAO.

————. 1997a. *Aerodrome Design Manual, Part 2: Taxiways, Aprons, and Holding Bays,* 3d ed., Corr. 1 (Doc 9157). Montreal: ICAO.

————. 1997b. *Aerodrome Design Manual, Part 3: Pavements,* 2d ed., Amdts. 1 and 2 (Doc 9157). Montreal: ICAO.

————. 1999. *Annex 14 to the Convention on International Civil Aviation, Aerodromes,* vol. 1, *Aerodrome Design and Operations,* 3d ed. Montreal: ICAO.

————. 2000. *Aerodrome Design Manual, Part 1: Runways,* 2d ed., Amdt. 1 (Doc 9157). Montreal: ICAO.

Transportation Research Board (TRB). 2002. *Aviation Demand Forecasting: A Survey of Methodologies.* Transportation Research Circular E-C040, National Research Council, TRB, Washington, DC.

CHAPTER 28
AIR TRAFFIC CONTROL SYSTEM DESIGN

Robert Britcher
Montgomery Village, Maryland

Aviation is a recent form of travel. During the 17th and 18th centuries, humans tried to build airplanes with wings that flapped like birds. The planes were called ornithopters. They did not succeed. In the 19th century, armies used lighter-than-air balloons to reconnoiter, while a few pioneers dreamed of and experimented with heavier-than-air machines. In 1804, George Cayley, often called the father of aviation, designed, built, and flew a small model glider. It was the first modern configuration airplane in history, with a fixed wing and a horizontal and vertical tail that could be adjusted. Powered flight awaited the German aviator Otto Lilienthal. Some 90 years after Cayley, Lilienthal constructed a glider with flapping wing tips that was to be powered by a small motor using compressed carbonic gas. His efforts were truncated. On August 9, 1896, he was killed when he stalled and crashed to the ground while gliding.

In 1903, the Wright brothers designed and built a flying craft that could be controlled while in the air. Every successful aircraft built since has had controls to roll the wings right or left, pitch the nose up or down, and yaw the nose from side to side. These three controls, roll, pitch, and yaw, let a pilot navigate an airplane in all three dimensions, making it possible to fly from place to place. The entire aerospace business, the largest industry in the world, depends on this simple but brilliant idea. More important, the Wright brothers changed the way we view our world. Before flight became commonplace, people traveled in two dimensions, crossing the borders that separate town from town and nation from nation. Seen from above, the artificial boundaries that divide us disappear. Distances shrink, the horizon stretches. The world seems grander and more interconnected.

Airplanes proved their worth in World War I. Soon after, commercial flying burgeoned. At first, civil aviation was limited to exploration and hauling cargo. But by the late 1920s, planes were carrying passengers. In 2000, U.S. commercial airlines flew over 600 million passengers a year.

In the United States, air traffic control began with the Air Commerce Act of 1926. An aeronautics branch of the Department of Commerce was established. It chartered safety standards, licensing, certification, rule-making, and the management of airways and navigational aids. These services remain the linchpin of civil aviation authorities around the world, including today's U.S. Federal Aviation Administration (FAA), now under the Department of Transportation.

This chapter emphasizes the design of systems to control commercial and general aviation in the United States. Our military and aviators of other nations also require air traffic control

(ATC) systems. But, in the main, they use the same technology and techniques as the U.S. civil system.

One of the first design principles to shape domestic air traffic control was the creation of airways. Early flights traversed airspace at the whim of the pilots. There were no airways. In 1926, that changed. Like the corners and turns of highways, airways would be marked. Navigational aids became known as fixes, and the leg between fixes as an *airway*. (Navigational aids, or *navaids,* consist of landmarks, lights, and radio signals—even bonfires in the early days. We can now add satellites to the list.) Planes would fly point to point, from a runway to a fix at a planned altitude, to the next fix, and so on until the plane landed. This procedural approach inherently constrains air travel, as do roads and railways on the ground. (Figure 28.1 shows how sector boundaries and airways are mapped onto the controller's display.)

Point-to-point control and radios enabled planes to be tracked from the ground. The nation's first air route traffic control center (ARTCC) was created in Newark, NJ, in 1935. (Figure 28.2 shows the 20 ARTCCs that monitor traffic across the continental United States.) The first controllers included Glen Gilbert and J. V. Tighe, who designed shrimp boats to track flights on a table map. Gilbert is often credited as the father of air traffic control.

FIGURE 28.1   Controller airspace.

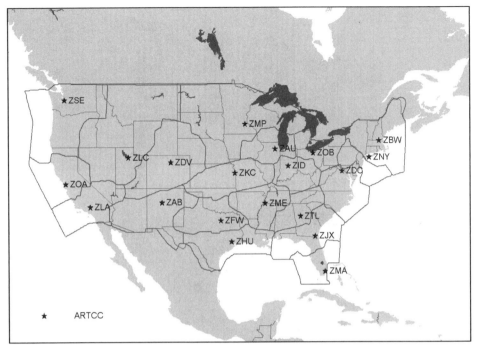

FIGURE 28.2 NAS en route centers.

Gilbert and others not only monitored flights, they issued voice clearances. A clearance gives pilots the go-ahead to depart from, arrive at, or hold at a *fix,* flight level (altitude), or runway. Ground control of airplanes soon became an FAA-regulated practice for flights flying instrument flight rules (IFR), such as commercial airliners. Flights using visual flight rules (VFR) are not considered "controlled"; the pilot is on his or her own: see and avoid. For controlled flights, the pilot must give priority to the controllers' directions, regardless of what he or she may see. The exception is weather. The pilot may use his or her judgment in negotiating weather.

Shortly before World War II a system coalesced. Regulations, maps, navigational aids, ground facilities, and communications began to interact in the pursuit of what the FAA calls separation assurance. Despite improvements in technology, the step-by-step, fix-by-fix, clearance-by-clearance approach, often called the board game, still dominates the design of air traffic control. All that is needed for the game to work is a flight plan (flight plans identifies an aircraft's call sign, intended speed, departure fix, proposed en route altitude, and route of flight, fix by fix, to the arrival fix; see Figure 28.6) and a calculator to determine the arrival time at each fix. Start the game. Planes take off. For every controlled flight, shortly before it reaches the next fix, a flight progress strip is printed that shows the time of arrival at that fix (see Figure 28.3).

World War II brought forth a plethora of technology and techniques. Radar and encryption are perhaps the most obvious. But the list includes computing, communications theory, information theory, game theory, and operations research. All would become important to modern living.

Radar proved crucial to ground control of aircraft. It gives controllers a view of the airspace situation in near real-time. After point-to-point routing, the situation display is the second great hallmark of air traffic control design.

FIGURE 28.3 En route controller flight strips.

In concept, radar is simple. A rotating signal bounces off airborne targets, planes among them. The time and angle of the return signal allows the location of the target to be calculated. One reply gives the position in terms of range and azimuth. From the second and succeeding replies the heading and velocity can be deduced. By the 1960s, many airplanes were equipped with transponders. The transponder intercepts the radar signal and adds a four-digit octal beacon code to the reply. This secondary return more clearly identifies the aircraft. What remains is to feed the return, with its position, velocity, heading, and maybe a beacon code, to a ground facility and the controllers' displays. This is done by land lines similar to telephone lines.

Radar helped shape the topology of the air traffic control system, i.e., how it is laid out. Long-range radars monitor wide-area en route airspace. In the United States, their reports feed the 20 ARTCCs. Figure 28.4 shows the "R" controller's sector display of weather and traffic, supported by the latest 20 × 20-inch console.

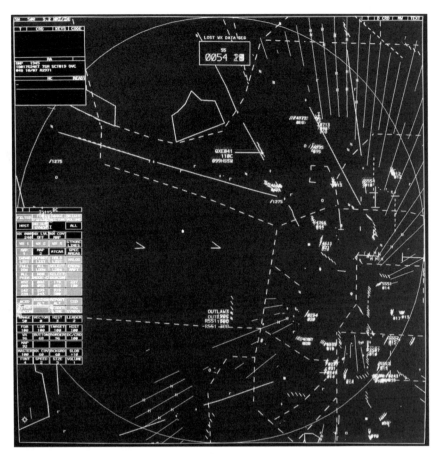

FIGURE 28.4 En route center radar controller.

At airports, short-range radars feed terminal radar (approach) control facilities (TRA-CONs), which control arrivals and departures in concert with airport towers. In the United States, there are slightly fewer than 200 TRACONs. They are placed near major airports, where they control departing and arriving traffic within a radius of about 60 nautical miles. There are over a thousand FAA-supported towers, and many more municipal and private airport towers.

A typical flight leaves the gate under the control of an airport tower. It takes off and ascends under TRACON control. About 60 nautical miles out, the flight transitions to en route airspace. A transcontinental flight will fly under the jurisdiction of several en route ARTCCs. It arrives as it left, but in reverse. The plane descends through approach control airspace, under control of a TRACON, and lands, guided by airport tower controllers. Along the way, controllers at each facility track the plane with radar. (See Figure 28.5.)

Controllers at TRACONs and ARTCCs manage only a sector of the approach or en route airspace. Therefore, control must be handed off from sector to sector, from controller to controller. The Chicago en route ARTCC watches over 60 sectors of airspace. A plane crossing Illinois will thus be handed off several times. As it is leaving the Chicago center's airspace, which spans a few hundred miles around the city's compass, control will be handed

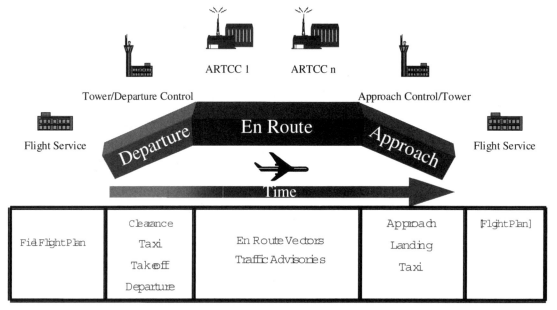

FIGURE 28.5   Gate-to-gate ATC.

off to the next ARTCC. If the flight is eastbound, for example, that will be the Indianapolis ARTCC.

The first computer used in air traffic control was located in Indianapolis. In the late 1950s, it stored flight plan data on a drum. The drum was connected to a printer that printed the flight progress strips. The computer calculated the times of arrival at each fix along the route of each flight, much like estimated times at rail stops along a train route.

The first automated system was developed by Burroughs and UNIVAC for the TRACONs. Unlike the en route ARTCCs, which are concerned with airway traffic and thus longer planning horizons, the TRACONs objective is to help guide airplanes through the tender process of departing and arriving. In this they rely mostly on radar and visual tracking, a more immediate, near-real-time job. The TRACON system is a radar tracking system. It was and is still called the Automated Radar Terminal System, or ARTS.

The ARTS computers, originally UNIVAC computers, are located in the TRACONs and provide (originally analog, now digital) displays of the departing and arriving traffic to the dozen or fewer approach controllers, who monitor the airspace around the airport. The ARTS also transmits the digital display data to the airport towers, which contain little in the way of computers, except for output devices like the displays attached to the remote ARTS TRACON, and flight strip printers, connected to the en route centers.

So much has changed since 1965, the year that the first ARTS sites went operational: technologies, social trends, space exploration, the geopolitical landscape. But not air traffic control automation. The ARTS equipment has been improved over the years, but its functions and algorithms have changed little.

The same can also be said for en route automation. IBM developed the en route automation system for the 20 ARTCCs. It was originally called the 9020 NAS En Route Stage A, for the 9020 multiprocessor IBM built for the FAA. The computer, twice upgraded, is now referred to simply as "the Host."

Over the decades in which automation has aided the controllers, the FAA has put more energy into safe and dependable automation than into rich automation. The 9020 multiprocessor is an excellent example of what the FAA calls its "failsafe" system. In the 1960s, the ARTS and the 9020 system used multiprocessors, so that if one computer failed, another would resume its processing. These machines were pioneering. Both UNIVAC and IBM built upon the concepts developed for the FAA to produce the first commercial multiprocessing, virtual storage computers. IBM's 9020 became the prototype for its System/370 virtual machines, which automatically mask storage, instruction fetch, cache memory, and channel faults while continuously computing. (That the IBM Enterprise System/390 today has a mean-time-to-failure of about 25 years is due largely to the objectives set by the FAA.)

The en route system combines radar tracking and flight planning, using algorithms developed by MIT, based on what had been learned at the early Indianapolis center. Thus, en route control is divided into two parts that the automation must reconcile, but that favor two different jobs: the "R" (radar) controller and the flight or "D" (data) controller. Each sector of airspace is under the control of a radar controller for near-real-time actions, and a flight controller, who is engaged in more long-term planning. (Figure 28.6 shows the fields of a filed flight plan.)

The ARTS required little more than 2 years to develop; the en route system took 9. Two aspects of the en route system design contributed to this: the complexities and size of en route airspace and the tricky business of pairing real-time radar tracking with intermediate- and long-term flight plans and their frequently amended altitudes and routes of flight. The en route software is about 10 times larger than the ARTS. Not the least of the automation challenges is automatically resynchronizing tracking and flight data during and after computer failures.

Flight planning is still the Achilles' heel of air traffic control. No flight or collection of flights can be planned perfectly. The exigencies of weather, dense traffic, and limited runways disrupt even careful planning. No airplane flies according to its projected plan, not for an hour, not for 10 minutes. The flight route estimation algorithms have not been precise or

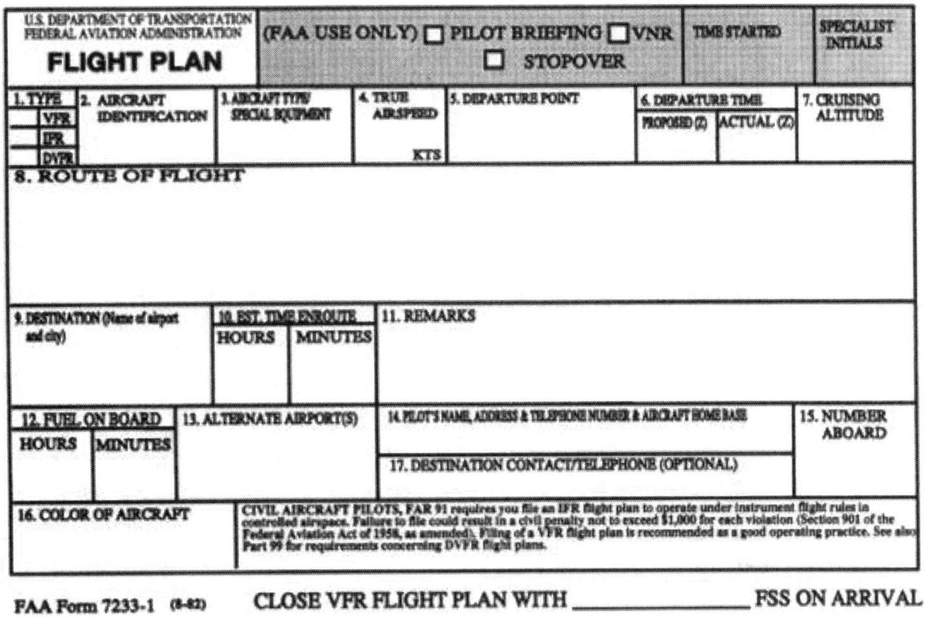

U.S. DEPARTMENT OF TRANSPORTATION
FEDERAL AVIATION ADMINISTRATION
FLIGHT PLAN

(FAA USE ONLY) ☐ PILOT BRIEFING ☐ VNR ☐ STOPOVER | TIME STARTED | SPECIALIST INITIALS

1. TYPE: VFR / IFR / DVFR
2. AIRCRAFT IDENTIFICATION
3. AIRCRAFT TYPE/SPECIAL EQUIPMENT
4. TRUE AIRSPEED KTS
5. DEPARTURE POINT
6. DEPARTURE TIME: PROPOSED (Z) / ACTUAL (Z)
7. CRUISING ALTITUDE
8. ROUTE OF FLIGHT
9. DESTINATION (Name of airport and city)
10. EST. TIME ENROUTE: HOURS / MINUTES
11. REMARKS
12. FUEL ON BOARD: HOURS / MINUTES
13. ALTERNATE AIRPORT(S)
14. PILOT'S NAME, ADDRESS & TELEPHONE NUMBER & AIRCRAFT HOME BASE
15. NUMBER ABOARD
16. COLOR OF AIRCRAFT
17. DESTINATION CONTACT/TELEPHONE (OPTIONAL)

CIVIL AIRCRAFT PILOTS, FAR 91 requires you file an IFR flight plan to operate under instrument flight rules in controlled airspace. Failure to file could result in a civil penalty not to exceed $1,000 for each violation (Section 901 of the Federal Aviation Act of 1958, as amended). Filing of a VFR flight plan is recommended as a good operating practice. See also Part 99 for requirements concerning DVFR flight plans.

FAA Form 7233-1 (8-82) CLOSE VFR FLIGHT PLAN WITH ______ FSS ON ARRIVAL

FIGURE 28.6 Flight plan.

accurate enough to keep up with the shifts in a flight's trajectory. In addition, individual flight strips are still the norm. As a flight progresses along its route, strips are printed at each fix and stacked along the side of the radar display. Several flights converging on a fix at various times and altitudes are manually sorted so that the controller can read and envision any conflicts. Needless to say, this nongraphical approach challenges even the most experienced en route flight controller, who must internalize the aircrafts' positions and velocities, convert hardcopy integer data into spatial relationships, and *imagine* the airspace picture.

In the early 1970s, the FAA added to the TRACON and en route automation altitude tracking and metering, which provides arrival controllers in both the centers and the TRACONs a list of flights arriving at destination fixes, organized by time of arrival and approach altitude. Before 1980, FAA had added to the ARTS and the 9020 en route systems two automation functions to enhance safety: conflict alert, which evaluates the position, velocity, and heading of tracks within a geographical are of interest and alerts the controller if separation standards will be violated; and minimum safe altitude warning (MSAW), which supports separation assurance between aircraft and terrain, such as tall buildings or mountains.

In 2001, Lockheed Martin, FAA's primary automation supplier, implemented an algorithm developed by MITRE to assist the flight ("D") controller in projecting aircraft routes. The tool has two advantages: it uses a more precise and accurate route estimation algorithm—called a trajectory modeler—and it shows the airplanes' trajectories graphically, aligned with airspace maps. The algorithm, packaged under the name of the user request evaluation tool (URET), where the user is the "D" controller and ultimately the pilot, also allows the controller to probe a flight beyond its current position, in order to evaluate where it and nearby flights will be in, say, 20 minutes. URET also allows the controller to proffer trial flight plans to investigate how conflicts might be avoided. Figure 28.7, shows, with red lines, two flights potentially converging as they approach the northwest geographical area of interest.

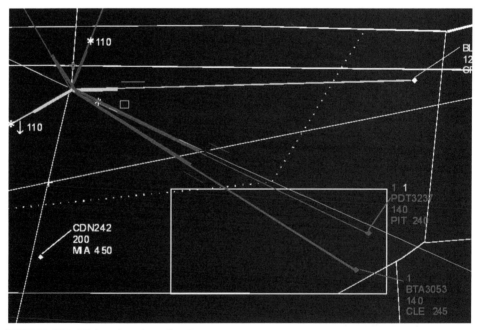

FIGURE 28.7   Flight trajectories and probe.

To recap, the air traffic control domains under FAA jurisdiction include the airport towers, with dominion over departures and arrivals, TRACONs, which guide a flight for about 60 miles in the approach areas, and the 20 national en route enters. The Department of Defense also owns and operates military airports and towers, as well as TRACONs. Military airspace is considered separate from civilian airspace. Thus, the civilian system under the FAA is marked by airspace that is not contiguous, with military and other special-use airspace fragmenting the national picture. There are automation links between the FAA and the military— mostly they exchange flight plans and voice alerts, but the two systems are run separately. This arrangement was sorely tested during the attack on America in 2001. That the FAA was able to land hundreds of flights within minutes of the attack on the World Trade Center is a testament to FAA-DoD coordination and air traffic controller proficiency.

Critical partners in commercial aviation are the airlines. Since deregulation, the airlines have been more active in using automation and information in concert with the FAA. In addition to the regional and local airline operation centers, major airlines operate ramp towers at our largest airports. The ramp towers manage traffic between and among the aprons, concourse, runways, ramps, and gates. (Some municipal airport authorities also use ramp towers airport-wide as part of their airport management.) It stands to reason that the FAA and the airlines would want to share what they know about flight conditions, including runway availability, ground delays, weather, special-use airspace restrictions, and flow control. To that end, the airlines and FAA have recently instituted collaborative decision-making (CDM). The practice was immediately effective. CDM is largely carried out by voice. Fully automatic sharing of data between airlines and the FAA remains an objective.

One of the areas of air traffic control that has emerged as a result of burgeoning traffic and crowded skies and runways is called traffic flow management (TFM). The FAA considers TFM an "essential," not a "critical," service. (Critical services involve the ground controller's actions to ensure aircraft-aircraft and aircraft-airspace separation.) Many believe that in the future long-range airspace analysis and flow control will be the most important function on the ground; that with on-board systems such as GPS, collision avoidance, conflict detection, advanced weather sensors, and automatic data link (automatically exchanging data computer to computer between cockpit and ground facilities), separation assurance will fall largely to the pilot, using the automation on the flight deck. (See Figure 28.8.)

Certainly the commercial airlines would benefit from improved TFM. The FAA can satisfy safety constraints simply by increasing flight separation standards and holding planes on the ground and, en route, at fixes until the competing traffic or weather dissipates. But this relies on air traffic controller intervention using rule-making and voice procedures more than automation. A more automated and collaborative TFM system would allow commercial and general aviators more flexibility in choosing routes and thus eliminate one-at-a-time delays.

The traffic flow management system is run by the FAA. It overlies the tactical control systems of the towers, TRACONs, and ARTCCs. Each of these facilities send proposed and active flight plans, second-order amendments, and tracking data to the Air Traffic Control System Command Center (ATCSCC) in Herdon, VA. The ATCSCC receives the data rather circuitously. The control facilities send their flight and tracking data to the FAA's support facility, the W. J. Hughes Technical Center, in Pomona, NJ, and from there to the Volpe National Transportation System Center in Cambridge, MA. The two support facilities transform the data from a local facility view to a regional and national view and integrate weather information from the National Weather Service. The picture the ATCSCC has to work with is then conducive to evaluating large volumes of airspace.

ARTSCC staff use traffic volume algorithms to predict and analyze traffic patterns throughout the United States. The TFM hub shares the results with traffic planners at the control facilities as well as the airlines, which can better plan for possible delays.

Oceanic air traffic is under the control of centers located in Hawaii, Alaska, off the west coast of the United States at the Oakland ARTCCs, and off the east coast of the United States at the New York ARTCC. The continental sites at Oakland and New York maintain oceanic control rooms separate from their en route counterparts. Virtually the same automation as used for en route control is used to ensure separation over the ocean: tracking and

FIGURE 28.8 Flight deck circa 2003.

flight planning and flight progress strips. However, over the ocean, instead of radar and land-based navigational aids being used, position, velocity, and heading are provided to the ground control facilities by radio, the coastal navigational system LORAN, inertial navigation, and GPS, which, like radar, uses signal distance and time—but from space—to derive position, velocity, and heading.

In 2003, Lockheed Martin tailored a New Zealand oceanic system that automates pilot reports, ground controller clearances, and supports automatic dependent surveillance (automatic position and velocity reports from the onboard GPS) over data link, allowing the FAA to reduce separation standards from 60 to 30 nautical miles over the sea.

The design of the U.S. air traffic control system has evolved conservatively. The FAA places safety first. United States air travel is the safest in the world. In operating one of most complex systems ever realized, the FAA has chosen to introduce changes deliberately. The system is largely procedural; humans are constantly in the loop. New technology is assimilated at a pace far slower than in the commercial sector and the military. This is the case in computing, telecommunications, navigation, graphical systems, and meteorology.

The FAA is organized to ensure that no advanced functions are introduced before they are deemed absolutely safe. Designs are ultimately concerned with reconciling the "new artificial" with the "old," both natural and artificial. As such, how people organize to design is part of the design. Two aspects of the FAA's organization are decisive in that regard.

The first is that the air traffic controllers, as well as the operations and maintenance staff and administrative employees, belong to unions. The unions throttle abrupt changes in technology, especially as it might disrupt the well-worn and effective habits of the air traffic controllers, without suffocating it. For technologists and those who wish to fly without any personal sacrifice, this arrangement must seem atavistic. But it provides a built-in governor.

The second aspect of the FAA organization also ensures against unwelcome change. It is the W. J. Hughes Technical Center (WJHTC), where FAA systems are tested and maintained. Regardless of the intensity of testing done by suppliers, all systems must pass a strict set of tests at the WJHTC, which reports directly to the administrator for research and acquisition, independent of systems development. As such, the WJHTC is the third-party quality assurance arm of the FAA.

The future of air traffic control design is certain: it will improve, but in specific areas and incrementally. In terms of major changes, economics and security concerns have reduced commercial air travel and the amount of money available for system upgrades. In the near term, improvements will be modest and directed.

In 1998, Mrs. Jane Garvey, the current FA Administrator, started the Free Flight initiative. The objectives for Free Flight Phase 1 were to use CDM, deploy URET, implement automated data link, and improve arrival and runway flow into the major airports. These objectives were met.

In 2003, the FAA is undertaking a new plan, the Operational Evolution Plan (OEP). The OEP increases the FAA's commitment to a modernization strategy that encapsulates the entire aviation community, including the airlines, airports, cargo carriers, the DOD, and NASA. No single objective or technique characterizes the OEP. It is a collection of specific objectives that will increase capacity and efficiency at airports, relieve en route congestion, and improve weather prediction. Airport efficiency, for example, will soon benefit from GPS and air navigation to help guide landings, augmenting the long-time instrument landing systems in use worldwide. Traffic management advisories will be used to smooth the arrival of many flights to few airports and runways, helping controllers find the optimum path from en route airspace to the final approach and runway, and ultimately the gate. To relieve en route congestion and reduce voice communications, automated data link—computer-to-computer—will add more messages to the core set.

Perhaps the most safety-enhancing element of the OEP falls in the weather domain. A quote from the OEP summarizes one such initiative: Cockpit surface movement maps have shown promise in improving crew situational awareness in low visibility. These tools supplement the pilot's out-the-window assessment of aircraft position, direction and speed. When coupled with positive identification of other surface traffic, procedures can be changed to

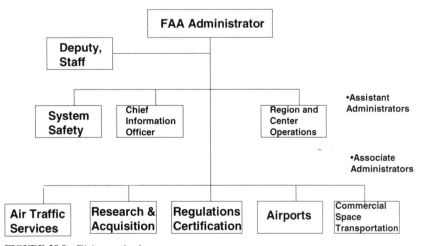

FIGURE 28.9 FAA organization.

direct one aircraft to follow another without visual references outside the cockpit. These changes may enhance pilot confidence and efficiency in moving about the airport surface. The key to success for this initiative as an OEP capacity enhancement is the ability to go beyond improvement in situational awareness to improved efficiency in surface movement.

Progress is not so much technological as geographical. The U.S. air traffic control design is being inserted into the developing nations around the world. Flight data, radar, oceanic control, voice clearances, digital maps, data link, GPS, runway management, flow control—all have their roots in the U.S. system, and all are being made available overseas, thanks to companies like Lockheed Martin. Already, the fundamentals of air traffic control design are at work in the United Kingdom, Scotland, Argentina, New Zealand, Taiwan, China, and Korea, and soon they will be in Africa and most of the developing nations in Europe and Asia.

To repeat from the introduction, the world is becoming smaller and therefore friendlier: Before flight became commonplace, people traveled in two dimensions, crossing the borders that separate town from town and nation from nation. Seen from above, the artificial boundaries that divide us disappear. Distances shrink, the horizon stretches. The world seems grander and more interconnected.

OPERATIONS AND ECONOMICS

CHAPTER 29
TRANSPORTATION PLANNING AND MODELING

Konstadinos G. Goulias
Department of Civil Engineering,
The Pennsylvania State University,
University Park, Pennsylvania

29.1. OVERVIEW

More than at any other time during the past decade, we have observed a clear shift in transportation policies away from construction of new facilities and toward more efficient management of an existing transportation system to meet the ever-increasing transportation demand. Contrary to a widespread belief, this change in direction has not eliminated projects for new highways and projects for major maintenance and reconstruction of existing highways. Every year billions of U.S. dollars are allocated to new transportation facilities, as the U.S. transportation appropriation Bills amply demonstrate (http://www.whitehouse.gov/omb/legislative/sap/107-1/HR2299-s.html, accessed December 2002).

There is, however, a general shift toward optimal management of existing facilities and strategic improvements at specific transportation system components that are considered to be critical interventions in a system of interconnected and mutually influencing facilities. In fact, in the United States some of the congressional appropriation recipients and their assets play a critical role in the functioning of the national transportation system and its interfacing state systems. Independent of role, location, and perceived need for investment, the overall goal of fund allocation is to maximize the performance of the transportation system in its entirety and avoid major new infrastructure building initiatives. In parallel, we also observe a movement toward policy actions and large investments that aim to protect the environment (see examples in Goulias 2003). These policy actions also view the world surrounding us as an ecosystem, placing more emphasis on its overall survival by examining direct and indirect effects of individual policy actions. This trend is not limited to transportation; as Lomborg (2001) explains, it encompasses the entire range of human activity and the entire ecosystem we live in. This good news, however, is countered by additional arguments in favor of higher human activity control targets (e.g., pollution controls) and a warning that human development pace contradicts biological pace, with unknown consequences (Tiezzi 2003). In fact, one of the most recent studies on research needs, addressing the transportation environment relationship (TRB 2002), expands the envelope to incorporate ecology and natural systems and addresses human health in a more comprehensive way than in the past, reiterating a variety of unresolved issues such as wasting land. As a result, in the past decade we have also experienced a clear shift to policy analysis approaches that have an expanded scope and domain and are characterized by explicit recognition of transportation system complexity and uncertainty, as the discussion later in this chapter illustrates. Reflecting all this, a relatively new term, *sustainable transportation,* is now often used to indicate a shift in the mentality of the transportation analyst community to represent a vision of a transportation system that attempts to provide services that minimize harm to the environment. In fact, one of the most comprehensive reviews of policies in North America (Meyer and Miller 2001) contrasts the nonsustainable and sustainable approaches and provides a compelling argument about the change in these policies and how we are moving toward a more sustainable path.

In the United States during the past 10 years, the need to examine these new and more complex policy initiatives has become increasingly pressing due to the passage of a series of legislative initiatives (Acts) and associated federal and state regulations on transportation policy, planning, and programming. The multimodal character of the new legislation, its congestion management systems, and the taxing air quality requirements for selected U.S. regions have motivated many new forecasting applications. In the early years these were predominantly based on the Urban Transportation Planning System and related processes, but during the last 5 years a shift to richer conceptual frameworks has occurred. Air quality mandates have motivated impact assessments of the so-called transportation control measures and the creation of statewide mobile source air pollution inventories (Stopher 1994; Loudon and Dagang 1994; Goulias et al. 1993) that require different analytical forecasting tools than in any pre-1990 legislative initiatives (Niemeier 2003). An added motivation is lack of substantial funding for transportation improvement projects and a shift to charging the firms that benefit the most from transportation system improvements, creating a need for impact fees assessment for individual private developers. These assessments create the need for higher resolution in the three dimensions of geography (space), time (time of day), and social space* (households and individuals), used in typical regional forecasting models but also the domain of jurisdictions where major decisions are made. They also create a pressing need for interfaces with traffic engineering simulation tools that are approved and/or endorsed in legislation (for examples see Paaswell, Rouphail, and Sutaria 1992). Another push for new tools is the assessment of technologies under the general name of intelligent transportation systems (i.e., bundles of technological solutions in the form of user services attempting to solve chronic problems such as congestion, safety, and air pollution).

These policies and assessments and the models developed to support their assessment should not be, and usually are not, studied in isolation from past initiatives. Humphrey (1990) is a demonstration of a widespread consensus about the inability of *traditional transportation solutions* to solve urban and suburban congestion and its counterpart, environmental degradation. In fact, the Clean Air Act Amendments of 1990 attempt to address these concerns with a series of actions/policy initiatives under the umbrella name of transportation control measures (TCMs):

1. Public transit

2. High-occupancy-vehicle (HOV) roads and lanes

3. Employer-based transportation management programs

4. Trip-reduction ordinances

5. Traffic flow improvement programs

6. Fringe and corridor parking for multiple occupancy vehicles and transit

7. Programs to limit and restrict vehicle use in downtown areas

---

*Social space here means the space where self-realization and relationships among individuals take place and where feelings, emotions, and experiences are shared. Golledge and Stimpson (1997) provide an alternative definition that includes geographic space.

8. Shared-ride and HOV programs

9. Roads designated for exclusive use by nonmotorized modes

10. Bicycle lanes and facilities (storage)

11. Control of vehicle idling

12. Programs to reduce cold-start vehicle emissions

13. Flexible work schedules

14. Programs and ordinances to reduce single-occupant-vehicle (SOV) travel, including provisions for special events and activity centers

15. Construction of paths for exclusive use by pedestrian and nonmotorized vehicles

16. Vehicle retirement/replacement (scrappage) programs

The framework thus defined also emphasized the market nature of these controls (a carrot-and-stick approach to implementation) and an air quality framework that in essence enriched our policy "tools" (see Meyer and Miller 2001; and Niemeier 2003). Today, these tools are becoming even more wide-ranging, as the list in Table 29.1 demonstrates.

As Garrett and Wachs (1996) discuss in the context of a lawsuit against a regional planning agency in the Bay Area, traditional four-step regional simulation models (Creighton 1970; Hutchinson 1974; Ortuzar and Willumsen 2001; and chapter 7 of this Handbook) are outpaced by the same legislation that defined many of the policies described above. Independently of the urgency and timeliness of modeling and simulation, one fundamental difference from past energy crises is a widespread recognition of the need to build integrated models. Under these initiatives, forecasting models, in addition to long-term land use trends and air quality impacts, need to address issues related to technology use and information provision to travelers in the short and medium term. Similarly, the European Union focuses on issues such as increasing citizen participation, intra-European integration, decentralization, deregulation, privatization, environmental concerns, mobility costs, congestion management by population segments, and private infrastructure finance (see van der Hoorn 1997).

These new policy initiatives place more complex issues in the domain of regional policy analysis and forecasting and amplify the need for methods that produce forecasts at the individual traveler and his or her household* levels instead of the traffic analysis zone** level. In addition to long-range planning activities and the typical traffic management activities, analysts and researchers in planning need to evaluate the following:

- Travel demand and supply impacts of new technologies (e.g., mobile voice and data transmission)

- Traveler and transportation system manager information provision and use (e.g., on-board traveler information systems)

- Pricing and financing strategies (e.g., congestion pricing)

- Combinations of transportation management actions and their impacts (e.g, parking fee structures and city center restrictions)

- Assessment of combinations of environmental policy actions (e.g., carbon taxes and information campaigns on the health effects of ozone)

The tools to perform all this need to have forecasting capabilities that are more accurate and detailed in space and time (e.g., we are moving toward parcel-by-parcel analysis and separate analyses for different seasons of a year and days of the week to capture seasonal and within-week variations of travel). Echoing all this and in the context of the Dutch reality,

---

*Viewed as a decision-maker.

**Convenient spatial subdivision that is often based on data availability from the U.S. Census.

TABLE 29.1 Typical Tools for the Planner and Operations Manager in Transportation

Type of tool	Brief description	Other source of information
Congestion pricing and toll collection programs.	A premium is charged to travelers who wish to travel during the most congested periods.	TRB 1999a.
Emission, vehicle miles traveled, and other fee programs (including carbon taxes and trading).	Programs that shift taxation from traditional sources towards pollutant emissions and natural-resource-depletion agents	An example can be found at: http://www.me3.org/projects/greentax/ (accessed January 2003).
Intelligent transportation systems (ITS).	Use of telecommunications and information technology to manage, control, and provide information about the transportation system to travelers and managers.	http://www.itsa.org/ and http://www.ertico.com/.
Accelerated retirement of vehicles programs.	Programs to eliminate high-emitting and older technology vehicles.	http://ntl.bts.gov/DOCS/SCRAP.html.
Telecommunications to substitute/complement travel.	The employment of telecommunications to substitute-complement-enhance travel.	http://www.vtpi.org/tdm/tdm43.htm.
Land use growth and management programs.	Legislation that controls for the growth of cities in sustainable paths.	http://www.awcnet.org/.
Land use design and attention to neighborhood design for nonmotorized travel.	Similar to the previous but with attention paid to individual neighborhoods.	http://www.sustainable.doe.gov/landuse/luothtoc.shtml.
Goods movements (freight) programs to improve operations.	A variety of programs to facilitate and minimize the damage for freight movement.	http://ntl.bts.gov/DOCS/harvey.html.
Highway system improvements in traffic operations and flow.	Improved data-collection, monitoring, and traffic management.	http://www.nawgits.com/icdn/.
Programs to increase the use of multirider modes.	Alternative transit service improvements and expansion and high-occupancy vehicle systems.	http://www.fhwa.dot.gov/operations/hovguide01.htm.
System improvements for nonmotorized travel.	Organized efforts to increase nonmotorized travel.	http://www.ci.missoula.mt.us/feetfirst/plan.html.
Special event planning and associated traffic management.	Enhanced procedures to handle the demands of a special event.	http://tmcpfs.ops.fhwa.dot.gov/cfprojects/new_detail.cfm?id=32&new=0.
Public involvement and education programs.	Programs aiming at defining goals based on the public's desires.	http://www.dot.state.pa.us/internet/secinet.nsf/ and search for PennPlan.
Individualized marketing techniques with improved information and communication with the customer.	Public programs to provide personal help in changing travel behavior in favor of environmentally friendly modes.	http://www.local-transport.dft.gov.uk/travelplans/index.htm.

Note: Many additional examples can be found at http://www.vtpi.org/tdm/ (accessed April 2003).

Borgers, Hofman, and Timmermans (1997) have identified five information need domains the new envisioned policy analysis models will have to address. Somewhat modified in format from the original list, they are:

1. Social and demographic trends that may produce a structural shift in the relationship between places and time allocation by individuals, invalidating existing travel behavior model systems

2. Increasing scheduling and location flexibility and degrees of freedom for individuals in conducting their everyday business, leading to the need to consider additional choices (e.g., departure time from home, work at home, shopping by the Internet, shifting activities to the weekend) in modeling travel behavior

3. Changing quality and price of transport modes based on market dynamics and not on policies external to the travel behavior (e.g., the effect of deregulation in public transport);

4. Shifting of attitudes and potential cycles in the population outlook about modes

5. Changing scales/jurisdictions (*scale* is the original term used to signify the different jurisdictions)—different policy actions in different sectors have direct and indirect effects on transportation, and different policy actions in transportation have direct and indirect effects in the other sectors (a typical U.S. example welfare-to-work program)

Based on this overview, in the next section we define the model attributes required to design major improvements over the more traditional systems engineering simulation models. The section defines a few key parameters for an improved model that contains both the flexibility and the potential to fill the information needs of today's policy initiatives. Also presented in this section a brief review of the many innovations in modeling and simulation that have advanced the field and are currently used to develop a variety of new behavioral paradigms and operations models. The final section provides a summary and an assessment of where we are today in transportation planning and modeling and where we are expected to go to deliver the policy analysis tool that is needed.

29.2 MODEL ATTRIBUTES AND INNOVATIONS IN MODELING

The policies we reviewed above have expanded the context of travel behavior models to entire life paths of individuals, and for this reason a more general modeling framework is emerging. Planning and modeling have achieved tremendous progress toward a comprehensive approach to, in essence, build simulated worlds on computer, enabling the study of policy scenarios. The emerging framework, however, is incomplete, containing many gaps. It is rich, however, in the directions taken and potential for scientific discovery and policy analysis. In this section we first review four dimensions of this framework and then discuss select innovations. We focus on and emphasize passenger travel, but similar innovations can be found in modeling of goods movement (Southworth 2003).

29.2.1 Model Dimensions

Four dimensions can be identified in building taxonomies of policies and models: *geographic space* and its conditional continuity; *temporal scale* and calendar continuity; interconnectedness of *jurisdictions;* and, most important, the set of relationships in *social space* for individuals and their communities. These four dimensions very often cannot be disengaged from each other. For example, when one considers issues associated with a change in propulsion fuels and energy consumption for transportation, the planning horizons are very often 25 or more years. This may also be accompanied by planning that expands beyond the borders of a single country because producers and consumers may be organized in market unions (OPEC, EU) and planned actions involve governments and private companies. Issues of this type are long-term, encompass larger geographic regions, and involve a complex network of organizations. These are in the realm of grand visions and policies. At the other end of the spectrum we find measures and actions aiming at resolving a very localized problem, such as access to facilities for disabled persons. The time horizon here is a few months, and the solutions range, to take the case of providing access to persons with disabilities, from installing ramps in buildings and cutting sidewalk curbs so that wheelchairs can access rooms in buildings and the street, to checking for compliance with the relevant rules and regulations of the fire protection code and the Americans with Disabilities Act.

The first dimension, *geographic space,* here means the physical space in which human action occurs. This dimension has played an important role in transportation planning and modeling because the first preoccupation of transportation system designers has been to move persons from one location to another. Initial applications considered the territory divided into large areas (traffic analysis zones), represented by a virtual center of gravity (centroid) and connected by facilities (higher-level highways). The centroids were connected to the higher-level facilities using a virtual connector summarizing the characteristics of all the local roads within the zone. As computational power increased and the types of policies/strategies required increased resolution, the zone became smaller and smaller. Today is not unreasonable to expect software to handle zones as small as a parcel of land and transportation facilities as low in the hierarchy as a local road (the centroid becomes the housing unit and the centroid connector the driveway of the unit, and they are no longer virtual). As we will see later in this chapter, in analyzing behavior we are interested in understanding human action. For this reason, in some applications geographic space needs to consider more than just physical features (Golledge and Stimpson 1997, 387), moving us into the notion of place and social space (see also below).

The second dimension, *time,* here means continuity of time, irreversibility of the temporal path, and the associated artificiality of the time period considered in many models. For example, models used in long-range planning applications use typical days (e.g., a summer day for air pollution). In many regional long-range models, the unspoken assumption is that we target a typical work weekday in developing models to assess policies. Households and their members, however, may not always (if at all) obey this strict definition of a typical weekday to schedule their activities and may follow very different decision-making horizons in allocating time to activities within a day, spreading activities among many days, including weekends, substituting out-of-home for in-home activities on some days but doing exactly the opposite on others, and using telecommunications only selectively (e.g., on Fridays and Mondays more often than on other days). Obviously, taking into account these scheduling activities is far more complex than what is allowed in existing transportation planning models.

The third dimension is *jurisdictions* and their interconnectedness. The actions of each person are regulated by jurisdictions with different and overlapping domains, such as federal agencies, state agencies, regional authorities, municipal governments, neighborhood associations, trade associations and societies, religious groups, and formal and informal networks of families and friends. The federal government defines many rules and regulations on environmental protection. These may end up being enforced by a local jurisdiction (e.g., a regional office of an agency within a city). On the one hand, we have an organized way of governance that clearly defines jurisdictions and policy domains (e.g., tax collection in the United States). On the other hand, however, the relationships among jurisdictions and decision-making about allocation of resources do not always follow this orderly governance principle of hierarchy. A somewhat different and more bottom-up relationship is found in the social network, and for this reason it requires a different dimension.

The fourth and final dimension is *social space* and the relationships among persons within this space. For example, individuals from the same household living in a neighborhood may change their daily time allocation patterns and location visits to accommodate or take advantage of changes in the neighborhood, such as elimination of traffic and the creation of pedestrian zones. Depending on the effects of these changes on the pedestrian network, we may also see a shift in social behavior within the neighborhood. In contrast, increase in traffic to surrounding places may create an outcry from other surrounding neighborhoods, thus complicating the relationships among the residents.

One important domain and entity within this social space is the household. This has been a very popular unit of analysis in transportation planning, since strong relationships within a household can be used to capture behavioral variation (e.g., the simplest method is to use a household's characteristics as explanatory variables in a regression model of travel behavior). In this way any changes in the household's characteristics (e.g., change in composition due to birth, death, children leaving the nest, or adults moving into the household) can be used to predict changes in travel behavior. New model systems are created to study this interaction within a household, looking at the patterns of using time in a day and the changes across days and years.

It is therefore very important in modeling and simulation as well as other types of policy analysis to incorporate in the models used for policy analysis not only the interactions described above but also interactions among these four fundamental dimensions. The typical example is long-range planning, which is usually defined for larger geographical areas (region, states, and countries) and addresses issues with horizons from 10 to 50 years. In many instances, we may find that large geographic scale means also longer time frames applied to wider mosaics of social entities and including more diverse jurisdictions. On the other side of the spectrum, issues that are relevant to smaller geographic scales are most likely to be accompanied by shorter-term time frames applied to a few social entities that are relatively homogeneous and subject to the rule of a very few jurisdictions. This is one important organizing principle but also an indicator of the complex relationships we attempt to recreate in our computerized models for decision support. In developing the blueprints of these models, we can choose from a variety of theories (e.g., neoclassical microeconomics) and conceptual representations of the real world that help us develop these models. At the heart of our understanding of how the world (as an organization, a household, or an individual human being) works are models of decision-making and conceptual representations of relationships among entities making up this world. The next two sections briefly review both.

29.2.2 Decision-Making Paradigms

Transportation planning applications are about judgment and decision-making of individuals and their organizations. There are different decision-making settings that we want to understand. Two of these settings are (1) the travelers and their social units from which motivations for their behavior and constraints to their behavior emerge; and (2) the transportation managers and their organizations that serve the travelers and their social units (note that we exclude from this discussion goods movement that contains a few additional actors; see Southworth 2003). Both settings have received considerable attention in transportation planning and its modeling of the decision-making process. Conceptual models of this process are transformed into computerized models of a city, region, or even state in which we utilize components, which are in turn human behavior models representing judgment and decision-making of travelers moving around the transportation network and visiting locations where they can participate in activities. Models of this behavior are simplified versions of strategies used by travelers when they select among options that are directly related to their desired activities. In some of these models, we also make assumptions about hierarchies of motivations, actions, and consequences. Some of these assumptions are explicit when deriving the functional forms of models, as in the typical disaggregate choice models, but in other models these assumptions are implicit. When designing transportation planning model interfaces for transportation planners and managers, we also implicitly make assumptions about the managers' ability to understand the input, agent representation, internal functioning, and output of these computerized models. Our objective is therefore not only to understand travel behavior and build models that describe and predict human behavior, but also to devise tools (e.g., decision support systems) that allow transportation managers to understand the assumed behavior in the models, study scenarios of policy actions, and define and explain policy implications to others. This, in essence, implies that we, the model system designers, create a platform for a relationship between planners and travelers. A similar but more direct relationship also exists between travelers and transportation managers when we design the observation methods that provide the data used for modeling but also the data used to measure attitudes and opinions, such as travel surveys. In fact, this relationship is studied in much more detail in the survey design context and linked directly to the image of the agency conducting the survey and the positive or negative impression of the travelers about the sponsoring agency (Dillman 2000). Most transportation research for modeling and simulation, however, has emphasized traveler behavior when building surveys and models neglecting the interface with the planners. The summary below, however, applies to individuals traveling in a network but also to organizations and planners in the sense used by H. A. Simon (1997).

Rational decision-making is a label associated with human behavior that follows a strategy in identifying the best course of action. In summary, a decision-maker solves an optimization problem and identifies the best existing solution to this problem. Within this more general strategy, when an operational model is needed and this operational model is required to provide quantitative predictions about human behavior, some kind of mathematical apparatus is needed to provide the predictions. One such apparatus is the *subjective expected utility* (SEU; Savage 1954) formulation of human behavior. In developing alternative models to SEU, Simon (1983) defines four theoretical components:

1. A person's decision is based on a utility function assigning a numerical value to each option—*existence and consideration of a cardinal utility function.*

2. The person defines an exhaustive set of alternative strategies among which just one will be selected—*ability to enumerate all strategies and their consequences*

3. The person can build a probability distribution of all possible events and outcomes for each alternative option—*infinite computational ability*

4. The person selects the alternative that has the maximum utility—*maximizing utility behavior.*

This behavioral paradigm has served as the basis for a rich production of models in transportation related choice analysis that include mode of travel, destinations to visit, as well as the household residence (see the examples in the seminal textbook by Ben-Akiva and Lerman 1985). It has also served as the theoretical framework for consumer choice models and for attempts to develop models for hypothetical situations (see the comprehensive book by Louviere, Hensher, and Swait 2000). It has also replaced the aggregate modeling approaches to travel demand analysis as the orthodoxy that many old and new theories and applications are compared to and compete with.

SEU can be considered to be a model from within a somewhat larger family of models under the label of weighted additive rule (WADD) models (Payne, Bettman, and Johnson 1993). Real humans, however, may never behave according to SEU or related maximizing and infinitely computational capability models (Simon 1983 labels this the Olympian model). Based on exactly this argument, different researchers in psychology have proposed a variety of decision-making strategies (or heuristics). For example, Simon created alternative model paradigms under the label of *bounded rationality*—the limited extent to which rational calculation can direct human behavior (Simon 1983, 1997) to depict a sequence of a person's actions when searching for a suitable alternative. The modeled human is allowed to make mistakes in this search, giving a more realistic description of observed behavior (see also Rubinstein 1998). Tversky (1969) is credited with another stream of decision-making models starting with the *lexicographic approach,* in which a person first identifies the most important attribute, compares all alternatives on the value of this attribute, and chooses the alternative with the best value on this most important attribute. Ties are resolved in a hierarchical system of attributes. Another Tversky model (1972) assumes a person selects an attribute in a probabilistic way and influenced by the importance of the attribute, all alternatives that do not meet a minimum criterion value (cutoff point) are eliminated. The process proceeds with all other attributes until just one alternative is left and that is the one chosen. This has been named the *elimination by aspects strategies* (EBA) model. Later, Kahneman and Tversky (1979) developed *prospect theory* and its subsequent version, *cumulative prospect theory* (Tversky and Kahneman 1992), in which a simplification step is first undertaken by the decision-maker editing the alternatives. Then a value is assigned to each outcome and a decision is made based on the sum of values multiplying each by a decision weight. Losses and gains are treated differently. All these alternatives to SEU paradigms did not go unnoticed in transportation research, with early significant applications appearing in the late 1980s. In fact, a conference was organized attracting a few of the most notable research contributors to summarize the state of the art in behavior paradigms (documented in Gärling, Laitila, and Westin 1998). One of the earlier examples using another of Simon's inventions, *satisficing behavior*—acceptance of viable choices that may not be optimal—is a series of transportation-specific applications described in Mahmassani and Herman (1990). Subsequent contributions continue along the path of more realistic models, and the most recent example, discussing a few models (Avineri and Prashker 2003), uses cumulative prospect theory, giving a preview of a movement toward more realistic travel behavior models. As Gärling (1998) and Avineri and Prashker (2003) point out, these paradigms are not ready for practical applications, contrary to the efforts by Mahmassani and colleagues that have been applied, and additional work is required to use them in a simulation framework for applications. In addition, Payne, Bettman, and Johnson (1993, 29–34) provide an excellent review of these models and a summary of the differentiating aspects among the paradigms. Most important, however, they provide evidence that decision-makers adapt, i.e., switch between decision-making paradigms, to the task and the context of their choices. They also make mistakes and may also fail to switch strategies. As Vause (1997) discusses of some length, transportation applications are possible using multiple decision-making heuristics within the same general framework and employing a production system approach (Newell and Simon 1972). A key consideration, however, that has received little attention in transportation is the definition of context within which decision-making takes place. Recent production systems (Arentze and Timmermans 2000) are significant improvements over past simulation techniques. However, travelers are still assumed to be passive in shaping the environment within which they decide to act (action space). This action space is viewed as largely made by constraints and not by travelers active shaping of their context. In contrast, Goulias (2001, 2003) reviews another framework from human development that is designed to treat decision-makers in their active and passive roles and explicitly accounts for mutual influence between an agent (active autonomous decision-maker) and her environment.

29.2.3 Innovations in Systems Modeling Approaches

In spite of the issues raised in the previous section, in transportation modeling and simulation we have experienced a few tremendously progressive steps forward. Interestingly, these key innovations are from nonengineering fields but they are very often transferred and applied to transportation systems analysis and simulation by engineers. These are listed here in a somewhat sequential chronological order, merging technological innovations and theoretical innovations.

Disaggregate Demand Models.   At the same time that the Bay Area Rapid Transit system was studied and evaluated in the 1960s, Dan McFadden (the 2000 Nobel Laureate in Economics) and a team of researchers produced practical mode choice regression models at the level of an individual decision-maker (see http://emlab.berkeley.edu/users/mcfadden/, accessed May 2003). The models are based on random utility maximization (of the SEU family), and their work opened up the possibility of predicting mode choice rates more accurately than ever before. These models were initially named *behavioral travel-demand models* (Stopher and Meyburg 1976), but later the more appropriate term of *discrete choice models* (Ben-Akiva and Lerman 1985) prevailed. Although restrictive in their assumptions, these models are still under continuous improvement and have become the standard tool in evaluating alternative transportation mode options. Some of the most notable and recent developments advancing the state of the art and practice are:

1. Better understanding of the theoretical and particularly behavioral limitations of these models (Gärling, Laitila, and Westin 1998; McFadden 1998; Golledge and Gärling 2003).

2. More flexible functional forms that resolve some of the problems raised in Williams and Ortuzar (1982), allowing for different choices to be correlated when using the most popular discrete choice regression models (Koppelman and Sethi 2000; Bhat 2000, 2003).

3. Combination of revealed preference, stated choices by travelers, with stated preferences and intentions, answers to hypothetical questions by travelers, availability of data in the same choice framework to extract in a more informative way travelers' willingness to use a mode and willingness to pay for a mode option (Ben-Akiva and Morikawa 1989; Louviere, Hensher, and Swait 2000). This latter improvement enables us to assess situations that are impossible to build in the real world.

4. Computer-based interviewing and laboratory experimentation to study more complex choice situations and the transfer of the findings to the real world (Mahmassani and Jou 2000). This direction, however, is also accompanied by a wide variety of research studies aiming at more realistic behavioral models that go beyond mode choice and travel behavior (Golledge and Gärling 2003).

5. Expansion of the discrete choice framework using ideas from *latent class models* with covariates that were first developed by Lazarsfeld in the 1950s and their estimation finalized by Goodman in the 1970s (see the review in Goodman 2002 and discrete choice applications in Bockenholdt 2002). This family of models was used in Goulias (1999) to study the dynamics of activity and travel behavior and in the study of choice in travel behavior (Ben-Akiva et al. 2002).

Constraints and Related Ideas.   As discussed earlier, the rational economic assumption of the maximum utility model framework (underlying many but not all of the disaggregate models) is very restrictive and does not appear to be a descriptive behavioral model, except for few special circumstances when the framing of decisions is carefully designed (something we cannot expect to happen every time a person travels on the network). Its replacement, however, requires conceptual models that can provide the types of outputs needed in regional planning applications. A few additional research paths, labeled *studies of constraints,* are also functioning as gateways into alternative approaches to replace or complement the more restrictive utility-based models. A few of these models also consider knowledge and information provision to travelers.

The first aspect we consider is the choice set in discrete choice models. Choice set is the set of alternatives from which the decision-maker selects one. These alternatives need to be mutually exclusive, exhaustive, and finite in number (Train 2003). Identification, counting, and issues related to the alternatives considered have motivated considerable research in choice set formation (Richardson 1982; Swait and Ben-Akiva 1987a,b; Horowitz 1991; Horowitz and Louviere 1995). The most important threat to misspecification of the choice set is the potential for incorrect predictions (Thill 1992). When this is a considerable threat, as in destination choice models where the alternatives are numerous, a model of choice set formation appears to be an additional burden (Haab and Hicks 1997). Other methods, however, also exist and may provide additional information about the decision-making processes.

Models of the processes can be designed to match the study of specific policies in specific contexts. One such example and a more comprehensive approach defining the choice sets is the situational approach (Brög and Erl 1989). The method uses in-depth information from survey respondents to derive sets of reasons for which alternatives are not considered for specific choice settings (individual trips). This allows separation of analyst-observed system availability from user-perceived system availability (e.g., due to misinformation and willingness to consider information). This brings us to the duality between objective choice attributes and subjective choice attributes. Most transportation applications, independently of the decision-making paradigm adopted, assume the analysts' (modelers) and the travelers' (modeled) measured attributes to be the same. Modeling the process of perceived constraints may be far more complex when one considers the influence of the context within which decisions are made. Golledge and Stimpson (1997, 33–34) describe this within a conceptual model of decision-making that has a cognitive feel to it. They also link the situational approach to the activity-based framework of travel extending, the framework further (Golledge and Stimpson 1997, 315–28).

From another viewpoint, the idea of constraints in the movement of persons was taken a step further by the time geography school in Lund. In that framework, the movement of persons among locations can be viewed as their movement in space and time. Movement in time is viewed as one-way (irreversible) movement in the path, while space is viewed as a three-dimensional domain. Practical calculations of the extreme points in such a domain use existing data in surveys (Pendyala 2003). Most important, however, in this framework is the idea of a project, which, according to Golledge and Stimpson (1997, 268–69), is a set of linked tasks that are undertaken somewhere at some time within a constraining environment. This idea of the project underlies one of the most exciting developments in travel behavior—the activity-based approaches to travel demand analysis and forecasting, which can be considered as a method of modeling time allocation.

Emphasis on Time Allocation.   Chapin's research (1974), providing one of the first comprehensive studies about time allocated to activity in space and time, is also credited with motivating the foundations of activity-based approaches to travel demand analysis. His focus has been on the propensity of individuals to participate in activities and travel, linking their patterns to urban planning. In about the same period, Becker developed his theory of time allocation from a household production viewpoint (Becker 1976), applying economic theory in a nonmarketing sector and demonstrating the possibility of formulating time-allocation models using economics reasoning (i.e., activity choice). In parallel, another approach was developing in geography. Hagerstrand's seminal publication on time-space geography (1970) presents the foundations of the approach. It provides the third base about *constraints* in human paths in time and space for a variety of planning horizons. These are *capability constraints* (e.g., physical limitations such as speed); *coupling constraints* (e.g., requirements to be with other persons at the same time and place); and *authority constraints* (e.g., restrictions due to institutional and regulatory contexts, such as the opening and closing hours of stores). Cullen and Dobson, in two papers in the mid-1970s as reviewed by Arentze and Timmermans (2000) and Golledge and Stimpson (1997), appear to be the first researchers to attempt to bridge the gap between the motivational (Chapin) approach to activity participation and the constraints (Hagerstrand) approach by creating a model that depicts a routine and deliberated approach to activity analysis. Most subsequent contributions to the activity-based approach emerge in one way or another from these initial frameworks, with important operational improvements (for reviews see Kitamura 1988; Bhat and Koppelman 1999; Arentze and Timmermans 2000; and McNally 2000).

The basic ingredients of an activity-based approach for travel demand analysis (Jones, Koppelman, and Orfeuil 1990; and Arentze and Timmermans 2000) are:

1. Explicit treatment of travel as derived demand (Manheim 1979), i.e., participation in activities such as work, shop, and leisure motivate travel, but travel can also be an activity (e.g., taking a drive). These activities are viewed as episodes (starting time, duration, and ending time) and are arranged in a sequence forming a pattern of behavior that can be distinguished from other patterns (a sequence of activities in a chain of episodes). In addition, these events are not independent and their interdependency is accounted for in the theoretical framework.

2. The household is considered to be the fundamental social unit (decision-making unit), and the interactions among household members are explicitly modeled to capture task allocation and roles within the household, as well as relationships and change in these relationships as households move along their life cycles and individuals' commitments and constraints change. These are depicted in the activity-based model.

3. Explicit consideration of constraints by the spatial, temporal, and social dimensions of the environment is given. These constraints can be explicit models of time-space prisms or reflections of these constraints in the form of model parameters and/or rules in a production system format (Arentze and Timmermans 2000).

The input to these models is the typical regional model data of social, economic, and demographic information of potential travelers and land use information to create schedules followed by people in their everyday life. The output is detailed lists of activities pursued, times spent in each activity, and travel information from activity to activity (including travel time, mode used, and so forth). This output is very much like a "day-timer" for each person in a given region. Figure 29.1 provides an example of time allocation to different activities from an application that collected activity participation data (Alam and Goulias 1998). Figure 29.2 shows the output from a model that combines the more traditional four-step model with an activity model in its trip-generation step. The output contains all the elements of the typical four-step model and predicted volumes of individuals at specific locations in the city (a more detailed description of this study can be found in Kuhnau and Goulias 2003).

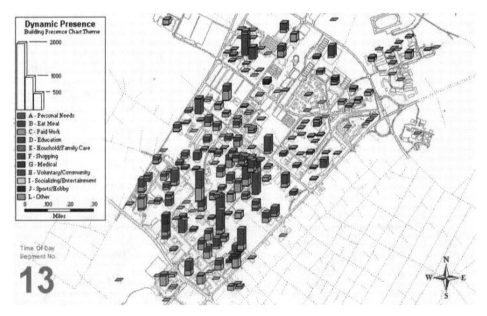

FIGURE 29.1   Building-by-building activity distribution at University Park (the Pennsylvania State University main campus) at 1:00 pm.

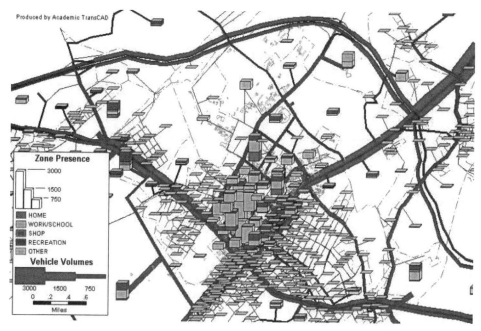

FIGURE 29.2   A combined presence at locations and travel demand model at a specific time of a day using GIS software.

Activity-based model systems that aim at becoming operational follow the three original traditions:

1. The microeconomics *Homo economicus* vision, based on Becker's formulation (Jara-Diaz 1998)
2. Production system/computational process modeling, following Newell and Simon (1972), such as Kitamura and Fujii (1998), Arentze and Timmermans (2000)
3. Statistical pattern recognition and transition probability approaches to create synthetic schedules (Ma 1997) or improve existing four-step models (Kuhnau and Goulias 2003)

Many planning and modeling applications, however, aim at forecasting. Inherent in forecasting are time changes in the behavior of individuals and their households and their response to policy actions.

Consideration of Behavioral Dynamics.   At the heart of behavioral change are questions about the process followed in shifting from a given pattern of behavior to another. In addition to measuring change and the relationships among behavioral indicators that change in their values over time, we are also interested in the timing, sequencing, and staging of these changes. Moreover, we are interested in the triggers that may accelerate desirable or delay undesirable changes and the identification of social and demographic segments that may follow one time path versus another in systematic patterns. Knowledge about all this is required to design policies but also to design better forecasting tools. Developments in exploring behavioral dynamics and advancing models for them have progressed in a few arenas. First is the data-collection arena, with panel surveys, repeated observation of the same persons over time, which are now giving us a considerable history in developing new ideas about data collection but also about data analysis (Golob, Kitamura, and Long 1997; Goulias and Kim 2003), and interactive and laboratory data-collection techniques (Doherty, 2003, Golledge 2002) that allow a more in-depth examination of behavioral process followed by decision-makers. The second arena is in the development of microeconomic dynamic formulations for travel behavior (Supernak 1990; Goodwin 1998) that challenge conventional assumptions and offer alternative formulations. The third arena is in the behavior from a developmental viewpoint as a single stochastic process (Kitamura 2000), a staged development process (Goulias 1999), or as the outcome from multiple processes operating at different levels (Goulias 2002). Experimentation with new theories from psychology emphasizing development dynamics is a potential fourth area that is just beginning to emerge (Goulias 2003).

Integration of Transportation with Other Models.   The examples of studies in the previous section focus more on the paths of persons in space and time within a somewhat short time horizon such a day, week, or maybe month. The consideration of behavioral dynamics has expanded the temporal horizons to a few years. However, regional simulation models are very often designed for long-range plans spanning 25 years or even longer time horizons. Within these longer horizons, changes in the spatial distribution of activity locations and residences (land use) are substantial, changes in the demographic composition and spatial distribution of demographic segments are also substantial, and changes in travel patterns, transport facilities, and quality of service offered can be extreme. Past approaches in modeling and simulating the relationship among land use, demographics, and travel in a region attempted to disengage travel from the other two, treating them as mutually exogenous. As interactions among them became more interesting and pressing for policy analysis, due to urban sprawl and suburban congestion, increasing attention was paid to their complex interdependencies. This led to a variety of attempts to develop integrated model systems that would enable the study of scenarios of change and mutual influence between land use and travel. An earlier review of these models with heavy emphasis on discrete choice models can be found in Anas (1982). Miller (2003) and Waddell and Ulfarsson (2003), 20 years later, provide comprehensive reviews of models that have integrated many aspects in the interdependent triad of demographics-travel-land use models. Both reviews trace the history of some of the most notable developments and link these models to the activity-based approach above. Both reviews also agree that a microeconomic and/or macroeconomic approach to modeling land and transportation interactions is not sufficient and more detailed simulation of the individuals and their organizations acting in a time-space domain needs to be simulated in order to obtain the required output for informed decision-making. They also introduce the idea of simulating interactive agents in a dynamic environment of other agents (multiagent simulation). Creation of integrated systems is further complicated by the emergence of an entire infrastructural system as another layer of human activity—telecommunication.

Telecommunication as an Infrastructure Component.   Today telecommunication and transportation relationships are absent from regional simulation planning and modeling as well from the most advanced land use and transportation integrated models (see previous section). Considerable research findings, however, have been accumulating since the 1970s. Telecommunication here means a much larger system of services and technologies, called information and (tele)communication technologies (ICT). The definition by Cohen, Salomon, and Nijkamp (2002) is very appropriate in this case: "[ICT is] a family of electronic technologies and services used to process, store and disseminate information, facilitating the performance of information-related human activities, provided by, and serving the institutional and business sectors as well as the public-at-large."

Consider, for example, advanced traveler information systems (ATIS). ATIS is a very good example of a direct impact on travel behavior because it may affect many aspects of daily life, including time allocation and related decisions by a household and its members and the type of goods a household considers and acquires. ATIS started as one of the many services offered by ITS and over time expanded beyond the roadside, migrating into vehicles, offices, and homes (Weilland and Purser 2000). This movement took place in all the four main media of television, radio, Internet, and telephone, expanding the interaction between transportation and telecommunications. For this reason, when assessing the effects of this technology we need a wider and more comprehensive framework than the single trip information acquisition and information use framework adopted by traffic simulation applications (for an overview of these applications see Mahmassani and Jou 1998) or the frameworks used in typical ITS deployment plans and evaluations (JHK et al. 1996; Patten and Goulias 2001; Patten et al. 2003).

Salomon (1986) sketched one such framework where he recognized four possible effects of ICT on travel: and they are: substitution, modification, enhancement, and neutrality. Substitution means that ICT can actually eliminate trips. Telecommuting, teleshopping, and teleconferencing are some examples. *Modification* indicates ICT can alter the travel behavior of individuals, changing the order of trips (sequencing), the travel mode, or the timing of the trip (e.g., departure time). From an operations standpoint, this is particularly important when a shift of commute trips to off-peak hours or a switching to public transportation and/or car-pooling happens because of ICT use. *Enhancement* reflects those trips that would not have been generated without ICT. For example, when there is more information available for particular activities, one would expect an increase in the desire to travel and participate in these activities. Also, people are able to save time by better planning of their schedules (thanks to ICT) and by communicating while traveling. The saved time is often used to make other trips. *Neutrality* reflects those instances of ICT that have no remarkable effect on travel behavior. There are, however, many gaps in our knowledge about ICT and transportation interactions that require additional research before policies can be defined in such a way that ICT can play a significant role in changing travel behavior. Ultimately, these changes should also benefit the level of service offered by the transportation system. Mokhtarian (1990) expanded the telecommunications-transportation framework further, and the ideas of substitution, modification, enhancement, and neutrality served as the basis for identifying new relationships. Her conceptual framework reflects the impacts of ICT on travel and contains the reverse causality of the effect of travel on ICT. In one of the most recent summaries about this interaction, Krizek and Johnson (2003) map the terrain of research, recognizing the many complexities of interaction. They agree with Mokhtarian (1990) that the four relationships discussed in Salomon (1986) are only a simplification. Krizek and Johnson (2003) expand the Salomon-Mokhtarian framework one step further, considering a triad of dimensions:

1. Nature of the activity pursued using ICT (borrowing the categorization of subsistence, maintenance, and leisure from activity analysis in travel behavior)
2. The effect of ICT on travel (using the four Salomon effects)
3. The effect of subtasks when pursuing an ICT action

This latter aspect of multiple activities at a specific destination is also becoming particularly interesting for research that goes beyond telecommunications and land use.

The example above demonstrates that key to understanding the interaction between telecommunications and transportation is also understanding the evolution of the technology, as discussed for example in Golob (2001), where evidence concerning the usage of personal computers, the Internet, mobile phones, and other new technologies may indicate generational differences among the users. This can be expanded to include awareness of technologies in which different segments of the population may approach the services in different ways (Goulias, Kim, and Pribyl 2003), due either to the time needed to gain familiarity with the new technology or to dismissal of usefulness due to generational and cohort effects. In a regional simulation model, in addition to the demographics, land use, and transportation facilities, we also need to consider telecommunications and information availability and use and their interactions with transportation. Frameworks are available and data collection schemas are also starting to emerge (Marker and Goulias 2000).

29.2.4 Enabling Technologies

The last two decades have produced a variety of enabling technologies for modeling and simulation, opening the door to tremendous improvements. A few of the most important are stochastic simulation, production systems, geographic information systems, interactive and technology-aided data-collection approaches, and more flexible data analysis techniques.

Stochastic microsimulation here refers to an evolutionary engine software that is used to replicate the relationships among social, economic, and demographic factors with land use, time use, and travel by people. As discussed above, the causal links among these groups of entities are extremely complex, nonlinear, and in many instances unknown or incompletely specified. This is the reason that no closed-form solution can be created for such a forecasting model system. An evolutionary engine, then, provides a realistic representation of person and household life histories (birth, death, marriages, divorces, birth of children, etc.), spatio-temporal activity opportunity evolution, and a variety of models that account for uncertainties in data, models, and behavioral variation (see Miller 2003 and Goulias 2002 for overviews and Sundararajan and Goulias 2003 for an application).

Production systems were first developed by Newell and Simon (1972) to depict explicitly the way humans go about solving problems. These are a series of condition-action (note the parallel with stimulus-response) statements in a sequence. From this viewpoint they are search processes that may never reach an absolute optimum, and they replicate (or at least attempt to replicate) human thought and action. Models of this kind are called *computational process models,* and through the use of If-Then rules they have made possible first the creation of a framework in the Rand Corporation (Hayes-Roth and Hayes-Roth 1979), which has also linked their framework to alternative planning strategies. Subsequently and based on a decade of work by Gärling, (Gärling, Kwan, and Golledge (1994) published an operational model for activity scheduling. This was followed by two applications named PCATS and AMOS (Kitamura and Fujii 1998) and the most complete of all these models, Albatross, in the Netherlands (Arentze and Timmermanns 2000).

Geographic information systems are software systems that can be used to collect, store, analyze, modify, and display large amounts of geographic data. They include layers of data that are able to incorporate relations among the variables in each layer and allow relationships to be built in data across layers. One can visualize a GIS as a live map that can display almost any kind of spatio-temporal information. Maps have been used by transportation planners and engineers for long time, and they are a natural interface to use in modeling and simulation. Figure 29.2 is in fact an image from a GIS software.

There are also two other technologies that merit a note, although not strictly for modeling. The first is about data collection, particularly using the Internet to build complex interviews that are interactive and dynamic (Doherty 2003). In the same line of development we also see the use of geographic positioning systems (GPS), which allow one to develop a trace of individual paths in time and space (Wolf et al. 2001; Doherty et al, 2001). In data analysis we see greater strides in using data mining and artificial intelligence-borne techniques to extract travel behavior patterns (Teodorovic and Vukadinovic 1998; Pribyl and Goulias 2003) and advanced and less restrictive statistical methods to discover relationships in travel behavior data (e.g., Kharoufeh and Goulias 2002).

29.3 SUMMARY AND NEXT STEPS

Policies are dictating the creation and testing of increasingly more sophisticated policy assessment instruments that account for direct and indirect effects of behavior, procedures for behavioral change, and provide finer resolution in the four dimensions of geographic space, time, social space, and jurisdictions. Tremendous progress has occurred in the past 20 years, but development requires faster pace to create the policy tools needed. These policy tools need to disentangle the actions of persons within projects–a sequenced series of activities in the four dimensions listed above. A rich collection of decision paradigms is already used, and a few new ideas are starting to migrate to practice. The plethora of advances includes:

1. Models and experiments to create computerized virtual worlds and synthetic schedules at the most elementary level of decision-making using microsimulation and computational process models

2. Data-collection methods and new methods to estimate, validate, and verify models using advanced hardware, software, and data analysis techniques

3. Integration of models from different domains to reflect additional interdependencies such as land use and telecommunications

However, much more work remains to be done in order to develop models that can answer many questions from policy analysis. For this reason, a few steps are presented in closing the chapter. Of the four dimensions above, the two consistently important dimensions are space and time. Since we are dealing with the behavior of persons, it is unavoidable to consider perceptions of time and space. To be convinced, run the following experiment. Identify two cities and ask 10 of your friends to tell you how far are they apart and how long will it take to travel between them. Then, armed with a precise watch and a well-calibrated speedometer, travel between the two cities and compare your findings with your friends' stated perceptions. The differences you will find have been studied from the perception and cognition viewpoints (Golledge and Stimpson 1997). In addition, moving to time allocation to develop travel demand models also opens the door to the multiple dimensions of time and associated perception of it, such as tempo, duration, and clock time (Levine 1997).

In social space and social networks particularly, research is needed on interactions among persons and group decision-making. For example, the relationships among people can be depicted by using indicators of their interaction to depict the exchange of resources in terms of a network of relationships. In this way one can build *social networks* that provide the structural environment within which opportunities and constraints to individual action (e.g., activity participation and travel) are depicted explicitly. While this conceptualization has not been used in data-collection and modeling, interaction among household members has been recognized as an important factor affecting behavior (Van Wissen 1989; Golob and McNally 1996; Chandrasekharan and Goulias 1999; Gliebe and Koppelman 2002), and building on these initial efforts will be an area of considerable advances.

Some time ago, Savas (1983, 379) said "In this idealistic scenario, decision-making is a neat, crisp act which follows quite automatically from the analysis. In fact, however, in the urban policy arena, policies are rarely *made;* at best they *emerge* from a vague, prolonged, diffuse, pluralistic and evolutionary process." This statement is indicative of our experience in planning, during both plan formation and implementation in programming, and, to a lesser extent, during project execution. Since integrated model systems require us to develop models of the effects of policy implementation (see the original Urban Transportation Planning System, which contains a few examples of implementation feedback, Roberts 1973), and since we would like them to be realistic, the need arises for a *metamodel* of jurisdictional power and implementation. Could Simon (1997) come again to our rescue?

29.4   *ACKNOWLEDGMENT*

Partial support for this chapter was offered by the Mid-Atlantic Universities Transportation Center at the Pennsylvania State University from the United States Department of Transportation.

29.5 *REFERENCES*

Anas, A. 1982. *Residential Location Markets and Urban Transportation.* New York: Academic Press.

Allam, B. S., and K. G. Goulias. 1999. "Dynamic Emergency Evacuation Management System Using GIS and Spatio-temporal Models of Behavior." *Transportation Research Record* 1660, 92–99.

Arentze, T., and H. Timmermans. 2000. *Albatross: A Learning Based Transportation Oriented Simulation System.* Eindhoven, NL: European Institute of Retailing and Services Studies, Technical University of Eindhoven.

Avineri, E., and Y. Prashker. 2003. "Sensitivity to Uncertainty: The Need for a Paradigm Shift." In *CD-TRB ROM Proceedings.* Paper presented at the 82nd Annual Transportation Research Board Meeting, January 12–16, 2003, Washington, DC.

Barret, C., K. Berkbigler, L. Smith, V. Loose, R. Beckman, J. Davis, D. Roberts, and M. Williams. 1995. *An Operational Description of TRANSIMS.* LA-UR-95-2393, Los Alamos, NM.

Becker, G. S. 1976. *The Economic Approach to Human Behavior.* Chicago: University of Chicago Press.

Ben-Akiva, M. E., and S. R. Lerman. 1985. *Discrete Choice Analysis: Theory and Application to Travel Demand.* Cambridge: MIT Press.

Ben-Akiva, M. E., and T. Morikawa. 1989. "Estimation of Mode Switching Models from Revealed Preferences and Stated Intentions." Paper presented at the International Conference on Dynamic Travel Behavior at Kyoto University Hall, Kyoto, July.

Ben-Akiva, M. E., J. Walker, A. T. Bernardino, D. A. Gopinath, T. Morikawa, and A. Polydoropoulou. 2002. "Integration of Choice and Latent Variable Models." In *In Perceptual Motion: Travel Behavior Research Opportunities and Application Challenges,* ed. H. S. Mahmassani. Amsterdam: Pergamon.

Bhat, C. R. 2000. Flexible Model Structures for Discrete Choice Analysis." In *Handbook of Transport Modelling,* ed. D. A. Hensher and K. J. Button. Amsterdam: Pergamon, 71–89.

———. 2003. "Random Utility-Based Discrete Choice Models for Travel Demand Analysis." In *Transportation Systems Planning: Methods and Applications,* ed. K. G. Goulias. Boca Raton: CRC Press, 10-1–10-30.

Bhat, C. R., and F. S. Koppelman. 1999. "Activity-Based Modeling of Travel Demand." In *Handbook of Transportation Science,* ed. R. W. Hall. Boston: Kluwer, 35–61.

Bockenholt, U. 2002. "Comparison and Choice: Analyzing Discrete Preference Data by Latent Class Scaling Models. In *Applied Latent Class Analysis,* ed. J. A. Hagenaars and A. L. McCutcheon. Cambridge: Cambridge University Press, 163–82.

Borgers, A. W. J., F. Hofman, and H. J. P. Timmermans. 1997. "Activity-Based Modelling: Prospects." In *Activity-Based Approaches to Travel Analysis,* ed. D. F. Ettema and H. J. P. Timmermans. Oxford: Pergamon, 339–51.

Bróg, W., and E. Erl. 1989. "Interactive Measurement Methods: Theoretical Bases and Practical Applications." *Transportation Research Record* 765.

Chandrasekharan, B., and K. G. Goulias. 1999. "Exploratory Longitudinal Analysis of Solo and Joint Trip Making in the Puget Sound Transportation Panel." *Transportation Research Record* 1676, 77–85.

Chapin, F. S., Jr. 1974. *Human Activity Patterns in the City: Things People Do in Time and Space.* New York: John Wiley & Sons.

Cohen, G., I. Salomon, and P. Nijkamp. 2002. "Information-Communications Technologies (ICT) and Transport: Does Knowledge Underpin Policy?" *Telecommunications Policy* 26:31–52.

Creighton, R. L. 1970. *Urban Transportation Planning.* Urbana: University of Illinois Press.

Dillman, D. A. 2000. *Mail and Internet Surveys: The Tailored Design Method,* 2nd ed. New York: John Wiley & Sons.

Doherty, S. 2003. "Interactive Methods for Activity Scheduling Processes." In *Transportation Systems Planning: Methods and Applications,* ed. K. G. Goulias. Boca Raton: CRC Press, 7-1-7.25.

Doherty, S. T., N. Noel, M. Lee-Gosselin, C. Sirois, and M. Ueno. 2001. *Moving Beyond Observed Outcomes: Global Positioning Systems and Interactive Computer-Based Travel Behavior Surveys.* Transportation Research Circular E-C026, National Research Council, Transportation Research Board, Washington, DC, March.

Garling T. 1998. "Behavioural Assumptions Overlooked in Travel-choice Modeling." In *Travel Behaviour Research: Updating the State of Play.* ed. J. de Dios Ortuzar, D. Hensher, and S. Jara-Diaz. Amsterdam: Pergamon, 3–18.

Gärling, T., M. Kwan, and R. Golledge. 1994. "Computational-Process Modeling of Household Travel Activity Scheduling." *Transportation Research B* 25:355–64.

Gärling, T., T. Laitila, and K. Westin. 1998. "Theoretical Foundations of Travel Choice Modeling: An Introduction." In *Theoretical Foundations of Travel Choice Modeling,* ed. T. Gärling, T. Laitila, and K. Westin. Oxford: Pergamon, 1–30.

Garrett, M., and M. Wachs. 1996. *Transportation Planning on Trial. The Clean Air Act and Travel Forecasting.* Thousand Oaks, CA: Sage.

Gliebe, J. P., and F. S. Koppelman. 2002. "A Model of Joint Activity Participation between Household Members." *Transportation* 29(1):49–72.

Golledge, R. G. 2002. "Dynamics and ITS: Behavioral Responses to Information Available from ATIS." In *In Perpetual Motion: Travel Behavior Research Opportunities and Application Challenges,* ed. H. Mahmassani. Amsterdam: Pergamon, 81–126.

Golledge, R. G., and T. Gärling. 2003. "Spatial Behavior in Transportation Modeling and Planning." In *Transportation Systems Planning: Methods and Applications,* ed. K. G. Goulias. Boca Raton: CRC Press, 3.1–3.27.

Golledge, R. G., and R. J. Stimson. 1997. *Spatial Behavior: A Geographic Perspective.* New York: Guilford Press.

Golob, T. F. 2001. "Travelbehaviour.com: Activity Approaches to Modeling the Effects of Information Technology on Personal Travel Behaviour." In *Travel Behavior Research: The Leading Edge,* ed. D. Hensher. Amsterdam: Pergamon, 145–84.

Golob, T. F., and M. G. McNally. 1996. "A Model of Household Interactions in Activity Participation and the Derived Demand for Travel." Paper presented at the 75th Annual Transportation Research Board Meeting, Washington, DC.

Golob, T. F, R. Kitamura, and L. Long. 1997. *Panels for Transportation Planning: Methods and Applications.* Boston: Kluwer.

Goodman, L. A. 2002. "Latent Class Analysis: The Empirical Study of Latent Types, Latent Variables, and Latent Structures." In *Applied Latent Class Analysis,* ed. J. A. Hagenaars and A. L. McCutcheon. Cambridge: Cambridge University Press, 3–55.

Goodwin, P. B. 1998. "The End of Equilibrium." In *Theoretical Foundations of Travel Choice Modelling.* In *Theoretical Foundations of Travel Choice Modeling,* ed. T. Gärling, T. Laitila, and K. Westin. Oxford: Pergamon, 103–32.

Goulias, K. G. 1999. "Longitudinal Analysis of Activity and Travel Pattern Dynamics Using Generalized Mixed Markov Latent Class Models." *Transportation Research B* 33:535–57.

———. 2001. "A Longitudinal Integrated Forecasting Environment (LIFE) for Activity and Travel Forecasting." In *Ecosystems and Sustainable Development III,* ed. Y. Villacampa, C.A. Brebbia, and J-L. Uso. Southampton: WIT Press, 811–20.

———. 2002. "Multilevel Analysis of Daily Time Use and Time Allocation to Activity Types Accounting for Complex Covariance Structures Using CorrelatedRandom Effects." *Transportation* 29(1):31–48.

———. 2003. "Transportation Systems Planning." In *Transportation Systems Planning: Methods and Applications,* ed. K. G. Goulias. Boca Raton: CRC Press, 1-1–1-45.

Goulias, K. G., and T. Kim. 2003. *Analysis of the Puget Sound Transportation Panel Survey Database in Waves 1-9.* Draft Final Report, Submitted to the Puget Sound Regional Council, Seattle, WA.

Goulias, K. G., and R. Kitamura. 1992. "Travel Demand Analysis with Dynamic Microsimulation." *Transportation Research Record* 1357, 8–18.

Goulias, K. G., and D. Szekeres. 1994. *Centre Region Transportation Demand Management Plan.* Final Draft Report. Prepared for the Centre Regional Planning Commission, University Park, PA.

Goulias, K. G., T. Kim, and O. Pribyl. 2003. "A Longitudinal Analysis of Awareness and Use for Advanced Traveler Information Systems." Paper to be presented at the European Commission Workshop on Behavioural Responses to ITS, April 1–3, Eindhoven, The Netherlands.

Goulias, K. G., T. Litzinger, J. Nelson, and V. Chalamgari. 1993. *A Study of Emission Control Strategies for Pennsylvania: Emission Reductions from Mobile Sources, Cost Effectiveness, and Economic Impacts.* PTI 9403. Final report to the Low Emissions Vehicle Commission. The Pennsylvania Transportation Institute, University Park, PA.

Haab, T. C., and R. L. Hicks. 1997. "Accounting for Choice Set Endogeneity in Random Utility Models of Recreation Demand." *Journal of Environmental Economics and Management* 34:127–47.

Hagerstrand, T. 1970. "What about People in Regional Science?" *Papers of the Regional Science Association* 10:7–21.

Hayes-Roth, B., and F. Hayes-Roth. 1979. "A Cognitive Model of Planning." *Cognitive Science* 3:275–310.

Horowitz, J. L. 1991. "Modeling the Choice of Choice Set in Discrete-Choice Random-Utility Models." *Environment and Planning A* 23:1237–46.

Horowitz, J. L., and J. J. Louviere. 1995. "What Is the Role of Consideration Sets in Choice Modeling?" *International Journal of Research in Marketing* 12:39–54.

Humphrey, T. F. 1990. "Suburban Congestion: Recommendations for Transportation and Land Use Responses. *Transportation* 16:221–40.

Hutchinson, B. G. 1974. *Principles of Urban Transport Systems Planning.* Washington, DC: Scripta.

Jara-Diaz, S. R. 1998. "Time and Income in Travel Choice: Towards a Microeconomic Activity-Based Theoretical Framework." In *Theoretical Foundations of Travel Choice Modeling,* ed. T. Gärling, T. Laitila, and K. Westin. Oxford: Pergamon, 51–73.

JHK & Associates, Clough, Harbour & Associates, Pennsylvania Transportation Institute, and Bogart Engineering. 1996. *Scranton/Wilkes-Barre Area Strategic Deployment Plan.* Final Report. Prepared for Pennsylvania Department of Transportation District 4-0, August 1996, Berlin, CT.

Jones, P., ed. 1990. *Developments in Dynamic and Activity-Based Approaches to Travel Analysis: A Compendium of Papers from the 1989 Oxford Conference.* Aldershot: Avebury, UK.

Jones P., F. Koppelman, and J Orfeuil. 1990. "Activity Analysis: State-of-the-Art and Future Directions." In *Developments in Dynamic and Activity-Based Approaches to Travel Analysis: A Compendium of Papers from the 1989 Oxford Conference.* Aldershot: Avebury, 4–55.

Kahneman, D., and A. Tversky. 1979. "Prospect Theory: An Analysis of Decisions under Risk." *Econometrica* 47(2):263–91.

Kasten, R. A., and F. J. Sammartino. 1990. "A Method for Simulating the Distribution of Combined Federal Taxes Using Census, Tax Return, and Expenditure Microdata." In *Microsimulation Techniques for Tax and Transfer Analysis,* ed. G. H. Lewis and R. C. Michel. Washington, DC: Urban Institute Press

Kenworthy, J. R., and F. B. Laube. 1999. "Patterns of Automobile Dependence in Cities: An International Overview of Key Physical and Economic Dimensions with Some Implications for Urban Policy." *Transportation Research A* 33:691–723.

Kharoufeh, J. P., and K. G. Goulias. 2002. "Nonparametric Identification of Daily Activity Durations Using Kernel Density Estimators." *Transportation Research B* 36:59–82.

Kitamura, R. 1988. "An Evaluation of Activity-Based Travel Analysis." *Transportation* 15:9–34.

———. 2000. "Longitudinal Methods." In *Handbook of Transport Modelling,* ed. D. A. Hensher and K. J. Button. Amsterdam: Pergamon, 113–28.

Kitamura, R., and S. Fujii. 1998. "Two Computational Process Models of Activity-Travel Choice." In *Theoretical Foundations of Travel Choice Modeling,* ed. T. Gärling, T. Laitila, and K. Westin. Oxford: Pergamon, 251–79.

Kitamura, R., E. I. Pas, C. V. Lula, T. K. Lawton, and P. E. Benson. 1994. "The Sequenced Activity Mobility Simulator (SAMS): An Integrated Approach to Modeling Transportation, Land Use and Air Quality." Mimeo.

Koppelman, F. S., and V. Sethi. 2000. "Closed-Form Discrete-Choice Models." In *Handbook of Transport Modelling,* ed. D. A. Hensher and K. J. Button. Oxford: Pergamon, 211–25.

Krizek, K. J., and A. Johnson. 2003. *Mapping of the Terrain of Information and Communications Technology (ICT) and Household Travel.* Transportation Research Board annual meeting CD-ROM, Washington, DC, January.

Kuhnau J., and K. G. Goulias. 2003. "Centre SIM: First-Generation Model Design, Pragmatic Implementation, and Scenarios." In *Transportation Systems Planning: Methods and Applications,* ed. K. G. Goulias. Boca Raton: CRC Press, 16-1–16-14.

Kwan, M. 1995. "GISICAS: An Activity-Based Spatial Decision Support System for ATIS." Paper presented at the conference AActivity Based Approaches: Activity Scheduling and the Analysis of Activity Patterns, Eindhoven, The Netherlands, May 25–29.

Law, A. M., and W. D. Kelton. 1991. *Simulation Modeling and Analysis.* New York: McGraw Hill.

Lawrence, M. F., and M. Tegenfeldt. 1997. "The Use of Economic and Demographic Forecasts by Metropolitan Planning Organizations." Paper presented at the 76th Annual Transportation Research Board Meeting, Washington, DC.

Lee, M., and K. G. Goulias. 1997. "Accessibility Indicators for Transportation Planning Using GIS." Paper presented at the 76th Annual Transportation Research Board Meeting, Washington, DC.

Levine, R. 1997. *A Geography of Time: On Tempo, Culture, and the Pace of Life.* New York: Basic Books.

Linzie, M. 2000. "Future of International Activities." In *Transportation in the New Millenium. State of the Art and Future Directions.* National Research Council, Transportation Research Board, Washington, DC.

Litman, T. 2001. "You Can Get There from Here: Evaluating Transportation Choice." Paper 01-3035, presented at the Transportation Research Board 80th Annual Meeting and included in the CD ROM proceedings, Washington, DC.

Loikanen, H. A. 1982. "Housing Demand and Intra-urban Mobility Decisions: A Search Approach." Mimeo.

Lomborg, B. 2001. *The Skeptical Environmnetalist: Measuring the Real State of the World.* Cambridge: Cambridge University Press.

Loudon, W. R., and D. A. Dagang. 1994. "Evaluating the Effects of Transportation Control Measures." In *Transportation Planning and Air Quality II,* ed. T. F. Wholley. New York: ASCE.

Loudon, W. R., J. L. Henneman, L. I. Hartnett, and M. J. Lawlor. 1997. "Integrating Transportation and Land Use Planning: Addressing the Requirements of Federal Legislation and Rule Making." Paper presented at the 76th Transportation Research Board Meeting, Washington, DC.

Louviere, J. J., D. A. Hensher, and J. D. Swait. 2000. *Stated Choice Methods: Analysis and Application.* Cambridge: Cambridge University Press.

Lowry, I. S. 1988. "Planning the Urban Sprawl." In *A Look Ahead: Year 2020.* National Research Council, Transportation Research Board, Washington, DC, 275–312.

Ma, J. 1997. "An Activity-Based and Micro-Simulated Travel Forecasting System: A Pragmatic Synthetic Scheduling Approach." Ph.D. dissertation, Department of Civil and Environmental Engineering, The Pennsylvania State University, University Park, PA.

Ma, J., and K. G. Goulias. 1997. "Multivariate Marginal Frequency Analysis of Activity and Travel Patterns in the First Four Waves of the Puget Sound Transportation Panel." *Transportation Research Record* 1566, 67–76.

Mackett, R. L. 1985. "Micro Analytic Simulation of Locational and Travel Behaviour." In *Proceedings of PTRC Summer Annual Meeting, Seminar L, Transportation Planning Methods.* London: PTRC, 175–88.

———. 1990. "Exploratory Analysis of Long Term Travel Demand Using Micro-analytical Simulation. In *New Developments in Dynamic and Activity-Based Approaches to Travel Analysis,* ed. P. M. Jones. Aldershot: Gower, 384–405.

Madder, G. H., and O. M. Bevilacqua. 1989. "Electric Vehicle Commercialization." In *Alternative Transportation Fuels. An Environmental and Energy Solution,* ed. D. Sperling. New York: Quorum Books.

Mahmassani, H. S., and R. Herman. 1990. "Interactive Experiments for the Study of Tripmaker Behaviour Dynamics in Congested Commuting Systems." In *Developments in Dynamic and Activity-Based Approaches to Travel Analysis: A Compendium of Papers from the 1989 Oxford Conference.* Aldershot: Avebury.

Mahmassani, H. S., and R.-C. Jou. 1998. "Bounder Rationality in Commuter Decision Dynamics: Incorporating Trip Chaining in Departure Time and Route Switching Decisions." In *Theoretical Foundations of Travel Choice Modeling,* ed. T. Gärling, T. Laitila, and K. Westin. Oxford: Pergamon.

Manheim, M. 1979. *Fundamentals of Transportation Systems Analysis.* Cambridge: MIT Press.

Marker, J. T., and K. G. Goulias. 2000. "Framework for the Analysis of Grocery Teleshopping." *Transportation Research Record* 1725, 1–8.

McFadden, D. 1998. "Measuring Willingness-to-Pay for Transportation Improvements." In *Theoretical Foundations of Travel Choice Modeling,* ed. T. Gärling, T. Laitila, and K. Westin. Oxford: Pergamon, 339–64.

McNally, M. G. 2000. "The Activity-Based Approach." In *Handbook of Transport Modelling,* ed. D. A. Hensher and K. J. Button. Amsterdam: Pergamon, 113–28.

Meyer, M. D., and E. J. Miller. 2001. Urban Transportation Planning, 2nd ed. Boston: McGraw-Hill.

Miller, E. J. 2003. "Land Use: Transportation Modeling." In *Transportation Systems Planning: Methods and Applications,* ed. K. G. Goulias. Boca Raton: CRC Press, 5-1–5-24.

Mokhtarian, P. L. 1990. "A Typology of Relationships Between Telecommunications and Transportation." *Transportation Research A* 24(3):231–42.

Newell, A., and H. A. Simon. 1972. *Human Problem Solving.* Englewood Cliffs: Prentice Hall.

Niemeier, D. A. 2003. "Mobile Source Emissions: An Overview of the Regulatory and Modeling Framework." In *Transportation Systems Planning: Methods and Applications,* ed. K. G. Goulias. Boca Raton: CRC Press, 13-1–13-28.

Ortuzar, J. de D., and L. G. Willumsen. 2001. *Modeling Transport,* 3rd ed. New York: John Wiley & Sons.

Paaswell, R. E., N. Rouphail, and T. C. Sutaria, eds. 1992. *Site Impact Traffic Assessment: Problems and Solutions.* New York: ASCE..

Patten, M. L., and K. G. Goulias. 2001. *Test Plan: Motorist Survey—Evaluation of the Pennsylvania Turnpike Advanced Travelers Information System (ATIS) Project, Phase III.* PTI-2001-23-I, University Park, PA, April.

Patten, M. L., M. P. Hallinan, O. Pribyl, and K. G. Goulias. 2003. *Evaluation of the Smartraveler Advanced Traveler Information System in the Philadelphia Metropolitan Area.* Technical memorandum, PTI 2003-33, University Park, PA, March.

Payne, J. W., J. R. Bettman, and E. J. Johnson. 1993. *The Adaptive Decision Maker.* Cambridge: Cambridge University Press.

Pendyala, R. 2003. "Time Use and Travel Behavior in Space and Time." In *Transportation Systems Planning: Methods and Applications,* ed. K. G. Goulias. Boca Raton: CRC Press, 2-1–2-37.

Pendyala, R., R. Kitamura, and D. V. G. P. Reddy. 1995. "A Rule-Based Activity-Travel Scheduling Algorithm Integrating Neural Networks of Behavioral Adaptation." Paper presented at the conference Activity Based Approaches: Activity Scheduling and the Analysis of Activity Patterns, Eindhoven, The Netherlands, May 25–29.

Pribyl, O., and K. G. Goulias. 2003. "On the Application of Adaptive Neuro-fuzzy Inference System (ANFIS) to Analyze Travel Behavior." Paper presented at the 82nd Transportation Research Board Meeting, January, and included in the CD ROM proceedings and accepted for publication in the *Transportation Research Record.*

Richardson, A. 1982. "Search Models and Choice Set Generation." *Transportation Research A* 16(5–6):403–16.

Roberts, P. O. 1973. "Resource Paper for Demand Forecasting for Long-Range and Contemporary Options." In *Urban Travel Demand Forecasting: Proceedings of a Conference at Williamsburg, Virginia.* Special Report 143, Highway Research Board, Washington, DC.

Rubinstein, A. 1998. *Modeling Bounded Rationality.* Cambridge: MIT Press.

Salomon, I. 1986. "Telecommunication and Travel Relationships: A Review." *Transportation Research A* 20(3):223–38.

Savage, L. J. 1954. *The Foundations of Statistics.* New York: John Wiley & Sons. Reprint, New York: Dover, 1972.

Savas, E. S. 1983. "Systems Analysis and Urban Policy." In *Systems Analysis in Urban Policy-Making and Planning,* ed. M. Batty and B. Hutchinson. NATO Conference Series 2, System Science 12. New York: Plenum Press.

Simon, H. A. 1997. *Administrative Behavior,* 4th ed. New York: Free Press.

———. 1983. "Alternate Visions of Rationality." In *Reason in Human Affairs,* ed. H. A. Simon. Stanford: Stanford University Press, 3–35.

Southworth, F. 2003. "Freight Transportation Planning: Models and Methods." In *Transportation Systems Planning: Methods and Applications,* ed. K. G. Goulias. Boca Raton: CRC Press, 4.1–4.29.

Stopher, P. R. 1994. "Predicting TCM Responses with Urban Travel Demand Models." In *Transportation Planning and Air Quality II,* ed. T. F. Wholley. New York: ASCE.

Stopher, P. R., and A. H. Meyburg. 1976. *Behavioral Travel-Demand Models.* Lexington, MA: Lexington Books.

Sundararajan, A., and K. G. Goulias. 2003. "Demographic Microsimulation with DEMOS 2000: Design, Validation, and Forecasting." In *Transportation Systems Planning: Methods and Applications,* ed. K. G. Goulias. Boca Raton: CRC Press, 14-1–14-23.

Supernak, J. 1990. "A Dynamic Interplay of Activities and Travel: Analysis of Time of Day Utility Profiles." In *Developments in Dynamic and Activity-Based Approaches to Travel Analysis: A Compendium of Papers from the 1989 Oxford Conference,* ed. P. Jones. Aldershot: Avebury, 99–122.

Swait, J., and M. Ben-Akiva. 1987a. "Empirical Test of a Constrained Choice Discrete Model: Mode Choice in São Paolo, Brazil." *Transportation Research B* 21(2):103–15.

———. 1987b. "Incorporating Random Constraints in Discrete Models of Choice Set Generation." *Transportation Research B* 21(2):91–102.

Teodorovic, D., and K. Vukadinovic. 1998. *Traffic Control and Transport Planning: A Fuzzy Sets and Neural Networks Approach.* Boston: Kluwer.

Thill, J. 1992. "Choice Set Formation for Destination Choice Modeling." *Progress in Human Geography* 16(3):361–82.

Tiezzi, E. 2003. *The End of Time.* Southampton: WIT Press.

Train, K. E. 2003. *Discrete Choice Methods with Simulation.* Cambridge: Cambridge University Press.

Transportation Research Board (TRB). 1999. *Curbing Gridlock: Peak Period Fees to Relieve Traffic Congestion.* Special Report 242, National Research Council, TRB, Washington, DC.

———. 1999b. *Transportation, Energy, and Environment. Policies to Promote Sustainability.* Transportation Research Circular 492, National Research Council, TRB, Washington, DC.

———. 2002. *Surface Transportation Environmental Research: A Long-Term Strategy.* National Research Council, TRB, Washington, DC.

Tversky, A. 1969. "Intransitivity of Preferences." *Psychological Review* 76:31–48.

———. 1972. "Elimination by Aspects: A Theory of Choice." *Psychological Review* 79:281–99

Tversky, A., and D. Kahneman. 1992. "Advances in Prospect Theory: Cumulative Representation of Uncertainty." *Journal of Risk and Uncertainty* 9:195–230.

van der Hoorn, T. 1997. "Practitioner's Future Needs." Paper presented at the conference Transport Surveys, Raising the Standard, Grainau, Germany, May 24–30.

Van Wissen, L. 1989. "A Model of Household Interactions in Activity Patterns." Paper presented at the International Conference on Dynamic Travel Behavior Analysis. Kyoto University Hall, Kyoto, Japan, July 18–19.

Vause, M. 1997. "A Rule-Based Model of Activity Scheduling Behaviour." *In Activity-Based Approaches to Travel Analysis,* ed. D. F. Ettema and H. J. P. Timmermans. Oxford: Pergamon, 73–88.

Waddell, P., and G. F. Ulfarsson. 2003. "Dynamic Simulation of Real Estate Development and Land Prices within an Integrated Land Use and Transportation Model System." Paper presented at the 82nd Annual Meeting of the Transportation Research Board, Washington, DC, January 12-16. Also available at http://www.urbansim.org/papers/, accessed April 2003.

Wegener, M., and F. Fürst. 1999. *TRANSLAND Integration of Transport and Land Use Planning, Deliverable D2a Land-Use Transport Interaction: State of the Art.* Final Draft, Institute of Spatial Planning, University of Dortmund, Dortmund, Germany.

Weiland, R. J., and L. B. Purser. 2000. "Intelligent Transportation Systems." In *Transportation in the New Millennium: State of the Art and Future Directions. Perspectives from Transportation Research Board Standing Committees.* National Research Council, Transportation Research Board, Washington, DC. Also available at http://nationalacademies.org/trb/.

Williams, H. C. W. L., and J. D. Ortuzar. 1982. "Behavioral Theories of Dispersion and the Misspecification of Travel Demand Models." *Transportation Research B* 16(3):167–19.

Wilson, F. D. 1979. *Residential Consumption, Economic Opportunity, and Race.* New York: Academic Press.

World Bank. 1996. *Sustainable Transport: Priorities for Policy Reform.* Washington, DC.

Wolf, J., R. Guensler, S. Washington, and L. Frank. 2001. *Use of Electronic travel diaries and Vehicle Instrumentation Packages in the Year 2000.* Atlanta Regional Household Travel Survey. Transportation Research Circular E-C026, National Research Council, Transportation Research Board, Washington, DC, March.

CHAPTER 30
TRANSPORTATION ECONOMICS

Anthony M. Pagano
Department of Managerial Studies,
University of Illinois at Chicago, Chicago, Illinois

30.1 INTRODUCTION

Transportation economics is a very broad field. It includes the application of economic principles to pricing, cost analysis, and regulatory issues. It also includes the analysis of transportation impacts on land use, economic development, and the environment. The field of transportation economics also includes the analysis of the costs and benefits of transportation improvement and initial construction projects. It is this latter aspect of transportation economics that is of interest to transportation engineers and is the subject of this chapter.

The chapter begins with a discussion of the methodology of project appraisal. This includes cost analysis, methods to deal with uncertainty, methods of benefit quantification and evaluation, rate of return analysis, net present value analysis, the appropriate discount rate to use, benefit-cost ratios, net benefit analysis, and suboptimization. The chapter then discusses transportation user benefits, intangible costs and benefits, and the externalities associated with transportation projects.

30.2 THE METHODOLOGY OF PROJECT APPRAISAL

The methodology of project appraisal involves the ascertainment of the costs and benefits of a transportation project. It utilizes the methodology of benefit-cost analysis to analyze alternative projects and make decisions as to which course of action is best. Accordingly, we will focus on benefit-cost analysis techniques in this chapter.

The first question is: What is benefit-cost analysis? Prest and Turvey (1966) state that it is "a practical way of assessing the desirability of projects, where it is important to take a long view (in the sense of looking at repercussions in the further, as well as the 'nearer,' future) and a wide view (in the sense of allowing for side effects of many kinds on many persons, industries, regions, etc.); i.e., it implies the enumeration and evaluation of all the relevant costs and benefits." Quade (1965) says that "it is any analytic study designed to assist a decision-maker identify a preferred choice from among possible alternatives." In a sense, it is a way to look at a problem, analyze it, and arrive at some type of solution. It involves the comparison of various alternatives to achieve a specific objective and essentially consists of the following six steps:

1. The statement of the desired objectives
2. A complete specification of all the relevant alternatives
3. An estimation of all the costs involved
4. An enumeration of all the benefits
5. Development of a model, either verbally or mathematically
6. Development of criteria for choice among the relevant alternatives

It must be cautioned that these steps are so interrelated that any attempt to discuss them as mutually exclusive parts is surely doomed to failure. This chapter, then, examines each of the steps while keeping in mind their mutual interdependence.

30.3 OBJECTIVES OF PROPOSED INVESTMENTS

Any project appraisal must start with an enumeration of the objectives that the projects or programs are designed to attain. There may be one objective or many. What is important is that the objectives be enumerated clearly and completely. There is no room for ambiguity. The objective function cannot be misspecified. The importance of the objective function is taken up elsewhere in this chapter. At present, it should suffice to say that the objective function need not be formulated mathematically. It may be either verbally or mathematically explained, or both. What is necessary is that it be clear.

30.4 ALTERNATIVES

All the alternative systems or programs that, in the eyes of the analyst, may possibly attain the stated objectives should be enumerated exactly and in great detail. Sometimes all the alternatives are not obvious. The analyst must explore all avenues to obtain a list of all the alternatives.

Another important aspect concerning the specification of the alternatives is that if all the initially listed alternatives are examined and no one of them achieves the desired objectives adequately, the analyst may be forced to design new alternatives. This is one of the primary advantages of benefit-cost analysis: to force the examination and exploration of new avenues where alternatives may lie.

30.5 COSTS

A good benefit-cost study requires the complete enumeration of all the relevant costs of each alternative project. Even if all the costs are known and can be quantified, this is still not an easy task. The costs internal to the project must be broken down by type. One approach is to break them down into three categories—fixed or sunk costs, investment or capital, and maintenance costs. Sunk costs are money that has already been expended on plant and equipment or resources. These are costs that have been made in the past. Economic analysis maintains that because these costs were expended in the past they have no relevance to future systems. In other words, they should not be counted as a cost of the various systems that the analyst is examining. Only the difference among alternatives is what is relevant in their comparison. Inasmuch as sunk costs are the same for all alternatives, there is no need to consider it in an analysis.

   Investment or capital costs are those costs that are outlays on plant and equipment for each alternative system. Maintenance costs are those costs that accrue over time and are expended to keep the various facilities at a suitable level of performance. These three categories, of course, can be broken down into various subcategories to enumerate more explicitly and completely the costs involved.

   One approach is to calculate the net average annual cost of a transportation improvement project. This is the amount by which the annual cost of the improvement exceeds the cost if no improvements were made. An application to highway improvements would result in the following cost categories:

1. Right-of-way
2. Grading, drainage, and minor structures
3. Major structures
4. Pavement and appurtenances

The following equation can be used to calculate the net average annual highway improvement project cost:

$$\Delta H = (C_1 K_1 + C_2 K_2 + C_3 K_3 + C_4 K_4) + \Delta M$$

where
ΔH = net average annual highway improvement project cost
C_1 = capital cost of right-of-way;
C_2 = capital cost of grading, drainage, and minor structures
C_3 = capital cost of major structures
C_4 = capital cost of pavement and appurtenances
ΔM = change in annual maintenance and operation cost for the project
K_1, K_2, K_3, K_4 = capital recovery factor for the known interest rate and service life of the respective item

   If the service life of an improvement is T years and the known interest rate is r, the capital recovery factor is given by:

$$K = \frac{r(1 + r)^T}{(1 + r)^T - 1}$$

The capital recovery factor distributes the capital costs over the T years of service equally, taking into account the time value of money

   The use of the capital recovery factor is typical of analyses in transportation. The rationale behind its use is that the transportation agency must finance its improvements with borrowed money or that, even if it does not borrow money, the inclusion of interest is an indication of the investment opportunities forgone by the taxpayers when the agency spends tax money.

   Another problem in cost estimation is establishing the planning or analysis period. The project life or planning period is usually highly subjective. It depends on personal judgments not only of the physical length of life of the project, but also of the likelihood of any changes that may make a particular project obsolete. The estimation of the planning period is further complicated in the analysis of transportation improvement projects due to varying physical lengths of life for each component of the improvement and the extreme uncertainty involved in the estimation of traffic growth in the more distant future. The estimation of traffic growth is an important part of an analysis of transportation improvements because the amount of benefit accruing from a given improvement is directly related to this factor. If the planning period is too long, the analyst may not be able to estimate this critical factor accurately. This can result in the wrong alternative being considered ''best'' or an unacceptable project being accepted.

There is not much agreement in transportation as to what constitutes the maximum length of time to use as the planning period. Periods of from 10 to 100 years have been mentioned by various authors as the maximum planning period. Some authors maintain, though, that the planning period should be either the physical length of life of the project or the useful life of the project, whichever is shorter.

Another problem in cost estimation is the estimation of the salvage value of the facility at the end of the analysis period. The salvage value can be positive in the case where the structures can be sold as scrap or negative in the case where there is a removal cost. Thus, the salvage value can either reduce or increase the cost of a given alternative and should be taken into account.

30.6   UNCERTAINTY

Rarely does the analyst have a chance to work on a problem where all the costs and benefits are known. Even when all the costs and benefits are quantifiable, the analyst usually is uncertain as to their actual values.

There are two types of uncertainty that need to be dealt with. The first is uncertainty about the world in the future. It includes items such as technological uncertainty and strategic uncertainty. This first type of uncertainty is difficult to deal with since these are unpredictable events. The second type is statistical uncertainty. It occurs because chance elements exist in the real world. This second type of uncertainty results from recurrent factors that can be modeled and estimated. Both types of uncertainty are usually present in long-run decision problems. Sensitivity analysis, contingency analysis, and a fortiori analysis are methods to treat the second type of uncertainty.

30.6.1   Contingency Analysis

Contingency analysis is most often used to evaluate military programs, and it may have some merit in transportation appraisal. The analysis must try to visualize the various changes that might occur that would have a significant impact on program outcomes and must take these contingencies into account in estimating the future worth of various investments.

30.6.2   A Fortiori Analysis

Another method to deal with uncertainty is a fortiori analysis. This method is used primarily in the military, but again may have some use in evaluating transportation improvement projects. To utilize this method, the analyst takes a set of circumstances unfavorable to a program and then compares that program with other possible programs. If the first program emerges as the winner in the unfavorable circumstances, the conclusion emerges, with high confidence, that that program is better than the others.

30.6.3   Sensitivity Analysis

Another approach to uncertainty is sensitivity analysis, which involves changing the values of certain parameters to see what effect this has on the final results. Given that the analyst is not sure exactly what the values of the parameters of various alternative systems will be, sensitivity analysis provides a means to examine the changes in the final results (i.e., which system should be accepted as best) if these parameters change. The analyst can vary each parameter and see how the final results change. If the final results are not appreciably affected by changes in certain parameters, it need not be considered further. If, on the other hand, the final result varies significantly with variations in other parameters, the analyst should spend more time, money, and energy on such parameters to try to achieve the best possible estimates of their actual values.

Sensitivity analysis can also involve finding a value for the uncertain parameter above or below which the optimal alternative may change. One example cited in the literature is the value of life. It can be helpful to know that below a certain value of life one project alternative is most desirable, but above that amount a different alternative would become optimal.

It might be worth mentioning that the finding of insensitivity is just as important as finding sensitivity in the estimates. If the results do not change much as key variables are changed, then the analyst can be more confident that the best project alternative is selected.

Sensitivity analysis, then, can be a helpful aid to the decision-maker either by indicating the risks involved in the decision or by reducing the number of parameters that need further consideration. It will not eliminate or even reduce the uncertainty surrounding the estimates of the various parameters. It will, however, indicate how uncertainty can or cannot affect the final results.

30.6.4 Monte Carlo Simulation

A final approach to dealing with uncertainty is to use Monte Carlo simulation. This involves specifying a probability distribution for each of the key parameters of the analysis. Each is varied according to the distribution and a distribution of outcomes developed for each alternative. Monte Carlo simulation has the advantage that all parameters and variables can be varied simultaneously. The disadvantage is that some knowledge of the probability distributions of the variables and parameters must be known.

30.7 BENEFITS—QUANTIFICATION AND EVALUATION

Probably the most conceptually difficult aspect of benefit-cost analysis is the measure of the benefits of each alternative project or system. The analyst would have a comparatively easy task if all the benefits were known and could be quantified. This is not the usual case. Most benefits are subject to a great deal of uncertainty, and the analyst may not be able to estimate even the likelihood that a benefit may be at any given level. Some benefits and costs are not even subject to quantification. How does the analyst measure the increased security and well-being of society? He cannot even measure, let alone place a value on, these benefits.

In view of these problems, the first point that should be noted is that all the benefits must be specified and enumerated. If not all benefits are taken into account, the wrong system may be considered best. This includes all benefits external as well as internal to the system. Also, the objectives should be kept in mind when the analyst decides what the appropriate measures of benefits or effectiveness will be. One example, cited in the literature, concerns the use of welfare payments as an alternative to training to help people out of poverty. One approach to evaluation is to discount the expected lifetime earnings of the trainees and match them against the cost of the program. If the earnings are greater than the costs, the training will be justified. Welfare payments could be justified as an alternative to training if the costs of training are greater than the discounted expected lifetime earnings of the trainees.

But the discounted expected lifetime earnings of the trainees substantially underestimate the benefits of the training programs. One of the major objectives should be to end poverty through the enhancement of personal opportunity. Clearly, the expected lifetime earnings of the trainees are not an adequate measure of the achievement of this objective. The alternative of welfare payments, if some measure of this personal opportunity objective were not included in the analysis, may be given more weight than it actually deserves. The analyst must examine carefully the objectives that he or she would like the alternatives to attain and use the measure or measures that most adequately indicate the level of attainment of the objectives. This is why the objectives must be clearly and distinctly defined. If the objectives are not precisely stated, it will be impossible to determine a suitable measure of the level of their attainment.

As mentioned previously, the most difficult task the analyst faces is what to do with benefits that cannot be quantified. There is a quantification fallacy that many analysts fall into, in which it is assumed that every factor pertinent to the analysis can be quantified. This cannot always be done. Certain factors, no matter how hard the analyst tries, cannot be quantified. If this is the case, the analyst should present his results to the decision-maker without the benefit of a quantitative analysis.

If the benefits can be quantified, the next step is to place some value on them. The usual analysis is to value benefits in a common measure so these can be compared. This common measure is in monetary units. Some authors have suggested the use of quasi- or shadow prices as a means to place some value on the various benefits. Shadow prices are those prices that would exist if a marketplace actually existed for these benefits. But some authors disagree with the use of shadow prices, arguing that the analyst should look elsewhere to place values on various benefits. For example, the use of the amount of money granted to individuals by courts of law is one means of placing value on the benefit received from reducing accidents or deaths.

30.8   EXPERT OPINION

One method of dealing with uncertainty and unquantifiables is to consult an expert or group of experts who are knowledgeable in the particular field of interest. One method is for the analyst to consult a single expert or groups of experts either individually or through direct fact-to-face confrontation of the experts. Another approach is to use the Delphi technique.

30.8.1   The Delphi Technique

The Delphi technique was developed by Olaf Helmer of the RAND Corporation in 1964. The Delphi technique eliminates the problems of group dynamics that are present in a face-to-face discussion and possibly permits a consensus. Direct discussion is replaced by a series of questionnaires, which can be administered through the mail or through e-mail.

As an example, suppose an analyst wants an estimate of the value of some number N. The analyst first asks each expert to place a value on N, independent of the others. The various responses are then arranged in order of magnitude and the quartiles Q_1, M, and Q_3 are determined.

N_1	N_2	N_3	N_4	N_5	N_6	N_7	N_8	N_9	N_{10}	N_{11}	N_{12}	N_{13}	N_{14}	N_{15}
1	1	1	1	1	1	1	1	1	1	1	1	1	1	1
			1				1				1			
			Q_1				M				Q_3			

Next, the values of Q_1, M, and Q_3 are given to the experts. Each is asked to revise the previous estimate, and if the revised estimate is outside the range Q_1 to Q_3, to explain why the estimate should be different than the 75 percent majority opinion. The results of the second round (that is, the revised values of Q_1, M, and Q_3) are given again to the experts along with the various reasons given in round 2 either to raise or lower the value of N. The reasons are given to the experts in such a way that the anonymity of the respondents is preserved. The experts are asked to evaluate the reasons given in round 2 and again revise their estimates. If their estimates still fall outside the range of Q_1 and Q_3 they are asked for the reasons why the estimate should be different. The estimation procedure can continue for an additional round if desired. The result is usually a consensus estimate that has less dispersion than the original estimates.

The various rounds reduce the dispersion of the estimates and increase the confidence that the analyst has in the estimate. The question remains, however, as to whether the resultant estimate is more accurate that the original.

The Delphi technique eliminates the problems of group dynamics that occur in face-to-face discussions, in which one or a few individuals can dominate the discussion. It also usually results in a consensus forecast that has less dispersion than the initial estimate. On the other hand, the Delphi technique can lead to ambiguous results, since a written questionnaire replaces face-to-face discussion. In addition, if there is too much time between rounds, the experts may have difficulty remembering the context in which the previous discussion was held. This last problem can be reduced through the use of e-mail.

30.9 THE MODEL

The next step in most benefit-cost studies is to design some sort of model to represent the system or systems that the analyst wishes to evaluate. This model is necessarily an abstraction, although it should be a reasonable representation of reality. The model can be either highly mathematical in form or merely a verbal description of reality. It can be a computer simulation or a written representation. The purpose of the model is to develop a *set of relationships* among the objectives, the alternatives available for attaining the objectives, the estimated cost of the alternatives, and the estimated benefits. The assumptions underlying the model should be made explicit.

After the model has been built, it should be checked to ascertain whether it is structured in such a way as to produce a reasonable representation of reality. Some possible questions that can be used to test the model are:

1. Can the model describe known facts and situations reasonably well?
2. When the principal parameters involved are varied, do the results remain consistent and plausible?
3. Can it handle special cases where we already have some indication as to what the outcome should be?
4. Can it assign causes to known effects?

30.10 DISCOUNTING

When the various costs and benefits appear as a stream over time, the analyst's job becomes more difficult. The costs or benefits may be larger in the first time period and decrease thereafter. Or the benefits and costs may grow larger through time, or the benefits may become larger while the costs become smaller. The important point is that both benefits and costs come in streams that are not necessarily equal, nor are they necessarily the same throughout the years.

If, in Figure 30.1, it is assumed that the two alternatives cost the same, which benefit stream is to be preferred? This is a question that the analyst often faces. Discounting can help answer it.

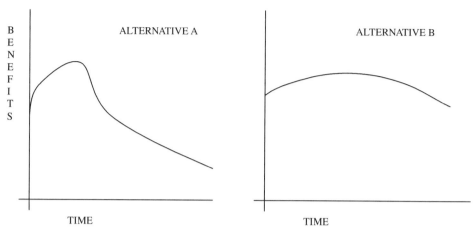

FIGURE 30.1   Time distribution for two alternatives.

One of the more controversial aspects of benefit-cost analysis is the use of the discount rate. It is controversial because there is little agreement as to which rate to use. Discount rates from 2 percent all the way up to 25 percent have been suggested by various authors. Some argue that discounting should not be allowed at all in benefit-cost studies. Of course, this argument boils down to the assumption of a zero rate of discount. What follows in this section is a brief discussion of some of the arguments of various authors as to what the proper discount rate should be.

Some authors believe there is no "correct" discount rate that can be used for government investments. It is argued that since the private investment decision can be changed or sold on the market, it makes sense to discount it. However, with regard to government investments, society cannot change its benefit stream after the decision to invest is made. That is, society cannot sell part of the benefits from an investment on the market. Thus, there is no discount rate that is valid.

Other authors tend to think that a higher interest rate should be used for riskier investments to reflect the risk involved. Alternatives that realize payoffs in the more distant future are considered more risky. The argument is that unforeseen technological change may eliminate the need for the alternative exhibiting payoffs in the more distant future. A higher discount rate is recommended to reflect the riskiness of these more distant payoffs.

Others believe that the total opportunity cost of government borrowing is the correct discount rate, since debt policy can be used to achieve preferred time profiles of both taxes and borrowing. They argue that additional government borrowing displaces an equal amount of private investment. The total opportunity cost is the sum of the interest rate on long-term government bonds plus the loss of tax revenues associated with society's forgone investment income.

Some economists argue that rather than an interest rate being used to discount benefit and cost streams, the real growth rate of the economy should be used instead. The real growth rate should be used if estimates of future costs and benefits are made in constant dollars. If the estimates take changes in future price levels into consideration, the unadjusted growth rate should be used.

If the analyst has examined all the issues involved and still cannot find the appropriate discount rate, one alternative is to use an upper bound rate and lower bound rate to see if it makes any difference in the final analysis. This is a good use of sensitivity analysis in benefit-cost studies.

30.11 CRITERIA FOR CHOICE AMONG ALTERNATIVES

The next step in the analysis involves choosing a criterion or criteria on which to base a decision as to which alternative system or project is "best." This is one of the more difficult steps in an analysis.

Some authors continually search for the alternative that yields the greatest amount of benefit for the least amount of cost. There is no such alternative. Other authors warn against the "sole criterion fallacy." This is the fallacy that a single criterion can be selected to evaluate all the alternatives. The real world is never this simple. To try to evaluate all of the alternatives on the basis of one criterion is meaningless. This occurs because of the complexities involved in each system and because the systems may differ in many respects.

In most benefit-cost studies, the costs and benefits accrue over time. The prices of both the costs and benefits can change over time. The question that arises is: Should the analyst take these changes in price into account when evaluating the various alternatives? Most economists feel that the analyst should take into account changes in the relative prices of costs and benefits, but not changes in the general price level. All prices should be in constant dollars, which for convenience is usually the initial year of the analysis.

Some of the more commonly used criteria are benefit-cost ratios, net benefit, incremental benefit-cost ratios, and the internal rate of return on investment. Each of these methods is discussed in turn.

30.11.1 Benefit-Cost Ratios

One method used to evaluate the various alternatives is to form a ratio of the benefits and the costs accruing to each alternative project or system and select the alternative that exhibits the highest benefit-cost ratio. This method of selection has produced a great deal of controversy.

Probably the most common criticism of the use of benefit-cost ratios is the argument that the ratios ignore the relative magnitude of the various costs and benefits involved. Some economists believe that benefit-cost ratios can be used as long as the level of cost is approximately the same for the alternatives under consideration. Another defect of the benefit-cost ratio method is that the definition of benefits and costs can affect the outcome. In the highway field, maintenance costs, for example, can be treated as either positive costs or negative benefits. If the maintenance costs are put in the numerator of the ratio, the ranking of alternatives may differ from the case where these costs are placed in the denominator of the fraction. This is not a problem for the net benefit or incremental benefit-cost ratio method.

The primary advantage of using the benefit-cost ratio method for ranking alternatives is that it can deal with cases in which the benefits and the costs are not expressed in the same units. This advantage is unique to the benefit-cost ratio method when compared to all the other methods previously mentioned.

In many transportation benefit-cost analyses, the benefits and the costs are expressed on an annual basis. This is done through the use of the capital recovery factor mentioned earlier. An example of the ratio most commonly used in the highway field is given as:

$$\frac{R_0 - R_1}{S_1 + M_1 - S_0 - M_0} = \frac{R_0 - R_1}{(S_1 - S_0) - (M_0 - M_1)}$$

where S = investment costs on an annual basis
M = maintenance costs on an annual basis
R = road user costs on an annual basis

The subscripts 0, 1 refer to the existing and the proposed facilities, respectively. It should be noted that the benefits of a given highway improvement as expressed in the numerator of the ratio are given in terms of a decrease in road user costs. Most studies of highway improvements calculate benefits in this manner.

Thus, the measure of benefits in the highway field is usually limited to the reduction in road user costs that would result from a proposed facility. Some studies incorporate other costs and benefit categories as well, such as externalities associated with highway use.

As mentioned previously, the benefits and the costs that accrue because of highway improvement are usually expressed on an annual basis. However, benefits and costs usually do not accrue in uniform annual streams. It is necessary to annualize benefits and costs by taking the present worth of each stream first, and then distribute these benefits and costs uniformly over the life of the project.

30.11.2 Net Benefit

The net benefit method requires the analyst to subtract the costs from the benefits of each alternative; the alternative that exhibits the largest net benefit (a difference between cost and benefit) is selected. Looking at this in terms of production theory, total benefits can be considered the same as total revenue, total costs have the same meaning, and net benefit can be considered as total profit. Maximizing net benefit then corresponds to the firm maximizing total profit.

Use of benefit-cost ratios to evaluate the various alternatives does not always lead to the same alternative being chosen, as would use of net benefits of the various alternatives. For example:

Alternative	Benefits	Costs	Benefit-cost ratio	Net benefit
A	200	100	2	100
B	30	10	3	20

In this example, alternative A will be chosen if the analyst uses net benefit as the criterion of choice, but alternative B will be chosen if the benefit-cost ratio method is used. If a firm used the principle of the benefit-cost ratio as the criterion to decide at which level of output it should produce, it is unlikely that the firm would be maximizing profit. The one defect in the net benefit method is that it cannot handle situations in which the costs and benefits cannot be expressed in the same units.

30.11.3 Incremental Benefit-Cost Ratios

A third method frequently mentioned in the literature as a criterion of choice is the incremental benefit-cost ratio. A simplified example illustrates how this method works. Assume that a DOT must pick 1 alternative from among 10 mutually exclusive alternatives. There is no budget limitation, because the DOT has funds available to cover the cost of any one of the 10 alternatives being considered. For convenience of illustration, it is assumed that none of the alternatives is dominated by any other (that is, is overshadowed by reason of another being any more effective while costing the same or less). The various alternatives are ranked by increasing cost in Table 30.1.

TABLE 30.1 Example of Costs and Benefits of
Alternative Projects

Alternative	Benefit, B	Cost, C	$B - C$
$A1$	10	2	8
$A2$	20	14	6
$A3$	50	25	25
$A4$	80	30	50
$A5$	90	45	45
$A6$	100	70	30
$A7$	140	75	65
$A8$	170	130	40
$A9$	220	160	60
$A10$	350	320	30

The essence of the method is to examine the ratio:

$$\frac{(B_{k+1} - B_k)}{(C_{k+1} - C_k)}$$

If this ratio is greater than 1, alternative A_{k+1} is accepted and compared with alternative A_{k+2} in the same manner. If the ratio is less than 1, A_k is accepted and compared with A_{k+2}. These comparisons are continued until the last acceptable alternative is reached. This alternative is accepted as best. Table 30.2 illustrates the method.

In this example, alternative $A7$ is considered best. In this situation, the incremental analysis leads to the same alternative being chosen as for the net benefit method. This can be readily seen, for, if

$$\frac{(B_j - B_i)}{(C_j - C_i)} > 1, \quad (B_j - B_i) > (C_j - C_i) \quad \text{and} \quad (B_j - C_j) > (B_i - C_i)$$

The incremental benefit-cost ratio method for ranking alternatives suffers from the same defect as the net benefit methods. That is, it cannot handle situations where the benefits and the costs are expressed in different units.

TABLE 30.2 Example of Incremental Method of Selection of Best Alternatives

Comparison	Incremental benefits	Incremental cost	Incremental benefits/cost	Decision in favor of
$A1$ vs. $A0$	10	2	5	$A1$
$A2$ vs. $A1$	10	12	5/6	$A1$
$A3$ vs. $A1$	40	23	40/23	$A3$
$A4$ vs. $A3$	30	5	6	$A4$
$A5$ vs. $A4$	10	15	2/3	$A4$
$A6$ vs. $A4$	20	40	1/2	$A4$
$A7$ vs. $A4$	60	45	4/3	$A7$
$A8$ vs. $A7$	30	55	6/11	$A7$
$A9$ vs. $A7$	80	85	16/17	$A7$
$A10$ vs. $A7$	210	245	6/7	$A7$

30.11.4   Ratios and Net Benefits—A Comparison

It might be helpful to examine the relationships that exist among the benefit-cost ratio method, the net benefit method, and the incremental benefit-cost ratio method for ranking the alternative projects or programs. The benefits and the costs of six alternatives are shown in Figure 30.2, in which the origin represents the existing situation. The benefits that accrue to alternative $A2$ can be represented by the line segment $A2–C_2$. The costs accruing to this alternative are represented by the line segment $A0–C_2$.

The benefit-cost ratio, then, is $(A2–C_2)/(A0–C_2)$; but this is nothing more than the tangent of the angle $A2–A0–C_2$. As this angle increases, the benefit-cost ratio increases. Conversely, as the benefit-cost ratio increases, the angle increases. The tangent function is ever increasing in the interval $0°$ to $90°$. Thus, choosing the alternative that exhibits the largest benefit-cost ratio corresponds to picking the alternative whose ray from the origin is the highest. The $45°$ line indicates where the benefits are equal to the costs. All along this line the benefit-cost ratio is 1.0, as should be expected, because $\tan 45° = 1$. For all the alternatives above the $45°$ line the ratio is greater than 1.0; for those below this line, less that 1.0. The benefit-cost ratio solution would indicate indifference between alternatives $A1$ and $A3$, because they both lie on the highest ray from the origin.

Net benefit in this figure is represented as the vertical difference between the $45°$ line and the alternative. As can be seen clearly, all the other alternatives are dominated by alternative $A3$ since this alternative exhibits the greatest benefit-cost ratio and has the largest net benefit of all the alternatives. Alternative $A5$ has a negative net benefit and a benefit-cost ratio less that 1.0.

The same six alternatives are shown in Figure 30.3. The incremental benefit-cost ratio in this figure is represented as the line segment that joins two alternatives; thus, the incremental benefit-cost ratio in going from project $A1$ to project $A2$ is the line segment $A1–A2$. Now,

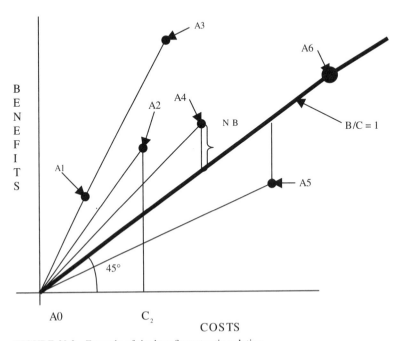

FIGURE 30.2   Example of the benefit-cost ratio solution.

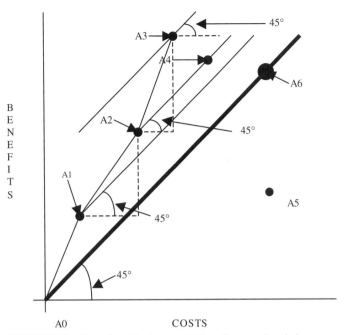

FIGURE 30.3   Example of the incremental benefit-cost ratio solution.

the criterion of choice using the incremental benefit-cost ratio procedure is that if the ratio is greater than 1.0, the alternative exhibiting the greater cost is accepted over the other. For alternative $A1$ as compared to alternative $A2$, a ratio of 1.0 can be displayed as the 45° line originating at point $A1$. So, if the incremental benefit-cost ratio is greater than 1.0 it will be above the 45° line; if it is less than 1.0 it will be below the 45° line. But the increment in going from $A0$ to $A1$ is greater than 45°, so $A1$ is considered better than $A0$. The increment in going from $A1$ to $A2$ is also greater than 45°, so $A2$ is accepted over $A1$. This procedure is continued for all the alternatives. As can be seen from the figure, alternative $A3$ is accepted as best by this criterion because a line segment joining $A3$ to any other alternative will be below the 45° line originating at point $A3$. Notice also that alternatives $A5$ and $A6$ are below the 45° line originating at alternative $A1$. There is no need to compare them to any other alternative because $A1$ is better than either of these two and any other alternative that is better than $A1$ will be better than either of them.

Again, the same six alternatives are shown in Figure 30.4. As is evident from the diagram, the 45° line extending from each alternative merely projects the net benefit for each alternative and compares it to the net benefit of the other alternatives. That is, the vertical distance from the 45° line coming out of the origin is the net benefit that accrues when alternative $A1$ is employed. When, for example, $A1$ is compared to $A2$ by incremental procedure, the 45° lines become the references to decide if one alternative is better than the other. If $A2$ were on the 45° line through $A1$, it would have the same net benefit as $A1$. Because $A2$ is better than $A1$ it lies on a higher 45° line than the corresponding parallel line through $A1$. This means that the net benefit for $A2$ is greater than that for $A1$. It should be clear from these three diagrams that the net benefit solution is equivalent to the incremental solution and the benefit-cost ratio solution need not necessarily be the same solution as the other two.

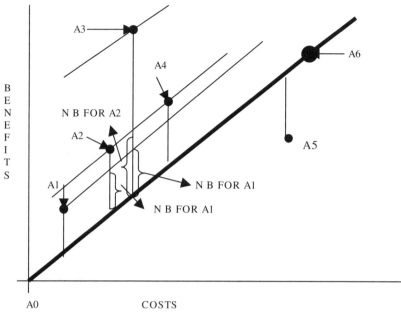

FIGURE 30.4   Example of the net benefit solution.

30.11.5 Rate of Return

The fourth criterion mentioned previously to evaluate the various alternatives is the rate of return method. This is that rate of interest that makes the annual benefits equal to the annualized costs.

Using an example from the highway field, annual benefits are given by $(R_0 - R_1) - (M_1 - M_0)$ and annual costs are given by $(I_1 - I_0) K(i, n)$. Thus, the rate of return is obtained by solving for i in:

$$(R_0 - R_1) - (M_1 - M_0) = (I_1 - I_0)K(i, n)$$

in which $K(i, n)$ = capital recovery factor,

$$K(i, n) = \frac{i(1 + i)^n}{(1 + i)^n - 1}$$

where R = road user costs on an annual basis
$\quad M$ = maintenance costs on an annual basis
$\quad I$ = initial investment
$\quad n$ = the life of the project, in years
$\quad i$ = the rate of return

The alternative that produces the highest rate of return is accepted as best.

Some economists take issue with the use of the rate of return method. They believe that this method does not provide an adequate basis for comparing one alternative to another. They propose using the incremental rate of return method instead. This method is similar to the incremental benefit-cost ratio method discussed previously.

30.12 SUBOPTIMIZATION

Another problem that occurs frequently in benefit-cost studies is that it is sometimes difficult for the analyst to relate how well the various alternatives attain the stated objectives. Suppose a hypothetical family wishes to perform a benefit-cost analysis to ascertain the optimal combination of goods and services to buy so as to maximize happiness. Happiness, then, is the family's stated objective. Of course, it is very difficult, if not impossible, to decide which combination of goods and services will maximize happiness. It may be impossible even to quantify the degree to which happiness is attained through the use of one good or another. In other words, although a utility function for the family may exist, in practice it is impossible for the analyst to decide its exact form.

To get out of this difficulty, the family might try to formulate new objectives whose attainment could be more easily calculated and whose attainment may be an indication of the attainment of happiness. The family, in this case, is said to be "suboptimizing" or optimizing on a lower level. Some of these new objectives that the family may try to attain are the satisfaction of hunger and thirst, the attainment of more leisure time, and the attainment of shelter from the extremes of temperature. These new objectives are referred to as lower-level objectives, and happiness can be referred to as the higher-level objective.

Suboptimization occurs when objectives are subclassified and apportioned to a lower level. For example, if the objective of government is to promote social welfare, the government suboptimizes when it grants its DOT the jurisdiction over highways. The goal of the DOT may be to reduce the cost of transportation along the highways in its jurisdiction. The higher-level objective then is to promote social welfare; the lower-level objective is to reduce the cost of highway transportation. It would be difficult indeed for the transportation department to relate various courses of action to the higher-level objective. This lower-level objective is then used to relate the decision of the DOT to this higher-level objective. Whenever the analyst suboptimizes, it is important for him or her to understand and state how the attainment of the lower-level objectives is an indication that the higher-level objectives are attained.

30.12.1 A Highway Example

It is assumed throughout this discussion that one of the objectives of transportation improvements is the promotion of social welfare. It is further assumed that transportation improvement promotes social welfare either through the reduction of the amount of real input required to obtain a stated level of transportation services or through an increase in transportation services for a stated level of inputs. What is measured then is how the productive capacity of the economy has increased through transportation improvement.

Many transportation improvements result not only in an increase in the productive capacity of the economy, but also in a redistribution of income from one group in the economy to another. Many highway studies fail to take this redistribution of income into account. An example of counting a transfer of income as a benefit occurs in the valuation of time savings that accrue to commercial vehicles. One method to value these time savings is to assume that these savings will permit the same level of transportation services to take place with fewer inputs. One of the inputs that is decreased in the number of trucks engaged in these services. Thus, the commercial vehicle owners need a smaller fleet of trucks to do the same amount of work. Included as benefits would be not only the cost to the owners of the trucks they did not have to buy, but also the decrease in interest charges on the trucks. However, interest charges in no way constitute a using up of real goods and services for the economy as a whole. They are merely a transfer of income from the commercial vehicle owners to the lenders of money. Thus, the inclusion of interest charges as a benefit is not consistent with the previously stated objective and would be subotimization at a lower level. Of course, if the goal of the transportation department is limited to the reduction of the cost of operation of commercial vehicle owners, a reduction in interest charges is a valid benefit to include in an analysis. This, however, is an example of the lower-level objective being inconsistent with the higher-level goal.

30.13   ADVANTAGES AND LIMITATIONS OF BENEFIT-COST ANALYSIS

A variety of authors over the years have documented the advantages and limitations of benefit-cost analysis as a method of project appraisal. One of the big limitations is there is less confidence in the estimates of benefits than costs. Another limitation is incompleteness. It is argued that limitations on both time and money place limits on an analysis. Imperfect information also places a limit on what can be quantified. It is argued that benefit-cost analysis can never look at all relevant factors. There is always something left for the decision-maker.

Many authors believe that, in view of its limitations, benefit-cost analysis should be used only as an aid to the decision-maker. It is argued that the main role of the analysis is to sharpen the intuition and judgment of the decision-maker. It should not be used to make the decision. Even so, it is an important part of the decision process. If used as an aid to the decision-maker, it is probably the best aid that he or she has. If nothing else, the analysis can eliminate the really bad alternatives and give the decision-maker a shorter list to choose from.

30.14   QUANTIFICATION AND EVALUATION OF BENEFITS OF TRANSPORTATION PROJECTS

As stated earlier, a good benefit-cost analysis takes into consideration all the relevant costs and benefits. This involves the complete enumeration and specification of all the benefits that accrue to users as well as nonusers, benefits that are easily quantified and those that do not lend themselves readily to quantification. It includes the enumeration of market as well as extra-market consequences. This does not imply, of course, that all benefits must be quantified, nor does it imply that all benefits must have some value attached to them. What it does imply is that no prospective source of benefit should go unnoticed in the analysis.

In the highway field, the beneficiaries of highway improvement are usually categorized as users and nonusers. User benefits can occur as a decrease in the cost of motor vehicle operation, a decrease in time spent on the highway, a decrease in accident rates, or a decrease in the strain and discomfort of nonuniform driving. Nonuser benefits can accrue as an increase in land values, stimulation of economic growth along a new roadway, decrease in the cost of transportation-intensive goods, or increase in the efficiency of delivery of transportation-intensive goods.

Transit improvement benefits include decreases in user costs, including decreases in travel and waiting times, increased comfort and convenience, and reduced accidents and crime. Many of the beneficiaries of transit improvements may seldom if ever use transit. These are the users of the road network who benefit from reduced congestion resulting from increased transit use. A variety of external benefits accrue to transit improvement, such as increases in land values, economic growth, etc., similar to those of highway improvements. In addition, there may be reductions in auto-related emissions resulting from increased transit use. One problem with the evaluation of nonuser external benefits is that double counting of benefits may occur. The analyst must be careful to understand how benefits actually result.

Another aspect of benefit estimation concerns understanding the distribution of benefits that occur. This involves determining which groups, areas, municipalities, etc benefit and which suffer a loss. One example concerns the provision of park-and-ride lots in suburban areas. A benefit-cost analysis may reveal that the social and regional benefits of such facilities far outweigh the costs. Increased transit use resulting from building of such facilities would result in decreased commuting costs for former auto commuters, reduced user costs for those that still commute by car, reduced air pollution, etc. However, there are a variety of cost and benefit impacts that accrue to the residents of the municipalities in which park-and-ride lots are built. These include increased congestion, pollution, and crime associated with increased traffic volumes and impacts on property values, tax base, and economic development in the vicinity of the park-and-ride lot. Unless municipalities are compensated for possible losses or the losses mitigated in some way, a great deal of opposition to such lots may result in these areas. An analysis of the distribution of benefits and costs may help to develop such compensation and mitigation measures.

Another complication in measuring the benefits of transit improvements is that there is an option value of transit that is not necessarily captured by the benefits associated with regular transit use. The argument is that some people benefit from the availability of transit in case the automobile cannot be used. For example, in bad weather conditions, or if the car breaks down or is not available, individuals benefit from the transit being there, even if they do not use it on a regular basis. Merely estimating the cost savings that these individuals experience when using transit underestimates the true benefits associated with this option value of transit. One way of attempting to estimate such option benefits is to value these using the Black and Scholes formula for valuing stock options. A discussion of this approach can be found in TCRP Report 78 (ECONorthwest and Parsons, Brinckerhoff, Quade & Douglas 2002).

30.15   ROAD USER BENEFITS

Since road user benefits accrue to both beneficiaries of road improvements and transit projects, it would be helpful to look at these in a little more depth. Much work has been done in attempting to quantify these benefits. The major work in this area is the AASHTO "Red Book" of 1977. This work is currently being updated by NCHRP Project 02-23, which should be released some time in 2003.

An important category of road user benefits is decreases in motor vehicle operating costs. These costs can be reduced by transportation improvements through a decrease in the consumption of fuel and oil, a reduction in tire wear, a decrease in maintenance and repairs, and a reduction in ownership costs. These operating costs differ for different types and conditions of roadway. Some of the more relevant factors that the analyst should take into consideration when measuring these operating costs are the number and arrangements of lanes, the type of roadway surface, the amount of grade along a road, the average running speed, the traffic volume, the alignment of the roadway, and whether the roadway is located in a rural or an urban area.

30.15.1   Fuel Consumption

One of the largest single components of vehicle operating cost is fuel consumption. Not all highway improvements will cause a decrease in this cost. Fuel consumption is affected by frequency of accelerations, grades, road roughness, horizontal curvature, operating speeds, and congestion, among other factors.

Generally, road-improvement projects that cause free-flowing average speeds to increase above 35 mph will cause fuel consumption to increase. This increase in speed will, however, produce time savings. Thus, there exists a trade-off between these two types of user costs. There exist many other trade-offs as well in transportation project appraisal. Thus, the analyst cannot really be certain that an improvement that reduces a few elements of cost will actually be the best improvement to make. It could be that other costs are increased so much that they far outweigh the savings produced by the given improvement. This example serves to emphasize that all the relevant costs and benefits should be taken into account in any project appraisal study. If only those elements of cost that are reduced are examined, the analyst may conclude that a given improvement is best or acceptable when in fact it is not.

30.15.2   Tire Wear, Oil Consumption, and Maintenance Costs

The effects of transportation improvements on tire wear, oil consumption, and vehicle maintenance costs have for the most part not been firmly established. The type of roadway surface, route shortening, and type of operation are the only factors that have been established as affecting one or more of these cost elements. Tire wear increases with surface roughness, wear being less for paved than for gravel surfaces. It also increases with increased stop and go driving. These same factors may also affect oil consumption and maintenance costs, although the effects are not well established.

30.15.3   Vehicle Ownership Costs

Vehicle ownership costs include garage rent, taxes, licenses, insurance, interest, and time depreciation. These costs are constant throughout the year regardless of vehicle use. If a transportation improvement, such as a light rail system, results in a reduction in the number of vehicles that are owned, then these costs will be affected.

Time depreciation is independent of vehicle use and is only a function of age. Another type of depreciation is that which is related to vehicle use. There is not much agreement in the literature as to all the factors that affect depreciation due to use, nor is there much agreement as to the percentage of total depreciation that should be attributed to time and use depreciation.

30.15.4   Time Savings

Another benefit that may accrue to road users is a reduction in travel time. Although it is no easy task in itself to measure the time savings in minutes or hours, there exists a much more conceptually difficult problem. This is the placing of some dollar value on the savings. Vehicles are usually divided into passenger cars and commercial vehicles. Separate values of time are calculated for each class of vehicle. Some authors also have tried to estimate the value of travel time to commuting motorists. The value of time varies with the purpose of use. Thus, the value of time for commuting motorists may differ from the value of time for leisure driving.

One problem with valuation of time savings concerns aggregation of such savings. Suppose a highway improvement results in a time savings of 1 minute per motorist. When the minute is then summed for all vehicles, 365 days a year, for the life of the project, it would seem that a very substantial savings would result. But what is the real value of a 1-minute savings per trip? For most vehicles, this savings has little or no value. However, if the 1-minute savings is then added to time savings from other transportation projects, a savings of a considerable block of time can result.

There are various ways that the analyst may go about placing a value on time savings. The revenue and the cost savings methods are two means frequently used to value the time savings of commercial vehicles. The willingness-to-pay method is a way used to value the times savings of private automobiles. Finally, there is the cost-of-time method, which is used not to value time savings but to ascertain the cost of providing such savings.

Revenue Method.   The revenue method, also referred to as the net operating profit method, is a technique used to value the time savings of commercial vehicles. It is assumed that time savings will permit an increase in revenue-miles driven by the commercial vehicle, as the vehicle will be driven more miles in the same amount of time. For every hour of time saved by a transportation improvement project, the owner of the commercial vehicle is assumed to have 1 additional hour of revenue operation at no additional cost to those factors that vary with hours of operation.

Cost Savings Method.   The cost savings method is also used for the evaluation of commercial vehicle time savings. While the revenue method assumes that time savings result in an increase in revenue-miles in the same amount of time, the cost savings method assumes that the time savings will permit the same amount of mileage to be driven as before, but with the use of fewer resources. Thus, according to the cost savings method, time savings will result in the use of fewer vehicles and drivers to do the same amount of work as before. The costs that are reduced are those costs that vary with hours of operation, or those associated with a reduction in the number of drivers and vehicles. These costs include interest, depreciation, and property tax on equipment and reduced wages and benefits.

Several of the costs that are assumed to be reduced are open to dispute. Interest charges are one of these. Transportation improvements are beneficial to society when they permit the expansion of the productive capacity of the economy through the use of fewer resources to obtain the same level of output or through the expansion of output with the use of the same level of resources. The cost to society in resources used to obtain a given level of output is not a money cost, but the cost in real goods and services whose consumption society must forgo to obtain this output. Money is only a means of measuring how many resources society must use to attain some stated level of output. Interest charges as a money cost do not use up any of society's resources. These are merely a transfer of income from highway users or highway-intensive goods users to bondholders. As such, reductions in these charges should not be counted as benefit to society.

The inclusion of property taxes in the cost savings method can be attacked on the same grounds as interest charges. That is, property taxes do not constitute a using up of real resources, and as such their reduction should not be calculated as a benefit of transportation improvement projects. These same arguments apply to the valuation of vehicle ownership costs mentioned earlier.

Willingness-to-Pay Method.   There are several willingness-to-pay methods used to value time savings. All the techniques, however, involve the same basic methodology. The essence of the method requires the calculation of the opportunity cost to estimate how much of other goods and services, at market prices, motorists are willing to forgo to obtain 1 hour of time savings. This opportunity cost is then the estimated value of time savings. The classic approach is to find situations where motorists save time, but must pay a toll, over another route that takes longer but is toll-free.

Cost-of-Time Method.   The cost-of-time method differs from the other methods because, instead of calculating the value of time savings itself, this method seeks to compare alternative transportation projects according to the cost of providing the savings. Because the various methods of valuing time savings can lead to different results, many analysts place little confidence in the estimated value of time savings. If little confidence can be placed in the estimate of the value of time, little confidence can be placed on the ranking of alternatives using this value. Instead, the cost-of-time method seeks to determine how the ranking of alternative projects will change for different assumed values of time.

30.15.5 Cost of Accidents

Another road user cost that can be reduced by transportation improvement projects is the cost of accidents. The analyst faces two rather difficult problems in attempting to use the cost of accidents in the analysis. The first problem involves the difficulty in obtaining statistically significant estimates of how much a given project will affect accident rates. The second problem involves attaching some dollar estimates to the cost of accidents. This latter problem is taken up in this chapter.

The usual method of placing a value on the cost of accidents involves enumeration of all costs associated with various types of accidents. The costs usually enumerated are the so-called direct costs of a given type of accident. These direct costs are then estimated for the various accidents that occur in the geographic area that the analyst studies during a given time period. Some average value is then formulated for the costs of each type of accident. Most studies classify accidents by severity. Thus, the usual estimates of the costs of accidents are presented as the cost of:

1. Property-damage-only accidents
2. Nonfatal injury accidents
3. Fatal accidents

Some approaches break down injury accidents into categories of severity.

The direct costs of accidents include:

- Damage to the vehicle itself
- Damage to property outside the vehicle
- Damage to other vehicles
- Cost of ambulance service
- Hospital and treatment services
- Funeral costs
- Value of work time lost
- Present value of the loss of future earnings by those fatally injured or impaired
- Other costs

Indirect costs are those costs incurred to deal with the accident problem as a whole. These include costs of police enforcement, driver licensing and education, overhead costs of automobile insurance, etc.

The direct and indirect costs of accidents can be lumped together and called the economic cost of accidents, since these involve the expenditure or loss of money. Noneconomic costs are those intangibles that do not involve money expenditure or loss. Intangibles include the pain, fear, and suffering of the victims due to death or personal injury.

Each of the elements of direct and indirect costs involves challenges of estimation. One approach is to use court damage awards as a proxy for these costs. This is an approach not without controversy since some analysts argue that these are merely income transfers that only indirectly relate to the true costs incurred. The court award of $7 million dollars, which was later reduced, for spilling hot coffee on a woman's lap is a case in point.

Estimation of expected lifetime earnings is also fraught with difficulties. The analyst must estimate the loss in potential earnings had the victim enjoyed a normal work life. Factors that can affect lifetime earnings include age, sex, employment status, level of education, etc. Thus, a male with a college degree will have a certain expected income, while a female without a degree will have another. An assumption must be made as to how many years the victim would have continued in active employment. Individuals who stay home to take care of children and retirees are difficult to deal with using this approach. Most studies assign arbitrary values to household and child-rearing services. Some studies assume the value of retiree lost earnings is zero, although this may lead to unacceptable conclusions, especially when maintenance costs are subtracted from income.

The intangible, noneconomic costs are even more difficult to ascertain. One approach is to carry the intangible costs through the analysis as a parameter, I. The analyst could then calculate all the economic costs of the project and arrive at a net benefit that includes the intangible cost. For example, suppose the net benefit of a given transportation improvement project is:

$$\text{Net benefit} = -\$100,000 + 0.1I$$

The analyst could then solve for the value of I that would make net benefit equal to zero. In this case, it would be:

$$I = \$1,000,000$$

That is, the intangible cost per accident must be at least $1 million to economically justify such a project.

This approach to dealing with intangibles can be used not only for accident costs but any intangible cost element in which the analyst does not have a good estimate of the possible costs or benefits.

30.15.6   External Benefits and Costs

External benefits and costs are those that accrue to the users of other facilities and nonusers. These are the externalities of transportation projects and should be taken into account in any benefit-cost analysis. These include air, water, and noise pollution effects, impacts on land values, economic development effects, decreases in the cost of transportation-intensive goods, and increases in the efficiency of delivery of transportation-intensive goods.

Some of these external benefits and costs are a using up of real resources or a real cost, such as increased levels of pollution, and should be taken into account. Some categories of external benefits and costs, however, may be either transferred user benefits or a transfer of benefits and costs from one group of nonusers to another. For example, if a new highway is built, land values in the vicinity of the new facility will rise. What is not considered in most analyses is that land values in another area may fall because of the new facility. This can occur to land along an older route that has traffic diverted from it to the new highway. It is important for the analyst to consider these impacts, to understand the distribution of benefits and costs of the improvement.

30.16   CONCLUSIONS

Transportation project appraisal using benefit-cost analysis is a powerful tool to prioritize transportation improvement projects in a systematic fashion. It attempts to substitute quantitative evaluations for the decisions based on intuition, judgment, and political considerations. However, such analyses cannot take all factors into account. There are many intangibles and unquantifiables that cannot be readily incorporated into an economic analysis. As such, benefit-cost analysis cannot be used to make decisions as to which projects to implement. The intuition and judgment of the decision-maker is still an important element in transportation project appraisal. Benefit-cost analysis, however, can shed light on important benefits and costs that can be quantified and valued. As such, it is an important aid to decision-making. It cannot ever replace the decision-maker. Its use in transportation project appraisal helps to ensure that wherever possible, such decisions are made in a businesslike basis.

30.17   REFERENCES

American Association of State Highway and Transportation Officials (AASHTO). 1977. *A Manual on User Benefit Analysis of Highway and Bus-Transit Improvements,* AASHTO, Washington, D.C.

Barnum, H. N., J.-P. Tan, J. R. Anderson, J. A. Dixon, and P. Belli. 2001. *Economic Analysis of Investment Operations: Analytical Tools and Practical Applications.* Washington, DC: World Bank, February.

Cambridge Systematics, Inc, R. Cervero, and D. Aschauer. 1998. *Economic Impact Analysis of Transit Investments: Guidebook for Practitioners.* TCRP Report 35, National Research Council, Transportation Research Board, Washington, DC.

ECONorthwest and Parsons, Brinckerhoff, Quade & Douglas, Inc. 2002. *Estimating the Benefits and Costs of Public Transit Projects: A Guidebook for Practitioners* TCRP Report 78, National Research Council, Transportation Research Board, Washington, DC.

Faiz, A., and R. S. Archondo-Callao. 1994. *Estimating Vehicle Operating Costs.* Washington, DC: World Bank, January.

Louis Berger & Associates. 1998. *Guidance for Estimating the Indirect Effects of Proposed Transportation Projects.* National Cooperative Highway Research Program Report 403, National Research Council, Transportation Research Board, Washington, DC.

Organization for Economic Co-operation and Development (OECD). 2001. *Assessing the Benefits of Transport.* OECD Code 752001091E1. http://www1.oecd.org/publications/e-book/7501091E.PDF Paris, April.

Prest, A. R., and R. Turvey. 1965. "Cost-Benefit Analysis: A Survey." *Economic Journal,* 75:683–735.

Quade, E. S. 1965. *Cost-Effectiveness: An Introduction and Overview.* The RAND Corp., P-3134, May.

Texas Transportation Institute. 1993. *Microcomputer Evaluation of Highway User Benefits.* National Cooperative Highway Research Program Report 7–12, National Research Council, Transportation Research Board, Washington, DC.

Tsunokawa, K. 1997. *Roads and the Environment: A Handbook.* Washington, DC: World Bank, November.

van der Tak, H. G., and L. Squire. 1975. *Economic Analysis of Projects.* Washington, DC: World Bank, January.

30.18   SOURCES OF INFORMATION THROUGH THE WEB

- World Bank publications can be found at: http://publications.worldbank.org/ecommerce/
- OECD publications can be found at: http://www1.oecd.org/publications
- TRB publications and the status of NCHRP projects and TCRP projects can be found at: Transportation Research Board, http://www.nas.edu/trb
- Two software tools are available to evaluate the benefits and costs of transportation improvement projects. One is called the Sketch Planning Analysis Spreadsheet Model (SPASM) and the other is the Surface Transportation Efficiency Analysis Model (STEAM). Both were developed for the Federal Highway Administration. For further information:
  - Federal Highway Administration, *STEAM User Manual,* http://www.fhwa.dot.gov/steam/users_guide.htm

- Federal Highway Administration, *SPASM User's Guide,* http://www.fhwa.dot.gov/steam/spasm.htm
- Federal Highway Administration, *Using SPASM for Transportation Decision-Making*, http://www.fhwa.dot.gov/steam/spasm.htm

CHAPTER 31
INNOVATIVE INFORMATION TECHNOLOGY APPLICATIONS IN PUBLIC TRANSPORTATION

John Collura
Northern Virginia Center,
Virginia Polytechnic and State University,
Falls Church, Virginia

31.1 INTRODUCTION

A major movement is taking place in the United States, Europe, and the Pacific Rim to improve public transportation facilities and services with the use of advanced information technologies (National Transit Institute 2001; Casey et al. 2000). These technologies include, for example, computer hardware and software, the Internet, satellite-based navigation and location systems, wire and wireless telecommunications, sensors, and advanced computational methods. Anticipated benefits associated with investments in these technologies are, for example, improvements in customer convenience, transit management and operations, and safety and security (Goodell 2000), and activities are underway in the United States to promote the evaluation of the actual benefits and costs in operational field tests using these technologies (Casey and Collura 1992).

The primary aim of this chapter is to review the major areas in which transit operators in the United States have invested in information technology and to examine their experiences and lessons learned. These areas include public transportation management and operations, traveler information services, transit signal priority, and electronic payment and fare collection.

A special effort will be made in this chapter to discuss the way in which use of systems engineering concepts are employed in the design and deployment of systems consisting of one or more of these information technologies. Attention will also be given to describing the relationship of these concepts to the National Intelligent Transportation System (ITS) Architecture developed by the U.S. Department of Transportation. More detailed reviews of systems engineering concepts, the National ITS Architecture, and their potential application to enhance public transportation services are provided elsewhere (Gonzalez 2002; FHWA 1997).

While this chapter may be of interest to transportation engineers and transit planners, it may also be useful to other members of the transportation community, including transportation managers and policy-makers, for the purpose of enhancing their understanding of the intent, expectations, and potential merits of information technology investments in public transportation. Such an understanding facilitates discussions with technology and equipment vendors and systems integrators and enables public sector transportation professionals and policy-makers to make more informed choices in information technology investments.

31.2 *SYSTEMS ENGINEERING CONCEPTS*

Systems engineering has been generally defined as an "approach to building systems that enhances the quality of the end result" (Gonzalez 2002). Such systems that employ one or more of the information technologies mentioned above to improve transportation facilities and services have been called intelligent transportation systems (ITS). Intelligent transportation systems designed to enhance public transportation services have also been referred to as advanced public transportation systems (APTS).

Central to the application of the systems engineering approach in the design of intelligent transportation systems are the concepts of system objectives, system functional requirements, and system architecture.

A system objective is the intent that an intelligent transportation system is designed to achieve. System functional requirements are the "what's" that a system must perform in order to accomplish intended objectives. For example, a system objective might be to reduce incident response times to address more quickly the needs of disabled transit vehicles, and a functional requirement might be, for example, the capability to identify the location of the vehicle with the use of one or more of the location technologies mentioned above.

System architecture depicts the structure of a system design and may be of the physical or logical type. A physical architecture includes a depiction of the subsystems in the system, and the logical architecture represents the flow of data in the system. The graphic in Figure 31.1 represents a very broad representation of the physical architecture and subsystems of an intelligent transportation system as defined in the National ITS Architecture, developed by the U.S. Department of Transportation.

Another concept employed in the National ITS Architecture is the ITS user service, which is a particular service provided to a user. The user may be a traveler, but could also be a public transportation dispatcher or vehicle operator. Table 31.1 provides a list of the seven general categories (or bundles) of ITS user services and the corresponding 31 individual ITS user services currently included in the National ITS Architecture.

The remainder of this chapter will review the four major application areas in which transit authorities in the U.S. have deployed intelligent transportation systems:

1. Public transportation management and operations (public transportation management, archived data function, ride matching and reservation, and public travel security)

2. Traveler information services (pretrip travel information and enroute transit information)

3. Transit signal priority (traffic control and public transportation management)

4. Electronic payments and fare collection (electronic payment services)

Noted in parentheses are the ITS user services that relate to these four application areas. These user services are highlighted in Table 31.1.

A special effort will be made to review these areas in terms of their systems, technologies, objectives, requirements, and associated ITS user services and architectures. In addition, selected ITS deployments will be reviewed in each of the four application areas and results and lessons learned will be presented.

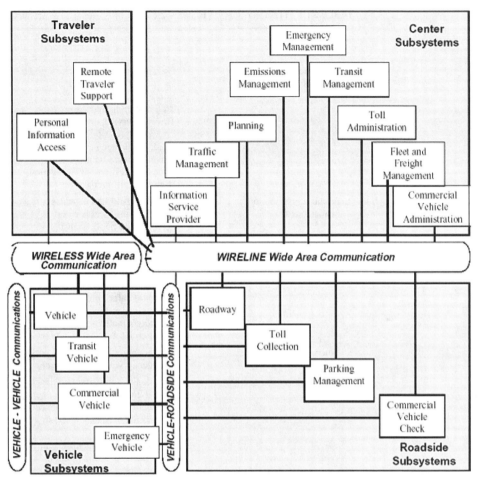

FIGURE 31.1 ITS Architecture subsystems and communication elements.

31.3 PUBLIC TRANSPORTATION MANAGEMENT AND OPERATIONS

Public transportation management and operations involve a variety of day-to-day activities performed by the public transportation agency personnel. Two major activities include fleet management and passenger security (Casey et al. 1998). While the two activities are often viewed as inseparable and interrelated, fleet management tends to focus on the efficient and safe operations of vehicles and passenger security attempts to reduce risk and harm to transit patrons. To facilitate the conduct of public transportation management activities related to fleet management and passenger security, many public transit providers have made investments in public transportation management systems, including the use of the following systems and technologies:

TABLE 31.1 ITS User Service Bundles Included in National ITS Architecture

Bundle	User services
1. Travel and transportation management	• Pretrip travel information • En route driver information • Route guidance • Ride matching and reservation • Traveler services information • Traffic control • Incident management • Travel demand management • Emissions testing and mitigation • Highway-rail intersection
2. Public transportation management	• Public transportation management • Enroute transit information • Personalized public transit • Public travel security
3. Electronic payment system	• Electronic payment services
4. Commercial vehicle operation	• Commercial vehicle electronic clearance • Automated roadside safety inspection • On-board safety monitoring • Commercial vehicle administrative processes • Hazardous material incident response • Commercial fleet management
5. Emergency management	• Emergency notification and personal security • Emergency vehicle management
6. Advanced vehicle safety systems	• Longitudinal collision avoidance • Lateral collision avoidance • Intersection collision avoidance • Vision enhancement for crash avoidance • Safety readiness • Pre-crash restraint deployment • Automated vehicle operation
7. Information management	• Archived data function

• Advanced communication systems
• Automatic vehicle location systems
• In-vehicle diagnostic systems
• Transit operations application software
• Automatic passenger counters

The aim of this section is to review these systems and technologies as they relate to public transportation fleet management and passenger security and their intended system objectives, system architectures, and anticipated benefits. In addition, several system deployments in the United States will be examined and the deployment results and lessons learned will be discussed.

31.3.1 Technologies

In order to understand better the capabilities and limitations of the technologies mentioned above, it is important to consider their functionalities and associated benefits as presented below.

Advanced communication systems (ACS) may have the functionalities of transmitting data, voice, and video. A popular technology still being used to transmit voice between vehicle operators and dispatchers in the public transportation industry in the United States and abroad is the conventional two-way radio.

New technology being employed by some transit operators in the United States to transmit data and text is the radio frequency-based mobile data terminals and cellular digital packet data (CDPD).

Anticipated benefits of these new technologies include easing the strain on the transit operator's existing communication network and the ability to accommodate increased telecommunications needs for both data and voice purposes and perhaps video.

Automated vehicle location (AVL) systems are used to monitor vehicles and track real-time location of vehicles and to transmit this information to a central location (Spring, Collura, and Black 1997). This information has been used for a variety of purposes in U.S. industry, including paratransit scheduling and dispatching (Hardin, Mathias, and Pietrzyk 1996; Stone 1993); passenger information systems; and locating vehicles in case of an emergency (Casey et al. 1998).

To facilitate the integration of AVL data into a comprehensive database for transit planning and operations, the SAE J 1708 standard has been used by the transit industry in the United States. This location referencing message specification (LRMS) describes a set of standard interfaces for the transmission of location references among different components of an advanced public transportation system. LRMS interfaces define standard meanings for the content of location reference messages, and standard, public domain formats for the presentation of location references to application software.

AVL technologies employed in the transit industry in the United States include global positioning systems (GPS), signpost and odometer, ground-based radio navigation and location, and dead reckoning. Each technology has a different set of operational characteristics associated with advantages and disadvantages (Casey et al. 2000). A general trend taking place in the United States is that the transit industry is moving toward the use of GPS-based AVL. In some cases a combination of GPS and dead reckoning has been considered for use where GPS signals cannot be picked up by buses due to the so-called canyon effect resulting from tall buildings in high-density downtown areas.

Anticipated benefits associated with AVL are safety- and service-related. Safety benefits include more timely decisions in emergency situations, quicker response to vehicle mechanical problems, and increases in driver and passenger safety and security. Service related benefits are improvements in dispatching efficiency, route and schedule planning, customer service, schedule adherence, and the collection of passenger information.

In-vehicle diagnostic (IVD) systems perform continuous measurement of vehicle components (e.g., oil pressure, engine temperature, status of electrical system, and tire pressure). In some IVD applications, out-of-tolerance conditions are sent to the dispatcher in real time without driver intervention. Examples of IVD functionalities include in cab diagnostic and fuel economy display, in-vehicle AVL linkage, and automatic data links to bus maintenance bays with the corresponding bus ID, odometer reading, and fuel level. Anticipated benefits of IVDs are quicker notification of mechanical problems and more rapid response to service diagnostics.

Transit operations software (TOS) is used in conjunction with AVC, AVL, and IVD technology and is employed by dispatchers and other transit staff in the delivery of fixed route and paratransit service. For example, in fixed-route bus service computer-aided dispatch (CAD), software is used to facilitate communication between the bus operator and the dispatch center, a function becoming more and more important due to the increase in data transmission that results from the use of AVL, AVC, IVD, and other advanced technologies. CAD applications reduce the amount of voice traffic. TOS applications also help in tracking the on-time status of each vehicle in a fleet, thus assisting operators, dispatchers, and customers by updating them about the location of vehicles. TOS may also be used to reduce bus bunching.

In rail transit, TOS applications may assist in integrating supervisory control and data acquisition systems (SCADA) with other control systems such as automatic train control (ATC), automatic vehicle identification (AVI), traffic signal loop detectors, and automated train dispatch. In addition, TOS systems in rail transit may aid in establishing vehicle location using signal block occupancy and/or AVI transmitters.

In paratransit operations, CAD software can be used for ride matching and assigning customers to demand responsive vehicles operating in a shared-ride, advanced reservation mode. Such CAD software is employed in combination with two-way data and voice communication systems, mapping software, and a global positioning system base station.

Anticipated benefits associated with the use of TOS applications and the other advanced technologies mentioned include improved scheduling and dispatching, more reliable service to the customer, improved efficiency of operations for the agency, increased productivity, and enhanced safety for drivers and passengers.

Automatic passenger counters (*APCs*) estimate the number of passengers boarding or alighting and thus provide data for service planning and operational purposes. Such data may be archived for long-term planning activities. APCs use two different types of technologies—infrared beams and treadle mats—and some APCs are integrated into AVL systems to the location of the passengers boarding and alighting. APC output is transferred in two ways: (1) in an off-line mode, in which data are stored on vehicle and downloaded at the end of the day at the transit center; and (2) in a real-time mode, in which a wireless communication system sends the data to a central computer at the transit center.

Anticipated benefits related to the use of APCs are reductions in data-collection costs, increase in the type and range of data available, decrease in the time to process collected data, and improved service planning.

31.3.2 Public Transportation Management and Operations System Objectives and Requirements

Primary objectives of public transportation management and operations systems designed to enhance fleet management and passenger security are:

1. To improve transit service effectiveness
2. To increase operating efficiency
3. To preserve passenger safety

Examples of system features directed at achieving such objectives include:

- A bus operator can activate a silent alarm discreetly and the dispatcher at the transit center can be alerted regarding a theft or robbery. With the use of a satellite-based automated location system (AVL), the dispatcher can determine vehicle location and then contact the appropriate authorities.
- With the aid of an AVL, a dispatcher knows when and where to send a replacement vehicle in the event another vehicle becomes disabled.
- With the help of in-vehicle diagnostic systems, the dispatcher may send a replacement vehicle before the vehicle becomes disabled.

- Computer-aided dispatch (CAD) software may be used to respond to real-time passenger requests for demand response door-to-door service or for fixed route/route deviation service.
- CAD software may also aid in the development of paratransit routes by assigning passengers and designing schedules, thus improving system efficiency by grouping riders and locating their origins and destinations

Examples of system requirements to achieve the above objectives and features are wireless transmission of either data or voice between the vehicle operator and the dispatcher; direct communication between the dispatcher and emergency response officials; and tracking of vehicles in real time.

31.3.3 Relationship to National System Architecture

Advanced public transportation systems designed for fleet management and passenger security relate to the ITS User service of public transportation management, public travel security, ride matching and reservations, and archived data function, as defined in the National ITS Architecture. It is also useful to think of such systems in terms of their subsystems, as reflected in the National ITS Architecture's so-called interconnect diagram as shown in Figure 31.2. As can be observed, the shaded subsystems include the vehicle, a transit management center, and the transit traveler. On board the vehicle are technologies including the AVL receiver and other components, the silent alarm, and vehicle diagnostic equipment. At the transit center are also AVL components and transit and paratransit operations software and associated computers. A third subsystem may include the traveler interested in scheduling a paratransit trip. Connecting the transit center and the vehicle is some form of wireless communication system. Either wireless or wireline communication provides the link between the traveler and the transit center.

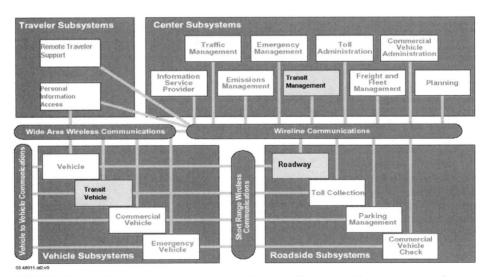

FIGURE 31.2 Architecture subsystem interconnect diagram: public transportation management and operations.

31.3.4 Deployments

Winston-Salem, North Carolina.   The Winston-Salem Transit Authority operates a 22-vehicle paratransit service called Trans-AID (Casey et al. 1998; Spring, Collura, and Black 1997). The service is designed to meet the special needs of elderly and handicapped individuals. The authority uses PC-based computer aided dispatch and scheduling software together with mobile data terminals (MDTs) in the vehicles and a satellite-based AVL system. Results associated with the deployment of these technologies are as follows:

- Operating expenses decreased by 8.5 percent per vehicle mile and by 2.4 percent per passenger trip.
- Customer satisfaction increased by 17.5 percent.
- Client base increased by 100 percent.
- Passenger waiting time decreased by 50 percent.

Santa Clara County, California.   Santa Clara County, located south of the San Francisco area, operates a relatively large paratransit service called Outreach (Casey et al. 1998). The county uses computer-aided scheduling and dispatch software, a digital geographic database, and a satellite-based automated vehicle location system. Results attributed to this technology deployment include:

- An increase in shared rides from 38 to 55 percent
- A decrease in fleet size from 200 to 130 vehicles

31.4   TRAVELER INFORMATION

Traveler information systems provide travelers with information to assist in making travel decisions prior to their departure and/or during their trip (Casey et al. 1998). While some traveler information systems are designed to assist only transit patrons, other systems are more comprehensive and provide information on transit and other local transportation facilities and services including highways, parking garages, and airports.

31.4.1 Technologies

The technologies employed in the design of a traveler information system typically include computer software and hardware and some form of wire or wireless communication system. A traveler information system might be static in that scheduled bus arrival times, for example, are available on a website and accessed via a touch screen in the lobby of a hotel. On the other hand, the system could be dynamic in that it provides such information via a website that provides up-to-date information on transit services with the use of an automated vehicle location (AVL) system, tracks transit vehicles in real time, and provides actual bus arrival times on the touch screen, rather than merely scheduled times. Pretrip information sources include schedules, routes, geographic information, and estimated times of arrival and departure (ETA and ETD). Pretrip access mechanisms include wireline telephone, WWW, wireless devices, kiosks, and display boards. En route information sources include ETA, ETD, and delay information. En route access mechanisms include, for example, cell phones and other wireless devices, bus stop displays, terminal displays, and in-vehicle display boards.

31.4.2 Traveler Information System Objectives and Requirements

A primary objective of traveler information systems is to improve traveler convenience. For example, if individuals learn at work, perhaps via a Web-based dynamic traveler information system, that their afternoon commuter bus is leaving on schedule, then they are able to determine when they must leave the office; if the bus has been delayed, commuters may then decide to stay at work later or perhaps run an errand before getting to the bus station.

Another objective of a traveler information system may be to improve passenger safety and security. For example, if transit riders late in the evening know that the bus from the shopping mall or downtown is leaving on time, then they can plan accordingly. If they know the bus is running behind schedule, they are possibly able to avoid waiting an extended period of time late at night at a bus stop in an insecure location. It should also be noted that traveler information systems may also improve safety and security by facilitating communication between bus and train operators and dispatchers regarding problematic passengers on board transit vehicles and in terminal areas.

Another, more long-term objective often associated with the use of transit traveler information systems is the potential of improving overall ridership (and possibly revenue) by increasing the trip frequency of current riders and attracting new patrons. The underlying assumption is that over time, enhancements in transit service quality, convenience, safety, and security will allow transit to become more competitive as a result of the use of traveler information systems coupled with other transit improvements.

Requirements of static traveler information systems may simply include providing schedule and route information on a website and providing access to the website via wire and wireless communication systems. Dynamic traveler information systems will likely have additional requirements such as tracking locations of vehicles in real time; calculating estimated times of arrivals and departures (ETA and ETD) and comparing these times to scheduled times, and providing forms of communication from the information source (e.g., AVL system), to the information processors, to the website, and ultimately to the user interface and display.

31.4.3 Relationship to National ITS Architecture

Traveler information systems for transit incorporate aspects related primarily to two of the 30 ITS user services included in the National ITS Architecture. One is en route traveler information, which is provided during the trip either in the transit vehicle or at the transit station/stop; the other is pretrip traveler information made available prior to departure to assist in the selection of a mode, route, and departure time.

As depicted in Figures 31.3 and 31.4, several transportation subsystems (e.g., traffic and transit management centers, the vehicle) can be employed to provide information regarding transit and other modes (FHWA 1997). For example, the transit management center might communicate with transit vehicles via a wireless communication system and, with the aid of an AVL system, provide travelers with actual vehicle locations and/or estimated vehicle arrival times, parking availability, and other information via the wireline telephone, cell phones, variable message signs, kiosks, and PCs with an Internet connection. In addition, a traffic management center can provide information on real-time roadway traffic conditions with the use of similar communication systems and devices.

31.4.4 Deployments

Seattle, Washington.   A traveler information system for transit has been designed as an integral part of the Seattle Metropolitan Model Deployment Initiative (MMDI), a comprehensive, multimodal effort to implement a variety of ITS user services in the Seattle regional

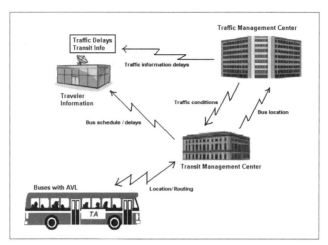

FIGURE 31.3 Traveler information systems.

area (Collura, Chang, and Carter 2002). The Seattle MMDI is a partnership involving government, industry, and academia. The major participants are the U.S. DOT, Washington State DOT (WSDOT), King County Metro, the University of Washington (UW), and industry partners. As presented in Figure 31.5, a distinguishing characteristic of the traveler information system in the Seattle MMDI is that it utilizes the Internet as a major communication medium between the transportation subsystems and the ITS information backbone (I2B).

The I2B, conceptually, is an information system design and specifies how participants (e.g., WSDOT, KC Metro, UW, private information service providers such as Fastline) organize information flows, consisting of gathering, processing, and dissemination steps, from the source to the actual users. From a functional standpoint, the I2B acts as a collector and distributor of information (i.e., it receives information from contributors and redistributes this information from processors, who in turn process the information, add value, and provide the information to travelers). The equipment components of the I2B consist of a series of geographically distributed computers located at the UW, WSDOT, and KC Metro. The com-

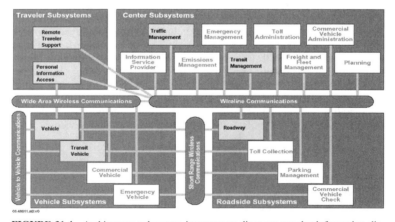

FIGURE 31.4 Architecture subsystem interconnect diagram: traveler information dissemination.

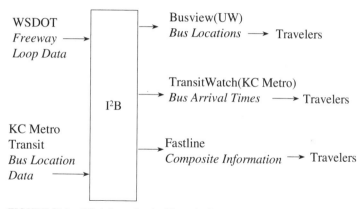

FIGURE 31.5 ITS information backbone in Seattle.

puters host processes that interact and collaborate with each other via the Internet. The administration of the I2B is performed by UW, which also provides software tools for connecting to the I2B.

As shown in Figure 31.6, UW also operates Busview, a service that uses the World Wide Web (WWW) to display KC Metro bus locations graphically. Travelers, prior to their departure, can connect to the Busview website and click on a selected location of KC Metro's service. Icons will appear showing the route and bus numbers, direction of travel, and the time a bus passed that location.

UW and KC Metro also jointly operate Transit Watch, another transit-specific software application accessible on the Web, which displays estimated bus arrival times at key locations as presented in Figure 31.7.

31.5 TRANSIT SIGNAL PRIORITY

Transit signal priority strategies provide buses and light rail vehicles with varying levels of preferential treatment at signalized intersections (ITS America 2002; Collura, Chang, and Gifford 2001). Priority may include an extension of the green interval and/or the truncation of a red interval. Preferential treatments are also provided to emergency vehicles (e.g., fire, ambulance, and/or police), usually in the form of an immediate green interval, often referred to as emergency vehicle signal preemption (McHale and Collura 2003). If such emergency preemption is provided, it is essential that its design and operation be considered in coordination with transit signal priority and that proper signal control transition strategies be included (Obenberger and Collura 2001).

This section deals primarily with the application of transit signal priority for buses on arterial streets. Transit priority system objectives, requirements, and architectures will be reviewed and the results and lessons learned in transit priority system deployments in the United States will be examined.

31.5.1 Technology

A relatively simple transit priority system, as depicted in Figure 31.8, might include a transmitter on board the vehicle that sends a signal to a receiver at the intersection via a short-range wireless communication link.

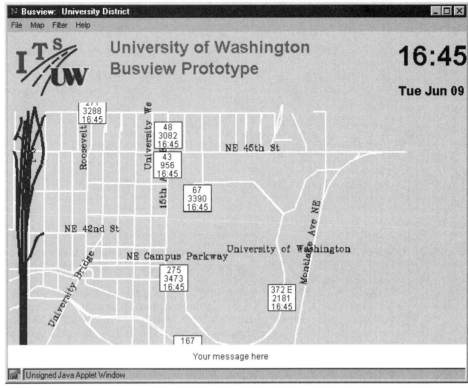

FIGURE 31.6   Busview screen.

The receiver passes on the message to the signal controller equipment, requesting priority. When priority is requested and a green interval exists in the direction of transit vehicle travel, the green interval may be held and extended as needed to allow the transit vehicle to clear the intersection. Should a red interval exist, the transit vehicle may be required to wait; alternatively, the red interval may be truncated by shortening green intervals of other signal phases while still maintaining proper clearance times.

Another, more technologically advanced transit priority system might include a transit management center that communicates via a wireless communication system with all fixed route buses to monitor their locations. Location data are then processed with the aid of an automated vehicle location (AVL) system and used by the transit management center to determine whether the bus is running behind schedule. If so, a message might then be sent to a central processing unit on board the bus to enable the bus then to request priority via a short-range wireless communications link as described above. A variation to this transit priority system might include determining bus lateness on board the vehicle. In addition, another variation might include the transit management center sending a message to the traffic management center for authorization. The traffic management center would accept and process the priority request via a centralized traffic signal control system.

31.5.2   Transit Signal Priority System Objectives and Requirements

An example of a transit priority system objective is to reduce vehicle running times for the purposes of increasing operational efficiency and maintaining scheduled times. Another ob-

Applet Viewer: buslink.tcserver.BusLinkApplet

Applet

Northgate TC

5:02 PM
Tue Jun 9

Route	Destination	Scheduled	At Bay	Depart Status
5	Downtown Seattle	5:12 PM	5	No Info Avail
16	Downtown Seattle	5:02 PM	6	On Time
16	Northgate	5:05 PM	2	1 Min Delay
41	Northgate	5:06 PM	2	No Info Avail
41	Northgate	5:12 PM	2	No Info Avail
62	Ballard	5:04 PM	6	On Time
67	Northgate P & R	5:03 PM	2	On Time
67	UW Campus	5:10 PM	5	On Time
75	University District	5:15 PM	1	On Time
302	Aurora Village	5:05 PM	4	No Info Avail
307	Woodinville P & R	5:00 PM	2	1 Min Delay
307	Downtown Seattle	5:00 PM	5	2 Min Delay

TransitWatch, by Noah and Dan, UW ITS Program

Last update: Tue Jun 09 17:02:07 PDT 1998

FIGURE 31.7   TransitWatch Buslink screen.

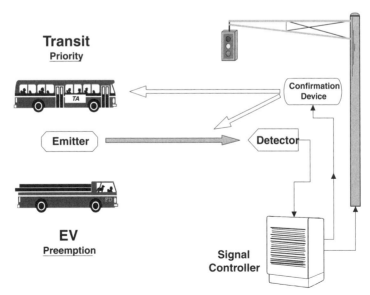

FIGURE 31.8   Transit signal priority systems.

jective might be to control vehicle flow when vehicles along the same route are following one another too closely. This phenomenon, referred to as bunching, is a common problem in some large urban areas along congested routes with short headways.

In order to meet one or more system objectives, transit priority systems are designed to meet certain requirements (Gifford, Pelletiere, and Collura 2001). For example, one requirement might be to detect buses prior to the intersection and to change signal phases in some prescribed manner (e.g., extending the green interval to allow a local bus to clear the intersection) for the purpose improving schedule adherence. Another requirement might be to grant priority selectively to certain buses to eliminate bus bunching and thus maintain proper headways. A third requirement might be to track actual vehicle location and time and compare them to scheduled time so that priority is granted only to buses running behind schedule by some predetermined threshold level. Other transit priority requirements might be to monitor traffic flow rates and/or to count passengers boarding and alighting so that bus priority can be granted depending on threshold levels of traffic congestion and/or bus occupancy.

31.5.3   Relationship to National ITS Architecture

Transit priority relates to two of the 30+ ITS user services as defined in the National ITS Architecture. One is public transportation management, a user service supporting transit operational functions. The other is traffic control, a user service related to managing the movement of traffic, including transit and other vehicles along roadways.

The subsystems for a transit priority system may be more complicated than presented above in Figure 31.8. For example, a transit management center might be required to request priority authorization from a traffic management center. In this instance, the transit management and traffic management centers would likely communicate with one another via a wire line connection. The transit management center subsystem communicates via a wide-area wireless communication system with transit vehicles (the vehicle subsystem); the vehicle communicates with the AVL system; and the transmitter on the vehicle communicates with the detector (the roadside subsystem) via a dedicated short-range communication (DSRC) link.

31.5.4   Deployments

Charlotte, North Carolina.   The City Department of Transportation deployed a transit priority strategy in the mid-1980s for express buses only along a 6-mile segment of a major arterial including some 14 signalized intersections (Casey et al. 2000). Priority requests include green extentions and red truncations in the inbound direction during the AM peak and in the outbound direction during the PM peak.

For safety purposes the priority phase ends after the bus clears the intersection. Also, to assist in reducing motorist confusion, no signal phases are skipped after the transit priority phase is completed. In addition, all nonpriority phases get a minimum green interval in the spirit of recognizing the importance of travel time as perceived by all travelers.

The physical system architecture of the priority system in place in Charlotte resembles the system in Figure 31.8. Essentially, an emitter mounted on the bus sends a strobe light signal to a detector at the intersection, which transmits the request to the signal controller. The emitter, detector, and controller equipment were provided by the 3M company.

Field studies in Charlotte indicate that bus travel times decrease by approximately 4 minutes and that there are no unacceptable cross-street delays. Anecdotal evidence suggests that other possible benefits include less wear and tear on braking systems, fewer rear-end collisions, decreases in idling time and localized emissions, reductions in driver stress, and improvements in traffic flow along the arterial.

The transit priority system in Charlotte is one of the earliest reported systems in the United States. A reason some cite for this early and relatively successful deployment relates to the institutional setting and decision-making structure in place in Charlotte, which includes the City Department of Transportation being responsible for both traffic and transit operations. Consequently, the centralized decision-making structure transportation authority made it easier to get the transit operator and the traffic engineering community to work together in a coordinated and cooperative effort. Such cooperation is essential in the planning and deployment of transit priority and other advanced public transit system applications.

Bremerton, Washington.   The transit priority system in Bremerton was deployed in the mid-1990s at 8 intersections on 40 buses (Casey et al. 2000; Williams, Haselkorn, and Alalusi 1993). The Bremerton system is very similar to the Charlotte system in terms of system objectives, requirements, and architecture. One difference is that phase-skipping is permitted in the Bremerton system.

Studies in Bremerton indicate that bus travel times were reduced by 10 percent, express bus travel times diminished by up to 16 percent, the recovery period back to signal coordination was about 19 seconds, and no significant changes in cross-street delays took place. Other anticipated benefits expected in Bremerton include reduced driver stress and improved morale and less stop-and-go thus leading potentially to an increase in brake life on the order of 20 percent.

Seattle, Washington.   The priority system in Seattle was deployed along two arterials, at 26 intersections, and on over 200 buses (Casey et al. 2000). While the system in Seattle is very similar to the systems in Bremerton and Charlotte in terms of system objectives, there are some differences in system requirements and architecture worth noting. For example, the Seattle system only provides a green extension and no red truncation. The system architecture in Seattle consists of the following subsystems: the in-vehicle subsystem includes a radio frequency (RF)-based tag; a roadside subsystem including pole-mounted antennas and readers which communicate with the vehicle via the RF signal and with the signal controller via a wireline connection; and center subsystems, including the King County transit center, which determine whether the priority request should be granted depending on time status and local traffic conditions established by the traffic management center staff.

Preliminary studies in Seattle indicate that signal-related stops encountered by buses in the AM peak in unsaturated flow conditions (i.e., level of service B) decreased by 50 percent, average stopped delay diminished by 57 percent, bus travel time dropped by 35 percent, and there were no significant increases in side street delays.

31.6   *ELECTRONIC PAYMENTS*

Electronic payment systems (EPSs) in transportation may apply to toll roads, parking, and transit services (Collura and Plotnikov 2001). EPSs may also serve broad nontransportation functions, such as retail applications, telephone services, access systems, medical records, and social programs (E-Squared Engineering 2000; Fleishman et al. 1998). EPSs may also be integrated with credit and debit cards in banking and other financial transactions.

EPSs in transportation are intended to address a variety of issues and problems that may be perceived by either the transit operators or the travelers (Dinning and Collura 1995). Issues and problems often seen by transit operators include costs and liability associated with coin and cash collection; the importance of accurate data collection and reporting; intermodal coordination; flexibility in fare policy implementation; and the need to reduce fare evasion and fraud. Issues and problems seen by transit riders include the need to have exact change; difficulties associated with intermodal transfer and multiple fares; and the desire for a single payment media accepted by various transit agencies, other transportation providers, and retail stores.

31.6.1 Technologies

Electronic payment systems being used on bus and rail transit employ a variety of technologies, such as card media, readers, vending and distribution equipment, computer hardware and software, wireless and wire telecommunication, and a clearinghouse (Hendy 1997).

Card media include magnetic stripe and microprocessor-based applications with varying levels of performance characteristics pertaining to storage capacity, processing speed, and security features. Microprocessor-based cards (also referred to as chip cards or smart cards) are both contact and contactless technologies and provide greater performance in terms of processing speed, storage capacity, and security, including accuracy, confidentiality, impersonality, data integrity, and repudiation. Finally, it should be noted that the card technology industry is currently considering the merits of a combination card that may employ two or more types of card media on one card.

Readers also vary in type. Some readers accommodate only magnetic stripe cards that are swiped manually through a slot in the reader. Other magnetic stripe cards are transported through the reader mechanically with a motorized unit, as is the case with the magnetic stripe card cards currently used on some rail transit systems such as the Washington, DC, METRO and Bay Area Rapid Transit (BART) in San Francisco. A technology trend taking place in the United States and abroad is to deploy contactless smart cards which are read by readers using a wireless communication link. The expectation is that such microprocessor-based and wireless reader technologies will be less costly to maintain than the motorized readers using multiple moving parts. The basic function of the reader is to provide communication between the card and host computer in the electronic payment system. In addition, the reader may validate the card and provide a data processing function.

Vending and distribution technologies are used to assist the transit operator in the sale and distribution of card media to transit riders. Vending machines may be located at rail stations or major bus stop/terminal locations. The initial sale of cards and the process of placing additional value on cards may be carried out with the use of credit cards in person at major locations or on-line via the Internet.

Computer hardware and software and telecommunication systems are also critical in the deployment of the electronic payment system for a number of purposes, including data processing and distribution, database management and recordkeeping, security, and accounting activities such as billing and reconciliation. In the case of electronic payment systems on rail transit, computer systems usually operate in a centralized, on-line mode and data are transferred and processed with wire telecommunication on a dedicated or leased arrangement. For bus transit the card transaction and payment data are captured and stored on readers and transferred to the computer system in a distributive, off-line mode, typically at the end of each day. A thought among designers of electronic payment systems on bus transit is that a wireless transmission link [e.g., wireless fidelity (WIFI)] may someday facilitate the transfer of data from the bus (as it enters the maintenance yard) to the central computer. While the off-line mode has worked to date, it is believed by some transit operators that it may create a greater potential for data loss and fraud and significantly complicate testing, maintenance, and repair of readers on board the bus.

In closing the discussion on EPS technologies, it should be mentioned that some public transit operators and other transportation providers are considering partnering with banks, credit card companies, and other financial institutions in the deployment of electronic payment systems. The primary role of the financial institution is to provide a so-called clearinghouse function, thus possibly reducing the need for the transit operator to make investments in some of the EPS technology and associated activities described above. The clearinghouse function might include, for example, managing the central computer system;

providing encryption-key and personal identification number (PIN) management; authenticating cards; validating transactions; and financial billing and reconciliation. It should also be noted that a third-party clearinghouse might be appropriate to operate an EPS in an open-system environment in which multiple transit operators and possibly other transportation providers (e.g., toll road agencies, parking authorities) are involved. An underlying aim of the inclusion of the clearinghouse is to maintain the integrity of the EPS and ensure its proper use by all the transportation operators and their users. Major questions surrounding the merits of a clearinghouse relate to the financial implications regarding costs and flow of revenues (Ghandforoush, Collura, and Plotnikov 2003; Lovering and Ashmore 2000).

31.6.2 Electronic Payment System Objectives and Requirements

Three commonly stated objectives of electronic payment systems in public transit are to enhance customer convenience and satisfaction, increase revenues, and reduce costs associated with payment collection and processing.

An example of an EPS requirement designed to achieve the above objective regarding customer convenience includes providing the ability to allow the transit customer to reload the value of the payment card with a credit card at a vending machine in the train station or via the Internet with a credit card. Requirements related to the other two objectives might include the use of a payment medium and reader to withstand certain temperature and precipitation levels, to satisfy certain minimum transaction processing speeds, to preserve customer confidentiality, and to ensure data security for fraud protection.

31.6.3 Relationship to National ITS Architecture

Electronic Payment Systems relate to the ITS user service included in the National ITS Architecture referred to as electronic payment services. Figure 31.9 suggests that the subsystems of a multimodal EPS might include the toll administration, the transit management center, and the fleet and freight management center. The toll administration is involved in collecting payments on the toll roads from transit vehicles, commercial vehicle operators, and private vehicles. The transit management center in responsible for collecting the pay-

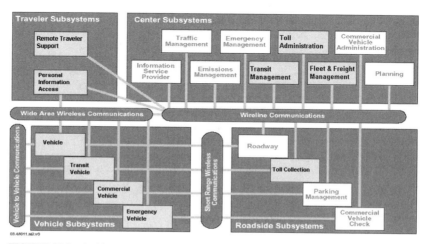

FIGURE 31.9 Architecture subsystem interconnect diagram: electronic payments.

ments on board the buses and in the train stations and park-and-ride parking lots. The fleet and freight management center includes private commercial vehicle operators that have contractual payment agreements with the toll road agencies.

31.6.4 Deployments

Washington, DC.   The Washington Metropolitan Area Transportation Authority (WMATA) deployed a regionwide EPS project in 2001. While the primary objective of the EPS project, called SmarTrip, is to improve customer convenience, other objectives include accommodating seamless intermodal passenger transfers, reducing maintenance and operating costs, increasing reliability and security of the system, preventing fare fraud, and facilitating better data collection.

The SmarTrip electronic payment system consists of a number of subsystems. One subsystem is the transit management center, including the card-issuing administrator (WMATA treasury) and the central database (WMATA host computer). Another subsystem includes roadside components such as the Rail Station Monitoring and Control System (SMACS), the parking lot access system, and the SmarTrip vending machines located in rail stations. A third subsystem includes the traveler who carries the smart card and uses it as a means to pay the rail fare and enter and exit the boarding platform area.

The project involves the use of a contactless smart card to pay for WMATA METRO rail and park-and-ride services. The card costs $5 and can be recharged up to $180 in value. Any purchase over $20 earns the cardholder a 10 percent bonus in fare. A pin number uniquely assigned to each card keeps track of the transactions. The value of the card is reduced according to its usage. Fare cards can also be used as daily, weekly, or monthly pass.

The central database host computer system, where customer accounts are maintained, communicates with a network of computers located at rail stations and parking lots, which in turn controls SmarTrip card readers, express vending machines, faregates, and other system access devices. The network of computers at the stations also maintains system performance log and transaction history prior to passing it to the host.

The SmarTrip card reader facilitates contactless communication between the card and the central computer for transactions. When the card is brought within close proximity (about 3 inches) to the reader, the reader initiates data exchange with the card over a low-power radio frequency (RF). The card contains a plastic body with integrated chips and coil antenna inside. Approximately 150,000 passengers, or more than 50 percent of WMATA's daily ridership, uses the SmarTrip system. WMATA plans to put SmarTrip readers in its buses and to extend SmarTrip coverage to other transit systems in suburban Washington, including DASH (Alexandria, VA), Ride-On (Montgomery County, MD), Virginia Railway Express (VRE), and the MARC trains. Efforts are also underway to explore the implications of including a third-party clearinghouse for the reasons cited above in the technology section.

Chicago, Illinois.   The major objectives of the Chicago Transit Authority's (CTA) Smart Card Customer Pilot Program, initiated in 2001, were to test the ease and acceptance of a smart card for transit customers and to explore economic and operational benefits of this technology (ITS 2000). Potential benefits of primary interest to the CTA related to the possible reduction in maintenance cost of fare-collection equipment and the durability and reliability of the new fare media. It was the nation's first operational, intermodal, multiagency smart card program. Smart cards can be used to pay fares on all CTA buses, at all CTA train stations, and on suburban buses operated by another transit agency.

Like the WMATA SmarTrip card, the smart card in Chicago is a permanent, rechargeable fare card. It is made of plastic and has an embedded computer chip that keeps track of the value of the card. The card can be charged in the vending machines at rail stations and other selected locations. The cost of the card is $5 and the card can be charged to a maximum value of $100. A bonus of $1 is earned for every additional $10. A pin number is assigned to each card and the transactions are kept track using the pin number. Users touch or wave their smart cards in close proximity to readers called Smart Card Touch pads mounted on rail turnstiles and bus fare boxes. The agency pays about $7 to $8 for a smart card versus 5 to 6 cents for a magnetic stripe card. In the long run, smart cards are more cost-effective because they can last for five or more years. If the card is lost or stolen, the CTA will issue a new card for a replacement fee of $5 with the value that was on the card at the time the customer reported the lost or stolen card. A cardholder is allowed to complete a trip even with insufficient value on the card. However, the negative balance will be adjusted during the next card-charging period.

The CTA is also considering partnerships with major financial institutions and is examining the potential of including multiple uses for the smart card, such as food, entertainment, and telecommunications purchases.

31.7   CHALLENGES AND FUTURE DIRECTIONS

While it is clear that some of the initial investments in information and communication technologies have benefited the public transit industry in the United States, there will be challenges along the way as more and more public transit providers consider the deployment of additional technologies in a widespread fashion.

One challenge will relate to securing the capital and operating funds associated with the investment of such technologies. Given that many transit providers operate under very tight fiscal constraints, there will be a continuing shortage of funds to invest in technologies such as new wire and wireless communication systems, computer hardware and software, automated vehicle-location systems, transit priority equipment, and electronic payment technologies.

A strategy transit providers should consider in light of the their fiscal situations is coordinating their plans to make these investments with other transit authorities and public agencies. For example, depending on the availability of capital and operating funds, a transit authority might collaborate with the local political jurisdiction (e.g., the county or city) to procure a radio system to meet the basic data and voice communication needs and provide the wireless communication link required for satellite-based AVL systems. The fixed cost of the radio system may require a smaller investment on the part of the transit authority since other agencies would share in the capital costs and the use of the system.

Other examples in which pooling financial resources might benefit the transit providers are in the areas of transit priority and electronic payment system planning and deployment. For example, transit providers might consider pooling funds with the fire and rescue community to design and invest in systems to deploy transit priority and emergency vehicle preemption strategies. Similarly, smaller suburban transit providers interested in a new electronic fare-collection system may benefit in collaborating with the larger metropolitan area transit provider, the net effect being both a reduction in the funds required from the suburban operators and the eventual establishment of a single, integrated fare-collection system region-wide.

In closing, it is important to stress that the deployment of information technologies is important to assist the transit agencies in the United States in providing their existing riders with continued improvements in transit service quality and safety. Such improvements are critical to the transit industry and, coupled with new and improved infrastructure investments, have the potential to aid the U.S. transit industry in maintaining stability in ridership and improving overall system efficiency and productivity.

31.8 REFERENCES

"CTA Begins Trial of Multi-Agency Smart Card." 2000. *ITS* 10(18).

Casey, R., and J. Collura. 1992. *Evaluation Guidelines for Advanced Public Transportation Systems.* Prepared for U.S. Department of Transportation, FTA, Washington, DC.

Casey, R., L. Labell, J. LoVecchio, R. Ow, J. Royal, J. Schwenk, L. Moniz, E. Carpenter, C. Schweiger, and B. Marks. 1998. *Advanced Public Transportation Systems: The State of the Art Update '98.* repared for U.S. Department of Transportation, FTA, Washington, DC.

Casey, R., L. Labell, L. Moniz, J. Royal, M. Sheehan, T. Sheehan, A. Brown, M. Foy, M. Zirker, C. Schweiger, B. Marks, B. Kaplan, and D. Parker. 2000. *Advanced Public Transportation Systems: The State of the Art Update 2000.* Prepared for U.S. Department of Transportation, FTA, Washington, DC.

Collura, J., and V. Plotnikov. 2001. "Evaluating Electronic Payment Systems in Public Transit." In *Proceedings of the 11th ITS American Annual Meeting,* Miami, June.

Collura, J., J. Chang, and M. Carter. 2002. "Seattle ITS Information Backbone." *Transportation Research Record* 1753.

Collura, J., J. Chang, and J. Gifford. 2000. "Traffic Signal Priority Strategies for Transit: A Review of Selected Experiences in the United States." In *Proceedings of the World Congress on ITS,* Torino, Italy.

Dinning, M., and J. Collura. 1995. "Institutional Issues Concerning the Implementation of Integrated Electronic Payment Systems in Public Transit." In *Proceedings of the Second World Conference on ITS,* Yokohama, Japan.

————. 1996. "Evaluating Payment Systems in Public Transit." Prepared for the 6th ITSA Annual Meeting, Houston.

E-Squared Engineering. 2000. *Introduction to Electronic Payment Systems and Transportation.* Primer, ITA America, September.

Federal Highway Administration (FHWA). 1997. *ITS Deployment Guidance for Transit Systems Technical Edition.* U.S. Department of Transportation, FHWA, Washington, DC.

Fleishman, D., C. Schweiger, D. Lott, and G. Pierlott. 1998. *Multipurpose Transit Payment Media.* TCRP Report 32, National Research Council, Transportation Research Board, Washington, DC.

Ghandforoush, P., J. Collura, and V. Plotnikov. 2003. "Developing a Decision Support System for Evaluating an Investment in Fare Collection Systems in Transit." *Journal of Public Transportation* 6(2).

Gifford, J., D. Pelletiere, and J. Collura. 2001. "Stakeholder Requirements for Traffic Signal Preemption and Priority in the Washington D.C. Region." *Transportation Research Record* 1748.

Goeddel, D. 2000. *Benefits Assessment of Advanced Public Transportation System Technologies, Update 2000.* Prepared for U.S. Department of Transportation, FTA, Washington, DC, November.

Gonzalez, P. 2002. *Building Quality Intelligent Transportation Systems through Systems Engineering.* Prepared for U.S. Department of Transportation, FHWA-OP-02-046, Washington, DC, April.

Hardin, J., R. Mathias, and M. Pietrzyk. 1996. *Automatic Vehicle Location and Paratransit Productivity.* National Urban Transit Institute, September.

Hendy, M. 1997. *Smart Card Security and Applications.* Norwood, MA: Artech House.

ITS America. 2002. *An Overview of Transit Signal Priority.* Washington, DC: ITS America.

Lovering, M., and D. Ashmore. 2000. "Developing the Business Case." *ITS International* (January/February).

McHale, G., and J. Collura. 2001. "Improving the Emergency Vehicle Signal Priority Methodology in the ITS Deployment Analysis System (IDAS)," In *Proceedings of the ITS Congress,* Sydney.

National Transit Institute. 2001. *Advanced Public Transportation Systems: Mobile Showcase One Day Workshop Notebook.* Brunswick, NJ: National Transit Institute.

Obenberger, J., and J. Collura. 2001. "Transition Strategies to Exit Preemption Control: State of the Practice." *Transportation Research Record* 1748.

Spring, C., J. Collura, and K. Black. 1997. "Evaluation of Automatic Vehicle Location Technologies for Paratransit in Small and Medium-Sized Urban Areas." *Journal of Public Transportation* 1(4).

Stone, J. 1993. "Paratransit Scheduling and Dispatching Systems: Overview and Selection Guidelines." In *Proceedings of IVHS America Third Annual Meeting,* April.

Williams, T., M. Haselkorn, and K. Alalusi. 1993. *Impact of Second Priority Signal Preemption on Kitsap Transit and Bremerton Travelers.* Seattle: University of Washington.

CHAPTER 32
PARKING MANAGEMENT

P. Buxton Williams
MPSA Partners, Oak Park, Illinois

Jon Ross
MPSA Partners, Chicago, Illinois

32.1 OVERVIEW

> Go to the village ahead of you, and at once you will find a donkey tied there, with her colt by
> her. Untie them and bring them to me.
> —Matthew 21:2, New International Version

Parking management—providing space where people can safely leave their form of transport (and link to other transportation) while they go about the business of working, shopping, and residing—has been a need of the common citizen for centuries. While the earliest documented reference to anything resembling a parking operation may be the excerpt above (it is unclear whether the village in question received any shekels from owners of donkeys who used the space), government has long faced the challenges posed by parking, and will continue to do so, no matter what vehicles and forms of transportation humans employ.

In modern times, government has discovered many ways to earn revenue for the use of parking spaces, a discovery that has, in turn, spawned a whole industry of hardware, software, management, and support services. Parking and related functions (such as transit, public works, and road/highway construction) are fundamental to the everyday activities of virtually every citizen. And, as the automobile is so central to the American way of life—and as the average American automobile becomes larger with SUVs dominating the landscape—parking as an industry and as a networked function of government faces new challenges.

To the average citizen, parking is like a trip to the dentist or buying an insurance policy. No one wants it, but when you need it, it had better be there for you. Parking is clearly taken for granted by those who use it. But parking, and the services and returns it can generate, are also undervalued and underutilized. Parking management in the bureaucracy of most American cities—including many of the largest cities—is isolated and marginalized, run by midlevel staff in police departments, municipal courts, and parking offices whose expertise is typically not shared with or leveraged across other agencies of government that could use their information to the economic and service benefit of the citizenry. As we will reaffirm later in this chapter, parking management has great potential to contribute to a host of key government activities, in areas from public works and transportation planning to economic development and tourism.

A look at industry statistics and economic indicators makes a solid case for elevating parking management out of its current stealth in most local governments. According to the International Parking Institute, the industry is responsible for more than $26 billion in U.S. economic activity—roughly evenly divided between the public and private sectors. More than a million Americans work in the field. Somewhere around 5 million parking meters are installed in the cities and towns of America. And, with well in excess of 100 million parking spaces in the United States—roughly two-thirds of them off-street—demand for parking is huge and growing.

Revenues from parking—even from parking fines alone—are significant. Even for cities with poor collection rates, there is a lot of money in parking—funds every city needs, regardless of the economic climate. To a megacity like New York, parking fines are worth hundreds of millions of dollars a year; to a top-50 city, revenues are typically in the tens of millions, and to a city of 75,000 people, parking revenues may fall under $1 million annually. But no matter the size of the city, optimizing parking management—in revenue generation and in terms of public safety and public service—is critical to the operation of local government.

For these reasons, this chapter will provide a highly strategic approach to parking management. It will discuss what a truly comprehensive parking management system includes and how it operates and will offer insight on how municipalities can develop and manage parking systems that control the use of on- and off-street parking spaces; optimize revenues, efficiencies, and customer service; and leverage the programs and information generated by a well-run parking management system into benefits across other levels of government.

32.2   SURVEY AFFIRMS MANY PARKING MANAGEMENT OPERATIONS ARE MARGINALIZED

To assess the state of the parking industry and its managers, and to confirm the authors' perspectives as delineated in this chapter, the authors conducted a survey. Surveys were sent to parking managers, traffic enforcement officers, and other city officials with responsibility for parking management in three categories of cities: the top 50 in population in the United States according to the 2000 Census, cities in the next tier of population (approximately 100,000–500,000), and cities in the 50,000–100,000 category. The questions spoke to issuance, enforcement, and collection practices; agencies/departments responsible for parking management and analytics; how parking revenues are generated and distributed; management and technology tools and practices (such as outsourcing); and quality assurance practices and metrics.

Surveys were completed on paper or via Internet, results were tabulated, and key findings were summarized below. Completed surveys were received from about 35 cities, distributed roughly evenly across the three categories of cities denoted above. While this number of responses is not sufficient to produce findings that are statistically significant, they do reveal and reaffirm many of the conclusions put forth by the authors.

32.2.1   Summary of Findings

- Police departments, courts, and public works are the primary or, in many cases, exclusive users and stewards of parking management systems across the United States. This practice generally mitigates against other departments using valuable parking and public safety data to enhance services to the general public and contribute to growth and new business initiatives.

- With few exceptions, revenues generated from parking management operations cover operating expenses and provide additional funds for other initiatives within the parking management system. This practice can stand in the way of directing resources toward other lucrative citation-related revenues (i.e., permitting, false alarm billings, and other fee/fine-based programs administered by municipal courts).

- Most parking management systems use total revenues and collection rates as the standard form of measurement of their efficiency. This suggests that these systems are not operating in a suitably sophisticated manner to provide sufficient analytics to gauge the impact of improvement initiatives.

- Parking management systems across the country are not yet taking full advantage of the Internet, credit cards, and other customer-driven alternatives as payment options. Comprehensive use of such measures is associated with improved collection rates, cash flow, and citizen/customer satisfaction.

- Municipalities are divided on outsourcing of parking management operations (i.e., ticket processing, collections, citation enforcement/issuance, and collections). However, most municipalities with a collection rate below 60 percent (below the minimal industry standard of 70 percent) do not outsource any of their parking management operations. This suggests that expanded outsourcing, when well managed, is likely to contribute to improved operations and efficiencies in parking systems.

- A major form of parking management outsourcing, automated ticket-writing, is on the rise. More cities today use hand-held ticket-writers than they did 3–5 years ago. Consistent with this trend, more municipalities are automating ticket tracking and processing. This translates into improved efficiencies, faster collections, and fewer voidances of citations due to human error.

- Municipalities are reluctant to turn parking operations management over to a parking authority. These entities typically have more independence and flexibility than traditional city-run parking operations and can translate to more efficient management and generation of new revenue sources through economic and business development initiatives.

- Municipalities are increasingly recognizing the value of providing more accessible multimedia information to citizens about parking and related services. This practice contributes to citizen participation and satisfaction, and therefore improved operation and efficiencies.

32.3 *WHAT IS COMPREHENSIVE PARKING MANAGEMENT?*

In a perfect world, citizens would adhere to the rules and regulations, thereby eliminating the need for a parking management system. People would use parking meters for the allotted duration, pay the required amount, and remove their vehicles when their time was finished, thus freeing up the space for other motorists. And there would be no need to develop a system around the issuance of parking tickets, regulation of parking spaces, enforcement of parking laws, and provision of adjudication procedures.

But we do not live in such a world, so we need parking management systems to ensure that citizens have sufficient parking facilities, are accountable for parking violations, and have the opportunity to appeal any real or perceived miscarriage of justice.

In this real world, parking management means more than just providing spaces for parking cars. Truly comprehensive parking management involves a good deal of social engineering and planning because it must thoughtfully address a host of resource management issues: public safety, traffic flow and management, urban planning and design, land use/environmental management, and how to build, manage, and draw adequate revenue from parking operations and services. Fundamentally, a comprehensive parking management system serves the citizen in all these ways. This chapter discusses the major building blocks of comprehensive parking management relevant to a midsize village or a megacity. It discusses key facets of parking management: hardware/software functionality, enforcement measures, customer relations programs, and others. The chapter will also discuss methods for validating and verifying parking management system functions and measuring the impact and return on investment of parking management programs. Using samples and case studies derived from experience and a companion survey (see attachment), this chapter delineates best practices in parking management, including how to integrate these programs into economic development and growth initiatives.

32.4 COMPREHENSIVE PARKING MANAGEMENT—SYSTEMS AND OPERATIONS

Any discussion of parking management systems would be incomplete without making reference to the full spectrum of components these systems include. Parking management involves parking lots, garages and on-street parking; it includes parking meters, signs, booting and towing equipment, lift gates, loop detectors, smart cards, cellular phones, paving equipment, striping and chalking equipment, lighting, and construction; it uses parking tickets, permits, hang tags, envelopes, automatic ticket-writing devices, booths, and kiosks; it requires hardware and software for ticket processing, ticket dispensers, revenue control, and traffic controllers, to name a few. Needless to say, the parking industry is dominated by a wide variety of specialized vendors, manufacturers and service providers. In such a fragmented market, it is difficult to try to determine the origins of parking management as an industry or service without encouraging a serious debate among industry experts.

Generally speaking, a state-of-the-art parking management system should be proactive, responsive, accountable, and effective. It must be built upon a solid foundation of technology and well-conceived business processes. An exemplary system should include several key features:

• A ticket-writing process that minimizes or eliminates data entry errors due to human error such as poor handwriting (if possible, all tickets should be written using automated ticket-writing devices)
• Speedy processing of tickets via data entry or downloading from automated ticket-writing devices
• On-line interface with secretary of state/department of motor vehicles in order to obtain critical information about violators (if possible, in a real-time interface)
• Ability to generate notices (statements of outstanding violations) to violators
• Clearly defined enforcement measures

Additionally, the system must include the capability to:

• Track tickets from issuance through to final resolution.
• Identify violators with multiple vehicles separately from rental or fleet owners.
• Track and associate correspondence directly to the ticket or vehicle owner in question.
• Generate statistical reports that help to measure the effectiveness of the system.
• Provide citizens with multiple payment methods, including telephone and Internet options.
• Allow citizens to appeal decisions in person or electronically.
• Provide interactive Internet access to citizens who wish to make inquiries, complete applications for permits, provide change of address, or make a complaint or comment.
• Encourage payment through enforcement measures such as booting, towing, license suspension, and plate suspension.
• Interface with other city departments (i.e., public works agencies) to track damaged meters and signs and some state departments (such as motor vehicle administrations).

- Generate standard and ad hoc correspondence letters to citizens regarding inquiries, complaints, and general ticket status.
- Maintain multiple addresses in addition to a default address for violators.
- Furnish citizens with information about available permit spaces in city-run garages.

An oversimplified view of a parking management system can be reflected in a small spreadsheet showing the number of tickets issued, the violation committed, and whether payment has been received. This approach—one put in place in many smaller cities at the outset of their parking management programs, and still in operation in a good number of jurisdictions—enables a small village to determine at least the number of tickets being used, when to reorder tickets, and how much is collected from parking offenders.

Technological advances, double-digit population growth, and the need to become proactive to citizen needs and more responsive to citizen demands have driven local governments to initiate more robust parking management system solutions. Some of these solutions include document imaging, automated ticket-writing devices, wireless interfaces, booting and towing, workflow processes, interactive voice response (IVR) applications, e-commerce, webcasting, and use of the Internet for queries, appeals, and payments. Moreover, these solutions are flexible and scalable and allow processing of millions of tickets in nanoseconds.

Figures 32.1a and 32.1b use a citation flow diagram to show at a glance sample parking management system activities over a period of time. In the diagram, the horizontal lines with their respective labels represent specific stages in the life cycle of a citation. The vertical arrows represent the movement of a citation from one stage to the next over a period of time. For example, the first arrow from ticket issuance to data entry represents the change in status of the ticket from being issued to being entered in the system. The arrowhead indicates the direction of flow of the activity.

Figures 32.1a and 32.1b represent these two extremes of parking management system solutions. (Most systems are, of course, somewhere in the middle in terms of their complexity.) Figure 32.1a is a simple approach where the need for high-volume processing, fast-paced decision-making, and intricate data interfaces is not present. In this scenario, tickets are issued and sent for data entry. At times, tickets are paid before they are entered in the system. Some tickets are paid after data entry. Within a specified timeframe, violation notices are sent out for those tickets remaining unpaid after data entry. Later still, tickets remaining unpaid are sent to a collection agent. (Of course, tickets may be paid at any time during the process, as depicted with the arrows going from all other lines/stages to the payment line/stage.) While this is a simplified process, it should be noted that there are municipalities that are not even at this level of processing and would need to mount a considerable effort to move up to this baseline.

Figure 32.1b provides a more sophisticated approach to parking management. This option is typically associated with larger cities that have a greater focus on parking as a function that enables economic development, improves public safety, and allows transportation engineers to gauge the impact of changes in traffic patterns. This type of system typically includes features such as ticket imaging, workflow process, interactive voice response (IVR), Internet interfaces, automated ticket-writing devices, e-commerce links, telephone inquiries, and citizen-friendly payment options. A system of this nature normally provides interactive access for citizens to make and track an appeal, request information, submit complaints or comments, and view transactions against their respective accounts. Also, a system with such a high level of functionality usually contains features and functions to allow easy management and control of city parking lots, garages, meters, and signs.

32.4.1 "Build-to-Suit" Parking Management Options

When trying to determine which parking management option or infrastructure is best for a given municipality, the public administrator has a variety of solution alternatives to consider.

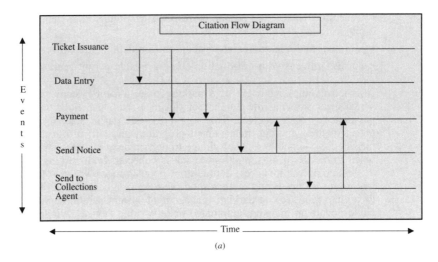

(a)

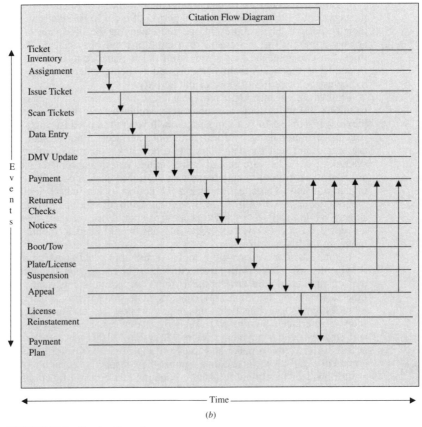

(b)

FIGURE 32.1 Citation flow diagram.

Typically, cities opt for an in-house alternative, which typically is cheaper but can be plagued by operational limitations and their associated costs. Some available alternatives worth considering include:

- *Full-scale facilities management:* In this approach, the parking management operation is turned over to a private operator, who assumes responsibility for all parking management functions and personnel, including ticket-writing officers. Police officers continue to issue tickets but on a smaller scale since the bulk of the ticketing is covered by the private operator's employees. In some cases, the ticket-writing officers are managed by a third party to avoid conflict of interest.
- *Functional outsourcing:* This is a variation of the full-scale facilities management approach. Instead of turning all functions over to a single operator, the municipality determines the parking management functions best suited for outsourcing and turns those over to private operators or even internal departments (such as an information technology department).
- *Software application purchase:* This option involves using technology and software to supplement an internal parking management operation and proves useful to many municipalities.
- *Private-public partnerships, revenue-sharing agreements, and privatization:* These options are becoming increasingly attractive to the public sector in the era of "reinventing government" programs, where privatization in particular has proven useful in delivering some infrastructure and public works functions. These three options have some significant application to parking and related areas, and because a private-public partnership usually includes a revenue-sharing agreement and a privatization element, they are necessarily linked for the purpose of this chapter. In a typical arrangement of this kind, an IT services company assumes the operation of the parking management system—possibly in a privatized or semiprivatized structure—for a fixed fee while the municipality handles all other functions. Although the private partner is responsible for the management of the technology solution, the citizenry hardly ever sees its staff, because municipal personnel handle all customer-facing activities. New revenues generated through greater efficiency of operation or from new programs may be split between the local government and the private contractor.

Of course, there are arguments for and against these approaches, depending on the municipality's perspectives and public/political sentiment. Full-scale facilities management is a boon to the municipality that stresses its core competencies of enhanced government-to-citizen services. All parking management issues are handled by the facilities management company, from a defective PC to a complaint about a parking meter malfunction. In a five-year agreement, for example (a standard term for a deal of this nature), the municipality has a fixed budget for this function, which means a consistent line item and a relief for many department heads. A properly negotiated contract will shield the municipality from the expenses of technology refreshment in order to keep pace with technology advancement, a responsibility of the facilities management contractor.

There are some potential downsides to this approach, and municipalities should perform careful due diligence before selecting a vendor for a facilities management arrangement. A local government should make sure the vendor is financially solid; if not, placing too much of the parking management operation in the hands of an unstable organization can be problematic. Also, cities should make sure they enforce sufficient checks and balances to prevent the contractor from abusing its position by harassing citizens with an inordinate amount of parking tickets. The municipality should also establish, through its contract, that timely and accurate analytical information will be provided. This information may be used to measure the performance of the contractor.

Functional outsourcing requires that the municipality conduct a thorough evaluation of its capabilities and expertise in order to determine which functions are best suited for outsourcing. With this approach, a municipality is able to nurture and develop its employees who will support the contractor responsible for the outsourced function. Also, this approach functions as a cost-containment measure because its cost is fixed over the contracted period. A negative impact of this approach is the possibility that a municipality may find itself overly dependent on a private contractor for key services. A typical example of functional outsourcing is the use of a company to manage meter installations, maintenance, and collections, with ticket processing provided in-house.

Similar to functional outsourcing, software application purchase requires an evaluation of the function for which the software application is required. Today, software application purchases with specific and refined procedures are commonplace in municipal purchasing departments. Following these procedures results in cost savings down the road. As in the other options addressed in this section, the right software application hinges on having an in-depth knowledge of the business process and its interfaces and impact on other departments. This approach puts responsibility for system and software availability onto the software contractor. Also, system upgrades and maintenance are left to the vendor, freeing municipal staff to focus on their duties and reducing software applications to just a tool. When considering this approach, cities need to account for additional training as the software vendor provides upgrades. Depending on the scope of the upgrade, training may require days as opposed to hours, and therefore demands coordination of schedules to limit service disruptions.

Private-public partnerships, revenue-sharing agreements, and privatization are innovative means of involving private-sector partners in efficiently and quickly finding cost-saving measures and alternate revenue streams. These options tend to move more quickly than other business models because the administrative and business processes are typically given legal authority to move faster and conduct business in a more enterprising way. (Parking authorities, for example, often operate in this fashion. Authorities are generally given more leeway in terms of taxing, land acquisition, and business transactions than traditional public-run governmental entities. The Miami Parking Authority, for example, is responsible for management of at least one mixed-use commercial building, in addition to public parking facilities throughout the city.)

In this approach, private partners receive additional payments based on a predefined percentage of savings and revenues they bring to the system. A plus is that contractors are paid for performance and therefore actively seek to increase revenues. For its part, the municipality typically has fewer burdens in terms of man-hours and hands-on activities and can take advantage of the economic benefits associated with privatized programs and public-private partnerships. City officials, though, should make sure not to give too much away in forging these kinds of business arrangements. Most business models of this type can be conducted so that local government maintains ownership of the land and infrastructure of its parking operation while turning over management of the system—and development of new programs, services, and revenue streams—to private partners.

32.4.2   System Planning

Planning for a parking management system is futile without first defining what and how much should be included. Depending on the size and complexity of the jurisdiction, a parking management system may be as simple as controlling and processing tickets issued for expired meters at local parking facilities. It may include controlling and processing tickets for expired meters, handicapped parking, and illegal parking; issuing residential permits; tracking citizen correspondence; conducting adjudication; performing collections activities; providing alternative payment methods; managing enforcement methods (booting, towing, and license/plate suspension; interfacing with the Internet; and providing alternative methods for citizens to obtain system information.

Whether the system is simple or complex, one critical area of concern is the municipality's responsibility to the citizen. The citizen, as the ultimate recipient of the service and its penalties, must be provided with every opportunity to provide input. Of course, the more complex the system becomes, the more need for closer management and control.

But handling a simple or complex parking management system is not necessarily a daunting task and, in fact, has proven to be quite enjoyable work, if you ask some parking directors. These leaders in the field are well informed about the value of the various industry associations, conventions, and publications, resources that are valuable to any parking professional and worth at least an initial review.

The most prominent association is the International Parking Institute (IPI; www.parking. org), which boasts members across the globe but holds its annual conventions primarily in North America. IPI publishes an annual directory of statistics and best practices about the industry (some are documented in this chapter). On a smaller scale is the Parking Industry Exhibition (PIE; www.parkingtoday.com), another group that brings together industry vendors and buyers. One other group, a network composed of industry vendors and service providers, is Expo1000 (www.expo1000.com/parking). All of these industry groups have regular conferences and seminars and publish reports and provide industry information for players throughout the parking field.

Other sources of useful information and networking include regional and state parking associations and municipal leagues, which offer memberships to parking industry leaders as well as other ranking municipal employees. Most hold an annual convention of some type. Many of their activities are tied in part to other government-related groups in public works, traffic engineering, planning/development, and other fields.

Most parking associations publish their own guides listing vendors and service providers, with *Parking Today* (published by PIE) and *The Parking Professional* (published by IPI) being the most reputable monthly industry publications. These provide interesting discussions on parking issues and general industry developments. While not specific to the parking industry, *Government Technology* magazine is also worth mentioning as a source for information on relevant industry issues. GT provides a broader spectrum of articles dealing with state and local governments and how they resolve social issues and enhance government-to-citizen relations.

32.4.3 Vendor Management—Tools of the Trade

An overview of parking management systems would not be complete without a discussion of vendor management, especially considering the large number of vendors who provide products and services to the industry. Municipal parking directors often find themselves unwittingly involved in partnerships with parking vendors of varying stripes. A successful parking management system requires willing participation from all vendors, along with consistent and reliable performance. To achieve this level of satisfactory performance, the parking director must pay more than lip service to vendor management. The parking director must nurture and develop a strong individual relationship with his or her vendors. He or she must include them in the initial stages of planning, seek their input on problems facing his or her department, know their product or service offerings, and be willing to trust their industry knowledge and expertise while still ensuring that they adhere to strict standards and milestones consistent with the jurisdiction and the contract.

32.5 *FUNCTIONAL PARKING MANAGEMENT REQUIREMENTS*

Parking management system solutions must take into account a wide variety of factors with direct impact on program success. These factors are numerous: placement and maintenance of signs and meters; coordination of on- and off-street parking facilities; partnering with private parking facility operators to help regulate traffic flow and enhance public safety; appropriate use of technology and supporting business practices; proper training of personnel to ensure proficiency in providing services to the general public; rigid audits and controls to ensure that checks and balances are performed as required to maintain a high degree of system transparency; and a continuous improvement program to keep pace with industry and technology advancements.

A comprehensive parking management system should begin with a solid infrastructure plan that details current operation components. This plan should identify areas of strengths and weaknesses, outstanding issues and concerns, departments that provide and receive system information, detailed documentation of the information in the system, the format and method of delivery of that information, and barriers to success. It should also project goals and objectives of the parking operation over a longer term (5 years minimally; 10 years optimally), including a general outline of initiatives that must be implemented to satisfy those goals and objectives. Also, clearly defined measurement criteria should be established in the plan so that the progress, improvements, and failures can be charted, evaluated, validated, verified, and acknowledged as lessons learned.

The process of developing the infrastructure plan gives the municipality a realistic view of the current program, a clear understanding of its services and how they affect the general public, its vendors, suppliers, and other municipalities or state departments with which it interfaces. Internally, this exercise reveals difficult issues and concerns among staff members and between departments. It uncovers artificial boundaries and needless activities that have lingered long after they have become obsolete due to procedural changes or system improvements. Upon completion of the infrastructure plan, the municipality will have a clear view of required steps to move forward. It may require major overhauling of the current system, including replacing applications, outsourcing some functions, business process reengineering, adding new technology, or turning over the entire operation to an outside contractor. Whatever the outcome, the municipality will have a thorough understanding and solid business process foundation in determining how to proceed in implementing its plan.

With an infrastructure plan that has the support and commitment of the core parking management officers and staff, what requirements must a comprehensive parking management solution now fulfill? Beginning with ticket control and issuance, it is imperative that the system be able to track all tickets from inventory through to resolution, whether tickets are written manually or by automated ticket-writing devices. Controls should be put in place to record and monitor the movement of tickets to and from the ticket inventory. Assignment of tickets to officers should be controlled to ensure that only the officer to whom the tickets were assigned can write those tickets. For manual tickets, this is accomplished by enforcing the requirement that the officer sign each ticket issued. (This is less of a problem for tickets written using automated ticket-writing devices, because the devices contain built-in features to facilitate ticket control and monitoring.) Tickets issued by the assigned officer should be entered into the system so that the officer data are retained as part of the profile, providing an internal mechanism to track the ticket and officer at every ticket-processing stage.

Once the ticket is issued and entered into the system, a robust ticket-processing application is needed to allow for proper tracking of tickets and correspondence with violators, aging of tickets, and generation of appropriate notices, fines and penalties. This application must also apply partial or full payments via Internet, telephone (by interactive voice response), mail-in, and in person; conduct on-line adjudication; provide interfaces to other departments and systems; and modify ticket-processing activities based upon external events and internal triggers. Also, this application must generate standard management, operations, and executive analysis reports as well as standard letters to inform citizens of the status of queries made against the system and to provide basic information. Additionally, the application must be able to generate ad hoc reports and customized letters.

Other features of the application should include the ability to:

- Obtain in-state and out-of-state license information.
- Furnish information about the parking management program to citizens via the Internet.
- Allow citizens to conduct inquiries, file complaints, and appeal against judgments via the Internet using ticket number and supporting data to ensure privacy and confidentiality.
- Manage information flow to departments such as public works, courts, city clerk, state department of motor vehicles, finance, and police.
- Accept payments for tickets issued but not yet entered into the system.
- Apply penalties for returned checks and halt, suspend or change processing of tickets in this category.
- Generate detailed transaction history by ticket, violator, plate number and account.
- Provide lists of violators whose vehicles are eligible for towing or booting.
- Facilitate suspension of license and denial of plate registration based upon a set level of outstanding fines.
- Monitor, control, and issue permits and renewals for city-owned garages via walk-in or Internet transactions.
- Assist the management of off- and on-street parking facilities.
- Provide adjudication officers with full view of ticket images and transaction history of each ticket.
- Generate court dockets and supporting correspondence to aid in the adjudication process.
- Manage maintenance of meters and signs.

An end-to-end view of the functional requirements of an automated parking management system for a mid- to large-sized city is depicted in Figure 32.2.

Typically, a system as depicted in Figure 32.2 would be implemented by a city with an annual ticket volume greater than 1½ million. Such a city would have the support infrastructure to facilitate the functions, features, and processes described without taking a large toll on other city departments. The automated workflow process enables the city to control ticket activities from inventory update through to data entry, payment, and on-line adjudication. Tickets and appropriate ticket images are moved to different work pools as ticket status and timing dictate. City personnel are able to move in and out of the different pools performing the necessary activities on the tickets in their pool. As this takes place, supervisors monitor the work pools and make corrections or reassign jobs to balance the workload and smooth out the process.

The ticket data entry process is necessary only for handwritten tickets. To improve ticket data processing accuracy and integrity, automated ticket-writing devices are recommended. Handwritten tickets should be kept to a bare minimum and eliminated whenever possible. To this end, a few cities have installed in-car devices for car patrol officers and handheld devices for foot patrol officers. This certainly increases the accuracy and effectiveness of the overall parking management system while reducing the need for ticket imaging.

The on-line adjudication process enables adjudicators to view all ticket activities, from inventory control up to and including the adjudication results. This puts appropriate information at the adjudicators' fingertips and expedites the process. Providing ticket information to users over the Internet gives citizens the opportunity to conduct interactive transactions without having to visit city hall or local parking management offices. Having online interfaces with external systems provides an added data security measure and in most cases reduces the need for redundant data input, increases data integrity, and expedites the overall process.

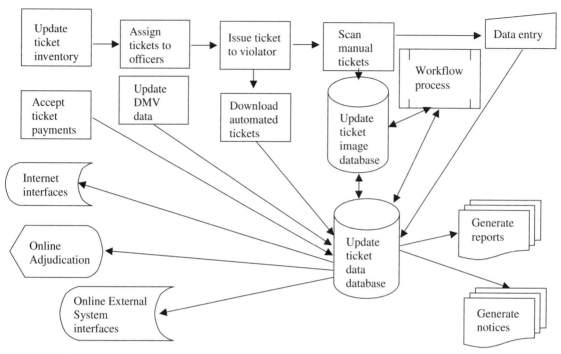

FIGURE 32.2

The reporting process in this system normally generates a large variety of analytical reports tailored to assisting managers and supervisors monitor the vital signs of the system to ensure increased productivity, improved data accuracy, and enhanced customer service. A few refined operational reports provide the necessary audit trails, system logs, and job statistics to ensure that the system achieves a high degree of transparency.

Of course, not all cities can afford to implement all the functions, features, and processes described in this approach. However, cities may use this example as a foundation for their plans and use the pieces that are feasible, then apply them to their specific situation. (This is beginning to take place at higher rates and in smaller cities. For example, a small Chicago suburb has eliminated handwritten tickets altogether by equipping all patrol cars with in-car ticket-writing devices and providing handheld ticket-writing devices to other officers.) Still, functions like interactive Internet processing, on-line adjudication, workflow process and online external system interfaces are deemed rather expensive measures for many cities to implement at this time.

Smaller municipalities with less need for automation, smaller ticket volume, and fewer interfaces will not have a need for such an extensive system. In fact, the smaller volume of tickets, and therefore revenue, precludes the use of expensive ticket-handling processes. It is important to remember that no city wishes to take on a parking management system that is going to lose money, no matter how noble the cause. Therefore, it is imperative that the municipality keep a tight rein on matters by sticking as closely as possible to its original infrastructure plan.

For most cities—those processing fewer than a million tickets annually—an efficient, citizen-responsive parking management system would maintain a number of core features:

- Ticket inventory
- Ticket assignment
- Ticket issuance procedures (for parking enforcement officers)
- Data entry of manual tickets
- Downloading of automated tickets
- DMV updates
- Payment processing
- Adjudication
- Internet interface
- Limited external system interfaces as warranted
- Small volume of analytical reports, audit-trail reports, system logs, and job statistics as necessary
- Notices
- Enforcement measures (i.e., booting, towing, plate denial, or license suspension)

Adherence to federal and state laws and regulations governing cities and municipalities is among the requirements that must be discussed for municipalities of any size. These requirements typically include regulations stipulating conditions that must be met in order to boot or tow a vehicle, suspend a driver's license, deny plate renewal registration, or escalate parking fines. Federal and state laws may also govern adjudication and determine what constitutes a proper mailing address and who must provide that address. Each municipality must conduct its own research to determine what federal and state laws and regulations apply to its particular parking management solution strategy, then enact appropriate laws to activate them.

32.6 GROWING THE PARKING MANAGEMENT SYSTEM INTO A COMPREHENSIVE CITATION MANAGEMENT SOLUTION

With a comprehensive infrastructure plan in place, a municipality can now begin to look at growing the parking management system into a centralized citation management system that incorporates processing activities for all citations and sets the stage for supporting economic development programs. The reasons why a city should integrate its parking and citation management programs are numerous and evident.

For one, IT integration of multiple, compatible government services and systems adds operational and bureaucratic efficiencies. There is less redundancy of effort and more integration of systems, leading to smoother operation and time and cost savings. Second, smart systems that unite back-office management of programs run by a common department (for example, a municipal court, which typically administers fine- or fee-based citations) streamline operation and enhance enterprise-wide systems management. Finally, it makes money—in raw revenues and by simplification and limiting duplication of effort.

The strategy for this approach should be established in the infrastructure plan so that the best suitable application and database designs are selected to accommodate this move. From the outset, a parking management system should be designed with flexibility, scalability, and a limited amount of proprietary software to allow ease of growth. Also, the system should be sufficiently robust to provide users with all the tools to perform operational responsibilities with an acceptable level of proficiency.

Certain functions are considered standard for any parking management operation and must be fully functional before moving into a centralized citation management system. These include enforcement measures such as:

- Ticket issuance, noticing, and warnings
- Booting and towing
- Driver's license and plate suspension
- Collection methods
- Adjudication procedures

Standard processing functions that must be accounted for in a basic parking management system being considered for expansion into a centralized citation management system include payment processing and interface with the department of motor vehicles and local police department. The parking management system should also accommodate changes resulting from federal and state regulations that must be taken into consideration. Again, the municipality may need to enact local laws to enforce these regulations. The municipality's legal counsel should be able to provide proper direction on this matter.

Once the groundwork has been laid to move forward with building a centralized citation management system, the municipality needs to evaluate the available programs in order to determine their suitability, the priority in which they will be added to the system, and how best to implement them. Basically all municipal programs that may generate a citation should be considered reasonable candidates. These include processing of:

- Permits
- False alarm violations
- City stickers
- Traffic violations
- Railroad at-grade crossing violations
- Railroad at-grade traffic interference violations
- Other city ordinances for which citations are issued

The City of Memphis and its Traffic Violations Bureau are a good example of how a parking management system can move into a comprehensive citation management program. The City of Memphis issues and manages 400,000-plus citations worth more than $12 million annually. To enhance this operation, it established a Traffic Violations Bureau (TVB) to manage and control all citations for traffic, parking, arrests, and other municipal ordinance violations. At the same time, the police department committed to eliminating handwritten tickets by placing automated ticket-writing devices in squad cars and using wireless technology to interface with the existing ticket processing system. The advanced nature of this initiative required that the existing ticket processing system be rewritten or replaced by a more comprehensive solution.

The City wanted to continue to maintain the operation and was not willing to enter into a transaction-based agreement with a ticket-processing vendor. Also, the City wanted to phase in the use of automated ticket-writing devices over time, starting with in-car units. Faced with a short implementation timeframe, limited resources, and a desire to implement a flexible and scaleable solution, the City embarked on a business process management solution, which included:

- Assessment of the current situation
- Identification, qualification, and selection of appropriate vendors
- Development of business requirements/objectives

- Review of entire business process and determination of implementation plan
- Scheduling and managing all phases of the request for proposal (RFP) process
- Preparation, development, and distribution of the RFP
- Review and selection of candidates for the bidding process
- Organization, definition, and facilitation of the bidders' conference
- Evaluation of bidder responses
- Development of recommendations for a final vendor

The City executed this strategy on a fast track over a 100-day period. The process was coordinated so that City operations were not unnecessarily interrupted or delayed. Extensive meetings were conducted with key stakeholders, including the City Court Clerk, senior officers of the City Court Clerk office, and supervisors and senior staff of the police department, in addition to the City's information technology outsource partner.

With this background and analysis in hand, the City then applied TVB's business requirements to development of the RFP and identified an initial field of vendors for consideration for bid. More than a dozen vendors were initially considered, but were quickly winnowed to a short list of candidates, who were forwarded to the RFP. From there, a preferred vendor was recommended and selected. The City plans to apply this business process approach to integrate all citation-related functions under one comprehensive citation management system.

32.6.1 Parking Management as a Foundation for Economic Development

As depicted in the hot-air balloon graphic in Figure 32.3, a comprehensive parking management system can serve as a foundation that enables the municipality to grow into a centralized citation management system and then use those roots to link with larger-scale planning and economic development programs. All of these can be incorporated into a properly developed parking management system without conducting a major overhaul of the program, assuming that the program was designed with sufficient modularity to allow existing components to remain unchanged. Therefore, adding permitting to a parking management system requires incorporating modules for a range of applications:

- Permit application, which should interface with the parking database to determine if there are outstanding parking fines and may interrogate other municipal databases to determine if the applicant has other outstanding debts with the municipality
- Permit tracking, which tracks permits issued and to whom and identifies renewals, which may be generated as part of the parking management system's noticing module
- Permit issuance, which prints permits for mailing to the approved applicant

Similarly, application modules are required for false alarm billing and city sticker registration/issuance. With careful planning these may be incorporated in the permit application module. The characteristics of false alarm billing citations—issuances applied to citizens or business whose home or office security systems improperly go off, unnecessarily requiring fire or police response—may be accommodated in the system by providing the necessary processing methods for citations that are marked as such. No special payment processing method is required for these. City stickers—annual fees levied by municipalities on vehicles whose owners reside in the municipality—are basically vehicle permits and may be treated the same way.

Traffic violations need very little explanation, as this is another citation with far more information than required for a parking ticket, and so it can be included in a centralized citation management system. Typically, only a separate data entry procedure is required for

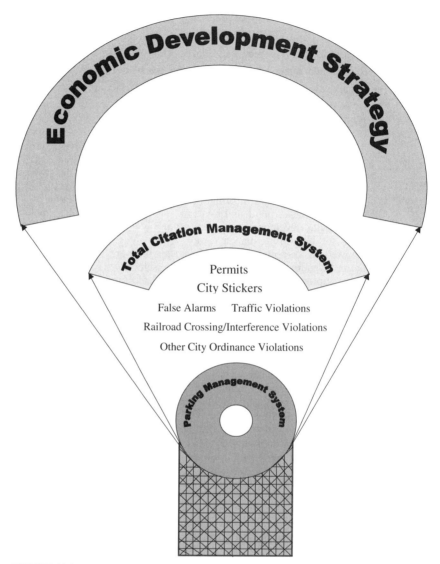

FIGURE 32.3

these violations because in most municipalities courts handle these violations. So the system must be able to generate appropriate court dockets and other court-specific documents to facilitate court processing. This does not mean the system should take on processing functions better handled by a courts or justice management system; the citation management system will restrict itself to accepting the traffic violations, generating appropriate court documents, and accepting payments as directed by the courts.

Adding other municipal ordinances for which citations are issued to the centralized citation management program should not require major efforts. For example, adding noise, dumping, or building code violations would require some modification in the data entry module to accommodate these citations and, similar to traffic violations, would require a module to facilitate court processing. Depending on municipal laws, additional processing may be required. However, in all situations, in-depth analysis of the business processes must be conducted in order to ensure that the new system accommodates all business requirements.

In all situations described above, no special equipment is necessary to facilitate the addition of the business processes. That is the primary difference between these processes and the railroad violations—at-grade crossing violations and at-grade traffic interference violations.

At-grade crossing violations require installation of unique equipment to detect when a vehicle evades level crossing gates and must be able to capture the license plate of the vehicle as well as the date and time of the infraction. Also, the system must be able to reprint pictures of the infraction for mailing to the violator.

At-grade traffic interference violations require similar technology to capture the date and time of infraction as well as identify the railroad company whose train was in violation of the statutes. At this time, there are not many vendors in the marketplace that provide this technology, but there is sufficient competition to keep the prices affordable. (It should be noted that untapped revenues from at-grade traffic interference violations are potentially huge; one small city in the Midwest, for example, by instituting a relatively low-technology citation management system for at-grade crossings, generated $8 million in citation revenues, even after negotiating with several freight companies to simplify issuance/dispensation of tickets.) With the technology in place, the centralized citation management system needs to have a module that accepts the citations from the railroad violations system and process them according to federal, state, and local statutes.

Municipalities wishing to proceed with this action must first determine whether state statute allows them the right to charge railroad companies for this infraction and what is the procedure by which to do so. This process is not uniform across all states. In Illinois, for example, law allows only municipalities that reside in counties of a million or more in population to issue such citations against freight companies.

As each new type of violation is added to the centralized citation management system, expertise and proficiencies should also improve reducing the degree of difficulty for each. One benefit of this approach is a more robust and maintainable system that allows the municipality to streamline operations, reduce costs, and improve services. In the end, the totality of these initiatives allows municipalities to build the groundwork for broadening the reach of citation management into larger-scale planning and economic development plans.

32.6.2 Parking Management, Infrastructure Planning, and Economic Growth

So how is parking connected to the bigger world of planning and economic development? The answer is one word: money. State and local governments are in a period of nearly unmatched deficits and debt. According to the U.S. Department of Commerce, 1 in every 10 dollars budgeted by state and local government in 2002 was borrowed. During a flat economy, their increase in spending—on welfare, health, and education as well as on public works, public safety, and homeland security, all high priorities—outpaced their increase in revenues by nearly a third in 2002. It is the highest proportion of debt in state and local governments since the high-growth 1950s, when infrastructure fought to keep pace with the Baby Boom.

High federal deficits—forecast at $300+ billion well into the mid-00s—mean pressure on state and local government to provide services and balance budgets without raising taxes. Tax increases are never popular, especially in a down economy, which is why local governments raised taxes only 1.2 percent in 2002. Many states and cities face big deficits, even as they must provide key social and public services that will likely require them to spend more, not less. As debt increases, access to capital becomes more difficult. The most critical indicator of public finance—credit rating—has shown a downward trend. Six states (including California, New Jersey, Colorado, and Wisconsin) have seen their credit rating (as measured by Standard & Poor's) drop since 2001, and many major cities are in the same position.

So that means local governments must be particularly innovative in bringing new revenues into their coffers—not just to cut debt and balance budgets, but to be competitive by maintaining services and infrastructure that prevent citizens and businesses from going elsewhere. Parking and related programs are a major area of opportunity for government to capture new revenues, be more citizen-responsive, spur economic development, and generate maximum return on investment without raising taxes.

A look at citation management alone proves the point. A typical growth city or suburb of 50,000 that has a collection rate on parking tickets below the industry standard of 70 percent realizes net returns of $250,000 or more if improvements in its parking management system increase ticket collections by 15–20 percent. Add in enhanced efficiencies and better cash flow that result from improved business processes, and the city can realize another $100,000 or more. For a larger city, management of a comprehensive parking operation can be worth millions; for a major city, tens of millions—from parking operations alone.

Optimized parking management and infrastructure programs do not only generate revenues to pay for expanded public safety, emergency, and public works programs. Infrastructure development also connects government to citizens and the private sector by supporting economic activity and boosting quality of life. And a number of tools and programs associated with parking bridge into other key government functions—economic development, tourism/conventions, and public works, to name a few—that make the public sector more competitive and attractive to business and private investment. There are several practical examples of parking-related programs that are used to stimulate economic activity and quality of life.

Zoned/Managed Parking in Residential Neighborhoods.   Many communities, particularly those in densely populated urban areas with a healthy mix of private homes and commercial/retail businesses, find zoned or managed parking (through the use of permits or meters) a vital tactic to provide peace of mind to homeowners. Typically, this device is used to prevent commuters or nonresidents from using residential streets from parking their cars during the business day or overnight and to ensure parking for property owners in the neighborhood.

Evidence from many cities has shown that blocks or entire neighborhoods made up of private homes that employ zoned or managed parking realize many benefits. For one, they encourage use of public transportation and therefore minimize auto traffic and associated environmental effects. Second, property owners in zoned neighborhoods tend to keep their homes longer; they reinvest in their properties, which increases their value and in turn produces higher property tax revenues to cities. And, because (especially in residential communities that use meters) they hold the threat of a costly parking ticket, they essentially dictate that the overwhelming majority of people who park in these neighborhoods are shoppers—people who feed meters and retailers' cash registers. This, of course, contributes to cities' bottom lines in a number of ways.

Shared or Multiuse Parking Programs.   These programs strike a balance between use of parking and transit for commercial needs during the weekday and retail/entertainment demands at night and on weekends. They are designed to ensure the most efficient use of available parking and public transit facilities throughout the day and to meet the multiple and sometimes conflicting demands of residents, commuters and businesses. Parking facilities that accommodate 9-to-5 workers convert to service theater and restaurant patrons after hours. Permitting can be used to enable people to use parking for whichever purpose or purposes they require, and communication tools—the same technologies used in stock tickers or airline scheduling services—can provide real-time information to the businesses and citizens that require up-to-the minute parking availability. The current proliferation of in-car GPS applications makes this service even more attainable in the near future.

Many cities employ these programs to optimize parking/transit systems and traffic flow. Park-and-ride and kiss-and-ride programs of this nature are common, especially around public transportation stations and airports. Often, these programs are complemented by circulator services that minimize congestion in central business districts by encouraging perimeter parking, then use buses to shuttle drivers en masse.

The San Francisco/Oakland Bay Area Rapid Transit system employed a variation of this model in late 2002. BART allowed commuters to purchase decals that ensure rush hour parking at some of its busiest and parking-dense stations. Not only do the new revenues help plug budget gaps. The program also minimizes congestion at targeted lots and saves many commuters the time and hassle associated with inability to leave their cars at desired stations. BART officials also report that many commuters are happy to pay for such a service because it allows them to spend extra time at home with their families before they leave for work.

Shared and multiuse parking is more than a convenience and a smart business move. In many cases, it enables cities to acquire much-needed public and private funding for large-scale development and revitalization programs. In many cases, the inability to forge shared and multiuse parking operations prevents cities from pressing forward with vital development initiatives. Because many cities' development departments maintain strict rules and methodologies governing parking and traffic flow attached to any new building project, cities and developers must work together to develop flexible parking management systems. Otherwise, they—and the citizenry—often lose money, jobs, and other economic benefits.

Magnet Infrastructure Programs.   Some cities use a combination of infrastructure/land management, parking/transit measures, and tax incentives to harness and redeploy existing public resources to increase retail, commercial, and residential development. These magnet programs can be applied to an entire business district or single building, and many are built around a financing mechanism known as TIF (tax increment financing).

One of the few true economic development incentives left for local governments, TIFs pool property tax revenues into infrastructure development, often in blighted or largely vacant areas. TIFs provide for land cleanup, road and building repairs, and other infrastructure improvements so that the land and properties in these districts generate revenues, typically without raising taxes to the community. Through their special designation, TIFs enable development projects to happen at all, and faster than they might otherwise. At least 44 states maintain some form of TIF program for redevelopment and revitalization.

Because of their unique status, TIF districts integrate transportation and parking in innovative, revenue-generating ways. One large southwestern city, for example, has tied parking infrastructure (including a downtown shuttle, smart-card meters, a special parking infrastructure fund, and encouragement of alternative forms of transportation such as bicycles) to the central business district TIF. As a result, developers have thrust themselves into new retail and commercial building projects in the city's ongoing downtown development plan, and developers are now working with the city (with the parking management staff playing a central role) to expand building projects into the residential sector. The city is also exploring semiprivatization of its parking and transit systems to facilitate longer-term development and growth of the downtown area.

Privatization and Partial Privatization Programs.   These initiatives, discussed earlier in this chapter, can be extremely useful in expediting public development activities. Entities such as parking, transit, airport, and development authorities typically enjoy powers regarding land use, zoning, and taxing (including, in some cases, TIFs) that spur development and create new services and revenues. The new entity can be public or quasi-public and can be managed by the public or private sector (or in partnership between the two).

While some cities have chosen to go the route of full privatization—in which they turn over management responsibility for these operations to private contractors in exchange for a guaranteed amount of revenue, typically up front—public-sector leaders should be cautious of this approach. Revenue- or profit-sharing—even in combination with an up-front payment—may be a better option, simply because cities should enjoy the full benefit of new revenues privatization can bring and not leave the lion's share of new revenues to private contractors. Nonetheless, privatization models can have significant benefits, as Figure 32.4 depicts.

Figure 32.4 represents an actual revenue-share privatization business model created for a small midwestern city. The city had no formal parking management operation or infrastructure, and required that a parking authority—owned by the city but managed by private partners—be created to this end. The plan created a governmental entity and parking authority appointed by the mayor and city council, with management and technical assistance provided by the private partner, which paid the city to turn over management responsibility, then split all profits with the city once that initial investment was recovered.

The parking management operation created by the new authority was also designed to form the foundation of the city's larger economic development plans. A regional hospital, the city's largest employer, was expanding and required more parking and the land to build it. Four commuter train stations carry thousands of people, many of them who drive their cars to the city to take the train, to and from work. A local community college was also building a magnet campus in the downtown. Additionally, the city was continuing a long-standing movement from an industrial-based economy to a service-driven economy, which was to include development of riverfront land just outside downtown. Finally, city leaders wanted to highlight and heighten the city's health care and retail base by establishing a medical district and arts district in and around the downtown area.

Figure 32.4 depicts the strategic planning behind the parking authority and subsequent economic development. The inner circle of the bullseye shows the core stakeholders needed

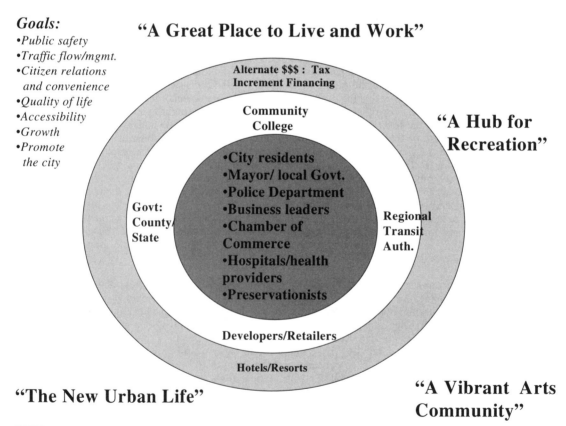

FIGURE 32.4   Parking authority privatization—stakeholder map.

to spur the initial phases of the project. The next rings depict the next phases of partners/ stakeholders and vehicles (such as financing) required to expand this infrastructure into broader-scale economic development. The phrases in quotes outside the bull's-eye suggest a few ways in which the city, citizens, and private sector could represent and market the plan within the city and to others outside of it, with an eye toward the goals listed in the upper left corner.

32.6.3 Bringing Parking Management's Role in Economic Development to the Table

While it is clear that parking management skills, technology, and business practices are potentially of infinite value, the function and discipline are still largely marginalized and fairly low on most cities' food chain. So how can those responsible for parking management and public-sector leaders better leverage the value parking management brings to the table and increase its return on investment? Several practical steps are required:

1. *Integrated management/decision-making:* Cities can do a much better job of aligning the people responsible for various parking-related functions—enforcement/ticketing officers, collections staff, courts officials, etc.—in overall planning and management. The integration of these staff, and simplified reporting relationships and lines of communication with the people to whom they report, streamline and therefore heighten the role of parking management in the bureaucracy.

2. *Better direct linkages between parking management and related infrastructure functions:* The economic, operational, and customer service benefits of parking management fall on deaf ears unless they are in front of decision-makers in other infrastructure-related public programs. Cities need to restructure to take full advantage. More direct roles, peer relationships, and bureaucratic linkages need to be made between the parking management function and those departments responsible for public works, transportation, housing, and public transit.

3. *Transform parking management skills, intelligence and business processes to larger government functions:* In a majority of cities today (including those surveyed in the material referenced in the accompanying sidebar), most of the information generated by the parking management function is rarely if ever shared with other departments related to infrastructure, much less those responsible for economic development and planning. The strategic value of parking and related infrastructure programs is too critical today for this trend to continue. Parking and infrastructure are central to revenue generation, public safety and citizen relations, and should be at the table in coordination of management of key agencies such as development authorities, planning/development departments, tourism/convention centers, and related entities.

32.7 CUSTOMER RELATIONS

Customer relations is an area that unfortunately has been largely neglected by purveyors of parking management systems and municipal employees responsible for these systems. Typically, the first encounter between the municipality and a citizen occurs upon a parking infraction and continues through the process of further notification and, finally, payment of fines and fees. During that time, the citizen is not considered a customer to whom certain civil courtesy should be extended, but rather as the perpetrator of a crime and one who must be harassed into compliance—even if, as in most cities, parking infractions are limited civil, not criminal offenses. This attitude has led to the use of negative approaches geared toward forcing citizens into compliance with local laws and payment requests. Most people realize that finding a parking ticket on their vehicle, or finding their vehicle immobilized by a boot or towed due to overdue fines, is a disturbing and costly experience. Parking managers should understand this and better recognize the need to ease citizen discomfort.

There are a number of methods that may be employed to change this interaction with the general public. But first there needs to be a change in the way that parking managers view the people who commit parking infractions. They should not be viewed as intransigent, recalcitrant, or even scofflaws. By changing the way they view the general public, parking managers can develop proactive and responsive strategies such as innovative public awareness programs, alternative payment methods, and citizen inquiry facilities. The primary focus of these strategies is on encouraging a more amicable dialogue with the general public.

For example, a brief description (even pictorially) of how the infraction will be treated by the municipality and the various options available to the individual should accompany the parking citation notice. Additionally, as is done in Austin, Texas, a statement of overdue ticket should be included with each new ticket notice received by the individual. This approach provides citizens with knowledge of their total outstanding debts to the municipality. Also, it removes the excuse of pleading ignorance to the process, because details of the consequence of not paying are always provided with the notice. A byproduct of this approach is the prevention of increased fines as a result of misapplied payments. For example, a citizen may have paid the fines but still be receiving a request for payment due to the municipality's poor bookkeeping.

Alternatively, municipalities may use Internet sites to provide details of how the parking management program works, updates to current practices, frequently asked questions, and guides to using parking services. Larger cities may look to provide regional service centers to disseminate information and accept in-person payments and inquiries from the general public, thus putting a friendlier face on the municipality in the eyes of the citizen.

Smaller municipalities may enter into agreements with small merchants to accept payments and hand out informative parking management brochures describing system updates and new procedures. Whatever the option or solution, the purpose of customer-service initiatives like these is to increase the public's willingness to comply with municipal laws through a greater understanding of policies, distribution of parking revenues, and public safety improvements. Other methods to improve public response and participation include:

- Improved maintenance of signs and meters
- Use of the Internet, TV, radio, newspapers, and frequent brochures to provide public information
- Warnings to motorists who are within weeks of being booted or towed or having their license suspended
- Special payment plan programs for motorists with very high outstanding balances (a variation of this is currently running in Memphis)
- Amnesty programs that allow citizens with high volumes of tickets to pay off citation debts without paying late fees or incurring additional civil penalties (Chicago instituted such a program in late 2002, with mixed results; the city did collect more than $10 million in unpaid parking tickets, however)

Of course, each municipality must devise its own public awareness and customer relations programs. Customer relations, though, should be an integral part of the parking management system designed to encourage public compliance and willingness to pay outstanding fines and penalties. Booting, towing, license suspension, and denial of plate registration renewal are sufficiently threatening measures in and of themselves and are almost always guaranteed to force compliance. Therefore, with these enforcement measures, municipalities would do well to encourage a friendlier interaction with citizens. Needless to say, criteria should be defined to determine the level of success of these programs so that educated decisions surrounding these programs can be made.

32.8   *SYSTEM VALIDATION AND VERIFICATION METHODS*

A comprehensive parking management solution must have validation and verification methods to promote and ensure accountability, efficiency, integrity, and accuracy. Today's parking management operations are singularly focused on maximizing enforcement measures and place little or no emphasis on monitoring manual and automated procedures for continuous improvements. As a result, systems tend to lack the functions and features necessary to identify faults and failures relating to accountability, efficiency, integrity, and accuracy.

The absence of these features can undermine confidence in the system and, over time, lead to general abuse of the system internally. This abuse manifests itself in increased public complaints, which can create animosity between government and citizens. To foster an environment of cooperation and harmony, the parking management system must be seen as unquestioned in terms of accountability, efficiency, integrity, and accuracy.

Citizens must be assured that when they follow the rules, they are dealt with fairly and equally. They should be confident that the municipality is operating the parking management system to provide sufficient availability of parking spaces (on-street, off-street, and in private and public lots) and to reduce operations costs by improving efficiencies, enhancing services to reduce complaints, and holding staff accountable.

The adage that "perception is reality" holds true for public services, and the general public must feel that government provides the highest level of service on a consistent basis. Anything less will give rise to apathy and general mistrust of the system. Therefore, it is imperative that system validation and verification methods be established as integral components of a comprehensive parking management solution. These are best suited to a continuous improvement strategy that solicits feedback from citizens, management, and operations staff. Mechanisms for obtaining feedback from the general public may include telephone surveys, focus groups, suggestion boxes in public locations, direct mail, and Internet questionnaires. Information obtained through these media should be carefully reviewed and evaluated to determine whether to apply the suggestions and how best to do so. Also, a mechanism should be in place to acknowledge appreciation for public participation.

32.9   *MEASURING THE IMPACT AND EFFICIENCY OF THE PARKING MANAGEMENT SYSTEM*

Many cities and municipalities are guilty of operating a parking management system that is separate and independent from other systems and departments and, in doing so, reducing or eliminating the opportunity to take advantage of successful business process innovations. In most cases these innovative measures generated cost savings and process efficiencies that would bring about exponential rewards to the parking management system. For example, a parking management system run by and for the police department is unlikely to share in the benefits gained from cost-saving and efficiency measures put in place by the city's IT department. Also, with very little sharing of information between departments, the limitations of the parking management system are not always visible. Notwithstanding, there are ways to ensure that the parking management system becomes a living, breathing organism and stays abreast of business process improvements and technology advancements.

There are some basic methods and procedures that are necessary to ensure the continuous relevance of any system. These include:

• A continuous improvement process
• System performance standards
• Business continuity strategy
• System backup and recovery procedures

Beyond these, there must be procedures and standards designed to measure the efficiency, impact, and progress of the program. Viewing the parking management program as a service to the general public, it stands to reason that regular public feedback should be sought to determine whether parking initiatives are having any effect. Also, it is important to have measures in place to ensure accountability of staff members, contractors, and managers. Too often the parking management function is treated as an isolated function or department with little or no interaction with other municipal departments, which in itself reduces the need for accountability and performance measurements. Measurements such as the revenues earned and revenues lost through negligence or inefficiencies are of equal importance to municipal governments. Other data, such as usage or lack of usage of parking facilities, may be of assistance to stakeholders in the planning, building, and licensing departments.

A properly run parking management operation should have procedures and standards that allow managers to gauge program progress. There should be standard response times for telephone and written inquiries with established procedures to measure staff responsible for these functions. The time for a citation to move through the system from initial issuance to final dispensation should be established and properly adhered to. Continuous improvement activities should be applied in order to ensure that these standards are met. Typically, it should take no more than ninety days (90) for a ticket to move from initial issuance to final dispensation. As well, telephone inquiries should be answered within 24 hours and written responses should be answered within 40 hours. While these are guidelines and may be adjusted according to each environment, lack of standards like these will only lead to abuse of the system as well as mistrust of the system by the general public.

Applying continuous improvement activities to the system without obtaining feedback will not help to determine the success of the initiative. Municipalities may use surveys or focus groups to help determine whether an implemented improvement plan is gaining acceptance or needs to be changed. And like the continuous improvement plan, the surveys and focus group should be a regular feature for getting feedback internally and externally. As a matter of fact, economic indicators, measures, and tools should be incorporated in the feedback surveys and focus groups to help determine return on investment for each improvement initiative implemented. Like corporate America, government must establish acceptable payback periods for its investments in order to determine what programs are best for the general public in terms of social impact, citizen satisfaction, financial outlay, and acceptance by the broadest cross-section of the population.

32.10   BENCHMARKS AND INDUSTRY TRENDS

Industry benchmarks cover a broad cross-section of analytics, ranging from hourly wages to hourly meter rates, and cover segments such as airports, hospitals, municipalities and universities. They also cover a variety of geographic regions across the United States and Canada. Such a multidimensional array of analytical variables makes the term *industry standards* almost meaningless. Notwithstanding, there are some notable observations worth mentioning. Uppermost among these is the 70 percent collection rate noted elsewhere in this chapter, which is considered a basic benchmark for the industry.

Other observations that are best stated as trends rather than standards include statistics such as (IPI):

- 12 percent of municipalities do not use outside contractors for any parking management service function.
- 64.2 percent of municipalities rely on in-house support to administer their parking management program.
- Disabled parking violation incurs the highest fine of all parking violations.

- Parking revenues are higher than expenses, making most U.S. parking management operations profitable.
- 77.8 percent of municipalities treat parking violations as a civil offense.
- 85.7 percent of municipalities use hand-held ticket-writing devices.
- 82 percent of municipalities assess penalties for overdue citations.
- 71.7 percent of municipalities issue citations via the police department.
- 92.6 percent of municipalities use towing as an enforcement measure. This number increases to 100 percent in Canada.
- 56 percent of municipalities use booting as an enforcement measure. This number is 0 in Canada.
- Most municipalities cite punishing scofflaws as the primary reason for booting and towing.

32.11 SUMMARY

Parking management is a much more complex and high-value area of operation than most cities currently give it credit for being. In most cases, cities treat parking systems as necessary evils rather than business and service operations. Greater strategic thinking and coordination are required to take better advantage of public safety, customer service, and economic benefits associated with parking management and related programs.

Through the combination of technology, financial management, business processes, and operations practices denoted in this chapter, public-sector leaders within multiple areas of government can build parking management systems that add service to the citizenry and dollars to the bottom line for many years to come.

32.12 REFERENCE

International Parking Institute (IPI). *Benchmarking the Parking Profession*. Fredericksburg, VA: IPI.

CHAPTER 33
TRUCKING OPERATIONS

Amelia Regan
Computer Science and
Civil and Environmental Engineering
University of California, Irvine,
Irvine, California

33.1 INTRODUCTION

33.1.1 The Importance of Trucking

Estimating the importance of the trucking industry is difficult, though all estimates point to its significance to our economy and the quality of our lives. In the latest data available from the U.S. Census Bureau, expenditures on transportation by truck in 2001 were over 318 billion dollars (U.S. Census 2003). Add in the costs of support facilities for road freight transportation and the number increases by 10–12 percent. Another way to look at it is that the nation's freight bill is about $1,200 per person per year. The Census Bureau data note that trucks moved almost 87 billion miles in 2001—nearly 80 percent of them loaded. Further, the trucking industry employed nearly 3 billion employees in 2001. No matter which numbers we use, its clear that the trucking industry is vital to our economy.

33.2 A BRIEF HISTORY OF THE TRUCKING INDUSTRY IN THE UNITED STATES

In 1900, there were 8,000 automobiles in the United States. By 1920 there were 8,131,522 automobiles and 1,107,639 trucks. In the early years of trucking, companies competed with short-haul railroad operations and managed to be significantly more efficient due to the lack of regulation in trucking relative to the heavily regulated and high-fixed-cost rail operations (Herbst and Wu 1973). The trucking industry was heavily regulated from the mid-1930s, when the Motor Carrier Act was adopted, until 1980, when it was dismantled. In 1937, the first year in which interstate commerce commission (ICC) received reports from trucking companies, there were 54 trucking companies with incomes in excess of $1 million. By 1955, that number had increased more than fifteenfold (Taff 1956). The ICC, created in 1887, was the first regulatory commission of the U.S. government. It gained control of the trucking industry in 1935 and essentially ran the industry like a public utility until 1980. It lost most of its power then and was completely dismantled in 1995, when its remaining functions were transferred to the National Surface Transportation Board.

33.3   *CLASSIFICATION OF TRUCKING OPERATIONS*

Trucking companies were historically placed in the following classes for the purposes of regulation. Private carriers are those that are owned by the company for whom they provide exclusive service. Common carriers (first called motor common carriers to distinguish them from rail common carriers) are companies that move general commodities for any customer in need of service. Under regulation, common carriers were required to provide service to any customer who asked, at a fixed and publicly available price set by the ICC. Today's common carriers have short-term or long-term contracts with shippers or third-party logistics providers. Contract carriers are carriers that provide service under long-term contracts with shippers. Many of these have highly specialized operations. Under regulation, contract carriers could negotiate the rates charged with shippers directly. Those rates, and their operational costs, tended to be lower than those charged by the for-hire common carriers (Taft 1956). Exempt carriers were those that hauled loads specifically exempted from regulation and nothing else. The largest group of exempt carriers were those moving agricultural commodities. Some interesting court cases related to the exception of various commodities (seafood, frozen food, etc.) made it all the way to the U.S. Supreme Court.

Today, common and contract motor carriers of property are grouped into the following three classes: class I carriers having annual carrier operating revenues of $10 million or more; class II carriers having annual carrier operating revenues of at least $3 million but less than $10 million; and class III carriers having annual carrier operating revenues of less than $3 million. When carriers change classes, they are required to notify the U.S. Bureau of Transportation Statistics.

Companies are further classified by the type of services they provide. The first system of classification breaks the industry down into full truckload carriers, less-than-truckload (LTL) carriers, and package delivery services. LTL carriers are those that handle freight weighing less than 10,000 pounds, while truckload carriers haul heavier freight in full truckloads. Note that LTL carriers typically also provide truckload services. A further classification might specify what type of equipment is used by the carrier, including van (sometimes called dry van), flatbed, refrigerated (known as reefers), and tanker trucks, to mention a few.

Carriers are typically further classified as providing local, regional, or long-distance operations. The operations of these different companies and the working conditions for their drivers differ significantly. Drivers in local operations tend to spend much of their time loading and unloading and waiting their turn at dock facilities; in return for such inconveniences, they sleep at home every night. Long-distance drivers tend to stay on the road for 3 to 6 weeks at a time, sleeping in their cabs or occasionally in inexpensive motels. Long-haul drivers experience quite a bit of autonomy, despite the emergence of on-board computers that keep them in continuous contact with dispatchers and managers. Long-distance truck drivers are the cowboys of the 21st century.

33.4   *THE EVOLUTION AND IMPACTS OF REGULATION*

It would be impossible to do proper justice to the fascinating history of the regulation and subsequent deregulation of the U.S. trucking industry in this chapter, which is mainly concerned with contemporary issues in the industry. Nonetheless, we discuss some of the basic ideas here. Under regulation, the ICC, with the backing of the railroad industry, the International Brotherhood of Teamsters, and the newly formed American Trucking Associations, severely limited both entry into the industry and competition with respect to prices charged to shippers. Railroads felt they were subject to unfair competition from the trucking companies of the day because as common carriers, they were required to serve all customers. In addition, the railroads felt that the fact that trucking companies were not required to pay for the construction of the highway network made them unfairly competitive. Under the ICC, a carrier wishing to provide service along a route it had not previously served had to obtain a "certificate of public convenience and necessity" and would further have to provide evidence that no carriers currently servicing the route would be negatively impacted by increased competition. The rules by which the private, for-hire and exempt carriers operated and the way those rules changed over time is an endlessly entertaining and interesting subject. Rates under regulation were set by rate bureaus, which were collaborative groups of carriers exempt from federal antitrust laws. At the time of deregulation in 1980, there were 10 large rate bureaus and 55 small ones (Hirsh 1988). The rates set by these bureaus were typically not competitive at all. Labor unions were able to extract much higher wages than they would otherwise have received in a competitive market because carriers were able to pass their costs directly on to shippers. In many cases, companies could only secure rights for one-way service, and had to return empty after each loaded move. The postderegulation industry is significantly more efficient than its regulated counterpart. There are many measures of these increases in efficiency. One is that empty miles in the last few years have remained under 20 percent (U.S. Census Bureau 2003). This inefficiency of the regulated industry and the efficiency gains achieved postderegulation has been the subject of numerous studies. See, e.g., Bailey (1986) and Ying (1990). By 1983, prices for freight transportation by truck had dropped 25 percent, and they continued to drop by as much as 35 pecent. Interestingly, the main argument for the deregulation of the industry was the fact that it lacked the economies of scale necessary to require regulation. Economists argued that deregulation would lead the industry to break into smaller more competitive firms (Spady and Friedlaender 1978). However, deregulation has led to a dichotomous industry in which 70 percent of all companies control fewer than seven vehicles but major market segments are dominated by a few large carriers that own tens of thousands of vehicles and effectively control, through spot markets and subcontracting agreements, as many as hundreds of thousands of others. Technically, the economies of scale arguments might in fact be correct, but clearly all segments of the industry are subject to significant economies of density and many of economies of scope. High densities in the parcel and LTL segments allow carriers to make better use of their terminal networks, and high densities in the truckload segment allows carriers to minimize waiting times and empty moves.

The postderegulation entrants into the industry were mainly truckload carriers because of the relative ease of entry into that market, which does not require the investments in infrastructure required to run LTL operations. Since deregulation, virtually no carriers entering the LTL market survived while many truckload entrants were successful (Beltzer 1995).

33.5   THE INDUSTRY AND UNIONS

Regulation made possible the rise of unionization in the industry, though there have always been a significant number of nonunion carriers. The height of the union movement in the trucking industry came in 1964 with the adoption of the International Brotherhood of Teamsters National Motor Freight Agreement. The Teamsters, led at the time by James R. Hoffa, achieved a long-time goal of effectively guaranteeing union wages nationwide. Unions had the largest impact on the intercity for-hire segment of the industry and were able to extract high wages in that segment in which companies were able to pass the higher costs directly off onto shippers because of industry-wide pricing agreements. Today, led by Hoffa's son, James P. Hoffa, the union is celebrating the hundredth anniversary of its creation in 1903. Union membership under regulation was around 60 percent of the industry, while by 1990, 10 years after deregulation, it had fallen to 25 percent (Hirsh 1993). According to the latest available data, in 2001 less than 20 percent of truck drivers were union members (Hirsh and MacPherson 2003).

33.6 TRUCKLOAD TRUCKING

The largest fraction of the trucking industry is the truckload sector, which moves full containers from shippers to their consignees. The largest long-distance truckload carriers use rail intermodal transportation whenever it is economically advantageous. Intermodal transportation is generally thought to be economically efficient for moves of more than 300 miles. The use of rail intermodal transportation became more popular during the mid- to late 1990s due to two simultaneous factors: a significant labor shortage in the over-the-road market and improvements in technologies that made intermodal operations more efficient. Long-haul truckload drivers, known as over-the-road drivers, typically spend 3 to 6 weeks on the road at one time. Many of these drivers drive sleeper cabs, and some work in teams of two known as doubles. Over-the-road drivers are typically paid by the loaded mile. A large portion of the local truckload market involves drayage operations—those operations in service to intermodal facilities. Dray operators pick up and deliver loads to railyards or ports and may move loads between such facilities. Dray operators typically move three to five loads per day and are alternatively paid per loaded distance or a flat fee for each load moved.

33.7 LESS-THAN-TRUCKLOAD

Less-than-truckload carriers typically haul loads that weigh less than 10,000 pounds. These carriers must operate consolidation terminals where multiple loads can be loaded onto vehicles en route to nearby destinations. Demands are picked up in a local area and delivered to a local terminal, where they are sorted and loaded into line-haul trailers. These line-haul trailers move loads to break-bulk transfer terminals, where the loads are unloaded and reloaded again. A load might be handled at several of these intermediate locations en route to its final destination. Figure 33.1, taken from Roy (2003), depicts the organization of these operations. Rates for LTL services are based on the class of product carried and the weight of the shipment. While the number of classes has varied over time, there are currently 18 different freight classifications. The tariffs charged for different classifications can be very different—base prices for the most expensive class (500) can be 10 times the price of goods in the least expensive class (50) of the same weight. Factors affecting the classification of goods include, at a minimum, the following: weight per cubic volume, value per pound, liability to loss damage or theft in transit, likelihood of injury to other freight, and risks due to hazard of carriage and expense of handling. The costs of LTL services are sufficiently high that for some shipments it is less expensive to move the goods as a truckload movement. For example, for many classes of goods, a 5,000-pound LTL shipment can be moved less expensively as a full truckload movement; the breakeven point for LTL versus truckload moves depends upon the distance traveled, and the class of goods carried. The shorter the distance and higher the classification of the goods, the more likely a truckload move will be competitive in price with an LTL move. Similarly, charges related to the weight of a shipment are broken into categories such as less than 500 pounds, between 500 and 1,000 pounds, between 1,000 and 2,000 pounds, between 2,000 and 5,000 pounds, and so on, but just as LTL freight might be better moved as a full truckload, sometimes it is more cost-effective to have a shipment classified in the next-higher weight category (these breakpoints are known as *break weights*). For certain items, the class rating structure based on weights is not a fair structure. Very light and bulky items are charged a rate based on the cubic size of the shipment rather than the weight. It should be clear that for companies with significant transportation needs, managing their freight transportation contracts can be quite complex. Most companies will have in-house transportation and logistics managers whose primary responsibility it is to manage their long and short-term contracts with trucking companies and to see that they pay the lowest fees possible. Much of their time is spent identifying ways to move LTL freight to truckload and move express freight to LTL.

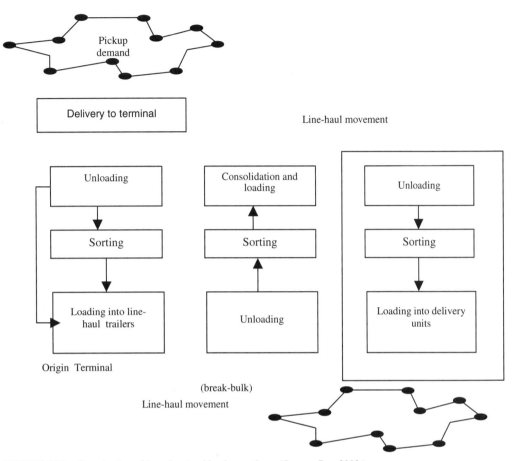

FIGURE 33.1 Organization of less-than-truckload operations. (*Source:* Roy 2003.)

33.8 EXPRESS SERVICES

There are two distinct types of express services: national and international package express services like United Parcel Service (UPS) and Federal Express (FedEx), and local express services, which handle both packages and irregular freight. The express package delivery companies got their start in the early 1900s with the emergence of the American Express Company, the United States Express Company, Wells Fargo Express Company, and Adams Express Company (Merritt 1908). These services were joined in 1912 by UPS, which, begun in 1908 as a courier service, began delivering goods from retailers to customers. In 1913, the United States Postal Service (USPS) extended its operations to include parcel post (Roper 1914). That move was not without its opponents (Merritt 1908) and remains controversial today. UPS gained the rights to expand its operations nationally in 1959, and from 1960 to 1970 its revenues grew from $4.9 million per year to $528 million; much of that growth at the expense of the USPS Parcel Post services (Nissen and Lago 1975). FedEx got its start in 1973 and was the first company to offer one- and two-day national express package delivery. FedEx's early operations benefited immensely from the deregulation of the air cargo industry in 1977 and relied heavily on a hub-and-spoke delivery system. Today UPS and

FedEX run both ground and air operations and no longer rely as heavily on their hub-and-spoke systems. Their ground operations are built around local sorting facilities, which serve pickup and delivery operations and send partial or full truckloads to other sorting facilities across the country. These local operations are typically no more than 5 hours' driving time apart, so that the truck drivers can make a single round trip move and return to their home locations within one work shift.

Local express services are dominated by courier companies that provide 2-, 4-, and 8-hour express delivery services within small geographic regions. Every city in the United States is served by several of these, and large cities are served by many. The larger and more successful of these companies typically also provide third-party logistics services—that is, they balance out their irregular operations with contracts to make deliveries for retailers who provide same-day or next-day deliveries within a local area (e.g., express office supply stores).

33.9   PRIVATE FLEETS

A significant fraction of the trucking industry is made up of private fleets. These fleets are the transportation and logistics units of large manufacturing or distribution companies (large food and beverage manufacturers, large groceries, toy manufacturers, etc.). According to a relatively recent study, shippers spend more money transporting freight by private fleets than by for-hire carriers (Roth 1995). Private fleets were estimated in that study to be responsible for 54 percent of trucking industry receipts in 1995. While it is quite likely that the fraction has gone down during the past few years as the result of increased outsourcing, that fraction remains considerable. There have been many debates over the years about when companies should outsource their logistics services and when they should keep them in house. In addition, many major shippers maintain a private fleet for their mission-critical transportation and outsource the rest. One issue for private fleets is the difficulty of keeping empty miles low. For example, the 1995 study mentioned above points out that standard private tractor-trailer fleets were running empty 24 percent of the time, significantly more than for-hire fleets. Similar increases were found in the dry van and tank truck market segments.

33.10   THIRD-PARTY LOGISTICS

Since deregulation, third-party logistics providers (or 3PLs, as they are commonly called) have emerged as a new industry, providing a wide variety of logistical and management services. By combining contemporary information technology with traditional freight-handling systems, an increasing number of intermediary firms are already managing freight transportation for companies that choose not to handle their own shipping and receiving (for example, medium and large manufacturers who find their goods movements too complex to handle themselves). Many of these manufacturers deal with multiple trucking, ocean, rail, and air cargo providers. The complexities of booking and moving their freight are enormous. Freight transportation intermediaries provide a bridge between shippers and carriers, facilitating the flow of information and goods. Lately, these companies have been benefiting from the communications capabilities of the Internet.

Some 3PLs evolved from the pre-deregulation freight brokerages and shipping agents. The term *freight broker* referred to ICC-licensed truck brokers that acted as marketing agents and load matchers for smaller trucking companies, while shipping agents bought capacity from railroads and sold it to shippers. Until recently, most 3PL companies were affiliated with a parent transportation or warehousing company. 3PL providers are of two types: asset-based or non-asset-based. Today, both asset-based and non-asset-based providers are transforming themselves to respond to new market demands. For example, both asset-based and non-asset-based companies are integrating themselves more deeply into manufacturers' operations rather than focusing exclusively on warehousing and distribution activities.

Many 3PLs are non-asset-based, meaning they do not own transportation equipment or warehouse facilities. These tend to be either management- or knowledge-based consulting companies, focusing mainly on strategic or tactical activities rather than handling physical distribution themselves—although many work closely with asset- based carriers and warehouse managers.

Recently, new types of freight-transportation intermediaries and new business models have emerged. On-line logistics providers are attempting to use the power of the Internet and new software tools to interact efficiently and simply with shippers, carriers, and traditional third-party logistics providers. Some firms provide on-line freight marketplaces, enabling the purchase and sale of freight transportation capacity. These range from simple load-posting boards to sophisticated on-line exchanges. Some firms develop software tools to optimize freight operations or to simplify complex shipping problems. Others supply information on container ports or other intermodal facilities, or organize and aggregate buying power for various companies.

These new intermediaries supplant 3PLs by providing many of the services previously handled by traditional 3PLs. But they also offer opportunities for 3PLs to operate more effectively and provide better services.

These new on-line freight transportation intermediaries and infomediaries promise to transform the industries by enabling companies to move beyond traditional business paradigms, profiting from the synergies of information. Whether or not they will succeed in doing so remains to be seen. Intuitively, it would appear that the freight transportation industry, made up of many small firms and with many existing levels of intermediation, is ideally suited to benefit from the promise of the Internet. However, as an industry it has been very slow to change and to adopt new technologies, because of both a reluctance to do so and a lack of resources needed to make such investments.

Better information about congestion, queues at intermodal facilities, and border crossings and attractive purchasing agreements can increase equipment utilization and network efficiencies and reduce operating costs. These reduced costs may make these systems attractive to trucking companies that see an opportunity to leverage the technology investments of others.

33.11 MAJOR INDUSTRY ISSUES

33.11.1 Technology Adoption

All sectors of the industry are moving towards increased real-time operations management. For large companies, this move has been gradual and steady and has been taking place for the past 15 years or so. The largest companies in the industry, especially UPS and FedEx, have led the charge to adopt new technologies, including trip recorders, on-board computers, tracking and tracing technologies, advanced vehicle location systems, electronic data interchange (EDI) systems, and optimization-based routing and scheduling systems. The large truckload and less-than-truckload companies followed the large express package carriers and adopted a wide variety of technologies and optimization tools in the last decade. Their adoption was due in part to pressure from shippers who were familiar with the service guarantees provided by the express carriers. Smaller companies have been slower to adopt advanced technologies, both because their operations need them less and because the costs to small companies can be prohibitive. Technology adoption may well transform the way companies do business, but to date, with the exception of the express trucking industry, the impacts of adoption have been incremental—while the net effects may be large, the impacts of adoption of any specific technology have been relatively small.

33.11.2   Operational Costs

Keeping operational costs low is of grave importance in the trucking industry, where profit margins are very slim. While labor costs are an importance factor in profitability, insurance costs and fuel costs can significantly impact firms' success. Diesel prices rose approximately 20 percent in 2002 and have continued to rise in 2003.

33.11.3   Hours of Service Rules

The hours of service rules in place in the trucking industry through 2003 were developed in 1937 and adopted in 1938. Those rules limit drivers to no more than 60 hours of work in any consecutive 7-day period, and no more than 70 hours of work in any consecutive 8-day period. Drivers were further restricted to 10 hours of driving time in any 24-hour period and at least 8 hours off in the same period. An amendment put in place in 1962 replaced the limit of 10 hours of driving per 24-hour period with a requirement that drivers have at least 8 consecutive off-duty hours after 10 hours of driving. After this off-duty period, the driver could resume driving so that a driver could legally use an 18-hour cycle: 10 hours of driving, 8 hours off-duty, etc. A driver on this cycle would be driving 16 hours in some 24-hour periods. After several years of discussion and after holding extensive public hearings, in April 2003 the Federal Motor Carrier Safety Administration (FMCSA) changed the hours of service rules. Beginning in January 2004 the new rules allow drivers to drive for 11 hours after 10 hours off duty and are restricted to drive those 11 hours within a 14-hour period. These rules were intended to help drivers to get onto a 24 hour work clock. An exception to the 14-hour period is that local drivers (who sleep at home every night) can increase that period to 16 hours once per work week. The weekly limits mentioned above remain in place, with a new rule allowing any driver who remains off-duty for 34 hours to restart the week. For an extensive examination of the history of the rules, and the costs and benefits of amending them, see Belzer and Saltzman (2002) and Crum, Morrow, and Daecher (2002).

There have also been several studies on the impact of operational structures imposed by shippers on driver safety and the ability of companies to adhere to HOS regulations. For example, shippers set up pick-up and delivery schedules that do not take into account HOS regulations and the need for rest. In addition, just-in-time distribution policies require drivers to arrive at a specific time to load/unload, but then may delay loading and unloading. These and other practices may cause drivers to violate the HOS regulations, leading to an increase in fatigue related accidents (Fleger et al. 2002).

33.11.4   Inadequate Parking/Rest Facilities

The lack of sufficient parking and rest facilities for truck drivers is reaching crisis proportions. Truck drivers are often forced to drive in peak passenger commuting hours due to delivery schedules. If there were adequate rest facilities in urban areas, they might be able to reduce some of their peak period travel by arriving at these rest facilities early and resting and taking care of their paperwork until just before their deliveries are due.

33.11.5   Safety

Truck safety has vastly improved over the last few decades, and if new technologies are adopted it will continue to improve. Between 1981 and 2001, there was a 37 percent increase in the number of registered large trucks and a 91 percent increase in miles driven by these trucks. At the same time, there was an 8 percent reduction in the number of heavy duty trucks involved in fatal accidents and a 52 percent reduction in participation of trucks in those accidents, as a fraction of vehicles involve in such accidents (FMCSA 2003). Despite the gains due to better overall equipment, improved roadway design, and a better-trained workforce, problems will continue to emerge, particularly on freeways exceeding practical saturation of truck traffic. In the most congested areas, state and local government officials are seriously talking about the addition of truck-only lanes (possibly tollways and possibly partially automated). Western states have considered developing long-distance truck-only toll lanes that would allow for longer combination vehicles to travel at relatively high speeds. Weigh-in-motion (WIM) systems, which are increasingly replacing manual weigh stations, can be used to measure the axle loads of a vehicle traveling at highway speed. WIM equipment provides information on wheel and axle loads, gross vehicle mass, axle spacing, vehicle speed, distribution of traffic movement through the day, headway between vehicles, freight movement, and traffic volumes and mix. These data can be used for a range of applications, including identifying overloaded vehicles, collecting weight-dependent tolls, verifying optimal load distribution, and providing inputs into pavement management systems. WIM systems usually use axle sensors to detect and classify vehicles and allow for complete electronic screening of compliance and safe carrier status. They can also be used to measure weight, height, and speed of a truck traveling at highway speed, and use this information to warn driver of unsafe conditions. Other technologies that will improve safety are various warning systems being built into new vehicles. These will surely see more use as time goes by.

33.11.6 Changes in Contracting

The development of electronic commerce has impacted the trucking industry just as it has impacted other industries. Its primary impact has been in the area of contracting. In addition to the development of on-line spot markets, where shippers and obtain quotes and drivers can find available loads, the industry is seeing the development of collaborative marketplaces and the use of combinatorial auctions. The collaborative marketplaces take many forms—some develop networks of shippers who manage their transportation needs as though they were a single, large company—others allow carriers to collaborate and for small companies to sub-contract loads from the large carriers or from groups of large carriers. Other collaborative models are likely to emerge in the near future (Song and Regan 2003b).

Combinatorial auctions are those in which the auctioneer places a set of heterogeneous items out to bid simultaneously and in which bidders can submit multiple bids for combinations of these items. Bids can be structured so that bidders can express their desire for a bundle of inseparable items (known as atomic bids), a collection of bids with additive values (known as OR bids), or a collection of atomic bids which are mutually exclusive (known as XOR bids). Shippers are increasingly using such auctions, in order to award contracts to trucking companies. The items put out to bid are lanes with associated volumes (for example an average of five loads per week on the lane from Los Angeles to Dallas). In order to bid effectively in such auctions, the carriers have to solve a number of very difficult problems. Large carriers can spend several days constructing their bids for major shippers, and will still not develop optimal bids. Recently, several researchers have begun to examine the optimization problems inherent in both the bid construction and winner determination in combinatorial auctions (Song and Regan 2003a).

33.12 CONCLUSION

The trucking industry is vibrant and active. The demands for movements of freight keep rising and, due to increasingly global manufacturing and distribution systems, they will continue to rise. Like all other industries, it is being impacted by emerging technologies and innovations in operating procedures. The past two decades have seen enormous productivity increases that are not likely to slow in the years ahead.

33.13   REFERENCES

Bailey, E. E. 1986. "Price and Productivity Change Following Deregulation: The US Experience." *Economic Journal* 96(3):1–17.

Belzer, M. H. 1995. "Collective Bargaining after Deregulation: Do the Teamsters Still Count?" *Industrial and Labor Relations Review* 48(4):636–55.

Belzer, M. H., and G. M. Saltzman. 2002. "The Case for Strengthened Motor Carrier Hours of Service Regulations." *Transportation Journal* 41(4):51–71.

Crum, M. R., P. C. Morrow, and C. W. Daecher. 2002. *Motor Carrier Scheduling Practices and their Influence on Driver Fatigue*, Federal Motor Carrier Safety Administration Report FMCSA-RT-03-005.

Fleger, S. A., N. Owens, L. Rice, and K. J Chen. 2002. "Assessment of Non-Carrier Encouraged Violations of Motor Carrier Regulations: Possible Solutions." In *Proceedings of the International Truck and Bus Safety Research and Policy Symposium*, 501–10.

Federal Motor Carrier Safety Administration (FMCSA). 2003. *Large Truck Crash Facts 2001*. FMCSA-R1-02-011, U.S. Department of Transportation, FMCSA, Washington, DC.

Herbst, A. F., and J. S. K. Wu. 1973. "Some Evidence of Subsidization: The U.S. Trucking Industry, 1900–1920." *Journal of Economic History* 33(2):417–33.

Hirsh, B. 1988. "Trucking Regulation, Unionization and Labor Earnings, 1973–1985." *Journal of Human Resources* 23(3):296–319.

Hirsh, B., and D. Macpherson. 2003. "Union Membership and Coverage Database from the Current Population Survey." *Industrial and Labor Relations Review* 56(2): 349–54.

Merritt, A. N. 1908. "Shall the Scope of Government Be Enlarged So as to Include the Express Business?" *Journal of Political Economy* 16(7):417–35.

Nissen, D. H., and M. Lago. 1975. "Price Elasticity of the Demand for Parcel Post Mail." *Journal of Industrial Economics* 23(4):281–99.

Roper, D. C. 1914. "Fundamental Principles of Parcel Post Administration." *Journal of Political Economy* 22(4):522–35.

Roy, J. 2003. *The Impact of New Supply Chain Management Practices on the Decision Tools Required by the Trucking Industry in Applications of Supply Chain Research and eCommerce in Industry*. Dordrecht: Kluwer, in press.

Song, J., and A. C. Regan. 2003a. "Approximation Algorithms for the Bid Valuation and Structuring Problem in Combinatorial Auctions for the Procurement of Freight Transportation Contracts." University of California, Irvine Working Paper.

———. 2003b. *An Auction Based Collaborative Carrier Network, Transportation Research E*, under review.

Spady, R. H., and A. F. Friedlaender. 1978. "Hedonic Cost Functions for the Regulated Trucking Industry." *Bell Journal of Economics* 9(1):159–79.

Taff, C. A. 1956. "The Competition of Long-Distance Motor Trucking: Farm and Industrial Productions and Supplies (in The Changing Patterns of Competition in Transportation and Other Public Utility Lines)." *American Economic Review* 46(2):508–20.

U.S. Census Bureau. 2003. *2001 Service Annual Survey: Truck Transportation, Messenger Services and Warehousing*.

Ying, J. S. 1990. "The Inefficiency of Regulating a Competitive Industry: Productivity Gains in Trucking Following Reform." *Review of Economics and Statistics* 72(2):191–201.

CHAPTER 34
THE ECONOMICS OF RAILROAD OPERATIONS: RESURGENCE OF A DECLINING INDUSTRY

Wesley W. Wilson
Department of Economics, University of Oregon,
Eugene, Oregon,
Upper Great Plains Transportation Institute

Mark L. Burton
Department of Economics, Marshall University,
Huntington, West Virginia

34.1 INTRODUCTION

The railroad industry is one of the oldest in the United States, dating from the inception of the Baltimore & Ohio in 1830 (Pegrum 1968; Wilner 1997). Very early in their history, railroads were relatively small firms that provided specialized services to local markets over small networks (Wilner 1997). At the time, rail transportation represented a tremendous advance over other available modes. Networks quickly connected, the mode became dominant and, in fact, spurred other industrial and agricultural developments. By the end of the nineteenth century, America's railroads stood as the centerpiece in a burgeoning industrial economy.*

As the railroads grew and firms consolidated through the latter half of the 1800s, abuses of market power, growing concentration, and rent-seeking led to passage of the Interstate Commerce Act in 1887. This legislation provided for the economic regulation of railroads. This statute was followed by a number of other regulatory initiatives that further tightened the federal government's control over the industry's economic decision-making.**

---

*The history and development of the railroad industry is available in many places. Locklin (1972), Pegrum (1968), and Wilner (1997) are examples of excellent reviews of the evolution of the railroad market.

**Between 1887 and 1920, there was a steady increase in the control exercised by the federal government, and, in fact, the government ceased control of the railroads during World War I, but returned them to private hands at that conflict's conclusion. The magnitude of regulatory oversight reached its zenith with the Transportation Act of 1920. Beginning with that Act, the railroad industry entered a 50-year period in which it had little or no control over its economic destiny.

In the first half of the 20th century, motor carriage and commercial aviation emerged as powerful competitors. By the end of World War II, railroad transportation had lost its dominance. The postwar decades saw a further decline in rail usage and in the industry's financial well-being. By the mid-1970s, it was clear that the highly regulated environment in which the railroads had operated was no longer consistent with a functioning and financially viable industry.* This realization was quickly followed by series of federal laws, the Regional Rail Reorganization Act (3-R Act), passed in 1973; the Railroad Revitalization and Regulatory Reform Act (4-R Act), passed in 1976; and the Staggers Rail Act of 1980. Each of these legislative actions was designed to help a threatened industry through a series of actions that created Conrail, provided subsidies to the industry, and, most importantly, provided for a substantial reduction of the economic regulations under which railroads operated for nearly 100 years.

In the wake of deregulation, the rail sector has reemerged as vital component of the nation's overall transportation system. Deregulation both induced and facilitated a number of cost-reducing activities.** Today, the industry is characterized by relatively few, very large firms. These firms provide services over large and complex networks, with substantially less fuel and labor usage than in the past.† Railroads actively participate in two types of markets—the general movement of high-volume, low-value bulk commodities and the transport of higher-valued goods in geographically limited, high-volume, and long-distance corridors. While U.S. carriers face measurable competition in most rail-served markets, there are circumstances in which large, bulk commodity shippers have very limited transportation alternatives. These "captive" shippers contend that postderegulation rail industry gains have been achieved at their expense. Consequently, further modifications to the regulatory environment are constantly under consideration.

Perhaps because of its pivotal economic role, the railroad industry has remained controversial. Historians have both lauded and lamented the role the railroads play and have played in economic development.‡ The industry has been chastised for noncompetitive behavior and inefficient production and is a frequently used example to illustrate the rationale for economic regulation. It has also been used to demonstrate the inefficiencies of regulation and the savings to society from deregulation.§ Regardless, however, of one's vantage point, few would argue that the railroad industry is unimportant as a means of transporting freight. To the contrary, the industry moves billions of tons of commerce each year and is the principal mode used to transport a number of bulk commodities such as coal and grain. Moreover, the railroads have emerged as an important component in the nation's system of intermodal transportation, a system that is vital to participation in international markets.

Due to their private ownership of infrastructure, spatial production technology, and historical importance, the economics of railroading tend to be somewhat more complicated than the economics of other transport industries. The purpose of the current chapter is to describe carefully the railroad industry, its operations, the markets it serves, the current regulatory environment, and the specific economic conditions that motivate firm decision-making. Accordingly, section 34.2 outlines railroad operating practices, section 34.3 describes rail-served

---

*See Friedlaender and Spady (1981), Gallamore (1999), Keeler (1983), Wilner (1997), Winston and Grimm (2000), and a variety of others.

**See Barnekov and Kliet (1990), Boyer (1977, 1981, 1987), Burton (1993), MacDonald (1987, 1989, 1999), MacDonald and Cavallusso (1996), Wilson (1994, 1996, 1997), and Winston and Grimm (2000).

†Bitzan (1999), Bitzan and Keeler (forthcoming), Ivaldi and McCullough (2001), and Wilson and Bitzan (2003).

‡See Gillen and Waters (1996), who introduce a special volume of the *Logistics and Transportation Review* dedicated to transport infrastructure and economic development. In addition, the reader is referred to Cain (1997), Nelson (1951), and Owen (1959). See also excellent discussions in Harper (1978), Locklin (1972), Pegrum (1968), and Winston (1985).

§See the classic books by Meyer et al. (1959) and Friedlaender and Spady (1981), along with more recent research by Barnekov and Kliet (1990), Boyer (1977, 1981), Burton (1993), MacDonald (1987, 1989a, b), MacDonald and Cavallusso (1996), McFarland (1989), Wilson (1994, 1996, 1997), Winston (1985, 1993), and Winston et al. (1990) and a litany of others underscoring the effects of regulation and the savings of partial deregulation. Winston (1993), in particular, provides a concise synopsis of the distortions created by regulation.

transport markets, section 34.4 provides a history of rail industry regulation, and section 34.5 reveals railroad pricing motivations and constraints through a series of economic models.

34.2 RAILROAD OPERATIONS

U.S. class I railroads are typically organized around three primary operating departments. Engineering departments are responsible for the construction and maintenance of track, signals, and other right-of-way structures. Mechanical departments oversee locomotives and freight cars. Transportation departments are responsible for both terminal and line-haul train operations.*

Common to all transport modes, railroads operate over networks composed of links (line-haul trackage) and nodes (terminals, yards, and junctions). The configuration and quality of any specific network element depends on local geography, topography, and the nature of the traffic that uses that element. There is also considerable variation in railroad vehicles. While diesel locomotives represent the standard source of power, freight cars vary widely in size and form, depending on the nature of the commodities they are designed to accommodate.** Generally, U.S. railroads own the network over which they operate. However, in some relatively isolated settings, carriers will share facilities or operate over the tracks of a connecting railroad.† Similarly, U.S. carriers typically own fleets of locomotives and freight cars. However, it is not at all unusual for a carrier to use equipment owned by another railroad or, in the case of freight cars, equipment that is owned by shippers.‡

Railroad traffic ranges from single-car shipments of relatively high-valued goods to multiple-car shipments of lower-valued bulk commodities. Typically, smaller movements are retrieved from shippers by switch crews who move the freight to classification yards, where it is consolidated or "blocked" with other shipments destined for the same general location.§ This process is repeated in reverse when the shipment is near its destination. Larger shipments, however, often move directly between shipper and receiver. These operations are referred to as "unit train" movements. By eliminating a substantial portion of terminal costs, unit trains can achieve substantially lower per-ton-mile costs.¶ Much of the growth in railroading discussed in section 34.3 resulted from growth in multiple car and unit train traffic stimulated by partial deregulation, the accompanying rate flexibility, and relaxed merger guidelines, which have dramatically increased the size of railroad networks.

In addition to carload and unit train service, intermodal traffic has emerged over the past two decades as an important source of railroad activity.¶¶ Intermodal service combines the line-haul movement of shipping containers or truck trailers by rail with traditional truck

---

*Our focus is on the last of these three operating departments. See Armstrong (1998) for a more complete discussion of all three areas.

**During the 20th century, the single most important change in locomotion was the switch from steam to diesel. Recently, more subtle developments, including AC traction and computerization of engine control systems, have produced significant advances in locomotive power and fuel efficiency.

†In some cases, shared facilities or trackage rights over a connecting carrier are the result of cooperative agreements between the railroads. In other cases, these outcomes have been imposed by the Interstate Commerce Commission or Surface Transportation Board as a condition for granting merger approval. The payment for the use of rail lines has not been well developed in the economics literature. Obviously, the granting of traffic rights provides for the possibility of greater competition on a route, but the effectiveness of the competition depends critically on the level of payment by the tenant carrier.

‡Of course, if the railroad does not own the equipment, there is a corresponding decrease in the rate as their costs are lower.

§For a more in-depth discussion of yard operations see Armstrong (1998).

¶There are a number of ways to measure railroad outputs. One standard measure is the movement of one ton of freight one mile. This measure is referred to as a ton-mile.

¶¶Trailer-on-flat-car (TOFC) services have been routinely available for nearly a half century. However, the rapid growth in intermodal traffic dates to the mid-1980s. Projections suggest that in 2003, the movement of containers and trailers will replace coal as the single largest source of railroad traffic.

service at the origin and destination of the movement. Typically, non-railroad-owned drayage firms retrieve containers or trailers from the shipper and deliver them to railroad intermodal facilities in advance of a prescribed cutoff time. The trailers and containers are then loaded by crane onto specially designed railroad equipment for the line-haul move.* The process is then reversed as the intermodal shipment nears its final destination. Most intermodal trains are operated as unit trains without being switched en route. Currently, Class I railroads operate approximately 200 intermodal terminals located in 42 states.**

Regardless of whether trains are assembled in classification yards, by switch crews, or dispatched directly as unit trains from an on-line shipper or intermodal facility, the transportation process between terminals is likely to be the same. Most line-haul route segments are divided into blocks of varying lengths. Generally, a centrally located train dispatcher controls a number of route segments and the blocks that comprise them. Trains progress from one location toward another as they are given permission to occupy successive blocks. This permission is conferred to the train crew through a variety of methods, including, but not limited to, written train orders, electronic lighted signals, and voice-issued track warrants. Regardless of which method is used, the dispatcher attempts to move each train from one location to another as quickly as possible while maintaining the requisite separation between all trains (and any track maintenance crews) that are present on the portion of the railroad he or she controls. In the earliest days of railroading, all train movements were controlled by strict timetables. However, throughout most of the 20th century, freight trains were run on irregular schedules as "extras" whenever a sufficient amount of traffic was gathered to form a train.† Within the past decade, however, the emergence of time-sensitive intermodal traffic, combined with labor demands, has caused a partial return to closely scheduled freight service in limited settings.

The capacity of a particular line-haul route segment depends on a variety of factors. First, this capacity is significantly affected by the number of mainline tracks. Route segments with two or more main tracks are vastly more efficient because they allow trains moving in opposing directions to meet and pass at any location along the route segment. Similarly, faster trains are easily able to overtake slower trains moving in the same direction. The same sort (but not degree) of operating flexibility is afforded by the presence of passing sidings, so that longer, more frequent sidings imply greater route capacity. Unfortunately, multiple main tracks and longer, frequent sidings impose significantly higher construction and maintenance costs. Consequently, they are used only on those locations where high traffic density justifies the additional expense.

Route link capacity and operating ease are also a function of the signaling system in place. "Dark" track segments with no signals, where dispatchers must control operations through track warrants or train orders, require a greater degree of train separation to ensure safety than do route segments with automatic block or dispatcher-controlled electronic signals.

Finally, the capacity of a specific route segment can be influenced by track alignment and grade. The ability to affect the efficient flow of trains is degraded as the variability of train speeds increases. To the extent that relatively sharp curves or steep grades reduce train speeds at specific locations along the route segment, the overall capacity of that segment is reduced.

From an operating perspective, labor and fuel usage constitute the two most important cost items. This is true with respect to both line-haul and terminal operations. Labor costs

---

*In the case of TOFC movements, the most modern equipment consists of articulated "spine" cars that minimize the ratio of equipment weight to lading. Currently, the most efficient form of container-on-flat-car (COFC) movement involves stacking the containers two-high in articulated "well" cars. This process is referred to as double-stacking.

**The number and location of intermodal facilities was determined by the authors based on railroad promotional materials.

†Train sizes vary considerably based on the nature of the commodity or commodities, the nature of the service, and the configuration of the trackage over which the train is to pass. Trains providing local service may be only a few hundred feet in length. Fast-moving intermodal trains are generally in the neighborhood of 7,000 feet long, whereas heavy unit coal trains may extend to a length of 10,000 feet or more.

accounted for 41 percent and fuel for 23 percent of total transportation expenditures in 2001.* However, the railroad industry has made tremendous strides in reducing both labor and fuel usage.** In 1980, the industry produced 932 billion ton-miles of freight transportation services with a work force of approximately 550,000. In 2001, class I railroads produced 1.53 trillion ton-miles of service with only 162,000 employees.† This represents a five-fold increase in output per worker over the span of two decades. Similarly, the average fuel efficiency of rail transport has improved dramatically. The average number of ton-miles of transport services per gallon of fuel consumed has increased from approximately 250 to nearly 400 over the past decade.‡

Unlike most freight modes, railroad traffic is routinely interchanged between different railroads. Although railroad mergers have reduced the frequency of interchange, in 2001, roughly 55 percent of all rail traffic was handled by two or more carriers.§ At most locations, interchange is handled directly between connecting carriers. However, in some larger locations, traffic is exchanged through the services of a terminal railroad that serves multiple carriers. From an operational standpoint, the existence of interchange means that a carrier's ability to efficiently utilize its own fleet of rail cars is very often dependent on the extent to which connecting railroads return interchanged cars in a timely fashion.¶

Like most transport modes, U.S. rail carriers are continually evaluating emerging technologies in search of ways to improve their operations. The railroads' search for improvement is generally driven by three factors. First, competition between carriers and from alternative modes forces a continual search for ways to reduce operating costs. Second, from a shipper's perspective, the predictability of transit times is increasingly essential. Thus, rail carriers routinely search for cost-effective ways to improve service reliability. Finally, because railroad accidents can be extraordinarily costly, the nation's railroads continually search for technologies that reduce the likelihood of such accidents. These motivations have led to the adoption of numerous new technologies over the past decade. Examples include, but are not limited to, new computer and electronic technologies that improve both the performance and fuel efficiency of locomotives, the use of global positioning systems (GPS) in train dispatching, the development of car identification systems that improve shipment tracking abilities, the development of new wayside detection devices that help eliminate the need for cabooses, and the use of remote-control locomotives in yard switching activities.

34.3 THE DEFINITION AND EVOLUTION OF RAILROAD MARKETS

In defining markets for analysis, economists do not have a generally accepted prescribed methodology. Various rules have been used in practice, but by and large, the definition of markets is, at best, a murky science. Stigler and Sherman (1986) begin by stating: "The role of the market is to facilitate the making of exchanges between buyers and sellers" (55). They define markets as "that set of suppliers and demanders whose trading establishes the

---

*See Surface Transportation Board (2001). Readers should note that this discussion focuses on operating costs to the exclusion of capital costs.

**See Davis and Wilson (1999) and Bitzan and Keeler (forthcoming).

† See www.aar.org. Davis and Wilson (1999) develop and estimate a model of railroad employment. As they discuss, industry employment has fallen dramatically during a time period in which industry output has been increasing. Labor has therefore become more productive. In their model, they find that innovations such as increasing use of unit trains, longer lengths of haul, and mergers have played significant roles in reducing the quantity of labor used by firms. Bitzan and Keeler (forthcoming) also examine productivity, focusing on labor saving devices such as the elimination of the caboose.

‡ See U.S. Army Corps of Engineers (1998).

§ This value was developed through the use of the Surface Transportation Board's 2001 Carload Waybill Sample.

¶ Initially, concerns regarding the timely return of rolling stock made many U.S. carriers hesitant to interchange traffic with Mexican carriers. However, reforms in the operation of Mexico's railroads have largely eliminated such issues.

price of a good'' (55). Historically and generally, there are two dimensions of focus in defining markets–the identification of products and the geographic locales of suppliers and demanders. However, neither the product nor geographic limits to a market are well defined. Indeed, as noted by Slade (1986), ''Because markets so frequently overlap, market definition is rarely an easy task'' (291).

Despite the difficulties encountered in practice, however, market definitions are necessary for economic analysis and for policy implementation. In economic analysis, market definitions are often arbitrarily chosen with occasional specification tests used to identify whether other products or locations not in the defined market affect results of the analysis. In antitrust policy, market definition is often a major issue, particularly in monopoly and merger cases. Given a court-determined market definition, a variety of summary measures, such as, concentration ratios, Herfindahl indices, etc., are then used to evaluate the structure of the market to infer some notion of market power. A narrowly defined market may give overstated degrees of market power, while broadly defined markets may yield underestimated degrees of market power.*

Given both the necessity and the importance of market definitions, a variety of approaches have evolved. Theoretically, substitution and complementary relationships among products interconnect markets together. These substitution or complementary relationships may be present for either demanders or suppliers, but the determination of the limits of market interconnections is not an easy task. Uri and Rifkin (1985) discuss the range of practices used, from a consideration of how ''reasonable'' is the interchangability of products to an examination of prices and their movements. But, as Slade (1986) states, ''In practice, market definition frequently involves rules of thumb and may rest on complex legal distinctions that bear little relationship to economic principles'' (292). Beginning in the early 1980s, the definition of a market used by the Department of Justice in antitrust proceedings changed to ''a product or group of products and a geographic area in which it is sold such that a hypothetical, profit-maximizing firm, not subject to rate regulation, that was the only present and future sell of those products in that area would impose a 'small but significant and nontransitory' increase in price above prevailing or likely future levels.'' ** Such a definition, as noted by Scheffman and Spiller (1985) and Spiller and Huang (1986), relies on counterfactual rather than observed behavior.

With network technologies, the complications of market definitions are more difficult. Connections of markets may emanate from the demand side of the market, e.g., wheat from North Dakota and wheat from Kansas may be substitute movements by a receiver in Portland. Connections may also emanate from the supply side, e.g., if economies of density and/or scope exist.† That is, the pricing of a commodity on a link in the network is affected by pricing of commodities either on the same link or, if economies of scope exist, by pricing on other links in the network.‡ Such interdependencies require recognition that railroad firms are multiproduct firms that operate over spatially separated nodes in a network. This recognition is very much consistent with Winston (1985), who points out that the output of a transportation firm ''is the movement of a commodity or passenger from a specific origin to a specific destination over a particular time period'' (60).

Following the discussion above and Winston (1985), railroads operate in many markets, hauling different commodities between lots of different origin-destination pairs. In aggregate,

---

*We note that the Department of Justice uses Herfindahl indices as one factor in considering whether to contest a merger between firms in a market. The Herfindahl index is simply the sum of squared firm market shares.

** The *Journal of Economic Perspectives* (1987) has a series of papers on the issues related to the definition of markets. See Schmalensee (1987), Fisher (1987a), and White (1987) for relatively detailed discussions of market definitions. In addition, Uri and Rifkin (1985) and Westbrook and Buckley (1990) have excellent discussions related to the rail market.

† Economies of density means that as output over a given link increase, per unit costs of providing the service fall. Economies of scope means that a single multiproduct firm can produce two or more outputs more cheaply than can separate specialized firms who together product the small level of outputs.

‡ Berry (1992) and DeVaney and Walls (1999) point out these and related interdependencies over a network.

however, railroads dominated U.S. freight transportation during the latter half of the 19th century and through the first few decades of the 20th century. In the early part of the 20th century, new modes developed and the dominance of the 19th century eroded. Today, although railroads continue to play an important role in the nation's overall transportation system, that role is much narrower. As reported in the 1997 Commodity Flow Survey, railroads provided 38.4 percent of all ton-miles moved, pointing to their continued importance. However, they provided only 4.6 percent of transportation services expressed in FOB value of shipments,* suggesting their service is larger for lower-valued commodities. Further, while they hauled 38.4 percent of ton-miles, this figure corresponds to only 14 percent of all tonnage moved. Together, these statistics suggest they haul relatively low-valued commodities over relatively longer distances.

In the remainder of this section, we describe the specific markets in which railroads operate. Our discussion initially focuses on identifying the products hauled and the extent of operations in each product category. We then describe the supply side of the market. In this regard, we describe the network over which railroads operate and how it has changed over time. We then turn to describing the number of firms, concentration levels, and how changes in regulation, firm behavior, and network characteristics have translated into changes in operating characteristics and costs.

34.3.1 Products Hauled

Railroad markets defined in terms of origins-destinations and commodities suffer from data availability and sheer magnitudes of the number of markets and potential markets served. Data at this level of disaggregation are extremely difficult to access and use in characterizing markets. However, as part of the reporting requirement to the Surface Transportation Board (formerly the Interstate Commerce Commission), railroads do provide aggregate measures of the products they haul. These data are summarized, by Standard Transportation Commodity Code at the two-digit level (STCC-2 code), in Table 34.1. In this regard, we provide, the commodities hauled (at the two-digit level), along with *originated* tonnages,** revenues, and the revenue per ton received from each product category.

In total, the industry hauled approximately 1.1 billion tons in 1983. By 1997, the amount hauled by the industry was about 1.6 billion, an increase of about 45 percent since 1983. These tons reflect movements of 36 different commodity (STCC-2) codes and encompass a wide array of different types of movements. The diversity of products hauled by railroads is large and includes coal, farm products, chemicals, machinery, freight forwarder traffic, etc. Despite the degree of diversity in the commodities that railroads haul, coal movements have dominated and continue to dominate railroad tonnage. Specifically, coal is far and away the leading commodity (in terms of originated tonnage) in both 1983 and 1997. Coal accounted for about 39 percent of all railroad in 1983 and increased to about 44.5 percent in 1997. When examining the contribution of coal to total revenues, however, the percentage is much lower. That is, in 1983 only 26 percent of railroad revenues were from coal ($5.693 billion), and that figure, despite the increase in market share, remained roughly the same, 26.6 percent, in 1997. After coal, there is a considerable drop-off in terms of percentage of tons originated and the revenue received from the various other commodities. In 1997, the next three largest commodities hauled (chemicals, farm products, and nonmetallic minerals) accounted for a total of 23.62 percent of tonnage originated and about 23 percent of total revenues.

Different commodities are and have been priced differently. These commodities have very different demand, cost, and competitive characteristics, and, as a result, it is not surprising

---

*FOB means free on board. It does not include the logistics costs of movements.

**Railroads often interline with other railroads to serve origin-destination pairs. The tonnage figures we report are the sum of tonnages originated on the railroad lines regardless of whether the service terminates on the associated railroad's lines or another railroad lines.

TABLE 34.1 Commodity Codes, Tonnage, Revenues, and Rates 1985 and 1997

		1985			1997		
STCC	Description	Tonnage (000,000)	Revenue (000,000)	Rev/Ton $/ton	Tonnage (000,000)	Revenue (000,000)	Rev/Ton $/Ton
stcc	tonssum revsum rpt						
1	Farm products	130.98	1,805.05	13.78	125.56	2,151.62	17.13
8	Forest products	0.37	10.45	28.19	0.51	9.09	17.98
9	Fresh fish or other marine products	0.02	0.65	28.05	0.17	5.15	30.93
10	Metallic ores	66.56	326.98	4.91	31.85	267.76	8.40
11	Coal	426.92	4,013.19	9.40	705.12	6,693.90	9.49
13	Crude petroleum, ntl gas or gasoline	1.40	16.62	11.83	2.15	17.94	8.34
14	Nonmetallic minerals	55.44	434.18	7.83	109.30	771.87	7.06
19	Ordance or accessories	0.27	9.93	37.15	0.26	8.32	32.00
20	Food or kindred products	67.94	1,302.71	19.17	85.71	1,745.73	20.36
21	Tobacco products	0.22	6.75	31.37	0.04	1.88	46.37
22	Textile mill products	0.32	11.34	35.37	0.15	6.24	41.45
23	Apparel	0.11	5.33	50.63	0.43	22.01	50.99
24	Lumber or wood products	48.60	813.75	16.74	48.14	893.20	18.55
25	Furniture or fixtures	0.45	29.43	65.60	0.41	29.26	70.86
26	Pulp, paper, or allied products	26.86	526.58	19.60	32.12	639.29	19.90
27	Printed matter	0.19	5.47	29.52	0.68	30.80	44.99
28	Chemicals or allied products	82.33	1,462.53	17.76	139.78	2,878.24	20.59
29	Petroleum or coal products	32.00	498.40	15.57	39.25	673.07	17.14
30	Rubber or misc. plastic products	1.30	48.20	37.18	1.31	50.99	38.97
31	Leather or leather products	0.05	2.02	41.36	0.06	2.02	35.67
32	Clay, concrete, glass, or stone	33.69	449.59	13.34	40.95	641.32	15.66
33	Primary metal products	26.73	414.75	15.51	49.56	833.68	16.82
34	Fabricated metal products	0.75	26.99	36.01	0.74	29.65	40.13
35	Machinery	0.83	36.47	43.86	0.96	49.53	51.49
36	Electrical machinery or equipment	1.09	58.18	53.61	1.15	82.68	71.95
37	Transportation equipment	21.13	1,033.00	48.87	31.40	1,967.14	62.64
38	Instruments or photographic goods	0.05	1.95	39.25	0.03	1.84	64.57
39	Misc. products or manufacturing	0.11	6.18	55.76	0.74	37.87	51.07
40	Waste or scrap metal	20.74	264.78	12.76	36.98	531.01	14.36
41	Misc freight shipments	1.16	47.36	41.00	1.94	75.96	39.15
42	Containers, shipping, returned empty	0.92	25.98	28.32	4.72	166.67	35.31
44	Freight forwarder traffic	1.29	50.40	39.04	4.92	290.29	58.98
45	Shipper asso. or similar traffic	7.89	429.83	54.45	0.85	29.96	35.42
46	Misc. mixed shipments	30.14	1,209.04	40.10	86.40	3,501.05	40.52
48	Waste hazardous or substanances	0.00	0.00	.	0.91	23.00	25.24
All	All Commodities	1,089.01	15,384.04	14.12	1,585.24	25,160.06	15.87

that rates vary across commodities. We summarize average rates per ton by STCC-2 code in Table 34.1. Overall, railroads earned about $14.12 per ton in 1983, which increased to $15.87 per ton in 1997 (weighted average by tonnage and expressed in nominal terms). In real terms (1992 base), the railroads earned 19.34 per ton in 1983, falling to 14.20 in 1997– nearly a 27 percent reduction in average revenues. The leading commodity hauled (coal) has an average rate per ton of about $9.49. This figure is well below the overall average rate per ton for all tonnage moved ($15.87). The rate per ton for the other major commodities hauled by the railroad (chemicals, farm products, and nonmetallic minerals) is $20.59,

$17.13, and $7.06, respectively. For other commodities, the rates tend to be much higher. For example, the average rates of apparel (23), furniture or fixtures (25), transportation equipment (37), instruments or photographic goods (38), and freight forwarder (44) traffic each earn in excess of $50 per ton. However, the combined market share (tonnage) for these relatively high-rated commodities is only 2.3 percent. In contrast, the combined market share of the leading five commodities (in tonnage), which yield lower rates is, 73.5 percent.

34.3.2 Shipment Characteristics

From Table 34.1 it is clear that railroads tend to haul a lot of different commodities but specialize in commodities that do not command high rates. There are several possible explanations. Some of these explanations relate to the characteristics of the commodities (and their associated demand and supply characteristics), and some relate to characteristics of how the service is provided. The characteristics we describe include the value of the commodity shipped, shipment sizes, density (weight) of the product, distances shipped, and the number of railroads involved in the movement.

Demand and supply characteristics of the commodity shipped have an important and obvious influence on the rate that can be charged. Table 34.2 displays commodity values per ton.* These data reinforce a number of points we have made throughout this chapter. In particular, railroads tend to haul low-value bulk commodities, which tend to be shipped over long distances with large shipment sizes. Indeed, coal (11) and nonmetallic minerals (14), commodities which ranked as number 1 and 4 in terms of tonnages in 1997, each have FOB values considerable less than the other commodity classes. Higher-valued goods, such as transportation equipment (37), tend to flow by other modes. Such higher-valued commodities have higher inventory costs, and as a result, motor carriers, which offer faster transit times, tend to dominate the movements of these commodities. This follows directly from the work

TABLE 34.2 Shipment Characteristics for Selected STCC Codes

Description	STCC	Cars/ ship	Wt/ car	Miles	RR	Rev/ RTM	Value
Farm products	1	5.3	70.7	1,118	1.20	5.32	223.74
Metallic ores	10	10.9	92.4	928	2.18	5.27	235.58
Coal	11	22.8	106.2	628	1.34	3.31	20.75
Nonmetallic minerals	14	5.8	93.0	708	1.26	5.04	11.58
Food & kindred prod.	20	1.1	65.8	1,120	1.19	6.39	996.65
Lumber & wood prod	24	1.0	76.4	1,032	1.34	6.75	190.94
Pulp, paper & prod.	26	1.0	59.3	1,054	1.63	9.20	897.66
Chemicals	28	1.1	83.9	931	1.42	10.42	977.08
Petrol & coal prod.	29	1.6	75.1	870	1.45	8.66	190.61
CCGS	32	1.3	85.4	772	1.52	6.16	114.28
Primary metal prod.	33	1.2	81.7	843	1.38	8.57	858.12
Transportation equip	37	1.0	21.4	887	1.13	36.43	7,446.89
Waste or scrap	40	1.2	64.7	584	1.24	16.39	139.49
Misc. mixed shipments	46	1.0	15.1	1,444	1.05	6.99	N.A.

Note: All information in the table except for Value was generated from the Waybill Sample for 2001. Value per ton is based on the 1993 Commodity Flow Survey.

---

*These data were calculated using data from the 1993 Commodity Flow Survey.

of Baumol and Vinod (1970) and other related research discussed in the surveys by Oum, Waters, and Yong (1992) and Winston (1985) which point to the notion that rate is just one element of the cost of alternative modes. Service characteristics do have a strong and important effect on the decision of shippers on which mode to use.*

In addition to demand considerations, how the service is provided also has an enormous influence on costs and through costs on rates. In this regard, there are a number of differences across commodity groupings. For example, coal movements tend to occur through larger shipment sizes, which allow railroads to use unit trains and multiple cars in servicing the demands. Such operations are much more efficient than are single-car movements and, as a result, point to lower costs and rates. Coal movements have the largest average shipment size, with more than 22 cars. Other major commodity groups such as farm products, metallic ores, and nonmetallic minerals also tend to be multiple car shipments. The remaining commodities are dominated by single-car movements, and we note that the average rate per ton-mile is larger than are the rates for coal, farm products, metallic ores and nonmetallic minerals.

Coal, farm products, metallic ores, and nonmetallic minerals also tend to have larger weights per car. Certainly, this likely represents differences in the type of car used but also reflected is the density of the product. For example, on a per-ton basis, feathers are far more expensive to ship than lead is on a per-ton basis. Thus, one would expect the rate attached to feathers to be much higher. Indeed, the rates per ton-mile are lower for higher car weights.

Overall, railroads ship commodities long distances. In all cases, shipment distances are in excess of 500 miles. In the 1997 Commodity Flow Statistics, the average shipment distance for all shipments (all modes) is 472 miles. For comparison purposes, for-hire truck, private truck, and shallow draft water are 485, 53, and 177 respectively. The provision of a service involves both variable costs as well as "quasi" fixed costs. Quasi fixed costs are costs that are incurred with the provision of service but do not vary with the extent of the service provided. It is well documented that as average lengths of haul increase, rates fall owing to the reduced per unit cost of providing the service as mileage increases. It is also well documented (e.g., Locklin 1972) that greater shipment distances generally favor rail movements vis-à-vis truck since rail per unit costs fall with distance faster than do truck per unit costs. Again, across commodities, there are considerable differences. Coal travels an average of 628 miles, miscellaneous mixed shipments travel the farthest (1,444 miles), while other primary products, including farm products (01), nonmetallic minerals (14), and chemicals (28), travel 1,118, 708, and 931 miles, respectively.

A final consideration, shown in Table 34.2, is the number of railroads involved in a movement. Specifically, railroads often work with other railroads to complete a movement from one location to another. An individual rail network may not be able to service a given origin-destination pair. But through interlining, railroads can extend the number of origin-destination pairs that can be serviced. In the table, a value of one means that one railroad both originates and terminates the movement on its own lines. While there is no discernible pattern across commodities, we do note that there is considerable interlining, particularly for metallic ores (10), pulp, paper, and products (26), and clay, concrete, glass, or stone (32). Interline movements are, of course, more costly than single-line movements and, as a result, should have an increasing effect on costs and through costs on rates.

In summary, the leading commodities of coal, chemicals, farm products, ores, etc. tend to be relatively low-valued commodities (except chemicals), which are hauled in large shipment sizes, have higher weights per car, are shipped long distances, and sometimes involve interlines–the railroads tend to operate in such markets wherein they have an advantage

---

*There are scores of demand studies that reinforce the notion that service characteristics play an important role in shipper demand decisions. See surveys by Winston (1985) and Oum, Waters, and Yon, (1992), and Small and Winston (1999). Also see Inaba and Wallace (1989) for an excellent example of such research.

relative to many of the other modes.* As we report in the ensuing sections, many of the changes occurring in the last 20 years reflect innovations, pricing, and consolidations that have occurred largely as a result of partial deregulation and that have yielded tremendous efficiency gains to railroads.

34.3.3 Network Characteristics and Firm Size

When the Staggers Rail Act was passed in 1980, railroads provided services over 180,000 miles of road. Over time, this figure has fallen to about 120,000 miles of road (Figure 34.1 contains this information from 1983 through 1997). Two primary factors explain the reduction in network size. Through the 1970s and '80s, there has been ongoing pressure on the railroads to abandon unprofitable low-density lines (branch lines). Under partial deregulation, some of the lines have indeed been abandoned, while others have been sold to form new short-line and regional railroads.** Overall, however, the size of the rail network has fallen. Corresponding to the fall in network size, however, there has been an increase in the ton-miles (the unit of output traditionally used in railroad economics). Indeed, industry ton-miles increased from about 800 billion in 1983 to about 1.3 trillion in 1997 (Figure 34.1) with a corresponding increase in network utilization (Figure 34.2).

Within firms, the changes are dramatic. Since partial deregulation, there has been an unprecedented consolidation of firm outputs and network through merger activity. The number of firms has fallen dramatically, from 28 in 1983 to just 9 in 1997 (Figure 34.3).† Associated with the consolidation of firms has been a tremendous increase in the size of firm networks and in output levels. While the size of the industry network has fallen, the average size of firm networks has increased from about 6000 miles per firm in 1983 to over 13,000 miles per firm in 1997 (Figure 34.4). In terms of output, the effects are even more dramatic. In 1983, Class I railroads operated an average of about 29 billion revenue ton-miles (Figure 34.4). By 1997, this figure had increased by more than a factor of 5 to nearly 150 billion revenue ton-miles (Figure 34.4). These figures together suggest that network utili-

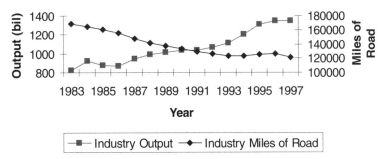

FIGURE 34.1   Industry output and network size.

---

*The primary source of competition for these commodities is generally either commercial navigation or local commodity usage, where truck transport is feasible such as local grain processing, local livestock operations, or local electricity generation

** Allen (1990), Klindworth (1983), and Tye (1990) describe some of these changes and effects.

† The decrease in firm numbers is due to the declassification of firms as Class I railroads. From 1983–1997, six railroads (BLE, BM, DH, DMIR, FEC, PLE) were declassified. In addition to these declassifications, there were 12 mergers, which accounted for remaining disappearance of firms. See Bitzan (1999) and Wilson and Bitzan (2003) for greater detail.

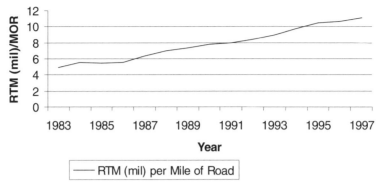

FIGURE 34.2 RTM (mil) per mile of road.

zation has been tremendously impacted by partial deregulation and continues to grow. The increases in outputs, networks, and utilization are very important in terms of realizing greater economies of density.

In addition to growing outputs, networks, and network utilization, there are a number of other sources of change in the industry. Specifically, economists have long examined the structure of costs in the railroad industry.* In so doing, they have provided a number of measures of operating and/or network characteristics, e.g., average length of haul, percentage of unit train traffic, percentage of interlined traffic, etc. Many of these have been impacted and have changed dramatically over the last two decades. By and large the primary variables are similar to those described earlier in Table 34.2.

As discussed earlier, longer lengths of haul are generally thought to be efficiency-enhancing. As we show in Figure 34.5, average length of haul has been increasing dramatically over the span of the data. In 1983, average length of haul across railroads was about 360 miles, but increased to nearly 500 miles by 1997 (arithmetic means across railroads). A change in average length of haul from 360 to 500 miles is a 39 percent increase, which points to dramatic reductions in costs.

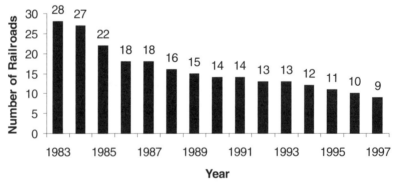

FIGURE 34.3 Number of railroads.

---

*See, for example, see Brown, Christensen, and Caves . (1979), Keeler (1974), Braeutigam and Daughety (1982), Caves et al. (1980; 1981, 1985), Tolliver (1984), Vellturo et al. (1992), Berndt et al. (1993), Friedlaender et al. (1993), Wilson (1997), Bitzan (1999), and Ivaldi and McCullough (2001).

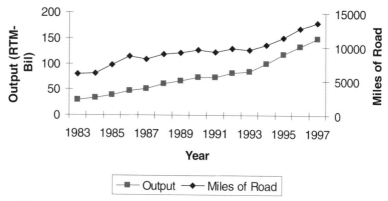

FIGURE 34.4 Average firm size.

Another dramatic change in railroad operations is the growing emphasis on unit train traffic. Unit train traffic follows a much different and much more efficient operation if the volumes are large enough, as we discussed in section 34.2. Such traffic occurs in large volumes between a specific shipper and receiver who have made investments, e.g., high speed loaders that allow large shipments. In Figure 34.6, we compare the use of unit train versus through train operations* and show that there has been an increased use of unit trains through the 1980s and '90s, while the more traditional use of through trains has fallen.

The increase in firm sizes, the changes in networks, the consolidation of firms, and the changes in traffic patterns have made firm inputs much more productive. In Figure 34.7, we have calculated the weighted (by revenue ton-mile) outputs per inputs. The inputs we use are labor, fuel, locomotives, and rail cars.** In all cases, we normalized the variables by

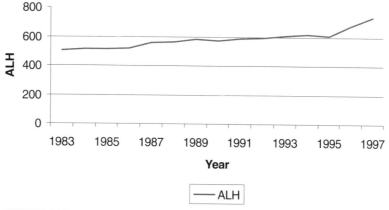

FIGURE 34.5 Average length of haul.

---

*Way train movements are the residual.

**We do note that the percentage of both types of equipment that are owned by the firms has fallen through time. In 1983, railroads owned 76 percent of the locomotives and 82 percent of the rail cars used in moving commodities (weighted averages by revenue ton-miles). By 1997, each these figures had fallen to 67 percent, reflecting a now longstanding trend to the use of leased equipment.

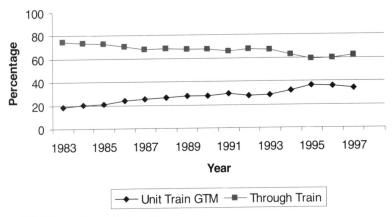

FIGURE 34.6 Type of train traffic.

1983 values. In evaluating these figures, it is clear that across the board, the primary inputs used by railroads have yielded efficiency gains. The production of both labor and cars are the largest, increasing output per unit input of over 150 percent. The other two inputs discussed here (network and miles of road were discussed earlier) have also increased, just not as dramatically. In particular, output per locomotive has nearly doubled, and output per gallon of fuel has increased by about 30 percent.

Obviously, when output per unit input is increasing *for all inputs,* costs per unit produced is falling. Indeed, econometric research (e.g., Wilson 1997) supports the fact that railroad costs have fallen and fallen dramatically since partial deregulation. In documenting the outcome of the efficiency enhancements, we simply plot average variable costs over time (Figure 34.8). Variable costs in this regard include fuel, labor, equipment and materials and supplies. To express the values in real terms, we used the GDP price deflator with a base year of 1992.* By any measure, the decreases in per unit cost have been large. In nominal terms,

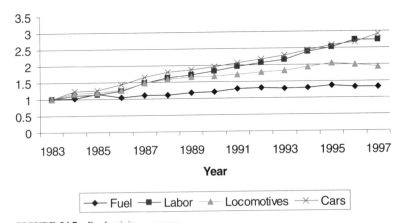

FIGURE 34.7 Productivity measures.

---

*They do not include way and structures. We also include an opportunity cost in the measurement of capital items. For a complete discussion of variable definitions, see Bitzan (1999) and Wilson and Bitzan (2003).

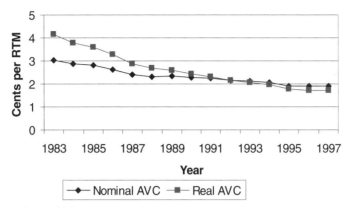

FIGURE 34.8 Average variable costs.

per unit costs fell from 3.03 cents per RTM in 1983 to 1.92 cents per RTM in 1997–a 36 percent reduction in per unit costs. In real terms, the change is much larger. In real terms, per unit costs fell from 4.14 cents per RTM in 1983 to 1.72 cents in 1997–a 58 percent reduction in per unit costs.

34.3.4 Changes in Market Structure

In addition to demand and cost (network) characteristics, the structure of the marketplace theoretically may have a large effect on explaining market outcomes. As noted earlier, there have been a number of mergers over the last two decades, which, by any standard, has led to a consolidation of market power. Industrial organization economists use a variety of measures to gauge the level of market power. Historically and most commonly, both concentration ratios and Herfindahl indices are used.* The employment of such measures rests heavily on the definition of the market, and as we discussed earlier, network markets are best characterized in terms of the flows of a commodity between two origin-destination pairs. We present these measures in this section, simply as descriptive measures of the consolidation of output among firms operating in the United States.

Presented in Figures 34.9 and 34.10 are four-firm concentration ratio (CR-4) and the Herfindahl (H) ratio in the United States from 1983–1997.** As seen in these figures, the consolidation of firms through merger activities has led to an astounding increase in concentration, particularly over the last 5 years of data. In 1983, the CR-4 was only 47 percent, but by 1997 it had risen to 87 percent. Hence, four firms provide 87 percent of ton-miles produced. By any standard, this level of concentration is large. The Herfindahl index yields a comparable conclusion. But, in examining the data, we note that most of the mergers in the latter period fall within DOJ anticompetitive guidelines. That is, the postmerger Herfindahl index is in excess of 1,800 and the effects of the merger are to increase the H by more than 100 points. Specifically, the UP-SP merger in 1995 and the BN-ATSF merger in 1996 would have fallen under DOJ merger guidelines. These measures point to increased market power in the railroad industry. While both costs and revenues per unit have been falling, the increase and growth in concentration point to potential inefficiencies in the market.

---

*These measures have been subject to some debate, and recent advances in industrial organization have led to more direct measures of market power. Nonetheless, these two measures, CR and H, continue to be used routinely to describe the structure of the marketplace.

**The CR-4 is the sum of the largest four firms' market shares. The H is the sum of the squares of market shares for the firms in the market.

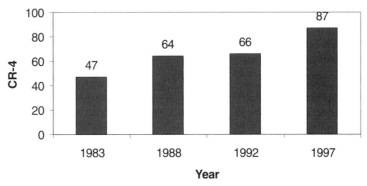

FIGURE 34.9 Four-firm concentration ratio (CR-4).

34.3.5 Railroad Markets and the Future

This section has attempted to summarize, albeit briefly, railroad markets. In this regard, we found that railroads operate in a lot of different commodity and geographic markets. However, the traffic is dominated by commodities that are low in value and can be shipped in large volumes. As a result, rates tend to be relatively low. This pattern of shipments that has been longstanding and will likely remain so. But, since deregulation, there have been a number of changes in railroading that point to considerable efficiencies. These include a sizable reduction in the size of the industry network, an increase in firm sizes, and improvements in operating characteristics such as output per route mile, average length of haul, and increases in shipment sizes, which explain tremendous efficiency gains. Some of these efficiencies have been gained as a result of pricing practices (e.g., unit train rates), but some are due to mergers and innovations geared to labor- and fuel-saving devices. In total, these effects explain the fact that per unit costs have fallen, and have fallen dramatically over the last 20 years.

Weighing against these cost reductions is growing levels of concentration in the U.S. railroad industry. Our data pertain to the whole of the United States as the market (all U.S. class I firms). However, all of the firms have limited geographic scope and the preponderance of shippers are captive shippers, i.e., served by one railroad. As a result, the overall level of concentration is likely understated and the extent of market power is likely larger than one

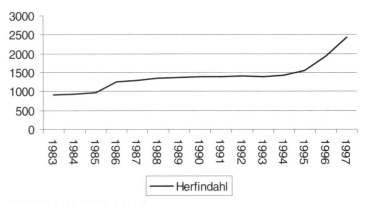

FIGURE 34.10 Herfindahl index.

might infer from these figures. Indeed, at the highest level of disaggregation, commodity movements between origin-destination pairs is probably the best level of analysis. It is at this level that the rigors of intramodal, intermodal, geographic, and product competition are best addressed.

The railroads have actively promoted larger shipment sizes through both pricing and operating practices. Bulk commodity shippers are given more favorable rates when shipments exceed specific size thresholds. Moreover, the railroads have worked to consolidate geographically dispersed shipping locations into more centralized, high-volume origins, and destinations. The result is that grain shipment size increased by 69 percent between 1995 and 2001, while coal shipment size increased by 410 over the same period (Waybill Statistics).

While railroads have generally conceded high-valued, low-volume traffic to motor carriers, they have worked hard to retain high-valued traffic where shipment volumes are sufficient to allow unit train operations. Thus, we observe a continued strength in automobile, chemical, and TOFC/COFC traffic. Higher-valued traffic will certainly never dominate overall tonnage, but it continues to grow in importance as a source of revenue.

Looking to the future, it is likely that both trends will continue. Other available transport modes do not have the capacity (trucking) or the route flexibility (commercial navigation) to compete effectively for most of the bulk commodity traffic that currently moves by rail. At the same time, general and significant constraints on the potential growth of truck traffic of any kind mean that the railroads have the potential to recapture an increasing share of higher-valued shipments when and where shipment volumes are relatively large and/or shipment distances are relatively long.* Whether or not railroads are successful in doing so depends heavily on whether they can provide a level of service that at least approaches that offered by motor carriers.

34.4   *REGULATION AND PARTIAL DEREGULATION***

Regulation of railroads, as noted by Keeler (1983), was rooted in British common law dating to the 16th century. From common law emanated common carrier obligations. That is, modes predating railroads were often granted operating certificates, and these carried into railroads. Keeler notes, "[T]he notion behind common carriage was that the government would grant the carrier certain powers and privileges, generally conferring an exclusive right to make a profit from transportation. In return for these privileges, the carrier was expected to assume certain obligations" (Keeler 1983, 20). Such obligations were that carriers may not refuse service, must provide service at a reasonable price, must provide all equally, and are responsible for safe delivery.† Owing to difficulties in implementing such regulations through the court system, due to growing and more complex railroad networks,‡ numerous states passed laws to regulate railroads during the 1870s and 1880s.§

---

*The nation's highway system is largely complete. Moreover, the ability to expand this system further is measurably constrained by both environmental concerns and fiscal realities. Thus, the availability of additional motor carrier capacity depends heavily on our ability to better use the network that is currently in place.

**The material covered in this section can be found in a number of sources. See, e.g., Hilton (1969), Keeler (1983), Locklin (1972), Pegrum (1968), and Winston et al. (1990, 1993).

†Daggett (1928) and Keeler (1983). In addition to common law principles, such obligations were manifested in corporate charters.

‡Such difficulties included differences in court decisions, differences in the conflicts in corporate charters, rate wars on competitive routes, and extremely high rates where competition was not present. Finally, in 1886, a Supreme Court ruling, *Wabash, St. Louis, and Pacific v. Illinois,* 118 U.S. 557 (1886), found that states were not "empowered" to regulate interstate commerce (Keeler 1983).

§In addition to Keeler (1983), Locklin (1972) and Pegrum (1968) provide excellent overviews of the history of railroad regulation and its roots.

The Interstate Commerce Act of 1887 was the first piece of federal regulation to regulate railroads.* Beginning with this legislation and various amendments, economic regulation by the Interstate Commerce Commission grew to include maximum rate regulation (Hepburn Act of 1906), minimum rate regulation (Transportation Act of 1920), entry and exit (Transportation Act of 1920), and promotion of mergers (Transportation Act of 1920).**

Throughout the 1920s and 30s, other modes began to develop and the railroads' dominance of transportation began to erode. During this time, other modes came under regulation. The Motor Carrier Act of 1935 and the Transportation Act of 1940 brought much of the motor carrier industry and some of the barge industry under ICC jurisdiction. By the end of World War II, virtually every aspect of railroad operations was tightly controlled by the Interstate Commerce Commission (ICC). In particular, the ICC had jurisdiction over rail rates, line abandonment, service discontinuities, rail mergers, car flows, and interchange rules.

The control, however, was subject to considerable criticism. Value-of-service pricing resulted in commodities captive to railroads paying higher prices than commodities where the railroads faced competition regardless of cost similarities. Small shippers tended to pay the same price as large shippers despite large cost differences in the provision of service. Innovations such as the Big John Hopper car by the Southern Railway and unit/multiple car movements were discouraged to protect other modes of competition and shippers. Rail lines could not be abandoned even though they were no longer profitable, and mergers were given much scrutiny, with proceedings lasting for years before a decision was reached.

By the early 1970s, it became clear to policy-makers that the then-current set of regulations and the associated regulatory practice were no longer serving its purpose. New innovations were not being adopted, market shares continued to fall, and there were a series of high-profile railroad bankruptcies.† In response, the federal government began the process of relaxing economic regulation of the railroad industry with passage of the Railroad Revitalization and Regulatory Reform (4R) Act of 1976. This legislation provided subsidies to ailing railroads and began the process of regulatory change. The regulatory changes included some rate flexibility for railroads, easing of merger restrictions, and the relaxation of some abandonment rules. However, a general assessment of that legislation and, in particular, its application by the ICC suggests that it did not offer the sort of meaningful regulatory relief needed to rescue the ailing industry.‡ Consequently, Congress acted again in 1980 to institute regulatory relief through the passage of the Staggers Rail Act.

The Staggers Act effectively ended the regulation of most railroad rates. Continuing to work toward the objective of the 4-R Act, a regulatory screening process was put in place. Specifically, the reasonableness of the rate could only be considered if the movement in question was found to be "market dominant." § The decision of whether a particular movement was the result of market dominance involved two stages. First, if the revenue-to-variable cost ratio (R/VC) was less than some given threshold, the rate could not be considered market dominant. If the R/VC ratio was in excess of the given threshold, then there was a qualitative evaluation performed to determine whether or not competitive pressures (i.e., intramodal, intermodal, product or geographic competitive elements) were present. If so,

---

*Since passage there have been a number of legislative bills enacted. Most notably, these include the Transportation Acts of 1920 and 1940, the 3-R Act in 1973, the 4-R Act in 1976, the Staggers Rail Act of 1980, and the ICC Termination Act of 1995.

**This statue allowed the ICC to force mergers to "save" railroad service from weak railroads.

† Such bankruptcies included the Penn Central, the Erie-Lackawanna, the Lehigh Valley, the Reading, the Central of New Jersey, the Katy, and later the Rock Island and the Milwaukee Road. The Regional Reorganization Act of 1973 created the Railway Association, an agency created to deal with the problems of railroads in the Northeast. One solution was to plan for public ownership.

‡ Winston et al. (1990).

§ See Eaton and Center (1985) and Wilson (1996) for a complete discussion of the rules, and Wilson (1996) for a discussion of railroad pricing behavior under such rules.

then the market was not considered market dominant and therefore not subject to reasonableness proceedings. As noted in Friedlaender and Spady (1981), the presence of competition was defined relatively liberally, except for coal markets. As a result, railroads had considerable latitude in the establishment of rates. The new law also allowed rail carriers to establish confidential contracts with shippers.* This provision effectively ended any requirement that rates be nondiscriminatory, thus opening the door to a number of revenue-enhancing and cost reducing innovations. Most notable among these practices was the use of multiple car shipments and unit trains. Finally, Staggers also radically reduced the regulatory hurdles formerly associated with approval of mergers and branch-line abandonment, and, as we documented above, the result has been a unprecedented consolidation of railroad outputs into larger firms.

Because of the railroads' ailing financial condition, there was a general expectation that deregulated railroad rates would increase. Indeed, initially after passage, rates did rise for a number of commodities, but it was not long before the rates for most commodities (in real terms) fell below prepartial deregulation time periods Wilson (1994). Under partial deregulation, the cost-cutting activities were implemented, and as a result, real operating costs per unit output fell, and fell dramatically.**

The Staggers Act also severely limited the activities of the rate bureaus through which railroads had historically gathered to recommend rates to the ICC. This provision, coupled with the ability to engage in confidential contracts, opened the door to rate competition between railroads, and often, where direct rail-to-rail competition did not exist, the competitive pressure exercised by motor carriage and commercial navigation assumed a disciplinary role.

The end result of all of these measures was a dramatic decrease in average operating costs that were rapidly transformed into lower rail rates. In addition, there were also tremendous improvements in the quality of service. During the first years after deregulation, transit times fell by 14 percent and the variability of transit times fell by 36 percent.† With pared-down route systems and marginally greater profits, the railroads began to invest more heavily in their remaining facilities. These investments helped sustain the continuing reduction in operating costs.

With the number of changes in the railroad operating environment, along with motor carrier deregulation, further change was implemented when President Clinton signed into law the ICC Termination Act of 1995. This statute replaced the ICC with a much smaller Surface Transportation Board (STB). Currently, it is the STB that is responsible for adjudicating shipper-based rate cases brought under the Staggers Act; it is the STB that now evaluates proposed mergers; and it is the STB that is more generally responsible for ensuring there is adequate competition in rail-served markets.

Given the market outcomes since 1980, it is hard to imagine any dissatisfaction with the deregulation process. However, there were other results of the Staggers Act that have drawn less favorable review. At the time of the statute's passage (1980) there were approximately 40 class I railroads operating in the United States. In 2000, this number had fallen to only 8. As noted by Wilson and Bitzan (2003) and others, Staggers streamlined the merger ap-

---

*See Hanson et al. (1989, 1990) and Barnekov and Kleit (1990) for further discussion.

**Richard C. Grayson, former president and CEO of Burlington Northern, once remarked, "We started with costs; we didn't know a thing about pricing, but we knew how to cut costs." Many of these activities are documented in Wilson (1997). MacDonald (1999) also provides a excellent synopsis of cost-cutting measures. Of course, industry consolidation has had an effect as well. Wilson and Bitzan (2003) document some of these effects. They find that the consolidation of output among firms in the industry account for a 20 percent reduction in industry costs over the time period 1983–1997. Finally, less viable but important innovations include the replacement of cabooses with way-side detectors and end-of-train (EOT) devices, the elimination of the fireman's position from locomotive service, increased mechanization of maintenance of way activities, and an across-the-board reduction in labor force. For a full description of the increase in railroad productivity see Bitzan and Keeler (forthcoming), Wilson (1996, 1999) and Davis and Wilson (1999, 2003).

† See Grimm and Smith (1986).

proval process and the rail carriers were quick to seize this new opportunity. The result of the sustained merger activity is that most rail shippers have very limited choices among rail options. When shipment and commodity characteristics are such that alternative modes are good substitutes and when those alternative modes are available, the paucity of rail competition is less confounding. However, for those shippers that must use rail, the reduction in the degree of rail-to-rail competition has lead to increasing captivity, i.e., shippers without meaningful competitive alternatives.

At the same time the ICC was being replaced with the STB, the railroads undertook another round of significant mergers.* Not only did this further reduce the degree of direct rail-to-rail competition, but the implementation of these mergers routinely led to severe service disruptions. While the STB generally approved proposed mergers, in many cases it also granted trackage rights to alternative carriers in order to mitigate partially the loss of direct rail competition. Moreover, when merger implementation led to service quality degradation, the STB imposed strict monitoring systems to ensure a timely service recovery.

34.5 MODELS OF RAILROAD DECISION-MAKING

With partial deregulation, the business of railroading has become more complicated. In the course of the business of railroading, firms make an enormous number of decisions. We categorize these decisions into network design, input, and output/pricing decisions. Network design decisions involve the size, configuration, and maintenance of the network. Input decisions are those related to the employment of economic resources, such as labor, fuel, and equipment, to produce some given set of outputs, i.e., movements between origin-destination pairs. Output and pricing decisions include decisions of whether or not to serve a market (participation decision) and how much to supply. In this section, we develop a series of relatively simple models to characterize the pricing and output decisions over a network. The network size, configuration, and maintenance decisions have received some attention in the economics literature, and the input decisions have often been addressed jointly with cost considerations.**

34.5.1 Captive Shippers and Railroad Pricing

We first describe the pricing of services to a single shipper, who ships a single commodity between two locations vis-à-vis Winston (1985) and section 34.3. This simple model illustrates how competition enters into railroad decision-making. In the ensuing sections, we generalize the model to capture flows over a more complicated network involving multiple commodities under conditions of decreasing average costs (economies of density) and ultimately render a model that allows for intramodal competition from a separate network. We conclude this section of the novelty of this pricing model from others that have appeared in the literature.

---

*In the west, the Union Pacific acquired the Southern Pacific and the Burlington Northern merged with the Santa Fe. In the east, rivals CSX and Norfolk Southern jointly acquired, then divided Conrail. In the central United States, Illinois Central, a north-south carrier, was acquired by the Canadian National.

**In contrast to the cost function literature, a series of relatively recent papers, e.g., MacDonald and Cavalluzzo (1996), Davis and Wilson (1999, 2003), and Bitzan and Keeler (forthcoming) examine labor decisions and the influence of partial deregulation on both employment and costs.

We begin by noting that there is an array of choices confronting an individual shipper.* If the producer of the good is the decision-maker, the decisions include how much to ship, where to ship, and by what mode to ship. If the receiver is the transportation decision-maker, choices may again include how much to ship and by what mode, but also from where to receive the product. In pricing services to a single shipper, these choices may or may not limit the pricing power of a railroad. In fact, such principles are present in the market dominance standards under partial deregulation to represent the adequacy of competition in limiting rates.**

We assume that shippers are price-takers in all markets and have a well-defined production technology for each of the choices, with appropriate regularity conditions. Shippers are assumed to be profit-maximizing, given the prices of inputs and outputs and the technology that maps inputs into outputs. Given these conditions, shippers maximize profits by choosing appropriate levels of inputs and outputs for each of the choices described above. Using these profit-maximizing input and output levels, the maximum profits associated with *each choice* can be expressed as a function of the associated input and output prices, i.e., $\pi_i = \pi_i(P, w, m)$, where P, w, m represent the output price (the price the shipper receives for the product shipped), the input prices (the price of labor, capital, etc.), and the price of transportation by mode m, respectively.† The shipper has a number of alternatives indexed by $i = 1, 2, \ldots, I$. The actual choice observed is the choice i that yields the maximal profits from the set of profit functions representing options to the shipper.

In pricing services to an individual shipper, railroads must remain the preferred mode of the shipper to provide the service. The railroad then chooses a price and output combination that maximizes profits subject to the constraint that it yields shippers a profit level that is greater than or equal to the shippers next best alternative. Formally, the railroad's problem is stated as:

$$\text{Max } \pi = R(r)r - C(R(r)) \qquad s.t. \ \pi_r^S(r) \geq \pi_m^S \qquad (34.1)$$

where r = rail rate
$R(r; P, w)$ = shipper's demand for railroad service,‡
$\quad\quad C(\cdot)$ = railroad's cost function
$\pi_r^S(r) \geq \pi_m^s$ = the constraint that railroad service to shipper (S), by the railroad offering rate (r), yields profits at least as great as profits by the next-best alternative mode (m).

The first-order conditions for this problem can be written as:§

$$\frac{r - mc}{e} = \frac{\lambda - 1}{\varepsilon} \qquad (34.2)$$

*In this section, the model follows from Wilson (1996). The ensuing sections represent modest extensions of this model. The models are fully consistent with a discrete choice models of shipper demands. For surveys of this literature see Winston (1985), Oum, Waters, and Yong (1992), and Small and Winston (1999). Specific models that are useful for understanding our model are Daughety and Inaba (1978) and Inaba and Wallace (1989).

**Essentially, partial deregulation introduced a screening mechanism into regulatory design. Before a rate could be considered for its reasonableness, a ruling that the movement was market dominant was necessary. Market dominance criteria included a revenue to variable cost threshold and a qualitative evaluation of whether intramodal, intermodal, geographic, and product competition was not significant in limiting railroad rates. Recently, the geographic and product competition standards were removed as criteria (Bitzan and Tolliver 1998). Also see Eaton and Center (1985) and Wilson (1996).

†While transportation is an input to the shipper, we represent it separately here because it is the focal point of the ensuing discussion. We further note that the price of transportation is not the rate but the rate and the associated handling charges, inventory costs, etc.

‡We note that under the assumptions of the model the railroad demand function(s) can be derived from the associated shipper profit functions as $\partial \pi_i(P, w, r)/\partial r = - R(P, w, r)$

§This condition can also be written as $r/mc = 1/(\lambda - 1)/\varepsilon)$, under the assumption that $mc \cong vc$ is the foundation for the r/vc threshold used in market dominance proceedings.

This is the Lerner index of market power (the percentage markup of price over marginal cost). In this expression, λ is the multiplier attached to the participation constraint and ε is the shipper's elasticity of demand. The basic point of this derivation is that the rate charged in a particular railroad market may or may not be constrained by alternative choices of the shipper. Specifically, λ may take a value from zero to one. If zero, the monopoly price obtains with the result that the railroad charges a monopoly price and is not constrained by competitive alternatives. This is the extreme case of a captive shipper, who, as discussed earlier, has lost more and more alternatives over time. If λ is in the range from zero to one, the railroad price in a market is limited by the alternative choice of the shipper, which may reflect alternative terminal markets, alternative modes, alternative products, or some combination thereof. The closer these alternatives are to the movement in question, the more constrained is the railroad rate. In the limiting case of $\lambda = 1$, the railroad prices at marginal costs.

In the development of these profit-maximizing rates, we note that marginal costs play an important role. The lower the marginal costs, the lower the rate due to standard economic theory. We also note that as marginal costs fall, the railroad markup may become larger as a lower railroad marginal cost may not affect the next best alternative. Events through the 1980s and 1990s that have led to consolidation, innovations afforded by pricing flexibility (e.g., contract rates, unit trains, shuttle trains, etc.), and greater freedom to rationalize the network by abandoning or selling railroad lines have given railroads lower costs and greater levels of market dominance.

We should finally discuss conditions of economies of density. At the individual shipper level in this model, economies of density simply mean that the cost of serving a large shipper are lower than the cost of serving a small shipper. To the extent that alternative modes such as trucking do not operate under economies, railroads would tend to offer large shippers lower rates and the level of dominance of railroads over competing alternatives (modes) would be larger.

34.5.2 Multiple Shippers and Network Externalities

In this section, we add a second shipper to the model. While the generalization to multiple shippers seems innocuous, it complicates the model except under very specific circumstances that are not generally thought to hold in railroad economics. This second shipper may ship the same commodity or a different commodity. In presenting the model, there are two specific changes. First, the railroad has a participation constraint for each shipper. That is, the shipper may produce a different commodity or may have differing characteristics—e.g., size of firm, loading facilities, etc.—that may affect the payoffs attached to using rail vis-à-vis other discrete alternatives. Second, in providing the services to two shippers, the railroad becomes a multioutput firm. Depending on the nature of the technology, the multiple outputs may or may not be aggregated into a single output term. A few different possibilities are described below.

The development of shipper alternatives and choices was outlined in the previous subsection. With appropriate indexing of the two shippers (1 and 2), the railroad's profit-maximization problem becomes (given that it chooses to serve both shippers):

$$\text{Max } \pi = R_1(r_1)r_1 + R(r_2)r_2 - C(R_1(r_1), R_2(r_2))$$
$$\text{s.t. } \quad \pi_r^{S_1}(r_1) \geq \pi_{m_1}^{S_1} \qquad (34.3)$$
$$\pi_r^{S_2}(r_2) \geq \pi_{m_2}^{S_2}$$

The associated first-order conditions can again be rearranged to form the Lerner index. These expressions are given by:

$$\frac{\partial L}{\partial r_1} = R_1'(r_1)[r_1 - C_{R_1}(R_1(r_1), R_2(r_2))] + R_1 - \lambda R_1 = 0$$

$$\frac{\partial L}{\partial r_2} = R_2'(r_2)[r_2 - C_{R_1}(R_1(r_1), R_2(r_2))] + R_2 - \lambda R_2 = 0$$

$$\frac{\partial L}{\partial \lambda_1} = \pi_r^{S_1}(r_1) \geq \pi_{m_1}^{S_1}$$

$$\frac{\partial L}{\partial \lambda_2} = \pi_r^{S_2}(r_2) \geq \pi_{m_2}^{S_2}$$

(34.4)

Similar to the previous treatment, these expressions can be rearranged to define Lerner indices for both shippers and can be solved to yield profit-maximizing price levels. A key point and the purpose of writing out the full set of first-order conditions is to note the interdependence of the rates paid by each shipper. This dependence enters into the expression through the cost function. There are a number of cases to consider.

First, if the cost function exhibits constant marginal costs (i.e., $C = c_1 R_1 + c_2 R_2$), then the prices do not marginally depend on one another and the optimization is separable.* In this particular case, there is no interdependence of the rates paid by each shipper. Over such cases are probably not observed in railroad economics.

Second, the cost function applies to outputs over the link, i.e., $C = C(R_1 + R_2)$, but costs are not constant. In this case, the interdependence of shipper 1 and 2 flows remains. Under conditions of decreasing marginal costs (i.e., $C'' < 0$), the more that is shipped on a route, the lower the marginal costs of service for each shipper. Thus, the larger is the shipper, the lower the rates to the other shipper. In addition, the greater the pressures of competitive alternatives in one market (the larger is λ), the lower the rate in that market. This added competition induces the associated shipper to ship more by rail (due to the lower rates), in turn making this shipper larger. But the effect spills over to the other shipper. Since there are greater flows over the network (since the volume of the second shipper has increased), the costs of serving the other shipper (now shipper 1) are lower. Thus, again rates are lower, inducing even larger flows. To our knowledge, such interdependencies have not been considered in the literature.

The final case is the most general case, which relates to the general first-order conditions. In this case, if outputs are complementary (that is, if added operations to one shipper reduce marginal costs to the other shipper), the end result is the same as described above under the decreasing marginal costs case. If, on the other hand, outputs are substitutes, then the results are opposite those described above. That is, increased operations for one shipper increase costs for the other shipper (e.g., congestion) and rates for the latter will increase. Added competitive pressures (increases in λ) on one shipper reduce rates to that shipper, thus increasing volumes of that shipper. Hence, the increased volume in turn increases the cost of providing service to the other shipper, increasing rates.

34.5.3   Multiple Shippers and Multiple Railroads

In this section, we extend the model in section 34.5.1 to reflect intramodal competition between different originations. Specifically, we assume there is one shipper (or groups of identical shippers) serviced by two railroads with different networks. The terminal point may

---

*Consider, for example, a case in which two shippers are served on entirely different segments of the route network.

be common with different cost characteristics across railroads, or, alternatively, the terminal points may also be different without loss of generality. In this context we examine the oligopolistic behavior of railroads competing for service to this shipper (or group of shippers).

In the model developed to this point, railroads choose the rate. The results obtained in the simple model apply directly. That is, the shipper chooses the railroad/terminal location that offers maximal profits. If all else is the same, the railroad offering the lowest rate wins the traffic. If there are cost differences across railroads, the low-cost railroad will offer the lowest rate, that being the rate at which the other railroad no longer wishes to provide the service. Such a framework is rich for explaining developments over the last several years. For example, when a merger offers a railroad greater network connectivity (more direct service at lower costs), it may have the effect of diverting traffic flows from another railroad and reducing rates to the shipper(s). Further, the granting of traffic rights to another railroad affords that railroad the opportunity to become the preferred alternative. Again, this results in a diversion of traffic from existing railroads. In cases in which traffic is not diverted, there are still important implications and price effects from this model. In particular, if the merger or the presence of traffic rights does indeed positively affect the nonparticipating carrier's ability to serve a given shipper but does not improve it enough to make it the low-cost firm, the railroad providing service may still need to lower price to remain the preferred mode.

In this chapter, the pricing of services to shippers has been developed in terms of discrete choices of a shipper. In this context, we have demonstrated that there is a trigger price at which the railroad may no longer provide the service, i.e., if the railroad prices a shipper too high it will lose the service priced. It may be that the shipper receives or sends the product from or to another location or ships by a different mode, or alternatively, in the case of a receiver, the product may be priced out of the market.

In a more richly developed model, a shipper may use two or more railroads, ship to more locations, etc. In this regard, we simply note that such a shipment plan remains a discrete choice from a menu of choices. We assume that all other options do not bind railroad rates (i.e., $\lambda = 0$) for all other options. A consequence of this framework is that in equilibrium, the shipper is indifferent between the two railroad services.

To illustrate various oligopoly models, we assume there is a demand for transportation service to a location. This demand is provided by two different railroads with rates r_1 and r_2. Both railroads provide service to the shipper, although again, the demand may be the serviced from two different locations and the railroads may have different networks and/or cost conditions. There are a large number of plausible assumptions of rivalry between firms.* One example of this is the product-differentiated demand model (i.e., the railroads provide service but they are differentiated in some regard).** In this model the demand functions for each railroad are now interdependent. That is, the shipper(s) use both railroads but the level of service from each is dependent on the railroads' choice variables (r_1 and r_2). Demands can then be written as $R_1 = R_1(r_1, r_2)$ and $R_2 = R_2(r_2, r_1)$, where the demands are decreasing in the first term and increasing in the second term (the railroads provide substitutes).† Within this framework, each railroad makes profit-maximizing decisions, wherein the first order conditions can be written in a general form as:‡

---

*Because our emphasis is on rates, we limit ourselves to assumptions of rivalry pertaining to prices. Thus, the homogeneous Cournot model does not apply.

**This can be horizontal product differentiation, i.e., at the same price and service attributes different shippers may chose different railroads or vertical product differentiation, i.e., one railroad provides a better service than the other so that at the same price, shippers would prefer the high-quality-of-service firm.

†Burton and Wilson (2003) describe a network situation in which an origin-destination pair is served by a sequence of different firms (rail-barge) in a network. Such a situation can allow for complementary effects in the demand model, i.e., the second term would also be negative.

‡Again, we note that all other possible combinations do not limit rates. The Lagrangian multipliers are equal to zero for all other choices that the shipper(s) can make.

$$\frac{\partial \pi_1}{\partial r_1} = \left(R_{r_1} + R_{r_2} \frac{dr_2}{dr_1} \right) (r_1 - MC_1) + R_1 = 0$$

$$\frac{\partial \pi_2}{\partial r_2} = \left(R_{r_2} + R_{r_1} \frac{dr_1}{dr_2} \right) (r_2 - MC_2) + R_2 = 0$$

(34.5)

In the literature, the terms dr_1/dr_2 and dr_2/dr_1 (from the above equation) are typically termed a conjectural variation, which can be interpreted as a firm's belief of how the other firm will react to its rate change (Martin 2002, 45). For our purposes, we simply include these terms as a way to index the nature of rivalry between firms. In the standard Nash framework, each firm takes the other firm's choice variable as given (the conjectural variation terms are zero) and these two equations are solved jointly for Nash equilibrium prices. In such a model, as the level of product differentiation rises, firms garner more market power.

There are a number of ways to proceed through the remainder of the oligopoly models. While not "vogue," a conjectural variation framework probably provides the best presentation for our purposes. In this regard, we follow Martin (2002) (who follows Bowley 1924 and Hicks 1935) and treat the conjectures as constants. If the conjectures are positive, then the railroads expect that as they increase their own price, their rival (the other railroad) will increase its price. The consequence of a price change, then, is an increase (decrease) in a firm's price, which induces a decrease (increase) in output originating from its own price increase (decrease). But, since the other firm's price also increases (decreases), there is a corresponding increase (decrease) in output. In effect, the greater the positive conjecture, the more intense the competition and the lower the prices and the markup.

Such a framework is often used to model origin-destination movements in railroad economics. For example, MacDonald (1987) used a markup model similar in spirit to the model described above.* In particular, he modeled confidential waybill data pertaining to corn, soybean, and wheat movements. His empirical model is based on a model wherein the "[f]irm-specific elasticity of demand depends on the market elasticity of demand for the commodity, M_k, the extent of competition at that location, C_i, and the nature of rivalry among sellers, which is represented by the conjectural variation term" (153). While such a framework is not identical to the model developed above, it is based on the same principles. In his model, he includes a variety of different measures to account for interfirm rivalry. These measures allow for intramodal (i.e., between railroads) and intermodal (i.e., between a railroad and barge modes) rivalry. The specific measures used were the distance a shipper was located to water and the inverse of the Herfindahl index. He found that both miles to water and the inverse Herfindahl had important effects on rates and that the effect of these measures differed across commodities. Specifically, his results showed that miles to water increase rates. In his base models, a one percent increase in miles to water increases rates by 0.0864, 0.0856 and 0.2576 percent for corn, soybeans, and wheat, respectively. Conversely, he found that a 1 percent change in the inverse Herfindahl index leads to a 0.28, 0.19, and 0.11 percent reduction in rates. These results provide substantial evidence for the role of both intramodal and intermodal competition effects, even at the individual shipper level. In the context of the models presented in this chapter there is much. For example, the attractiveness of an alternative (i.e., water) falls as that shipper is located further from water. In the context of the first subsection, the railroad becomes more market dominant as this distance increases.** Further, the attractiveness and rivalry between railroads developed in this subsection relates to the inverse Herfindahl measure. In particular, as the number of railroads increases, accompanied by a decrease in market shares in a region (i.e., competition

---

*Wilson (1994) used a similar type of model as well. In his case, he used a markup model to examine railroad rates over a wide range of STCC level 2 codes to estimate the effects of partial deregulation across commodities.

**There are other empirical studies that link rail rates to the availability of commercial navigation. See, e.g., Burton (1996).

in a region increases), there is a corresponding decrease in railroad rates. In the context of the model presented here, the notion is that the conjecture term is positive and increasing in the Herfindahl index.

34.6   SUMMARY AND CONCLUSIONS

The postderegulation productivity gains achieved by U.S. railroads effectively reversed a trend that would have otherwise signaled their elimination as an important transport mode. Fortunately, 20 years after the implementation of the Staggers Rail Act, a healthy railroad industry is well positioned to respond to the capacity constraints that challenge other transport modes. However, whether U.S. freight railroads play an increasingly important role within the overall transportation landscape depends on many factors, some of which are well beyond the railroads' control.

First, policy-makers must realize that the productivity gains that restored the vitality of the railroad industry were the direct result of competition. As section 34.5 demonstrates, rail carriers have every incentive to exercise market power when that power exists. This simultaneously reduces the incentive to engage in innovative cost-reducing activities. In those settings where competing transport modes cannot supply adequate competitive pressure, policy-makers must stand ready to ensure that there is sufficient rail-to-rail competition. This may require regulatory mechanisms that are not currently in evidence.

Shippers (even of bulk commodities) continue to increase the level of service they demand from transportation providers. Thus, if the currently observable rail renaissance is to continue, it will be necessary for U.S. railroads to continually improve the level of service they offer. This, in turn, will require the railroads to invest in improving the quality of both vehicles and infrastructure. Currently, there is a trend toward the pooling of private and public funds to support rail infrastructure projects that benefit both rail customers and other affected constituencies. Such partnerships may well be necessary to achieve the required level of future investment.

The explosion in intermodal traffic is directly traceable to the tremendous growth in international trade. If this trade growth declines, so will the importance of increased intermodal capacity. It is likely, however, that the ability to increase the role of international economic activity depends on the ability to ensure secure container shipments. Thus, there is little the rail industry can do to affect such an outcome.

34.7   ACKNOWLEDGMENTS

The authors gratefully acknowledge the editorial and research assistance of Chris Clark in the preparation of this chapter.

34.8   REFERENCES

Abdelwahab, W. M. 1992. "Modeling the Demand for Freight Transport." *Journal of Transport Demand and Policy* 26(1):49–70.

———. 1998. "Elasticities of Mode Choice Probabilities and Market Elasticities of Demand: Evidence from a Simultaneous Mode Choice/Shipment-Size Freight Transport Model." *Transportation Research E* 34(4):257–66.

Allen, R. A. 1990. "Railroad Line Sales: Their Uncertain Legal Status after Pittsburgh & Lake Erie." *Transportation Practitioners Journal* 57(3):255–80.

Armstrong, J. H. 1998. *The Railroad, What It Is, What It Does,* 4th ed. Omaha: Simmons-Boardman.

Barnekov, C. C., and N. Kleit. 1990. "The Efficiency Effects of Railroad Deregulation in the United States." *International Journal of Transport Economics* 17:21–36.

Berndt, E. R., A. F. Friedlaender, J. S. W. Chiang, M. Showalter, and C. A. Vellturo. 1993. "Cost Effects of Mergers and Deregulation in the U.S. Railroad Industry." *Journal of Productivity Analysis* 4(1–2): 127–44.

Berry, S. 1992. "Estimation of a Model of Entry in the Airline Industry." *Econometrica* 60(4):889–917.

Bitzan, J. D. 1999. "The Structure of Railroad Costs and the Benefits/Costs of Mergers." *Research in Transportation Economics* 5:1–52.

Bitzan, J. D., and T. E. Keeler. Forthcoming. "Productivity Growth and Some of its Determinants in the Deregulated U.S. Railroad Industry." *Southern Economic Journal.*

Bitzan, J. D., and D. T. Tolliver. 1998. "Market Dominance Determination and the Use of Product & Geographic Competition." A Verified Statement before the Surface Transportation Board.

Bowley, A. L. 1924. *The Mathematical Groundwork of Economics.* Oxford: Oxford University Press.

Boyer, K. D. 1977. "Minimum Rate Regulation, Modal Split Sensitivities, and the Railroad Problem." *Journal of Political Economy* 85:493–512.

———. 1981. "Equalizing Discrimination and Cartel Pricing in Transport Rate Regulation." *Journal of Political Economy* 89:270–86.

———. 1987. "The Cost of Price Regulation: Lessons from Railroad Deregulation." *Rand Journal of Economics* 18:408–16.

Braeutigam, R. R., and A. F. Daughety. 1982. "The Estimation of a Hybrid Cost Function for a Railroad Firm." *Review of Economics and Statistics* 64(3):394–404.

Brown, R. S., L. R. Christensen, and D. W. Caves. 1979. "Modeling the Structure of Cost and Production for Multiproduct Firms." *Southern Economic Journal* 46:256–73.

Bureau of Transportation Statistics (BTS). *1993 Commodity Flow Survey.* U.S. Department of Transportation, BTS, Washington, DC.

———. *1997 Commodity Flow Survey.* U.S. Department of Transportation, BTS, Washington, DC.

———. 2001. *National Transportation Statistics.* U.S. Department of Transportation, BTS, Washington, DC.

Burton, M. L. 1993. "Railroad Deregulation, Carrier Behavior, and Shipper Response: A Disaggregated Analysis." *Journal of Regulatory Economics* 5:417–34.

Burton, M. L. 1996. *Rail Rates and the Availability of Barge Transportation: The Missouri River Basin.* U.S. Army Corps of Engineers, Omaha, Nebraska.

Cain, L. P. 1997. "Historical Perspective on Infrastructure and US Economic Development. *Regional Science and Urban Economics* 27(2):117–38.

Caves, D. W., L. R. Christensen, and J. A. Swanson. 1980. "Productivity in the US Railroads, 1955–74." *Bell Journal of Economics* 11:166–81.

———. 1981. "Productivity Growth, Scale Economies, and Capacity Utilization in the U.S. Railroads, 1955–1974." *American Economic Review* 71(5):994–1002.

Caves, D. W., L. R. Christensen, M. W. Trethaway, and R. J. Windle. 1985. "Network Effects and the Measurement of Returns to Scale and Density for U.S. Railroads." In *Analytical Studies in Transport Economics,* ed. A. F. Daugherty. New York: Cambridge University Press.

Daggett, S. 1928. *Principles of Inland Transportation.* New York: Harper Press.

Daughety, A. F., and F. S. Inaba. 1978. "Empirical Aspects of Service-Differentiated Transport Demand." In *Proceedings of the Workshop on Motor Carrier Economic Regulation.* Washington, DC: National Academy of Sciences, 329–55.

Davis, D. E. and W. W. Wilson. 1999. "Deregulation, Mergers, and Employment in the Railroad Industry." *Journal of Regulatory Economics* 15:5–22.

———. 2003. "Wages in Rail Markets: Deregulation, Mergers, and Changing Networks Characteristics." *Southern Economic Journal* 69(4):865–85.

DeVany, A. S., and D. W. Walls. 1999. "Price Dynamics in a Network of Decentralized Power Markets." *Journal of Regulatory Economics* 15(2):123–40.

Eaton, J. A., and J. A. Center. 1985. "A Tale of Two Markets: The ICC's Use of Product and Geographic Competition in the Assessment of Rail Market Dominance." *Transportation Practitioners' Journal* 53: 16–35.

Fisher, F. M. 1987a. "Horizontal Mergers: Triage and Treatment." *Journal of Economic Perspectives* 1(2):13–40.

———. 1987b. "On the Misuse of the Profits-Sales Ratio to Infer Monopoly Power." *Rand Journal of Economics* 18(3):384–96.

Friedlaender, A. F., and R. H. Spady. 1980. "A Derived Demand Function for Freight Transportation." *Review of Economics and Statistics* 62(3):432–41.

———. 1981. *Freight Transportation Regulation: Equity, Efficiency, and Competition in the Rail and Truck Industries.* Cambridge: MIT Press.

Friedlaender, A. F., E. R. Berndt, J. S. W. Chaing, M. Showalter, and C. A. Vellturo. 1993. "Rail Costs and Capital Adjustments in a Quasi-regulated Environment." *Journal of Transport Economics and Policy* 27(2):131–52.

Gallamore, R. E. 1999. "Regulation and Innovation: Lessons from the American Railroad Industry." In *Essays in Transportation Economics and Policy.* Washington, DC: Brookings Institution Press, 493–529.

Gillen, D., and W. G. Waters. 1996. "Transportation Infrastructure and Economic Development: A Review of the Recent Literature." *Transportation Research: Part E: Logistics and Transportation Review* 32(1):39–62.

Grimm, C. M., and K. G. Smith. 1987. "The Impact of Rail Regulatory Reform on Rates, Service Quality, and Management Performance: A Shipper Perspective." *Logistics and Transportation Review* 22:57–68.

Hanson, S. D., C. P. Baumel, and D. Schnell. 1989. "Impact of Railroad Contracts on Grain Bids to Farmers." *American Journal of Agricultural Economics* 71(3):638–46.

Hanson, S. D., S. B. Baumhover, and C. P. Baumel. 1990. "Characteristics of Grain Elevators that Contract with Railroads." *American Journal of Agricultural Economics* 72(4):1041–46.

Harper, D. V. 1978. *Transportation in America.* Englewood Cliffs: Prentice-Hall.

Hicks, J. R. 1935. "Annual Survey of Economic Theory: The Theory of Monopoly." *Econometrica* 3(1): 1–20.

Hilton, G. 1969. *The Transportation Act of 1958.* Bloomington: Indiana University Press.

Inaba, F. S., and N. E. Wallace. 1989. "Spatial Competition and the Demand for Freight Transportation." *Review of Economics and Statistics* 71(3):614–25.

Ivaldi, M., and G. J. McCullough. 2001. "Density and Integration on Class I U.S. Freight Railroads." *Journal of Regulatory Economics* 19(2):161–82.

Keeler, T. E. 1974. "Railroad Costs, Returns to Scale and Excess Capacity." *Review of Economics and Statistics* 56(2):201–08.

———. 1983. *Railroads, Freight, and Public Policy.* Washington, DC: The Brookings Institute.

Klindworth, K. A. 1983. "Impact of Staggers Rail Act on the Branchline Abandonment Process." *Proceedings of the Twenty Fourth Annual Meetings of the Transportation Research Forum* 24(1):451–60.

Koo, W. W., D. D. Tolliver, and J. D. Bitzan. 1993. "Railroad Pricing in Captive Markets: An Empirical Study of North Dakota Grain Rates." *Logistics and Transportation Review* 29(2):123–37.

Locklin, P. D. 1972. *Economics of Transportation.* Irwin Series in Economics. Homewood, IL: Richard D. Irwin.

MacDonald, J. M. 1987. "Competition and Rail Rates for the Shipment of Corn, Soybeans, and Wheat." *Rand Journal of Economics* 18(1):151–63.

———. 1989. "Railroad Deregulation, Innovation, and Competition: Effects of the Staggers Act on Grain Transportation." *Journal of Law and Economics* 32:63–96.

———. 1998. "Railroad Deregulation, Innovation, and Competition: Effects of the Staggers Act on Grain Transportation." *Foundations of Regulatory Economics* 3:274–306.

MacDonald, J. M., and L. C. Cavalluzzo. 1996. "Railroad Deregulation: Pricing Reforms, Shipper Responses, and the Effects on Labor." *Industrial Labor Relations Review* 50:80–91.

Martin, S. 2002. *Advanced Industrial Economics,* 2nd ed. Malden, MA: Blackwell.

McFarland, H. 1989. "The Effects of United States Railroad Deregulation on Shippers, Labor, and Capital." *Journal of Regulatory Economics* 1:259–70.

Meyer, J. R., M. J. Peck, J. R. Stenason, and C. Zwick. 1959. *The Economics of Competition in the Transportation Industries.* Cambridge: Harvard University Press.

Nelson, J. C. 1951. "Changes in National Transportation Policy: Highway Development, the Railroads, and National Transport Policy." *American Economic Review* 41(2):495–505.

Oum, T. H. 1979a. "A Cross Sectional Study of Freight Transport Demand and Rail-Truck Competition in Canada." *Bell Journal of Economics* 10(2):463–82.

———. 1979b. "Derived Demand for Freight Transport and Inter-Modal Competition in Canada." *Journal of Transport Economics and Policy* 13(2):149–68.

Oum, T. H., W. G. Waters II, and J.-S. Yong. 1992. "Concepts of Price Elasticities of Transport Demand and Recent Empirical Estimates." *Journal of Transport Economics and Policy* (May): 139–54.

Owen, W. 1959. "Special Problems Facing Underdeveloped Countries: Transportation and Economic Development." *American Economic Review* 49(2):179–87.

Pegrum, D. F. 1968. *Transportation: Economics and Public Policy.* Irwin Series in Economics. Homewood, IL: Richard D. Irwin.

Scheffman, D. T., and P. T. Spiller. 1985. "Geographic Market Definition under the DOJ Guidelines." Discussion Paper, U.S. Federal Trade Commission, Bureau of Economics.

Schmalensee, R. 1987. "Horizontal Merger Policy: Problems and Changes." *Economic Perspectives* 1(2): 41–54.

Slade, M. E. 1986. "Exogeneity Tests of Market Boundaries Applied to Petroleum Products." *Journal of Industrial Economics* 34(3):291–303.

Small, K., and C. Winston. 1999. "The Demand for Transportation: Models and Applications." In *Essays in Transportation Economics and Policy: A Handbook in Honor of John R. Meyer.* Washington, DC: The Brookings Institution Press, 11–55.

Spiller, P. T., and C. J. Haung. 1986. "On the Extent of the Market: Wholesale Gasoline in the Northeastern United States." *Journal of Industrial Economics* 35:131–46.

Stigler, G. J., and R. A. Sherwin. 1986. "The Extent of the Market." *Journal of Law and Economics* 28:555–85.

Surface Transportation Board. 2001. R-1 Reports, Industry Composite.

Tolliver, D. D. 1984. "Economies in Density in Railroad Cost Finding: Applications to Rail Form A." *Logistics and Transportation Review* 20(1):3–24.

Tye, W. B. 1990. "Regulatory Financial Tests for Rail Abandonment Decisions." *Transportation Practitioners Journal* 57(4):385–403.

Uri, N. D., and E. J. Rifkin. 1985. "Geographic Markets, Causality, and Railroad Deregulation." *Review of Economics and Statistics* 67(3):422–28.

U.S. Army Corps of Engineers. 1998. *Available Navigation, Fuel Consumption and Pollution, Abatement: The Missouri River Basin.* Omaha, NE.

U.S. Department of Commerce, Economics and Statistics Administration. 1997. *1997 Economic Census—Transportation.*

Vellturo, C. A., E. R. Berndt, A. F. Friedlaender, J. S.-E. W. Chiang, and M. H. Showalter. 1992. "Deregulation, Mergers, and Cost Savings in the Class I U.S. Railroads, 1974–86." *Journal of Economics and Management Strategy* 1(2):339–69.

Westbrook, M. D., and P. A. Buckley. 1990. "Flexible Functional Forms and Regularity: Assessing the Competitive Relationship Between Truck and Rail Transportation." *Review of Economics and Statistics* 72(4):623–30.

White, L. J. 1987. "Antitrust and Merger Policy: A Review and Critique." *Journal of Economic Perspectives* 1(2):13–22.

Wilner, F. N. 1997. *Railroad Mergers: History, Analysis, Insight.* Omaha, NE: Simmons-Boardman.

Wilson, W. W. 1994. "Market-Specific Effects of Rail Deregulation." *Journal of Industrial Economics* 42:1–22.

———. 1996. "Legislated Market Dominance." *Research in Transportation Economics* 4(1):33–48.

———. 1997. "Cost Savings and Productivity Gains in the Railroad Industry." *Journal of Regulatory Economics* 11:21–40.

Wilson, W. W., and J. Bitzan. 2003. "Industry Costs and Consolidation: Efficiency Gains and Mergers in the Railroad Industry." Mimeo.

Wilson, W. W., and M. Burton. 2003. "Network Pricing and Vertical Exclusion in Railroad Markets." Mimeo.

Winston, C. 1981. "A Disaggregate Model of the Demand for Intercity Freight Transportation." *Econometrica* 49(4):981–1006.

———. 1985. "Conceptual Developments in the Economics of Transportation: An Interpretive Survey." *Journal of Economic Literature* 23(1):57–94.

———. 1993. "Economic Deregulation: Days of Reckoning for Microeconomists." *Journal of Economic Literature* 31(9):1263-90.

Winston, C., T. M. Corsi, C. M. Grimm, and C. A. Evans. 1990. *The Economic Effects of Surface Freight Deregulation.* Washington, DC: The Brookings Institute.

Winston, C., and C. Grimm. 2000. "Competition in the Deregulated Railroad Industry: Sources, Effects, and Policy Issues." In *Deregulation of Network Industries: What's Next?* Washington, DC: The Brookings Institution Press, 41–71.

CHAPTER 35
AIRLINE MANAGEMENT AND OPERATIONS

Saad Laraqui
Business Administration Department,
Embry-Riddle Aeronautical University,
Daytona Beach, Florida

35.1 EXECUTIVE SUMMARY

The overall structure of firm-level management infrastructures in the airline industry is important to understand when analyzing airline operations management. More integrated relationships between strategic, financial, marketing, and other functions of management and operations and its management are increasingly evident in increasing cooperation among airlines to include cross-cultural integration among different regions and markets. Integrating mechanisms include a wide variety of strategic approaches, such as alliances, bilateral agreements, code-sharing, and other techniques. Integrating management and operations has been found to be a basis for maintaining a competitive advantage and a robust financial portfolio. Driven by adverse conditions in the competitive environment, airlines must persistently seek new approaches to create synergy between the management approach and the operations that they manage.

Collaboration on a global level is the best way to achieve a more cohesive relationship within an airline's operational structure. In this chapter, several variables will be analyzed relative to the successful combination of management structures and airline operations. Anecdotal evidence presented here suggests that there is a relationship between management style and function and operational utility and performance. No longer operating independently of each other, airlines seek to be part of larger networks and integrate their operations in the global environment. Therefore, airlines must consider not only the effects of competitors within the same region, but also of those that operate in other regions. The airline environment is a truly global enterprise, where interactions among airlines in different regions are crucial for survival.

35.2 INTRODUCTION

Management infrastructures and dynamics have a pronounced effect on the successful execution and flow of airline operations. Airlines fail regularly even in times of unprecedented growth, suggesting that failure is not entirely driven by external conditions such as pressures on operating economics due to high fuel prices. Tides of financial success are not simply reflections of favorable forecasts and stout economic indicators. Success in the airline industry appears to be better correlated with the functions of management, particularly management's ability to rationalize and synchronize strategies and maintain good working relationships with labor to ensure success of the organization's business model.

Several startup airlines have demonstrated an understanding of the importance of acquainting all levels of management with actual operations, building a solid link with the operating employees. Southwest, JetBlue, and a host of other dynamic airlines have clearly expressed this relationship as fundamental to survival. But the benefits of linking management strategy with airline operations are not exclusive to newer, smaller airlines. This strategy to diminish the seemingly enormous fissures that exist between upper management and those involved in everyday operations can work for any airline. Certainly, the financial turnarounds of SAS in the 1980s under Jan Carlzon and of Continental Airlines in the late 1990s under Gordon Bethune are testimony to this.

Concurrent with this trend is the evolution of cooperative alliances, which have enabled airlines to rationalize their route structures and adjust their market strategies to compete with other newly formed global alliances and aggressively capitalized startups. While it is not evident at the present time due to union contract issues between carriers in an alliance, in time the trend toward globalization may cause greater integration of flight crews and maintenance personnel in the actual day-to-day operations of alliance-wide flight schedules. Already Continental Airlines and Northwest Airlines are coordinating the scheduling of Northwest's Asian hub at Narita, where a fleet of Continental B-737-700s and Northwest A-320s are based to feed the Northwest-operated B-747 long-haul fleet. In the recovery of the airline industry after the September 11, 2001, terrorist attacks, the level of coordination is likely to push levels that could not have been contemplated by managers or union leaders in the 1970s–90s.

Prior to the Airline Deregulation Act of 1978, airlines in the United States faced severe restrictions on their ability to expand under the Civil Aeronautics Board (CAB) policies of the day. Managements trying to improve the efficiency of their route networks or to add new service had to petition the CAB under carefully promulgated rules of administrative procedure. Even getting authorization to introduce a new fare type took up management time. In its attempt to manage a 12 percent return on equity to airline shareholders, following strict public utility doctrine, the CAB carefully controlled the growth of each carrier and the overall level of capacity being added, as well as fares, subsidy levels to small communities, and even the levels of flights offered between major cities pairs. Following the economic havoc caused by the 1973–74 energy crisis, several of the local service carriers and the newly emerging cargo carriers like Federal Express began to lobby for deregulation. Starting with modest liberalization during the Ford administration on fare and route regulation, most notably on dropping service to subsidized small communities where jet economics no longer worked at fuel prices 10 times higher than in 1972, the momentum built under the Carter administration to deregulate completely. On October 1, 1978, the Kennedy-Cannon Airline Deregulation Act became law and the CAB began to phase itself out of existence. Airlines operating in the environment that evolved after the 1978 Deregulation Act faced new challenges that stifled many airlines and their management.

The Carter administration embarked on a mission to reduce regulatory controls and ensure that they were kept to a minimum. One of his election pledges was to support the interests of the consumers: "In air transport, as in other industries, this meant less regulation and more choice" (Doganis 2001, 23). In a short time, there was pressure from the newly freed airlines to extend the domestic freedoms to the international sphere. To achieve this, negotiation of bilateral air service agreements was necessary. The aim was to promote the following (Doganis 2001, 24):

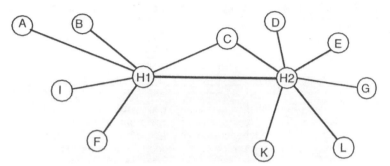

FIGURE 35.1   Hub-and-spoke airport structure. (*Source:* Button 2002, 30.)

- Greater opportunities for innovative and competitive pricing
- Elimination of restrictions on capacity, frequency, and route operating rights
- Elimination of discrimination and unfair competitive practices faced by U.S. airlines abroad
- Flexibility for multiple designation of U.S. airlines
- Authorization of more U.S. cities as international gateways
- Liberalization of rules regarding charter flights

In time, in light of the rapid expansion of the numbers of passengers transported in the United States from 250 million in 1978 to nearly 600 million in 2000, other governments around the world began to abandon the public utility theory of airline regulation and adopted the U.S. approach toward airline deregulation.

Today, most countries have liberalized domestic and international air operations with much the same result in terms of expansion of airline fleets and passengers transported as fares generally come down and frequency over an expanded city pair matrix increases. One unexpected benefit to many governments has been the substantial increase in revenues from landing fees, air navigation charges, gate rentals, and concession participations. The upward slope of revenue growth has permitted the self-funding expansion of infrastructure such as new runway construction and passenger terminal additions. At many airports around the world, the terminal has become a shopping mall rather than a sterile place where passengers are processed to and from flights. This is partly due to the structure changes after the deregulation in 1978. With the hub-and-spoke structure, passengers have longer waits between their connections, and this encourages terminals to generate more nonoperational revenue. Figure 35.1 shows the current airport structure.

Unfortunately, the industry is not immune to geopolitical problems and economic recession. However, with the long-term trend towards globalization, air transportation will continue to play a vital role in air commerce. The thin margins of a global commodity business will require nimble management working closely with operating employees under their own control and those who are nominally under control of the alliance partners as the networks become more finely meshed.

In the next section, an historical view of airline operations will be presented with a focus on Europe and the United States. Notable successes and failures will be compared and contrasted. Afterward, the new and continuously changing airline business environment will be discussed, suggesting that structural changes need to be met with a strategic paradigm that integrates strategic, financial, and operations management into a cohesive whole.

35.3 *OPERATING ENVIRONMENT AMONG AIRLINES IN DIFFERENT REGIONS*

35.3.1 The Growth Initiative

The operating environment of an airline should dictate the type of organization that it is or will become. Levels of management and functions such as planning and marketing and the structure of operating departments such as flight, maintenance, and station operations should reflect the dictates of the operating environment. Although elements of the airline business are constantly changing, airlines must develop and implement specific strategic plans in order to survive, focusing on their most important asset: customers.

While international service introduces additional complexities to which domestic operators are not fully exposed, the global nature of air commerce affects the manner in which all management strategies are articulated. In addition to the challenges of filling seats and running airplanes on time, managers with international exposure have to deal with foreign regulations, labor practices, and a varying mix of cultures, customs, and political meddling. As the industry globalizes, many of the daily challenges will resolve themselves. Outside of the United States, the ICAO and IATA approaches to operational standardization have facilitated the process of acquainting operating personnel and line managers with airline procedures, supplemented by customer service standards developed by the alliances. Of course, the level of difficulty in achieving a harmonized approach varies between regions. Some regions present less hostile environments than others. Several examples will be discussed, illustrating the challenge of operating across dissimilar environments that can present difficulties to airline management.

35.3.2 Defragmenting Europe

Prior to the beginning of the 1990s, the European airline industry was heavily regulated, with the goal of protecting the large national carriers that had evolved, country by country, since the end of World War II. With the creation of the European Union, the dropping of the bilateral restrictions between member states led to progressive liberalization of capacity controls and the procedures to obtain new routes. New domestic and intra-EU markets were opened, and a number of European carriers capitalized on the opportunities for new growth by reconfiguring their route systems and improving their marketing strategies. Some of the changes to the route network and fleet mix affected operations and caused new labor issues to emerge. Moreover, pressure from the United States on "open skies," and the great success of the U.S.-Netherlands open skies agreement and the near merger of Northwest Airlines and KLM, opened management eyes to the opportunities of adopting U.S.-style hub-and-spoke operations, feeding short-haul domestic, intra-EU into long-haul transatlantic and other global destinations. Figure 35.2 illustrates this type of network structure.

Table 35.1 summarizes revenue and traffic effects of major code sharing alliances in 1998 values. As can be seen from the table, Northwest/KLM benefits from their alliance a lot more than others do.

The EU operating environment changed as a result of the integration of several different operating environments, prompting a more unified and allied approach than was at first thought possible due to linguistic differences, labor practice, and operating regulation. A key element of success was the adoption of the Joint Airworthiness Authority (JAA) between the member states, which served to consolidate a mix of national operating regulations into a cohesive suite of rules for operating EU aircraft throughout Europe and the world. The adoption of certification rules for large and small carriers, as well as aircraft maintenance organizations, suddenly made it easier than ever in the past to start, grow, or restructure an airline in Europe.

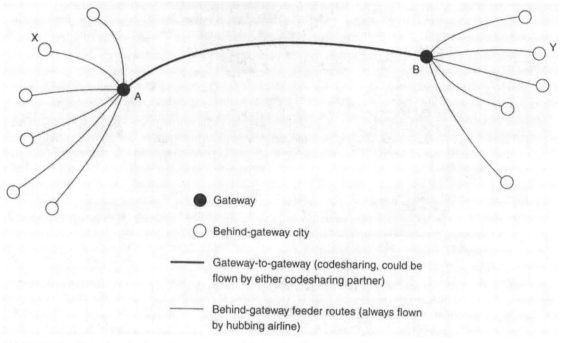

FIGURE 35.2 Global long-haul and short-haul network hub-and-spoke structure. (*Source:* Spitz 1998, 493.)

The opening of new markets created challenges for both major carriers and new low cost carriers (LCCs) that entered the market. The new LCCs implemented strategies never seen in the European airline industry, designed to compete for passengers long transported by the largely state-owned carriers, which were also under great pressure from the EU to privatize. The LCCs developed a "lean and mean" management organization that had never been seen before in Europe, where the state-owned carriers had become top-heavy in a strictly regulated

TABLE 35.1 Revenue and Traffic Effects of Code-Sharing Alliances

Alliance	Annualized impacts	
	Revenue ($millions)	Traffic (000 pax)
Northwest/KLM	$125–175 (NW) $100 (KL)	200 (NW) 150 (KL)
BA/USAir	$100 (BA) $ 20 (US)	150
United/Lufthansa	n/a	219
Delta/Virgin Atlantic	$100	250

Source: Spitz 1998, 493.

environment that approved most requests for increases in fares to match requests for higher wages paid to growing workforces. These new operators took advantage of growing price and cost pressures, using lower fares as a way of luring passengers away from established, well-known carriers (Daly 1995). Efficiency was the key to their success, and a new game plan was fashioned to accommodate the need to develop their cost structures against long-established charges for landings, handling, and navigation by the European civil aviation authorities.

Airlines in different regions have shared a similar evolution irrespective of their location. The period of heavy regulation is followed by a gradual liberalization, which in most cases leads to complete deregulation of the operating environment. This change in the regulatory environment triggers a number of startup airlines. Most startups usually begin flying early in the newly deregulated environment. Soon the economic realities of the thin-margined airline business serve to eliminate the poorly capitalized or poorly conceptualized carriers. In the bigger countries, the market becomes saturated and a round of consolidation occurs, eliminating many of the startups. Those that are well capitalized or capture enough market share to continue to grow usually survive, particularly if their management is nimble.

The Italian airline market serves as an example. Italy's industry historically consisted of mainly one major carrier, Alitalia, which emerged as a result of consolidation of a number of carries started in the post-World War II era. This government-owned entity controlled the domestic market and maintained hegemony in all international markets to and from Italy. Deregulation of the Italian marketplace opened up new opportunities, and new operators started to appear. The postderegulation Italian industry shared a pattern similar to that of the U.S. industry, where price wars between the new entrants and Alitalia eliminated all opportunities for profit and, more importantly, returns to shareholders who had invested in the startups. With much lower fares between the larger cities, the volume of Italian domestic traffic has greatly increased over prederegulation levels. Alitalia has fought off the new entrants very aggressively by developing an extensive code-share network with the Italian regional operators. Italian labor law makes it hard for a startup to have a major labor advantage, and the airport ground-handling monopolies make it difficult for a new operator to gain competitive advantage through lower boarding costs. Under the pressure of higher fuel prices since the summer of 2000, the Italian industry has consolidated into two large and two smaller groups. Some of the independent carriers have shut down or shifted their focus to increased charter activity. The shake-out among the jet operators has created some opportunities for regional turboprop operators, who soon become affiliated with Alitalia, which has been successful in maintaining market dominance, albeit at the price of profitability of its domestic route system. With its international network under severe economic pressure, Alitalia is struggling to right-size itself and find a long-term alliance partner. The EU rules make it very difficult for the Italian government to orchestrate an economic bailout of the carrier as it has done in the past.

The first challenge to European airline managements was to cope with the need of the industry to undergo changes of ownership from state governments to privatized entities. The second major challenge was to restructure operational procedures, oftentimes requiring renegotiating longstanding labor practices with unions. The third challenge was the need to integrate multicountry operating environments into successful operational units, a challenge the U.S. managers did not have to face when the United States began deregulating. It took some time for approaches to the legal, economic, and political problems that arose to be developed. Moreover, since the formation of the EU, airlines have found themselves in the midst of an arduous process to privatize carriers that now operate in a highly competitive, unrelenting air transport market.

Closely related to deregulation is privatization. The issue of privatization has deeply affected the airline environment in Europe. After partial and then full deregulation, European airlines were able to test their strategic visibility through privatization and, consequently, through carrier alliances, some involving cross-ownership or other financial participation. Privatization in the EU occurred in two phases: (1) the privatization of British Airways in

1987 via a public stock flotation, and (2) the phased reduction of state ownership of KLM Royal Dutch Airlines, Lufthansa German Airlines, and ultimately most of the other European state-owned airlines, which "reflected an ideological belief in the contraction of the state" (Standiland 1999).

Privatization was facilitated by a strong European economy and booming stock markets and a period of low fuel prices, ensuring most of the carriers good earnings as they made their hub-and-spoke networks work around fortress hubs in the major cities of their original origin. The early success of some airlines like British Airways with its London hub and KLM with its hub at Amsterdam and business relationship with Northwest, which gave it near complete access to the U.S. and Asian market, created a climate of great expectations in the stock markets of Europe. Tides of economic success usually facilitate expansion, whereas uncertainty about the economic climate forces many airlines to contract. Many other newly liberated European airlines adopted a "me too" strategy, trying to improve earnings as best they could to effect an offering of stock to the investing public. Some carriers were more successful than others at coping with all of the challenges of operating profitably to satisfy shareholders and other stakeholders, knowing that under the new EU rules they could not go back to the national government for cash or a fare increase when they failed to meet budget projections.

Olympic Airways: Finding Interface between Management and Operations.   Like some other ailing airlines in Europe at the time, Greece's national flag carrier, Olympic Airways, faced symptoms of what was referred to as the "distressed state airline syndrome," common during the aftermath of the Gulf War in the early 1990s when the world-wide industry was nearly put out of business by the steep increase in fuel prices and subsequent economic recession (Doganis 2001). Once controlled by Aristotle Onassis, Olympic had become a ward of the state during the period of unprofitabilty that commenced after the 1973–74 energy crisis, and the changes in the traffic makeup of its long-haul network to Australia, Canada, and the United States. Olympic's politically appointed executives had to deal with the indecisiveness of its indifferent operating managers, and its board of directors consistently failed to face the challenges to the structural integrity of the organization. In fact, during the 1-year period between February 1995 and March 1996, Olympic's board of directors was changed three times (Doganis 2001). The focus was more on entertaining political objectives than on the airline's fiscal and operational strategy. The problems that the airline frequently faced, extensively impaired its ability to successfully operate, were:

- It was largely undercapitalized, with huge debts.
- It had serious financial difficulties securing credit.
- It was frequently overpoliticized.
- It had overweight, poorly planned labor networks.

After years of dire financial trauma, executives and closely thereafter the government began to probe the extent of the problem. After months of investigation, it was concluded that yield and operating expenses were aspects of the business that were simply ignored. Unusually low utilization was damaging Olympic's ability to maintain its competitive edge against other airlines in Greece and Europe. Specifically, the economic perspective of operating the airline was visibly disregarded. According to a senior economist for the MITRE Corporation, "[E]conomic analysis provides more precise estimates of the causal relationship among airline operating expenses, yield, and demand" (Homan 1999, 506).

Economic analysis requires management organizations to locate problem areas within the operational structure and focus their efforts on reducing unnecessary utilization of assets and finding ways to redeploy them. Its own aircraft heavily burdened Olympic Airways. For example, the airline used an Airbus A300-600R, an aircraft with a capacity of 260 passengers, on early morning routes from Athens to the northern Greek city of Thessaloniki. The

TABLE 35.2 How Revenue Management Is Impacting Airline Thinking

Conventional Thinking	Revenue Management Thinking
Airlines control the price consumers will pay for their products and services.	Consumers determine price through buying behaviors; successful airlines respond to changing consumer signals.
Consumers pay for products and services.	Consumers pay for the opportunity to buy at the price and conditions acceptable at a particular moment in time.
Airlines segment consumers by standard demographic and psychographic factors (age, sex, income, and so forth).	Airlines segment consumers by variations in buying behaviors based on price sensitivity and time sensitivity.
Prices are set on a cost-plus profit margin basis.	Airlines rework costs to accommodate consumer-driven prices and still achieve profits.
Discounts are determined by salespeople to close sales and meet volume objectives.	Revenue management objectives determine discounts in terms of revenue maximization goals.

Source: Cross 1998, 307.

route's main occupants were Athens newspapers ferried for distribution in Thessaloniki, the second city of Greece. Rather than managing the aircraft and the route for profit, Olympic used it on this route for four years simply because the marketing department had once decided that most peak time service between Athens and Thessalonki would be flown in a wide-body, irrespective of experienced demand patterns. New management eventually realized the effect this scheduling practice had on the value-generating capacity of the aircraft: "Aircraft cost plays an important role in operating economics . . . and this dictates the profitability of individual transactions" (Vella 1999, 385).

The process of transforming the airline came at a time when airlines worldwide were facing their most pronounced financial crisis ever. Although new Airbus A-340s were delivered to alleviate economic losses being incurred by operating aging Boeing 747s on many long-haul routes, the airline failed to regain its competitive position and serve its entire customer base. During this period, the Greek government decided to allow the startup of new carriers to serve the domestic market in response to demands by regional politicians for an alternative to moribund Olympic. Ultimately management discovered a valuable lesson in its evolution, namely that the concept of "the one-size-fits-all airline is dead" (Gertzen 2002). How Olympic learned that valuable lesson, and how revenue management is impacting airlines' thinking process, is shown in Table 35.2.

As a consequence, Olympic continues to restructure its European operations and is struggling to salvage the few parts of its long-haul operations that are economically viable, given the fundamental facts that fewer Greeks are emigrating to countries like Australia and that third-generation immigrants do not return to the motherland in the same numbers as did the first and second generations.

35.3.3 The North American Precedent: Operational Optimization

The airline industry in the United States is one of the largest industries in terms of sales, employment, capital requirements, and importance to the overall economy. Figure 35.3 shows the impact that the aviation industry has on the U.S. economy.

North American airlines have long been exposed to challenges of integrating a strategic approach to the business by developing route networks across a broad swath of geography

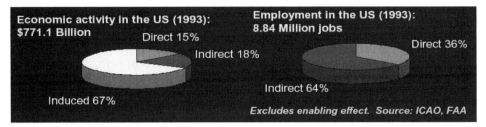

FIGURE 35.3   Aviation industry's impact on U.S. economy. (*Source: ICAO.*)

while meeting the demands of public private and institutional shareholders who financed the industry almost from infancy, labor, and government policies. While the scheduled business was initially more developed in pre-war Europe, the United States and Canada were pioneers in approaching the airline industry from the perspective of dynamic management and intelligent operations. Management dynamics is the cornerstone of American and Canadian business, and the degree to which airlines develop their management organizations has a significant impact on the way an airline operates. The management structures at the major carriers like American Airlines and Air Canada evolved in a manner reflecting the regulatory requirements of the time, while dealing with the problems of running a continent-wide airline operation in a period when telecommunications was primitive. This pioneering class of carriers had to invent solutions ranging from a standardized approach to air traffic control (initially without significant government involvement) to demanding that manufacturers develop specific airliners such as the Douglas DC-3 and later the DC-8 as well as the Boeing 707 to meet growing requirements of their customer base. In the early 1970s, competitive pressures forced Southwest Airlines to develop management tactics that got everyone involved in the overall operation of the aircraft from station to station. For instance, flight attendants assisted in the boarding of passengers in the gate area, which was unheard of at such well-established competitors as Braniff International. Braniff tried to fend off the incursion of pesky Southwest into the Texas marketplace by selling below its cost, anticipating falsely that the unheard-of ticket prices were also below Southwest's internal costs, which were further refined to deal with the lowered ticket yields. The success of this approach provided the benchmark for many of the world's airlines, which have tried to emulate the unusually strong bond among the people at Southwest.

Airline management is responsible for ensuring that the industry does not collapse on technicalities such as safety of flight, gross management incompetence, and clumsy business decisions. Over the years, many of the surviving U.S. airlines have taken the initiative and implemented several strategies aimed at bridging the gap between management and operations. For example, American Airlines pioneered yield management and focused its strategy on developing a solid relationship with the customer. Subsequently, the airline established what is referred to as customer relations management (CRM). This conceptual approach to operations management is described as "how one can maximize or create the right sort of experience every time" (Ott 2000, 52).

The ultimate goal of U.S. airlines has been to optimize operations and integrate them with the goals of upper-level management. The objective of this approach, as in the case of the American Airlines initiative, was to create an interface of cooperation among organizational structures. In turn, this "should be tailored to match specific measurable corporate objectives of the company based on a well-defined mission statement" (Asi 2002). Another approach U.S. airlines have advanced is compensating travelers for inconveniences: "on a specific consumer level, airlines must use information on a customer's operational history to set the stage for ways an airline could compensate for negative experiences . . . knowing a customer's wants and needs in greater detail will help develop 'the right product for the consumer'" (Ott 2000, 51).

People Express: Pioneering Leadership Development.   People Express began operations in April of 1981 and quickly emerged as a model airline with an innovative management team and an aggressive pricing and frequency strategy. According to Donald Burr, founder and CEO of People Express, the most predominant reason for establishing the new company "was to try and develop a better way for people to work together . . . that's where the name People Express came from" (People Express 1990). Having cut his teeth with Frank Lorenzo, in the turnaround of Texas International Airlines in the early 1970s, Mr. Burr called the traditional style of management a "deadening grind and a lack of vision" (People Express 1990).

From its inception, the airline focused on combining the strategies of management with the interests of its customers. Profit was no longer the core focus; instead, serving people and customers were precursors to profit, and that became the focus of management (People Express 1990). People Express channeled its energies into finding a crossover point between the organization's design and the successful execution of that design. By assembling "management teams," People Express effectively improved the performance of its operations. The airline was one of the first to pioneer leadership development in its long-term strategy. However, its pioneering labor approach and its rapid expansion strategy failed to produce earnings, and in 1986 Burr surrendered control of the People Express experiment to Continental Airlines, then run by Frank Lorenzo, who went on to integrate the People Express Newark hub successfully into the dominance of the New York market. In the aftermath of the Gulf War, Mr. Lorenzo's hard-nosed approach with the labor unions in wrestling cost savings forced Continental into bankruptcy for the second time in a little over a decade, which provided an opportunity for new management to focus the Continental strategy narrowly around three hubs at Cleveland, Houston, and Newark, and set about to simplify and rejuvenate the fleet.

35.3.4   Operations Management Among Airlines in an Alliance

Overview.   The emergence of worldwide airline alliances over the past decade has instigated a gradual reevaluation of airline operations among participating carriers. The need to become a more cohesive marketing organization providing a seamless experience to alliance customers has prompted airline management to consider new ways of integrating joint scheduling, gate operations, and reservations data. According to industry sources, half of the world's jet fleet is controlled by only 17 airlines, and almost half of world passenger traffic flies on one of the four largest alliances: Star Alliance, Wings, OneWorld, or SkyTeam (Barry 1998). The fifth worldwide airline alliance, which was recently restructured, is Qualiflyer. Table 35.3 summarizes these five alliances and their member airlines.

The environment in which an alliance operates should be clarified. A *strategic alliance* is one where present or potential competitors commingle their assets in order to pursue a single or joint set of business objectives. Such commingled assets may include terminal facilities, maintenance bases, aircraft, staff, traffic rights, or even capital resources. The Northwest/KLM alliance is an example of a successful crossborder, crosscultural strategic partnership, originating in KLM's financial participation in the 1988 leveraged buyout (LBO) of Northwest Airlines in a hostile takeover by an LBO group. *Marketing alliances*, on the other hand, are not strategic alliances, in that partners use their assets independently of one another and pursue their individual objectives. Alliances of this sort are quite common: code-sharing agreements, joint frequent-flyer programs, and other agreements are widespread, but the degree of cooperation is limited by the degree to which each airline offers its assets to another. Table 35.4 illustrates different phases of airline alliances common between international carriers.

The Star Alliance spearheaded by Air Canada, Lufthansa, and United Airlines is perhaps the best example of a strong marketing alliance that has expanded to encompass a global network of more than a dozen carrriers (Doganis 2002).

TABLE 35.3 Five Worldwide Airline Alliances and Their Member Airlines

Alliance	Members	
	Before 2001	After 2001
SkyTeam	Delta Air France Alitalia Korean Air Aero Mexico TSA-Czech Airlines	SAME
Qualiflyer	Air Littoral Austrian Airlines AOM French Airlines LOT Polish Airlines Air Europe Sabena Crossair Swissair PGA Portugalia Airlines Turkish Airlines Volare Airlines TAP Air Portugal	Air Littoral Air Liberte LOT Polish Airlines Swiss Air Lines SN Brussels Airlines TAP Air Portugal
Wings	Northwest Airlines KLM Continental	SAME
Star	United Airlines Tyrolean Airways Thai Airways International Singapore Airlines Scandinavian Airlines Varig Brazilian Airlines Asiana Airlines Austrian Airlines Air New Zealand Lauda Air Mexicana Airlines LOT Polish Airlines Lufthansa German Airlines Spainair All Nippon Airways BMI British Midland Air Canada	SAME
OneWorld	American Airlines British Airways Air Lingus Cathay Pacific Finnair Iberia LanChile Qantas	SAME

Source: Oster 2001, 22.

TABLE 35.4   Phases of Airline Alliances to Full Merger

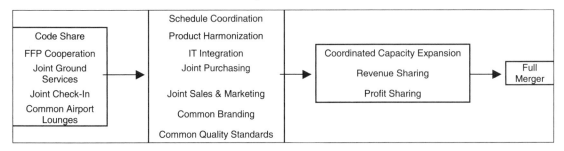

Table 35.5 summarizes the features of alliances and the advantages of being in an alliance in various ways.

Why Alliances Fail.   Despite the success of many global alliances, several have failed. The lack of clear objectives and a changing economic marketplace, compounded by conflicting management objectives, have been the major causes of alliance disruption and disintegration.

In order to avoid failure airlines should consider the process described in Figure 35.4 before they choose their global partners. This will ensure the alliance to last for long time periods.

No greater example of such a failure exists than the collapse of the Swiss-based Qualiflyer Group, in which Belgium's SABENA, France's Air Liberte, and Switzerland's SWISS-AIR all collapsed within a very short time of each other. To avoid such operational and financial malfunctions, management's role is to define, implement, and measure its short- and long-term strategic goals.

A difficulty in assessing the viability of integrating airlines in alliances lies in the complexity of maintaining streamlined operations in the regulatory environment. For example, the integration of airlines in Europe and North America has been characterized as complicated and convoluted. A number of factors have conspired to stifle many attempts of international airlines to cooperate effectively and, to a lesser extent, merge their operational units into one. While in 1987–89 Northwest and KLM successfully lobbied the U.S. government for the most integrated approach from a financial, marketing, and operational standpoint, the U.S. administration, favorably disposed to unions, placed limitations on KLM's role. North-

TABLE 35.5   Alliance Features

Network extension	Operational craft reduction	Competition reduction
Complementary code share	Shared airport lounge and gate facilities	Parallel code sharing
Schedule coordination	Ground handling	Joint pricing and yield management
Special prorate	Catering	Revenue pooling
Through baggage	Maintenance	Joint flight
Single check-in	Joint purchasing	Joint venture
Shared or proximate gates	Joint insurance purchasing	Multiple listing of code-shared flights
Marketing alliance	Joint marketing	
Frequent flyer	Management contract	
Block seat	Wet-lease	

Source: Oster 2001, 24.

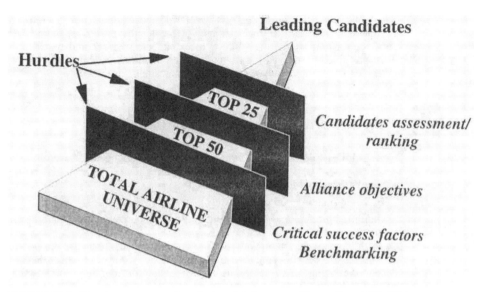

FIGURE 35.4   The rigorous process for selecting global partners. (*Source:* Baur 1998, 537.)

west needed more financial involvement from KLM in the period after the Gulf War and before the public markets allowed Northwest to go public in 1993, and the U.S. government allowed KLM to take up to a 49 percent equity position in Northwest, provided KLM did not violate the 25 percent voting stock limitation on foreign ownership. Conversely, when American Airlines and British Airways attempted to unify transatlantic operations in the United States and Britain, the principal competitors, United Airlines and Virgin Atlantic Airways, mounted a lobbying effort aimed at triggering a change in the U.S.-U.K. bilateral agreement. Slots at London Heathrow airport were reallocated, which ultimately led to the abandonment of most features of the service agreement, except for the OneWorld alliance. Had the deal overcome regulatory hurdles, the alliance would have given the two airlines 60 percent control of the transatlantic market (Barry 1998). Moreover, the longer-term consequences would have fundamentally challenged foreign ownership limitations placed on air carriers by nationalistic governments, as well as how flight crews are flowed through an integrated global route network. In short, it was too much too soon for politicians on both sides of the Atlantic.

Converging Strategies: Successful Allied Cooperation.   The success of many alliances is attributed to the ways that operational utility and flexibility are managed. As evidenced by the increasing number of alliances, the operational environment and management strategy play significant roles in their successful implementation. Again, one of the most longstanding major alliances is Northwest and KLM. Beginning with KLM's participation in the 1998 leveraged buyout of Northwest, the alliance quickly adopted code-sharing as a means of flowing traffic, and hence money, between themselves. By late 1992, the U.S. Department of Transportation granted them antitrust immunity. This allowed unobstructed entry and capacity privileges between the United States and the Netherlands, especially considering the open skies agreement between the two governments (Wang and Evans 2002). With the broadened agreement, they were able successfully to implement a strategy that would facilitate a mutual opportunity for expansion and market access. While flowing some Northwest airplanes into Amsterdam from cities such as Boston and Detroit, Northwest was able to use the KLM fleet to gradually establish a Northwest-branded hub at Amsterdam. Meanwhile,

KLM mimicked the Northwest flight offerings to extend its reach into the United States beyond the principal Northwest hubs. As the results proved positive, the two companies developed a joint marketing strategy to extend the network and reach farther into Europe. This was aided by the U.S. and Dutch governments by establishing less restrictive bilateral agreements with the newly emerging CIS states and Eastern Europe.

35.4   THE CHANGING AIRLINE BUSINESS ENVIRONMENT

During the past decade, full-service airlines have responded to several challenges:

- New entrants and low-cost competition
- Increased pressures on costs
- Simplified product and process function with strategies that average down labor cost

With the wave of attempts to average down labor cost, managers hope to include changes to labor contracts. This approach to defend and protect market share has not been effective, despite the laying off of numerous employees and outsourcing maintenance and airport operations (Collings 2002).

In the United States, established labor law makes it difficult to achieve significant changes to labor contracts and work practice, except where bankruptcy restructurings force a renegotiation. In Europe, managers have had greater success in affecting changes to the restructuring of work, but have found less success in changing compensation rates, in many cases established by national government decrees. Further, in Europe, national carriers have long contracted out ground handling at locations outside their countries and have used third-party maintenance contractors. The average U.S. major, in contrast, has more integrated station and maintenance operations, making it harder to change work practices by contracting to third parties.

Where the approach of replacing airline workers with third-party contractors has been attempted, airline managers expect or imagine a company where employees understand what it takes for the airline to make money. Improvements from a streamlined airline would improve cash flow so that greater investment can be made in customer service, attracting prospective clients and turning them into loyal customers. Improved customer service would include features that passengers want at prices they are willing to pay. Unfortunately, to date such a scenario or understanding is unrealistic or rare in the full-service airlines. Few managers can state clearly or view the business model of their airlines, much less how their companies make money or protect market share: "This is clearly visible with respect to the competition with new entrants, comprised of operators such as Jet Blue, and low-cost carriers like EasyJet" (Collings 2002).

However, there are indications that this is changing in the United States. Carriers have tried to reinvent themselves in the wake of the September 11 terrorist attacks, made worse by the subsequent recession triggered by instability in energy markets. Of the U.S. majors, Continental has had the greatest success with this approach. Begun in the turmoil of the acquisition of Eastern Air Lines by the then-parent of Continental Air Lines, Texas Air Corp., in the mid-1980s, the new management headed by Gordon Bethune, installed after the Gulf War, was able to build on the approaches of the previous managements to improve the product. The fleet was streamlined from more than 10 aircraft types to 3, most of which were relatively new, reducing maintenance costs and improving reliability. Hubs were carefully analyzed route by route, and in some cases service was abandoned in order to ease self-induced air traffic control delays, improving the ability of the carrier to run on time. This was a major factor in the refusal of prospective customers to fly the carrier. Above all, employees at the operating level were empowered to solve customer service issues as they arose. The internal slogan of "Work Hard, Fly Right" was extended to become the marketing theme pitched to Continental customers, who have responded enthusiastically to the changes at the carrier, greatly improving earnings to shareholders and workers.

Full-service airlines have begun to change their approaches in the implementation and adoption of more realistic business models where the core logic lies in creating value. Some approaches have included:

- A set of values that the airline offers to its stockholders and employees
- An operating process to deliver these values
- Arrangement as a coherent system (point to point, hub-and-spoke, international)
- A system that relies on building assets, capabilities, and relationships

By carrying out these steps, airlines have begun to realize the goal of creating value. For example, "[Managers at] American airlines [have] finally announced [that] they recognize that they have to change their business model, and cannot focus on maximizing revenues; they must focus on maximizing efficiency" (Collings 2002).

Managers of successful airlines find that no single business model is guaranteed to produce financially superior results; however, successful models do share certain characteristics. Successful airlines such as JetBlue offer unique value, sometimes in the form of a new idea where the carrier started up with new equipment instead of used aircraft. This has been coupled with adopting a distinctive approach to customer service at a time when major carriers were cutting back on inflight service. Often, a good formula is a combination of product and service, features that offer more value or lower price for the same benefit, or deliver more benefit for the same price.

Southwest and JetBlue have radically different approaches to creating value, even though they exploit the same low-fare market segment. Southwest offers a simple, reliable product favored by business clients and leisure passengers who like the frequency of its short-haul services. JetBlue has developed a lower-frequency, longer-haul product aimed more narrowly at the leisure market, providing the low-fare filler for the U.S. majors in coach.

A winning business model can be difficult to imitate. This is often achieved by establishing a key differentiator, such as customer attention or superb execution. True differentiation creates a barrier to entry that protects the main profit stream. Successful business models are grounded in reality and are based on accurate assumptions about customer behavior, financial structures, and day-to-day operations. As they have become globe-circling giant network carriers, it appears that many full-service airlines lack understanding of where they make money, why customers prefer their offerings, and which customers only drain resources.

Since airlines compete for customers and resources, a business model must highlight what is distinctive about the firm, how it wins customers, attracts investors, and earns profits. Effective business models are rich and detailed, and components reinforce each other. It is important to remember, however, that regulatory and economic changes, such as increased fuel prices, can make a once-effective business model less capable of generating earnings to the carrier, without which it slowly dies.

35.4.1 Airline Pricing—A Historical Perspective

A key objective of any company that produces and sells a commodity product is to gain as much control as possible over the supply of that product. The organization can thus avoid selling at commodity price levels, which are always just marginally above the production costs of the lowest-cost producer (Greenslet 2002). Historically, airline strategy has followed this fundamental rule, resulting in attempts to control the supply of seats in markets. This has led to the creation of primary hubs for all the major carriers, allowing them to offer customers a vast network of routes and markets. This tactic, when combined with the enticement of frequent flyer programs, has, however, led to the narrowing of options for many consumers who are forced to select the airlines that serve their hub cities.

Generally, network airlines have learned that there is no earnings leverage in competing aggressively against each other at major hubs. This has led to the emergence of patterns of regional dominance through "fortress" hubs, where a single carrier may have as much as an 80 percent share of the hub airport. Over the years, airlines have been able to employ this strategy most effectively in maintaining a yield premium servicing the business passenger, who historically has proved willing to pay a higher fare for the convenience of frequency. Business travel has always been the foundation of airline revenue generation. A United Airlines study in the 1990s stated that 9 percent of its passengers accounted for 43 percent of total revenue (Greenslet 2002, 1). Since the availability of frequent and reliable service has stimulated business travel over many years, business travelers have developed very short travel planning horizons, even on international trips in a multinational work environment. One consequence is that the airlines have been better able to segment passengers by fare type and offer an array of fares attractive to many different passenger segments on the same aircraft.

Generally, the shorter the booking horizon, the higher the fare. With short travel horizons, business travelers have become accustomed to paying higher fares. Additionally, restrictions on discount fares that required a Saturday night stay made them unattractive to business travelers. The full fare in the 1980s was about twice the average discount fare, and this disparity increased threefold during the economic boom of the late 1990s. While perhaps only 5 percent of the inventory of the major airlines sold at the highest fare levels as the price segmentation became more and more refined, many of the major corporations contested this pricing disparity by negotiating company discounts. Moreover, the carriers competed aggressively for identifiable major corporate travel accounts. This allowed the biggest customers greater access to a wide range of hidden discounts. But with the trend to smaller firms and self-employed people in the changing information age economy, a growing segment of customers traveling at the last minute were forced to pay high fares.

At the same time, more and more conferences and conventions, which account for as much as 33 percent of all annual business travel, began to be scheduled in such a manner as to allow access to the discounted fares. Airlines attempted to manipulate their yield-management systems in such a way as to preserve revenue streams. This had the effect of pushing some short-notice coach and business class fares in key markets up at a very aggressive rate in the late 1990s, just as the information technology boom, which had fueled traffic growth, was ending.

Figure 35.5 shows the number of enplanements and business traveler enplanements between years 1992 and 2010. From this graph it can be seen that the growth rate for business travelers is slightly less than the total enplanement growth rate.

35.4.2 Impact of the Global Economic Downturn on Operations Management

The business traveler has often complained about the disparity that exists between full fare prices and the many discount prices that are offered to entice the leisure traveler. Until December 2000, due to a strong global economy, the aviation industry was able to extract a yield premium in exchange for schedule flexibility to the business passenger.

However, as soon as the information technology sector began to decline, particularly as the e-commerce/telecom bubble collapsed in late 2000, airline revenue from full-fare travel fell drastically from a $10 billion annual rate to about $7 billion. It appears that in 2001 this may have declined by as much as 30 percent more, to $3.7 billion (Greenslet 2002). The primary reasons for this drastic impact on the bottom line of airline financial statements are that:

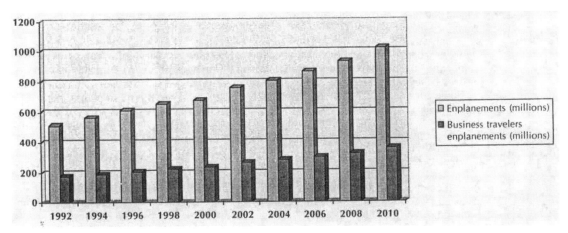

FIGURE 35.5   Enplanements and business travelers enplanement growth rates between 1992 to 2010. (*Source:* Baur 1998, 537.)

1. The business traveler is no longer a captive customer of the large full-service airlines. He or she is aware of the price and service options that are now available and easily accessible due to the World Wide Web. Moreover, travelers are willing to change lifestyles sufficiently to travel in periods when low fares are available, as small and large companies are more open to compensatory flextime in exchange for holding a weekend meeting or traveling through the weekend.

2. In the wake of September 11, major carriers reduced many of the restrictions on discount fares in attempts to build market share and regain customer loyalty. Initially, airlines that attempted to reinstate these restrictions, such as US Airways, appeared to have induced passengers to switch to flying on other carriers. More recently, carriers have developed an approach to increased revenues through surcharges at airports rather than at the time of booking.

In the United States, established network carriers have had to face a challenge not faced by any generation of airline managers since the tenfold increase in fuel prices from 1972 to 1975. U.S. managers are under intense pressure to develop innovative solutions to restore earnings. First, they must rebuild their traffic bases and encourage customers to fly again, as opposed to driving or taking other modes of transportation. Managers at U.S. major airlines are beginning to realize that they may not be able to restore the most important element of the traditional business model—the business traveler. A permanent loss of this revenue stream may prove detrimental to the financial health of airlines, since it will not allow them to offset the increases to their cost structures resulting from schedule cutbacks.

In the United States, in the period from September 11, 2001, until this writing, yield per passenger mile has decreased by about 1 cent while costs have increased by about 1 cent. With load factors trending slightly down, this is an unbearable condition in an industry that considers a 5 percent net operating margin to be exceptional. This implies that a structural change in the prevailing full-service airline business model will be ultimately necessary in order to ensure success in the long-term: "Mr. Schenk, president of the New York-based firm Airline Associates, said if the airline business wanted to be profitable again it had to ditch many of the strategies which had brought rich reward in the past" (Collings 2002). Even the best-designed business model cannot last forever. Business models must change to keep pace with shifting customer needs, markets, and competitive threats.

Second, full-service airlines may have to modify their existing business models by expanding geographically into other markets in order to add customers, adjust prices, and extend products and services. All these are incremental changes that can boost the returns from an existing model.

Another business approach being adopted is to renew the distinctiveness of an existing business model. This focuses on revitalizing value to counteract the forces that encourage competition based solely on price. Decades after the creation of Southwest, many full-service airlines believe that their future success lies in cutting costs.

Some successful airlines replicate their business model in new domains by taking new products into new markets. For instance, EasyJet, the successful U.K. low-cost carrier, has extended the underlying pricing concept into rental cars and Internet cafés. Airlines adopting their existing advantages to new models will grow by building around unique skills, assets, and capabilities.

Operations managers will have to clarify their roles in improving their airlines' focus. They need to understand the organization's business model and how it contributes to the overall company strategy. Flight operations must understand what makes the airline distinctive and how the airline makes money. Likewise, management needs to establish a team-wide framework to allow the carrier to be a more nimble competitor. The business model becomes a series of strategic building blocks, expanded in strategic range by experimentation with new blocks, then mixed and matched to create profitable new combinations.

35.5   AN ERA OF CHANGE IN THE AIRLINE INDUSTRY

Air transport is regarded as an initiator of economic development and has traditionally experienced greater growth than most economic sectors (ICAO 2002). The economic stimuli of airlines, airports, and their direct affiliates contributes more than 4.5 percent of world output and accounts for 15.4 million jobs. The ramifications of the recent global economic downturn that has been magnified by terrorist attacks serve to remind us of the crucial role played by civil aviation.

In the weeks following September 11, 2001, airlines, manufacturers, and analysts attempted to envision the long-term implications of the terrorist attacks (Morrison 2002). Their analysis yielded the following two worst-case scenarios:

1. The world economy would go into a meltdown prompted by a collapse in consumer spending, rising oil prices, and plummeting stock market.

2. Passengers, worried about airline security and safety, would refuse to fly again.

It has two years since those ghastly attacks took place, and both of the above scenarios are still hanging. The U.S. economy has largely weathered the recession and is trying hard to begin a recovery phase, prompting European economies to try hard to gain lost ground and regain investor confidence. The military actions in Afghanistan, the threat of war in the Middle East, and the implementation of security measures on board aircraft and at airports have not helped the economic situation. These steps have won back some customer confidence, and there has been resurgence in passenger numbers, both domestic and international. Airlines, however, face a deeper crisis resulting from the fact that there are too many airlines chasing too few customers. This problem is particularly acute in Europe and the United States, where there is a dire need for consolidation. The economic boom of the late 1990s had masked the problems of inefficient operations and incompetent management at most major carriers. These inadequacies were revealed once the economic boom collapsed and airlines were forced to compete for lucrative business customers.

There presently exist many barriers to structural reform in the airline industry that must be overcome before real change occurs and a healthier industry results (Morrison 2002). First and foremost, airlines must switch their focus from maintaining market share to building more profitable operations. The obsession of major carriers with size is essential to maintaining a hub-and-spoke network and supporting giant infrastructures. However, airlines need to identify and retain their core operations while outsourcing all other functions to external contractors.

Second, the outdated regulatory framework that plagues airlines globally needs to be replaced by transparency and trust among member states. Ownership restrictions that ban foreign nationals from owning majority stakes in domestic airlines are antiquated and do not reflect changes that have been implemented in various other industries such as auto manufacturing. Bilateral agreements between the United States and individual European countries restrict competition and promote inefficiency. As former International Air Transport Association (IATA) director-general Pierre Jeanniot insisted, the airline sector needs to become like a normal industry.

The third area that requires immediate change is pricing strategy. As discussed, airlines have realized the folly of complex business models and are now striving to achieve simplicity in their pricing strategies. A risk that full-service airlines face is the evolution of air travel as a commodity. By offering customers a unique experience, airlines can compete on value and profit rather than price and revenue.

35.5.1 Risk Management at Airlines

Mercer Management Consulting recently analyzed aviation industry risks for the 10-year period from April 1991 to April 2001 (Zea 2002). The primary risks facing the industry can be broadly classified into the following four categories:

1. *Strategic risks* are defined by business design choices and their interaction with external factors. Challenges from new competitors, shifts in customer preferences and industry consolidation are examples of strategic risks.

Southwest Airlines has mitigated these risks successfully by building a business model that is simple and operationally cost-effective (Zea 2002). Its use of secondary airports insulates from direct competitive pressures and reduces aircraft turnaround times. Profit sharing with employees and introducing "fun" in the workplace has been an integral part of Southwest's corporate culture and its recipe for success.

Lufthansa's diversification into nonflying businesses began in 1994 with four companies being created in engineering, cargo, services, and systems (Zea 2002). This decision was made to reduce strategic risk and the volatility of its earnings base from the passenger airline business. Thus, Lufthansa is now placed in a better position than its competitors, due to reduced earnings volatility while simultaneously maintaining its core airline segment.

2. *Financial risks* involve the management of capital and cash. External environmental factors that affect the variability and predictability of cash flow such as general economic conditions or currency exchange rates are also taken into consideration.

Ryanair, Europe's low-cost carrier, pursues an aggressive cash-management strategy. It made history in January 2002 by placing a record-breaking order with Boeing for 100 737-800s along with 50 options. Ryanair thus overcame the airline crisis by adding capacity rather than retrenching and downsizing company personnel. Chief Executive Mike O'Leary's perspective on this issue was, "When many of our high-fare competitors were grounding aircraft and canceling flights, we did the opposite and put a million tickets on sale at £9 ($13) each—our forward bookings went through the roof" (Campbell & Kingsley-Jones 2002, 35).

Fuel hedging is also a common way to manage the financial risk of volatile price changes. Cathay Pacific relies on long-term fuel hedging contracts and has saved approximately $80 million over the last five years. In 2000, 15 percent of Cathay's operating profits were directly attributable to savings from fuel hedging.

3. *Operational risks* arise from the tactical aspects of running airline operations. Examples include crew scheduling, accounting and information systems, and e-commerce systems. Although airlines address major risks such as business interruption quite effectively, they often fail to recognize subtle issues such as managing relations with government agencies. These regulatory issues cost the industry millions in operational inefficiencies and legal actions.

Operations managers need to reshape their units by incorporating familiar concepts such as process reengineering, contingency planning, and improved communication at all levels of the organization. The challenge facing airlines today is that they need to look at risks holistically, and evaluate each potential response, by gauging the impact on shareholder value (Zea 2002).

4. *Hazard risks* include risks arising from unforeseen events such as terrorist activities, war, and environmental hazards. Ironically, it seems that hazard risks were the least likely to result in loss of value to the aviation industry (Zea 2002).

Airlines have turned towards the world of technology to help deal effectively with increased security measures at airports and optimize internal operations. Self-service kiosks are now available at many ticket counters, and airlines strongly encourage passengers to use them to help speed up the boarding process (Gilden 2002). This is done through publicity campaigns, having ticket agents at check-in counters who advise travelers to use the automated service, and awarding frequent flier bonus mileage. Some carriers have gone a step further and now allow customers to check in at home using the Internet, a boon for those who are traveling without bags or plan to use the curbside baggage check. James Lam, former chief risk officer at General Electric, aptly summed up the issue: "Leaders recognize that over the long-term, the only alternative to risk management is crisis management" (Zea 2002, 3).

35.6   CONCLUSION

The convergence of management and airline operations is a symptom of the trend to integrate strategy and everyday performance. The strengthening of this relationship ensures a synthesis among management and business processes and creates an interrelated working environment. The changing face of international aviation management and the volatile environment of today make it essential that these processes be implemented. Although the crisis of an ever-widening financial burden threatens the economic conditions and fiscal viability of the airlines, management organizations are well positioned to offset the consequences.

As the tide of consolidation and increasing cooperation among airlines develops, the future of operations management and the industry as a whole will coevolve. Consequently, passengers will find an increasing role in the future development of the air transport paradigm as new measures to ensure safety and security are employed. The realization of these measures has already been seen and effectively employed by airlines and airports around the world. However, while management's initiative to accelerate the implementation of strategies has been slowed by waves of unpleasant economic indicators, by no means has it been halted. In time, the recovery of the industry is inevitable, although the environment in which these business processes are carried out will be considerably changed.

35.7   REFERENCES

Asi, M. 2002. "Institution of 'Corporate Culture' in Jordanian Air Transport." *Middle East News Online* (May 19), http://www.middleeastwire.com.

Barry, J. 1998. "EU Fears Alliances Foster High Fares." *The International Herald Tribune* (September 7): 17.

Baur, U. 1998. "Winning Strategies in a Changing Global Airline Environment." In *Handbook of Airline Marketing,* ed. G. F. Butler and M. R. Keller. New York: McGraw Hill, 533–44.

Button, K. 2002. "Airline Network Economics." In *Handbook of Airline Economics,* ed. D. Jenkins. New York: McGraw Hill, 27–34.

Campbell, A., and M. Kingsley-Jones. 2002. "Rebel Skies." *Flight International* (April 9): 30–39.

Chrystal, P., and S. LeBlanc. 2001. "Alliances: Beyond Marketing Facing Reality." In *Handbook of Airline Strategy,* ed. G. F. Butler and M. R. Keller. New York: McGraw Hill, 269–94.

Collings, R. 2002. "U.S. Air Industry 'Must Change.'" BBC News Online (August 19). Accessed October 13, 2002, from the World Wide Web: http://news.bbc.co.uk/1/low/business/2202442.stm.

Cross, R. G. 1998. "Trends in Airline Revenue Management." In *Handbook of Airline Marketing,* ed. G. F. Butler and M. R. Keller. New York: McGraw Hill, 303–18.

Daly, K., ed. 1995. "Networkers of the Future." *Air Transport Intelligence (ATI) Online* (May).

———. 2002. "Consolidation Time in Italy." *Air Transport Intelligence (ATI) Online* (August).

Doganis, R. 2001. *The Airline Business in the 21st Century.* New York: Routledge.

Gertzen, J. 2002. "Airline Industry Faces Fear of Failing: Inevitable Shakeout Expected to Lead to Higher Fares, Fewer Flights." *Business Newsbank* (September 8).

Gilden, J. 2002. "The World of Travel Keeps on Changing." *Orlando Sentinel* (May 12). Accessed September 26, 2002, from the World Wide Web: http://www.orlandosentinel.com/travel/la-sourcebook-security.story.

Greenslet, E. 2002. Executive Summary. *The Airline Monitor.*

Homan, A. C. 1999. "Charges in Airline Operating Expenses: Effects on Demand and Airline Profits." In *Handbook of Airline Finance,* ed. G. F. Butler and M. R. Keller. New York: McGraw Hill, 503–10.

ICAO Secretariat. 2002. "11 September's Negative Impact on Air Transport Is Unparalleled in History." *ICAO Journal* 57(2):6–8.

Laney, E., Jr. 2002. "The Evolution of Corporate Travel Management: Reacting to the Stresses & Strains of Airline Economics." In *Handbook of Airline Economics,* ed. D. Jenkins. New York: McGraw Hill, 477–86.

Morrison, M. 2002. "Time to Worry." *Flight International* (April 23): 3.

Oster, C. V., and J. S. Strong. 2001. "Competition and Antitrust Policy." In *Handbook of Airline Strategy,* ed. G. F. Butler and M. R. Keller. New York: McGraw Hill, 3–34.

O'Toole, K. 2002. "Keeping It Simple." *Airline Business* (September).

Ott, J. 2000. "Customer Service Drive Yields More Coach Space." *Aviation Week and Space Technology* (February 28): 50–51.

People Express Airlines: Rise and Decline. 1990. Harvard University Business School Case Study.

Spitz, W. H. 1998. "International Code Sharing." In *Handbook of Airline Marketing,* ed. G. F. Butler and M. R. Keller. New York: McGraw Hill, 489–502.

Vella, S. L. 1999. "Aircraft Asset Value Management." In *Handbook of Airline Finance,* ed. G. F. Butler and M. R. Keller. New York: McGraw Hill, 405–22.

Wang, Z. H., and M. Evans. 2002. "The Impact of Market Liberalization on the Formation of Airline Alliances." *Journal of Air Transportation* 7(2):26–30.

Zea, M. 2002. "Is Airline Risk Manageable?" *Airline Business* (April).

CHAPTER 36
THE MARINE TRANSPORTATION SYSTEM

James J. Corbett
Department of Marine Studies,
University of Delaware, Newark, Delaware

36.1   INTRODUCTION

This chapter describes the role of the marine transportation system (MTS) within a context of more familiar transportation modes, such as automobile, trucking, and rail. Within this context, the role and complexity of the MTS can be better understood in overview. The waterway network may not have rail track or asphalt defining its limits and extent, but navigable waterways are more like highways than an uncharted horizon. Specific features unique to the MTS will be summarized, including fleet characteristics for vessels engaged in cargo transportation, fishing, or other service. Like any broad system of different vehicles and destinations, the MTS cannot be summarized by only one or two images of ships. Diversity in vessel technologies, national fleet ownership and registration, and concentration of trade activity all affect the way the MTS is managed by industry and regulated by government. This diversity is driven by trends in the transportation system toward increasing global trade in an era where governments and businesses are requiring greater transparency in the MTS to foster commerce and satisfy security and environmental objectives.

36.2   OVERVIEW OF MARINE TRANSPORTATION

The marine transportation system (MTS) as a primary mode of trade transportation is global in nature and is older than any other modern form of transportation except perhaps the dirt road. Before locomotives crossed the continents, shipping enabled long-distance exploration and trade. Before roads linked metropolitan centers on a global scale, major cities arose near coastlines to take advantage of marine transportation. And before people knew the earth was not flat, sailors used ships to trade with distant cultures as civilization (both western and eastern) used the MTS to explore and expand their horizons.

Today, the MTS is key to global and multimodal transportation of energy resources, raw materials and bulk goods, manufactured products, and even people (through tourism and ferry transit). While land and air transportation is most familiar to the traveling public, the MTS is nearly invisible to the average person, who nonetheless benefits from its service. In its broadest sense, the marine transportation system is a network of specialized vessels, the ports they visit, and transportation infrastructure from factories to terminals to distribution centers to markets. In this regard, the marine transportation system depends upon the land-based modes for point-to-point movement of goods and people.

Maritime transportation is a necessary complement to and occasional substitute for other modes of freight transportation (see Figure 36.1). For many commodities and trade routes, there is no direct substitute for waterborne commerce. (Air transportation has replaced most ocean liner passenger transportation, but it carries only a small fraction of the highest-value and lightest cargoes.) On other routes, such as some coastwise or short-sea shipping or within inland river systems, marine transportation may compete with roads and rail, depending upon cost, time, and infrastructure constraints. Other important marine transportation activities include passenger transportation (ferries and cruise ships), national defense (naval vessels), fishing and resource extraction, and navigational services (vessel-assist tugs, harbor maintenance vessels, etc.).

36.2.1   Role of Marine Transportation in Global Trade and Regional Transportation

The role of marine transportation in commerce and society today is often overlooked because of the public visibility of roads and rail. However, marine transportation in the 21st century is an integral, if sometimes less publicly visible, part of the global economy. In the United States, more than 95 percent of imports and exports are carried by ships (EC 2001; MARAD 2000). On a worldwide basis, some 35,000 oceangoing vessels move cargo more than 13 billion tonne-km, annually. International trade by water modes connects international trade with all modes of transportation in nearly every nation. In the United States, the Maritime Administration (MARAD) reports that waterborne commerce within the United States moves cargo more than 2.1 billion tonne-km annually (see Figures 36.2 and 36.3). In the European Union, marine transportation moves more than 70 percent (by volume) of all cargo traded with the rest of the world; in the United States, more than 95 percent of imports and exports are carried by ships (EC 2001; MARAD 2000).

The MTS provides two basic functions in the multimodal transportation system: first, the MTS is the only mode capable of efficiently transporting large volumes of high-density cargo between continents and therefore provides a necessary complement to other modes. This

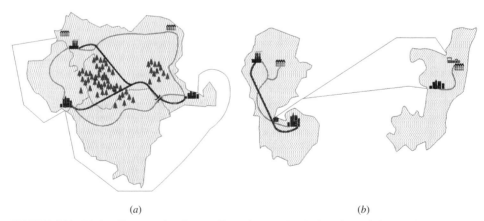

(a) (b)

FIGURE 36.1   Marine Transportation System illustration as (a) substitute for and (b) complement to land modes.

FIGURE 36.2   United States waterway network. (*Source:* NDC Publications and U.S. Waterway Data CD, U.S. Army Corps of Engineers, 2001, latest version available at http://www.iwr.usace.army.mil/hdc/.)

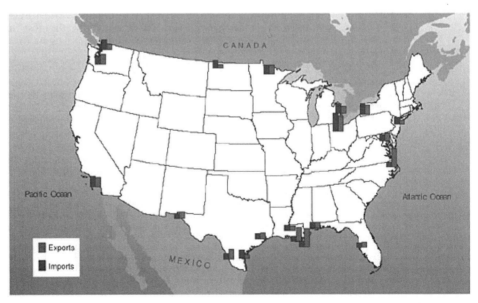

FIGURE 36.3   Top 20 international freight gateways by cargo tonnage in the United States (1998). (*Source:* U.S. Department of Transportation, Federal Highway Administration, Freight Analysis Framework, 2002.)

includes most of our energy resources (coal and oil), raw materials (ore, aggregate sands and stone), food products (grains, fruit, vegetables, and meat), and manufactured goods (machinery, shoes, apparel, furniture). (Air transportation provides higher-speed service, but it generally moves only small volumes of low-density, high-value cargoes like people, electronics, and last-minute surge deliveries of apparel or other goods.) Second, the coastal waterway and inland river network enables the MTS to substitute in many areas for many land-based modes such as automobile, heavy-duty truck, and rail. This includes passenger ferry service as part of commuter networks in New York, Seattle, and San Francisco, inland-river barges providing connections between rail and highway networks, and coastwise ships moving cargo between major trade centers.

Marine transportation is connected with the other modes of transportation so intrinsically that some studies include the road and rail networks as part of the system to deliver marine cargoes. The U.S. Department of Transportation reported to Congress in 1999 that the MTS included "the network of railroads, roadways, and pipelines" (Slater and MTS 1999). In terms of the work performed—measured in tonne-km or ton-miles—the marine transportation system rivals the other modes. The relative share of cargo moved by water compared with truck and rail modes varies from year to year, but U.S. waterborne commerce represents between 22 and 24 percent of the total tonne-km of cargo movements in recent years. Truck and rail modes in the United States each account for about 25 to 29 percent of annual cargo tonne-km. In Europe, waterborne commerce between nations (called shortsea shipping) accounts for nearly 41 percent of the goods transport market, compared to 44 percent for trucking and only 8 percent for rail and 4 percent for inland rivers (EC 2001).

One of the reasons that marine transportation moves so many tonne-km of cargo is cost efficiency. Choice of mode is a function of cost (shipping is usually cheaper), time (shipping is often slower), and other quality-of-service factors. Only pipelines move goods at lower average cost than shipping, because of their fixed capital and ability to move fluids (liquids and gases) with very low energy requirements. Bulk shipment of liquid and dry goods such as oil and grain costs about two to three times as much per tonne-km as pipelines, averaging between 2.5 and 3.5 cents per tonne-km, according to U.S. national transportation statistics (BTS 1996, 1998, 1999). These average costs are generally less than but similar to rail, whereas marine transportation of bulk goods can cost much less than trucking. Except for high-value containerized goods that are shipped intermodally, the cost per tonne-km for trucking can be more than an order of magnitude greater than the cost for marine transportation.

Of course, intermodal shipping of market-ready items can be more time-sensitive than cost-sensitive, which makes the price (freight-rate) to ship a container by sea more similar to the price to ship a container by road or rail. These differences in the demand for cargo transportation that is low-cost, timely, and reliable are the primary determinants in modal choice. Another factor includes the volume of material required, where just-in-time (JIT) inventories may favor single container deliveries or even partial loads that trucks are best suited to move. However, when the distances are large and demand is high, water transportation is often the mode of choice. Examples include primary commodity flows where the market for goods is either dominated by a few closely located firms or other receivers with large demand for cargo (like crude oil destined for major port refineries in Texas, New York, and California) or consumer-ready commodities where the market is large and can be centrally supplied via ports on navigable waterways (such as East Coast markets for juice or fruit via the Port of Wilmington, Delaware, or for home furnishings via the Port of New York/New Jersey).

For most transportation engineers, planners, and policy-makers, it is this intermodalism that demands attention. Where a transportation network can handle cargo in similar packets, the ability to transfer cargo between modes benefits from greater capacity, greater security, and greater speed. These characteristics enabled a relatively well-defined transportation infrastructure to accommodate much greater numbers of vehicles moving freight. In the next section, containerized vessel designs that facilitated this intermodal uniformity will be discussed. Containerization of ship cargoes, in its infancy in the 1950s, revolutionized ocean trade and now accounts for over 70 percent of nonbulk ocean cargo. This shift to containerization accompanied the boom in global trade during the later part of the 20th century that is associated with increased wealth and consumption and also with increased road congestion and environmental problems.

Regionally, passenger ferry service and local freight movements by barges or small vessels have become increasingly important to the marine transportation system. As automobile transportation and the construction of bridges and highways enabled greater personal mobility, many regions shifted away from ferry transportation. Similarly, as trucking became more efficient and subsidized highways made freight movement more cost-competitive, coastwise shipping declined. Today, road congestion in many regions exceeds infrastructure capacity, and adding highway lanes is not always a good choice. Building new bridges into New York City, for example, and adding parking garages if bridges were added would require exorbitant costs and consume precious real estate. However, passenger ferry service across the waterfront business centers in New York, San Francisco, and other major cities provides a much more viable and attractive solution to congested passenger transportation. As a result, commuter ferry service has emerged in the United States as one of the fastest-growing domestic markets for marine transportation. Similar efforts are underway in Europe to increase overall transportation capacity through better integration of marine transportation with road and rail. Transportation planners and policy-makers have begun advocating shortsea shipping, inland-river barge service, and passenger ferry transportation as a necessary part of a successful transportation network.

36.2.2 Complex Marine Transportation System

The MTS is a complex element of global trade. For many, the most visible elements of the MTS are the vessels. Complexity of marine transportation may be seen most easily by examining the complex mix of ships and boats that comprise the system's vehicles. Ships, barges, ferries, towboats, and tugboats are among the most recognized and romanticized parts of the MTS, perhaps second only to the water itself. Unlike automobiles, where many thousands of each model are built and sold each year and a popular car design may remain largely the same for years, each ship design is unique to only one or a few vessels.

Because of different service requirements, ship hull designs, power plants, cargo loading and discharge equipment, and other features vary greatly. Inland-river and coastal towboats range in power from 0.5 megawatts (MW) to 4 MW (600 Hp to ~5000 Hp), and can push from 1 to more than 30 barges in a tow. Passenger ferries often have capacity to carry 150 to 300 passengers and range in power from 2 MW to 4 MW (~2600 Hp to ~5000 Hp), depending upon speed requirements; some ferries, like the Staten Island ferries in New York, have capacity for 6,000 passengers. Roll-on, roll-off ships (or RoRos, so called because vehicles and other rolling cargo are driven onto the vessel rather than loaded by crane) can carry between 200 and 600 automobiles and have installed power between 15 and 25 MW (~20,000 and ~33,000 Hp). Tankers and dry bulk carriers can carry more than 250,000 tons of cargo, with installed power often in the range of 25 to 35 MW (~33,000 to ~46,000 Hp). And while container ships are not often as large as tankers, they have much larger power plants to accommodate greater vessel speeds. Average container ships carrying between 1,750 and 2,000 standard shipping containers have installed power of 20 to 25 MW (~26,000 to ~33,000 Hp); larger containerships with capacity for more than 4,000 containers can have installed power of 35 to 45 MW (46,000 to 60,000 Hp), and the largest container ships can carry more than 6,000 standard containers with engines rated as high as 65 MW (86,000 Hp).

Generally speaking, the MTS is concerned with cargo or passenger transport vessels—ships that move cargo or passengers from one place to another in trade. These vessels account for almost 60 percent of the internationally registered fleet and are analogous to on-road vehicles because they generally navigate well-defined (if unmarked) trade routes similar to a highway network. Other vessels are primarily engaged in extraction of resources (eg., fishing, oil or other minerals) or primarily engaged as support vessels (vessel-assist tugs, supply vessels). Fishing vessels are the largest category of nontransport vessels and account for more than one-quarter of the total fleet. Fishing vessels and other nontransport ships are more analogous to nonroad vehicles, in that they do not generally operate along the waterway network of trade routes. Rather, they sail to fishing regions and operate within that region, often at low power, to extract the ocean resources.

Cargoes, vessels, ports, and terminals also vary greatly. The primary common goals at ports and terminals are to achieve rapid vessel loading and unloading and to maximize the rate of cargo transport through the port facilities. Many ports are designed to facilitate container movements between ship, truck, and rail. Recently, transportation planners and engineers have been working to speed this port transfer with on-dock rail, larger cranes, and improved queuing schedules for trucks. Some ports are megahubs that attract much of the world trade (by value and volume), where other ports are networked ports that handle much of the niche cargoes or serve as feeders for megaports. And each of the world's ports is unique by location (proximity to market), waterway configuration, and ecosystem, which makes general characterizations difficult.

36.2.3   Cargo Networks: Shortsea Shipping and Multimodal Networks

In terms of the marine transportation system, the network of ports, waterways, and associated rail and highway provides the linkages between commodity supplies and their markets. In addition to global trade, regional transportation—both domestic and international—relies upon water modes for movement of bulk goods. In fact, regional trade of containerized and general cargoes by water is gaining attention, especially in Europe. This is termed shortsea shipping.

In general, shortsea shipping applies to certain maritime transport services that do not involve an ocean crossing. This includes shipping along coastlines or inland rivers within the North American and European continents, although politically the term applies more to Europe because so much of European trade among nations involves shipment of this type. As defined politically by the European Commission (EC 1999), shortsea shipping "means the movement of cargo and passengers by sea between ports situated in geographical Europe or between those ports and ports situated in non-European countries having a coastline on the enclosed seas bordering Europe." However, the United States has begun considering its coastwise and inland shipping as shortsea shipping as well, and there is no fundamental difference since shortsea shipping can refer to both domestic and international cargoes.

One of the primary advantages of shortsea shipping is that these networks may enable reverse cargo flows (backhauls) and higher average cargo capacities. In this sense, shortsea shipping may improve the economics of certain feeder services. This may be more complementary than competitive in most transportation logistics, because the average distance of a metric ton transported in the 1990s has been 100 kilometers for road, 270 kilometers for inland waterways, 300 kilometers for rail, and 1,385 kilometers for shortsea shipping (EC 1999). In other words, shortsea shipping may be most economic when moving cargo longer distances, complementing transport efficiencies of shorter routes by other modes.

Similar statistics exist for shipments in the United States, although less containerized cargo is carried by water modes because cross-country highways are favored over longer coastwise routes. However, the north-south coastal (and perhaps inland river) corridors could relieve more highway congestion in the Western, Eastern, and (perhaps) Central United States. Recent studies sponsored by the U.S. Department of Transportation (Federal Highway Administration) have produced a Freight Analysis Framework that evaluates how cargoes are moved in the United States. This important work may be motivated by the need to understand highway traffic and congestion, but it reveals important information about the maritime transportation system as well. Importantly, there is a strong cross-country link to

East Coast states; this suggests that trucking is an important link for cargoes between the major coastlines and that Pacific Rim shipping routes carrying cargo destined for New York and other states may use California as the port of call. (These flows appear very similar whether one picks New York or California, showing symmetry in the Freight Analysis Framework.)

However, heavy freight flows along north-south highway networks illustrate the potential for waterborne commerce to complement current truck transportation routes along the major highways between Seattle and Los Angeles in the west and between Maine and Florida in the east. The potential for the marine transportation system to assist in the movement of cargoes on shortsea routes will depend upon three things: policy action to make sustainable shipping a priority and remove barrier legislation; business practices to demonstrate that coastwise and shortsea shipping can be reliable and administratively efficient; and transportation planning that emphasizes and develops the multimodal network.

36.3 VESSELS

Vessels are the most numerous and active link in the marine transportation system. Whether one considers international oceangoing trade, coastwise shipping, or inland river and harbor activity, ships are the means by which transportation between nodes is accomplished. This section discusses the international fleet, and United States domestic fleets of vessels. By way of these examples, it can be seen that vessels in domestic service are not simply smaller versions of international fleets.

36.3.1 International Fleet Profile (Fleet Characteristics by Vessel Type, Nation of Registry, Geographic Operation)

Despite the variation among vessels, general types of internationally registered ships can be categorized according to their service and design characteristics. A profile of the world oceangoing fleet is presented in Table 36.1; vessels larger than 100 gross registered tons (grt) are generally considered capable of ocean or coastal commercial transit.

TABLE 36.1 Profile of World Oceangoing Fleet[a]

Ship type	Number of ships in world fleet	Percent of world fleet
Transport ships		
Container vessels	2662	3.0%
General cargo vessels	23,739	26.8%
Tankers	9098	10.3%
Bulk/combined carriers	8353	9.4%
Miscellaneous passenger	8370	9.4%
Non-transport ships		
Fishing vessels	23,371	26.4%
Tugboats	9348	10.5%
Other (research, supply)	3719	4.2%
Total	88,660	100.0%

[a] Vessels greater than 100 gross registered tonnes.

Container ships are specialized cargo vessels that carry standard-sized boxes stacked together in cargo holds and on deck. These vessels represented a revolutionary innovation in shipping because they were designed to enable cargo to be packaged into a standard container suitable for intermodal transport by ship, truck, and rail (even some aircraft are equipped to carry containers). The standard unit of measure for container ships is the 20-foot-equivalent-unit (TEU), which represents a container that is 20 feet long, 8 feet wide, and 8.5 feet high. Today, most shipping containers are twice as long and are referred to as 40-foot-equivalent-units (FEUs). However, container ships are uniformly referred to by the number of TEUs they can carry when fully loaded. Container ships account for only about 3 percent of the world fleet, but over the decades since their introduction the rate of growth for these vessels has been about twice as great as for other vessel types. This is because of their versatility in carrying relatively high-value cargoes, from refrigerated fruits and meats to furniture and shoes, and because the container is so well suited to the other modes of freight transportation. Container ships are typically faster than other vessel types, with average cruise speeds greater than 18 nautical miles per hour (knots); recently built container ships often have speeds greater than 24 knots. These higher speeds enable container ships to provide scheduled reliable service, whereas much of the world fleet sails between ports without published schedules, following the cargoes and markets. Also, because of economies of scale and trade growth, recent container ships have been very large. The fleet average container capacity is around 2,000 TEUs per vessel, but recently built container ships have exceeded 6,000 TEU capacity and new container ship designs exceed 10,000 TEU. While these vessels are getting larger and faster, there may be an upper bound to these economies of scale—perhaps not in vessel design, but because very large container ships may be larger than most world ports can accept without significant investment in channel deepening or terminal equipment.

General cargo ships include a large variety of vessel types, but all are capable of carrying cargoes in various package configurations. These vessels were the predecessors to modern container ships and still constitute the majority of cargo vessels in the world fleet. However, the average age for these vessels is more than 10 years old, their average speeds are slower (~12 knots), and their cargo capacities are typically smaller than those of container ships. Noncontainerized nonbulk cargoes, referred to break-bulk, typically require more labor to load and discharge because the configurations for cargo packaging are not standard. Unlike the container ships, which rely on terminal cranes that can offload dozens of TEUs or FEUs per hour, general cargo ships may have cargo-handling equipment on deck or rely on smaller terminal equipment. Slower cargo handling tends to limit the economic value of larger capacities because transport vessels are most productive when time in port is minimized. Their size, speed, and capacity make general cargo ships better suited to regional and coastwise transport.

For the purposes of this chapter, certain specialized vessels are included in the general cargo ship category although they serve more specialized cargo markets. This includes vehicle carriers, refrigerated cargo ships, and others. It is worth noting that some of these vessels are more like container ships than general cargo ships in terms of design standardization, speed, and specialized port handling. However, the markets for their cargoes is smaller and they account for only about 1 percent of all general cargo ships (less than 0.4 percent of the world fleet).

Liquid and dry bulk carriers are similar in terms of their service, sizes, and speeds. Liquid bulk carriers, or tankers, are familiar to many because of their service in moving energy resources, particularly crude oil. Cargoes carried by these vessels include mostly crude oil and petroleum products, but tankers also are used to carry bulk juices, wines, and beer. (It is much less expensive to move juice, wine, or beer in bulk and bottle it near the consumer market than to move these products in bottles, cans, or crates.) Dry bulk vessels carry cargoes such as grain and other agricultural products, ores and minerals, and coal. Advantages of moving liquid and dry cargoes in bulk are that loading and discharge can be faster and less labor-intensive and that unrefined cargoes can be transported from ports near their point of harvest or extraction to other ports that are nearer to refineries, factories, or markets. These cargoes can then be processed, packaged, or used to manufacture products for shipment to consumers. On average, tankers are smaller than bulk carriers, although the larger tankers have more than twice the capacity of the largest bulk carriers; this is primarily because many smaller tankers serve regional or feeder markets. Average speeds for bulk carriers are around 13–15 knots, with liquid tankers typically a bit slower. The average fleet ages for both liquid and dry bulk ships are roughly the same, and older than container or general cargo vessels.

Passenger vessels in the international fleet represent both ferry transport and cruise excursion vessels. By way of analogy, ferry transport vessels are more like scheduled bus service on highways, while cruise excursion vessels are more like tour buses. Both move passengers along the waterway network, but cruise ships carry the same passengers both to and from destinations during a given voyage. Ferries tend to serve markets where passengers in one location want to go to another location for work, tourism, or other purposes. Passenger vessels include many specialized designs, but two general characteristics dominate. Some passenger vessels are conventional monohull designs that operate at slower speeds (~13 knots on average), while others are twin-hull (catamaran) designs that operate at higher speeds (greater than 22 knots on average). Certain routes have enough demand for fast service to justify the significantly higher power and fuel requirements of faster vessels; twin hulls, or catamarans, can achieve these speeds efficiently with adequate stability and passenger comfort in typical sea conditions.

36.3.2 U.S. Domestic Cargo Fleet Profile

The U.S. domestic cargo fleet can be seen simply as a subset of vessels in the world fleet, but this would be misleading. As in other major nations with active commercial coastlines and inland river networks, many ships in the U.S. cargo fleet are smaller than internationally registered vessels. Moreover, nonpropelled vessels such as barges are counted in U.S. inventories that would not be reported in global inventories. Tables 36.2 and 36.3 present the summary of cargo vessels in the U.S. fleet, by those of 1,000 gross tons and greater and by those less than 1,000 gross tons, respectively. Passenger vessels and military craft are not included.

Of the 462 self-propelled vessels in Table 36.2, those 127 in foreign trade are most comparable to the world fleet profile for cargo vessels, although some of the 164 vessels in domestic trade may also be oceangoing. The government ships are typically operated in dedicated civilian service to meet military sealift demands and do not compete directly for general commercial cargoes. The self-propelled vessels in Table 36.3 are typically not oceangoing vessels, as shown by their service area description as coastal, inland waterway, or Great Lakes.

It is interesting to note that the U.S. inventory of ships includes so many non-self-propelled vessels, mostly barges on inland rivers. This is partly an artifact of the inventory methodology, but the information can be important to transportation planners and engineers, particularly in terms of designing waterway locks and dams or when considering port terminal designs.

Not shown in these tables are some 1,330 commercial passenger vessels (mostly ferries carrying less than 150 passengers each), more than 5,400 tugs and towboats that assist ships in U.S. ports or push barge tows along rivers and coastlines, and some 1,600 workboats, including crewboats, supply, and utility vessels. And another significant element of the domestic MTS is recreational boating. These small vessels are rarely registered internationally and do not typically transit oceans, operating most of the time in domestic waters on a weekend or seasonal basis.

36.4 PORTS

Perhaps the most visible element of the marine transportation system, ports are the MTS nodes where cargo and passengers transition from vessel to on-road vehicle or rail. These

TABLE 36.2 U.S.-Flag Cargo Carrying Fleet (Vessels of 1,000 Gross Tons and Over) by Area of Operation[a]

	Total		Liquid carriers		Dry bulk carriers		Containerships		Other freighters[b]	
	No.	Tons	No.	Tons	No.	Tons	No.	Tons	No.	Tons
Grand total	**3,869**	**30,495**	**2,196**	**15,714**	**759**	**5,889**	**123**	**3,108**	**791**	**5,784**
Foreign trade	**268**	**5,319**	**53**	**946**	**116**	**1,115**	**61**	**2,510**	**38**	**748**
Oceanborne	263	5,214	51	927	113	1,029	61	2,510	38	748
Great Lakes	5	105	2	19	3	86	0	0	0	0
Domestic trade	**3,430**	**21,921**	**2,116**	**13,887**	**643**	**4,774**	**57**	**512**	**614**	**2,748**
Coastal	1,344	13,299	567	8,770	355	2,146	57	512	365	1,871
Inland waterway	2,000	6,454	1,542	5,075	227	576	0	0	231	803
Great Lakes	86	2,168	7	42	61	2,052	0	0	18	74
Government	**171**	**3,255**	**27**	**881**	**0**	**0**	**5**	**86**	**139**	**2,288**
Total self-propelled	**462**	**14,914**	**114**	**6,230**	**69**	**2,600**	**90**	**2,898**	**189**	**3,186**
Foreign trade	**127**	**4,588**	**17**	**771**	**12**	**579**	**61**	**2,510**	**37**	**728**
Oceanborne	127	4,588	17	771	12	579	61	2,510	37	728
Great Lakes	0	0	0	0	0	0	0	0	0	0
Domestic trade	**164**	**7,071**	**70**	**4,578**	**57**	**2,021**	**24**	**302**	**13**	**170**
Coastal	105	5,063	68	4,559	2	71	24	302	11	131
Inland waterway	0	0	0	0	0	0	0	0	0	0
Great Lakes	59	2,008	2	19	55	1,950	0	0	2	39
Government	**171**	**3,255**	**27**	**881**	**0**	**0**	**5**	**86**	**139**	**2,288**
Total non-self-propelled[c]	**3,407**	**15,581**	**2,082**	**9,484**	**690**	**3,289**	**33**	**210**	**602**	**2,598**
Foreign trade	**141**	**731**	**36**	**175**	**104**	**536**	**0**	**0**	**1**	**20**
Oceanborne	136	626	34	156	101	450	0	0	1	20
Great Lakes	5	105	2	19	3	86	0	0	0	0
Domestic trade	**3,266**	**14,850**	**2,046**	**9,309**	**586**	**2,753**	**33**	**210**	**601**	**2,578**
Coastal	1,239	8,236	499	4,211	353	2,075	33	210	354	1,740
Inland waterway	2,000	6,454	1,542	5,075	227	576	0	0	231	803
Great Lakes	27	160	5	23	6	102	0	0	16	35

Source: U.S. Maritime Administration (MARAD), current July 2002. Adapted by MARAD from Corps of Engineers, Lloyd's Maritime Information Service, U.S. Coast Guard and Customs Service Data. Self-propelled vessels ≥1,000 gross tons; excludes one domestic coastal passenger vessels of 3,988 dwt and eleven other passenger vessels of 96,474 dwt.

[a] Carrying capacity expressed in thousands of metric tons.

[b] Includes general cargo, Ro-Ro, multipurpose, LASH vessels, and deck barges; excludes offshore supply vessels.

[c] Integrated tug barges of 1,000 grt and greater are contained in non-self-propelled categories as follows: foreign trade—1 dry bulk (36,686 tons), 1 other freighter (20,000 tons); domestic coastal—11 liquid (449,370 tons), 3 dry bulk (111,000 tons); U.S./Translakes—Great Lakes 2 liquid (18,955 tons) and 3 dry bulk (85,514 tons).

ports range from megahubs of world trade to fishing or recreational centers of activity. Trends indicate that port activity overall will increase dramatically as cargo volumes double or triple over the next decades and as more of the world's population uses coastal waters for commerce, fishing, and recreation. These trends create conflicts over waterfront real estate and environmental resources and provide strong incentive for new technology and modernization. To ensure that marine transportation continues to meet our needs, planners must recognize, understand, and project the intermodal nature of the MTS and its inherent interfaces and complexities.

36.4.1 World Port Ranking

Table 36.4 presents a summary of the top 40 world ports, ranked according to the volumes of cargoes they handle annually, of total cargoes and containerized cargoes handled annually,

TABLE 36.3 U.S.-Flag Cargo Carrying Fleet (Vessels less than 1,000 Gross Tons) by Area of Operation[a]

	Total		Liquid carriers		Dry bulk carriers		Containerships		Other freighters	
	No.	Tons	No.	Tons	No.	Tons	No.	Tons	No.	Tons
Grand Total	**32,229**	**46,381**	**2,214**	**3,965**	**23,010**	**36,438**	**4**	**2**	**7,001**	**5,976**
Foreign trade	**109**	**50**	**3**	**1**	**106**	**49**	**0**	**0**	**0**	**0**
Oceanborne	109	50	3	1	106	49	0	0	0	0
Great Lakes	0	0	0	0	0	0	0	0	0	0
Domestic trade	**32,120**	**46,331**	**2,211**	**3,964**	**22,904**	**36,389**	**4**	**2**	**7,001**	**5,976**
Coastal	3,930	3,562	241	982	573	741	1	1	3,115	1,838
Inland waterway	27,890	42,394	1,961	2,975	22,211	35,478	3	1	3,715	3,940
Great Lakes	300	375	9	7	120	170	0	0	171	198
Total self-propelled	**384**	**948**	**77**	**797**	**4**	**2**	**0**	**0**	**303**	**149**
Domestic trade	**384**	**948**	**77**	**797**	**4**	**2**	**0**	**0**	**303**	**149**
Coastal	256	902	71	795	0	0	0	0	185	107
Inland waterway	109	27	2	0	0	0	0	0	107	27
Great Lakes	19	19	4	2	4	2	0	0	11	15
Total non-self-propelled	**31,845**	**45,433**	**2,137**	**3,168**	**23,006**	**36,436**	**4**	**2**	**6,698**	**5,827**
Foreign trade	**109**	**50**	**3**	**1**	**106**	**49**	**0**	**0**	**0**	**0**
Oceanborne	109	50	3	1	106	49	0	0	0	0
Great Lakes	0	0	0	0	0	0	0	0	0	0
Domestic trade	**31,736**	**45,383**	**2,134**	**3,167**	**22,900**	**36,387**	**4**	**2**	**6,698**	**5,827**
Coastal	3,674	2,660	170	187	573	741	1	1	2,930	1,731
Inland Waterway	27,781	42,367	1,959	2,975	22,211	35,478	3	1	3,608	3,913
Great Lakes	281	356	5	5	116	168	0	0	160	183

Source: U.S. Maritime Administration (MARAD), current July 2002. Adapted by MARAD from Corps of Engineers, Lloyd's Maritime Information Service, U.S. Coast Guard and Customs Service Data.
[a] Carrying capacity expressed in thousands of metric tons.

in tons and TEUs, respectively. As shown, less than half (17) of the top 40 world ports appear in the top 40 for both tonnage and container TEUs. This is important for transportation engineering because it means that the transportation infrastructure is very different when considering the needs for large ports handling bulk cargoes versus containerized cargoes. Moreover, the value of cargo (per ton) will be very different and may influence the economic importance of different ports more than the volume of cargo transported. This observation will be repeated at the national level for most large trading nations.

36.4.2 U.S. Port Ranking

There are over 300 ports in the United States, but most of the cargo and ship traffic is handled by fewer than 100 ports. Table 36.5 summarizes statistics for the top 30 ports in the United States, cargo and containerized cargo, respectively. As with the world port profile, less than half (10) of the top 30 U.S. ports appear in the top 30 for both tonnage and container TEUs. Illustrations of principal U.S. ports are shown in Figures 36.4 and 36.5.

36.4.3 Major Fishing Ports

Major fishing ports look quite different than the major cargo ports. Unless supported by factory or mother ships or fitted with freezing equipment, fishing boats need to get to port

TABLE 36.4   World Port Ranking[a]

Total cargo volume 2000 Metric tons (000s), except where noted					Container traffic 2000 Twenty foot equivalent units (TEUs)			
Rank	Port	Country	Unit[b]	Tons	Rank	Port	Country	TEUs
1	Singapore	Singapore	FT	325,591	1	Hong Kong	China	18,098,000
2	Rotterdam	Netherlands	MT	319,969	2	Singapore	Singapore	17,090,000
3	South Louisiana	U.S.A.	MT	197,680	3	Pusan	South Korea	7,540,000
4	Shanghai	China	MT	186,287	4	Kaohsiung	Taiwan	7,426,000
5	Hong Kong	China	MT	174,642	5	Rotterdam	Netherlands	6,274,000
6	Houston	U.S.A.	MT	173,770	6	Shanghai	China	5,613,000
7	Chiba	Japan	FT	169,043	7	Los Angeles	U.S.A.	4,879,000
8	Nagoya	Japan	FT	153,370	8	Long Beach	U.S.A.	4,601,000
9	Ulsan	South Korea	RT	151,067	9	Hamburg	Germany	4,248,000
10	Kwangyang	South Korea	RT	139,476	10	Antwerp	Belgium	4,082,000
11	Antwerp	Belgium	MT	130,531	11	Tanjung Priok	Indonesia	3,369,000
12	New York/New Jersey	U.S.A.	MT	125,885	12	Port Kelang	Malaysia	3,207,000
13	Inchon	South Korea	RT	120,398	13	Dubai	U.A.E.	3,059,000
14	Pusan	South Korea	RT	117,229	14	New York/New Jersey	U.S.A.	3,050,000
15	Yokohama	Japan	FT	116,994	15	Tokyo	Japan	2,899,000
16	Kaohsiung	Taiwan	RT	115,287	16	Felixstowe	U.K.	2,793,000
17	Guangzhou	China	MT	101,521	17	Bremer Ports	Germany	2,712,000
18	Quinhuangdao	China	MT	97,430	18	Gioia Tauro	Italy	2,653,000
19	Ningbo	China	MT	96,601	19	San Juan	U.S.A.	2,334,000
20	Marseilles	France	MT	94,097	20	Yokohama	Japan	2,317,000
21	Osaka	Japan	FT	92,948	21	Manila	Philippines	2,289,000
22	Richards Bay	South Africa	HT	91,519	22	Kobe	Japan	2,266,000
23	Kitakyushu	Japan	FT	87,346	23	Yantian	China	2,140,000
24	Qingdao	China	MT	86,360	24	Qingdao	China	2,120,000
25	Hamburg	Germany	MT	85,863	25	Laem Chabang	Thailand	2,105,000
26	Dalian	China	MT	85,053	26	Algeciras	Spain	2,009,000
27	Kobe	Japan	FT	84,640	27	Keelung	Taiwan	1,955,000
28	Tokyo	Japan	FT	84,257	28	Nagoya	Japan	1,905,000
29	New Orleans	U.S.A.	MT	82,400	29	Oakland	U.S.A.	1,777,000
30	Dampier	Australia	MT	81,446	30	Colombo	Sri Lanka	1,733,000
31	Vancouver	Canada	MT	76,646	31	Tianjin	China	1,708,000
32	Corpus Christi	U.S.A.	MT	75,461	32	Charleston	U.S.A.	1,629,000
33	Beaumont	U.S.A.	MT	75,032	33	Genoa	Italy	1,501,000
34	Newcastle	Australia	MT	73,871	34	Seattle	U.S.A.	1,488,000
35	Tubarao	Brazil	MT	73,182	35	Le Havre	France	1,486,000
36	Tianjin	China	MT	72,980	36	Tacoma	U.S.A.	1,376,000
37	Port Hedland	Australia	MT	72,914	37	Barcelona	Spain	1,364,000
38	Hay Point	Australia	MT	69,379	38	Cristobal	Panama	1,354,000
39	Le Havre	France	MT	67,492	39	Hampton Roads	U.S.A.	1,347,000
40	Port Kelang	Malaysia	FT	65,227	40	Melbourne	Australia	1,328,000

[a]Primary source: American Association of Port Authorities, www.aapa-ports.org; AAPA Sources: *Shipping Statistics Yearbook 2001*, 359; U.S. Army Corps of Engineers, *Waterborne Commerce of the United States CY 2000; AAPA Advisory,* May 21, 2001; various port authority Internet sites.

[b]Abbreviations: MT = metric ton; HT = harbor ton; FT = freight ton; RT = revenue ton.

Note: The cargo rankings based on tonnage should be interpreted with caution since these measures are not directly comparable and cannot be converted to a single, standardized unit.

TABLE 36.5 United States Port Ranking[a]

		Total cargo volume 2000 Short tons (000s)					Container traffic 2002 Twenty foot equivalent units (TEUs)	
Rank	Port name	Total	Domestic	Foreign	Rank	Port name	Total	
1	South Louisiana, Port of	217,757	119,141	98,615	1	Los Angeles	4,060,000	
2	Houston	191,419	62,617	128,802	2	Long Beach	3,184,000	
3	New York	138,670	72,273	66,397	3	New York	2,627,000	
4	New Orleans	90,768	38,316	52,452	4	Charleston	1,197,000	
5	Corpus Christi	83,125	23,989	59,136	5	Savannah	1,014,000	
6	Beaumont	82,653	16,043	66,609	6	Norfolk	982,000	
7	Huntington	76,868	76,868	0	7	Oakland	979,000	
8	Long Beach	70,150	17,400	52,750	8	Houston	851,000	
9	Baton Rouge	65,631	42,505	23,126	9	Seattle	850,000	
10	Texas City	61,586	20,330	41,256	10	Tacoma	769,000	
11	Port of Plaquemine	59,910	38,864	21,046	11	Miami	752,000	
12	Lake Charles	55,518	20,476	35,042	12	Port Everglades	370,000	
13	Mobile Harbor	54,157	24,232	29,925	13	Baltimore	302,000	
14	Pittsburgh	53,923	53,923	0	14	New Orleans	216,000	
15	Los Angeles	48,192	6,065	42,127	15	Portland	185,000	
16	Valdez	48,081	46,409	1,672	16	San Juan PR Harbor	159,000	
17	Tampa Bay	46,460	31,662	14,798	17	Palm Beach	142,000	
18	Philadelphia	43,855	14,066	29,788	18	Wilmington, DE	133,000	
19	Norfolk	42,377	10,505	31,872	19	Gulfport	132,000	
20	Duluth-Superior, MN & WI	41,678	28,165	13,512	20	Philadelphia	115,000	
21	Baltimore	40,832	14,535	26,297	21	Jacksonville	114,000	
22	Portland, OR	34,334	16,357	17,977	22	Boston	80,000	
23	St Louis	33,338	33,338	0	23	Wilmington, NC	71,000	
24	Freeport	30,985	5,599	25,386				
24	Chester	59,000						
25	Portland, ME	29,330	2,337	26,993	25	Newport News	57,000	
26	Pascagoula	28,710	10,454	18,256	26	Freeport	54,000	
27	Paulsboro	26,874	9,186	17,688	27	Port Bienville	41,000	
28	Seattle	24,159	8,718	15,441	28	Richmond	36,000	
29	Chicago Harbor	23,929	20,063	3,866	29	Honolulu	32,000	
30	Marcus Hook	22,584	8,872	13,712	30	Ponce	29,000	

[a]Primary source: U.S. Army Corps of Engineers Navigational Data Center (for tonnage data) and American Association of Port Authorities, www.aapa-ports.org (for containerized data).
Note: The cargo rankings by tonnage are for a different year than the rankings by TEU and should be considered representative of relative port ranking, which may vary from one year to another.

and send their cargo on to the market quickly. Table 36.6 presents the top 40 fishing ports, both by volume and value of fish landed. Note that these ports are generally different from the top cargo ports. Most of these ports are smaller and more clearly dedicated to fishing activities and related industry (including tourism).

36.4.4 Port Infrastructure Development

While these data are important indicators of vessel traffic and other modal traffic to major ports, transportation planners must also consider the number of vehicles (ships, trucks, trains) in addition to the cargoes transported. In this regard, port calls are an important direct

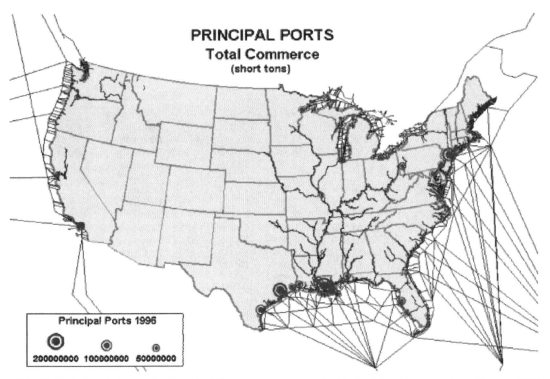

FIGURE 36.4 Principal ports in the United States waterway network, identified by total cargo tons. (*Source:* NDC Publications and U.S. Waterway Data CD, U.S. Army Corps of Engineers, 2001, latest version available at http://www.iwr.usace.army.mil/ndc/.)

measure of marine transportation infrastructure demand. Table 36.7 presents a summary of the top 25 world ports by the number of ship visits, or port calls; this table also provides estimates of the observed capacity of these ports in terms of the total deadweight tonnage of the ships that called on each port. This may qualitatively indicate relative demand for intermodal infrastructure to carry cargoes into or out of the port facilities.

From this table, it is clear that frequent calls to port are typically associated with general cargo and containerized cargoes, more so that with bulk cargoes. Containerized cargo vessels calling on the 25 ports shown in Table 36.4 account for some 46 percent of all container ship port calls worldwide. Moreover, about half of the ports receiving the most port calls are also listed as ports with the greatest cargo volumes for both total and containerized cargoes. These ports may qualify as true megaports, which are primary hubs for all types of marine transportation. Another interesting note is that there are more ports in Table 36.4 that also appear as major container ports. Lastly, 4 of the busiest 25 ports in terms of port calls are not listed among the top world ports by total cargo volume or by container volume. This suggests that on a global basis, at least, transportation planning and engineering needs to consider the total system when developing new infrastructure for maritime ports.

36.5 TRENDS, OPPORTUNITIES, AND CHALLENGES

The marine transportation system is distinct in many ways from land-based modes, but it has much in common with truck, rail, and even air transportation systems. In particular, the

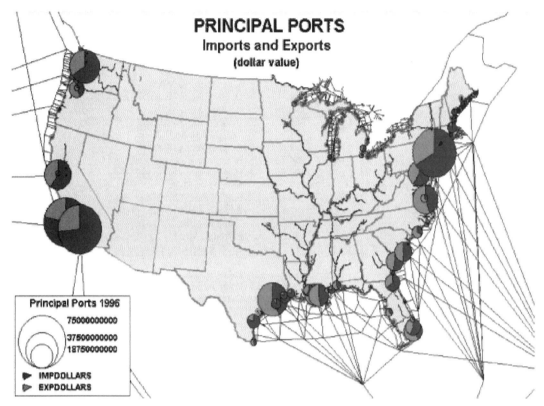

FIGURE 36.5 Principal ports in the United States waterway network, identified by cargo value. (*Source:* NDC Publications and U.S. Waterway Data CD, U.S. Army Corps of Engineers, 2001, latest version available at http://www.iwr.usace.army.mil/ndc/.)

MTS must respond to increasing trade over the longer term, meet emerging and stricter environmental regulation and safety requirements, and adopt innovative technologies. Responding to these trends presents opportunities and challenges to all modes of transportation, but for the marine transportation system the environmental challenges are relatively new and unfamiliar (compared to trucking, for example, which has been more heavily regulated than shipping, especially in the United States and Europe). And unlike other modes, where vehicle life is measured in years and/or where periodic overhauls involve reengining with newer designs, ships last for decades without major design changes to hull or power train.

36.5.1 Increasing Trade Volumes

Global trade has increased dramatically over the past century, and projections continue to forecast a doubling of trade volume for major nations, and as much as a tripling in cargo volumes to major world ports. Carriage of these increasing trade volumes will by necessity include water transportation. Three possible ways exist for the marine transportation system to accommodate these forecasted cargoes. First, the existing fleet might simply carry more through better logistics and more fully loaded ships to maximize efficiency. Second, additional ships may be added to carry the additional cargoes resulting in an increase in ship traffic overall. Third, the fleet might transition to larger vessels without increasing the overall

TABLE 36.6 Top 40 Fishing Ports, by Commercial Landings in Dollars and Pounds 1999

Rank	Port	Pounds (millions)	Port	Dollars (millions)
1	Dutch Harbor-Unalaska, AK	699.8	New Bedford, MA	146.3
2	Cameron, LA	414.5	Dutch Harbor-Unalaska, AK	124.9
3	Empire-Venice, LA	396.2	Kodiak, AK	94.7
4	Reedville, VA	366.8	Brownsville-Port Isabel, TX	88.6
5	Intracoastal City, LA	321.7	Dulac-Chauvin, LA	68.1
6	Kodiak, AK	289.6	Empire-Venice, LA	61.6
7	Los Angeles, CA	254.7	Honolulu, HI	56.0
8	Pascagoula-Moss Point, MS	199.9	Hampton Roads Area, VA	52.8
9	Port Hueneme-Oxnard-Ventura, CA	162.2	Key West, FL	50.6
10	Astoria, OR	130.1	Port Arthur, TX	49.3
11	Newport, OR	102.3	Bayou La Batre, AL	48.9
12	Sitka, AK	95.5	Cameron, LA	47.6
13	New Bedford, MA	89.0	Portland, ME	45.4
14	Beaufort-Morehead City, NC	68.4	Golden Meadow-Leeville, LA	44.9
15	Naknek-King Salmon, AK	63.1	Sitka, AK	44.6
16	Portland, ME	62.8	Palacios, TX	41.8
17	Cape May-Wildwood, NJ	59.9	Point Judith, RI	41.4
18	Point Judith, RI	59.3	Galveston, TX	40.8
19	Ketchikan, AK	57.3	Los Angeles, CA	38.1
20	Moss Landing, CA	50.5	Naknek-King Salmon, AK	37.3
21	Atlantic City, NJ	50.4	Seward, AK	35.8
22	Dulac-Chauvin, LA	48.2	Gulfport-Biloxi, MS	35.5
23	Gloucester, MA	40.1	Homer, AK	30.9
24	Point Pleasant, NJ	38.2	Delcambre, LA	30.8
25	Westport, WA	37.5	Gloucester, MA	30.0
26	Wanchese-Stumpy Point, NC	33.3	Reedville, VA	29.5
27	Petersburg, AK	32.3	Cape May-Wildwood, NJ	28.6
28	Brownsville-Port Isabel, TX	29.2	Astoria, OR	28.0
29	Seward, AK	28.6	Grand Isle, LA	26.4
30	Rockland, ME	28.3	Atlantic City, NJ	26.1
31	Honolulu, HI	27.0	Newport, OR	24.6
32	Golden Meadow-Leeville, LA	26.9	Intracoastal City, LA	24.5
33	Bayou La Batre, AL	23.0	Wanchese-Stumpy Point, NC	24.0
34	Hampton Roads Area, VA	22.7	Freeport, TX	22.8
35	Morgan City-Berwick, LA	20.2	Tampa Bay-St. Petersburg, FL	20.4
36	Ilwaco-Chinook, WA	19.8	Port Hueneme-Oxnard-Ventura, CA	20.2
37	Coos Bay-Charleston, OR	19.2	Delacroix-Yscloskey, LA	20.1
38	Grand Isle, LA	18.2	Ketchikan, AK	20.0
39	Homer, AK	18.1	Petersburg, AK	19.4
40	Bellingham, WA	18.0	Stonington, ME	18.0

Source: Fisheries Statistics & Economics Division of the National Marine Fisheries Service; www.st.nmfs.gov/st1/commercial/

number of ships in service in order to provide greater capacity within the same network of routes. There are advantages and challenges to the MTS under each alternative. In practice, each of these alternatives may be applied in some combination.

On certain routes and for certain cargoes, it may be possible to carry more cargo with the current fleet. However, there is an upper bound to this alternative. Most shipping routes are practically unidirectional, meaning that most of the cargo flows in one direction. For example, oil is primarily imported into the United States, and grain is primarily exported.

TABLE 36.7 Top 25 World Port Calls and Observed Capacity by Vessel Type 2000[a]

Port	Tanker[b]		Dry bulk		Containership		Other general cargo[c]		Total	
	Calls	Capacity	Calls	Capacity	Calls	Capacity	Calls	Capacity	Calls	Capacity
Singapore	5,351	436,844	4,581	242,709	11,286	354,686	3,232	63,656	24,450	1,097,895
Hong Kong	637	26,774	1,040	34,262	12,462	412,264	1,360	24,240	15,499	497,540
Kaohsiung	773	48,032	1,387	69,756	5,808	199,284	692	12,504	8,660	329,576
Pusan	150	4,555	1,181	51,191	5,217	164,795	1,009	18,681	7,557	239,222
Rotterdam	2,112	121,957	900	73,730	2,528	110,192	1,579	33,243	7,119	339,122
Antwerp	990	34,071	921	41,747	2,111	76,312	2,183	44,795	6,205	196,924
Yokohama	505	36,129	530	17,725	3,298	103,399	1,663	30,212	5,996	187,465
Keelung	256	10,350	491	13,545	4,344	94,522	542	8,704	5,633	127,121
Port Kelang	425	10,480	522	18,797	3,950	109,883	668	12,035	5,565	151,195
Los Angeles & Long Beach	911	66,045	783	37,568	2,955	124,281	677	15,057	5,326	242,951
Nagoya	265	28,669	814	51,991	2,699	91,331	1,374	24,596	5,152	196,587
Houston	2,988	134,809	748	28,342	614	19,799	779	24,881	5,129	207,831
New Orleans	1,371	81,956	2,676	119,270	388	10,853	655	21,957	5,090	234,036
Kobe	301	9,012	381	14,049	3,325	116,447	660	11,149	4,667	150,657
New York & New Jersey	1,271	65,965	301	10,099	2,172	87,463	861	23,104	4,605	186,631
Taichung	668	25,561	1,228	54,158	1,998	33,604	513	9,276	4,407	122,599
Laem Chabang	207	15,027	495	20,057	2,600	49,820	442	7,860	3,744	92,764
San Francisco & Oakland	787	50,653	626	22,619	1,936	82,958	226	6,841	3,575	163,071
Santos	637	17,342	727	31,262	1,547	42,749	637	14,336	3,548	105,688
Hamburg	440	14,349	565	32,753	1,745	74,067	764	16,210	3,514	137,379
Tokyo	1	260	222	7,692	2,987	102,198	238	4,547	3,448	114,697
Durban	442	23,604	809	27,354	1,043	29,088	1,115	21,028	3,409	101,074
Shanghai	180	6,208	782	44,157	1,763	47,449	582	10,718	3,307	108,532
Le Havre	699	53,308	104	6,681	2,013	82,329	433	9,768	3,249	152,086
Osaka	95	5,244	478	17,638	2,030	57,659	475	9,424	3,078	89,966
Top 25 world ports	22,462	1,327,204	23,292	1,089,151	82,819	2,677,433	23,359	478,822	151,932	5,572,609
Total world port	138,296	8,751,934	126,246	5,917,050	180,766	5,406,073	115,127	2,344,277	560,435	22,419,335
Top 25 as percent of world total	16.2%	15.2%	18.4%	18.4%	45.8%	49.5%	20.3%	20.4%	27.1%	24.9%

[a] Primary source: U.S. Maritime Administration, office of statistics, 3/8/2002; MARAD source: Lloyd's Maritime Information Services, Vessel Movements. *Observed capacity* is estimated by multiplying the number of vessel calls by vessel deadweight tonnage capacity; it does not represent total capacity for a given port. Excludes calls by vessels under 10,000 dwt tons.
[b] Includes petroleum, chemical, and gas carriers.
[c] Includes Roll-on/Roll-off (Ro/Ro), Ro/Ro container, vehicle carriers, general cargo, partial containership, refrigerated, barge carrier, livestock carrier, and combination carriers.

Bulk carriers typically transit empty (or under ballast) in one direction and return with a load of cargo; this results in a practical upper bound of 50 percent capacity overall on routes for bulk carriers (e.g., averaging 100 percent one way and 0 percent the other, or more realistically, averaging 80 to 90 percent one way and 10 to 20 percent on the return). For container and general cargoes, some routes are able to secure a backhaul; however, there is limited ability for ships to realize this idealized goal. Where a strong backhaul market can be established, the backhaul freight rates usually are much lower and may not directly cover the costs of the voyage; here, as for bulk carriers, the freight rate for inbound cargoes must cover most of the cost of the round trip voyage. For inland-river and shortsea shipping, backhauls are more practical because vessel capacities are smaller. Even so, backhauls on inland rivers are relatively few and are commodity specific. In fact, the industry appears to behave as though there is an upper bound. Most ships carry loads that average 50 to 65 percent capacity or less. When cargo capacities exceed 70 percent, it can be an indication that too few ships are available for the route (Abrams 1997); under these conditions, ships are added to the route.

Adding ships to increase route capacity is easier than it might be for onroad vehicles in the current transportation system. This is primarily because ocean routes are unconstrained highways. For most route distances, few modifications will be needed to enable more vessel traffic. However, in and near port regions this may not hold. The number of berths at terminals and cargo handling rates provide at least two important limits on how many vessels a port can accommodate in a given period. As discussed above, no consistent metric to evaluate port capacity exists, but clearly the capacity of the marine transportation system will be limited if ships have to wait for a berth at a port.

36.5.2 Environmental and Safety Constraints

Environmental impacts from shipping have been the focus of increasing attention over the past decades, primarily focused on waterborne discharges and spills of oil, chemical, and sewage pollution. A number of international treaties and national laws have been adopted, along with industry best practices, to prevent these pollution releases through accident or substandard operation. However, the shipping industry is being seen as an important hybrid between land-based transportation and large stationary power systems. On the one hand, these are non-point-source vehicles operating on a transportation network of waterways; on the other hand, these vehicles operate ship systems that compare to small power-generating plants or factories. As the industry itself and policy-makers at state, federal, and international levels recognize these facts, environmental performance is being measured and regulated. Recent efforts to mitigate environmental impacts from shipping, including invasive species in ballast water, toxic hull coatings, and air emissions, are relatively new for the industry and will take decades to address.

Environmental regulation of the shipping industry typically lags behind regulation of similar industrial processes (engines, etc.) on land. At the international level, the International Maritime Organization has effectively developed a set of treaties addressing much of these ship pollution issues. Collectively called the International Convention for the Prevention of Pollution from Ships (MARPOL 73/78), these regulations apply global standards to environmental protection practices aboard ship. However, national and state jurisdictions can apply stricter regulations to certain vessel operations and to the ports that receive these ships. For example, the U.S. Environmental Protection Agency recently regulated shipping under authority mandated by the Clean Air Act. Another example includes efforts by port states in Sweden and Norway and by individual ports in California to apply market-based or voluntary regulations to reduce air pollution from ships entering their ports.

In terms of ship safety, increasing regulations are also seen as a result of international security concerns, not to mention efforts to standardize best practices for crew training and safe vessel navigation. Specific efforts include automated ship identification systems, more transparent cargo tracking processes, and additional shipboard and corporate responsibilities for ship security to reduce the risk of theft, terrorism, or smuggling.

These increasing regulations are imposing changes in both ship operation and ship equipment. For many vessels this will simply mean retrofitting the minimum necessary equipment to comply with regulations. However, as seen in shoreside stationary systems, there may be a point at which retrofits may become more costly than replacement. At this point, emerging regulations may speed up the normal process of fleet technology and modernization.

36.5.3 Technology and Modernization

As growth in trade and stricter regulations continue over the coming decades, vessel technology will adapt through the two processes of expansion and modernization. Technology and modernization affect both ports terminals and vessels. Internationally, the trend toward larger vessels—as large as 10,000 TEUs—is advancing the concepts of megaports and networked ports. Additionally, the trend continues in certain ship types for increased speed. More fundamental than these leading trends in certain segments of the MTS are crosscutting efforts. This modernization affects different industries in the marine transportation system very differently, depending on the rate of technology capitalization and expected working life.

In perspective, efforts to innovate in marine systems are not at all new. The efficient use of energy in marine transportation has been important since before Greek and Roman times. Shipping itself has been an important human activity throughout history, particularly where prosperity depended primarily on commerce with colonies and interregional trade with other nations (Mokyr 1990). From ancient times when ships sailed mostly in sight of land using wind-driven sails or hand-driven oars to today's oceangoing fleet of steel ships powered by the world's largest internal combustion engines, the history of shipping includes many examples of technical innovations designed to increase ship performance while conserving three factors: energy, capital, and labor. Whether the primary energy for ship propulsion was wind and sail or engines and petroleum, all major innovations in marine transportation throughout history involve balancing these factors.

However, fleets typically have built up during brief periods (typically during and following major wars) and then aged without much replacement. Some nations have instituted planned replacement policies for over-age ships, but most rely on industry to modernize when vessel economics indicate. Other nations have implemented unsuccessful policies that protect domestic shipbuilding and/or other policies that remove incentives to replace older ships. For these nations' fleets, shipboard technologies tend to be older than the state-of-the-art.

In this regard, the U.S. fleet has modernized more slowly than the world fleet. Over the past decades, the U.S. fleet has not built new ships at the same rate as the world fleet. Considering all vessels greater than 100 gross tons, the U.S. fleet construction rate has been less than 0.5 percent annually (MARAD 1996), while the average rate of construction for the world fleet of ships has been about 2 percent (LMIS 2002; Lloyd's Register). Under current market and policy conditions (e.g., U.S. versus foreign labor rates for merchant vessel construction and operation), there are limited economic incentives to modernize the U.S. fleet at the pace of the world fleet. This results in an average fleet age for the U.S. (>23 years old) that is equal to or greater than the average age that most ships are scrapped (UNCTAD 1995).

Many ship operators, especially those using ships purchased on the second-hand market or in nations like the United States, currently face the alternatives of life extension for these vessels (either with or without technology and efficiency improvements) or new construction. This choice will continue to conflict with efforts to improve fleets for greater trade, stricter environmental demands, and safer operation. Ports and transportation systems that connect the MTS to the general transportation network will need to plan for the emerging MTS at the same time as they accommodate less-preferred technologies that will be replaced at some point in the future.

36.6 REFERENCES

Abrams, A. 1997. "Ship Cargo Capacity Tightens in 2nd Quarter." *Journal of Commerce.*

Bureau of Transportation Statistics (BTS). 1996. *1993 Commodity Flow Survey.* U.S. Department of Transportation, BTS, Washington, DC.

———. 1999. *1997 Commodity Flow Survey.* U.S. Department of Transportation, BTS, Washington, DC.

————. 1998. *National Transportation Statistics.* U.S. Department of Transportation, BTS, Washington, DC.

Corbett, J. J., and P. S. Fischbeck. 1997. "Emissions from Ships." *Science* 278:823–24.

————. 2000. "Emissions from Waterborne Commerce in United States Continental and Inland Water." *Environmental Science and Technology* 34(15): 3254–60.

————. 2001. "Commercial Marine Emissions and Life-Cycle Analysis of Retrofit Controls in a Changing Science and Policy Environment." In *Marine Environmental Engineering Technology Symposium (MEETS) 2001.* Arlington, VA: ASNE/SNAME.

Corbett, J. J., P. S. Fischbeck, and S. N. Pandis. 1999. "Global Nitrogen and Sulfur Emissions Inventories for Oceangoing Ships." *Journal of Geophysical Research* 104:3457–70.

Energy Information Administration (EIA). 2000. *World Energy Database and International Energy Annual 1998.* Washington, DC: EIA.

————. 2001. *World Energy Database and International Energy Annual 2001.* Washington, DC: EIA.

European Commission (EC). *The Development of Short Sea Shipping in Europe: A Dynamic Alternative in a Sustainable Transport Chain, Second Two-yearly Progress Report.* 1999, European Commission: Brussels, Belgium.

————. 2001. *White Paper: European Transport Policy for 2010, Time to Decide.* Luxembourg: Office for Official Publications of the European Communities.

Farrell, A. E., D. W. Keith, and J. J. Corbett. 2003. "A Strategy for Introducing Hydrogen into Transportation." *Energy Policy* 31(13): 1357–67.

Harrington, R. L., ed. 1992. *Marine Engineering.* Jersey City: Society of Naval Architects and Marine Engineers.

Lloyds Maritime Information System (LMIS). 2002. *The Lloyds Maritime Database.* Lloyd's Register-Fairplay Ltd.

Lloyd's Register of Shipping. 1970–1994. *Merchant Shipbuilding Return.* London.

Maloney, M. J. 1996. *World Energy Database.* Washington, DC: Energy Information Administration.

Maritime Administration (MARAD). 1996. *Outlook for the U.S. Shipbuilding and Repair Industry.* U.S. Department of Transportation, Maritime Administration, Office of Ship Construction, Washington, DC.

————. 2000. Website pages. U.S. Maritime Administration, Department of Transportation.

Mokyr, J. 1990. *The Lever of Riches.* New York, Oxford University Press.

Skjølsvik, K. O., et al. 2000. *Study of Greenhouse Gas Emissions from Ships (MEPC 45/8 Report to International Maritime Organization on the outcome of the IMO Study on Greenhouse Gas Emissions from Ships).* MARINTEK Sintef Group, Carnegie Mellon University, Center for Economic Analysis, and Det Norske Veritas, Trondheim, Norway.

Slater, R., and MTS Task Force. 1999. *An Assessment of the U.S. Marine Transportation System: A Report to Congress.* U.S. Department of Transportation, Washington, DC.

UNCTAD. 1995. *Review of Maritime Transport 1994.* United Nations, New York and Geneva.

CHAPTER 37
FREIGHT TRANSPORTATION PLANNING

Kathleen Hancock
Civil and Environmental Engineering Department,
University of Massachusetts at Amherst,
Amherst, Massachusetts

37.1 INTRODUCTION

The transportation system consists of a vast network of multiple modes that carry both people and goods locally, regionally, nationally, and globally. The goals of transportation professionals to maintain mobility, improve safety, and ensure sustainability are increasingly challenged as demand increases.

As part of the transportation system, the effective movement of freight also plays a large economic role. When congestion on the transportation system increases, businesses and customers are affected in two ways. First, movement of freight becomes less productive, causing the price of moving goods to increase and, second, more freight transportation must be consumed to meet the needs of an expanding economy. To ensure that freight movement remains viable and competitive, transportation planners must include freight in planning activities, and ideally, they must incorporate it in an integrated manner into the overall planning process.

37.1.1 Legislative Requirements

A primary driving force behind today's freight planning activities occurred with the passage of the Intermodal Surface Transportation Efficiency Act (ISTEA) of 1992. Within ISTEA, both the metropolitan and statewide planning requirements were modified, and states were mandated to create a series of management systems, two of which address freight planning: the intermodal management system (IMS) and the congestion management system (CMS). This mandate was changed in 1995 by the National Highway System Designation Act to an operational recommendation, and many states that had begun the process under the mandate continued to develop systems and planning techniques to improve planning within their states. The Transportation Equity Act for the 21st Century (TEA-21), passed in 1998, continued an emphasis on implementing and integrating freight planning. In addition to these highway acts, the 1990 Clean Air Act Amendments (CAAA), followed by the Environmental Protection Agency's (EPA) 1993 General Conformance Regulations, have also influenced the need for transportation planning.

For metropolitan areas, section 1024 of ISTEA required the inclusion of the following factors into the transportation planning process (FHWA and FTA 1993):

• Methods to enhance the efficiency of freight
• International border crossings and access to ports, airports, intermodal transportation facilities, major freight distribution routes, national parks, recreations areas, monuments, historic sites, and military installations

For statewide planning, the law required that the following factors be considered (FHWA and FTA 1993):

• International border crossings and access to ports, airports, intermodal transportation facilities, and major distribution routes
• Methods to enhance the efficiency of commercial motor vehicles

Although the management systems are no longer mandated by the federal government, many states are continuing to follow their philosophy. Understanding their intent is therefore important to understanding the basis of current planning activities. The CMS is a "systematic process that provides information on transportation system performance and alternative strategies to alleviate congestion and enhance the mobility of people and goods" (FHWA 1994). Because of this definition, freight movement becomes a major factor, even though many argue that freight represents only a small portion of vehicle flow. As one measure of success or failure, planners are supposed to analyze their actions in terms of the enhancement of the mobility of goods.

The IMS is a systematic process that (FHWA and FTA 1994):

• Identifies key linkages between one or more modes of transportation where the performance or use of one mode will affect another
• Defines strategies for improving the effectiveness of those modal interactions
• Evaluates and implements these strategies

Unlike the CMS, which has a clearly defined goal of reducing congestion, the IMS requires the measurement of performance of certain critical parts of the total transportation system, most of which have not previously been measured. Defining these performance measures has taken two different paths: (1) measuring the efficiency of the entire freight system, and (2) measuring the efficiency of specific points of intermodal connection.

Guidelines, with associated funding from FHWA and FTA, have been in place for passenger transportation and, indirectly, for freight moving on our highways. However, major freight facilities, including seaports, rail lines, and airports, were not considered until the CAAA as clarified by the 1993 EPA General Conformity Regulations was established. The EPA requires that conformity for carbon monoxide and ozone include both direct on-site emissions and indirect emissions caused by vehicles coming and going from freight facilities (EPA 1993). The inclusion of indirect emissions requires planners to document impacts of freight facilities at a much greater level of detail.

37.1.2 Practical Considerations

Transportation is essential for development and growth and the primary need for transportation is economic. For people, this need is to have access to work and food. For freight, this need is to supply the goods by moving raw materials to processing to distribution to the point of final consumption to disposal. People and goods, for the most part, share transportation facilities and compete for services and resources.

Today's society is moving more freight than ever before, and the total is projected to grow by nearly 70 percent by 2020 (FHWA 2002). In 1997, the United States moved, on average, 41 million tons of freight per day valued at $23 billion. This represents 14.8 billion tons and $8.6 trillion dollars and translates to 3.9 billion ton-miles of freight movement per year (BTS 2001). Many components of the transportation infrastructure are already under stress due to heavy traffic congestion, safety and environmental concerns, and the recent emphasis on national security. With the projected growth and increasingly limited available resources, planning for freight will play an increasingly important role in the future of transportation and economic systems.

Congestion and Capacity. Because freight typically moves in large vehicles with reduced operating characteristics, particularly on roads, increases in freight traffic have a greater impact on the capacity of the transportation system. With congestion increasing due to increases in both passenger and freight demand, conflict between these two groups is exacerbating the problem. Growth in international trade is creating greater pressures on ports, airports, and border crossings in addition to creating and expanding high-growth trade routes over facilities that are already congested by domestic traffic. When demand exceeds supply, resulting congestion impacts speed and reliability. The Texas Transportation Institute estimated that in 2000 3.6 billion person-hours of delay occurred on the highways of 75 urban areas (TTI 2002). This system delay has increased the value of the travel time of freight between $25 and $200 per hour, depending on the product being shipped. Unexpected delay for trucks can increase this amount 50 to 250 percent (FHWA 2001). Therefore, congestion increases the cost of goods, which has a direct effect on the U.S. economy.

Financing. Traditionally, investments in freight transportation facilities, whether public or private, have met the needs of moving freight. With existing conditions and expected growth, this is no longer the case and neither sector has the necessary resources to meet these needs. Both sectors must look for creative ways to finance projects jointly. In addition, improvements that will maximize productivity must be identified and targeted for these limited resources. Planners must have adequate tools and information to identify which improvements to invest in.

Safety. Safety is the top priority for public and private sector transportation professionals. For public sector professionals, safety relates to reduced crashes and associated societal costs on publically maintained facilities. For private sector professionals, safety relates directly to the cost of doing business. As freight movement increases across all modes, the interaction and competition of moving people and goods increases, and consequently the concern about safety increases. Although separation of freight and people has been considered, the practicality of doing so in the foreseeable future is nonexistent. This means that planners will need to continue to consider and plan for the very different characteristics of each.

National Security. Recent events have changed concerns about security from controlling theft and reducing contraband to preventing attacks and enhancing security while keeping commerce moving. This involves protecting assets—facilities and vehicles as well as the supporting communications and power supply—both from direct attack and from use as instruments to deliver an attack. Although most of the focus by the profession is on screening goods and tracking their movement, consideration for facility enhancements is also important, particularly as space and specialized equipment are needed for this screening and other security activities.

Environment. Since the passage of the National Environmental Policy Act of 1969, all major transportation projects have been required to include environmental impact assessments. Because of increased awareness of the impact that transportation, particularly freight transportation, has on air and water quality and land use and development patterns, transportation professionals must ensure that transportation projects are environmentally sound and economically sustainable. Air pollution, dredging, and noise are key issues facing the freight transportation professionals. An understanding of these issues when planning facility improvements or expansions will reduce their impacts and reduce the planning times required for conducting environmental reviews.

37.2 FREIGHT TRANSPORTATION OPERATIONS TODAY

Freight operations consist of everything required to move an item of freight from its origin or shipper to its destination or receiver. In the United States, most freight operations are handled by the private sector and are considered part of the supply chain management of business operations. The public sector owns and manages many of the facilities, including the highway system, that are required to move freight. It also regulates and taxes freight movement. This division in ownership and responsibility creates some unique challenges for freight planning.

37.2.1 Public Versus Private Sector

Movement of freight is unique because it is predominantly managed by the private sector across facilities owned and maintained by the public sector. The most telling distinction between how these entities view transportation is in their respective definitions of transportation. The public sector defines transportation as the effective and efficient movement of people and goods from one place to another. The private sector defines transportation as the creation of place and time utility, where *place utility* means that goods or people are moved to places of higher value and *time utility* means that this service occurs when it is needed.

In the public sector, planning for transportation improvements occurs from 6 months in the short term to 20 years or more in the long term. In the private sector, short term planning is delivery of tomorrow's goods, while long-term is considered 6 months to a year in the future. This difference in time frames has made partnering between public and private stakeholders difficult, particularly when the players do not understand this difference.

Transportation goals also vary between the public and private sector. Goals of the public sector are to provide safe, reliable, and sustainable transportation to all users. Because transportation is considered a public utility in the sense that it is vital to the overall public interest, it is the responsibility of the public sector to ensure that it operates effectively and fairly. Goals of the private sector are to provide reliable, cost-effective service to specific customers in a competitive environment of providing a better service than the competitors.

37.2.2 Logistics

Movement of freight is a derived demand, meaning that goods are moved only in response to a need. A firm needs a commodity, either as input to a product or as an item to sell to a consumer. The provider of the commodity ensures delivery of the product to the customer. Several players can be involved in accomplishing this, including transportation firms, wholesalers, and third-party logistics firms that specialize in providing logistics management services.

Logistics management determines how and where freight moves, and because the goal of logistics management is to minimize costs, an understanding of these movements and the associated costs is important to the planning process. Movement of a product from point of production to point of consumption consists of several separate movements, each of which has costs associated with it. Factors to consider include the number and location of storage sites, storage time at each site, transport modes used between sites, and shipment sizes. Associated costs include building and operating storage facilities costs, inventory costs, shelf-life costs, transport costs, loading and unloading costs, loss and damage, order costs, and stock-out costs resulting from late deliveries (Cambridge Systematics et al. 1997).

Freight transport decisions are based on cost and also on customer satisfaction. Therefore, reliability and the ability to deliver goods undamaged, on time, and when needed is as important as minimizing the cost associated with that delivery.

37.2.3 Influences

Evolution of the movement of freight since World War II has occurred quickly due primarily to the influences of four exogenous factors:

*   Globalization of business
*   Deregulation of transportation and a changing governmental infrastructure
*   Organizational changes in business
*   Rapidly changing technology

Globalization has impacted freight movement in many ways, from foreign sourcing of procurement to selling goods to other countries to multifaceted international distribution, manufacturing, and marketing. With improved transportation services, use of land bridges—moving goods across countries without those countries being either the origin or destination for those goods—has increased.

Deregulation of air, motor, and rail carriers in the 1980s dramatically changed the way freight is transported. Overall, the cost and/or quality of transportation service has improved for shippers. Additional changes in governmental infrastructure that have impacted freight movement include deregulation of banking and communications, deregulation of motor carrier transportation in Canada, and changes in the European economic community. The opening of Eastern Europe and the dissolution of the U.S.S.R. have changed freight transportation patterns. The North American Free Trade Agreement (NAFTA) has redefined the north-south freight transportation corridors in the United States.

The restructuring of business, including mergers, acquisitions, and related activities, has changed how businesses ship goods. In some instances, logistics functions have been consolidated; in others, they have been outsourced to third-party logistics organizations. This has directly impacted the sizes and modes of shipments.

Improvements in technology, particularly related to computers and communication, have allowed businesses to improve inventory control and scheduling leading to increased use of just-in-time (JIT) delivery of goods and improved service quality. Motor carrier companies are now able to meet narrowly defined windows for pickup and delivery. Effects of the growth of the internet and e-commerce are just beginning to be seen and will change how business is conducted.

37.2.4 Commodities

Freight is made up of commodities that vary from raw materials to finished goods. Most raw materials are shipped as either dry bulk, such as coal or grain, or liquid bulk, such as oil or milk. Processed goods can be shipped in almost any form, either containerized or noncontainerized. The diversity of types of freight can be seen from examples like mail, automobiles, machine parts, scrap, garbage, hazardous materials, computers, pressurized liquid natural gas (LNG), clothing, and fresh flowers. How this incredibly diverse population of goods is grouped for transportation planning analysis is still an area of discussion and research.

For accounting, either to keep track of where goods are going or what goods are being manufactured and exported or imported, commodities have been grouped and classified by various agencies in the U.S. government. The U.S. Census Bureau used the Standard Industrial Classification (SIC) until 2002, when the SIC was replaced by the North American Industrial Classification System (NAICS) (http://www.census.gov/epcd/www/naics.html). The Federal Railroad Administration uses the Standard Transportation Commodity Classification System (STCC) and the Bureau of Transportation Statistics, in cooperation with Canada, uses the Standard Classification of Transported Goods (SCTG) (http://www.bts.gov/cfs/sctg/advants.htm). The U.S. Department of Commerce uses Export Control Classification Numbers (ECCN) (http://www.bxa.doc.gov/licensing/facts2.htm). When working with commodities, one or more of these commodity classifications have been and may be used to group them for analysis purposes or for determining national and international flows of goods.

37.2.5 Modes

Freight, more so than people, travels by many modes including rail, water, air, and pipeline as well as truck. Figure 37.1 shows growth by mode of freight movement over the last ten years of the 20th century. For almost all movements by modes other than truck, freight trips must also be multimodal. In other words, freight must change modes during a shipment from its origin to destination.

Each mode has unique operating characteristics that require that planners understand and account for these differences.

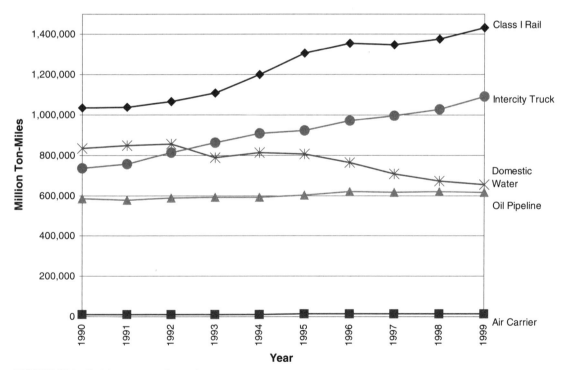

FIGURE 37.1   Freight movement by mode.

Highway.   Highway transportation is the dominant mode for passenger transportation and is a rapidly growing mode for moving freight. The highway network includes approximately 4 million miles of roads, ranging from freeways to unpaved rural roads. The majority of highways are paid for and maintained by public funds and are under the jurisdiction of public agencies. The highway system carries approximately 30 percent of intercity ton-miles of freight, which represents about 75 percent of intercity freight revenue.

The trucks using these highways vary in size, ownership, and use. About 9 percent of all trucks are used for over-the-road intercity freight and are classified by gross vehicle weight. These trucks are required to conform to standards for height, width, length, gross weight, weight per axle, and number of axles, with the current maximum weight at 80,000 lb. The trucking industry is a major force in moving freight but it is fragmented and diverse. Privately owned trucks transport their own company's products and are not regulated except for meeting federal and state safety requirements. For-hire trucks transport goods under contract and are generally regulated with some exceptions, such as those carrying exempt agricultural products or operating in a single state. For-hire trucking is also divided between truckload and less-than-truckload (LTL) services, with the latter including package carriers. Truckload carriers generally move directly from the shipper to the consignee. Less-than-truckload shipments are usually picked up by one truck, taken to a nearby terminal, reloaded on the line-haul truck, taken to a terminal near the consignee, reloaded onto a delivery truck, and delivered to the consignee.

Costs associated with trucking are primarily variable costs. These include labor; depreciation of vehicles; and repair and maintenance of vehicles, including tires, user fees, and fuel and lubrication, including taxes. Fixed costs, which account for about 25 percent of trucking costs, include license fees, property taxes, management, and terminals (Wood and Johnson 1996). Direct costs associated with highways, except for tolls and some user fees, are provided by the public and are not part of trucking costs.

The diversity in size and configuration of trucks provides the trucking industry with greater flexibility in moving goods, since the size and characteristics of the vehicle can be more easily tailored to the size and nature of a shipment. Motor carriers also have the reputation of providing a high level of customer service since they have more flexibility in scheduling and the ability to reach most origins or destinations. Also, as mentioned earlier, almost all freight shipped by other modes requires a truck movement at one or both ends of the shipment.

Highway operations, including traffic control and an understanding of congestion, are well known to transportation planners, and the incorporation of trucks into models used in the planning process is relatively straightforward. The biggest challenges come from predicting the number of trucks to be included and determining their impact on the capacity of the roadways, since capacity calculations include a fixed percentage of trucks in the vehicle mix and the purpose of freight planning is determining the number of trucks to include.

Rail.   U.S. railroads are privately owned, although the original development of the rail system was supported by land grants that allowed for the quick expansion and dominance of rail as the freight transporter in the 19th and early 20th centuries. Track mileage has been reduced to approximately 113,000 miles, less than half of its peak. Individual rail companies own their track and have contractual agreements with other companies, including the federal government, to use this track.

The rail industry is grouped into three classes based on revenues: class I carriers with greater than $253.7 million in revenues, class II railroads with revenues between $20.3 and $253.7 million, and class III railroads with less than $20.3 million in revenues. Class I carriers, now less than seven companies, provide the majority of line-haul services and account for approximate three-quarters of total rail mileage and 90 percent of freight revenue (Muller 1999). Class II railroads are regional and usually connect to class I or other class II railroads. Class III include switching and terminal railroads.

Costs associated with use of railroads are approximately 50 percent fixed and do not vary with the volume of business. These include rights-of-way, classification yards, general management, and maintenance. Variable costs include purchase and maintenance or usage charges of equipment, usage charges of rights-of-way, and labor and fuel and lubrication costs. Because of the high fixed cost, rail can recognize significantly increased economies of scale: as volume increases, total cost of production decreases on a per-unit basis.

Railroads are especially suited to carrying large, bulky shipments that are not time-sensitive, such as milk, over long distances. Raw materials and other bulk commodities, motor vehicles, intermodal containers or containers on flat cars (COFC), and piggyback truck-trailers or trailer on flat car (TOFC) are the most common items shipped by rail. The biggest limitation to rail service is the fixed track, which limits the routes available, and the locations of the beginning and ending points of a rail movement.

Planners face a major challenge when incorporating rail movements into existing planning models. Most public sector planners do not have the same understanding of rail operations that they do of highways. Rail capacity is based on the number of trains that can safely traverse a section of rail, which is based on the type of signaling and control on that segment and whether more than one set of tracks is available or where and how often sidings are available to allow for passing in either direction. The speed of a train is dependent on the type of track control, the geometrics of the route, and the makeup and weight of the train. Another complicating factor when considering rail movements is that freight moves in cars that are combined to make up trains. From the beginning of a rail trip, a given car can change trains at classification yards any number of times before reaching the termination of that trip.

Pipelines.   For the most part, planners do not consider pipelines in their planning activities. Pipelines are privately owned and maintained, and freight that uses pipelines flows in bulk without containerization. Truck, rail, or tanker movements to or from pipeline terminals are generally modeled in those modes.

Domestic Water Carriers.   Moving freight by water falls into two categories, domestic and international. This section summarizes the characteristics of domestic movements.

Domestic waterways used for moving freight include navigable rivers, canals, the Great Lakes, and the Gulf and Atlantic Intracoastal Waterways. The U.S. Army Corps of Engineers maintains navigable rivers and harbors, including dams and locks. The U.S. Coast Guard also plays a role in the safety of waterways, including providing navigational aids such as buoys to guide vessels on their course. Terminals very greatly and can be as basic as a privately owned dock on a waterway up to a complex maritime port that serves both domestic and international shipments. Ports are usually public/private partnerships that are managed by a publicly based authority overseeing multiple public or private terminals consisting of several docks. Movement of domestic freight can also occur from port to port using open water, although this has substantially declined over time.

Carrier vessels consist of barges controlled by tugboats and towboats on inland rivers and canals, lakers or vessels that are not considered seaworthy for operation on ocean waters used on the Great Lakes, and both barges and seagoing vessels on the intracoastal waterways.

Maintenance and management of waterways and ports are generally publicly funded, while specific terminals may be privately or publicly funded. Costs for using these facilities are covered by shippers/carriers through user fees based on the services provided. Variable costs for shipping goods by water include maintenance or usage charges of vessels, usage charges, and labor and fuel costs.

Like railroads, domestic water carriers are suited to carrying large, bulky shipments that are not time-sensitive over long distances. Petroleum products, raw and construction materials and other bulk commodities, and intermodal containers are the most commonly shipped items. The biggest limitations to domestic water service are the geographically limited net-work and terminal locations, shipment time, and seasonal and weather effects.

The domestic waterway is primarily a long-distance mode that is of interest in regional and corridor planning. However, ports and terminals are of interest to local planners since most of the freight is offloaded into trucks, with some freight going to rail and pipeline depending on connectivity and commodity type.

Domestic Aviation.   The fastest-growing mode for moving freight is domestic aviation. Although expensive, the ability of aviation to deliver goods across long distances faster than any other mode has increased demand, particularly for parcel delivery and just-in-time delivery backup. Freight can travel in the belly of a passenger aircraft, on a dedicated cargo aircraft, and on a super cargo aircraft. Security concerns resulting from September 11, 2001, have affected this mode more than any other, particularly for the first option. Facilities associated with air freight include freight terminal(s) at the airport and access to the airplane. These are usually managed by a private entity, either the carrier or a freight forwarder. Individual airlines, including Federal Express and UPS, own and maintain aircraft used to move freight.

Like water, air is primarily a long-distance mode that is of interest to national planners while airports are of interest to local and regional planners. For this mode, landside freight transportation is almost exclusively truck.

Intermodal Transportation.   Over the past decade, the distinction between intermodal and multimodal transportation has become blurred. *Intermodal transportation* is commonly used to refer to movement of freight containers across one or more modes. Where planners need this distinction is at intermodal terminals that transfer containers from truck to rail or rail to truck. Although container transfer can occur at maritime ports or, less commonly, at airports, these activities occur within the port itself, are generally considered to be part of port operations and are usually represented in local planning as a generator of truck traffic. Truck/rail intermodal activity has a more direct impact on local planning. Characteristics of intermodal terminals that are of interest include transfer capacity and rate, storage capacity, and hours of operation along with associated costs for normal operation and for delayed operation.

37.3 PLANNING

Freight transportation improvements are planned and implemented by both the public and private sectors. This has traditionally occurred independently. In general, publicly funded improvement planning follows a lengthy structured process with prescribed involvement by many stakeholders. In contrast, private sector planning has a short response time without external involvement and is driven by market trends. The freight planning process presented in this section focuses on planning performed by the public sector. However, effective planning cannot ignore the latter, since the business aspects of moving freight play an important role in how decisions are made.

The extent to which freight is considered in public planning both at the state and local metropolitan planning organization (MPO) level varies from being integrated into the prioritization and funding process to not being considered at all. This section presents the state of the practice in public sector planning for freight.

37.3.1 Purpose

The purpose of public sector planning is to determine systematically the use of available resources to meet the goals of the public being served. Some factors used to justify selection of projects include improvements to traffic flow and safety, savings in energy consumption, economic growth, increased accessibility, employment stimulation, competition with other cities or states, politics, or personal benefit. Although planning has focused in the past on moving people, the public sector must now include freight in the process and, ideally, allocate resources to the projects that most closely meet the goals of the overall system, irrespective of the user—people or freight—or mode. This implies having a planning process that considers all of the components of the transportation system related to those goals. However, this ideal has yet to be achieved in all but very unusual cases, and the state of the practice is to plan independently for freight and people.

37.3.2   Process

Because of the complexities outlined early in this chapter and the limited availability of data, freight planning procedures for the public sector remain extremely challenging. This section attempts to summarize some of the procedures currently available and accepted for use by planners.

The planning process consists of seven basic steps, which are interrelated and not necessarily sequential: defining the situation, defining the problem, exploring solutions, analyzing performance, evaluating alternatives, choosing the project, and constructing or implementing the selected project.

Situation and Problem Definition.   Because of competing needs of the public and private sectors for freight, defining the situation and problem usually requires interaction between these sectors. This can occur through the establishment of a formal relationship, usually through a freight advisory group. Members of this group could include staff from the local planning agencies, port authorities, major carriers from rail and trucking companies, package delivery companies, and the region's major shippers. Private sector associations, such as chambers of commerce or economic development agencies, are also important players. FHWA's *Public-Private Freight Planning Guidelines* provide more extensive information about establishing and using these groups to assist with freight planning activities (FHWA 2002). If the group is structured appropriately, it can have an active role in all parts of the planning process.

After identifying the problem, planners must then quantify it, which consists of inventorying facilities and collecting related information. This information can include existing geometry or facility configuration, traffic control, traffic counts, land use, employment, economic activity, and origins and destinations. This is generally readily available for traditional planning but does not necessarily include details required for freight planning. The freight advisory group can be very beneficial in assisting with this activity.

Exploring Solutions, Analyzing Performance, and Evaluating Alternatives.   As with problem definition, freight advisory groups can provide valuable information for exploring solutions by identifying lists of cost-effective efforts that can be implemented quickly and easily to improve freight movement. This is often as simple as retiming a signal to account for left-turn movements by large trucks. For larger-scale studies, these groups can assist with providing the expertise to correctly model freight activities.

Once all possible solutions have been narrowed to those that will be considered in more detail, planners begin to analyze and evaluate these alternatives. The analysis of the performance of possible alternatives is often referred to as the planning process, but it is really the process that integrates system supply on a network with travel demand forecasts to show equilibrium travel flows. Data required for this process and models that are currently used are presented in the following sections.

The evaluation phase considers the effectiveness of each alternative under consideration to achieve the objectives of the project as defined by performance measures. The performance data from the analysis phase are used to determine costs and benefits for each alternative.

Performance measures for freight facilities and for evaluating freight on the transportation system are presented in the Performance section.

37.3.3 Data

Probably the biggest issue for freight planners is the availability of data about freight movements that can be used for planning activities. Data collection is expensive and resource-intensive and most planning activities do not include the time or money to obtain these data. In addition, data on freight movements are usually considered proprietary and are held closely by shippers and carriers to maintain their competitive advantage. As a result, planners often select analysis techniques based on available data instead of determining the best techniques to meet their planning goals and acquiring the necessary data to support those techniques. This will remain an issue for planners, at least in the near future, and the models and methods discussed in the following section reflect this limitation.

Publicly available sources for freight data are outlined in the following paragraphs. Web pages for each data set are included after the name of the data set (Meyberg and Mbwana 2001).

Commodity Flow Survey (*CFS*) (**http://www.bts.gov/ntda/cfs/**). Data were collected on shipments by domestic establishments in manufacturing, wholesale, mining, and selected other industries for 1993 and 1997 by the U.S. Census Bureau in partnership with the Bureau of Transportation Statistics. This is an origin shipper survey and does not have detailed origin-destination data for commodity by mode. Locational aggregation is by state or by National Transportation Analysis Region (NTAR). Data in the CFS are best used to determine what is sent out by states or NTAR to the rest of the country, the mode used, and the distance traveled. Commodities coming into a state or NTAR and market area dynamics are not effectively presented. Other limitations include lack of geographical detail, lack of industry coverage for industries that are moving very fast, absence of several kinds of movements, lack of vehicular flow information, focus only on domestic movements, and difficulty with matching with other data sources.

Transborder Surface Freight Data (*TSFD*) (**http://www.bts.gov/ntda/tbscd/**). North American merchandise trade data by commodity type, by surface mode of transportation (rail, truck, pipeline, mail, and other), and with geographic detail for U.S. exports to and imports from Canada and Mexico have been collected and available monthly since 1993 by the U.S. Customs. This data set is a subset of official U.S. international merchandise trade data. The TSFD has been used to monitor freight flows and changes since NAFTA and for trade corridor studies. This is a customs data set, not a transportation survey data set, and is focused primarily on revenues.

Vehicle Inventory and Use Survey (*VIUS*) (**http://www.bts.gov/ntda/tius/**). Information on trucks domiciled within a state and owned by businesses and individuals, ranging from multitrailer combination vehicles to pickups, vans, and minivans, is collected and available every 5 years by the U.S. Census Bureau. These data provide limited information about where trucks are used by zones or ranges of motion. No information is provided on what commodity trucks carried.

Motor Carrier Financial and Operating Statistics (**http://www.bts.gov/ntda/mcs/**). Annual and quarterly data are collected on industry, financial, employee, and operating for motor carriers of property and passengers.

***Waterborne Transportation Data* (http://www.wrsc.usace.army.mil/ndc/).**   A series of databases and statistics pertaining to waterborne commodity and vessel movements are collected by the U.S. Army Corps of Engineers and made available annually.

Additional information includes domestic commercial vessel characteristics, prot and waterway facilities, lock facilities, lock operations, and navigation dredging projects.

***State Freight Transportation Profiles* (http://www.bts.gov/ntda/sftp).**   State profiles present information of freight transportation for each of the 50 states, combining major federal databases into tables. Reports give a description of the database and contact points.

***TRANSEARCH Data* (*Reebie Associates*) (http://www.reebie.com/).**   TRANSEARCH is an integrated, multimodal freight flow database constructed from public and proprietary data sources by Reebie Associates. Market-to-market freight movements are provided for the United State, Canada, and Mexico for the year 2000. Tonnage moved by market pair, by commodity, and by seven modes of transportation are available at the county, business economic area (BEA), metropolitan area, state, or province level. Other information includes secondary traffic and freight rehandled by truck from warehouse and distribution center. Modal coverage includes for-hire truckload, for-hire less-than-truckload, private truck, rail carload, rail/truck, intermodal, air, and water. A cost is associated with using these data.

***Rail Waybill* (http://www.ntis.gov/fcpc/cpn8441.htm).**   An annual sample of freight movements terminating on railroads in the United States is collected by the Surface Transportation Board (STB). Sample size is approximately 2.5 percent of all rail traffic. Two files are created. The master waybill file contains confidential information on specific stations, railroads, and revenue and is not available to the public. The public-use file provides information on freight movements at the BEA level at five-digit Standard Transportation Commodity Code level. Origin, destination, intermediate railroads and junctions, commodity, type of car, number of cars, tons, and revenue are provided.

***National Roadside Survey—Canada* (http://strategis.ic.gc.ca/SSG/ti01101e.html).**   A 1991 survey of commercial vehicles was performed by the Canadian Conference of Motor Transport Administrators to provide information to assist with identifying the impact of changes within the Canadian trucking industry. Complementary surveys were performed in 1995 and 1999. Information includes carrier type, vehicle configuration, trailer configuration, capacity utilization, and drive category for interprovincial movements.

***Employer Database* (http://www.doleta.gov/alis/edbnew1.asp).**   The State Employment Security Agencies collects information about employers, including who they are, where they are, what activities they do, and how much freight they generate or attract. The America's Labor Market Information System (ALMIS) Employer Database is available to the general public. Information about accessing this information is available at the website.

Many local planning agencies have collected information related to freight at different levels of aggregation for specific needs. Private companies are often willing to share information if confidentiality is ensured and they are convinced that the result will provide a benefit to their operations, i.e., improvements in their movement of freight. Also, cost-effective data collection on a smaller scale can be accomplished for a localized project. Contacting the local planning organization can often identify what data may be available for use on planning initiatives.

37.3.4 Planning Models and Methods

Analysis of the performance of planning options requires valid models for forecasting the demands for freight movement and how these demands change over time. Traditional efforts have focused on growth in types of commodities based on time series data to predict future commodity flows or in using the traditional four-step urban transportation planning model adjusted appropriately to consider the unique characteristics of freight. Although criticism exists of using the four-step process, current practice for public sector planning is to use some form of this approach (Maze 1994). The basic four-step approach, as it relates to freight, consists of (1) determining how much and where goods are generated and how much and where goods are used, (2) estimating volumes of freight moving from each origin to each destination, (3) selecting the mode or modes moving the freight, and (4) determining the route to be used. Either in step (2) or (3) the volume of goods must be converted to numbers of vehicles.

When determining freight demand, several factors make it unique and more complex than determining passenger demand. *Units of measure* can be easily determined for passengers. However, freight may be measured in units, by weight, by value, or by volume. The *value of time* is very different where the value of commodities being shipped can be directly impacted by time, such as fresh cut flowers. Passengers generally require no assistance for *loading and unloading,* while effective movement of freight is highly dependent on being loaded and unloaded both at the origin and destination and at transfer points along the trip. *Types of vehicles* also vary substantially for freight. Passengers, for the most part, only require seats in vehicles. Freight requires a wide variety of accommodations, from refrigerated containers to dry bulk hoppers to liquid bulk tank cars to flat beds to specialty vehicles. These differences should be included in considerations as planners use the models presented in the next sections.

Freight Generation and Attraction.   Base data of the amount of freight generated from or attracted to an area or location are often difficult to acquire or collect. When these data are available, the units are in average numbers of tons or average dollars of value and are usually based on nationally or regionally based samples. In addition to the data sets listed in the previous section, Fischer and Han (2001) have compiled other major sources and types of truck trip generation data.

If the planner has the option to perform data collection as part of the planning process, the following methods are in use today (Southworth 2003):

- Vehicle classification counts: traffic loop counters or videos or other traffic sensors
- Vehicle intercept and special traffic generator surveys: counting classifying or surveying vehicles as they enter and leave a specific location or area
- Truck trip travel diaries: daily travel surveys completed by drivers or dispatchers
- Carrier activity surveys: regulated surveys for safety or user fee legislation
- Commodity flow surveys: shipment inventories completed by shippers or establishments

These data have been used to develop commodity- or vehicle-based freight trip generation rates in different ways. The most straight-forward is to combine vehicle traffic counts or tons with employment or land use values to develop simple trip rates or tons moved per employee or unit of land (Cambridge Systematics 1995; Fisher and Han 2001). Care should be used in transferring these rates to other applications.

Another relatively simple procedure is to assume that demand is directly related to economic indicator variables such as those that measure goods output or demand in physical units. If these measures are not available, constant-dollar measures of output or demand or employment can be used. Indicator variables are used to derive either annual growth rates or growth factors of ratios of forecast-year to base-year values (Cambridge Systematics et al. 1997).

A more robust method is to fit least squares regression models to traffic counts or tons by commodity. This approach consists of identifying one or more independent variables, usually measures of economic activity, which can be used to predict the dependent variable, usually a measure of freight activity such as tons or ton-miles. For forecasting, regression techniques are applied to time-series data for both the independent and dependant variables. The resulting model of this relationship is then applied using forecasts of independent variables to future time periods. The *Quick Response Freight Manual* (Cambridge Systematics 1997) summarizes several models that have been used for different planning applications.

With all forecasts, several factors should be considered in their development, including the level of aggregation of geographic area, the level of aggregation of commodity, and whether the planning activity is for a new or existing facility.

Freight Trip Distribution.   Once supply (generation) and demand (attraction) values have been developed, planners must determine the pattern of freight movements. One method for modeling commodity flows is to develop commodity-specific spatial interaction (SIA) models (Southworth 2002). The volume of freight for commodity c, V_{ci}, originating from area i is allocated to destination j using the general SIA model (Wilson 1970):

$$T_{cij} = V_{ci} * A_{ci} * W_{cj} * B_{cj} * f(c_{cij}) \qquad (37.1)$$

where T_{cij} = volume of freight of commodity c allocated moving from origin i to destination j
  V_{ci} = volume of freight of commodity c in area i
  W_{cj} = volume of freight of commodity c of interest in area j
  $f(c_{cij})$ = inverse function of costs, c_{cij}, of transporting a unit of commodity c from i to j
  A_{ci} and B_{cj} = balancing factors that ensure compliance with observed activities

Specifically,

$$A_{ci} = [\Sigma_{cj} * W_{cj} * B_{cj} * f(c_{cij})]^{-1} \qquad \forall_{ci} \qquad (37.2)$$

and

$$B_{cj} = [\Sigma_{ci} * V_{ci} * A_{cj} * f(c_{cij})]^{-1} \qquad \forall_{cj} \qquad (37.3)$$

These factors are solved using an iterative proportional fitting procedure that ensures

$$\Sigma_{cj} S_{cij} = V_{ci} \quad \text{for all } i \text{ and } \quad \Sigma_{cj} S_{cij} = W_{cj} \quad \text{for all } j \qquad (37.4)$$

This is a doubly constrained SIA model (Wilson 1970). The origin-to-destination freight costs, c_{cij}, are derived directly from empirical data or by econometric modeling. These models are applied to zonally aggregated data for zones ranging from traffic analysis zones (TAZs) for urban areas to counties for regional or statewide planning.

More detailed analyses can be performed using advanced models such as a logit choice model (see Southworth 2002).

Several issues should be considered when applying trip distribution models to freight movements. First, a significant differences exists between modes and the associated cost values need to be carefully thought out to represent these differences effectively. In addition, backhauling of empty vehicles/containers and the associated costs should be included. If more than one mode is used for an individual shipment, the costs associated with changing modes, as well as mode-specific costs, should be included in the cost fuction used for distribution.

Freight Mode Choice.   Moving freight short distances tends to be accomplished by a single mode, truck, without any real competition from other modes. Also, incorporating transportation improvements on the landside of maritime ports, airports, or intermodal terminals focuses on highway planning. For these types of planning activities, mode choice is not a consideration. However, for longer-distance freight movements, expanding terminal activities, and moving high-valued time-sensitive freight, planning across modes is important and mode choice is a part of the planning process.

This very quickly becomes an involved and difficult process because mode choice usually relates to the type of service required for moving freight which is not independent from a mode. Examples of questions to consider include whether the freight be moved on scheduled service or whether the commodity being shipped needs special packaging or handling.

The most common approach to analyzing mode split is use of a discrete choice model such as a logit model (Cambridge Systematics 1997). The selection is then based on the mode or combination of modes that provides the lowest cost. The general formulation for selecting mode K from k available modes between the origin i and destination j is:

$$P_{K/ij} = \exp(-\lambda c_{ijk})/\Sigma_k \exp(-\lambda c_{ijk}) \tag{37.5}$$

With a given modal cost sensitivity parameter, λ, the averaged modal cost is:

$$c_{ij} = -1/\lambda \ln(\Sigma_k \exp(-\lambda c_{ijk})) \tag{37.6}$$

The cost terms determine the effectiveness of the model, and it is therefore crucial that the modeler understand them and the context in which they are used. One cost that is difficult to include in the cost formulation is the value of reliability, which is a major factor from the business perspective in moving freight. In some cases, costs may consist of trade-off values instead of actual costs. Wigan et al. (2000) provide some quantitative insight into this. Additional methods have been proposed for determining mode split (for example Jiang 1999), although these are currently more academic than applied.

Because of the complexity and interrelated nature of selecting between modes, which is based not only on cargo type, type of firms involved, and the nature of the geography, but also on values of service measures and perceptions of those involved, mode choice modeling is still very much in its infancy for freight planning.

Converting Volume to Vehicles.   For many planning activities, freight movements must be converted from volume or economic value to number of vehicles or vehicular equivalents. Although this sounds straightforward, several factors complicate this process, including variability in vehicle sizes within and across modes, variability in shipment sizes, variability in the relationship between volume and weight of different commodities, variability in packaging, and the amount of backhauling (returning an empty vehicle to its origin) involved. Estimated backhauling in the United States with current operating practices ranges from 15 to 50 percent of truck miles traveled (BTS 2001).

Freight Traffic Assignment.   Assigning vehicles to the transportation network provides the information required for determining most transportation-related impacts, such as congestion, air quality, and physical deterioration of facilities. For multiple origins and destinations several models are available, from the straightforward all-or-nothing assignment to comprehensive nonlinear programming optimized assignment.

Truck assignments on highways are the most common analysis currently performed. Several methods have been used by planners, depending on the purpose and the complexity of the problem. Many MPOs have included truck movements as a specific vehicle size class which is represented as a passenger car equivalent (PCE). Large single-unit trucks are considered as 1.5 PCEs while a tractor-trailer combination may be 3 PCEs for impacts on travel speeds and congestion.

The Wardrop equilibrium assignment model has been used for mixed passenger and freight highway traffic (Southworth et al. 1983). This approach assumes that all routes that are used have the same travel cost or impedance while routes that are not used cost more. Logit route choice models can be used for small networks with identifiable routes, while least-cost routing models can be used for long-distance movements. Because of the trip-chaining behavior of many truck trips, adjusting the "cost" using circuity factors may be required to adequately model truck flows. Validating assignment models can be problematic depending on the availability of track traffic counts for the analysis region.

37.3.5 Performance

Performance measures must be established to assess the current freight transportation system and to evaluate planning alternatives. This section presents several different performance measures that have evolved or been developed to address the needs of freight planning in the public sector.

Performance of a transportation network is evaluated based on the mobility of its users. In the past, traffic has been measured by the movement of people, rather than freight, which was consistent with the focus of older analytical procedures. As the awareness of the importance of goods movement increased, freight-based performance measures have been developed. Because both freight and people should be considered when evaluating a transportation network, freight-related measures should also relate to measures designed for the flow of people.

Traditional measures have been based on the level of service (LOS) concept, which was related to the ratio of volume to capacity (V/C). This was used as a measure of the success or deficiency of a transportation segment. One measure that came out of this approach is *delay*, which measures congestion conditions.

Accessibility has gained support as a measure that more effectively measures performance. This is the achievement of travel objectives within an acceptable time limit and can be applied at any level of evaluation. The advantage to this measure is that it focuses on the goal of the transportation system instead of its symptoms.

Additional measures have been proposed by different studies for specific applications. Hagler Bailly Services, Inc. developed a set of measures (interchangeably called indicators) for the FHWA to evaluate productivity and efficiency improvements in goods movement by motor vehicles on highways (Hagler Bailly 2000). Indicators compiled from previous efforts were ranked in terms of descriptive value and technical appropriateness. Descriptive value indicates the comprehensibility of the indicator for general audiences, and technical appropriateness relates to the usefulness of the indicator as a measure of freight movement productivity. Consideration was also given to data availability and costs. Thirteen indicators were identified as first-tier, meaning they are potentially valuable measures of the quality or cost of freight service or of the performance of facilities important for goods movement. The first four indicators relate to the quality or cost of freight service to shippers. The others address travel time and reliability of system performance significant to freight movements.

- Freight service:
  - Cost of highway freight per ton-mile
  - Cargo insurance rates
  - Fuel consumption of heavy trucks per ton-mile
  - On-time performance for highway-freight deliveries
- Highway system performance:
  - Crossing time at international border crossings
  - Point-to-point travel times for selected freight-significant highways
  - Hours of delay per 1,000 vehicle-miles on freight-significant highways
  - Ratio of peak period travel time to off-peak travel time at freight-significant nodes
  - Ratio of variance to average minutes per trip in peak periods at freight significant nodes
  - Hours of incident-based delay on freight-significant highways

- Annual miles per truck
- Customer satisfaction
- Conditions on connectors between the National Highway System and intermodal terminals (pavement and traffic conditions)

Czerniak, Gaiser, and Gerard (1996) analyzed performance measures used by 15 state departments of transportation in their intermodal management systems. The measures are classified into six goals.

- Goal 1: Accessibility of intermodal facilities
  - Level of service
  - Conditions of a transportation route
  - Bridge restrictions (e.g., weight restrictions)
  - Queuing of vehicles
  - Turning radius into facility
  - Deficiencies of the facility (e.g., number of structures lacking 21-foot vertical clearance)
- Goal 2: Availability of intermodal facilities
  - Volume-to-capacity ratios
  - Railroad track capacity
  - Storage capacity
- Goal 3: Cost and economic efficiency
  - Cost per ton-mile by mode
  - Revenue ton-miles
  - Expenditures (e.g., for freight rail, to retire deficiencies)
- Goal 4: Safe intermodal choices
  - Number of accidents
  - Cost of accidents
  - Number of fatalities
- Goal 5: Connectivity between modes (ease of intermodal connection)
  - Number of intermodal facilities
  - Delay of trucks at facilities
  - Travel times
- Goal 6: Time
  - Total transfer time
  - Freight transfer time between modes
  - Average travel time

These are internal and external performance measures. Internal measures, such as those listed under goals 2, 3, and 4, focus on the operations of an intermodal facility. External measures address the overall system performance. Goals ranked below goal 6 include reliability of facility and environmental protection with measures of air quality and fuel usage.

The Volpe National Transportation Systems Center identified three levels of activity for defining performance measures for intermodal planning needs: system level standards relating to network connectivity, operational standards relating to service delivery, and facility level standards relating to terminal accessibility (Norris 1993) as shown in Table 37.1.

Depending on the application, planners should select the appropriate measures of effectiveness to evaluation options under consideration or to measure current performance of the system to be evaluated.

TABLE 37.1   Performance Standards by Analysis Level

System	Components	Measures
Network level	Access	Vertical clearance
		Capacity characteristics
		Bridge rehabilitation
	Safety	Grade crossings
		Waiting time at drawbridges
	Transferability	Cross-modal intersections
		Prioritization of track usage
		Legal/regulatory restrictions
Operations level	Service delivery efficiency	Line haul speed
		Door-to-door delivery time
		Customer service
		Real-time cargo information
		Percent on-time performance
	Economic efficiency	Cost per ton-mile
		Revenue per ton-mile
		Operating ratio
		Market share
	Environmental resource use efficiency	Fuel use per ton-mile
		Emission per ton-mile
Facility level	Terminal accessibility	Access time to major transportation link
	Pickup delivery system	Gate queue length/time
		Container parking/storage capacity
	Interchange/transfer	Intermodal transfer time
		Average drayage length
		Average drayage cost
	Economic efficiency	Average annual revenue

37.3.6   Case Studies

This section is meant not to provide detailed information about planning case studies, but to provide links to studies that have been performed with enough of a description for the reader to decide whether to pursue it further.

Chicago (**http://www.fhwa.dot.gov/freightplanning/chicago.html**).   Chicago has maintained a freight component in its regional transportation planning since the 1970s. In the years immediately before and after ISTEA, Chicago included freight sector input in its planning process through both formal data collection efforts and industry outreach.

Chicago is the hub of the nation's freight transportation system, with the largest intermodal freight market in the nation. Class I railroads, as do a number of class IIs and class IIIs, operate within the region, which features 27 major intermodal yards, two waterborne freight facilities, three clusters of lesser-sized water terminals, and three auto transloaders. In 1986, trucks accounted for 12.5 percent of regional vehicle traffic (measured in vehicle equivalents).

Puget Sound (**http://www.fhwa.dot.gov/freightplanning/puget.html**).   The focus of this case study is to (1) describe the environment for freight mobility in the Puget Sound region, (2) describe the structure and efforts of the Puget Sound Regional Council and the Regional Freight Mobility Roundtable, (3) chronicle the achievements of these organizations, and (4) capture key lessons that transportation planners might apply to other locales.

Puget Sound is a large U.S. port with growing containerized international trade coupled with increasing regional vehicle miles of travel and population growth. The metropolitan planning organization for the area, the Puget Sound Regional Council (PSRC), is addressing freight mobility issues and is using private freight sector input for decision making. In 1992, the PSRC and the Economic Development Council of Seattle and King County formed an advisory panel, the main purpose of which was to capture private freight sector input on freight issues. The Regional Freight Mobility Roundtable has developed into a model example.

Delaware Valley Regional Planning Commission (**http://www.dvrpc.org/transportation /freight.htm**).   Moving freight and stimulating economic development are appropriate and worthwhile goals for transportation planning. A region's vitality and businesses, jobs, and consumers all rely on a transportation system that can handle goods efficiently and safely. The Delaware Valley Regional Planning Commission (DVRPC) has committed significant resources and technical capabilities to examining freight issues in the Philadelphia-Camden region because of their strong support for economic development. This site provides abstracts to several of their freight planning studies.

The Delaware Valley region is a freight transportation gateway with a large freshwater port; freight service from three large class I railroads and 12 smaller short lines; an airport with expanding international cargo services; an excellent highway network, including intermodal connectors; and numerous rail and port intermodal terminals that are equipped to handle all types of cargo. To capitalize on these resources, DVRPC has integrated these assets and placed freight directly in the transportation planning process.

Iowa Statewide Planning (**http://www.ctre.iastate.edu/Research/statmod/planning. htm**).   This study develops a matrix for the identification and development of tools and databases to support freight planning and modeling. Dimensions of the matrix include selected freight planning issues and scenarios and a prioritized list of commodity types for Iowa. Issues are identified by public and private sector practitioners and stakeholders and tools and databases are developed using GIS and Internet technologies.

This project investigates and establishes guidelines for the development of analytical methods to support the Iowa Department of Transportation's (Iowa DOT) intermodal statewide planning process. The Iowa DOT has been involved in freight transportation modeling efforts since the early 1970s, primarily on the development of grain forecasting models. The Iowa DOT is now developing multimodal techniques to mode movements of several commodities. The objective of the Department is to simulate the impacts of changes in transportation and nontransportation service variables on freight movements to investigate industry location decisions, the rationale behind commodity movements, and public policy impacts on freight movements.

37.4 *THE FUTURE*

With the rapidly changing environment for moving goods both in terms of business practices including integrated supply chain management and e-commerce, and transport practices including intelligent transportation systems, planning for freight will potentially undergo many changes in terms of both available data and modeling techniques.

37.4.1 Advanced Technologies

Advanced technologies will impact freight planning activities primarily in two ways. First will be the ability to obtain and make available data related to freight movement and traffic operations that impact its movement. Second will be the increasing ability of businesses to improve supply chain management using improved communications and inventorying technologies on the private sector side and the increasing ability of the public sector to manage traffic flow.

The potential for unobtrusive freight data-collection methods is high given the expansion of advanced technologies, including electronic data interchange (EDI), intelligent transportation systems for commercial vehicle operation (ITS-CVO), global positioning systems (GPS), Web-based data retrieval and assembly, and automated freight handling activities.

Private sector professionals are moving quickly to the use of real-time electronic information for managing and optimizing freight handling and movement to minimize costs, maximize service, and meet regulatory requirements. This same information could provide data required for public evaluation and planning purposes. Information from EDI and GPS includes what and how much is being moved, who is moving it, how it is being moved, why it is being moved, when it is moving, and where it is moving. The challenge is to harvest and disseminate this information in a way that does not compromise the confidentiality of individual firms and consequently their viability, competitiveness, and existence. Similarly, publicly gathered information, such as weigh-in-motion, ITS-CVO monitoring, and travel-time/volume data would also be collected and available to users (Hancock 2000).

Use of advanced technologies to link private and public sectors for information exchange and communications will increasingly affect where, when, and how freight is moved over transportation networks as vehicle and product tracking, in combination with traffic travel time, weigh-in-motion, and incident monitoring, provide on-the-fly intelligent decision-making and self-regulation. How and when the interaction develops will play an important role in freight planning.

37.4.2   Advanced Planning Techniques

Forecasting freight requires an understanding of freight demand and supply chain logistics. In addition to the traditional multistep model presented in this chapter, several additional approaches have been proposed, including Boarkamps, Binsbergen, and Bovy's GoodTrip model (2000), Nagurney et al.'s multilevel spatial price equilibrium modeling (2002), and Hancock, Nagurney, and Southworth's microsimulation modeling (2001). All three approaches emphasize multimodal network-based models and require an underpinning in decision-making of shippers, carriers, and receivers. The GoodTrip model expands the multistep process by explicitly including transactional stages and associated players involved in the supply chain before determining mode and route. The methodology proposed by Nagurney et al. models the components of the supply chain as a network of commodity flows, information flows, and prices for a given industry. Supply and demand equilibrium is obtained by iteratively solving across demands, production functions, and transactions costs for each movement in the supply chain. Transaction costs are dynamically updated to include costs of physically moving that freight as well as associated business costs.

High-speed desktop computing and more available data have allowed planners to consider detailed microsimulation of individual vehicle- or trip-based movements proposed by Hancock et al. (2001) as a feasible option. Individual movements are aggregated to obtain traffic volumes and modal information. This approach, as well as the multi-level network model, are both in the initial stages of prototyping. Although each shows promise, they are still a long way from practical applicability.

37.5   CONCLUSION

As presented in this chapter, planning for freight is a complex and difficult task. Because of the differences between public and private sector operations and planning and the variability of commodities and the means for shipping commodities, the ability of a public planning agency to grasp the breadth of this activity is problematic at best. However, if transportation planners are to make effective decisions in the current political and economic environment, a method for incorporating freight into the planning process is crucial. This chapter has attempted to provide a backbone on which to build a process for doing this while alerting the planner to the potential limitations and problems that will be encountered.

37.6 REFERENCES

Boarkamps, J. H. K., A. J. Binsbergen, and P. H. L. Bovy. 2000. "Modeling Behavioral Aspects of Urban Freight Movements in Supply Chains." *Transportation Research Record* 1725, 17–25.

———. 2002. *National Transportation Statistics 2002.* U.S. Department of Transportation, BTS, Washington, DC. Available at http://www.bts.gov/publications/national_transportation_statistics/index.html.

Bureau of Transportation Statistics (BTS). 2001. *Transportation Statistics Annual Report 2000.* U.S. Department of Transportation, BTS, Washington, DC.

Cambridge Systematics, Inc. 1995. *Characteristics and Changes in Freight Transportation Demand: A Guidebook for Planners and Policy Analysis.* NCHRP Report 8-30. National Research Council, Transportation Research Board, Washington, DC.

Cambridge Systematics, Inc. 1997. *Quick Response Freight Manual.* Report DOT-T-97-10, U.S. Department of Transportation and U.S. Environmental Protection Agency, Washington, DC. Available at http://tmip.fhwa.dot.gov/clearinghouse/docs/quick/.

Cambridge Systematics, Inc., Leeper, Cambridge & Campbell, Inc., Sydec, Inc, T. M. Corsi, and C. M. Grimm. 1997. *A Guidebook for Forecasting Freight Transportation Demand.* Report 388, National Research Council, National Cooperative Highway Research Program, Transportation Research Board, Washington, DC.

Coogan, M. A. 1996. *Synthesis of Highway Practice 230: Freight Transportation Planning Practices in the Public Sector.* National Research Council, National Cooperative Highway Research Program, Transportation Research Board, Washington DC.

Czerniak, R., S. Gaiser, and D. Gerard. 1996. *The Use of Intermodal Freight Performance Measures by State Departments of Transportation.* Report DOT-T-96-18, U.S. Department of Transportation, Washington, DC, June.

Meyburg, A. H., and J. R. Mbwana. 2001. "Data Needs in the Changing World of Logistics and Freight Transportation." Conference Synthesis, New York State Department of Transportation and the Transportation Research Board Freight Transportation Data Committee.

Environmental Protection Agency (EPA). 1993. 40 CFR Parts 6, 51, and 93, "Determining the Conformity of General Federal Actions to State or Federal Implementation Plans."

Federal Highway Administration (FHWA). 2001. Creating a Freight Sector within HERS. White paper prepared for FHWA by HLB Decision Economic, Inc. U.S. Department of Transportation, FHWA, Washington, DC, November.

———. 2002. *The Freight Story: A National Perspective on Enhancing Freight Transportation.* U.S. Department of Transportation, FHWA, Washington, DC, November. Available at http://ops.fhwa.dot.gov/freight/publications/freight%20story/.

———. Public-Private Freight Planning Guidelines. U.S. Department of Transportation, FHWA, http://www.fhwa.dot.gov/freightplanning/guide12.html, accessed January 29, 2002.

Federal Highway Administration (FHWA) and Federal Transit Administration (FTA). 1993. 23 CFR Part 450 and 49 CFR Part 613, "Regulations—Statewide Planning; Metropolitan Planning."

———. 1994. 23 CFR Parts 500 and 626 and 49 CFR Part 614, "Regulations—Management and Monitoring Systems.

Jiang, F., P. Johnson, and C. Calzada. 1999. Freight Demand Characteristics and Mode Choice: An Analysis of the Results of Modeling with Disaggregate Revealed Preference Data, 158 *Journal of Transportation and Statistics,* December 1999.

Fischer, M. J., and M. Han. 2001. *Truck Trip Generation Data.* Synthesis 298, National Research Council, National Cooperative Highway Research Program, Transportation Research Board, Washington, DC.

Hagler Bailly Services, Inc. 2000. *Measuring Improvements in the Movement of Highway and Intermodal Freight.* Report DTFH61-97-C-00010, U.S. Department of Transportation, Federal Highway Administration, Washington, DC, March 20.

Hancock, K. L. 2000. *Freight Transportation Data in the New Millennium—a Look Forward.* National Research Council, Transportation Research Board, Washington, DC. Available at http://gulliver.trb.org/publications/millennium/00043.pdf.

Hancock, K. L., A. Nagurney, and F. Southworth. 2001. "Enterprise-Wide Simulation and Analytical Modeling of Comprehensive Freight Movements." National Science Foundation Workshop on Engineering the Transportation Industries, Washington, DC, August 13–14.

Maze, T. H. 1994. *Freight Transportation Planning Process (or Chaos).* Second Annual National Freight Planning Conference Report, February.

Muller, G. 1999. *Intermodal Freight Transportation,* 4th ed. Washington, DC: Eno Transportation Foundation, Inc., and Greenbelt, MD: Intermodal Association of North America.

Nagurney, A., K. Ke, J. Cruz, K. Hancock, and F. Southworth. 2002. "Dynamics of Supply Chains: A Multilevel (Logistical/Information/Financial) Network Perspective." *Environment and Planning B* 29: 795–818.

Norris, B. B. 1993. "Intermodal Performance Standards." Volpe National Transportation Systems Center, U.S. Department of Transportation, Integrating Transportation Management Systems into Transportation Planning and Operations National Conference, Nashville, TN, November..

Peyrebrune, H. L. 2000. *Synthesis of Highway Practice 286: Multimodal Aspects of Statewide Transportation Planning.* National Research Council, National Cooperative Highway Research Program, Transportation Research Board, Washington, DC.

Southworth, F. 2003. "Freight Transportation Planning: Models and Methods." In *Transportation Systems Planning: Methods and Applications,* ed. K. G. Goulias. Boca Raton: CRC Press, 4.1–4.29.

Southworth, F., et al. 1983. "Strategic Freight Planning for Chicago in the Year 2000", *Transportation Research Record* 920.

Texas Transportation Institute (TTI). 2002. *2002 Urban Mobility Report.* College Station, TX: TTI.

Transportation Research Board (TRB). 2002. *Freight Transportation Research Needs Statements.* Transportation Research Circular, Number E-C048, National Research Council, Transportation Research Board, Washington, DC, December. Available at http://gulliver.trb.org/publications/circulars/ec048.pdf.

Wigan, M., N. Rockliffe, T. Thoresen, and D. Tsolakis. 2000. "Valuing Long-Haul and Metropolitan Freight Travel Time and Reliability." *Journal of Transportation and Statistics* 3(3):83–89.

Wilson, A. G. 1970, Entropy in Urban and Regional Modelling, London, Pion.

Wood, D. F., and J. C. Johnson. 1996. *Contemporary Transportation,* 5th ed. Upper Saddle River: Prentice Hall.

CHAPTER 38
TRANSPORTATION MANAGEMENT

George L. Whaley

Organization and Management Department,
College of Business, San José State University,
San José, California

38.1 DEFINITION OF TRANSPORTATION

Pouliot (2002) indicates that transportation represents one of the most important human activities worldwide. Furthermore, he maintains that transportation is a multidimensional service that affects many aspects of our daily lives. While transportation is complex and multifaceted, one might also view transportation as simply the movement between two or more points. According to Wood and Johnson (1996), transportation is the physical movement of people and goods between points. Therefore, transportation is a very broad, generic term. It consists of three major industries: air, water, and surface transportation. Wood and Johnson (1996) even include pipelines as another category of transportation. Within these three major industries, different carrier modes are regarded as subindustries. One major subindustry is transit within the area of surface transportation. Within these major industries, different carrier modes are regarded as subindustries. One major subindustry is transit within the area of surface transportation. Within the transit subindustry there are many carrier modes, like *local* passenger transportation by means of buses, trolleys, light rail, and so on. Sometimes carrier modes such as ferry boats cross subindustries. Ferry boats fit within both water transportation and transit because they are usually oriented toward local passenger service. These differences in carrier modes foster unique industry characteristics such as public versus private ownership, degree of government regulation, type of organizational structure, degree of unionization, and organizational culture. To make meaningful generalizations about transportation management, it is often necessary to distinguish between industries within the industry.

38.2 INFLUENCE OF MIXED TRANSPORTATION SECTORS

Transportation differs from many other industries in that there are multiple subindustries and organizational boundaries that influence the effectiveness of management. One important distinction is that a considerable number of publicly owned and operated carriers are found within transportation. While almost all airlines, railroads, and maritime operations are privately owned, this is not true of the transit industry. Many transit facilities are owned and operated by a government agency (often city or county), while others contract out to private firms and are privatized in a limited sense. Some public transit authorities subcontract city or county transit services out to privately owned transit companies under the Memphis formula. Under this formula, the local government agency contracts transit services out to a private enterprise and continues to retain broad control over the top management of that private company. Therefore, the transportation industry is partly in the private sector and partly in the public sector and allows the organization to qualify for federal funding. Oestreich and Whaley (2001) point out that this means day-to-day management activities are partly governed by private sector laws and partly by public sector laws. A wide number of parties are interested in these management activities. Vitale and Giglierano (2002) call these interested parties stakeholders and indicate they are individuals and organizations that have an interest in the organization, its operation and its performance. From the perspective of the total organization, departments within the organization and individual management activities, simultaneous involvement in both the public and private sectors increases the number of potential stakeholders who should be considered in any management decision. Consider for a moment that a strike by unionized bus drivers occurs in a city bus agency that is managed by a private management company, how many stakeholders exist? Obviously, the union, bus drivers, bus agency, management company and bus customers are stakeholders in this labor–management impasse. However, local politicians, business and the non bus riding public are likely stakeholders. Politicians that dislike privatization will most likely question the ineffectiveness of the management company compared to a completely public sector bus company. Both the local businesses and the nonbus riding public who are impacted financially and convenience-wise will also most likely question whether the management activities contribute to the strike. The operational and political realities related to survival of these organizations today and in the future are more likely to cause these organizations to cross sectors and industry boundaries than in the past. This alone makes management of transportation organizations more complex and difficult.

38.3   *THE EXTERNAL ENVIRONMENT*

Pouliot (2002) indicates that the external environment of an organization provides constant and pervasive changes that impact all transportation organizations. Some of the external forces fostering internal and external change are:

1. Economic
2. Geographic
3. Environmental
4. Social
5. Political

For example, the decision by a transportation organization to use an alternative fuel such as liquid gas may be based on a federal grant. Initially, the federal grant may be viewed as a new market and source of revenue for the organization. However, the manager may find out there are other unintended consequences related to both the internal and external environments. The use of an alternative fuel may require engineers to redesign existing company equipment and develop new techniques for vehicle maintenance that result in employee resistance to the change. The federal grant may require new regulations that impact external governmental reports, external employee safety reports, and other work procedures. In order to implement the change to liquid gas technology successfully, management needs to be aware and understand the viewpoints of many internal and external stakeholders.

   External factors that increase management challenges in transportation more than other industries are unionization, government regulation, safety, and the economic environment.

38.3.1 Unionization Issues

Oestreich and Whaley (2001) suggest that the degree of unionization impacts many facets of management. The transportation industry is more highly unionized than many other industries, and public sector employees are more highly unionized than private sector employees. Government statistics (U.S. Department of Commerce, Bureau of the Census, October 1998) show that in 1997, total union membership for *private* sector employees in this country was less than 10 percent but for *public* sector employees it was more than 27 percent. Yet in the private sector of the transportation industry, union membership amounted to close to 25 percent, which is almost equal to the total percentage union membership level in the public sector in the United States in general. The high rate of unionization in the transportation field, regardless of subindustries and carrier modes, means that supervisors and managers need to encompass extensive knowledge of labor relations to realize effective management. Individual managers and supervisors will be required to have extensive knowledge of work rules covered in the collective bargaining agreement and may also be called upon to serve on a joint labor-management committee. The topics covered by the joint labor-management committee could range in complexity from how to handle a current budget crisis to how to implement a new federal law. Oestreich and Whaley (2001) point out this level of labor relations knowledge is not only true for the legal aspects of traditional union-management relations, but it is also true for the increasing trend of union-management cooperation in the transportation area.

38.3.2 Government Regulation and Promotion Issues

The federal government seeks to promote and regulate transportation more than most other industries (Moffat and Blackburn 1996; Oestreich and Whaley 2001; and Pouliot 2002). Federal legislation such as the Transportation Equity Act for the 21st Century (TEA-21) requires all metropolitan regions in the United States to have a regional transportation plan that looks at 20 or more years into the future. On the regional and local level, the country's economic development and the quality of life for most individuals depend on the development of effective transportation systems. Throughout history, the country's economic development on the regional and local level and the quality of life for most individuals depend on the development of effective transportation systems. You may recall the Erie Canal was considered a great engineering feat two centuries ago; however, its greatest impact may have been on the early economic and social development of the state of New York. The Erie Canal made New York the only state at the time that was linked to both the Atlantic Ocean and the Great Lakes region. More recently, development of U.S. transportation systems has been credited with successes on a national level and beyond its borders. For example, Wood and Johnson (1996) indicate that national defense also depends to some extent on the transportation system of the country. They indicate during the Desert Storm and Desert Shield military operations, for example, 63% of the military cargo to the Gulf region was shipped by vessels of private maritime companies, and 64% of the military personnel were flown to the region by commercial airplanes. The new emphasis on rapid deployment of people and equipment overseas in the case of war and internal homeland security extends this view. More recent U.S. military campaigns in Afghanistan, Liberia and Iraq underscore dependence on private sector and military transportation systems. Transportation availability, head-way, average trip length and cost are part of the decision to locate, relocate, expand and reduce organizations as well as determine the success of a military campaign or the desirability of a location as a "good place to work." Therefore, the federal government not only has a compelling interest in promoting transportation, but it also has an interest in regulating it.

Currently, laws regulate federal funding of transportation services, and different laws regulate different modes of transportation, such as railroads and domestic water carriers. Federal and state government agencies monitor and control the various modes of transportation. The key federal agencies are the Federal Aviation Administration, the Federal Highway Administration, the Federal Railroad Administration, the Maritime Administration, and the Federal Transit Administration. Most of these agencies are under the control of a superagency, the U.S. Department of Transportation (DOT), headed by the U.S. Secretary of Transportation, which is a cabinet-level position. One notable exception is the U.S. Coast Guard, which transitioned from DOT to the newly formed Department of Homeland Security in March 2003. As these examples show, transportation is a highly regulated industry. Managers, particularly in the top echelons, must have a high degree of legal expertise and considerable political skill in dealing with the direct influences of federal and state agencies on the viability of their own organizations. One recent example is the federal government decision to require all baggage inspection employees to be federal employees rather than contractor employees. What started out as a strategic, top management legal and political concern quickly became a day-to-day operational issue. In today's environment, the skills, knowledge, and abilities (SKAs) requirements quickly trickle down the management hierarchy to the first-level managers or supervisor.

38.3.3 Safety Issues

A wide range of employment activities in transportation, ranging from employment testing to training and discipline, are regulated to some extent. There tends to be a higher concern, perhaps even a preoccupation, with safety in transportation than in most other industries. This is particularly true for passenger safety and the safety of the general public, where carriers' vehicles have greater than normal potential for injuring the general public. Ethical, legal and political pressures on these carriers to maintain a solid safety record are enormous and these pressures often lead to crisis management. Vitale and Giglierano (2002) suggest crisis management can result from ethical choices made by stakeholders in a situation. They offer as an example the ethical lapses of stakeholders such as the aircraft manufacturer, the airline management and the investigating government agency in the aftermath of the TWA Flight 800 crash. They report the pressure to reassure the public that flying is safe while attempting to deal with the tragedy in a professional and compassionate way resulted in ethical compromises by several levels of management inside and outside TWA. Among others, legal pressures include strict control of substance abuse by employees in safety-sensitive positions. Oestreich and Whaley (2001) report that one of the most severe constraints upon transportation facilities is the legally mandated drug and alcohol testing of employees in safety-sensitive positions. There are pitfalls for both union and management in transportation in this area of testing. Ed Wytkind (1997), Executive Director of the Transportation Trades Department of the AFL-CIO, states:

> The Teamsters and the ATU and the TWU and others know all too well that there's not a day in a week when one of their members aren't subjected to some form of drug testing. Post-accident, pre-employment, return to service, random—I don't care what it is.

The Omnibus Transportation Employee Testing Act of 1991 requires drug and alcohol testing in all areas of transportation. The DOT has published detailed rules that require employers in the transportation industry to have certain programs in place that control drug and alcohol abuse. These rules include mandatory procedures for urine drug testing and breath alcohol testing. For union officers, supervisors, and managers in the transportation industry who deal with employees in safety-sensitive positions, it is important to be familiar with the government's requirements of drug and alcohol testing. Even minor violations of these regulations can easily cause formal grievances for organizations with unions. Oestreich and Whaley (2001) indicate that while most unions in the transportation field strongly support safety measures, some are opposed to random drug and alcohol testing. Indeed, it is random testing that is the most controversial aspect of this issue, as opposed to not testing for "just cause." The political pressure inside and outside the federal agencies and pressure on elected officials are magnified in such crisis situations. These external factors affect all transportation subindustries and sectors and complicate the role of all managers, especially supervisors because they are the front-line managers.

38.3.4 Economic Issues

One constant in transportation is the changing external economic environment. The economic environment alternates between cycles of growth, contraction, and stability. In periods of economic expansion, transportation organizations experience more than normal external pressure to expand service levels. Since the expansion of service levels often require capital investment, legislation, and other long-term impacts on the transportation infrastructure, the economic environment often changes before the expanded service levels can be implemented. In periods of extended contraction, transportation organizations tend to reduce service levels, increase prices, and reduce infrastructure. Timing of management activities is critical because the underlying economics often change toward growth while management raises prices and contracting. This leads to criticism of transportation managers as poor planners and providers of inadequate service. Stability is seldom recognized because organizations tend to remember only the recently experienced impact of boom or bust. Often this is further complicated by the U.S. economy and the local economy where the transportation organization is located moving in opposite cycles of expansion, contraction, and stability. If the current operating environment is one of fiscal austerity and contraction, responsible management needs to have three different plans: further contraction, future growth, and stability. The more successful transportation organizations utilize contingency management techniques over time and throughout the management hierarchy.

38.4 THE INTERNAL ENVIRONMENT: ORGANIZATIONS

Organizations of all kinds are a key part of our daily lives because they produce the majority of our goods and services. An organization is two or more persons engaged in a systematic effort to achieve their collective goals. Therefore, organizations involve the behavior of individuals who have a common purpose. In order to achieve this common purpose, it requires the effective utilization of critical resources available to the organization. This involves the utilization of three key internal factors: people, financial, and technical resources. The characteristics of the transportation industry and key external forces on these organizations influence the internal management of transportation organizations. The internal focus is equally critical to successful management of transportation organizations as the aforementioned external forces. Section 38.5 will define management and emphasize the role that managers play in the allocation of the internal resources across both internal and external boundaries.

38.5 WHAT IS MANAGEMENT?

Management is the process of achieving organizational goals by engaging in functions that transcend the roles, levels, and industries of those involved in these ongoing activities. These activities or functions of management include planning, organizing, directing, decision-making, controlling, and leadership. Since these functions tend to overlap, Cook and Hunsaker (2001) reduce the larger set of ongoing activities to four basic functions: planning, organizing, directing, and controlling. A brief overview of the four management functions and how they relate to each other, different aspects of managerial activities in general, and management in the transportation industry follows.

38.5.1 Planning

Planning is the management function that focuses on goal-setting and how best to achieve goals. There are various levels of goals, ranging from the vision and mission at the top of the organization down to rules and procedures at the bottom. For example, the vision of a transportation agency might be "To be the agency with the highest ridership and fare-box recovery in our region by year 2005." An example of a procedure for an individual coach operator might be "Bus operators are required to call out all stops in a manner they can be understood by all riders." A procedure for an engineering manager at the same transportation agency might be "All engineering hours charged to a project must have a valid charge number to be approved by an engineering supervisor." Goals at all levels serve to guide the actions of employees and managers, motivate them to achieve the plans of the organization, help in resource allocation, act as constraints, and help to measure results.

Management researchers and practitioners such as Cook and Hunsaker (2001) indicate that planning is considered the most important management function. Planning is a more recent functional addition to transportation organizations than many organizations in other industries. In transportation organizations, short-run and operational activities often take priority over both long-run and strategic activities. Moffat and Blackburn (1996) indicates a common shortcoming of bus field supervisors in transportation is their focus on what is urgent and short range, and there is a propensity to ignore items that are goal oriented and future oriented. This suboptimization extends to other management levels, functions, and other types of transportation organizations. If the goal of the city traffic department is "to reduce traffic congestion and increase the flow of traffic by 20 percent within one year without increased cost and reduced benefit," why does suboptimization occur? Consider the following traffic control scenario: "A budgeted city project to replace an outdated traffic signal over the next 90 days that could improve traffic flow receives less priority than a request to install a new traffic signal where pedestrian jaywalking has increased and one death at the intersection has occurred." Urgency, internal and external politics, and lack of training and direction may lead the traffic department supervisor to ignore the goals of the organization before making a decision. Some cities such as the City of San José, California (2001 Annual Transportation Report) have developed level of service (LOS) metrics to guide such transportation management decisions and align the decisions with the City General Plan. Wood and Johnson (1996) suggest that lack of a profit motive, the number of stakeholders, and the structure and plethora of laws involved in goal-setting in the public sector make planning a longer, more involved process than in the private sector. If a transportation organization is hierarchical, runs by the book, and communicates strictly through the chain of command, supervisors may feel a disconnection from the planning function. This feeling of disconnection often leads to going-alone behavior and cutting corners rather than improving the planning process. The internal need to prioritize activities as well as the previously mentioned external forces on the organization such as TEA-21 serve to align local, regional, and national transportation goals and increase the importance of planning at all levels of transportation organizations. As in other industries, planning is now considered the most important management function in transportation because it attempts to integrate the goals of all stakeholders at all levels of the organization. After all, without planning, any direction the organization would take at any time would be okay.

38.5.2 Organizing

Management effectiveness also involves organizing or activities related to the structure of the organization. At the smaller, local or micro-level of the organization, organizing involves effective grouping of tasks and placing the right people into the right jobs. This may involve conducting task or job analyses to write appropriate job descriptions and specification statements. The larger, higher or macro-level activities focus on arranging jobs and people into larger units that are efficient and that are most likely to help achieve the goals of the organization. When these structural arrangements are displayed in a picture or chart format, it is called an organizational chart. Wood and Johnson (1996) and Moffat and Blackburn (1996) suggest that the typical structural units at the top of most transportation organizations are operations, maintenance, and administration. As previously mentioned, planning is increasing being considered a key top-level function in transportation organizations. The administrative unit is often divided into subunits such as human resources and finance. Information technology and marketing are more recent subunits found inside administrative units, and they usually report to the top manager in larger transportation organizations. Typically, engineering is either a separate unit reporting to the top manager of the organization, disbursed throughout the four aforementioned units at the top, or placed within the operations and maintenance units. These previously mentioned organizational structures are considered functional structures because common jobs inside units are based on similar disciplines, activities, or outputs such as marketing, engineering, finance, and information technology. Other common organizational structures are product, process, territory, customer, and project structures. If a transit organization's structure at the top is based solely on bus service and light rail service, this is a form of product organization structure. A trip- or route-based transit organization is a form of territory or customer organization structure. Hybrid structures are combinations of these common organizational structures.

The matrix structure is a popular hybrid structure. A matrix structure exists when temporary projects are combined with another organizational structure. Marketing and transportation planning activities that include several time-sensitive projects and where the balance of the organization is structured along functional or product lines should consider a pure project or hybrid matrix organizational structure. Since matrix structures require department employees to work on several temporary projects simultaneously that are managed by employees outside the department and employees on the projects report to their own department managers, there is less tendency to overstaff the organization. Although this structure can create a resource savings to the organization, employees who work in matrix organizations often feel they actually report to more than one manager and do not feel committed to any single project. As a result, self-motivated and self-directed employees are usually more productive in these roles and become known as problem-solvers. Each organizational structure has advantages and disadvantages and the management challenge is to find the structure that best suits and assists the organization in meeting its goals. If a city transportation department wants to start a "campaign" to make its streets safer, how should it organize the activities? Should the activities belong to an existing administrative department such as community relations, be assigned to a project office along with other projects, be organized as a temporary project as part of a matrix structure, or serve as a vehicle to start a separate marketing department as a new function?

To be effective, organizations need to change the structure of jobs, departments, and the entire enterprise as the situation dictates. Often the structure of jobs, departments, and the entire enterprise is altered or redesigned to incorporate these changes. The four common forms of job redesign found in organizations are job enrichment, job enlargement, job rotation, and job simplification. Cook and Hunsaker (2001) suggest job enrichment usually provides more employee motivation than the other three methods because it accommodates the task complexity without changes in the amount of work required and fulfills the higher-level needs of the employee. The degree of work motivation is based on each jobholder's perception of the change involved with the job redesign. Would a design engineer who works for a state transportation agency on a highway improvement project perceive increased task complexity in earthquake retro-fit design work as job enrichment or simply more work to do (job enlargement)? A private sector consulting engineer may view the complexity of the same retro-fit design work differently. The difference in perception might exist as a result of the external and internal forces faced by the organization and the individual. For example, changes in federal regulations such as Americans with Disabilities Act (ADA) or recent advances in information technology such as the Internet impact on most jobs in transportation. Whatever the reason, the perception of the task is the key to job redesign and buy-in of the employees.

Reengineering is a different form of redesign at the enterprise level that focuses on business processes. Champy (1995) defines reengineering as a radical redesign of business processes to achieve dramatic improvements in critical, contemporary measures of performance such as cost, quality, service, and speed. The key to adding value to the organization based on reengineering is connected to how tasks relate to each other in order to meet the organizational goal. A single state highway overpass design department may undertake reengineering with the goal of providing higher-quality service to customers at a lower cost and find the reengineering effort involves the interrelated tasks of many departments and jobs within the organization. We can examine the impact of reengineering on a different state highway design department, the purchasing department, and one document, a manual parts catalogue. The analysis of anticipated changes in business processes across the state highway agency may identify the manual parts catalogue as a problem area. The engineers may want an automated parts catalogue to increase the speed and accuracy of buying parts for the organization while lowering cost. In order to implement this change, they discover the effects on activities such as purchasing, information technology, finance, accounting, and facilities. In this example, the workflow in the form of changed work rules, roles, and responsibilities may be more important for achieving the goals than the change in job tasks. If the cost of implementing this change is less than the benefits, a cost-benefit analysis suggests the change should be implemented. On the other hand, if the cost exceeds the benefits, the change should not be implemented in the manner suggested by the reengineering workflow analysis. Each form of redesign considered should be carefully analyzed to make sure it fits the needs of the organization.

38.5.3 Directing

This management function is sometimes referred to as leading (Hersey, Blanchard, and Johnson 2001). Directing or leading involves influencing others through motivation, innovation, change, and communication, and it involves more than simply giving instructions to employees. Directing often involves interpretation of the organizational vision, plans and policies provided to others. This management activity may range from explaining how access to fewer resources will impact an employee's job to how to implement a new provision in the union-management labor agreement. This could involve a change in job, resources, and skill level and employees may resist the required change. Employees can be motivated to change temporarily based on management tactics such as intimidation and fear, but more lasting changes are usually based on items such as trust, self-interest, and education. Union and nonunion employees alike may suggest that money is a motivator, but, modern motivation research (Hersey, Blanchard, and Johnson, 2001) suggests that money is seldom a motivator. This research suggests that generally speaking, money is really a maintenance factor and can easily demotivate employees, especially when it is not allocated, administered, and communicated properly. Experienced managers note that employees do not usually work harder nor smarter after a salary increase or promotion. However, they usually complain and produce less output and quality when these positive employee actions do not occur. In highly unionized settings such as transportation, wages for many employees are set by the collective bargaining process and most raises and promotions are determined by seniority. This should reduce the chance of money being a motivator for many transportation employees. A manager should not conclude money is unimportant to employees, but it is just not always a primary motivator. This is especially true in situations where employees are unionized because the manager does not always have the discretion to grant monetary raises.

Any successful change strategy requires effective organizational, group, and individual communications. Communication skills such as persuasion and conflict resolution are required in the directing function. If your city's engineering department is accountable for the new "traffic calming" project, what communication techniques might you use, as the department supervisor, to explain the goals of the project to engineers with more traditional traffic control experiences? Good ideas often fail due to poor communications. Traditional labor-management relationships are based on short-run interpersonal items such as power and fear. As Oestreich and Whaley (2001) indicate, interest based bargaining or negotiations (IBN) is based on quite different interpersonal items and is gaining acceptance in transportation settings. They conclude that IBN is most likely to succeed when the survival of both union and management is at stake and when the aforementioned values of trust, self-interest, and education exist between the parties. The bottom line is that directing assists the planning, organizing, and controlling functions and requires effective leadership skills.

38.5.4 Controlling

This functional management approach involves comparing the goals and standards of the organization to actual results. Key controlling activities are regulating and monitoring organizational activities and taking corrective action where necessary. A common monitoring tool is the budget, which is the organizational plan expressed in monetary terms. The actual expenses are usually compared to the budget in the form of a variance report. Managers are expected to monitor results, explain variances to the budget, and make the necessary changes to meet the budget. Sometimes effective controlling requires advocating changes to the existing budget or development of a different type of budget. For example, should you advocate increasing construction expenses in the current year and exceed the current budget in order to reduce the total project costs over several years? In this situation, what impact does the external economic environment have on the controlling function? If a transportation organization considers downsizing the organization based on an unfavorable economic environment, what should be the budget impact? Should the organization reduce the budget of each department on a fixed percentage, reduce the budget based on reduced headcount or some other resource needs, or simply start from scratch with a zero-based budget approach? Many public transit agencies today are faced with reduced financial resources stemming from unfavorable economic conditions. For example, what should a local transportation organization do when it is dependent upon federal funding to complete a long term project and it receives notice that federal funding is dependent on scaling back the project? Should the local transportation organization scale back the project, stretch out the completion time or look for substitute funding sources? These dwindling resources result from many factors such as reduced fare-box recovery, lower ridership, less income due to less taxing authority, less income based on lower public priority of services provided or less need because funded projects have been completed. Whatever economic climate exists in the future, transportation managers will increasingly be asked to do more with fewer resources. This current austere economic environment requires higher levels of budget/financial skills than previously required for all levels of transportation managers. When a more favorable economic environment returns, the same high level of financial skill will be needed to manage expansion. Other significant controlling activities in transportation are maintenance schedules, accident prevention, safety training, and numerous human resource reports such as sick leave usage, workers' compensation claims, overtime, and vacation. These increased skill requirements in the controlling function are quickly moving down to the supervisory level.

38.6 ARE MANAGEMENT FUNCTIONS AND ACTIVITIES GENERIC?

Although classical era management scholars and practitioners maintained that all managers should use management functions and principles in precisely the same manner, contemporary management views suggest that effective organizations depend on situational or contingency approaches to management. Cook and Hunsaker (2001) list situations that make a difference to organizational effectiveness: role of the manager, level of the manager, goals of the organization, structure of the organization, technology of the organization, and external environment of the organization. In addition, the unique characteristics of the industry require consideration.

38.7 TRANSPORTATION MANAGEMENT

The effectiveness of transportation organizations depends on an intricate set of relationships among people, materials, machines and technology. For example, transportation engineers focus on these four relationships from an equipment, materials, and systems perspective, and planners focus on these same items from an organizational goal, revenue, and time perspective. It is common to hear the term *management* applied to purely engineering and operational activities. However, engineering and operations departments usually only focus on the transportation system. The management of transportation systems involves the inanimate components, such as roads, bridges, materials, machines, vehicles, demand, price, and information systems. Therefore, management of the systems is not synonymous with management of the people in order to accomplish the work. The focus of transportation management is thus the interrelationship between the people within the organization, with less emphasis on the inanimate, economic, or infrastructure components.

38.8 ORGANIZATIONAL GOALS

The goals of an organization are critical to the management functions that need emphasis. Efficiency or productivity is a basic organization goal. Efficiency is the ability to make the best use of available resources. In most cases, efficiency is measured as the ratio of outputs to inputs of the organization. Efficiency is easier to measure in private sector transportation firms because quantifiable marketplace results such as profit are used as metrics. However, efficiency is measured in the public sector in terms of quantifiable metrics such as service levels, stakeholder complaints and commendations, and bond ratings. Five scenarios are demonstrated in Figure 38.1 concerning outputs and inputs that describe efficiency. The plus symbol + indicates outputs or inputs are increasing, the negative symbol − indicates they

EFFICIENCY = OUTPUT / INPUT

1. OUTPUT INCREASES FASTER THAN INPUT (+/+)

2. INPUT DECREASES MORE THAN OUTPUT (−/−)

3. PRODUCE SAME OUTPUT WITH LESS INPUT (0/−)

4. PRODUCE MORE OUTPUTS WITH SAME INPUTS (+/0)

5. OUTPUT INCREASE AS INPUT DECREASES (+/−)

FIGURE 38.1 Efficiency metrics.

are decreasing, and the zero 0 symbol indicates no change in inputs or outputs. For example, efficiency is doing more with less, but doing less with even less resources is also a form of efficiency.

Drucker (1967) has described efficiency as "doing things right." Efficiency is important but deceptive at the same time. For example, an organization can be efficient and yet un-successful, that is, the organization can efficiently pursue the wrong goal. According to Drucker, it is important to "do things right" as well as "do the right thing." Norman Au-gustine, former CEO of Martin Marietta, tells the following story concerning an organization that is efficient but is not doing the right thing (Dess and Lumpkin 2003):

> I am reminded of an article I once read in a British newspaper which described a problem with the local bus service between the towns of Bagnall and Greenfields. It seemed that, to the great annoyance of customers, drivers had been passing long queues of would-be passengers with a smile and a wave of the hand. This practice was however, clarified by a bus company official who explained, it is impossible for drivers to keep their timetables if they must stop for passengers.

In order for an organization to be successful over time, both efficiency and effectiveness are needed. Usually competing goals exist in transportation. Metrics such as percentage sick outs, ridership, farebox recovery, and time schedule require a critical balance for a bus company to survive and thrive. Balance of competing internal goals is just as important to the success of other types of transportation organizations.

38.9 MANAGERIAL ROLES

Cook and Hunsaker (2001), Hersey, Blanchard, and Johnson (2001), and Mintzberg (1990) define role as an organized set of expected behaviors of a particular position. It was pointed out above that management approaches need to be altered based on different situations such as a change in job or role within the organization. Moffat and Blackburn (1996) indicate that most bus field supervisors are promoted from bus operator jobs, and that very little is usually done by the bus companies to prepare and support these supervisors in their new role. Moreover, the bus field supervisor study suggests these new supervisors lack leadership as well as management skills. Kotter (1990) describes the differing roles of nonmanagement, managers, and leaders in organizations and the skill, knowledge, and abilities that make employees successful in those roles. The challenges and dilemmas inherent in changes in roles are illustrated in Figure 38.2.

Since success in one role does not automatically prepare employees for success at another role, both individuals and organizations need to recognize additional preparation is needed as employees transition from one organizational role to another role. The roles that managers play in the organization affect whether the four managerial functions are effective. Mintzberg (1990) identified three general roles that describe what managers typically do and what makes them effective. The three general managerial role categories are: Interpersonal, In-formational and Decisional. These three role categories are further divided into ten mana-gerial roles: Figurehead, Leader, Liaison, Monitor, Disseminator, Spokesperson, Entrepre-neur, Disturbance Handler, Resource Allocator, and Negotiator. Each role is placed under the appropriate category in Figure 38.3. In some organizations, employees are required to play all three roles while performing the same job.

Effectiveness in each role contributes to the overall effectiveness of the organization. Effective transportation managers are able to recognize the role required for different situ-ations and fulfill that role within the resource constraints of the organization.

INDIVIDUAL CONTRIBUTOR	SUPERVISORS AND MANAGERS	LEADERS
1. Task-oriented	1. Results-oriented	1. Goal-oriented
2. Perform task	2. Establish mission	2. Establish vision
3. Follow rules	3. Thrive on order	3. Tolerate ambiguity
4. Avoid mistakes	4. Correct failures	4. Make failure success
5. Self motivated	5. Motivate by systems	5. Motivate by inspiration
6. Resist change	6. Adjust to change	6. Create change

DILEMMAS?

FIGURE 38.2 Transition in roles.

38.10 MANAGERIAL LEVELS

Managers need the critical skill, knowledge, and abilities (SKAs) to carry out the four management functions: planning, organizing, directing, and controlling. Supervisors or first-level managers are the lowest level in the management hierarchy and have only nonmanagement employees reporting to them. Typically, first-line managers in the transportation industry are: bus supervisor, dispatch supervisor, project engineering manager, transportation planning department manager, and financial department manager. Middle-level managers usually have

❏ INTERPERSONAL
1. FIGUREHEAD
2. LEADER
3. LIAISON

❏ INFORMATIONAL
4. MONITOR
5. DISSEMINATOR
6. SPOKESPERSON

❏ DECISIONAL
7. ENTREPRENEUR
8. DISTURBANCE HANDLER
9. RESOURCE ALLOCATOR
10. NEGOTIATOR

FIGURE 38.3 Mintzberg's management roles.

other managers reporting to them. Since most medium-sized and large transportation organizations are composed of at least four major functions—transportation planning, operations, maintenance, and administration—which report to top management, middle-level managers are typically the heads of these functions. In small transportation organizations, the middle-level managers may have management and nonmanagement employees reporting to them and they in turn usually report to the top manager. In larger transportation organizations, the top management level is usually composed of vice-presidents of the major functions and the top manager of the organization. When the top manager has the title chief executive officer (CEO), the typical titles of managers reporting to the top are: chief operating officer (COO), chief financial officer (CFO), chief information officer (CIO), and so on. In addition to role, different levels of management emphasize a different set of skills, knowledge, and abilities that are needed to be effective. Hersey, Blanchard, and Johnson (2001) and Katz (1974) identify three key skills that make a difference in managerial performance: technical, human and conceptual (see Figure 38.4).

Although managers use all three skills, Katz's (1974) research indicates that effective top managers emphasize conceptual skills, middle-level managers emphasize human skills, and lower-level managers emphasize technical skills. One could conclude that technically trained employees such as engineers tend to have more difficulty with the less technical, higher levels of management than the lower levels of management, where they can rely on their technical expertise. If transportation managers are promoted from engineer to supervisor and higher levels of management, they need to prepare for these changes in skill level through education or planned job experiences. Small transportation organization and transportation consulting firms do not usually have these three distinct levels of management and SKA flexibility, and training is thus even more important to them. Moffat and Blackburn (1996) indicate bus companies have reduced their support for training over time. This trend toward lower support for internal training requires bus companies and bus supervisors to seek training from outside sources such as local universities, the National Transit Institute (NTI), and internal training through on-the-job training (OJT). When economic times are difficult, the trend toward less supervisory training and the delay of management training in transportation is evident.

Management Levels

Top Level Managers

Mid Level

Low Level Managers

Skill Emphasis

Conceptual: Ability to grasp the big picture and analyze abstract issues

Human: Ability to demonstrate positive interpersonal skills

Technical: Ability to demonstrate competence and expertise in a particular field

FIGURE 38.4 Managerial roles, skills, and levels.

38.11   LEADERSHIP IN TRANSPORTATION MANAGEMENT

It was mentioned above that directing as a managerial function requires effective leadership. Mintzberg (1990) identifies leadership as one of the ten basic management roles that focuses on interpersonal roles. However, is leadership based on the formal authority a manager acquires due to positional power? Does positional power alone guarantee effectiveness? New managers often assume that power, influence, authority, and leadership all automatically come with their new title. Consider for a moment the promotion of an individual contributor in an engineering position in transportation to a supervisory position in the same organization. Since leadership is not identified as critical to success for most individual contributor jobs, organizations have cut back on training new supervisors in these skills. Where does the new supervisor acquire this new skill set? A few external sources for training new supervisors were mentioned above, these external training sources are not always available when needed by the supervisor. Bennis and Thomas (2002) indicate that leaders are made and not born. They report that every leader, regardless of age, had at least one intense transformational experience. Therefore, organizations need to identify leaders, convince new managers that leadership is a necessary skill, provide job opportunities to develop and demonstrate leadership, and provide the training necessary to develop new managers as leaders.

Covey (1990), Kotter (1990), and Hersey, Blanchard, and Johnson (2001) identify leadership as a key success criterion for managers in all industries and all level of management. Kotter (1990) identifies leadership as a distinctly different role and set of activities from management. Blanchard and Johnson (2001) define leadership as an influence system. Moffat and Blackburn (1996) identify leadership as a key success criterion for the bus supervisor. Do these findings apply to other transportation settings? Let us suppose you are a new supervisor who was formerly a traffic engineer with the same organization. Furthermore, let us assume you are a brilliant traffic engineer but you had poor peer group rapport before promotion to supervisor of the traffic department. Most likely you will not work out as an effective leader in this department. You may have appropriate content knowledge for the traffic supervisor job but lack the skill, knowledge and ability to influence subordinates because of poor interpersonal relationships. Although it helps to have good relationships with subordinates, it is not necessary that they like you for you to influence them in a positive manner. Critical interpersonal skills such as effective communications emphasize problem-solving, trust, and follow-up instead of personalities. Harvey and Lucia (1995) identify follow-up behavior such as "walking the talk" to provide positive influence. Perhaps, in the earlier scenario, you had excellent interpersonal skills but had a leadership style that did not match the task situation of individuals within the department. The task-leadership skill match is needed as well as the resources to implement the activities involved in order to be effective (Hersey, Blanchard, and Johnson 2001). Therefore, both management and leadership are important success criteria, and we will focus on the value that leadership provides in what follows.

It is difficult to demonstrate only one leadership style in transportation and become a successful manager. A government-owned and operated carrier may require different leadership characteristics in its top management than a private carrier. A small transportation organization may call for a different type of leadership than a large organization. In the past, the dominant management approaches of many transportation organizations were bureaucratic, mechanistic, traditional, and autocratic in leadership style. This meant that management was most likely hierarchy-oriented and valued stability more than change. Managers might have leaned toward centralization more than decentralization of authority and lacked the flexibility needed to launch innovative programs. Moffat and Blackburn (1996) suggest this is still particularly true in repair and maintenance operations, and in bus and trucking operations first-line supervisors often lack the leadership training needed to relate well to the people they supervise. The organization expects first-line supervisors to be control-oriented, to enforce rules, and to punish noncompliance. Bus operators frequently complain that the only time they hear from their supervisor is when they are in trouble. Engineers in the same environment frequently complain that their supervisors micromanage their work and get in the way of their productivity. This internal organizational environment or culture makes innovative practices and changes in leadership styles very difficult.

There are numerous theories of leadership that are supported by research. These theories or models are based on items such as traits, styles, power, and the situation. With a few exceptions, trait- or characteristic-based theories are not generally supported by modern research. A plethora of credible leadership style theories exists. These include multiple style categories, but they almost always include two basic styles, "autocractic" and "democratic." The autocratic style is similar to theory X and the democratic style is similar to theory Y, identified by McGregor (1960). As Figure 38.5 shows, assumptions that people in general make about human behavior are related to theory X and Y styles.

Traditional style theories of leadership suggest that a manager uses the same approach to influencing others in all situations. Pfeiffer and Jones (1972) provide an exercise (shown in Figures 38.6 and 38.7) that allows the supervisor to assess their leadership style based on assumptions the supervisor generally makes about human behavior in work settings. You can

THEORY X ASSUMPTIONS

1. The average human being has an inherent dislike of work and will avoid it if he can.

2. Because of this human characteristic of dislike for work, most people must be coerced, controlled, directed, and threatened with punishment to get them to put forth adequate effort toward the achievement of organizational objectives.

3. The average human being perfers to be directed, wishes to avoid responsibility, has relatively little ambition, and wants security above all.

THEORY Y ASSUMPTIONS

1. The expenditure of physical and mental effort in work is as natural as play or rest.

2. External control and the threat of punishment are not the only means of bringing about effort towrd organization objectives. Man will exercise self-direction and self-control in the service of objectives to which he is committed.

3. Commitment to objectives is a function of the rewards associated with their achievement.

4. The average human being learns under proper conditions not only to accept but also to seek responsibility.

5. The capacity to exercise a high degree of imagination, ingenuity and creativity in the solution of organizational problems is widely, not narrowly, distributed in the population.

FIGURE 38.5   Assumptions about human behavior.

Indicate on the scale below ✓ where you would classify your own basic attitudes toward your subordinates in terms of McGregor's Theory X and Theory Y assumptions.

Theory X _____**Theory Y**

        10   20   30   40

FIGURE 38.6   Management style.

participate in this exercise by placing a mark (X) somewhere from 10–40 points on the line provided in Figure 38.6 indicating whether you believe your style is theory X or theory Y.

The exercise in Figure 38.7 provides 10 workplace scenarios, and the exercise requires you to respond to each situation as you would actually respond on the job in your role as a supervisor. You are provided with a score between 10–40 based on your responses to each scenario. You are provided with a score between (1–4) based on your responses to each scenario.

You are asked to add your score on each scenario to produce a total score. A total score between 10–20 points suggests you will use a theory X style across all situations and a score between 30–40 points suggests you will use a theory Y style consistently. Is there a large gap in your two scores on this exercise? Whaley (2003) indicates supervisors who have responded to the exercise in Figures 38.6 and 38.7 find their style based on the ten questions in Figure 38.7 is much closer to their actual style at work than the initial "guess" provided in Figure 38.6.

Directions: The following are various types of behavior which a supervisor (manager, leader) may engage in relation to subordinates. Read each item carefully and then put a check mark in one of the columns to indicate what you would do.

	Make a Great Effort to Do This	**Tend to Do This**	**Tend to Avoid Doing This**	**Make a Great Effort to Avoid This**
IF I WERE A SUPERVISOR :				
Points:	**(1)**	**(2)**	**(3)**	**(4)**
1. Closely supervise my subordinates in order to get better work from them.				
2. Set the goals and objectives for my subordinates and sell them on the merits of my plans.				
3. Set up controls to assure that my subordinates are getting the job done.				
4. Encourage my subordinates to set their own goals and objectives.				
5. Make sure that my subordinates' work is planned out for them.				
6. Check with my subordinates daily to see if they need any help.				
7. Step in as soon as reports indicate that the job is slipping.				
8. Push my people to meet schedules if necessary.				
9. Have frequent meetings to keep in touch with what is going on.				
10. Allow subordinates to make important decisions.				

**Award the number of points (1–4) for each response as indicated above, depending on your response to each scenario. Reverse the point scale for scenario #4, #10. Add the total points for all ten scenarios.

FIGURE 38.7   Supervisory style: the *x-y* scale.

Job-related factors suggest that it is impractical and ineffective to use the same leadership styles in all situations if the goal is to become a successful transportation manager. Hersey, Blanchard, and Johnson (2001) indicate that one factor that requires consideration is the nature of the task. If you are the supervisor of bus maintenance and the mechanics perform simple brake repairs, a more directive, task-oriented, authority-oriented, theory X style is more effective. On the other hand, if the fuel used for the coaches is a new fuel such as liquid gas, this usually suggests more task complexity and a more supportive, people-oriented, participative, theory Y style is more effective. Moreover, if the mechanics performing the simple brake job are inexperienced in performing the task, a theory X style is more effective, while if the mechanics are highly experienced, a theory Y style is more effective. Researchers tend to agree that the situation is critical to the selection of an appropriate leadership style, but they disagree concerning which aspects of the situation to take into consideration. A popular situational model of leadership is the Blanchard Situational Leadership Model (Hersey, Blanchard, and Johnson 2001). This model, shown in Figure 38.8, suggests that the appropriate leadership style is based on the development level of each employee on each specific task performed. In earlier versions of this model, the developmental level was called the "readiness" level and the "maturity" level.

The Situational Leadership Model is attractive for the transportation supervisor because it places the emphasis on the development level of the subordinate and the specific task rather than the style of the manager. Therefore, if there is not a match between the development level of the subordinate and the style of the leaders, several approaches are shown in Figure 38.9 that may work in your organization.

Many other credible situational leadership models exist. Some situational models emphasize the role or style of the leader (Fielder 1967; House 1971; Kouzes and Posner 1996), while others emphasize leadership in terms of the interaction of the team (Blake and Mouton 1984; Bales and Cohen 1994). The newer SYMLOG model (Bales and Cohen 1994) provides a very flexible tool for measuring the effectiveness of a leader to each team member, a most effective person (MEP) score, and the effectiveness of the entire team across different internal and external settings. Additionally, SYMLOG is more metrics-oriented and can be used to determine a team effectiveness score and compare individual scores to other important or-

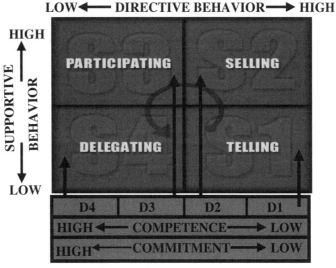

FIGURE 38.8 Situational model.

1. The leader can change styles.
2. The leader can focus on developing the employee around a specific task.
3. The leader can temporarily use others in organization to influence the employee.
4. The leader can focus on changing the content of the task to match the employee.
5. The leader can select a different employee or different situation.

FIGURE 38.9 Situational leadership approaches.

ganizational concepts. Some of these comparisons are the perception of current organizational culture, the perception of future organizational culture, the perception of values employees are rewarded for in the organizational culture, and the perception of values for which your customers believe you demonstrate in the culture. Teams that are not co-located are called virtual teams. They are a new organizational structure trend in many organizations as well as transportation. As the work in transportation organizations becomes more team-oriented, these team-based leadership models and assessment instruments become more helpful.

38.12 TYPICAL CHALLENGES FACED BY NEW MANAGERS IN TRANSPORTATION

The new managers within the transportation industry are beset by more challenges than their counterparts decades ago. These challenges come from outside the organization, from within the organization itself, and from the individual. The results of interviews and discussions with new engineering managers are shown in Figure 38.10 indicating there are eight typical challenges faced by new managers in transportation (Whaley 2003).

New transportation managers need to be especially aware of the eight common themes listed in Figure 38.10 and to become proactive. For example, when employees remind you, as a new manager, that you made mistakes in the past as a nonmanagement employee, you can take this negative communications as an opportunity to discuss your new role as supervisor and admit your past behavior was not appropriate. Perhaps you can turn lemons into lemonade by asking these same employees for help in aligning their behavior with the goals of the organization. As a new manager, you need to understand your new role as supervisor and become comfortable with the role before you can expect your employees to be com-

1. Subordinates do not respect your authority because they remember you broke the work rules as a nonmanagement employee.
2. Managers are not familiar with the transportation or organizational culture.
3. Several employees have more task knowledge and experience than the manager and are vocal about it.
4. Managers are asked to enforce organization rules and regulations and they do not personally agree with the organization's policies and procedures.
5. Managers do not have sufficient breadth of management training or experience.
6. New managers do not identify with management because they were union members in the past.
7. What the manager does and says is usually interpreted as organizational policy.
8. Constraints do not allow the manager to treat everything and everyone individually.

FIGURE 38.10 Eight typical challenges faced by new managers in the transportation industry (1994–2003).

fortable with you in this new role. When you are not comfortable with organizational policies and procedures, we suggest you do not undermine the company's goals by telling employees about your disagreement with the policy or procedure. First, we suggest you seek to understand the reasons and benefits of the policy from your management and then explain these points to the employees in terms of their self-interest and in a neutral manner. The new supervisor is often shocked to find out that most things they say at work are interpreted by employees as company policy. This situation can be taken by the new supervisor as an opportunity to explain to employees the multiple roles you play at work and provide examples to demonstrate when you are speaking for the company. This clarification builds trust and is especially important in employee discipline situations. Oestreich and Whaley (2001) point out that it is important in discipline situations where the employee is covered by a collective bargaining agreement for the employee to know when you are giving a direct order. Although the specific contexts where each of the eight situations arises are unique, they happen often enough to new managers that they should be anticipated and a unique proactive strategy to deal with them should be developed. However, the most important thing for any manager to remember concerning building employee trust is that the manager needs to "walk the talk."

Many of these eight challenges are related to external pressures on the organization, and organizations need to consider the impact of internal pressures through organizational culture. Many transportation organizations are changing their business models and management approaches based on a shift in the values or culture of the organization. The formal, top-down, "do it by the book" culture in transportation is being slowly replaced with more open, bottom-up, "do it by the planned goals" culture. This transition in values should make it easier to change the organization irrespective of whether the changes are based on external or internal forces. One noteworthy change is the previously mentioned growth of teams in transportation organizations. If an organization embraces the values related to teams, it should be easier for the new supervisor to make use of the critical SKAs of each employee and team and not feel the need to be the expert on each task the subordinate employees perform. As previously mentioned, the team approach to work should also change the leadership style approach of each manager.

Individual employees provide much of the impetus for change today because they expect different things from the organization than previous generations of managers (Hersey, Blanchard, and Johnson, 2001). Wellner (2002) provides a breakdown of the U.S. into seven generational categories based on age and percentage of the population in year 2000. Generation X employees and transportation managers alike who were between ages 24–35 in year 2000, expect to participate in organizational decision-making. If participation is also an organizational cultural value, employees are even more inclined to demand participation. The individual focus on participation helps to balance the group focus when group-oriented values are also important organizational values. This makes employee empowerment easier to implement. Figure 38.11 shows the four conclusions Moffat and Blackburn (1996) formed based on their survey of transit agencies, directors, managers and supervisors of operations concerning the role and responsibilities of the bus field supervisor.

These findings reinforce the other findings (Whaley 2003) that led to the list of eight typical challenges for new transportation managers shown in Figure 38.10.

---

1. Almost all supervisors are hired from the ranks of bus operators.
2. Knowledge of transportation systems and experience with them is critical, but little is done to prepare supervisors for this new role.
3. Little or no formal training is provided to supervisors in people management.
4. Increased interaction is required of supervisors with bus operators and the public.

FIGURE 38.11 Four conclusions regarding role and responsibilities of supervisors.

38.13   *THE FUTURE OF TRANSPORTATION MANAGEMENT*

Hersey, Blanchard, and Johnson (2001), Moffat and Blackburn (1996), Whaley (2003), and Wood and Johnson (1996) indicate that a quicker pace of change is expected for all organizations in the future. Jacob (1995) and Kanter (1990) suggest that change is the major management challenge. Kanter (1990) states: "The major challenge management faces today is living in a world of turbulence and uncertainty where new competitors arrive on the scene daily and competitive conditions change. We can no longer count on a stable world that is unchanging and unvarying and manage accordingly." Some expected changes in transportation are:

1. Fewer resources available and expanded service requirements
2. Expanded privatization and emphasis on technology, safety, and security
3. Change in the organization culture from authoritarian to participative
4. Increase use of lower cost and nonunion labor
5. Less reliance on gasoline and diesel fuel and increased exploration of alternative fuels
6. Increased use of technology across the board from route planning and communications to intelligent highways and computerized services
7. Increased workforce diversity in the United States and abroad
8. More interest in labor-management cooperation processes such as IBN
9. Shift of power from transportation companies to stakeholders
10. Knowledge becoming the primary source of power within the organization
11. More participation of stakeholders and quicker expected responses due to expanded use of the Internet
12. Increased links among transportation, the environment, and local economic vitality and planning

Wood and Johnson (1996) suggest that a comprehensive national transportation policy is needed. Parts of a national policy exist today, but the parts need to be integrated for overall effectiveness. For example, NAFTA is not a transportation policy per se, but it impacts the movement of carriers across national borders in North America. Federal legislation in 1991 increased support to mass transit and is not integrated very well with environmental laws. The Intermodal Surface Transportation Efficiency Act of 1991 (as amended by TEA-21) provides more support for management training in the transportation industry but it is not well coordinated with training under existing transportation legislation and organizations. Airline deregulation has brought important changes to that industry, and several airlines are in bankruptcy today. Publicly supported transportation organizations across the country are suffering large budget cuts. There is an urgent need to integrate sound economic policy at the national, state, and local levels with effective regulatory policies in transportation. The Transportation Equity Act (TEA-21) requires regional transportation plans. Closer alignment of city, county, and state planning with federal goals is required to achieve the goals of each in an environmentally sensitive manner. For example, it will not be uncommon in the future for a city agency to perform a transportation study to convince a private firm in a different industry to locate in their city or not to leave their city. This trend could also lead to greater marketing of services provided by local transportation agencies. Lastly, no stakeholders should be overlooked in formulating governmental policy. Osterman et al. (2001) indicate that unions tend to be an overlooked stakeholder in the planning process. Many transportation organizations have enlisted their labor unions to help survive financial crunches. Osterman et al. (2001) suggest that unions can play an even larger and positive role in public policy-making at the national level and that this role is important for the survival and future growth of unions.

Communication is a key component of interpersonal skills, but effective listening, diversity management, impression management, and emotional intelligence are at the top of the list. Cook and Hunsaker (2001) indicate that effective listening is one of the most important communication skills for managers. Increased diversity will be the future norm, and effective listening will be a key element for managers and organizations to address differences effectively. Diversity based on legally protected groups such as age, gender, race, national origin, religion, and disability will no doubt continue to be covered under federal law in the future. Employers that value diversity should also consider the role of other types of diversity, such as learning styles and leadership styles. Not only is considering other forms of diversity a good thing to do, it also provides the organizations with a competitive advantage when their customers are also diverse.

A new measurement of interpersonal skill is emotional intelligence (EQ or EI). Unlike IQ, which seeks to measure raw intelligence, emotional intelligence seeks to measure interpersonal and communications skills. Goleman (1998) suggests EQ is the ability to recognize and manage the emotions of others and yourself and EQ can be used to enhance the performance of the manager and the organization. Although most professionals are trained to think and behave rationally, they tend to have emotional reactions before their rational reactions. Have you caught yourself wishing you never expressed something or behaved in an inappropriate manner? On the other hand, do you recall that workplace outcomes were more productive when you recognized the feelings of others and exercised empathy and self-control? Goleman (1998) and other advocates of EQ go further and indicate that close to 90 percent of success in leadership is attributable to emotional intelligence. The five components of EQ that transportation managers need to keep in mind are:

1. Self-awareness
2. Managing emotions
3. Motivating oneself
4. Empathy
5. Social skill

Self-regulating emotional control is helpful so that constant change does not get the manager off-track. If doing more with less resources is the wave of the future, then expecting employee performance at a rate higher than stated in the job description is the next logical step. Coleman and Borman (2000) identify this performance expectation as organizational citizenship (OC). The level of job performance expectation involved in the OC concept requires organizational support, and implementation of the OC concept is vulnerable to inappropriate self-promotion by employees. However, self-promotion and impression management is not automatically synonymous with dysfunctional organizational politics. If employee performance is high, impression management is usually perceived by others as positive and the use of persuasion skills, not organizational politics. Change agent and change management skills will be emphasized for new transportation managers in the future because change will be faster, pervasive, expected, and more valued. An increased emphasis on self-learning, speed of learning, e-learning, and interpersonal skills by individual managers and their organizations is needed. Employees and organizations expect more from managers today because they see behavior in this area as lowering cost and increasing revenue or other metrics that represent increased organizational value. These future trends in transportation raise the expectations and performance bar for managers. The required management skill set for new transportation managers will most likely transform what was a luxury skill set to a minimally expected skill set.

38.14 REFERENCES

American Public Transit Association (APTA). 1999. *Transit Fact Book,* 50th ed. Washington DC: APTA.

Bales, R. F., and S. P. Cohen.1994. *SYMLOG: A System for the Multiple Level Observation of Groups.* New York: Free Press.

Bennis, W. G., and R. J. Thomas. 2002. *Geeks and Geezers: How Era, Values, and Defining Moments Shape Leaders.* Cambridge: Harvard Business School Press.

Blake, R. R., and J. S. Mouton. 1984. *The Managerial Grid III.* Houston: Gulf.

Champy, J. 1995. *Reengineering Management: The Mandate for New Leadership.* New York: Harper-Collins.

Coleman, V. I., and W. C. Borman. 2000. "Investigating the Underlying Structure of the Citizenship Performance Domain." *Human Resource Management Review* 10:25–44.

Cook, C. W., and P. L. Hunsaker. 2001. *Management and Organizational Behavior.* New York: McGraw-Hill Higher Education.

Covey, S. R. 1990. *Principle-Centered Leadership.* New York: Simon & Schuster.

Dess, G. G., and G. T. Lumpkin. 2003. *Strategic Management: Creating Competitive Advantages.* New York: McGraw-Hill Higher Education.

Drucker, P. F. 1967. *The Effective Executive.* New York: Harper & Row

Fatt, J. P. T. 2002. "Emotional Intelligence: For Human Resource Managers." *Management Research News* 57–74.

Fiedler, F. E. 1967. *A Theory of Leadership Effectiveness.* New York: McGraw-Hill.

Goleman, D. 1998. Calculating the Competencies of Stars. *Working with Emotional Intelligence,* New York: Bantam Books.

Goleman, D. 1998. *Working with Emotional Intelligence.* New York: Bantam Books.

Harvey, E. C., and A. Lucia. 1995. *Walk the Talk and Get the Results You Want.* Dallas: Performance.

Hersey, R. E., K. H. Blanchard, and D. E. Johnson. 2001. *Management of Organizational Behavior: Leading Resources.* Upper Saddle River: Prentice-Hall.

House, R. J. 1971. "A Path-Goal Theory of Leadership." *Administrative Science Quarterly* 16:321–38.

Jacob, R. 1995. "The Struggle to Create an Organization for the 21st Century." *Fortune* (April 3): 90.

Kanter, R. M. 1990. *The Planning Forum Network* 2(1): 1.

Katz, R. L. 1974. "Skills of an Effective Administrator." *Harvard Business Review* 52(5): 90–102.

Kotter, J. 1990. "What Leaders Really Do." *Harvard Business Review* 68(3):103–11.

Kouzes, J. M., and B. Z. Posner. 1996. *The Leadership Challenge.* San Francisco: Jossey-Bass.

McGregor, D. 1960. *The Human Side of Enterprise.* New York: McGraw-Hill.

Mintzberg, H. 1990. "The Manager's Job: Folklore and Fact." *Harvard Business Review* 90(2):163–176.

Moffat, G. K., and D. R. Blackburn. 1996. *Changing Roles and Practices of Bus Field Supervisors.* TCRP Synthesis 16. National Research Council, Transportation Research Board, Washington, DC.

Oestreich, H. H., and G. L. Whaley. 2001. *Transit Labor Relations Guide.* Mineta Transportation Institute Report 01-02. San José: Mineta Transportation Institute, San José State University.

Osterman, P., T. A. Kochan, R. M. Locke, and M. J. Piore. 2001. *Working in America: A Blueprint for the New Labor Market.* Cambridge: MIT Press.

Pfeiffer, J. W., and J. E. Jones. 1972. *The 1972 Handbook for Group Facilitators.* Iowa City: University Associates.

Pouliot, M. 2002. *Transport Geography on the Web.* Department of Economics and Geography, Hofstra University, Hempstead, NY.

San José Department of Transportation. 2001. City of San José Transportation Annual Plan.

U.S. Department of Commerce, Bureau of the Census. 1998. *Statistical Abstracts of the United States 1998,* 444.

Vitale, R. P. and Giglierano, J. J. 2002. *Business to Business Marketing: Analysis & Practice in a Dynamic Environment.* Mason: South-Western.

Wellner, A. S. 2000. Generational Divide, *American Demographics,* October, pp. 53–58.

Whaley, G. L. 2003. Interviews of New Transportation Managers, 1994–2003.

Wood, D. F., and J. C. Johnson. 1996. *Contemporary Transportation.* Upper Saddle River: Prentice Hall.

Wytkin, E. 1997. *Toward a Cooperative Future.* Mineta Transportation Institute Report 97-2. San José: Mineta Transportation Institute.

INDEX

ABOUT THE EDITOR

Myer Kutz has been president of Myer Kutz Associates, Inc., a publishing and information services consulting firm, since 1990. He was vice president in charge of scientific/technical publishing at John Wiley & Sons for 5 years. He has been a member of the board of the Online Computer Library Center (OCLC) and chair of the ASME Publications Committee. He has a BS in mechanical engineering from MIT and a master's degree from RPI. He is the editor of *Standard Handbook of Biomedical Engineering and Design,* also published by McGraw-Hill.